Element	Symbol	Atomic Number	Atomic Mass
Manganese	Mn	25	54.9380
Mendelevium	Md	101	(256)[a]
Mercury	Hg	80	200.59
Molybdenum	Mo	42	95.94
Neodymium	Nd	60	144.24
Neon	Ne	10	20.179
Neptunium	Np	93	237.0482[c]
Nickel	Ni	28	58.71
Niobium	Nb	41	92.9064
Nitrogen	N	7	14.0067
Nobelium	No	102	(254)[a]
Osmium	Os	76	190.2
Oxygen	O	8	15.9994
Palladium	Pd	46	106.4
Phosphorus	P	15	30.97376
Platinum	Pt	78	195.09
Plutonium	Pu	94	(242)[a]
Polonium	Po	84	(210)[a]
Potassium	K	19	39.0983
Praseodymium	Pr	59	140.9077
Promethium	Pm	61	(147)[a]
Protactinium	Pa	91	231.0359[c]
Radium	Ra	88	226.0254[c]
Radon	Rn	86	(222)[a]
Rhenium	Re	75	186.2
Rhodium	Rh	45	102.9055
Rubidium	Rb	37	85.4678

Element	Symbol	Atomic Number	Atomic Mass
Ruthenium	Ru	44	101.07
Samarium	Sm	62	150.4
Scandium	Sc	21	44.9559
Selenium	Se	34	78.96
Silicon	Si	14	28.0855
Silver	Ag	47	107.868
Sodium	Na	11	22.98977
Strontium	Sr	38	87.62
Sulfur	S	16	32.06
Tantalum	Ta	73	180.9479
Technetium	Tc	43	98.9062[c]
Tellurium	Te	52	127.60
Terbium	Tb	65	158.9254
Thallium	Tl	81	204.37
Thorium	Th	90	232.0381[c]
Thulium	Tm	69	168.9342
Tin	Sn	50	118.69
Titanium	Ti	22	47.90
Tungsten	W	74	183.85
Uranium	U	92	238.029
Vanadium	V	23	50.9415
Xenon	Xe	54	131.30
Ytterbium	Yb	70	173.04
Yttrium	Y	39	88.9059
Zinc	Zn	30	65.38
Zirconium	Zr	40	91.22

[a]Mass number of most stable or best known isotope.
[b]Tentative name.
[c]Mass of most commonly available, long-lived isotope.

Chemistry

A Life Science Approach

Stuart J. Baum
State University of New York
College at Plattsburgh

Charles W. J. Scaife
Union College
Schenectady

Chemistry

A Life Science Approach

2nd Edition

Macmillan Publishing Co., Inc.
NEW YORK
Collier Macmillan Publishers
LONDON

MACMILLAN PUBLISHING CO., INC.
866 Third Avenue, New York, New York 10022

COLLIER MACMILLAN CANADA, LTD.

Library of Congress Cataloging in Publication Data

Baum, Stuart J
Chemistry, a life science approach

Includes index.
1. Chemistry. I. Scaife, C. W. J., joint author. II. Title.
QD31.2.B39 1980 540 79-53
ISBN 0-02-306610-5

Printing: 1 2 3 4 5 6 7 8 Year: 0 1 2 3 4 5 6

Preface

This book, written for a two-semester or a three-quarter course, presents an integrated treatment of general chemistry along with organic chemistry and biochemistry. It provides the necessary background for students entering biological and biomedical fields of study including nursing, medical technology, dental hygiene, and nutrition. The level of the book is such that it assumes no high school chemistry course as a prerequisite. Students with a strong high school chemistry background might exempt one quarter (approximately the first eleven chapters) and begin study using the second two thirds of the book. It is anticipated that this book will be used for the terminal course in chemistry for most of these students. Our aim in writing this text is to assist the student in acquiring a sound background in the principles and concepts of general chemistry, organic chemistry, and biochemistry and a clear understanding of the functioning of biological systems at the molecular level. An attempt is made to achieve a logical progression in thinking so that chemical formulas and equations become meaningful symbols of the chemical language. Students need to understand the principles of general and organic chemistry before they can comprehend fully the complex transformations that occur in the cell. It is not sufficient to be able to draw the structure of a carbohydrate, a protein, or a lipid. Students must also understand clearly the interrelationships among these foodstuffs. They must be given an understanding and an appreciation of the superb chemical mechanisms that the body employs in transforming and conserving energy. Students who merely memorize the Krebs cycle and the electron transport chain are not learning biochemistry. It is necessary that they understand the significance of these processes.

This book is divided into three main sections. The first eleven chapters are devoted to establishing the fundamental principles of general chemistry. A balance of historical perspective, observations, models, and practical applications is given. Simple experimental observations are normally presented first, and models are deduced from them. Descriptive inorganic chemistry, organic chemistry, and biochemistry are used liberally for illustrative purposes. Chapter 1 can be covered in class or utilized for outside reading. Chapters 2–4 are complete in themselves and can be covered in any order. Parts of Chapters 2 and 3 are indicative of the fact that this book is not written in the fashion of the times, but is directed instead toward the need by certain students for fairly specific backgrounds. Certain terms or topics are consciously omitted when they are not absolutely necessary for the understanding of later topics. Chemical calculations are presented early (Chapter 4) so that they can be used as tools in later chapters or in a concurrent laboratory. Only simple algebra is utilized in the calculations, and the use of moles is stressed. Macroscopic properties of gases, liquids, solids, and solutions (Chapters 5–7) and reaction pathways (Chapter 8) are explained primarily on the basis of the microscopic bonding picture. Factors that affect the rates of reactions are discussed in Chapter 9. Chapter 10, "Chemical Equilibrium," emphasizes solution chemistry, pH, and buffers, particularly as these topics relate to biochemistry. The medical aspects of nuclear chemistry are stressed in Chapter 11.

Chapters 12–21 deal with the structure and properties of the different classes of organic compounds, with emphasis on the characteristic reactions of the various functional groups. A few selected reaction mechanisms have been presented in order to familiarize the student with such terms as free radical, carbonium ion, electrophile, and nucleophile. Chapter 22 deals with stereoisomerism, and its importance to an understanding of enzyme specificity is stressed.

The remaining chapters present the fundamental concepts of biochemistry in a form that will be understandable to students having only a limited background in organic chemistry. The major emphasis is placed on the dynamic nature of biochemistry and the interrelationships of the various metabolic pathways. Chapters 23–25 discuss the chemistry of the three major classes of foodstuffs—carbohydrates, lipids, and proteins. Enzymes (Chapter 26) are presented in a separate chapter as a special class of proteins. Because of the limited mathematics background of most of the students using this text, statements have to be qualitative rather than mathematical. Thus, for example, no extensive treatment of enzyme kinetics is attempted. In Chapter 27, the molecular basis of life is sketched by a clear and concise discussion of nucleic acid structure and replication, and by the role of nucleic acids in protein synthesis and in the action of viruses. Chapters 28–30 present the basic metabolic reaction sequences that occur within cells from the point of view of their interrelationships and integration into the fundamental whole. Metabolism is concerned with the production and utilization of energy, and it is here that the student is able to tie together the chemical principles learned in general chemistry and organic chemistry with biochemistry. The human body is viewed as an intricate machine that utilizes the energy of foods to run chemical reactions so as to meet its own needs. Chapter 31 discusses the blood, and it is especially well suited to those students who are majoring in one of the health sciences. The final chapter on nutrition and health appeared in an issue of *Chemical and Engineering News*. We are particularly pleased that the author, Howard J. Sanders, and the editor of that journal granted us permission to reprint the article. It is certain to be of interest and value to all readers of this text.

Many of the chapters in this second edition have been revised to reflect suggestions made by reviewers and users of the text. There is a much greater integration of biochemical examples throughout the discussions of organic compounds. Some added topics include aging, octane number, toxic compounds, vision, odor, laetrile, recombinant DNA, and the **R-S** convention for absolute configuration. At the end of each chapter we have included a selection of study questions and problems, ordered approximately according to topical coverage in each chapter. Students should be encouraged to work some of these exercises even while studying appropriate sections of each chapter. It is our contention that students learn chemistry only through repeated practice. The more questions that they are able to answer, the more confident they will be about their understanding of the subject. A supplementary Answer Book is available for students to check their answers.

In preparing this text, we received invaluable aid from many sources. We are indebted to one of our students, Mrs. Deborah S. Miller, for checking many of the numerical problems. We are particularly grateful for the suggestions and critical reviews provided by James A. Campbell, El Camino College; Andrew J. Glaid, Duquesne University; Truman A. Jordan, Cornell College; Lee Pike, East Tennessee State University; Thomas I. Pynadath, Kent State University; William H. Voige, James Madison University; and Ron Widera and Cornelia Lupash, Long Beach City College. We appreciate the efforts of Gregory W. Payne, Elisabeth Belfer, and other members of the staff of Macmillan who encouraged the writing of this book and saw it through production. Finally, we acknowledge the proofreading assistance and encouragements of our wives and the patience of our children during the preparation of the manuscript and the production of this book.

S. J. B.
C. W. J. S.

Contents

1

Some Fundamental Concepts

Chemistry is a study of the *composition, structure, and properties of matter.* What elements are present in water and with what abundance? How is water constructed on a microscopic and a macroscopic scale? What characteristics does water exhibit on a microscopic and a macroscopic scale?

Chemistry is a study of *the changes that matter undergoes.* What happens when water freezes, or when water reacts violently with sodium metal to yield bubbles of gas and a clear solution?

Chemistry is a study of *the related energy changes that accompany material changes.* Is energy released or absorbed when ice is formed from water, or when sodium nitrate is dissolved in water?

Chemistry is a study of *the reasons for particular compositions, structures, properties, material changes, and energy changes.* What is the relationship among these? How is the structure of water on a microscopic scale related to its macroscopic properties? How does the structure of water change upon the addition of ethyl alcohol to water?

Chemistry is concerned with matter and energy. The purpose of this chapter is to discuss the states and nature of matter and the various forms of energy.

1.1 Matter

Matter is *anything that has mass and occupies space.* Our entire universe is composed of matter. Familiar examples include wood, glass, paper, sugar, and air.

a. Mass and Weight

Mass is *a measure of the amount of matter.* The mass of a body is a fixed property of that body. It is *constant* everywhere, independent of other objects. The mass of a sample of moon rock is the same on the moon, in space, or at any altitude on the earth's surface. Mass is the property that gives matter the tendency to stay at rest when already at rest or to stay in motion when already in motion. Thus, riding on an express elevator as it begins its downward motion, one finds his body wanting to stay at rest. On the other hand, when the elevator suddenly stops descending, his body wants to continue in motion and to travel through the floor of the elevator.

Weight is *a measure of the pull of gravity on an object.* The magnitude of the gravitational pull is directly proportional to the mass of the object and the mass of the attracting body (usually the earth). Thus, the weight or pull of gravity on a moon rock is less the smaller the mass of the moon rock. Likewise, the weight of a given moon rock determined on the moon is smaller than the weight of the same moon rock determined in your laboratory because the less massive moon exerts less gravitational force on the rock than does our more massive earth. Moreover, the gravitational pull is inversely proportional to the distance between the centers of the attracting bodies; that is, the gravitational pull is larger when the bodies are closer together and is smaller when the bodies are farther apart. Thus, the gravitational pull and the weight vary depending upon where the measurement is made. The weight of the moon rock is smaller when determined at a higher altitude on earth or when determined eighty miles out in space than when determined in your laboratory. The distinction between mass and weight may be understood in terms of an astronaut in space. The astronaut is weightless whenever he is so far away from the earth or other planetary bodies as to be unaffected by their force of gravity, but the astronaut is never massless since mass is constant everywhere.

Mass and weight can be distinguished by the method of their determination. The mass of an object is determined by balancing it against standard masses, as shown in Figure 1.1a. The chemical balances in your laboratory (common types shown in Figure 1.1c and 1.1d) use this principle. By convention the mass of an object equals the weight of the same object only at sea level on the earth. The weight of an object is frequently determined by means of a spring balance (shown in Figure 1.1b). No standard masses are involved and the weight will depend only on the force of gravity at the point of measurement. Bathroom scales and grocer's scales use this principle. Unfortunately, determining mass and determining weight are both commonly called weighing.

You should recognize the difference between mass and weight and realize that you are using mass in laboratory experiments and calculations when you compare *amounts of matter.* Many texts use mass and weight interchangeably, but this one will try to use them correctly. In fact, mass will be the appropriate term in almost all cases in this book.

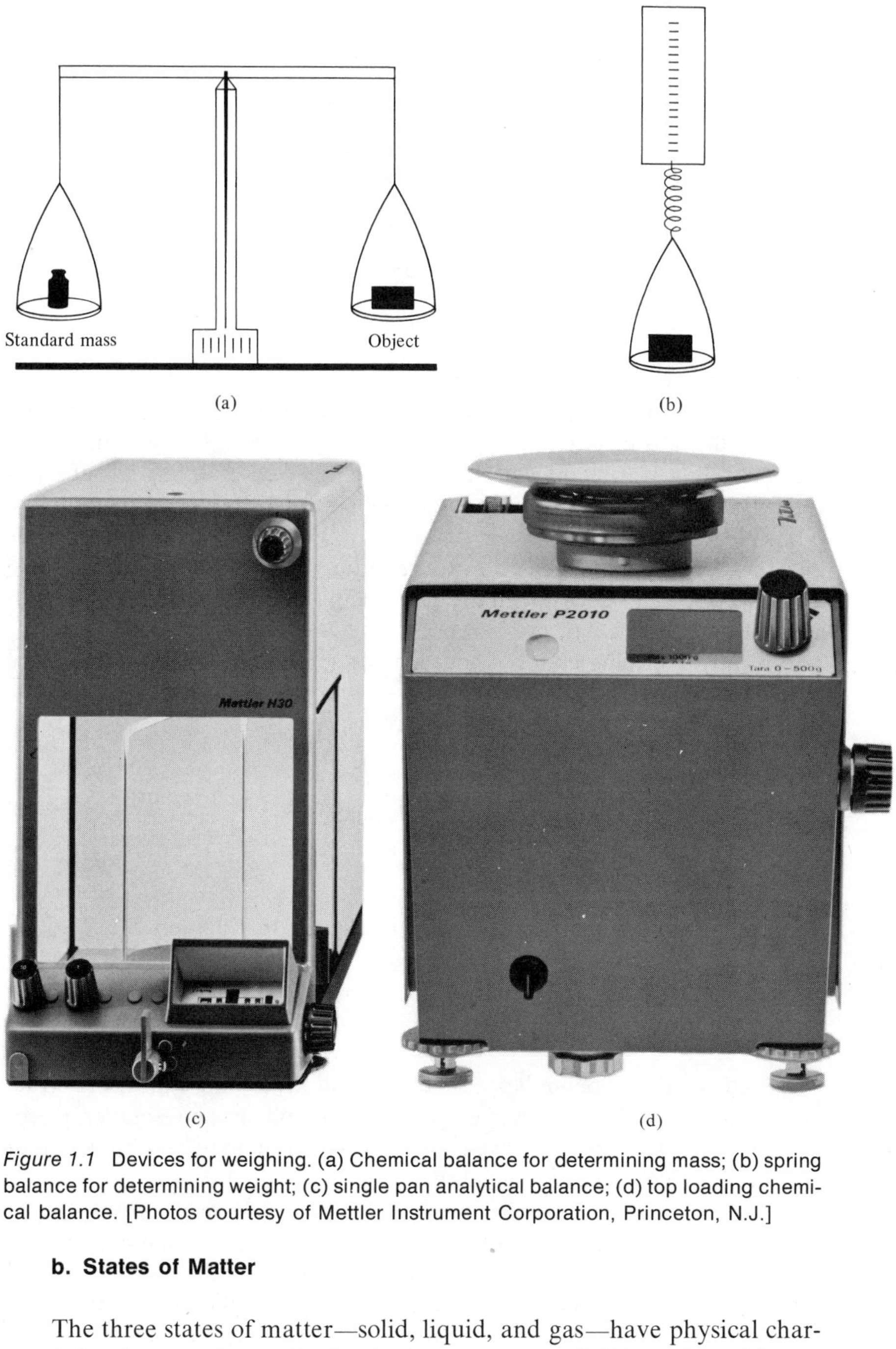
Standard mass
Object
(a)
(b)
Mettler H30
Mettler P2010
(c)
(d)

Figure 1.1 Devices for weighing. (a) Chemical balance for determining mass; (b) spring balance for determining weight; (c) single pan analytical balance; (d) top loading chemical balance. [Photos courtesy of Mettler Instrument Corporation, Princeton, N.J.]

b. States of Matter

The three states of matter—solid, liquid, and gas—have physical characteristics that are detectable by the human senses. Solids and liquids can usually be identified by sight and touch. A piece of iron has a definite shape and is hard, whereas liquid water takes the shape of its container and is soft

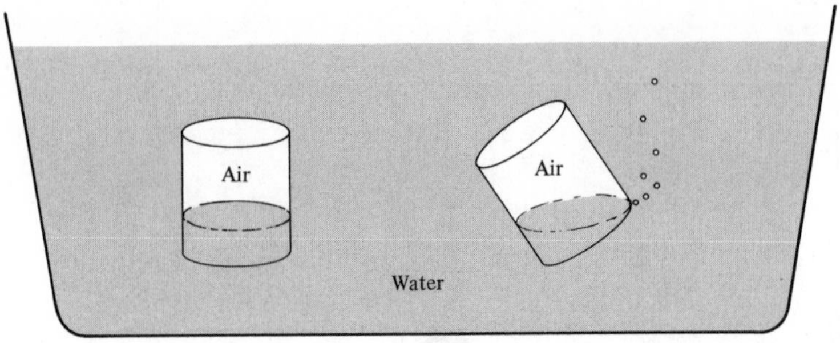
Air
Air
Water

Figure 1.2 Displacement of air from a glass by water.

to the touch. Gases are not so easily identified unless they are colored. Thus, reddish brown nitrogen dioxide (a common air pollutant from automobile exhaust) fills a transparent bottle and is easily detected in the bottle by its color. Air, however, being colorless, may be detected by trying to put other matter in its place and obtaining evidence that the space is already occupied. If one places an inverted glass into water as in Figure 1.2, water partially fills the glass because the increased pressure reduces the volume required by the air; however, water does not fill the glass completely until the glass is tilted to allow the trapped air to bubble out.

Terms describing changes between the solid, liquid, and gas state are illustrated for water.

$$\underset{\text{(Solid)}}{\text{Ice}} \underset{\text{freeze}}{\overset{\text{melt}}{\rightleftharpoons}} \underset{\text{(Liquid)}}{\text{Water}} \underset{\text{condense}}{\overset{\text{evaporate}}{\rightleftharpoons}} \underset{\text{(Gas)}}{\text{Water vapor}}$$

Gases, liquids, and solids are discussed in detail in Chapters 5 and 6.

c. Nature of Matter

A clean copper penny appears to be continuous and uniform in structure. If you could cut a copper penny into smaller and smaller pieces indefinitely, would you still have copper, or would there be an ultimate limit to which you could carry the subdivision?

Matter, on the bulk scale in which we can observe it, appears to be continuous. However, Democritus, an ancient Greek philosopher, proposed by 400 B.C. that matter was composed of small indivisible particles called atoms. He had no experimental means of testing his theory. Experimental data was eventually forthcoming that led John Dalton, a British chemist and science teacher in the Manchester schools, to a restatement in 1808 that *all matter is composed of tiny, discrete particles called atoms.* The particle nature of matter is fundamental to chemistry.

d. Pure Substances

A cube of sugar has a constant composition, structure, and physical and chemical behavior throughout the cube and cannot be further resolved into other materials by simple changes of state. Such *a material with an unchanging composition, structure, and behavior* is a **pure substance.** Pure substances may be composed of elements or compounds.

1. Elements. Mercury in a thermometer is an **element** because it is *a substance that is composed of only one kind of atom.* The various chemical elements are often indicated by a shorthand notation called **chemical symbols,** decided upon by international agreement. These symbols consist of a capitalized initial letter, sometimes followed by a lower case second letter to distinguish between elements with the same initial letter. Thus, the symbols for beryllium, boron, bromine, barium, bismuth, and berkelium are Be, B, Br, Ba, Bi, and Bk, respectively. Many symbols—including those starting with B already noted—begin with the same letter as the name of the element they represent. Others, derived from the Latin name for the element (Ag for silver from *argentum*) or the German name for the element (W for tungsten from *Wolfram*), have no correspondence with their current names. However, extensive use in the past has led to their continued use. Symbols of some common elements are given in Table 1.1. A complete table of symbols of the chemical elements is given inside the front cover.

The natural distribution by mass of the ten most common elements in the earth's crust, oceans, and atmosphere is also shown in Table 1.1. The elements are distributed very unequally. Oxygen, the most abundant element, comprises nearly 50% of the total; oxygen and silicon combined constitute nearly 75% of the total; and the ten most abundant elements combined make up nearly 99% of the total.

The chemical elements are distributed in the human body in very different proportions from their natural abundances, as shown by the data in Table 1.1. Oxygen, the most abundant element naturally, comprises nearly 65% by mass of the human body; oxygen and carbon combined constitute nearly 84% of the total; and oxygen, carbon, hydrogen, and nitrogen make up about 96% of the total. Each of these four elements is present in a much higher percent in living organisms than in nature.

Elements are often classified further as metals or nonmetals. *Metals* such as iron, aluminum, magnesium, and copper are solids at room temperature (only mercury is an exception), shiny, good conductors of heat and electricity, and are easily hammered into sheets or drawn into wire. Metals in Table 1.1 and on the periodic table inside the back cover are unshaded. Metals are used in such things as cooking ware, tools, cars, and bridge and building frameworks. *Nonmetals* such as carbon, sulfur, and iodine are brittle solids; bromine is a liquid; and some nonmetals are gases at room temperature. All are poor conductors of heat and electricity. Nonmetals in Table 1.1 and on the periodic table inside the back cover are shaded. Nonmetals are present in

Table 1.1 Symbols of Some Common Elements

Element	Symbol	Percent by Mass Natural	Percent by Mass Human	Element	Symbol	Percent by Mass Natural	Percent by Mass Human
Aluminum	Al	7.5		Lead	Pb		
Argon	Ar			Magnesium	Mg	1.9	
Arsenic	As			Mercury	Hg		
Barium	Ba			Nickel	Ni		
Bromine	Br			Nitrogen	N		3.1
Calcium	Ca	3.4	2.0	Oxygen	O	49.2	64.6
Carbon	C		18.1	Phosphorus	P		1.1
Chlorine	Cl		0.45	Potassium	K	2.4	0.37
Chromium	Cr			Silicon	Si	25.7	
Copper	Cu			Silver	Ag		
Fluorine	F			Sodium	Na	2.6	0.11
Gold	Au			Sulfur	S		0.25
Hydrogen	H	0.9	10.0	Tin	Sn		
Iodine	I			Titanium	Ti	0.6	
Iron	Fe	4.7		Zinc	Zn		

wood, glass, plastics, and clothing. Metals are found to the left and nonmetals to the right of the dark solid line through the periodic table inside the back cover. Four fifths of the elements are metals.

2. Compounds. A **compound** is *a pure substance that is formed when two or more atoms of the same element or of different elements combine.* Diatomic oxygen (Section 5.5-g) is a compound formed from two atoms of the *same* element. Like many compounds made up of atoms of a single element, diatomic oxygen is also the common form of the element oxygen. Table salt is a compound made up of a fixed proportion of atoms of two *different* elements and exhibits properties unlike those of its constituent elements. A pure substance composed of atoms of two or more different elements can be shown to be a compound by decomposing it to its constituent elements or by synthesizing it from its constituent elements. When an electric current is passed through molten table salt, the two elements, sodium and chlorine

(Section 5.5-i), are separated. Likewise, sodium and chlorine, when brought together, combine to produce sodium chloride (table salt—Section 3.4 and Figure 3.2). Compounds are classified in a great variety of ways including the nature of their constituent elements, their physical and chemical properties, and the nature of their chemical bonding.

Many compounds are composed of small distinguishable units called molecules. A **molecule** is *an electrically neutral collection of atoms bonded together strongly enough to be considered a recognizable unit.* Two nitrogen atoms bind together to form a molecule of nitrogen (Section 5.5-d). Two hydrogen atoms and two oxygen atoms bind together to form a molecule of hydrogen peroxide (a common antiseptic). One sulfur atom and two oxygen atoms bind together to form a molecule of sulfur dioxide (a common air pollutant that arises from the burning of sulfur-containing fuels—Section 5.5-h). Fifty-seven carbon atoms, 110 hydrogen atoms, and 6 oxygen atoms bind together to form a molecule of stearin (the main component of beef fat and a precursor in the preparation of soaps). A molecule exhibits distinctive characteristics that may not be related in any obvious fashion to the properties of the atoms comprising the molecule.

Just as chemical symbols are used to signify elements, chemical formulas are used as abbreviations for compounds. A **formula** *conveys information about the composition and structure of a compound.* An empirical formula is the simplest kind of formula. It gives only a minimum of information about the composition of a substance and says nothing about its structure. An **empirical formula** states only *the* ***relative*** *numbers of atoms in a compound.* By convention an empirical formula includes symbols of the elements making up a compound and subscripts affixed to the symbols to designate the relative numbers of atoms. Thus, the empirical formula of hydrogen peroxide is HO, meaning that there is one atom of hydrogen for each atom of oxygen in the compound.

A **molecular formula** states *the* ***actual*** *numbers of atoms in a molecule,* but says nothing about its structure. A molecular formula includes symbols of the elements making up a molecule with subscripts affixed to the symbols to designate the actual numbers of atoms. The molecular formula of hydrogen peroxide is H_2O_2 in contrast to its empirical formula, HO. The molecular formula of sulfur dioxide is SO_2 and of stearin is $C_{57}H_{110}O_6$, the same as their respective empirical formulas.

A **structural formula** gives a maximum amount of information about a molecule. It gives *the actual numbers of atoms in a molecule, indicates which atoms are bonded to each other,* and *may indicate the three-dimensional configuration of the molecule.* Figure 1.3a indicates which atoms are bonded to each other in the hydrogen peroxide molecule; Figure 1.3b represents the three-dimensional configuration of the hydrogen peroxide molecule, which is illustrated in the space-filling models of Figure 1.3c and 1.3d. In Figure 1.3b a wedge is used to indicate a bond coming out of the paper toward you. Details of bonding and molecular structure are discussed in Chapter 3.

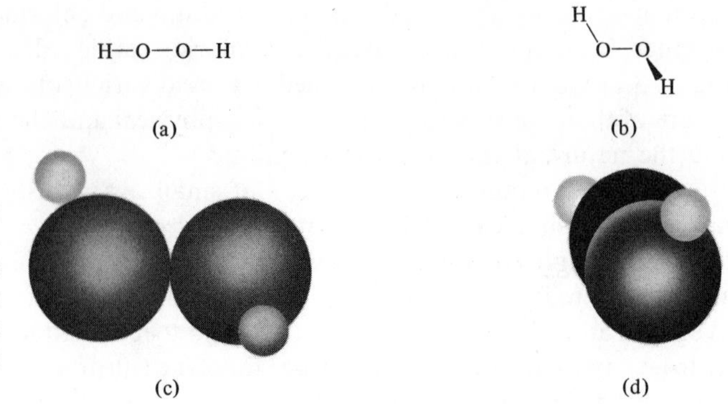
H—O—O—H
(a)
(b)
(c)
(d)

Figure 1.3 Structural formulas (a, b) and space-filling models (c, d) for hydrogen peroxide, H_2O_2. The O—H groups form an angle of about 95° when viewed along the O—O bond (d). Larger spheres represent oxygen atoms; smaller spheres represent hydrogen atoms.

e. Mixtures

Many of the materials that we encounter are **mixtures** of substances; that is, they *contain two or more elements and/or compounds* not bound to each other and not in any fixed proportions. Air is a mixture consisting primarily of nitrogen gas and oxygen gas, but also containing small amounts of water vapor, carbon dioxide, and other substances. The composition of a mixture may be variable, sometimes over a wide range. The amount of water vapor in air changes over a relatively narrow range, whereas the composition of a mixture of sand and table salt can be varied continuously in any ratio desired to produce an infinite number of compositions.

A given mixture can be studied in great detail and can be described further as a solution or as a heterogeneous mixture. Any given **solution** has *a constant composition, structure, and physical and chemical behavior.* Given quantities of solid sugar dissolved in water, liquid ethyl alcohol dissolved in water, or hydrogen chloride gas dissolved in water form three different solutions of constant composition.

Portions of **heterogeneous mixtures** *exhibit different compositions, structures, and physical and chemical properties, so that abrupt discontinuities or boundaries may be observed.* Thus, a glass filled with iced tea and ice cubes (Figure 1.4a) contains a heterogeneous mixture since there is an observable boundary between the liquid solution and solid ice. There may also be undissolved solid sugar in the bottom of the glass (Figure 1.4b). Heterogeneous mixtures may be formed by composites of elements and/or compounds. Thus, a mixture of magnesium turnings and powdered copper (Figure 1.4c) is heterogeneous and the distinct fragments of each metal can be perceived easily. Likewise, a sealed flask containing sand, water, and air (Figure 1.4d) holds a heterogeneous solid-liquid-gas mixture.

The term **phase** is often applied to *a mechanically separate, homogeneous portion of matter. Any solution is,* therefore, *a one-phase system.* Natural gas, a solution of a number of compounds containing carbon and hydrogen, is a one-phase system. All gas mixtures are homogeneous. *A heterogeneous mixture contains two or more phases.* Thus, a sealed wine bottle contains two phases: the wine which is itself a solution, and the air above the wine which is also a solution. If solid cream of tartar precipitates in the bottom of the bottle, then a third phase is present. If a piece of the cork stopper tears loose and falls into the bottle, then a fourth phase is present (two of the phases are *separate* solid phases).

f. Separation of Substances

When carrying out a chemical reaction in the laboratory, one frequently encounters a mixture of two or more substances. The task then is to separate the mixture into its components, hopefully without converting the substances into new ones and losing the desired product.

Two-component mixtures are easiest to resolve into their components but the same principles and techniques apply to mixtures with more than two components. If a two-component mixture is heterogeneous, that is, has two phases, the separation may be relatively easy. One of the components can be simply removed physically from the other by methods such as decantation

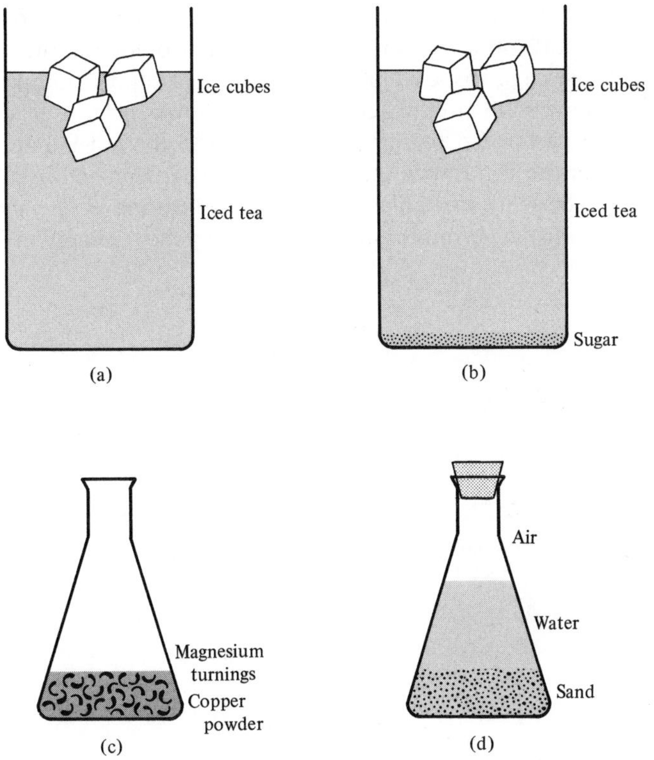
Ice cubes
Iced tea
(a)
Ice cubes
Iced tea
Sugar
(b)
Magnesium
turnings
Copper
powder
(c)
Air
Water
Sand
(d)

Figure 1.4 Examples of heterogeneous mixtures. (a) Ice cubes and iced tea (two phases); (b) ice cubes, iced tea, and solid sugar (three phases); (c) magnesium turnings and powdered copper (two phases); (d) air, water, and sand (three phases).

and filtration. If a two-component mixture is homogeneous, that is, has only one phase, somewhat more sophisticated techniques such as crystallization, distillation, sublimation, extraction, or chromatography are required.

1. Decantation. When one cleans his fish tank, he first removes the fish, then carefully pours the water from the tank, leaving behind most of the sand, shells, and plants. When you drink your coffee, you sip more carefully near the bottom of the cup, leaving the small residue of coffee grounds behind in the cup. These are examples of decantation. **Decantation** is *a gentle pouring off of a liquid without disturbing a solid sediment.* Decantation is useful for separating a liquid from a solid that has settled to the bottom of a container. A special form of decantation is utilized to separate two immiscible liquids (for example, motor oil and water) using a separatory funnel (Figure 1.5). After the separatory funnel has been left standing and the two immiscible liquids have separated, the lower layer (the liquid having the greater density or mass per unit volume) is drawn off through a stopcock into another container, thus providing a separation.

2. Filtration. Primary treatment in a sewage disposal plant involves passage of sewage through coarse screens to remove pieces of wood, metal, rags, garbage, and other sizable materials that would otherwise damage pumps in the sewage disposal plant. In some coffeepots boiling water is forced up a hollow stem after which it percolates down through ground coffee contained in a perforated pan and then returns to the pot below. In many vacuum cleaners air is pumped through a porous bag that acts as a filter and collects dust. These are examples of filtration. **Filtration** is *the passage of a liquid or gas through a porous substance for the purpose of removing suspended*

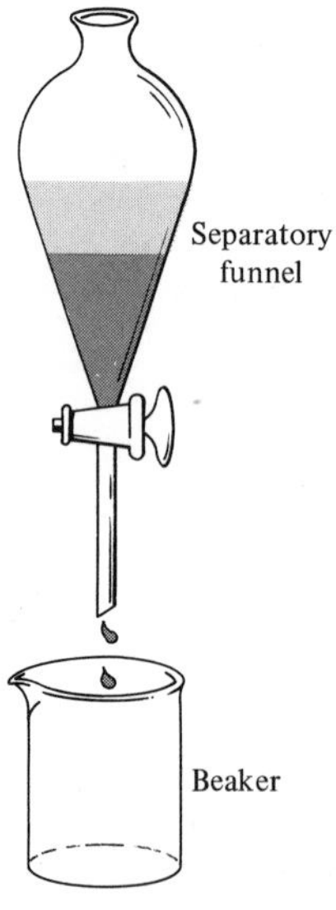
Separatory
funnel
Beaker

Figure 1.5 Separating two immiscible liquids. The more dense liquid forms the lower layer and can be drawn off through the stopcock.

solids. Filtration is useful for separating solids from liquids as in the first two cases above or solids from gases as in the third case above.

*3. **Crystallization.*** Many minerals in the earth's crust were probably formed by crystallization under appropriate conditions from liquid mixtures. Calcite (calcium carbonate) occurs as limestone, a common sedimentary rock, and as marble, a common metamorphic rock. Calcite is relatively insoluble in water, but is appreciably soluble in water containing carbon dioxide that is picked up as water percolates through soil. Stalactites and stalagmites in limestone caves form due to the deposition of calcite from ground water as carbon dioxide escapes into the atmosphere. This is an example of crystallization, and takes advantage of the change in carbon dioxide concentration as ground water reaches the earth's surface.

*4. **Distillation.*** Ordinary tap water should not be used in a steam iron because as the water is heated and converted to steam, deposits of salt are left behind. These salts react chemically with the interior, corroding the iron and eventually closing the steam holes so that the iron no longer has any steaming capability. This is an *undesirable* separation of substances in a solution by distillation. **Distillation** *involves evaporation and subsequent condensation of a substance,* and is useful for separating two substances (gases, liquids, or solids) in a solution that have quite different boiling points. Steam iron manufacturers suggest purchasing distilled water that has already been freed from trace quantities of dissolved salts as a way to avoid the above problem. The basic apparatus and techniques utilized for distilling water are illustrated in Figure 1.6. The water is heated and boiled, the vapor is then condensed to the liquid state by passing it through a condenser in which cold water is circulated, and the distilled water is collected in a receiver flask. Meanwhile, the nonvolatile salts originally in the natural water remain in the distillation flask.

On ships distillation is the method for converting readily available seawater to fresh water for drinking and for the operation of steam boilers. Similar techniques are being utilized for production of fresh water from seawater for use by coastal cities. This technique is not new: Aristotle (384–322 B.C.) wrote that fresh water was made by the distillation of seawater.

Nature provides a cycle involving distillation that is very important to us. Water evaporates from the oceans into the air. When moist air cools, the water vapor condenses. Fog or stratus clouds (high fog sheets) form if condensation occurs close to the earth. Altostratus or altocumulus clouds form if condensation occurs higher in the atmosphere. Cirrus clouds composed of ice crystals form if condensation occurs high enough that the temperature is below freezing. The cycle is completed when moisture returns to the oceans as dew, rain, or snow.

*5. **Sublimation.*** If dry ice (solid carbon dioxide) is left exposed at room

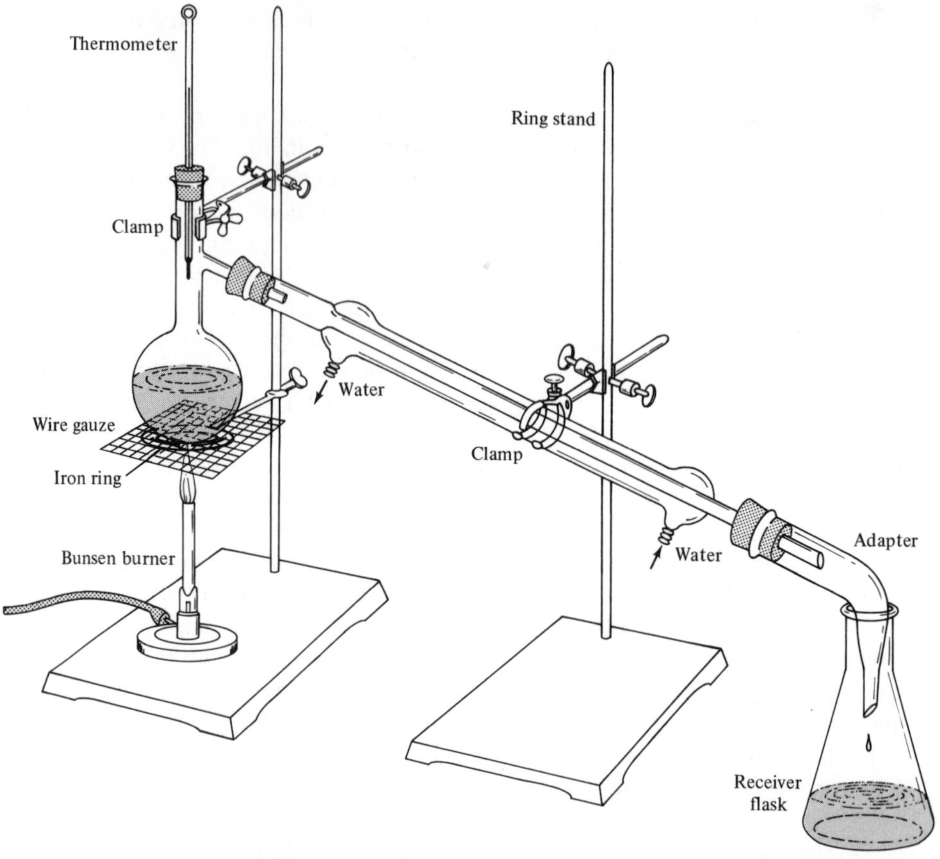
Thermometer
Ring stand
Clamp
Water
Wire gauze
Iron ring
Clamp
Bunsen burner
Water
Adapter
Receiver
flask

Figure 1.6 The distillation of water. Water in the round bottom flask is heated and boiled; water vapor is condensed to liquid water by passing it through a condenser in which cold water is circulated; and the distilled water is collected in a receiver flask.

temperature, it seems to disappear. Actually it has been converted directly from the solid to the gas state. If a few purple crystals of iodine are placed in an evacuated bottle at room temperature, the purple color of iodine vapor soon forms above the crystals and eventually fills the bottle. These examples illustrate **sublimation,** which is *the direct conversion of a solid to a gas without passing through the liquid state.* If an ice cube is held against the side of the evacuated bottle containing iodine crystals, the purple crystals in the bottom of the bottle begin to disappear and purple crystals form on the side of the bottle at the ice cube. This illustrates sublimation followed by crystallization. This is a common method for separating a volatile solid from a nonvolatile solid, for example, separating a mixture of iodine and table salt. The mixture is placed in an evacuated bottle and an ice cube is held against the side of the bottle. Iodine crystals will sublime and crystallize on the side of the bottle whereas the less volatile table salt will remain on the bottom of the bottle. Thus, a separation is achieved.

6. Extraction. Cookbooks recommend adding a pared and cut potato to a stew that is too salty. The potato absorbs salt from the stew and makes the dish more palatable. This is an example of extraction. **Extraction** is *the withdrawal or forcing out of a substance by chemical or mechanical action.* Extraction is one method of removing perfumes from roses. Rose petals are immersed in hot petroleum ether. Heat ruptures the petal cells and the discharged perfumes are absorbed by the petroleum ether. The remains of the petals are removed by filtration; then the petroleum ether is distilled leaving behind the perfume components.

7. Chromatography. If a solution of color pigments from leaves is passed through a vertical glass tube containing activated charcoal, and if more solvent is poured slowly through the column, a separation of the pigments into bands, called a chromatogram, is obtained by the time they pass from the bottom of the column. This is an example of liquid *chromatography.* Interpretation of the results of such an experiment in 1903 allowed Michael Tswett, a Russian botanist, to elucidate the basic principles of adsorption chromatography. A solution is poured into a chromatographic column containing a packing that has a greater affinity for one or more components in the solution. This packing material is referred to as the stationary phase. The mixture is adsorbed by the stationary phase and held at the top of the column. When a second moving phase is passed slowly through the column, the components of the mixture are subjected repeatedly to extraction by the mobile phase and adsorption on the stationary phase. The more strongly a component is adsorbed to the stationary phase and the less soluble it is in the mobile phase, the slower it moves through the column. Thus, a separation is achieved and pure zones of substances are collected at the bottom of the column. Various types of chromatography have been developed: adsorption, gas or vapor phase (VPC), ion exchange, partition or paper, and thin-layer (TLC). All utilize the same basic principles, but differ in the nature of their stationary and mobile phases and in the kinds of substances they can separate most effectively. Both gas and thin-layer chromatography require only micrograms of a sample (1 microgram = 0.000001 gram, or about $\frac{1}{100}$ of the mass of a grain of sugar). Gas chromatography is more capable of separating similar substances than any other method presently known.

Various methods for separating substances from mixtures have been discussed individually. In fact, these methods are often used in conjunction with each other as illustrated in the basic procedures for production of sugar from sugar beets. The beets are first placed in a perforated, tumbling washer that selectively *filters* dirt and small stones from the larger beets. The beets are then sliced into chips and placed in a heated tank where hot water *extracts* sugar from the beet chips. The sugar solution is purified by the addition of lime which precipitates impurities so that they can be separated by *filtration.* Sulfur dioxide is added to bleach residual coloring matter and to precipitate excess lime so that it can be separated by *filtration.* Water is *distilled* under vacuum from the sugar solution so that the sugar slowly *crystallizes*. The sugar crystals are then *filtered* in porous centrifugal baskets. Even at this point adsorption *chromatography* indicates the presence of a number of sugars (with sucrose being the dominant one) as well as small amounts of other impurities.

g. Identification of Substances

Obtaining a pure substance is of little use if one cannot be certain what the substance is. A known substance that has been characterized previously may be relatively easy to identify. One can compare certain of its properties (its fingerprints) with those of a known substance just as the FBI tracks down criminals by comparing observed fingerprints with known fingerprints in its files.

Most substances are *analyzed* to determine what chemical elements are present and how much of each element is present. An empirical formula is then calculated from the analytical data (Section 4.11-c). A solid is often identified by its *melting point*—that temperature at which the solid converts to a liquid. A liquid is often identified by its normal *boiling point*—that temperature at which the liquid converts to a gas at one atmosphere pressure (a pressure equal on the average to the atmospheric pressure at sea level, and marginally greater than our usual atmospheric pressure). A liquid may also be identified by its *refractive index*—a measure of the angle to which visible light is bent or refracted on entering the liquid. A gas is frequently identified by measuring the mass of a given volume of the gas and calculating its *vapor density* (mass divided by volume).

1.2 Energy

Energy (Greek, *en,* in; *ergon,* work) is *the capacity to do work* or to move an object through a distance against a force. Energy changes always accompany chemical changes; thus, the concept of energy is fundamental in chemistry. When your leg muscles act as you pedal a bicycle up a hill against both frictional forces and the force of gravity, energy is expended. Your leg muscles possess energy and are able to do work causing the bicycle pedals to move; the pedals move the crank and chain, causing the wheels to move; the wheels possess energy and are able to do work causing the bicycle to move; and the bicycle possesses energy and is able to do work causing your body to move.

a. Kinetic Energy

When a bicycle is in motion, it possesses energy that it does not have when it is standing still. *Energy of motion* is called **kinetic energy.** Any mass in motion possesses kinetic energy. Particularly in relation to equipment with moving parts, kinetic energy is often called *mechanical energy*. Kinetic energy is manifested in a number of forms that differ only in the nature of the particles in motion.

1. Electrical energy. If the matter in motion is a stream of electrons flowing through a conductor and giving rise to an electric current, the energy is called electrical energy. Electrical energy causes some chemical reactions to occur, and electrical energy can be produced by other chemical reactions. Electrical energy can be converted easily to light as in the case of the incandescent bulb. The electrical nature of nerve impulses is of particular interest to biochemists.

2. Heat. If the movement involves random thermal motion of atoms or groups of atoms in a substance, the energy is called heat. Heat increases as matter becomes hotter and its atoms move more rapidly. Heat and temperature are related, but should be distinguished clearly. **Temperature** is *a measure of the* ***intensity*** *of heat,* but specifies nothing about quantity of heat. Thus, a burning match and a bonfire may both be at the same temperature, but the bonfire certainly gives off more heat.

Temperatures near room temperature are usually measured by following the expansion of mercury with increased temperature. The mercury is contained in a glass thermometer (Figure 1.7a) consisting of a terminal bulb surmounted by a calibrated capillary tube from which the temperature is read.

Temperature is measured in units called *degrees* using one of three temperature scales. The *Fahrenheit* scale (°F), used commonly in the United States, is based upon 0°F as the freezing point of a solution containing a maximal amount of table salt dissolved in water and is based upon 100°F as body temperature. On this scale, the freezing point of water is 32°F and the boiling point of water is 212°F. Each Fahrenheit degree is $\frac{1}{180}$ of the temperature interval between these two points. The *Celsius* scale (°C), used commonly in chemistry and currently being promoted in the United States, is based upon the freezing point of water as 0°C and the boiling point of water as 100°C.[1] Each Celsius degree is $\frac{1}{100}$ of the temperature interval between these two points. It follows that a Celsius degree is 1.80 times as large as a Fahrenheit degree and that the Celsius freezing point of water is 32° lower than the Fahrenheit value. Thus, Fahrenheit temperatures (°F) and Celsius temperatures (°C) can be converted to each other using the following equations.

$$°F = 1.80°C + 32° \qquad °C = 0.556(°F - 32°)$$

Certain chemical calculations, particularly those involving gases, require the

[1] This is also frequently called the centigrade scale because of the 100-degree interval.

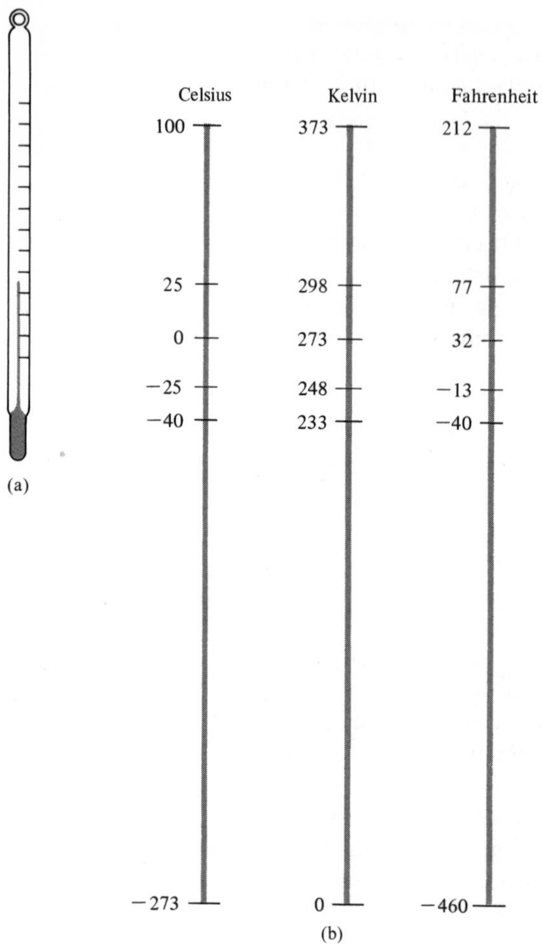
(a)
Celsius
Kelvin
Fahrenheit
100
373
212
25
298
77
0
273
32
−25
248
−13
−40
233
−40
−273
0
−460
(b)

Figure 1.7 (a) Mercury thermometer (−10° to 110°, Celsius scale). (b) Relationship of three temperature scales.

use of *absolute* temperatures (suffixed with a K for Kelvin that does *not* require a superscript degree sign before it). The size of each degree on the absolute scale is identical to a degree on the Celsius scale, but each absolute temperature is 273.16° higher than its corresponding Celsius temperature. Thus, Celsius temperatures (°C) and absolute temperatures (K) can be converted to each other using the following equations.

$$K = {}^\circ C + 273.16^\circ \qquad {}^\circ C = K - 273.16^\circ$$

The relationship of the three temperature scales is shown in Figure 1.7b. Unless stated otherwise, Celsius temperatures will be used in this book and frequently the C will be omitted.

Example 1.1 Calculate the Celsius and absolute temperatures that are equivalent to 64°F, room temperature in many homes.

First convert °F to °C.

$$°C = 0.556(°F - 32°) = 0.556(64° - 32°) = 18°$$

Then convert °C to K.

$$K = °C + 273.16° = 18° + 273.16° = 291 \text{ K}$$

The unit used most commonly as a measure of the *quantity* of heat is the calorie (cal). One **calorie** is *the amount of heat required to raise the temperature of one gram of water one degree Celsius,* or more specifically, from 14.5° to 15.5°. This particular temperature range is specified because the heat required is slightly different at different temperatures. Heat must be added to raise the temperature of a substance whereas heat must be withdrawn to lower the temperature of a substance. Heat flows from a warmer body to a colder body. Thus, when you dip your hand into cold water, heat is added to the water and its temperature increases while heat is simultaneously removed from your hand and its temperature decreases. Since the calorie is a very small unit of energy, the kilocalorie (equal to 1000 calories and abbreviated kcal, or Cal in nutrition) is frequently used to indicate energy lost or gained in chemical reactions or to indicate energy available in food products.

3. Radiant energy. If the movement involves photons, or packets of electromagnetic radiation, the energy is called radiant energy. If the temperature of the heating coils in an oven is increased gradually, invisible radiation called *infrared radiation* (of lower energy than visible light) is emitted. As the temperature is increased further, visible rays called *light* are emitted. At very high temperatures, as in the sun, invisible *ultraviolet radiation* (of higher energy than visible light) is emitted. Infrared radiation, light, and ultraviolet radiation are forms of radiant energy.

4. Sonic energy. If the movement involves the cooperative motion of molecules producing audible vibrations in air, the energy is called sonic energy. The electrical discharge of lightning causes sudden heating and expansion of air around it. Cooling and contraction follow quickly, creating vibrations in the air that result in thunder. Sound waves, caught by your external ear flaps, set your eardrums in vibration and pass the oscillatory movements through the intermediate segments of your middle and inner ears. This example is a manifestation of *sound.*

Sound will not travel in a vacuum because of the absence of molecules to undergo cooperative vibration. Thus, if a ringing doorbell is placed in a container that is gradually evacuated, the ringing becomes less intense as evacuation proceeds until finally you cannot hear the bell ringing even though you can still see the clapper beating against the bell. It should be noted that few of the sounds in the battles in the vacuum of outer space in the movie "Star Wars" would actually be audible; however, the battles would be dull and unemotional scenes without the added sound effects since we are not accustomed to life in a vacuum.

b. Potential Energy

When an automobile is at rest near the top of a hill, it has no kinetic energy, but it does possess energy of position that it does not have at the bottom of the hill. This energy is converted easily to kinetic energy by releasing the brake. When a bow is held motionless in its bent form, it has no kinetic energy, but it possesses energy of configuration that it does not have in its normal shape. Releasing the string imparts kinetic energy to an arrow. A mixture of hydrogen gas and oxygen gas possesses energy because, under appropriate conditions, these gases react explosively to produce sound, heat, light, and mechanical energy. These examples are manifestations of potential energy. **Potential energy** is *that energy which matter has stored in it by reason of its relative position, configuration, or chemical composition.*

Chemical energy and nuclear energy result when there is a difference in the potential energy of reactants and products in a chemical reaction or in a nuclear reaction. When an electric current is passed through water, hydrogen gas and oxygen gas are liberated in a chemical process called electrolysis. When calcium hydroxide is dehydrated by heating, water and calcium oxide are formed. These are endothermic processes. An **endothermic** reaction is a *reaction that absorbs energy* or one in which the potential energy of the reactants is less than that of the products. Many endothermic reactions are responsible for the building of tissues in the human body. Through an appropriate reaction a lightning bug emits light. If charcoal is burned in air, carbon dioxide is formed and heat is liberated. These are exothermic processes. An **exothermic** reaction is a *reaction that releases energy* or one in which the potential energy of the reactants is greater than the potential energy of the products.

Various forms of energy have been discussed individually under the classifications of kinetic energy and potential energy. The importance of these forms in everyday life arises primarily from the relatively easy transformation of one form of energy into another. For example, the potential energy of water held behind a dam is converted to the mechanical energy of rushing water. The rushing water moves blades in a generator and allows conversion of mechanical energy to electrical energy. Electric current is passed through a coil in a toaster to produce heat, through a filament in an incandescent bulb to produce light, through an electrolytic cell to produce chemical energy in the electrolysis of water, through an electric motor to convert the electrical energy back to mechanical energy, or through a circuit of a doorbell that under appropriate conditions operates a mechanical clapper that produces sound. Another example of energy transformation involves our primary source of energy, the sun. Heat and light are radiated from the sun. Heat provides the necessary warmth for proper functioning of our bodies. Light is absorbed by green plants and is transformed to stored chemical energy by photosynthesis, involving the rearrangement of atoms to form new chemical compounds such as sugar. We ingest this stored chemical energy through the foods we eat. Over centuries plant life is converted to natural fuels such as coal, petroleum, and natural gas. The chemical energy of these natural fuels is transformed by combustion to heat for our homes. Heat converts water to steam that drives turbines to generate electrical energy.

1.3 Measurements

The basic principles of chemistry have been deduced from the interpretation of results from experiments carried out under well-defined conditions. These results became meaningful only as they became reproducible. Results can be reproduced only when the experimental conditions have been described in sufficient detail so that they can be repeated exactly. Experimental conditions and results require both qualitative and quantitative description, that is, the questions as to both *what* and *how much* must be answered.

Accurate and precise measurements as well as a standard numerical form and system of units are required to describe the properties of matter quantitatively. Exponential numbers, the metric system, and the significance of the measurement of selected properties will be described.

a. Exponential Numbers

Numbers that are either very large or very small are encountered frequently in chemical measurements and problems. It is most convenient to express such numbers in exponential form involving a coefficient, a number between one and ten, times an exponential, a whole-number power of ten.

To write a number greater than one in exponential form, place a decimal point to the right of the first digit of the number and multiply the result by 10 raised to a positive power equal to the number of digits to the right of this decimal point. Thus, $83{,}417 = 8.3417 \times 10^4$ since there are four digits (3417) to the right of the decimal after the decimal is placed to the right of the first digit in the number. As an additional example, $538{,}000{,}000{,}000 = 5.38 \times 10^{11}$.

To write a number less than one in exponential form, place a decimal point to the right of the first nonzero digit of the number and multiply the result by 10 raised to a negative power equal to the number of digits to the left of this decimal point. Thus, $0.000639 = 6.39 \times 10^{-4}$ since there are four digits (0006) to the left of the decimal after the decimal is placed to the right of the first nonzero digit in the number. As an additional example, $0.0000000003468 = 3.468 \times 10^{-10}$.

Reversing the process and converting numbers from exponential form to nonexponential form is accomplished by moving the decimal point to the right a number of places equal to the positive exponent of 10 or to the left a number of places equal to the negative exponent of 10, adding zeroes where necessary. Thus, $2.48 \times 10^{7} = 24{,}800{,}000$ after moving the decimal point seven places to the right, and $2.48 \times 10^{-7} = 0.000000248$ after moving the decimal point seven places to the left.

b. Metric System

Commercial operations and the general public in the United States use the unwieldy British foot-pound system (Great Britain no longer uses it). However, the metric system is used commonly in many other countries, is currently being promoted in the United States, and has become the standard system for scientists. In the metric system the common units of length, volume, and mass are the meter, the liter, and the gram, respectively. The main advantage of the metric system is that subdivisions or multiples of its basic units are in factors of ten and are expressed using suitable prefixes attached to the basic unit. Common prefixes are

mega-	one million (1,000,000 or 1×10^{6})
kilo-	one thousand (1000 or 1×10^{3})
centi-	one-hundredth (0.01 or 1×10^{-2})
milli-	one-thousandth (0.001 or 1×10^{-3})
micro-	one-millionth (0.000001 or 1×10^{-6})
nano-	one-billionth (0.000000001 or 1×10^{-9})
pico-	one-trillionth (0.000000000001 or 1×10^{-12})

You are already familiar with conversion from one unit to another by division or multiplication by ten or powers of ten through use of our decimal monetary system. Thus, \$.32 = 32 cents, and \$2.32 = 232 cents. Similar simple procedures are required with the metric system.

1. Length. The fundamental unit of length in the metric system is the *meter* (m). The meter was defined originally as the length between two marks on an inert platinum–iridium reference stick kept at Sèvres, France. A meter is slightly longer than three feet. Conversion factors between various units of length in the metric system and the British system are given below.

1 kilometer (km)	= 1000 meters (m)	= 0.6214 mile (mi)
	1 meter	= 39.37 inches (in.)
1 centimeter (cm)	= 0.01 meter	= 0.3937 inch
1 millimeter (mm)	= 0.001 meter	= 0.03937 inch
1 micrometer (μm)	= 0.000001 meter or 1×10^{-6} meter	
1 nanometer (nm)	= 0.000000001 meter or 1×10^{-9} meter	
1 Ångström (Å)	= 0.0000000001 meter or 1×10^{-10} meter	

Another commonly used conversion factor is 2.54 cm = 1 in. Lengths commonly measured in the laboratory using a meter stick are expressed in millimeters or centimeters.

2. ***Volume.*** The common unit of volume in the metric system is the *liter*. It is slightly larger than a quart (1 liter = 1.057 quarts). The milliliter (ml), equal to 0.001 liter, is used for small volumes, especially for liquids and gases. The cubic centimeter (cm^3), exactly equal to 1 ml, is also used frequently, particularly for solids.

Volumes of sizable solids, depending on their shapes, can be determined by measuring lengths or radii, or by displacement of measured volumes of liquid. Volumes of liquids are measured in graduated cylinders, burets, transfer pipets, measuring pipets, or volumetric flasks (Figure 1.8). These vessels are designed to contain or deliver known, calibrated volumes. Volumes of gases are measured in known, calibrated, closed volumes.

3. ***Mass.*** The standard unit of mass in the metric system is the kilogram; however, the *gram* (g) is used more commonly. The gram is defined as one thousandth of the mass of an inert, platinum–iridium, standard kilogram block kept at Sèvres. A gram is slightly less than half the mass of a dime. The common conversion factor between the British and metric systems is 1 pound = 453.6 grams. Mass is always used to specify the amount of a solid because the volumes of solids vary markedly with shape, state of subdivision, and packing of the particles. Thus, you would not describe the size of your

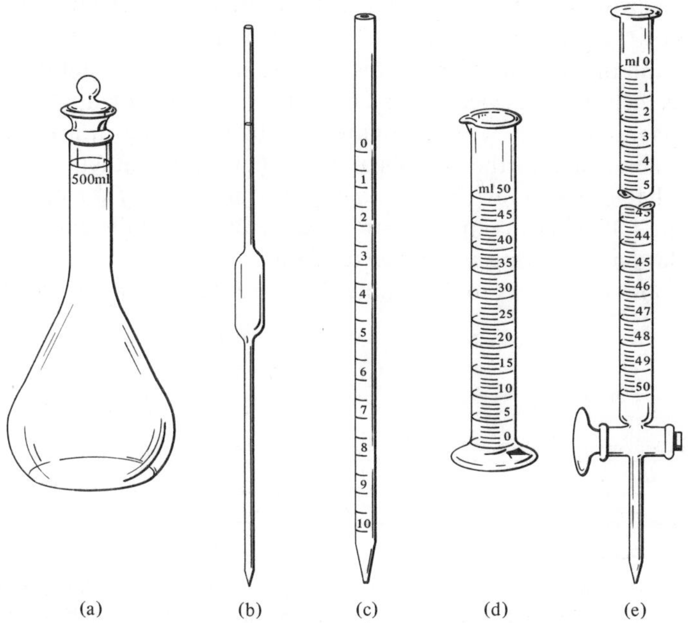

500ml
0
1
2
3
4
5
6
7
8
9
10
ml 50
45
40
35
30
25
20
15
10
5
0
ml 0
1
2
3
4
5
44
45
46
47
48
49
50
(a)
(b)
(c)
(d)
(e)

Figure 1.8 Vessels for measuring volumes of liquids. (a) Volumetric flask; (b) transfer pipet; (c) measuring pipet; (d) graduated cylinder; (e) buret.

baby sister in liters because her irregular shape makes it difficult to determine her volume. A fine powder occupies less volume than the same mass of chunks of the same solid because the powder leaves much less unfilled air space. Degree of packing is illustrated by the prevalence of the following statement on cereal boxes: "This package is sold by weight, not by volume; some settling of contents may have occurred during shipping and handling." Mass is measured by means of a chemical balance (Figure 1.1c and 1.1d).

c. Significant Figures

Some numbers are pure numbers and are *exact.* Thus, one can speak of 7 pencils or 3 cats. These numbers are not subject to the uncertainties of measurement. Such is not the case of many experimental observations.

The concept of significant figures is designed to indicate the accuracy or reliability of a measurement, and to provide the maximum amount of information with no misinformation. By convention, **significant figures** in a number include *all digits known with certainty and one doubtful or approximated digit.* Significant figures can be illustrated by the example depicted in Figure 1.9. If one measures the width of a piece of business-sized paper ($8\frac{1}{2}$ inches) using a stick one meter in length but with no calibrations, even the first digit is an approximation and one estimates a length of 0.2 meter. If one measures the same width with a meter stick calibrated in 0.1 meter intervals, one observes that the page is between 0.2 and 0.3 meter wide. Thus, 0.2 is no longer doubtful and one can estimate one more digit. The width is then 0.22 meter. Likewise, if one uses a meter stick calibrated in 0.01 meter intervals, the width is 0.222 meter. The first value has one significant figure, the second has two, and the third has three. This discussion also shows that the calibration of a measuring device is very important in terms of how accurate a measurement is and how many figures can be expressed in a result.

A general rule for the use of significant figures in mathematical operations is that *calculations cannot improve the precision of an experimental result.* The following example illustrates some of the pitfalls that often arise in calculations involving experimental data.

Suppose one wants to determine the density of a cube of wood. Since

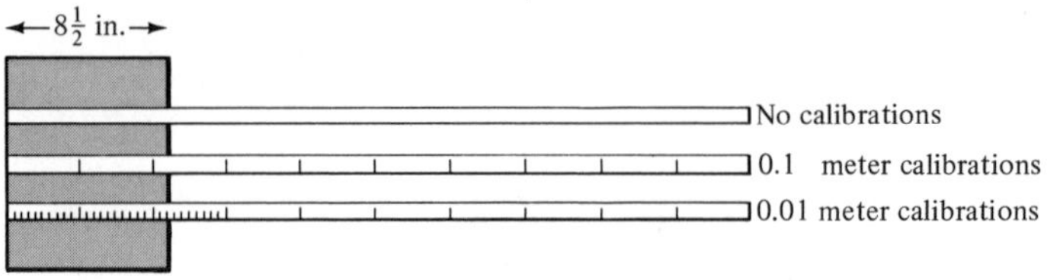

$8\frac{1}{2}$ in.
No calibrations
0.1 meter calibrations
0.01 meter calibrations

Figure 1.9 Measurement of length illustrating significant figures. Lengths from top to bottom are 0.2, 0.22, and 0.222 meter, representing one, two, and three significant figures, respectively.

density is equal to mass divided by volume, one must determine the mass of the cube, and then measure a side of the cube so that he can calculate its volume. One weighs the cube and finds it has a mass of 0.6448 gram. One measures a side of the cube and finds a length of 9.8 mm or 0.98 cm. The volume of a cube of side s can be expressed by the following equation.

$$V = s^3 = (0.98 \text{ cm})^3 = 0.9604 \text{ cm}^3$$

This is the answer one obtains by exact multiplication. Since density = mass/volume,[2]

$$\text{density} = \frac{0.6448 \text{ g}}{0.9604 \text{ cm}^3} = 0.671386922\ldots \text{ g/cm}^3$$

Suppose that one stops after obtaining nine digits. This implies that one can calculate a density with very high precision by using data of only moderate precision. This implication is contrary to the general rule that calculations cannot improve the precision of an experimental result, and is incorrect. The result is correctly reported as 0.67 g/cm^3.

1. Addition and Subtraction. By applying the above rule to addition and subtraction, one finds that the last digit obtained in a sum or difference should correspond to the first doubtful decimal place in any of the added or subtracted numbers.

Example 1.2 What is the mass of a flask and stopper if the flask, weighed on a platform balance, is 187.84 grams and the stopper, weighed on an analytical balance, is 7.3843 grams?

Determine the sum using all digits. Then round off to two places to the right of the decimal since one of the added numbers has its doubtful digit in the second place to the right of the decimal.

```
187.84|  g
  7.38|43 g
195.22|43 g rounded off to 195.22 g
```

Note how the units of the quantities used are involved in the mathematical operation. When grams (of flask) are added to grams (of stopper), the sum is expressed in grams.

[2] A diagonal slash / is used throughout this book preceding a unit in the denominator (the reciprocal of a unit). Thus, *grams per cubic centimeter* (grams per centimeter cubed) is written *g/cm*^{3}.

Example 1.3 Dry ice sublimes at 195 K. Convert 195 K to degrees Celsius, recalling that °C = K − 273.16°.

Perform the subtraction and round off to one place to the left of the decimal since one of the numbers has its doubtful digit one place to the left of the decimal.

$$\begin{array}{r|l} 195 & ^\circ \\ -273. & 16^\circ \\ \hline -\ \ 78. & 16^\circ \end{array} \text{ rounded off to } -78^\circ\text{C}$$

Note how the units of the quantities used are involved in the mathematical operation. When degrees are subtracted from degrees, the difference is expressed in degrees. Note also that any Celsius temperature below the freezing point of water will be a negative number, whereas an absolute temperature is always a positive number.

2. ***Multiplication and Division.*** By applying the above rule to multiplication and division, you will find that the result contains the same number of significant figures as the least number in the data used in the calculation.

Example 1.4 How many grams of breakfast cereal are in a box that contains 0.50 pound?

The conversion factor required is 1 lb = 453.6 g. Since one of the numbers entering the calculation has only two significant figures, the final result may have only two significant figures. Therefore, perform the multiplication and round off to two significant figures.

$$(0.50\ \cancel{\text{lb}})(453.6\ \text{g}/\cancel{\text{lb}}) = 226.8\ \text{g rounded off to } 230\ \text{g}$$

Note how the units of the quantities used are involved in the mathematical operation. When pounds, lb (of cereal), is multiplied by grams per pound, g/lb, lb cancels 1/lb (shown by cancellation lines here and in future examples), and the product is expressed in grams.

Note that the zero in 230 (to the left of the decimal) is not a significant figure, but only indicates the placement of the decimal. Use of exponential notation eliminates this ambiguity since the decimal point is indicated by a power-of-ten term. Thus, 230 g = 2.3×10^2 g. The exponential result shows clearly that there are two significant figures. If there were three significant figures, the exponential result would be written as 2.30×10^2 g.

As shown in the previous example, zeroes at the end of a number and at the *left* of the decimal may be significant or may only indicate the place of the decimal. However, zeroes at the end of a number and at the right of the decimal are always significant because they contribute to the precision of the number. Thus, 0.04300 has four significant figures. The zero to the left of the 4 is not significant because it only indicates the place of the decimal.

Example 1.5 What will be the volume of 40.1 g of rubbing alcohol if the density of rubbing alcohol is 0.7851 g/ml?

Density equals mass divided by volume, or $d = m/v$. Rearranging, volume equals mass divided by density, or $v = m/d$. Perform the division, and round off to three significant figures since one of the numbers entering the calculation (40.1) has only three significant figures.

$$\frac{40.1\ \cancel{g}}{0.7851\ \cancel{g}/\text{ml}} = 51.08\text{ ml rounded off to }51.1\text{ ml}$$

Note how the units of the quantities used are involved in the mathematical operation. When grams (of rubbing alcohol) are divided by grams per milliliter, g in the numerator cancels g in the denominator, and the quotient (volume) is expressed in reciprocal of 1/ml, or ml. *Always write down units and carry them through your mathematical operations.* Incorrect units in an answer tell you that you performed an incorrect mathematical operation. Thus, volume units of g^2/ml or 1/ml in Example 1.5 would indicate a wrong mathematical procedure because you know from its definition that volume must be expressed in milliliters.

Exercises

1.1 Define clearly the following terms.

(a) matter
(b) mass
(c) weight
(d) weightlessness
(e) atom
(f) pure substance
(g) element
(h) chemical symbol
(i) metal
(j) nonmetal
(k) compound
(l) molecule
(m) formula
(n) empirical formula
(o) molecular formula
(p) structural formula
(q) mixture
(r) solution
(s) heterogeneous
(t) phase
(u) decantation
(v) filtration
(w) crystallization
(x) distillation
(y) sublimation
(z) extraction
(aa) chromatography
(ab) melting point
(ac) boiling point
(ad) energy
(ae) kinetic energy
(af) mechanical energy
(ag) radiant energy
(ah) heat
(ai) temperature
(aj) Fahrenheit temperature scale
(ak) Celsius temperature scale
(al) absolute temperature scale
(am) calorie
(an) electrical energy
(ao) sonic energy
(ap) potential energy

(aq) chemical energy
(ar) endothermic
(as) exothermic
(at) exponential numbers
(au) metric system
(av) meter
(aw) centimeter
(ax) micrometer
(ay) liter
(az) milliliter
(ba) gram
(bb) kilogram
(bc) significant figures
(bd) density

1.2 Give two examples of matter in each of the physical states.

1.3 The Commission on Chemical Nomenclature (France, 1787) classified silica as a simple chemical substance (element) because early chemists could not decompose it. However, silicon, when it was discovered, was found to react with oxygen to form silica as the only product. Is silica an element or a compound? Why?

1.4 Classify each of the following as an element, a compound, or a mixture.
(a) air
(b) glass
(c) mercury
(d) table salt
(e) sugar
(f) sulfur
(g) water

1.5 Describe what each of the following formulas tells you.
(a) the molecular formula CO_2
(b) the empirical formula CH
(c) the molecular formula $C_2H_5NO_2$
(d) the structural formula 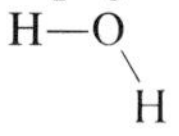

1.6 Suggest a specific example of each of the following.
(a) a one-phase system
(b) a two-phase system (solid and solid)
(c) a two-phase system (solid and liquid)
(d) a two-phase system (solid and gas)
(e) a two-phase system (liquid and liquid)
(f) a two-phase system (liquid and gas)
(g) a three-phase system (all solids)

1.7 Considering the general properties of metals and nonmetals, would you select a metal or a nonmetal for the following uses?
(a) a saucepan for cooking
(b) a mirror
(c) insulation for electrical wiring
(d) filler for a balloon
Explain your choices.

1.8 Distinguish between each pair.
(a) mass and weight
(b) heat and temperature
(c) element and compound
(d) distillation and sublimation
(e) centimeter and millimeter

1.9 State whether each of the following is a solution or is heterogeneous, and suggest an appropriate procedure for separation.
(a) the various hydrocarbon components in natural gas
(b) the components in a water solution of potassium bromide
(c) an immiscible mixture of ether and water
(d) the components of air
(e) a mixture of sand and water

1.10 Some of the flavor that peas have fresh out of the pod is lost on cooking. Comment on the separation processes that may take place.

1.11 Cite two specific examples for each conversion.
(a) kinetic energy to potential energy
(b) potential energy to kinetic energy
(c) one form of kinetic energy to another form of kinetic energy
(d) one form of potential energy to another form of potential energy

1.12 Express the following numbers in exponential form.
(a) 16,800,000
(b) 134,600
(c) 0.000000742
(d) 3,400,000,000,000
(e) 0.0001389

1.13 Express the following numbers in exponential form.
(a) 1,368
(b) 0.00635
(c) 0.0000000019
(d) 17,830,000
(e) 0.0000513

1.14 Convert the following exponential numbers to nonexponential form.
(a) 8.32×10^{9}
(b) 1.6×10^{-4}
(c) 5.48×10^{-7}
(d) 2.1×10^{13}
(e) 6.19×10^{-2}

1.15 How many significant figures are in each of the following numbers?
(a) 17.430
(b) 1.83×10^{-18}
(c) 0.00007
(d) 1064
(e) 0.068000

1.16 How many significant figures are in each of the following numbers?
(a) 68.4
(b) 0.0006
(c) 2.11×10^{6}
(d) 0.01300
(e) 1083

1.17 Perform the following mathematical operations noting carefully the correct number of significant figures and the units.
(a) (17.38 g) + (0.643 g) + (21.6 g)
(b) $(1.37 \times 10^{4}\ \text{g}) \div (168\ \text{cm}^3)$
(c) (348 ml) − (6.37 ml) + (18.34 ml)
(d) $(27\ \text{g/ml}) \times (16.4\ \text{ml}) \times (1.00 \times 10^{-3}\ \text{kg/g})$
(e) $(3.0\ \text{lb}) \times (453.6\ \text{g/lb})$

1.18 Perform the following mathematical operations with numbers in exponential form noting carefully the correct number of significant figures.
(a) $(7.38 \times 10^{4}) + (6.2 \times 10^{5}) + (4.1265 \times 10^{3})$
(b) $(1.317 \times 10^{1}) - (4.8 \times 10^{-1})$
(c) $(3.87 \times 10^{6}) \times (1.649 \times 10^{-8})$
(d) $(2.73 \times 10^{-7}) \div (3.1 \times 10^{-11})$
(e) $\dfrac{(6.14 \times 10^{4}) - (9.06 \times 10^{3})}{(5.16 \times 10^{-6})}$

1.19 Calculate the following.
(a) 84.1 mm = ? m
(b) 1.8 in. = ? cm
(c) 1.63×10^{7} m = ? mi
(d) 0.161 liter = ? cm^3
(e) 721 ml = ? pt
(f) 1.4 lb = ? mg
(g) 2.74×10^{6} mg = ? kg
(h) −196°C = ? °F
(i) 112°F = ? K
(j) 323 K = ? °C

1.20 How many Celsius degrees are there between 74 K and 74°F?

1.21 Magnesium melts at 651°C. What are the Fahrenheit temperature and the absolute temperature for the melting point of magnesium?

1.22 The diameter of a human hair is about 0.04 mm. What is this diameter in inches?

1.23 After you cross the Mexican border, you see a speed limit sign indicating 70 km/hr. How fast may you travel in miles per hour?

1.24 Calculate the densities of **(a)** diamond if a 0.386 gram crystal displaces 0.11 ml and **(b)** mercury if 20.0 ml weighs 272 grams.

1.25 What will be the total mass of a 58.36 gram beaker and contents after the addition of 12 ml of concentrated sulfuric acid (density 1.84 g/ml)?

2

Atomic Structure and the Periodic Table

Although there was much prior indirect evidence for the existence of atoms, nobody had ever seen an atom until a few years ago. Albert Crewe and Michael Isaacson, two American physicists, observed atoms and produced films showing their motion using an electron microscope that was capable of magnification by fifteen million times, enough to magnify a grapefruit to the size of the earth. These experiments provided additional evidence relating to the postulates of early Greek philosophers. As early as the first century before Christ, Lucretius, a Roman, described the discrete particle nature of matter, according to the theories of the Greek philosophers, such as Leucippus, Epicurus, and Democritus, in a poem, *De Rerum Natura* ("Concerning the Nature of Things"). Although later philosophers tried to discredit the atomic theory, the writings of René Descartes, Pierre Gassendi, Robert Hooke, and John Locke in the seventeenth century brought relatively wide acceptance of this idea. Still basing his thoughts only on intuition, Isaac Newton wrote in 1718:

> . . . it seems probable to me that God, in the beginning, formed matter in solid, massy, hard, impenetrable, movable particles, of such sizes and figures, and with such other properties, and in such proportion, as most conduced to the end for which he formed them. . . .

The eighteenth century saw the beginning of quantitative experiments and a change from chemistry as an art to chemistry as a science. The purpose of this chapter is to discuss some of the experiments that served as a basis (1) for the quantitative laws that led John Dalton to his atomic theory, (2) for developing the periodic table of the chemical elements, and (3) for defining atomic structure.

2.1 Law of Conservation of Mass

By 1774 Antoine Lavoisier, a French chemist who is regarded as the founder of modern chemistry, had performed careful quantitative experiments involving the reaction of tin with oxygen and had shown that within experimental error the masses of reactants and products were the same. These experiments are of little more than historical significance to you, in part because you can perform a simple experiment that is even more precise than the ones performed by Lavoisier. Consider the reaction of a magnesium filament (Mg) with oxygen gas (O_2) in a flashbulb to produce solid magnesium oxide (MgO) and light. This reaction can be represented by the following chemical equation.

$$2\ Mg_{(s)} + O_{2(g)} \longrightarrow 2\ MgO_{(s)} + \text{light}$$

The formulas for the reactants, written on the left, are separated by an arrow from the formula for the product, written on the right. The subscript notation (s) indicates the solid phase and (g) the gas phase. Since light is evolved, this is an exothermic reaction (Section 1.2-b). This is a convenient system to study because both reactants and product are sealed within the flashbulb. Careful weighing of a flashbulb before and after flashing on an analytical balance indicates no detectable change in mass. Lavoisier's and many other similar quantitative experiments led to the **law of conservation of mass,** which states that *there is no detectable change in mass in an* ***ordinary*** *chemical reaction.*

Note that the law states that there is no *detectable* mass change in an *ordinary* chemical reaction. Albert Einstein recognized the quantitative interconversion of mass and energy in chemical reactions, resulting in small mass changes. Thus, mass is converted to heat and light energy in the combustion of charcoal or in the flashbulb reaction. However, using Einstein's equation, it can be shown that the mass change in the typical flashing of a flashbulb is only about 5×10^{-9} gram, well below the sensitivity of the best analytical balance. On the other hand, in nuclear reactions (Sections 11.2 and 11.3), which are not considered as ordinary chemical reactions, conversions of mass to energy or vice versa are often large enough to detect experimentally. To encompass both ordinary chemical reactions and nuclear reactions, we should speak of the law of conservation of mass and energy.

2.2 Law of Constant Composition

When carbon (C) is burned in excess oxygen (O_2), carbon dioxide (CO_2) is formed. This reaction can be represented by the following equation.

$$C_{(s)} + O_{2(g)} \longrightarrow CO_{2(g)}$$

Mass data for the amount of C and O_2 used and the amount of CO_2 formed in several such experiments are given in Table 2.1.

As expected, all sets of data in Table 2.1 verify the law of conservation of mass; that is, the sum of the masses of reactants equals the mass of product. In addition, the data in runs 1 and 2 indicate that the results are reproduci-

Table 2.1 Mass Data for Combustion of Carbon in Excess Oxygen

Run	*Mass of C (g)*	*Mass of O_2 (g)*	*Mass of CO_2 (g)*
1	1.00	2.67	3.67
2	1.00	2.67	3.67
3	10.00	26.70	36.70
4	3.75	10.00	13.75

ble. However, you, like early chemists, can use your imagination and look for other relationships in the mass data. For example, consider the carbon/oxygen mass ratio and the oxygen/carbon mass ratio.

$$\frac{\text{Mass of C}}{\text{Mass of O}} = \frac{1.00\text{ g}}{2.67\text{ g}} = \frac{10.00\text{ g}}{26.70\text{ g}} = \frac{3.75\text{ g}}{10.00\text{ g}} = 0.374$$

$$\frac{\text{Mass of O}}{\text{Mass of C}} = \frac{2.67\text{ g}}{1.00\text{ g}} = \frac{26.70\text{ g}}{10.00\text{ g}} = \frac{10.00\text{ g}}{3.75\text{ g}} = 2.67$$

A pattern emerges immediately. You find that all of the carbon/oxygen mass ratios are identical and all of the oxygen/carbon mass ratios are identical. On the basis of similar analyses of a great deal of data from quantitative experiments of this nature, Louis Proust, a French chemist, postulated in 1779 the **law of constant composition.** This law states that when two elements combine to form a given compound, they always do so in a fixed proportion by mass or, more simply, *a given compound always shows a fixed proportion of elements by mass.*

The data can be analyzed in another fashion, namely, through the calculation of percent carbon and percent oxygen. Percent carbon is obtained by dividing the mass of carbon in the compound by the total mass of the compound and multiplying by 100%. A similar calculation is performed for oxygen.

$$\text{Percent C} = \left(\frac{1.00\text{ g}}{3.67\text{ g}}\right)(100\%) = \left(\frac{10.00\text{ g}}{36.70\text{ g}}\right)(100\%) = \left(\frac{3.75\text{ g}}{13.75\text{ g}}\right)(100\%)$$
$$= 27.2\%\text{ C}$$

$$\text{Percent O} = \left(\frac{2.67\text{ g}}{3.67\text{ g}}\right)(100\%) = \left(\frac{26.70\text{ g}}{36.70\text{ g}}\right)(100\%) = \left(\frac{10.00\text{ g}}{13.75\text{ g}}\right)(100\%)$$
$$= 72.8\%\text{ O}$$

All percents of carbon are identical, and all percents of oxygen are identical. Many similar analyses led to another statement of the law of constant composition—*a chemical compound always contains the same elements in the same percents by mass.*

2.3 Dalton's Atomic Theory

In 1808 John Dalton published his *New System of Chemical Philosophy*. In it he proposed an atomic theory based on the eighteenth-century quantitative laws discussed above and on numerous subsequent experiments. The basic postulates of his theory along with illustrative examples follow.

1. *Elements are composed of extremely small particles called atoms.* The element lead is made up of lead atoms.
2. *Atoms of a given element exhibit identical chemical properties; atoms of different elements exhibit different chemical properties.* All lead atoms exhibit identical chemical properties and all copper atoms exhibit identical chemical properties; however, lead atoms and copper atoms can be differentiated by their distinctly different chemical properties.
3. *No atom of one element is changed into an atom of another element in an ordinary chemical reaction.* When iron atoms are heated with sulfur atoms, only iron atoms and sulfur atoms are present in the iron(II) sulfide (FeS) product that is formed.
4. *Compound substances are formed when an integral number of atoms of one element combine with an integral number of atoms of one or more other elements. The relative numbers of atoms of different elements will be fixed in a given pure compound.* In the compound iron pyrites there is always one atom of copper for every atom of iron and every two atoms of sulfur.

A test of any theory is how well it explains observed facts or scientific laws derived from these facts. We can understand the two quantitative laws discussed previously on the basis of Dalton's atomic theory.

1. *Law of conservation of mass.* According to Dalton's third postulate, atoms are conserved in an ordinary chemical reaction; thus, there can be no change in mass.
2. *Law of constant composition.* According to Dalton's fourth postulate, the relative numbers of atoms of different elements are constant. The relative masses or percents by mass must also be constant.

2.4 Development of the Periodic Table from Properties

The empirical development of the periodic classification of the chemical elements on the basis of physical and chemical properties was an important stimulus for the further development of chemistry and was a remarkable illustration of the role of creativity in science. Nobel laureate Melvin Calvin has said that it is no trick for someone to reach a correct conclusion when he has all of the data; rather, real creativity is involved if one can reach a correct conclusion when he has only half of the data, realizing that part of the data he has is wrong and not knowing which part it is. These statements certainly characterize the efforts of Dmitri Mendeleev, a Russian chemist, whose work was first published in 1869 and who could be considered the father of our modern periodic table.

A number of scientists had tried to develop systematic relationships among the known chemical elements by plotting various physical properties of the elements as a function of atomic mass (Sections 4.3 and 4.4). (A table of the accepted values for the atomic masses of the elements is given inside the front cover.) Unaware of the efforts of most of these earlier workers, Mendeleev developed the modern periodic table by arranging the elements in order of increasing atomic mass and also in columns of elements with similar chemical characteristics. His insight in interpreting even the rather inaccurate data available to him was phenomenal. His ideas concerning the relationship of properties of groups of elements and the logical patterns into which these elements fall have provided the basis for chemical progress ever since.

Four aspects of primary importance stand out in Mendeleev's work. First, he stated a periodic law—the properties of elements are periodic functions of their atomic masses. Second, he recognized (as is reflected in our present periodic tables) that hydrogen was somehow unique. Third, he suggested that chemical properties were more important in determining periodicity than were atomic masses. Thus, as shown in the classical short Mendeleev table (Table 2.2), he realized that Sb, Te, and I should fall in that order under As, Se, and Br, even though order of atomic masses suggested Sb, I, and Te. This was because the chemical properties of Te were more like those of Se than those of Br, and the chemical properties of I were more like

Table 2.2 Table of Mendeleev

I	*II*	*III*	*IV*	*V*	*VI*	*VII*	*VIII*
H							
Li	Be	B	C	N	O	F	
Na	Mg	Al	Si	P	S	Cl	
K	Ca	—	Ti	V	Cr	Mn	Fe Co Ni Cu
(Cu)	Zn	—	—	As	Se	Br	
Rb	Sr	Y	Zr	Nb	Mo	—	Ru Rh Pd Ag
(Ag)	Cd	In	Sn	Sb	Te	I	
Cs	Ba	—	Ce	—	—	—	— — — —
—	—	—	—	—	—	—	
—	—	—	—	Ta	W	—	Os Ir Pt Au
(Au)	Hg	Tl	Pb	Bi	—	—	
—	—	—	Th	—	U	—	

those of Br than those of Se. Likewise, Fe, Co, and Ni should be placed in that order, even though Co had a greater atomic mass than Ni, because the chemical properties of Co were more like those of Fe than were the chemical properties of Ni. Fourth, he made predictions on the basis of his table. Using the *unoccupied* positions in his table as a basis, not only did he predict the existence of at least a dozen undiscovered elements, but he also predicted properties of some of these elements. He realized that the properties of an undiscovered element would be intermediate between those of the four elements surrounding it in his table. With the discovery of gallium in 1875, scandium in 1879, and germanium in 1886, the accuracy of Mendeleev's predictions was quickly verified. A comparison of predicted and experimentally measured properties of germanium, shown in Table 2.3, illustrates Mendeleev's remarkable insight into the relationships among properties of elements, particularly considering the fact that only three (Si, As, and Sn) out of the four elements surrounding Ge were known at the time.

Mendeleev's predictions convinced even the most skeptical of the value of his periodic classification. It only remained for Henry Moseley, a British physicist, to find the explanation for the several atomic mass irregularities in Mendeleev's table and to develop a totally consistent periodic law. In order to appreciate Moseley's work more clearly, a number of experiments that indicated that atoms are divisible and electrical in nature should be examined.

Table 2.3 Predicted and Observed Properties of Germanium

Property	*Predicted*	*Observed*
Atomic mass (amu)	72	72.59
Atomic volume (cm^3)	13	13.6
Density (g/cm^3)	5.5	5.35 (20°)
Specific heat	0.073	0.076
Color	dirty gray	grayish white
Product of heating in air	white GeO_2	white GeO_2
GeO_2 properties	refractory solid	refractory solid
Molar volume (cm^3)	22	22.16
Density (g/cm^3)	4.7	4.228 (25°)
$GeCl_4$ properties	colorless liquid	colorless liquid
Molar volume (cm^3)	113	113.35
Density (g/cm^3)	1.9	1.844 (30°)
Boiling point (°C)	100	84

2.5 Electrons

Electrolysis studies by Humphrey Davy and Michael Faraday, two British chemists, in the early part of the nineteenth century had indicated that matter was electrical in nature. However, only in the latter part of that century did gas discharge tube experiments provide concrete evidence for the existence of charged particles.

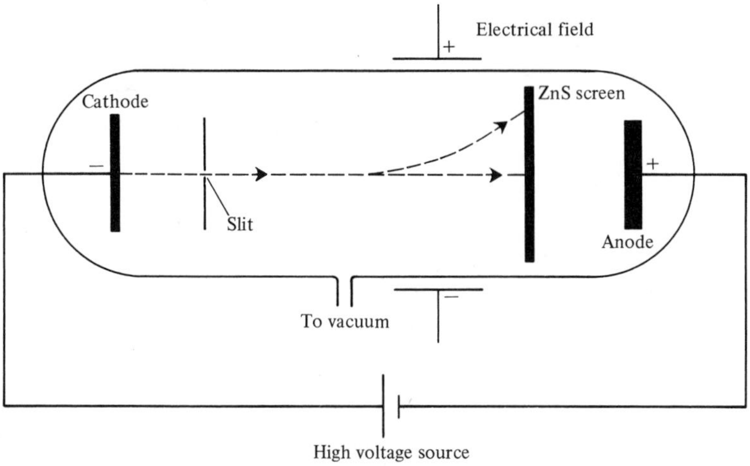
Electrical field
+
ZnS screen
Cathode
–
+
Slit
Anode
To vacuum
–
High voltage source

Figure 2.1 Gas discharge tube for studying electrons.

When a high voltage was applied across electrodes contained in a partially evacuated tube (Figure 2.1), a glowing of the residual gas in the tube was observed. Various characteristics of this gas discharge could be studied experimentally by using additional apparatus with the tube. If a zinc sulfide screen was placed between the cathode and the anode, the rays impinged only on the cathode side of the screen and discrete flashes could be observed on the screen. These two observations indicated that the rays were emanating from the cathode and that they possessed a particle nature. If an electrical field was placed around the discharge tube, the rays were shown to deviate toward the positively charged plate (dashed curve). This observation indicated that the rays were negatively charged since unlike charges attract each other. Moreover, the degree of deviation was independent of what residual gas was in the tube. These cathode rays are what we now call electrons, and they are identical no matter what atoms are ionized to form them.

In somewhat more sophisticated experiments, Joseph Thomson, a British physicist, in 1897 and Robert Millikan, an American nuclear physicist, in 1909 determined the charge/mass ratio for an electron and the charge of an electron, respectively. From these results it was determined that the mass of an electron is 9.110×10^{-28} gram. It is assigned a relative charge of -1.

2.6 Protons

Gas discharge tubes of slightly different construction (Figure 2.2) led to the discovery of protons. When a high voltage was passed across such a tube, flashes were noted on the zinc sulfide screen placed on the *opposite* side of the cathode from the anode. Furthermore, the rays were deflected toward the *negative* plate in the presence of an electrical field and the degree of deflec-

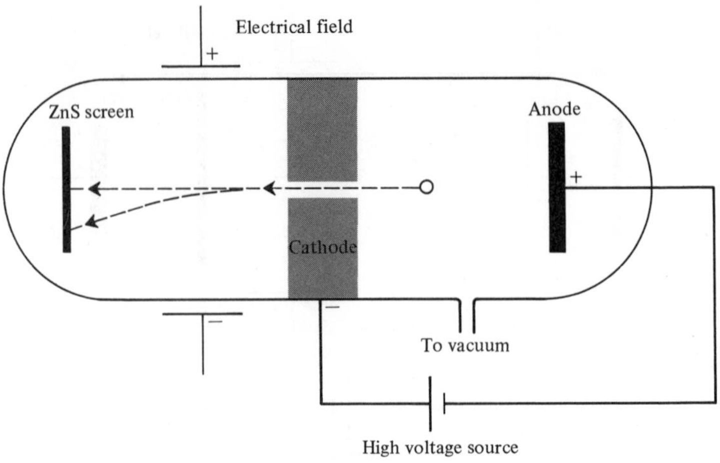
Electrical field
+
ZnS screen
Anode
+
Cathode
−
−
To vacuum
High voltage source

Figure 2.2 Gas discharge tube for studying protons.

tion *depended* on what residual gas was present in the tube. When hydrogen gas was present in the tube, there was significant deflection, whereas the deflection was markedly less with oxygen gas or nitrogen gas. Such behavior can be understood as follows. Electrons, emanating from the cathode, collided with residual gas molecules on the way to the anode. Some of these electrons had high enough energy to eject electrons from the gas molecules, thereby forming positive ions. Any positive ions would be accelerated toward the cathode since unlike charges attract. However, some positive ions would pass through the perforated cathode and impinge on the zinc sulfide screen. The positive ions were deflected on their way to the screen if an electrical field were present, the deflection being greater for positive ions of smaller mass.

Experiments similar to Thomson's experiment, requiring the use of both electrical and magnetic fields, could be used to determine the charge/mass ratio for the positive ions. Further, if they were formed by ejection of an electron from a neutral species, the positive ions could be assumed to have a charge of the *same* magnitude as, but *opposite* in sign to, that of the electron, and the mass of these positive ions could be determined. Protons (H^+ ions), which resulted if hydrogen were the residual gas in the tube, were found to be very much heavier than the electron (about 1836 times, or 1.673×10^{-24} gram).

2.7 Neutrons

Although the existence of neutrons was postulated very much earlier, it was not until 1932 that James Chadwick, a British physicist, was able to find direct evidence for such particles. The appropriate apparatus is shown in Figure 2.3. When Li, Be, or B sheets were bombarded with dipositively charged alpha particles (helium atoms from which the electrons have been

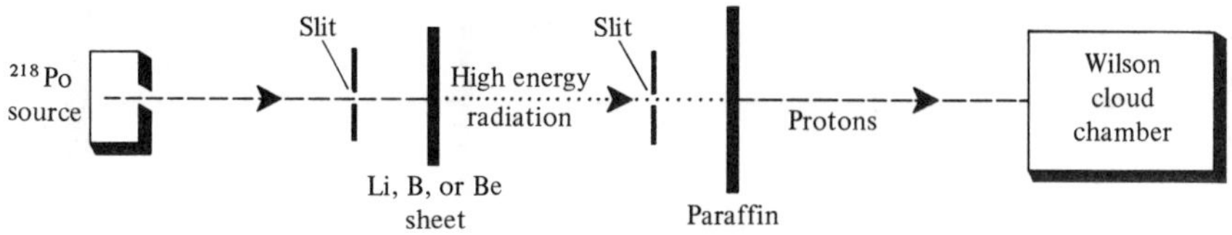
Slit
Slit
^{218}Po
source
High energy
radiation
Wilson
cloud
chamber
Protons
Li, B, or Be
sheet
Paraffin

Figure 2.3 Apparatus for studying neutrons.

removed), high energy particles were emitted that could penetrate a 1 inch lead sheet and lose only half of their intensity. The path of these particles was unaffected by an electrical field, and thus the particles were uncharged. These particles were originally assumed by several workers to be a form of gamma radiation; however, they were much too energetic. It was observed that these particles ejected protons from a block of paraffin wax placed in their path. Protons were indicated by the fact that the ejected particles were deflected toward the negative plate of an electrical field and left a chain of water droplets in a Wilson cloud chamber. Chadwick suggested the name "neutron" and postulated that the mass of a neutron was similar to the mass of a proton. If this were the case, a neutron could eject a proton from the paraffin by direct collision. Francis Aston, a British scientist, later determined the mass of a neutron (1.675×10^{-24} gram) by means of a mass spectrometer.

2.8 The Nuclear Atom

In 1898, following his measurement of the charge/mass ratio for electrons, Thomson proposed a structure for an atom. According to his model the atom was a sphere of positive electricity containing negative electrons much as seeds are embedded within the pulp of a grape. Since protons and other cations had been found to be much more massive than electrons, Thomson postulated that most of the mass of an atom was associated with the positive electricity. Positively charged ions could presumably be formed by stripping away one or more electrons, thereby leaving most of the mass of the atom and a resultant positive charge.

An alpha-particle-scattering experiment, performed in 1911 by Ernest Rutherford, a British physicist, disproved the Thomson model and placed the concept of a nuclear atom on a firm basis. The important features of Rutherford's experiment are diagramed in Figure 2.4. A beam of alpha particles resulting from the radioactive decay of polonium (Section 11.2-a) was directed onto a thin metal foil by collimation through a pinhole in a lead plate. Most of the very energetic alpha particles easily penetrated the thin metal foil and continued to a zinc sulfide screen with little or no deflection. However, some alpha particles were deflected at quite large angles and a few were even *reflected* back from the foil. Rutherford was astounded by these results—"it was almost as incredible as if you fired a 15 inch shell at a piece of tissue paper and it came back and hit you."

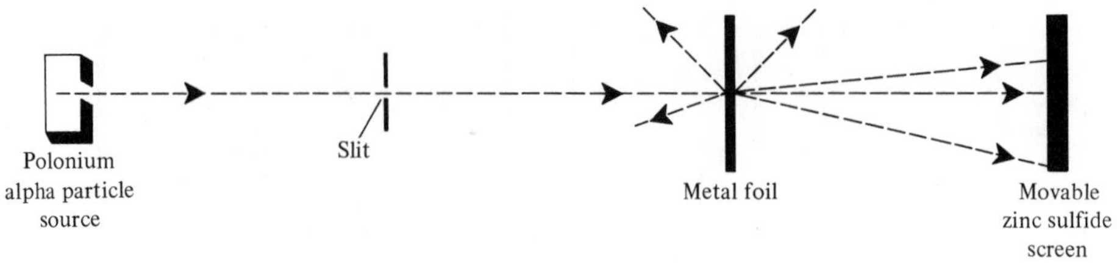

Polonium
alpha particle
source
Slit
Metal foil
Movable
zinc sulfide
screen

Figure 2.4 Rutherford's alpha-particle-scattering experiment.

Rutherford recognized that an atom according to the Thomson model, with its mass and positive charge spread uniformly, would not provide significant repulsion for positively charged alpha particles and that, contrary to his experimental results, all alpha particles should pass through the metal foil with little deviation from their original path. However, Rutherford further recognized that if an atom's *positive electricity and mass* were concentrated in a very small region, most of the alpha particles would pass through with no deflection, but a few would directly encounter a concentration of positive charge (immovable because of its high mass) and would be repelled strongly and deflected considerably from their original path. These ideas are illustrated in Figure 2.5. Rutherford thus postulated that an atom has a very small nucleus in which positive charge and mass are concentrated and a relatively empty extranuclear region containing electrons. Quantitative alpha-particle-scattering studies indicated that the nucleus had a diameter of about 10^{-13} cm, whereas the diameter of an atom was close to 10^{-8} cm.

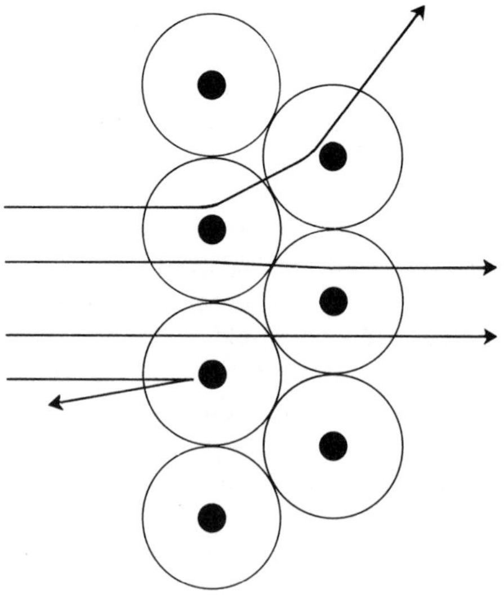

Figure 2.5 Deflection of alpha particles by a nuclear atom.

2.9 Atomic Number

Moseley, a student in Rutherford's laboratory, further clarified the concept of a nuclear atom and, at the same time, suggested a rational periodic law that is the basis of our modern periodic table. Using vacuum tubes like those used previously in gas discharge tube experiments, Moseley bombarded replaceable anodes (positive electrodes) with electrons and determined the frequencies (number of pulses passing a given point per second) of x-rays emitted from the target anodes. He found a linear relationship between the square root of the frequency (for the highest-frequency x-ray of a series emitted for each element) and the order in which the elements appeared in Mendeleev's periodic table. His results are plotted in Figure 2.6. Two significant observations emerge from his plot. First, gaps appeared corresponding to elements still undiscovered, but predicted earlier by Mendeleev. Second, just as in Mendeleev's periodic table, Moseley found that iron, cobalt, and nickel and antimony, tellurium, and iodine (also chlorine, argon, and potassium) appeared in those orders on his graph in opposition to the order of increasing atomic mass.

On the basis of these observations, Moseley postulated, "there is in the atom a fundamental quantity which increases by regular steps as we pass from one atom to the next. This quantity can only be the charge on the central positive nucleus." Moseley called this quantity the atomic number. The **atomic number** is *the number of protons in the nucleus of an atom and* is also *the number of extranuclear electrons in a neutral atom.* On the basis of Moseley's work, a new periodic law can be stated—*properties of atoms are periodic functions of their atomic numbers.*

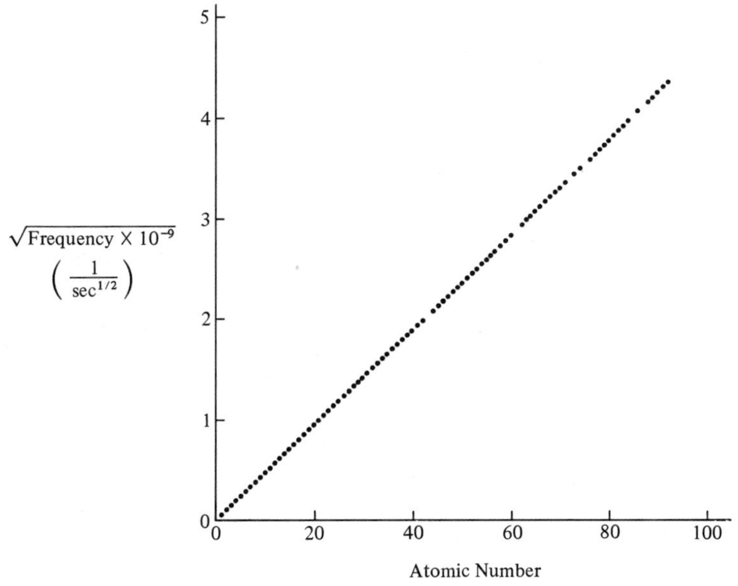

$\sqrt{\text{Frequency}} \times 10^{-9}$
$\left(\frac{1}{\sec^{1/2}}\right)$
0
1
2
3
4
5
0
20
40
60
80
100
Atomic Number

Figure 2.6 Moseley's plot of x-ray frequencies versus atomic number.

2.10 Isotopes

Isotopes are *atoms that have the same atomic number but different masses.* Most of the mass of an atom is in its nucleus which is composed of protons and neutrons. *Isotopes of a given atom have the same number of protons but different numbers of neutrons.* Isotopes are often designated using an *atomic number* (number of protons) subscript and a *mass number* (number of protons and neutrons) superscript, followed by the chemical symbol for the atom of that atomic number. The number of neutrons equals the mass number minus the atomic number. For example, ${}^{31}_{15}P$ indicates an atom with 15 protons, that is, phosphorus, and (31 − 15) or 16 neutrons. Likewise, ${}^{238}_{92}U$ indicates an atom with 92 protons, that is, uranium, and (238 − 92) or 146 neutrons.

All the isotopes of an element undergo similar chemical reactions although the rates of these chemical reactions (especially those involving bonds of hydrogen to another element) may differ. Only the isotopes of hydrogen have been given specific names—${}^{1}_{1}H$, protium; ${}^{2}_{1}H$, deuterium; and ${}^{3}_{1}H$, tritium.

Isotopes are responsible for the fact that many elements have nonintegral atomic masses. For example, copper has two naturally occurring isotopes. Copper's atomic mass of 63.546 amu results from the average of ${}^{63}_{29}Cu$ and ${}^{65}_{29}Cu$ weighted appropriately for their natural abundances of 69.09% and 30.91%, respectively. (See Example 4.3 and Exercise 4.2 for calculations of this type.)

Isotopes are also responsible for the fact that atomic mass is not directly related to atomic number, for the fact that chlorine, argon, and potassium; iron, cobalt, and nickel; and antimony, tellurium, and iodine appear in the periodic table in order of increasing atomic number, but not in the order of increasing atomic mass. An overabundance of a higher mass isotope or a lower mass isotope of an element (a function of nuclear stability) can occasionally cause an average atomic mass that is higher or lower than expected.

2.11 Atomic Spectra

The relationship of electron configurations in the extranuclear region of an atom and properties of the atom were discovered largely through interpretation of atomic spectra. A simple spectroscope, having the components shown in Figure 2.7, can be used for illustration. A *continuous spectrum* or a rainbow of colors is obtained on the film in Figure 2.7a if a beam from a white-light source such as the sun is dispersed by passage through a glass prism. Lower-energy or lower-frequency components (red) are bent the least

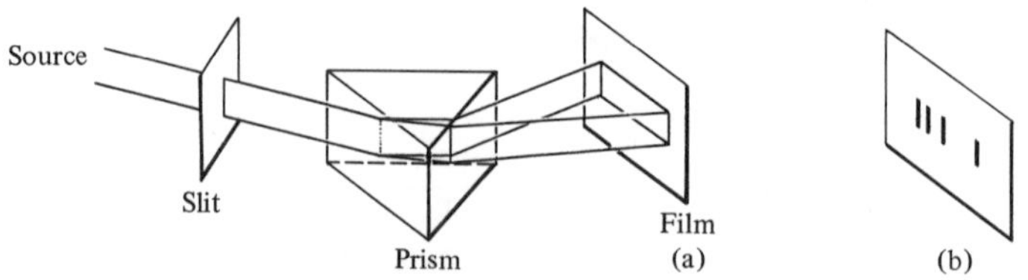

Source
Slit
Prism
Film
(a)
(b)

Figure 2.7 Components of a simple spectroscope.

and appear at the lower right edge of the film in Figure 2.7a; higher-energy or higher-frequency components (blue) are bent the most and appear at the upper left edge of the film in Figure 2.7a. If, on the other hand, the source is colored light emitted by a heated gaseous sample, a series of narrow lines of color (called a *line spectrum*) rather than a continuous blend of colors is obtained on the film. Thus, hydrogen atoms give four lines in the visible region of the spectrum as shown in Figure 2.7b. Furthermore, different elements can be shown to have their own characteristic pattern of lines. Interpretation of such line spectra resulted in models for the extranuclear region of the atom. Building such a model was no easy task. It could be likened to your trying to construct a harp when you had never seen one and your only knowledge of such an instrument was that you had heard tones emitted from it when its strings were plucked by a three-year-old child living in the apartment above yours.

In 1913 Niels Bohr, a Danish physicist, developed the first reasonable model for the electronic structure in the extranuclear region of the hydrogen atom. He explained four basic atomic spectral observations as follows.

1. *Heated gaseous atoms emitted a spectrum.* The gaseous atoms absorbed heat energy and were raised to higher energy states. Upon their return to lower energy states light energy was emitted. It was postulated that electrons in atoms were responsible for the spectra, that electrons were excited to higher energy states and then emitted light upon returning to lower energy states. A continuous spectrum (Figure 2.7a) would be expected by this mechanism assuming a continuous range of energy states existed.

2. *Line spectra were observed* (Figure 2.7b). Bohr postulated that electrons were present only in certain allowed (quantized) energy levels and that no electrons could have energies lying between energy levels. These energy levels represented the energies of electrons that were pictured as moving in circular paths called orbits around the nucleus. The frequencies of the emitted spectral lines were a measure of the spacing of these energy levels.

3. *Line spectra were characteristic for each element.* The number of protons (equal to the total positive charge in the nucleus and to the number of electrons in a neutral atom) differs for atoms of unlike elements. The attractive forces between the nucleus and electrons would then vary and could be expected to alter the spacings of the energy levels for atoms of different elements.

4. *Energy levels were closer together at higher energies.* If energy levels were equally spaced, only one spectral line would be observed rather than a series of spectral lines. Bohr developed a mathematical relationship that accounted for energy level spacings in the hydrogen atom.

An energy level diagram for the hydrogen atom that is consistent with the Bohr model, but not to scale, is shown in Figure 2.8a. The individual energy levels labeled *K*, *L*, *M*, *N*, *O*, *P*, and *Q* are called *shells*. Although the Bohr model accurately described atomic spectra for the hydrogen atom, it proved unsatisfactory for atoms with more than one electron, and has since been abandoned. However, the basic ideas of the Bohr model, coupled with the interpretation of more refined atomic spectra obtained under special experimental conditions, leads to a more realistic energy level diagram.

If the gaseous atoms emitting a line spectrum are placed in an electrical

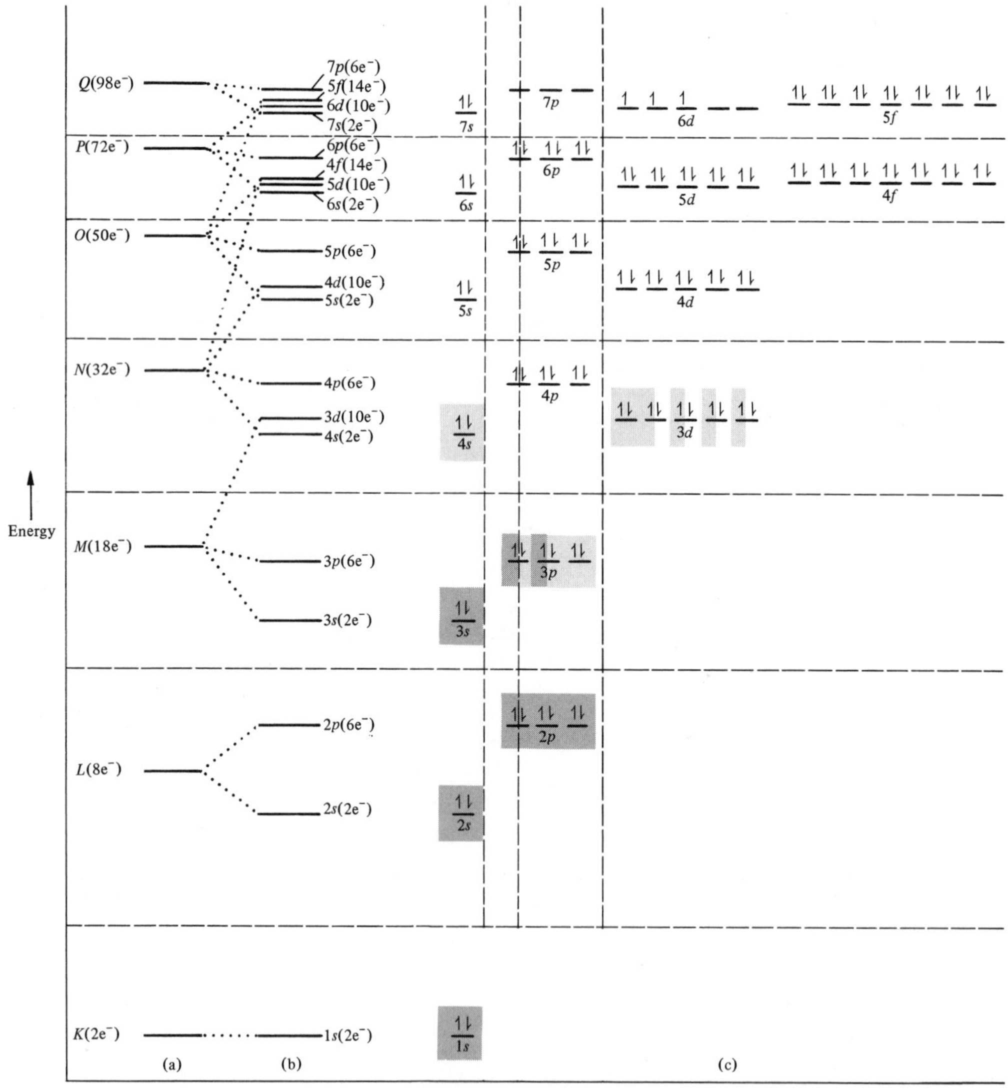

Energy
Q(98e⁻)
P(72e⁻)
O(50e⁻)
N(32e⁻)
M(18e⁻)
L(8e⁻)
K(2e⁻)
7p(6e⁻)
5f(14e⁻)
6d(10e⁻)
7s(2e⁻)
6p(6e⁻)
4f(14e⁻)
5d(10e⁻)
6s(2e⁻)
5p(6e⁻)
4d(10e⁻)
5s(2e⁻)
4p(6e⁻)
3d(10e⁻)
4s(2e⁻)
3p(6e⁻)
3s(2e⁻)
2p(6e⁻)
2s(2e⁻)
1s(2e⁻)
7s
7p
6d
5f
6s
6p
5d
4f
5s
5p
4d
4s
4p
3d
3s
3p
2s
2p
1s
(a)
(b)
(c)

Figure 2.8 Energy level diagram.

field (or for certain atoms where the electrons present produce their own field), many of what were single spectral lines are split into two or more lines. In terms of an energy level diagram two or more sublevels called *subshells* arise from the shells in Figure 2.8a in order to account for the new spectral lines. The subshells that result are shown schematically in Figure 2.8b. Note that subshells from different shells overlap at higher energy. The letter designations *s*, *p*, *d*, and *f* were originally chosen on the basis of observations of the line spectra (sharp, principal, diffuse, fundamental). It can be shown by calculations from quantum mechanics (a study of the laws of motion that describe the behavior of very small particles such as electrons) that the *s*, *p*, *d*, and *f* subshells hold 2, 6, 10, and 14 electrons, respectively, as indicated in Figure 2.8b.

If the gaseous atoms emitting a line spectrum are placed in a magnetic field, many of what were single spectral lines in the electrical field experiment above are further split into three, five, or seven lines. In terms of an energy level diagram several sublevels arise from the subshells in Figure 2.8b. From quantum mechanical calculations it can be shown that these sublevels, called *orbitals*, represent regions in space in which an electron with a specific energy is most likely to be found. The energy levels of the orbitals that result are shown schematically in Figure 2.8c. The orbital energy levels are pictured side by side rather than split in energy so as not to complicate the diagram further and because the sets of orbitals are of the same energy under usual experimental conditions, that is, in the absence of a magnetic field.

Finally, it should be noted that any spinning charge such as an electron behaves like a tiny magnet. An electron can spin in either a clockwise or a counterclockwise direction about its axis. Two spin states, which correspond to two magnets whose fields are oriented in opposite directions, arise for each orbital. Otto Stern and Walther Gerlach showed in 1921 that these two spin states can be separated in the presence of an inhomogeneous magnetic field. This means that each of the orbitals in Figure 2.8c can be split into two sublevels. This results in a simple statement of the *Pauli exclusion principle*—each orbital can contain a maximum of two electrons, and the two electrons must have opposite spin. Electron spin is usually indicated by arrows: ↿ and ⇂.

2.12 Development of the Periodic Table from Electron Configurations

You can now use the energy level diagram in Figure 2.8c and predict the ground state (most stable or lowest energy) electron configuration for any atom.[1] Three basic rules are followed when filling the energy level diagram.

[1]Note that the relative energies of orbitals in the energy level diagram depend on the nuclear charge (number of protons) and number of electrons present. Therefore, no single diagram is *quantitatively* appropriate for all elements although it may be qualitatively correct. Note also that energy levels may shift and in some cases even the relative order of levels may change as nuclear charge increases. These problems, however, will not seriously jeopardize the predictive power of such a diagram.

1. Electrons occupy the lowest energy orbitals first to obtain the most stable configuration.
2. A maximum of *two* electrons may occupy an orbital to conform to the Pauli exclusion principle.
3. Electrons in orbitals of the same energy (as in the $2p$ subshell in Figure 2.8c) occupy *separate* orbitals (all electrons having the same spin) until each orbital has one electron. Only then are two electrons of opposite spin paired in each orbital. This is necessary because although electrons of opposite spin attract each other as would two tiny magnets, the repulsive effect resulting from two electrons having the same negative charge and in close spatial proximity is significantly greater. Thus, they spread over as much space as possible by occupying separate orbitals that have different spatial distributions. After a subshell is half full, pairing of electrons occurs because the pairing repulsion energy is significantly less than the energy required to place an electron in the empty orbital next higher in energy.

Some examples will illustrate the use of the energy level diagram in Figure 2.8c and the preceding three rules. The one electron of hydrogen occupies the $1s$ orbital. A shorthand notation $1s^1$ is used to indicate this electron configuration since it is inconvenient to construct an energy level diagram each time you want to write an electron configuration. The $1s$ notation indicating a specific orbital is followed by an exponent indicating the number of electrons occupying that orbital. The second electron of helium occupies the $1s$ orbital, but has *opposite* spin ($1s^2$). The $1s$ orbital is now filled and the third electron of lithium occupies the $2s$ orbital ($1s^22s^1$). The fourth electron of beryllium occupies the $2s$ orbital, but has opposite spin ($1s^22s^2$). The fifth electron of boron occupies a $2p$ orbital ($1s^22s^22p^1$). The sixth electron of carbon occupies a second $2p$ orbital to conform with rule 3 ($1s^22s^22p^2$). Note that the shorthand electron configuration does not indicate whether the two electrons in the $2p$ orbitals are paired. A more specific alternate notation sometimes used is $1s^22s^2\underset{2p}{\underline{\uparrow}\ \underline{\uparrow}\ \underline{}}$. The seventh electron for nitrogen enters the third $2p$ orbital giving three unpaired electrons ($1s^22s^2\underset{2p}{\underline{\uparrow}\ \underline{\uparrow}\ \underline{\uparrow}}$). The eighth electron for oxygen enters the first $2p$ orbital leaving two unpaired electrons ($1s^22s^2\underset{2p}{\underline{\uparrow\downarrow}\ \underline{\uparrow}\ \underline{\uparrow}}$). The ninth electron for fluorine enters the second $2p$ orbital leaving one unpaired electron ($1s^22s^2\underset{2p}{\underline{\uparrow\downarrow}\ \underline{\uparrow\downarrow}\ \underline{\uparrow}}$). The tenth electron for neon enters the third $2p$ orbital leaving no unpaired electrons ($1s^22s^2$ $\underset{2p}{\underline{\uparrow\downarrow}\ \underline{\uparrow\downarrow}\ \underline{\uparrow\downarrow}}$). The electron configurations for the next eight elements result by filling the $3s$ and $3p$ orbitals in the same way that the $2s$ and $2p$ orbitals were filled for the elements lithium through neon. The electron configuration for silicon ($1s^22s^22p^63s^23p^2$) is shaded darkly in Figure 2.8c. Similarly, the electron configuration for iron ($1s^22s^22p^63s^23p^64s^23d^6$) is shaded lightly in Figure 2.8c. Except for minor irregularities that occur because of apparent extra stabilities of filled and half filled shells (for example, Cr $1s^22s^22p^63s^23p^64s^13d^5$ and Cu $1s^22s^22p^63s^23p^64s^13d^{10}$) and because of shifting of energy levels with changing nuclear charge, the electron configurations of all known elements can be predicted. The probable electron configuration for a recently discovered man-made element (hahnium, atomic number 105) is shown unshaded in Figure 2.8c. The ground state electron configurations for all known elements, based on observations of atomic spectra and magnetic properties, are given in Table 2.4. Shaded configurations are still in doubt.

A device that can help you remember the order of filling of orbitals is shown in Figure 2.9. First set up the orbitals and draw the diagonal lines. Starting with $1s$, fill toward the head of each arrow, then back to the tail of the next arrow. Thus, the order of filling is $1s$, $2s$, $2p$, $3s$, $3p$, $4s$, $3d$, $4p$, $5s$, $4d$, $5p$, $6s$, $4f$, $5d$, $6p$, $7s$, $5f$, and $6d$, the same order predicted by Figure 2.8c. Noble gases, the elements appearing at the end of each period, occur just before you cross the dashed line from right to left in Figure 2.9, or just before beginning to fill an s orbital. Using Figure 2.9, the electron configuration for arsenic (atomic number 33) is $1s^22s^22p^63s^23p^64s^23d^{10}4p^3$. Note that the sum of the exponents in a shorthand electron configuration equals the atomic number of the element that the shorthand notation represents.

Careful use of the energy level diagram in Figure 2.8c leads directly to the periodic table shown inside the back cover. It can be seen that the end of a *period* (a horizontal row in the periodic table) results at each point in the

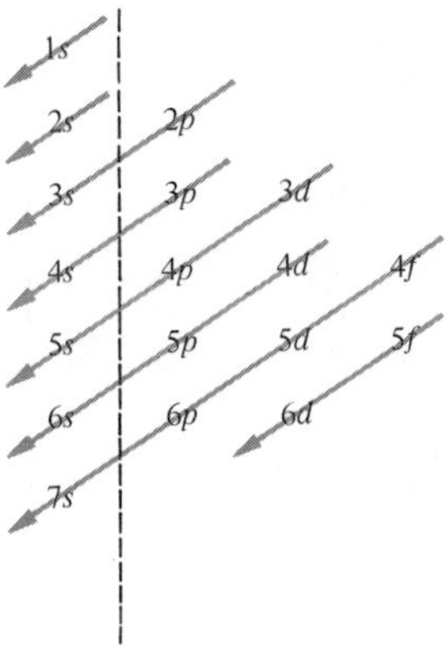
1s
2s
2p
3s
3p
3d
4s
4p
4d
4f
5s
5p
5d
5f
6s
6p
6d
7s

Figure 2.9 Order of filling of orbitals.

Table 2.4 Ground State Electron Configurations of the Elements

		1	2		3			4				5				6				7
Z	*Element*	*s*	*s*	*p*	*s*	*p*	*d*	*s*	*p*	*d*	*f*	*s*	*p*	*d*	*f*	*s*	*p*	*d*	*f*	*s*
1	H	1																		
2	He	2																		
3	Li	2	1																	
4	Be	2	2																	
5	B	2	2	1																
6	C	2	2	2																
7	N	2	2	3																
8	O	2	2	4																
9	F	2	2	5																
10	Ne	2	2	6																
11	Na	2	2	6	1															
12	Mg	2	2	6	2															
13	Al	2	2	6	2	1														
14	Si	2	2	6	2	2														
15	P	2	2	6	2	3														
16	S	2	2	6	2	4														
17	Cl	2	2	6	2	5														
18	Ar	2	2	6	2	6														
19	K	2	2	6	2	6		1												
20	Ca	2	2	6	2	6		2												
21	Sc	2	2	6	2	6	1	2												
22	Ti	2	2	6	2	6	2	2												
23	V	2	2	6	2	6	3	2												
24	Cr	2	2	6	2	6	5	1												
25	Mn	2	2	6	2	6	5	2												
26	Fe	2	2	6	2	6	6	2												
27	Co	2	2	6	2	6	7	2												
28	Ni	2	2	6	2	6	8	2												
29	Cu	2	2	6	2	6	10	1												
30	Zn	2	2	6	2	6	10	2												
31	Ga	2	2	6	2	6	10	2	1											
32	Ge	2	2	6	2	6	10	2	2											
33	As	2	2	6	2	6	10	2	3											
34	Se	2	2	6	2	6	10	2	4											
35	Br	2	2	6	2	6	10	2	5											
36	Kr	2	2	6	2	6	10	2	6											
37	Rb	2	2	6	2	6	10	2	6			1								
38	Sr	2	2	6	2	6	10	2	6			2								
39	Y	2	2	6	2	6	10	2	6	1		2								
40	Zr	2	2	6	2	6	10	2	6	2		2								
41	Nb	2	2	6	2	6	10	2	6	4		1								
42	Mo	2	2	6	2	6	10	2	6	5		1								
43	Tc	2	2	6	2	6	10	2	6	6		1								
44	Ru	2	2	6	2	6	10	2	6	7		1								
45	Rh	2	2	6	2	6	10	2	6	8		1								
46	Pd	2	2	6	2	6	10	2	6	10										
47	Ag	2	2	6	2	6	10	2	6	10		1								
48	Cd	2	2	6	2	6	10	2	6	10		2								
49	In	2	2	6	2	6	10	2	6	10		2	1							
50	Sn	2	2	6	2	6	10	2	6	10		2	2							
51	Sb	2	2	6	2	6	10	2	6	10		2	3							
52	Te	2	2	6	2	6	10	2	6	10		2	4							
53	I	2	2	6	2	6	10	2	6	10		2	5							
54	Xe	2	2	6	2	6	10	2	6	10		2	6							

Table 2.4 *(Continued)*

		1	2		3			4				5				6				7
Z	*Element*	*s*	*s*	*p*	*s*	*p*	*d*	*s*	*p*	*d*	*f*	*s*	*p*	*d*	*f*	*s*	*p*	*d*	*f*	*s*
55	Cs	2	2	6	2	6	10	2	6	10		2	6			1				
56	Ba	2	2	6	2	6	10	2	6	10		2	6			2				
57	La	2	2	6	2	6	10	2	6	10		2	6	1		2				
58	Ce	2	2	6	2	6	10	2	6	10	2	2	6			2				
59	Pr	2	2	6	2	6	10	2	6	10	3	2	6			2				
60	Nd	2	2	6	2	6	10	2	6	10	4	2	6			2				
61	Pm	2	2	6	2	6	10	2	6	10	5	2	6			2				
62	Sm	2	2	6	2	6	10	2	6	10	6	2	6			2				
63	Eu	2	2	6	2	6	10	2	6	10	7	2	6			2				
64	Gd	2	2	6	2	6	10	2	6	10	7	2	6	1		2				
65	Tb	2	2	6	2	6	10	2	6	10	9	2	6			2				
66	Dy	2	2	6	2	6	10	2	6	10	10	2	6			2				
67	Ho	2	2	6	2	6	10	2	6	10	11	2	6			2				
68	Er	2	2	6	2	6	10	2	6	10	12	2	6			2				
69	Tm	2	2	6	2	6	10	2	6	10	13	2	6			2				
70	Yb	2	2	6	2	6	10	2	6	10	14	2	6			2				
71	Lu	2	2	6	2	6	10	2	6	10	14	2	6	1		2				
72	Hf	2	2	6	2	6	10	2	6	10	14	2	6	2		2				
73	Ta	2	2	6	2	6	10	2	6	10	14	2	6	3		2				
74	W	2	2	6	2	6	10	2	6	10	14	2	6	4		2				
75	Re	2	2	6	2	6	10	2	6	10	14	2	6	5		2				
76	Os	2	2	6	2	6	10	2	6	10	14	2	6	6		2				
77	Ir	2	2	6	2	6	10	2	6	10	14	2	6	7		2				
78	Pt	2	2	6	2	6	10	2	6	10	14	2	6	9		1				
79	Au	2	2	6	2	6	10	2	6	10	14	2	6	10		1				
80	Hg	2	2	6	2	6	10	2	6	10	14	2	6	10		2				
81	Tl	2	2	6	2	6	10	2	6	10	14	2	6	10		2	1			
82	Pb	2	2	6	2	6	10	2	6	10	14	2	6	10		2	2			
83	Bi	2	2	6	2	6	10	2	6	10	14	2	6	10		2	3			
84	Po	2	2	6	2	6	10	2	6	10	14	2	6	10		2	4			
85	At	2	2	6	2	6	10	2	6	10	14	2	6	10		2	5			
86	Rn	2	2	6	2	6	10	2	6	10	14	2	6	10		2	6			
87	Fr	2	2	6	2	6	10	2	6	10	14	2	6	10		2	6			1
88	Ra	2	2	6	2	6	10	2	6	10	14	2	6	10		2	6			2
89	Ac	2	2	6	2	6	10	2	6	10	14	2	6	10		2	6	1		2
90	Th	2	2	6	2	6	10	2	6	10	14	2	6	10		2	6	2		2
91	Pa	2	2	6	2	6	10	2	6	10	14	2	6	10	2	2	6	1		2
92	U	2	2	6	2	6	10	2	6	10	14	2	6	10	3	2	6	1		2
93	Np	2	2	6	2	6	10	2	6	10	14	2	6	10	4	2	6	1		2
94	Pu	2	2	6	2	6	10	2	6	10	14	2	6	10	6	2	6			2
95	Am	2	2	6	2	6	10	2	6	10	14	2	6	10	7	2	6			2
96	Cm	2	2	6	2	6	10	2	6	10	14	2	6	10	7	2	6	1		2
97	Bk	2	2	6	2	6	10	2	6	10	14	2	6	10	8	2	6	1		2
98	Cf	2	2	6	2	6	10	2	6	10	14	2	6	10	10	2	6			2
99	Es	2	2	6	2	6	10	2	6	10	14	2	6	10	11	2	6			2
100	Fm	2	2	6	2	6	10	2	6	10	14	2	6	10	12	2	6			2
101	Md	2	2	6	2	6	10	2	6	10	14	2	6	10	13	2	6			2
102	No	2	2	6	2	6	10	2	6	10	14	2	6	10	14	2	6			2
103	Lr	2	2	6	2	6	10	2	6	10	14	2	6	10	14	2	6	1		2
104	Ku	2	2	6	2	6	10	2	6	10	14	2	6	10	14	2	6	2		2
105	Ha	2	2	6	2	6	10	2	6	10	14	2	6	10	14	2	6	3		2
106	—	2	2	6	2	6	10	2	6	10	14	2	6	10	14	2	6	4		2

energy level diagram where there is a set of filled orbitals and a relatively large energy gap before the next empty orbital. Horizontal dashed lines show such gaps in Figure 2.8c. Thus, the first period has only 2 elements, the second has 8, the third has 8, the fourth has 18, the fifth has 18, the sixth has 32, and the seventh should have 32 before it is finally completed.

Groups (vertical columns) of the periodic table can be developed by drawing vertical lines in Figure 2.8c, filling electrons in orbitals until you reach the vertical lines in successive attempts, and identifying the element at each attempt. If you fill successively to the left-hand vertical dashed line, you obtain first Be, then Mg, Ca, Sr, Ba, and Ra, the elements of group II. These elements having the same number of electrons in the outermost shell are chemically similar. Likewise, filling successively to the middle vertical dashed line yields B, Al, Ga, In, and Tl, the elements of group III, and filling successively to the right-hand vertical dashed line yields He, Ne, Ar, Kr, Xe, and Rn, the noble gas elements which have filled-shell electron configurations. Care must be exercised as to where vertical lines are drawn to obtain the elements of groups IV, V, VI, and VII, and elements of the various subgroups in the transition elements due to the spreading of electrons in equal energy orbitals before pairing in accordance with rule 3 on page 44.

Two common classifications of elements are used on the basis of what orbitals are being filled. **Representative elements** (main group elements) are *those in which **s** and **p** orbitals are being filled.* This includes groups IA, IIA, IIIA, IVA, VA, VIA, VIIA, and VIIIA on the periodic table inside the back cover. **Transition elements** (subgroup elements) are *those in which **d** and **f** orbitals are being filled.* This includes groups IB, IIB, IIIB, IVB, VB, VIB, VIIB, and VIIIB as well as the lanthanide series and the actinide series on the periodic table inside the back cover.

2.13 Electronic Symbols

Electrons in the outermost shell of an atom usually determine the chemical properties of that atom. Therefore, it is convenient to specify the electron configuration of the outermost shell of representative elements by electron dot symbols. **Electronic symbols** consist of *the symbol of the element* (representing the nucleus and electrons in inner shells) *surrounded by dots* (representing electrons in the outermost shell). The dots are only a means of keeping track of the outermost electrons and they do not represent actual spatial distributions of the outermost electrons.

The electron configuration for a phosphorus atom (atomic number 15) is $1s^2 2s^2 2p^6 3s^2 3p^3$. The electronic symbol for phosphorus, representing only the electrons in the outermost energy level ($3s^2 3p^3$), is shown along with electronic symbols for magnesium, carbon, chlorine, and neon.

$$\cdot\overset{\cdot\cdot}{\underset{\cdot}{\mathrm{P}}}\cdot \qquad \mathrm{Mg}: \qquad \overset{\cdot\cdot}{\underset{\cdot}{\mathrm{C}}}\cdot \qquad \cdot\overset{\cdot\cdot}{\underset{\cdot\cdot}{\mathrm{Cl}}}: \qquad :\overset{\cdot\cdot}{\underset{\cdot\cdot}{\mathrm{Ne}}}:$$

Note that paired and unpaired electrons are represented by a pair of dots and a single dot, respectively.

Exercises

2.1 Define clearly the following terms.

(a) mass
(b) chemical equation
(c) reactants
(d) products
(e) empirical formula
(f) molecular formula
(g) element
(h) atom
(i) compound
(j) atomic mass
(k) periodic law
(l) periodic table
(m) electron
(n) proton
(o) neutron
(p) alpha particle
(q) nuclear atom
(r) atomic number
(s) isotope
(t) mass number
(u) line spectrum
(v) energy level
(w) shell
(x) subshell
(y) orbital
(z) Pauli exclusion principle
(aa) electron spin
(ab) ground state
(ac) energy level diagram
(ad) electron configuration
(ae) period
(af) group
(ag) representative elements
(ah) transition elements
(ai) electronic symbols

2.2 Define clearly, giving specific examples.
(a) the law of conservation of mass **(b)** the law of constant composition

2.3 Describe briefly the contributions of the following scientists toward the development of the periodic table of the elements.
(a) Mendeleev
(b) Moseley
(c) Bohr

2.4 **(a)** State briefly the basic postulates of Dalton's atomic theory.
(b) Describe how one or more of these postulates explain (1) the law of conservation of mass and (2) the law of constant composition.
(c) Are any modifications of Dalton's atomic theory required in light of our present knowledge of atomic structure? State clearly your reasoning.

2.5 Consider the elements chlorine, argon, and potassium with atomic numbers 17, 18, and 19, respectively, and atomic masses 35.453, 39.948, and 39.102 amu, respectively. Show in what order you would have placed these three elements in your periodic table if you had been **(a)** Mendeleev, **(b)** Moseley, and **(c)** Bohr. State clearly the knowledge available to, and the thinking of, these men in order to defend your answers.

2.6 Explain clearly the following statements.
(a) The atomic mass of chlorine (35.453 amu) disproves an early hypothesis by Prout, a British chemist, that hydrogen (1.0079 amu) was the basic element from which all others could be derived.
(b) The atomic numbers and atomic masses of tellurium (52, 127.60 amu) and iodine (53, 126.905 amu) illustrate the manner in which we determine the order of elements in the periodic table.

2.7 Outline clearly the experiment you would perform, the apparatus you would use, the results you would expect, and the conclusions you would draw while trying to convince your roommate that the following existed.

(a) electrons
(b) protons
(c) neutrons
(d) atomic nuclei
(e) quantized electronic energy levels

2.8 Describe briefly the structure of an atom according to models put forth by Thomson and by Rutherford and show how alpha-particle-scattering experiments provide evidence opposing the Thomson model but supporting the Rutherford model.

2.9 Describe the number of protons and neutrons in the nucleus and the number of electrons in the extranuclear region for each of the following.

(a) $^{19}_{9}F$
(b) $^{64}_{29}Cu$
(c) $^{90}_{38}Sr$
(d) $^{127}_{53}I$
(e) $^{235}_{92}U$

2.10 Write the electron configuration and electronic symbol for each of the following atoms in the ground state.

(a) O
(b) Ar
(c) N
(d) Cl
(e) C

2.11 Find a block of three consecutive elements in Table 2.4 that illustrate the third sentence of the footnote in Section 2.12.

2.12 Describe the meaning of each of the following symbols.

(a) $1s^22s^22p^2$
(b) $\dot{A}l:$
(c) $:\dot{\underset{\cdot}{N}}\cdot$
(d) $1s^22s^22p^63s^1$
(e) $1s^22s^22p^63s^23p^64s^23d^7$

2.13 How would you describe the relationship between the electron configurations of atoms and the placement of atoms in the periodic table?

2.14 Account for the fact that Si has one electron in each of two $3p$ orbitals rather than two electrons of opposite spin in the same $3p$ orbital.

2.15 Referring only to Figure 2.8c, determine the number of electrons in the neutral atoms of elements that make up the following.

(a) period 2
(b) group IV
(c) group VII
(d) period 5

3

Chemical Bonding and Molecular Architecture

Glycine is the simplest amino acid and is the major amino acid found in gelatin. The glycine molecule consists of two carbon atoms, five hydrogen atoms, one nitrogen atom, and two oxygen atoms. Physically, glycine is a sweet-tasting, colorless, crystalline compound that is readily soluble in water. Chemically, glycine can be obtained by the hydrolysis of proteins and undergoes reactions characteristic of both amino groups and carboxylic acids. The physical and chemical properties of glycine are critically dependent upon how its various atoms are bonded together to form a three-dimensional structure.

Thus far we have dealt with the structure of individual atoms. The purposes of this chapter are to examine the forces (ionic, covalent, dipolar, hydrogen bonding, metallic, and dispersion) that cause atoms to bind together into molecules or into larger aggregates, and to study several factors that determine the shapes of simple molecules.

3.1 Chemical Bonds

Chemical bonds are *the forces or interactions that cause atoms to be held together as molecules or cause atoms, ions, and molecules to be held together as more complex aggregates.* If no such forces existed, no atoms or molecules would condense to liquids or solids; all atoms would exist as monatomic gases, even at temperatures approaching absolute zero. Chemical bonds range from strong ionic and covalent bonds to weaker metallic bonds to very much weaker dipole–dipole interactions, hydrogen bonds, and dispersion forces. All of these types of chemical bonds can be understood at least partially in terms of electrostatic interactions (either attractive or repulsive) between charged bodies such as positively and negatively charged ions, electrons and nuclei containing protons, polar molecules, or instantaneous dipoles formed by atoms or nonpolar molecules.

3.2 Ions

An **ion** is *an atom or distinguishable group of atoms bearing one or more positive or negative charges due to the loss or gain of one or more electrons.* Common table salt, sodium chloride, is composed of Na^+ and Cl^- ions. These ions are formed from single atoms. Ammonium nitrate consists of ammonium ions (NH_4^+) and nitrate ions (NO_3^-). These ions are formed from distinguishable groups of atoms. *A positively charged ion* such as Na^+ or NH_4^+ is called a **cation.** *A negatively charged ion* such as Cl^- or NO_3^- is called an **anion.**

The formation of a cation requires a *loss* or removal of one or more electrons from a neutral atom. The **ionization energy** is *the energy required to remove an electron from a gas phase atom or ion.* It is a measure of the hold that an atom or cation has on a given electron. This hold results from the electrostatic attraction between the positive nuclear charge and the negative charge of the electron. The first ionization energy for Mg is the energy required for the following process.

$$Mg_{(g)} \longrightarrow Mg^+{}_{(g)} + e^-$$

This step involves removal of an electron from a neutral gas phase Mg atom to form a Mg^+ cation. A second step, involving removal of an electron from a positively charged gas phase cation, requires energy equivalent to the second ionization energy, and can be represented as

$$Mg^+{}_{(g)} \longrightarrow Mg^{2+}{}_{(g)} + e^-$$

With few exceptions, first ionization energies increase across periods and decrease down groups of the periodic table. Such trends are illustrated in Figure 3.1. In addition, second, third, and higher ionization energies are successively greater than first ionization energies. Thus, the first five ionization energies for magnesium are 176, 347, 1848, 2521, and 3256 kcal/mole, respectively. Note that it is much more difficult to remove electrons from an inner shell—as indicated by the high third, fourth, and fifth ionization energies—than it is to remove electrons from the outer shell—as indicated by the lower first and second ionization energies. Therefore, in ordinary chemical reactions atoms only lose outer-shell electrons.

The formation of an anion requires a *gain* of one or more electrons by a neutral atom. **Electron affinity** is *the energy released when an electron is added to a gas phase atom or ion.* It is a measure of the hold that an atom or anion

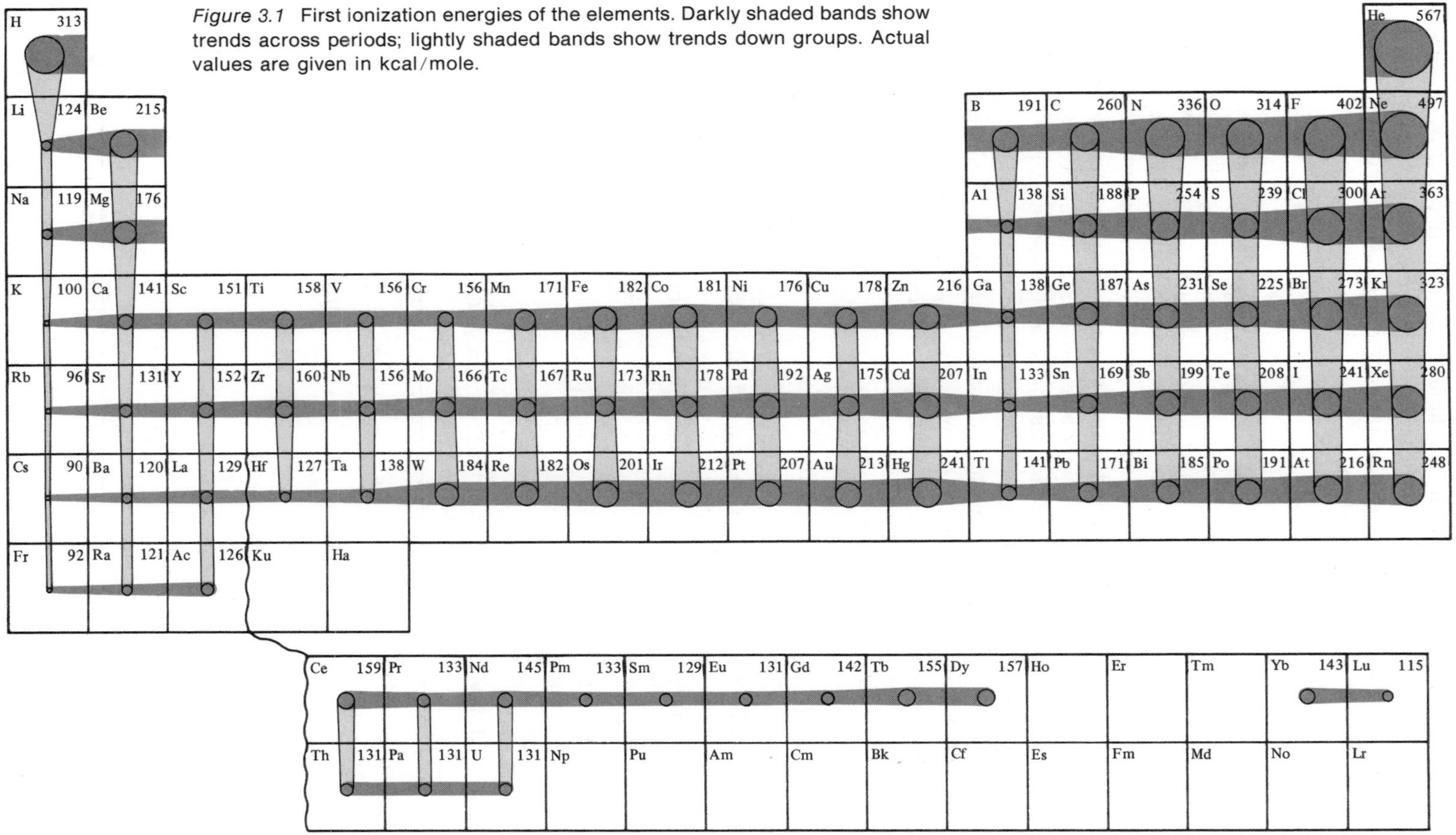
H 313
He 567
Li 124
Be 215
B 191
C 260
N 336
O 314
F 402
Ne 497
Na 119
Mg 176
Al 138
Si 188
P 254
S 239
Cl 300
Ar 363
K 100
Ca 141
Sc 151
Ti 158
V 156
Cr 156
Mn 171
Fe 182
Co 181
Ni 176
Cu 178
Zn 216
Ga 138
Ge 187
As 231
Se 225
Br 273
Kr 323
Rb 96
Sr 131
Y 152
Zr 160
Nb 156
Mo 166
Tc 167
Ru 173
Rh 178
Pd 192
Ag 175
Cd 207
In 133
Sn 169
Sb 199
Te 208
I 241
Xe 280
Cs 90
Ba 120
La 129
Hf 127
Ta 138
W 184
Re 182
Os 201
Ir 212
Pt 207
Au 213
Hg 241
Tl 141
Pb 171
Bi 185
Po 191
At 216
Rn 248
Fr 92
Ra 121
Ac 126
Ku
Ha
Ce 159
Pr 133
Nd 145
Pm 133
Sm 129
Eu 131
Gd 142
Tb 155
Dy 157
Ho
Er
Tm
Yb 143
Lu 115
Th 131
Pa 131
U 131
Np
Pu
Am
Cm
Bk
Cf
Es
Fm
Md
No
Lr

Figure 3.1 First ionization energies of the elements. Darkly shaded bands show trends across periods; lightly shaded bands show trends down groups. Actual values are given in kcal/mole.

has on an *additional* electron. This hold results from the electrostatic attraction between the positive nuclear charge and the negative charge of the electron. The first and second electron affinities for oxygen are the energies released in the following steps.

$$O_{(g)} + e^- \longrightarrow O^-_{(g)}$$
$$O^-_{(g)} + e^- \longrightarrow O^{2-}_{(g)}$$

Being dependent upon the same factors, electron affinities show approximately the same trends as ionization energies (Figure 3.1)—they increase across periods and decrease down groups. However, they are not easily measured experimentally, and accurate values for only a few elements are known. Second, third, and higher electron affinities are successively much less favorable than first electron affinities, and atoms gain at most two or three electrons in ordinary chemical reactions.

Many commonly occurring ions have noble gas or filled-shell electron configurations (Section 2.12). Atoms frequently gain or lose sufficient electrons to attain noble gas configurations, as illustrated by the third period elements in Table 3.1.

3.3 Ionic Bonds

The simplest kind of electrostatic attraction, that between two oppositely charged ions, gives rise to an ionic bond. An **ionic bond** results from *an electrostatic attraction between a positively charged ion* (cation) *and a negatively charged ion* (anion). We can imagine the formation of an ionic bond as involving three *hypothetical* steps:

1. The loss of one or more electrons by one element to form a cation (requires energy).

Table 3.1 Formation of Ions with Noble Gas Configurations for Third Period Elements

Cations	*Anions*
$Na\cdot \longrightarrow Na^+ + e^-$	
$Mg: \longrightarrow Mg^{2+} + 2\,e^-$	
$\dot{Al}: \longrightarrow Al^{3+} + 3\,e^-$	
$\cdot\dot{Si}: \longrightarrow Si^{4+} + 4\,e^-$	
$\cdot\dot{P}: \longrightarrow P^{5+} + 5\,e^-$	$\cdot\dot{P}: + 3\,e^- \longrightarrow :\ddot{P}:^{3-}$
$\cdot\ddot{S}: \longrightarrow S^{6+} + 6\,e^-$	$\cdot\ddot{S}: + 2\,e^- \longrightarrow :\ddot{S}:^{2-}$
	$:\ddot{Cl}: + e^- \longrightarrow :\ddot{Cl}:^-$

2. The gain of one or more electrons by another element to form an anion (releases energy).
3. The attraction of the cation and anion formed (releases energy).

As a result of wanting to minimize the energy in step 1 and to maximize the energy in step 2, an ionic bond is favored when an atom with low ionization energy (Section 3.2) and an atom with high electron affinity (Section 3.2) combine. Sodium chloride, formed by the reaction of Na atoms and Cl atoms, is an example of an ionic compound. Little energy is required to remove an electron from a gaseous Na atom to form a Na^+ cation in step 1 since Na has a very low ionization energy for the process

$$Na_{(g)} \longrightarrow Na^+_{(g)} + e^-$$

Significant energy is released when an electron is added to a gaseous Cl atom to form a Cl^- anion in step 2 since Cl has a high electron affinity for the process

$$Cl_{(g)} + e^- \longrightarrow Cl^-_{(g)}$$

Considerable energy is also released when the two ions are brought together in step 3 and their opposite charges attract each other to form an ionic bond.

$$Na^+_{(g)} + Cl^-_{(g)} \longrightarrow Na^+Cl^-_{(g)}$$

The net result in the formation of a fully ionic bond is complete electron transfer from the atom with low ionization energy to the atom with high electron affinity.

Charge must be conserved in any chemical reaction; that is, there must be a balance of electrons gained and lost in the electron transfer causing formation of an ionic bond. This results in a product that is electrically neutral. When magnesium atoms react with fluorine atoms, each Mg atom loses two electrons

$$Mg_{(g)} \longrightarrow Mg^{2+}_{(g)} + 2\,e^-$$

and each F atom gains only one electron.

$$F_{(g)} + e^- \longrightarrow F^-_{(g)}$$

As a result two F atoms are needed to accept the two electrons lost by the Mg atom, and the net reaction can be written as

$$Mg_{(g)} + 2\,F_{(g)} \longrightarrow MgF_{2(g)}$$

Whether atoms gain or lose one, two, or more electrons in the formation of ionic compounds depends upon the relative energies released for each case in the three steps on page 54. The compound that has the greatest release of energy during its formation will be the favored one and the most stable one. Thus, energy calculations can predict that MgF_2 is formed, but that Mg_2F, MgF, and MgF_3 are not. Frequently, as in this case, the stable salt formed is one whose ions have noble gas or filled-shell electron configurations (Section 2.12).

Ionic compounds are favored between elements on the far left of the periodic table, having low ionization energies, and elements on the far right of the periodic table, having high electron affinities. Examples include potassium chloride (KCl), cesium bromide (CsBr), and magnesium oxide (MgO). Relatively high electron affinities also result from the formation of some of the common oxyanions such as hydroxide (OH^-), nitrate (NO_3^-), carbonate (CO_3^{2-}), sulfate (SO_4^{2-}), and phosphate (PO_4^{3-}). Therefore, these oxyanions form ionic compounds with cations derived from elements on the far left of the periodic table. Examples include rubidium nitrate ($RbNO_3$), calcium carbonate ($CaCO_3$), and barium sulfate ($BaSO_4$). Formation of some transition metal cations with *low* charge requires relatively low ionization energies. Therefore, these cations form ionic compounds with anions derived from elements on the far right of the periodic table. Examples include manganese(II) chloride ($MnCl_2$) and iron(II) fluoride (FeF_2).

3.4 Nondirectional Nature of Ionic Bonds

Electrostatic interactions are not specific for one direction; they occur in all directions. Therefore, ionic bonds are nondirectional. A given cation may be attracted simultaneously by a number of anions and vice versa. Thus, sodium chloride has a crystal lattice (Figure 3.2) consisting of Na^+ ions, each surrounded by six Cl^- ions, and Cl^- ions, each surrounded by six Na^+ ions. Sodium chloride, like all ionic compounds, consists of an infinite array of positively and negatively charged ions in such ratios that the total number of positive charges is just balanced by the same total number of negative charges. No isolated NaCl unit can be identified within this infinite array. However, we can still consider NaCl as the simplest formula describing sodium chloride and can define a bond length for NaCl as the center-to-center distance between adjacent Na^+ and Cl^- ions.

Since electrostatic forces of attraction between oppositely charged ions are strong in all directions, ionic compounds are usually quite hard, fairly dense, high melting, solids. However, many ionic compounds dissolve readily in water, a property responsible for their importance in biological processes. Salts of ions such as Na^+, K^+, and Cl^-, ingested daily by each of us, dissolve readily in and are easily transported by our blood and other body fluids.

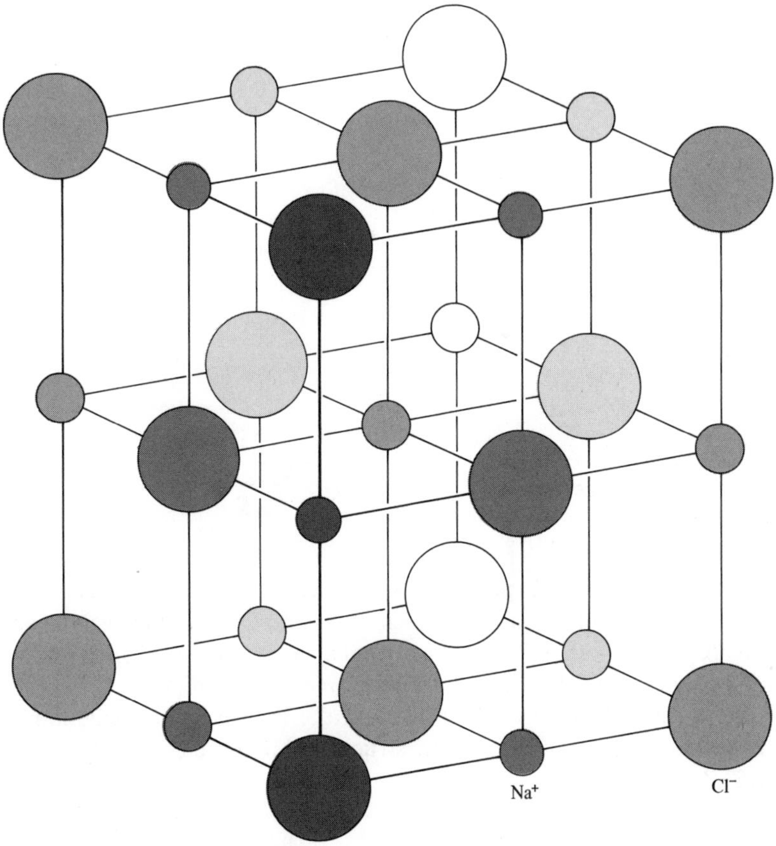
Na⁺
Cl⁻

Figure 3.2 Crystal lattice of sodium chloride. Darkly shaded ions are close to you and lightly shaded ions far from you. Na^+ and Cl^- ions essentially touch each other but have been reduced in size for clarity.

3.5 Covalent Bonds

When elements with a similar hold on outer shell electrons combine, a complete transfer of outer shell electrons to form ions is unreasonable. Both atoms try to maintain a hold on the electrons involved. A **covalent bond** arises *from the attraction between two atoms as a result of sharing one or more outer shell electrons by the two atoms.* A single covalent bond results when two atoms are held together by the sharing of a pair of electrons with each atom normally contributing one electron to the pair.

The hydrogen molecule is the simplest case involving a covalent bond. Figure 3.3 presents an electrostatic picture of what happens when two H atoms, each with a single 1*s* electron, form a H_2 molecule. This is only a fixed, *hypothetical* picture because the protons and electrons are constantly in motion. Attractive forces arise because each electron is attracted by both

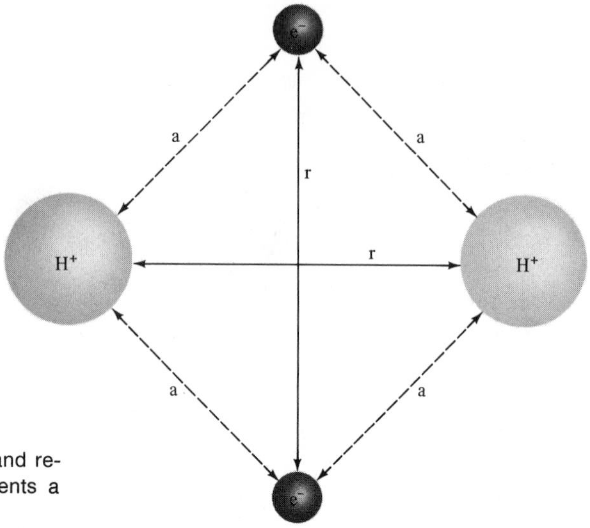
e⁻
a
a
r
H⁺
r
H⁺
a
a
e⁻

Figure 3.3 Representations of the attractions (a) and repulsions (r) in the hydrogen molecule. H^+ represents a proton and e^- an electron.

positively charged nuclei (protons). Repulsive forces arise because the two negatively charged electrons repel each other and the two positively charged protons repel each other. There are a total of four attractive forces and only two repulsive forces, and the attractive forces predominate.

The formation of a covalent bond in the hydrogen molecule is represented more simply using electronic structures composed of electronic symbols (Section 2.13).

$$\mathrm{H}\cdot + \cdot\mathrm{H} \longrightarrow \mathrm{H}:\mathrm{H} \quad \text{or} \quad \mathrm{H{-}H}$$

The shared electron pair forming the covalent bond is shown as two dots (or as a short line) *between* the two H atoms.

The covalent bond in the diatomic H_2 molecule has specific properties. It is a *single* bond because the two H atoms share *one* electron pair. Its *bond length* is the center-to-center distance between the two nuclei. Its *bond energy* is the energy required to break a gaseous H_2 molecule apart into two gaseous H atoms (103 kcal/mole, about enough energy to raise the temperature of one liter of water from 0° to 100°). It is a *nonpolar* covalent bond because both atoms are identical, both have the same attraction for the electron pair, and both share the electron pair equally. There is no separation of positive and negative charge between the two H atoms; the center of positive charge coincides with the center of negative charge.

Chlorine gas consists of nonpolar, covalent, diatomic, molecules. Each Cl atom originally has a $[\mathrm{Ne}]3s^23p^5$ electron configuration. Each Cl atom shares one of its $3p$ electrons with another Cl atom to form a shared-pair, single, covalent bond.

$$:\ddot{\underset{..}{\mathrm{Cl}}}\cdot + \cdot\ddot{\underset{..}{\mathrm{Cl}}}: \longrightarrow :\ddot{\underset{..}{\mathrm{Cl}}}:\ddot{\underset{..}{\mathrm{Cl}}}: \quad \text{or} \quad :\ddot{\underset{..}{\mathrm{Cl}}}{-}\ddot{\underset{..}{\mathrm{Cl}}}:$$

The other three pairs of electrons on each Cl atom do not take part in bonding and are called nonbonding electron pairs or unshared electron pairs.

Electronic structures are only bookkeeping devices and incorrectly imply that certain pairs of electrons, shown between two nuclei, are the only ones that are involved in a covalent bond. On the contrary, all outer shell electrons are generally associated with the molecule as a whole rather than the individual atoms and contribute partially to bond formation. Inner shell electrons reside almost completely on individual atoms and contribute little or nothing to bond formation.

The Cl—Cl bond is nonpolar because each Cl atom has the same attraction for the shared electron pair. The Cl—Cl single covalent bond is weaker than an H—H single covalent bond, illustrating the general rule that larger atoms are not as close to each other as smaller atoms, have less effective sharing of electron density, and thus form weaker bonds.

Nonpolar covalent bonding usually results in the formation of *discrete* molecules with strong bonding within molecules, but only very weak dispersion forces (Section 3.13) between molecules. The molecules are broken apart into their constituent atoms only with large expenditures of energy; however, relatively little energy is required to separate the molecules from each other. Thus, nonpolar covalent compounds are typically low melting solids, low boiling liquids, or even gases at room temperature, and they exhibit low densities and high vapor pressures (Section 6.4). The odors of a number of covalent solids and liquids (for example, gasoline) can be detected at considerable distances from the samples themselves because of their high vapor pressures, even near room temperature. Nonpolar compounds are insoluble in water and in other polar (Section 3.10) liquids, but dissolve readily in nonpolar liquids.

Some substances have strong covalent bonds between *all* atoms resulting in the formation of giant molecules. These so-called *covalent solids* have lattices that, like ionic lattices, have no discrete molecules. Covalent bonds in fixed directions create an infinite interlocking framework. Silicon dioxide (of which flint, opal, quartz, and sand are common forms) is an example of a substance with strong covalent bonds in three dimensions. Figure 3.4 depicts the crystal structure of quartz. Diamond (Sections 6.7 and 6.11) is another example of a covalent solid. Such substances are usually very hard, high melting solids. They are insoluble in almost all solvents and are poor electrical conductors.

3.6 Multiple Covalent Bonds

Multiple covalent bonds occur when more than one electron pair is shared between two atoms. When ***two*** *electron pairs are shared between two atoms,* a **double bond** is formed. It is stronger than a single bond, but not twice as strong, because the two electron pairs repel each other and cannot become fully concentrated between the two atoms. Its bond length is shorter than that

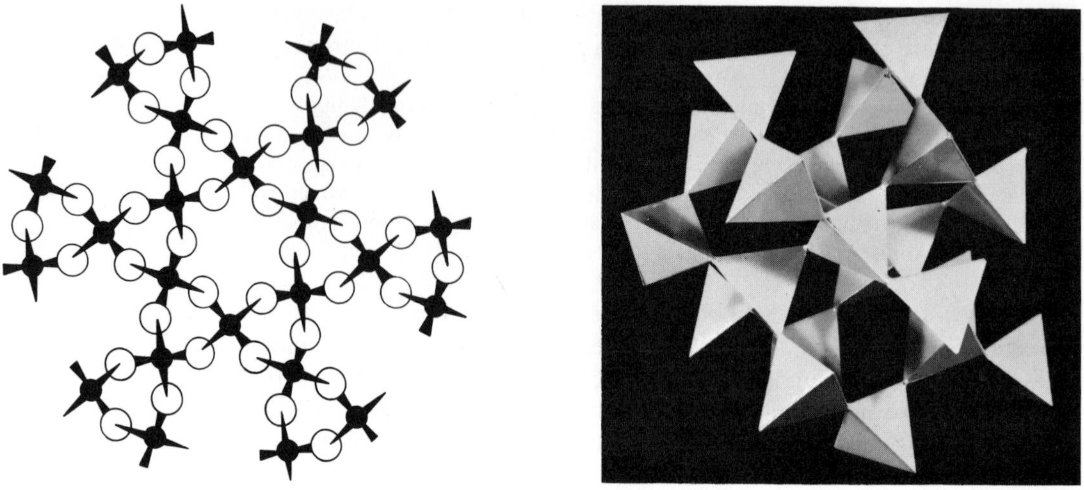

(a) (b)

Figure 3.4 (a) Structure of quartz. Small black circles represent Si atoms; open circles represent O atoms. (b) Picture of Si tetrahedra in quartz; O atoms not shown. [Reproduced with permission from A. F. Wells, *Structural Inorganic Chemistry,* The Clarendon Press, Oxford, 1962.]

of a single bond. When ***three*** *electron pairs are shared between two atoms,* a **triple bond** is formed. It is stronger than a single bond and a double bond, but not three times as strong as a single bond. Its bond length is shorter than that for a double bond. A quadruple bond, involving the sharing of four electron pairs, is known for $Re_2Cl_8^{2-}$. *Saturated* molecules possess only single bonds; *unsaturated* molecules possess one or more double or triple bonds.

Nitrogen gas consists of nonpolar, diatomic, covalent molecules having triple bonds. Each N atom has a $1s^22s^22p^3$ electron configuration. Each N atom shares three $2p$ electrons with another N atom to form a triple bond.

$$:\dot{\underset{\cdot}{N}}\cdot + \cdot\dot{\underset{\cdot}{N}}: \longrightarrow :N:::N: \quad \text{or} \quad :N{\equiv}N:$$

It is a nonpolar bond because both atoms are identical and have the same attraction for the three electron pairs.

Table 3.2 gives examples of some carbon compounds which illustrate electronic structures and the increase in carbon–carbon bond energy and decrease in carbon–carbon bond distance upon the formation of multiple bonds.

Multiple bonds are formed commonly by carbon, nitrogen, and oxygen atoms and occasionally by boron, silicon, phosphorus, sulfur, and other atoms. Multiple bonds arise whenever there are too few outer shell electrons to provide each atom with a stable noble gas electron configuration (Section 2.12) using a completely single-bonded structure. Electronic structures can be used to predict whether or not multiple bonds exist in simple molecules (Section 3.8).

Table 3.2 Compounds Illustrating Single and Multiple Carbon–Carbon Bonds

Name	*Electronic Structure*	*Bond Energy (kcal/mole)*	*Bond Distance (Å)*
Ethane	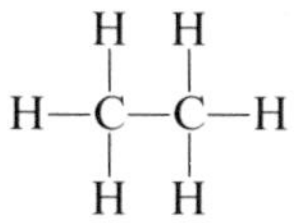	83	1.54
Ethylene	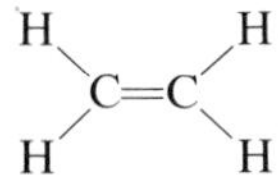	147	1.33
Acetylene	H—C≡C—H	194	1.20

3.7 Polar Covalent Bonds

A nonpolar covalent bond (Section 3.5) always arises between two *identical* atoms or between two *different* atoms having *identical* electronegativities. **Electronegativity** is *a measure of the tendency of an atom in a molecule to attract shared electrons to itself.* It indicates the relative tendency of an atom to attain a negative condition when forming a bond. It is a measure of the hold that an atom has on *shared* electrons. This hold arises from electrostatic attraction between the shared electrons and the protons in the nucleus of an atom. Electronegativities are dependent upon the same factors as ionization energies (Section 3.2) and show the same trends—increase across periods and decrease down groups (Figure 3.5).

Hydrogen chloride gas consists of polar, diatomic, covalent molecules. Each H atom shares a 1*s* electron with a Cl atom, which in turn shares a 3*p* electron with a H atom. A single covalent bond is formed. It is a *polar* bond

$$H\cdot + \cdot\ddot{\underset{..}{Cl}}: \longrightarrow H:\ddot{\underset{..}{Cl}}: \qquad \text{or} \qquad H{-}\ddot{\underset{..}{Cl}}:$$

because Cl has a greater electronegativity than H (Figure 3.5) and has a stronger attraction for the shared electron pair than H. The electron pair is shared unequally. A charge separation or *dipole* is produced. Experimental observations of the orientation of HCl molecules between two charged plates indicate that the H end of the molecule has a partial positive charge (symbolized, using the Greek delta, by $\delta+$) and the Cl end a partial negative charge (symbolized by $\delta-$); thus: $^{\delta+}H{-}Cl^{\delta-}$.

The degree of polarity in a covalent bond can be predicted from differences in electronegativities. The greater the difference in electronegativities, the greater the polarity of the bond. Consider the series of molecules HF, HCl, HBr, and HI. Electronegativities decrease through F (4.0), Cl (3.0), Br (2.8), and I (2.5), and approach the value for H (2.1). The difference in electronegativities then decreases through the series HF, HCl, HBr, and HI, and there is a trend toward equal sharing of electrons. Thus, a trend of decreasing polarity through HF, HCl, HBr, and HI is predicted. Such a trend has been verified experimentally. Already you can see that *there is no clear*

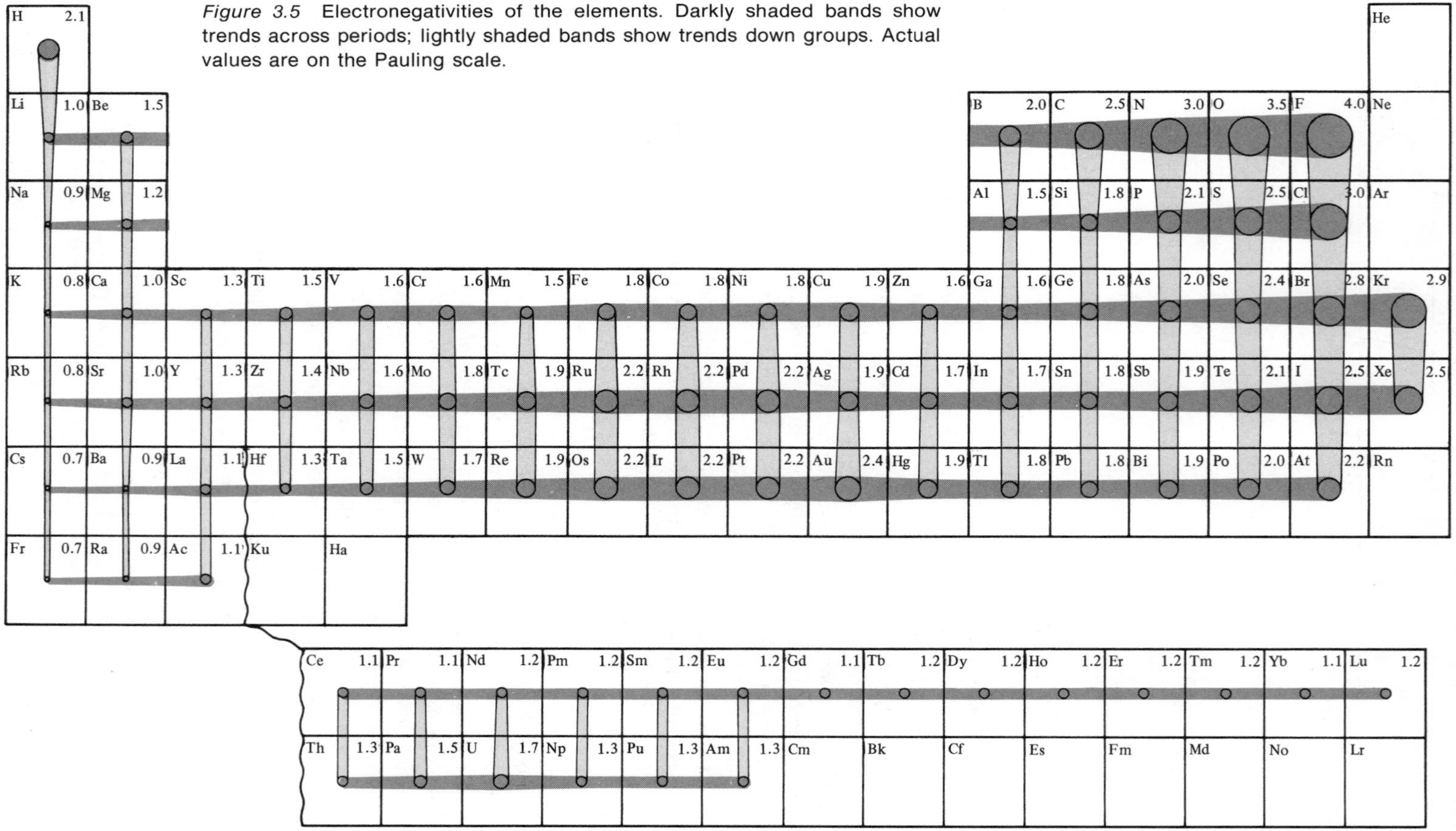
H 2.1
He
Li 1.0
Be 1.5
B 2.0
C 2.5
N 3.0
O 3.5
F 4.0
Ne
Na 0.9
Mg 1.2
Al 1.5
Si 1.8
P 2.1
S 2.5
Cl 3.0
Ar
K 0.8
Ca 1.0
Sc 1.3
Ti 1.5
V 1.6
Cr 1.6
Mn 1.5
Fe 1.8
Co 1.8
Ni 1.8
Cu 1.9
Zn 1.6
Ga 1.6
Ge 1.8
As 2.0
Se 2.4
Br 2.8
Kr 2.9
Rb 0.8
Sr 1.0
Y 1.3
Zr 1.4
Nb 1.6
Mo 1.8
Tc 1.9
Ru 2.2
Rh 2.2
Pd 2.2
Ag 1.9
Cd 1.7
In 1.7
Sn 1.8
Sb 1.9
Te 2.1
I 2.5
Xe 2.5
Cs 0.7
Ba 0.9
La 1.1
Hf 1.3
Ta 1.5
W 1.7
Re 1.9
Os 2.2
Ir 2.2
Pt 2.2
Au 2.4
Hg 1.9
Tl 1.8
Pb 1.8
Bi 1.9
Po 2.0
At 2.2
Rn
Fr 0.7
Ra 0.9
Ac 1.1
Ku
Ha
Ce 1.1
Pr 1.1
Nd 1.2
Pm 1.2
Sm 1.2
Eu 1.2
Gd 1.1
Tb 1.2
Dy 1.2
Ho 1.2
Er 1.2
Tm 1.2
Yb 1.1
Lu 1.2
Th 1.3
Pa 1.5
U 1.7
Np 1.3
Pu 1.3
Am 1.3
Cm
Bk
Cf
Es
Fm
Md
No
Lr

Figure 3.5 Electronegativities of the elements. Darkly shaded bands show trends across periods; lightly shaded bands show trends down groups. Actual values are on the Pauling scale.

distinction between covalent bonds and ionic bonds. The charge separations in the hydrogen halide molecules suggest the presence of covalent bonds, but with superimposed ionic character.

3.8 Writing Electronic Structures

Correct electronic structures (often referred to as Lewis electron dot structures) are essential for determining whether or not multiple bonding exists in a molecule or complex ion, and more important, for determining the three-dimensional spatial configuration of a molecule or complex ion. For many molecules for which you do not know the bonding picture and structure, several possible electronic structures can be written. It is relatively easy to decide which of the several electronic structures is most valid. At this stage we are not concerned with the specific geometry of several atoms around another atom.

Electronic structures in which there are two electrons around hydrogen, eight electrons around second period elements, and frequently (but not always) *eight electrons around representative elements in the third through seventh periods, are preferred.* Shared electrons forming covalent bonds are counted as being around *both* of the atoms that share them when applying this rule. This rule results from the fact that atoms tend to share sufficient electrons to attain the very stable electron configuration of the next noble gas. They do this by forming a number of covalent bonds equal to the number of *additional* electrons required. A hydrogen atom, $\mathrm{H}\cdot$, shares *one* additional electron to achieve the electron configuration of $\mathrm{He}:$ and forms *one* covalent bond. A fluorine atom, $:\ddot{\underset{\cdot\cdot}{\mathrm{F}}}\cdot$ (and usually atoms of other elements of group VII), shares *one* additional electron and forms *one* covalent bond; an oxygen atom, $:\ddot{\underset{\cdot}{\mathrm{O}}}\cdot$ (and usually atoms of other elements of group VI), shares *two* additional electrons and forms *two* covalent bonds; a nitrogen atom, $:\dot{\underset{\cdot}{\mathrm{N}}}\cdot$ (and usually atoms of other elements of group V), shares *three* additional electrons and forms *three* covalent bonds; and a carbon atom, $:\dot{\mathrm{C}}\cdot$ (and usually atoms of other elements of group IV), shares *four* additional electrons and forms *four* covalent bonds—all to achieve the $:\ddot{\underset{\cdot\cdot}{\mathrm{Ne}}}:$ electron configuration.

This rule also leads to the formation of the greatest possible number of covalent bonds—*one* for hydrogen and group VII elements, *two* for group VI elements, *three* for group V elements, and *four* for group IV elements. The maximum number of covalent bonds provides the greatest possible bond energy and results in the most stable molecule or the one with the lowest energy. This maximum number of covalent bonds can be achieved by any combination of single and/or multiple bonds. Thus, each carbon atom in the molecules in Table 3.2 forms four bonds—by means of four single bonds in ethane, two single bonds and one double bond in ethylene, and one single bond and one triple bond in acetylene.

Example 3.1 Determine the correct electronic structures for the following molecules: (a) CH_3I, (b) CO_2, and (c) C_2H_6O.

Although electrons are indistinguishable, it is helpful to designate outer shell electrons of different atoms with different symbols until you have determined the correct electronic structure.

(a) :Ċ· shares 4 additional electrons and forms 4 covalent bonds.

H∘ shares 1 additional electron and forms 1 covalent bond.

:I× (with ×× above and below) shares 1 additional electron and forms 1 covalent bond.

Usually the atom capable of forming the most covalent bonds is the central atom. With the exception of hydrogen, which is always a terminal atom, the central atom is also the least electronegative atom. Try a structure in which 3 H atoms and 1 I atom are bonded to C to provide C with 4 covalent bonds.

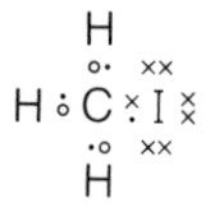

Note that each H atom has 2 electrons around it and forms 1 bond, the C atom has 8 electrons around it and forms 4 bonds, and the I atom has 8 electrons around it and forms 1 bond, according to the previous rules. This is the correct electronic structure for CH_3I.

The following structure is an incorrect one.

It obeys the general rule of having 2 electrons around H and 8 around C and I, and it has 4 covalent bonds like the previous structure. However, it has one too few bonds for C and one too many bonds for I.

(b) :Ċ· shares 4 additional electrons and forms 4 covalent bonds.

:O× (with ×× above and × below) shares 2 additional electrons and forms 2 covalent bonds.

The C atom can form more covalent bonds, is the less electronegative atom, and is probably the central atom. Therefore, try a structure with the C atom bonded to 2 O atoms and with single bonds. Always start with a single-bonded structure first unless you know that double bonds exist.

This is a poor structure because the O atoms do not have an octet of electrons and the C atom forms only 2 bonds and the O atoms only one bond (none of the atoms form the maximum number of bonds of which they are capable).

The following structure suffers from the same two difficulties except that it is the C atom rather than the O atoms that lacks an octet of electrons.

The preceding difficulties can be alleviated by the use of multiple bonding. The following structures are only marginally better than previous ones because they still

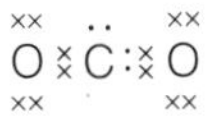

suffer partially from the two problems already mentioned. However, introduction of a second C=O double bond solves both problems.

Note that each atom has 8 electrons around it and that each atom has formed its maximum number of bonds. This is the correct electronic structure for CO_2.

Perhaps you can tell simply by looking at previous structures that the following one is incorrect even though all atoms have an octet of electrons and 4 covalent

bonds are formed. It is incorrect because one O atom forms 1 bond whereas the other forms 3 bonds; both normally form 2 bonds.

It is left to you as an exercise to show that any electronic structures having O as the central atom are also incorrect. Note that no previous knowledge other than the molecular formula was required to obtain the correct electronic structure for CO_2.

(c) :Ċ· shares 4 additional electrons and forms 4 covalent bonds.

H∘ shares 1 additional electron and forms 1 covalent bond.

ẍO× shares 2 additional electrons and forms 2 covalent bonds.

A longer chain of atoms is required to write a correct electronic structure for C_2H_6O. Use the C atom and the O atom as central atoms in the chain since they are capable of forming more covalent bonds. An H atom can only be a terminal atom in an electronic structure because it can only form one bond. Two chains are possible: C—C—O or C—O—C.

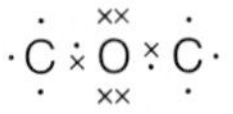

After writing the chains (including outer shell electrons), simply place the H atoms in terminal positions wherever there remain unpaired electrons.

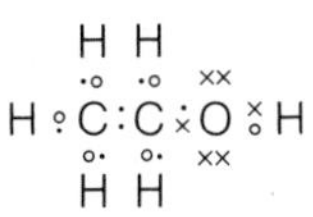

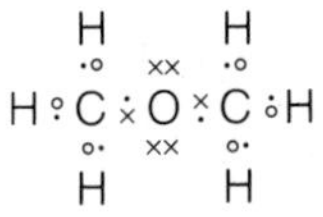

Note that in both structures there are 2 electrons around each H atom and 8 electrons around each C and O atom. Each H atom forms 1 bond, each O atom forms 2 bonds, and each C atom forms 4 bonds. Both structures are correct for C_2H_6O. These are two isomers (Section 12.3), ethyl alcohol and dimethyl ether (see Table 12.2).

Electronic structures that exemplify the preceding rules cannot be written for some molecules. Three types of exceptions will be discussed.

a. Atoms with Less Than Eight Electrons

An atom with fewer than four outer shell electrons may have less than eight electrons around it after sharing its outer shell electrons to form covalent bonds. The boron atom is an example. Boron trichloride (BCl_3) is a colorless gas with a boiling point slightly below room temperature. It fumes from reaction with moisture in air and is a useful starting material for the preparation of many interesting boron compounds. It can be represented as forming from one B atom and three Cl atoms. The BCl_3 molecule is stable

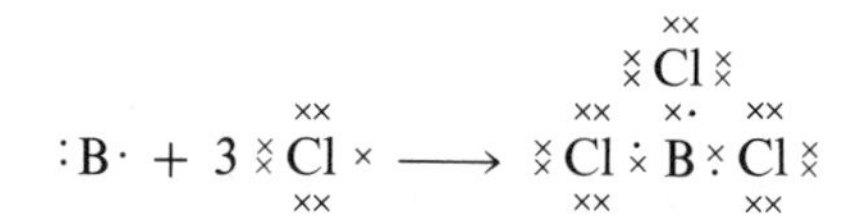

even though B has only six electrons associated with it. Other boron compounds including BF_3, BBr_3, and $B(CH_3)_3$ have similar electronic structures.

An atom like boron in the above stable molecules readily achieves an octet of electrons when the molecules containing boron react further with other molecules having an unshared pair of electrons. Thus, boron trichloride reacts with ammonia (NH_3) to form a compound in which the nitrogen donates both electrons that form the shared-pair covalent bond joining the two originally stable molecules.

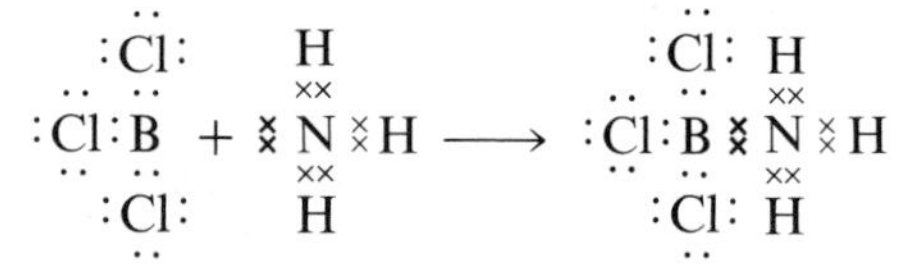

Similar shared-pair covalent bonds to which one atom donates both electrons are formally constituted between transition metal atoms or cations that are capable of accepting electron pairs and anions or molecules that are capable of donating electron pairs. Metal complexes, which contain a central metal atom or cation covalently bound to a cluster of two, four, or six anions or molecules, are commonly formed. For example, in the presence of ammonia, silver(I) forms a colorless, soluble, complex ion, $[Ag(NH_3)_2]^+$, that is a part of Tollens' reagent for the oxidation of aldehydes (Section 18.5-c-2).

$$Ag^+ + 2\,{:}NH_3 \longrightarrow [H_3N{:}Ag{:}NH_3]^+$$

Likewise, in basic solution in the presence of citrate ion, copper(II) forms a blue, soluble, complex ion, $[Cu(citrate)_2]^{6-}$, that is a part of Benedict's solution for the oxidation of aldehydes (Section 18.5-c-1) and to test for the presence of reducing sugars (Section 23.9-b-1).

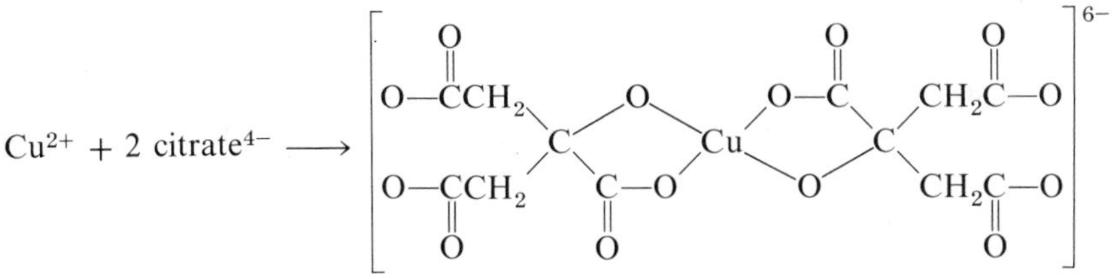

Many metal complexes (for example, heme; Figure 25.1) are of vital importance in biological systems.

A number of molecules possess unpaired electrons, and one or more atoms in these molecules will not have eight electrons associated with it. Electronic structures are inadequate for determining the nature of bonding and the shape of many such molecules. For example, the best electronic structure for nitric oxide (a colorless gas formed when dilute nitric acid, HNO_3, is used as an oxidizing agent) is :N: :O, but NO can be shown experimentally to have a bond energy and bond distance appropriate for $2\frac{1}{2}$ covalent bonds rather than just a double bond. Likewise, the best electronic structure for the common oxygen molecule is O: :O, but O_2 is found experimentally to have two *unpaired* electrons. Thus, we represent O_2 as :O:O:, although this structure disobeys the generalizations discussed previously (eight electrons around O and O forming two bonds) and does not predict observed double bond character.

b. Atoms with More Than Eight Electrons

Representative elements in the third through seventh periods may have more than eight electrons associated with them, and transition elements almost always have more than eight electrons associated with them in typical molecules. Phosphorus pentafluoride (PF_5, a colorless gas), thionyl chloride ($SOCl_2$, a colorless liquid), xenon tetrafluoride (XeF_4, a white solid), and chromium hexacarbonyl [$Cr(CO)_6$, a white solid and a useful starting material for preparing compounds with carbon bonded to Cr] illustrate this point. Third through seventh period elements need not always restrict themselves to an octet of electrons because *d* orbtals can be utilized to hold additional electrons.

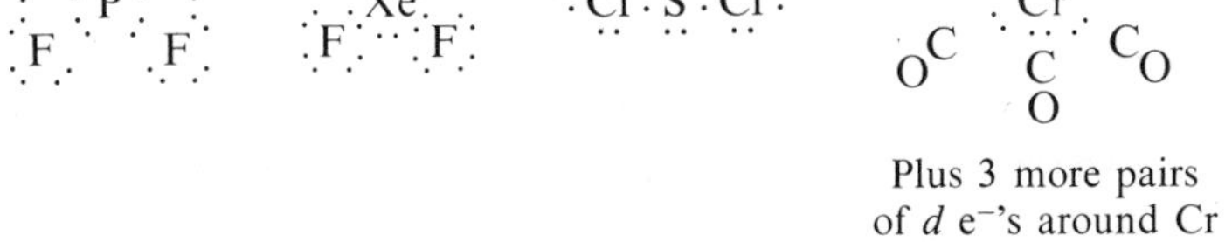

Plus 3 more pairs of *d* e⁻'s around Cr

c. Resonance

The phenomenon in which no single classical electronic structure adequately describes experimentally observed bonding parameters (bond energy and bond distance) is called **resonance.** Ozone, O_3 (a gas with a sharp, penetrating odor that is formed in significant amounts in the atmosphere during electrical storms), requires the use of resonance structures. Structure I is a valid electronic structure according to guidelines given in this section.

I II

However, structure I incorrectly implies one single O—O bond (longer) and one double O=O bond (shorter). Experimentally the two bonds are found to be identical and equivalent to about $1\frac{1}{2}$ bonds. The actual electron distribution is *intermediate* between those depicted by two contributing resonance structures I and II and is never like that depicted by either contributing structure.

Resonance is a misnomer because it implies oscillation from one structure to another or oscillation of an electron pair from one bond to another. *This is not true.* A mule, the offspring of a male donkey and a mare, can be considered a hybrid of a donkey and a horse. However, it is not a horse at one instant and a donkey at another; it is always a mule. Likewise, a molecule or complex ion exhibiting resonance has *one* real structure, which is never any of the rather unsophisticated electronic structures used to describe it.

Example 3.2 The nitrate ion, NO_3^-, exhibits resonance. Write the contributing electronic structures for NO_3^-. What information about bonding do these structures convey?

 shares 3 additional electrons.

 shares 2 additional electrons.

The charge, -1, provides one additional electron.

Add to N with its 5 outer shell electrons the one additional electron from the charge. Then let N share an electron pair with each of 3 O atoms. The resulting structure leaves N with only 6 electrons around it whereas 8 would provide a more

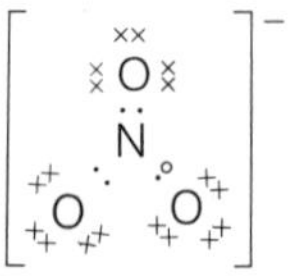

stable structure. This problem can be alleviated by introducing one double bond. This can be accomplished in any of the three bonds producing three equivalent resonance structures. These three resonance structures taken together imply that all three N—O bond energies and bond distances are identical and correspond to about $1\frac{1}{3}$ bonds.

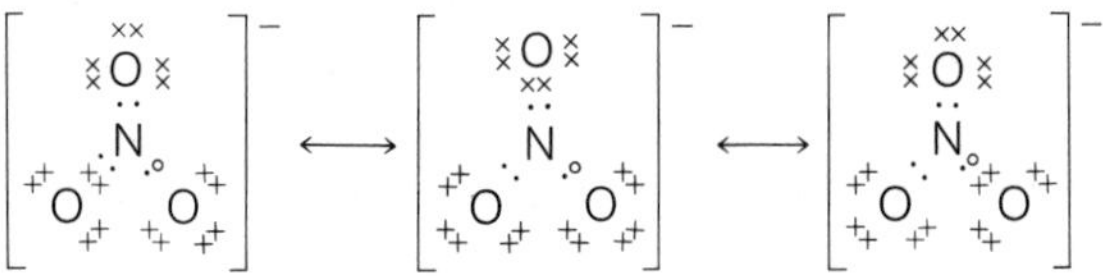

3.9 Shapes of Simple Molecules

Even simple molecules show a great variety of three-dimensional geometries. Figure 3.6 shows a few examples. Molecules containing two atoms, such as I_2 and HBr, are necessarily linear (two points determine a straight line). Molecules containing three atoms are necessarily planar (three points determine a plane); however, they may be linear like CO_2 or nonlinear (bent) like H_2S. Molecules containing four atoms may be planar or nonplanar. The planar molecules may be linear like $(CN)_2$, nonlinear like HN_3, triangular like BBr_3, or T-shaped like BrF_3. The nonplanar molecules may be pyramidal like PH_3 or nonlinear like H_2O_2. As the number of atoms in a molecule increases to five and higher, the variety of configurations and their complexity increase. Electronic structures (Section 3.8) provide useful representations of the nature of bonding in atoms, but provide little direct informa-

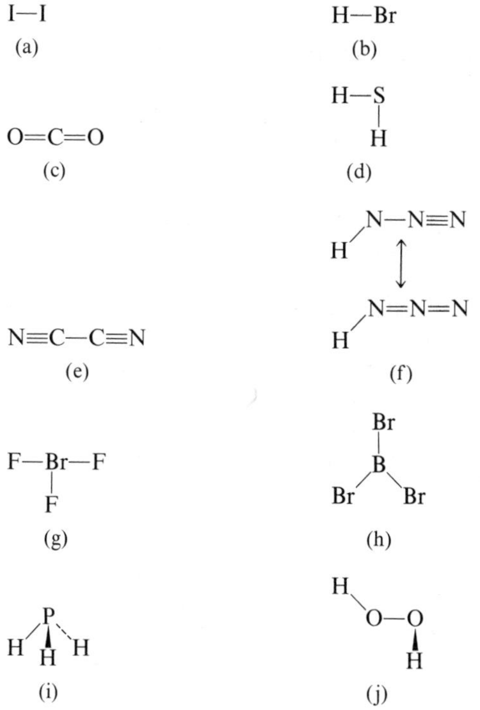

Figure 3.6 Geometries of a few simple molecules. (a) Iodine; (b) hydrogen bromide; (c) carbon dioxide; (d) hydrogen sulfide; (e) cyanogen; (f) hydrazoic acid; (g) bromine trifluoride; (h) boron tribromide; (i) phosphine; (j) hydrogen peroxide. All atoms and bonds are in the plane of the paper except those shown with a dashed line, which are back of the paper (away from the reader), and those shown with a darkened wedge, which are out of the paper (toward the reader).

tion about the distribution of electron density within molecules or the geometry of atoms within molecules.

A number of bonding theories can be used for predicting shapes of simple molecules. We shall use the valence shell electron pair repulsion (VSEPR) theory. This model has the advantages of being relatively simple and nonmathematical while still providing accurate qualitative shapes for most simple molecules and ions involving representative elements and, with a minor modification, for most simple molecules and ions involving transition elements.

The basic premise of VSEPR theory is that electron pairs, having like charge, repel each other. Other factors being equal, outer shell or valence shell electron pairs around an atom will try to spread as far apart as possible. Furthermore, interactions between nonbonding electron clouds of atoms bonded to a central atom are assumed to be negligible. Therefore, *orientation of several atoms around a given central atom is dependent only on interactions between electron pairs around the central atom.*

a. Central Atoms with Identical Substituent Atoms and Only Shared Electron Pairs

Under these conditions all electron pairs around the central atom are equivalent because they are all shared between the same two atoms. The electron pairs have the same charge distribution and density and thus have identical repulsive properties. To get as far apart as possible, they can orient symmetrically on the surface of a sphere around the central atom.

Example 3.3 Determine the shapes of the following isolated molecules: (a) $BeCl_2$, (b) BF_3, (c) CH_4, (d) PF_5, (e) SF_6.

It is left to you as an exercise to determine that the correct electronic structures for these molecules are the left-hand structures of Figure 3.7.

(a) There are 2 shared electron pairs around Be. They can be kept as far apart as possible by placing them on opposite sides of Be, or 180° away, thus forming a linear molecule. The **bond angle,** which is *the angle formed by two covalent bonds,* is 180°. These ideas are illustrated schematically in the right three structures of Figure 3.7.

(b) There are 3 shared electron pairs around B. They can be kept as far apart as possible by placing them at the corners of an equilateral triangle with the B atom at the center and thus in the plane of the equilateral triangle (bond angle = 120°; Figure 3.7).

(c) There are 4 shared electron pairs around C. They can be kept as far apart as possible by placing them at the corners of a tetrahedron with the C atom at the center of the tetrahedron (bond angle = 109.5°; Figure 3.7). This is the approximate configuration around all C atoms in molecules in which the C atoms are bonded to 4 other atoms, as in alkanes (Chapter 13), alkyl halides (Chapter 16), alcohols and ethers (Chapter 17), and amines (Chapter 20).

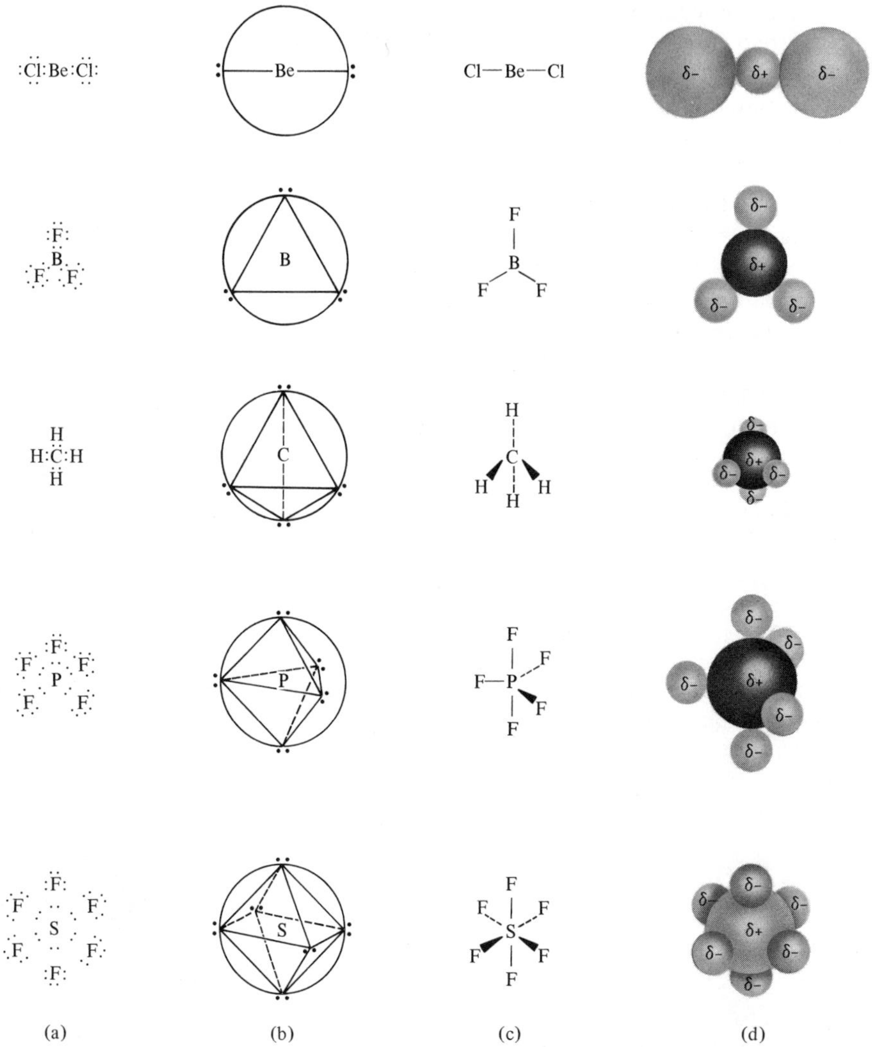

Figure 3.7 Electronic structures (a), structures showing electron pair distributions (b), three-dimensional geometric structures (c), and space-filling models (d) for beryllium dichloride, boron trifluoride, methane, phosphorus pentafluoride, and sulfur hexafluoride.

(d) There are 5 shared electron pairs around P. They can be kept as far apart as possible (but not quite at equivalent positions) by placing them at the corners of a trigonal bipyramid with the P atom at the center of the trigonal bipyramid. There are 2 kinds of positions on the trigonal bipyramid and 2 bond distances and bond angles (90° and 120°) in the resulting structure (Figure 3.7).

(e) There are 6 shared electron pairs around S. They can be kept as far apart as possible by placing them at the corners of an octahedron with the S atom at the center of the octahedron (bond angle = 90°; Figure 3.7).

b. Central Atoms with Identical Substituent Atoms and Both Shared and Unshared Electron Pairs

A shared electron pair is attracted by two nuclei, and its charge cloud (the region in space where the electrons of a pair are most likely to be found) is relatively localized. An unshared electron pair is attracted to only one nucleus, and its charge cloud is more diffuse and tends to occupy a larger volume in space around a central atom. Stated simply, unshared electron pairs require more space than shared electron pairs. Alternatively, unshared-pair-unshared-pair repulsions are greater than unshared-pair-shared-pair repulsions, which are greater than shared-pair-shared-pair repulsions. Thus, when shared pairs around a central atom are replaced by unshared pairs, the bond angles (angles between shared pairs) decrease compared to the ideal angles for the various geometries indicated in Example 3.3.

Example 3.4 Compare qualitatively the bond angles in CH_4, NH_3, and H_2O.

It is left to you as an exercise to determine that the correct electronic structures for these molecules are those in Figure 3.8a.

The central atom in each case has 4 electron pairs around it. They can be kept as far apart as possible by placing them at the corners of a tetrahedron with the central atom at the center of the tetrahedron. (Enlarged dots indicate unshared electron pairs in Figure 3.8b.)

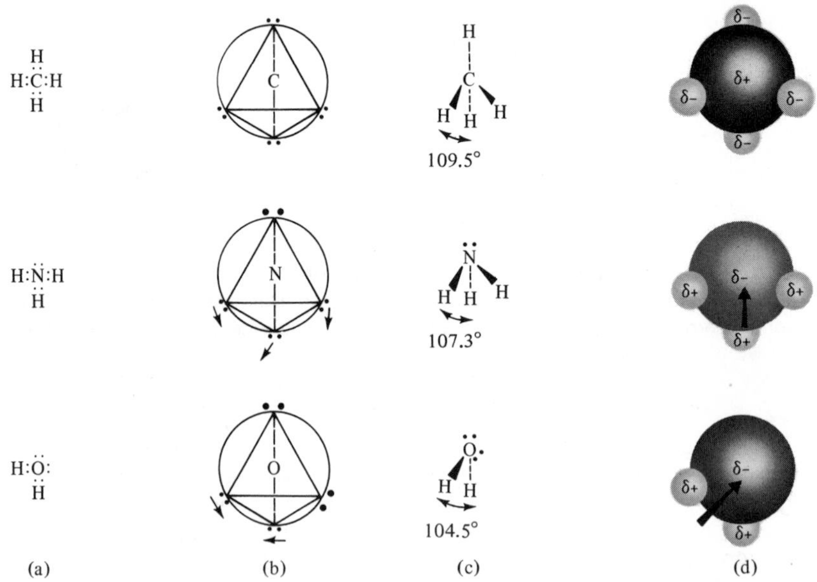

109.5°
107.3°
104.5°
δ+
δ−
(a)
(b)
(c)
(d)

Figure 3.8 Electronic structures (a), structures showing electron pair distributions (b), three-dimensional geometric structures (c), and space-filling models (d) for methane, ammonia, and water. Heavy arrows in (d) indicate net molecular polarity (Section 3.10).

Addition of 1 unshared pair requiring more space in NH_3 compresses the H—N—H angles somewhat as illustrated by the arrows (Figure 3.8b). Addition of a second unshared pair in H_2O compresses the H—O—H angle even farther. Thus, you predict that the H—X—H bond angle decreases from CH_4 to NH_3 to H_2O. Experimental data agree with this conclusion (Figure 3.8c).

c. Central Atoms Forming Multiple Bonds

Double or triple bonds involving two or three shared electron pairs can be considered as just another region of electron density and don't affect the gross shape of a molecule. However, they have larger electron clouds than a single shared electron pair. Therefore, like unshared electron pairs, double and triple shared electron pairs cause greater repulsion and occupy more space than single shared electron pairs.

Example 3.5 Compare qualitatively the two kinds of bond angles in ethylene, C_2H_4.

The electronic structure for C_2H_4 shows 3 regions of electron density around each C atom (Figure 3.9a). They can be kept as far part as possible by placing them at the corners of an equilateral triangle with a C atom in the center of the equilateral triangle and thus in the plane of the equilateral triangle (Figure 3.9b). However, the double shared electron pair occupies more space than the single shared electron pairs, thus opening the H—C—C bond angles and closing the H—C—H bond angle. These predictions are verified by the experimental bond angles (Figure 3.9c). Similar bond angles around a doubly-bonded C atom are typical for all alkenes (Chapter 14) as well as aldehydes and ketones (Chapter 18), acids and esters (Chapter 19), and amides (Chapter 20).

The effects on geometry around a central atom of shared electron pairs, unshared electron pairs, and multiple bonding have been treated separately. In many interesting molecules these factors and others more subtle act simultaneously. Nevertheless, with practice, you can quickly get a feeling for the relative importance of these factors and can predict the structures of relatively complicated molecules and ions.

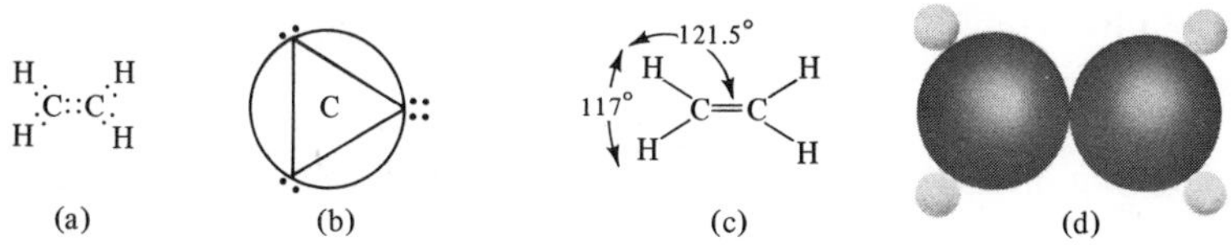
H. .H
:C::C:
H˙ ˙H
(a)
C
(b)
121.5°
117°
H
H
C=C
H
H
(c)
(d)

Figure 3.9 Electronic structure (a), structure showing electron pair distributions (b), three-dimensional geometric structure (c), and space-filling model (d) for ethylene. All six atoms of ethylene lie in the same plane.

3.10 Polar Molecules

A **polar molecule** is *one in which there is a net separation of positive and negative charge,* as in $^{\delta+}H{-}Cl^{\delta-}$. We saw in Sections 3.5 and 3.7 that it was a simple matter to determine whether or not diatomic molecules are polar. A polar diatomic molecule arises from two covalently bound atoms having different electronegativities, whereas a nonpolar diatomic molecule results from two covalently bound atoms that are identical or have identical electronegativities.

For a molecule composed of more than two atoms, it is more difficult to predict whether it is polar or nonpolar. You must consider both the polarity of individual bonds and the orientation of these bonds in space with respect to each other. You must examine the shape of the molecule and determine how the dipoles of individual polar bonds either add to or cancel each other.

Symmetrical molecules have a structure exhibiting a regular repeating pattern of bonds and are nonpolar even though individual bonds within the molecules are polar. For example, in carbon dioxide the bond dipoles are equal in magnitude of charge separation and opposite in orientation (indicated by arrows). Thus, the bond dipoles cancel each other, leading to a nonpolar molecule.

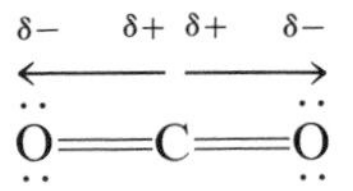
δ− δ+ δ+ δ−
O═C═O

Likewise, all of the molecules in Figure 3.7 are nonpolar because of their symmetry. You can understand the cancellation of bond dipoles by imagining a tug-of-war in which an appropriate number of teams placed at each terminal atom pull on a ball representing the central atom. If each team pulls equivalently (two teams for $BeCl_2$, three teams for BF_3, four teams for CH_4, five teams for PF_5, and six teams for SF_6), their effects will cancel, and the ball representing the central atom will not move. In a similar fashion shown by the charge separations in the right-hand structures of Figure 3.7, the two bond dipoles in $BeCl_2$ cancel each other; the three bond dipoles in BF_3 cancel each other; the four bond dipoles in CH_4 cancel each other; the five bond dipoles in PF_5 cancel each other; and the six bond dipoles in SF_6 cancel each other. There is no net charge separation for each molecule taken as a whole, and the molecules are nonpolar.

Unless the individual bond dipoles for unsymmetrical molecules cancel fortuitously, such molecules are polar. Unlike CH_4 in Figure 3.8d, NH_3 and H_2O are not symmetrical, and have net dipoles for which the directions are indicated by arrows with the arrowhead toward the negative end of the net dipoles according to the usual convention.

Example 3.6 Decide which of the molecules in Figure 3.10 are polar.

You can see from Figure 3.10 that PF_5 and XeF_2 are symmetrical molecules. Using the tug-of-war analogy it can be seen that the five polar bonds in PF_5 cancel

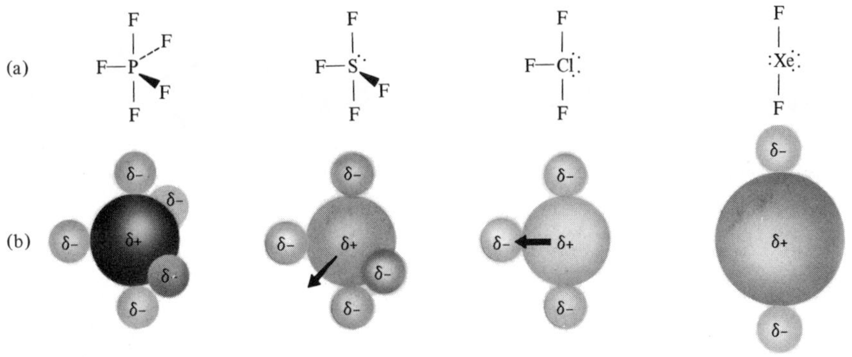
(a)
F
F
F—P
F
F
F
F—S
F
F
F
F—Cl
F
F
:Xe:
F
(b)
δ−
δ+
δ−
δ+
δ−
δ+
δ−
δ+

Figure 3.10 Three-dimensional geometric structures (a) and space-filling models (b) for phosphorus pentafluoride, sulfur tetrafluoride, chlorine trifluoride, and xenon difluoride.

each other, and the two polar bonds in XeF_2 cancel each other (Figure 3.10b). They are nonpolar molecules.

You can see from Figure 3.10 that SF_4 and ClF_3 are unsymmetrical molecules. Using the tug-of-war analogy it can be seen that the polar bonds in these two molecules do not cancel. They have net dipoles in directions indicated by arrows (Figure 3.10b).

Polar molecules can be distinguished experimentally from nonpolar molecules when placed in an electrical field. Nonpolar molecules show little tendency to orient themselves in an electrical field. As depicted in Figure 3.11, polar molecules tend to orient with their positive poles toward the negative plate and their negative poles toward the positive plate of an electrical field.

Polar molecules have weak electrostatic forces between molecules resulting from attraction between opposite ends of individual dipoles. These forces determine certain physical properties of aggregates of polar molecules. These forces are much weaker than electrostatic attractions between ions

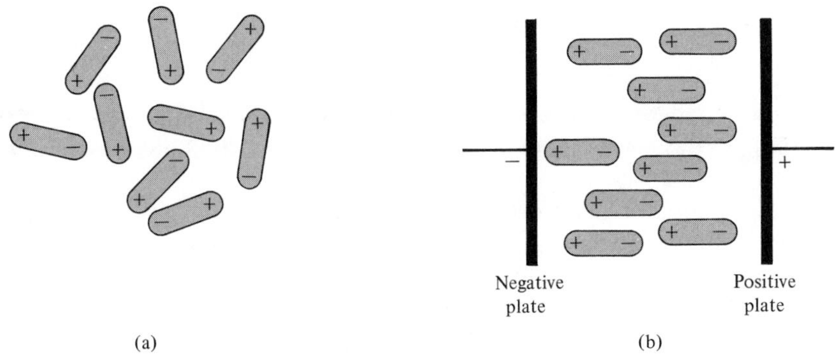
Negative
plate
Positive
plate
(a)
(b)

Figure 3.11 (a) Random orientation of polar molecules in the absence of an electrical field; (b) specific orientation of polar molecules in the presence of an electrical field.

(Section 3.3), but somewhat stronger than dispersion forces (Section 3.13) between nonpolar molecules. Consequently, melting points and boiling points for polar molecules tend to be higher than those for nonpolar substances, but significantly lower than those for ionic substances. Such a trend is illustrated in Table 3.3 for three compounds, the molecules of which have comparable masses and thus comparable dispersion forces (Section 3.13).

Table 3.3 Physical Properties of Br_2, ICl, and RbBr

Compound	*Molar Mass (g/mole)*	*Nature of Bonding*	*Melting Point (°C)*	*Boiling Point (°C)*
Br_2	159.82	nonpolar covalent	−7.2	58.78
ICl	162.36	polar covalent	27.2	97.4
RbBr	165.38	ionic	682	1340

3.11 Hydrogen Bonds

A primarily electrostatic attractive force between a positively charged hydrogen atom in one molecule and a small, highly electronegative atom in a second molecule is called a **hydrogen bond.** An H atom covalently bound to a small, electronegative atom as in HF, has such a small share of the bonding electrons that it behaves almost like a bare proton. As such, it is easily attracted to the electrons of a nearby electronegative atom like F, O, or N. Furthermore, it is so small that it only has room for two atoms around it and allows a very close approach of these two atoms. In effect, a hydrogen bridge consisting of a proton shared between two atoms is formed. Hydrogen bonds can be represented as in HF shown below.

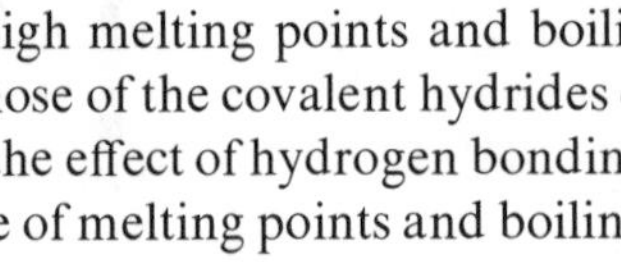

Hydrogen bonds are much weaker than normal ionic bonds (Section 3.3) but are somewhat stronger than ordinary dipole–dipole interactions between polar molecules (Section 3.10).

The abnormally high melting points and boiling points of NH_3, H_2O, and HF compared to those of the covalent hydrides of other group V, VI, and VII elements manifest the effect of hydrogen bonding. In Figure 3.12 you can see the gradual increase of melting points and boiling points due to increased dispersion forces (Section 3.13) through the nonpolar covalent hydrides of group IV elements (CH_4, SiH_4, GeH_4, and SnH_4). A similar parallel trend is exhibited by the last three members of the other three series. However, NH_3,

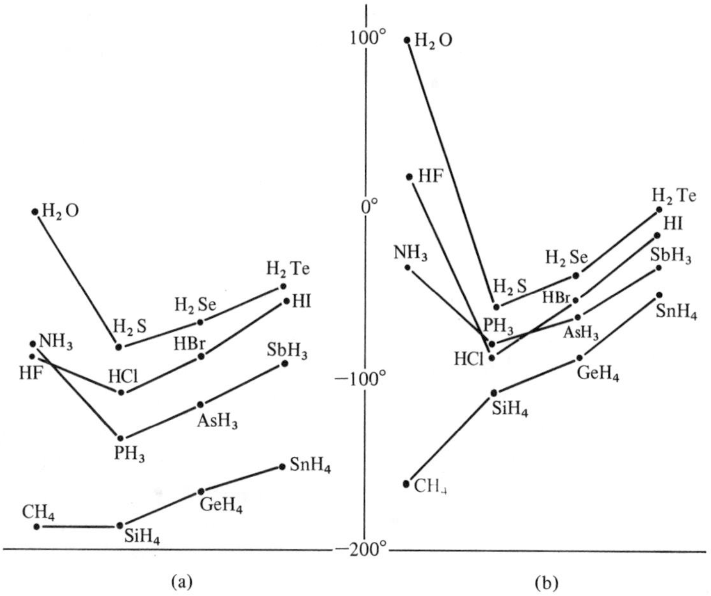

100°
0°
−100°
−200°
H₂O
NH₃
HF
H₂S
HCl
PH₃
H₂Se
HBr
AsH₃
H₂Te
HI
SbH₃
SnH₄
GeH₄
SiH₄
CH₄
(a)
(b)

Figure 3.12 Melting points (a) and boiling points (b) in °C versus mass of molecules for compounds of hydrogen and elements of groups IV, V, VI, and VII.

H_2O, and HF have significantly higher melting points and boiling points than expected due to stronger forces between these molecules as a result of hydrogen bonding.

Hydrogen bonding is of primary importance in accounting for the structure and seemingly abnormal properties of water (Section 6.6). Likewise, hydrogen bonding is responsible for the unexpected melting points, boiling points, and solubilities of alcohols (Section 17.3) and for the unusual structures and properties of many proteins (Sections 25.5 and 25.6) and nucleic acids (Section 27.2).

3.12 Metallic Bonds

As in ionic bonding and covalent bonding, outer shell electrons are responsible for bonding between metal atoms. However, it is unreasonable to assume that ionic bonds occur between metal atoms since all the atoms are alike and no single atom would give up electrons to another atom. Covalent bonding between metal atoms is almost as unreasonable because not enough outer shell electrons are available for as many shared-pair bonds as each metal atom seems to form. Instead, a **metallic lattice** consists of a *regular array of positive ions immersed in a cloud of highly mobile outer shell electrons.* Metals have relatively low ionization energies (Section 3.2) or relatively loose holds on their outer shell electrons. These electrons are free to move throughout the metallic lattice. Metallic bonding results from attraction between the positive ions and the cloud of negative electrons. Such attractive forces are weaker than ionic or covalent bonding forces. Thus, many metals are soft and fairly low melting. Potassium is soft enough to be cut with a knife and melts at 68.7°. On the other hand, some of the transition metals, where significant covalent character is superimposed on the metallic lattice, are hard and high melting. Tungsten is very hard and melts at about 3410°.

Other properties generally exhibited by most metals can be explained on the basis of the bonding model presented above. Metals have high electrical conductivity. The highly mobile outer shell electrons are free to move through the metallic lattice under the influence of an electrical field, and thus conduct a current. Copper wire is used commonly in house circuits. Metals have high thermal conductivity. Collisions between the highly mobile electrons cause a transfer of heat energy through the metal. Cooking ware is constructed from copper, aluminum, or iron. Metals appear shiny. The mobile electrons absorb and re-emit light of various energies, thus making the metal a very good reflector. Metals are malleable (capable of being hammered into sheets) and ductile (capable of being drawn into wire). It is relatively easy to move planes of positive ions over one another without producing any change in the crystal structure. Many metals eject electrons upon exposure to light and upon sufficient heating. The mobile electrons are loosely held and can escape from the metal entirely given sufficient energy. Photocells operate as a result of light-induced emission of electrons. Some vacuum tubes depend upon heat-induced emission of electrons for their operation.

3.13 Dispersion Forces (van der Waals Forces)

None of the bonding interactions discussed thus far account for bonding *between* nonpolar molecules like H_2 or atoms like Ar and thus the existence of these species in the liquid or solid phase. One further type of interaction called dispersion forces or van der Waals forces provides the explanation.

Dispersion forces *arise from attraction between temporary or instantaneous dipoles.* On the average an Ar atom has a spherical distribution of electron density and no separation of charge (Figure 3.13a). However, at any given instant, the electron cloud may be more concentrated at one side of the atom producing an instantaneous dipole as shown in Figure 3.13b. Electrons

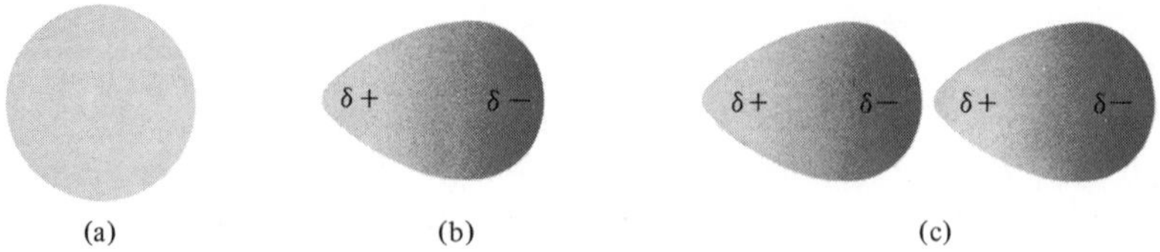

δ+
δ−
δ+
δ−
δ+
δ−
(a)
(b)
(c)

Figure 3.13 Model of dispersion forces. (a) Spherical distribution of electron density of an isolated atom; (b) instantaneous dipole produced by distortion of the electron cloud; (c) attraction of two instantaneous dipoles.

of one atom repel electrons of adjacent atoms and cause a correlation of their motions. As a result, the first instantaneous dipole induces a similar dipole in an adjacent molecule. These two dipoles oriented in the same direction (Figure 3.13c) produce an attractive force.

Dispersion forces are very weak compared to ionic or covalent interactions and depend upon the ease with which the electron cloud can be distorted or polarized. As a first approximation, dispersion forces increase with increasing numbers of electrons in the atom or molecule. Consequently, properties depending primarily upon interatomic or intermolecular dispersion forces show a similar trend. Thus, the melting points and boiling points of CH_4, SiH_4, GeH_4, and SnH_4 (Figure 3.12) increase almost linearly with the increasing number of electrons in these molecules (10, 18, 36, and 54, respectively). Since the number of protons and neutrons increases as the number of electrons in a series of molecules increases, the masses in that series must also increase. Thus, as a second approximation, one can say that dispersion forces increase with increasing mass in a series of atoms or molecules.

Exercises

3.1 Define clearly the following terms.

(a) chemical bond
(b) ion
(c) cation
(d) anion
(e) ionization energy
(f) electron affinity
(g) ionic bond
(h) bond length
(i) covalent bond
(j) single bond
(k) bond energy
(l) nonpolar covalent bond
(m) covalent solids
(n) double bond
(o) triple bond
(p) saturated molecule
(q) unsaturated molecule
(r) polar covalent bond
(s) electronegativity
(t) electronic structures
(u) resonance
(v) valence shell electron pair repulsion theory
(w) bond angle
(x) polar molecule
(y) hydrogen bond
(z) metallic bond
(aa) dispersion forces

3.2 **(a)** Write the empirical formula for the compound that would be formed between Ca and Cl.
(b) Indicate the primary kind of bonding found in this substance, and describe how the bonding forces arise.
(c) What will be the standard physical state of this substance at room temperature? Why?

3.3 Referring to Figure 3.5, arrange the following bonds in order of decreasing polarity: H—F, C—C, I—Br, B—N, Al—Cl.

3.4 Compare qualitatively the bond energies and bond distances of each pair.
(a) N—N and N≡N
(b) O—O and O═O
(c) C═O and C—N
(d) C═C and C≡C
(e) Cl—Cl and I—I

3.5 Write electronic structures for the following. Note that (d) and (e) may have greater than eight electrons around the central atom.
(a) C_2H_6
(b) BI_3
(c) AsH_3
(d) SO_2
(e) SO_3^{2-}

3.6 Write electronic structures for the following. Note that (d) and (e) may have greater than eight electrons around the central atom.
(a) NO^+
(b) OH^-
(c) CN^-
(d) XeF_4
(e) ClO_4^-

3.7 Describe in words and by means of figures the geometries of the molecules in Exercise 3.5. Show clearly your reasoning.

3.8 Describe in words and by means of figures the geometries of the following. Show clearly your reasoning.
(a) PO_4^{3-}
(b) NF_3
(c) PCl_5
(d) C_3H_6
(e) C_2H_4O

3.9 Compare qualitatively the bond angles in each pair. Show clearly your reasoning.
(a) NH_3 and H_2O
(b) BF_3 and BF_4^-
(c) CCl_4 and $SiCl_4$
(d) CO_2 and NO_2
(e) NH_3 and NH_4^+

3.10 Compare qualitatively the various kinds of bond angles, showing clearly your reasoning.
(a) $COCl_2$
(b) CH_4O
(c) ClF_3
(d) N_2H_4
(e) SNF_3

3.11 **(a)** Describe how the configuration around B changes in going from BF_3 to F_3BNH_3 according to the following reaction.

$$BF_3 + NH_3 \longrightarrow F_3BNH_3$$

(b) Does the H—N—H bond angle increase, decrease, or show no change in going from NH_3 to F_3BNH_3? Why?

3.12 The electronic formula for acetic acid is

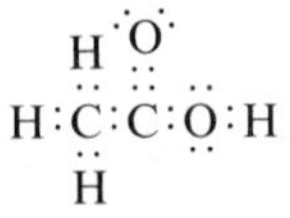

(a) What is the configuration of atoms around each C atom? Why?
(b) What estimate would you give for the C—O—H bond angle? Why?
(c) Compare qualitatively the two carbon–oxygen bond distances.

3.13 Write electronic formulas for each of the following.

(a) a linear molecule consisting of three or more atoms
(b) a planar anion
(c) a molecule with an octahedrally coordinated central atom
(d) a nonplanar molecule consisting of four atoms
(e) a molecule having a double bond at a C atom

3.14 Which of the following molecules are polar? Show clearly your reasoning.

(a) HN_3
(b) C_2H_2
(c) CH_2O
(d) CO_2
(e) BrI

3.15 Which of the molecules in Exercise 3.5 are polar? Show clearly your reasoning.

3.16 Describe three properties usually exhibited by metals, and explain on the basis of metallic bonding why metals exhibit these properties.

3.17 Indicate the mechanism of bonding and a specific chemical example that illustrates each of the following.

(a) ionic bonding
(b) covalent bonding
(c) hydrogen bonding
(d) metallic bonding
(e) dispersion forces

3.18 Predict which substance in each of the following pairs will have the higher boiling point. Show clearly your reasoning.

(a) HF or HCl
(b) $MgCl_2$ or BrCl
(c) C_5H_{12} or C_3H_8
(d) PH_3 or SiH_4
(e) K or KI

3.19 Explain in terms of interatomic and intermolecular forces

(a) why polar molecules boil at higher temperatures than nonpolar molecules of similar masses
(b) why it is unreasonable to talk about NaCl molecules as discrete species existing at room temperature
(c) why water has a higher melting point than H_2S
(d) why diamond and diatomic hydrogen, H_2, both exhibit covalent bonding, but diamond is a high melting solid whereas H_2 is a very low boiling gas

3.20 Comment critically on each of the following.

(a) A bond between two different atoms must necessarily be polar.
(b) Boiling points of molecules increase with increasing mass of the molecules.
(c) The PF_3 molecule is expected to have a planar triangular structure and thus will be nonpolar.
(d) Dispersion forces are so weak that they have no effect on properties of molecules.

4

Stoichiometry

In 1948 cyanocobalamin, isolated from liver, was found to have therapeutic activity in the treatment of pernicious anemia, a disease involving deficient maturation of red blood cells. By 1955 the structure of cyanocobalamin, now better known as Vitamin B_{12}, was elucidated. A single ion of this complex species contains 1 cobalt atom, 1 phosphorus atom, 63 carbon atoms, 90 hydrogen atoms, 14 nitrogen atoms, and 14 oxygen atoms. Vitamin B_{12} has a percent composition of 4.34% cobalt, 2.28% phosphorus, 55.75% carbon, 6.68% hydrogen, 14.45% nitrogen, and 16.50% oxygen. It is represented by the formula, $CoPC_{63}H_{90}N_{14}O_{14}^{+}$. One mole of this ion weighs 1357.41 grams. When reduced with hydrogen molecules in the presence of a platinum catalyst, one mole of Vitamin B_{12} is converted to one mole of hydroxocobalamin. This species is represented by the formula $CoPC_{62}H_{91}N_{13}O_{15}^{+}$ and has a molar mass of 1348.40 g/mole.

The purpose of this chapter is to discuss percent composition, atoms, molecules, formulas, ions, moles, and molar masses, terms used to indicate quantities of elements or compounds, and to present the mathematical operations for converting one to another. These units, along with balanced chemical equations, will be utilized to determine the stoichiometry (Greek, *stoicheion,* element; *metron,* measure) of chemical reactions, or the amounts of materials consumed and produced in chemical reactions.

4.1 Percent Composition

The percent composition by mass of a compound is obtained by dividing the mass of each element present in the compound by the total mass of the compound, then converting to percent by multiplying by 100%. The masses of each element present must be determined experimentally by chemical analysis.

Example 4.1 If 2.44 grams of titanium combines with 3.60 grams of chlorine to form a compound, what is the percent composition of this compound?

The total mass of the compound equals 2.44 grams + 3.60 grams, or 6.04 grams.

$$\text{Percent Ti} = \left(\frac{\text{mass of Ti}}{\text{mass of compound}}\right)(100\%) = \left(\frac{2.44\ \text{g}}{6.04\ \text{g}}\right)(100\%) = 40.4\%\ \text{Ti}$$

$$\text{Percent Cl} = \left(\frac{\text{mass of Cl}}{\text{mass of compound}}\right)(100\%) = \left(\frac{3.60\ \text{g}}{6.04\ \text{g}}\right)(100\%) = 59.6\%\ \text{Cl}$$

Note that there are three significant figures in each result according to the rules in Section 1.3-c. The sum of the percents of all elements present must equal approximately 100% if the calculation is correct.

If you know the percent composition of a compound, you can calculate the mass of any element present in a given mass of the compound by converting the percent to a decimal and multiplying the decimal by the given mass of the compound.

Example 4.2 Titanium forms two other compounds with chlorine in which the percents of titanium are 31.05% and 25.25%, respectively. What mass of titanium is present in 6.04 grams of each of these compounds?

Change the percents to decimals by dividing by 100%. Thus, 31.05% = 0.3105 and 25.25% = 0.2525. Then multiply by the given mass of the compound.

$$\text{Mass of Ti} = (6.04\ \text{g})(0.3105) = 1.88\ \text{g of Ti}$$

$$\text{Mass of Ti} = (6.04\ \text{g})(0.2525) = 1.53\ \text{g of Ti}$$

Note that there are three significant figures in each result acording to the rules in Section 1.3-c-2.

4.2 Atom

An **atom** (Greek, *atomos,* indivisible) is *the smallest particle of a chemical element that possesses all the inherent characteristics of that element and that remains unchanged in an ordinary chemical reaction* (Sections 1.1-c and 2.3). Atoms can be further subdivided in nuclear processes, but in such processes the original chemical identity of the atom is destroyed. Atoms have a nuclear region (Section 2.8) made up of protons (Section 2.6) and neutrons (Section 2.7) and an extranuclear region containing electrons (Section 2.5). The masses of protons (1.00728 amu) and neutrons (1.00867 amu) are comparable (Section 4.3). However, the mass of an electron is only 1/1835 that of a proton, and mass changes due to gain or loss of electrons in chemical reactions are assumed to be negligible.

4.3 Atomic Mass Unit

Two units of mass are convenient for chemical calculations, the *gram* and the *atomic mass unit*. The gram is the most commonly used unit of mass (Section 1.1-a) in the metric system (Section 1.3-b-3) and is defined as 0.001 kilogram. In 1961 the carbon-12 isotope (^{12}C, Section 2.10), containing six protons and six neutrons in its nucleus and six electrons in its extranuclear region, was designated as the standard for atomic masses. Using this standard, now common in chemistry, the gram is equal to $\frac{1}{12}$ of the mass of 1 mole (Section 4.9) of ^{12}C. Using the same standard the atomic mass unit, abbreviated amu, is defined as $\frac{1}{12}$ of the mass of *one atom* of ^{12}C. Since an atom is extremely small and has very little mass, it follows that the atomic mass unit is extraordinarily small compared to the gram (Section 4.5).

4.4 Atomic Mass

The **atomic mass** of an element is *the average mass for an aggregate of atoms of that element compared to the mass of a* ^{12}C *atom having a mass of* 12.000 *amu*. The atomic mass of helium is 4.00260 amu; thus, the average mass of a helium atom is about one third of the mass of a ^{12}C atom. The atomic mass of titanium is 47.90 amu; thus, the average mass of a titanium atom is about four times the mass of a ^{12}C atom. A table of the accepted values for the atomic masses of the elements is given inside the front cover.

The average atomic mass for an aggregate of atoms of the same kind can be determined by taking a weighted average of the stable isotopes (Section 2.10) of that atom.

Example 4.3 If boron has naturally occurring isotopes with mass numbers 10 and 11 amu with relative percentage abundances of 19.6 and 80.4, respectively, calculate the average atomic mass of boron.

The relative contribution of each isotope must be considered to obtain the average atomic mass. Each contribution is determined by multiplying the mass number by the percent abundance changed to a decimal.

$$\begin{aligned}(10\ \text{amu})(0.196) &= 1.96\ \text{amu}\\(11\ \text{amu})(0.804) &= \underline{8.84\ \text{amu}}\\&\ 10.80\ \text{amu}\end{aligned}$$

Note that there are four significant figures in the result according to the rules in Section 1.3-c.

4.5 Avogadro Number

The Avogadro number, usually denoted by *N*, is defined in a wide variety of ways; however, it is simply the conversion factor between the *gram* and the *atomic mass unit*. It has been determined by electrochemical and crystallographic techniques to have a value of 6.02252×10^{23}. Thus, 1 gram = 6.02252×10^{23} amu. This number is very important in numerical problems to be discussed shortly.

4.6 Molecule

A **molecule** (Section 1.1-d-2) is *an electrically neutral collection of atoms bonded together strongly enough to be considered as a recognizable unit.* Two carbon atoms and four hydrogen atoms bind together to form a molecule called ethylene (Tables 14.1 and 14.3), a reactive gaseous compound. Six carbon atoms and twelve hydrogen atoms bind together to form a molecule called cyclohexane (Section 13.8), a liquid in which many organic compounds dissolve readily. A molecule exhibits distinctive charcteristics that may not be related in any obvious fashion to the properties of the atoms that comprise the molecule.

4.7 Formula

A **formula** (Section 1.1-d-2) *conveys information about the composition and structure of a compound.* An **empirical formula** states only *the **relative** numbers of atoms in a compound* and says nothing about its structure. By convention an empirical formula includes symbols of the elements making up a compound and subscripts affixed to the symbols to designate the relative numbers of atoms. Thus, the empirical formula for both ethylene and cyclohexane is CH_2, meaning that there is one carbon atom for every two hydrogen atoms in the compound.

A **molecular formula** states *the **actual** numbers of atoms in a molecule,* but says nothing about its structure. A molecular formula includes symbols of the elements that make up a molecule with subscripts affixed to the symbols to designate the actual numbers of atoms. The molecular formula of ethylene is C_2H_4, and the molecular formula of cyclohexane is C_6H_{12}, both simple multiples of their empirical formula, CH_2.

A **structural formula** gives *the actual numbers of atoms in a molecule, indicates which atoms are bonded to each other, and may indicate the three-dimensional configuration of the molecule.* Thus, to indicate which atoms are bonded to each other in ethylene and cyclohexane, we write

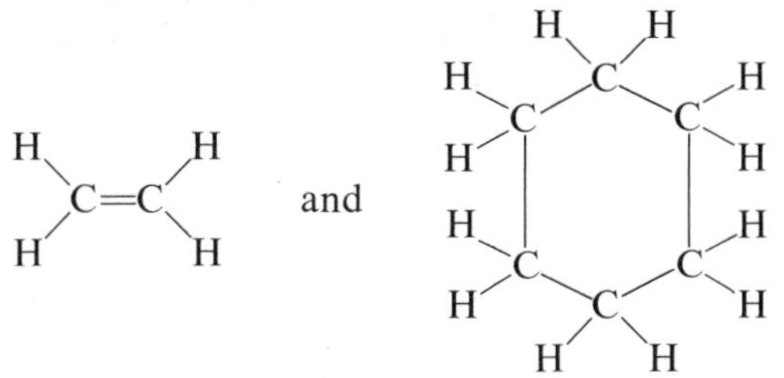

In cyclohexane there are six C atoms forming a six-membered ring. In both ethylene and cyclohexane each C atom has two H atoms bonded to it. More detailed structural formulas are shown for ethylene in Figure 3.9 and for cyclohexane in Figures 13.7 and 13.8.

4.8 Ion

An **ion** (Section 3.2) is *an atom or distinguishable group of atoms bearing one or more positive or negative charges due to the loss or gain of one or more* *electrons.* Titanium(II) chloride, the compound in Example 4.1, is composed of Ti^{2+} and Cl^- ions. These ions are formed from single atoms. Ammonium sulfate consists of NH_4^+ and SO_4^{2-} ions, both of which are formed from distinguishable groups of atoms. *A positive ion* such as Ti^{2+} or NH_4^+ is called a **cation.** *A negative ion* such as Cl^- or SO_4^{2-} is called an **anion.**

4.9 Mole

Atoms and most molecules, formulas, and ions are extremely small and cannot be counted directly by chemists who want to keep track of quantities of these particles. The process of weighing is used instead to count these particles indirectly. Since individual atoms, molecules, formulas, and ions have very little mass, any sample massive enough to be weighed in the laboratory must involve an extremely large number of these particles. Chemists have devised the concept of the mole to express large numbers of these particles.

One **mole** *of atoms, molecules, formulas, or ions* is *the Avogadro number of atoms, molecules, formulas, or ions,* respectively. One mole of titanium atoms is the Avogadro number or 6.023×10^{23} titanium atoms. One mole of sulfur dioxide molecules is the Avogadro number of sulfur dioxide molecules. One mole of H_2O_2 formulas is the Avogadro number of H_2O_2 formulas. One mole of nitrate ions is the Avogadro number of nitrate ions.

When using the mole concept, you must exercise great care in keeping track of what particle you are counting. You must specify the particle in words or by formula. For example, one mole of sulfur might refer to either the Avogadro number of sulfur atoms or the Avogadro number of S_8 molecules (a sulfur molecule consists of eight sulfur atoms arranged in the form of a crown-shaped ring). The former weighs 32.06 grams; the latter weighs 256.48 grams.

4.10 Molar Mass

For the mole to be a useful quantity that can be weighed in the laboratory, it must be related to mass. The **molar mass** *of any particle* (atom, molecule, formula, or ion) is *the sum of the average atomic masses of all atoms forming that particle.* (Many texts refer to this quantity as molecular weight.) The molar mass, frequently symbolized by M, can be computed if the formula of the particle is known or can be determined by a number of experimental methods if the formula of the particle is unknown.

The molar mass of an atom is its average atomic mass expressed in grams. It is the mass of one mole of such atoms. One mole of titanium atoms weighs 47.90 grams and consists of the Avogadro number of titanium atoms.

The molar mass of a molecule is the sum of the average atomic masses of all the atoms comprising the molecule. It is the mass of one mole of such molecules. One mole of SO_2 molecules is 32.06 grams of sulfur plus 2(15.9994 grams) of oxygen, or 64.06 grams, and consists of the Avogadro number of

SO_2 molecules. It is left to you as an exercise to show that 64.06 has the correct number of significant figures.

Many substances in the solid and liquid state are not constructed of discrete molecules, but rather the crystal or the liquid may be considered as one giant molecule. Table salt, sodium chloride, a part of the crystal lattice of which is given in Figure 3.2, is an example of an ionic substance in which an infinite array of positively charged ions is associated with an infinite array of negatively charged ions. The ions are in such ratios that the total number of positive charges is just balanced by the same total number of negative charges. No discrete molecules can be discerned; thus, molecular formula has no meaning in this case. The empirical formula, NaCl, shows what ions are present and states the ratio of these ions as simply as possible. The molar mass is the empirical-formula mass in calculations involving such substances.

The molar mass of a formula is the sum of the average atomic masses of all atoms comprising the formula. It is the mass of one mole of such formulas. One mole of $TiCl_2$ formulas weighs 47.90 grams of titanium plus 2(35.453 grams) of chlorine, or 118.81 grams, and consists of the Avogadro number of $TiCl_2$ formulas.

The molar mass of an ion is the sum of the average atomic masses of all atoms comprising the ion. It is the mass of one mole of such ions. One mole of nitrate ions, NO_3^-, weighs 14.0067 grams of nitrogen plus 3(15.9994 grams) of oxygen, or 62.0049 grams, and consists of the Avogadro number of nitrate ions. Note that the mass of one additional electron which gives rise to the -1 charge is negligible and is neglected.

4.11 Sample Calculations

a. Atoms

Example 4.4 For a 250 gram sample of titanium metal, determine (a) the number of moles of titanium atoms present, (b) the number of titanium atoms present, and (c) the average mass of one titanium atom.

(a) The molar mass of a Ti atom is 47.90 g/mole. Determine the number of moles of Ti atoms by dividing the mass of Ti by the molar mass of a Ti atom.[1]

$$\text{Number of moles of Ti atoms} = \frac{\text{mass of Ti}}{\text{molar mass of Ti}} = \frac{250\ \not{g}}{47.90\ \not{g}/\text{mole}} = 5.22 \text{ moles of Ti atoms}$$

[1]When taking data from tables, always use at least as many significant figures as you have in other data if they are available.

(b) One mole of Ti atoms consists of the Avogadro number of Ti atoms. Determine the number of Ti atoms by multiplying the number of moles of Ti atoms by the Avogadro number.

$$\begin{aligned}\text{Number of Ti atoms} &= (\text{number of moles of Ti})(\text{Avogadro number})\\ &= (5.22\ \cancel{\text{moles}})(6.023 \times 10^{23}\ \text{atoms}/\cancel{\text{mole}})\\ &= 3.14 \times 10^{24}\ \text{Ti atoms}\end{aligned}$$

(c) One mole of Ti atoms weighs 47.90 grams and consists of the Avogadro number of Ti atoms. Determine the mass of one Ti atom by dividing the molar mass of a Ti atom by the Avogadro number.

$$\begin{aligned}\text{Mass of a Ti atom} &= \frac{\text{molar mass of Ti}}{\text{Avogadro number}}\\ &= \frac{47.90\ \text{g}/\cancel{\text{mole}}}{6.023 \times 10^{23}\ \text{atoms}/\cancel{\text{mole}}}\\ &= 7.953 \times 10^{-23}\ \text{g/atom}\end{aligned}$$

Alternatively, it was calculated above that 250 grams of Ti consists of 3.14×10^{24} Ti atoms. Therefore,

$$\frac{250\ \text{g}}{3.14 \times 10^{24}\ \text{atoms}} = 7.96 \times 10^{-23}\ \text{g/atom}$$

Note that dimensions attached to the various quantities are carried through the same mathematical operations as the numbers themselves. Also note that there are (after calculations in accordance with the rules in Section 1.3-c-2) four significant figures in the answer in (c) using the method requiring only constants from tables but only three significant figures in the method using experimental data.

b. Molecules

Example 4.5 **For a 250 gram sample of sulfur dioxide, determine (a) the number of moles of SO_2 molecules present, (b) the number of moles of sulfur atoms and oxygen atoms present, (c) the number of SO_2 molecules present, (d) the number of sulfur atoms and oxygen atoms present, and (e) the average mass of one SO_2 molecule.**

(a) The molar mass of a SO_2 molecule is 64.06 g/mole. Determine the number of moles of SO_2 molecules by dividing the mass of SO_2 by the molar mass of a SO_2 molecule.

$$\begin{aligned}\text{Number of moles of } SO_2 \text{ molecules} &= \frac{\text{mass of } SO_2}{\text{molar mass of } SO_2}\\ &= \frac{250\ \cancel{\text{g}}}{64.06\ \cancel{\text{g}}/\text{mole}}\\ &= 3.90 \text{ moles of } SO_2 \text{ molecules}\end{aligned}$$

(b) The molecular formula, SO_2, indicates that there are one mole of S atoms and two moles of O atoms in each mole of SO_2 molecules.

$$\begin{aligned}\text{Number of moles of S atoms} &= \text{number of moles of } SO_2 \text{ molecules}\\ &= 3.90 \text{ moles of S atoms}\\ \text{Number of moles of O atoms} &= (2)(\text{number of moles of } SO_2 \text{ molecules})\\ &= 7.80 \text{ moles of O atoms}\end{aligned}$$

(c) One mole of SO_2 consists of the Avogadro number of SO_2 molecules. Determine the number of SO_2 molecules by multiplying the number of moles of SO_2 molecules by the Avogadro number.

$$\begin{aligned}\text{Number of } SO_2 \text{ molecules} &= (\text{number of moles of } SO_2)(\text{Avogadro number})\\ &= (3.90 \cancel{\text{moles}})(6.023 \times 10^{23} \text{ molecules}/\cancel{\text{mole}})\\ &= 2.35 \times 10^{24}\ SO_2 \text{ molecules}\end{aligned}$$

(d) The molecular formula, SO_2, indicates that there are one S atom and two O atoms in each SO_2 molecule.

$$\begin{aligned}\text{Number of S atoms} &= \text{number of } SO_2 \text{ molecules}\\ &= 2.35 \times 10^{24} \text{ S atoms}\\ \text{Number of O atoms} &= (2)(\text{number of } SO_2 \text{ molecules})\\ &= 4.70 \times 10^{24} \text{ O atoms}\end{aligned}$$

(e) One mole of SO_2 molecules weighs 64.06 grams and consists of the Avogadro number of SO_2 molecules. Determine the mass of one SO_2 molecule by dividing the molar mass of a SO_2 molecule by the Avogadro number.

$$\begin{aligned}\text{Mass of a } SO_2 \text{ molecule} &= \frac{\text{molar mass of } SO_2}{\text{Avogadro number}}\\ &= \frac{64.06 \text{ g}/\cancel{\text{mole}}}{6.023 \times 10^{23} \text{ molecules}/\cancel{\text{mole}}}\\ &= 1.064 \times 10^{-22} \text{ g/molecule}\end{aligned}$$

Alternatively, it was calculated above that 250 grams of SO_2 consists of 2.35×10^{24} SO_2 molecules. Therefore,

$$\frac{250 \text{ g}}{2.35 \times 10^{24} \text{ molecules}} = 1.06 \times 10^{-22} \text{ g/molecule}$$

c. Formulas

The empirical formula of a compound can be determined using molar masses of constituent atoms and analytical data for the compound, either in the form of masses of each element present or percent composition. Both methods involve calculation of the ratio of moles of constituent atoms as illustrated in the following three examples.

Example 4.6 Determine the empirical formula of the compound in Example 4.1 formed from 2.44 grams of titanium and 3.60 grams of chlorine.

$$\text{Number of moles of Ti atoms} = \frac{\text{mass of Ti}}{\text{molar mass of Ti}}$$

$$= \frac{2.44\ \cancel{g}}{47.90\ \cancel{g}/\text{mole}}$$

$$= 0.0509 \text{ mole of Ti atoms}$$

$$\text{Number of moles of Cl atoms} = \frac{\text{mass of Cl}}{\text{molar mass of Cl}}$$

$$= \frac{3.60\ \cancel{g}}{35.453\ \cancel{g}/\text{mole}}$$

$$= 0.102 \text{ mole of Cl atoms}$$

The ratio of moles of Ti atoms to moles of Cl atoms is 0.0509:0.102. Divide both numbers by the smaller one to reduce to a simple whole-number ratio.

$$\text{Ti:Cl} = \frac{0.0509}{0.0509}:\frac{0.102}{0.0509} = 1.00:1.99$$

Therefore, the empirical formula is $TiCl_2$.

Example 4.7 Determine the empirical formula of an alcohol that has a percent composition by mass of *60.0*% carbon, *13.4*% hydrogen, and *26.6*% oxygen.

Assume 100 grams of the compound. The percent composition indicates the number of grams of each element present in 100 grams of the compound.

$$\text{Mass of C} = (100 \text{ g})(0.600) = 60.0 \text{ g of C}$$

$$\text{Mass of H} = (100 \text{ g})(0.134) = 13.4 \text{ g of H}$$

$$\text{Mass of O} = (100 \text{ g})(0.266) = 26.6 \text{ g of O}$$

$$\text{Number of moles of C atoms} = \frac{\text{mass of C}}{\text{molar mass of C}}$$

$$= \frac{60.0\ \cancel{g}}{12.011\ \cancel{g}/\text{mole}}$$

$$= 5.00 \text{ moles of C atoms}$$

$$\text{Number of moles of H atoms} = \frac{\text{mass of H}}{\text{molar mass of H}}$$

$$= \frac{13.4\ \cancel{g}}{1.0079\ \cancel{g}/\text{mole}}$$

$$= 13.3 \text{ moles of H atoms}$$

$$\text{Number of moles of O atoms} = \frac{\text{mass of O}}{\text{molar mass of O}}$$

$$= \frac{26.6\ \cancel{g}}{15.9994\ \cancel{g}/\text{mole}}$$

$$= 1.66 \text{ moles of O atoms}$$

$$\text{C:H:O} = \frac{5.00}{1.66} : \frac{13.3}{1.66} : \frac{1.66}{1.66} = 3.01:8.01:1.00$$

Therefore, the empirical formula is C_3H_8O, and the alcohol is 1-propanol (Tables 17.1 and 17.2).

Frequently fractional ratios of moles of constituent atoms are obtained. In such cases multiply by a number that will reduce the ratio to small whole numbers. This process is illustrated in the following example.

Example 4.8 Hematite, an iron-containing ore, is composed of 69.9% iron and 30.1% oxygen. What is its empirical formula?

Assume 100 grams of the compound. The percent composition indicates the number of grams of each element present in 100 grams of the compound.

$$\text{Mass of Fe} = (100\text{ g})(0.699) = 69.9\text{ g of Fe}$$

$$\text{Mass of O} = (100\text{ g})(0.301) = 30.1\text{ g of O}$$

$$\text{Number of moles of Fe atoms} = \frac{\text{mass of Fe}}{\text{molar mass of Fe}}$$

$$= \frac{69.9\ \cancel{g}}{55.847\ \cancel{g}/\text{mole}}$$

$$= 1.25 \text{ moles of Fe atoms}$$

$$\text{Number of moles of O atoms} = \frac{\text{mass of O}}{\text{molar mass of O}}$$

$$= \frac{30.1\ \cancel{g}}{15.9994\ \cancel{g}/\text{mole}}$$

$$= 1.88 \text{ moles of O atoms}$$

$$\text{Fe:O} = \frac{1.25}{1.25} : \frac{1.88}{1.25} = 1.00:1.50$$

Reduce the ratio of moles of constituent atoms, 1.00:1.50, to small whole numbers by multiplying by two. This gives a simplest ratio of 2.00:3.00, and the empirical formula for hematite is Fe_2O_3.

If the empirical formula of a compound is known, the molar mass of that empirical formula and the percent composition of the compound can be calculated.

Example 4.9 What are the molar mass and the percent composition of titanium(III) sulfate, $Ti_2(SO_4)_3$?

Molar mass of a $Ti_2(SO_4)_3$ formula

$$= 2(47.90\text{ g})\text{ of Ti} + 3(32.06\text{ g})\text{ of S} + 12(15.9994\text{ g})\text{ of O}$$
$$= 383.97\text{ g/mole}$$

$$\text{Percent Ti} = \left[\frac{\text{mass of Ti}}{\text{molar mass of } Ti_2(SO_4)_3}\right](100\%) = \left[\frac{2(47.90\text{ g})}{383.97\text{ g}}\right](100\%)$$
$$= 24.95\%\text{ Ti}$$

$$\text{Percent S} = \left[\frac{\text{mass of S}}{\text{molar mass of } Ti_2(SO_4)_3}\right](100\%) = \left[\frac{3(32.06\text{ g})}{383.97\text{ g}}\right](100\%)$$
$$= 25.05\%\text{ S}$$

$$\text{Percent O} = \left[\frac{\text{mass of O}}{\text{molar mass of } Ti_2(SO_4)_3}\right](100\%) = \left[\frac{12(15.9994\text{ g})}{383.97\text{ g}}\right](100\%)$$
$$= 50.002\%\text{ O}$$

In the above examples molar masses of constituent atoms and analytical data in the form of masses of each element present or percent composition were required to determine the empirical formula. To determine the molecular formula, an experimentally determined molar mass of the molecule is also required.

Example 4.10 Ascorbic acid (vitamin C), found in citrus fruits, tomatoes, and green vegetables, is composed of 40.9% carbon, 4.6% hydrogen, and 54.5% oxygen. Determine (a) the empirical formula of ascorbic acid and (b) the molecular formula if the experimentally determined molar mass of an ascorbic acid molecule is 176.13 g/mole.

(a) Assume 100 grams of the compound. The percent composition indicates the number of grams of each element present in 100 grams of the compound.

$$\text{Mass of C} = (100\text{ g})(0.409) = 40.9\text{ g of C}$$
$$\text{Mass of H} = (100\text{ g})(0.046) = 4.6\text{ g of H}$$
$$\text{Mass of O} = (100\text{ g})(0.545) = 54.5\text{ g of O}$$

$$\text{Number of moles of C atoms} = \frac{\text{mass of C}}{\text{molar mass of C}}$$
$$= \frac{40.9\text{ g}}{12.011\text{ g/mole}}$$
$$= 3.41\text{ moles of C atoms}$$

$$\text{Number of moles of H atoms} = \frac{\text{mass of H}}{\text{molar mass of H}}$$

$$= \frac{4.6\ \cancel{g}}{1.0079\ \cancel{g}/\text{mole}}$$

$$= 4.6 \text{ moles of H atoms}$$

$$\text{Number of moles of O atoms} = \frac{\text{mass of O}}{\text{molar mass of O}}$$

$$= \frac{54.5\ \cancel{g}}{15.9994\ \cancel{g}/\text{mole}}$$

$$= 3.41 \text{ moles of O atoms}$$

$$\text{C:H:O} = \frac{3.41}{3.41} : \frac{4.6}{3.41} : \frac{3.41}{3.41} = 1.00{:}1.3{:}1.00$$

$$= 3.00{:}3.9{:}3.00$$

Note that it is reasonable to round off 3.9 to 4 when converting to whole numbers in this example because the fact that there are two significant figures in 3.9 implies that the error is at least ± 0.1. Therefore, the empirical formula of ascorbic acid is $C_3H_4O_3$.

Note additionally that it would have been unreasonable to round off 1.50 to 2 in Example 4.8 because three significant figures imply that the error is in the hundredths place.

(b) The molecular formula is some integral multiple of the empirical formula. Compute the molar mass of the empirical formula and determine what multiple the experimental molar mass of the molecule is of the molar mass of the empirical formula.

Molar mass of the empirical formula

$$= 3(12.011 \text{ g}) \text{ of C} + 4(1.0079 \text{ g}) \text{ of H} + 3(15.9994 \text{ g}) \text{ of O}$$

$$= 88.063 \text{ g/mole}$$

$$\frac{\text{Molar mass of the molecule}}{\text{Molar mass of the empirical formula}} = \frac{176.13\ \cancel{g}/\cancel{\text{mole}}}{88.063\ \cancel{g}/\cancel{\text{mole}}}$$

$$= 2.0000$$

Therefore, two empirical formulas comprise the molecular formula, and the molecular formula of ascorbic acid is $C_6H_8O_6$. (See Section 24.6 for the structural formula of ascorbic acid.)

d. Ions

Example 4.11 For a 250 gram sample of titanium(III) sulfate, $Ti_2(SO_4)_3$, determine (a) the percent composition of SO_4^{2-} in this compound, (b) the mass of SO_4^{2-} ions present, (c) the number of SO_4^{2-} ions present, and (d) the average mass of one SO_4^{2-} ion.

(a) Molar mass of SO_4^{2-} ion = 32.06 g of S + 4(15.9994 g) of O
= 96.06 g/mole

Note that the mass of two additional electrons that give rise to the −2 charge is negligible and is neglected.

The molar mass of a $Ti_2(SO_4)_3$ formula, calculated in Example 4.9, is 383.97 g/mole.

$$\text{Percent } SO_4^{2-} = \left[\frac{3(\text{molar mass of } SO_4^{2-})}{\text{molar mass of } Ti_2(SO_4)_3}\right](100\%)$$
$$= \left[\frac{3(96.06 \text{ g/mole})}{383.97 \text{ g/mole}}\right](100\%)$$
$$= 75.05\% \text{ of } SO_4^{2-}$$

Note that this same result can be obtained by adding the percent sulfur and the percent oxygen of Example 4.9 since electrons have negligible mass compared to atoms.

(b) Determine the mass of SO_4^{2-} in 250 g of $Ti_2(SO_4)_3$ by multiplying the mass of $Ti_2(SO_4)_3$ by the percent SO_4^{2-} converted to a decimal.

$$\text{Mass of } SO_4^{2-} = (250 \text{ g})(0.7505) = 188 \text{ g of } SO_4^{2-}$$

(c) The molar mass of a SO_4^{2-} ion is 96.06 g/mole. Determine the number of moles of SO_4^{2-} ions by dividing the mass of SO_4^{2-} ions by the molar mass of a SO_4^{2-} ion.

$$\text{Number of moles of } SO_4^{2-} \text{ ions} = \frac{\text{mass of } SO_4^{2-}}{\text{molar mass of } SO_4^{2-}}$$
$$= \frac{188 \text{ g}}{96.06 \text{ g/mole}}$$
$$= 1.96 \text{ moles of } SO_4^{2-} \text{ ions}$$

One mole of SO_4^{2-} ions consists of the Avogadro number of SO_4^{2-} ions. Determine the number of SO_4^{2-} ions by multiplying the number of moles of SO_4^{2-} ions by the Avogadro number.

$$\text{Number of } SO_4^{2-} \text{ ions} = (\text{number of moles of } SO_4^{2-})(\text{Avogadro number})$$
$$= (1.96 \text{ moles})(6.023 \times 10^{23} \text{ ions/mole})$$
$$= 1.18 \times 10^{24}\ SO_4^{2-} \text{ ions}$$

(d) One mole of SO_4^{2-} ions weighs 96.06 g and consists of the Avogadro number of SO_4^{2-} ions. Determine the mass of one SO_4^{2-} ion by dividing the molar mass of a SO_4^{2-} ion by the Avogadro number.

$$\text{Mass of a } SO_4^{2-} \text{ ion} = \frac{\text{molar mass of } SO_4^{2-}}{\text{Avogadro number}}$$
$$= \frac{96.06 \text{ g/mole}}{6.023 \times 10^{23} \text{ ions/mole}}$$
$$= 1.595 \times 10^{-22} \text{ g/ion}$$

Alternatively, it was calculated above that 188 g of SO_4^{2-} consists of 1.18×10^{24} SO_4^{2-} ions. Therefore,

$$\frac{188\ \text{g}}{1.18 \times 10^{24}\ \text{ions}} = 1.59 \times 10^{-22}\ \text{g/ion}$$

A pictorial summary of the mathematical relationships of the various units used to specify quantities of chemical substances follows.

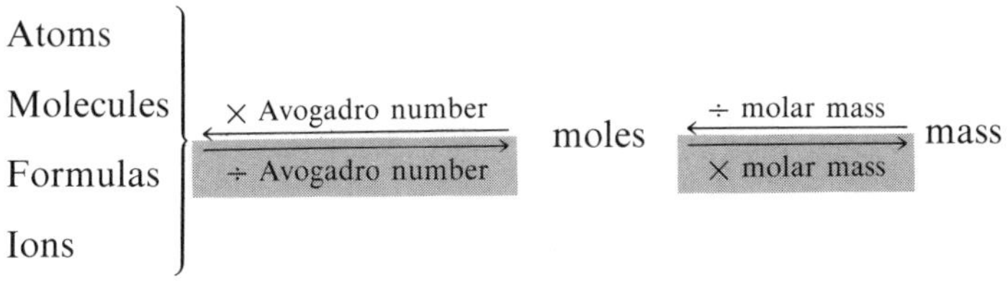
Atoms
Molecules
Formulas
Ions
× Avogadro number
÷ Avogadro number
moles
÷ molar mass
× molar mass
mass

It should be emphasized that the *mole concept* can be used for working all problems involving quantities of substances provided you keep track of what particles (atoms, molecules, formulas, or ions) you are considering. Remaining problems in this book will be worked using the mole concept. An additional example is given for practice.

Example 4.12. The molecular formula of cream of tartar (potassium hydrogen tartrate), one of the components in baking powder, is $KC_4H_5O_6$. For a cream of tartar sample containing 8.01×10^{23} potassium atoms, determine (a) the number of carbon atoms present, (b) the number of cream of tartar molecules present, (c) the number of moles of cream of tartar molecules present, (d) the number of moles of hydrogen atoms present, (e) the mass of cream of tartar present, and (f) the mass of hydrogen present.

(a) The molecular formula, $KC_4H_5O_6$, indicates that there are four C atoms for each K atom in a $KC_4H_5O_6$ molecule. Therefore,

$$\begin{aligned}\text{Number of C atoms} &= 4(\text{number of K atoms})\\ &= 3.20 \times 10^{24}\ \text{C atoms}\end{aligned}$$

(b) There will be the same number of cream of tartar molecules as potassium atoms since there is one K atom in each $KC_4H_5O_6$ molecule.

$$\begin{aligned}\text{Number of } KC_4H_5O_6 \text{ molecules} &= \text{number of K atoms}\\ &= 8.01 \times 10^{23}\ KC_4H_5O_6 \text{ molecules}\end{aligned}$$

(c) One mole of cream of tartar molecules consists of the Avogadro number of $KC_4H_5O_6$ molecules. Determine the number of moles of $KC_4H_5O_6$ molecules by dividing the number of $KC_4H_5O_6$ molecules by the Avogadro number.

$$\text{Number of moles of } KC_4H_5O_6 \text{ molecules} = \frac{\text{number of } KC_4H_5O_6 \text{ molecules}}{\text{Avogadro number}}$$

$$= \frac{8.01 \times 10^{23} \text{ molecules}}{6.023 \times 10^{23} \text{ molecules/mole}}$$

$$= 1.33 \text{ moles of } KC_4H_5O_6 \text{ molecules}$$

(d) The molecular formula, $KC_4H_5O_6$, indicates that there are five moles of hydrogen atoms in each mole of $KC_4H_5O_6$ molecules.

$$\text{Number of moles of H atoms} = 5(\text{number of moles of } KC_4H_5O_6 \text{ molecules})$$
$$= 6.65 \text{ moles of H atoms}$$

(e) The molar mass of a cream of tartar molecule is 188 g/mole. Determine the mass of cream of tartar by multiplying the number of moles of cream of tartar molecules by the molar mass of a cream of tartar molecule.

$$\text{Mass of } KC_4H_5O_6 = (\text{number of moles of } KC_4H_5O_6)(\text{molar mass of } KC_4H_5O_6)$$
$$= (1.33 \text{ moles})(188 \text{ g/mole})$$
$$= 250 \text{ g of } KC_4H_5O_6$$

(f) Determine the mass of H by multiplying the number of moles of H atoms by the molar mass of a H atom.

$$\text{Mass of H} = (\text{number of moles of H})(\text{molar mass of H})$$
$$= (6.65 \text{ moles})(1.0079 \text{ g/mole})$$
$$= 6.70 \text{ g of H}$$

Alternatively, solve (f) by determining the percent H in $KC_4H_5O_6$ using the molecular formula and molar masses of a H atom and a $KC_4H_5O_6$ molecule. Then use the percent composition to determine what part of the 250 grams calculated in (e) is H. This is left to you as an exercise.

4.12 Chemical Equations

A **chemical equation** *specifies the substances used up* (*reactants*) *and the substances formed* (*products*) *in a chemical reaction,* as established by experimentation. By convention, formulas for the reactants, written on the left side of the equation, and formulas for the products, written on the right side of the equation, are separated by a single arrow $\longrightarrow$, a double arrow $\rightleftharpoons$, or an equal sign $=$. Thus, an *unbalanced* chemical equation describing the reaction between solid lead(IV) oxide, PbO_2, and gaseous hydrogen, H_2, to form solid lead, Pb, and liquid water, H_2O, is written as follows.

$$PbO_{2(s)} + H_{2(g)} \longrightarrow Pb_{(s)} + H_2O$$

The subscript notation (s) indicates the solid phase and (g) the gas phase. The subscript notations (l) for the liquid phase and (aq) for an aqueous solution may be used, but are frequently omitted.

***Balancing** a chemical equation makes it consistent with the conservation of mass and electrical charge,* neither of which can be created or destroyed. A balanced equation (other than one describing a nuclear reaction) has the same number of respective atoms on the left and right sides and has the same *net* electrical charge on the left and right sides. In the above equation none of the chemical species is charged. The net electrical charge on both sides of the equation is zero, and the electrical charge is balanced or conserved. However, the mass is not balanced or conserved. Although there are the same numbers of lead and hydrogen atoms on the left and right sides, there are two oxygen atoms on the left side and only one on the right side. *Since the ratios of atoms in the formulas are fixed, only smallest whole-number coefficients before the chemical species may be used to balance chemical equations.*

Many chemical equations are simple enough to balance by inspection. Consider, for example, the previous equation involving the reaction of PbO_2 and H_2 to form Pb and H_2O. Placing a coefficient of two before H_2O balances the oxygen atoms, but leaves the hydrogen atoms unbalanced. However, placing a coefficient of two before H_2 balances the hydrogen atoms. The balanced equation for the reaction is

$$PbO_{2(s)} + 2\,H_{2(g)} \longrightarrow Pb_{(s)} + 2\,H_2O$$

Another example involving the reaction of calcium ions, Ca^{2+}, with phosphate ions, PO_4^{3-}, to form insoluble calcium phosphate, $Ca_3(PO_4)_2$, illustrates the association of ions in aqueous solution.

$$Ca^{2+} + PO_4^{3-} \longrightarrow Ca_3(PO_4)_{2(s)}$$

Neither the charge nor the mass is balanced in this equation. The net charge on the left is -1 whereas it is zero on the right. There are more calcium, phosphorus, and oxygen atoms on the right than on the left. The formula of the product indicates that three Ca^{2+} ions are required for every two PO_4^{3-} ions. Placing a coefficient of three before Ca^{2+} and a coefficient of two before PO_4^{3-} on the left provides the number of calcium, phosphorus, and oxygen atoms required on the right. In addition, this balances the electrical charge leaving a net charge of $(3)(+2) + (2)(-3) =$ zero on the left and also zero on the right. The balanced equation for the reaction is

$$3\,Ca^{2+} + 2\,PO_4^{3-} \longrightarrow Ca_3(PO_4)_{2(s)}$$

All equations required in problems at the end of this chapter can be balanced by inspection. The *half-reaction method,* used to balance more complicated equations involving electron transfer, is discussed in Section 8.6.

A balanced chemical equation provides a concise summary of much information concerning a given chemical reaction. It specifies the substances involved and the elements comprising such substances. In addition, *a balanced chemical equation describes the relative numbers of moles of particles* (atoms, molecules, formulas, and ions) *and the relative numbers of particles taking part in a chemical reaction.* Furthermore, the relative masses of atoms, molecules, formulas, or ions reacted and formed can be determined easily by multiplying the relative numbers of moles of these particles by appropriate molar masses.

The quantitative information readily deduced from two balanced chemical equations is illustrated below.

	$PbO_{2(s)}$ +	$2\ H_{2(g)}$ ⟶	$Pb_{(s)}$ +	$2\ H_2O$
Relative number of moles involved	1 mole	2 moles	1 mole	2 moles
Relative number of particles involved	1 formula	2 molecules	1 atom	2 molecules
Relative masses involved	1(239.2 g)	2(2.0158 g)	1(207.2 g)	2(18.0152 g)

	$3\ Ca^{2+}$ +	$2\ PO_4^{3-}$ ⟶	$Ca_3(PO_4)_{2(s)}$
Relative number of moles involved	3 moles	2 moles	1 mole
Relative number of particles involved	3 ions	2 ions	1 formula
Relative masses involved	3(40.08 g)	2(94.9714 g)	1(310.18 g)

4.13 Mass Relations in Chemical Reactions

A balanced chemical equation, in conjunction with the various mathematical operations described previously in this chapter, is used to determine the masses of materials consumed and produced in a given chemical reaction. One assumption is also implicit in the following examples—the chemical reaction goes to completion so that all of the reactants are used up entirely to form products.

Example 4.13 The enzyme mixture called zymase, present in bread yeast, catalyzes the fermentation of glucose, $C_6H_{12}O_6$, to ethyl alcohol, C_2H_5OH, and gaseous carbon dioxide, CO_2, as represented by the following balanced chemical equation. (See Section 28.3.)

$$C_6H_{12}O_6 \xrightarrow{\text{zymase}} 2\ C_2H_5OH + 2\ CO_{2(g)}$$

Determine how many molecules of CO_2 and what mass of C_2H_5OH are produced from 8.01×10^{20} $C_6H_{12}O_6$ molecules.

The balanced equation indicates that two CO_2 molecules are produced for each $C_6H_{12}O_6$ molecule consumed.

$$(8.01 \times 10^{20}\ \cancel{C_6H_{12}O_6\ \text{molecules}})\left(\frac{2\ CO_2\ \text{molecules}}{1\ \cancel{C_6H_{12}O_6\ \text{molecule}}}\right)$$

$$= 1.60 \times 10^{21}\ CO_2\ \text{molecules}$$

The same number of C_2H_5OH molecules, 1.60×10^{21}, is produced. Determine the number of moles of C_2H_5OH molecules produced by dividing the number of C_2H_5OH molecules by the Avogadro number.

$$\text{Number of moles of } C_2H_5OH \text{ molecules} = \frac{1.60 \times 10^{21}\ \cancel{\text{molecules}}}{6.023 \times 10^{23}\ \cancel{\text{molecules}}/\text{mole}}$$

$$= 2.66 \times 10^{-3} \text{ mole of } C_2H_5OH \text{ molecules}$$

Determine the mass of C_2H_5OH produced by multiplying the number of moles of C_2H_5OH molecules by the molar mass of a C_2H_5OH molecule.

$$\text{Mass of } C_2H_5OH = (2.66 \times 10^{-3}\ \cancel{\text{mole}})(46.069\ \text{g}/\cancel{\text{mole}})$$

$$= 0.123 \text{ g of } C_2H_5OH$$

Example 4.14 Solid potassium nitrate, KNO_3, a common constituent in chemical fertilizers, decomposes on heating to solid potassium nitrite, KNO_2, and gaseous oxygen, O_2. (a) Write a balanced chemical equation for this reaction. (b) Determine how many moles of oxygen molecules and what mass of potassium nitrite are formed from complete decomposition of 18.0 grams of potassium nitrate.

(a) Write an unbalanced chemical equation.

$$KNO_{3(s)} \longrightarrow KNO_{2(s)} + O_{2(g)}$$

By inspection the equation is balanced if a coefficient of one-half is placed before O_2. However, since whole-number coefficients are desired, multiply coefficients through by two and obtain

$$2\ KNO_{3(s)} \longrightarrow 2\ KNO_{2(s)} + O_{2(g)}$$

(b) Determine the number of moles of KNO_3 in 18.0 g since problems are most easily solved in terms of moles.

$$\text{Moles of } KNO_3 = \frac{\text{mass of } KNO_3}{\text{molar mass of } KNO_3}$$
$$= \frac{18.0\ \cancel{g}}{101.105\ \cancel{g}/\text{mole}}$$
$$= 0.178 \text{ mole of } KNO_3$$

The balanced equation indicates that one mole of O_2 forms for each two moles of KNO_3 consumed.

$$(0.178\ \cancel{\text{mole of } KNO_3})\left(\frac{1 \text{ mole of } O_2}{2\ \cancel{\text{moles of } KNO_3}}\right) = 0.0890 \text{ mole of } O_2$$

The balanced equation also indicates that one mole of KNO_2 forms from one mole of KNO_3. Therefore, 0.178 mole of KNO_2 is produced. Determine the mass of KNO_2 produced by multiplying the number of moles of KNO_2 produced by the molar mass of KNO_2.

$$\text{Mass of } KNO_2 = (0.178\ \cancel{\text{mole}})(85.106\ \text{g}/\cancel{\text{mole}})$$
$$= 15.1 \text{ g of } KNO_2$$

4.14 Theoretical Yield, Actual Yield, and Percent Yield

The assumption implicit in Secton 4.13, that reactants are converted *completely* to desired products in a chemical reaction, frequently is not valid. There are several reasons for this. A state of equilibrium (Section 10.1) may develop such that significant amounts of reactants remain at the end of the reaction time. Side reactions may occur to produce undesired products. An excess of one reactant may be used purposely in order to convert a second more expensive or difficultly prepared reactant quantitatively to products.

In the last step in the commercial preparation of aspirin (Section 19.11), excess acetic anhydride, $C_4H_6O_3$, along with sulfuric acid as a catalyst, is added to salicylic acid, $C_7H_6O_3$, to produce aspirin, $C_9H_8O_4$, and acetic acid, $C_2H_4O_2$.

$$C_7H_6O_3 + C_4H_6O_3 \longrightarrow C_9H_8O_4 + C_2H_4O_2$$

The **theoretical yield** of aspirin in this reaction is *the maximum amount that can be produced based on the* ***limiting*** *reactant,* the one not in excess.

Example 4.15 What is the theoretical yield of aspirin, $C_9H_8O_4$, if 5.00 ml of acetic anhydride, $C_4H_6O_3$ (density = 1.082 g/ml), is stirred with 3.00 grams of salicylic acid, $C_7H_6O_3$?

Determine the number of moles of each of the reactants and which is the limiting reactant.

$$\text{Mass of } C_4H_6O_3 = (\text{volume})(\text{density}) = (5.00\ \cancel{\text{ml}})(1.082\ \text{g}/\cancel{\text{ml}}) = 5.41\ \text{g}$$

$$\text{Number of moles of } C_4H_6O_3 = \frac{\text{mass of } C_4H_6O_3}{\text{molar mass of } C_4H_6O_3} = \frac{5.41\ \cancel{\text{g}}}{102.090\ \cancel{\text{g}}/\text{mole}} = 0.0530 \text{ mole of } C_4H_6O_3$$

$$\text{Number of moles of } C_7H_6O_3 = \frac{\text{mass of } C_7H_6O_3}{\text{molar mass of } C_7H_6O_3} = \frac{3.00\ \cancel{\text{g}}}{138.123\ \cancel{\text{g}}/\text{mole}} = 0.0217 \text{ mole of } C_7H_6O_3$$

The balanced equation indicates that $C_7H_6O_3$ and $C_4H_6O_3$ react in a 1 : 1 mole ratio. Thus, $C_7H_6O_3$, with the smaller number of moles, is the limiting reactant, and the theoretical yield of aspirin that can be produced must be based on 0.0217 mole of $C_7H_6O_3$. The balanced equation also indicates that one mole of aspirin is produced for each mole of $C_7H_6O_3$ consumed. Therefore, 0.0217 mole of aspirin is produced.

$$\text{Mass of } C_9H_8O_4 = (\text{number of moles of } C_9H_8O_4)(\text{molar mass of } C_9H_8O_4) = (0.0217\ \cancel{\text{mole}})(180.161\ \text{g}/\cancel{\text{mole}}) = 3.91 \text{ g of aspirin}$$

The **actual yield** of a product is *that recovered experimentally*. It is almost invariably less than the theoretical yield because of an unfavorable equilibrium, because of side reactions, or because losses may occur in separation and purification. Percent yield, defined as

$$\textbf{Percent yield} = \left(\frac{\textit{actual yield}}{\textit{theoretical yield}}\right)(100\%)$$

is used to express the degree of conversion of reactants to products.

Example 4.16 Determine the percent yield of aspirin in Example 4.15 if 3.42 grams of aspirin is recovered.

$$\text{Percent yield} = \left(\frac{\text{actual yield}}{\text{theoretical yield}}\right)(100\%) = \left(\frac{3.42\ \cancel{\text{g}}}{3.91\ \cancel{\text{g}}}\right)(100\%) = 87.5\%$$

Exercises

4.1 Define clearly the following terms.

(a) percent composition
(b) atom
(c) atomic mass unit
(d) gram
(e) atomic mass
(f) isotope
(g) Avogadro number
(h) molecule
(i) formula
(j) empirical formula
(k) molecular formula
(l) structural formula
(m) ion
(n) cation
(o) anion
(p) mole
(q) molar mass
(r) chemical equation
(s) balanced chemical equation
(t) theoretical yield
(u) actual yield
(v) percent yield
(w) limiting reactant

4.2 Calculate the average atomic mass of the indicated element, given the naturally occurring isotopes and their relative percentage abundances.

(a) lithium: ${}^{6}_{3}Li$, 7.42%; ${}^{7}_{3}Li$, 92.58%
(b) silicon: ${}^{28}_{14}Si$, 92.21%; ${}^{29}_{14}Si$, 4.70%; ${}^{30}_{14}Si$, 3.09%
(c) chlorine: ${}^{35}_{17}Cl$, 75.53%; ${}^{37}_{17}Cl$, 24.47%
(d) copper: ${}^{63}_{29}Cu$, 69.09%; ${}^{65}_{29}Cu$, 30.91%
(e) germanium: ${}^{70}_{32}Ge$, 20.52%; ${}^{72}_{32}Ge$, 27.43%; ${}^{73}_{32}Ge$, 7.76%; ${}^{74}_{32}Ge$, 36.54%; ${}^{76}_{32}Ge$, 7.76%

4.3 Determine the molar mass for each of the following.

(a) a phosphorus atom
(b) a P_4 molecule
(c) a $Co(NH_3)_6{}^{3+}$ ion
(d) a C_2H_5 empirical formula
(e) a $C_6H_4(NO_2)_2$ molecule

4.4 Determine the percent composition for each of the following.

(a) $Ni(CO)_4$
(b) $C_2H_5NO_2$
(c) $HgBr_2$
(d) $B_{10}C_2H_{12}$
(e) $C_4H_{10}Cl$

4.5 Determine the number of moles of appropriate particles (atoms, molecules, formulas, or ions).

(a) 3.61 grams of iron
(b) 3.61×10^{21} NH_3 molecules
(c) 9.27×10^{19} $NO_3{}^-$ ions
(d) 48.7 grams of $Mn(CO)_5$
(e) 1.83×10^{24} C_2H_4 empirical formulas

4.6 Determine the number of iodine atoms in each of the following.

(a) 13.0 grams of KIO_3
(b) 4.06 moles of I_2 molecules
(c) 1.23×10^{22} $I_3{}^-$ ions
(d) 167 grams of IF_5
(e) 3.00×10^{-3} mole of ICl molecules

4.7 Determine the empirical formula for a compound containing each of the following.

(a) 5.31×10^{20} Sr atoms, 5.31×10^{20} S atoms, 2.12×10^{21} O atoms
(b) 0.154 mole of K atoms, 0.154 mole of Cr atoms, 0.539 mole of O atoms
(c) 1.73 grams of B, 2.24 grams of N, 3.84 grams of C, 0.97 gram of H
(d) 0.606 gram of Ni, 6.21×10^{21} S atoms
(e) 3.83×10^{21} C atoms, 0.0159 mole of H atoms, 0.0254 gram of O

4.8 Determine each mass, in grams.
(a) one aluminum atom
(b) 2.35 moles of $C_6H_{10}O_4$ molecules
(c) the sulfur in 18.2 grams of $KHSO_4$
(d) one atomic mass unit
(e) the carbon in 8.34×10^{20} CO_2 molecules

4.9 In the introduction to Chapter 25 it is stated that the elemental composition of most proteins is remarkably constant at about 53% carbon, 7% hydrogen, 23% oxygen, 16% nitrogen, 1% sulfur, and less than 1% phosphorus. What are the ratios of C, H, O, and N atoms, respectively, to S atoms in most proteins?

4.10 Determine the empirical formulas of compounds with the following percent compositions.
(a) 74.3% Cs, 7.8% N, 17.9% O
(b) 49.8% Zr, 15.3% Si, 34.9% O
(c) 89.9% C, 10.1% H
(d) 51.3% C, 9.5% H, 12.0% N, 27.3% O
(e) 4.54% Mg, 71.82% C, 11.68% H, 11.96% O

4.11 Determine the mass of chromium in 1.00 gram of each of the following.
(a) Cr_2O_3
(b) $FeCr_2O_4$
(c) $Cr(ClO_4)_3$
(d) $Cr(NH_3)_6Br_3$
(e) $KCr(SO_4)_2 \cdot 12\,H_2O$ (the dot implies that the waters of hydration are integrally bound in the crystal lattice of this compound)

4.12 Determine the percent composition and empirical formula of the oxide of nitrogen formed by reacting completely 1.00 gram of nitrogen gas, N_2, with the following masses of oxygen gas, O_2.
(a) 0.57 gram
(b) 1.14 grams
(c) 2.28 grams
(d) 1.71 grams
(e) 2.86 grams

4.13 Determine the molecular formulas of compounds with the following empirical formulas and the molar mass for one molecule of each.
(a) HgI, 655 g/mole
(b) $C_7H_6O_2$, 122 g/mole
(c) P_2O_5, 284 g/mole
(d) C_3H_2Cl, 147 g/mole
(e) $PtH_8N_2Cl_6$, 444 g/mole

4.14 Nitrous oxide is composed of 63.65% N and 36.35% O. In a deficiency of oxygen 1.00 gram of nitrogen reacts with 1.14 grams of oxygen to produce 2.14 grams of nitric oxide. In an excess of oxygen 0.0357 mole of N_2 molecules reacts with 0.0714 mole of O_2 molecules. Show by approximate calculations or reasoning how the above data illustrate the following laws.
(a) conservation of mass
(b) constant composition

4.15 The human adult daily requirements for the trace metals, copper, magnesium, manganese, and zinc, are 2.5 mg, 300 mg, 4 mg, and 0.3 mg, respectively. Determine the number of atoms of each of these metals required daily.

4.16 Epinephrine is a hormone of the adrenal glands, which are just above the kidney, that provides a rapid physiological response to emergencies such as heat, fatigue, and shock. It has a molecular formula of $C_9H_{13}NO_3$. Determine the percent composition of epinephrine and the number of carbon atoms in 0.00712 gram of epinephrine.

4.17 Electrolysis of water, H_2O, yields hydrogen gas, H_2, and oxygen gas, O_2.

(a) Write a balanced chemical equation for this reaction.

(b) Determine how many moles of water molecules must be electrolyzed to give 18.2 grams of oxygen.

(c) Determine how many hydrogen molecules will be formed.

4.18 Lithium hydroxide, LiOH, cannisters were used as part of the life-support system in the Apollo 13 spacecraft. The solid LiOH absorbed gaseous carbon dioxide, CO_2, from the cabin atmosphere forming solid lithium carbonate, Li_2CO_3, and liquid water, H_2O, as products.

(a) Write a balanced chemical equation for this reaction.

(b) Determine what mass of LiOH was required each day to absorb the 933 grams of CO_2 exhaled on breathing by each astronaut and what mass of water was formed in the process.

4.19 LSD with a chemical formula of $C_{24}H_{30}N_3O$ liberates gaseous carbon dioxide, CO_2, gaseous nitrogen dioxide, NO_2, and water vapor, H_2O, on combustion with gaseous oxygen, O_2.

(a) Write a balanced chemical equation for this reaction.

(b) Determine the mass of oxygen required to react completely with 0.248 gram of LSD.

(c) If a column designed to absorb water vapor from the gaseous products weighed 2.0053 grams before the reaction, what would its mass be after the reaction?

4.20 Magnesium reacts with phosphorus to form magnesium phosphide, Mg_3P_2, as the only product.

(a) Write a balanced chemical equation for this reaction.

(b) Determine the number of moles of Mg_3P_2 expected when 0.143 gram of magnesium reacts with excess phosphorus according to the above reaction.

(c) Determine the percent yield if 0.230 gram of Mg_3P_2 forms.

4.21 An important reaction of amino acids, the building blocks for proteins, is decarboxylation on heating with barium hydroxide to form barium carbonate, a primary amine, and water. Thus, alanine, $C_3H_7NO_2$, forms ethylamine, C_2H_7N, according to the reaction

$$C_3H_7NO_{2(s)} + Ba(OH)_{2(s)} \longrightarrow BaCO_{3(s)} + C_2H_7N + H_2O$$

Determine the following.

(a) the mass of $Ba(OH)_2$ required to react completely with 1.63 grams of alanine

(b) the mass of $BaCO_3$ expected

(c) the number of moles of C_2H_7N molecules expected

(d) the number of water molecules expected

(e) the percent yield of solid $BaCO_3$ if 3.08 grams of $BaCO_3$ is recovered

4.22 You have prepared 2.35 grams of compound A in the laboratory by reacting metallic silver, Ag, with fluorine gas, F_2.

(a) If compound A contains 91.9% Ag, determine the minimum mass of silver with which you started the preparation and the empirical formula of compound A.

(b) If the 2.35 grams of compound A, on heating with water, decomposes into 1.08 grams of metallic silver and a water solution of Ag^+ and F^- ions, determine (1) the number of moles of Ag^+ ions in solution, (2) the number of F^- ions in solution, and (3) the empirical formula for compound B, which forms if the Ag^+ and F^- ions combine when the solution is evaporated to dryness.

(c) Write balanced chemical equations for the two reactions in (a) and (b).

5

Gases

Chlorine is a yellow-green gas with an odor that causes choking. Chlorine shows many of the properties typical of gases. It has mass, takes the shape and volume of a container, exerts a pressure, is easily compressible, diffuses readily, is miscible with other gases, and shows specific behavior with changes of pressure, volume, and temperature.

The purpose of this chapter is to give you an understanding of the behavior of gases, behavior that is simpler on the whole than that of liquids or solids, and to illustrate the use of stoichiometric calculations involving gases.

5.1 Variables for Describing Gases

Gases are matter and have **mass** (Section 1.1-a), illustrated by the fact that an evacuated glass bulb weighs less than the same bulb filled with air. However, owing to their very low densities and difficulties in handling, gases are not commonly measured quantitatively in terms of mass. They must be contained in sealed vessels, which are often difficult to weigh on ordinary chemical balances, in amounts that are so small as to be near the sensitivity limits of many chemical balances. Instead, quantities of gases are described more conveniently in terms of three interrelated variables—volume, temperature, and pressure.

a. Volume

The **volume** (Section 1.3-b-2) *of a substance* is *the space occupied by that substance.* We have illustrated vessels for measuring volume (Figure 1.8) and have described units for specifying volume. A gas occupies the *total* volume of its container, unlike a liquid, which occupies a fixed volume regardless of its container.

Volume is usually sufficient to specify the amount of a solid or liquid because the volumes of solids or liquids change very little with changes of temperature and pressure. Such is not the case for gases; temperature and pressure must also be specified. Thus, 1 liter of oxygen at a certain temperature and pressure has a different mass and contains a different number of moles from 1 liter of oxygen at the same temperature but a different pressure or at the same pressure but a different temperature.

b. Temperature

We have already noted (Section 1.2-a) that temperature is a measure of the intensity of heat. We have illustrated a thermometer for measuring temperature (Figure 1.7a) and units for specifying temperature (Figure 1.7b). Temperature determines the direction of heat flow. Thus, heat flows spontaneously from a region of higher temperature to a region of lower temperature.

c. Pressure

Pressure is *the exertion of a force on one body by another.* The pressure of a gas is the force that the gas exerts on the walls of its container. Pressure is a property that determines the direction in which mass flows. When you pump air into a deflated bicycle tire, the tire expands because the pressure on the inside of the tire is greater than the pressure on the outside. When you open the valve and allow air to escape from the tire, the air moves from a region of higher pressure to a region of lower pressure.

The pressure at any given point in a gas or a liquid is the same in all directions. Thus, atmospheric pressure exerted by air on a mountain climber is the same in all directions. Likewise, at any given depth in a quiet pond, the pressure exerted by water on an underwater swimmer is the same in all directions. For the mountain climber or the swimmer, there is no net push in any direction. However, in the presence of a strong wind in the air or a strong current in the water, the mountain climber and the swimmer experience a push from the direction of high pressure to the direction of low pressure.

Atmospheric pressure and water pressure both vary with depth. As the mountain climber ascends to higher altitudes, the atmospheric pressure decreases and his body is compressed by a lesser column of air above it. At high enough altitudes the pressure of oxygen gas becomes too small to support respiration. As the swimmer dives deeper, the water pressure increases and his body is compressed by a greater column of water above it acted on by the force of gravity. At great enough depths the increased pressure on his eardrums creates a sensation of pain.

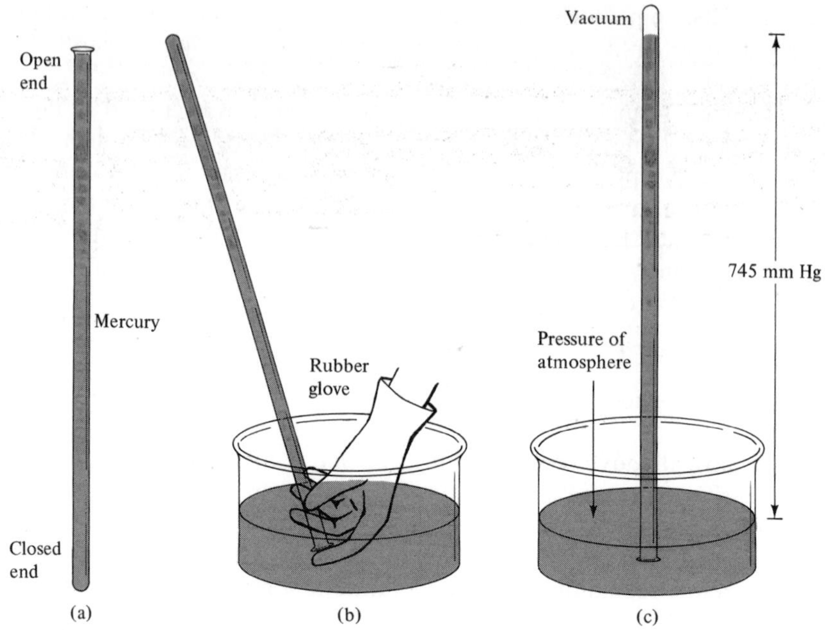
Open
end
Mercury
Closed
end
(a)
Rubber
glove
(b)
Vacuum
745 mm Hg
Pressure of
atmosphere
(c)

Figure 5.1 Construction of a mercury barometer.

A *barometer* is used to measure atmospheric pressure. Preparing a simple working model is a good way to understand the principles of operation of a barometer. A long glass tube, sealed at one end, is filled with mercury[1] (Figure 5.1a). The open end of the tube is covered long enough to invert it in a dish of mercury (Figure 5.1b). When the end is uncovered, some of the mercury flows out of the tube forming a vacuum (essentially zero pressure) at the sealed end (Figure 5.1c). However, some mercury remains in the tube because the atmospheric pressure pushing down on the surface of the mercury in the dish supports the column of mercury in the tube. Stated alternatively, the pressure exerted by the column of mercury in the tube is equal to the pressure exerted by the air (atmospheric pressure). The difference in the two levels of mercury (the height of the mercury column, 745 mm in this case) is a measure of atmospheric pressure.

Pressure changes with altitude and temperature; therefore, a standard reference point for measuring gas pressures has been defined. **One atmosphere** is *the pressure that supports a 760 mm column of mercury at 0°*. Since pressures are commonly measured using mercury barometers and manometers, pressures are frequently expressed in millimeters of mercury (1 atm = 760 mm Hg). For example, the atmospheric pressure in Figure 5.1c is 745 mm Hg and places the same pressure on the surface of mercury in the

[1]Other liquids would suffice, but mercury has the advantages of not requiring too long a tube, of having a very low rate of evaporation, and of being a liquid over a wide temperature range. However, mercury is poisonous, and a rubber glove is used to protect the skin.

dish as does the 745 mm column of mercury. A mm Hg is frequently referred to as a *torr* (named after Evangelista Torricelli, the inventor of the barometer). The weatherman reports prevailing atmospheric pressures in inches of mercury. Conversion factors between these various units of pressure are

$$1 \text{ atm} = 760 \text{ mm Hg} = 760 \text{ torr} = 29.9 \text{ in. Hg}$$

A *manometer* is used to measure the pressure of an enclosed sample of gas. Preparing a simple working model is the best way to understand the operation of a manometer. A liquid (usually mercury) is placed in the bottom of a U-tube in the sealed arm of which is trapped the gas sample of interest. The unsealed arm is either open to the atmosphere (Figure 5.2a) or connected to a vacuum pump (Figure 5.2b). At the same height in each arm (for example, at each dashed line in Figure 5.2) the pressures are equal; otherwise there would be a net flow of mercury from one arm to the other.

In Figure 5.2a the pressure at the dashed line in the left arm is the sum of the pressure of the gas sample (P_{gas}) and the pressure of the mercury column (P_{Hg}, the difference in height between the mercury columns in the left and right arms). The pressure at the dashed line in the right arm is atmospheric pressure (P_{atm}). Therefore,

$$P_{atm} = P_{gas} + P_{Hg} \quad \text{or} \quad P_{gas} = P_{atm} - P_{Hg}$$

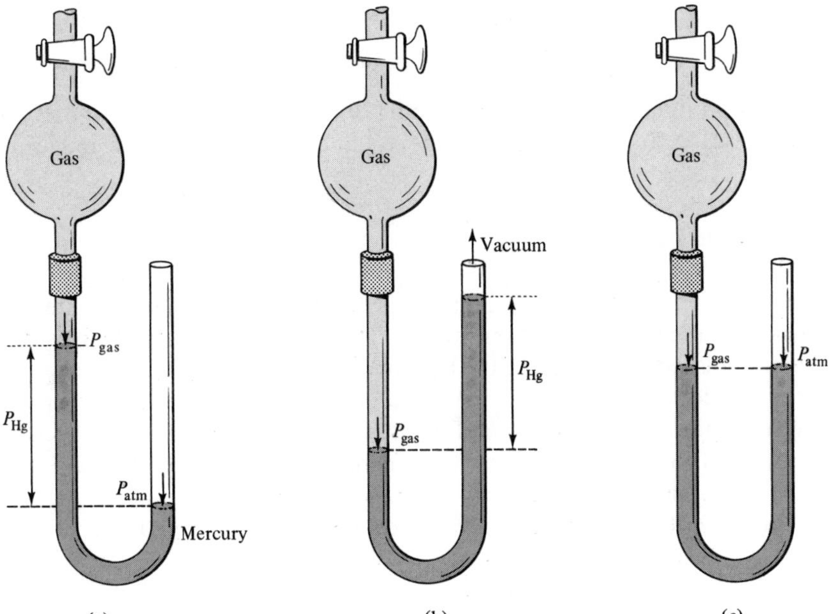
Gas
Gas
Gas
Vacuum
P_{gas}
P_{Hg}
P_{atm}
Mercury
P_{Hg}
P_{gas}
P_{gas}
P_{atm}
(a)
(b)
(c)

Figure 5.2 Types of manometers.

Note that atmospheric pressure must be known (from a barometer) when measuring pressure by this method. Note further that if the bottom of the U-tube is constructed of flexible rubber tubing, the right arm of Figure 5.2a can be raised with respect to the left arm until the two mercury levels are at the same height (Figure 5.2c). Then,

$$P_{Hg} = 0 \qquad \text{and} \qquad P_{gas} = P_{atm}$$

In Figure 5.2b the pressure at the dashed line in the left arm is the pressure of the gas sample (P_{gas}). The pressure at the dashed line in the right arm is simply the pressure of the mercury column (P_{Hg}, the difference in height between the mercury columns in the right and left arms) since the vacuum contributes essentially zero pressure. Therefore,

$$P_{gas} = P_{Hg}$$

Note that atmospheric pressure is not required when measuring pressure by this method.

Manometers of the kind discussed are useful for quantitative measurements involving large volumes of gases, for example, the analysis of alveolar air. However, clinical measurements, for example, of partial pressures of dissolved oxygen or carbon dioxide in blood, require much smaller samples than can be determined accurately in manometers. Modern blood gas analyzers make use of specifically designed electrodes that measure voltage differences from oxidation-reduction reactions (Section 8.7).

5.2 The Gas Laws

Consider an inflated balloon the opening of which has been tied so that no air can escape. In what ways can the volume of the balloon be reduced?

a. Boyle's Law

One way to reduce the volume of the balloon is to increase the external pressure on the balloon (and thus the internal air pressure as well). This can be accomplished by squeezing the balloon with your hands or by forcing the balloon under several feet of water. The decrease in volume in this experiment indicates that the volume of a gas is dependent on pressure. Specifically, other factors remaining constant, increasing the pressure decreases the volume; decreasing the pressure increases the volume.

In 1662 Robert Boyle, a British chemist, physicist, and theologian, proposed that *at constant temperature the volume of a given sample of a gas is* ***inversely*** *proportional to the pressure exerted on the gas.* This inverse relationship is illustrated schematically in Figure 5.3. When the initial pressure is

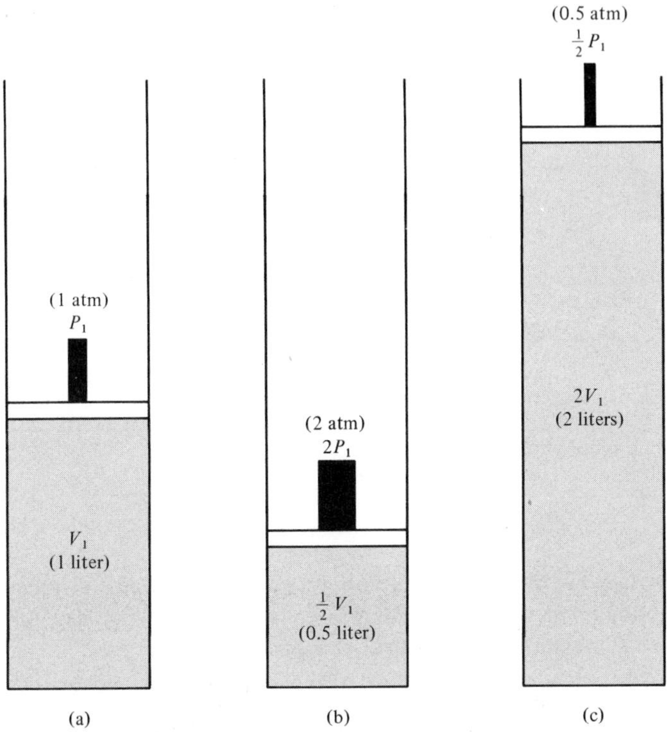
(0.5 atm)
$\frac{1}{2}P_1$
(1 atm)
P_1
$2V_1$
(2 liters)
(2 atm)
$2P_1$
V_1
(1 liter)
$\frac{1}{2}V_1$
(0.5 liter)
(a)
(b)
(c)

Figure 5.3 The variation of the volume of a gas in a cylinder fitted with a movable piston and held at constant temperature. P = pressure; V = volume. (a) Initial conditions; (b) V_1 halves when P_1 is doubled; (c) V_1 doubles when P_1 is halved.

doubled, the initial volume is halved. When the initial pressure is halved, the initial volume is doubled. This inverse relationship is illustrated graphically in Figure 5.4. Volume plotted against the reciprocal of pressure $\left(\frac{1}{\text{pressure}}\right)$ yields a straight line.

If the original volume and pressure of a given sample of gas are known, the volume that the gas will occupy at a different pressure (but at constant temperature) can be calculated.

Example 5.1 A sample of oxygen gas occupies a volume of 400 ml at a pressure of 720 torr. What will be its volume at a pressure of 760 torr, assuming the temperature remains constant?

The pressure is increasing. The volume of a gas varies inversely with the pressure of the gas; therefore, the volume decreases as the pressure increases. Set up an appropriate ratio of the initial and final pressures such that when multiplied by the initial volume the ratio causes the initial volume to decrease.

$$\text{Volume of sample of oxygen at 760 torr} = (400 \text{ ml})\left(\frac{720 \cancel{\text{torr}}}{760 \cancel{\text{torr}}}\right) = 379 \text{ ml}$$

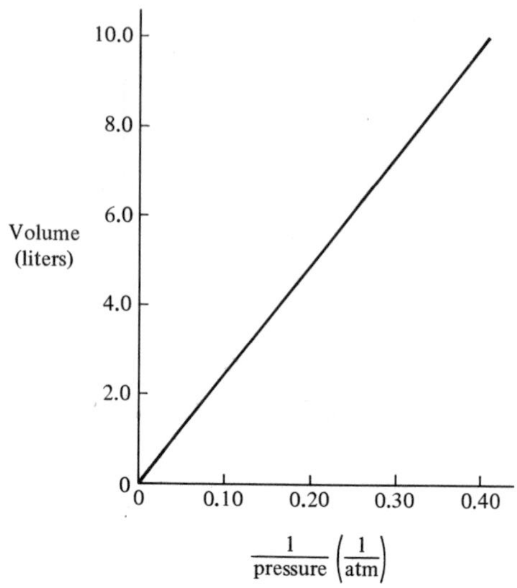
10.0
8.0
6.0
4.0
2.0
0
Volume
(liters)
0
0.10
0.20
0.30
0.40
$\frac{1}{\text{pressure}}\left(\frac{1}{\text{atm}}\right)$

Figure 5.4 Variation of volume of one mole of an ideal gas at 25°C as a function of reciprocal pressure. Temperature and number of moles remain constant.

Alternatively, if the original volume and pressure of a given sample of gas are known, the pressure that the gas will exert when its volume has changed (at constant temperature) can be calculated.

Example 5.2 At what pressure will the initial volume of oxygen in Example 5.1 occupy 0.600 liter, assuming the temperature remains constant?

Convert volumes to the same units: 0.600 liter = 600 ml.

The volume is increasing. The volume of a gas varies inversely with the pressure of the gas; therefore, the pressure decreases as the volume increases. Set up an appropriate ratio of the initial and final volumes such that when multiplied by the initial pressure the ratio causes the initial pressure to decrease.

$$\text{Pressure of sample of oxygen occupying 0.600 liter} = (720\ \text{torr})\left(\frac{400\ \cancel{\text{ml}}}{600\ \cancel{\text{ml}}}\right)$$
$$= 480\ \text{torr}$$

b. Charles' Law

Another way to reduce the volume of the inflated balloon is to place the balloon in a freezer for a few minutes. The decrease in volume upon cooling indicates that the volume of a gas is dependent on temperature. Specifically, other factors remaining constant, increasing the temperature increases the volume; decreasing the temperature decreases the volume.

In 1787 Jacques Charles, a French physicist, proposed that *at constant pressure the volume of a given sample of a gas is* ***directly*** *proportional to the* [*absolute*] *temperature of the gas.* This direct relationship is illustrated schematically in Figure 5.5. When the initial absolute temperature is doubled, the

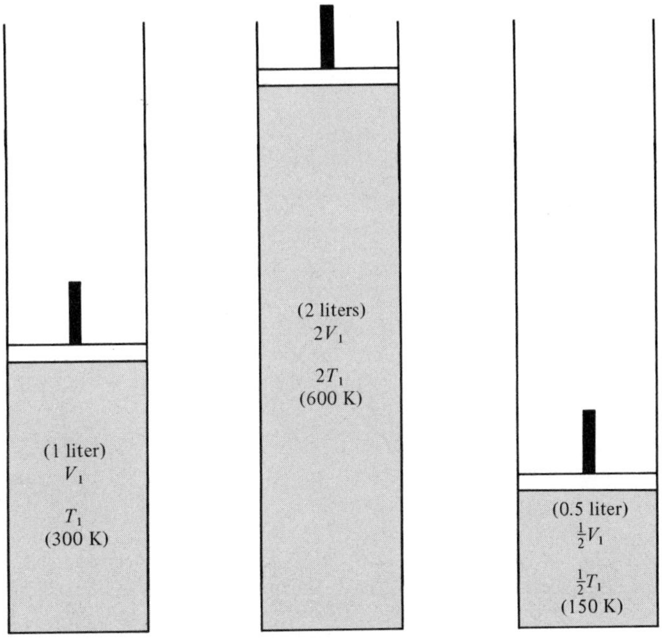
(1 liter)
V_1
T_1
(300 K)
(2 liters)
$2V_1$
$2T_1$
(600 K)
(0.5 liter)
$\frac{1}{2}V_1$
$\frac{1}{2}T_1$
(150 K)
(a)
(b)
(c)

Figure 5.5 The variation of the volume of a gas in a cylinder fitted with a movable piston to which is applied a constant pressure. V = volume; T = temperature. (a) Initial conditions; (b) V_1 doubles when T_1 is doubled; (c) V_1 halves when T_1 is halved.

initial volume is doubled. When the initial absolute temperature is halved, the initial volume is halved. This direct relationship is illustrated graphically in Figure 5.6. Volume plotted against absolute temperature yields the solid straight line.

If the original volume and temperature of a given sample of gas are known, the volume that the gas will occupy at a different temperature (but constant pressure) can be calculated.

Example 5.3 A sample of oxygen gas occupies a volume of 2.10 liters at 25°C. What volume will this sample occupy at 150°C assuming the pressure remains constant?

All quantitative calculations involving temperature changes of gases require the use of *absolute* temperatures.

Initial temperature: $25°\text{C} + 273° = 298\ \text{K}$

Final temperature: $150°\text{C} + 273° = 423\ \text{K}$

The temperature is increasing. The volume of a gas varies directly with the absolute temperature of the gas; therefore, the volume increases as the temperature increases. Set up an appropriate ratio of the initial and final absolute temper-

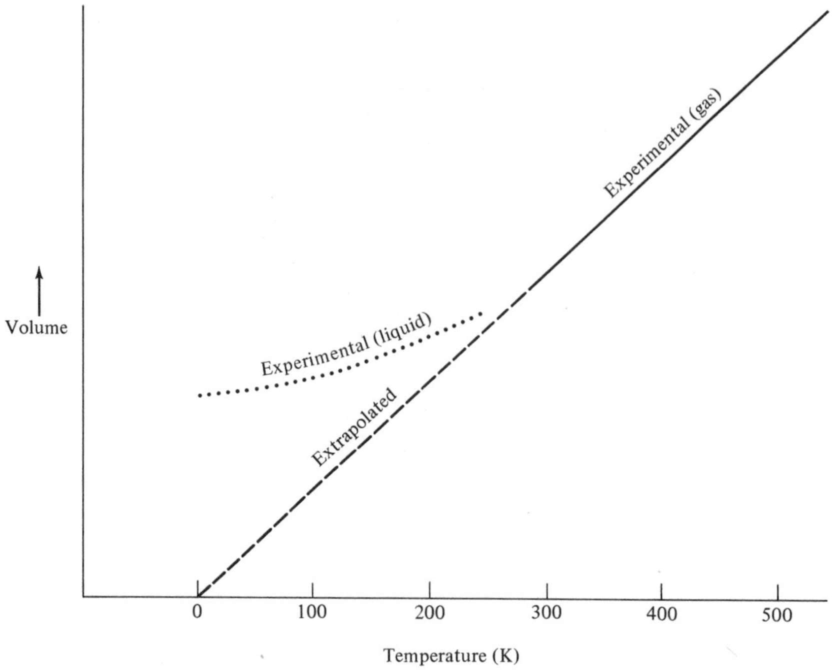

Volume
Experimental (gas)
Experimental (liquid)
Extrapolated
0
100
200
300
400
500
Temperature (K)

Figure 5.6 The variation of volume of a gas with absolute temperature, pressure and number of moles remaining constant.

atures such that when multiplied by the initial volume the ratio causes the initial volume to increase.

$$\text{Volume of sample of oxygen at } 150\,°\text{C} = (2.10 \text{ liters})\left(\frac{423\ \cancel{K}}{298\ \cancel{K}}\right) = 2.98 \text{ liters}$$

Alternatively, if the original volume and temperature of a given sample of gas are known, the temperature of the gas when its volume has changed (at constant pressure) can be calculated.

Example 5.4 At what Celsius temperature will the initial volume of oxygen in Example 5.3 occupy 0.750 liter, assuming the pressure remains constant?

The volume is decreasing. The volume of a gas varies directly with the absolute temperature of the gas; therefore, the temperature decreases as the volume decreases. Set up an appropriate ratio of the initial and final volumes such that when multiplied by the initial temperature the ratio causes the initial temperature to decrease.

$$\begin{aligned}\text{Temperature of sample of oxygen occupying 0.750 liter} &= (298 \text{ K})\left(\frac{0.750\ \cancel{\text{liter}}}{2.10\ \cancel{\text{liter}}}\right)\\ &= 106 \text{ K}\end{aligned}$$

Convert to °C.

$$
\begin{aligned}
{}^\circ C &= K - 273.16^\circ \\
&= 106^\circ - 273.16^\circ \\
&= -167^\circ C
\end{aligned}
$$

The temperature of a gas cannot be reduced indefinitely. At a sufficiently low temperature any given gas will liquefy. For example, boron trichloride (BCl_3) liquefies at 12°C (285 K) and fits the experimental data in Figure 5.6. Below 285 K the volume change of liquid BCl_3 with decreasing temperature is quite small as shown by the dotted curve in Figure 5.6. However, the solid line (gas phase) can be extended until it crosses the temperature axis. This intersection corresponds to the temperature at which the volume of the BCl_3 would become zero if BCl_3 behaved as an *ideal* (Section 5.3) gas at such low temperatures. This intersection is the same for all gases and provides an experimental determination of absolute zero (0 K). Any temperature below 0 K would imply the unrealistic situation of *negative* volume.

c. Avogadro's Law

The most obvious way to reduce the volume of the inflated balloon is to prick it with a pin or to untie the opening of the balloon and release some of the molecules composing the air. This indicates that the volume of a gas is dependent on the number of molecules present. Specifically, other factors being equal, increasing the number of molecules increases the volume; decreasing the number of molecules decreases the volume.

In 1811 Amadeo Avogadro, an Italian physicist, proposed that equal volumes of gases at the same temperature and pressure contain equal numbers of molecules. Stated alternatively, the volume of a gas at constant temperature and pressure is directly proportional to the number of molecules of the gas. Since a mole (Section 4.9) is related to the number of molecules only by a constant (Avogadro number, Section 4.5), it can also be stated that *at constant temperature and pressure the volume of a gas is **directly** proportional to the number of moles of the gas.* This direct relationship is illustrated schematically in Figure 5.7. When the initial number of moles is doubled, the initial volume is doubled. When the initial number of moles is halved, the initial volume is halved. This direct relationship is illustrated graphically in Figure 5.8. Volume plotted against number of moles yields a straight line.

d. Dalton's Law of Partial Pressures

The inflated balloon at the beginning of this section contained air, a mixture of gases composed primarily of nitrogen and oxygen. In 1801 John Dalton summarized the behavior of two or more *nonreacting* gases in the

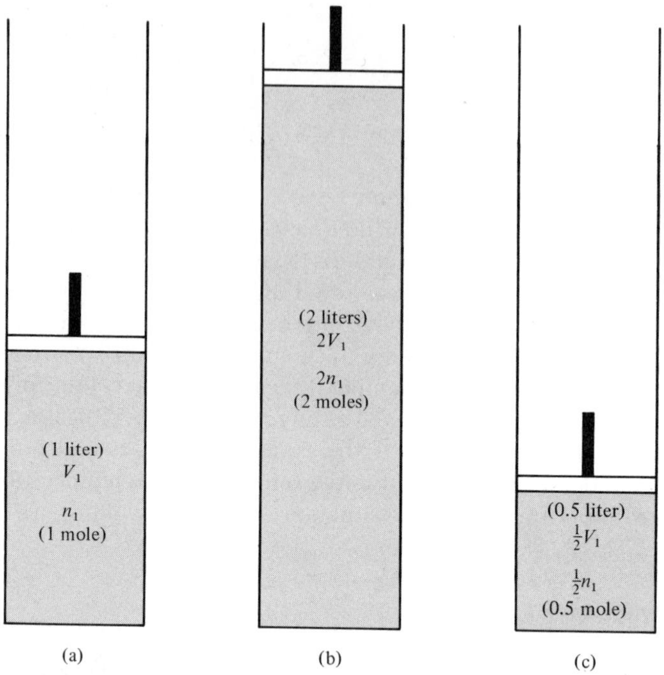
(1 liter)
V_1
n_1
(1 mole)
(2 liters)
$2V_1$
$2n_1$
(2 moles)
(0.5 liter)
$\frac{1}{2}V_1$
$\frac{1}{2}n_1$
(0.5 mole)
(a)
(b)
(c)

Figure 5.7 The variation of the volume of a gas at constant temperature and pressure in a cylinder fitted with a movable piston. V = volume; n = number of moles. (a) Initial conditions; (b) V_1 doubles when n_1 is doubled; (c) V_1 halves when n_1 is halved.

same container—*the total pressure exerted by a mixture of gases is equal to the sum of the partial pressures of the various gases.* Partial pressure is the pressure a gas would exert if it were the only gas in the container. For the inflated balloon Dalton's law of partial pressures can be written

$$P_{\text{total}} = P_{N_2} + P_{O_2} + P_{CO_2} + P_{H_2O} + P_{Ar} + \cdots$$

where the subscripts denote the various gases occupying the same volume of the balloon and at the same temperature.

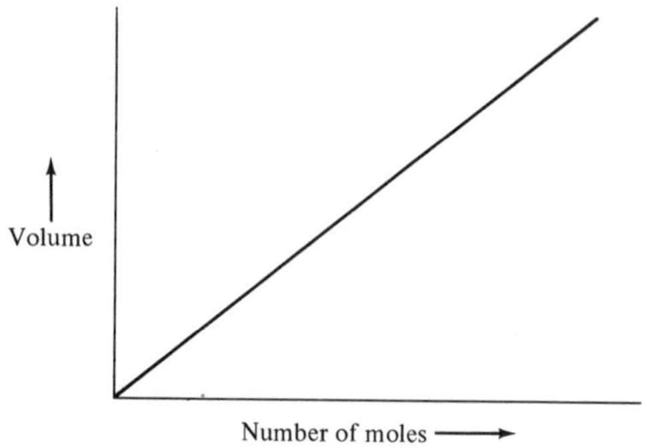
Volume
Number of moles

Figure 5.8 The variation of volume of a gas with number of moles of the gas, temperature and pressure remaining constant.

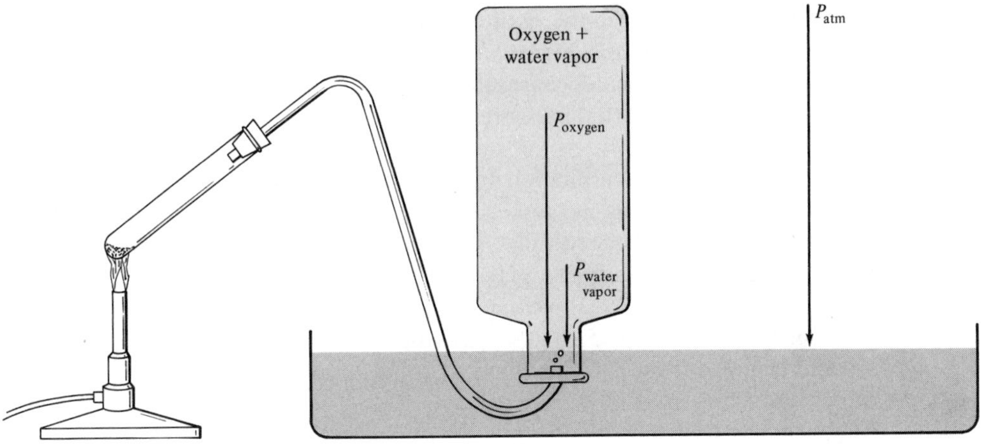
Oxygen +
water vapor
P_{atm}
P_{oxygen}
$P_{water\ vapor}$

Figure 5.9 Collecting oxygen over water.

Dalton's law of partial pressures is applied in many laboratory experiments in which gases are collected over water (Figure 5.9), for example, in the laboratory preparation of oxygen by thermal decomposition of potassium chlorate ($KClO_3$) in the presence of a manganese dioxide (MnO_2) catalyst.

$$3\ KClO_{3(s)} \xrightarrow[\text{heat}]{MnO_2} 3\ KCl_{(s)} + 2\ O_{2(g)}$$

Evolved oxygen displaces water in the initially filled collection bottle. At the end of the experiment the bottle contains both oxygen and water vapor. If the water levels inside and outside the bottle are made equal,

$$P_{atm} = P_{oxygen} + P_{water\ vapor}$$

The partial pressure of oxygen can be determined after P_{atm} is obtained from a barometer and $P_{water\ vapor}$ is obtained from a table showing the variation of water vapor pressure with temperature (see Table 6.1).

$$P_{oxygen} = P_{atm} - P_{water\ vapor}$$

Example 5.5 What is the partial pressure of oxygen collected over water at 23°C when the atmospheric pressure is 741 torr?

From Table 6.1, water has a vapor pressure of 21.1 torr at 23°C. The partial pressure of oxygen is equal to the atmospheric pressure minus the vapor pressure of water, or 741 torr − 21.1 torr = 720 torr.

e. Diffusion

When a bottle of deodorizer is opened in the bathroom, vapor spreads sufficiently to odorize the entire room. This *spontaneous spreading of a gas throughout any volume accessible to it* is called **diffusion.** The rapid random motion of gas molecules causes them to move from a region of relatively high concentration or partial pressure to a region of lower concentration or partial pressure.

To a first approximation the rate of diffusion is dependent upon the mass of the diffusing gas molecules and upon the partial pressure gradient. This is illustrated in the following examples.

If gaseous ammonia (NH_3) and hydrogen chloride (HCl) are injected simultaneously at equal partial pressures into opposite ends of a long tube filled with air (Figure 5.10), the gases diffuse through the tube until they meet and react to form a white cloud of solid ammonium chloride (NH_4Cl).

$$NH_{3(g)} + HCl_{(g)} \longrightarrow NH_4Cl_{(s)}$$

The position of the white cloud shows that the lighter NH_3 molecules diffuse more rapidly than the heavier HCl molecules. It has been found experimentally that the rate of diffusion (R) of gas molecules at constant temperature and partial pressure is inversely proportional to the square root of the molar mass (M) of the molecules.

$$R \propto \frac{1}{\sqrt{M}}$$

Many biochemical processes are dependent upon the passage of gas molecules across semipermeable membranes (Section 7.5-d). It has been shown experimentally that for a given gas at constant temperature the rate of diffusion is proportional to the partial pressure gradient across the membrane; that is, the greater the partial pressure difference, the faster will be the net passage of molecules across the membrane. A given gas diffuses in response to its own partial pressure gradient independent of what other gases are present, and continues to diffuse until its partial pressures are equal on both sides of the membrane.

The gas laws in this section represent the behavior of ideal gases (Section 5.3). Any real gas shows deviations from Boyle's law, Charles' law, Avogadro's law, Dalton's law, and the diffusion rate particularly at high pressures and at temperatures near the point of liquefaction. However, these deviations are small for many gases and will be assumed to be negligible in our discussion.

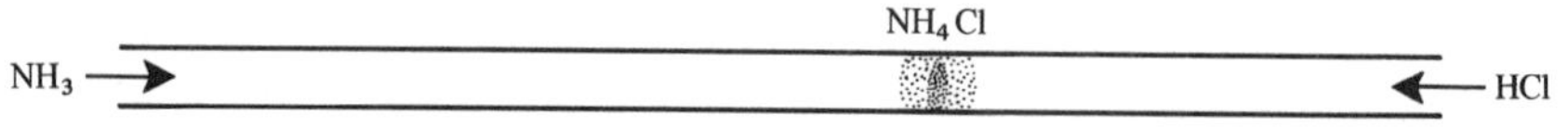

NH₄Cl
NH₃
HCl

Figure 5.10 Rate of diffusion of ammonia versus hydrogen chloride.

5.3 Kinetic Molecular Theory Applied to Gases

When dust particles in a room are illuminated from the side by sunlight from a window, it is observed that the particles do not settle to the floor but rather undergo irregular zigzag motion (called *Brownian motion*) in space. Furthermore, the motion of the particles becomes more rapid as the temperature of the room increases. Apparently the minute dust particles are bombarded constantly by molecules comprising air. Such observations provide evidence that the molecules comprising air are constantly in motion and that the velocities of these molecules are related to temperature.

On the basis of the behavior of gases presented above and earlier in this chapter, James Maxwell, a Scottish physicist, and Ludwig Boltzmann, an Austrian physicist, proposed in 1859 the kinetic molecular theory. The basic assumptions of this model as it applies to gases and reasons for the validity of these assumptions are:

1. *Gases consist of tiny, discrete molecules.* The very small size of atoms (Section 2.8), and thus the molecules that they make up, have already been discussed in Chapters 2 and 3.
2. *The average distance between gas molecules is so large compared to the size of the molecules themselves that the volume of a gas consists mostly of empty space.* Calculations show that in air, for example, more than 99.9% of the total volume is empty space. Thus, air can be compressed significantly upon the exertion of pressures much greater than atmospheric pressure.
3. *There are essentially no attractive forces between gas molecules so that they are completely independent of each other.* Gases expand spontaneously to fill any volume accessible to them so that there can be no significant binding forces between one molecule and another.
4. *Gas molecules are in perpetual, rapid, random, straight-line motion, colliding frequently with each other and with the walls of the container.* Gas molecules bombard minute particles giving rise to Brownian motion and bombard the walls of a container resulting in the exertion of pressure. Gas molecules expand into larger volumes and diffuse from one container to another.
5. *Collisions between molecules and between molecules and container walls are elastic; thus, there is no loss of kinetic energy* (Section 1.2-a) *on collision.* Gases do not change temperature spontaneously on standing, indicating that gas molecules do not lose energy by collisions. In the course of a collision, one molecule may transfer energy to another molecule, but the average energy of the pair of molecules remains constant.
6. *The* ***average*** *kinetic energy of all gas molecules is directly proportional to the absolute temperature.* Temperature is a measure of the intensity of heat and rises when additional heat is added. Heat is a form of energy which imparts increased velocity and higher kinetic energy to the gas molecules. Molecules will exhibit a range of kinetic energies

at any temperature because collisions continually change the velocity and kinetic energy of any given molecule.

Gases whose behavior is consistent with the postulates of the kinetic molecular theory are said to be **ideal gases.**

The success with which the kinetic molecular theory can explain the observed behavior of gases is the true test of its validity.

a. Boyle's Law

Refer to Figure 5.3. The pressure exerted by a gas depends only on the number of molecular impacts on a unit wall area per second. When the volume is reduced, the molecules collide more frequently with a unit wall area, and the pressure increases (Figure 5.3b). The converse is true when the volume is increased (Figure 5.3c).

b. Charles' Law

Refer to Figure 5.5. When the temperature is increased, the molecules move with greater velocities, collide harder and more frequently with the container walls, and would produce a greater pressure. To compensate and maintain the pressure constant, the gas expands the volume (Figure 5.5b) and maintains the number of molecular impacts on a unit wall area per second at a constant value. The converse is true when the temperature is decreased (Figure 5.5c).

c. Avogadro's Law

Refer to Figure 5.7. When the number of moles (and thus molecules) is increased, the number of molecular impacts on a unit wall area increases, and the pressure would increase. To compensate and maintain the pressure constant, the gas expands the volume (Figure 5.7b) and maintains the number of molecular impacts on a unit wall area per second at a constant value. The converse is true when the number of moles is decreased (Figure 5.7c).

d. Dalton's Law of Partial Pressures

If attractive forces between molecules are negligible, each molecule of a gas mixture hits the container walls the same number of times per second and with the same force as it would if no other gas molecules were present. Thus, the partial pressure of a given gas is unaffected by the presence of other gases.

e. Diffusion

The average kinetic energy for a collection of molecules is fixed for a given temperature and is expressed quantitatively by

$$\text{Average kinetic energy } (KE) = \tfrac{1}{2}\text{M}v^2$$

where M is the molar mass and v is the average velocity of the molecules. The rate of diffusion of gas molecules is related to their average velocity. Solving from above,

$$v = \sqrt{\frac{2KE}{\text{M}}}$$

Thus, velocity and rate of diffusion are inversely proportional to the square root of the molar mass of the molecules.

5.4 Stoichiometric Calculations Involving Gases

a. Standard Temperature and Pressure

Volumes of gases change with changes of temperature and pressure. Therefore, **standard temperature and pressure** (abbreviated **STP**) have been chosen for comparing volumes of gases. Standard temperature is *0 °C* and standard pressure is *1 atm or 760 torr*. Volumes of gases are corrected to STP prior to most stoichiometric calculations by performing a combined Boyle's law–Charles' law calculation.

Example 5.6 A sample of hydrogen gas has a volume of 1.10 liters at −40°C and 0.520 atm. What will be its volume at STP?

Convert the required temperatures to K.

$$\text{K} = {}^\circ\text{C} + 273^\circ = -40^\circ + 273^\circ = 233 \text{ K}$$
$$\text{K} = {}^\circ\text{C} + 273^\circ = 0^\circ + 273^\circ = 273 \text{ K}$$

The pressure is increasing and should cause a corresponding decrease in the volume. Set up an appropriate ratio of initial and final pressures such that when multiplied by the initial volume the ratio causes the initial volume to decrease.

The temperature is increasing and should cause a corresponding increase in the volume. Set up an appropriate ratio of initial and final temperatures such that when multiplied by the initial volume the ratio causes the initial volume to increase.

Both calculations are performed simultaneously.

$$\text{Volume of } H_2 \text{ at STP} = (1.10 \text{ liters})\left(\frac{0.520 \cancel{\text{atm}}}{1.00 \cancel{\text{atm}}}\right)\left(\frac{273 \cancel{\text{K}}}{233 \cancel{\text{K}}}\right) = 0.670 \text{ liter}$$

b. Molar Volume

The number of moles of a gas is required to perform stoichiometric calculations using the mole concept. It has been found experimentally that *one mole of any ideal gas at STP occupies 22.4 liters* (a volume about three times that of a basketball). Thus, 1.00 mole H_2 (2.02 grams) at STP occupies 22.4 liters; 0.500 mole H_2 (1.01 gram) at STP occupies 11.2 liters.

Example 5.7 Calculate the number of moles of hydrogen collected in Example 5.6.

The volume of H_2 at STP is 0.670 liter. Determine the number of moles of H_2 by dividing the volume by the molar volume of an ideal gas.

$$\text{Number of moles of } H_2 = \frac{0.670 \cancel{\text{liter}}}{22.4 \cancel{\text{liters}}/\text{mole}} = 0.0299 \text{ mole of } H_2$$

Stoichiometric calculations involving gases are performed using the concepts outlined in this section in combination with those presented in Chapter 4.

Example 5.8 What volume of ammonia (NH_3) at STP is required to react with excess hydrogen chloride (HCl) to produce 1.78 grams of ammonium chloride (NH_4Cl)?

The balanced chemical equation representing the reaction is

$$NH_{3(g)} + HCl_{(g)} \longrightarrow NH_4Cl_{(s)}$$

Determine the number of moles of NH_4Cl formed by dividing the mass of NH_4Cl formed by the molar mass of NH_4Cl.

$$\text{Number of moles of } NH_4Cl = \frac{1.78 \cancel{g}}{53.491 \cancel{g}/\text{mole}} = 0.0333 \text{ mole of } NH_4Cl$$

From the balanced equation, 1 mole of NH_3 is required for each mole of NH_4Cl formed. Therefore, 0.0333 mole of NH_3 is required to produce 0.0333 mole of NH_4Cl.

Determine the volume of NH_3 at STP required by multiplying the number of moles of NH_3 by the molar volume at STP.

$$\begin{aligned} \text{Volume of } NH_3 \text{ at STP} &= (0.0333 \cancel{\text{mole}})(22.4 \text{ liters}/\cancel{\text{mole}}) \\ &= 0.745 \text{ liter or } 745 \text{ ml} \end{aligned}$$

Example 5.9 What mass of zinc must be reacted with excess hydrochloric acid solution to produce 2.00 liters of hydrogen collected over water (Figure 5.9) at 25°C when the atmospheric pressure is 745 torr?

The balanced equation representing the reaction is

$$Zn_{(s)} + 2\ HCl \longrightarrow ZnCl_2 + H_{2(g)}$$

The pressure in the collection vessel is due to both H_2 and water vapor. From Table 6.1, water has a vapor pressure of 23.8 torr at 25 °C. The partial pressure of H_2 is equal to the atmospheric pressure minus the vapor pressure of water, or 745 torr − 23.8 torr = 721 torr.

Convert the required temperatures to K.

$$K = °C + 273° = 25° + 273° = 298\ K$$
$$K = °C + 273° = 0° + 273° = 273\ K$$

The volume of dry H_2 (2.00 liters) is then corrected to STP. The pressure is increasing and should cause a corresponding decrease in the volume. The temperature is decreasing and should also cause a corresponding decrease in the volume.

$$\text{Volume of dry } H_2 \text{ at STP} = (2.00 \text{ liters}) \left(\frac{721\ \cancel{\text{torr}}}{760\ \cancel{\text{torr}}}\right)\left(\frac{273\ \cancel{K}}{298\ \cancel{K}}\right)$$
$$= 1.74 \text{ liters}$$

Determine the number of moles of H_2 produced by dividing the volume of dry H_2 at STP by the molar volume at STP.

$$\text{Number of moles of } H_2 = \frac{1.74\ \cancel{\text{liters}}}{22.4\ \cancel{\text{liters}}/\text{mole}} = 0.0777 \text{ mole of } H_2$$

From the balanced equation, 1 mole of Zn is required to produce 1 mole of H_2. Therefore, 0.0777 mole of Zn is required to produce 0.0777 mole of H_2.

Determine the mass of Zn needed by multiplying the number of moles of Zn by the molar mass of Zn.

$$\text{Mass of Zn} = (0.0777\ \cancel{\text{mole}})(65.38\ \text{g}/\cancel{\text{mole}}) = 5.08\ \text{g}$$

5.5 Important Inorganic Gases

a. Hydrogen, H_2

Hydrogen gas is prepared commercially by passing steam over hot carbon

$$H_2O_{(g)} + C_{(s)} \xrightarrow{1000\,°C} H_{2(g)} + CO_{(g)}$$

or as a by-product when hydrocarbons are broken down thermally in refining crude oil for the production of gasoline. Hydrogen is also a by-product from the electrolysis (Section 8.8) of aqueous salt solutions for the production of chlorine (Section 5.5-i) and sodium hydroxide (Section 8.2-e). Hydrogen is a colorless, odorless, tasteless gas. It is used in large quantities for the preparation of ammonia (Section 5.5-e) and for the hydrogenation of fats and oils (Section 24.5-c). Hydrogen reacts with most elements under appropriate conditions of temperature and pressure.

b. Carbon Monoxide, CO

Carbon monoxide results from the incomplete combustion of carbon or carbon-containing compounds and is produced whenever carbon or its compounds are burned in a deficiency of oxygen.

$$2\ C_{(s)} + O_{2(g)} \xrightarrow{\text{heat}} 2\ CO_{(g)}$$

One of the primary sources of this gas is from incomplete combustion of hydrocarbons in automobile engines. Carbon monoxide is poisonous, a problem made all the more troublesome because carbon monoxide is difficult to detect as a result of being colorless, odorless, tasteless, and nonirritating in any obvious way. Many people die each year as a result of exposure to lethal amounts of carbon monoxide from automobile exhaust in a closed garage, from a faulty automobile exhaust system, or from poorly ventilated furnaces or space heaters. The toxicity of carbon monoxide results from its strong affinity for hemoglobin and the formation of a stable complex that can no longer carry oxygen (Section 31.4). Early symptoms of carbon monoxide poisoning include headache, dizziness, difficulty in breathing, and muscle weakness.

c. Carbon Dioxide, CO_2

Carbon dioxide results from the complete combustion of carbon or carbon-containing compounds and arises whenever carbon or its compounds are burned in an excess of oxygen.

$$C_{(s)} + O_{2(g)} \xrightarrow{\text{heat}} CO_{2(g)}$$

$$\underset{\text{methane}}{CH_{4(g)}} + 2\ O_{2(g)} \xrightarrow{\text{heat}} CO_{2(g)} + 2\ H_2O$$

$$\underset{\text{glucose}}{C_6H_{12}O_{6(s)}} + 6\ O_{2(g)} \xrightarrow{\text{heat}} 6\ CO_{2(g)} + 6\ H_2O$$

Carbon dioxide is a colorless, odorless, tasteless gas. Its primary uses are as a refrigerant in the form of dry ice (solid CO_2) and for the carbonation of beverages. Carbon dioxide and oxygen in the atmosphere are part of a cycle involving animal and plant life. Animals take up oxygen during respiration and use it to convert food into energy, water, and carbon dioxide. The carbon dioxide, discharged during respiration, is then taken up by plants and used in photosynthesis. Oxygen, produced in photosynthesis, is liberated by plants to complete the cycle.

d. Nitrogen, N_2

Air is about 78% nitrogen by volume, and thus nitrogen is prepared commercially by fractional distillation (Section 1.1-f) of liquid air. The industrial importance of nitrogen is indicated by the fact that it was the fifth chemical in terms of total production in the United States in 1978. Nitrogen is a colorless, odorless, tasteless gas. It is fairly inert chemically, and the N≡N triple bond is cleaved only at high temperatures in the presence of a catalyst (Section 9.5) or by certain bacteria that can convert nitrogen to ammonia, NH_3, at room temperature. The primary industrial use of nitrogen is in the production of ammonia (Section 5.5-e).

e. Ammonia, NH_3

Ammonia is produced commercially in the Haber process in which high pressures of nitrogen and hydrogen are heated in the presence of an iron–alumina catalyst.

$$N_{2(g)} + 3\,H_{2(g)} \xrightarrow[200\text{ atm}]{500\,°\text{C}} 2\,NH_{3(g)}$$

Ammonia was the fourth chemical in terms of total production in the United States in 1978. Ammonia is a colorless gas with a characteristic pungent odor. It is a weak base and it turns moist red litmus paper blue. Ammonia is very soluble in water to form what are commonly called ammonium hydroxide solutions, but what are really aqueous ammonia solutions (Section 8.2-e) since there is no chemical evidence for the presence of NH_4OH. The primary use for ammonia is in the production of nitrogenous fertilizers such as ammonium sulfate, but large quantities are also used in the preparation of nitric acid (Section 8.2-e) and the manufacture of synthetic fibers and plastics.

f. Nitrogen Dioxide, NO_2

The commercial production of nitrogen dioxide occurs primarily through the oxidation of nitric oxide, NO, as an intermediate step in the production of nitric acid (Section 8.2-e).

$$2\,NO_{(g)} + O_{2(g)} \longrightarrow 2\,NO_{2(g)}$$

Nitrogen dioxide is a brown gas that is both poisonous and corrosive. It is at least partially responsible for the brown haze over metropolitan areas as a result of rush hour traffic. Its precursor, nitric oxide, is formed whenever air is utilized in high temperature combustion processes that are followed by rapid cooling of exhaust gases, conditions common to automobiles. Nitric oxide is slowly oxidized to nitrogen dioxide which may then be photochemically dissociated upon interaction with sunlight. Further reactions with organic compounds produce additional pollutants that cause your eyes to tear and that cause difficulty in breathing. These various interactions of pollutants with sunlight give rise to what is called photochemical smog.

g. Oxygen, O_2

Air is about 21% oxygen by volume, and thus oxygen is produced commercially by the fractional distillation (Section 1.1-f) of liquid air. Oxygen was the third chemical in terms of total production in the United States in 1978. Oxygen is a colorless, odorless, tasteless gas. It is very reactive and capable of combining with most elements to form oxides. Industrially oxygen is used primarily in metal refining and especially for oxygen enrichment in blast furnaces utilized in steel production. Some oxygen is used in wastewater treatment as well as in hospitals to aid the respiration of patients. The oxygen taken up during respiration combines with hemoglobin (Section 31.4) and is carried to various parts of the body where it is used in the conversion of food to energy.

h. Sulfur Dioxide, SO_2

Large quantities of sulfur dioxide are produced industrially for the manufacture of sulfuric acid (Section 8.2-e) by burning sulfur in air.

$$S_{(s)} + O_{2(g)} \xrightarrow{\text{heat}} SO_{2(g)}$$

Sulfur dioxide is a colorless gas with a sharp, pungent odor. It is both poisonous and corrosive. Large amounts of sulfur dioxide are also generated in burning fossil fuels that contain sulfur. Sulfur dioxide, combined with soot, fly ash, and smoke from partially oxidized hydrocarbons, was primarily responsible for the severe London smogs in 1952, 1956, and 1962. More than 4000 deaths occurred in 1952, and nearly 900 people died in 1956. Stricter controls on the allowable sulfur content in fossil fuels has made sulfur dioxide a less serious air pollutant although there have been recent moves to allow the burning of high sulfur coal to ease the energy crisis.

i. Chlorine, Cl_2

Chlorine is produced by the electrolysis (Section 8.8) of aqueous salt solution,

$$2\,Na^+ + 2\,Cl^- + 2\,H_2O \xrightarrow{\text{electrolysis}} 2\,Na^+ + 2\,OH^- + H_{2(g)} + Cl_{2(g)}$$

and it is also a by-product in the Downs process involving the production of sodium by electrolysis of molten sodium chloride (Section 8.8). Chlorine was the seventh chemical in terms of total production in the United States in 1978. Chlorine is a dense, greenish yellow gas with a pungent, choking odor. It is both poisonous and an irritant of the mucous membranes of the nasal passages. It is very reactive and capable of combining with most elements to form chlorides. It is used extensively for water purification and for bleaching wood pulp and textiles.

j. Argon, Ar

Air is 0.934% argon by volume, and thus argon is produced commercially by the fractional distillation (Section 1.1-f) of liquid air. Argon is a colorless, odorless, monatomic gas. It is very stable chemically as a monatomic species because it has a filled-shell electron configuration (Section 2.12) like the other noble gases of group VIIIA. No compounds of argon are known, but a number of compounds of krypton and xenon have been prepared in the last twenty years. The chemical inertness of argon accounts for its uses as an inert atmosphere for welding and in electric light bulbs and radio tubes.

Exercises

5.1 Define clearly the following terms.
(a) gas
(b) volume
(c) temperature
(d) pressure
(e) atmospheric pressure
(f) barometer
(g) 1 atm pressure
(h) 740 mm Hg pressure
(i) torr
(j) manometer
(k) vacuum
(l) Boyle's law
(m) Charles' law
(n) absolute zero of temperature
(o) Avogadro's law
(p) partial pressure
(q) Dalton's law of partial pressures
(r) diffusion
(s) diffusion rate
(t) kinetic molecular theory
(u) kinetic energy
(v) ideal gas
(w) standard temperature and pressure
(x) molar volume at STP

5.2 Express each of the following pressures in torr.
(a) 0.213 atm
(b) 384 mm Hg
(c) 29.67 in. Hg
(d) 6.48 atm

5.3 Describe what would happen to the level of the mercury column in the barometer in Figure 5.1c under the following circumstances.
(a) Atmospheric pressure decreased.
(b) More mercury was added to the dish.
(c) Atmospheric pressure increased.
(d) A pinhole leak occurred at the sealed end of the tube.

5.4 Describe what would happen to the mercury levels in the manometers in Figure 5.2a *and* b if the following occurred.
(a) Atmospheric pressure increased.
(b) The flask was cooled.
(c) More gas was added to the flask.
(d) The flask was heated.
(e) Gas was removed from the flask.
(f) Atmospheric pressure decreased.
(g) The stopcock above the flask developed a slow leak.

5.5 What would happen to the level of the piston in Figure 5.3a with these changes?
(a) The gas in V_1 is heated.
(b) More gas is added to V_1.
(c) Additional force is applied on the top side of the piston.
(d) Gas is removed from V_1.
(e) The gas in V_1 is cooled.

5.6 Explain, on the basis of Charles' law, why absolute zero represents a minimum attainable temperature.

5.7 Explain, on the basis of Boyle's and Charles' laws, why the pressure of automobile tires increases during prolonged high speed driving.

5.8 A sample of nitrogen gas occupies 80.0 ml at 25°C and 1.00 atm pressure. What volume will the nitrogen occupy at the same temperature, but at each of the following pressures?
(a) 730 torr
(b) 0.64 atm
(c) 945 mm Hg
(d) 300 torr
(e) 12.3 atm

5.9 A sample of argon gas occupies 1.40 liters at 25°C and 1.00 atm pressure. What volume will the argon occupy at the same pressure, but at each of the following temperatures?
(a) 0°C
(b) 480 K
(c) 298 K
(d) 100°C
(e) −80°C

5.10 Correct the following volumes of helium to STP.
(a) 6.4 liters at 40°C and 350 mm Hg
(b) 38.4 ml at 300 K and 680 torr
(c) 400 ml at 200 K and 720 torr
(d) 0.400 liter at 23°C and 0.750 atm
(e) 280 ml at 80°C and 390 mm Hg

5.11 Outline the basic assumptions of the kinetic molecular theory as applied to gases. Describe experimental results which indicate that these assumptions are valid.

5.12 Explain the following in terms of the kinetic molecular theory.
(a) Boyle's law
(b) Charles' law
(c) Avogadro's law
(d) Dalton's law of partial pressures
(e) diffusion rate

5.13 Determine the volume that 0.20 mole of nitrogen gas occupies at each of the following.
(a) STP
(b) 100°C and 1.00 atm
(c) 308 K and 308 torr
(d) 273 K and 0.30 atm
(e) 25°C and 740 torr

5.14 Rank the following gases in order of decreasing rate of diffusion, assuming that each gas has the same partial pressure gradient.

(a) N_2
(b) He
(c) Xe
(d) CO
(e) CO_2

5.15 What would be the volume of dry oxygen at STP if 250 ml of oxygen were collected over water (Figure 5.9 and Table 6.1) at the following conditions of temperature and atmospheric pressure?

(a) 20°C and 752 mm Hg
(b) 28°C and 734 torr
(c) 25°C and 0.92 atm

5.16 A steel tank has a capacity of 800 liters. Calculate the resulting pressure if 3200 liters of nitrogen at 40°C and 745 torr pressure and 1800 liters of oxygen at 10°C and 750 torr pressure are pumped into the tank and brought to 25°C.

5.17 "Standard" conditions for clinical measurements are frequently different from STP conditions. Clinical "standard" conditions for blood gas measurements are body temperature and pressure fully saturated, labeled BTPS. Body temperature is 37.0°C; the pressure is the atmospheric pressure to which the body is exposed; and saturated refers to maximum water vapor pressure at 37°, which is 47 torr.

(a) If p_{O_2} and p_{CO_2} electrodes indicate that the partial pressures of oxygen and carbon dioxide are 99 torr and 39 torr, respectively, in alveolar air that is saturated with water vapor, what is the partial pressure of nitrogen (the only other major constituent) under BTPS conditions given that atmospheric pressure is 745 torr?

(b) Note the importance of specifying your conditions by calculating the partial pressure of nitrogen in water vapor saturated alveolar air if the results in (a) were obtained under STP conditions. (See Table 6.1 for p_{H_2O}.)

5.18 What volume of hydrogen at STP will combine with 12.0 grams of oxygen to form water?

5.19 How many liters of carbon dioxide, measured at 25°C and 740 torr, could be obtained upon thermal decomposition of 10.0 grams of pure limestone according to the following reaction?

$$CaCO_{3(s)} \longrightarrow CaO_{(s)} + CO_{2(g)}$$

5.20 What mass of pure octane (C_8H_{18}) is burned by reaction with the oxygen in air at 740 mm pressure and 200°C and filling a 0.95 liter cylinder of an automobile engine? (Air is one-fifth oxygen by volume.) The equation for combustion is

$$2\,C_8H_{18(g)} + 25\,O_{2(g)} \longrightarrow 16\,CO_{2(g)} + 18\,H_2O_{(g)}$$

5.21 If equal masses of nitrogen and argon are injected into separate flasks of equal volume and at the same temperature, which of the following are true? Show clearly your reasoning.

(a) More molecules are present in the nitrogen flask.
(b) The pressure is greater in the argon flask.
(c) The molecules are moving faster in the argon flask.
(d) More atoms are present in the argon flask.
(e) The molecules collide more frequently with the walls in the nitrogen flask.

5.22 Write a balanced chemical equation to describe each of the following.
(a) the commercial preparation of NH_3
(b) the formation of NO_2 from NO
(c) a commercial preparation for Cl_2

5.23 Does burning carbon in a deficiency of oxygen or in an excess of oxygen give the same product as the preparation of H_2 by passing steam over carbon? Write three balanced chemical equations to show clearly your reasoning.

5.24 Write a balanced chemical equation to describe the complete combustion of each of the following carbon compounds upon heating in an excess of oxygen.
(a) CH_4
(b) C_6H_6
(c) C_8H_{18}
(d) $C_7H_{16}O$
(e) $C_{12}H_{22}O_{11}$

6

Liquids and Solids

Like most solids, ice has a highly ordered structure composed of regular patterns of atoms and molecules. Oxygen atoms of neighboring water molecules, attracted by hydrogen bonds to a given central O atom, are located at the corners of a regular tetrahedron around the given central O atom. This basic structure is extended in three dimensions to form a regular crystal lattice with hexagonal channels as in a honeycomb. When ice melts, its highly ordered structure is partially broken down. Liquid water maintains short-range order around any given O atom, but the overall arrangement of tetrahedra is more random and is constantly changing. The structure of the liquid is between the extremes of complete disorder characteristic of gases and precise order characteristic of solids. Structural changes between liquids and solids give rise to differences in properties.

The purpose of this chapter is to discuss some of the properties of liquids and solids, how these properties differ from each other and from those of gases, and how these properties can be understood in terms of the kinetic molecular theory.

6.1 General Properties of Liquids

Like gases, *liquids have no characteristic shape*. Liquids assume the shape of whatever volume of a vessel is required to contain the liquid, whereas gases assume the shape of an entire closed container.

Unlike gases, *liquids maintain their volume,* regardless of the shape or size of the container. Whereas a gas occupies the total volume accessible to it, 100 ml of water occupies a 100 ml volume whether it is placed in a 100 ml beaker or a quart (943.4 ml) bottle.

Unlike gases, *liquids are essentially incompressible.* The volume of a gas at constant temperature halves when its pressure is doubled, whereas the volume of a sample of water decreases only $5 \times 10^{-3}\%$ when the pressure on the sample is doubled from 1 atm to 2 atm.

Unlike gases, *liquids show little change in volume with changes in temperature.* The volume of a gas at constant pressure doubles when its absolute temperature is doubled, whereas the volume of a sample of water increases only 4% with an increase in temperature from 273 K to 373 K.

Unlike gases, *liquids held at constant volume show a spectacular change in pressure with changes in temperature.* The pressure of a gas at constant volume doubles when the absolute temperature is doubled, but simply heating mercury contained in a fixed volume by 1 °C increases the pressure from 1 atm to 46 atm. Thus, you can't ignore pressure changes due to temperature fluctuations when constructing apparatus to contain fixed volumes of liquids!

Like gases, *liquids increase in volume as additional moles of the sample are added.* The volume of water at constant temperature and pressure doubles as the number of moles of water is doubled.

Liquids diffuse much more slowly than gases. Unless coffee is stirred when cream is added to it, several minutes are required for the cream to spread spontaneously throughout the cup. No such stirring is required for rapid mixing of two gases.

Liquids exhibit internal resistance to flow called **viscosity.** It is easier to pour water out of a narrow-mouth bottle than to pour molasses from the same bottle because the molasses has a higher viscosity.

The surface of a liquid experiences a net inward pull, called **surface tension,** that tends to draw the surface molecules into the body of the liquid. In the absence of gravity and other forces, the stable form of a mass of liquid is a sphere. This is nearly exhibited by a free-falling drop of rain. More commonly visible is the tear-shaped drop acted upon by gravity as it hangs from a medicine dropper.

The magnitude of surface tension is responsible for the phenomenon called **capillary action,** *the rise or fall of liquids in very small diameter tubes.* When a very small diameter tube is placed vertically in a beaker of water, the water rises in the capillary tube well above the level of water in the beaker. This climbing up the walls occurs because the attraction between water and the glass surface is stronger than the surface tension between water molecules. In the same manner, plant roots get water through capillaries in the soil even though the water table may be several feet below the roots of the plants. There are extended networks of very fine capillaries in many types of soil. Water molecules have a stronger attraction to the capillary walls of the soil than to other water molecules and thus rise by capillary action well above the water table to plant roots.

Liquids evaporate to form gases. Water placed in an open beaker gradually evaporates so that the beaker will be empty within a few days.

6.2 Kinetic Molecular Theory Applied to Liquids

Assumptions 1, 4, 5, and 6 (Section 5.3), explaining the behavior of gases, are also valid for liquids. Thus, liquids consist of tiny, discrete atoms or molecules that are in perpetual random motion, that suffer elastic collisions with each other and with container walls, and whose average kinetic energies are directly proportional to absolute temperature. However, explanation of the observed properties of liquids by means of the kinetic molecular theory requires several postulates in direct contrast to assumptions 2 and 3 (Section 5.3) for gases.

1. *Individual atoms or molecules composing a liquid must be close together, essentially in contact, so that free space between them is very small.* Four general properties can be explained on the basis of this assumption.
   a. Liquids are practically incompressible because, with free space reduced to a minimum, electron clouds of adjacent molecules repel immediately upon attempting to force molecules closer together.
   b. Liquids at constant volume show drastic changes in pressure with changes in temperature because liquid molecules are much closer together than gas molecules and a much greater number of collisions per unit wall area results for liquids than for gases when higher temperature imparts higher velocities and kinetic energies to the molecules.
   c. The volume of a liquid increases as additional moles (or molecules) of liquid are added because, with a minimum of free space, more molecules cannot be forced into the same volume.
   d. Liquids diffuse more slowly than gases because the average distance between molecules is small, and a molecule suffers great hindrance from billions of collisions with neighboring molecules while migrating from one side of a container to another.
2. *Attractive forces between atoms or molecules in a liquid prevent them from separating spontaneously from each other, but they are not strong enough to hold the positions of liquid molecules fixed.* Five general properties can be explained on the basis of this assumption.
   a. Liquids have no characteristic shape because the positions of liquid molecules are not fixed, but are free to slip past each other and fill from the bottom of a container upward so as to occupy positions of lowest potential energy.
   b. Liquids maintain their volume regardless of the shape or size of the container.
   c. Liquids show little change in volume with changes in temperature.
   d. Liquids exhibit surface tension and resistance to flow because attractions between molecules are strong enough to keep them clustered together.
   e. Liquids evaporate because, in a collection of molecules, a finite

number may be near the surface of the liquid and possess kinetic energy great enough to overcome attractive forces and escape into the gas phase.

6.3 Evaporation

Evaporation is *the conversion of a liquid to a gas.* Kinetic energy must be applied to break attractive forces between liquid molecules in order that molecules may escape to the gas phase. In the process of boiling (Section 6.5), kinetic energy is applied in the form of heat; but how do many liquids vaporize at ambient temperatures without the addition of heat?

The average kinetic energy of a collection of liquid molecules is directly proportional to the absolute temperature. Nevertheless, all molecules do not have the same energy; any given molecule may have a high or a low kinetic energy. A collection of molecules has a distribution of kinetic energies at any given temperature as illustrated in Figure 6.1. If the dashed line represents the minimum kinetic energy required to overcome attractive forces of the liquid and escape into the gas phase, it can be seen that, even at the lower temperature, a few molecules have the necessary kinetic energy. At the higher temperature, more molecules have this requisite kinetic energy.

In addition to decreasing the amount of liquid, evaporation causes cooling of the remaining liquid. Cooling occurs because highly energetic molecules overcome attractive forces and pass from the liquid. They leave behind molecules of *lower* average kinetic energy, giving a corresponding decrease in the temperature of the liquid. See, for example, the use of ethyl chloride as a temporary anesthetic in Section 16.3.

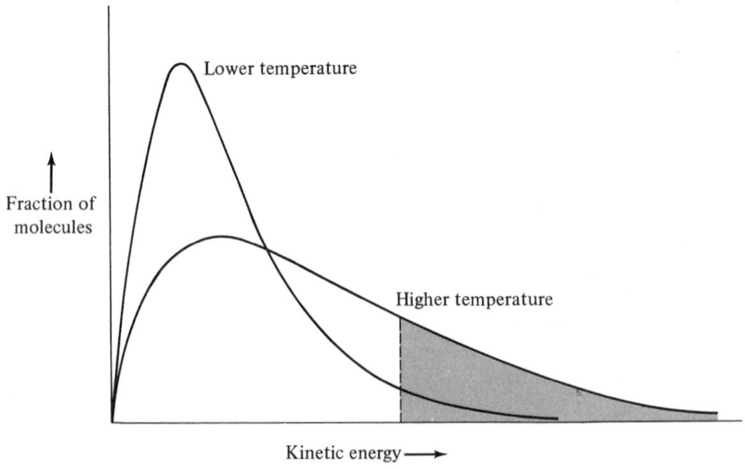
Lower temperature
Fraction of molecules
Higher temperature
Kinetic energy

Figure 6.1 Kinetic energy distributions of molecules of a liquid at two different temperatures. The dashed line represents the minimum kinetic energy required for molecules of the liquid to overcome attractive forces and escape into the vapor phase.

The decrease in temperature is actually observed experimentally only when the liquid is held in an *insulated* container. When water evaporates from a *noninsulated* beaker, there is sufficient heat flow from the surroundings to compensate for the loss of energy in the escaping highly energetic molecules. The temperature of the liquid is maintained at or near room temperature even though evaporation continues. However, when water evaporates from an *insulated* open container such as a thermos bottle, the heat flow from the surroundings is not sufficient to compensate for the loss of energy in the highly energetic escaping molecules. Thus, the temperature of the liquid drops, the curve in Figure 6.1 shifts to the left, the number of highly energetic molecules decreases, and the rate of evaporation is less. The net result is that water evaporates less rapidly from an insulated thermos bottle than from an open beaker. This phenomenon is used to advantage when evaporation of liquid nitrogen ($-196°$), used as a coolant, is decreased by containing it in a thermos bottle.

6.4 Equilibrium Vapor Pressure

Water placed in an open beaker evaporates freely into the air. If a bell jar is sealed over the beaker, all the water does not evaporate. The liquid level drops for a time, but then remains constant. Initially, highly energetic molecules pass into the vapor state, the amount of liquid decreases, heat is absorbed by the water from the surroundings to maintain a nearly constant temperature, and evaporation continues. However, as an increasing number of water molecules convert to water vapor, the probability that gaseous water molecules, diffusing throughout the bell jar in their rapid, random motion, will collide with and condense to the liquid phase, also increases. Eventually condensation and evaporation occur at the same rate; that is, water molecules enter the liquid at the same rate at which water molecules leave the liquid. A condition of *dynamic equilibrium,* in which two changes oppose each other exactly, is attained. Neither process stops; they simply occur at the same rates, with the result that the number of molecules in each phase remains *constant* (but not equal).

The gaseous water molecules, in their rapid, random motion, constantly bombard the walls of the bell jar and thus exert a pressure. *The pressure exerted by a vapor in equilibrium with its liquid* is called *the* **equilibrium vapor pressure** (frequently termed *vapor pressure* with equilibrium assumed).

The magnitude of the vapor pressure depends on the nature of the liquid and on its temperature. Liquids with relatively large forces of attraction, such as the dipole–dipole and hydrogen bonding forces in water, have small tendencies to escape into the vapor phase and have low vapor pressures. Liquids with relatively small forces of attraction, such as the weak dispersion forces present in butane, have large tendencies to escape into the vapor phase and have high vapor pressures. *Substances having high vapor pressures at room temperature* are said to be **volatile.** As the temperature of any liquid is increased, the average kinetic energy of its molecules and the number of highly energetic molecules increase. Hence, more molecules can escape to the vapor phase, and vapor pressure increases. Conversely, the vapor pressure decreases with decreasing temperature. The vapor pressures of *n*-butane,

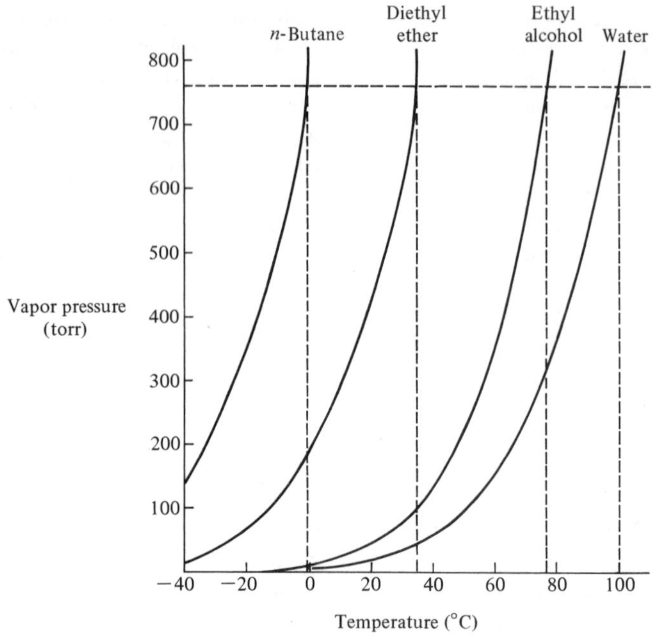

n-Butane
Diethyl ether
Ethyl alcohol
Water
Vapor pressure (torr)
800
700
600
500
400
300
200
100
−40
−20
0
20
40
60
80
100
Temperature (°C)

Figure 6.2 Vapor pressures of *n*-butane, diethyl ether, ethyl alcohol, and water as a function of temperature. The horizontal dashed line indicates one atmosphere pressure, and the vertical dashed lines indicate normal boiling points of the respective compounds.

diethyl ether, ethyl alcohol, and water, plotted as a function of temperature in Figure 6.2, illustrate these ideas. Table 6.1 shows the variation of the vapor pressure of water with temperature.

Our physical well-being is determined partially by humidity, or the presence of water vapor in the air. Air can hold a great quantity of water at high temperatures, but very little at low temperatures. On cold winter days the vapor pressure of water in the air may be less than 4 torr. Since water vapor from inside condenses on cold windows and since there are air passages to the outside despite our best construction efforts, the vapor pressure inside may also approach this value. This is much too low for comfort and health. Dry air causes rapid evaporation of moisture from mucous membranes in our breathing passages, causing irritation and increasing the chance of infection. Thus, humidifiers are used to add water vapor to the air. The reverse problem frequently occurs in summer when very moist air prevents the necessary elimination of excess heat from our bodies by evaporation. Then dehumidifiers are used.

6.5 Boiling

Evaporation occurs for highly energetic molecules close to the liquid–vapor interface. **Boiling** is *a special form of evaporation in which conversion from liquid to vapor occurs through formation of bubbles within the liquid.* These

Table 6.1 Equilibrium Vapor Pressure of Water

Temperature (°C)	*Pressure (torr)*	*Temperature (°C)*	*Pressure (torr)*
0.0	4.58	20.0	17.54
1.0	4.93	21.0	18.65
2.0	5.29	22.0	19.83
3.0	5.69	23.0	21.07
4.0	6.10	24.0	22.38
5.0	6.54	25.0	23.76
6.0	7.01	26.0	25.21
7.0	7.51	27.0	26.74
8.0	8.05	28.0	28.35
9.0	8.61	29.0	30.04
10.0	9.21	30.0	31.82
11.0	9.84	35.0	42.18
12.0	10.52	40.0	55.32
13.0	11.23	45.0	71.88
14.0	11.99	50.0	92.51
15.0	12.79	60.0	149.4
16.0	13.63	70.0	233.7
17.0	14.53	80.0	355.1
18.0	15.48	90.0	525.8
19.0	16.48	100.0	760.0

bubbles increase markedly the area of the liquid–vapor boundary and allow evaporation even from the interior of the liquid. Boiling occurs when the vapor pressure of a liquid increases with rising temperature until, at some characteristic temperature that is different for different liquids, bubbles form and push the surrounding air aside. The boiling point of a liquid is the temperature at which the vapor pressure of the liquid equals the prevailing atmospheric pressure.

The boiling point of a liquid depends on the nature of the bonding forces in the liquid. Liquids with relatively small forces of attraction between molecules have low boiling points; liquids with relatively large forces of attraction have high boiling points.

The boiling point of a liquid also depends on pressure. The **normal boiling point** is defined as *the temperature at which the vapor pressure of a liquid is equal to 1 atm or 760 torr* (100° for water). The boiling point of water decreases under the lower atmospheric pressures prevalent at higher altitudes, being 97.7° at a pressure of 700 torr. On the other hand, water boils at

120.1° in a pressure cooker if the pressure is 2 atm. The boiling point at any pressure may be determined from vapor pressure curves like those of Figure 6.2 by finding the temperature that corresponds to the desired pressure. Thus, the boiling point of ethyl alcohol at 400 torr is 63°.

The boiling point of a pure liquid is constant at a given pressure. Addition of heat to a liquid at its boiling point only gives more molecules sufficient kinetic energy to escape from the liquid to the gas phase and does not change the temperature of the sample until all of the liquid has vaporized. The boiling point at a given pressure is a commonly used criterion for identifying a liquid and characterizing its purity. A liquid containing impurities boils at a higher temperature than the corresponding pure liquid (Section 7.5-b).

6.6 Water

Water is an extraordinary liquid. It covers about three fourths of the earth's surface and is present in the atmosphere and in cloud formations. It is by far the most abundant compound in living organisms, constituting 60% by mass of the human body and about 90% by mass of certain foods such as lettuce, turnips, and apples. Water is of major importance to chemists because of its ability to dissolve a great variety of substances. In addition, it is a reactant or product in many chemical reactions (Chapter 8), a coolant in distillations, and a medium for heat transfer or a coolant for maintaining reactions at constant temperature.

The bent molecular structure of water (Example 3.4 and Figure 3.8) is responsible for its unique properties. Water has polar covalent bonds between a highly electronegative oxygen atom and two hydrogen atoms. It is a polar molecule (Section 3.10) with a partial negative charge residing on the oxygen and a partial positive charge shared by two hydrogen atoms (Figure 3.8d). The polar nature of the water molecule accounts for its high solvent ability and for a number of properties arising from strong intermolecular hydrogen bonding (Section 3.11) acting between adjacent H_2O molecules in the liquid.

Water is the most common dissolving medium in nature and in the laboratory, but is not a universal solvent. It dissolves many polar molecules such as hydrogen chloride (HCl) and ionic compounds such as sodium chloride (NaCl). The polar ends of the water molecules interact with dissolved ions or with oppositely charged ends of dissolved molecules. Water does not dissolve nonpolar substances such as oxygen (O_2), sulfur (S_8), bromine (Br_2), carbon tetrachloride (CCl_4), or hydrocarbons. These nonpolar molecules are repelled by water molecules and tend to cluster about each other rather than dispersing in water. A detailed discussion of the factors influencing solubility of substances in water is given in Section 7.3.

The melting and boiling points of water are unusually high compared to those of similar compounds. Figure 3.12, which shows graphically the melting points and boiling points of H_2O, H_2S, H_2Se, and H_2Te, illustrates this fact. Unlike water, the other three are gases at room temperature. This anomalous behavior of water results from the strong intermolecular hydrogen bonding present in both ice and water. The hydrogen atoms, though covalently bonded to an oxygen in one H_2O molecule, are capable of bridging adjacent water molecules together into three-dimensional aggregates throughout the bulk of the liquid. The energy necessary to break a hydrogen bond (ca 5 kcal/mole) is only about one tenth of the energy required to disrupt a covalent bond. Nevertheless, the cumulative effects of all of the hydrogen bonds within the water structure necessitate the expenditure of a considerable amount of energy and result in the abnormally high melting and boiling points.

Solid water (ice) is less dense than liquid water (0.917 versus 1.00 g/ml). Very specific hydrogen bonding in ice gives rise to open hexagonal channels as in a honeycomb, as shown in Figure 6.3. Liquid water has a randomly changing, collapsed structure with less empty space, and thus is more dense.

The density difference between ice and water has important consequences. One good consequence is that ice floats on water and forms on the

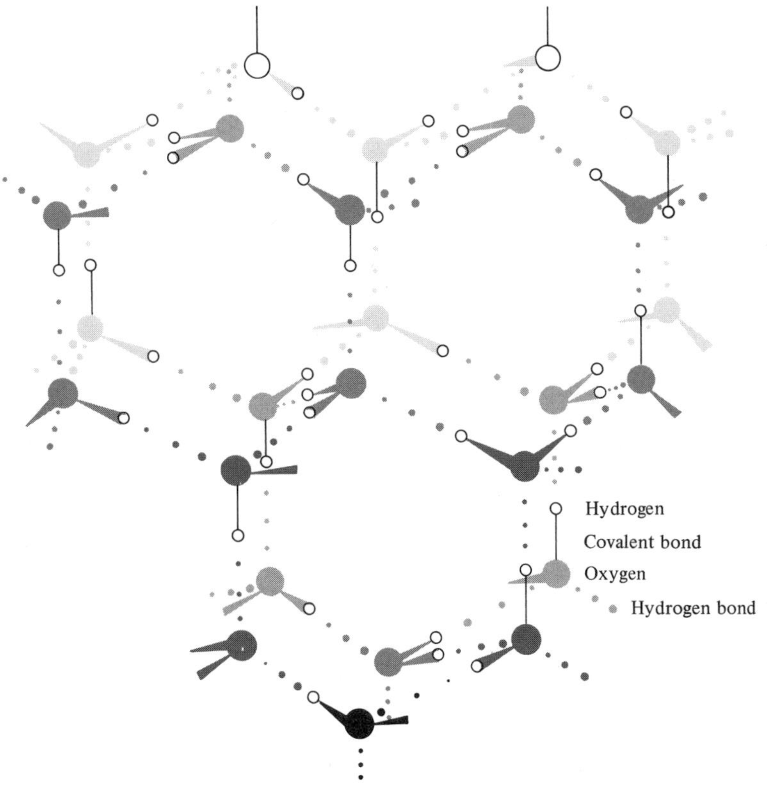
Hydrogen
Covalent bond
Oxygen
Hydrogen bond

Figure 6.3 The structure of ice showing the open hexagonal channels and tetrahedral coordination around oxygen caused by hydrogen bonding between H_2O molecules. Darker O Atoms are closer to the reader.

top surface of a pond in winter. The ice acts as an insulator between cold air and the remaining water below so that the pond does not freeze solid. Furthermore, the insulating ice moderates the temperature of the underlying water so that fish and frogs can survive the rigors of winter. This would not be so if ice were more dense than water and the pond froze from the bottom upward. Bad features that result from ice being less dense than water include the need to drain water pipes and toilet bowls in summer cottages that are closed for the winter and the need to maintain minimal heating in your home when you are away for a winter vacation. Since the same amount of water will occupy a larger volume as a solid than as a liquid, on freezing the increased pressure from ice requiring a larger volume will burst the pipes and could cause extensive flooding and water damage.

6.7 General Properties of Solids

Unlike gases and liquids, *solids have characteristic shapes, geometric patterns, or crystal structures that are distinctive for particular substances.* Solid sodium chloride forms a lattice of cubes as shown in Figure 3.2. The plane surfaces or faces within the crystal always intersect at an angle of 90°. When a sodium chloride crystal is broken, or even when ground into a fine powder, it cleaves in preferred directions so that the characteristic faces and angles are retained. Solids also crystallize in the variety of other forms depicted in Figure 6.4. Furthermore, there are several modifications of some of these forms. Some solids, which form two or more different crystalline modifications, are said to exhibit *polymorphism.* Diamond and graphite are different modifications of solid carbon. Different modifications are favored under different conditions. Thus, graphite can be converted to the more dense diamond under very high pressure.

The actual crystal structure of a solid can be determined experimentally by x-ray diffraction. X-rays, like those used for chest x-rays, are passed through a solid. They are deflected by the electron density of atoms or ions in the solid and produce characteristic patterns on a photograph. A detailed analysis of these patterns can yield an exact crystal structure.

Unlike gases, but like liquids, *solids maintain their volume.* A cube of sugar has an identical volume to the same sugar recrystallized in the form of a sphere or some other regular or irregular shape.

Unlike gases, but like liquids, *solids are essentially incompressible.* The volume of an iron sample decreases negligibly when the pressure on the sample is doubled from 1 atm to 2 atm.

Unlike gases, but like liquids, *solids show little change in volume with changes in temperature.* The volume of an iron sample held at constant pressure increases only 0.35% with an increase in temperature from 0° to 100°C.

Unlike gases, but like liquids, *solids held at constant volume show a tremendous change in pressure with changes of temperature.* Expansion joints

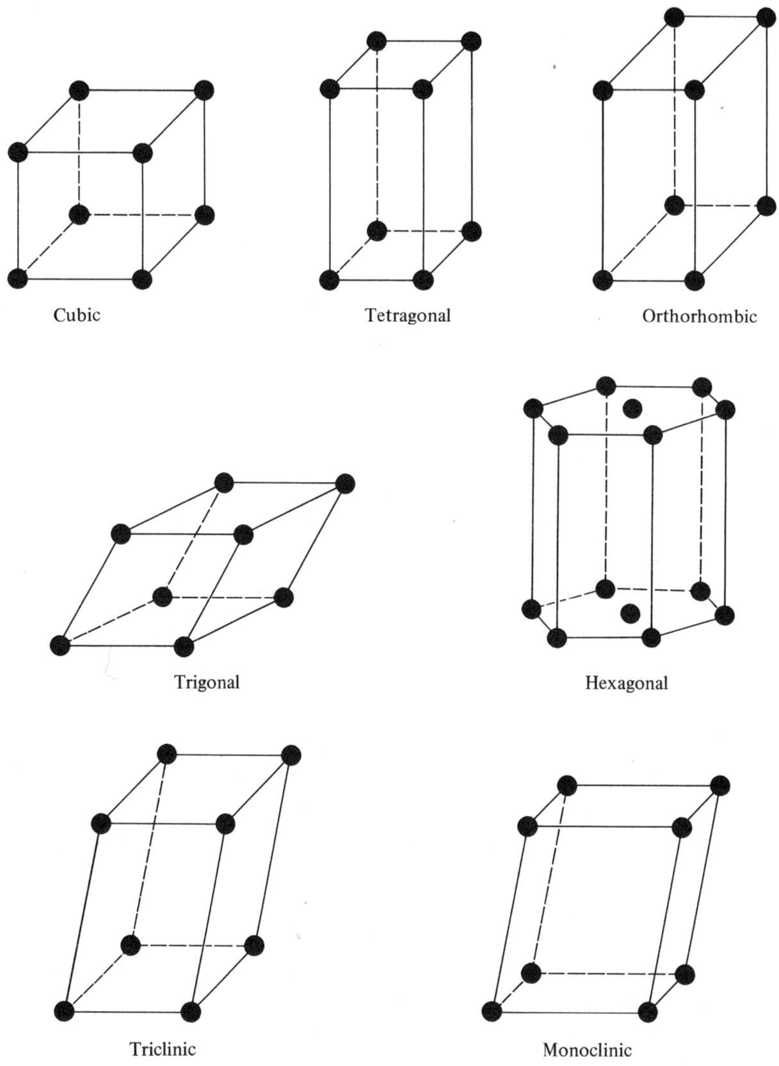
Cubic
Tetragonal
Orthorhombic
Trigonal
Hexagonal
Triclinic
Monoclinic

Figure 6.4 The simplest units in different crystal systems.

are used in heavy construction so that solid materials are not forced to remain in a constant volume through temperature fluctuations.

Like gases and liquids, *solids increase in volume as additional moles of a sample are added.* A 0.6 carat (a unit of mass equal to 0.2 gram) diamond has twice the number of moles and also twice the volume of a 0.3 carat diamond at the same temperature and pressure.

Solids diffuse even more slowly than liquids. A gold plate in contact with a lead plate shows only slight evidence for diffusion between the two plates after five years. Rock layers in contact with each other in the earth's crust still retain their boundaries even on a time scale of millions of years.

Under appropriate conditions, *solids melt to form liquids and sublime to form gases*. At room temperature and atmospheric pressure, ice melts to form liquid water whereas dry ice (solid carbon dioxide) sublimes, or converts directly from the solid to CO_2 vapor.

6.8 Kinetic Molecular Theory Applied to Solids

Assumptions 1, 4, 5, and 6 (Section 5.3), explaining the behavior of gases, are valid for solids as well as liquids. Thus, solids consist of tiny, discrete atoms, molecules, or ions, that are in perpetual random motion, that suffer elastic collisions with each other, and whose average kinetic energies are directly proportional to absolute temperature. However, explanation of the observed properties of solids by means of the kinetic molecular theory requires, as it did for liquids, modification of assumptions 2 and 3 (Section 5.3) for gases.

1. *Atoms, molecules, or ions in a solid are in contact with each other.* Free space is minimal. Four general properties of solids can be explained on the basis of this assumption.
   a. Solids are essentially incompressible because, with minimal free space, electron clouds of adjacent atoms, molecules, or ions repel immediately on attempting to force them closer together.
   b. Solids at constant volume show drastic changes in pressure with changes of temperature because the directly contacting atoms, molecules, or ions cause an even greater number of collisions per unit wall area than those in liquids when higher temperature imparts higher velocities and kinetic energies to the molecules.
   c. The volume of a solid increases as additional moles are added because, with minimal free space, more molecules cannot be forced into the same volume.
   d. Solids diffuse even more slowly than liquids because the average distance between atoms, molecules, or ions is so small that they suffer great hindrance from billions of collisions with neighboring atoms, molecules, or ions in a crystal.
2. *Attractive forces between atoms, molecules, or ions in a solid are strong enough to hold them in fixed positions.* Three general properties of solids can be explained on the basis of this assumption.
   a. Solids have characteristic shapes or geometric patterns because of the strong attractive forces holding atoms, molecules, or ions in well-defined positions.
   b. Solids maintain their volume and show little change in volume with changes in temperature because the strong attractive forces keep the atoms, molecules, or ions clustered together.
   c. Solids melt or sublime because a finite number of atoms, molecules, or ions may be near the surface of the solid and possess kinetic energy great enough to overcome attractive forces and escape into the liquid or vapor phase.

6.9 Melting

Ice on the sidewalk melts when the temperature rises above 0°. **Melting** is *the conversion of a solid to a liquid.* Melting occurs when a solid is warmed sufficiently so that atoms, molecules, or ions at the lattice points in a crystal overcome attractive forces and are free to slide past each other and exchange neighbors.

The melting point of a solid depends on the nature of the bonding forces in the solid. Solids with relatively small forces of attraction have low melting points; solids with relatively large forces of attraction have high melting points (see Table 6.2).

Melting points also change slightly with pressure. *The* **normal melting point** is defined as *the temperature of the solid-to-liquid transition at a pressure of* 1 *atm or* 760 *torr.* Melting points increase marginally with increasing pressure for most solids. However, the opposite is true for ice.

This uniqueness of ice is dependent upon the differences of density of water and ice (Section 6.6). It is very important to us in ways that we take for granted. For example, narrow runners on ice skates are designed to apply pressure on ice. This pressure lowers the melting point of the underlying ice and favors melting so as to form a film of liquid between the skates and the ice. This film acts as a lubricant and allows us to glide smoothly over the ice. However, if you skate outside on a very cold day, even the applied pressure of your skates does not lower the melting point enough to make the ice melt and provide a lubricating film. As a result your skates catch, and you may even trip occasionally.

The melting point of a pure solid is constant at a given pressure. Addition of heat to a solid at its melting point only gives more molecules sufficient kinetic energy to form the liquid and does not change the temperature of the sample until all of the solid has melted. The melting point is a commonly used criterion for identifying a solid and characterizing its purity. A solid containing impurities melts at a lower temperature than the corresponding pure solid (Section 7.5-c).

6.10 Sublimation

Ice and snow disappear gradually in the winter even when the temperature does not rise above 0°. No melting occurs; instead the ice and snow are converted directly to water vapor. **Sublimation** is *the direct conversion of a solid to a gas* without passing through the liquid state. Just as in a liquid (Figure 6.1), the atoms, molecules, or ions composing a solid have a distribution of kinetic energies at any given temperature. At any given time a few of them have the requisite kinetic energy to escape into the gas phase (to sublime). As with evaporation of liquids, an *equilibrium vapor pressure* arises from sublimation of solids. At higher temperatures more atoms, molecules, or ions have the necessary kinetic energy to sublime. Thus, vapor pressures of solids increase with increasing temperature, having lower magnitudes but the same trends as illustrated for liquids in Figure 6.2. Solids with high vapor pressures at room temperature are said to be *volatile.*

Vapor pressures of solids depend on the nature of the bonding forces in a solid. Solids with relatively weak forces of attraction have high vapor pressures; solids with relatively strong forces of attraction have low vapor pressures (see Table 6.2).

6.11 Types of Solids

We have indicated how the specific properties of solids are related to the nature of the bonding forces in such solids. Bonding forces in turn depend upon what particles occupy the lattice sites in a crystal. The bonding forces may be very strong covalent forces as between C atoms in the diamond lattice, or may be very weak dispersion forces as between Ar atoms in solid argon.

Six types of solids can be distinguished on the basis of particles that occupy lattice sites. Table 6.2 summarizes these six types in terms of the nature of bonding forces between particles, the general properties which each type exhibits, and specific examples of each type. Many solids can be classified into one of these six general categories.

Most of the aspects summarized by Table 6.2 have been discussed in this chapter or in Chapter 3. However, one of the types, polymeric solids, deserves further comment. Polymeric solids have polymers at the lattice points. **Polymers** are *very large molecules,* frequently consisting of long chains of atoms, two-dimensional sheets of atoms, or three-dimensional networks of atoms. The formation of some specific polymers is discussed in Section 8.3. Polymeric solids like Teflon are very important industrially. Others like cellulose and proteins are important biologically.

6.12 Amorphous Solids

Many polymeric materials have a particularly strong tendency to cool below their freezing point without crystallizing. Under appropriate conditions, a number of liquids can be made to do this. As the temperature is decreased, such liquids become less mobile (more viscous) and eventually become quite rigid. *Rigid, noncrystalline materials* are called **amorphous solids** *or* **glasses.** Amorphous solids resemble liquids in so far as they have short-range order rather than the long-range order characteristic of crystalline solids. However, they differ from liquids and resemble solids in being rigid and immobile.

Ordinary window glass is a good example of an amorphous solid. It does not have the highly ordered structure of quartz (SiO_2), illustrated in Figure 3.4. Instead, it has Si and O atoms arranged in *irregular* tetrahedral networks, each Si surrounded by four O atoms and each O atom shared between two Si atoms. Glass has reasonable strength and remains quite rigid except at high temperatures because of strong Si—O bonds. However, if it does break, due to its disorderly structure it will fracture into irregular shell-like chips rather than cleaving with the formation of plane faces.

Various types of glass, rubber, and plastic are popular commercial

Table 6.2 Types of Solids

Type	*Particles Occupying Lattice Sites*	*Forces Between Particles*	*Properties of Solids*	*Examples*
Atomic solids	Atoms	Covalent bonds between atoms	Very high melting points Very hard Nonvolatile Insulators	Diamond, $C_{(s)}$ $SiO_{2(s)}$ $Ge_{(s)}$ $B_{(s)}$ $Si_{(s)}$ Boron nitride
Ionic solids	Cations and anions	Electrostatic attraction between oppositely charged ions	High melting points Nonvolatile Hard, but shatter Insulators	$NaCl_{(s)}$ $CaF_{2(s)}$ $MgSO_{4(s)}$ $(NH_4)_3PO_{4(s)}$ $KBr_{(s)}$ $KNO_{3(s)}$
Polar molecular solids	Polar molecules	Dipole–dipole attraction between molecules	Moderate melting points Moderately volatile and hard Insulators	$HCl_{(s)}$ $H_2O_{(s)}$ $SO_{2(s)}$ $NO_{(s)}$ $NH_{3(s)}$ $HF_{(s)}$ $PCl_{3(s)}$
Nonpolar molecular solids	Nonpolar molecules (or atoms)	Dispersion forces between molecules (or atoms)	Low melting points Volatile Soft Insulators	$I_{2(s)}$ $CO_{2(s)}$ $CH_{4(s)}$ $SiCl_{4(s)}$ $Ar_{(s)}$
Metallic solids	Cations in an electron gas	Attraction between cations and electrons	Variable melting Variable hardness Low volatility Good conductors	$Na_{(s)}$ $Cu_{(s)}$ $Mg_{(s)}$ $Al_{(s)}$ $Fe_{(s)}$ $Cr_{(s)}$ $Ni_{(s)}$ $Zn_{(s)}$
Polymeric solids	Polymers	Dispersion forces and hydrogen bonding	Variable melting Variable hardness Low volatility	Polyethylene Polystyrene Nylon Orlon Starch Protein

amorphous solids because they can be fabricated into almost any shape by molding or blowing. Their constituents can be chosen to meet a variety of requirements: reasonable strength per unit mass, good insulation of heat or electricity, good resistance to chemical attack, or good resistance to mechanical abrasion.

6.13 Important Inorganic Solids

a. Ammonium Nitrate, NH_4NO_3

Ammonium nitrate is prepared commercially by an acid-base reaction (Section 8.2) between ammonia and nitric acid.

$$NH_{3(g)} + HNO_3 \longrightarrow NH_4NO_{3(s)}$$

Ammonium nitrate was the twelfth chemical in terms of total production in the United States in 1978. Ammonium nitrate is a white crystalline solid that is readily soluble in water. It is a rather dangerous solid and must be handled with care because it decomposes explosively when confined to limited space or exposed to elevated temperatures.

$$2\,NH_4NO_{3(s)} \xrightarrow{\text{heat}} 2\,N_{2(g)} + O_{2(g)} + 4\,H_2O_{(g)}$$

Because of this property ammonium nitrate is used extensively in the manufacture of explosives. A mixture of ammonium nitrate and trinitrotoluene (TNT, Section 15.6) is nearly as powerful an explosive as TNT alone, but is much less expensive. Ammonium nitrate is also used in the production of fertilizers.

b. Ammonium Sulfate, $(NH_4)_2SO_4$

Ammonium sulfate is prepared commercially by an acid-base reaction (Section 8.2) between ammonia and sulfuric acid.

$$2\,NH_{3(g)} + H_2SO_{4(l)} \longrightarrow (NH_4)_2SO_{4(s)}$$

Ammonium sulfate is a white crystalline solid that is readily soluble in water. Its primary use is in the production of fertilizers.

c. Calcium Oxide (lime or quicklime), CaO

Calcium oxide is prepared commercially by the thermal decomposition of calcium carbonate (limestone), $CaCO_3$.

$$CaCO_{3(s)} \xrightarrow{850\,^{\circ}C} CaO_{(s)} + CO_{2(g)}$$

Calcium oxide was the second chemical in terms of total production in the United States in 1978. Calcium oxide is a white solid that reacts with water in a process called slaking to produce calcium hydroxide (slaked lime), $Ca(OH)_2$.

$$CaO_{(s)} + H_2O \longrightarrow Ca(OH)_{2(s)}$$

This reaction is accompanied by a threefold expansion in volume; thus, building contractors must be careful that their lime supplies to be used for cement do not get wet. Cement is about one part lime to three parts sand and is a primary use for calcium oxide. In addition, lime is spread on fields to reduce the acidity of soil, and it is used for furnace linings both because it withstands high temperatures and because it counteracts acidic impurities especially found in furnaces used for steel production or in processing beet sugar. Lime is also used as a fluxing material to carry away impurities in steel production.

d. Sodium Hydrogen Carbonate (Sodium Bicarbonate, Baking Soda), $NaHCO_3$

Sodium hydrogen carbonate is separated commercially from the mineral trona ($Na_2CO_3 \cdot NaHCO_3 \cdot 2\ H_2O$), which is found particularly in Wyoming and California. Sodium hydrogen carbonate is a white crystalline solid that is readily soluble in water. It is used to neutralize acids, it is the leavening agent in baking powder, and it is used in fire extinguishers by virtue of its liberation of CO_2 with acids (Section 8.2).

$$NaHCO_{3(s)} + H^+ \longrightarrow Na^+ + CO_{2(g)} + H_2O$$

It is used in the production of sponge rubber because the heat of vulcanization produces thousands of tiny bubbles of CO_2 that are trapped in the rubber.

$$2\ NaHCO_{3(s)} \xrightarrow{\text{heat}} Na_2CO_{3(s)} + CO_{2(g)} + H_2O$$

It is also an important ingredient in oral hygiene preparations and a preservative for yeast and dairy products.

e. Sodium Carbonate (Soda Ash), Na_2CO_3

Sodium carbonate is obtained commercially from the mineral trona ($Na_2CO_3 \cdot NaHCO_3 \cdot 2\ H_2O$). The ore is first heated to convert sodium hydrogen carbonate, $NaHCO_3$, to sodium carbonate,

$$2\ NaHCO_{3(s)} \xrightarrow{\text{heat}} Na_2CO_{3(s)} + CO_{2(g)} + H_2O$$

and then the sodium carbonate is purified. Sodium carbonate was the eleventh chemical in terms of total production in the United States in 1978. Sodium carbonate is a white crystalline solid that is readily soluble in water. It is used primarily in the manufacture of glass and less extensively in the production of soap and in the pulp and paper industry.

f. Sodium Chloride, NaCl

Sodium chloride is produced commercially in a pure form by crystallization from brine solutions, either from sea water or brine wells, and in an impure form by direct mining from salt beds. Sodium chloride is a white crystalline solid that is readily soluble in water. Preparation of other chemicals such as chlorine (Section 5.5-i), hydrogen (Section 5.5-a), sodium, and sodium hydroxide (Section 8.2-e) accounts for about 70% of the sodium chloride produced. About 4% is blended with traces of other salts to produce table salt, and much of the rest is used in ice cream manufacture and the salting of animal hides.

Exercises

6.1 Define clearly the following terms.

(a) diffusion
(b) viscosity
(c) surface tension
(d) evaporation
(e) equilibrium vapor pressure
(f) dynamic equilibrium
(g) volatile
(h) boiling
(i) normal boiling point
(j) crystal structure
(k) polymorphism
(l) melting
(m) normal melting point
(n) sublimation
(o) atomic solid
(p) ionic solid
(q) polar molecular solid
(r) nonpolar molecular solid
(s) metallic solid
(t) polymeric solid
(u) amorphous solid

6.2 Outline the basic assumptions of the kinetic molecular theory as applied to liquids. Describe experimental results which indicate that these assumptions are valid.

6.3 Explain the following in terms of the kinetic molecular theory as applied to liquids.

(a) Liquids maintain their volume, but assume the shape of their container.
(b) Liquids diffuse more slowly than gases.
(c) Liquids are essentially incompressible.
(d) Liquids show little change in volume with changes in temperature.
(e) Liquids exhibit surface tension.

6.4 List some practical uses that depend upon the fact that liquids are essentially incompressible.

6.5 Use Figure 6.2 to determine the following.
(a) the boiling point of ethyl alcohol at a pressure of 200 torr
(b) the normal boiling point of diethyl ether
(c) the pressure at which water will boil at 60°

6.6 Outline the basic assumptions of the kinetic molecular theory as applied to solids. Describe experimental results that indicate that these assumptions are valid.

6.7 Explain the following in terms of the kinetic molecular theory as applied to solids.
(a) Solids maintain characteristic shapes or geometric patterns.
(b) Solids show little change in volume with changes in temperature.
(c) Solids maintain their volume.
(d) Solids increase in volume as additional moles of sample are added.
(e) Solids melt to form liquids and sublime to form gases.

6.8 Explain clearly why each of the following occurs.
(a) The vapor pressure of water increases with increasing temperature.
(b) Bubbles of gas in a boiling liquid generally increase in volume as they approach the top surface of the liquid.
(c) Boiling points vary with geographic location.
(d) Your arm cools when alcohol, having been spilled on it, evaporates.
(e) On a warm summer day, the outside walls of a glass of iced tea become covered with water droplets.
(f) Molasses becomes almost impossible to pour if it is accidently placed in the refrigerator.

6.9 Compare the solid, liquid, and gas states of water with respect to the following.
(a) rate of diffusion
(b) bonding forces acting between molecules
(c) distance between molecules
(d) energy of the molecules
(e) compressibility
(f) velocity of the molecules
(g) freedom of movement of the molecules
(h) bonding forces acting within molecules

6.10 Explain clearly why each of the following occurs.
(a) A liquid in an insulated container cools on evaporation.
(b) RbBr and Br_2 have about the same molar mass, but RbBr melts at 682° whereas Br_2 melts below room temperature.
(c) A mixture of ice and water remains at 0° until all the ice is melted, even though heat is applied to it.
(d) Alcohol spilled on the bathroom floor evaporates even though its boiling point is 82°.
(e) An automobile engine turns over only with great difficulty on a very cold morning.

6.11 A water pipe in your basement that has a 0.75-inch inside diameter and is 22 feet between valves is allowed to freeze. Recalling that the formula for the volume of a cylinder is $V = \pi r^2 h$, use data from Section 6.6, and calculate the change in volume between the liquid water and the ice that forms in the pipe. The consequences are obvious!

6.12 Consider the substances Br_2, HCl, elemental Si, and H_2O. Describe the intermolecular forces present in each of these substances, and predict the order of decreasing melting points. Show clearly your reasoning.

6.13 Explain clearly why each of the following occurs.

(a) A freely falling drop of water is spherical in shape.
(b) A lake freezes from the top down.
(c) Ice disappears from a sidewalk even when the temperature is constantly below freezing.
(d) Your hands gradually feel colder if you do not dry them after washing.
(e) A car radiator or engine block may crack if the cooling water in them is allowed to freeze.

6.14 Consider the substances F_2, Cl_2, Br_2, and I_2. Describe the intermolecular forces present in these substances, and predict the order of increasing boiling points. Show clearly your reasoning.

6.15 Give an example of each of the following types of solids. For each example tell what particles occupy the lattice sites and what kinds of bonding forces hold the particles in their positions.

(a) atomic solid
(b) ionic solid
(c) polar molecular solid
(d) nonpolar molecular solid
(e) metallic solid
(f) polymeric solid

6.16 Indicate the type of solid you would expect each of the following to form. For each of the following tell what particles occupy the lattice sites and what kinds of bonding forces hold the particles in their positions.

(a) carbon dioxide, CO_2
(b) calcium fluoride, CaF_2
(c) ammonia, NH_3
(d) polystyrene, $\left(\underset{\displaystyle C_6H_5}{\underset{|}{CH}}-CH_2 \right)_n$
(e) copper, Cu

6.17 How do amorphous solids differ from crystalline solids in terms of (a) structure and (b) properties?

6.18 Write a balanced chemical equation to describe each of the following.

(a) thermal decomposition of NH_4NO_3
(b) thermal decomposition of $CaCO_3$
(c) thermal decomposition of $NaHCO_3$
(d) reaction between NH_3 and HNO_3
(e) reaction of $NaHCO_3$ with H^+

7

Solutions

We have previously considered gases, liquids, and solids in a pure state. However, many samples that we encounter are not pure substances, but rather are solutions. Antifreeze is a solution of ethylene glycol in water. Blood plasma is a complex solution of proteins, other organic compounds, and salts in water. Sterling silver is an alloy composed of a solid solution of copper in silver. Many household chemicals such as cleaning fluids are solutions. The gasoline we use in our cars is a solution of hydrocarbons. Solutions exhibit characteristic properties that are often different from those of the pure substances composing the solutions.

The purpose of this chapter is to discuss the general properties of solutions and why these properties arise, to consider solubility and the factors affecting solubility, and to present means for describing concentrations of solutions.

7.1 General Properties of Solutions

A **solution** is *a system in which one or more substances is uniformly or homogeneously mixed or dissolved in another substance.* (See Sections 1.1-d and 1.1-e for a review of pure substances and solutions.) The components of a solution are referred to as solute and solvent. The **solute** is *the substance that is dissolved;* the **solvent** is *the substance that does the dissolving.* When a solution is prepared from two substances of different phases, the solvent is the substance that is of the same phase as the resulting solution, and the solute is the substance dissolved in the solvent. If a solution is prepared from two substances of the same phase, the solvent is conventionally the substance present in the larger amount. When water is the solvent, the resulting solutions are called *aqueous solutions.* Liquid solutions not involving water are called *nonaqueous solutions.*

The properties of a typical solution are best derived from observations of specific substances. A few purple iodine (I_2) crystals dropped into a flask containing colorless, liquid carbon tetrachloride (CCl_4) gradually dissolve. The solution around the dissolving crystals takes on a violet color. Eventually the violet color disperses evenly through the entire volume of the solution. A solution is a homogeneous mixture of solute and solvent. Any given solution has the same chemical composition and physical and chemical properties throughout its volume. A solution may be colored, but is usually transparent. Spectroscopic evidence on the above solution indicates that I_2 molecules are distributed uniformly throughout the CCl_4. The dissolved solute in a solution is either molecular or ionic in nature. The properties of the above solution remain unaltered after passage through filter paper. No further changes occur if the flask is sealed and left standing. The solute in a solution remains uniformly distributed throughout the solution and does not settle out with time. Addition of more I_2 only deepens the violet color; the solution remains transparent. The composition of a solution may be varied within limits. On mild heating, the CCl_4 can be distilled, leaving behind purple iodine crystals. The solute in a solution may frequently be separated from the solvent by techniques such as crystallization, distillation, or chromatography (Section 1.1-f).

7.2 Kinetic Molecular Theory Applied to Several Types of Solutions

In principle it is possible to have nine different types of two-component solutions: gas dissolved in solid, liquid, or gas; liquid dissolved in solid, liquid, or gas; and solid dissolved in solid, liquid, or gas. Five of these are encountered frequently.

Air is an example of a gas dissolved in a gas. In air O_2 and N_2 molecules, as well as other trace gaseous constituents such as CO_2, H_2, H_2O vapor, He, Ne, Ar, Kr, and Xe, are distributed uniformly and move independently of each other. *Gases mix readily in all proportions to form solutions.* The tiny, discrete gas molecules have large distances between them and thus vast amounts of empty space available for additional molecules. The gas molecules have essentially no attractive forces between them to inhibit mixing during the course of their rapid, random motion.

Alloys have metallic properties and result from the combination of two or more metals. Alloys are examples of solids dissolved in solids. Brass, an alloy of copper and zinc, is a solid solution in which some Cu atoms of the cubic structure of pure copper are displaced by Zn atoms. *The formation of solid solutions is relatively difficult.* The tiny, discrete atoms, molecules, or ions in a solid are in direct contact. Free space for additional particles is minimal. The particles forming a solid have strong attractive forces between them. This inhibits mixing and results in distinct packing arrangements in the crystal. An incoming solute particle must be neither too large nor too small so that it does not disrupt the orderly packing arrangement of the solvent particles. In an ionic solid the solute particles must also be of appropriate charge so as to maintain electroneutrality in the crystal.

Hydrogen chloride gas dissolved in water (forming hydrochloric acid) is an example of a gas dissolved in a liquid. The HCl molecules break into H^+ and Cl^- ions upon dissolving. These ions are distributed uniformly in the aqueous solution. Ethyl alcohol (C_2H_5OH) dissolved in water is an example of a liquid dissolved in a liquid. The colorless solution has C_2H_5OH molecules distributed randomly in the aqueous solution. Yellow powdered sulfur dissolved in carbon disulfide (CS_2) is an example of a solid dissolved in a liquid. This nonaqueous solution is composed of S_8 molecules distributed uniformly in the CS_2. *The ease of formation of liquid solutions varies considerably.* A particular liquid solvent may dissolve one solute readily, another solute only with difficulty, and a third solute not at all. The tiny, discrete particles composing liquids are in close contact, but have more space for additional particles than do solids. The particles forming a liquid have attractive forces between them, but the forces are weaker than in solids. There is thus less inhibition of mixing in liquid solutions.

When discussing the mixing of two liquids, the terms miscible and immiscible are often used. Ethyl alcohol and water mix readily with each other in all proportions. *Two liquids that are capable of mixing to form a solution* are said to be **miscible.** Conversely, oil and water separate into two layers (the less dense oil on top) when they are placed in the same container. *Two liquids that do not mix in each other and do not form solutions* are said to be **immiscible.**

7.3 Solubility

Methyl alcohol (CH_3OH) and water are miscible with each other in *all* proportions, whereas only a certain amount of table salt (NaCl) will dissolve in a given amount of water. In general, most solutes dissolve in any given solvent only to a certain extent. **Solubility** is *a measure of the amount of solute that will dissolve in a given amount of solvent.* Solubility can be expressed qualitatively or quantitatively. Qualitatively we can say that at room temperature calcium chloride ($CaCl_2$) is very soluble in water, mercury(II) chloride ($HgCl_2$) is moderately soluble, lead(II) chloride ($PbCl_2$) is only slightly soluble, and silver chloride (AgCl) is relatively insoluble. Quantitatively we can say that at 20 °C $CaCl_2$ dissolves to the extent of 74.5 grams per 100 ml of water, $HgCl_2$ to the extent of 6.9 grams per 100 ml, $PbCl_2$ to the extent of 0.99 gram per 100 ml, and AgCl to the extent of 2.9×10^{-4} gram per 100 ml. Solubility can also be expressed quantitatively in terms of molar concentrations (Section 7.7-a) and solubility product constants (Section 10.2-b).

If 20 grams of solid $CaCl_2$ is added to a beaker containing 100 ml of water, an unsaturated solution results. An **unsaturated solution** *contains less than the maximum amount of dissolved solute that the solution can hold under the existing conditions of temperature and pressure.* A great number of unsaturated solutions of $CaCl_2$ can be prepared by adding a few crystals at a time and allowing them to dissolve. However, eventually the added $CaCl_2$ does not dissolve, but remains as a solid in the bottom of the beaker. A limit to the amount of $CaCl_2$ that can be dissolved has been reached. A **saturated solution** *contains the maximum amount of dissolved solute that the solution can hold under the existing conditions of temperature and pressure.* Both dissolving and precipitation of $CaCl_2$ continue in the saturated solution, but they occur at the same rate. The number of solute particles dissolving equals the number of solute particles precipitating. The mass of $CaCl_2$ dissolved in solution and the mass of excess $CaCl_2$ present as a solid remain constant. The mass of $CaCl_2$ dissolved in a saturated solution is the solubility of $CaCl_2$. The solubility of $CaCl_2$ increases as the temperature of the solution is raised. Upon heating, the solution which was formerly saturated becomes unsaturated, and more $CaCl_2$ can be dissolved.

For certain salts it is possible to prepare **supersaturated solutions** that *have a greater amount of dissolved solute than a saturated solution.* A water solution of sodium acetate ($NaC_2H_3O_2$), which is saturated at some sufficiently high temperature, can be cooled slowly and carefully without having any solute precipitate from the solution. A supersaturated solution, holding more dissolved solute than a saturated solution can hold, results. Supersaturated solutions are unstable with respect to the separation of excess solute to form a saturated solution. This separation of excess solute can be made to occur by mechanical shock, by scratching the inside container wall, by the presence of dust particles, or by the addition of a solute crystal on which the excess solute can crystallize.

Unsaturated, saturated, and supersaturated solutions of $NaC_2H_3O_2$ can be distinguished from each other by adding a small crystal of $NaC_2H_3O_2$. If the solution is unsaturated, the crystal dissolves. If the solution is saturated, the mass of the crystal does not change (although its shape may change). If the solution is supersaturated, additional crystallization onto the crystal occurs until the amount of dissolved solute is reduced to the level of a saturated solution.

a. Why Dissolving Occurs

Whether or not a solute dissolves in a given solvent depends on a balance between two factors: (1) the bonding interactions that are broken and formed during the dissolving process and (2) the changes in disorder or randomness that occur during the dissolving process. These factors are discussed sequentially, and then the interaction of the two factors is considered for the mixing of a few solutes and solvents.

When a solute and solvent are mixed, bonding interactions in the solute—called *solute-solute interactions*—and bonding interactions in the solvent—called *solvent-solvent interactions*—are broken. These two processes require energy. During mixing of solute and solvent, bonding interactions between solute and solvent—called *solute-solvent interactions*—are formed. This process liberates energy and is one driving force for dissolving. Considering only changes of bonding interactions and assuming no change of disorder or randomness, dissolving of a solute in a given solvent is favored if the solute–solvent interactions are great enough to overcome the solute–solute and solvent–solvent interactions whereas dissolving is not favored if the solute–solvent interactions are small compared to the solute–solute and solvent–solvent interactions.

The effect of a change in disorder or randomness is best understood by considering an example. When a valve between two flasks, one filled with an ideal gas (Section 5.3) and the other evacuated, is opened, gas rushes into the empty flask until the pressures in the two flasks are equal. Since the attractive forces between the molecules of an ideal gas are negligible, the formation and destruction of bonding interactions could not have provided the driving force for the escape of gas from one flask into the other. Instead, the driving force is an increase in disorder or randomness of the gas molecules indicated by the fact that they are more disordered in the two flasks than they were in one. We have less knowledge about the positions of individual gas molecules in the two flasks than we did in one. An increase of disorder or randomness also occurs when a solute is mixed with a solvent, thus favoring the dissolving of a solute in a given solvent.

It might appear that predicting solubilities would be an easy task that could be accomplished simply by balancing solute–solute and solvent–solvent interactions against solute–solvent interactions and the change in disorder and seeing which set dominates. On the contrary, predictions are very complex, primarily because the same factors that cause increased solute–solute or solvent–solvent interactions also cause stronger solute–solvent interactions. If, however, we know the extent to which a given solute is soluble in a particular solvent (in other words, if we measured the solubility in the laboratory), we can explain the solute's solubility in the solvent on the basis of solute–solute interactions, solvent–solvent interactions, solute–solvent interactions, and the change in disorder as shown in the following examples.

Methyl alcohol and water mix readily because the favorable increase in disorder combined with hydrogen bonding and dipole–dipole interactions between alcohol and water compensate for breaking the same types of interactions originally present in the two pure liquids. Iodine dissolves in carbon tetrachloride because the favorable increase in disorder combined with dispersion forces between I_2 and CCl_4 molecules compensate for breaking the same types of interactions originally present in the two pure substances. Water does *not* dissolve in carbon tetrachloride because, even with the favorable increase in disorder, dispersion forces that can form between H_2O and CCl_4 molecules are not strong enough to overcome the significant hydrogen bonding and dipole–dipole interactions initially present in pure water.

Hydrogen chloride gas dissolves readily in water because the favorable increase in disorder combined with strong ion–dipole interactions between each H^+ and the negative end of a number of H_2O molecules (Figure 7.1a) and between each Cl^- and the positive end of a number of H_2O molecules

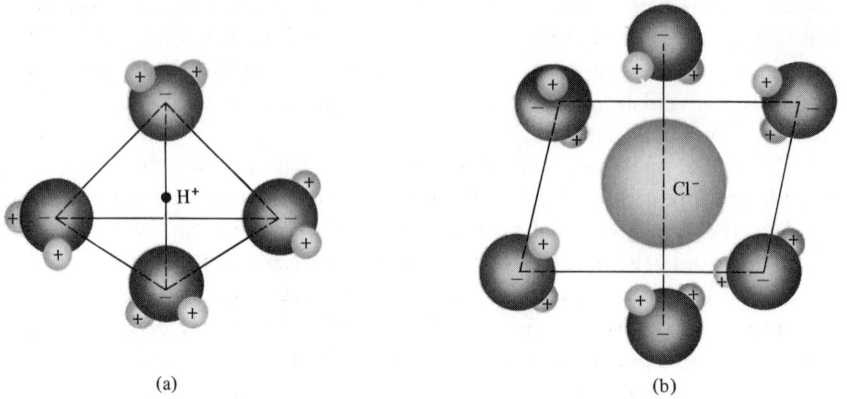

H⁺
Cl⁻
(a)
(b)

Figure 7.1 Ion-dipole interactions beween H^+ and a number of H_2O molecules (a) and between Cl^- and a number of H_2O molecules (b). All charges indicated on H_2O molecules are partial charges.

(Figure 7.1b) compensate for breaking covalent bonds in HCl molecules and hydrogen bonding and dipole–dipole interactions initially present in pure water. Similarly, sodium chloride dissolves in water because the favorable increase in disorder combined with the strong ion–dipole interactions between each Na^+ and the negative end of a number of H_2O molecules and between each Cl^- and the positive end of a number of H_2O molecules compensate for breaking the ion–ion interactions in NaCl crystals and hydrogen bonding and dipole–dipole interactions initially present in pure water. Conversely, silver chloride is insoluble in water because, even with the favorable increase in disorder, ion–dipole interactions between each Ag^+ and a number of H_2O molecules and between each Cl^- and a number of H_2O molecules are too small to overcome the ionic and covalent interactions initially present in AgCl crystals. Some general solubility rules for salts in water are given in Table 7.1.

b. Factors Affecting Solubility

Fortunately, solubilities are fairly easy to determine experimentally. On the basis of experimental evidence, it has been determined that solubilities are affected mainly by three factors.

1. Nature of the Solute and the Solvent. Generally, *polar solvents dissolve ionic and polar solutes, and nonpolar solvents dissolve nonpolar solutes* (like dissolves like). A polar solvent such as water is a good solvent for a polar gas like HCl, a polar liquid like CH_3OH, or an ionic solid like NaCl. The polar or ionic nature of these substances creates particularly strong dipole–dipole or ion–dipole solute–solvent interactions, and thus favors solubility. On the other hand, water is a poor solvent for a nonpolar gas like O_2, a nonpolar

Table 7.1 General Solubility Rules for Salts in Water[a]

1. Nitrates are generally soluble.
2. Chlorides, bromides, and iodides are generally soluble, except for the halides of Ag(I), Hg(I), and Pb(II), and HgI_2.
3. Sulfates and chromates are generally soluble, except for those of Ag(I), Hg(I), Pb(II), Sr(II), and Ba(II), and also $CaSO_4$ and $CuCrO_4$.
4. Salts of the alkali metal cations (Li^+, Na^+, K^+, Rb^+, Cs^+, and Fr^+) and of the ammonium ion (NH_4^+) are generally soluble.
5. Carbonates, fluorides, hydroxides, oxides, phosphates, sulfides, and sulfites are generally insoluble, except for those of cations in 4, as well as fluorides of Ag(I), Al(III), Hg(II), and Sn(II); hydroxides and oxides of Ca(II), Sr(II), and Ba(II); and sulfides of Mg(II), Ca(II), Sr(II), Ba(II), Al(III), and Cr(III).

[a] Roman numerals in parentheses indicate oxidation numbers (Section 8.5).

liquid like salad oil, or a nonpolar solid like naphthalene. The only solute–solvent interactions in these examples are weak dispersion forces, not strong enough to overcome the stronger hydrogen bonding and dipole–dipole interactions initially present in water. Conversely, a nonpolar solvent like benzene is a reasonably good solvent for O_2, salad oil, and naphthalene, but a poor solvent for HCl, CH_3OH, or NaCl.

*2. **Temperature.*** The usual tendency to heat a solution in order to dissolve more solute works for some solutes, but not for others. *Gases usually become less soluble in water as the temperature of the solution is increased;* however, this generalization does not hold for the solubility of gases in a number of other solvents. The decreasing solubility of gases in water with increasing temperature is illustrated for HCl and SO_2 in Figure 7.2 and for O_2 in Table 7.2. Data in Table 7.2 indicate that trout can reach higher concentrations of oxygen by swimming to deeper, colder parts of a lake on a hot summer day.

Water that has been in contact with air normally contains small amounts of dissolved gases. The tiny gas bubbles can be driven off by heating the

Table 7.2 Solubility of Oxygen in Water[a]

Temperature (°C)	*Solubility (g O_2 per 100 g H_2O)*
0	0.0069
20	0.0043
40	0.0031
60	0.0023
80	0.0014

[a] $p_{oxygen} + p_{water\ vapor} = 760$ torr.

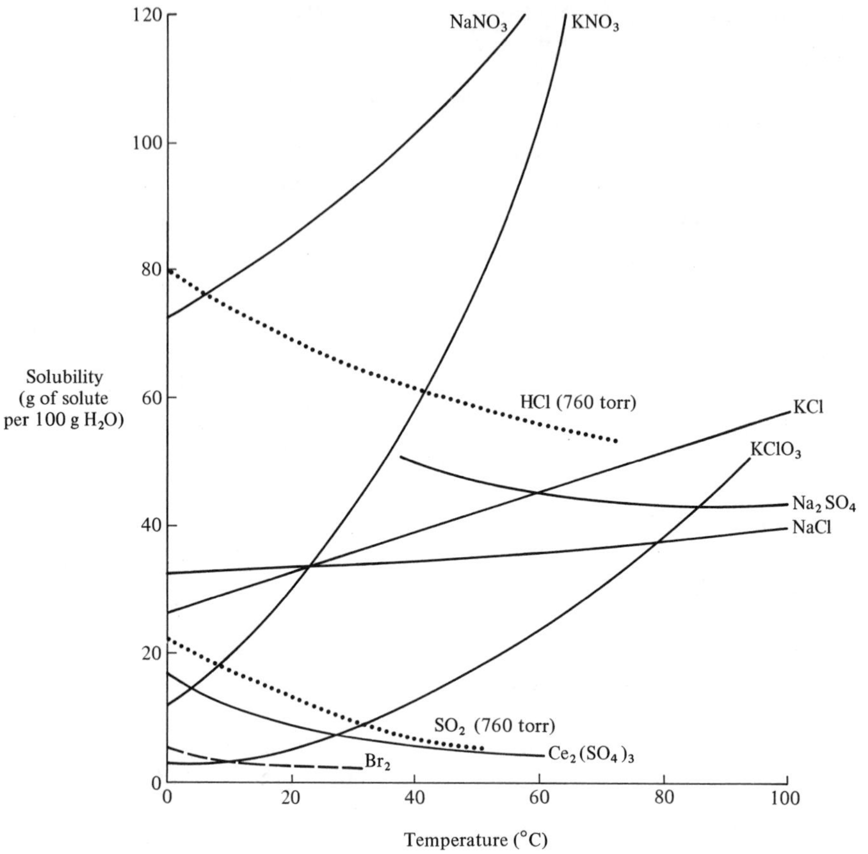

120
100
80
60
40
20
0
Solubility
(g of solute
per 100 g H2O)
NaNO3
KNO3
HCl (760 torr)
KCl
KClO3
Na2 SO4
NaCl
SO2 (760 torr)
Ce2 (SO4)3
Br2
0
20
40
60
80
100
Temperature (°C)

Figure 7.2 Variation of solubilities of various substances in water as a function of temperature. Solids ———, liquid ––––, gases ········ .

solution sufficiently. These bubbles are the ones you see rising before the water starts to boil. Water that has been boiled has a characteristic flat taste because dissolved air has been expelled.

There is no general rule for the temperature dependence of solubility of liquids and solids in a liquid solvent. Nicotine in water shows a strange variation of solubility with temperature—an increase in solubility with increasing temperature in one temperature range, and the opposite effect in another temperature range. Figure 7.2 shows solubility–temperature curves for a number of solids in water at atmospheric pressure. It can be seen that with increasing temperature, KNO_3 becomes more soluble; NaCl shows little change in solubility; and $Ce_2(SO_4)_3$, a rare case, becomes less soluble.

3. ***Pressure.*** *At constant temperature the solubility of all gases in any solvent is increased as the partial pressure of the gas above the solution is increased.* The linear increase in solubility with increasing partial pressure of several gases over a solution at constant temperature is presented graphically in Figure 7.3. Carbonated beverages are bottled under a high pressure of CO_2 so that a significant amount of CO_2 is dissolved in solution. When the bottle is opened, the partial pressure of CO_2 above the solution diminishes rapidly giving rise to a "pop," the solubility of CO_2 in the solution decreases, and bubbles of excess CO_2 escape to provide effervescence. The escaping CO_2 also provides the tingling taste in the beverage.

Unlike gases, liquids and solids exhibit practically no change of solubility with changes in pressure.

7.4 Electrolytes

We found in Section 7.1 that the dissolved solute in a solution may be either molecular or ionic in nature. When sugar dissolves in water, the sugar molecules retain their identity in solution and are simply solvated or surrounded by a sheath of water molecules. However, when NaCl dissolves in water, the solution process is accompanied by a disruption of the NaCl crystal structure into Na^+ and Cl^- ions. *The breaking apart of a solute into its ions in aqueous solution* is called **dissociation.** Each ion then interacts with a number of water molecules to produce a hydrated ion (Figure 7.1). It is not surprising that ionic solids, having ions at their lattice sites, are capable of dissociating into ions in aqueous solution. In addition, certain molecular substances may also dissolve in water to produce ions. Discrete gaseous HCl molecules dissociate into H^+ and Cl^- ions when dissolved in water. These ions then interact with a number of water molecules to produce hydrated

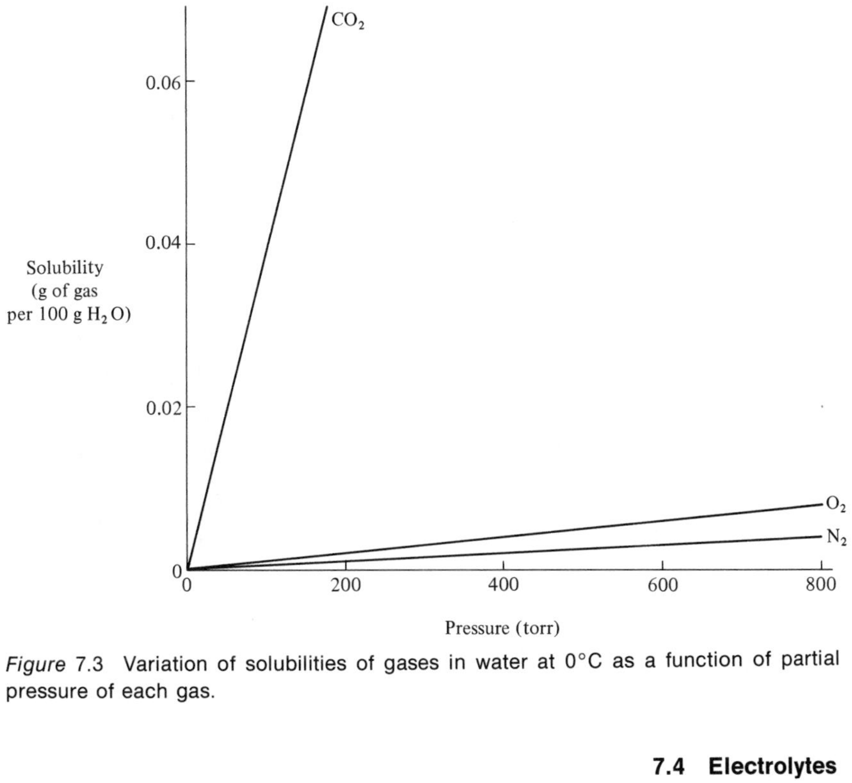

CO2
0.06
0.04
Solubility
(g of gas
per 100 g H2O)
0.02
O2
N2
0
0
200
400
600
800
Pressure (torr)

Figure 7.3 Variation of solubilities of gases in water at 0°C as a function of partial pressure of each gas.

ions. For simplicity, the water of hydration is often omitted from chemical equations. Thus we represent the dissociation of HCl in aqueous solution by the equation

$$HCl \longrightarrow H^+ + Cl^-$$

keeping in mind that all species are hydrated.

It is a simple matter to determine whether a solute is dissociated into ions in solution by measuring the electrical conductivity of the solution using the apparatus shown in Figure 7.4. A pair of electrodes is connected in series to a light bulb and to a source of electricity. As long as the electrodes are separated in air (a nonconductor), no electric current flows through the circuit, and the bulb does not light. However, if the two electrodes are touched to each other, the circuit is completed, and the bulb lights. If the electrodes are dipped into a beaker containing pure water, the bulb does not light, indicating that water is not a good conductor of electricity and is incapable of completing the circuit. Likewise, a sugar solution does not conduct electricity, and the bulb does not light when dipped into a sugar solution. On the other hand, the bulb does light when the electrodes are dipped into a solution of NaCl. Movement of Na^+ and Cl^- ions in solution constitutes an electric current and thus is capable of completing the circuit. Solutions of other substances can be tested similarly. *Substances producing solutions that conduct electricity* are called **electrolytes;** *substances producing solutions that do not conduct electricity* are called **nonelectrolytes.** Examples of electrolytes and nonelectrolytes are given in Table 7.3.

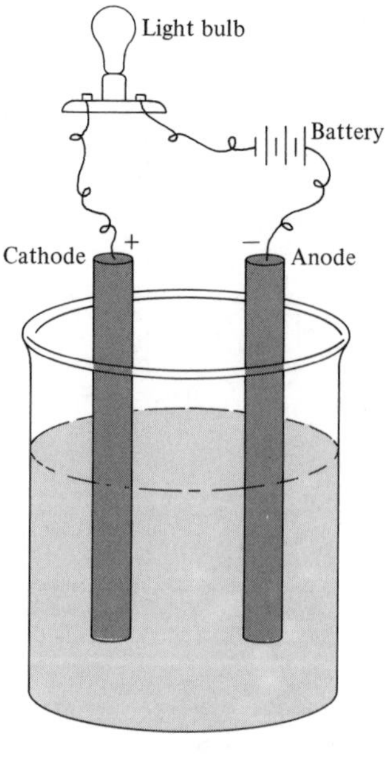
Light bulb
Battery
Cathode
+
−
Anode

Figure 7.4 Apparatus for determining the electrical conductivity of a solution.

Table 7.3 Some Electrolytes and Nonelectrolytes

Strong Electrolytes	
Sulfuric acid, H_2SO_4 [$H^+ + HSO_4^-$][a]	Potassium chloride, KCl [$K^+ + Cl^-$]
Hydrochloric acid, HCl [$H^+ + Cl^-$]	Ammonium carbonate, $(NH_4)_2CO_3$ [$NH_4^+ + CO_3^{2-}$]
Nitric acid, HNO_3 [$H^+ + NO_3^-$]	Silver nitrate, $AgNO_3$ [$Ag^+ + NO_3^-$]
Sodium hydroxide, NaOH [$Na^+ + OH^-$]	Copper(II) sulfate, $CuSO_4$ [$Cu^{2+} + SO_4^{2-}$]
Calcium hydroxide, $Ca(OH)_2$ [$Ca^{2+} + OH^-$]	Iron(III) bromide, $FeBr_3$ [$Fe^{3+} + Br^-$]
Weak Electrolytes	*Nonelectrolytes*
Acetic acid, CH_3COOH [CH_3COOH]	Methyl alcohol, CH_3OH [CH_3OH]
Hydrogen sulfide, H_2S [H_2S]	Ethyl alcohol, C_2H_5OH [C_2H_5OH]
Ammonia, NH_3 [NH_3]	Sucrose (sugar), $C_{12}H_{22}O_{11}$ [$C_{12}H_{22}O_{11}$]
Mercury(II) chloride, $HgCl_2$ [$HgCl_2$]	Acetone, CH_3COCH_3 [CH_3COCH_3]

[a]Brackets surround the predominant species in solution. All are hydrated.

A number of substances dissociate only partially in aqueous solution and cause the bulb to be only dimly lit in the electrical conductivity experiment described above. On the basis of such experimental observations, electrolytes may be subdivided into two groups. *Substances that dissociate almost completely and produce solutions that are good conductors of electricity* are called **strong electrolytes;** *substances that dissociate only slightly and produce solutions that are only weak conductors of electricity* are called **weak electrolytes.** Examples of weak electrolytes are included in Table 7.3.

7.5 Colligative Properties

We have examined the general properties of solutions in terms of the differences between a pure solvent and a solution on a microscopic scale. We have considered the distinctive species composing various solutions and have discussed the interactions between these species. We now need to examine those properties, caused by the presence of a solute, that differentiate a pure solvent and a solution on a macroscopic scale. Such properties are called *colligative properties.*

Colligative properties of a solution depend only on the ***number*** *of solute particles dissolved in a given amount of solvent and are independent of the* ***nature*** *or the* ***identity*** *of the solute particles.* An equal number of atoms, molecules, or ions dissolved in a given amount of solvent will have the same effect on the colligative properties of the resulting solution. Five examples of colligative properties will be discussed.

a. Vapor Pressure Lowering

If a beaker containing pure water and an identical beaker containing the same volume of a NaCl solution are sealed under a bell jar (Figure 7.5a), the

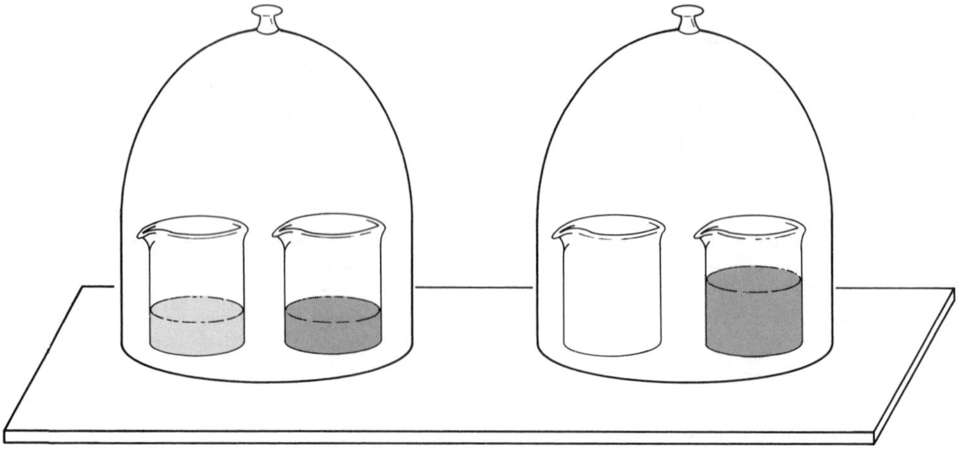
Pure
water
NaCl
solution
Empty
beaker
NaCl
solution
(a)
(b)

Figure 7.5 Change of levels of pure water and a NaCl solution when beakers are sealed under a bell jar. (a) Initial levels; (b) levels some time later.

level of pure water gradually drops and the level of the NaCl solution rises. Water is transferred from the beaker of pure water, through the vapor phase, to the NaCl solution. Given enough time, the beaker of pure water will empty, and all the water will end up in the NaCl solution (Figure 7.5b). This process occurs because the vapor pressure of the pure water is greater than the vapor pressure of the NaCl solution; that is, water has a greater tendency to escape to the vapor phase from the pure water than from the NaCl solution. *Adding a solute to a solvent lowers the escaping tendency of solvent molecules or lowers the vapor pressure of the solvent.* This lowering of vapor pressure with added solute occurs commonly in nature. For example, evaporation from the oceans, which could be very extensive, is diminished due to the large amounts of dissolved salts present.

b. Boiling Point Elevation

If the boiling points of pure water and of a NaCl solution are determined, it is found that the NaCl solution boils at a higher temperature than pure water. *Adding a solute to a solvent raises the boiling point of the resulting solution.* This is in agreement with the fact that adding a solute to a solvent lowers the tendency of the solvent molecules to escape into the vapor phase, or lowers the vapor pressure at any given temperature. A higher temperature is required for the vapor pressure of the solution to reach atmospheric pressure, the definition of boiling point (Section 6.5).

c. Freezing Point Depression

If the temperatures at which pure water and a NaCl solution freeze are determined, it is found that the NaCl solution freezes at a lower temperature than pure water. *Adding a solute to a solvent lowers the freezing point of the resulting solution.* The tendency of H_2O molecules to convert from the liquid phase into the solid phase is decreased by the addition of solute. We take advantage of the lowering of the freezing point of a solution as solute is added when we spread salt on icy sidewalks and roads. The ice first melts in localized areas because the salt–ice mixture has a freezing point that is lower than ambient temperature. The ice that melts forms a highly concentrated salt solution so that the process can continue until the entire layer of ice is melted. The freezing point of a concentrated NaCl solution may be lower than $-20\,°C$.

The net result of boiling point elevation and freezing point depression is that the liquid range of a solvent is extended in both directions when a solute is dissolved in the solvent. We take advantage of both freezing point depression and boiling point elevation when we add a solute such as ethylene glycol (antifreeze) to water in the cooling system of a car. The resulting solution does not freeze to ice and expand to crack the engine block or radiator even at temperatures well below freezing. Moreover, we can operate the engine at temperatures in excess of 100° (without boiling the water away), thus increasing its efficiency.

d. Osmotic Pressure

Certain membranes made of an animal bladder, a slice of vegetable tissue, or a piece of parchment, act as a barrier between two solutions, and simultaneously allow specific types of molecules to pass through the membrane while prohibiting the passage of other types of molecules. These are called *semipermeable membranes.* Semipermeable membranes that allow passage of solvent molecules but do not allow passage of solute molecules or ions are called *osmotic membranes.* If a NaCl solution is separated from pure water by an osmotic membrane, H_2O molecules spontaneously penetrate the membrane from both directions; however, passage across the membrane from the pure water side is faster than passage across the membrane from the solution side. The net result is exactly like that illustrated already in Figure 7.5 and involves a net transfer of H_2O from the pure water side of the membrane to the solution side of the membrane (Figure 7.6a). The passage of solvent molecules from a region with little or no dissolved solute, through an osmotic membrane, to a region with more dissolved solute is called *osmosis.*

Osmosis can be stopped by applying a pressure to the side with more dissolved solute. *The pressure required to just counteract the osmosis process and prevent the net passage of solvent* is called the **osmotic pressure** (Figure

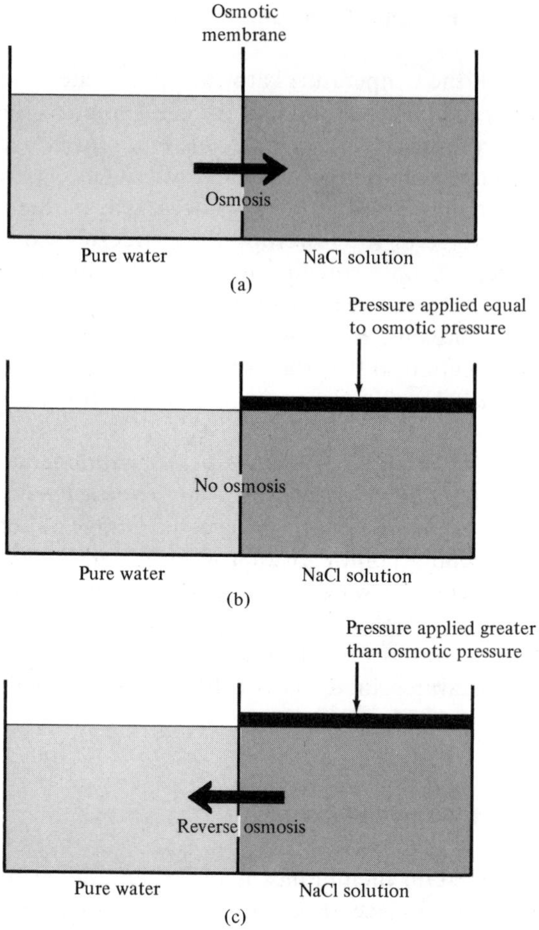
Osmotic membrane
Osmosis
Pure water
NaCl solution
(a)
Pressure applied equal to osmotic pressure
No osmosis
Pure water
NaCl solution
(b)
Pressure applied greater than osmotic pressure
Reverse osmosis
Pure water
NaCl solution
(c)

Figure 7.6 Illustration of (a) osmosis, (b) osmotic pressure, and (c) reverse osmosis.

7.6b). Osmotic pressure varies directly with absolute temperature, just as does the pressure of a gas. Osmotic pressure is very important in many biological processes (Section 31.2) since plant and animal cell walls act as semipermeable membranes. Water is taken up by trees as a result of passage of water from ground water containing relatively little dissolved solute, through root membranes, into the concentrated nutrient solution inside the roots. Solutions used for intravenous feeding must have carefully regulated amounts of solute so as to have the same osmotic pressure as the solution inside our cells. If too little solute is present, water passes through the cell walls and into the cell, swelling the cells and eventually bursting them. If too much solute is present, water passes out of the cells, causing them to shrivel up.

If pressures greater than osmotic pressure are applied, solvent molecules can be forced from the side with more dissolved solute to the pure water side (Figure 7.6c). This process is called *reverse osmosis*. It is used in the pulp and paper industry to remove impurities from water that are present after the pulping process. It is also being tried as a means for removing salt from seawater to make it drinkable.

e. Decrease of Solubility of Gases

If a beaker containing pure water and a beaker containing the same volume of a NaCl solution are sealed under a bell jar that has an atmosphere of CO_2, it is found that more CO_2 is dissolved in the pure water than in the NaCl solution. *Adding a solute to a solvent lowers the solubility of a gas in that solvent* (assuming the solute doesn't react with the gas). This creates a problem for us biologically because blood and body fluids act like aqueous solutions of various salts and do not dissolve oxygen readily. However, nature has provided us with a mechanism for transporting O_2 that does not require significant dissolving, through the reversible binding of O_2 to hemoglobin (Section 31.4).

7.6 Colloids

When NaCl dissolves in water, discrete Na^+ and Cl^- ions that formed the crystal become uniformly dispersed among the water molecules. The solute has been broken down as far as possible without destroying the ions by a chemical change. The dispersed ions do not separate out on standing, they cannot be seen with a microscope, and they cannot be removed by filtration. At the other extreme, when powdered lead(II) sulfate ($PbSO_4$) is shaken with water, the small particles of $PbSO_4$ become dispersed in the water only momentarily to form a *suspension*. On standing, the suspended $PbSO_4$ particles separate so that two distinct phases are apparent, usually even to the naked eye. It is easy to separate the $PbSO_4$ particles from the water by decantation or filtration.

Intermediate between a homogeneous solution and a suspension of separate phases is a *colloid*. Colloids have one phase fairly uniformly distributed in a second phase; however, the dispersed particles are neither so large that they form an apparent second phase, nor so small that they can be said to be in solution. Typically, colloids are made up of globular particles with diameters between about 10^{-7} and 10^{-4} cm. They may in some cases be in the form of filaments or thin films in which at least one dimension is very small. Colloidal particles may be composed of aggregates of atoms, as in the case of colloidal gold; aggregates of molecules, as in the case of S_8 molecules composing colloidal sulfur; or very large molecules or polymers, as in the case of hemoglobin.

There are a variety of kinds of colloids. Some, such as the proteins that are responsible for carrying out vital body functions, occur naturally. Others, including food products such as cheese and marshmallow, are manufactured. Colloids are usually classified according to the phase of the colloidal particles, called the *dispersed phase,* and the phase in which the colloidal particles are scattered, called the *continuous phase.* Table 7.4 lists some types of colloids with examples of each.

The particles of a colloid do not separate out on standing, they cannot be separated by filtration, and except for the case of high polymer particles, they cannot be seen with a microscope. Nevertheless, these particles are responsible for the unique properties exhibited by colloids—among them, light scattering, adsorption, and dialysis.

A flashlight beam, passed through a solution of sodium thiosulfate ($Na_2S_2O_3$), cannot be observed from the side (Figure 7.7a). The dissolved Na^+ and $S_2O_3^{2-}$ ions in the solution are too small to scatter light. If a few drops of hydrochloric acid (HCl) are added to the $Na_2S_2O_3$ solution and it is left standing, the flashlight beam passing through the solution can be seen from the side (Figure 7.7b). Colloidal sulfur particles, produced by a chemical reaction, are large enough to scatter the light even though the particles cannot be seen by the naked eye. Light scattering by colloidal particles is called the *Tyndall effect.*

Colloidal particles, because of their degree of subdivision, have many surface atoms and thus a very high surface area. Moreover, these surface atoms cannot bond in three dimensions as most atoms do, and thus bonding forces are not completely satisfied. These two factors are reponsible for the considerable force of attraction that causes other substances to adhere to the surface of colloids. This surface adsorption is usually highly specific. Some colloids adsorb neutral molecules. The adsorption of polar poisonous gases, such as H_2S, with simultaneous passage of nonpolar gases, such as O_2, makes charcoal an excellent air purification medium for a gas mask. Various oils are good lubricants because they are adsorbed strongly to metal surfaces and remain between the bearing surfaces to reduce friction. Other colloids selectively adsorb certain cations or anions (but not both) on their surface

Table 7.4 Kinds of Colloids

Type	*Dispersed Phase*	*Continuous Phase*	*Common Examples*
Liquid foam	Gas	Liquid	Soap suds, meringue, whipped cream
Solid foam	Gas	Solid	Pumice, marshmallow
Liquid aerosol	Liquid	Gas	Mist, fog, clouds
Liquid emulsion	Liquid	Liquid	Milk, cream, mayonnaise
Solid emulsion	Liquid	Solid	Butter, cheese
Solid aerosol	Solid	Gas	Smoke, dust in air
Gel or liquid sol	Solid	Liquid	Gelatin, jellies, gelatinous precipitates, agar, paints
Solid sol	Solid	Solid	Black diamonds, alloys

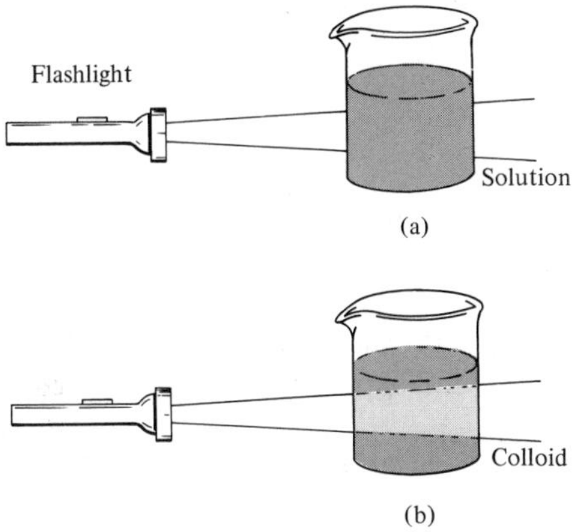
Flashlight
Solution
(a)
Colloid
(b)

Figure 7.7 Illustration of the Tyndall effect. (a) No light scattering by a solution; (b) light scattering by a colloid.

and become electrically charged. They are stabilized as colloids because charge–charge repulsions prevent them from sticking together and eventually precipitating to destroy the colloid. Many proteins in our blood are in an electrically charged, colloidal state.

Dialysis is *the separation of colloids from a solution by diffusion of the solvent and dissolved solute through a semipermeable membrane.* Certain semipermeable membranes (Section 7.5-d) pass dissolved molecules and ions in addition to solvent molecules, but do not pass colloidal particles. Such membranes, called *dialyzing membranes,* are slightly more permeable than osmotic membranes. Most of the membranes in our bodies are dialyzing membranes, for example, the walls of blood vessels. These walls normally pass water, O_2, CO_2, ions, and small organic molecules, but are usually not permeable to large protein molecules such as hemoglobin, albumin, globulins, lipoproteins, and fibrinogen. As a result of the influence of colloids, substances that normally diffuse through a dialyzing membrane become unequally distributed on either side of the dialyzing membrane. This resultant unequal distribution of ions and small molecules as a result of hindrance of dialysis by colloids is called the *Donnan effect.* The Donnan effect is involved in absorption and secretion and in the maintenance of differential concentrations of ions and small molecules between various compartments of the body. The human kidney is a place where the differences of concentrations of ions and small molecules on the two sides of dialyzing membranes are both very large and strictly controlled.

Artificial kidneys, which have been in use for about 25 years, have dialyzing membranes that form minute channels for blood (Section 31.1) on one side and a dialyzing fluid on the other side. The thin cellulose membranes allow all constituents of blood except plasma proteins and cellular components to diffuse freely in both directions. The net transfer of constituents across the membranes depends upon the concentration differential of each constituent and on the time of contact of blood with the membranes and the dialyzing fluid. Thus, the concentrations in the dialyzing fluid and the time of contact can be tailored to the needs of each individual using the artificial kidney.

7.7 Concentration of Solutions

An essentially infinite number of unsaturated solutions (Section 7.3) can be prepared for any given solute in any given solvent. Properties of these solutions such as the salty taste of a NaCl solution, the intensity of the blue color of a $CuSO_4$ solution, or the freezing point or osmotic pressure of whole blood, depend on the quantitative composition of the solutions. The **concentration** of a solution specifies *the composition in terms of the amount of solute dissolved in a given amount of solvent or in a given amount of solution.* Concentrations are commonly stated in terms of molarity.

a. Molarity

The **molarity** *of a solute* is *the number of moles of solute per liter of solution.* Molarity is designated by an italicized capital *M*. A sugar solution prepared by dissolving one mole of sucrose in enough water to make one liter of *solution* is 1.00 molar and is labeled 1.00 *M*. A 0.32 *M* sucrose solution has 0.32 mole of sucrose per liter of solution.

The molarity of a solute can be determined if we know how much solute was dissolved in a specific volume of solution.

Example 7.1 What is the molarity of a sugar solution prepared by dissolving 85.5 grams of sucrose ($C_{12}H_{22}O_{11}$) in enough water to give 1.00 liter of solution?

It is left to you as an exercise to determine that the molar mass of a $C_{12}H_{22}O_{11}$ molecule is 342 g/mole.

Calculate the number of moles of $C_{12}H_{22}O_{11}$ molecules dissolved.

$$\text{Number of moles of } C_{12}H_{22}O_{11} \text{ molecules} = \frac{85.5 \not{g}}{342 \not{g}/\text{mole}}$$

$$= 0.250 \text{ mole of } C_{12}H_{22}O_{11} \text{ molecules}$$

This is the number of moles of sucrose molecules in 1.00 liter; therefore, the sucrose solution is 0.250 *M*.

Example 7.2 What is the molarity of a sodium chloride solution prepared by dissolving 11.7 grams of NaCl in enough water to give 80 ml of solution?

Calculate the number of moles of NaCl formulas dissolved.

$$\text{Number of moles of NaCl formulas} = \frac{11.7 \not{g}}{58.5 \not{g}/\text{mole}}$$

$$= 0.200 \text{ mole of NaCl formulas}$$

Calculate the molarity of NaCl by dividing the number of moles of NaCl formulas by the volume of solution in liters.

$$\text{Molarity of NaCl} = \frac{0.200 \text{ mole}}{0.080 \text{ liter}} = 2.5 \text{ mole/liter} = 2.5\ M$$

In many cases, including the one above, the molarity indicates what the solution was made from, but does not actually specify what the solution contains. If the nature of the dissolved solute is known, the molarities of actual species in solution can be determined.

Example 7.3 **Recognizing that sodium chloride is a strong electrolyte and dissociates essentially 100% in solution, determine the molarities of the species in solution in Example 7.2.**

Sodium chloride dissociates in solution according to the following equation.

$$NaCl \longrightarrow Na^+ + Cl^-$$

One mole of NaCl formulas dissociates to one mole of Na^+ ions and one mole of Cl^- ions. Thus, 2.5 M NaCl produces 2.5 M Na^+ and 2.5 M Cl^-. A formula in brackets, for example $[Na^+]$, is often used to represent the concentration of the indicated species in solution. Therefore, we can write $[Na^+] = [Cl^-] = 2.5\ M$.

These results also could have been obtained from data in Example 7.2 by calculating the percent composition of Na^+ and Cl^- in NaCl, the mass of Na^+ ions and Cl^- ions used, the number of moles of Na^+ ions and Cl^- ions used, and the molarities of Na^+ and Cl^-. This is left to you as an exercise. Review Example 4.11 if you experience difficulty.

The mass of solute required to prepare a given volume of solution of a given concentration can be deduced by reversing the order of calculations in Examples 7.1 and 7.2.

Example 7.4 **How many grams of potassium sulfate are required to prepare 200 ml of a 0.225 M K_2SO_4 solution?**

Calculate the number of moles of K_2SO_4 formulas required by multiplying the molarity of the solution by the volume of the solution.

$$\begin{aligned}\text{Number of moles of } K_2SO_4 \text{ formulas} &= (0.225 \text{ mole}/\cancel{\text{liter}})(0.200\ \cancel{\text{liter}}) \\ &= 0.0450 \text{ mole of } K_2SO_4 \text{ formulas}\end{aligned}$$

Calculate the mass of K_2SO_4 required.

$$\begin{aligned}\text{Mass of } K_2SO_4 &= (0.0450\ \cancel{\text{mole}})(174 \text{ g}/\cancel{\text{mole}}) \\ &= 7.83 \text{ g } K_2SO_4\end{aligned}$$

Many solutes that will later be used in solution are stored conveniently as solutions. When a certain amount of the solute is needed, a portion of the sample can be measured out using any of the vessels shown in Figure 1.8. As long as the molarity of a solution is known, a specific number of moles of a solute can be obtained by measuring out an appropriate volume.

Example 7.5 What volume of a 15.0 *M* aqueous ammonia (NH_3) solution should be measured out so that the resulting volume contains 0.45 mole of dissolved NH_3 molecules?

Calculate the volume of aqueous NH_3 required by dividing the number of moles of NH_3 molecules required by the molarity of the aqueous NH_3 solution.

$$\text{Volume of aqueous NH}_3\text{ required} = \frac{0.45\ \cancel{\text{mole}}}{15.0\ \cancel{\text{moles}}/\text{liter}} = 0.030\text{ liter or }30\text{ ml}$$

Dilution is *the process in which more solvent is added to a solution so as to reduce the concentration of solute in the original solution.* Often it is necessary to prepare a solution of a required molarity by dilution of a specific volume of a more concentrated solution. From the definition of molarity, the molarity of a solution is inversely proportional to the volume. This relationship is utilized when performing calculations involving dilution.

Example 7.6 To what volume should 30.0 ml of 12.0 *M* acetic acid, CH_3COOH, be diluted to produce a 1.50 *M* acetic acid solution?

Decreasing the molarity of the solution requires an increase in the volume of the solution, or a dilution. Set up an appropriate ratio of the initial and final molarities such that when multiplied by the initial volume the ratio causes the initial volume to increase.

$$\text{Final volume of acetic acid} = (30.0\text{ ml})\left(\frac{12.0\ \cancel{M}}{1.50\ \cancel{M}}\right) = 240\text{ ml}$$

Our calculations indicate that to prepare a 1.50 *M* acetic acid solution from a 12.0 *M* acetic acid solution, enough water must be added to 30.0 ml of the 12.0 *M* solution to make the final *total* volume equal to 240 ml. In other words, approximately 210 ml of water must be added, only approximately because volumes of solutions on mixing are usually not precisely additive.

Example 7.7 What volume of 15.7 *M* nitric acid (HNO_3) should be used to prepare 400 ml of 6.00 *M* nitric acid?

The final molarity of the known final volume of solution is less than the initial molarity. Volume is inversely proportional to molarity; therefore, the known final volume is greater than the unknown initial volume. Conversely, the unknown initial volume is less than the known final volume. Set up an appropriate ratio of the initial and final molarities such that when multiplied by the final volume the ratio causes the final volume to decrease.

$$\text{Initial volume of nitric acid} = (400\text{ ml})\left(\frac{6.00\ \cancel{M}}{15.7\ \cancel{M}}\right) = 153\text{ ml}$$

Thus, 153 ml of 15.7 *M* nitric acid should be diluted with enough water to make the final *total* volume equal to 400 ml in order to prepare 400 ml of 6.00 *M* nitric acid.

The concentrations of solutions for which exact molarities are not important are frequently expressed in terms of mass percent or volume percent.

b. Mass Percent

Mass percent is *the percent of the total mass of a solution contributed by the solute.* It is obtained by dividing the mass of the solute by the total mass of the solution and multiplying the result by 100% to convert to percent.

$$\text{Mass percent} = \frac{\text{mass of solute}}{\text{total mass of solution}} \times 100\%$$

Hydrogen peroxide sold in drugstores as an antiseptic is typically 3% H_2O_2 by mass, meaning that there are 3 grams of dissolved H_2O_2 per 100 grams of solution.

$$\text{Mass percent} = \left(\frac{3\,\cancel{g}}{100\,\cancel{g}}\right)(100\%) = 3\%$$

c. Volume Percent

Volume percent is *the percent of the total volume of a solution contributed by the solute.* It is obtained by dividing the volume of the solute by the total volume of the solution and multiplying the result by 100% to convert to percent.

$$\text{Volume percent} = \frac{\text{volume of solute}}{\text{total volume of solution}} \times 100\%$$

Rubbing alcohol that is 8% by volume represents a solution prepared by adding 8 ml of alcohol to sufficient water to bring the total volume to 100 ml. Note that the volumes of solute and solvent usually are not additive.

$$\text{Volume percent} = \left(\frac{8\,\cancel{ml}}{100\,\cancel{ml}}\right)(100\%) = 8\%$$

d. Parts per Million (ppm)

Parts per million is a unit for expressing very small concentrations. It is designed to avoid the writing of very small decimal numbers. It is useful, for example, in expressing the concentrations of pollutants in air or water. Very small concentrations in gas mixtures are usually expressed in parts per million by volume, defined as

$$\textbf{ppm} = \frac{\textit{volume of pure gas}}{\textit{total volume of gas mixture}} \times 10^6\,\textit{ppm}$$

Thus, if 100 liters of a gas mixture over a metropolitan area contains 0.0040 liter of carbon monoxide, the CO is present at 40 ppm by volume.

$$\text{ppm} = \left(\frac{0.0040\ \cancel{\text{liter}}}{100\ \cancel{\text{liters}}}\right)(10^6\ \text{ppm}) = 40\ \text{ppm}$$

Very small concentrations of ions in water are more commonly expressed in parts per million by mass, defined similarly as

$$\textbf{ppm} = \frac{\textit{mass of pure ion}}{\textit{total mass of water solution}} \times 10^6\,\textit{ppm}$$

Thus, if 1000 grams of Hudson River water contain 0.0054 gram of calcium ion, the Ca^{2+} is present at 5.4 ppm by mass.

$$\text{ppm} = \left(\frac{0.0054\ \cancel{\text{g}}}{1000\ \cancel{\text{g}}}\right)(10^6\ \text{ppm}) = 5.4\ \text{ppm}$$

Exercises

7.1 Define clearly the following terms.

(a) solution
(b) solvent
(c) solute
(d) aqueous solution
(e) nonaqueous solution
(f) alloy
(g) miscible
(h) immiscible
(i) solubility
(j) soluble
(k) insoluble
(l) unsaturated solution
(m) saturated solution
(n) supersaturated solution
(o) dissolve
(p) solute–solute interactions
(q) solute–solvent interactions
(r) solvent–solvent interactions

(s) polar solvent
(t) dissociation
(u) hydrated ion
(v) electrolyte
(w) nonelectrolyte
(x) strong electrolyte
(y) weak electrolyte
(z) colligative properties
(aa) vapor pressure lowering
(ab) boiling point elevation
(ac) freezing point depression
(ad) semipermeable membrane
(ae) osmotic membrane
(af) osmosis
(ag) osmotic pressure
(ah) reverse osmosis
(ai) suspension
(aj) colloid
(ak) dispersed phase
(al) continuous phase
(am) liquid foam
(an) solid foam
(ao) liquid aerosol
(ap) liquid emulsion
(aq) solid emulsion
(ar) solid aerosol
(as) gel
(at) solid sol
(au) Tyndall effect
(av) adsorption
(aw) dialysis
(ax) dialyzing membrane
(ay) Donnan effect
(az) concentration
(ba) molarity
(bb) dilution
(bc) mass percent
(bd) volume percent
(be) parts per million

7.2 Pure water and an unsaturated solution of NaBr are both homogeneous and colorless. Cite several methods by which you could experimentally distinguish between them.

7.3 Explain the following on the basis of the kinetic molecular theory applied to solutions.
(a) A gas dissolves more readily in another gas than in a liquid.
(b) A solid dissolves in a liquid more readily than it dissolves in another solid.

7.4 How could you determine whether a given clear green solution of nickel(II) sulfate ($NiSO_4$) is saturated or unsaturated?

7.5 Describe what processes occur, in terms of interactions, when these events take place.
(a) NaI, a strong electrolyte, dissolves in water.
(b) Acetone, CH_3COCH_3, a nonelectrolyte, dissolves in water.
(c) Carbon tetrachloride, CCl_4, dissolves in the nonpolar solvent benzene, C_6H_6.

7.6 Predict the order of solubility of the substances N_2, KBr, and CH_3OH in each of the following solvents. Show clearly your reasoning.
(a) H_2O
(b) CH_3CH_2OH
(c) CCl_4

7.7 Referring to Figure 7.2, describe two different methods by which you could obtain some crystals of KNO_3 if you had a KNO_3 solution at room temperature that contained 30 grams of KNO_3 per 100 grams of water.

7.8 Classify each of the following as a strong electrolyte, a weak electrolyte, or a nonelectrolyte, and indicate for each what species are predominant in aqueous solution.
(a) $CoCl_2$
(b) H_2S
(c) acetone (CH_3COCH_3)
(d) HNO_3
(e) $CuSO_4$

7.9 Explain clearly why each of the following occurs.

(a) A warm bottle of a soft drink effervesces more vigorously than a cold bottle.
(b) Flowers placed in a sugar solution quickly wilt due to dehydration, but just as quickly regain their freshness when transferred to pure water.
(c) Certain colloids are ideal as selective adsorbers of poisonous gases in gas masks.
(d) Salt water can be converted to pure water by reverse osmosis.
(e) A nonvolatile solute is required to produce vapor pressure lowering or boiling point elevation of a solvent.

7.10 Calculate the molarity of solute in a solution prepared by dissolving the following.

(a) 0.450 mole of NaBr formulas in enough water to produce 100 ml of solution
(b) 114 grams of $C_{12}H_{22}O_{11}$ in enough water to produce 750 ml of solution
(c) 20.2 grams of KNO_3 in enough water to produce 400 ml of solution
(d) 0.60 mole of NH_3 molecules in enough water to produce 0.25 liter of solution
(e) 14.3 grams of $CuSO_4$ in enough water to produce 110 ml of solution

7.11 Calculate the mass percent of solute in the following solutions.

(a) 5.0 grams of C_2H_5OH in 80.0 grams of water
(b) 0.45 gram of K_2SO_4 in 100 grams of water
(c) 0.0382 gram of I_2 in 50.0 grams of CCl_4
(d) 0.050 mole of $C_{12}H_{22}O_{11}$ molecules in 200 grams of water
(e) 2.00×10^{-4} mole of LiCl formulas in 75 grams of CH_3OH

7.12 Calculate the volume percent of solute in each of the following solutions.

(a) 5.0 ml of CH_3OH in enough water to produce 40.0 ml of solution
(b) 9.00 ml of H_2O_2 in enough water to produce 40 ml of solution
(c) 2.48 ml of Br_2 in enough CCl_4 to produce 120 ml of solution
(d) 1.0 liter of ethylene glycol (CH_2OHCH_2OH) in enough water to produce 7.5 liters of solution
(e) 18 ml of C_2H_5OH in enough water to produce 120 ml of solution

7.13 Calculate the number of moles of solute molecules (or formulas) in each of the following.

(a) 0.250 liter of a 6.0 *M* NH_3 solution
(b) 20 ml of a 0.200 *M* Na_2SO_4 solution
(c) 120 ml of a 3.00 *M* $HC_2H_3O_2$ solution
(d) 35 grams of a 12% by mass NaCl solution
(e) 20 ml of a 0.0250 *M* sucrose solution

7.14 Calculate the mass of sodium hydroxide (NaOH) in each of the following.

(a) 50 grams of a 10% by mass NaOH solution
(b) 2.25 liters of a 6.00 *M* NaOH solution
(c) 30.5 ml of a 0.100 *M* NaOH solution
(d) 24 grams of a 7.0% by mass NaOH solution
(e) 0.35 liter of a NaOH solution prepared by dissolving 40 grams of NaOH in enough water to produce 400 ml of solution

7.15 Calculate the mass of solute required to prepare each of the following solutions.

(a) 200 ml of a 0.500 *M* NaOH solution
(b) 45 ml of a 6.0 *M* KOH solution
(c) 4.0 liters of a 0.300 *M* $ZnCl_2$ solution
(d) 80 grams of a 20% by mass NaCl solution
(e) 500 ml of a 0.100 *M* $AgNO_3$ solution

7.16 Making use of Figure 7.2, determine the following.
- **(a)** the approximate molarity of a NaCl solution that is saturated at 25°
- **(b)** the approximate temperature at which saturated solutions of KCl and KNO_3 have the same molarity
- **(c)** the mass percent of a $NaNO_3$ solution that is saturated at 20°
- **(d)** the approximate temperature at which saturated solutions of NaCl and $KClO_3$ have the same mass percent
- **(e)** the approximate molarity of a Na_2SO_4 solution that is saturated at 80°

7.17 Determine the volume to which each of the following solutions should be diluted to produce the solution indicated.
- **(a)** a 6.0 *M* HCl solution from 35 ml of a 12.0 *M* HCl solution
- **(b)** a 0.250 *M* $MnCl_2$ solution from 40.0 ml of a 1.50 *M* $MnCl_2$ solution
- **(c)** a 0.100 *M* HCl solution from 50.0 ml of a 12.0 *M* HCl solution
- **(d)** a 0.12 *M* KI solution from 23.0 ml of a 4.0 *M* KI solution
- **(e)** a 1.5 *M* $HC_2H_3O_2$ solution from 100 ml of a 6.0 *M* $HC_2H_3O_2$ solution

7.18 Determine what volume of a 12.0 *M* HCl solution is required to produce each of the following solutions.
- **(a)** a solution that contains 2.50 moles of HCl molecules
- **(b)** 300 ml of a 0.100 *M* HCl solution
- **(c)** 40 ml of a 2.0 *M* HCl solution
- **(d)** a solution that contains 7.2 grams HCl
- **(e)** 4.0 liters of a solution that contains 0.18 gram HCl per liter

7.19 Assuming that all of the following are strong electrolytes, calculate the molarity of Cl^- ion in a solution prepared by mixing the following.
- **(a)** 100 ml of a 12.0 *M* HCl solution with enough water to produce 2.5 liters of solution
- **(b)** 4.5 grams of $CaCl_2$ in enough water to produce 100 ml of solution
- **(c)** 25 ml of 0.100 *M* NaCl solution with enough water to produce 0.70 liter of solution
- **(d)** 0.350 mole of RbCl formulas in enough water to produce 50.0 ml of solution
- **(e)** 4.5 ml of a 0.100 *M* $CrCl_3$ solution and 20.5 ml of a 0.025 *M* $BaCl_2$ solution (assume the volumes are additive)

7.20 Calculate each of the following.
- **(a)** the parts per million by volume of NO_2 in 15 liters of an air sample containing 0.24 ml of NO_2
- **(b)** the parts per million by volume of neon in the atmosphere if 100 liters of an air sample contains 0.182 ml of neon
- **(c)** the volume of CO required to produce a 50-liter air test sample that is 20 ppm by volume CO
- **(d)** the parts per million by volume of O_2 in a mixture of 1.0 ml of O_2 in 68 liters of argon
- **(e)** the volume of SO_2 required to produce a 12-liter air test sample that is 75 ppm by volume SO_2

7.21 Calculate each of the following.
- **(a)** the parts per million by mass of O_2 in 1.6 kilograms of lake water containing 0.015 gram of O_2
- **(b)** the parts per million by mass of I_2 if 0.0030 gram of I_2 is dissolved in enough CCl_4 to produce 6.0 kilograms of solution

(c) the mass of CO_2 in 100 grams of river water containing 613 ppm by mass CO_2
(d) the parts per million by mass of Ba^{2+} in 500 grams of river water containing 0.0043 gram of $BaSO_4$
(e) the mass of ethylene dichloride required in a 150-gram corn oil mixture that is to be 5.0 ppm by mass ethylene dichloride for feeding rats in a carcinogenesis bioassay

7.22 Arrange letters for the following solutions in order of decreasing Cl^- concentrations. Show clearly your calculations and reasoning.
(a) a solution originally *half*-saturated in KCl at 10° (see Figure 7.2)
(b) a 0.238 *M* $CaCl_2$ solution
(c) a solution prepared by dissolving 32.0 grams of NaCl in enough water to produce 100 ml of solution
(d) a solution that is 1.0% by mass HCl in water
(e) a solution that contains 485 ppm by mass Cl^- in water

8

Chemical Reactions

In Chapter 4 we used the reduction of lead(IV) oxide by hydrogen

$$PbO_{2(s)} + 2\,H_{2(g)} \longrightarrow Pb_{(s)} + 2\,H_2O$$

to illustrate the meaning of a balanced chemical equation (Section 4.12), we used the fermentation of glucose ($C_6H_{12}O_6$)

$$C_6H_{12}O_6 \xrightarrow{\text{zymase}} 2\,C_2H_5OH + 2\,CO_{2(g)}$$

to present stoichiometric calculations involving chemical reactions (Section 4.13), and we used the preparation of aspirin ($C_9H_8O_4$) from salicylic acid ($C_7H_6O_3$) and acetic anhydride ($C_4H_6O_3$)

$$C_7H_6O_3 + C_4H_6O_3 \longrightarrow C_9H_8O_4 + C_2H_4O_2$$

to discuss the calculation of theoretical and actual yields (Section 4.14). In Chapter 7 we used the dissociation of hydrochloric acid

$$HCl \longrightarrow H^+ + Cl^-$$

to illustrate the meaning of electrolytes (Section 7.4). These are examples of chemical reactions. In this book we encounter only a very small fraction of the known chemical reactions. However, by studying various classes of chemical reactions we can often predict the probable products of and the feasibility of a particular new reaction by grouping it with reactions about which we already know.

The purposes of this chapter are to classify chemical reactions into (1) reactions in which there is no electron transfer and (2) reactions in which there is a transfer of electrons from one atom to another atom, and to give numerous examples of each type. Other classifications (for example, acid-base reactions) will be presented within the context of the above two types. A method in addition to that of inspection (Section 4.12) will be presented for balancing equations in which electron transfer occurs.

Reactions Involving No Electron Transfer

8.1 Dissociation and Precipitation

Many reactions involve simple formation or breakage of ionic or covalent bonds with no net transfer of electrons between chemical species. One example is the formation of ions in aqueous solution by dissolving a soluble salt such as white magnesium sulfate.

$$MgSO_{4(s)} \longrightarrow Mg^{2+} + SO_4{}^{2-}$$

Another example is the removal of ions from water by precipitating an insoluble salt such as white barium carbonate.

$$Ba^{2+} + CO_3{}^{2-} \longrightarrow BaCO_{3(s)}$$

Like $MgSO_4$, nearly all salts that are soluble in water, both inorganic and organic, dissociate into ions upon dissolving in water. On the other hand, like $BaCO_3$, all salts that are insoluble in water, both inorganic and organic, can be formed by the precipitation of appropriate ions from solution. The driving force for such a reaction is the formation of a stable insoluble product. Therefore, it is important that you know the solubility rules given in Table 7.1 and use them as a guide for determining when dissociation or precipitation reactions occur. For example, iron(II) chloride is soluble in water, and dissociates completely

$$FeCl_{2(s)} \longrightarrow Fe^{2+} + 2\ Cl^-$$

whereas iron(II) sulfide is an insoluble salt and precipitates when the two ions are present in solution.

$$Fe^{2+} + S^{2-} \longrightarrow FeS_{(s)}$$

Recall that ions such as Mg^{2+}, Ba^{2+}, Fe^{2+}, $SO_4{}^{2-}$, $CO_3{}^{2-}$, Cl^-, and S^{2-}, appearing alone in a chemical equation, undergo ion–dipole interactions with water (Figure 7.1) and are assumed to be hydrated. Other examples of dissociation reactions were presented in the discussion of electrolytes (Section 7.4) and will be introduced again in Section 8.2 and in Chapter 10. Two additional important classes of chemical reactions that involve no electron transfer are acid-base reactions and polymerization reactions.

8.2 Acid-Base Reactions

One method of defining acids and bases is according to characteristic properties exhibited by water solutions of acids and bases. Acidic solutions have a sour taste as imparted by the acetic acid component in vinegar or the oxalic acid component in rhubarb. Acidic solutions change the color of vegetable dye litmus from blue to red. Absorbent paper impregnated with vegetable dye litmus can thus serve as an *indicator* for the presence of acids. On the other hand, basic solutions have a bitter taste and the slippery, soapy feeling characteristic of liquid detergents or household ammonia. Basic solutions change the color of vegetable dye litmus from red to blue. Absorbent paper impregnated with vegetable dye litmus can thus also serve as an indicator for the presence of bases.

a. Arrhenius Acids and Bases

Another method of defining acids and bases was suggested by Svante Arrhenius, a Swedish chemist, in 1884. It is based upon the species produced when acids or bases are added to water. **Acids** are *substances that on addition to water produce H^+ ions or increase the H^+ concentration.* Nitric acid (HNO_3) is a soluble strong electrolyte (Section 7.4) or **strong acid** (Section 10.3) that *dissociates completely in water to produce H^+ ions.*

$$HNO_3 \longrightarrow H^+ + NO_3^-$$

Nitrous acid (HNO_2) is a soluble weak electrolyte (Section 7.4) or **weak acid** (Section 10.3 and Table 10.3) that *dissociates only partially to produce H^+ ions.*

$$HNO_2 \rightleftharpoons H^+ + NO_2^-$$

A single arrow $\longrightarrow$ is used for reactions that go completely to the right. A double arrow $\rightleftharpoons$ is used for reactions that go only partially to the right and are reversible reactions in dynamic equilibrium (Section 10.1). It should be emphasized that the bare proton (H^+) does not exist in solution as such; we only write it this way for simplicity. The H^+ ion undergoes significant ion–dipole interactions with water and is therefore hydrated (Figure 7.1).

Bases, according to the Arrhenius definition, are *substances that on addition to water produce OH^- ions or increase the OH^- concentration.* Sodium hydroxide (NaOH) is a **strong base** that *dissociates completely to produce OH^- ions.*

$$NaOH \longrightarrow Na^+ + OH^-$$

Aqueous ammonia is a **weak base** that *reacts partially to produce OH^- ions.*

$$NH_3 + H_2O \rightleftharpoons NH_4^+ + OH^-$$

Acids, according to the Arrhenius definition, need not contain hydrogen; they only need to produce H^+ ions when added to water. Likewise, Arrhenius bases need not contain a hydroxyl group (OH), but must produce OH^- ions when added to water. Hydrolysis reactions illustrate these statements. **Hydrolysis** is *the reaction of a cation, anion, or molecule with water to produce H^+ or OH^- and a weak electrolyte or weakly dissociated species* (either a complex ion or a molecule). The formation of the weak electrolyte or weakly dissociated species is the driving force for the reaction. Hydrolysis of cations and certain molecular species produces H^+ and tends to make a solution more acidic. For example, ground water can become acidic by reacting with carbon dioxide (CO_2) from the atmosphere. Hydrogen carbonate ion (HCO_3^-) is formed in addition to H^+ ions.

$$CO_{2(g)} + H_2O \rightleftharpoons HCO_3^- + H^+$$

Moreover, an aqueous solution prepared by dissolving white iron(III) nitrate, $Fe(NO_3)_3$, in water is acidic by virtue of the following reaction.

$$Fe^{3+} + H_2O \rightleftharpoons FeOH^{2+} + H^+$$

Hydrolysis of an anion always produces OH^- and tends to make a solution more basic. For example, aqueous solutions prepared by dissolving white sodium fluoride (NaF) or white potassium phosphate (K_3PO_4) in water are basic by virtue of the following reactions.

$$F^- + H_2O \rightleftharpoons HF + OH^-$$
$$PO_4^{3-} + H_2O \rightleftharpoons HPO_4^{2-} + OH^-$$

It is often not obvious from a chemical formula whether a substance is an acid or a base. Only experimental behavior can define with certainty whether a substance is an acid, a base, neither, or even both as illustrated by possible reactions for the HS^- ion.

$$\underset{\text{(as an acid)}}{HS^-} \rightleftharpoons H^+ + S^{2-}$$

$$\underset{\text{(as a base)}}{HS^-} + H_2O \rightleftharpoons H_2S + OH^-$$

In general, formulas for inorganic acids are written so that available protons appear first in the formulas (Tables 7.3 and 10.3); however, even this can be deceiving. For example, the formula for phosphorous acid is written as

H_3PO_3 rather than as $P(OH)_3$ or some other way. Unfortunately the H_3PO_3 formula suggests three available acidic protons; experimentally there are only two, a fact easily understood only by examining the structure and bonding in the H_3PO_3 molecule. Formulas for organic acids may or may not be written in such a way that available protons appear first in the formulas (Tables 7.3 and 10.3). Similar difficulties persist for inorganic and organic bases.

b. Neutralization

One of the simplest and most important reactions of acids and bases is called neutralization. It is frequently stated that **neutralization** involves *reaction of an acid and a base to produce water and a salt.* This statement is true for reactions not run in aqueous solution, for example, the reaction of gaseous hydrogen chloride and powdered sodium hydroxide to produce water and solid sodium chloride.

$$HCl_{(g)} + NaOH_{(s)} \longrightarrow H_2O + NaCl_{(s)}$$

Note that the salt consists of the cation of the base and the anion of the acid. In this case the NaCl is actually isolated because there is too little water formed in the reaction to dissolve it. However, when *solutions* of any strong acid like HCl and any strong base like NaOH are brought together, the net reaction is quite different. All of the substances in the above equation except water are strong electrolytes, and the species present in solution (except for water) are the respective dissociated ions. Therefore, the equation is written as

$$H^+ + Cl^- + Na^+ + OH^- \longrightarrow H_2O + Na^+ + Cl^-$$

where all ions are assumed to be hydrated. Moreover, Na^+ and Cl^- cancel because they appear on both sides of the equation and do not take part in the reaction. The net equation becomes

$$H^+ + OH^- \longrightarrow H_2O$$

and the driving force for this neutralization reaction is the formation of the weak electrolyte, H_2O. Note that the salt, NaCl, is not isolated from the neutralization reaction; it can be obtained only by further efforts after the neutralization reaction, such as removal of the excess water by evaporation. Neutralization reactions are exothermic (Section 1.2-b). The above reaction represents the neutralization in aqueous solution of any strong acid by a strong base as can be shown by the experimental observation that the amount of heat evolved per mole of H^+ or OH^- reacting is the same, 13.4 kcal, regardless of the identity of the strong acid or strong base.

Acid-base reactions involving a weak acid and a strong base or a strong acid and a weak base are represented by net equations in which the weak acid or base is written in its undissociated form. For example, reaction of the weak acid, hydrocyanic acid (HCN), with the strong base, sodium hydroxide, can be written as

$$HCN + Na^+ + OH^- \longrightarrow H_2O + Na^+ + CN^-$$

or, after canceling Na^+, the net equation is

$$HCN + OH^- \longrightarrow H_2O + CN^-$$

In this reaction H_2O and CN^- are weaker electrolytes than HCN and OH^-; thus, there is a driving force from the side of the stronger electrolytes toward the side of the weaker electrolytes. Moreover, the reaction of the strong acid, hydrochloric acid, with the weak base, trimethylamine [$(CH_3)_3N$], can be written as

$$H^+ + Cl^- + (CH_3)_3N \longrightarrow (CH_3)_3NH^+ + Cl^-$$

or, after canceling, the net equation is

$$H^+ + (CH_3)_3N \longrightarrow (CH_3)_3NH^+$$

Acid-base reactions are used frequently to determine quantitatively the concentration or amount of acid or base present in a mixture. For example, in the United States it is unlawful to sell oranges until they have ripened enough to have about an 8 : 1 ratio of total soluble solids to acids in the fruit juice. Thus, the amount of acid in the orange juice must be determined periodically near picking time. This is accomplished by a process called *titration*. A previously standardized sodium hydroxide solution is added dropwise to a given volume of orange juice until an appropriate indicator in the orange juice shows that excess base has been added. Simple stoichiometric calculations then give the concentration of acid.

Example 8.1 Assume that the only acid present in orange juice is citric acid ($C_6H_8O_7$) with three acidic protons. Calculate the molar concentration of citric acid in an orange juice sample if 35.0 ml of 0.120 *M* NaOH is required to neutralize the acid in 50.0 ml of the orange juice.

The net equation for the reaction of NaOH with citric acid is

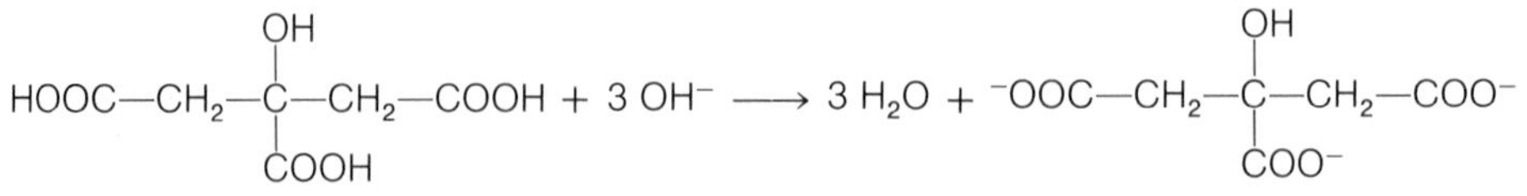
OH
HOOC—CH2—C—CH2—COOH + 3 OH− ⟶ 3 H2O + −OOC—CH2—C—CH2—COO−
COOH
OH
COO−

It can be seen from the equation that one mole of citric acid reacts for every three moles of OH^-

Calculate the number of moles of OH^- ions used by multiplying the molarity of NaOH by the volume in liters of NaOH used.

$$\text{Number of moles of } OH^- \text{ ions} = (0.120 \text{ mole}/\cancel{\text{liter}})(0.0350 \cancel{\text{liter}})$$
$$= 0.00420 \text{ mole of } OH^- \text{ ions}$$

One third as many moles of citric acid, or 0.00140 mole, is required.

Calculate the molarity of citric acid by dividing the number of moles of citric acid by the volume of orange juice in which the citric acid is present.

$$\text{Concentration of citric acid} = \frac{0.00140 \text{ mole}}{0.0500 \text{ liter}} = 0.0280\ M \text{ citric acid}$$

For many problems involving acid-base reactions, including most of those of biochemical interest, the Arrhenius definition of acids and bases in terms of the production of H^+ or OH^- in water solution is satisfactory. However, two more general models that expand the range of acid-base reactions are also encountered. These models were introduced almost simultaneously in 1923 by Johannes Brønsted, a Danish chemist, and Thomas Lowry, a British chemist, and by Gilbert Lewis, an American chemist.

c. Brønsted–Lowry Acids and Bases

The Brønsted–Lowry method of defining acids and bases depends upon the way acids and bases react with each other. An acid-base reaction is one in which there is a proton transfer from one species to another. **Acids** are *species that give up or donate a proton;* **bases** are *species that accept a proton.* Stated simply, an acid is a proton donor, and a base is a proton acceptor. A glance at previous examples in this section will show that all Arrhenius acids are also Brønsted–Lowry acids and all Arrhenius bases are also Brønsted–Lowry bases; however, the converse of these statements may not be true. Reactions involving water provide illustrations. In the dissociation of benzoic acid (C_6H_5COOH),

$$\underset{\text{Acid}}{C_6H_5COOH} + \underset{\text{Base}}{H_2O} \rightleftharpoons C_6H_5COO^- + H_3O^+$$

water acts as a Bronsted–Lowry base even though no OH^- is produced and water is not an Arrhenius base. In the reaction of methylamine (CH_3NH_2) with water,

$$\underset{\text{Acid}}{H_2O} + \underset{\text{Base}}{CH_3NH_2} \rightleftharpoons CH_3NH_3^+ + OH^-$$

water acts as a Brønsted–Lowry acid even though no H^+ is produced and water is not an Arrhenius acid. The two previous reactions do not proceed very far toward the products on the right because the reactants are weaker electrolytes than the products, thus providing a driving force toward the left.

It is not necessary that a water molecule be one of the reactants in a Brønsted–Lowry acid-base reaction. One important application of Brønsted–Lowry theory is to gas phase reactions, for example, the reaction of hydrogen chloride and ammonia resulting in the formation of ammonium chloride.

$$\underset{\text{Acid}}{HCl_{(g)}} + \underset{\text{Base}}{NH_{3(g)}} \longrightarrow NH_4Cl_{(s)}$$

Another very important application of Brønsted–Lowry theory is to reactions occurring in nonaqueous solvents, for example, the reaction of the hydride ion (H^-) with liquid ammonia.

$$\underset{\text{Acid}}{NH_3} + \underset{\text{Base}}{H^-} \longrightarrow H_{2(g)} + NH_2^-$$

In the first reaction NH_3 acts as a Brønsted–Lowry base by accepting a proton from HCl to form the NH_4^+ ion of NH_4Cl. In the second reaction NH_3 acts as a Brønsted–Lowry acid by giving up a proton to form the H_2 molecule. The driving forces for the two previous reactions are the formation of a crystalline solid in the NH_4Cl case and the formation of an insoluble gas in the H_2 case. Note that none of the species would be classified as Arrhenius acids or bases in these two particular reactions.

d. Lewis Acids and Bases

The Lewis method of defining acids and bases depends upon the way acids and bases react with each other, but is not limited to proton transfer reactions. Thus, it is a further extension of the Brønsted–Lowry model. The only restriction according to the Lewis concept is that acid-base reactions involve an electron pair transfer. Thus, any reaction in which one atom donates both electrons that form the shared-pair covalent bond (Section 3.8-a) is an acid-base reaction. **Lewis acids,** then, are *species that accept an electron pair;* **Lewis bases** are *species that give up or donate an electron pair.* Stated simply, an acid is an electron pair acceptor, and a base is an electron pair donor. Examination of previous examples in this section will show that all Arrhenius acids and Brønsted–Lowry acids are also Lewis acids and that all Arrhenius bases and Brønsted–Lowry bases are also Lewis bases; however, the converse of these statements may not be true. Consider the reaction of boron trifluoride (BF_3) with ammonia to form the aminotrifluoroborane adduct (F_3BNH_3).

$$\underset{\text{Acid}}{BF_3} + \underset{\text{Base}}{:NH_{3(g)}} \longrightarrow F_3BNH_{3(s)}$$

The BF_3 molecule acts as a Lewis acid in this reaction because boron has only six electrons in its outer shell (Section 3.8-a) and is a good electron pair acceptor. It is not, however, a Brønsted–Lowry acid or an Arrhenius acid in this reaction. On the other hand, NH_3 functions as a Lewis base in the above reaction, but is not a Brønsted–Lowry base or an Arrhenius base in this reaction.

A very important application of Lewis theory is to reactions of complex ions (Section 3.8-a). For example, if aqueous ammonia is added to a solution containing Cu^{2+}, deep blue tetraamminecopper(II) ion is formed.

$$\underset{\text{Acid}}{Cu^{2+}} + \underset{\text{Base}}{4\ :NH_3} \rightleftharpoons Cu(NH_3)_4{}^{2+}$$

The Cu^{2+} ion acts as an electron pair acceptor, and NH_3 acts as an electron pair donor. Moreover, neutral complexes can also be formed. Metal carbonyls, composed of a transition metal atom bonded to molecules of carbon monoxide, provide the most interesting examples. For example, tetracarbonylnickel(0) is formed by passing carbon monoxide over elemental nickel at 60°.

$$Ni_{(s)} + 4\ :CO_{(g)} \longrightarrow Ni(CO)_4$$

Reactions of complex ions are intimately involved in carbon monoxide poisoning. In oxyhemoglobin a central Fe^{2+} ion is bonded to six other groups, one of which is O_2. Unfortunately, hemoglobin forms a considerably more stable complex with CO than with O_2, and the complex with CO is formed preferentially even at CO concentrations as low as one part per thousand. Oxygen is not carried to the tissues, and muscular paralysis and death can eventually result (Section 31.4).

O2
Fe
+ CO(g)
CO
Fe
+ O2(g)
Oxyhemoglobin Carboxyhemoglobin

e. Important Inorganic Acids and Bases

A discussion of acids and bases would not be complete without mention of the preparations and properties of a number of important inorganic acids and bases.

Hydrochloric acid, an aqueous solution of hydrogen chloride gas, is produced by first forming hydrogen chloride by reaction of sodium chloride with sulfuric acid

$$2\ NaCl_{(s)} + H_2SO_4 \xrightarrow{\text{heat}} 2\ HCl_{(g)} + Na_2SO_4$$

or by burning hydrogen (Section 5.5-a) in chlorine (Section 5.5-i)

$$H_{2(g)} + Cl_{2(g)} \longrightarrow 2\ HCl_{(g)}$$

followed by bubbling hydrogen chloride into water to form a concentrated aqueous solution. Hydrochloric acid is a colorless liquid that fumes in moist air and has the sharp, pungent odor of HCl gas. It is a strong acid (Section 8.2-a) and is used to prepare chloride salts, to manufacture aniline dyes, and to clean iron surfaces before galvanizing or tin plating.

Nitric acid is prepared commercially by the Ostwald process involving the catalytic oxidation of ammonia (Section 5.5-e)

$$4\ NH_{3(g)} + 5\ O_{2(g)} \longrightarrow 4\ NO_{(g)} + 6\ H_2O_{(g)}$$

followed by further oxidation of nitric oxide, NO,

$$2\ NO_{(g)} + O_{2(g)} \longrightarrow 2\ NO_{2(g)}$$

and hydration of nitrogen dioxide, NO_2 (Section 5.5-f).

$$3\ NO_{2(g)} + H_2O \longrightarrow 2\ HNO_3 + NO_{(g)}$$

Nitric acid was the tenth chemical in terms of total production in the United States in 1978. It is a colorless liquid, but often becomes pale yellow on standing from the light-catalyzed formation of NO_2.

$$4\ HNO_3 \xrightarrow{\text{light}} 4\ NO_{2(g)} + 2\ H_2O + O_{2(g)}$$

Nitric acid fumes in moist air and has a characteristic choking odor. It is both a strong acid (Section 8.2-a) and a good oxidizing agent (Section 8.4). It stains proteins (Chapter 25), for example, woolen fabrics and animal tissues, a bright yellow. It is used in the manufacture of dye intermediates, explosives, nitrogen compounds for fertilizers, and many different organic chemicals.

Phosphoric acid is prepared commercially by reaction of calcium phosphate with sulfuric acid.

$$Ca_3(PO_4)_{2(s)} + 3\ H_2SO_4 \longrightarrow 2\ H_3PO_4 + 3\ CaSO_{4(s)}$$

Phosphoric acid was the ninth chemical in terms of total production in the United States in 1978. It is a colorless liquid and a medium strength acid. It is used primarily in the manufacture of fertilizers and of polyphosphates and phosphate salts.

Sulfuric acid is prepared commercially by the catalytic oxidation of sulfur dioxide (Section 5.5-h)

$$2\ SO_{2(g)} + O_{2(g)} \longrightarrow 2\ SO_{3(g)}$$

followed by reaction between SO_3 and H_2SO_4

$$SO_{3(g)} + H_2SO_4 \longrightarrow H_2S_2O_7$$

and a hydration step to form sulfuric acid.

$$H_2S_2O_7 + H_2O \longrightarrow 2\ H_2SO_4$$

Sulfuric acid was the top chemical in terms of total production in the United States in 1978. It is a colorless, odorless, oily liquid. It is a strong acid (Section 8.2-a) and has a great affinity for and reacts violently and exothermically with water. It is very corrosive toward all body tissues. It is used in the production of fertilizers, explosives, dyestuffs, and other acids.

Sodium hydroxide is prepared commercially by the electrolysis (Section 8.8) of aqueous sodium chloride solutions.

$$2\ Na^+ + 2\ Cl^- + 2\ H_2O \xrightarrow{\text{electrolysis}} H_{2(g)} + Cl_{2(g)} + 2\ Na^+ + 2\ OH^-$$

Sodium hydroxide was the eighth chemical in terms of total production in the United States in 1978. Sodium hydroxide is a white solid that is readily soluble in water with the generation of considerable heat. It is a strong base (Section 8.2-a) and is corrosive to animal and vegetable tissues. It is used to neutralize acids in many chemical processes as well as to hydrolyze fats and form soaps. Sodium hydroxide is used in products like Drano to dissolve proteins (Section 25.9-c) such as hair.

Aqueous ammonia is prepared by dissolving ammonia (Section 5.5-e) in water. It is a colorless liquid with an intense, pungent, suffocating odor. It is a weak base (Section 8.2-a) that is used for bleaching and removing stains and for manufacturing ammonium salts.

8.3 Polymerization Reactions

Polymers (Greek, *poly,* many; *meros,* parts) are *compounds that consist of many small repeating structural units*. For example, nylon 6–10, the name for one of a whole series of high-molar-mass polyamides, consists of a large number of $C_{16}H_{30}N_2O_2$ units.

$$\begin{array}{cc} \parallel & \parallel \\ +C(CH_2)_8CNH(CH_2)_6NH + C(CH_2)_8CNH(CH_2)_6NH + C(CH_2)_8CNH(CH_2)_6NH + \end{array}$$

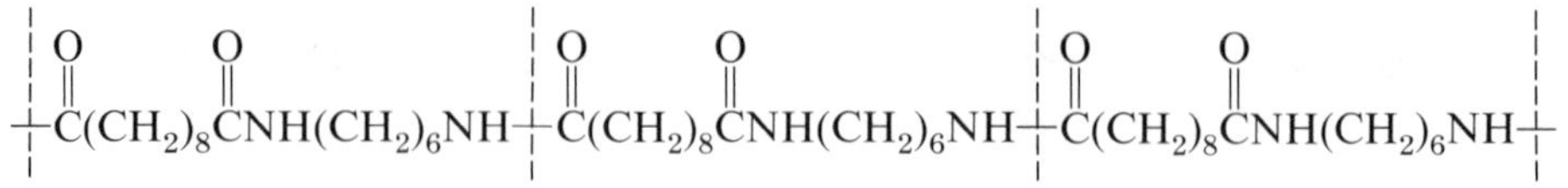

$C_{16}H_{30}N_2O_2$
Repeating unit

Synthetic polymers can be prepared from small molecules called monomers. In some cases, except for the shift of a bond, the repeating unit in the polymer is identical in chemical composition to the small molecule from which it is derived. In other cases, the repeating unit in the polymer resembles the monomers only in the sense that it can be constructed from the monomers. Thus, nylon 6–10, so named because one monomer has six carbon atoms and the other has ten carbon atoms, is prepared from sebacoyl chloride and hexamethylenediamine. The repeating unit is constructed from one of each monomer.

$$n\ ClC(CH_2)_8CCl\ +\ n\ \begin{array}{c} H \\ N(CH_2)_6N \\ H \end{array} \begin{array}{c} H \\ H \end{array} \longrightarrow 2n\ HCl\ +\ \left[\begin{array}{cc} O & O \\ \parallel & \parallel \\ C(CH_2)_8CNH(CH_2)_6NH \end{array} \right]_n$$

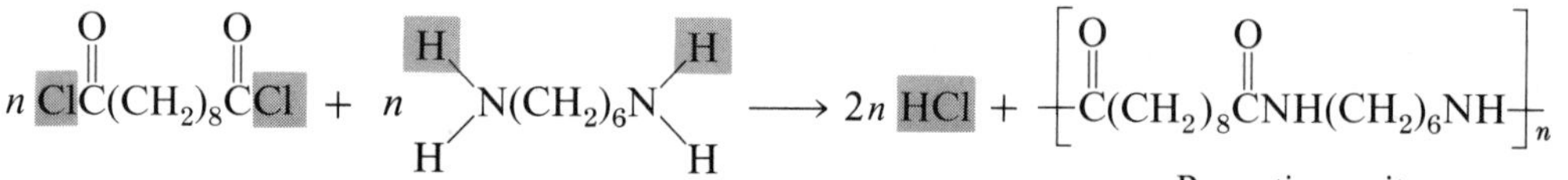

Sebacoyl chloride   Hexamethylenediamine   Repeating unit

Polymers are readily distinguishable from the small molecules from which they are synthesized. Besides the fundamental differences in molar mass, structure, and bonding, polymers usually exhibit different physical and chemical properties. The small molecules tend to be volatile liquids of low viscosity, or even gases at room temperature. Polymers, on the other hand, tend to be nonvolatile, highly viscous liquids, glasses, or solids that soften only at high temperatures.

Polymers have great practical importance because of the useful properties that they exhibit. For example, nylon 6–10 can be extruded from a melt as monofilaments, or spun from a solution of formic acid and phenol. The resulting fibers have low density, are elastic and lustrous, and mass for mass, are stronger than steel. However, the fibers have the disadvantages of being low melting and difficult to dye. Natural materials such as proteins (Chapter 25), natural rubber, cellulose, starch (Chapter 23) and complex silicate minerals are polymers. Artificial materials such as fibers, films, plastics, semisolid resins, synthetic rubbers, and silicate glasses are also polymers. Included among these artificial materials are Teflon, Styrofoam, Melmac,

Orlon, Kodel, Mylar, and Saran Wrap, as well as others that will be discussed.

Polymers are formed by two general methods: addition polymerization and condensation polymerization.

a. Addition Polymerization

Addition polymerization is *a process in which small molecules join together to form polymers that have the same percent composition as the original small molecules.* For example, under appropriate conditions, one molecule of ethylene ($CH_2{=}CH_2$) adds to another, and the addition process continues to produce a very long chain of $—CH_2—CH_2—$ units called polyethylene (Section 14.4-d). A double bond in the original small molecule is converted to two single bonds in the product. Two single bonds have greater bond strength and bond energy than one double bond.

$$n\,CH_2{=}CH_2 \xrightarrow{\text{catalyst}} \underbrace{\left[CH_2CH_2 \right]_n}_{\text{Repeating unit}}$$

As a result, the above reaction, typical of addition polymerizations, is exothermic. Considerable heat is evolved in some addition polymerizations.

Polyethylene is an example of a *linear* polymer (in the form of a long chain). Such polymers are usually soluble in suitable solvents. Since they also tend to soften on heating, linear polymers are called *thermoplastic polymers.*

Lucite, or Plexiglas, is a linear polymer resulting from the addition polymerization of methyl methacrylate in the presence of benzoyl peroxide initiator.

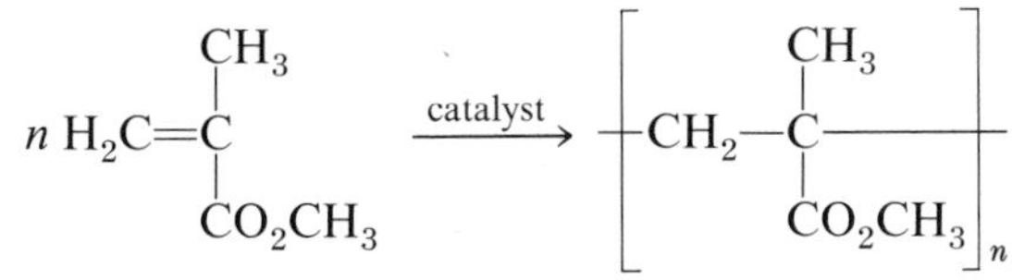

CH3
n H2C=C
CO2CH3
catalyst
CH3
CH2—C
CO2CH3
n

Methyl methacrylate Repeating unit

The oxygen–oxygen bond in benzoyl peroxide is weak, and bond rupture by heating or by ultraviolet light yields a *species having an unpaired electron* and called a **free radical** (Sections 13.4–13.5). Such a free radical

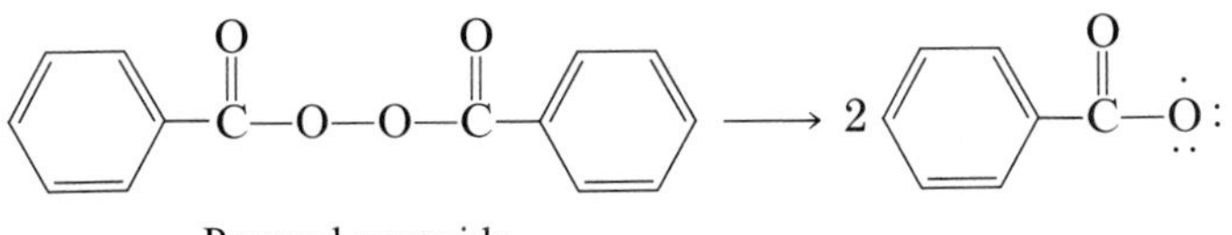

O
O
C—O—O—C
2
O
C—O:

Benzoyl peroxide

adds rapidly to a carbon–carbon double bond and forms a new highly reactive free radical intermediate that is capable of adding to another monomer unit. In this way the polymerization reaction is catalyzed by free radicals such as those produced by decomposition of benzoyl peroxide.

If addition polymerization is carried out using a mixture of two different small molecules, *copolymers* containing two kinds of monomers in every polymer molecule are formed. For example, Dynel is a copolymeric fiber resulting from addition polymerization of 50 percent acrylonitrile ($CH_2{=}CHCN$) and 50 percent vinyl chloride ($CH_2{=}CHCl$). No simple chemical formula can be used to represent the copolymer because the acrylonitrile and vinyl chloride molecules add together randomly to form the polymer.

b. Condensation Polymerization

Condensation polymerization is *a process in which polymers are produced as a result of the elimination of small, volatile molecules in each step of the process.* For example, under appropriate conditions, one molecule of terephthalic acid (see Section 19.2-a for synthesis) and one molecule of ethylene glycol condense to form a molecule of water and a species that can react

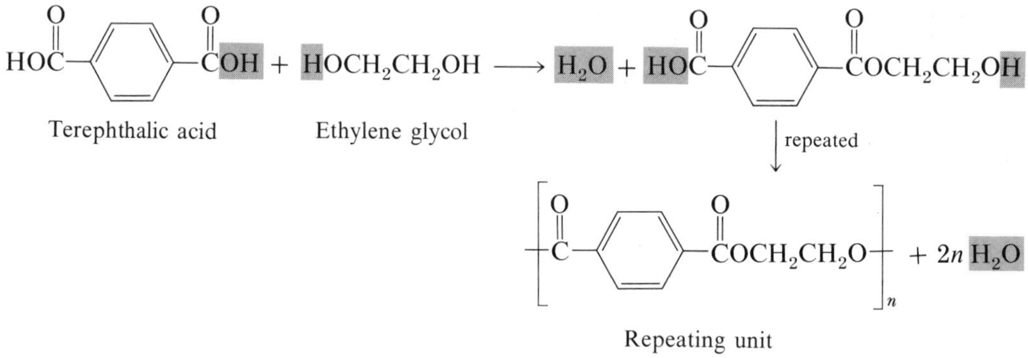
Terephthalic acid
Ethylene glycol
repeated
Repeating unit

repeatedly in the same fashion to form a linear polymer. The very important synthetic fiber Dacron is produced from this linear polymer. For condensation polymerization to occur, small molecules containing more than one reactive site (shaded in the above equation) are required so that simple coupling reactions can take place at several sites on each reactant. Condensation reactions are usually endothermic and are forced to completion by removing the volatile condensation product (water in the above example) from the reaction mixture.

Glyptal resin is a polymer resulting from condensation polymerization of polyalcohols such as glycerol with anhydrides such as phthalic anhydride.

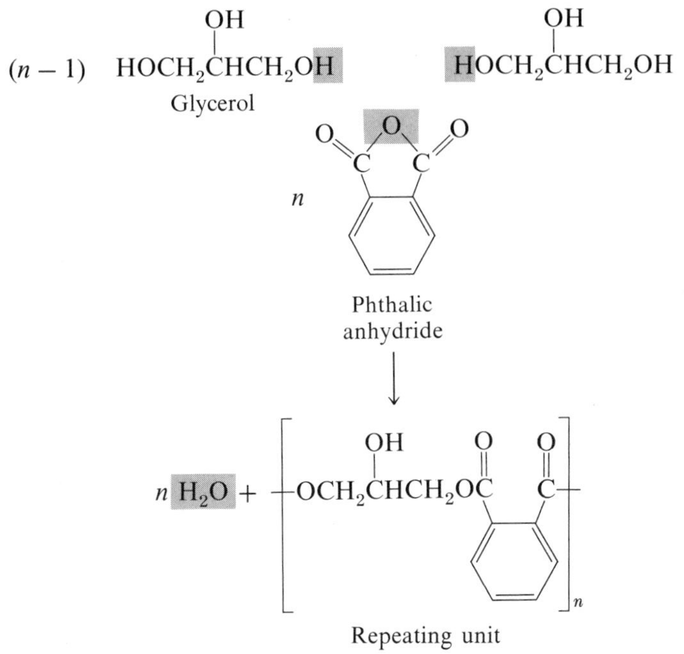

OH
(n − 1) HOCH2CHCH2OH
Glycerol
OH
HOCH2CHCH2OH
n
Phthalic
anhydride
n H2O +
OH
OCH2CHCH2OC
O
O
C
n
Repeating unit

Such resins are often used in the manufacture of paints and enamels.

Silicones are polymers resulting from condensation polymerization of alkyl- and aryl-substituted silicon dihydroxides. For example, dimethylsilicondihydroxide undergoes rapid, spontaneous condensation polymerization to form a methyl-substituted silicone. Silicones have unique properties

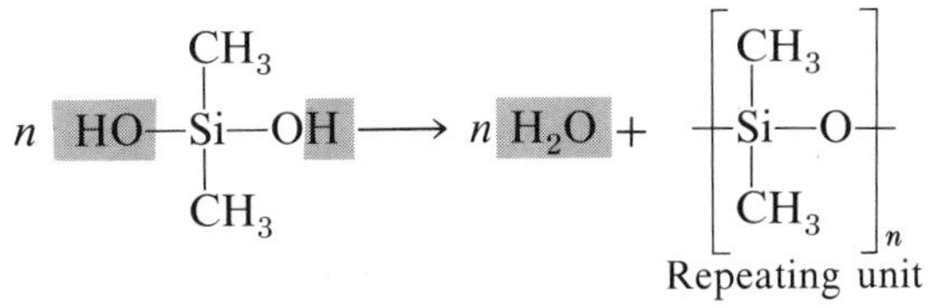

CH3
n HO—Si—OH
CH3
n H2O +
CH3
Si—O
CH3
n
Repeating unit

as a result of their silicon–oxygen backbone coupled with carbon atoms only in substituent groups. The strong Si—O bonds make silicones resistant to oxidation over a wide range of temperatures. Organic substituents at the surface of the polymer cause silicones to be oils that are highly water repellent. These properties make them useful for waterproofing leather, various fabrics, and paper products, as well as for lubricants, particularly in engines that operate at both low and high temperatures.

The primary structure of proteins (Section 25.4), a term that covers a great variety of biological macromolecules, can be understood in terms of condensation polymerization. Polyglycine can be made by condensation

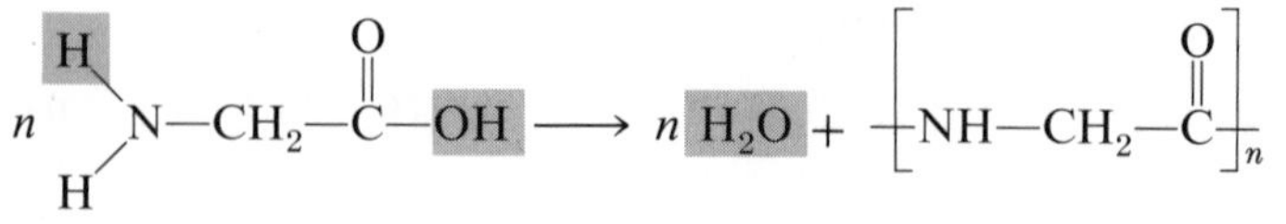

$$n \quad \overset{H}{\underset{H}{\diagdown}}N-CH_2-\overset{O}{\overset{\|}{C}}-OH \longrightarrow n\ H_2O + \left[NH-CH_2-\overset{O}{\overset{\|}{C}} \right]_n$$

Glycine   Repeating unit

polymerization of the simplest amino acid, glycine. Proteins are not strictly polymers and differ from polyglycine in that rather than being composed of a sequence of a single repeated amino acid, they are composed of many different amino acids linked together in a single macromolecule. About twenty different amino acids have been isolated from the various proteins. The structure and function of any given protein is greatly dependent upon which amino acids are present and their specific sequence.

All of the previous examples have been linear polymers. *Cross-linked polymers* may be produced by addition polymerization if reactant molecules have more than one double bond, or by condensation polymerization if reactant molecules have more than two reactive sites. The properties exhibited by the resulting polymers depend upon the degree of cross-linking, as illustrated by natural rubber. Natural rubber, formed from the latex of certain trees or shrubs, is a polymer composed of a large number of isoprene

$$\left[CH_2-\overset{CH_3}{\overset{|}{C}}=\overset{H}{\overset{|}{C}}-CH_2 \right]_n$$

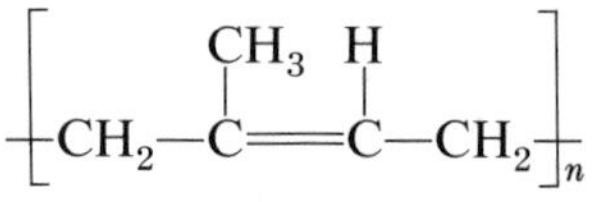

Repeating unit

(C_5H_8) units (Section 14.5). When unvulcanized (simply the above linear polymer), it has an irregular, noncrystalline structure. The molecules are coiled and intertwined, and unvulcanized rubber is very sticky. Examination of the above repeating unit shows that a double bond is still available for addition polymerization. On heating with a carefully controlled amount of sulfur in the presence of appropriate catalysts, the three-dimensional, cross-linked structure of vulcanized rubber (a portion of which is shown below) is formed.

$$-CH_2-\overset{CH_3}{\overset{|}{C}}=CH-CH_2-CH_2-\overset{CH_3}{\overset{|}{\underset{|}{C}}}-\overset{S}{\overset{|}{\underset{|}{CH}}}-CH_2-$$

$$-CH_2-\overset{CH_3}{\overset{|}{\underset{|}{C}}}-\overset{S}{\overset{|}{\underset{|}{CH}}}-CH_2-CH_2-\overset{CH_3}{\overset{|}{C}}=CH-CH_2-$$

$$\overset{S}{|}$$

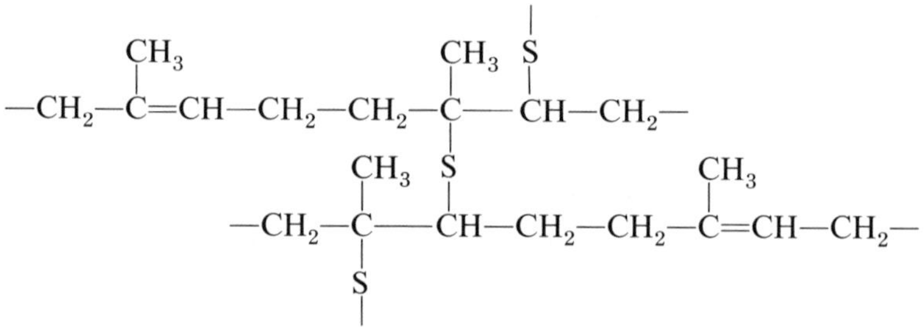

This polymer is only partially cross-linked; some double bonds still remain. However, it has the desirable properties of stretch and bounce that we attribute to most forms of rubber. If, on the other hand, additional sulfur is introduced, further cross-linking occurs with the formation of an extensive three-dimensional network. The resulting material no longer has the typical rubber-like properties; instead it is a hard, insoluble, high melting solid. Such network polymers that have extensive cross-linking are called *thermosetting polymers*.

8.4 Oxidation and Reduction

Reactions Involving Electron Transfer

Historically, *oxidation* meant reaction of a substance with oxygen, and *reduction* meant removal of oxygen from a substance. Oxidation included combustion of an element such as yellow sulfur to produce colorless sulfur dioxide (SO_2),

$$S_{(s)} + O_{2(g)} \longrightarrow SO_{2(g)}$$

or roasting of a compound such as white zinc sulfide (ZnS) to produce white zinc oxide (ZnO) and SO_2.

$$2\ ZnS_{(s)} + 3\ O_{2(g)} \longrightarrow 2\ ZnO_{(s)} + 2\ SO_{2(g)}$$

Reduction included reaction of an oxide with another substance, for example, reduction of black copper(II) oxide (CuO) by hydrogen to produce copper metal and water,

$$CuO_{(s)} + H_{2(g)} \longrightarrow Cu_{(s)} + H_2O$$

or thermal decomposition of an oxide such as red mercury(II) oxide (HgO) to produce metallic mercury and oxygen.

$$2\ HgO_{(s)} \xrightarrow{\text{heat}} 2\ Hg + O_{2(g)}$$

More recently, it has been recognized that most of the changes caused by combination of a substance with O_2 or by removal of O_2 from a substance need not involve O_2 at all, but can be accomplished by treatment of the same substances with other reagents. Thus, comparing with two reactions given above, zinc oxide can be produced by heating zinc sulfide with nitric acid rather than O_2, yielding yellow sulfur, reddish brown nitrogen dioxide (NO_2), and water as additional products,

$$ZnS_{(s)} + 2\ HNO_3 \longrightarrow ZnO_{(s)} + S_{(s)} + 2\ NO_{2(g)} + H_2O$$

and copper metal can be made by reduction with hydrogen of yellow copper(II) chloride ($CuCl_2$) rather than copper(II) oxide.

$$CuCl_{2(s)} + H_{2(g)} \longrightarrow Cu_{(s)} + 2\,HCl_{(g)}$$

Therefore, the terms oxidation and reduction have taken on broader meanings. **Oxidation** involves *a loss of electrons by a chemical species,* as illustrated in the following half-reactions (Section 8.6).

$$\begin{aligned} Ca_{(s)} &\longrightarrow Ca^{2+} + 2\,e^- \\ Fe^{2+} &\longrightarrow Fe^{3+} + e^- \\ UO_2{}^+ &\longrightarrow UO_2{}^{2+} + e^- \\ Mn^{2+} + 2\,H_2O &\longrightarrow MnO_{2(s)} + 4\,H^+ + 2\,e^- \end{aligned}$$

The species that loses electrons is oxidized and is the **reducing agent** (also called the reductant). **Reduction** involves *a gain of electrons by a chemical species,* as illustrated in the following half-reactions.

$$\begin{aligned} Sn^{4+} + 2\,e^- &\longrightarrow Sn^{2+} \\ Br_2 + 2\,e^- &\longrightarrow 2\,Br^- \\ MnO_4{}^- + 8\,H^+ + 5\,e^- &\longrightarrow Mn^{2+} + 4\,H_2O \\ O_{3(g)} + 2\,H^+ + 2\,e^- &\longrightarrow O_{2(g)} + H_2O \end{aligned}$$

The species that gains electrons is reduced and is the **oxidizing agent** (also called the oxidant).

Oxidation and reduction are complementary processes. They always occur simultaneously. One chemical species loses electrons, and a second chemical species gains those electrons. This simultaneous occurrence results in an oxidation-reduction reaction. Thus, an **oxidation-reduction reaction** involves *a transfer of electrons from one chemical species to another.* It may involve a complete transfer of electrons resulting in the formation of ionic bonds (Section 3.3). Reaction of potassium with bromine can be regarded as resulting from the transfer of an electron from potassium to bromine.

$$2\,K\cdot_{(s)} + :\ddot{\underset{..}{Br}}:\ddot{\underset{..}{Br}}:_{(g)} \longrightarrow 2\,K^+ \left[:\ddot{\underset{..}{Br}}:\right]^-$$

Alternatively, an oxidation-reduction reaction may involve only partial transfer of electrons resulting in the formation of covalent bonds. For example, the reaction of hydrogen with bromine can be regarded as resulting from the partial transfer of an electron from hydrogen toward bromine. Since

$$H:H_{(g)} + :\ddot{\underset{..}{Br}}:\ddot{\underset{..}{Br}}:_{(g)} \longrightarrow 2\,H:\ddot{\underset{..}{Br}}:_{(g)}$$

bromine is more electronegative, the shared electron pair in HBr is not shared equally, and a polar covalent bond (Section 3.7) results.

8.5 Oxidation Numbers

An oxidation number (sometimes called oxidation state) is a convenient device for keeping track of electron transfer in oxidation-reduction reactions. An **oxidation number** is the *charge that an atom **appears** to have* when electrons are counted according to two arbitrary rules: (1) electrons shared between two like atoms are divided equally between the two atoms, and (2) electrons shared between two unlike atoms are assigned to the more electronegative atom. An oxidation number may be zero, or a whole or fractional positive or negative number. It should be emphasized that oxidation numbers are only *arbitrary* bookkeeping devices that do *not* represent actual charges of atoms. A distortion, a polarization, or a redistribution of electron density occurs in all chemical species so that the *effective* charge of any atom is seldom greater than ±1.

If electrons are assigned according to the above rules, the oxidation number of a given atom is calculated by taking the number of outer shell electrons of that neutral atom minus the number of outer shell electrons actually surrounding that atom. For example, dividing the shared electrons of N_2 equally, the oxidation number of each N is $5 - 5 = \mathbf{0}$; assigning all the

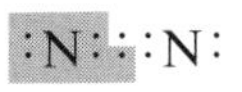

shared electrons in NH_3 to the more electronegative N atom, the oxidation number of N in NH_3 is $5 - 8 = \mathbf{-3}$, and the oxidation number of H in NH_3

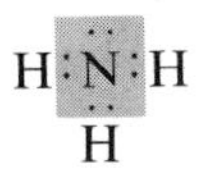

is $\mathbf{+1}$. Application of these ideas to many chemical species leads to the following less laborious conventions for assigning oxidation numbers.

1. *The oxidation number of an atom in a free element is* **zero,** no matter how complicated the formula of the element—for example, zero for C in $C_{(s)}$, Cl in $Cl_{2(g)}$, P in $P_{4(s)}$, and S in $S_{8(s)}$.
2. *The oxidation number of a monatomic ion is* **its charge**—for example, $+1$ for Li in Li^+, $+2$ for Ni in Ni^{2+}, $+3$ for Al in Al^{3+}, $+4$ for Th in Th^{4+}, -1 for Br in Br^-, and -2 for S in S^{2-}.
3. *The oxidation number of combined hydrogen is* $\mathbf{+1}$, except that it is -1 in hydrides of the more active metals and boron (LiH, CaH_2,

$LiAlH_4$, and KBH_4), in which H is bonded to a less electronegative atom.

4. *The oxidation number of combined oxygen is* **−2,** except that it is −1 in peroxides (for example, Na_2O_2 and H_2O_2), $-\frac{1}{2}$ in superoxides (for example, KO_2), and +1 and +2 in O_2F_2 and OF_2, respectively, in which O is bonded to a more electronegative atom.
5. *The most common oxidation number of each atom in a compound is* **+ the group number** for elements in groups I, II, and III—for example, +1 for K in KBr, +2 for Mg in Mg_3N_2, and +3 for Al in Al_2S_3.
6. Oxidation numbers of any elements that do not fit the above rules are assigned by keeping in mind that *all oxidation numbers must be consistent with the conservation of charge.* The sum of the oxidation numbers of all atoms in a neutral molecule must be zero, and the sum of the oxidation numbers of all atoms in a complex ion must equal the charge on that ion—for example, −2 for O and +2 for C in CO, and thus 0 for Mn in $Mn(CO)_{5(s)}$; −1 for F and thus +2 for Mn in $MnF_{2(s)}$; +4 for C and −5 for N in CN^-, and thus +3 for Mn in $Mn(CN)_6^{3-}$; −2 for O and thus +4 for Mn in $MnO_{2(s)}$; −2 for O and thus +6 for Mn in MnO_4^{2-}; and −2 for O and thus +7 for Mn in MnO_4^-.

The oxidation number of an atom is written *underneath* the atom in a formula so as to avoid confusion with the charge on an ion. Thus, oxidation numbers of Mn in the formulas above can be represented as follows.

$Mn(CO)_5$	MnF_2	$Mn(CN)_6^{3-}$	MnO_2	MnO_4^{2-}	MnO_4^-
0	+2	+3	+4	+6	+7

a. Oxidizing and Reducing Agents

Oxidizing and reducing agents can also be described in terms of changes of oxidation number. The **oxidizing agent** *contains the atom that shows a decrease in oxidation number.* The oxidizing agent is the substance that *does* the oxidizing and *is* reduced. The **reducing agent** *contains the atom that shows an increase in oxidation number.* The reducing agent is the substance that *does* the reducing and *is* oxidized. These concepts can be illustrated using the primary reaction that occurs when gasoline is burned in a limited amount of air, the combustion of octane (C_8H_{18}) with oxygen to produce colorless carbon monoxide (CO) and water vapor.

$$\underset{-\frac{9}{4}\ \ +1}{2\,C_8H_{18(g)}} + \underset{0}{17\,O_{2(g)}} \longrightarrow \underset{+2\ -2}{16\,CO_{(g)}} + \underset{+1\ -2}{18\,H_2O_{(g)}}$$

Oxidation numbers for each of the atoms are assigned in the conventional manner. It can be seen that C atoms increase in oxidation number from $-\frac{9}{4}$ to $+2$. Thus, octane is oxidized, and is the reducing agent. At the same time, O atoms decrease in oxidation number from 0 to -2. Thus, oxygen is reduced, and is the oxidizing agent.

In some chemical reactions a single chemical species acts as both the oxidizing agent and the reducing agent. The chemical species contains in some instances one atom that shows both an increase and a decrease in oxidation number. For example, in the spontaneous oxidation and reduction of pale blue nitrous acid (HNO_2) solutions to produce colorless nitric oxide (NO), red-brown nitrogen dioxide (NO_2), and water, N both increases and decreases in oxidation number.

$$\underset{+3}{2\,H\mathbf{N}O_2} \longrightarrow \underset{+2}{\mathbf{N}O_{(g)}} + \underset{+4}{\mathbf{N}O_{2(g)}} + H_2O$$

In effect, one molecule of HNO_2 is oxidized and the other is reduced. Likewise, in the spontaneous oxidation and reduction of Cu^+ in aqueous solution to produce copper metal and a blue solution of Cu^{2+}, one Cu^+ gains an electron that is lost by the other Cu^+.

$$\underset{+1}{2\,\mathbf{Cu}^+} \longrightarrow \underset{0}{\mathbf{Cu}_{(s)}} + \underset{+2}{\mathbf{Cu}^{2+}}$$

In other instances the chemical species may contain one atom that increases in oxidation number and another atom that decreases in oxidation number. For example, in the thermal decomposition of white ammonium nitrate (NH_4NO_3) to produce colorless nitrous oxide (N_2O) and water, one N increases in oxidation number whereas the other N decreases in oxidation number.

$$\underset{-3\ \ +5}{\mathbf{N}H_4\mathbf{N}O_{3(s)}} \xrightarrow{\text{heat}} \underset{+1}{\mathbf{N}_2O_{(g)}} + 2\,H_2O$$

Likewise, in the thermal decomposition of orange ammonium dichromate [$(NH_4)_2Cr_2O_7$] to produce the appearance of a miniature volcano during formation of green chromium(III) oxide (Cr_2O_3), nitrogen, and water, N increases in oxidation number and Cr decreases in oxidation number.

$$\underset{-3\qquad +6}{(\mathbf{N}H_4)_2\mathbf{Cr}_2O_{7(s)}} \xrightarrow{\text{heat}} \underset{+3}{\mathbf{Cr}_2O_{3(s)}} + \underset{0}{\mathbf{N}_{2(g)}} + 4\,H_2O$$

b. Nomenclature of Inorganic Compounds

Inorganic compounds, with some exceptions, normally include compounds of elements other than carbon. Since two of the common conventions for naming inorganic compounds utilize oxidation numbers, it is useful to discuss inorganic nomenclature at this point.

The various kinds of chemical formulas and the meaning of each kind have already been discussed (Sections 1.1-d-2 and 4.7). A basic rule in writing formulas of inorganic compounds is that elements making up a given compound are normally written in order of *increasing electronegativity* (Section 3.7), as illustrated by the following formulas.

HCl	NO	MnO_2	$Ca_3(PO_4)_2$
NaH	SrH_2	Li_2O	$Zn(SCN)_2$
MgO	CuF_2	$KClO_3$	$NiSO_4$

The only common exceptions to this rule involve compounds containing a complex cation *and* a complex anion, for example, NH_4NO_3 and $(NH_4)_2Cr_2O_7$, discussed in Section 8.5-a.

Names of inorganic compounds composed of only two elements (called binary compounds) are derived directly from the elements. The name of the less electronegative element is written first followed by the stem of the name of the more electronegative element and the suffix *-ide*. Thus, we call NaCl sodium chloride, KBr potassium bromide, LiH lithium hydride, and SrS strontium sulfide. A number of compounds involve only two elements, but have more than one atom per formula of one or both elements. If the compounds are composed of a metal having only a single common positive oxidation number and a nonmetal, the above simple rule generally applies. Thus, we call CaH_2 calcium hydride, $BaCl_2$ barium chloride, AlF_3 aluminum fluoride, and Al_2O_3 aluminum oxide. However, for cases involving only nonmetals, Greek prefixes are used to indicate the number of atoms per formula. Thus, we call CO carbon monoxide, CO_2 carbon dioxide, N_2O_4 dinitrogen tetroxide, and OF_2 oxygen difluoride.

Although the method of Greek prefixes would normally suffice for naming compounds involving elements with varying oxidation numbers, two additional methods are encountered more frequently. The older method applies when only *two* oxidation numbers are possible for the element in question. For the compound involving the element in its lower oxidation number, the stem of the name of the element and the suffix *-ous* is used. For the compound involving the element in its higher oxidation number, the stem of the name of the element and the suffix *-ic* is used. Thus, we call CoF_2 cobaltous fluoride and CoF_3 cobaltic fluoride. Frequently the stem of the old Latin name is used, for example, in ferrous chloride ($FeCl_2$) and ferric chloride ($FeCl_3$). The newer method, recommended by The International Union of Pure and Applied Chemistry (IUPAC), applies for *any* number of possible oxidation numbers for the element in question. It uses the name of the element followed by its oxidation number written in Roman numerals and placed in parentheses. Thus, we call CoF_2 cobalt(II) fluoride (read "cobalt two fluoride"), CoF_3 cobalt(III) fluoride, $FeCl_2$ iron(II) chloride, and $FeCl_3$ iron(III) chloride. Using this method to name the oxides of tungsten, we call W_2O_3 tungsten(III) oxide, WO_2 tungsten(IV) oxide, W_2O_5 tungsten(V) oxide, and WO_3 tungsten(VI) oxide.

Unfortunately, the naming of compounds of only two elements is further complicated by three other factors. One of the problems is the fact that several methods of nomenclature may be utilized within the same series of compounds. Thus, the oxides of nitrogen are generally named as follows.

N_2O	nitrous oxide
NO	nitric oxide
N_2O_3	dinitrogen trioxide
NO_2	nitrogen dioxide
N_2O_4	dinitrogen tetroxide
N_2O_5	dinitrogen pentoxide

A second problem is the fact that for a few compounds common names are used rather than systematic names developed from the above rules. Most prevalent among such compounds are water (H_2O) and ammonia (NH_3). A third problem is the fact that aqueous acid solutions are named using the prefix *hydro-* followed by the stem of the name of the nonmetal plus the suffix *-ic* and the additional word *acid.* Thus, we call aqueous solutions of HCl hydrochloric acid and of HBr hydrobromic acid.

Compounds composed of more than two elements must have a complex cation, a complex anion, or both. The names of such compounds involve the names of the complex cations and/or anions. Names of complex cations end in *-ium*. The most common complex cation is the ammonium ion (NH_4^+). Its compounds are named according to rules mentioned previously for binary compounds. Thus, we call NH_4Cl ammonium chloride and $(NH_4)_2S$ ammonium sulfide. Names of complex anions end in *-ide* or *-ate*. Common complex anions and salts of these anions are named below.

Anion		*Compound*	
OH^-	hydroxide	$Mg(OH)_2$	magnesium hydroxide
O_2^-	peroxide	BaO_2	barium peroxide
NO_3^-	nitrate	$Cr(NO_3)_3$	chromium(III) nitrate (chromic nitrate)
CO_3^{2-}	carbonate	$CaCO_3$	calcium carbonate
SO_4^{2-}	sulfate	$FeSO_4$	iron(II) sulfate (ferrous sulfate)
PO_4^{3-}	phosphate	$(NH_4)_3PO_4$	ammonium phosphate
CN^-	cyanide	$NaCN$	sodium cyanide
SCN^-	thiocyanate	$KSCN$	potassium thiocyanate
CrO_4^{2-}	chromate	K_2CrO_4	potassium chromate
$Cr_2O_7^{2-}$	dichromate	$K_2Cr_2O_7$	potassium dichromate

Names of the aqueous acids of *-ide* complex anions use the prefix *hydro-* followed by the stem of the name of the anion plus the suffix *-ic* and the additional word *acid*. Thus, we call HCN hydrocyanic acid.

Names of the aqueous acids and the corresponding salts of *-ate* complex anions utilize the stem of the name of the anion and depend upon the oxidation number of the central element in the anion. If the central element has only a single oxidation number, the stem of the name of the central element is used with the suffix *-ic* for the acid and the suffix *-ate* for the salt. Thus, we call H_3BO_3 boric acid and Na_3BO_3 sodium borate. If the central element has only two possible oxidation numbers, the acid corresponding to the lower oxidation number adds the suffix *-ous* to the stem and the acid corresponding to the higher oxidation number adds the suffix *-ic* to the stem. Thus, we call H_2SO_3 sulfurous acid and H_2SO_4 sulfuric acid. Salts derived from these acids are named by replacing the suffix *-ous* by *-ite* and the suffix *-ic* by *-ate*. Thus, we call Na_2SO_3 sodium sulfite and $MgSO_4$ magnesium sulfate. If the central element has more than two possible oxidation numbers, the prefix *hypo-* may be used to indicate an oxidation number lower than that of the *-ous* acid, and the prefix *per-* may be used to indicate an oxidation number higher than that of the *-ic* acid. These ideas are illustrated in naming the oxyacids and oxyanion salts of chlorine.

Acid		Salt	
$HClO$	hypochlorous acid	$NaClO$	sodium hypochlorite
$HClO_2$	chlorous acid	$NaClO_2$	sodium chlorite
$HClO_3$	chloric acid	$KClO_3$	potassium chlorate
$HClO_4$	perchloric acid	$KClO_4$	potassium perchlorate

8.6 Balancing Oxidation-Reduction Equations

Chemical equations describing oxidation-reduction reactions are often too complicated to be balanced simply by inspection (Section 4.12), but are readily balanced by one of two systematic methods. The method described here is called the "half-reaction" method. It is based on splitting the reaction into an oxidation half-reaction and a reduction half-reaction, balancing the two half-reactions including electrons separately, and then adding the two half-reactions to eliminate the electrons from the final balanced equation. The following steps are used.

1. Separate the given chemical reaction into two half-reactions so that atoms of the same kind (usually other than H or O) occur in the same half-reaction. When both are balanced one will be an oxidation half-reaction, and the other will be a reduction half-reaction.
2. Balance each half-reaction separately by:
   a. Using appropriate coefficients to balance all atoms except H and O by inspection.

b. Balancing O by adding H_2O to the side deficient in O.
c. Balancing H in acid solution by adding H^+ to the side deficient in H, or, alternatively, balancing H in basic solution by adding H_2O molecules equal in number to the deficiency in H to the side deficient in H and an equal number of OH^- to the opposite side. Reactions run in neutral solution or not in solution at all are balanced as if run in acid solution providing H appears in the skeletal equation; otherwise, this step is unnecessary for such reactions.
d. Balancing electrical charge by adding electrons to the side deficient in negative charge.
e. Subtracting any species duplicated on the left and right.

3. Multiply the two half-reactions by appropriate integers necessary to make the number of electrons lost in one half-reaction equal to the number of electrons gained in the other half-reaction. Then add the two half-reactions.
4. Subtract any species duplicated on the left and right.

Example 8.2 Complete and balance the following equation describing the oxidation of oxalic acid ($H_2C_2O_4$), discussed in Section 19.6-d, by potassium permanganate ($KMnO_4$) in sulfuric acid (H_2SO_4) solution to produce manganese(II) sulfate ($MnSO_4$), potassium sulfate (K_2SO_4), carbon dioxide (CO_2), and water.

$$H_2C_2O_4 + KMnO_4 + H_2SO_4 \longrightarrow MnSO_4 + K_2SO_4 + CO_2 + H_2O$$

What is the oxidizing agent? What is the reducing agent? Which species is oxidized? Which species is reduced?

When reactants and products are given in molecular form, soluble salts and strong acids and bases should be written in dissociated form first, and then H^+, OH^-, H_2O, and all species that remain unchanged (shaded in the following equation) should be neglected when setting up the half-reactions. (See Table 7.1 for solubilities.)

$$H_2C_2O_4 + K^+ + MnO_4^- + H^+ + SO_4^{2-} \longrightarrow Mn^{2+} + SO_4^{2-} + K^+ + SO_4^{2-} + CO_2 + H_2O$$

Set up and balance the first half-reaction using the steps outlined above.

Step 1 $\quad MnO_4^- \longrightarrow Mn^{2+}$

This half-reaction shows the chemical species containing Mn.

Step 2a MnO_4^- $\longrightarrow$ Mn^{2+}

The Mn atoms are already balanced, and no coefficients are required.

Step 2b MnO_4^- $\longrightarrow$ $Mn^{2+} + \mathbf{4\ H_2O}$

Four H_2O molecules added to the right side eliminate the deficiency of O on that side.

Step 2c $MnO_4^- + \mathbf{8\ H^+}$ $\longrightarrow$ $Mn^{2+} + 4\ H_2O$

Eight H^+ added to the left side eliminate the deficiency of H on that side.

Step 2d $MnO_4^- + 8\ H^+ + \mathbf{5\ e^-}$ $\longrightarrow$ $Mn^{2+} + 4\ H_2O$

Five electrons added to the left side make the net charge on the left and right equal to +2.

Step 2e $MnO_4^- + 8\ H^+ + 5\ e^- \longrightarrow Mn^{2+} + 4\ H_2O$

No species are duplicated on both sides.

The second balanced half-reaction is obtained by the same procedure.

Step 1 $H_2C_2O_4 \longrightarrow CO_{2(g)}$

This half-reaction shows the chemical species containing C.

Step 2a $H_2C_2O_4 \longrightarrow \mathbf{2}\ CO_{2(g)}$

A coefficient of two is required before CO_2 in order to balance the C atoms.

Step 2b $H_2C_2O_4 \longrightarrow 2\ CO_{2(g)}$

The O atoms are already balanced, and no H_2O molecules are required.

Step 2c $H_2C_2O_4 \longrightarrow 2\ CO_{2(g)} + \mathbf{2\ H^+}$

Two H^+ added to the right side eliminate the deficiency of H on that side.

Step 2d $H_2C_2O_4 \longrightarrow 2\ CO_{2(g)} + 2\ H^+ + \mathbf{2\ e^-}$

Two electrons added to the right side make the net charge on the left and right equal to zero.

Step 2e $H_2C_2O_4 \longrightarrow 2\ CO_{2(g)} + 2\ H^+ + 2\ e^-$

No species are duplicated on both sides.

The two half-reactions are then added after the first one is multiplied by two and the second one is multiplied by five.

Step 3

$$
\begin{array}{rl}
2\ MnO_4^- + 16\ H^+ + 10\ e^- & \longrightarrow 2\ Mn^{2+} + 8\ H_2O \\
5\ H_2C_2O_4 & \longrightarrow 10\ CO_{2(g)} + 10\ H^+ + 10\ e^- \\
\hline
2\ MnO_4^- + 16\ H^+ + 10\ e^- + 5\ H_2C_2O_4 & \longrightarrow 2\ Mn^{2+} + 8\ H_2O + 10\ CO_{2(g)} + 10\ H^+ + 10\ e^-
\end{array}
$$

The balanced equation is obtained by subtracting duplicated species.

Step 4 $\quad 2\ MnO_4^- + 5\ H_2C_2O_4 + 6\ H^+ \longrightarrow 2\ Mn^{2+} + 10\ CO_{2(g)} + 8\ H_2O$

A balanced molecular equation is obtained from the balanced net equation by adding back the cations and anions that were neglected in the beginning. Appropriate numbers are chosen to balance charge. Thus, 2 K^+ are required to balance the charge of 2 MnO_4^-; then 2 K^+ must be added to the right side. Moreover, 3 SO_4^{2-} are required to balance the charge of 6 H^+; then 3 SO_4^{2-} must be added to the right side.

$$5\ H_2C_2O_4 + 2\ K^+ + 2\ MnO_4^- + 6\ H^+ + 3\ SO_4^{2-} \longrightarrow 2\ Mn^{2+} + 3\ SO_4^{2-} + 2\ K^+ + 10\ CO_{2(g)} + 8\ H_2O$$

The ions are now combined to form the molecular equation. Note that the SO_4^{2-} on the right side is used to form two different products.

$$5\ H_2C_2O_4 + 2\ KMnO_4 + 3\ H_2SO_4 \longrightarrow 2\ MnSO_4 + K_2SO_4 + 10\ CO_{2(g)} + 8\ H_2O$$

From the half-reactions in step 3, it can be seen that MnO_4^- gains electrons and $H_2C_2O_4$ loses electrons. Thus, MnO_4^- (or $KMnO_4$) is reduced and is the oxidizing agent, and $H_2C_2O_4$ is oxidized and is the reducing agent.

Example 8.3 Complete and balance the following equation describing the reduction of the deep blue tetraamminecopper(II) cation [$Cu(NH_3)_4^{2+}$] by the dithionite anion ($S_2O_4^{2-}$) in *basic* solution to produce copper metal, the sulfite anion (SO_3^{2-}), and ammonia (NH_3).

$$Cu(NH_3)_4^{2+} + S_2O_4^{2-} \longrightarrow Cu_{(s)} + SO_3^{2-} + NH_{3(g)}$$

What is the oxidizing agent? What is the reducing agent? Which species is oxidized? Which species is reduced?

First half-reaction

Step 1	$Cu(NH_3)_4^{2+}$	$\longrightarrow$	$Cu_{(s)} + NH_{3(g)}$
Step 2a	$Cu(NH_3)_4^{2+}$	$\longrightarrow$	$Cu_{(s)} + \mathbf{4}\ NH_{3(g)}$
Step 2b	$Cu(NH_3)_4^{2+}$	$\longrightarrow$	$Cu_{(s)} + 4\ NH_{3(g)}$
Step 2c	$Cu(NH_3)_4^{2+}$	$\longrightarrow$	$Cu_{(s)} + 4\ NH_{3(g)}$
Step 2d	$Cu(NH_3)_4^{2+} + \mathbf{2\ e^-}$	$\longrightarrow$	$Cu_{(s)} + 4\ NH_{3(g)}$
Step 2e	$Cu(NH_3)_4^{2+} + 2\ e^-$	$\longrightarrow$	$Cu_{(s)} + 4\ NH_{3(g)}$

Second half-reaction

Step 1	$S_2O_4^{2-}$	$\longrightarrow$	SO_3^{2-}
Step 2a	$S_2O_4^{2-}$	$\longrightarrow$	$\mathbf{2}\ SO_3^{2-}$
Step 2b	$S_2O_4^{2-} + \mathbf{2\ H_2O}$	$\longrightarrow$	$2\ SO_3^{2-}$
Step 2c	$S_2O_4^{2-} + 2\ H_2O + \mathbf{4\ OH^-}$	$\longrightarrow$	$2\ SO_3^{2-} + \mathbf{4\ H_2O}$
Step 2d	$S_2O_4^{2-} + 2\ H_2O + 4\ OH^-$	$\longrightarrow$	$2\ SO_3^{2-} + 4\ H_2O + \mathbf{2\ e^-}$
Step 2e	$S_2O_4^{2-} + 4\ OH^-$	$\longrightarrow$	$2\ SO_3^{2-} + 2\ H_2O + 2\ e^-$

Combination of the two half-reactions

Step 3

$$Cu(NH_3)_4^{2+} + 2\ e^- \longrightarrow Cu_{(s)} + 4\ NH_{3(g)}$$

$$S_2O_4^{2-} + 4\ OH^- \longrightarrow 2\ SO_3^{2-} + 2\ H_2O + 2\ e^-$$

$$Cu(NH_3)_4^{2+} + 2\ e^- + S_2O_4^{2-} + 4\ OH^- \longrightarrow Cu_{(s)} + 4\ NH_{3(g)} + 2\ SO_3^{2-} + 2\ H_2O + 2\ e^-$$

Step 4

$$Cu(NH_3)_4^{2+} + S_2O_4^{2-} + 4\ OH^- \longrightarrow Cu_{(s)} + 4\ NH_{3(g)} + 2\ SO_3^{2-} + 2\ H_2O$$

From the half-reactions in step 3, it can be seen that $Cu(NH_3)_4^{2+}$ gains electrons and $S_2O_4^{2-}$ loses electrons. Thus, $Cu(NH_3)_4^{2+}$ is reduced and is the oxidizing agent, and $S_2O_4^{2-}$ is oxidized and is the reducing agent.

Example 8.4 Complete and balance the following equation describing the complete oxidation of the colorless oil, nicotine ($C_{10}H_{14}N_2$), an organic, nitrogen-containing, natural product found in tobacco.

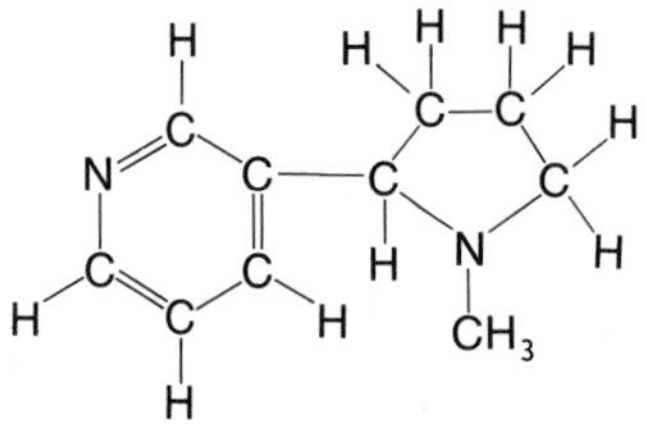

$$+ O_{2(g)} \longrightarrow CO_{2(g)} + N_{2(g)} + H_2O$$

What is the oxidizing agent? What is the reducing agent? Which substance is oxidized? Which substance is reduced?

First half-reaction

Step 1	$C_{10}H_{14}N_2$	$\longrightarrow$	$CO_{2(g)} + N_{2(g)}$
Step 2a	$C_{10}H_{14}N_2$	$\longrightarrow$	$\mathbf{10}\ CO_{2(g)} + N_{2(g)}$
Step 2b	$C_{10}H_{14}N_2 + \mathbf{20\ H_2O}$	$\longrightarrow$	$10\ CO_{2(g)} + N_{2(g)}$
Step 2c	$C_{10}H_{14}N_2 + 20\ H_2O$	$\longrightarrow$	$10\ CO_{2(g)} + N_{2(g)} + \mathbf{54\ H^+}$

Note that reactions run in neutral solution or not in solution at all are balanced as if run in acid solution providing H appears anywhere in the skeletal equation.

Step 2d	$C_{10}H_{14}N_2 + 20\ H_2O$	$\longrightarrow$	$10\ CO_{2(g)} + N_{2(g)} + 54\ H^+ + \mathbf{54\ e^-}$
Step 2e	$C_{10}H_{14}N_2 + 20\ H_2O$	$\longrightarrow$	$10\ CO_{2(g)} + N_{2(g)} + 54\ H^+ + 54\ e^-$

Second half-reaction

Step 1	$O_{2(g)}$	$\longrightarrow$	H_2O
Step 2a	$O_{2(g)}$	$\longrightarrow$	**2** H_2O
Step 2b	$O_{2(g)}$	$\longrightarrow$	$2\ H_2O$
Step 2c	$O_{2(g)}$ + **4 H^+**	$\longrightarrow$	$2\ H_2O$
Step 2d	$O_{2(g)} + 4\ H^+$ + **4 e^-**	$\longrightarrow$	$2\ H_2O$
Step 2e	$O_{2(g)} + 4\ H^+ + 4\ e^-$	$\longrightarrow$	$2\ H_2O$

Combination of the two half-reactions

Step 3

$$2\ C_{10}H_{14}N_2 + 40\ H_2O \longrightarrow 20\ CO_{2(g)} + 2\ N_{2(g)} + 108\ H^+ + 108\ e^-$$

$$27\ O_{2(g)} + 108\ H^+ + 108\ e^- \longrightarrow 54\ H_2O$$

$$2\ C_{10}H_{14}N_2 + 40\ H_2O + 27\ O_{2(g)} + 108\ H^+ + 108\ e^- \longrightarrow 20\ CO_{2(g)} + 2\ N_{2(g)} + 108\ H^+ + 108\ e^- + 54\ H_2O$$

Step 4

$$2\ C_{10}H_{14}N_2 + 27\ O_{2(g)} \longrightarrow 20\ CO_{2(g)} + 2\ N_{2(g)} + 14\ H_2O$$

From the half-reactions in step 3, it can be seen that $C_{10}H_{14}N_2$ loses electrons and O_2 gains electrons. Thus, $C_{10}H_{14}N_2$ is oxidized and is the reducing agent, and O_2 is reduced and is the oxidizing agent.

8.7 Reduction Potentials

The previous examples have illustrated how an oxidation-reduction reaction can be separated into two half-reactions for the purpose of obtaining a balanced equation. However, nothing has been said about the relative tendency for each of the half-reactions or for any net oxidation-reduction reaction to take place. Reduction potentials provide this information.

Many chemical reactions are exothermic and result in a decrease of potential energy. Usually this energy appears as heat evolved; however, by appropriate design of the system in which the reaction occurs, the energy can be made to appear as electrical energy. Oxidation-reduction reactions that occur spontaneously can be made to produce an electric current. The net voltage resulting from a pair of oxidation and reduction half-reactions can be measured experimentally. On the contrary, there is no way that the voltage for an *individual* half-reaction can be measured experimentally. Therefore, by international agreement, the reduction potential for the standard hydrogen electrode

$$2\ H^+\ (1\ \text{mole per}\ 1000\ \text{g water}) + 2\ e^- \xrightarrow{25°} H_{2(g)}\ (1\ \text{atm})$$

is arbitrarily assigned a value of 0.00 volt, and all other half-reaction reduction potentials are compared to this. Thus, a **standard reduction potential,** frequently labeled E_0, is *a measure of the* ***relative*** *tendency of a substance to gain one or more electrons compared to the standard hydrogen electrode.* Table

8.1 lists a number of half-reactions with their standard reduction potentials. The numerical values of these potentials apply only for standard conditions, that is, for gases at 25°C and 1 atm pressure and for aqueous solutions at 25°C and having a concentration of dissolved species of 1 mole per 1000 grams of water. If reactant pressure or concentration is greater than the standard value, the nonstandard potential will be more favorable or more positive. For example, if the Ag^+ concentration is 10 moles per 1000 grams of water, the reduction potential for the half-reaction $Ag^+ + e^- \longrightarrow Ag_{(s)}$ is +0.86 volt as compared to the standard reduction potential of +0.80 volt. The opposite is true if reactant pressure or concentration is less than the standard value. If product pressure or concentration is greater than the standard value, the nonstandard potential will be less favorable or less positive. The opposite is true if the product pressure or concentration is less than the standard value. The potentials given are for half-reactions proceeding as written to the right. For reverse reactions, the sign of the voltages must be changed.

As a result of the definition of reduction potential, Table 8.1 is arranged in such a way that reducing agents are listed in order of *decreasing* strength to the right of the arrows. A negative reduction potential indicates that the

Table 8.1 Some Half-reactions and Their Standard Reduction Potentials

Half-reaction	*Standard Reduction Potential, E_o (volts)*
$K^+ + e^- \longrightarrow K_{(s)}$	−2.93
$Ca^{2+} + 2\,e^- \longrightarrow Ca_{(s)}$	−2.87
$Na^+ + e^- \longrightarrow Na_{(s)}$	−2.71
$Mg^{2+} + 2\,e^- \longrightarrow Mg_{(s)}$	−2.37
$U^{3+} + 3\,e^- \longrightarrow U_{(s)}$	−1.80
$Al^{3+} + 3\,e^- \longrightarrow Al_{(s)}$	−1.66
$Zn^{2+} + 2\,e^- \longrightarrow Zn_{(s)}$	−0.76
$Fe^{2+} + 2\,e^- \longrightarrow Fe_{(s)}$	−0.44
$Pb^{2+} + 2\,e^- \longrightarrow Pb_{(s)}$	−0.13
$2\,H^+ + 2\,e^- \longrightarrow H_{2(g)}$	0.00
$Cu^{2+} + 2\,e^- \longrightarrow Cu_{(s)}$	+0.34
$I_2 + 2\,e^- \longrightarrow 2\,I^-$	+0.54
$Fe^{3+} + e^- \longrightarrow Fe^{2+}$	+0.77
$Ag^+ + e^- \longrightarrow Ag_{(s)}$	+0.80
$Br_2 + 2\,e^- \longrightarrow 2\,Br^-$	+1.07
$O_{2(g)} + 4\,H^+ + 4\,e^- \longrightarrow 2\,H_2O$	+1.23
$Cr_2O_7^{2-} + 14\,H^+ + 6\,e^- \longrightarrow 2\,Cr^{3+} + 7\,H_2O$	+1.33
$Cl_{2(g)} + 2\,e^- \longrightarrow 2\,Cl^-$	+1.36
$MnO_4^- + 8\,H^+ + 5\,e^- \longrightarrow Mn^{2+} + 4\,H_2O$	+1.51
$F_{2(g)} + 2\,e^- \longrightarrow 2\,F^-$	+2.87

species to the right of the arrow is a stronger reducing agent than H_2; a positive reduction potential indicates that the species to the right of the arrow is a weaker reducing agent than H_2. Among the chemical species listed to the right of the arrow, $K_{(s)}$ is the best reducing agent and F^- is the poorest reducing agent. In addition, oxidizing agents are listed in order of *increasing* strength to the left of the arrows. Among the chemical species listed to the left of the arrows, K^+ is the poorest oxidizing agent and F_2 is the best oxidizing agent.

The tendency for a given oxidation-reduction reaction to proceed can be determined by adding potentials for the two half-reactions composing that oxidation-reduction reaction. If the net voltage is positive, the oxidation-reduction reaction has a tendency to take place as written. If the net voltage is negative, the reverse reaction is favored. It should be emphasized, however, that the net voltage gives no indication as to whether the reaction proceeds fast enough to be observed.

Example 8.5 Determine whether iron(III) cation (Fe^{3+}) can oxidize iodide anion according to the following equation.

$$2\ Fe^{3+} + 2\ I^- \longrightarrow 2\ Fe^{2+} + I_2$$

Break the oxidation-reduction reaction into two appropriate half-reactions, find potentials for the half-reactions in Table 8.1, and add the potentials algebraically. Note that the sign of the voltage for the I_2 half-reaction is changed since the reverse reaction from that given in Table 8.1 is being considered.

$$\begin{array}{rll} Fe^{3+} + e^- \longrightarrow & Fe^{2+} & +0.77\ V \\ 2\ I^- \longrightarrow & I_2 + 2\ e^- & -0.54\ V \\ \hline & & +0.23\ V \end{array}$$

The net voltage for the reaction is positive; therefore, the reaction is predicted to (and does) proceed spontaneously to the right under standard conditions.

Many oxidation-reduction reactions can be utilized to produce electric current. They are designed into a *cell* that separates the oxidizing and reducing agents from each other so that electron transfer must occur through a wire in the form of an electric current. For practical use two or more cells are generally connected in series to form a battery. The lead storage battery used in automobiles is a very important example. Electrodes of lead and lead(IV) oxide (PbO_2), dipping into an aqueous solution of sulfuric acid (H_2SO_4), comprise the battery. The half-reactions and net reaction

$$\begin{array}{rll} Pb_{(s)} + HSO_4^- \longrightarrow & PbSO_{4(s)} + H^+ + 2\ e^- & +0.36\ V \\ PbO_{2(s)} + HSO_4^- + 3\ H^+ + 2\ e^- \longrightarrow & PbSO_{4(s)} + 2\ H_2O & +1.69\ V \\ \hline Pb_{(s)} + PbO_{2(s)} + 2\ HSO_4^- + 2\ H^+ \longrightarrow & 2\ PbSO_{4(s)} + 2\ H_2O & +2.05\ V \end{array}$$

indicate that Pb and PbO_2 are used up and that $PbSO_4$ is formed on the electrodes during use or discharge. In addition, H_2SO_4 is used up and the concentration of H_2SO_4 decreases during discharge. Since the density of an aqueous H_2SO_4 solution is dependent mainly on the H_2SO_4 concentration, a gas station attendant can measure density with a hydrometer and thus determine the H_2SO_4 concentration and how far the battery is "down" or discharged. The battery is "charged up" to its original condition simply by reversing the electrode reactions in a process called electrolysis.

8.8 Electrolysis

The net discharge reaction for the lead storage battery has a positive voltage and proceeds to the right as written above. Conversely, the reverse reaction, carried out on recharging, has a negative voltage, is unfavorable energetically, and will not proceed spontaneously. In order to make this reaction occur, energy must somehow be applied. A battery charger utilizes electrical energy from a house circuit to supply the driving force. *A process in which electrical energy is applied to cause a chemical change* is called **electrolysis.**

Electrolysis is used in a wide variety of industrial processes to effect chemical reactions that are unfavorable energetically. In the Downs process sodium is produced commercially by electrolysis of a molten mixture of NaCl and $CaCl_2$ (added to reduce the melting point of the mixture) at about 600°.

$$2\ NaCl \xrightarrow[CaCl_2]{\text{heat, electrolysis}} 2\ Na + Cl_{2(g)}$$

The chlorine produced in this reaction is also an important industrial chemical. In the Hall–Héroult process aluminum is produced commercially by electrolysis of bauxite [$Al(OH)_3$], dissolved in a molten mixture of fluorides, at about 1000°.

$$Al(OH)_3 \xrightarrow[Na_3AlF_6,\ CaF_2,\ NaF]{\text{heat, electrolysis}} Al + O_2,\ F_2,\ \text{and carbon compounds}$$

Very pure oxygen and hydrogen can be produced by electrolysis of water.

$$2\ H_2O \xrightarrow{\text{electrolysis}} 2\ H_{2(g)} + O_{2(g)}$$

Power consumption in such processes is high; consequently, they are economically feasible only near cheap sources of electricity.

Exercises

8.1 Define clearly the following terms.

(a) chemical reaction
(b) chemical equation
(c) indicator
(d) Arrhenius acid
(e) Arrhenius base
(f) strong electrolyte
(g) weak electrolyte
(h) strong acid
(i) strong base
(j) weak acid
(k) weak base
(l) hydrolysis
(m) neutralization
(n) salt
(o) acid-base reaction
(p) titration
(q) Brønsted–Lowry acid
(r) Brønsted–Lowry base
(s) Lewis acid
(t) Lewis base
(u) complex ion
(v) polymer
(w) monomer
(x) repeating unit
(y) addition polymerization
(z) condensation polymerization
(aa) linear polymer
(ab) cross-linked polymer
(ac) thermoplastic polymer
(ad) thermosetting polymer
(ae) copolymer
(af) silicones
(ag) oxidation
(ah) reduction
(ai) oxidation-reduction reaction
(aj) electron transfer
(ak) oxidation number
(al) oxidizing agent
(am) reducing agent
(an) half-reaction
(ao) reduction potential
(ap) standard hydrogen electrode
(aq) cell
(ar) battery
(as) electrolysis

8.2 Write balanced chemical equations describing the dissociation in aqueous solution of each of the strong electrolytes in Table 7.3.

8.3 Using the general solubility rules in Table 7.1, write balanced chemical equations describing the precipitation of five insoluble salts from aqueous solution. How would you carry out each of these reactions experimentally starting with appropriate soluble salts?

8.4 Using the data in Table 7.3, write balanced *net* ionic equations for the acid-base reactions that occur when solutions of the following are mixed. Do not include ions that do not take part in the reaction.

(a) HNO_3 and NaOH
(b) H_2S and NaOH
(c) CH_3COOH and aqueous NH_3
(d) H_2SO_4 and aqueous NH_3
(e) CH_3COOH and $Ca(OH)_2$

8.5 Write balanced net ionic equations illustrating why aqueous ammonia (NH_3) could be classified as each of the following.

(a) Brønsted–Lowry base
(b) Arrhenius base
(c) Brønsted–Lowry acid
(d) Lewis base

8.6 Write balanced net ionic equations that explain why solutions prepared by adding the following species to water are acidic.

(a) H_2S
(b) $H_2PO_4^-$ as NaH_2PO_4
(c) HBr
(d) CO_2
(e) Fe^{3+} as $Fe(NO_3)_3$

8.7 Write balanced net ionic equations that explain why solutions prepared by adding the following species to water are basic.
(a) KOH
(b) NH_3
(c) CH_3COO^- as $NaOOCCH_3$
(d) CH_3NH_2
(e) CO_3^{2-} as K_2CO_3

8.8 Calculate the concentration of acetic acid (CH_3COOH) in vinegar if 44.2 ml of 0.115 *M* NaOH is required to neutralize the acid in a 50.0 ml sample of vinegar.

8.9 Calculate the concentration of household ammonia if 23.5 ml of 0.150 *M* HCl is required to neutralize a 25.0 ml sample of the ammonia.

8.10 Explain how the following reactions can be classified as acid-base reactions.
(a) $H^+ + OH^- \longrightarrow H_2O$
(b) $H_3PO_4 + OH^- \longrightarrow H_2PO_4^- + H_2O$
(c) $Al^{3+} + H_2O \rightleftharpoons AlOH^{2+} + H^+$
(d) $Ag^+ + 2\,CN^- \rightleftharpoons Ag(CN)_2^-$
(e) $CO_3^{2-} + H_2O \rightleftharpoons HCO_3^- + OH^-$

8.11 Describe in detail how one could prepare the following.
(a) NH_4Cl from NH_3
(b) KBr from KOH
(c) Na_3PO_4 from H_3PO_4
(d) NH_3 from NH_4Cl
(e) $CaBr_2$ from HBr

8.12 Write balanced chemical equations to show how each of the following can be prepared.
(a) Hydrochloric acid
(b) Nitric acid
(c) Phosphoric acid
(d) Sulfuric acid
(e) Solid sodium hydroxide

8.13 Comment critically on each of the following statements.
(a) The polymer resulting from the condensation polymerization of the amino acid alanine, $CH_3-\underset{\displaystyle NH_2}{\underset{|}{C}H}-COOH$, has the same percent composition as alanine itself.
(b) Alanine can form a linear polymer, but not a cross-linked polymer.
(c) A polymer resulting from addition polymerization has greater total bond strength than the sum of the bond strengths in all the small molecules used to form the polymer.

8.14 Orlon is a copolymeric fiber resulting from the catalyzed addition copolymerization of 90% acrylonitrile, $CH_2{=}CHCN$, and 10% 2-vinylpyridine,

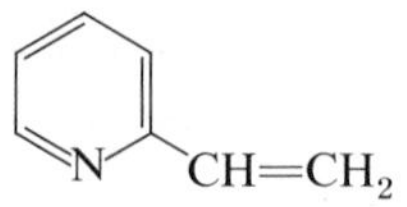
N
CH=CH2

Use Orlon as an example for describing clearly the meaning of each term.
(a) addition polymerization
(b) copolymer
(c) linear polymer
(d) percent composition of the polymer

8.15 Melmac plastic dishes are made from a polymer resulting from the condensation polymerization of melamine and formaldehyde. Sketch a partial structure

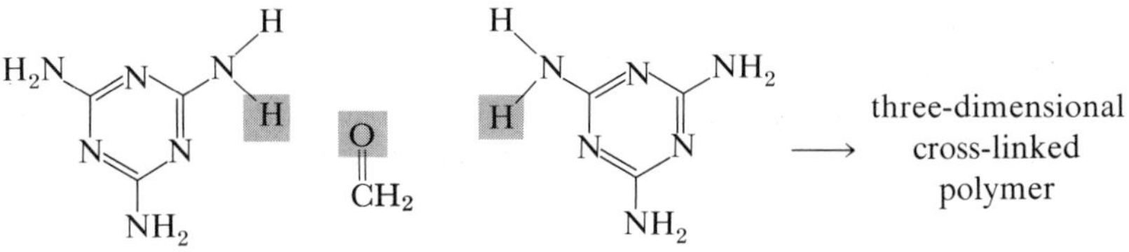

H
H
H_2N
N
N
NH_2
H
O
H
N
N
N
N
CH_2
NH_2
NH_2
three-dimensional
cross-linked
polymer

Melamine Formaldehyde

for the three-dimensional, cross-linked, polymeric product. What differences in properties would you expect between this polymer and one formed using the diamino derivative having only two reactive sites, in place of melamine in the above reaction?

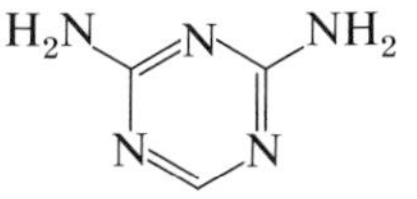

H_2N
N
NH_2
N
N

8.16 The structural formula for natural rubber, represented by the repeating unit given in Section 8.3-b, does not show the true directions of the bonds around the carbon atoms. On the basis of the real directions of the bonds for $-\overset{|}{C}=$ and $-\underset{|}{\overset{|}{C}}-$ when carbon has coordination numbers of three and four, respectively, account for the ability of rubber molecules to coil and uncoil.

8.17 Polypropylene is formed by the addition polymerization of propylene ($CH_3-CH=CH_2$). Sketch a portion of the polypropylene polymer. Calculate the percent carbon by mass in polypropylene.

8.18 Determine the oxidation number of Cr in each of the following species.

(a) $CrCl_2$
(b) Cr_2O_3 and $Cr(NH_3)_6{}^{3+}$
(c) $Cr(CO)_6$
(d) CrO_2Cl_2 and $Cr_2O_7{}^{2-}$
(e) $Na_2Cr_2(CO)_{10}$

8.19 Determine the oxidation number of all atoms in each of the following species.

(a) $BaCl_2$
(b) $KClO_3$
(c) CH_3OH
(d) N_2H_4
(e) H_2SO_4

8.20 Write the correct chemical formula for each of the following compounds.

(a) barium iodide, aluminum nitrate, vanadium(III) chloride, nitrous oxide, zinc carbonate

(b) silver nitrate, magnesium fluoride, sulfur trioxide, iron(II) sulfide, hydrogen bromide

(c) lead(II) carbonate, lithium hydride, silicon dioxide, nickel(II) sulfate, chromium(III) hydroxide

8.21 Write the correct chemical name for compounds represented by the following formulas.

(a) RbH, $CrCl_3$, $BaSO_4$, ClF_3, K_3PO_4
(b) $CuBr$, $MgCO_3$, $NaSCN$, NO_2, MnO
(c) BF_3, $TiCl_4$, SO_2, $Sr(NO_3)_2$, Na_2O_2

8.22 Complete and balance the following equations. In each case label the oxidizing agent, the reducing agent, the substance oxidized, and the substance reduced.

(a) $MnO_4^- + Fe^{2+} \xrightarrow{\text{acid}} Mn^{2+} + Fe^{3+}$

(b) $ClO^- \xrightarrow{\text{base}} Cl^- + ClO_3^-$

(c) $Si_3H_{8(g)} + O_{2(g)} \longrightarrow SiO_{2(s)} + H_2O$

8.23 Complete and balance the following equations. In each case label the oxidizing agent, the reducing agent, the substance oxidized, and the substance reduced.

(a) $H_2SO_3 + IO_3^- \xrightarrow{\text{acid}} HSO_4^- + I^-$

(b) $Cl_{2(g)} + S_2O_3^{2-} \xrightarrow{\text{base}} SO_4^{2-} + Cl^-$

(c) $KClO_{3(s)} + C_{12}H_{22}O_{11(s)} \longrightarrow KCl_{(s)} + CO_{2(g)} + H_2O_{(g)}$

8.24 Complete and balance the following equations.

(a) $WO_{3(s)} + CN^- \xrightarrow{\text{base}} W(CN)_8^{4-} + O_{2(g)}$

(b) $NO_3^- + Bi_2S_{3(s)} \xrightarrow{\text{acid}} Bi^{3+} + NO_{(g)} + S_{(s)}$

(c) $Sn^{2+} + IO_3^- \xrightarrow{\text{acid}} Sn^{4+} + I^-$

(d) $Re_2O_{8(s)} + H_2S_{(g)} \longrightarrow S_{(s)} + Re_2S_{7(s)} + H_2O_{(g)}$

(e) $H_2O_2 + CrO_{3(s)} \xrightarrow{\text{acid}} Cr^{3+} + O_{2(g)}$

8.25 Write a balanced chemical equation in molecular form for each of the following. In each case label the oxidizing agent, the reducing agent, the substance oxidized, and the substance reduced. (Review Section 8.5-b if you have difficulties with nomenclature.)

(a) oxidation of iron(II) sulfate by potassium dichromate in sulfuric acid solution to produce chromium(III) sulfate, iron(III) sulfate, potassium sulfate and water
(b) reduction of nitric acid by zinc metal to produce zinc nitrate, ammonium nitrate, and water
(c) simultaneous oxidation and reduction of chlorine gas in sodium hydroxide solution to produce sodium chloride, sodium chlorate, and water

8.26 **(a)** Complete and balance the equation describing the reduction of hematite (Fe_2O_3) by coke (C).

$$Fe_2O_{3(s)} + C_{(s)} \longrightarrow Fe_{(s)} + CO_{2(g)}$$

(b) How much coke would be needed to reduce 75 tons of hematite?
(c) How much iron would be produced?

8.27 Using the data in Table 8.1, predict which of the following reactions will proceed spontaneously. Show clearly your reasoning.
(a) $Cu_{(s)} + 2\,H^+ \longrightarrow Cu^{2+} + H_{2(g)}$
(b) $Cl_{2(g)} + 2\,Br^- \longrightarrow 2\,Cl^- + Br_2$
(c) $Zn_{(s)} + 2\,H^+ \longrightarrow Zn^{2+} + H_{2(g)}$
(d) $2\,Ag_{(s)} + Pb^{2+} \longrightarrow 2\,Ag^+ + Pb_{(s)}$
(e) $Zn_{(s)} + Cu^{2+} \longrightarrow Zn^{2+} + Cu_{(s)}$

8.28 The first really workable fuel cell for the direct conversion of chemical energy to electrical energy utilized the reaction between hydrogen gas and oxygen gas.

$$2H_{2(g)} + O_{2(g)} \longrightarrow 2\,H_2O$$

(a) Write the two half-reactions that would take place in this fuel cell operating under basic conditions.
(b) Calculate the mass of water that could be obtained if 16 liters of O_2 at 25° and 4.0 atm pressure reacted with excess H_2.

9

Chemical Kinetics

Some chemical reactions performed in the laboratory and many chemical reactions in biological systems take place almost instantaneously. They frequently occur in less time than that required for mixing two solutions. For example, an acid-base neutralization between H^+ and OH^- ions is complete in a fraction of a microsecond (10^{-6} sec). Under appropriate conditions, other chemical reactions, such as the decomposition of hydrogen peroxide (H_2O_2) to form water and O_2 gas, take place at moderate speeds and can be studied by conventional laboratory techniques. Additional reactions, such as those involving geological changes in nature, proceed so slowly that they cannot be detected in one's lifetime, but may still be very important over a time scale of the order of millions of years. The speed of many chemical reactions may be altered simply by changing the conditions under which the reaction is allowed to take place. For example, a mixture of H_2 gas and O_2 gas can be kept almost indefinitely at room temperature; their rate of reaction is infinitesimally small. However, passage of a spark causes the reaction to proceed explosively, resulting in the formation of water. Finally, it is often possible to delineate intermediate species through which a reaction passes and to elucidate the pathway by which a reaction proceeds (for example, see the reaction mechanism for the chlorination of alkanes in Section 13.4). The elucidation of chemical and biochemical reaction mechanisms is a frontier of chemistry in which much work remains to be done. For example, drugs are responsible for both many benefits and many detrimental effects; yet we understand very little about the reaction pathways that are involved.

Chemical kinetics is *the study of the rates of chemical reactions and the pathways by which chemical reactions proceed.* The purpose of this chapter is to illustrate the factors that influence the rates of chemical reactions and to examine a model that explains the influences of these factors on the rates of chemical reactions.

9.1 Rate of a Chemical Reaction

The **rate of a chemical reaction** is *the speed at which reacting substances are used up or the speed at which products are formed.* To express a reaction rate quantitatively we first need the overall stoichiometry of the reaction in the form of a balanced equation. For the decomposition of hydrogen peroxide according to the following equation,

$$2\,H_2O_2 \longrightarrow 2\,H_2O + O_{2(g)}$$

the reaction rate can be expressed as

$$\begin{aligned}\text{Rate} &= \tfrac{1}{2}\text{ rate of consumption of } H_2O_2\\ &= \tfrac{1}{2}\text{ rate of formation of } H_2O\\ &= \text{rate of formation of } O_2\end{aligned}$$

The coefficients in these equations arise from consideration of the overall stoichiometry of the reaction as expressed by the balanced equation.

The rate of a reaction is more commonly expressed as the change in concentration of a reactant or product per unit time. The rate of the above reaction can be expressed in terms of concentrations as

$$\begin{aligned}\text{Rate} &= -\tfrac{1}{2}\frac{\text{change in concentration of } H_2O_2}{\text{elapsed time}} = -\tfrac{1}{2}\frac{\Delta[H_2O_2]}{\Delta t}\\ &= \tfrac{1}{2}\frac{\text{change in concentration of } H_2O}{\text{elapsed time}} = \tfrac{1}{2}\frac{\Delta[H_2O]}{\Delta t}\\ &= \frac{\text{change in concentration of } O_2}{\text{elapsed time}} = \frac{\Delta[O_2]}{\Delta t}\end{aligned}$$

where the symbol Δ indicates *change of* a quantity and bracketed formulas represent molar concentrations of those species. The sign conventions in the above equations are chosen so that the rate of the forward reaction is a positive quantity even though the concentration of H_2O_2 is decreasing with time.

Expressed in this fashion, *reaction rates are dependent upon the concentrations of reactants but are independent of the amount of reactants under consideration.* For example, suppose that we carry out the iodide ion-catalyzed decomposition of hydrogen peroxide using a 100 ml sample of 1.0 M H_2O_2 and a 10 ml sample of 1.0 M H_2O_2. Although a larger volume of gaseous O_2 is produced in any given time from the decomposition of the 100 ml sample than from decomposition of the 10 ml sample, the H_2O_2 concentrations of both solutions are initially equal and remain equal at any given time after decomposition begins. The rate of the reaction, dependent only on concentration, is the same in both instances.

The rate of a reaction can be determined experimentally by measuring the rate of consumption of reactants or formation of products, or the change of concentrations of reactants or of products with time, providing appropriate techniques exist for measuring these quantities. This is easily accomplished for the decomposition of H_2O_2 by monitoring the volume of O_2 evolved with time when measured at constant pressure.

When this is done, results leading to the heavy solid line in Figure 9.1 are obtained. The reaction rate at any given time can be calculated from the slope of a tangent drawn to the smooth curve at the point in question. The slopes of the dotted tangents in Figure 9.1 are approaching zero as the reaction proceeds. Therefore, the rate of the reaction is decreasing with time.

If the temperature at which the decomposition of an identical H_2O_2 solution is carried out is increased, the dashed line in Figure 9.1, showing generally faster rates of reaction, is obtained. The same dashed line representing O_2 evolution can be obtained alternatively by adding iodide ion (I^-) as a catalyst to an identical H_2O_2 solution, by increasing sufficiently the initial concentration of H_2O_2, or by evolving O_2 instead from decomposition of dinitrogen pentoxide (N_2O_5) in CCl_4 according to the following equation (NO_2 is soluble in CCl_4 and is not evolved as a gas).

$$2\,N_2O_5 \longrightarrow 4\,NO_2 + O_{2(g)}$$

Similar simple qualitative observations for a wide variety of chemical reactions show that reaction rates depend, in general, on four factors: nature of the reactants, concentration of the reactants, temperature, and the presence of catalysts. Each of these factors will be examined individually in this chapter and will be discussed again in connection with biochemical reactions in Section 26.7.

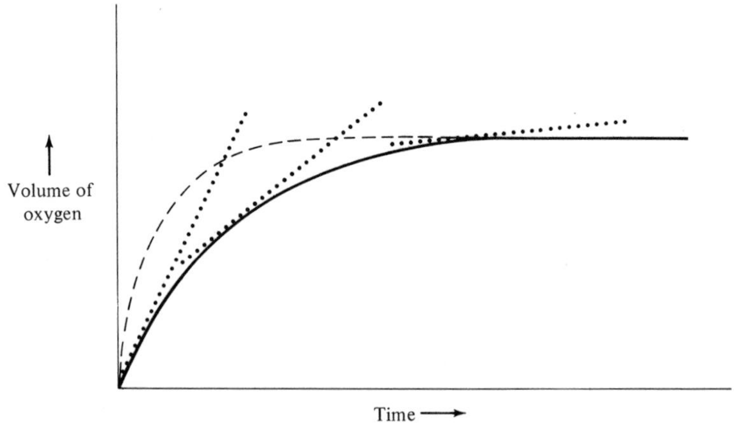

Volume of oxygen
Time

Figure 9.1 Curves showing the changes in volume of O_2 with time for the reaction

$$2\,H_2O_2 \longrightarrow 2\,H_2O + O_{2(g)}$$

The heavy solid line represents room temperature. The dashed line represents a higher temperature. Slopes of dotted lines give rates of reaction at several points.

9.2 Effect of Nature of Reactants on Reaction Rates

The observed rates of reactions between different reactants vary widely. At room temperature white phosphorus undergoes spontaneous combustion in air

$$P_{4(s)} + 5\,O_{2(g)} \longrightarrow P_4O_{10(s)}$$

whereas red phosphorus can be exposed to air without appreciable oxidation. At room temperature sodium reacts violently with water

$$2\,Na_{(s)} + 2\,H_2O \longrightarrow 2\,Na^+ + 2\,OH^- + H_{2(g)}$$

whereas magnesium is fairly unreactive toward water. Neutralization of oxalic acid ($H_2C_2O_4$) by NaOH is instantaneous

$$H_2C_2O_4 + 2\,OH^- \longrightarrow 2\,H_2O + 2\,C_2O_4^{2-}$$

whereas oxidation of $H_2C_2O_4$ by permanganate ion (MnO_4^-) proceeds slowly as indicated by the persistence of the purple color of MnO_4^- long after the reagents are mixed.

$$5\,H_2C_2O_4 + 2\,MnO_4^- + 6\,H^+ \longrightarrow 2\,Mn^{2+} + 10\,CO_{2(g)} + 8\,H_2O$$

The rate of a reaction is specific for each different chemical reaction under a particular set of conditions. This is not surprising since, in a chemical reaction, bonds are broken and bonds are formed. The rate of a reaction is thus expected to be a function of the kinds of atoms in the reactants, how these atoms are bonded, and how easily these bonds can be broken and reformed to produce new substances.

9.3 Effect of Concentration of Reactants on Reaction Rates

An iron file rusts very slowly in air at room temperature whereas the same file oxidizes more rapidly in pure oxygen.

$$4\,Fe_{(s)} + 3\,O_{2(g)} \longrightarrow 2\,Fe_2O_{3(s)}$$

Moreover, although the iron file oxidizes very slowly in air, very finely powdered iron, prepared by reducing powdered iron(III) oxide (Fe_2O_3) with H_2, oxidizes so rapidly in air that it may become red-hot and burst into flame.

$$Fe_2O_{3(s)} + 3\,H_{2(g)} \longrightarrow 2\,Fe_{(s)} + 3\,H_2O$$

Cigarettes will do the same thing and can provide a spark for the combustion of other materials; thus, smoking is prohibited in hospital rooms when oxygen is being used. These observations lead to the conclusion that *the rate of a chemical reaction increases as effective concentrations of the reacting particles increase*. This conclusion is not as straightforward as it appears because the effective concentrations of reacting particles are partially determined by whether the reactants are homogeneous or heterogeneous.

For homogeneous reactions, in which the reactants are in the same gas or liquid phase, *the rate depends on the concentrations of reactants in solution*. The concentration of a reactant can be changed in several ways. It can be increased by adding more reactant, or by decreasing the volume of the system (compressing if a gas solution or distilling solvent if a liquid solution). It can be decreased by removing some reactant, or by increasing the volume (expanding if a gas solution or adding solvent if a liquid solution).

A balanced equation provides no information about how the rate of a reaction is altered by changing the concentration of reactants. Such information can only be detemined by experiment. For example, it can be determined for the decomposition of H_2O_2 that doubling the concentration of H_2O_2 doubles the rate; tripling the concentration of H_2O_2 triples the rate; and halving the H_2O_2 concentration halves the rate. This evidence can be summarized by saying that the rate of decomposition of H_2O_2 is directly proportional to the concentration of H_2O_2. Expressed mathematically,

$$\text{Rate} \propto [H_2O_2] \qquad \text{or} \qquad \text{rate} = k[H_2O_2]$$

where brackets around H_2O_2 represent the molar concentration of H_2O_2 and k is a proportionality constant called the *specific rate constant*. A specific rate constant is characteristic for a given reaction at a given temperature. It will be different for a different chemical reaction or for the same reaction at a different temperature.

The above equation is known as the *rate law* for the decomposition of H_2O_2. A **rate law** *describes quantitatively the effect of concentration on the rate of a reaction*. The rate law is a direct consequence of the rate of the step or steps (called the mechanism) by which a reaction proceeds. Determination of a rate law is important because the rate law, along with actual isolation of intermediates, is the primary information from which a plausible mechanism for a reaction can be postulated.

The general form of a rate law is

$$\text{Rate} = k[\text{A}]^a[\text{B}]^b \cdots$$

where the exponents a, b, . . . may be zero, whole, fractional, positive, or negative numbers. The overall order of a reaction is given by the sum of the exponents in the rate law. The order with respect to a given reagent is the exponent of the concentration term for that reagent in the rate law. For the rate law describing the decomposition of H_2O_2, the exponent of $[H_2O_2]$ is one, and the reaction is said to be first order with respect to H_2O_2. The decomposition of H_2O_2 is also first order overall since $[H_2O_2]$ is the only concentration term in the rate law. It should be emphasized that *there is no necessary correlation between a rate law and the stoichiometry of a reaction.*

The exponents in a rate law are *not* determined by the coefficients in a balanced chemical equation. All terms in a rate law must be determined by experiment.

Example 9.1 Determine the rate law for the reaction

$$2\ H_{2(g)} + 2\ NO_{(g)} \longrightarrow 2\ H_2O_{(g)} + N_{2(g)}$$

if doubling the concentration of H_2 doubles the rate whereas doubling the concentration of NO quadruples the rate, and tripling the concentration of H_2 triples the rate whereas tripling the concentration of NO multiplies the rate by nine times.

It can be seen that the rate is directly proportional to the concentration of H_2, but directly proportional to the square of the concentration of NO. Therefore, the rate law is

$$\text{Rate} = k[H_2][NO]^2$$

where k is the specific rate constant. The reaction is first order with respect to H_2, second order with respect to NO, and third order (the sum of the exponents) overall.

For heterogeneous reactions, in which the reactants are in more than one phase, *the rate depends upon the area of contact between the phases.* Mossy zinc, which has a relatively small surface area, oxidizes only at very high temperatures, whereas powdered zinc burns readily in air at 300° to form white zinc oxide (ZnO).

$$2\ Zn_{(s)} + O_{2(g)} \longrightarrow 2\ ZnO_{(s)}$$

The greater the degree of subdivision of solids, the greater will be the surface area and the total number of particles that will be at the surface where they can react. Dissolving occurs as a result of interaction of solvent with surface particles of solute. Therefore, effective grinding of a solid salt increases its rate of dissolving in water.

9.4 Effect of Temperature on Reaction Rates

It was noted in the introduction of this chapter that there is no observable reaction between H_2 and O_2 when these gases are held at room temperature. However, with an increase of temperature of the mixture, the rate of reaction increases rapidly. This indicates that *raising the temperature of a reaction mixture increases the rate of reaction.* Conversely, lowering the temperature of a reaction mixture decreases the rate of reaction. We can easily understand then why reactions are heated to induce more rapid chemical change and why explosive reactions are cooled to reduce their violence. An important interest of chemical engineers is the design of methods for the addition of or removal of heat for the purpose of controlling chemical reactions.

The magnitude of the change of rate with temperature varies considerably for different reactions and for different temperature ranges. A crude rule of thumb is that the rate of a reaction approximately doubles for every 10 °C rise in temperature. In a rate law the change of rate with temperature is reflected in the magnitude of the specific rate constant k. Specific rate constants increase with increases in temperature and decrease with decreases in temperature.

9.5 Effect of Catalysts on Reaction Rates

Although reaction upon mixing of H_2 and O_2 at room temperature is not observable, these two gases react spontaneously upon the addition of platinum gauze to the mixture. Similarly, the laboratory preparation of O_2 by thermal decomposition of potassium chlorate ($KClO_3$)

$$2\,KClO_{3(s)} \longrightarrow 2\,KCl_{(s)} + 3\,O_{2(g)}$$

is effective only when a trace of manganese(IV) oxide (MnO_2) is added to speed up the reaction. Finally, an aqueous solution of maltose, a sugar that is stable toward hydrolysis indefinitely, hydrolyzes rapidly in our bodies without heating in the presence of the enzyme maltase (Section 23.7-b). Of further interest is the fact that after each of these reactions the platinum gauze, the MnO_2, and the maltase can be recovered unchanged. *Substances that increase the rate of a chemical reaction, but themselves remain unchanged after the reaction,* are called **catalysts.**

The amount of a catalyst required to accelerate a chemical reaction varies from one reaction to another. In many reactions, as in the thermal decomposition of $KClO_3$, a trace of the catalyst is sufficient. In other reactions, including many in our bodies that are catalyzed by enzymes, the rate of the reaction is proportional to some power of the concentration of the catalyst. When this occurs the catalyst concentration to the appropriate power actually appears in the rate law just as do reactant concentrations to appropriate powers.

Catalysts can be classified as homogeneous catalysts and heterogeneous catalysts. Homogeneous catalysts are in the same phase as the reactants. Homogeneous catalysts undergo rapid reaction with reactant particles to form active intermediates which then react further to free the catalyst. Enzymes (Chapter 26), the complex substances that accelerate biochemical processes, are homogeneous catalysts. Heterogeneous catalysts are present as an additional phase. They are usually solids to which reactant particles are adsorbed, thereby weakening bonds within the reactant molecules and rendering them more susceptible to reaction. For example, the effectiveness of a given mass of Pt gauze in catalyzing the reaction of H_2 and O_2 is directly proportional to its surface area. Certain substances, called inhibitors, can cover the surface of heterogeneous catalysts so as to prevent adsorption and reduce the rate of a catalyzed reaction. Inhibitors also interact with enzymes to cause a slowing down or cessation of biochemical reactions (Section 26.8).

Catalysts are very important both industrially and biologically. The refining of petroleum to obtain fuel oils and gasoline, the hydrogenation of oils to produce synthetic fats such as margarine or shortening, the manufacture of most of our important industrial chemicals, and the myriad of essentially instantaneous reactions occurring in all living organisms, require the use of catalysts. Catalysts are particularly useful in cases where higher temperatures cannot be used to speed up a reaction because the higher temperatures would also cause thermal decomposition of the desired products.

9.6 Collision Theory of Reaction Rates

Many of the experimental observations concerning the rates of chemical reactions can be accounted for in terms of the collision theory. **Collision theory** assumes that *reactions occur as a result of collisions between reacting particles,* whether they be atoms, molecules, or ions. Atoms and electrons rearrange during such collisions resulting in the rupture of some chemical bonds and the formation of others.

Only a very small percent of the total number of collisions leads to a chemical reaction. In a one liter, equimolar mixture of H_2 and O_2 molecules at STP, more than 10^{28} collisions occur per second; nevertheless, reaction between these molecules at room temperature is insignificant. Such observations have led to amplifying assumptions of collision theory—*the rate of any step in a reaction is directly proportional to (1) the number of collisions per second between reacting particles in that step and (2) the fraction of these collisions that are effective.*

The reasoning behind these assumptions can be understood by examining what happens on a molecular scale during the chemical reaction

$$CO_{(g)} + H_2O_{(g)} \longrightarrow CO_{2(g)} + H_{2(g)}$$

Molecules of CO and H_2O, initially at a distance, have no interaction between each other. As they approach each other and collide, their electron clouds, being similarly charged, interact repulsively with each other. They may only graze each other and separate again, retaining their original identity. Similarly, they may collide head-on, but with small enough kinetic energies that they simply bump gently and bounce apart as a result of repulsion between their electron clouds. On the other hand, they may have high kinetic energies and approach with high enough velocities so that upon collision their electron clouds interpenetrate to form a short-lived complex particle. This transient, high energy intermediate is called an *activated complex*. This activated complex may simply split back into the starting CO and H_2O molecules. However, atomic and electronic rearrangement leading to new chemical species, CO_2 and H_2, is equally likely.

The potential energy (Section 1.2-b) of this reaction system may be plotted against a reaction coordinate describing the extent to which the reaction has occurred as shown in Figure 9.2. In the initial state, where CO and H_2O molecules are far enough apart so as not to interact with each other, the total potential energy is simply the sum of the potential energies of CO and H_2O. As CO and H_2O molecules collide, their electron clouds interact repulsively, and the potential energy of the system increases. In the activated complex, where the repulsive interaction of the electron clouds is at a maximum, the potential energy of the system also reaches a peak. As the activated complex splits into CO_2 and H_2 molecules, and as they separate causing lessening interaction of their electron clouds, the potential energy decreases. When the CO_2 and H_2 molecules are far enough apart so as not to interact with each other, the potential energy reaches a value which is the sum of the potential energies of CO_2 and H_2.

The potential energy curve in Figure 9.2 provides several pieces of information. The difference between the potential energy of the reacting molecules, CO and H_2O, and the activated complex (illustrated by a black arrow in Figure 9.2) is called the activation energy. The **activation energy** for a reaction is *the minimum energy that colliding molecules must have for reaction to occur.* It is kinetic energy (possessed jointly by two reacting molecules) that is available for conversion into potential energy through the breaking of bonds or the freeing of electrons for redistribution from one atom to another. Several kinetic energy distributions were shown in Figure 6.1 to indicate that some molecules in a liquid sample have enough kinetic energy to overcome attractive forces and escape into the vapor phase. Similarly, in

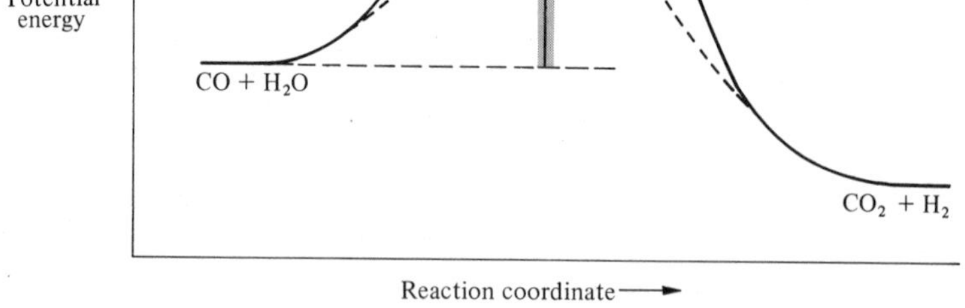

Activated complex
Activation energy
Potential energy
CO + H₂O
CO₂ + H₂
Reaction coordinate

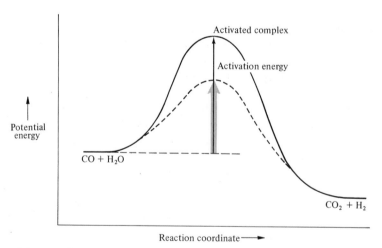

Figure 9.2 Potential energy diagrams for catalyzed (dashed curve) and uncatalyzed (solid curve) reactions.

a collection of molecules at a given temperature some molecules already possess sufficient kinetic energy (equal to the activation energy) to undergo reaction whereas others will not.

The reacting molecules can be likened to a ball rolling up a hill. If the ball is rolled slowly and has little kinetic energy, it goes only part way up the hill before stopping and rolling back down the hill. Likewise, molecules with little kinetic energy collide, interact, and fly apart unchanged. If the ball is rolled rapidly and has significant kinetic energy, it reaches the top of the hill and may roll down the other side. Likewise, molecules with the requisite kinetic energy may collide, interact, rearrange, and separate into new species. Other factors being equal, a large activation energy results in a slow rate of reaction and a small specific rate constant; a small activation energy leads to a rapid rate of reaction and a large specific rate constant.

Although it has no bearing on the activation energy or on the rate of a reaction, it should be noted that the difference between the potential energy of the reactants and the products indicates whether a reaction is exothermic or endothermic (Section 1.2-b). If there is a net decrease in potential energy as the reaction proceeds, as in the curve in Figure 9.2, excess energy is liberated, and the reaction is exothermic. Conversely, if there is a net increase in potential energy as the reaction proceeds, the reaction is endothermic.

The four factors that influence the rates of reactions can be accounted for by the collision theory and can be explained in terms of a potential energy diagram like that in Figure 9.2. Two of the factors, nature of the reactants and catalysts, cause changes of the activation energy. The other two factors, concentration of the reactants and temperature, affect the total number of collisions and the fraction of these collisions which lead to reaction.

The nature of the reactants influences the rates of reactions because each particular reaction has a characteristic activation energy. The repulsive interaction of the electron clouds of two colliding particles, and thus the height of the energy barrier that is the activation energy, depends upon the identity of the colliding particles and differs from one reaction to another.

Catalysts influence the rates of reactions by permitting easier formation of an activated complex. Catalysts enable a reaction to proceed by a different pathway (dashed curve in Figure 9.2) having a lower activation energy (shaded arrow in Figure 9.2) than the uncatalyzed pathway. With a lower activation energy, collisions between even less energetic particles are capable of inducing a reaction.

Temperature influences the rates of reactions because, as the temperature is increased, molecules move faster and have higher kinetic energies (Figure 6.1). They experience both more frequent collisions and more violent collisions. A greater fraction of the molecules has sufficient energy to do more than simply rebound unchanged. A greater fraction actually interpenetrates to form an activated complex from which new stable products can be formed. Thus, the rate of reaction increases with increased temperature.

The concentration of the reactants influences the rates of reactions because a greater number of collisions occurs as reactant concentrations increase. At any given temperature a certain fraction of these collisions produces a reaction. The greater the total number of collisions the greater the number producing a reaction.

Exercises

9.1 Define clearly the following terms.

(a) chemical kinetics
(b) reaction rate
(c) reactants
(d) products
(e) concentration
(f) homogeneous reaction
(g) rate law
(h) specific rate constant
(i) order of a reaction
(j) heterogeneous reaction
(k) catalyst
(l) homogeneous catalyst
(m) heterogeneous catalyst
(n) inhibitors
(o) collision theory
(p) activated complex
(q) potential energy
(r) activation energy
(s) exothermic reaction
(t) endothermic reaction

9.2 **(a)** Which of the factors that influence the rates of chemical reactions cannot be altered by the experimenter?
(b) Which of the factors that influence the rates of chemical reactions can be varied by the experimenter?
(c) Explain on the basis of collision theory how each of the above factors affects the rates of chemical reactions.

9.3 For the reaction

$$NO_{2(g)} + CO_{(g)} \longrightarrow CO_{2(g)} + NO_{(g)}$$

it is found that doubling the NO_2 concentration doubles the rate, halving the CO concentration halves the rate, tripling the NO_2 concentration triples the rate, and quadrupling the CO concentration quadruples the rate. Determine the following.
(a) the rate law for the reaction
(b) the order of the reaction with respect to NO_2
(c) the overall order of the reaction

9.4 Given rate laws for the following reactions, state the order of the reaction with respect to each reactant and the overall order of the reaction.
(a) $NH_4^+ + CNO^- \longrightarrow H_2NCONH_2$ rate $= k[NH_4^+][CNO^-]$
(b) $2\,NO_{(g)} + O_{2(g)} \longrightarrow 2\,NO_{2(g)}$ rate $= k[NO]^2[O_2]$
(c) $CHCl_{3(g)} + Cl_{2(g)} \longrightarrow CCl_{4(g)} + HCl_{(g)}$ rate $= k[CHCl_3][Cl_2]^{1/2}$
(d) $2\,HI_{(g)} \xrightarrow{Au} H_{2(g)} + I_{2(g)}$ rate $= k$
(e) $5\,Br^- + BrO_3^- + 6\,H^+ \longrightarrow 3\,Br_2 + 3\,H_2O$ rate $= k[Br^-][BrO_3^-][H^+]^2$

9.5 **(a)** Write the rate law for the reaction

$$NO_{(g)} + O_{3(g)} \longrightarrow NO_{2(g)} + O_{2(g)}$$

which has been shown to be first order with respect to both NO and O_3.

(b) What effect should the following changes have on the initial rate of the reaction, the specific rate constant for the reaction, and the activation energy for the reaction? (1) a decrease in the initial partial pressure of NO, (2) an increase in the temperature, (3) the addition of a catalyst, (4) an increase in the initial concentration of O_3.

9.6 Explain clearly the following.

(a) Diluting reaction mixtures with solvent of the same temperature does not decrease the rate of every reaction.

(b) Two gases, capable of reacting chemically, do not react instantaneously when the gases are mixed.

(c) The rate of a reaction increases more rapidly with an increase in temperature than does the frequency of molecular collisions.

9.7 Explain clearly the following.

(a) An increase in pressure accelerates the rate of a homogeneous gas phase reaction.

(b) Oxygen and hydrogen react explosively at room temperature when an electric spark is produced within the mixture even though they are inert toward each other in the absence of the spark.

(c) Lead shot exposed to air tarnishes very slowly on its surface, whereas finely powdered (pyrophoric) lead bursts into flame in air.

9.8 Explain clearly the following.

(a) A certain exothermic reaction has to be heated initially to get the reaction started, but then has to be cooled during the balance of the reaction to keep it under control.

(b) Thermal decomposition of potassium chlorate ($KClO_3$) yields potassium chloride (KCl) and potassium perchlorate ($KClO_4$) in the absence of catalyst, but yields potassium chloride (KCl) and oxygen gas (O_2) in the presence of MnO_2 as a catalyst.

(c) Diluting a reaction mixture with hot solvent may either increase or decrease the rate of the reaction.

9.9 Sketch a potential energy diagram similar to that in Figure 9.2 for an endothermic reaction.

9.10 When the temperature of a reaction system is increased, which factor is more important in causing the reaction rate to increase, an increased frequency of molecular collisions or an increased fraction of the effective collisions? Show clearly your reasoning.

9.11 **(a)** What is the meaning of the activation energy for a reaction?

(b) Cite physical evidence supporting the concept of an activation energy for a reaction.

9.12 **(a)** Sketch a potential energy diagram for the exothermic reaction

$$2\,N_{2(g)} + 3\,H_{2(g)} \longrightarrow 2\,NH_{3(g)}$$

Label the energies of the reactants, product, and activated complex on your diagram.

(b) The above reaction is reversible (occurs readily in both the forward and reverse directions). Which has the greater activation energy, the forward reaction or the reverse reaction?

9.13 (a) Complete and balance the following equation.

$$HCrO_4^- + HSO_3^- \xrightarrow{\text{acid}} Cr^{3+} + SO_4^{2-}$$

(b) Doubling the concentration of either $HCrO_4^-$ or H^+ doubles the rate of the reaction whereas doubling the concentration of HSO_3^- quadruples the rate of the reaction. Indicate (1) the rate law for the reaction, (2) the order of the reaction with respect to HSO_3^- and H^+, and (3) the overall order of the reaction.

(c) What effect should the following changes have on the initial rate of the reaction, the specific rate constant for the reaction, and the activation energy for the reaction? (1) an increase in the initial concentration of $HCrO_4^-$, (2) the addition of a catalyst, (3) a decrease in the temperature, (4) a simultaneous equivalent increase of the concentration of HSO_3^- and decrease of the concentration of H^+.

10

Chemical Equilibrium

Most of the stoichiometric calculations in Chapter 4 were based on the assumption that chemical reactions go to completion; that is, all of the reactants are converted to products. The chemical reactions of Chapter 8 were treated primarily as reactants going completely to products. In Chapter 9 we assumed that if a reaction occurred at all, it went to completion. However, it can be shown experimentally that, for many reactions, conversion of reactants to products is incomplete, regardless of the time allowed for the reactions to take place. These reactions are reversible. Reactants form products, but the products in turn react to produce the original reactants. Dynamic chemical equilibria persist.

Our bodies depend for their proper functioning on a great variety of chemical equilibria in aqueous solutions. Human blood, for example, loses its capability for transporting oxygen from the lungs to the cells if the pH is not maintained at about 7.4. The following equilibria are responsible for maintaining this pH (Section 31.6).

$$CO_2 + H_2O \rightleftharpoons H^+ + HCO_3^-$$

$$H_2PO_4^- \rightleftharpoons H^+ + HPO_4^{2-}$$

The purposes of this chapter are to describe and illustrate the nature of dynamic chemical equilibrium, to show how concentrations of species present at equilibrium can be specified by an equilibrium constant, to discuss factors that affect chemical equilibria, and to perform simple equilibrium calculations.

10.1 Dynamic Chemical Equilibrium

If phosphorus pentachloride (PCl_5) is injected into a closed container at 250°, reversible reactions involving phosphorus trichloride (PCl_3) and chlorine (Cl_2) take place.

$$PCl_{5(g)} \rightleftharpoons PCl_{3(g)} + Cl_{2(g)}$$

It is instructive to ascertain how the concentrations of reactant and products vary quantitatively as these reversible reactions proceed. Concentrations of the three compounds can be measured experimentally as a function of time. Results of these measurements are plotted in Figure 10.1. Suppose that at the time of injection the PCl_5 concentration is 1.20 M (1.20 mole/liter). The concentrations of PCl_3 and Cl_2 are both zero. The PCl_5 concentration decreases rapidly at first, then less rapidly, and eventually becomes constant. Conversely, the concentrations of PCl_3 and Cl_2 increase rapidly at first, then less rapidly, and eventually become constant. The PCl_3 and Cl_2 concentrations are identical at any given time because they are formed in a 1:1 mole ratio when the above reversible reaction proceeds to the right. In time it is found that 0.98 M PCl_5, 0.22 M PCl_3, and 0.22 M Cl_2 are present in the container. A state of **chemical equilibrium** is attained at *the point at which the*

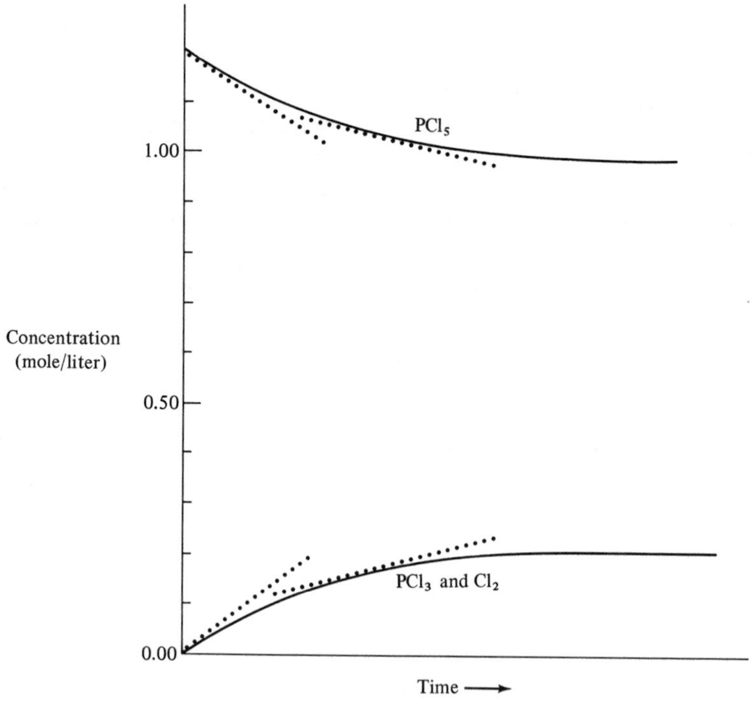
PCl₅
1.00
Concentration
(mole/liter)
0.50
PCl₃ and Cl₂
0.00
Time

Figure 10.1 Changing concentrations of PCl_5, PCl_3, and Cl_2 on approach to equilibrium. Dotted slopes show rates of change of concentrations.

concentrations of all species remain constant, even though reactions in *both* directions continue. Once equilibrium is attained under a given set of conditions, the equilibrium concentrations persist indefinitely unless the initial conditions are changed.

The slopes of the dotted tangents to the curves in Figure 10.1 depict the rates of change of concentration of the indicated species at two different times along each curve. The slopes of the tangents to the upper curve depict the rate of conversion of PCl_5 to PCl_3 and Cl_2, and the slopes of the tangents to the lower curve depict the rate of formation of PCl_3 and Cl_2 from PCl_5. The *change* of slopes indicates the *change* in the rate of destruction of PCl_5 on the upper curve and the *change* in the rate of formation of PCl_3 and Cl_2 on the lower curve. Thus, the *net* destruction of PCl_5 is rapid at first, then decreases and eventually reaches zero. Likewise, the *net* formation of PCl_3 and Cl_2 is rapid at first, then decreases and eventually reaches zero. At equilibrium there is no net conversion in either direction. It is a necessary consequence that *at equilibrium the rate of the forward reaction and the rate of the reverse reaction are equal.*

There are two possible kinds of equilibrium, *static* and *dynamic.* Static equilibrium is common in mechanical systems. For example, two children on a seesaw can find equilibrium positions such that the seesaw is balanced and does not move about its central support. We have already discussed examples of dynamic equilibrium between phases (Section 6.4). Liquid water in a closed container is in dynamic equilibrium with its vapor. As long as the temperature remains constant, the amounts of water in the liquid and vapor phase remain constant although there is continual passage of water molecules from the liquid to the vapor and vice versa.

Chemical equilibria are dynamic equilibria, not static, and reactions continue at the same rate in both directions after equilibrium is attained. That chemical equilibria are dynamic can be illustrated for the PCl_5–PCl_3–Cl_2 system by an isotope-scrambling experiment. Suppose that an equilibrium is set up by injecting PCl_3 and Cl_2 into the closed container instead of PCl_5.

$$PCl_{5(g)} \rightleftharpoons PCl_{3(g)} + Cl_{2(g)}$$

Specifically, PCl_3, composed only of isotopically pure ^{35}Cl, and Cl_2, composed only of isotopically pure ^{37}Cl, are injected. By the time equilibrium is achieved, the two Cl isotopes are scrambled through all three species. The $^{37}Cl_2$ has combined with $P^{35}Cl_3$ (reaction to the left) to form PCl_5 containing both Cl isotopes. This, in turn, has dissociated (reaction to the right) in a different manner from which it was formed to give Cl_2 containing the ^{35}Cl isotope and PCl_3 containing the ^{37}Cl isotope. The net result of this occurring many times is a completely statistical scrambling of the two Cl isotopes. All species will contain ^{35}Cl and ^{37}Cl in a 3:2 ratio since that is the ratio in which they were injected initially.

10.2 Law of Chemical Equilibrium

a. Gaseous Equilibria

Suppose that a series of experiments are performed on the PCl_5–PCl_3–Cl_2 system and that equilibrium concentrations, attained at the same temperature but from different initial concentrations, are measured. Results of several such experiments are shown in Table 10.1. The question arises whether there is any simple relation applying to each of the systems. Manipulating the numbers from Table 10.1 in a variety of ways shows, among other things, that the product of the concentrations of PCl_3 and Cl_2 divided by the concentration of PCl_5 yields about 0.050 in all cases. A simple relationship does exist, and such a relationship can be found for all chemical equilibria.

Each particular reversible reaction has its own specific state of dynamic equilibrium; however, a general equation defining the relation between concentrations of reactants and products can be written. Such an equation is called the **law of chemical equilibrium.** This law states that *for a reversible reaction at equilibrium, the product of the molar concentrations of the products divided by the molar concentrations of the reactants, each concentration raised to that power which is the coefficient of the substance in the balanced chemical equation, is a constant for a given temperature.* For the reversible reaction

$$PCl_{5(g)} \rightleftharpoons PCl_{3(g)} + Cl_{2(g)}$$

the law of chemical equilibrium leads to the following equilibrium constant expression.

$$K_t = \frac{[PCl_3][Cl_2]}{[PCl_5]}$$

Formulas enclosed in brackets represent the molar concentrations of the indicated species in moles per liter. All molar concentrations are to the first power since all coefficients in the balanced equation are one. The constant K_t is called the equilibrium constant for the reversible reaction. It has a fixed value for a given reversible reaction at a given temperature t, independent of the absolute values of the molar concentrations, as verified by the data in Table 10.1.[1]

Some reversible reactions include substances whose concentration is not able to change. In such cases the constant concentrations are combined mathematically with the equilibrium constant to form a new constant and leave a simplified equilibrium constant expression that omits the substance

[1]Note that equilibrium constants defined according to molar concentrations have units of moles per liter to some power (where that power may be zero if there is the same power of concentration terms in the numerator and denominator). However, equilibrium constants defined more rigorously according to activities are dimensionless; therefore, no units will be used for equilibrium constants in calculations.

Table 10.1 Equilibrium Concentrations at 250° for PCl_5–PCl_3–Cl_2 Systems

	Concentrations (mole/liter)		
Experiment	*PCl_5*	*PCl_3*	*Cl_2*
1	1.71	0.29	0.29
2	0.98	0.22	0.22
3	0.050	0.050	0.050
4	0.014	0.026	0.026
5	0.87	0.13	0.33

whose concentration is invariant. This happens for heterogeneous equilibria involving a gaseous phase and one or more condensed phases. For the reversible reaction

$$C_{(s)} + CO_{2(g)} \rightleftharpoons 2\ CO_{(g)}$$

the equilibrium constant expression is

$$K_t = \frac{[CO]^2}{[C_{(s)}][CO_2]}$$

However, it can be shown that the concentration of any pure condensed phase (for example, solid carbon) is a constant as long as *any* of the condensed phase is present. The addition of more solid carbon increases its number of moles and its volume by the same factor so that the concentration of carbon in moles per liter remains unchanged. Since $[C_{(s)}]$ is constant, the equilibrium constant expression for this reaction can be simplified by defining a new constant K_t'.

$$K_t' = [C_{(s)}]K_t = \frac{[CO]^2}{[CO_2]}$$

Equilibrium constants for heterogeneous equilibria are normally written in this fashion unless stated otherwise.

b. Ionic Equilibria

The law of chemical equilibrium also applies to ionic equilibria. When a weak electrolyte (Section 7.4) such as acetic acid (CH_3COOH) is dissolved in aqueous solution, it partially dissociates into ions. This dissociation is a reversible change, and equilibrium between the undissociated species and its ions is, unlike many gaseous equilibria, attained almost instantaneously.

$$CH_3COOH \rightleftharpoons H^+ + CH_3COO^-$$

The equilibrium constant expression for this reversible reaction is

$$K_t = \frac{[H^+][CH_3COO^-]}{[CH_3COOH]} = K_a$$

When used to describe the dissociation of weak acids, K_t is usually labeled K_a. When used to describe the dissociation of weak bases, K_t is usually labeled K_b.

Equilibria involving the ionization of water and equilibria involving slightly soluble salts contain substances whose concentration is not able to change. In these cases the constant concentrations are combined mathematically with the equilibrium constant to form a new constant and leave a simplified equilibrium constant expression that omits the substance whose concentration is invariant.

Water is a very weak electrolyte that dissociates slightly to hydrogen ions and hydroxide ions.

$$H_2O \rightleftharpoons H^+ + OH^-$$

The equilibrium constant expression for the reversible dissociation of water is

$$K_t = \frac{[H^+][OH^-]}{[H_2O]}$$

The concentration of H_2O in pure water is constant at 55.5 *M*. Since $[H_2O]$ is a constant, the equilibrium constant expression can be simplified by defining a new constant K_w.

$$K_w = K_t[H_2O] = [H^+][OH^-]$$

K_w is called the dissociation constant or ion product of water and has a value of 1.0×10^{-14} at 25°.

In dilute aqueous solutions, the concentration of H_2O is also about 55.5 *M* and remains essentially constant. Its concentration is so large compared to the concentrations of H^+ and OH^- ions that it does not change appreciably with changes in the concentration of H^+ and OH^-. Thus, the ion product of pure water is also a valid expression for dilute aqueous solutions.

For similar reasons the concentration of H_2O is not incorporated in the equilibrium constant for any reversible reactions in which liquid water is both the solvent *and* a reactant or product. Such is the case for hydrolysis equilibria. It was noted in Section 8.2 that hydrolysis is the reaction of a cation or anion with water to produce H^+ or OH^- and a weakly dissociated species. Hydrolysis of a cation always produces H^+ and tends to make a solution more acidic. Hydrolysis of an anion always produces OH^- and tends to make a solution more basic. For the reversible hydrolysis reactions

$$Al^{3+} + H_2O \rightleftharpoons H^+ + AlOH^{2+}$$

and

$$S^{2-} + H_2O \rightleftharpoons OH^- + HS^-$$

the equilibrium constant expressions are

$$K_h = K_t[H_2O] = \frac{[H^+][AlOH^{2+}]}{[Al^{3+}]}$$

and

$$K_h = K_t[H_2O] = \frac{[OH^-][HS^-]}{[S^{2-}]}$$

There are a number of reactions involving organic compounds in which water is a reactant or product but *not* the solvent for the reactions (Section 19.8). The concentration of H_2O can change over a large range in such reactions, and the concentration of H_2O must appear in the equilibrium constant expression. For example, when glucose ($C_6H_{12}O_6$) is treated with methanol (CH_3OH) in the presence of an acid catalyst, an equilibrium mixture of α- and β-methylglucoside ($C_7H_{14}O_6$) is formed (see Figure 23.6 for structural formulas).

$$C_6H_{12}O_6 + CH_3OH \rightleftharpoons C_7H_{14}O_6 + H_2O$$

The equilibrium constant expression for this reversible reaction is

$$K_t = \frac{[C_7H_{14}O_6][H_2O]}{[C_6H_{12}O_6][CH_3OH]}$$

The process of dissolving a sparingly soluble salt is reversible, and the solid solute tends to come to equilibrium with its saturated solution.

$$CaCO_{3(s)} \rightleftharpoons CaCO_{3(aq)}$$

Since most salts are strong electrolytes, the small amount of the dissolved salt in the saturated solution is practically 100% dissociated. Thus, the reversible reaction of $CaCO_3$ in water can be written

$$CaCO_{3(s)} \rightleftharpoons Ca^{2+} + CO_3^{2-}$$

and the equilibrium constant expression is

$$K_t = \frac{[Ca^{2+}][CO_3^{2-}]}{[CaCO_{3(s)}]}$$

However, since the concentration of any pure condensed phase is a constant as long as any of the condensed phase is present, $[CaCO_{3(s)}]$ is constant, and the equilibrium constant expression can be simplified by defining a new constant K_{sp}.

$$K_{sp} = K_t[CaCO_{3(s)}] = [Ca^{2+}][CO_3^{2-}]$$

K_{sp} is called the solubility product and is equal to the product of the molar concentrations of the ions in the saturated solution taken to powers equal to the coefficients of the ions in the reversible reaction. Thus, for the reversible solution of silver phosphate

$$Ag_3PO_{4(s)} \rightleftharpoons 3\ Ag^+ + PO_4^{3-}$$

the solubility product expression is

$$K_{sp} = K_t[Ag_3PO_{4(s)}] = [Ag^+]^3[PO_4^{3-}]$$

Equilibrium constant expressions for other representative reversible reactions are given in Table 10.2.

10.3 Equilibrium Calculations

The equilibrium constant for a given reversible reaction at a given temperature is determined by allowing the system to come to equilibrium, measuring *experimentally* the molar concentrations of each substance present, and substituting the measured concentrations into the equilibrium constant expression.

Example 10.1 Calculate the equilibrium constant for the dissociation of acetic acid at 25° if the equilibrium concentrations of CH_3COOH, H^+, and CH_3COO^- are 0.50 *M*, 0.0030 *M*, and 0.0030 *M*, respectively.

Write the equilibrium constant expression for the reversible dissociation of acetic acid, $CH_3COOH \rightleftharpoons H^+ + CH_3COO^-$.

$$K_a = \frac{[H^+][CH_3COO^-]}{[CH_3COOH]}$$

Substitute the molar concentrations and calculate $K_{25°}$.

$$K_{25°} = \frac{(0.0030)(0.0030)}{0.50} = 1.8 \times 10^{-5}$$

Equilibrium constants may be very large, near unity, or very small. A small equilibrium constant ($K \ll 1$), as in Example 10.1, implies that the reaction does not proceed far from left to right. The experimental equilib-

Table 10.2 Reversible Reactions and Equilibrium Constant Expressions

Reaction	Equilibrium Constant Expression
$2\ NO_{(g)} + O_{2(g)} \rightleftharpoons 2\ NO_{2(g)}$	$K_t = \frac{[NO_2]^2}{[NO]^2[O_2]}$
$CaO_{(s)} + SO_{3(g)} \rightleftharpoons CaSO_{4(s)}$	$K_t' = \frac{1}{[SO_3]}$
$HCN \rightleftharpoons H^+ + CN^-$	$K_a = \frac{[H^+][CN^-]}{[HCN]}$
$HgI_4^{2-} \rightleftharpoons Hg^{2+} + 4\ I^-$	$K_t = \frac{[Hg^{2+}][I^-]^4}{[HgI_4^{2-}]}$
$CH_3COOH + CH_3CH_2OH \rightleftharpoons CH_3COOCH_2CH_3 + H_2O$	$K_t = \frac{[CH_3COOCH_2CH_3][H_2O]}{[CH_3COOH][CH_3CH_2OH]}$
$Bi_2S_{3(s)} \rightleftharpoons 2\ Bi^{3+} + 3\ S^{2-}$	$K_{sp} = [Bi^{3+}]^2[S^{2-}]^3$
$Cr^{3+} + H_2O \rightleftharpoons H^+ + CrOH^{2+}$	$K_h = \frac{[H^+][CrOH^{2+}]}{[Cr^{3+}]}$

rium concentrations in Example 10.1 verify that little H^+ and CH_3COO^- are present compared to CH_3COOH when the system is at equilibrium. A large equilibrium constant ($K \gg 1$) implies that the reaction proceeds extensively from left to right.

A dissociation constant calculated as in Example 10.1 holds for all dilute solutions of that weak electrolyte at the same temperature. Other representative dissociation constants for weak acids are given in Table 10.3. Similar data are available in handbooks for weak bases. Strictly speaking these values for the equilibrium constants are only applicable to dilute solutions, in which ions are essentially independent of each other. In concentrated solutions, ions have significant interactions with each other even when separated by intervening water molecules.

The equilibrium constants in Table 10.3 are a quantitative indication of acid strength. A larger value of K_a implies a stronger acid; a smaller value of K_a implies a weaker acid. Common inorganic acids such as perchloric ($HClO_4$), nitric (HNO_3), sulfuric (H_2SO_4, first dissociation step only), hydrochloric (HCl), hydrobromic (HBr), and hydriodic (HI) acid have K_a values greater than 10 and are essentially 100 percent dissociated in all except very concentrated solutions. On the other hand, the majority of organic acids and some inorganic acids are only slightly dissociated and have low concentrations of hydrogen ion and high concentrations of undissociated acid in solution. Boric acid is weak enough, for example, that it is used commonly as an eyewash. The effect of structure on the acidity of organic acids will be discussed in Section 19.4.

Table 10.3 Dissociation Constants for Weak Acids

Name	*Reaction*[a]	$K_{25°}$
Oxalic	$HOOCCOOH \rightleftharpoons H^+ + HOOCCOO^-$	2.0×10^{-1}
	$HOOCCOO^- \rightleftharpoons H^+ + OOCCOO^{2-}$	6.1×10^{-5}
Hydrogen sulfate ion	$HSO_4^- \rightleftharpoons H^+ + SO_4^{2-}$	1.2×10^{-2}
Phosphoric	$H_3PO_4 \rightleftharpoons H^+ + H_2PO_4^-$	7.3×10^{-3}
	$H_2PO_4^- \rightleftharpoons H^+ + HPO_4^{2-}$	6.2×10^{-8}
	$HPO_4^{2-} \rightleftharpoons H^+ + PO_4^{3-}$	1.7×10^{-12}
Nitrous	$HNO_2 \rightleftharpoons H^+ + NO_2^-$	4.5×10^{-4}
Formic	$HCOOH \rightleftharpoons H^+ + HCOO^-$	1.8×10^{-4}
Benzoic	$C_6H_5COOH \rightleftharpoons H^+ + C_6H_5COO^-$	6.5×10^{-5}
Acetic	$CH_3COOH \rightleftharpoons H^+ + CH_3COO^-$	1.8×10^{-5}
Carbonic	$CO_{2(aq)} + H_2O \rightleftharpoons H^+ + HCO_3^-$	4.3×10^{-7}
	$HCO_3^- \rightleftharpoons H^+ + CO_3^{2-}$	5.6×10^{-11}
Boric	$H_3BO_3 + H_2O \rightleftharpoons H^+ + B(OH)_4^-$	6.0×10^{-10}
Water	$H_2O \rightleftharpoons H^+ + OH^-$	1.0×10^{-14}

[a] All species are assumed to be in aqueous solution and are therefore hydrated.

Equilibrium constants can be used for calculating the concentrations of ions present in solutions of any total molar concentration. Whatever the total quantity of electrolyte in solution, the respective molar concentrations of ions and undissociated molecules must obey the requirements of the equilibrium constant expression. The $K_{25°}$ values in Table 10.3 indicate that hydrogen sulfate ion (HSO_4^-) is a stronger acid than nitrous acid. If we calculate the hydrogen ion concentrations of equimolar solutions of hydrogen sulfate ion and nitrous acid, we should expect that the solution of hydrogen sulfate ion should have the higher concentration of H^+. This expectation is shown to be correct in Examples 10.2 and 10.3.

Example 10.2 Calculate the equilibrium concentrations of all species and the percent dissociation of a 0.10 *M* solution of nitrous acid at 25°.

Write the equation for the reversible dissociation of nitrous acid. Determine the before-reaction concentrations and the equilibrium concentrations using the procedures outlined below, and place them under the equation.

	HNO_2	$\rightleftharpoons$ H^+	+ NO_2^-
Before reaction	0.10 *M*	0	0
Equilibrium	$(0.10 - y)$ *M*	y *M*	y *M*

Let y equal the concentration of nitrous acid that dissociates in 0.10 *M* solution to establish equilibrium. The equilibrium concentration of undissociated nitrous acid is thus the before-reaction concentration minus that dissociated, or $(0.10 - y)$ *M*. There is no H^+ (neglecting the very slight dissociation of water) or NO_2^- before the reaction, but according to the balanced equation, dissociation of y *M* HNO_2 yields y *M* H^+ and y *M* NO_2^-.

Write the equilibrium constant expression for the reversible dissociation of HNO_2, and substitute the equilibrium concentrations of H^+, NO_2^-, and HNO_2.

$$K_{25°} = \frac{[H^+][NO_2^-]}{[HNO_2]} = \frac{(y)(y)}{0.10 - y} = 4.5 \times 10^{-4} \quad \text{(from Table 10.3)}$$

Sometimes it is necessary to solve an equation like this one using the quadratic formula (see Example 10.3). However, careful analysis of the problem often may lead to approximations that eliminate the need for such laborious mathematical manipulations. Such is the case in this example. Since nitrous acid is a weak electrolyte as indicated by the small value of $K_{25°}$, dissociation or reaction to the right is very slight, and y is small compared to 0.10. It is sufficiently accurate to neglect y in $(0.10 - y)$ and to consider the concentration of undissociated HNO_2 to be 0.10 M. The above equation then becomes one that is solved easily.

$$\frac{(y)(y)}{0.10} = 4.5 \times 10^{-4}$$

$$y^2 = (0.10)(4.5 \times 10^{-4}) = 45 \times 10^{-6}$$

$$y = [H^+] = [NO_2^-] = 0.0067\ M$$

$$[HNO_2] = 0.10\ M$$

Note that 0.0067 M is essentially negligible compared to 0.10 M; therefore, our approximation was a valid one. As a general rule, such an approximation is valid if the calculated quantity added to or subtracted from the known quantity is *less than* 10% of the known quantity. Thus, the above approximation would have been valid for any y less than 0.010 M. If the approximation proves invalid, then you must go back and solve the quadratic equation (see Example 10.3).

The percent dissociation is given by the concentration of either of the dissociated ions (for example, $[NO_2^-]$) divided by the initial concentration of HNO_2, and multiplied by 100% to convert to percent.

$$\text{Percent dissociation} = \left(\frac{0.0067\ \cancel{M}}{0.10\ \cancel{M}}\right)(100\%) = 6.7\%$$

This means that less than seven of every one hundred molecules of nitrous acid are dissociated at any given time.

Example 10.3 **Calculate the equilibrium concentrations of all species and the percent dissociation of a 0.10 M HSO_4^- solution.**

Write the equation for the reversible dissociation of HSO_4^-. Determine the before-reaction concentrations and the equilibrium concentrations using the procedures outlined below, and place them under the equation.

	HSO_4^-	$\rightleftharpoons$	H^+	+	SO_4^{2-}
Before reaction	0.10 M		0		0
Equilibrium	$(0.10 - y)\ M$		$y\ M$		$y\ M$

Let y equal the concentration of HSO_4^- that dissociates in the 0.10 M solution to establish equilibrium. The equilibrium concentration of undissociated HSO_4^- is thus the before-reaction concentration minus that dissociated, or $(0.10 - y)$ M. There is no H^+ (neglecting the very slight dissociation of water) or SO_4^{2-} before the reaction, but according to the balanced equation, dissociation of y M HSO_4^- yields y M H^+ and y M SO_4^{2-}.

Write the equilibrium constant expression for the reversible dissociation of HSO_4^-, and substitute the equilibrium concentrations of H^+, SO_4^{2-}, and HSO_4^-.

$$K_{25^\circ} = \frac{[H^+][SO_4^{2-}]}{[HSO_4^-]} = \frac{(y)(y)}{0.10 - y} = 1.2 \times 10^{-2} \quad \text{(from Table 10.3)}$$

Because it requires so little effort, one should first try the assumption that y is negligible compared to 0.10, as in Example 10.2, and then solve the resulting equation.

$$\frac{(y)(y)}{0.10} = 1.2 \times 10^{-2}$$

$$y^2 = (0.10)(1.2 \times 10^{-2}) = 12 \times 10^{-4}$$

$$y = 0.035\ M$$

It can be seen readily that y (0.035 M) is greater than 10% of 0.10 and that our approximation is not valid. This might have been expected since K_{25° is not very small. It is close enough to unity (within a couple powers of ten) that significant concentrations of H^+ and SO_4^{2-} are produced by dissociation of HSO_4^-. Therefore, the original equation with no approximations must be solved rigorously using the quadratic formula.

$$\frac{(y)(y)}{0.10 - y} = 1.2 \times 10^{-2}$$

$$y^2 = (0.10 - y)(1.2 \times 10^{-2}) = 0.0012 - 0.012\,y$$

$$y^2 + 0.012\,y - 0.0012 = 0$$

This is a quadratic equation because its variable y is raised to the second power, but no higher. The solution to a quadratic equation written in the form $ay^2 + by + c = 0$ is

$$y = \frac{-b \pm \sqrt{b^2 - 4ac}}{2a}$$

The $\pm$ sign indicates that there are two roots; however, only one of them will be physically meaningful in equilibrium calculations. The roots for our equation above can be obtained using the quadratic formula ($a = 1$, $b = 0.012$, $c = -0.0012$).

$$y = \frac{-0.012 \pm \sqrt{(0.012)^2 - (4)(1)(-0.0012)}}{2}$$

$$= \frac{-0.012 \pm \sqrt{0.000144 + 0.0048}}{2}$$

$$= \frac{-0.012 \pm 0.070}{2}$$

$$= 0.029 \quad \text{or} \quad -0.041$$

The negative concentration is physically meaningless and can be discarded. Thus, $y = [H^+] = [SO_4^{2-}] = 0.029\ M$, and $[HSO_4^-] = (0.10 - y)$ or $0.071\ M$.

$$\text{Percent dissociation} = \left(\frac{0.029\ \cancel{M}}{0.10\ \cancel{M}}\right)(100\%) = 29\%$$

This means that twenty-nine of every one hundred HSO_4^- ions are dissociated at any given time.

The hydrogen ion concentration of $0.10\ M\ HSO_4^-$ ($0.029\ M$) is, as predicted, greater than that for $0.10\ M\ HNO_2$ ($0.0067\ M$). Parallel to this, the percent dissociation of HSO_4^- (29%) is greater than that for $0.10\ M\ HNO_2$ (6.7%).

Calculations can be performed in a similar manner for solutions of weak bases.

Example 10.4 Calculate the equilibrium concentrations of all species in a $0.20\ M$ solution of hydroxylamine at 25° if $K_b = 1.07 \times 10^{-8}$.

The equation for the reversible dissociation of hydroxylamine is given below. Determine the before-reaction concentrations and the equilibrium concentrations using the procedures outlined previously, and place them under the equation.

	$HONH_2$	$+\ H_2O \rightleftharpoons$	$HONH_3^+$	$+\ OH^-$
Before reaction	$0.20\ M$		0	0
Equilibrium	$(0.20 - y)\ M$		$y\ M$	$y\ M$

Write the equilibrium constant expression for the reversible dissociation of $HONH_2$, and substitute the equilibrium concentrations of $HONH_3^+$, OH^-, and $HONH_2$.

$$K_{25^\circ} = \frac{[HONH_3^+][OH^-]}{[HONH_2]} = \frac{(y)(y)}{0.20 - y} = 1.07 \times 10^{-8}$$

Neglect y in the denominator compared to 0.20 since K_{25° is very small (as outlined in Example 10.2), and solve the resulting equation.

$$\frac{(y)(y)}{0.20} = 1.07 \times 10^{-8}$$

$$y^2 = (0.20)(1.07 \times 10^{-8}) = 21 \times 10^{-10}$$

$$y = [HONH_3^+] = [OH^-] = 4.6 \times 10^{-5}\ M$$

$$[HONH_2] = 0.20\ M$$

10.4 Factors Affecting Equilibria

When a reversible chemical reaction at equilibrium is disturbed, chemical reaction in one direction or the other dominates until equilibrium is re-established. The reversible reaction is said to shift to overcome the disturbance. An equilibrium can be disturbed by changing the concentration of a reactant or product, by changing the temperature, or by changing the volume.

a. Change of Concentration of a Reactant

Suppose that additional $H_2PO_4^-$ (in the form of solid NaH_2PO_4) is added to an aqueous solution in which $H_2PO_4^-$, H^+, and HPO_4^{2-} are in equilibrium. What effect does this have on the following reversible reaction?

$$H_2PO_4^- \rightleftharpoons H^+ + HPO_4^{2-}$$

A principle put forth by Henri Le Châtelier, a French chemist, in 1884 states that *if a stress is applied to a system at equilibrium, the system readjusts to reduce the stress.* Applied to the problem of $H_2PO_4^-$ addition to the $H_2PO_4^-$-H^+-HPO_4^{2-} system, the system readjusts to reduce the stress of an increased $H_2PO_4^-$ concentration. The increased $H_2PO_4^-$ concentration causes the rate of the forward reaction ($H_2PO_4^- \rightarrow H^+ + HPO_4^{2-}$) to increase relative to the rate of the reverse reaction ($H^+ + HPO_4^{2-} \rightarrow H_2PO_4^-$). The net result is that some of the added $H_2PO_4^-$ dissociates to form H^+ and HPO_4^{2-}, and we say that the reversible reaction is *shifted to the right.* However, the imbalance in the rates of the forward and reverse reactions is only temporary. Eventually the system returns to equilibrium, and the rates of the forward and reverse reactions are again equal. When this new equilibrium is attained, the concentrations of H^+, HPO_4^{2-}, and $H_2PO_4^-$ are different from what they were before the stress was applied (all increased in this particular case). However, their ratio (K_a) is the same as before the stress was applied.

Analysis of the equilibrium constant expression

$$K_a = \frac{[H^+][HPO_4^{2-}]}{[H_2PO_4^-]}$$

leads to a similar conclusion. Adding $H_2PO_4^-$ and increasing the $H_2PO_4^-$ concentration increases the denominator. Other concentrations remaining the same, the fraction becomes too small, and the system is no longer at equilibrium. The system can establish a new equilibrium, and thus obey the constancy of K_a, by simultaneously increasing the numerator and decreasing the denominator. This increases the concentrations of H^+ and HPO_4^{2-} while decreasing the $H_2PO_4^-$ concentration and corresponds to a shift of the reversible reaction to the right.

Quantitative calculations verify these predictions. It is left to you as an exercise to determine that in a 0.10 *M* $H_2PO_4^-$ solution the equilibrium concentrations of $H_2PO_4^-$, H^+, and HPO_4^{2-} are 0.10 *M*, 7.9×10^{-5} *M*, and 7.9×10^{-5} *M*, respectively. If 1.00 mole of solid NaH_2PO_4 is added to 1.00 liter of this particular $H_2PO_4^-$-H^+-HPO_4^{2-} system, the reversible reaction

$$H_2PO_4^- \rightleftharpoons H^+ + HPO_4^{2-}$$

shifts to the right, and the final equilibrium concentrations of $H_2PO_4^-$, H^+, and HPO_4^{2-} are 1.10 *M*, 8.3×10^{-4} *M*, and 8.3×10^{-4} *M*, respectively.

By converse reasoning, it can be argued that selective removal of $H_2PO_4^-$ from a solution in which $H_2PO_4^-$, H^+, and HPO_4^{2-} are in equilibrium shifts the reversible reaction to the left.

b. Change of Concentration of a Product

Suppose that additional SO_4^{2-} (in the form of solid Na_2SO_4) is added to an aqueous solution in which HSO_4^-, H^+, and SO_4^{2-} are in equilibrium. What effect does this have on the following reversible reaction?

$$HSO_4^- \rightleftharpoons H^+ + SO_4^{2-}$$

According to Le Châtelier's principle, the system tries to reduce the stress of added SO_4^{2-}. This can be accomplished by shifting the reversible reaction to the left, increasing the HSO_4^- concentration and decreasing the concentrations of H^+ and SO_4^{2-}. Note, however, that the final equilibrium concentration of SO_4^{2-} is still greater than its initial equilibrium concentration. The final equilibrium concentration of HSO_4^- is also larger than its initial equilibrium concentration, but the converse is true for H^+.

Analysis of the equilibrium constant expression

$$K_a = \frac{[H^+][SO_4^{2-}]}{[HSO_4^-]}$$

leads to a similar conclusion. Adding SO_4^{2-} and increasing the concentration of SO_4^{2-} increases the numerator. Other concentrations remaining the same, the fraction becomes too large, and the system is no longer at equilibrium. The system can establish a new equilibrium and thus obey the constancy of K_a by simultaneously decreasing the numerator and increasing the denominator. This decreases the concentrations of H^+ and SO_4^{2-} while increasing the HSO_4^- concentration and corresponds to a shift of the reversible reaction to the left.

Quantitative calculations verify these predictions. For instance, in Example 10.3 we calculated the equilibrium concentrations in a 0.10 *M* HSO_4^- solution to be 0.071 *M* HSO_4^-, 0.029 *M* H^+, and 0.029 *M* SO_4^{2-}. If 0.050 mole of solid sodium sulfate (Na_2SO_4) is added to 1.00 liter of a 0.10 *M* HSO_4^- solution, the reversible reaction

$$HSO_4^- \rightleftharpoons H^+ + SO_4^{2-}$$

shifts to the left, and the final equilibrium concentrations are 0.085 *M* HSO_4^-, 0.015 *M* H^+, and 0.065 *M* SO_4^{2-}.

By similar reasoning, it can be argued that addition of H^+ to a solution in which HSO_4^-, H^+, and SO_4^{2-} are in equilibrium also shifts the reversible reaction to the left. On the other hand, by converse reasoning, it can be argued that selective removal of either H^+ (for example, by neutralization with NaOH) or SO_4^{2-} (for example, by precipitation with $BaCl_2$) from a solution in which HSO_4^-, H^+, and SO_4^{2-} are in equilibrium shifts the reversible reaction to the right.

In addition, this example indicates that at equilibrium, molar concentrations need not necessarily be in the same ratio as the mole ratio indicated by the equation. The only requirement is that the law of chemical equilibrium be followed so that the total equilibrium constant expression is equal to the equilibrium constant. In dilute solutions we can assume strong electrolytes to be 100% dissociated. Therefore, as was done in this example, we can increase the concentration of one of the ions involved in the dissociation of a weak electrolyte by adding a soluble strong electrolyte that will furnish that ion. The resultant shift of a reversible reaction when one of the ions of that reaction is added and its concentration is increased is frequently called the *common-ion effect*. Buffer solutions (Section 10.6) function as a result of the common-ion effect.

c. Change of Temperature

Suppose that the temperature of a container in which gaseous H_2O, H_2, and O_2 are in equilibrium is increased. What effect does this have on the reversible reaction

$$H_2O_{(g)} \rightleftharpoons H_{2(g)} + \tfrac{1}{2}O_{2(g)}$$

if it is known that reaction to the right is an endothermic reaction?

According to Le Châtelier's principle, the system tries to reduce the stress of increased temperature and added heat. This can be accomplished by shifting the reversible reaction to the right since reaction in that direction requires heat or uses heat.

$$H_2O_{(g)} + 57.79\text{ kcal} \rightleftharpoons H_{2(g)} + \tfrac{1}{2}O_{2(g)}$$

This decreases the H_2O concentration and increases the concentrations of H_2 and O_2.

Analysis of the equilibrium constant expression

$$K_t = \frac{[H_2][O_2]^{1/2}}{[H_2O]}$$

leads to a similar conclusion. The equilibrium constant has a specific value at a given temperature. Increasing the temperature causes K_t to decrease for exothermic reactions and to increase for endothermic reactions. The reaction under consideration is an endothermic reaction; from experiment, K_t increases from 7.9×10^{-40} at 300 K to 9.1×10^{-10} at 1000 K to 18.3 at 5000 K. Thus, the concentrations of H_2 and O_2 increase and the H_2O concentration decreases, or alternatively, the reversible reaction shifts to the right as the temperature increases. Most decomposition reactions, in which a larger molecule breaks down to smaller molecules, are endothermic. They proceed to a greater extent at higher temperatures.

By converse reasoning, it can be argued that decreasing the temperature of a container in which H_2O, H_2, and O_2 are in equilibrium shifts the reversible reaction to the left.

d. Change in Volume

Suppose that the volume of a container in which N_2, H_2, and NH_3 are in equilibrium is decreased. What effect does this have on the following reversible reaction?

$$N_{2(g)} + 3\,H_{2(g)} \rightleftharpoons 2\,NH_{3(g)}$$

According to Le Châtelier's principle, the system tries to reduce the stress of the increased crowding of molecules resulting from decreasing the volume of the container. This can be accomplished by decreasing the total number of molecules in the system as a result of shifting the reversible reaction toward the side of fewer molecules—toward the right in the N_2–H_2–NH_3 system.

Analysis of the equilibrium constant expression

$$K_t = \frac{[NH_3]^2}{[N_2][H_2]^3}$$

leads to a similar conclusion. The concentrations of all species increase as the volume of the container is decreased. In this case two concentration terms are increasing in the numerator (since $[NH_3]$ is squared) and four concentration terms are increasing in the denominator (since $[H_2]$ is cubed). Other factors remaining the same, the fraction becomes too small, and the system is no longer at equilibrium. The system can establish a new equilibrium, and thus obey the constancy of K_t, by simultaneously increasing the numerator and decreasing the denominator. This increases the NH_3 concentration while decreasing the concentrations of N_2 and H_2 and corresponds to a shift of the reversible reaction to the right.

By converse reasoning, it can be argued that increasing the volume of a container in which N_2, H_2, and NH_3 are in equilibrium shifts the reversible reaction to the left.

For reversible reactions having the same number of molecules on both sides of the equation, there is no net change when the volume of the system is increased or decreased. For each of the reversible reactions

$$2\,HI_{(g)} \rightleftharpoons H_{2(g)} + I_{2(g)} \qquad K_t = \frac{[H_2][I_2]}{[HI]^2}$$

$$H_2O_{(g)} + CO_{(g)} \rightleftharpoons H_{2(g)} + CO_{2(g)} \qquad K_t = \frac{[H_2][CO_2]}{[H_2O][CO]}$$

Le Châtelier's principle predicts that neither the forward nor the reverse reaction can reduce the stress resulting from a decreased volume of the system. Likewise, analysis of each equilibrium constant expression shows that a decrease of volume causes an increase of two concentration terms in each numerator and two concentration terms in each denominator. These increases cancel each other, the constancy of each K_t is obeyed, the systems are still at equilibrium, and no shifts in either reversible reaction take place.

In addition, it should be noted that, since liquids and solids are quite incompressible, volume changes are inconsequential for reactions involving *only* pure liquids and pure solids. However, dilution with a solvent can result in volume changes for solutions.

e. Addition of a Catalyst

Suppose that maltase, an effective catalyst for the hydrolysis of maltose to glucose (Section 23.7-b), is added to a solution in which maltose ($C_{12}H_{22}O_{11}$), water, and glucose ($C_6H_{12}O_6$) are in equilibrium. What effect does this have on the following reversible reaction?

$$C_{12}H_{22}O_{11} + H_2O \rightleftharpoons 2\,C_6H_{12}O_6$$

Le Châtelier's principle ignores the possibility of a catalyst and cannot answer this question.

The equilibrium constant expression

$$K_t = \frac{[C_6H_{12}O_6]^2}{[C_{12}H_{22}O_{11}]}$$

involves only the equilibrium concentrations of the substances in the net equation. A catalyst affects intermediates that do not appear in the net equation and increases the rates of opposing reactions. *A catalyst does not,* however, *change equilibrium concentrations,* the constancy of K_t is obeyed, the system is still at equilibrium, and no net change in the reversible reaction takes place.

10.5 Hydrogen Ion Concentration and pH

It was noted in Section 10.2 that the ion product for the reversible dissociation of water, $H_2O \rightleftharpoons H^+ + OH^-$, is $K_w = [H^+][OH^-] = 1.0 \times 10^{-14}$ at 25°. In pure water, all H^+ and OH^- come from the dissociation of water, and H^+ and OH^- are produced in a 1:1 mole ratio. Therefore, $[H^+] = [OH^-] = 1.0 \times 10^{-7}$ *M*. Pure water is said to be **a neutral substance** because it *contains equal concentrations of H^+ and OH^-.*

Although the ion product, $[H^+][OH^-]$, must equal 1.0×10^{-14} in any aqueous solution at equilibrium at 25°, the concentrations of H^+ and OH^- need not be equal. If the H^+ concentration is greater than 1.0×10^{-7} *M*, the solution is acidic; if the H^+ concentration is less than 1.0×10^{-7} *M*, the solution is basic.

Example 10.5 What is the H^+ concentration in an aqueous solution in which the OH^- concentration is 1.0×10^{-5} *M*? Is the solution acidic or basic?

$$K_w = [H^+][OH^-] = 1.0 \times 10^{-14}$$

Solving for $[H^+]$

$$[H^+] = \frac{K_w}{[OH^-]} = \frac{1.0 \times 10^{-14}}{1.0 \times 10^{-5}} = 1.0 \times 10^{-9}\ M$$

The H^+ concentration is less than 1.0×10^{-7} *M*; therefore, the solution is basic.

Since it is cumbersome to work with the powers of ten involved with the small H^+ concentrations in most aqueous solutions, the term pH has been devised to express H^+ concentrations. The **pH** is defined as *the negative logarithm (base 10) of the hydrogen ion concentration.*

$$\mathbf{pH} = -\log [H^+]$$

Example 10.6 Calculate the pH of a neutral solution ($[H^+] = 1.0 \times 10^{-7}$ *M*).

$$\text{pH} = -\log [H^+] = -\log (1.0 \times 10^{-7})$$

The logarithm of a product ($a \times b$) equals the sum of the logarithms of a and b, or $\log (a \times b) = \log a + \log b$. Therefore,

$$\text{pH} = -\log (1.0 \times 10^{-7}) = -\log 1.0 - \log 10^{-7}$$

The logarithm of 10 raised to a power n equals n times the logarithm of 10, or $\log 10^n = n \log 10$. Therefore,

$$\text{pH} = -\log 1.0 + 7 \log 10$$

Now since log 1.0 = 0.0 and log 10 = 1.0,

$$pH = -0.0 + 7(1.0) = 7.0$$

All neutral solutions have a pH of 7. Acidic solutions have pH values less than 7; basic solutions have pH values greater than 7. Figure 10.2 shows pH values for some common substances. Note that most ingestible substances have pH values below 7.

It can be generalized from Example 10.6 that for all H^+ concentrations that are integral powers of 10, their pH values are always equal to minus the exponent of 10. Thus, if $[H^+] = 1.0 \times 10^{-3}$ *M*, the pH is 3.0; if $[H^+] = 1.0 \times 10^{-6}$ *M*, the pH is 6.0; if $[H^+] = 1.0 \times 10^{-10}$ *M*, the pH is 10.0.

However, many H^+ concentrations are not integral powers of 10. Their pH values are not obvious and must be determined by the use of a logarithm table like that in Table 10.4.

Example 10.7 It is left to you as an exercise to determine that the H^+

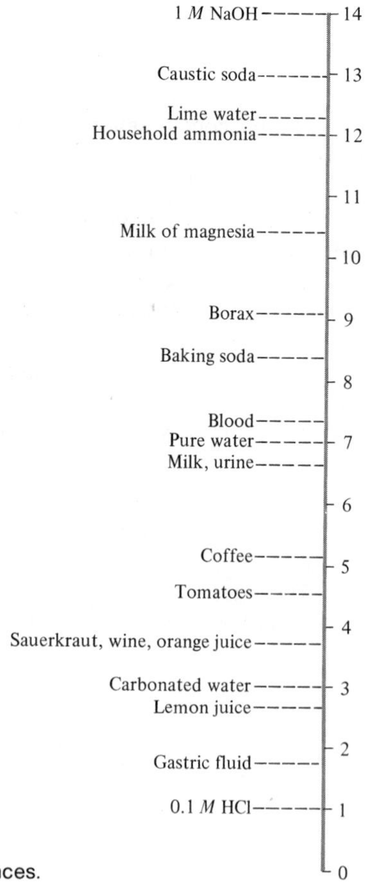
1 M NaOH
Caustic soda
Lime water
Household ammonia
Milk of magnesia
Borax
Baking soda
Blood
Pure water
Milk, urine
Coffee
Tomatoes
Sauerkraut, wine, orange juice
Carbonated water
Lemon juice
Gastric fluid
0.1 M HCl
14
13
12
11
10
9
8
7
6
5
4
3
2
1
0

Figure 10.2 pH values of some common substances.

Table 10.4 Two-Place Common Logarithms

	0.0	*0.1*	*0.2*	*0.3*	*0.4*	*0.5*	*0.6*	*0.7*	*0.8*	*0.9*
1	.00	.04	.08	.11	.15	.18	.20	.23	.26	.28
2	.30	.32	.34	.36	.38	.40	.42	.43	.45	.46
3	.48	.49	.51	.52	.53	.54	.56	.57	.58	.59
4	.60	.61	.62	.63	.64	.65	.66	.67	.68	.69
5	.70	.71	.72	.72	.73	.74	.75	.76	.76	.77
6	.78	.79	.79	.80	.81	.81	.82	.83	.83	.84
7	.85	.85	.86	.86	.87	.88	.88	.89	.89	.90
8	.90	.91	.91	.92	.92	.93	.94	.94	.95	.95
9	.95	.96	.96	.97	.97	.98	.98	.99	.99	1.00

concentration of a 0.10 *M* formic acid solution (see Table 10.3) is 0.0042 *M*. What is the pH of 0.10 *M* formic acid?

Convert the H^+ concentration to an exponential number.

$$[H^+] = 0.0042\ M = 4.2 \times 10^{-3}\ M$$
$$pH = -\log [H^+] = -\log (4.2 \times 10^{-3})$$

The logarithm of a product equals the sum of the logarithms. Therefore,

$$pH = -\log 4.2 - \log 10^{-3}$$

To get the logarithm of the number 4.2, look for 4 in the first vertical column of Table 10.4 and then move horizontally to the vertical column headed by 0.2. The logarithm of 4.2 is 0.62. The logarithm of 10^{-3} equals -3.00. Therefore,

$$pH = -0.62 + 3.00 = 2.38$$

Sometimes the reverse procedure is required to calculate an H^+ concentration from a pH.

Example 10.8 Calculate the concentrations of H^+ and OH^- in a 0.10 *M* pyridine solution for which the pH is 9.08.

$$pH = -\log [H^+] = 9.08 \qquad \text{or} \qquad \log [H^+] = -9.08$$

Split the logarithm into a sum of a positive number and the next higher negative integer.

$$\log [H^+] = 0.92 - 10$$

Solve this equation for $[H^+]$ by finding an antilog through the use of Table 10.4 in the reverse of the way it was used in Example 10.7. Find 0.92 in the body of Table 10.4 and determine the number from the first column and top row which corresponds to it. The antilog of 0.92 is 8.3. The antilog of an integer is 10 raised to the power equal to that integer. Therefore, the antilog of -10 is 10^{-10}. Thus, $[H^+] = 8.3 \times 10^{-10}\ M$. You can easily check this calculation by taking the negative logarithm of this result and seeing that you get back to a pH of 9.08.

Use the ion product of water to solve for $[OH^-]$.

$$K_w = [H^+][OH^-] = 1.0 \times 10^{-14}$$

$$[OH^-] = \frac{K_w}{[H^+]} = \frac{1.0 \times 10^{-14}}{8.3 \times 10^{-10}} = 1.2 \times 10^{-5}\ M$$

10.6 Buffer Solutions

In a large number of chemical reactions (especially those carried out in aqueous solution), it is important that the pH remains relatively constant. Practically all biological and physiological processes occur most readily at some unique pH value. The pH of human blood, for example, varies only slightly between the values of 7.35 and 7.45, and for any given person, the pH of venous and arterial blood differs by only 0.02 pH unit despite the production of numerous acids and bases in the cells. Blood would lose its capability for transporting oxygen if the pH were to fall outside this range.

Any solution that maintains a near constancy of pH is referred to as a **buffer solution.** Such a solution contains a combination of two chemical species that work as a pair in resisting marked changes of pH. Solutions containing an excess of a weak acid and the anion of that weak acid, or an excess of a weak base and the cation of that weak base, serve as excellent buffer solutions.

Maintenance of a nearly constant pH is achieved by the buffering action of an acid-base equilibrium. The buffering phenomenon is best understood by examining the processes that occur when a strong acid or a strong base is added to a buffer solution. We shall consider the bicarbonate buffer pair $[CO_{2(aq)}/HCO_3^-]$, prepared by bubbling CO_2 gas through an aqueous solution containing HCO_3^- ion, but the general discussion is applicable to all buffer pairs. The principal reversible reaction in solution is

$$[CO_{2(aq)} + H_2O \rightleftharpoons H^+ + HCO_3^-]^2$$

According to the arguments in Section 10.4-b, if a strong acid such as HCl is added to this buffer solution, the reversible reaction shifts to the left. The added H^+ ions (from the complete dissociation of HCl: $HCl \longrightarrow H^+ + Cl^-$) combine with excess HCO_3^- ions, present in the solution, to form hydrated CO_2 and water. Nearly all of the added H^+ ions are used up, and the H^+ concentration of the resulting solution is not appreciably altered from that of the original solution. A net conversion of HCO_3^- to CO_2 results, but since these species are present in excessive amounts in the original buffer solution, the change in their concentrations is insignificant.

Conversely, by arguments in Section 10.4-b, if a strong base such as

[2] An aqueous solution of CO_2 is often referred to as carbonic acid (H_2CO_3); however, there is no evidence that H_2CO_3 actually exists. Instead, hydrated CO_2 is probably the species present.

NaOH is added to this buffer solution, the reversible reaction shifts to the right. The added OH^- ions (from the complete dissociation of NaOH: $NaOH \longrightarrow Na^+ + OH^-$) neutralize H^+ ions to form water, and remove H^+ ions from solution. As this occurs, excess $CO_{2(aq)}$, present in the solution, reacts with water to form H^+ and HCO_3^- and re-establish equilibrium. Nearly all of the added OH^- ions are used up, and the H^+ ions are replenished so that the H^+ concentration in the resulting solution is not appreciably altered from that in the original solution. A net conversion of CO_2 to HCO_3^- results, but since these species are present in excessive amounts in the original buffer solution, the change in their concentrations is insignificant.

Like all buffer solutions, this one has a finite capacity. It is most effective when the numbers of moles of $CO_{2(aq)}$ and HCO_3^- are nearly equal and it ceases functioning when the number of moles of added acid or base approaches the number of moles of $CO_{2(aq)}$ and HCO_3^- present in the solution.

By starting with the equilibrium constant expression for the above reversible reaction, we can easily derive an equation that permits us to calculate the pH of the given buffer solution. The derivation is applicable to solutions containing any buffer pairs.

$$K_a = \frac{[H^+][HCO_3^-]}{[CO_{2(aq)}]}$$

Rearranging,

$$[H^+] = K_a \frac{[CO_{2(aq)}]}{[HCO_3^-]}$$

Taking the logarithm of both sides of the equation,

$$\log [H^+] = \log K_a + \log \frac{[CO_{2(aq)}]}{[HCO_3^-]}$$

Multiplying both sides of the equation by -1,

$$-\log [H^+] = -\log K_a - \log \frac{[CO_{2(aq)}]}{[HCO_3^-]}$$

Since $-\log [H^+] = pH$, $-\log K_a = pK_a$, and

$$-\log \frac{[CO_{2(aq)}]}{[HCO_3^-]} = +\log \frac{[HCO_3^-]}{[CO_{2(aq)}]}$$

we can obtain the expression

$$pH = pK_a + \log \frac{[HCO_3^-]}{[CO_{2(aq)}]}$$

This expression is referred to as the Henderson–Hasselbach equation. This equation is valid only when the ratio of the concentrations in the log term above is near unity, that is, when $[HCO_3^-]$ is approximately equal to $[CO_{2(aq)}]$.

Since pK_a is the negative logarithm of the dissociation constant of an acid, it too is a constant, and thus the pH of any particular buffer solution varies with the ratio of the concentration of the anion to that of the acid. If the weak acid and its anion are present in equal concentrations, the hydrogen ion concentration of the buffered solution will be equal to the dissociation constant of the weak acid; thus, $pH = pK_a$. Reference to a table of dissociation constants for weak acids allows us to choose a system that will buffer at almost any pH we desire. A number of common buffer systems involving weak acids and anions of those weak acids are indicated in Table 10.5. Similar data are available in handbooks for buffer systems involving weak bases and cations of those weak bases.

Using the values in Table 10.5 and the Henderson–Hasselbach equation, we can calculate the relative concentrations of weak acid and anion for any given buffer at an appropriate pH.

Example 10.9 Calculate the ratio of concentrations of HCO_3^- to $CO_{2(aq)}$ required to maintain the pH of blood plasma at 7.4.

$$pH = pK_a + \log \frac{[HCO_3^-]}{[CO_{2(aq)}]}$$

Rearranging,

$$\log \frac{[HCO_3^-]}{[CO_{2(aq)}]} = pH - pK_a = 7.4 - 6.37 = 1.0$$

Taking the antilog as outlined in Example 10.8,

$$\frac{[HCO_3^-]}{[CO_{2(aq)}]} = 10$$

The HCO_3^- concentration in the blood is ten times that of hydrated CO_2. The disproportionate concentrations of the two components of the buffer system would seem to indicate that this particular buffer pair has an extremely narrow buffering capacity. However, as we shall see in Section 31.6, the respiratory removal of carbon dioxide keeps this buffer ratio nearly constant.

It is left to you as an exercise to determine that, at a pH of 7.4, the concentration of HPO_4^{2-} in the blood is about 1.6 times that of $H_2PO_4^-$.

A similar derivation and calculation can be performed for a buffer system composed of a weak base and a cation of that base.

Example 10.10 Determine the pH of a solution containing 0.10 *M* NH_3 and

Table 10.5 Common Buffer Systems

Weak Acid	*Anion*	K_a	*Buffer pH* $= pK_a$ *when* $\frac{[Anion]}{[Acid]} = 1$
HSO_4^-	SO_4^{2-}	1.2×10^{-2}	1.92
H_3PO_4	$H_2PO_4^-$	7.3×10^{-3}	2.14
HF	F^-	6.7×10^{-4}	3.17
HNO_2	NO_2^-	4.5×10^{-4}	3.35
HCOOH	$HCOO^-$	1.8×10^{-4}	3.74
$CH_3CHOHCOOH$	$CH_3CHOHCOO^-$	1.4×10^{-4}	3.85
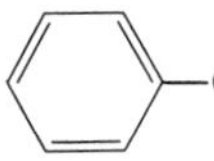—COOH	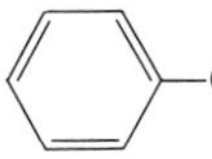—COO^-	6.5×10^{-5}	4.19
CH_3COOH	CH_3COO^-	1.8×10^{-5}	4.74
$CO_{2(aq)}$	HCO_3^-	4.3×10^{-7}	6.37
$H_2PO_4^-$	HPO_4^{2-}	6.2×10^{-8}	7.21
HSO_3^-	SO_3^{2-}	5.6×10^{-8}	7.25
H_3BO_3	$B(OH)_4^-$	6.0×10^{-10}	9.22
HCN	CN^-	4.9×10^{-10}	9.31
HCO_3^-	CO_3^{2-}	5.6×10^{-11}	10.25
HPO_4^{2-}	PO_4^{3-}	1.7×10^{-12}	11.77

0.20 *M* NH_4^+ if K_b for the reversible dissociation of aqueous ammonia is 1.8×10^{-5}.

The equation for the reversible dissociation of aqueous ammonia is

$$NH_3 + H_2O \rightleftharpoons NH_4^+ + OH^-$$

The equilibrium constant expression is

$$K_b = \frac{[NH_4^+][OH^-]}{[NH_3]}$$

Solving for $[OH^-]$, then setting equal to $\frac{K_w}{[H^+]}$

$$[OH^-] = K_b\frac{[NH_3]}{[NH_4^+]} = \frac{K_w}{[H^+]}$$

Solving for $[H^+]$,

$$[H^+] = \frac{K_w}{K_b} \times \frac{[NH_4^+]}{[NH_3]}$$

Taking the negative logarithm of both sides and solving,

$$-\log [H^+] = -\log K_w + \log K_b - \log \frac{[NH_4^+]}{[NH_3]}$$

$$pH = -\log (1.0 \times 10^{-14}) + \log (1.8 \times 10^{-5}) - \log \left(\frac{0.20}{0.10}\right)$$

$$pH = 14.00 - 4.74 - 0.30 = 8.96$$

Exercises

10.1 Define clearly the following terms.

(a) reversible reaction
(b) chemical equilibrium
(c) static equilibrium
(d) dynamic equilibrium
(e) law of chemical equilibrium
(f) equilibrium constant
(g) weak electrolyte
(h) dissociation
(i) heterogeneous equilibria
(j) ion product of water
(k) hydrolysis
(l) solubility product
(m) acid strength
(n) quadratic formula
(o) Le Châtelier's principle
(p) strong electrolyte
(q) common-ion effect
(r) catalyst
(s) neutral solution
(t) acidic solution
(u) basic solution
(v) pH
(w) buffer solution
(x) Henderson–Hasselbach equation

10.2 State the law of chemical equilibrium in your own words, and describe how it applies to the reversible reaction

$$2\,NO_2Cl_{(g)} \rightleftharpoons 2\,NO_{2(g)} + Cl_{2(g)}$$

10.3 What effect, if any, would the following changes cause in the equilibrium concentrations of *all* species in the reversible reaction of Exercise 10.2? Show clearly your reasoning.

(a) NO_2 is added.
(b) N_2 is added.
(c) The container volume is doubled.
(d) Cl_2 is selectively removed.
(e) NO_2Cl is added.

10.4 Write an equilibrium constant expression for each of the following reversible changes.

(a) $H_{2(g)} + Br_{2(g)} \rightleftharpoons 2\,HBr_{(g)}$
(b) $MgCO_{3(s)} \rightleftharpoons MgO_{(s)} + CO_{2(g)}$
(c) $Cu^{2+} + 4\,NH_3 \rightleftharpoons Cu(NH_3)_4^{2+}$
(d) $HS^- \rightleftharpoons H^+ + S^{2-}$
(e) $Hg^{2+} + H_2O \rightleftharpoons HgOH^+ + H^+$

10.5 Write an equilibrium constant expression for each of the following reversible changes.

(a) $CH_{4(g)} + H_2S_{(g)} \rightleftharpoons CS_{2(g)} + 4\,H_{2(g)}$
(b) $Au(CN)_4^- \rightleftharpoons Au^{3+} + 4\,CN^-$
(c) $PO_4^{3-} + H_2O \rightleftharpoons OH^- + HPO_4^{2-}$
(d) $Ag_2SO_{4(s)} \rightleftharpoons 2\,Ag^+ + SO_4^{2-}$
(e) $2\,NO_{2(g)} \rightleftharpoons N_2O_{4(g)}$

10.6 What would happen to the number of moles of CO_2 in each of the following systems at equilibrium if the volume were decreased? Show clearly your reasoning.

(a) $C_{(s)} + O_{2(g)} \rightleftharpoons CO_{2(g)}$
(b) $H_{2(g)} + CO_{2(g)} \rightleftharpoons H_2O_{(g)} + CO_{(g)}$
(c) $CaCO_{3(s)} \rightleftharpoons CaO_{(s)} + CO_{2(g)}$
(d) $2\,CO_{(g)} + O_{2(g)} \rightleftharpoons 2\,CO_{2(g)}$
(e) $CO_{2(s)} \rightleftharpoons CO_{2(g)}$

10.7 What would happen to the number of moles of H_2 in each of the following systems at equilibrium if the temperature were decreased? Show clearly your reasoning.

(a) $N_{2(g)} + 3\,H_{2(g)} \rightleftharpoons 2\,NH_{3(g)}$ (exothermic)
(b) $2\,H_2O_{(g)} \rightleftharpoons 2\,H_{2(g)} + O_{2(g)}$ (endothermic)
(c) $H_2O_{(g)} + CO_{(g)} \rightleftharpoons H_{2(g)} + CO_{2(g)}$ (exothermic)
(d) $2\,HI_{(g)} \rightleftharpoons H_{2(g)} + I_{2(g)}$ (endothermic)
(e) $H_{2(g)} + Cu_2O_{(s)} \rightleftharpoons H_2O_{(g)} + 2\,Cu_{(s)}$ (endothermic)

10.8 (a) Write the equilibrium constant expression for each of the reversible reactions in Exercise 10.7.
(b) Would the equilibrium constant increase or decrease for each of these equilibrium systems if the temperature were decreased? Show clearly your reasoning.

10.9 If a 0.010 *M* solution of HCN is dissociated to the extent of 0.020%, calculate the dissociation constant of HCN.

10.10 Using data from Tables 10.3 and 10.5, arrange the following compounds in order of increasing acid strength.

(a) acetic acid
(b) hydrocyanic acid
(c) formic acid
(d) oxalic acid
(e) phosphoric acid

10.11 Calculate the OH^- concentration of a solution having the indicated H^+ concentration.

(a) $1.0 \times 10^{-9}\,M$
(b) $6.4 \times 10^{-2}\,M$
(c) $3.6 \times 10^{-8}\,M$
(d) $2.0 \times 10^{-4}\,M$
(e) $9.8 \times 10^{-11}\,M$

10.12 Calculate the pH of each of the solutions in Exercise 10.11.

10.13 If a 0.14 *M* benzoic acid solution dissociates to the extent of 2.1%, calculate

(a) the dissociation constant of benzoic acid
(b) the pH of a 0.14 *M* benzoic acid solution

10.14 Using data from Table 10.3, calculate the concentrations of H^+ and OH^- and determine the pH of the following solutions.
(a) 0.050 *M* formic acid
(b) 0.22 *M* acetic acid
(c) 0.10 *M* aqueous CO_2 solution
(d) 0.50 *M* boric acid
(e) 0.10 *M* $H_2PO_4^-$ (neglect second dissociation)

10.15 Using data from Table 10.3, calculate the concentrations of H^+ and OH^- and determine the pH of the following solutions.
(a) 0.10 *M* oxalic acid (neglect second dissociation)
(b) 0.025 *M* HSO_4^- solution
(c) 0.10 *M* phosphoric acid (neglect second dissociation)
(d) 0.010 *M* nitrous acid
(e) 0.0010 *M* formic acid

10.16 Calculate the percent dissociation of each solution in Exercise 10.14.

10.17 A 0.50 *M* solution of aqueous ammonia dissociates to the extent of 0.6% according to the reaction

$$NH_3 + H_2O \rightleftharpoons NH_4^+ + OH^-$$

(a) What is the dissociation constant for this base?
(b) Calculate the pH of a 0.50 *M* solution of aqueous NH_3 and also the pH of a 0.010 *M* solution of aqueous NH_3. Account for the pH differences.

10.18 Show by appropriate chemical equations what happens when acids and bases are added to a buffer solution containing 1.0 *M* HSO_3^- and 1.0 *M* SO_3^{2-}.

10.19 Using data from Table 10.3, calculate the pK_a for the following.
(a) oxalic acid
(b) the hydrogen oxalate anion ($HOOCCOO^-$)
(c) the monohydrogen phosphate anion (HPO_4^{2-})
(d) benzoic acid
(e) formic acid

10.20 Calculate the H^+ concentration of **(a)** a 0.10 *M* benzoic acid solution and **(b)** a solution that contains 0.10 *M* benzoic acid and 0.10 *M* sodium benzoate ($NaC_6H_5COO \rightleftharpoons Na^+ + C_6H_5COO^-$). Account for the differences in the H^+ concentrations in (a) and (b) by Le Châtelier's principle and by equilibrium constant arguments.

10.21 Using data from Table 10.5 and the Henderson–Hasselbach equation, calculate the pH of the following buffer solutions.
(a) 0.10 *M* HF and 0.050 *M* F^-
(b) 0.50 *M* $H_2PO_4^-$ and 2.0 *M* HPO_4^{2-}
(c) 0.10 *M* HCOOH and 0.25 *M* $HCOO^-$
(d) 0.15 *M* HNO_2 and 0.080 *M* NO_2^-
(e) 0.48 *M* HCO_3^- and 0.63 *M* CO_3^{2-}

10.22 Using data from Table 10.5 and the Henderson–Hasselbach equation, calculate the ratio of concentrations of anion to weak acid for each of the following buffer pairs at the given pH.
(a) $HCOOH/HCOO^-$ at pH 3.40
(b) CH_3COOH/CH_3COO^- at pH 5.10
(c) $CO_{2(aq)}/HCO_3^-$ at pH 7.10
(d) $H_2PO_4^-/HPO_4^{2-}$ at pH 6.60
(e) HCO_3^-/CO_3^{2-} at pH 10.70

10.23 Calculate the pH of the following.

(a) a solution of 0.10 *M* aqueous ammonia (NH_3) if K_b for the reversible dissociation of NH_3 is 1.8×10^{-5}

(b) a solution of 0.10 *M* hydroxylamine ($HONH_2$) if K_b for the reversible dissociation of $HONH_2$ is 1.07×10^{-8}

(c) a solution that is 0.20 *M* in hydroxylamine ($HONH_2$) and 0.10 *M* in hydroxylammonium ion ($HONH_3^+$)

11

Nuclei and Radioactivity

It was originally believed that atoms were immutable entities, that chemical changes were caused only by rearrangements of electron configurations while nuclei retained their identity. Ordinary chemical reactions presented in previous chapters were interpreted in this light. However, in 1896 Henri Becquerel, a French physicist, observed the first case of natural radioactivity when he discovered that uranium atoms emitted unusual radiations and disintegrated into new atoms, now called thorium. Through the twentieth century, reactions involving transformations of nuclei have taken on great importance—for example, in utilization of nuclear fission for production of electric power, in bombardment of atoms with accelerated particles to synthesize new elements of high atomic numbers, and in use of radioisotopes for diagnosis and therapy or for determining mechanisms of simple chemical reactions and complex biological processes.

The purposes of this chapter are to examine properties of nuclei that result in some being radioactive, to study some of the decay products, to consider methods for inducing nuclear reactions artificially, and to indicate some medical applications of radioactivity.

11.1 Nuclear Stability

Recall from Section 2.10 that nuclei are composed of protons and neutrons, and, in addition, that any given isotope can be represented by an atomic number (number of protons) subscript and a mass number (number of protons and neutrons) superscript followed by the chemical symbol for the atom of that atomic number. Thus, $^{214}_{82}Pb$ indicates an isotope composed of 82 protons and (214 − 82) or 132 neutrons, and $^{206}_{82}Pb$ indicates an isotope composed of 82 protons and (206 − 82) or 126 neutrons. These isotopes may also be designated as lead-214 and lead-206.

It is somewhat perplexing that certain isotopes, such as $^{214}_{82}Pb$, are radioactive and disintegrate spontaneously to different isotopes whereas other isotopes of the same element, such as $^{206}_{82}Pb$, are not radioactive. Perhaps even more puzzling is the question of why *any* isotopes are not radioactive. What forces are responsible for a number of protons (positive charges) being packed together in nuclei having radii similar to 10^{-13} cm without being forced apart by electrostatic repulsion?

Nuclear binding forces are not well understood, and an analysis of nuclear forces is beyond the scope of this book. It is sufficient for us to recognize three characteristics that contrast nuclear forces with electrostatic forces: nuclear forces are charge-independent and arise between proton and proton, proton and neutron, and neutron and neutron; nuclear forces operate only over very short distances of the order of nuclear diameters; and nuclear forces are much stronger than electrostatic forces.

Experimentally it is found that the more protons there are in an isotope (causing greater electrostatic repulsion), the more neutrons are required per proton for stability. More neutrons cause increased attractive nuclear forces without simultaneously causing more electrostatic repulsion. Moreover, there are particular ratios of neutrons to protons that provide maximum stability. These ideas are illustrated by the belt of stability shown in Figure 11.1. The straight line represents isotopes having the *same* number of neutrons and protons or having a neutron/proton ratio of one. The black points represent known stable, nonradioactive isotopes containing given numbers of neutrons and given numbers of protons. It can be seen that isotopes of low atomic numbers (few protons) are stable when they contain approximately equal numbers of protons and neutrons whereas stable isotopes of high atomic numbers (many protons) contain increasingly more neutrons than protons. Isotopes lying within the belt of stability are nonradioactive; isotopes lying above, below, or beyond the belt of stability are radioactive and decay to form product isotopes that have neutron/proton ratios that are within or closer to the belt of stability.

11.2 Natural Radioactivity

Natural radioactivity is *the spontaneous decay of naturally occurring unstable isotopes* (called radioisotopes). Some of these unstable isotopes were present when the earth was created and are still decaying very slowly. For example, $^{238}_{92}U$ decays according to the following equations leading to a series of

$$^{238}_{92}U \longrightarrow {}^{4}_{2}He + {}^{234}_{90}Th\ (\text{unstable}) \longrightarrow \cdots$$

$$\longrightarrow {}^{0}_{-1}\beta + {}^{234}_{91}Pa\ (\text{unstable})$$

radioactive products. The half-life (Section 11.4-b) of $^{238}_{92}U$, or time required for half of a given number of $^{238}_{92}U$ atoms to disintegrate, is very long, 4.5×10^9 years. Other unstable isotopes are decaying very rapidly and would be used up were it not for the fact that they are being constantly replenished

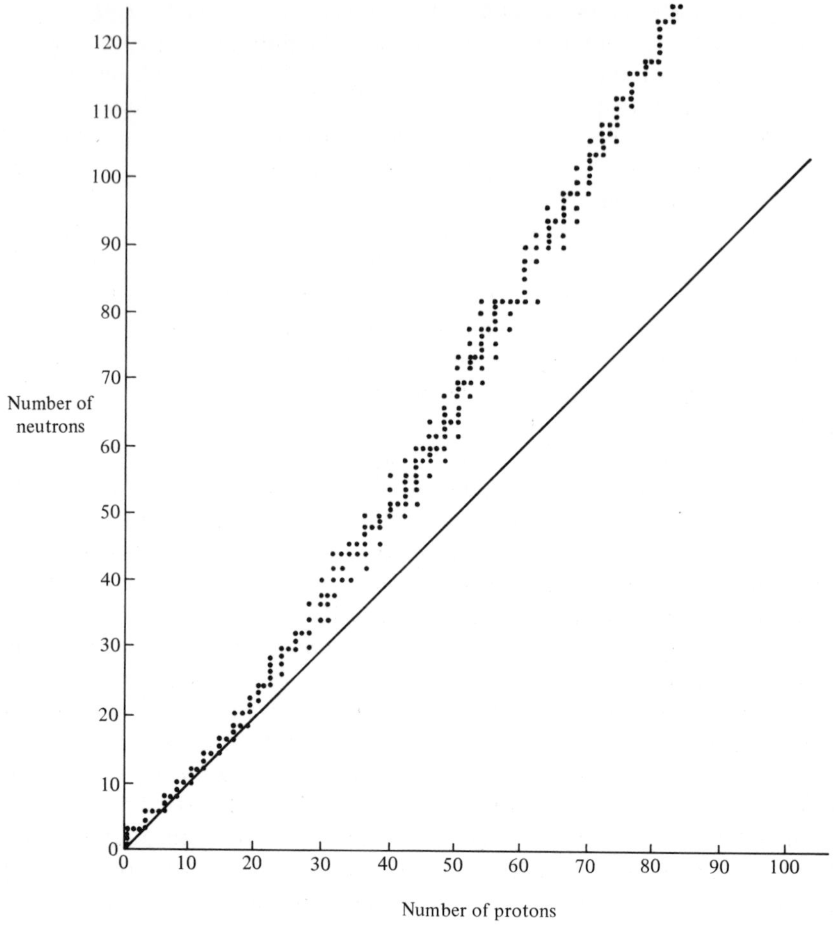
120
110
100
90
80
70
Number of neutrons
60
50
40
30
20
10
0
0
10
20
30
40
50
60
70
80
90
100
Number of protons

Figure 11.1 A plot of stable isotopes (black dots) as a function of number of neutrons and number of protons. The solid line represents a neutron/proton ratio of unity.

by decay of longer-lived isotopes. Such is the case for $^{234}_{90}Th$ in the decay scheme above. It has a half-life of only 24.1 days, but is constantly regenerated by the decay of $^{238}_{92}U$.

Radioactive decays are frequently classified according to the types of small particles that unstable nuclei eject or in one case accept. The most common types are outlined.

a. Alpha Decay

An alpha particle is an isotope of helium, $^{4}_{2}He$, composed of two protons and two neutrons. Alpha decay involves the spontaneous ejection of an alpha particle by an isotope. As illustrated in the following examples, alpha decay results in a product nucleus having two fewer protons and four less mass units.[1]

$$^{210}_{84}\text{Po} \longrightarrow {}^{4}_{2}\text{He} + {}^{206}_{82}\text{Pb (stable)}$$

$$^{226}_{88}\text{Ra} \longrightarrow {}^{4}_{2}\text{He} + {}^{222}_{86}\text{Rn (unstable)}$$

Alpha decay occurs for all nuclei having 84 or more protons. These nuclei lie beyond the belt of stability and have so many protons that they cannot be stable no matter how many neutrons are present. They successively split off small particles (frequently a combination of both alpha and beta particles) until they reach $^{206}_{82}\text{Pb}$ and the belt of stability. Some lighter isotopes such as $^{144}_{60}\text{Nd}$ and $^{147}_{62}\text{Sm}$ also undergo alpha decay.

A primary difference between ordinary chemical reactions and nuclear reactions is the amount of energy involved. It was noted in Section 2.1 that the approximate mass change in the chemical reaction in a flashbulb

$$2\ \text{Mg}_{(s)} + \text{O}_{2(g)} \longrightarrow 2\ \text{MgO}_{(s)}$$

was 5×10^{-9} gram. Using an equation developed in 1905 by Albert Einstein for the interconversion of mass and energy, it can be shown that this is equivalent to an energy change of about 105 kcal/mole. On the other hand, energy changes in most nuclear reactions are very much larger, typically about 10^6 times as large.

Alpha particles are emitted with discrete energies of about 10^8 kcal/mole, or with velocities of the order of 9000 to 15,000 miles/sec. In spite of these high velocities, since alpha particles carry a +2 charge and are relatively massive, they cannot penetrate far into matter. A few cm of air, a thick sheet of paper, all but the thinnest films of metal, or several mm of water are sufficient to stop alpha particles emitted in radioactive decay. Alpha particles are harmless on striking our skin because they cannot penetrate the outer layer of dead cells on the skin; however, they can cause serious tissue damage (up to 20 times as much as an equivalent quantity of x-rays) if they get inside the body by our breathing radioactive dust or swallowing radioactive liquids.

b. Beta Decay

Beta particle is simply another name for an electron. It carries a −1

[1]Note that since there is a transformation of nuclei in nuclear equations, they cannot be balanced by conventional methods used for chemical equations. Instead, balancing subscripts (proton numbers) results in conservation of charge, and balancing superscripts (mass numbers) results in conservation of mass.

charge and has so little mass that its emission does not appreciably change the mass of a nucleus. Beta decay involves the spontaneous ejection of a beta particle by a nucleus. As illustrated in the following examples, beta decay results in a product nucleus of the next higher atomic number (having one more proton) and the same mass.

$$^{87}_{36}\text{Kr} \longrightarrow {}^{0}_{-1}\beta + {}^{87}_{37}\text{Rb (unstable)}$$

$$^{116}_{49}\text{In} \longrightarrow {}^{0}_{-1}\beta + {}^{116}_{50}\text{Sn (stable)}$$

Beta decay decreases the neutron/proton ratio as illustrated in the following example,

$$^{24}_{11}\text{Na} \longrightarrow {}^{0}_{-1}\beta + {}^{24}_{12}\text{Mg (stable)}$$

Neutron/proton ratio: 1.18 1.00

and occurs for isotopes lying above the belt of stability or having neutron/proton ratios that are too high.

Beta particles are emitted with a distribution of energies ranging from zero up to a maximum of nearly 10^8 kcal/mole, corresponding to velocities of about 100,000 miles/sec. Having half the charge of an alpha particle and almost no mass, a beta particle can be nearly 100 times as penetrating as an alpha particle. About 2 cm of water or thin sheets of metal such as aluminum are required to stop the most energetic beta particles emitted in radioactive decay. On striking our skin, beta particles can penetrate into the living cells of our skin and can give the appearance of burns; however, they cannot penetrate to internal organs. Once inside the body, beta particles can cause tissue damage comparable to that from an equivalent quantity of x-rays.

c. Positron Emission

A positron is identical to an electron or beta particle except that it has a positive charge. Positron emission involves the spontaneous ejection of a positron by a nucleus. As illustrated in the following examples, positron emission results in a product nucleus of the next lower atomic number (having one less proton) and the same mass.

$$^{22}_{11}\text{Na} \longrightarrow {}^{0}_{+1}\beta + {}^{22}_{10}\text{Ne (stable)}$$

$$^{30}_{15}\text{P} \longrightarrow {}^{0}_{+1}\beta + {}^{30}_{14}\text{Si (stable)}$$

Positron emission increases the neutron/proton ratio as illustrated in the following example,

$$^{14}_{8}\text{O} \longrightarrow {}^{0}_{+1}\beta + {}^{14}_{7}\text{N (stable)}$$

Neutron/proton ratio: 0.75 1.00

and occurs for isotopes lying below the belt of stability or having neutron/proton ratios that are too low. In addition, positron emission occurs primarily for isotopes having less than 30 protons. Electron capture is favored by isotopes lying below the belt of stability and having atomic numbers greater than 30.

The energies, velocity distribution, penetration characteristics, and resultant tissue damage of positrons are comparable to those of beta particles.

d. Electron Capture

Electron capture involves absorption by a nucleus of a *K* or *L* shell electron of the atom resulting in the conversion of a proton to a neutron. As illustrated in the following example, electron capture, like positron emission, results in a product nucleus of the next lower atomic number (having one less proton) and the same mass.

$$^{195}_{79}\text{Au} + {}^{0}_{-1}\beta \text{ (electron capture)} \longrightarrow {}^{195}_{78}\text{Pt (stable)}$$

Like positron emission, electron capture increases the neutron/proton ratio as illustrated in the following example,

$$^{83}_{37}\text{Rb} + {}^{0}_{-1}\beta \text{ (electron capture)} \longrightarrow {}^{83}_{36}\text{Kr (stable)}$$

Neutron/proton ratio: 1.24 1.28

and occurs for isotopes lying below the belt of stability or having neutron/proton ratios that are too low. Unlike positron emission, electron capture is favored for isotopes having more than 30 protons.

e. Gamma Emission

A gamma ray is a bundle of energy that has no charge or mass. It is a form of electromagnetic radiation of higher energy than an x-ray. The types of radioactive decay described previously often leave a nucleus in an excited nuclear state. Gamma emission frequently accompanies these other types of radioactive decay and is the primary means by which excited nuclei can release their excess energy. Since gamma rays have no charge and no mass, they are omitted from balanced nuclear equations. Gamma emission causes no change in the neutron/proton ratio and does not help a nucleus reach the belt of stability.

Having no charge or mass and traveling at the speed of light (186,000 miles/sec), gamma rays do not readily interact with matter. They have very high penetrating power, of the order of 10,000 times that of alpha particles.

About 10% pass through 70 cm of water or even through 30 cm of lead. A combination of lead shielding and concrete is commonly used to stop gamma rays. Gamma rays cause tissue damage comparable to that of an equivalent quantity of x-rays.

11.3 Induced Nuclear Reactions

Induced nuclear reactions are unfavorable energetically and do not occur spontaneously, but can be made to take place by bombarding appropriate nuclei with other accelerated, high energy particles. The kinetic energy of the accelerated particles is converted to potential energy of reaction and overcomes electrostatic repulsion on close approach of the reacting nuclei.

Alpha-induced reactions were observed first because of the availability of energetic alpha particles from natural radioactivity. In 1919 Ernest Rutherford noticed that high energy alpha particles from decay of $^{214}_{83}Bi$,

$$^{214}_{83}Bi \longrightarrow {}^{4}_{2}He + {}^{210}_{81}Tl \text{ (unstable)}$$

on passing through nitrogen gas, produced new *stable* products.

$$^{14}_{7}N + {}^{4}_{2}He \longrightarrow {}^{17}_{8}O \text{ (stable)} + {}^{1}_{1}H \text{ (stable)}$$

In 1934 Irène Joliot-Curie and Jean-Frédéric Joliot bombarded $^{27}_{13}Al$ with alpha particles to produce a *radioactive* isotope ($^{30}_{15}P$) *artificially* for the first time.

$$^{27}_{13}Al + {}^{4}_{2}He \longrightarrow \begin{cases} {}^{30}_{15}P \text{ (unstable)} + {}^{1}_{0}n \text{ (neutron)} & 5\% \\ {}^{30}_{14}Si \text{ (stable)} + {}^{1}_{1}H \text{ (proton)} & 95\% \end{cases}$$

This alpha-induced reaction illustrates that nuclear reactions, like ordinary chemical reactions, can lead to several sets of products.

In addition to alpha particles, nuclear reactions have been induced by gamma rays, protons ($^{1}_{1}H$), neutrons ($^{1}_{0}n$), and deuterons ($^{2}_{1}H$), as well as by atoms heavier than alpha particles. Light-particle-induced nuclear reactions are very important for producing artificially many of the radioisotopes used for diagnosis and therapy in medicine. For example, gold-198 is formed by a neutron-induced reaction on gold-197.

$$^{197}_{79}Au + {}^{1}_{0}n \longrightarrow {}^{198}_{79}Au \text{ (unstable)}$$

Technetium-99 is formed in two steps involving a neutron-induced reaction on molybdenum-98

$$^{98}_{42}Mo + {}^{1}_{0}n \longrightarrow {}^{99}_{42}Mo \text{ (unstable)}$$

followed by beta decay.

$$^{99}_{42}Mo \longrightarrow {}^{0}_{-1}\beta + {}^{99}_{43}Tc \text{ (unstable)}$$

Heavy-particle-induced nuclear reactions have been useful in producing certain isotopes of the 14 man-made transuranium elements. One example for each follows.

$$^{238}_{92}\text{U} + ^{2}_{1}\text{H} \longrightarrow ^{239}_{93}\text{Np (unstable)} + ^{1}_{0}\text{n (neutron)}$$
$$^{238}_{92}\text{U} + ^{4}_{2}\text{He} \longrightarrow ^{239}_{94}\text{Pu (unstable)} + 3\,^{1}_{0}\text{n}$$
$$^{239}_{94}\text{Pu} + ^{4}_{2}\text{He} \longrightarrow ^{240}_{95}\text{Am (unstable)} + ^{1}_{1}\text{H} + 2\,^{1}_{0}\text{n}$$
$$^{239}_{94}\text{Pu} + ^{4}_{2}\text{He} \longrightarrow ^{242}_{96}\text{Cm (unstable)} + ^{1}_{0}\text{n}$$
$$^{241}_{95}\text{Am} + ^{4}_{2}\text{He} \longrightarrow ^{243}_{97}\text{Bk (unstable)} + 2\,^{1}_{0}\text{n}$$
$$^{238}_{92}\text{U} + ^{12}_{6}\text{C} \longrightarrow ^{246}_{98}\text{Cf (unstable)} + 4\,^{1}_{0}\text{n}$$
$$^{238}_{92}\text{U} + ^{14}_{7}\text{N} \longrightarrow ^{247}_{99}\text{Es (unstable)} + 5\,^{1}_{0}\text{n}$$
$$^{238}_{92}\text{U} + ^{16}_{8}\text{O} \longrightarrow ^{249}_{100}\text{Fm (unstable)} + 5\,^{1}_{0}\text{n}$$
$$^{253}_{99}\text{Es} + ^{4}_{2}\text{He} \longrightarrow ^{256}_{101}\text{Md (unstable)} + ^{1}_{0}\text{n}$$
$$^{246}_{96}\text{Cm} + ^{12}_{6}\text{C} \longrightarrow ^{254}_{102}\text{No (unstable)} + 4\,^{1}_{0}\text{n}$$
$$^{252}_{98}\text{Cf} + ^{10}_{5}\text{B} \longrightarrow ^{257}_{103}\text{Lr (unstable)} + 5\,^{1}_{0}\text{n}$$
$$^{249}_{98}\text{Cf} + ^{12}_{6}\text{C} \longrightarrow ^{257}_{104}\text{Ku (unstable)} + 4\,^{1}_{0}\text{n}$$
$$^{249}_{98}\text{Cf} + ^{15}_{7}\text{N} \longrightarrow ^{260}_{105}\text{Ha (unstable)} + 4\,^{1}_{0}\text{n}$$
$$^{249}_{98}\text{Cf} + ^{18}_{8}\text{O} \longrightarrow ^{263}_{106}\text{X (unstable)} + 4\,^{1}_{0}\text{n}$$

11.4 Quantitative Aspects of Radioactivity

The presence of radioactivity is of little use unless we can detect what type of radiation is being emitted and how much is being emitted per unit time.

a. Detection of Radiation

The presence of radioactivity is easily determined through the use of a photographic film or plate or of a fluorescent screen. The decay particles mentioned previously cause spotty exposure of a photographic film or plate. We have already noted in Sections 2.5 and 2.6 that electrons and protons produce scintillations on a fluorescent ZnS screen. Alpha particles, positrons, and gamma rays produce similar flashes.

We can determine the nature of radiation by observing its deflection in an electric field as shown in Figure 11.2. A radioactive material is placed in a lead block both to protect the experimenter and to obtain a collimated beam of emitted particles. If a variety of radioactive sources are used, an electrical field placed near the path of the particles separates the beam into four distinct parts. The negatively charged beta particles ($^{0}_{-1}\beta$) or electrons are deflected toward the positive electrode. Two kinds of positively charged particles—alpha particles ($^{4}_{2}$He) and positrons ($^{0}_{+1}\beta$)—are deflected toward the negative electrode. Positrons show an equal, but opposite, deflection from beta particles on the ZnS screen. The more highly charged, but much more

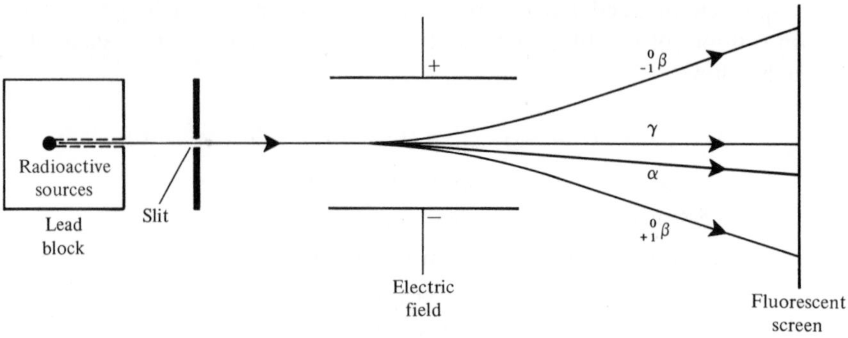

+
−
Radioactive sources
Lead block
Slit
Electric field
$_{-1}^{0}\beta$
γ
α
$_{+1}^{0}\beta$
Fluorescent screen

Figure 11.2 Separation of alpha particles ($_2^4He$), beta particles ($_{-1}^{0}\beta$), positrons ($_{+1}^{0}\beta$), and gamma rays (γ) by an electric field. Gamma rays are unaffected. Heavier alpha particles are deflected considerably less than beta particles.

massive, alpha particles are deflected to a lesser extent than positrons. Neutral gamma rays are undeflected by the electric field.

Counting the number of particles emitted per unit time requires an additional piece of apparatus. A Geiger–Müller counter, illustrated in Figure 11.3, is commonly used. This device consists of a partially evacuated tube containing coaxial cylindrical electrodes. Between them a voltage too small to produce a flow of current is maintained. An emitted particle passing into the tube is energetic enough to eject electrons from gaseous atoms such as argon that are present in the tube. The resultant positively charged ions then conduct an electric current between the two electrodes. This electric current can be detected on a calibrated meter or can produce light flashes, clicks on a loudspeaker, or counts on a digital tape or counter.

New apparatus and new techniques for counting radioactive disintegration are

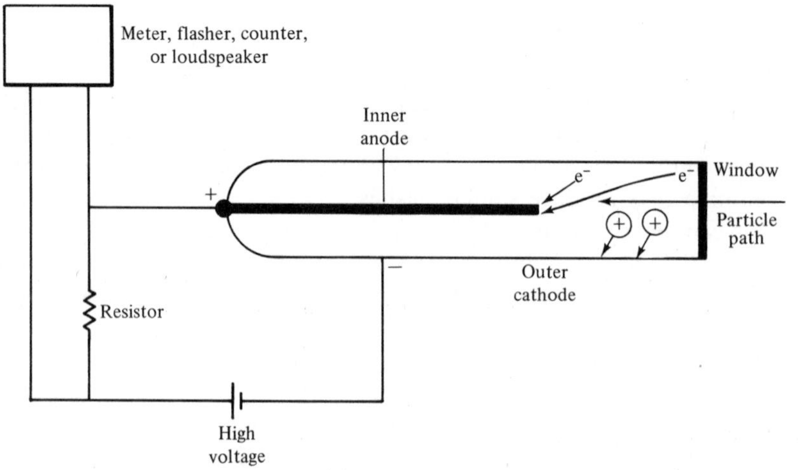

Meter, flasher, counter, or loudspeaker
Inner anode
+
e^-
e^-
Window
Particle path
−
Outer cathode
Resistor
High voltage

Figure 11.3 Geiger–Müller counter.

being developed continually. For example, it has been shown that a transistorized silicon diode detector about the size of a grain of sand can be constructed and placed within the human body with little discomfort.

b. Units of Measurement

The rate of disintegration of unstable nuclei varies significantly from one element to another and among isotopes of the same element. However, the rate of decay of any given isotope is constant and this activity is designated by its half-life. A **half-life** is *the time required for half the nuclei in a sample of a specific isotope to undergo radioactive decay*. For example, the half-life of $^{239}_{93}Np$ toward beta decay is 2.3 days. This means that, for 1.00 mole of $^{239}_{93}Np$ nuclei, half of it or 0.50 mole will have decayed in 2.3 days, and half of the 0.50 mole remaining, or 0.25 mole will have decayed in another 2.3 days. Short half-lives indicate rapid decay rates; long half-lives indicate slow decay rates. These concepts are illustrated by the data in Table 11.1 for isotopes of uranium.

Radiocarbon dating provides an example of decay rate calculations. Radioactive $^{14}_{6}C$ has a constant concentration in the atmosphere as a result of an equilibrium between its formation by cosmic ray neutron bombardment of $^{14}_{7}N$

$$^{14}_{7}N + ^{1}_{0}n \longrightarrow ^{14}_{6}C \text{ (unstable)} + ^{1}_{1}H$$

and its destruction by beta decay.

$$^{14}_{6}C \longrightarrow ^{0}_{-1}\beta + ^{14}_{7}N \text{ (stable)}$$

Thus, the $^{14}CO_2/^{12}CO_2$ ratio in the atmosphere is also constant at about 1.20×10^{-12}. Living materials have the same $^{14}C/^{12}C$ ratio as long as they remain alive as a result of plants absorbing CO_2 from the atmosphere during photosynthesis or of animals eating foodstuffs tracing back to plants. Living materials exhibit the characteristic decay rate for that $^{14}C/^{12}C$ ratio of 13.7 beta disintegrations per minute per gram of carbon. However, because of the decay of ^{14}C and no further replenishment of ^{14}C, both the $^{14}C/^{12}C$ ratio and the beta decay rate diminish after the plant or animal dies (by a factor of two over each half-life for ^{14}C decay, 5730 years). The age in years of any material that was once living can be determined by measuring only the $^{14}C/^{12}C$ ratio or the beta decay rate in that material and utilizing a formula derived from first order kinetics (Section 9.3).

$$\text{Age} = 1.904 \times 10^4 \log\left(\frac{1.20 \times 10^{-12}}{r}\right) = 1.904 \times 10^4 \log\left(\frac{13.7}{R}\right)$$

where r is the measured $^{14}C/^{12}C$ ratio or R is the measured beta decay rate.

Table 11.1 Decay Data for Uranium Isotopes

Isotope	*Radiation*	*Half-life*	*Decay Rate (disintegrations/sec) for 1 gram sample*	*Intensity (curies) for 1 gram sample*
$^{227}_{92}U$	Alpha Electron capture	1.3 min	2.4×10^{19}	6.4×10^{8}
$^{228}_{92}U$	Alpha Electron capture	9.3 min	3.3×10^{18}	8.9×10^{7}
$^{229}_{92}U$	Electron capture Alpha	58 min	5.1×10^{17}	1.4×10^{7}
$^{230}_{92}U$	Alpha	20.8 days	1.0×10^{15}	2.7×10^{4}
$^{231}_{92}U$	Electron capture Alpha	4.2 days	5.1×10^{15}	1.3×10^{5}
$^{232}_{92}U$	Alpha	73.6 yr	7.6×10^{11}	2.1×10^{1}
$^{233}_{92}U$	Alpha	1.62×10^{5} yr	3.5×10^{8}	9.5×10^{-3}
$^{234}_{92}U$	Alpha	2.48×10^{5} yr	2.3×10^{8}	6.2×10^{-3}
$^{235}_{92}U$	Alpha	7.13×10^{8} yr	7.9×10^{4}	2.1×10^{-6}
$^{236}_{92}U$	Alpha	2.39×10^{7} yr	2.3×10^{6}	6.3×10^{-5}
$^{237}_{92}U$	Beta	6.75 days	3.0×10^{15}	8.2×10^{4}
$^{238}_{92}U$	Alpha	4.51×10^{9} yr	1.2×10^{4}	3.3×10^{-7}
$^{239}_{92}U$	Beta	23.5 min	1.2×10^{18}	3.3×10^{7}
$^{240}_{92}U$	Beta	14.1 hr	3.4×10^{16}	9.3×10^{5}

Example 11.1 Linen wrappings for the Dead Sea scrolls were found to have a beta decay rate of 10.9 disintegrations per minute per gram of carbon. What is the approximate age of these linen wrappings?

$$t = 1.904 \times 10^4 \log\left(\frac{13.7}{R}\right)$$

$$= 1.904 \times 10^4 \log\left(\frac{13.7}{10.9}\right)$$

$$= 1890 \text{ years}$$

The intensity of radiation can be expressed in both a physical sense and a biological sense. Physical measurement of the intensity of radiation is related to the number of particles emitted by a source per second. The primary unit used is the curie. A **curie** is *the quantity of radioactive material that yields* the same number of disintegrations per second as 1 gram of radium, namely, *3.70* $\times$ *10*^{10} *disintegrations/sec.* The number of curies in 1 gram of the various isotopes of uranium is given in Table 11.1.

Biological measurement of the intensity of radiation is accomplished by examining tissue damage that results from ionizations of molecules caused by absorption of radiation. It is defined on the basis of the number of ion pairs produced by the radiation. Three different units are used: a roentgen for describing intensities of x-rays or gamma rays only, a rad for describing intensities of other types of radiation, and a rem for describing intensities of radiation absorbed specifically by humans. A **roentgen** is *the quantity of ionizing radiation that produces approximately* 2×10^9 *ion pairs per cubic centimeter of dry air at STP.* The number of ion pairs produced by 1 roentgen varies with different materials and corresponds to about 1.8×10^{12} ion pairs per gram of living tissue. A **rad** is approximately equal to a roentgen and corresponds to *the quantity of energy absorbed from ionizing radiation equal to* 2.4×10^{-8} *kcal per gram of irradiated material.* A 500-rad dose can cause enough random ionizations among key body molecules to be lethal. A **rem** is based on a roentgen and is *the quantity of ionizing radiation that must be absorbed by man to produce the same biological effect as 1 roentgen of high penetration x-rays.*

11.5 Biological Effects of Radiation

Radiation represents a potential danger to biological systems. Ignorance of the properties of radiation and its biological effects only heightens the danger. However, if we know the properties of various types of radiation and understand the biological effects of radiation, we can minimize the danger by taking proper safety precautions.

As ionizing radiation penetrates a biological system, it loses energy during bombardment of the complex molecules in the tissues. These collisions are energetic enough to eject electrons and rupture chemical bonds. Molecular fragments that may be either ions or radicals are produced. The radicals and organic ions are often unstable and quickly recombine in different ways from which they were formed and produce new molecules that are foreign to the original cell. Frequently the electrons that are ejected have enough energy to bombard neighboring molecules and cause secondary ionization and radical formation. Thus, one particle can potentially cause more than one reaction. In addition, the ionizing radiation can react with water either inside or outside the cell. This would seem to be an ideal means for de-energizing the ionizing radiation until one recognizes that free radicals (Sections 8.3-a and 13.4–13.5) formed from water can do additional damage by causing oxidation-reduction reactions on complex molecules comprising the tissues.

The relatively immediate outward symptoms from overexposure to radioactivity are many and varied. The most common symptoms are a decrease in the white blood cell count and gastrointestinal disturbances manifested as diarrhea, nausea, and general weakness. In addition, there may be internal bleeding, loss of hair, ulceration of the mouth, and the appearance of ugly skin blemishes that may not heal.

The body can frequently recover from the outward effects of radiation even though residual effects such as cancer, genetic mutations, or accelerated aging may occur much later. Damaged cells are often involved with processes such as the growing of hair or fingernails; the replacement of skin layers damaged by abrasion; the replacement of mucous linings in the mouth, throat, stomach, or intestines; or the replacement of red blood cells. If only some of these cells are damaged, those that remain intact can divide and eventually replace and do the work of those that were damaged. A gradual recovery then results. On the other hand, if too many cells are damaged, the symptoms become more severe and death results.

The extent of radiation damage depends on a variety of factors. We have already discussed the energy of and penetrating ability of the common *types of radiation* and have noted that increasing either the *energy of the radiation* or the *intensity of the radiation* causes greater tissue damage. In addition, the extent of radiation damage depends on the *organ of the body that is irradiated* and on the *age* and *general health* of the individual.

The outward symptoms previously described as well as long-term residual effects arise primarily as a result of radiation damage to the genetic apparatus of cells. Very intense exposure to ionizing radiation breaks chromosomes and kills cells outright. Weaker exposure to ionizing radiation leaves chromosomes intact, but alters delicate gene structures, for example, that of a DNA molecule (Section 27.2). When this happens, several possibilities arise. Changes in gene structures may be so great that cells cannot undergo replication. They are not killed themselves, but are reproductively dead. Cancer cells are deliberately exposed to radiation to halt replication by this means. Alternatively, cells may still be able to divide, but can only yield mutant daughter cells that will lead to future erratic behavior, perhaps in the form of cancer or perhaps in the form of genetic defects.

11.6 Radioisotopes in Medicine

The use of radioisotopes in medicine is based on the fact that radioisotopes are chemically indistinguishable from other atoms of the same kind and thus can be directed to various specific parts of the body where other atoms of the same kind are known to go. Actual procedures differ depending on whether the radioisotopes are being used for diagnosis or for therapy.

a. Diagnosis

A radiologist must consider at least three criteria when choosing a radioisotope to be used for diagnosis. In many diagnoses radioisotopes are deliberately placed in the body to serve as tracers in order to determine where in the body a given chemical species goes. Therefore, a radiologist must choose a radioisotope of an atom contained in the chemical species under study. The presence of the tracer is usually determined by an external counter. Therefore, radioisotopes emitting alpha particles are not generally useful in diagnosis because alpha particles have too low a penetrating power to be detected outside the body. On the other hand, radioisotopes emitting gamma rays are ideal because the highly penetrating gamma rays are readily detected outside the body. Finally, radiation damage must be minimized. To do this, a radiologist chooses a radioisotope that has a short half-life so that it will give the maximum number of disintegrations during the short time required for the test. He uses this radioisotope in the form of a compound that will be eliminated from the body shortly after its tracer function is completed. He also uses the smallest possible amount of the radioisotope that can be detected.

More than 100 different radioisotopes have been used for diagnosis. Among the most important ones are $^{51}_{24}Cr$, $^{198}_{79}Au$, $^{131}_{53}I$, $^{59}_{26}Fe$, $^{32}_{15}P$, and $^{99}_{43}Tc$. None of these occur naturally; all are produced artificially (Section 11.3). Several applications of chromium-51, a radioisotope that undergoes electron capture

$$^{51}_{24}\text{Cr} + {}^{0}_{-1}\beta \text{ (electron capture)} \longrightarrow {}^{51}_{23}\text{V (stable)}$$

and simultaneous emission of gamma rays with a half-life of 27.8 days, will illustrate the use of radioisotopes for diagnosis. Chromium, in the form of sodium chromate (Na_2CrO_4) binds readily to red blood cells. This makes chromium-51 useful for a number of blood tests.

One test is the determination of total volume of red blood cells (V_{rbc}). A sample of a patient's blood, stabilized with heparin after withdrawal, takes up chromium-51 on incubation with radioactive Na_2CrO_4. The intensity of radioactivity R of a given volume of blood sample V is then determined prior to reinjection into the patient. After the short time required for complete mixing throughout the blood stream, the same volume V of blood is withdrawn and its intensity of radioactivity R' is measured. A simple radioactivity dilution calculation (similar to molar dilution calculations in Section 7.7-a and a minor correction for $^{51}_{24}Cr$ that decayed during mixing in the bloodstream) gives the total volume of red blood cells.

$$V_{\text{rbc}} = \frac{R}{R'} V$$

Frequently red blood cells die prematurely and lead to certain types of anemia. This condition can also be diagnosed using chromium-51. A patient's blood sample, already incubated with radioactive Na_2CrO_4, is injected back into the patient and into a compatible normal individual. If on later withdrawal of blood from both individuals, the red blood cell survival time is too short in both individuals, the red blood cells of the patient were abnormal. If, however, only the red blood cell survival time in the patient is too short, then the red blood cells are normal, but the patient's blood contains some component that destroys red blood cells.

Thallium-201, a radioisotope that also undergoes electron capture

$$^{201}_{81}\text{Tl} + {}^{0}_{-1}\beta \text{ (electron capture)} \longrightarrow {}^{201}_{80}\text{Hg (stable)}$$

as well as simultaneous emission of gamma rays with a half-life of 73 hours, has been approved recently by the Food and Drug Administration for clinical diagnosis of heart disease. Thallium-201 mimics potassium ion behavior in its uptake from the bloodstream into normal cells of the heart. Following injection of thallium-201, photo images reveal any damaged heart muscle cells because thallium-201 will not have entered malfunctioning sections of heart muscle. Thus, the extent of damage even from mild heart attacks can be pinpointed in this fashion.

There are also some diagnostic procedures (for example, radioimmunoassays) that use radioactive materials, but do not involve injection of the isotope into the patient.

b. Therapy

A radiologist has different goals when using radioisotopes for therapy than he has when using them for diagnosis. The tracer capabilities of radioisotopes are of no use to him. Therefore, he need not be so concerned about determining the presence of a radioisotope by an external detector. This means that he can utilize an alpha-particle-emitting radioisotope as well as those that emit beta particles and gamma rays. The main objective of the radiologist is to destroy selectively cells or tissues that are abnormal for the purpose of bringing a particular disease under control. This requires a more intense dose of radiation, but it also must be highly localized so as not to damage nearby healthy organs. Radioisotopes that emit alpha particles or beta particles are better for this purpose because their penetrating power is less and their effect is limited to the immediate area in which they are implanted.

Radiation is particularly damaging to cells while they are in the process of dividing. Cancer is a rapid, uncontrolled division of abnormal cells to produce more abnormal cells; therefore, cancer cells are particularly vulnerable to radiation. Radiation thus provides a useful method for cancer therapy especially for cases where surgery is not practical. A number of radioisotopes have been used for cancer therapy. Among the most common have been $^{226}_{88}Ra$ that occurs naturally and $^{198}_{79}Au$, $^{60}_{27}Co$, $^{131}_{53}I$, $^{132}_{53}I$, and $^{90}_{39}Y$ that must be produced artificially.

Radium salts were first used for cancer therapy. Radium-226 undergoes alpha decay

$$^{226}_{88}\mathrm{Ra} \longrightarrow {}^{4}_{2}\mathrm{He} + {}^{222}_{86}\mathrm{Rn}\ \text{(unstable)}$$

and simultaneous emission of gamma rays with a half-life of 1620 years. Pinpoint application is achieved by inserting tiny hollow gold or platinum needles containing a radium salt directly into a tumor. Radium treatment has several disadvantages. Highly radioactive radon gas that is formed as a product must be contained and not allowed to pass through and cause radiation damage in other parts of the body. In addition, radium salts are extremely expensive, and radium has a very long half-life. Therefore, the radium salts must be implanted in a tumor temporarily and be removed when the tumor is destroyed. This prevents objectionable radiation damage and makes the radium sample available for further use.

Some of the disadvantages of radium are overcome using yttrium-90 that undergoes beta decay

$$^{90}_{39}\text{Y} \longrightarrow {}^{0}_{-1}\beta + {}^{90}_{40}\text{Zr (stable)}$$

with a far shorter half-life of 64.2 hours. Tiny yttrium oxide ceramic beads can be implanted directly in a gland or tumor. The beta particles have little enough penetrating power that they affect only the immediate area of the implant. For example, yttrium oxide beads have been implanted directly into the pituitary gland on the reasoning that destruction of this gland, that stimulates cell reproduction, slows down the growth of a tumor elsewhere in the body.

High intensity treatments of perhaps a thousand curies or more are accomplished with cobalt-60 that is produced in a nuclear reactor by neutron irradiation of cobalt-59.

$$^{59}_{27}\text{Co} + {}^{1}_{0}\text{n} \longrightarrow {}^{60}_{27}\text{Co (unstable)}$$

Cobalt-60 undergoes beta decay

$$^{60}_{27}\text{Co} \longrightarrow {}^{0}_{-1}\beta + {}^{60}_{28}\text{Ni (stable)}$$

and simultaneous emission of gamma rays with a half-life of 5.24 years. Teletherapy, using a finely collimated beam from a tiny cobalt-60 source, can bring radiation to bear on a small area of the body.

The energy evolved in the decay of radioisotopes can be utilized therapeutically rather than the decay products. Thus, the heat generated from the alpha decay of $^{238}_{94}\text{Pu}$ is converted to electricity in a tiny nuclear powered battery that operates a heart pacemaker. The pacemaker is implanted in the body and need be replaced only about every ten years.

Exercises

11.1 Define clearly the following terms.

(a) nucleus
(b) proton
(c) neutron
(d) isotope

(e) atomic number
(f) mass number
(g) electrostatic forces
(h) nuclear forces
(i) belt of stability
(j) neutron/proton ratio
(k) natural radioactivity
(l) radioisotope
(m) nuclear equation
(n) alpha particle
(o) alpha decay
(p) beta particle
(q) beta decay
(r) positron
(s) positron emission
(t) electron capture
(u) gamma ray
(v) gamma emission
(w) induced nuclear reaction
(x) proton-induced nuclear reaction
(y) neutron-induced nuclear reaction
(z) alpha-induced nuclear reaction
(aa) Geiger–Müller counter
(ab) half-life
(ac) curie
(ad) roentgen
(ae) rad
(af) rem
(ag) radioactive tracer

11.2 Describe the composition of the nucleus of the six radioisotopes listed in Section 11.6-a as being used for medical diagnosis.

11.3 Write balanced nuclear equations for alpha decay of the following.
(a) $^{226}_{88}Ra$
(b) thorium-230
(c) americium-241
(d) $^{218}_{84}Po$
(e) $^{260}_{105}Ha$

11.4 Write balanced nuclear equations for beta decay of the following.
(a) carbon-14
(b) $^{234}_{91}Pa$
(c) $^{210}_{82}Pb$
(d) indium-116
(e) $^{210}_{83}Bi$

11.5 Write balanced nuclear equations for positron emission of the following.
(a) nitrogen-13
(b) $^{120}_{51}Sb$
(c) $^{20}_{11}Na$
(d) fluorine-17
(e) aluminum-26

11.6 Write balanced nuclear equations for electron capture of the following.
(a) $^{65}_{31}Ga$
(b) nickel-59
(c) iron-55
(d) $^{75}_{34}Se$
(e) $^{85}_{38}Sr$

11.7 Using the belt of stability in Figure 11.1, predict what mode of natural radioactivity will occur for the following.
(a) $^{222}_{86}Rn$
(b) $^{10}_{6}C$
(c) $^{191}_{80}Hg$
(d) $^{19}_{8}O$
(e) $^{25}_{13}Al$

11.8 Uranium-238 decays by fourteen steps (alpha, beta, beta, alpha, alpha, alpha, alpha, alpha, beta, beta, alpha, beta, beta, alpha) on the way to formation of stable lead-206. Write balanced nuclear equations for the fourteen steps.

11.9 Complete the following nuclear equations.
(a) $^{18}_{8}O + ^{1}_{1}H \longrightarrow \ ? \ + ^{1}_{0}n$
(b) $^{14}_{7}N + \ ? \ \longrightarrow ^{17}_{8}O + ^{1}_{1}H$
(c) $? \longrightarrow ^{0}_{+1}\beta + ^{30}_{14}Si$
(d) $^{238}_{92}U + \ ? \ \longrightarrow ^{247}_{99}Es + 5\,^{1}_{0}n$
(e) $^{249}_{98}Cf + ^{12}_{6}C \longrightarrow ^{257}_{104}Ku + \ ?$

11.10 Suggest a reasonable induced nuclear reaction for the synthesis of an isotope of element 108 from fermium-249.

11.11 Iodine-131 undergoes beta decay with a half-life of 8.0 days.
(a) Write a balanced nuclear equation describing this decay.
(b) How many ions of this isotope would remain in a thyroid gland after 24 days if 1.0 micromole (10^{-6} mole) of I^- ion was injected and taken up by the gland initially?
(c) How many curies of radiation intensity would still be present after 24 days if the iodine-131 exhibited 1.5×10^{17} disintegrations/sec?

11.12 If a sample of a radioisotope is half disintegrated in 6.0 hours, what fraction of it remains after 48 hours?

11.13 **(a)** Planking from what was suspected to be a vessel used in the Norwegian crossing to England just prior to the Battle of Hastings (1066 A.D.) was found to have a $^{14}C/^{12}C$ ratio of 1.07×10^{-12}. What is the approximate age of this planking?
(b) Another piece of planking had a beta decay rate of 12.7 disintegrations per minute per gram of carbon. Could it have been from the same vessel?

11.14 Calculate the $^{14}C/^{12}C$ ratio expected for the linen wrappings of the Dead Sea scrolls in Example 11.1 on page 266.

11.15 Compare the charges, masses, energies, speeds, penetrating power, and relative capability of damaging tissue for the various particles emitted by radioactive decay.

11.16 Describe the outward symptoms of overexposure to ionizing radiation.

11.17 Explain clearly the following.
(a) It is impossible to have a pure sample of any radioactive element.
(b) Ionizing radiations can both cause cancer and arrest cancer.
(c) Beta particles are emitted from nuclei consisting originally of neutrons and protons only.
(d) Radioactive decay of various isotopes of an element can produce new elements having both higher and lower atomic numbers.

11.18 Discuss the various criteria that a radiologist must consider when using a radioisotope for **(a)** diagnosis and **(b)** therapy.

11.19 Explain clearly the following.
(a) Isotopes having short half-lives are used for medical diagnosis.
(b) Greater intensity of radiation is required for therapy than for diagnosis.
(c) Radiation damage to certain cells causes only temporary symptoms from which the body can recover whereas radiation damage to other cells causes long-range residual effects.

12

Introduction to Organic Chemistry

In 1685 Nicolas Lémery, a French chemist, published a book entitled *Cours de Chyme* in which he classified substances as animal, vegetable, or mineral on the basis of their origin. This was probably the first attempt made to distinguish substances derived from plant or animal sources from those obtained from mineral constituents. The term **organic** was later applied to those compounds derived from living matter; substances that originated from nonliving sources were accordingly referred to as **inorganic.** Furthermore, it was universally believed that living organisms contained some mysterious vital force necessary to the formation of organic substances. Throughout the ensuing century, scientists were thwarted in all their attempts to synthesize organic substances from inorganic materials. Their failures served to entrench more firmly the vital force theory, which eventually achieved a status akin to religious dogma.

In 1814, a Swedish chemist, J. J. Berzelius, dealt the vital force theory a serious blow when he proved that the basic laws of chemical change (the law of definite composition and the law of multiple proportions) applied both to organic and inorganic compounds. Fourteen years later, the erroneous theory suffered a crippling blow by a stroke of chemical serendipity.

The birth of modern organic chemistry is generally placed in the year 1828. It was in that year that Friedrich Wöhler, a professor of chemistry at Germany's University of Göttingen, attempted to prepare ammonium cyanate by heating a mixture of two inorganic salts, potassium cyanate and ammonium chloride. To his surprise, instead of ammonium cyanate he obtained crystals of the well-known organic compound urea. (Urea is a

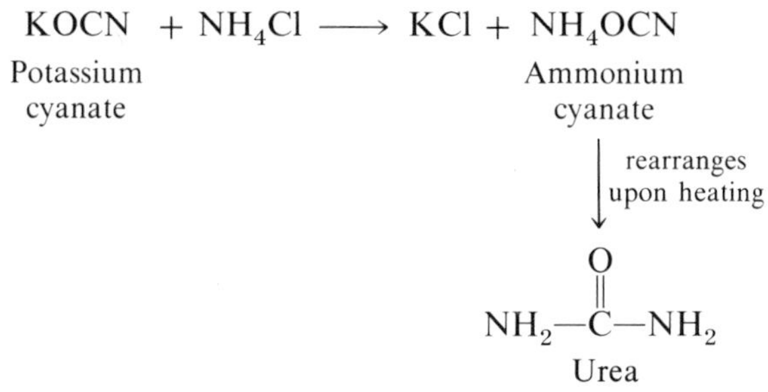
KOCN + NH4Cl ⟶ KCl + NH4OCN
Potassium cyanate
Ammonium cyanate
rearranges upon heating
NH2—C(=O)—NH2
Urea

substance excreted in the urine of mammals.) Wöhler correctly concluded that ammonium cyanate is first formed, but then rearranges under the influence of heat to yield urea. Urea contains the same number and kind of atoms as ammonium cyanate, but these atoms are arranged differently.

The next several decades witnessed a renewed effort on the part of chemists to synthesize organic compounds from inorganic starting materials. As a result of the enlightenment of Wöhler's discovery, many other organic compounds were synthesized in chemical laboratories. Moreover, while many of these compounds were identical to compounds found in nature, many others were entirely new, having no known counterpart in nature.

12.1 The Nature of Organic Compounds

By 1850 the vital force theory was essentially dead, and the relationship between the two branches of chemistry was clearly recognized. Table 12.1 contrasts the general properties of organic and inorganic compounds. It must be understood, however, that there are exceptions to every entry in this table (see Sections 3.3–3.7 and 3.10–3.13).

The one constituent common to all organic compounds is the element carbon. Today, the term organic chemistry, although no longer descriptive, implies the study of carbon-containing compounds.[1] There are approximately 90,000 known inorganic compounds and this number is not rapidly increasing. On the other hand, there are over one million known organic compounds (isolated from nature or synthesized in the laboratory) and several thousand new compounds are synthesized and described each year. Over 95% of all known chemical compounds are compounds of carbon.

What is so unique about carbon that differentiates it from all of the other elements in the periodic table? Carbon has the ability to bond successively to other carbon atoms to form chains and rings of varying sizes. As the number of carbon atoms in a chain increases, the number of ways that these

[1]This definition is not strictly adhered to. In Section 8.5-b we have mentioned several of the following compounds of carbon, which properly belong to the domain of inorganic chemistry.

Carbon monoxide, CO	Carbonates, e.g., Na_2CO_3	Thiocyanates, e.g., $NaSCN$
Carbon dioxide, CO_2	Bicarbonates, e.g., $NaHCO_3$	Cyanates, e.g., $KOCN$
Carbon disulfide, CS_2	Cyanides, e.g., KCN	Carbides, e.g., CaC_2

Table 12.1 Contrasting Properties of Organic and Inorganic Compounds

Organic	*Inorganic*
1. Low melting points	1. High melting points
2. Low boiling points	2. High boiling points
3. Low solubility in water; high solubility in nonpolar solvents	3. High solubility in water; low solubility in nonpolar solvents
4. Flammable	4. Nonflammable
5. Solutions are nonconductors of electricity	5. Solutions are conductors of electricity
6. Chemical reactions are usually slow	6. Chemical reactions are rapid
7. Exhibit isomerism	7. Isomers are limited to a few exceptions (e.g., the transition elements)
8. Exhibit covalent bonding	8. Exhibit ionic bonding
9. Exist as gases, liquids, and solids at room temperature	9. Exist predominantly as solids at room temperature

atoms may arrange themselves increases, yielding compounds with the same chemical composition but with different structures. Finally, carbon can form equally strong bonds with a number of different elements. Those elements most frequently encountered in organic compounds are hydrogen, oxygen, nitrogen, sulfur, phosphorus, and the halogens.

12.2 Chemical Formulas

In Sections 1.1-d and 4.7 we defined and illustrated the terms empirical formula, molecular formula, and structural formula. A fourth formula, the **condensed structural formula,** is perhaps the most widely employed by organic chemists. As the name implies, it is a shorthand (and less descriptive) method of representing the structural formula. The convention is to omit the bonds between each carbon and the hydrogens attached to it. Very often, carbon–carbon single bonds are also omitted. Table 12.2 illustrates the differences that exist among all four types of chemical formulas, using as examples specific organic compounds that will be studied later in this text.

12.3 Isomers

For the most part, the empirical and molecular formulas contain enough information for the inorganic chemist, since two different inorganic compounds do not usually have the same formula. They are, however, of limited use in describing organic compounds. This is clearly seen by reference to Table 12.2. Recall that ethyl alcohol and methyl ether both have the molecular formula C_2H_6O (Section 3.8), and the molecular formulas for acetic acid and methyl formate are both $C_2H_4O_2$. These are not exceptional cases.

Table 12.2 Chemical Formulas of Some Organic Compounds

Name of Compound	*Class of Compound*	*Empirical Formula*	*Molecular Formula*	*Structural Formula*	*Condensed Structural Formula*
Ethane	Alkane	CH_3	C_2H_6	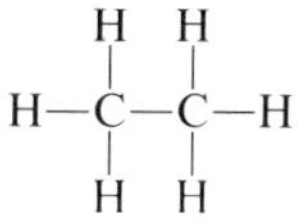	CH_3CH_3
Ethylene	Alkene	CH_2	C_2H_4	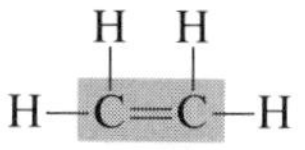	$H_2C{=}CH_2$
Acetylene	Alkyne	CH	C_2H_2	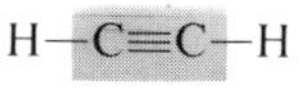	$HC{\equiv}CH$
Ethyl alcohol	Alcohol	C_2H_6O	C_2H_6O	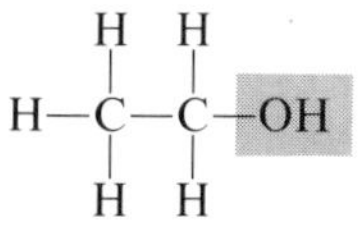	CH_3CH_2OH
Dimethyl ether	Ether	C_2H_6O	C_2H_6O	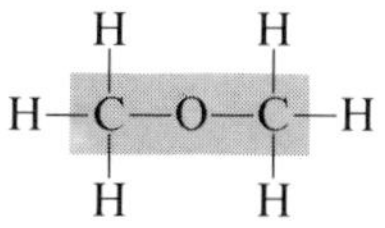	CH_3OCH_3
Acetaldehyde	Aldehyde	C_2H_4O	C_2H_4O	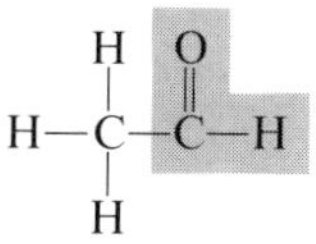	CH_3CHO
Acetone	Ketone	C_3H_6O	C_3H_6O	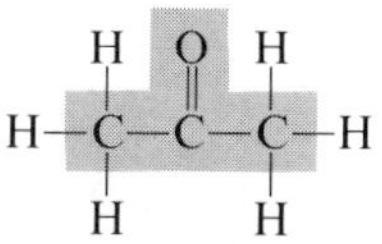	CH_3COCH_3
Acetic acid	Carboxylic acid	CH_2O	$C_2H_4O_2$	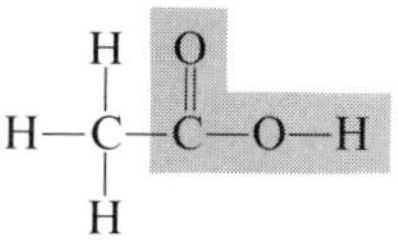	CH_3COOH
Methyl formate	Ester	CH_2O	$C_2H_4O_2$	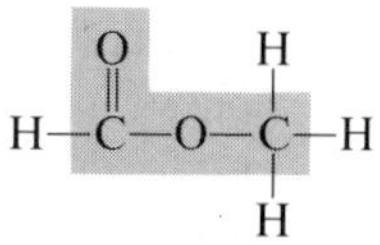	$HCOOCH_3$
Acetamide	Amide	C_2H_5ON	C_2H_5ON	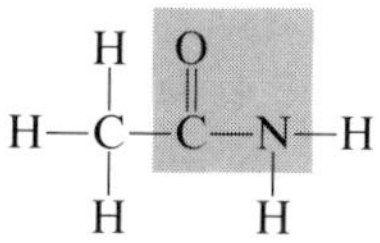	CH_3CONH_2
Ethylamine	Amine	C_2H_7N	C_2H_7N	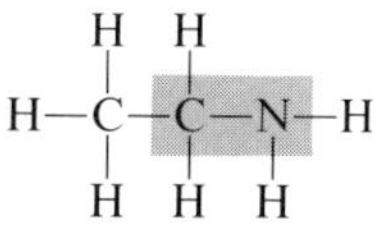	$CH_3CH_2NH_2$

Commonly, several different organic compounds will have identical chemical composition but widely different chemical and physical properties. (This fact was cited in Section 12.1 as being partly responsible for the great abundance of organic compounds.) *Compounds that have the same molecular formula but different structural formulas* are **isomers** (Greek, *iso,* same; *meros,* part). The phenomenon that describes the existence of such compounds is **isomerism.** In our study of organic and biological compounds we will refer to isomerism on many occasions.

12.4 Homologous Series and Functional Groups

The total number of possible organic compounds is infinite. However, these compounds may be divided into a comparatively small number of classes on the basis of similarities in structure. Therefore, thorough knowledge of a few members of each class familiarizes one with the chemical and physical properties of almost all the other members of that class. *A family of similar compounds* is a **homologous series** (*homos,* same). *The individual members of the series* are **homologs.** Characteristics of a homologous series include

1. All compounds in the series contain the same elements and can be represented by a single general formula.
2. The molecular formula of each homolog differs from the one above it and the one below it by a "—CH_2—" increment.
3. There is a gradual variation in physical properties with increasing molar mass.
4. All compounds in the series have similar chemical properties.

Examination of Table 12.2 reveals that each of the classes of compounds (i.e., alkene, alkyne, aldehyde, ketone, etc.) is distinguished from each of the other classes by some unique structural feature. Such a structural feature is a **functional group.** A functional group is defined as *any particular arrangement of a few atoms that bestows characteristic properties on an organic molecule.* (The functional group of each class of compounds is shown as the shaded portion of the structural formulas.) Each member of a homologous series contains the same functional group, and the characteristic differences between the families of organic compounds are a result of differing functional groups. In the following chapters, we will study each of the important homologous series of organic compounds. In particular, we will be concerned with the special properties and reactions associated with each functional group in order to understand better the chemical reactions involving these same characteristic groupings.

Exercises

12.1 What was the major deterrent to the development of organic chemistry prior to the nineteenth century? What was the significance of Wöhler's experiment?

12.2 The number of different carbon-containing compounds appears to be unlimited, whereas there are relatively many fewer compounds of the other elements. Explain.

12.3 Briefly discuss how organic and inorganic compounds differ.

12.4 Suggest a test that one might use to determine if a newly discovered compound is organic or inorganic.

12.5 Distinguish between the terms ionic bond and covalent bond. Give an example of each.

12.6 What property of the carbon atom accounts for the occurrence of so many carbon compounds?

12.7 Represent the structure of propane, $CH_3CH_2CH_3$, using an electron dot formula.

12.8 Write structural formulas to represent the different isomers of the following.

(a) C_3H_8O
(b) $C_2H_4Cl_2$
(c) $C_2H_3Cl_3$
(d) C_3H_7Cl
(e) C_3H_6BrCl

12.9 Which of the following pairs of compounds are isomers?

(a) $CH_3CH_2CH_3$ and $CH_3CH_2CH_2CH_3$

(b) $CH_3CH{=}CH_2$ and 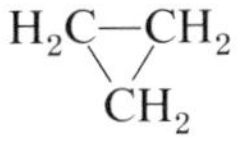

(c) $CH_3CH{=}CH_2$ and $CH_3C{\equiv}CH$

(d) 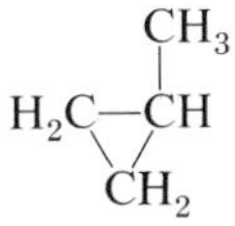 and 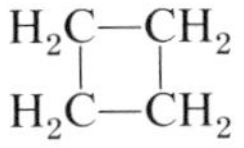

(e) CH_3CH_2OH and 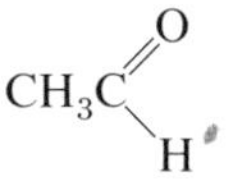

(f) 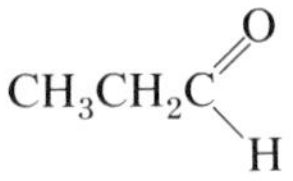 and 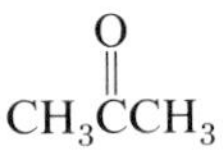

(g) 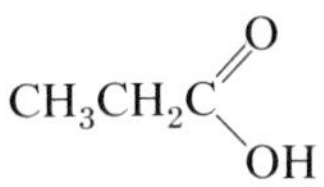 and 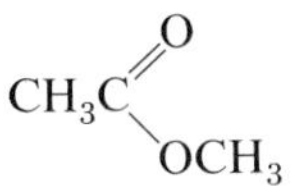

(h) 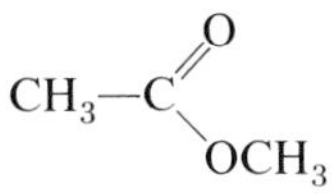 and 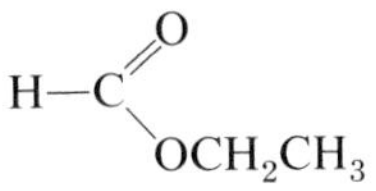

12.10 Can homologs be isomers? Explain.

12.11 Define the following terms and illustrate each with a suitable example.

- **(a)** empirical formula
- **(b)** molecular formula
- **(c)** structural formula
- **(d)** isomers
- **(e)** homologous series
- **(f)** functional group

12.12 A hydrocarbon was shown by analysis to contain 80% carbon and 20% hydrogen. At STP a liter of the hydrocarbon weighed 1.34 grams. What is the molecular formula of the hydrocarbon?

12.13 Quantitative combustion of an organic compound indicated that it contained 82.8% carbon and 17.2% hydrogen.

- **(a)** What is the empirical formula of this compound?
- **(b)** If the molar mass of the compound is 58, what is the molecular formula of the compound?
- **(c)** Write structural formulas for the two isomers of this compound.

13

The Saturated Hydrocarbons

The saturated hydrocarbons are often referred to as the *parent* compounds of organic chemistry since all other known organic compounds can be considered to be derivatives of them. **Hydrocarbons,** as the name implies, are *compounds composed only of carbon and hydrogen atoms.* The adjective, **saturated,** describes the type of bonding within the hydrocarbon molecule. It signifies that *each carbon atom is covalently bonded to four other atoms by* ***single*** *bonds.* The compounds are also referred to as paraffin hydrocarbons or alkanes. The name *paraffin* (Latin, *parum affinis,* little activity) alludes to their unreactive nature. The name *alkane* is the generic name for this class of compounds based upon the IUPAC system of nomenclature.

The saturated hydrocarbons belong to the *aliphatic* series of compounds. (The two other main series, aromatic and heterocyclic, will be mentioned later in the text.) Since many of the first hydrocarbons to be studied were derived from fatty acids, the name aliphatic (Greek, *aleiphatos,* fat) was applied to them to indicate their source. Today, aliphatic compounds comprise all the open-chain hydrocarbons and cyclic hydrocarbons as well as their derivatives.

13.1 Nomenclature of Alkanes

The general formula for the alkane homologs is C_nH_{2n+2}. Table 13.1 contains the names and formulas for the first ten saturated hydrocarbons. Also included is a listing of the number of possible isomers that correspond to each molecular formula. The number of possible isomers increases extremely

Table 13.1 Straight-Chain Alkanes

Name	Molecular Formula	Condensed Structural Formula	Number of Possible Isomers
Methane	CH_4	CH_4	1
Ethane	C_2H_6	CH_3CH_3	1
Propane	C_3H_8	$CH_3CH_2CH_3$	1
Butane	C_4H_{10}	$CH_3(CH_2)_2CH_3$	2
Pentane	C_5H_{12}	$CH_3(CH_2)_3CH_3$	3
Hexane	C_6H_{14}	$CH_3(CH_2)_4CH_3$	5
Heptane	C_7H_{16}	$CH_3(CH_2)_5CH_3$	9
Octane	C_8H_{18}	$CH_3(CH_2)_6CH_3$	18
Nonane	C_9H_{20}	$CH_3(CH_2)_7CH_3$	35
Decane	$C_{10}H_{22}$	$CH_3(CH_2)_8CH_3$	75

rapidly with each addition of a carbon atom to the chain. For example, addition of four carbon atoms and eight hydrogen atoms to hexane, to form decane, causes a fifteen-fold increase in the number of possible isomers. Even more striking is the calculation that there are 366,319 possible isomers of the alkane whose molecular formula is $C_{20}H_{42}$ and there are 62,491,178,805,831 possible isomers of the alkane whose molecular formula is $C_{40}H_{82}$. These numbers are given just by way of illustration, and in no way are meant to imply that all of the isomers are known or have been synthesized.

a. Common System

There is no difficulty in naming the first three alkanes, since there is only one structure that can be written for each of these compounds. However, it is possible to write two (and only two) structural formulas for butane. Both structures correspond to the molecular formula C_4H_{10} (Figure 13.1). The important point here is that not only can two separate structures be drawn that correspond to this formula but two distinct compounds actually exist. Each is found in nature and each can be synthesized in the laboratory. They are structural isomers of one another and thus have different physical properties.

Compound I is a straight-chain hydrocarbon. Its four carbon atoms are arranged in one continuous chain. Compound II is a branched-chain hydrocarbon. Three of its carbon atoms are arranged in a straight chain, while the fourth is branched off from the middle carbon atom. The straight-chain hydrocarbon is designated by prefixing the letter *n* (for normal) to the name of the compound. Thus, compound I is *n*-butane. The branched-chain compound is designated by the prefix *iso*-; compound II is isobutane.

There are only three different structural formulas that can be drawn to represent the compound with the molecular formula C_5H_{12} (Figure 13.2).

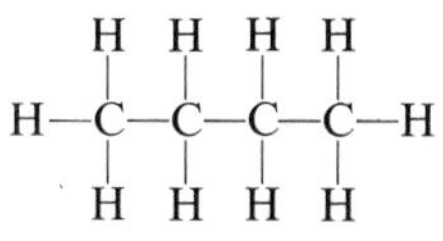

$CH_3CH_2CH_2CH_3$

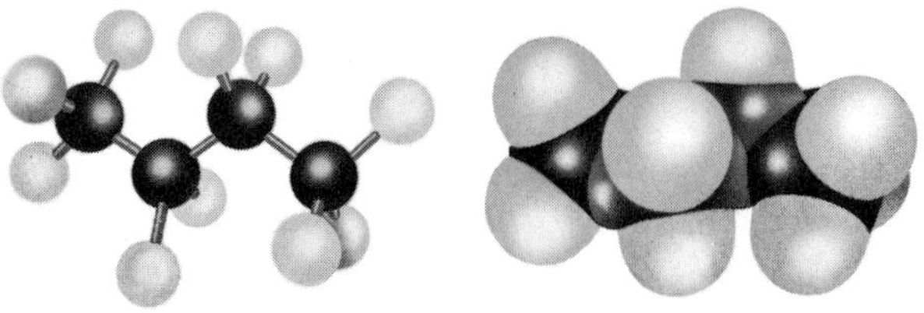

I
n-Butane

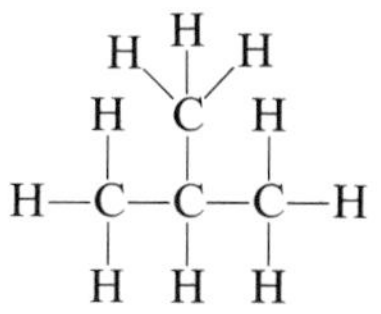

$CH_3CH(CH_3)CH_3$

II
Isobutane

Figure 13.1 Structures of the butane isomers.

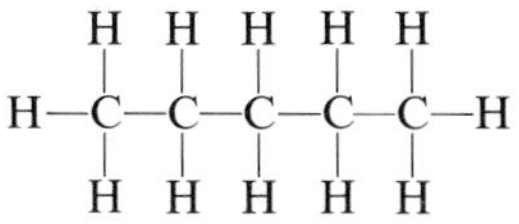

$CH_3CH_2CH_2CH_2CH_3$

n-Pentane

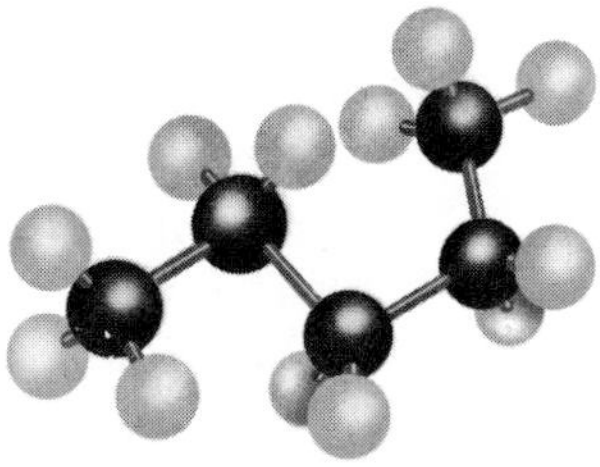

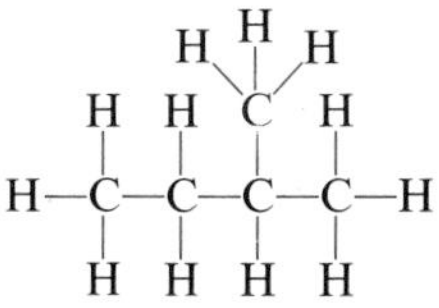

$CH_3CH_2CH(CH_3)CH_3$

Isopentane

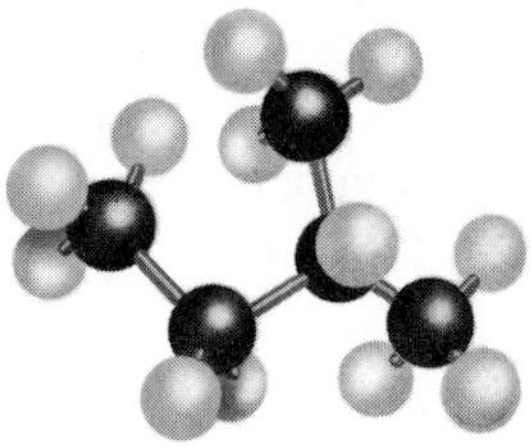

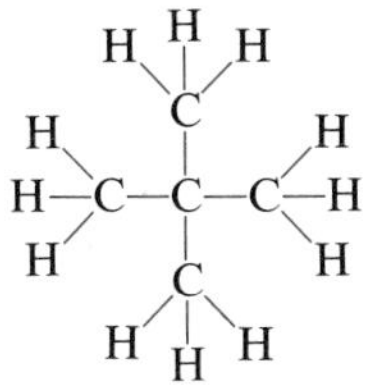

$C(CH_3)_4$

Neopentane

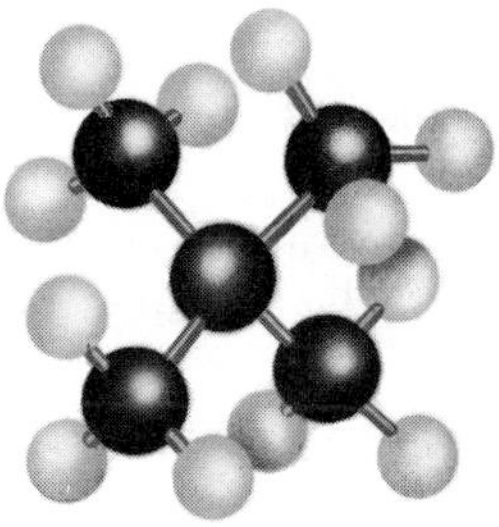

Figure 13.2 Structures of the pentane isomers.

b. IUPAC System

For the higher homologs of this series, as well as for other classes of organic compounds, the system of using prefixes is unsatisfactory. It would overtax the memory of anything short of a computer, and language simply cannot provide meaningful prefixes. To tackle this problem, an international meeting of chemists was held at Geneva, Switzerland, in 1892. The meeting resulted in a simple, unequivocal system for naming organic compounds. The system was modified by the International Union of Chemists and is kept up to date by its successor, the International Union of Pure and Applied Chemistry (IUPAC). What was evolved is a set of rules referred to as the Geneva System, or the IUPAC System of Nomenclature. The rules for the alkanes are as follows.

1. Saturated hydrocarbons are named according to the longest continuous chain of carbons in the molecule. This is the parent compound.
2. The suffix *-ane* indicates that the molecule is a saturated hydrocarbon.
3. The name of the parent hydrocarbon is modified by noting what *alkyl groups* (see below) are attached to the chain.
4. The chain is numbered and the position of each substituent alkyl group is indicated by the number of the carbon atom to which it is attached. The chain is numbered in such a way that the substituents occur on the carbon atoms with the lowest numbers.
5. Names of the substituent groups are placed in alphabetical order before the name of the parent compound.
6. If the same alkyl group appears more than once, the numbers of the carbons to which it is attached are both expressed. If the same group appears more than once on the same carbon, the number of the carbon is repeated as many times as the group appears.
7. Hyphens are used to separate numbers from names of substituents; numbers are separated from each other by commas. The number of identical groups is indicated by the Greek prefixes *di-*, *tri-*, *tetra,* etc.
8. The last alkyl group to be named is prefixed to the name of the parent alkane forming one word.

c. Alkyl Groups

The group of atoms that results when a hydrogen atom is removed from an alkane is called an **alkyl group.** Thus, the general formula for an alkyl group is C_nH_{2n+1}. The group is named by replacing the *-ane* suffix of the parent hydrocarbon with *-yl* (Table 13.2).

Note that not all of the hydrogen atoms of the propane molecule are

Table 13.2 Common Alkyl Groups

Parent Hydrocarbon		*Alkyl Group*		*Condensed Formulas*		
Methane	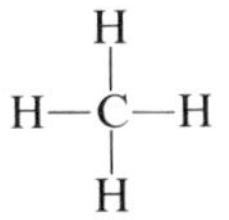	Methyl	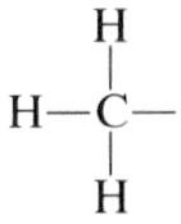	CH_3—		
Ethane	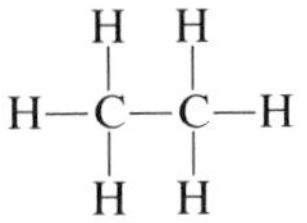	Ethyl	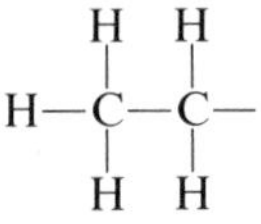	CH_3CH_2—	or	C_2H_5—
Propane	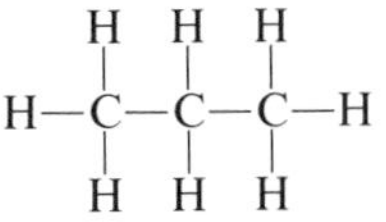	*n*-Propyl	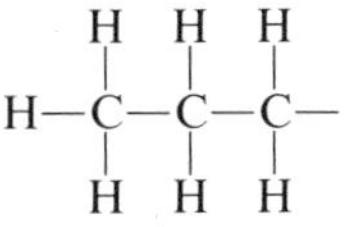	$CH_3CH_2CH_2$—	or	C_3H_7—
		Isopropyl	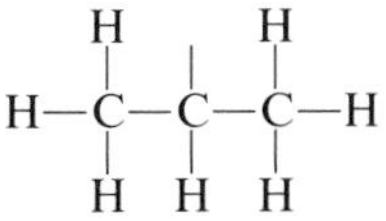	$CH_3\overset{\vert}{C}HCH_3$	or	$(CH_3)_2CH$—

equivalent. Six of them, three attached to each terminal carbon, are equivalent. The two attached to the middle carbon, however, are different. In Section 17.1 this difference is more fully explained. Let it suffice for now to say that if either of the hydrogen atoms on the middle carbon is removed, a different alkyl group results.

For convenience, the organic chemist often represents an alkyl group by the symbol **R.** This symbol is used whenever it is desirable to represent a general class of compounds. Thus, the symbol for the family of saturated hydrocarbons is R—H.

The application of these rules may appear to be very difficult, but, in reality, learning the systematic nomenclature is probably the easiest part of organic chemistry. All that is required is a good deal of practice. (The student is encouraged to do many nomenclature problems early in the semester. It has been the author's experience that through repeated practice the student quickly learns the application of these rules and organic nomenclature becomes a game rather than a chore.)

13.2 Application of the IUPAC Rules

Table 13.3 shows the isomers of butane and pentane. Examination of the table raises the following questions: If compounds I and II are both isomers of butane, why is one called butane and the other methylpropane? Similarly, if compounds III, IV, and V are all isomers of pentane, C_5H_{10}, why is one designated as pentane, one as a derivative of butane, and the other as a derivative of propane? The answer is that under the IUPAC system, a

Table 13.3 Isomers of Butane and Pentane

	Formula	*IUPAC Name*
I	$CH_3CH_2CH_2CH_3$	*n*-Butane
II	CH_3 | CH_3CHCH_3	2-Methylpropane
III	$CH_3CH_2CH_2CH_2CH_3$	*n*-Pentane
IV	CH_3 | $CH_3CH_2CHCH_3$	2-Methylbutane
V	CH_3 | $CH_3—C—CH_3$ | CH_3	2,2-Dimethylpropane

compound may be an isomer of another compound without having the same parent name. The only criterion that two compounds must meet in order to be isomers is that they have the same molecular formula. It is to be emphasized that according to the IUPAC system, the name of the parent hydrocarbon corresponds to the number of carbons in the *longest continuous chain*, not to the total number of carbon atoms in the molecule. This point is well illustrated by the IUPAC names for the five isomers of hexane, C_6H_{14} (Table 13.4).

There are several aspects of nomenclature that invariably trouble the

Table 13.4 Isomers of Hexane

Isomer	*Formula*	*IUPAC Name*
1	$CH_3CH_2CH_2CH_2CH_2CH_3$	*n*-Hexane
2	CH_3 | $CH_3CH_2CH_2CHCH_3$	2-Methylpentane
3	CH_3 | $CH_3CH_2CHCH_2CH_3$	3-Methylpentane
4	CH_3 | $CH_3CH_2CCH_3$ | CH_3	2,2-Dimethylbutane
5	CH_3 | $CH_3CHCHCH_3$ | CH_3	2,3-Dimethylbutane

beginning student in organic chemistry. Repeated practice will insure that the student will avoid the following common pitfalls.

1. *Thinking that a substituent positioned above the chain is different from the substituent positioned below the chain.* There is free rotation about any of the single bonds in the molecule, so above and below have no meaning. The significant thing about the position of a substituent group is the number of the carbon atom to which it is attached.

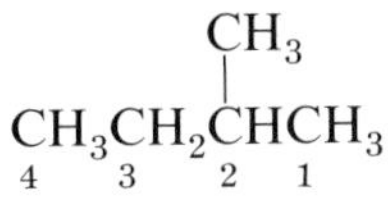

$$\underset{4\quad 3\quad 2\quad 1}{CH_3CH_2\overset{\overset{\displaystyle CH_3}{|}}{C}HCH_3} \text{ is the same as } \underset{4\quad 3\quad 2\quad 1}{CH_3CH_2\underset{\underset{\displaystyle CH_3}{|}}{C}HCH_3}$$

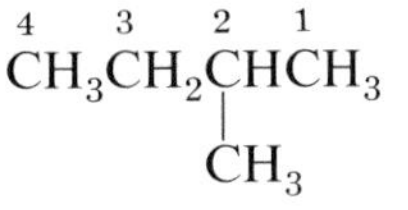

2. *Failure to identify the longest continuous chain.* Nothing but convention dictates that an organic molecule must be represented in the form of a linear chain. Examine curiously branched structures and seek out the longest continuous chain.

$$\underset{\substack{3 \\ CH_2-\overset{4}{C}H-CH_3 \\ | \\ ^5CH_2 \\ | \\ ^6CH_2 \\ | \\ ^7CH_3}}{\overset{1}{C}H_3-\overset{\overset{\displaystyle CH_3}{|}}{\underset{\displaystyle |}{\overset{2}{C}}}-CH_3}$$

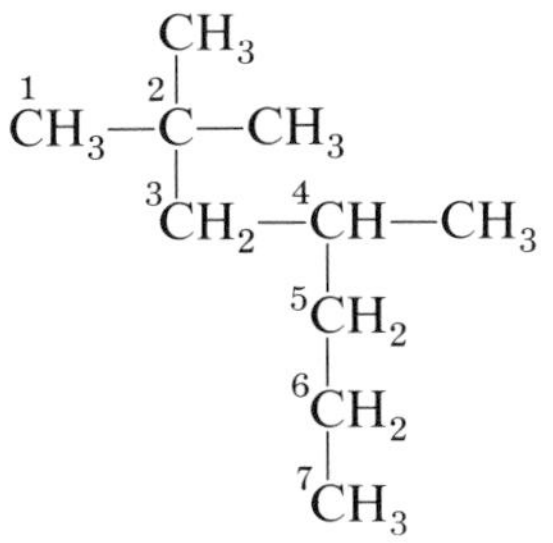

is the same as 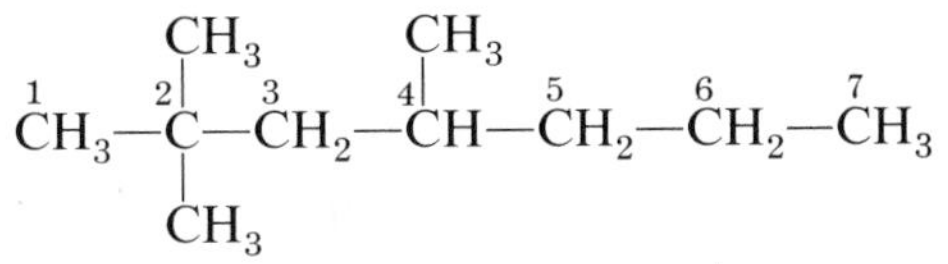

$$\underset{CH_3}{\overset{\overset{\displaystyle CH_3 \qquad CH_3}{| \qquad\quad |}}{\overset{1}{C}H_3-\overset{2}{C}-\overset{3}{C}H_2-\overset{4}{C}H-\overset{5}{C}H_2-\overset{6}{C}H_2-\overset{7}{C}H_3}}$$

2,2,4-Trimethylheptane

3. Don't let an *apparent* alkyl substituent on a terminal carbon fool you.

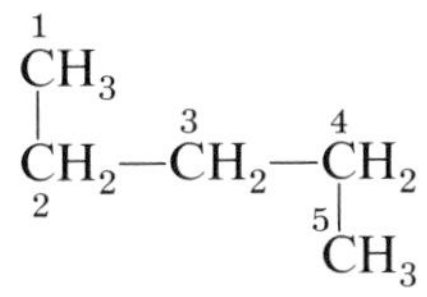

$$\underset{2}{\overset{\overset{\displaystyle ^1CH_3}{|}}{C}H_2}-\overset{3}{C}H_2-\overset{\overset{4}{C}H_2}{\underset{\underset{\displaystyle ^5CH_3}{|}}{}} \text{ is the same as } \overset{1}{C}H_3-\overset{2}{C}H_2-\overset{3}{C}H_2-\overset{4}{C}H_2-\overset{5}{C}H_3$$

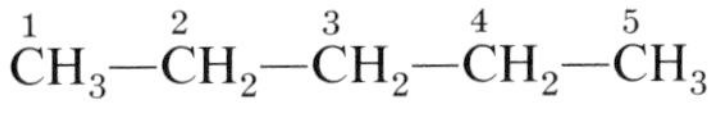

4. *Numbering from the wrong end.*

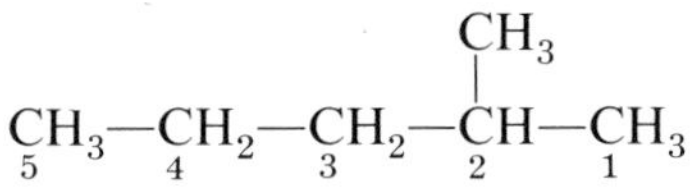

$$\underset{5}{C}H_3-\underset{4}{C}H_2-\underset{3}{C}H_2-\underset{2}{\overset{\overset{\displaystyle CH_3}{|}}{C}}H-\underset{1}{C}H_3 \quad \begin{array}{c}\text{is the}\\\text{same as}\end{array} \quad \underset{1}{C}H_3-\underset{2}{\overset{\overset{\displaystyle CH_3}{|}}{C}}H-\underset{3}{C}H_2-\underset{4}{C}H_2-\underset{5}{C}H_3$$

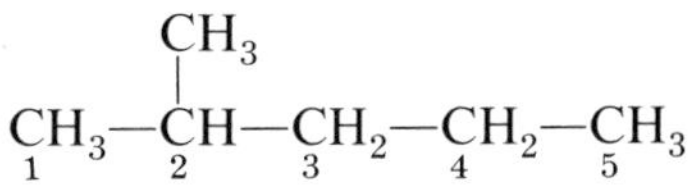

5. *Failing to write the same number twice and/or failing to use the prefix di-, tri-, etc., when two identical substituents are bonded to the same carbon atom.*

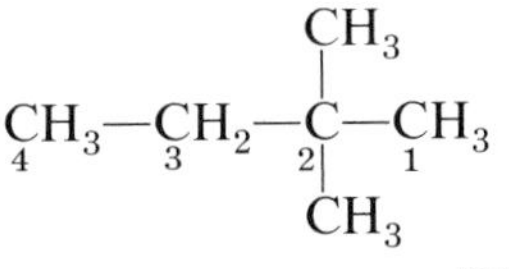

$$\underset{4}{C}H_3-\underset{3}{C}H_2-\underset{2}{\overset{\overset{\displaystyle CH_3}{|}}{\underset{\underset{\displaystyle CH_3}{|}}{C}}}-\underset{1}{C}H_3 \quad \text{is 2,2-dimethylbutane}$$

and *not* 2-dimethylbutane or 2,2-methylbutane

Table 13.5 Examples of IUPAC Nomenclature

Condensed Structural Formula	*Rewritten and Numbered*	*IUPAC Name*
$CH_3CH_2CH(CH_3)CH_2CH(CH_3)_2$	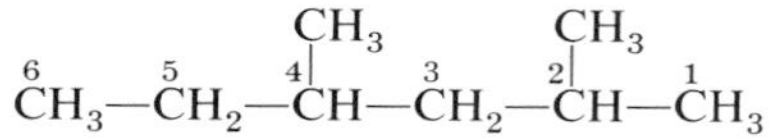	2,4-Dimethylhexane *not* 3,5-Dimethylhexane
$(C_2H_5)_2CHCH(CH_3)CH_2CH_2CH_3$	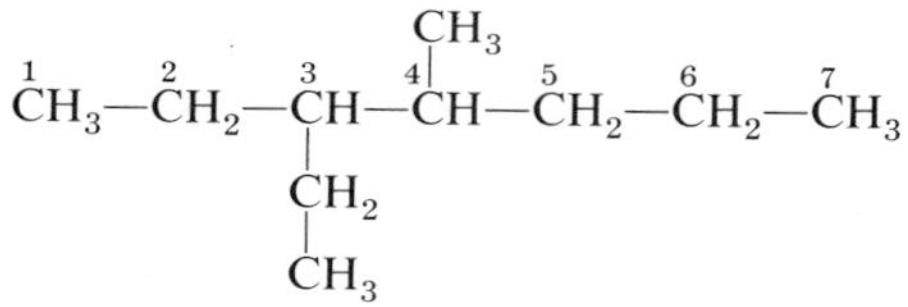	3-Ethyl-4-methylheptane *not* 4-Methyl-5-ethylheptane
$(CH_3)_2CHCH(C_3H_7)_2$	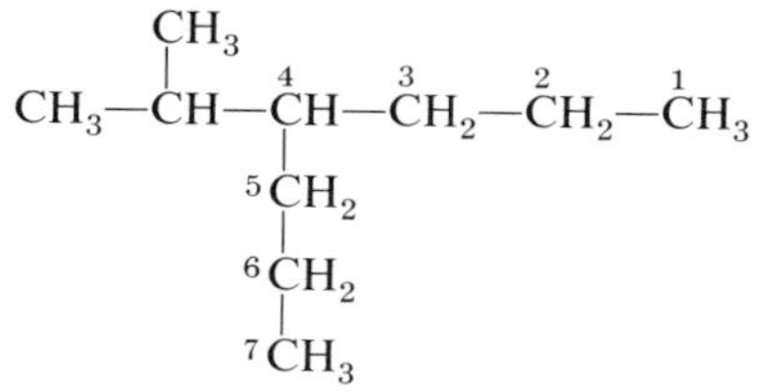	4-Isopropylheptane *not* 2-Methyl-3-*n*-propylhexane
$(CH_3)_2CHCH_2CH_2C(CH_3)_3$	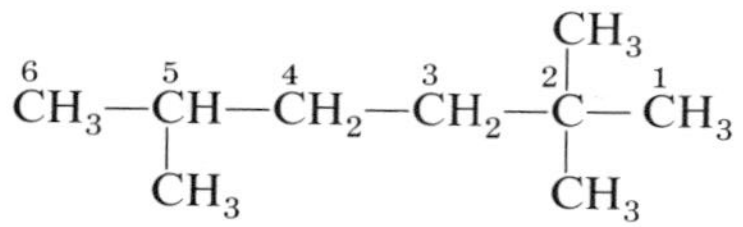	2,2,5-Trimethylhexane *not* 2,5-Trimethylhexane
$(C_2H_5)_2CHC(CH_3)(C_2H_5)_2$	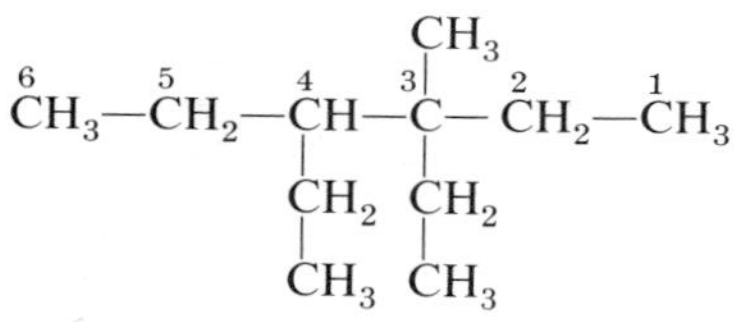	3,4-Diethyl-3-methylhexane *not* 3,4-Ethyl-3-methylhexane

Table 13.5 contains some further examples of the application of the IUPAC system for naming organic compounds.

13.3 Physical Properties of Alkanes

The alkanes are nonpolar molecules. They are, therefore, insoluble in water, but are soluble in organic solvents such as benzene and carbon tetrachloride. Carbon tetrachloride is the principal constituent of cleaning fluid because it readily dissolves grease and oil, which are composed chiefly of hydrocarbons. The hydrocarbons are generally less dense than water, their densities being less than 1.0 g/ml. They are colorless, tasteless, and odorless. The odor associated with natural gas is not that of methane or any of the other akanes, but rather is due to an odorant purposely added to the gas in sufficient quantities to allow for the detection of leaks. Table 13.6 summarizes the properties of the first ten straight-chain alkanes.

Throughout the chapters on organic chemistry and, in fact, in all organic chemistry textbooks, the student will find tables of physical properties. The figures in these tables are not meant to be memorized but are given in order to present a numerical description of the physical characteristics of organic molecules and to serve as standards of purity. The physical states of substances at room temperature are obtained directly from a knowledge of their melting points and boiling points. Room temperature is considered to be approximately 67 °F or about 20 °C. If the melting point of a substance is above 20 °C, the substance will exist as a solid at room temperature. (For example, the first member of the alkane series that is a naturally occurring solid is *n*-octadecane, $C_{18}H_{38}$, mp 28 °C.) If the boiling point of a substance is above 20 °C, the substance will exist as a liquid at room temperature. (Note that *n*-pentane, bp 36 °C, is the first liquid alkane.) Similarly, knowledge of the density of a substance indicates whether the substance is lighter or heavier than water. (The density of water is 1.00 g/ml.) We observe that oil

Table 13.6 Physical Properties of Some Alkanes

Name	*Molecular Formula*	*MP (°C)*	*BP (°C)*	*Density (g/ml)*	*Normal State*
Methane	CH_4	−182	−161	—	Gas
Ethane	C_2H_6	−183	−89	—	Gas
Propane	C_3H_8	−188	−42	—	Gas
n-Butane	C_4H_{10}	−138	−0.5	—	Gas
n-Pentane	C_5H_{12}	−130	36	0.63	Liquid
n-Hexane	C_6H_{14}	−95	69	0.66	Liquid
n-Heptane	C_7H_{16}	−91	98	0.68	Liquid
n-Octane	C_8H_{18}	−57	125	0.70	Liquid
n-Nonane	C_9H_{20}	−54	151	0.72	Liquid
n-Decane	$C_{10}H_{22}$	−30	174	0.73	Liquid

or grease do not mix with water, but rather float on the surface. They do not mix with water because they are insoluble in water; they float on top of the water because they are less dense than water. Densities are also useful in calculations that call for the conversion of a certain mass of a liquid into the corresponding volume of that liquid, or vice versa. (Recall that density equals the mass of a substance divided by its volume.)

As indicated in Table 13.6, the first four members of the alkane series are gases. A comparison of the densities of these gases with the density of air allows one to predict whether or not they will rise in the air. Natural gas is composed chiefly of methane, which has a density of about 0.65 g/liter. The density of air is about 1.29 g/liter. Natural gas, then, is much lighter than air, and it will rise in a room in which there may be a natural gas leak. Once the leak is detected and eliminated, the gas can be removed from the room by opening an upper window. On the other hand, the two constituents of bottled gas[1] are much heavier than air. Propane has a density of 1.6 g/liter, and butane has a density of about 2.0 g/liter. If bottled gas escapes into a room, it collects near the floor and a much more serious situation develops because it is then more difficult to rid the room of the gas.

As shown in Table 13.6, the boiling points of the straight-chain alkanes increase with increasing molar mass. This general rule holds true within all the families of organic compounds whenever the straight-chain homologs of the family are considered. Larger molecules are able to wrap around and interact with one another, and thus more energy is required to separate them. In general, the straight-chain isomers have higher boiling points than the branched-chain isomers. The former compounds can be likened to strands of spaghetti. The molecules can be very closely packed together, resulting in relatively strong intermolecular van der Waals forces of attraction. Chain branching increases the distance between adjacent molecules; hence, the attractive forces are decreased and molecules can more easily escape from the liquid. Table 13.7 lists the melting points and boiling points of the isomers of butane and pentane.

Notice that there is no obvious trend that enables us to predict melting points. In general, the more symmetrical isomers tend to have higher melting

[1]Bottled gas is chiefly a mixture of propane and butane that, through pressure, has been condensed to a liquid so that it may be compactly stored and shipped in metal containers.

Table 13.7 Physical Properties of the Isomers of Butane and Pentane

IUPAC Name	*Condensed Structural Formula*	*MP (°C)*	*BP (°C)*
n-Butane	$CH_3(CH_2)_2CH_3$	−138	−0.5
2-Methylpropane	$CH_3CH(CH_3)_2$	−145	−11
n-Pentane	$CH_3(CH_2)_3CH_3$	−130	36
2-Methylbutane	$CH_3CH_2CH(CH_3)_2$	−160	28
2,2-Dimethylpropane	$(CH_3)_4C$	−20	9.5

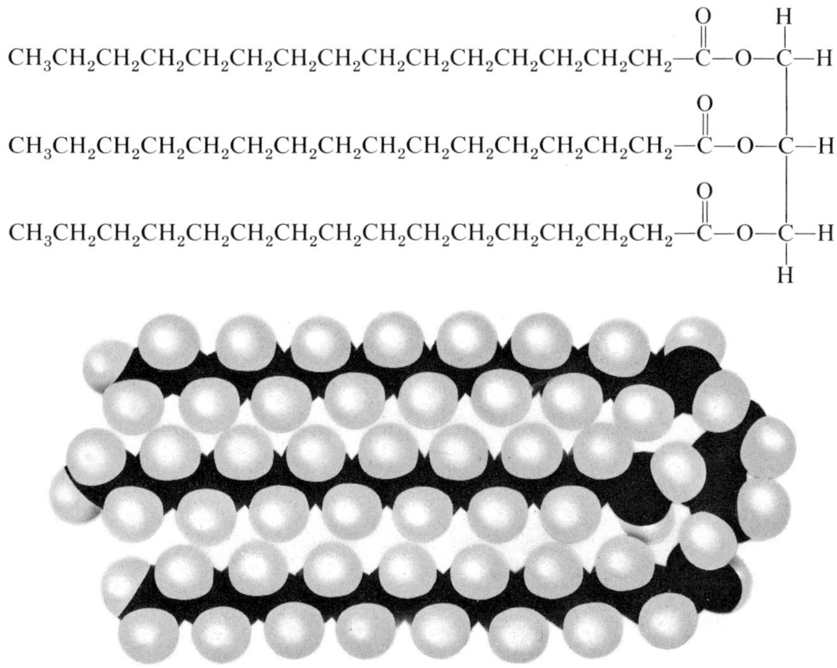

$$CH_3CH_2CH_2CH_2CH_2CH_2CH_2CH_2CH_2CH_2CH_2CH_2CH_2CH_2CH_2-\overset{\overset{\textstyle O}{\|}}{C}-O-\overset{\overset{\textstyle H}{|}}{\underset{\textstyle |}{C}}-H$$

$$CH_3CH_2CH_2CH_2CH_2CH_2CH_2CH_2CH_2CH_2CH_2CH_2CH_2CH_2CH_2-\overset{\overset{\textstyle O}{\|}}{C}-O-\overset{|}{\underset{|}{C}}-H$$

$$CH_3CH_2CH_2CH_2CH_2CH_2CH_2CH_2CH_2CH_2CH_2CH_2CH_2CH_2CH_2-\overset{\overset{\textstyle O}{\|}}{C}-O-\overset{|}{\underset{\textstyle H}{C}}-H$$

Figure 13.3 Tripalmitin, a typical lipid molecule.

points than the less symmetrical isomers. Contrast 2,2-dimethylpropane with the less symmetrical pentanes in Table 13.7. Contrast, also, the very symmetrical octane isomer $(CH_3)_3CC(CH_3)_3$ with *n*-octane. The former is a solid and melts at $103\,°C$; the latter is a liquid whose melting point is $-57\,°C$.

An extensive review of the physical properties of the alkanes has been given here, not because these compounds are so important in and of themselves, but rather because of their importance in discussing the other families of organic and biological compounds. A knowledge of alkane chemistry is also vital to an understanding of the functions of lipids because large portions of their structures consist of segments of alkane moieties (Figure 13.3). One of the major functions of lipids is to serve as structural components of living tissues. Their biological importance here is dependent upon the presence of both polar and nonpolar groups, which enable them to bridge the gap between water-soluble and water-insoluble phases. This is a requisite in maintaining the selective permeability of cell membranes (Figure 13.4). We shall discuss the dual nature of lipid molecules in Chapter 24.

13.4 Chemical Properties of Alkanes

The alkanes are the most unreactive of all classes of organic compounds. Their lack of reactivity accounts for their widespread use as solvents. They are normally unreactive toward such strong reagents as concentrated sulfuric acid, concentrated sodium hydroxide, sodium metal, and potassium

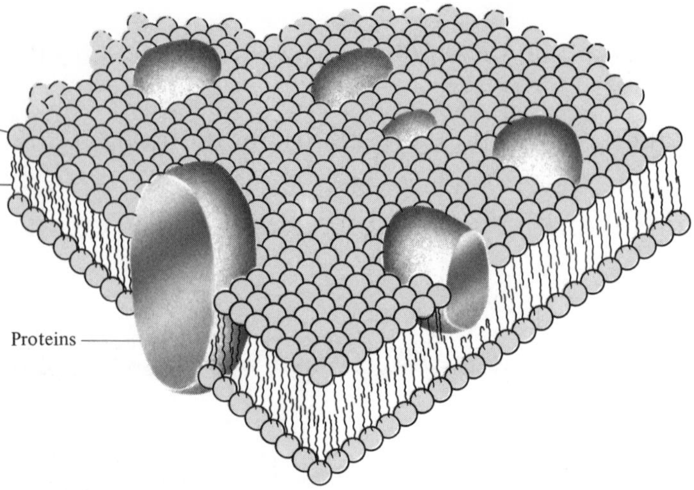
Polar portions
of lipids
Nonpolar portions
of lipids
Proteins

Figure 13.4 Fluid mosaic model of membrane structure. [Adapted from S. J. Singer and G. L. Nicolson, *Science,* **175,** 723 (1972).]

permanganate. Under favorable reaction conditions, they can be made to react with oxygen (combustion), nitric acid, and with chlorine and bromine. The reactions with nitric acid and with the halogens occur by a process known as **substitution.** In a substitution reaction, a part of the reagent molecule substitutes for one of the hydrogen atoms of the alkane.

***Halogenation of Alkanes* (*Substitution of Halogen for Hydrogen*).** The student is probably aware that a mixture of alkanes and air can exist together indefinitely without entering into any reaction. As soon as a match is lit, however, the gas begins to burn. Similarly, mixtures of alkanes and either chlorine or bromine are quite stable when kept in dark containers. When the reaction mixtures are heated or exposed to the sunlight, rapid exothermic reactions occur. Therefore, heat and light must be involved in activating an otherwise unreactive mixture. Through considerable research, this reaction, as well as many of the others we shall study, has been elucidated. When we have arrived at that point in research where we know the pathway by which the reactants are converted to the products, we say that we know the *reaction mechanism* for the reaction under study (see Chapter 9).

The first step in the reaction mechanism for the chlorination or bromination of an alkane involves the splitting of a halogen molecule into two highly energetic halogen atoms. This step is followed by attack of the energetic halogen atom upon a molecule of the alkane. The result of this attack is the formation of a molecule of hydrogen halide and a new highly reactive species, referred to as a **free radical.** (Free radicals are species having odd numbers of electrons; they are important intermediates in many organic reactions.) This newly formed free radical then attacks an unreacted halogen molecule to form the halogenated alkane. At the same time, another energetic halogen atom is formed, and the process can then repeat itself. (This reaction is not applicable to iodine or fluorine. Iodine is unreactive, whereas fluorine is too reactive to control).

This particular type of reaction mechanism is called a **chain mechanism** because once it is initiated, it is self-perpetuating. The process is terminated only by the chance combination of two radicals. The following is a general mechanism for the chlorination of alkanes:

Initiation

$$Cl:Cl \xrightarrow{\text{sunlight}} 2\ Cl\cdot$$

Propagation

$$R-\overset{\overset{\displaystyle H}{|}}{\underset{\underset{\displaystyle H}{|}}{C}}:H + Cl\cdot \longrightarrow R-\overset{\overset{\displaystyle H}{|}}{\underset{\underset{\displaystyle H}{|}}{C}}\cdot + H:Cl$$

$$R-\overset{\overset{\displaystyle H}{|}}{\underset{\underset{\displaystyle H}{|}}{C}}\cdot + Cl:Cl \longrightarrow R-\overset{\overset{\displaystyle H}{|}}{\underset{\underset{\displaystyle H}{|}}{C}}:Cl + Cl\cdot$$

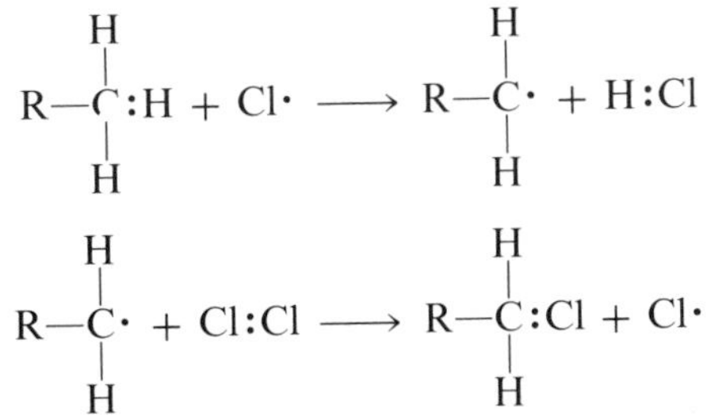

Termination

$$R-\overset{\overset{\displaystyle H}{|}}{\underset{\underset{\displaystyle H}{|}}{C}}\cdot + Cl\cdot \longrightarrow R-\overset{\overset{\displaystyle H}{|}}{\underset{\underset{\displaystyle H}{|}}{C}}:Cl$$

$$R-\overset{\overset{\displaystyle H}{|}}{\underset{\underset{\displaystyle H}{|}}{C}}\cdot + R-\overset{\overset{\displaystyle H}{|}}{\underset{\underset{\displaystyle H}{|}}{C}}\cdot \longrightarrow R-\overset{\overset{\displaystyle H\ H}{|\ \ |}}{\underset{\underset{\displaystyle H\ H}{|\ \ |}}{C:C}}-R$$

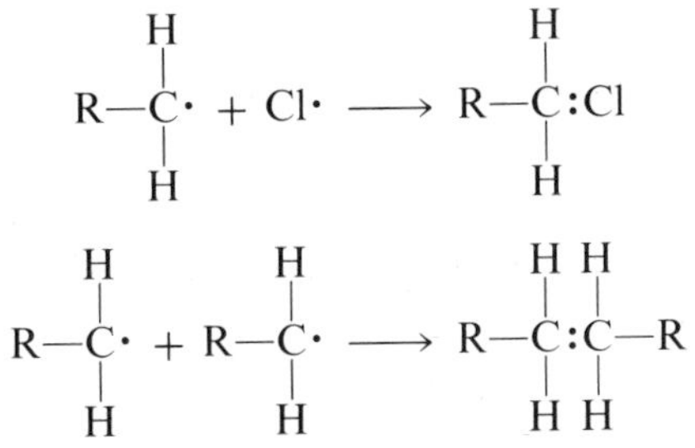

$$Cl\cdot + Cl\cdot \longrightarrow Cl:Cl$$

Free radical halogenation reactions are quite difficult to control. The halogenation of an alkane usually results in a mixture of products, which can be separated from one another by distillation.

$$CH_4 + Cl_2 \xrightarrow[\text{light}]{\text{heat or}} CH_3Cl + CH_2Cl_2 + CHCl_3 + CCl_4 + C_2H_6 + \text{other products}$$

13.5 Free Radical Theory of Aging

Over the years numerous theories have been advanced to explain the causes of human aging. The free radical theory is just one of these, and although there is supportive evidence based upon experimental observations, this theory, like a half dozen others, is controversial and speculative.

The basic concept of the free radical theory of aging is that a steady flux of radicals exists in the body cells and these radicals can react with critical cellular components. Free radicals may be especially harmful when they attack the unsaturated fatty acids present in the lipids of cell membranes (recall Figure 13.4). The resulting altered lipids may seriously impede the diffusion of substances through the membrane—the flow of nutrients and oxygen into the cell and the flow of waste products out of the cell—and thus lead to death of the cell. Some of the damage to the membranes is probably repaired by cellular processes. Aging, according to this theory, could involve either an increased rate of membrane damage with time, or a decrease in the efficiency of repair, or both.

What is the source of these biological free radicals? Since radiation causes a cleavage of chemical bonds, one possibility is the dosage received from the background radiation of cosmic rays and radioactive elements. Although this radiation is partially responsible for cell damage and the low level of natural mutations, it probably has no effect on the aging process per se.[2] It is more likely that free radicals arise from the hydrogen peroxide (HOOH) produced in the oxidative deamination of amino acids (Section 30.3-b) and from the organic hydroperoxides (ROOH) formed by the oxidation of unsaturated lipids (Section 24.6).

It is well known that normal cleavage of a peroxide molecule to form radicals is an extremely slow process at the relatively low temperatures (37 °C) of living systems.

$$\text{H—O:O—H} \longrightarrow 2\,\text{OH}\cdot \qquad \text{extremely slow}$$

However, cellular fluid contains several transition metal ions (e.g., iron, copper) that can readily undergo one-electron oxidation-reduction reactions. Ferrous ions, for example, react with hydrogen peroxide or with organic hydroperoxides by a redox process to produce radicals at an exceedingly rapid rate.

$$\text{HO—OH} + Fe^{2+} \longrightarrow \text{HO}\cdot + \text{HO}^- + Fe^{3+}$$
$$\text{RO—OH} + Fe^{2+} \longrightarrow \text{RO}\cdot + \text{HO}^- + Fe^{3+}$$

Ozone is another source of radicals. It reacts with unsaturated fatty acids to produce organic radicals. The normal ozone content of air is about 0.02 ppm. (Ozone levels in smog normally reach 0.2 to 0.3 ppm and a value as high as 0.6 ppm has been reached.) If 0.02 ppm of ozone is entirely converted to radicals, then about 10^{-6} mole of radicals per human body per day would be formed.

The free radical theory of aging received its impetus from reports that several inhibitors of free radical reactions had been found to increase the average life span of mice when added to their daily diet. The synthetic

[2] William A. Pryor, *Chem. and Eng. News,* June 7, 1971, p. 34.

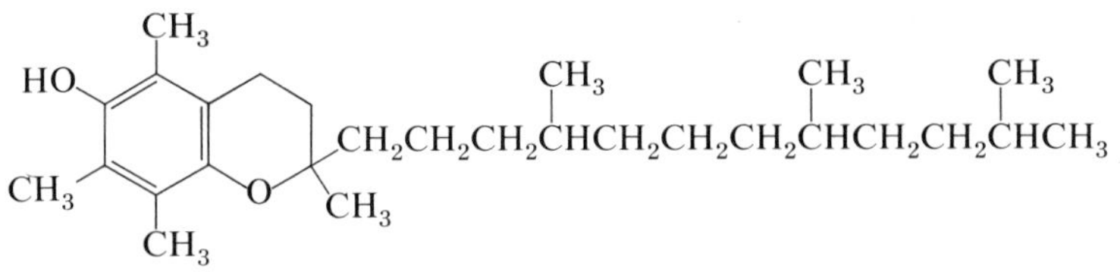

Figure 13.5 Structure of the natural antioxidant vitamin E.

antioxidants 1-amino-2-mercaptoethane and 3,5-di-*tert*-butyl-4-hydroxytoluene (butylated hydroxytoluene) were found to be the most effective.

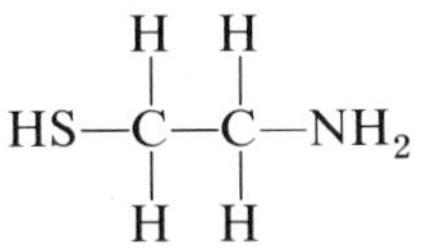

1-Amino-2-mercaptoethane
(2-Mercaptoethylamine)

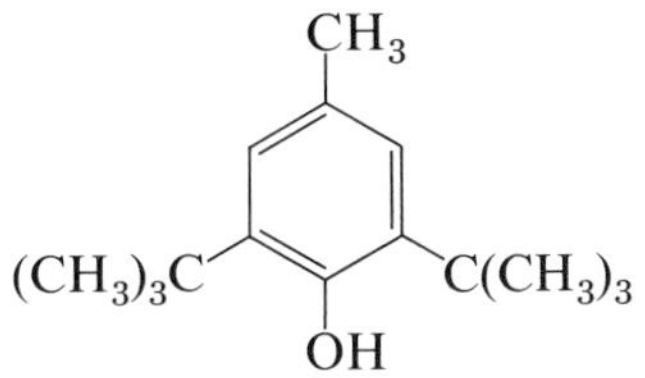

3,5-Di-*tert*-butyl-4-hydroxytoluene
(Butylated hydroxytoluene, BHT)

More recently, the natural antioxidant vitamin E (Figure 13.5) has been found to extend the life of fruit flies by about 13%. However, this does not prove that the antioxidant property of these compounds is responsible for their life-extending action. The following alternative explanations have been postulated.

> . . . these antioxidants induce chemical stress. Certain kinds of stress itself lengthens life. Another possibility is that these antioxidants prolong life because they are enzyme inducers; that is, they increase the formation of microsomal enzymes. The final possibility is that these compounds restrict either appetite or food absorption; it is known that restriction of calories prolongs life.[3]

In summation, then, experimental observations suggest that radicals may be involved in several pathological conditions and that they play a role in the aging of mammals. There is, however, no evidence that synthetic antioxidants or additional vitamin E (beyond that required for nutritional needs) increases the life span of humans. It is probable that human aging is caused by a number of different mechanisms operating simultaneously. Clearly, much more extensive research is required in this area.

13.6 Sources and Uses of Alkanes

Petroleum is a complex mixture of hydrocarbons produced by the decomposition of animal and vegetable matter that has been entrapped in the earth's crust for long periods of time. Refining of petroleum involves distillation into fractions of different boiling ranges (Table 13.8). Crude oil and

[3] Ibid.

Table 13.8 Approximate Boiling Ranges of Various Petroleum Fractions

Fraction	*Boiling Range (°C)*
Natural gas (C_1 to C_4)	<20
Petroleum ether (C_5 to C_6)	30–60
Gasoline (C_6 to C_{12})	85–200
Kerosene (C_{12} to C_{15})	200–300
Fuel oil (C_{15} to C_{18})	300–400
Lubricating oil, paraffin wax, greases, asphalt, mineral oil (C_{16} to C_{40})	>400

natural gas are the liquid and gaseous components of petroleum. Natural gas contains about 80% methane and 10% ethane, the remaining 10% being a mixture of the higher alkanes. Methane, sometimes referred to as marsh gas, is a product of the bacterial decay of plant and marine organisms that have become buried beneath the earth's surface. It is the chief constituent of cooking fuel. When methane burns in an ample supply of oxygen, the following reaction takes place.[4]

$$CH_4 + 2\,O_2 \longrightarrow CO_2 + 2\,H_2O + 210\ \text{kcal/mole}$$

Notice that the reaction is exothermic. The basic source of this released energy is the capacity of these carbon-containing compounds to form stable chemical bonds with oxygen. The larger the alkane molecule, the greater the amount of heat produced. Approximately 150 kcal/mole of energy is produced with each additional carbon atom. The general equation for the reaction may be written as follows.

$$C_nH_{2n+2} + \frac{3n+1}{2}O_2 \longrightarrow n\,CO_2 + (n+1)\,H_2O + \text{energy}$$

The hydrocarbons have numerous commercial and industrial uses. For example, the combustion of hydrocarbons provides the energy that operates the engine of a car. The fuel in this case consists of a mixture of hydrocarbons composed predominantly of isomers of octane. Also present in gasoline are

[4] A limited supply of oxygen results in the incomplete combustion of hydrocarbons (e.g., octane) producing the poisonous air pollutant carbon monoxide or elemental carbon (which forms deposits on automobile pistons). Unburned hydrocarbons (and reaction by-products such as aldehydes) are also released into the atmosphere, significantly contributing to the smog problem.

$$2\,C_8H_{18} + 17\,O_2 \longrightarrow 16\,CO + 18\,H_2O + \text{energy}$$

$$2\,C_8H_{18} + 9\,O_2 \longrightarrow 16\,C + 18\,H_2O + \text{energy}$$

other hydrocarbons having from nine to twelve carbons. The equation for the energy-releasing combustion of hydrocarbons is similar to the equation for the burning of food in our body to produce energy (see Chapter 28).

$$C_6H_{12}O_6 + 6\ O_2 \longrightarrow 6\ CO_2 + 6\ H_2O + 686\ \text{kcal/mole}$$

13.7 Octane Rating

In rating the various hydrocarbon fuels, a standard one-cylinder test engine was employed. It was discovered that pure isooctane burned very smoothly (without vibration or "knocking") and it was assigned an octane number of 100. *n*-Heptane caused very severe knocking and was given an octane number of 0. Mixtures of these two compounds give knocking properties between the two extremes. Therefore the octane number of any fuel is determined by the percentage of isooctane in the heptane-isooctane mix. For example, if a fuel has the same knocking properties as a 6% heptane–94% isooctane mix, it would have an octane number of 94. Technological advances have resulted in engines that have greater power and compression ratios. This has necessitated a higher quality fuel. Some fuels containing more highly branched hydrocarbons or aromatic hydrocarbons may have octane numbers greater than 100 if they burn more smoothly than isooctane. Most regular grades of gasoline have octane ratings of 92–94, and premium grades of gasoline average 98–100. A method used to minimize engine knocking is to add small quantities of a so-called antiknocking ingredient such as tetraethyllead, $(C_2H_5)_4Pb$. Its function is to control the concentration of free radicals and prevent the premature explosions that are characteristic of knocking. Tetraethyllead adds another pollutant (lead) to those already produced by the automobile. The catalytic converters found in most of today's cars are deactivated by lead, and therefore some other additive had to be found. The major oil companies now add aromatic compounds such as benzene, toluene, and xylene (Chapter 15) to their unleaded gas to increase the octane number. The addition of these compounds has substantially increased the cost of unleaded gas compared to the leaded variety. Recent research investigations suggest that rare earth metals (e.g., lanthanum and cerium) and their compounds could replace tetraethyllead as antiknock additives for motor fuels.

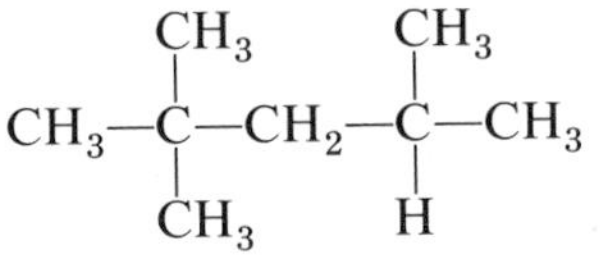

"Isooctane"
2,2,4-Trimethylpentane

$CH_3CH_2CH_2CH_2CH_2CH_2CH_3$

n-Heptane

13.8 Cycloalkanes

The general formula for cycloalkanes is C_nH_{2n}; they contain two less hydrogen atoms than the open-chain hydrocarbons. The cycloalkanes can be thought to arise from the removal of one hydrogen atom from each of the terminal carbons of the corresponding open-chain hydrocarbon. These carbons are then joined together.

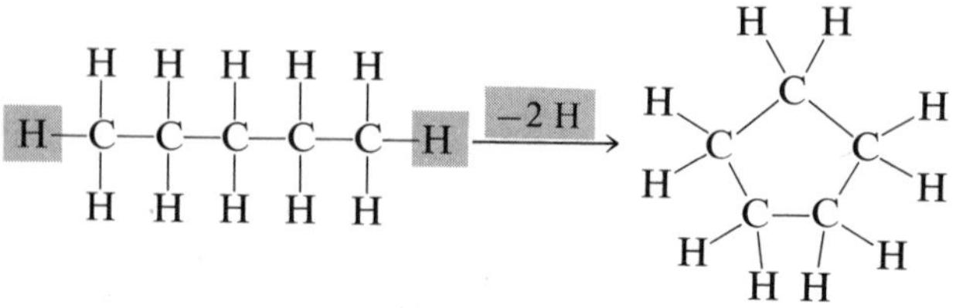

The simplest cycloalkane, cyclopropane, contains three carbons. The names of the cycloalkanes are obtained by attaching the prefix *cyclo-* to the name of the parent alkane. Again, substituents are indicated by number, and the ring carbons are numbered so that the carbons bearing the substituents have the lowest numbers. It is a convention in the writing of cycloalkane formulas that the hydrogen atoms attached to the ring carbons need not be shown. The appropriate geometrical figure is drawn assuming that there is a carbon atom at each corner of the figure, and substituent groups are shown attached to the proper carbon atoms of the ring.

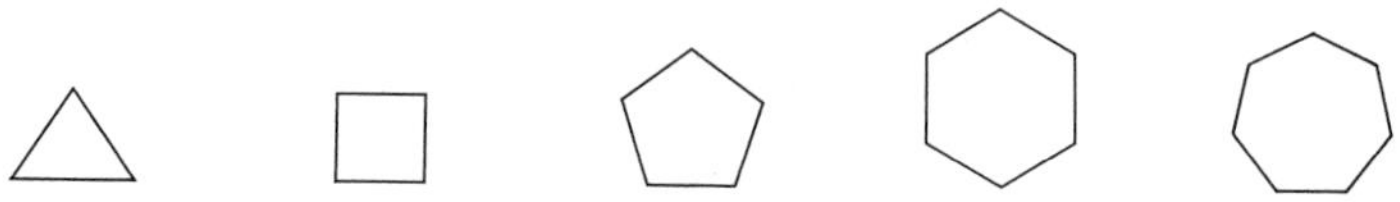

Cyclopropane Cyclobutane Cyclopentane Cyclohexane Cycloheptane

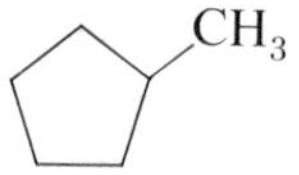

Methylcyclopentane

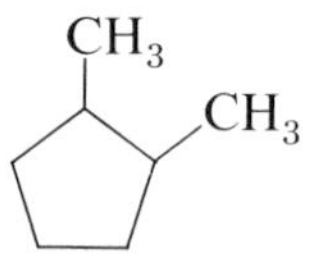

1,2-Dimethylcyclopentane

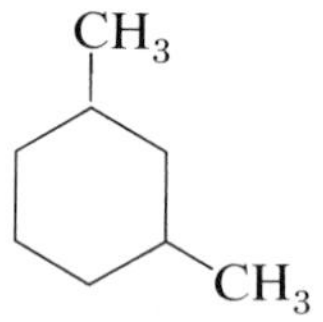

1,3-Dimethylcyclohexane

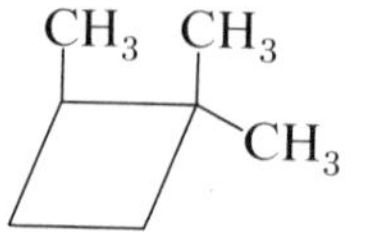

1,1,2-Trimethylcyclobutane

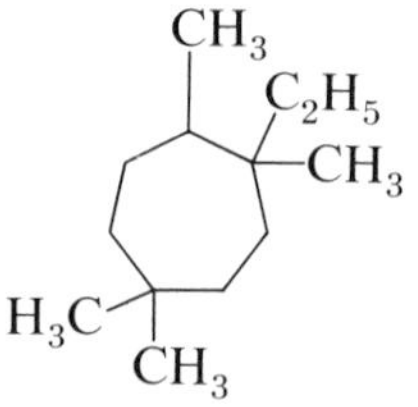

1-Ethyl-1,2,5,5-tetramethylcycloheptane

In most respects, the chemical and physical properties of the cycloalkanes are similar to those of their noncyclic counterparts. There are, however, some variations which deserve special consideration.

The major distinction between the cyclic and noncyclic alkanes arises from the differences in the geometric configuration of the two types of compounds. As mentioned previously, there is free rotation about all carbon–carbon bonds of open-chain alkanes. This is not the case with the cycloalkanes. Here the carbons are held rigidly in place (Figure 13.6). Free rotation is impossible without disruption of the ring structure. For these compounds, the position of any substituent relative to the ring becomes extremely important. (The substituent will be situated either above or below the ring.) A new form of isomerism is possible, which we shall discuss in Chapter 22. For now, let it suffice to say that

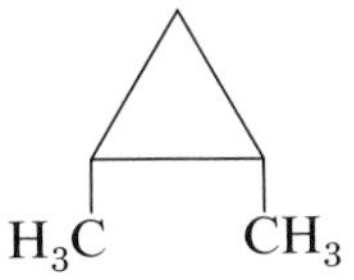

H3C
CH3

is not the same compound as, but rather is an isomer of,

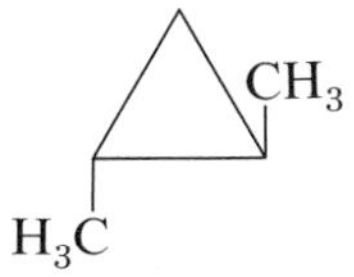

CH3
H3C

A second distinction concerns the compound cyclopropane. Cyclopropane is a sweet-smelling, colorless gas that is sometimes used as an anesthetic. (It is the most potent of the inhalation anesthetics, producing deep unconsciousness in a matter of seconds.) Its unique character lies in the fact

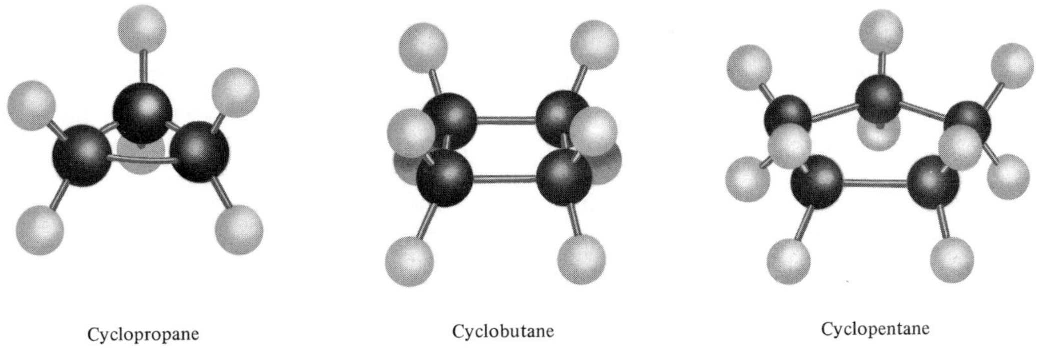

Cyclopropane
Cyclobutane
Cyclopentane

Figure 13.6 Ball-and-stick models of some cycloalkanes.

that it is chemically reactive. Construction of a model of cyclopropane makes this easy to understand for it is quite difficult to connect the three carbon atoms in the form of a ring. A great strain must be applied to the model since, to join the two terminal carbons, one must compress the carbon–carbon bond angle from 109° to 60°. The stress thus placed upon the molecule can be relieved if the ring opens. In order for this to occur, two atoms must be added to the molecule. Cyclopropane then, has a great tendency to undergo such a reaction. It will react with hydrogen bromide to form 1-bromopropane; it will react with hydrogen to form *n*-propane; it will react with chlorine in the dark by *addition* of two clorine atoms to form 1,3-dichloropropane. In the sunlight, it reacts to give the expected monohalo substitution product via a free radical chain mechanism. Cyclopropane does not react, however, with aqueous potassium permanganate.

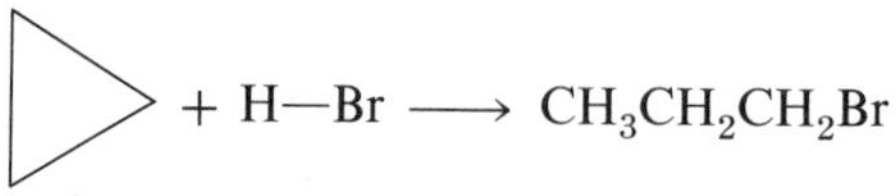
+ H—Br
CH3CH2CH2Br

1-Bromopropane

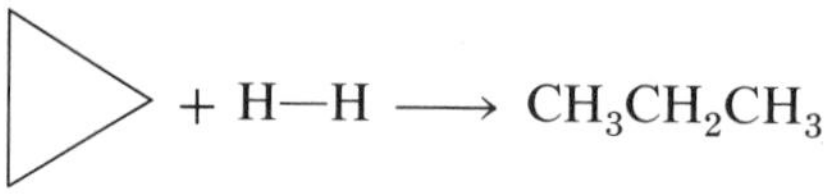
+ H—H
CH3CH2CH3

Propane

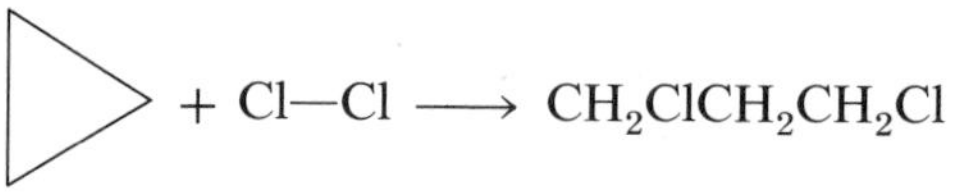
+ Cl—Cl
CH2ClCH2CH2Cl

1,3-Dichloropropane

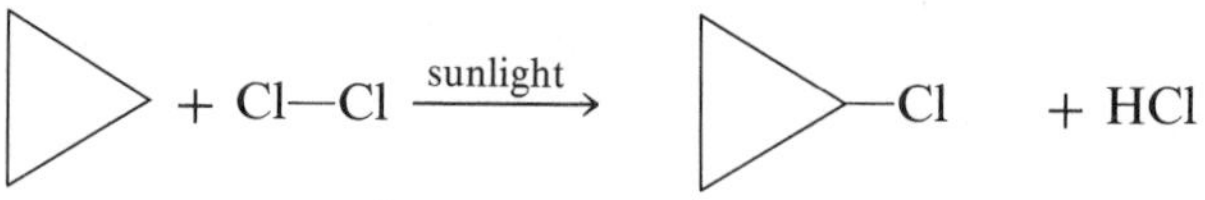
+ Cl—Cl
sunlight
—Cl
+ HCl

Chlorocyclopropane

We now know that cyclic compounds containing more than four carbons in the ring are not planar molecules. Four of the carbon atoms in cyclopentane are coplanar, but the fifth carbon is puckered slightly out of the plane. If cyclohexane were planar, the internal bond angles would have to be 120° and the molecule would be strained. By assuming a bent geometric shape, cyclohexane and the higher cycloalkanes can achieve the unstrained tetrahedral angle. Molecular models indicate that a six-membered ring may assume two distinct nonplanar **conformations** (*the molecular geometry of a compound*). One form is shaped like a chair and is called the **chair conformation;** the other form resembles a boat and is called the **boat conformation** (Figure 13.7). In the boat form the hydrogen atoms on carbons 1 and 4 are

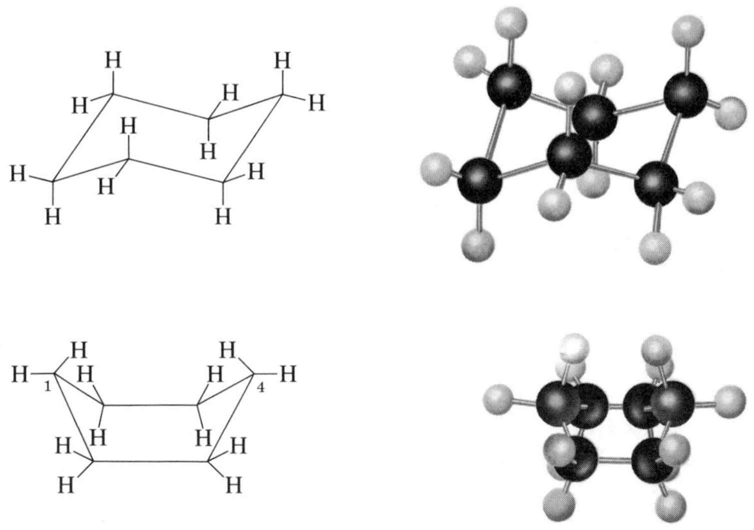

(b)

Figure 13.7 Conformations of cyclohexane. (a) "Chair" form; (b) "boat" form.

close enough to repel each other, whereas all of the hydrogen atoms are staggered and are maximally separated in the chair form. Therefore, the chair form is more stable than the boat form and is the preferred conformation.

The carbon atoms in the puckered cyclohexane ring approximate a plane. Six of the twelve hydrogen atoms lie in this plane and they are referred to as **equatorial** hydrogens. The other six hydrogen atoms, which lie roughly at right angles above or below this plane, are called **axial** hydrogens (Figure 13.8). Substituent groups usually occupy equatorial positions so that they are as far away as possible from ring hydrogens and/or other groups.

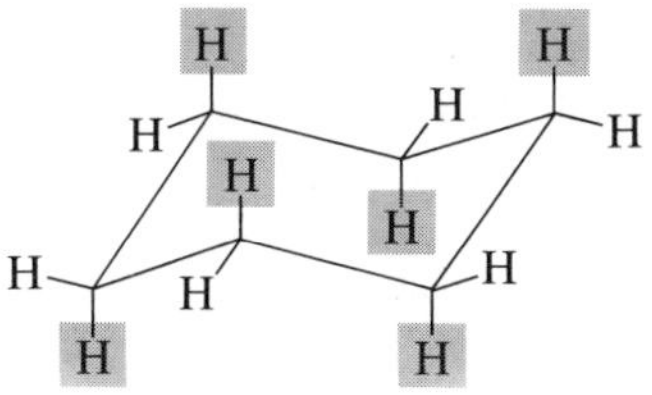

Figure 13.8 Axial and equatorial positions in cyclohexane (shaded hydrogens are axial).

In summary, it should be emphasized that the alkanes as a class are generally unreactive toward most chemicals. This characteristic is usually imparted to the alkane portion of more complex molecules. In other words, although a compound may contain one or more highly reactive functional groups, the alkane portion of that molecule will generally remain unchanged despite transformations elsewhere in the molecule.

Exercises

13.1 Write structural formulas and give IUPAC names for all the isomeric heptanes, C_7H_{16}.

13.2 Name the following compounds.

(a) $(CH_3)_2CHCH_2CH_2CH(CH_3)CH_2CH_3$
(b) $CH_3CH_2CH_2CH_2C(CH_3)_3$
(c) $(CH_3)_2CHC(CH_3)_2C(C_2H_5)_3$
(d) 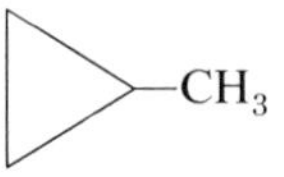
(e) $(CH_3)_2CHCH(CH_3)_2$
(f) $(CH_3)_2CHC(C_2H_5)_3$
(g) $(CH_3)_2CHCH(n\text{-}C_3H_7)CH_2CH(CH_3)_2$
(h) 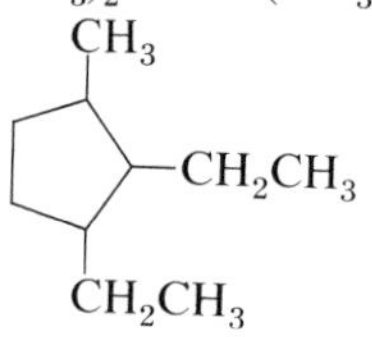

13.3 Write structural formulas for the following compounds.

(a) isobutane
(b) neopentane
(c) 4-isopropyloctane
(d) 4-methylheptane
(e) 3-ethyl-2,2-dimethylhexane
(f) 3,4-dimethyl-4-*n*-propylheptane
(g) 2,2,4-trimethylpentane
(h) 3,3-dimethylpentane
(i) 3,4-diethyl-4,6-dimethylnonane
(j) 1,4-diethyl-2-methylcycloheptane
(k) 1,1-dimethylcyclobutane
(l) 1-ethyl-3-*n*-propylcyclohexane

13.4 What is wrong with each of the following names? Give the structure and the correct name for each compound.

(a) 4,4-dimethylhexane
(b) 2-*n*-propylpentane
(c) 1,1-diethylbutane
(d) 2,8-dimethyloctane
(e) 5-*n*-butyloctane
(f) 1,1,2-trimethylbutane
(g) 1,2-dimethyl-3-ethylpropane
(h) 2-methyl-3-ethylbutane
(i) 3-isopropyl-4-methylpentane
(j) 3,5,5-trimethylhexane
(k) 1,4-dimethylcyclobutane
(l) 2,3,3-trimethylbutane

13.5 Give another name for 3-ethyl-2,4-dimethylpentane.

13.6 Using appropriate examples, distinguish between the following terms.

(a) alkyl group and alkane
(b) alkyl group and alkyl free radical
(c) straight and branched chains
(d) cyclic and noncyclic alkanes
(e) chair and boat forms
(f) axial and equatorial positions

13.7 What generalizations can you make about solubilities in water, densities, boiling points, and melting points of the alkanes?

13.8 Of the three isomers, octane, 2-methylheptane, and 2,2,3,3-tetramethylbutane, predict the one that has the highest boiling point, and the highest melting point.

13.9 How does natural gas differ from bottled gas?

13.10 How can you account for the fact that some CH_3CH_2Cl is formed during the chlorination of methane?

13.11 Complete the following equations.

(a) 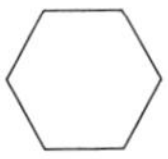 $+ Cl_2 \xrightarrow{\text{light}}$

(b) $CH_3CH_2CH_2CH_2CH_3 + O_2 \longrightarrow$

(c) 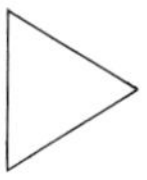 $+ H_2 \longrightarrow$

13.12 An alkane having a molar mass of 72 forms only one monosubstituted bromine derivative upon its bromination in the presence of sunlight.

(a) What is the structural formula of the alkane?

(b) Bromination of an isomer of this alkane yields a mixture of four isomeric monobromo derivatives. What is the structural formula of the isomer?

13.13 The density of gasoline is 0.69 g/ml. On the basis of the complete combustion of octane, calculate the amount of carbon dioxide and water formed per gallon of gasoline used in an automobile.

13.14 Air is approximately 20% oxygen by volume. How many liters of air would be needed to supply just enough oxygen for the complete combustion of 2.24 liters of ethane at STP?

13.15 A hydrocarbon was found to contain 85.7% carbon and 14.3% hydrogen and it had a density of 1.25 g/liter (at STP). Write a structural formula for the compound.

14

The Unsaturated Hydrocarbons

Two families of homologous compounds are included under the general classification of unsaturated hydrocarbons. In both classes each compound contains fewer hydrogen atoms than the corresponding alkane. Because these compounds are deficient in hydrogen, they are said to be **unsaturated** and must therefore contain multiple bonds. One family of compounds contains carbon–carbon double bonds; that is, two carbon atoms are joined together by two pairs of shared electrons. These compounds are called **alkenes** or **olefins.**[1] The other family of compounds contains carbon–carbon triple bonds (three pairs of shared electrons) and are known as **alkynes** or **acetylenes.**

Alkenes

In the following sections, we shall deal mainly with those compounds that contain only one double bond per molecule. The general formula for such a homologous series is C_nH_{2n}.[2] Since there can be no alkene with only one carbon atom, the first member of the series has the molecular formula C_2H_4, and its structural formula, superimposed on its space-filling model, is shown in Figure 14.1.

[1]The action of chlorine gas on compounds that contain double bonds produces an oily liquid. The name olefin comes from the Latin *oleum,* an oil, and *ficare,* to make.

[2]Notice that the alkenes and the cycloalkanes have the same general formula. Therefore, alkenes and cycloalkanes that contain the same number of carbon atoms are isomers (for example, $CH_3{-}CH{=}CH_2$ is an isomer of cyclopropane).

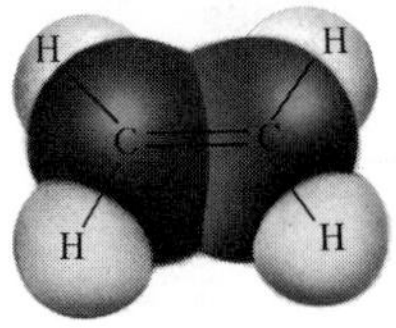

Figure 14.1 The structure of ethylene.

14.1 Nomenclature of Alkenes

The IUPAC rules for alkane nomenclature have the following modifications when applied to the alkenes:

1. Locate the longest chain of carbon atoms that contains the functional group (in this case the carbon–carbon double bond). Again, the longest chain determines the parent hydrocarbon.
2. The name of the parent hydrocarbon is modified by replacing the *-ane* suffix with the ending *-ene*. The *-ene* ending signifies that the molecule is an alkene.
3. Number the parent hydrocarbon from the end of the chain that will confer the smallest number upon the carbon atom containing the functional group (for example, $CH_3CH_2CH{=}CH_2$ is 1-butene and not 3-butene).

Table 14.1 Nomenclature of Alkenes

Molecular Formula	*Structural Formula*	*IUPAC Name*	*Common Name*
C_2H_4	$H-\overset{H}{\overset{\vert}{C}}=\overset{H}{\overset{\vert}{C}}-H$	Ethene[a]	Ethylene
C_3H_6	$CH_3-\overset{H}{\overset{\vert}{C}}=\overset{H}{\overset{\vert}{C}}-H$	Propene[a]	Propylene
C_4H_8	$CH_3-CH_2-\overset{H}{\overset{\vert}{C}}=\overset{H}{\overset{\vert}{C}}-H$	1-Butene	α-Butylene
C_4H_8	$CH_3-\overset{H}{\overset{\vert}{C}}=\overset{H}{\overset{\vert}{C}}-CH_3$	2-Butene	β-Butylene
C_4H_8	$CH_3-\overset{CH_3}{\overset{\vert}{C}}=\underset{H}{\underset{\vert}{C}}-H$	2-Methylpropene	Isobutylene

[a]These compounds need not be named 1-ethene and 1-propene since it is understood that the double bond can only be between carbon one and carbon two. For all higher homologs, the position of the double bond must be indicated by number.

Table 14.2 Application of the IUPAC Rules to the Naming of Alkenes and Cycloalkenes

Condensed Structural Formula	*Rewritten and Numbered*	*IUPAC Name*
$CH_3(CH_2)_2CH{=}CH_2$	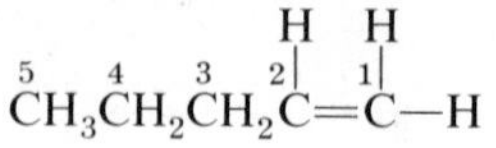	1-Pentene
$CH_3CH{=}CHCH_2CH_3$	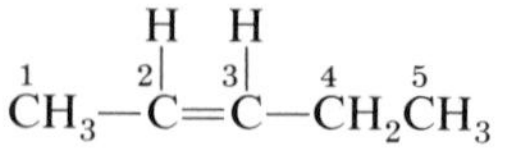	2-Pentene
$CH_3CH_2CH{=}C(CH_3)_2$	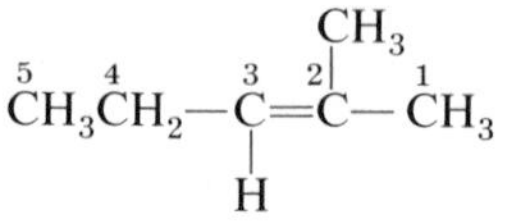	2-Methyl-2-pentene
$(CH_3)_2CHC(CH_3){=}CHCH_3$	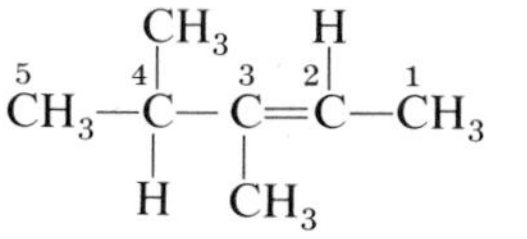	3,4-Dimethyl-2-pentene
$CH_3CH_2CH(CH_3)C(C_3H_7){=}CH_2$	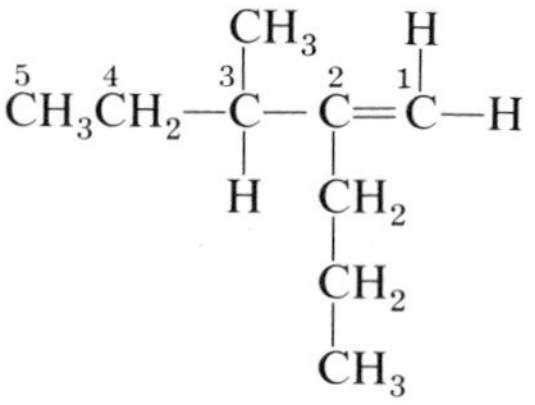	3-Methyl-2-*n*-propyl-1-pentene
		Cyclopentene
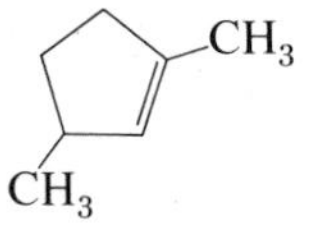	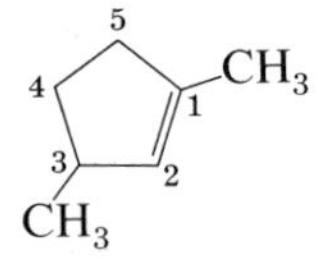	1,3-Dimethylcyclopentene
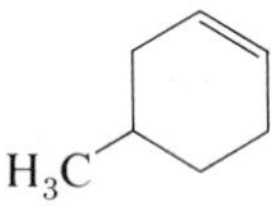	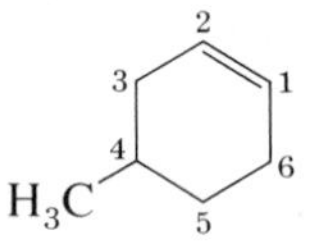	4-Methylcyclohexene
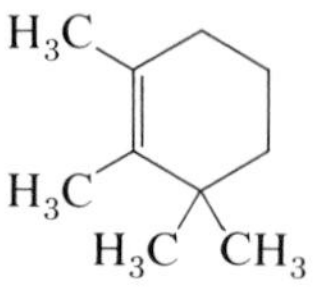	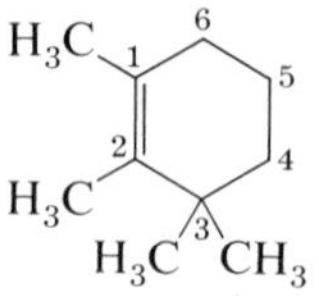	1,2,3,3-Tetramethylcyclohexene
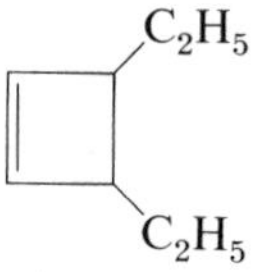	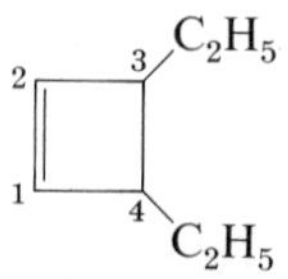	3,4-Diethylcyclobutene

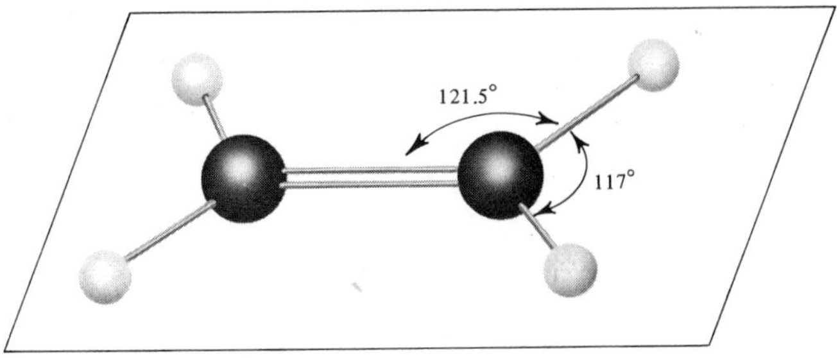
121.5°
117°

Figure 14.2 Planar configuration of the double bond.

Regardless of all the rules of nomenclature that have been agreed upon by chemists throughout the world, there still exists a problem of nomenclature of which the student should be aware. This problem was alluded to in the previous chapter, when it was stated that many of the common names of the more familiar organic compounds still persist in the chemical literature. Chemists often use these common names, especially when referring to the lower homologs of each series of organic compounds. The student must be familiar with these common names as well as with the IUPAC names. Table 14.1 contains the names and formulas of some of the lower members of the alkene series. Table 14.2 contains some examples of the application of the IUPAC rules of nomenclature to the alkenes and cycloalkenes.

Before leaving the subject of alkene nomenclature, one additional aspect must be mentioned. The double bond of alkene molecules, like the ring structure of cycloalkanes, imposes geometrical restrictions on the molecules. Free rotation is not possible about a double bond as it is about a single bond. *This restriction bestows a flatness or planar configuration upon the double bond* (Figure 14.2). Therefore, the relative positions of substituent groups located above or below the double bond become very significant. The nomenclature employed in situations such as this, as well as the consequences of restricted rotation, will be discussed in Chapter 22.

14.2 Preparation of Alkenes

Ethylene and, to a lesser extent, the higher homologs of the alkene series can be obtained from the cracking of petroleum. *Cracking* is a process in which the saturated hydrocarbons are subjected to very high temperatures in the presence of a catalyst. The result is the elimination of hydrogen and the smaller hydrocarbons, such as methane and ethane. However, the process always yields a mixture of products and is thus of limited use in a chemical laboratory.

$$CH_3CH_2CH_2CH_3 \xrightarrow[\text{catalyst}]{>400^\circ} \begin{cases} CH_2{=}CH_2 + CH_3CH_3 \\ CH_3CH{=}CH_2 + CH_4 \\ CH_3CH{=}CHCH_3 + H_2 \end{cases}$$

a. Dehydrohalogenation of Alkyl Halides

Dehydrohalogenation is not a widely used preparative reaction because the alkenes that would be formed are more readily available by the dehydration of an alcohol. Dehydrohalogenation involves the removal of the elements of hydrogen halide (HX) from adjacent carbon atoms. The action of a strong base upon the alkyl halide is required to bring about the dehydrohalogenation. Since the reaction is carried out in solution, it is necessary to find a solvent that will dissolve both the alkyl halide and the strong base. Water, the common solvent in inorganic chemistry, is of no value here since alkyl halides are water insoluble. The best solvent for this reaction is alcohol. (See Section 16.4-a for a discussion of the mechanism of the dehydrohalogenation reaction.)

General Equation

$$\underset{\underset{\text{H X}}{|}}{\overset{\overset{\text{R R}}{|\ |}}{R-C-C-R}} \xrightarrow[\text{alcohol}]{\text{KOH}} \overset{\overset{\text{R R}}{|\ |}}{R-C=C-R} + KX + HOH$$

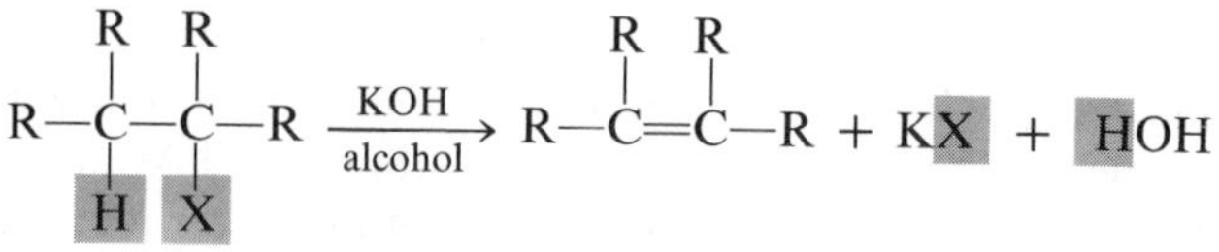

Specific Equation

$$\underset{\underset{\text{Br H}}{|}}{\overset{\overset{\text{H}_3\text{C H}}{|\ |}}{CH_3-C-C-H}} \xrightarrow[\text{alcohol}]{\text{KOH}} \overset{\overset{\text{H}_3\text{C H}}{|\ |}}{CH_3-C=C-H} + KBr + HOH$$

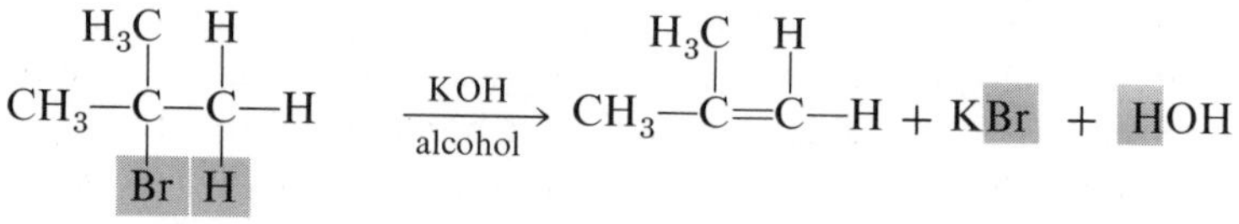

2-Bromo-2-methylpropane   2-Methylpropene

b. Dehydration of Alcohols

Dehydration reactions are extremely common in both organic chemistry and biochemistry. As the name implies, dehydration involves the removal of a molecule of water from a reactant molecule. (See Section 17.4-b for a discussion of the mechanisms of the dehydration reaction.)

With alcohols the reaction can be brought about by passing the alcohol through a heated tube containing alumina, Al_2O_3, or by heating the alcohol in the presence of concentrated sulfuric acid.

General Equation

$$\underset{\underset{\text{H OH}}{|\ \ |}}{\overset{\overset{\text{H H}}{|\ |}}{R-C-C-H}} \xrightarrow[\Delta]{\substack{\text{conc H}_2\text{SO}_4 \\ \text{or Al}_2\text{O}_3}} \overset{\overset{\text{H H}}{|\ |}}{R-C=C-H} + HOH$$

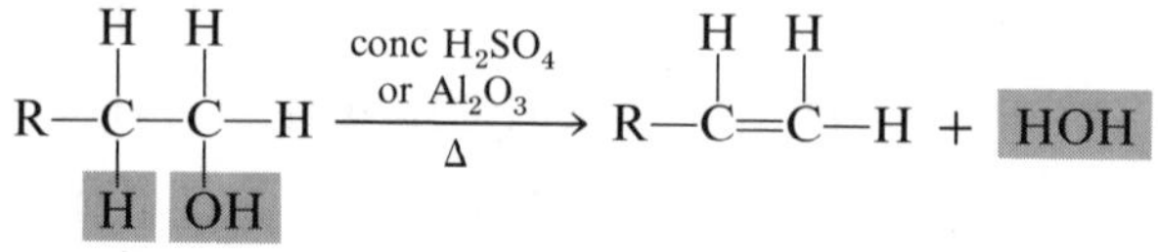

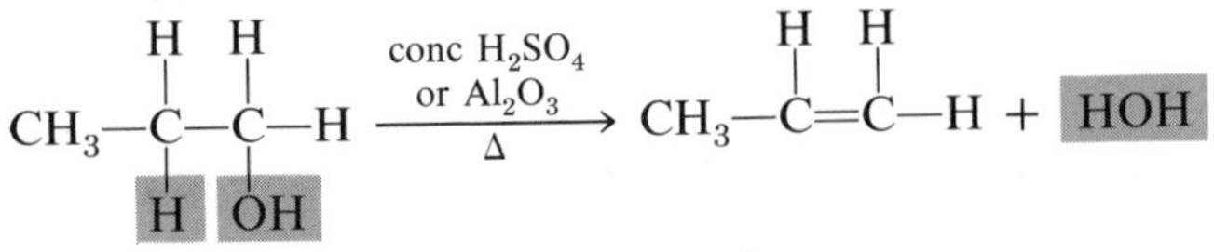

$$CH_3-\underset{\underset{\displaystyle H}{|}}{\overset{\overset{\displaystyle H}{|}}{C}}-\underset{\underset{\displaystyle OH}{|}}{\overset{\overset{\displaystyle H}{|}}{C}}-H \xrightarrow[\Delta]{\substack{\text{conc } H_2SO_4 \\ \text{or } Al_2O_3}} CH_3-\underset{}{\overset{\overset{\displaystyle H}{|}}{C}}=\underset{}{\overset{\overset{\displaystyle H}{|}}{C}}-H + \boxed{HOH}$$

Propanol   Propene

14.3 Physical Properties of Alkenes

Table 14.3 lists the physical constants for several members of the alkene series. Since an alkene is only two atomic mass units lighter than its corresponding alkane, it is not surprising to find similarities in their melting points and boiling points. Like all of the hydrocarbons, the alkenes are insoluble in water, but soluble in organic solvents.

Table 14.3 Physical Constants of Some Alkenes

Molecular Formula	Structural Formula	IUPAC Name	MP (°C)	BP (°C)
C_2H_4	$CH_2\!=\!CH_2$	Ethene	−169	−104
C_3H_6	$CH_3CH\!=\!CH_2$	Propene	−185	−47
C_4H_8	$CH_3CH_2CH\!=\!CH_2$	1-Butene	−130	−6.5
C_4H_8	$(CH_3)_2C\!=\!CH_2$	2-Methylpropene	−141	−6.9
C_5H_{10}	$CH_3(CH_2)_2CH\!=\!CH_2$	1-Pentene	−138	30
C_6H_{12}	$CH_3(CH_2)_3CH\!=\!CH_2$	1-Hexene	−141	64

14.4 Chemical Properties of Alkenes

The major distinction between the alkanes and the alkenes is apparent from a comparison of their chemical properties. Alkanes are inert. The few reactions that they do undergo occur by substitution. The alkenes, on the other hand, are very reactive and the majority of their reactions are characterized by addition to the double bond.

$$\overset{}{\underset{}{C}}=\overset{}{\underset{}{C} + \boxed{YZ}} \longrightarrow -\underset{\underset{\displaystyle Y}{|}}{\overset{\overset{\displaystyle |}{}}{C}}-\underset{\underset{\displaystyle Z}{|}}{\overset{\overset{\displaystyle |}{}}{C}}-$$

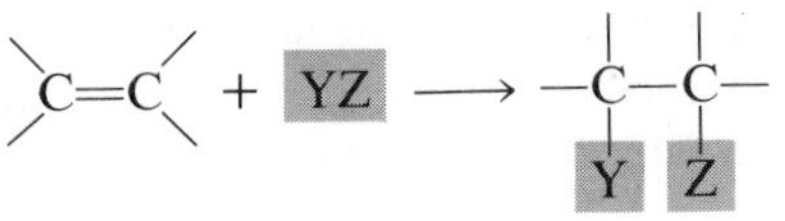

a. Addition of Symmetrical Reagents

1. Addition of Halogen (Halogenation). Alkenes react with chlorine or bromine. With the less reactive iodine molecule, the reaction is very slow. Illumination speeds up this latter reaction considerably. Reactions between the halogens and the alkenes are usually carried out in some inert solvent such as carbon tetrachloride. In fact, this reaction is the basis for a very useful test for the presence of an unsaturated compound. The unknown is added to a solution of bromine in carbon tetrachloride. If the red color of the solution is dissipated, the bromine is being consumed in the reaction. This is a strong indication that the unknown compound contains an unsaturated linkage.

General Equation

$$\underset{\substack{|\ \ \ |\\ \text{H}\ \text{H}}}{\text{R}-\text{C}=\text{C}-\text{H}} + \boxed{\text{X:X}} \longrightarrow \underset{\substack{|\ \ \ |\\ \boxed{\text{X}}\ \boxed{\text{X}}}}{\text{R}-\text{C}-\text{C}-\text{H}}$$

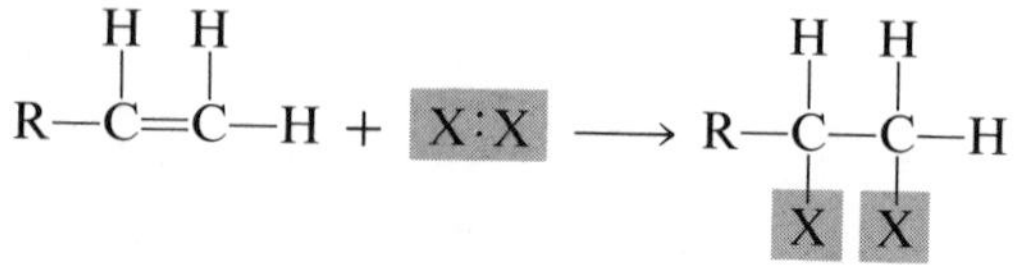

Specific Equation

$$\underset{\substack{|\ \ \ |\\ \text{H}\ \text{H}}}{\text{H}-\text{C}=\text{C}-\text{H}} + \boxed{\text{Br}_2} \longrightarrow \underset{\substack{|\ \ \ |\\ \boxed{\text{Br}}\ \boxed{\text{Br}}}}{\text{H}-\text{C}-\text{C}-\text{H}}$$

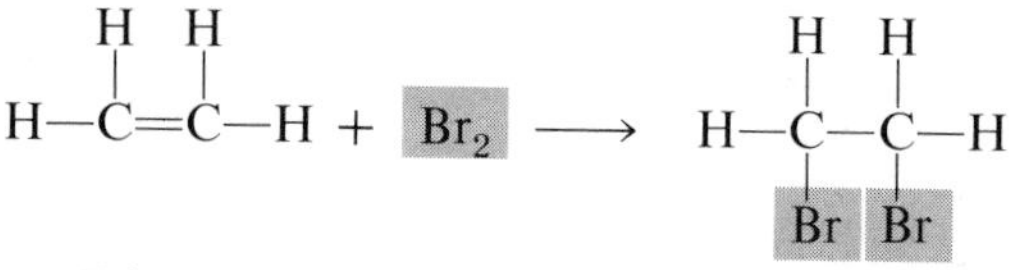

Ethene   1,2-Dibromoethane

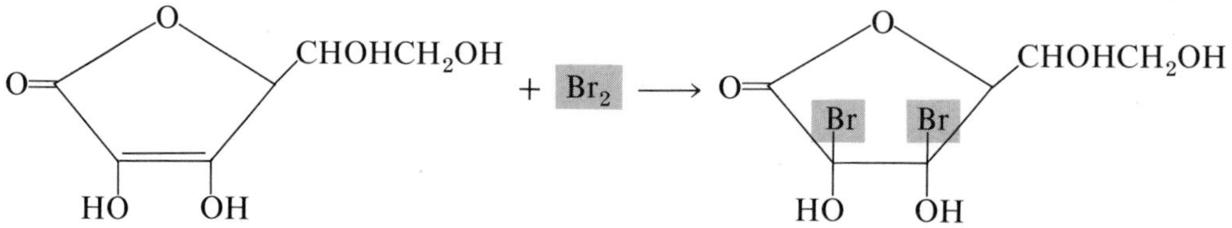

Ascorbic acid
(Vitamin C)   2,3-Dibromoascorbic acid

2. Addition of Hydrogen (Hydrogenation). When hydrogen adds to an alkene, the corresponding alkane is produced. The process requires a catalyst and is thus termed catalytic hydrogenation. The best catalysts for the reaction are platinum and palladium, but because of the high cost of these metals finely divided nickel is often used in the laboratory. The reaction takes place on the surface of the catalyst, and thus increasing the surface area of the catalyst increases the rate of the reaction.

General Equation

$$\underset{\substack{|\ \ \ |\\ \text{H}\ \text{H}}}{\text{R}-\text{C}=\text{C}-\text{H}} + \boxed{\text{H:H}} \xrightarrow[\text{catalyst}]{\text{Pt, Pd, or Ni}^*} \underset{\substack{|\ \ \ |\\ \boxed{\text{H}}\ \boxed{\text{H}}}}{\text{R}-\text{C}-\text{C}-\text{H}}$$

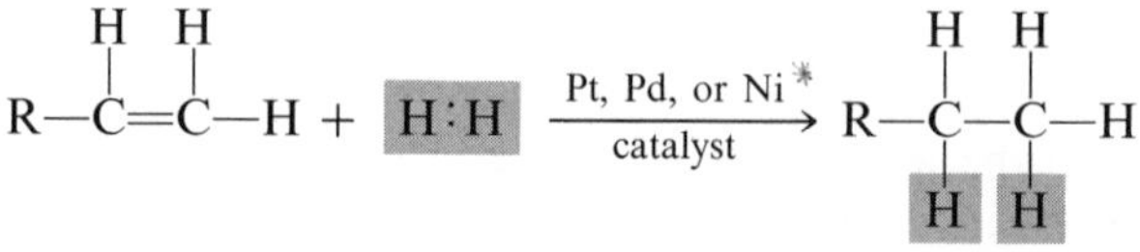

Specific Equation

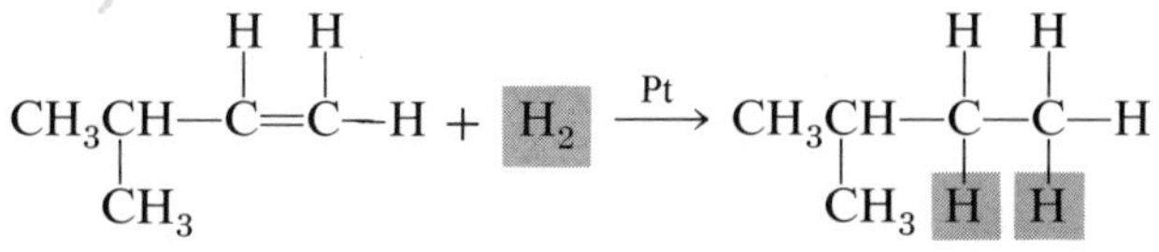

3-Methyl-l-butene   2-Methylbutane

The catalytic hydrogenation of the alkenes provides a method for the preparation of pure alkanes. The reaction is also of considerable commercial importance. Margarine and cooking shortenings are produced by the catalytic hydrogenation of polyunsaturated vegetable oils (see Section 24.5-c).

b. Addition of Unsymmetrical Reagents

The majority of alkene addition reactions are initiated by the attack of a proton or a positive ion on the electrons of the double bond. The electron-seeking reagents that are capable of forming a covalent bond with carbon are referred to as **electrophilic reagents** or **electrophiles** (electron-loving). The electrophile reacts with the alkene to form an intermediate carbonium ion, which subsequently combines with the anionic species to yield the addition product.

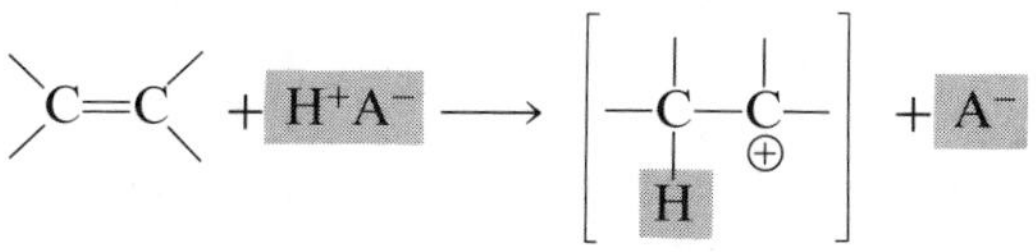

Carbonium ion

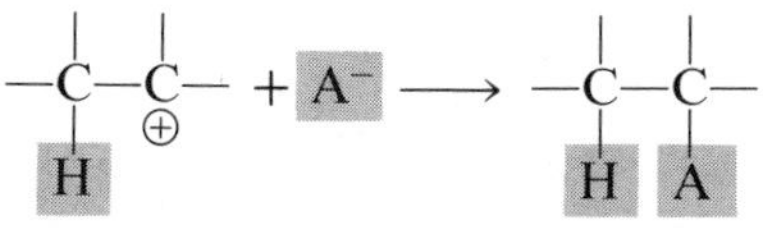

Carbonium ions are important intermediates in many organic and biochemical reactions. A **carbonium ion** is defined as *an electron-deficient carbon with a sextet of electrons.* The following are Lewis structures for some carbonium ions.

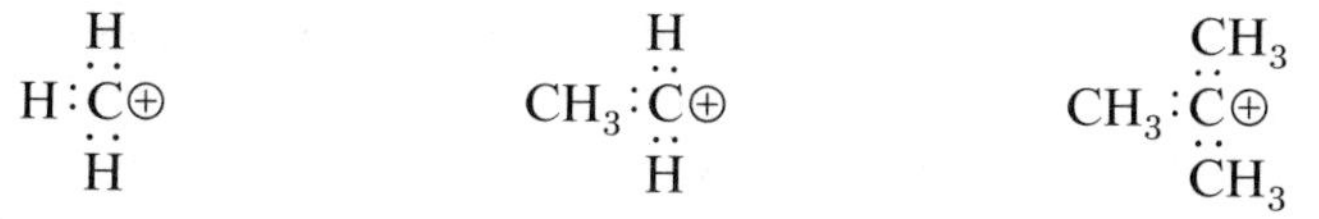

Methyl carbonium ion   Ethyl carbonium ion   *tert*-Butyl carbonium ion

Reaction between a symmetrical alkene (such as $H_2C{=}CH_2$) and an unsymmetrical reagent (HA) presents no problem since only one product results no matter which carbon atom accepts the H and which accepts the A. For example, consider the reaction of a symmetrical alkene with hydrogen chloride.

$$\underset{\text{}}{H-\overset{\overset{\displaystyle H}{|}}{C}=\overset{\overset{\displaystyle H}{|}}{C}-H} + \boxed{HCl} \longrightarrow H-\overset{\overset{\displaystyle H}{|}}{\underset{\underset{\displaystyle \boxed{H}}{|}}{C}}-\overset{\overset{\displaystyle H}{|}}{\underset{\underset{\displaystyle \boxed{Cl}}{|}}{C}}-H \quad \text{or} \quad H-\overset{\overset{\displaystyle H}{|}}{\underset{\underset{\displaystyle \boxed{Cl}}{|}}{C}}-\overset{\overset{\displaystyle H}{|}}{\underset{\underset{\displaystyle \boxed{H}}{|}}{C}}-H$$

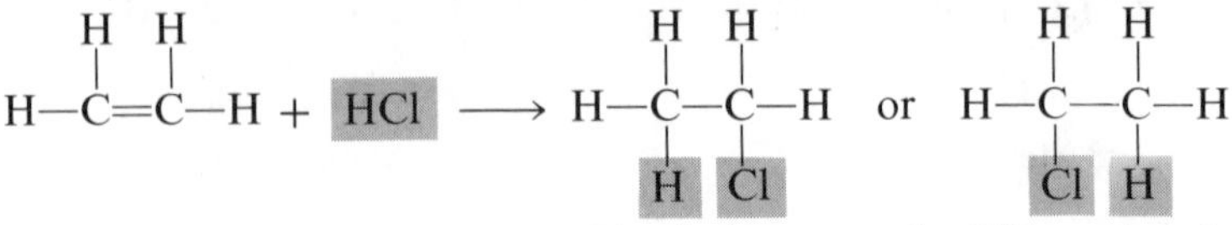

Identical compounds (Chloroethane)

However, in the case of a reaction between an unsymmetrical alkene and an unsymmetrical reagent, one might predict that two possible products would be formed. For example, the addition of hydrogen bromide to propene might be expected to yield both 1-bromopropane and 2-bromopropane. If the reaction were governed solely by the statistical addition of the fragments H and Br, a mixture of the two isomeric alkyl halides would be expected.

$$CH_3-\overset{\overset{\displaystyle H}{|}}{C}=\overset{\overset{\displaystyle H}{|}}{C}-H + \boxed{HBr} \longrightarrow CH_3-\overset{\overset{\displaystyle H}{|}}{\underset{\underset{\displaystyle \boxed{H}}{|}}{C}}-\overset{\overset{\displaystyle H}{|}}{\underset{\underset{\displaystyle \boxed{Br}}{|}}{C}}-H \quad \text{and/or} \quad CH_3-\overset{\overset{\displaystyle H}{|}}{\underset{\underset{\displaystyle \boxed{Br}}{|}}{C}}-\overset{\overset{\displaystyle H}{|}}{\underset{\underset{\displaystyle \boxed{H}}{|}}{C}}-H$$

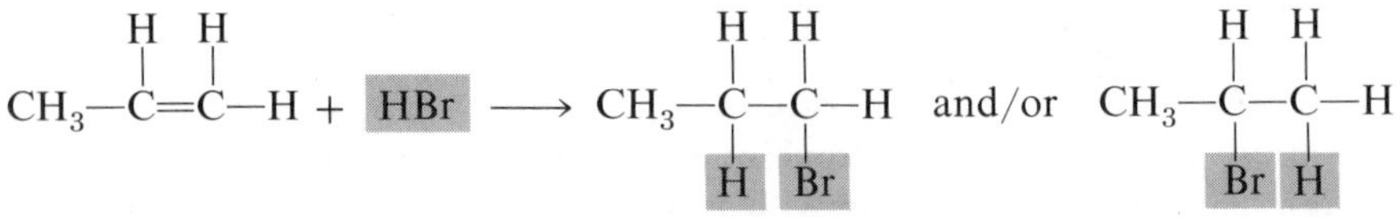

Propene   1-Bromopropane   2-Bromopropane
            I   II

Actually, when this reaction is carried out in the laboratory, only one product is formed, and that product is compound II. This phenomenon was observed in all reactions between unsymmetrical reagents and alkenes, and in 1871 a Russian chemist, Vladimir Markownikoff, suggested a generalization that enables us to predict what the outcome of such a reaction will be. This empirical generalization is known as **Markownikoff's rule** and can be stated as follows: *In the addition of unsymmetrical reagents to olefins, the positive portion of the reagent (which is usually hydrogen) adds to that carbon atom which already has the most hydrogen atoms.*

  The direction of addition according to Markownikoff's rule can be rationalized on the basis of the relative electronegativities between alkyl groups and hydrogen atoms. Alkyl groups are less electronegative than hydrogen and thus they repel electrons by comparison with hydrogen. In an unsymmetrical alkene the electrons are shifted in a direction away from the carbon bonded to the alkyl group(s). The electrophile (usually H^+) will then seek out that portion of the alkene having the greatest electron density. This electronic distribution can be represented by the symbols δ^+ and δ^-, which designate relatively low and high electron density, respectively, but do not imply a formal positive and negative charge.

$$\underset{\underset{\displaystyle \delta^+ \quad \delta^-}{}}{CH_3-\overset{\overset{\displaystyle H_3C}{|}}{C}=\overset{\overset{\displaystyle H}{|}}{C}-H} + \boxed{H^+Br^-} \longrightarrow CH_3-\overset{\overset{\displaystyle H_3C}{|}}{\underset{\underset{\displaystyle \oplus}{}}{C}}-\overset{\overset{\displaystyle H}{|}}{\underset{\underset{\displaystyle \boxed{H}}{|}}{C}}-H + \boxed{Br^-}$$

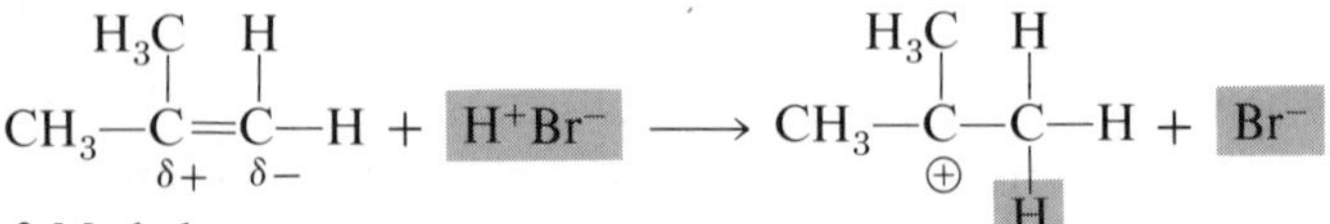

2-Methylpropene

The intermediate carbonium ion is stabilized by the electron-repelling effect of the adjacent methyl groups.

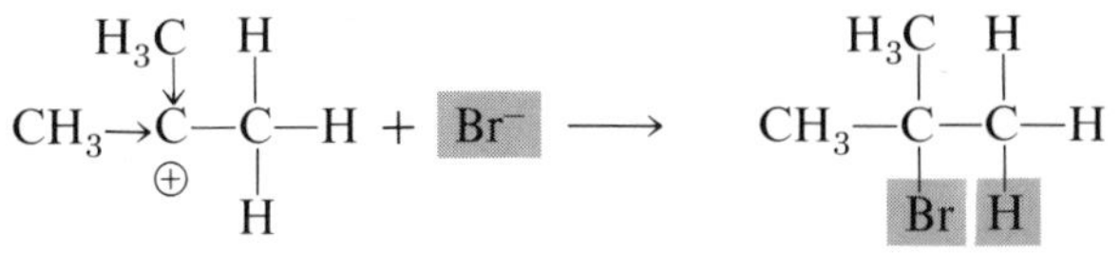

2-Bromo-2-methylpropane
(t-Butyl bromide)

General Equation

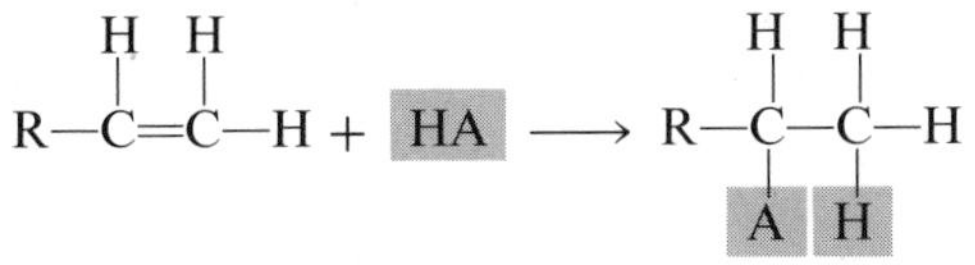

where R is not hydrogen and A = Cl, Br, I, OSO_3H, etc.

Specific Equations

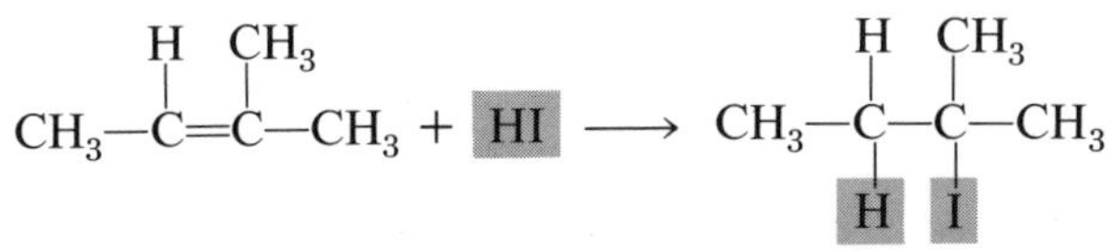

2-Methyl-2-butene — 2-Iodo-2-methylbutane

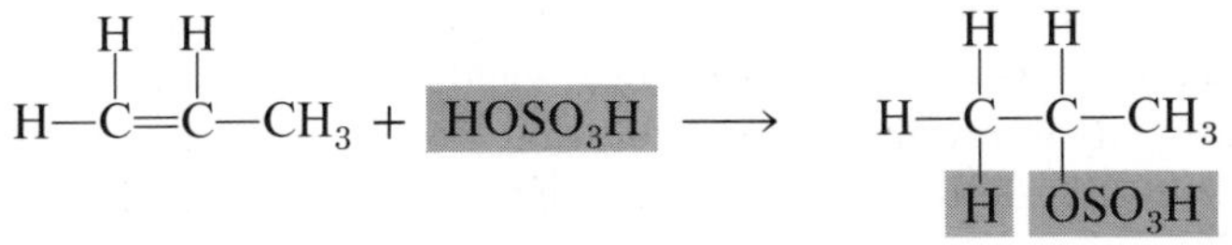

Propene — Isopropyl hydrogen sulfate

Although the direct hydration of alkenes is an important industrial process for the preparation of alcohols (see Section 17.2-a), it is seldom used as a laboratory procedure because of the reaction conditions that are required. The hydration reaction, however, is very common in biochemistry; many of the hydroxy compounds that occur in living systems are formed in this manner. The following reaction, for example, occurs as one of the steps in the Krebs cycle (see Figure 28.5).

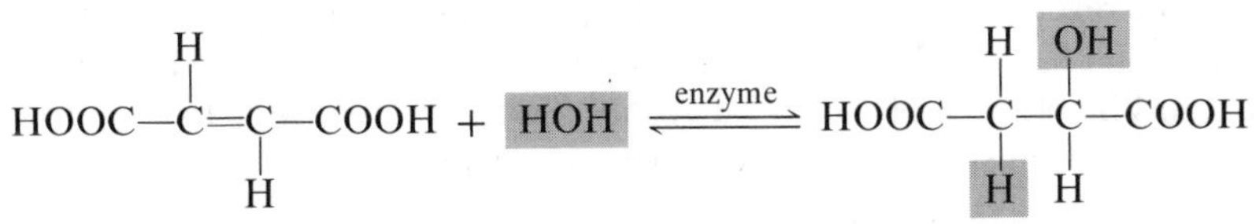

Fumaric acid — Malic acid

c. Oxidation

Alkenes, like alkanes, can be completely oxidized to carbon dioxide and water. They release slightly less energy than the corresponding alkanes.

$$CH_2{=}CH_2 + 3\,O_2 \longrightarrow 2\,CO_2 + 2\,H_2O + 332\ kcal$$

Alkenes are readily oxidized by a solution of potassium permanganate. The products of this reaction are dependent upon reaction conditions. This is not at all unusual in organic chemistry. Therefore it is very important that the reaction conditions be specified as part of the equation.

1. Oxidation with Cold, Dilute, Neutral Potassium Permanganate Solution. Under these conditions, alkenes are oxidized to **glycols** (compounds containing two hydroxyl groups).

General Equation

$$3\ R{-}\overset{\displaystyle H}{\underset{\displaystyle }{C}}{=}\overset{\displaystyle H}{\underset{\displaystyle }{C}}{-}H + 2\ KMnO_4 + 4\ H_2O \longrightarrow 3\ R{-}\overset{\displaystyle H}{\underset{\displaystyle HO}{C}}{-}\overset{\displaystyle H}{\underset{\displaystyle OH}{C}}{-}H + 2\ MnO_2 + 2\ KOH$$

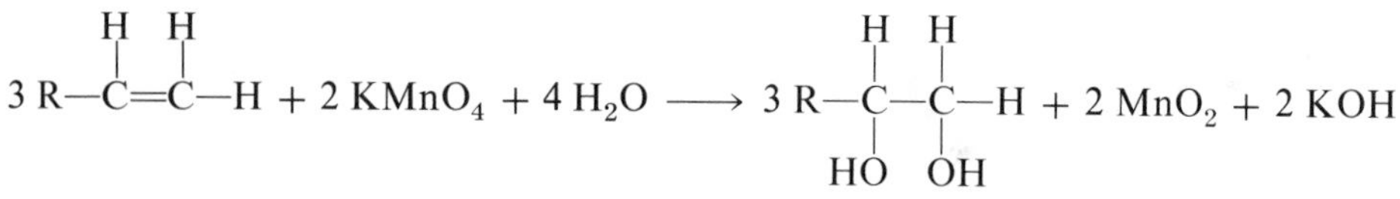

A glycol

Note that the above represents a balanced net equation. It is necessary to write the complete equation whenever the reaction is to be carried out in the laboratory. A knowledge of the stoichiometry of a reaction enables one to calculate the necessary quantities of each reagent and the expected mass of the product to be formed (the theoretical yield—Section 4.14). Often, however, we are only interested in knowing what organic products are obtained. In these cases, it is sufficient just to indicate the inorganic reagents above the arrow.

Specific Equation

$$H{-}\overset{\displaystyle H}{\underset{\displaystyle }{C}}{=}\overset{\displaystyle H}{\underset{\displaystyle }{C}}{-}H \xrightarrow{\text{dil } KMnO_4} H{-}\overset{\displaystyle H}{\underset{\displaystyle HO}{C}}{-}\overset{\displaystyle H}{\underset{\displaystyle OH}{C}}{-}H$$

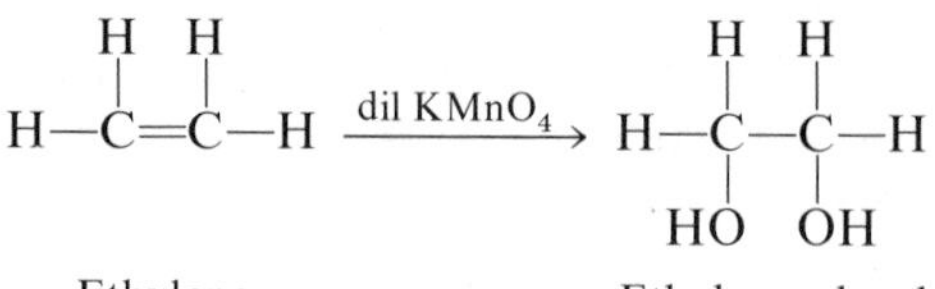

Ethylene   Ethylene glycol

The reaction of alkenes with dilute potassium permanganate is the basis of a second test (the Baeyer test) for unsaturation. If an unknown compound contains no other easily oxidized groups, then a color change from purple to brown (from purple $KMnO_4$ to brown MnO_2) indicates the presence of an unsaturated linkage in the compound.

2. Oxidation with Hot Concentrated KMnO₄. Vigorous oxidation of alkenes with concentrated $KMnO_4$ at elevated temperatures results in a cleavage of the carbon–carbon double bond. The products of the reaction will depend upon the structure of the reacting alkene. This reaction is employed in the identification of the structure of an unknown alkene by determining what fragments (CO_2, ketones, or acids) are obtained as products of the oxidation. The following equation is illustrative of this type of reaction.

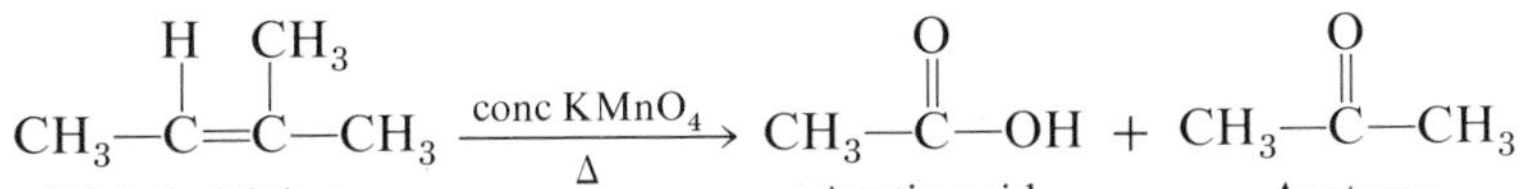

$$
\underset{\text{2-Methyl-2-butene}}{CH_3-\overset{\overset{\displaystyle H}{|}}{C}=\overset{\overset{\displaystyle CH_3}{|}}{C}-CH_3} \xrightarrow[\Delta]{\text{conc } KMnO_4} \underset{\substack{\text{Acetic acid} \\ \text{(An acid)}}}{CH_3-\overset{\overset{\displaystyle O}{||}}{C}-OH} + \underset{\substack{\text{Acetone} \\ \text{(A ketone)}}}{CH_3-\overset{\overset{\displaystyle O}{||}}{C}-CH_3}
$$

d. Polymerization (see Section 8.3)

Perhaps the most familiar polymer is polyethylene, which is made from the polymerization of billions of ethylene monomers. Annual production of polyethylene in the United States is well over one million tons. It is used for electrical insulation and for making plastic cups, refrigerator trays, squeeze bottles, etc.

The polymerization reaction requires the presence of a small amount of an initiator. The most common initiators are peroxide molecules, which break down to produce free radicals. The radical adds to an ethylene molecule (monomer) to form a new free radical. This radical then adds to another molecule of ethylene to produce a larger radical. The addition is repeated many times until a long chain of atoms is obtained. The polymerization process, once begun, continues until a termination step occurs, such as the combination of two radicals.

Initiation

$$ R-O\!:\!O-R \longrightarrow 2\,R-O\cdot $$

$$ R-O\cdot + \underset{\underset{H\ \ \ H}{|\ \ \ |}}{\overset{\overset{H\ \ \ H}{|\ \ \ |}}{C=C}} \longrightarrow R-O-\underset{\underset{H\ \ H}{|\ \ |}}{\overset{\overset{H\ \ H}{|\ \ |}}{C-C}}\cdot $$

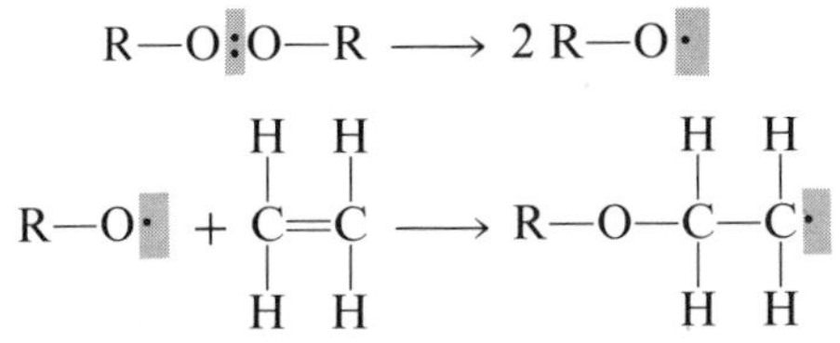

Propagation

$$ R-O-\underset{\underset{H\ \ H}{|\ \ |}}{\overset{\overset{H\ \ H}{|\ \ |}}{C-C}}\cdot \xrightarrow{\ \ H-C=C-H\ \ } R-O-\underset{\underset{H\ \ H\ \ H\ \ H}{|\ \ |\ \ |\ \ |}}{\overset{\overset{H\ \ H\ \ H\ \ H}{|\ \ |\ \ |\ \ |}}{C-C-C-C}}\cdot \xrightarrow{\ \ H-C=C-H\ \ } $$

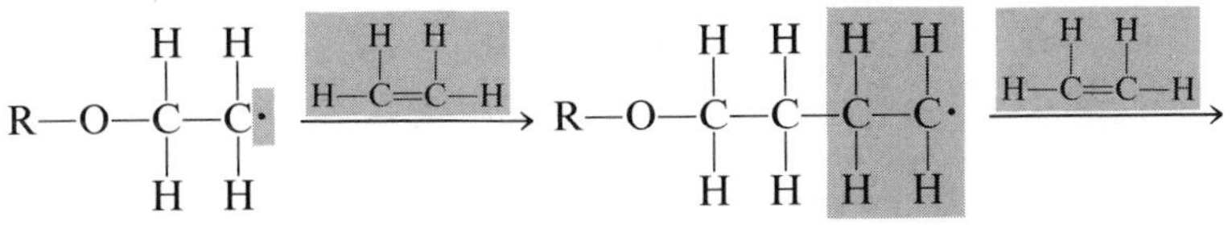

Termination

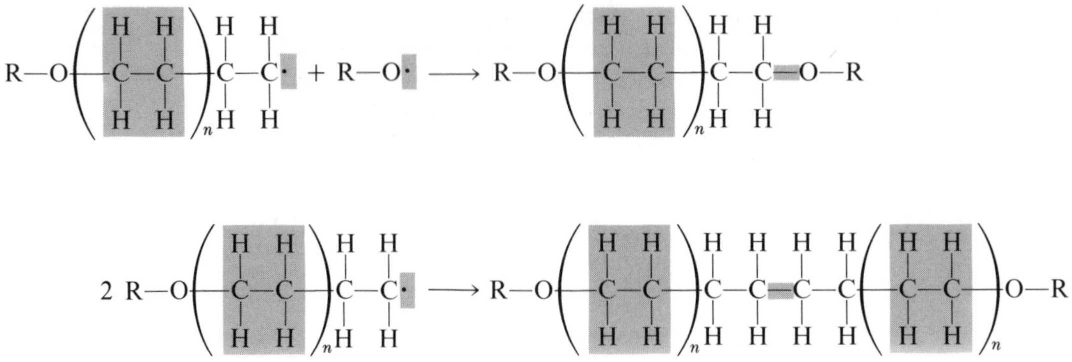

Notice that the repeating unit of the polymer is

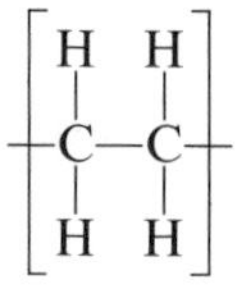

Therefore the equation for the overall process can be written as

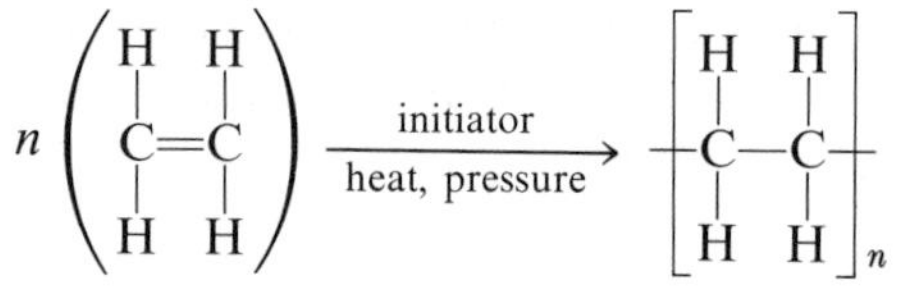

initiator
heat, pressure

$n = 100\text{–}1000$

Many other olefins serve as excellent monomer units to provide many of the following well-known polymers.

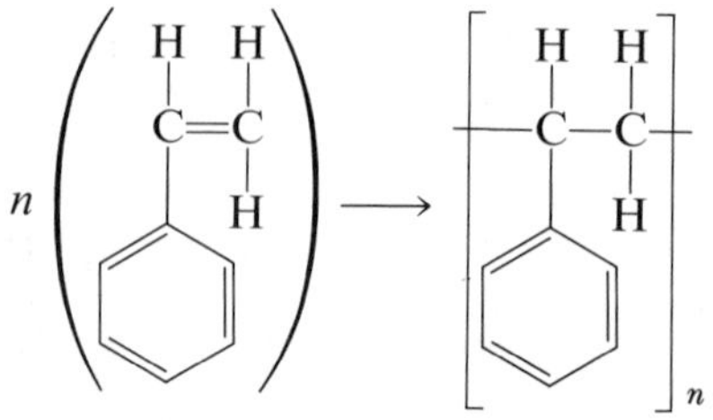

Styrene Polystyrene

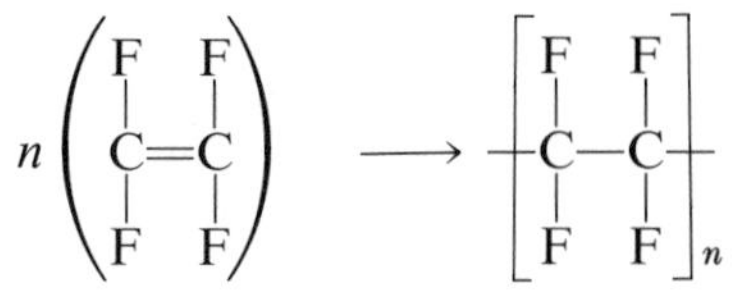

Tetrafluoroethylene Teflon

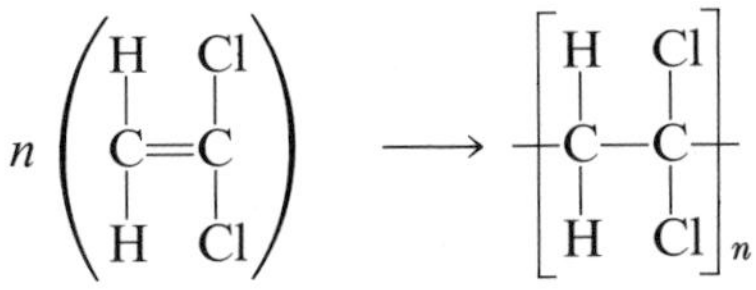

1,1-Dichloroethylene Saran

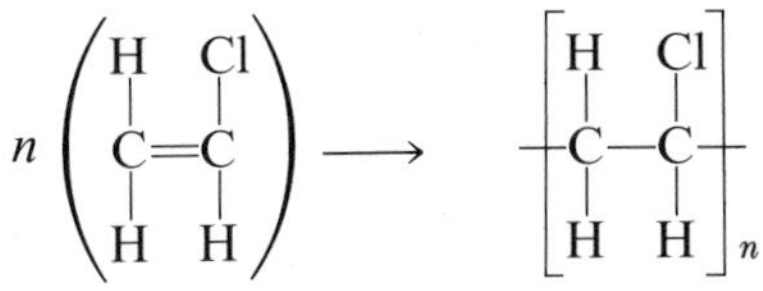

Vinyl chloride Polyvinyl chloride

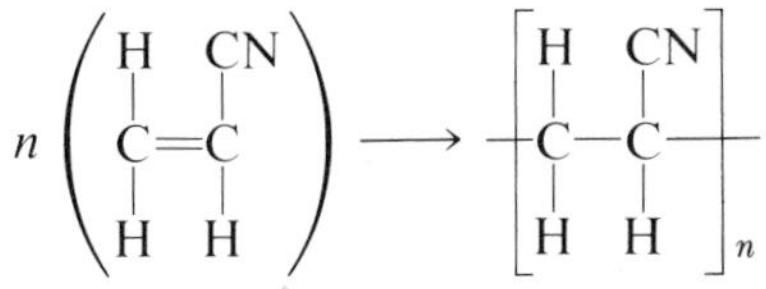

Cyanoethylene Orlon

14.5 The Isoprene Unit

Complex alkenes containing multiple double bonds within the molecules occur widely in nature. One of our important commodities, natural rubber, is a linear polymer of the diolefin 2-methyl-1,3-butadiene, *isoprene.*

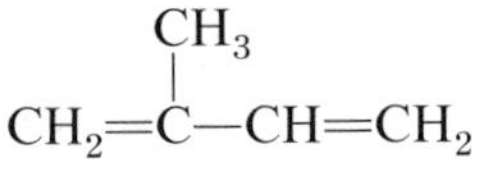

Isoprene
(2-Methyl-1,3-butadiene)

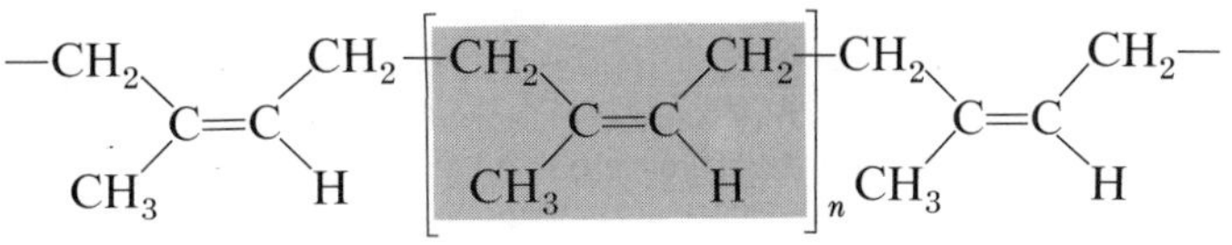

Natural rubber

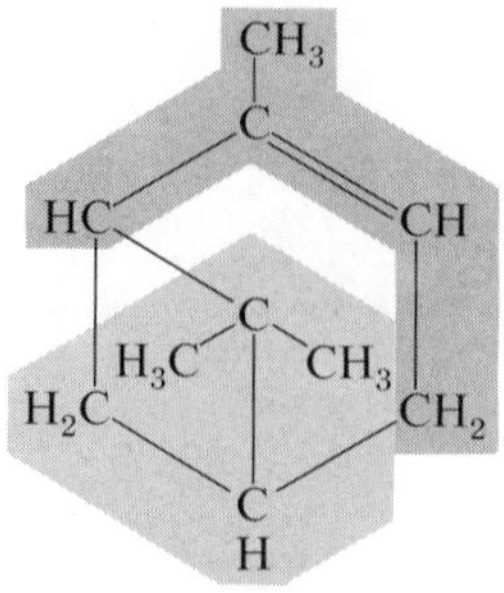

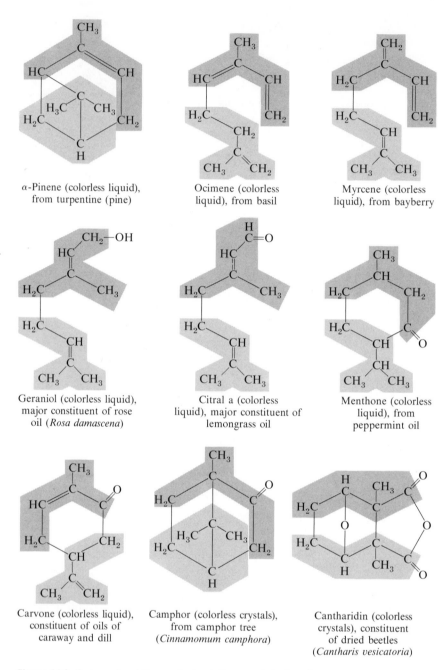

α-Pinene (colorless liquid), from turpentine (pine)

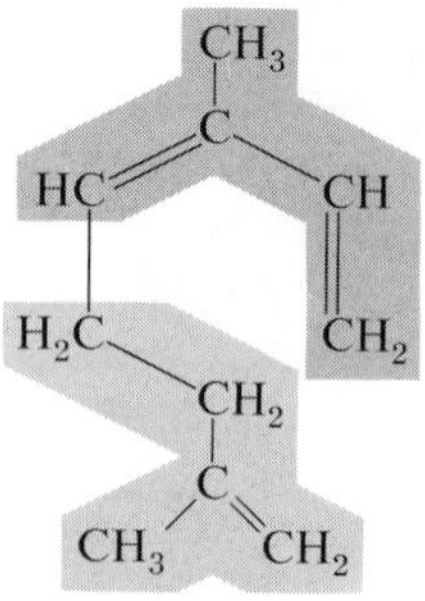

Ocimene (colorless liquid), from basil

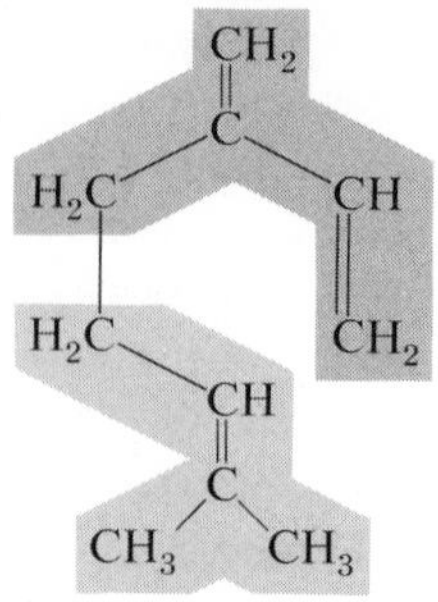

Myrcene (colorless liquid), from bayberry

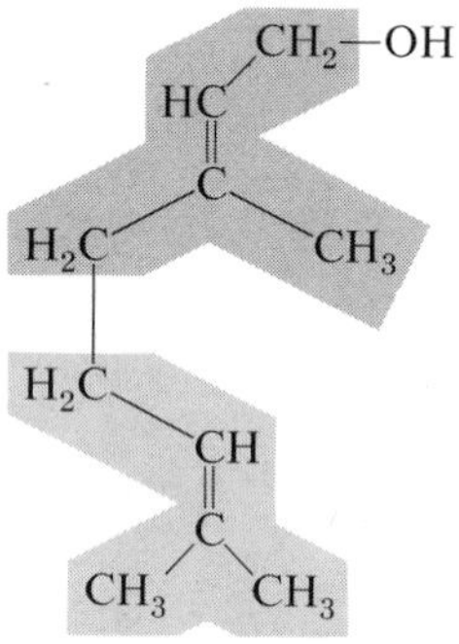

Geraniol (colorless liquid), major constituent of rose oil (*Rosa damascena*)

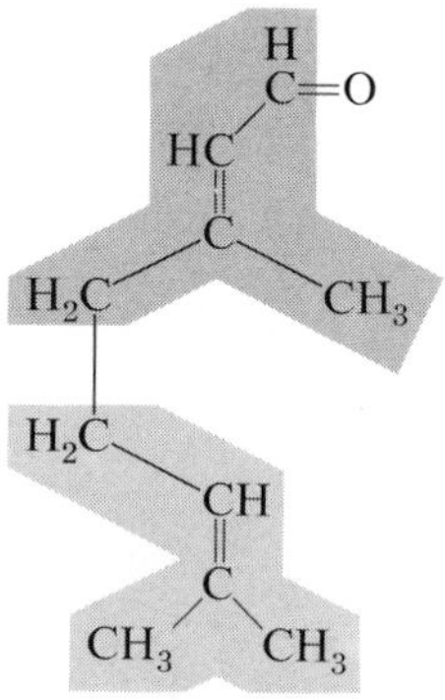

Citral a (colorless liquid), major constituent of lemongrass oil

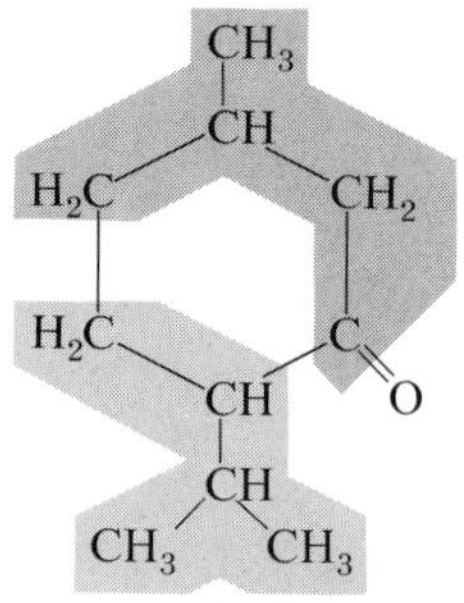

Menthone (colorless liquid), from peppermint oil

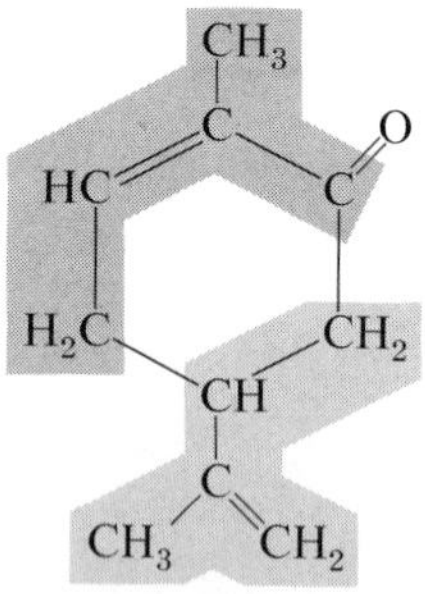

Carvone (colorless liquid), constituent of oils of caraway and dill

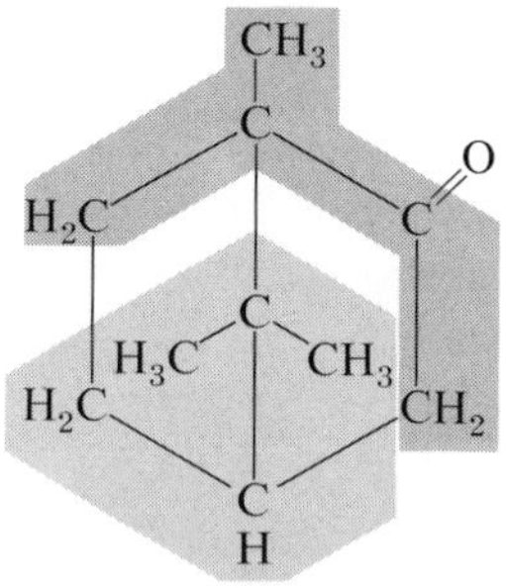

Camphor (colorless crystals), from camphor tree (*Cinnamomum camphora*)

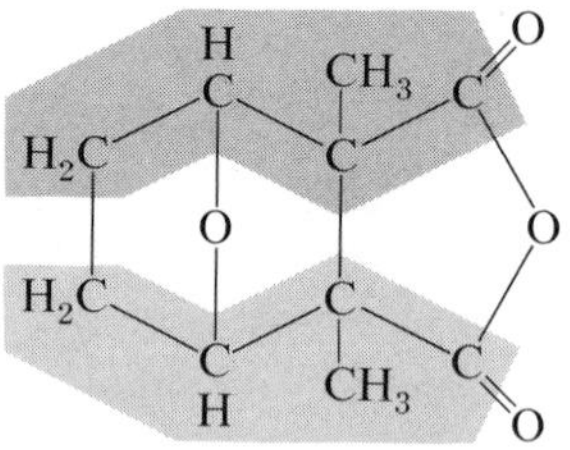

Cantharidin (colorless crystals), constituent of dried beetles (*Cantharis vesicatoria*)

Figure 14.3 Some natural C_{10} products of diverse origin that obey the isoprene rule. The plant origin of each substance is given, and the isoprene units are distinguished by the light and dark grey areas.

The odoriferous constituents of many plants (e.g., cedar, clove, eucalyptus, peppermint, rose) are volatile compounds referred to as *essential oils*. These oils are widely used in food flavorings, medicines, and perfumes. They all contain isomeric hydrocarbons of the composition $C_{10}H_{16}$, which are called *terpenes*. The structural feature that is common to nearly all of the terpenes is the five-carbon isoprene unit.[3] The compound responsible for the color of carrots and autumn leaves is a tetraterpene called β-carotene. The carotenes are important intermediates in photosynthesis, in the biosynthesis of vitamin A (the reason that carrots are a good source of vitamin A), and in other cellular processes.

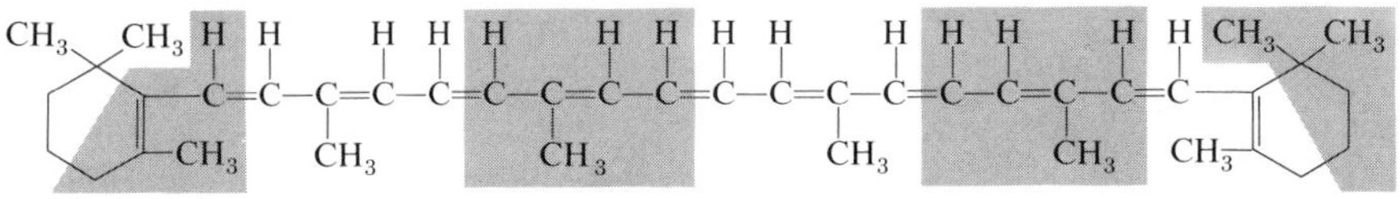

β-Carotene
(8 isoprene units)

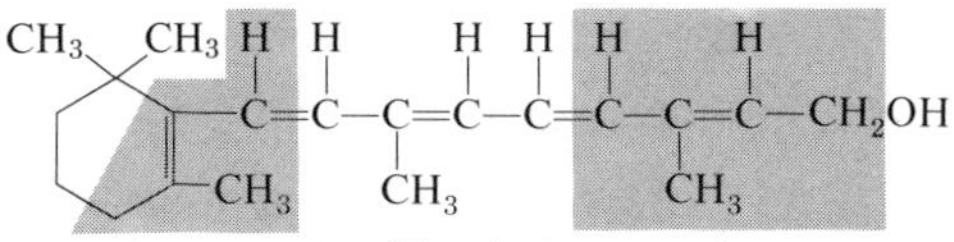

Vitamin A
(4 isoprene units)

The great abundance of naturally occurring plant products whose carbon skeletons contain isoprene units has resulted in the formulation of the **isoprene rule:** *All terpenes should be formally divisible into isoprene units.* Some examples of terpenes that obey the rule are shown in Figure 14.3.

Alkynes

The **alkynes** are unsaturated hydrocarbons that *contain a triple bond between adjacent carbon atoms*. The general formula for this class of compounds is C_nH_{2n-2}. The first and most important member in the series is called acetylene, and has the molecular formula C_2H_2 (Figure 14.4). Because of the

[3]The terpenes are not actually formed in nature from isoprene, which itself has never been detected in a natural product. The true precursor of all terpenes is mevalonic acid, a compound that arises from acetyl CoA.

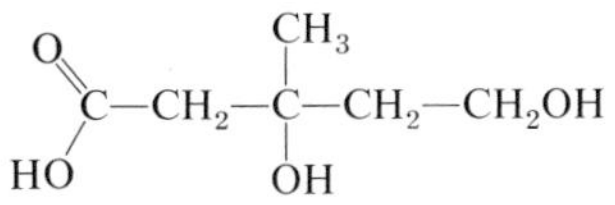

Mevalonic acid

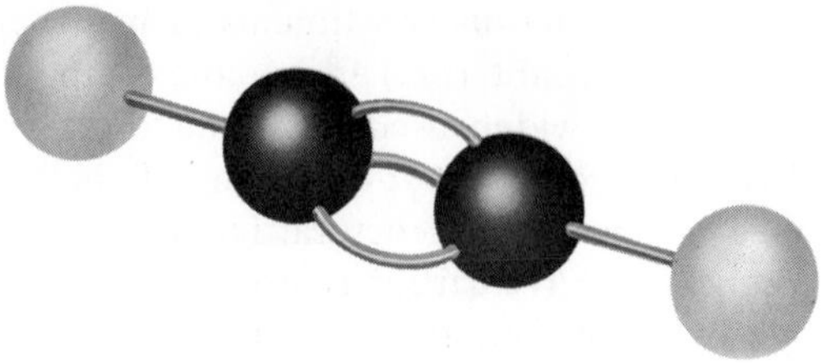

Figure 14.4 Ball-and-stick model of acetylene showing linear conformation of the triple bond.

importance of this simple compound, the higher homologs are often referred to as acetylenic hydrocarbons or acetylenes.

14.6 Nomenclature of Alkynes

The procedure used in naming the alkynes is identical to that for naming the alkenes, but the generic ending for this class of compounds is, of course, different. The suffix *-yne* is used to indicate the presence of the triple bond in the compound, and must be added to the stem name of the parent alkane. Once again, we find that the lower homologs of the series are very often referred to by their common names rather than by their IUPAC names. Table 14.4 lists formulas and names for some of the alkynes.

When double and triple bonds exist in the same molecule, the nomenclature becomes a bit more complex. The situation is handled by the insertion of numbers within the name of the parent compound to indicate the position of each multiple bond. By convention, *-ene* precedes *-yne.*

$$\overset{8}{C}H_3\overset{7}{C}H_2\overset{\overset{H}{|}}{\underset{6}{C}}{=}\overset{\overset{H}{|}}{\underset{5}{C}}\overset{4}{C}H_2\overset{3}{C}H_2\overset{2}{C}{\equiv}\overset{1}{C}H$$

Oct-5-en-1-yne

$$\overset{1}{C}H_3\overset{\overset{H}{|}}{\underset{2}{C}}{=}\overset{\overset{CH_3}{|}}{\underset{3}{C}}\overset{4}{C}H_2\overset{\overset{CH_3}{|}}{\underset{\underset{CH_3}{|}}{\overset{5}{C}}}\overset{6}{C}H_2\overset{7}{C}{\equiv}\overset{8}{C}\overset{9}{C}H_3$$

3,5,5-Trimethylnon-2-en-7-yne

this is # backwards

Table 14.4 Formulas and Nomenclature for Some Alkynes

Molecular Formula	*Structural Formula*	*Common Name*	*IUPAC Name*
C_2H_2	$H{-}C{\equiv}C{-}H$	Acetylene	Ethyne
C_3H_4	$CH_3{-}C{\equiv}C{-}H$	Methylacetylene	Propyne
C_4H_6	$CH_3CH_2{-}C{\equiv}C{-}H$	Ethylacetylene	1-Butyne
C_4H_6	$CH_3{-}C{\equiv}C{-}CH_3$	Dimethylacetylene	2-Butyne
C_5H_8	$CH_3CH_2CH_2{-}C{\equiv}C{-}H$	*n*-Propylacetylene	1-Pentyne
C_5H_8	$CH_3CH_2{-}C{\equiv}C{-}CH_3$	Methylethylacetylene	2-Pentyne
C_5H_8	$CH_3{-}\underset{\underset{CH_3}{\vert}}{CH}{-}C{\equiv}C{-}H$	Isopropylacetylene	3-Methyl-1-butyne (*not* 2-Methyl-3-butyne)

14.7 Preparation of Alkynes

Despite the reactivity of the triple bond, there is little difficulty in synthesizing the alkynes. The C≡C bond is found in birth control pills (see Section 24.11) and in a variety of living organisms, particularly molds. Acetylene is an abundant and important industrial organic raw material. It can be prepared easily and inexpensively from limestone, coal, and water. The first step in the process involves the breakdown of limestone in an electric furnace. The products are lime (calcium oxide) and carbon dioxide. The lime then reacts with the coal, while still in the furnace, to form calcium carbide. Finally, by means of a low temperature reaction, the calcium carbide is treated with water to form acetylene.

$$\underset{\text{Limestone}}{CaCO_3} \xrightarrow{\Delta} \underset{\text{Lime}}{CaO} + CO_2$$

$$CaO + 3\,C \xrightarrow{2000\,°C} \underset{\text{Calcium carbide}}{CaC_2} + CO$$

$$CaC_2 + 2\,H_2O \longrightarrow \underset{\text{Acetylene}}{HC{\equiv}CH} + Ca(OH)_2$$

Acetylene is also prepared by the cracking of methane in an electric arc. This process was perfected and used extensively by Germany during World War II. The acetylene produced in the process was employed as the starting material for the manufacture of synthetic rubber.

14.8 Physical Properties of Acetylene

Acetylene is a colorless gas, insoluble in water, and it has a rather inoffensive odor when pure. The very disagreeable odor usually associated with acetylene is due to the presence of phosphine impurities. The gas can be condensed to a liquid at $-84\,°C$ and a pressure of one atmosphere. In the liquid state, acetylene is very sensitive to shock and is highly explosive. In order to transport the gas safely, it must be dissolved in acetone at moderate pressure. Acetylene burns with a highly luminous flame; one of its earliest uses was a fuel for miners' lamps. It was later utilized in bicycle and automobile lamps. The combustion of acetylene in oxyacetylene torches generates enough heat to cut and weld metals.

$$2\,C_2H_2 + 5\,O_2 \xrightarrow{3000\,°C} 4\,CO_2 + 2\,H_2O + 620\text{ kcal}$$

14.9 Chemical Properties of Alkynes

The chemical reactions of the alkynes are, for the most part, strictly analogous to the reactions of the alkenes. For example, two moles of the reagents that add to the alkenes will add to the alkynes. The reactions may be visualized as occurring in two steps.

Symmetrical Reagents

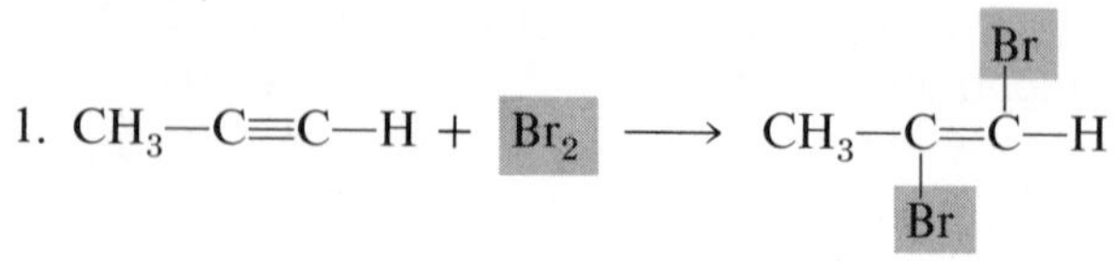

Propyne — 1,2-Dibromopropene

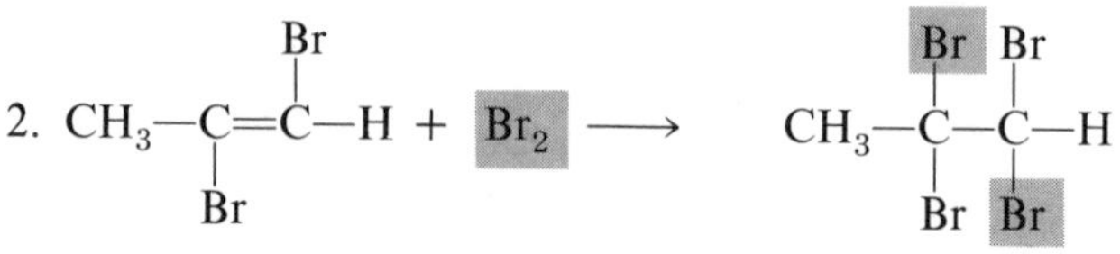

1,1,2,2-Tetrabromopropane

Unsymmetrical Reagents (follow Markownikoff's Rule)

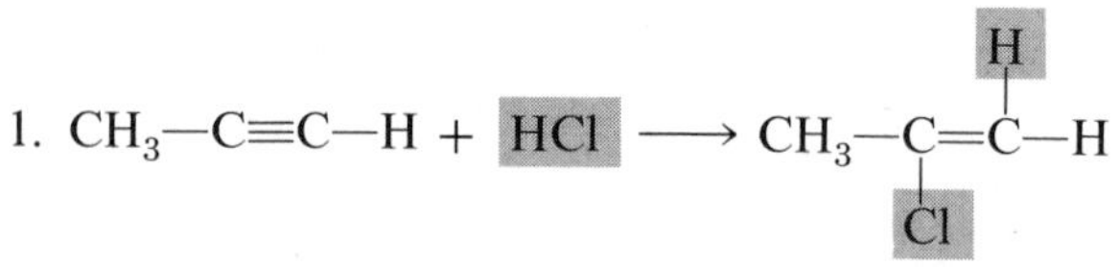

2-Chloropropene

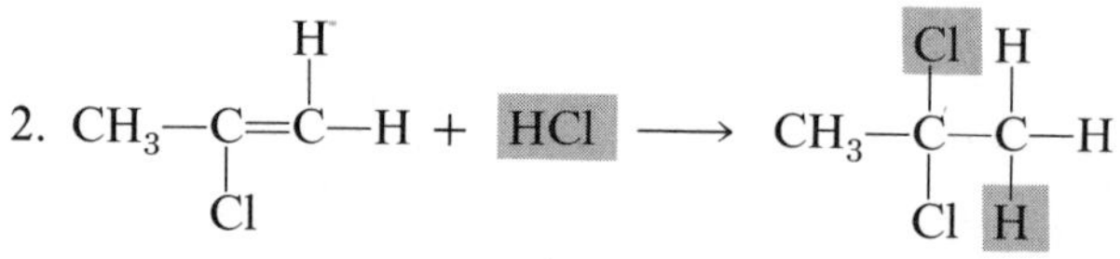

2,2-Dichloropropane

Because the alkynes are more unsaturated than the alkenes, they are also more reactive. Thus, under the same conditions, they will react with several reagents toward which the alkenes are inert. Acetylene, in the presence of certain catalysts, will react with hydrogen cyanide, while ethylene will not. The product of the reaction is cyanoethylene and is an important industrial chemical from which Orlon is produced (Section 14.4-d).

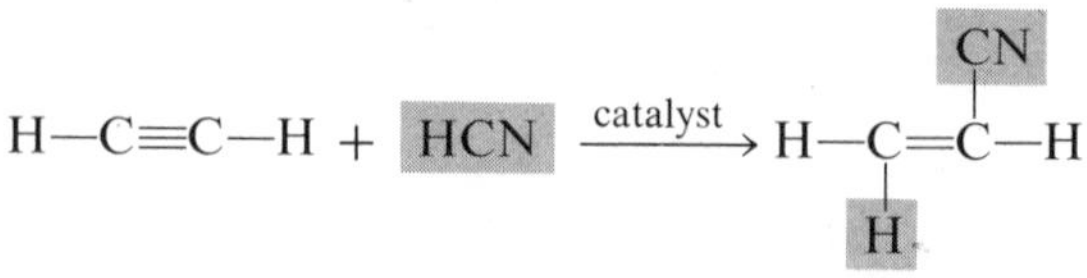

Acetylene — Cyanoethylene (Acrylonitrile)

Acetylene also reacts, by addition, with both water and acetic acid. The hydration reaction leads to an unstable product that rearranges to form acetaldehyde. The vinyl acetate that is obtained from the reaction of acetylene and acetic acid is used in the production of vinyl resins.

$$H-C\equiv C-H + HOH \xrightarrow[H_2SO_4]{HgSO_4} \left[\begin{array}{c} OH \\ | \\ H-C=C-H \\ | \\ H \end{array} \right] \longrightarrow \begin{array}{c} H \quad O \\ | \quad // \\ H-C-C \\ | \quad \backslash \\ H \quad H \end{array}$$

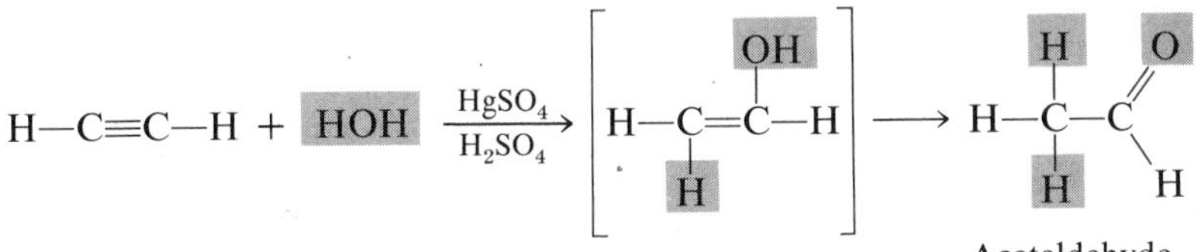

Acetaldehyde

$$H-C\equiv C-H + CH_3-C\overset{O}{\underset{OH}{\diagup\!\!\!\diagdown}} \longrightarrow \begin{array}{c} \quad\quad O \\ \quad\quad || \\ \quad\quad C-CH_3 \\ \quad\quad / \\ H-C=C-O \\ | \quad | \\ H \quad H \end{array}$$

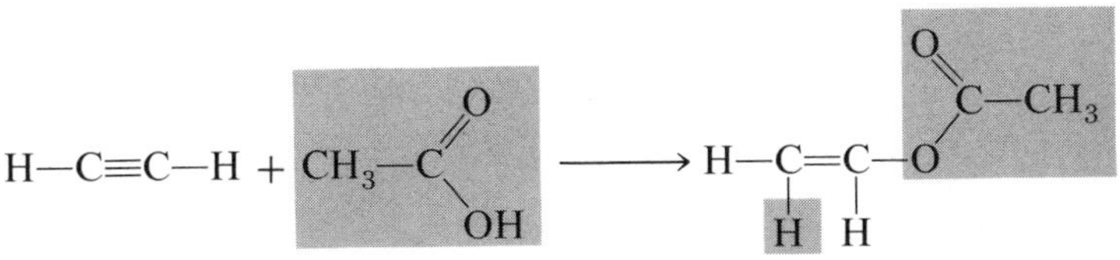

Acetic acid　　　　　　　　　Vinyl acetate

(Note that $H-\underset{\underset{H}{|}}{C}=\underset{\underset{H}{|}}{C}-$ is called the *vinyl group*)

Acetylene also undergoes dimerization to form vinylacetylene, a key intermediate in the production of the synthetic rubber neoprene. Neoprene, unlike natural rubber, is resistant to attack by lubricants such as grease and oil, and it therefore has wide industrial use.

$$H-C\equiv C-H + H-C\equiv C-H \xrightarrow[NH_4Cl]{CuCl} \begin{array}{c} H-C=C-C\equiv C-H \\ | \quad | \\ H \quad H \end{array}$$

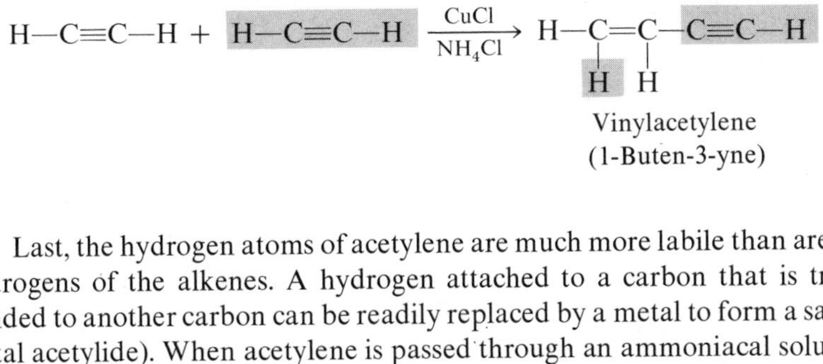

Vinylacetylene
(1-Buten-3-yne)

Last, the hydrogen atoms of acetylene are much more labile than are the hydrogens of the alkenes. A hydrogen attached to a carbon that is triply bonded to another carbon can be readily replaced by a metal to form a salt (a metal acetylide). When acetylene is passed through an ammoniacal solution of cuprous chloride or silver nitrate, insoluble metal acetylides are produced. Sodium acetylide can be produced by passing acetylene gas through molten sodium. Copper and silver acetylides differ from sodium acetylide in that they are extremely sensitive to shock. In the dry state they are highly explosive. The disubstituted alkynes of the type $R-C\equiv C-R'$ do not have acetylenic hydrogens and are incapable of forming metal acetylides. This difference in behavior is used as a diagnostic test to distinguish terminal from nonterminal alkynes.

$$\mathrm{H{-}C{\equiv}C{-}H} \xrightarrow[\mathrm{NH_4OH}]{\mathrm{CuCl\ (or\ AgNO_3)}} \underset{\text{Copper acetylide}}{\mathrm{Cu{-}C{\equiv}C{-}Cu}} \ (\text{or } \underset{\text{Silver acetylide}}{\mathrm{Ag{-}C{\equiv}C{-}Ag}})$$

$$\mathrm{H{-}C{\equiv}C{-}H} \xrightarrow[180\,^{\circ}\mathrm{C}]{\mathrm{Na}} \underset{\text{Sodium acetylide}}{\mathrm{Na{-}C{\equiv}C{-}Na}}$$

$$\mathrm{CH_3CH_2{-}C{\equiv}C{-}CH_3} \xrightarrow[180\,^{\circ}\mathrm{C}]{\mathrm{Na}} \text{No reaction}$$

Exercises

14.1 Draw and name all the noncyclic structural isomers of hexene, C_6H_{12}.

14.2 Draw and name all the cyclic isomers of C_6H_{12}.

14.3 Draw and name all the noncyclic isomers of C_5H_8.

14.4 Name the following compounds.
- **(a)** $CH_2{=}CHCH(CH_3)_2$
- **(b)** $CH_3CH_2C{\equiv}CCH_3$
- **(c)** $CH_3CH{=}CHCH_2C{\equiv}CH$
- **(d)** $(CH_3)_2C{=}CHCH_2CH(CH_3)CH_2CH_3$
- **(e)** $CH_3CH_2CH{=}CHCH_2CH(C_2H_5)_2$
- **(f)** $(C_2H_5)_2C{=}C(CH_3)_2$
- **(g)** $CH_2{=}C(CH_3)C(CH_3){=}CHCH_3$
- **(h)** $CH_2{=}C{=}CHCH_2CH_3$
- **(i)** $CH_3CH{=}CHCH{=}CHCH_3$
- **(j)** $CH_3CH_2C(CH_3)(C_2H_5)CH_2C{\equiv}CCH_3$
- **(k)** 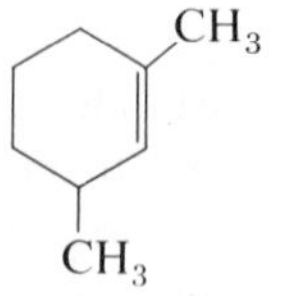
CH3
CH3

14.5 Write structural formulas for the following compounds.
- **(a)** 2-pentene
- **(b)** 4-octyne
- **(c)** 3-*n*-propyl-2-heptene
- **(d)** cyclopentene
- **(e)** 3-ethyl-3-hexene
- **(f)** 2-chloro-1,3-butadiene
- **(g)** 3,3-dimethylcyclobutene
- **(h)** 3-ethyl-2-methyl-1-hexene
- **(i)** 5-ethyl-4-methyl-1-heptene
- **(j)** 3,3-dimethyl-1-butyne
- **(k)** 1,2-dimethylcyclopentene
- **(l)** 1,3,5,7-cyclooctatetraene

14.6 What is wrong with each of the following names? Give the structure and correct name for each compound.
(a) 3-methyl-2-butene
(b) 5-ethylcyclopentene
(c) 2-*n*-propyl-1-propene
(d) 2-isopropyl-3-heptene

14.7 Complete the following equations.

(a)  $-CH_2CH_3 + HBr \longrightarrow$

(b) $(CH_3)_2C{=}CHCH_3 + H_2SO_4 \longrightarrow$
(c) $CH_3C{\equiv}CH + Cl_2$ (excess) $\longrightarrow$
(d) $CH_3C{\equiv}CH + 2HI \longrightarrow$
(e) $CH_3C{\equiv}CH + H_2O \xrightarrow[H^+]{Hg^{2+}}$
(f) $CH_3C{\equiv}CH + CuCl \xrightarrow{NH_4OH}$
(g) $CH_3CH_2CH{=}CH_2$ + dil $KMnO_4 \longrightarrow$
(h) $CH_2{=}CHCH_2Br + Br_2 \longrightarrow$

14.8 By means of equations distinguish between each of the following.
(a) an addition reaction and a substitution reaction
(b) a dehydration reaction and a dehydrohalogenation reaction

14.9 Define and illustrate each of the following terms.
(a) electrophile
(b) carbonium ion
(c) polymerization
(d) isoprene
(e) isoprene rule
(f) terpene

14.10 What are the chief uses of acetylene? How is it produced industrially?

14.11 Write equations for two reactions of acetylene that are similar to those of ethene and for two reactions that are unique to acetylene.

14.12 Describe visual chemical tests that would enable you to distinguish among the following compounds.
(a) *n*-hexane, 1-hexene, 1-hexyne, 2-hexyne
(b) propane, propene, cyclopropane

14.13 A hydrocarbon whose molecular formula is C_4H_6 reacts with two moles of chlorine to yield $C_4H_6Cl_4$. Is this information sufficient to determine the structure of the original hydrocarbon? Explain.

14.14 A hydrocarbon of molecular formula C_5H_8 combines with two moles of hydrogen to yield 2-methylbutane. What is the structural formula of the original compound?

14.15 A colorless compound having the molecular formula C_7H_{14} decolorizes bromine water, adds one mole of hydrogen, and reacts with concentrated $KMnO_4$ to give the following products:

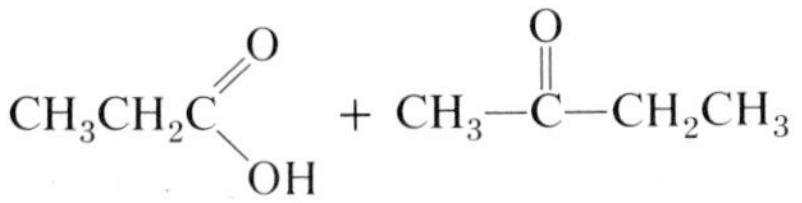
CH3CH2C
O
OH
+ CH3—C—CH2CH3
O

What is the structural formula of the original compound?

14.16 An unknown compound has the molecular formula, C_6H_{10}, yet it only adds one mole of hydrogen or of bromine per mole of compound. What is a possible structural formula?

14.17 A hydrocarbon having a molar mass of 82 contains 88% carbon and 12% hydrogen. It reacts with ammonical silver nitrate to form a silver containing precipitate. What is a possible structure of this compound?

14.18 Three isomeric pentenes, X, Y, and Z, can be hydrogenated to 2-methylbutane. Addition of chlorine to Y gives 1,2-dichloro-3-methylbutane, where 1,2-dichloro-2-methylbutane is obtained from Z. Write the structural formulas for the three isomers.

14.19 How many grams of HBr are necessary to convert 13 grams of acetylene into 1,1-dibromoethane?

14.20 Excess bromine was added to one mole of ethene and 150 grams of product was formed. What percent of the theoretical yield was obtained?

14.21 The complete combustion of 11.2 ml of a four-carbon hydrocarbon at STP requires 67.2 ml of oxygen. Is the hydrocarbon an alkane, alkene, or alkyne?

15

The Aromatic Hydrocarbons

Previously, distinctions between classes of hydrocarbons were made according to the type of bonding that occurred between the carbon atoms in the molecule. The alkanes were characterized by the occurrence of single bonds, the alkenes by double bonds, and the alkynes by triple bonds.

It is not so easy to characterize the aromatic hydrocarbons in this manner. The compounds of this family were called aromatic originally because many of them have spicy or sweet-smelling odors. However, not all aromatic compounds are odorous, and not all fragrant organic compounds are aromatic. The name is used today to denote a particular type of chemical structure and not to refer to any particular odor.

15.1 Structure of the Aromatic Ring

Benzene, the simplest member of the aromatic family, exhibits the common structural feature of this class of compounds. It was first isolated from illuminating gas by the English chemist Michael Faraday in 1825. Later, the compound was given the name of benzene because it could be obtained by distilling benzoic acid with calcium oxide.

The molecular formula of the compound (C_6H_6) indicates that there is only one hydrogen atom for each carbon atom in the molecule. This implies a high degree of unsaturation, and a very high reactivity would therefore be predicted for the molecule. Chemists soon discovered, however, that benzene is chemically stable and that it behaves more like an alkane than an alkene or alkyne. It is unreactive with respect to both the bromine and permanganate

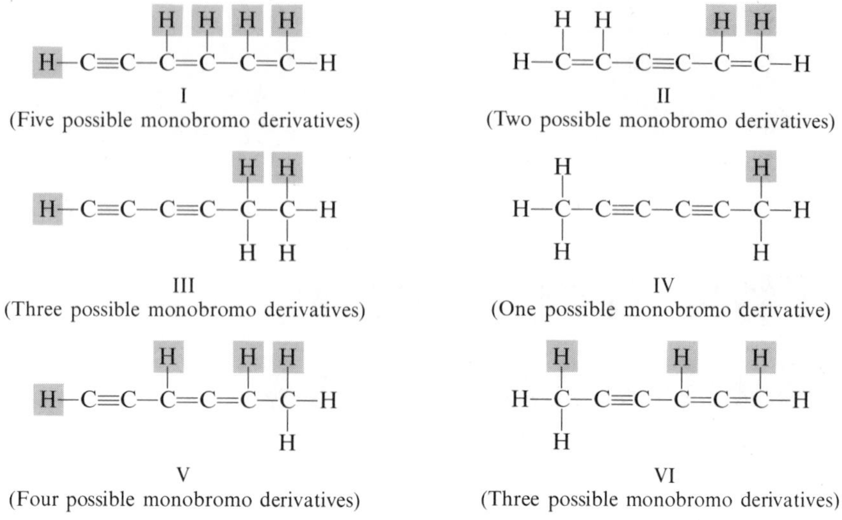
I
(Five possible monobromo derivatives)
II
(Two possible monobromo derivatives)
III
(Three possible monobromo derivatives)
IV
(One possible monobromo derivative)
V
(Four possible monobromo derivatives)
VI
(Three possible monobromo derivatives)

Figure 15.1 Noncyclic isomers of C_6H_6.

tests for unsaturation (Section 14.4). Under special conditions, it can be made to react with bromine, but the reaction is a substitution reaction and not an addition.

$$C_6H_6 + Br_2 \xrightarrow[\text{catalyst}]{FeBr_3} \underset{\text{Bromobenzene}}{C_6H_5Br} + HBr$$

This reaction proved significant in elucidating the structure of benzene. It was found that one and only one monobromobenzene is formed in the reaction. There are no isomers of this compound, and therefore the six hydrogens of benzene must be equivalent. This evidence thus ruled out all but one of the possible straight-chain structures for benzene. Compound IV (Figure 15.1) could be discarded, however, because it too would be expected to react vigorously with bromine and potassium permanganate.

Finally, in 1865, the German chemist Friedrich August Kekulé proposed a structure that could account for all of the known chemical properties of the compound.[1] His theory was that the benzene molecule consisted of a cyclic, hexagonal, planar structure of six carbon atoms with alternate single and double bonds. Each carbon atom was bonded to only one hydrogen atom. He accounted for the equivalence of all six carbon atoms by suggesting that the double bonds are not static but rather are mobile, and that they oscillate from one position to another.

[1]There is a story told that one evening Kekulé fell asleep while sitting in front of a fire. He dreamt about chains of atoms having the forms of twisting snakes. Suddenly one of the snakes caught hold of its own tail, forming a whirling ring. He awoke, freshly inspired, and spent the remainder of the night working on his now famous hypothesis. Kekulé is said to have written the following, "Let us learn to dream gentlemen, and then perhaps we shall learn the truth."

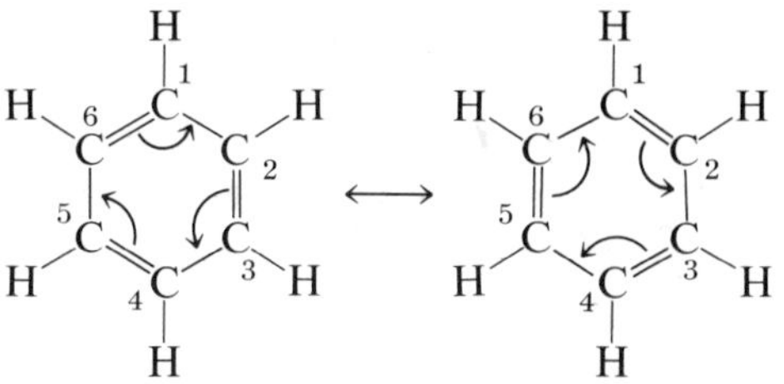
H
1
H
6
C
H
C
C
2
5
C
C
H
4
C
3
H
H
H
1
H
6
C
H
C
C
2
5
C
C
H
4
C
3
H
H

These two structures differ from each other only in the positions of the double and single bonds, and while they satisfy the requirements for equivalent hydrogens, they do not explain why benzene does not behave as an unsaturated hydrocarbon. To explain this one remaining discrepancy, chemists have postulated that the bonds between the carbon atoms of benzene are neither single bonds nor double bonds, but rather are some hybrid of the two. We say that benzene exhibits **resonance**[2] (Section 3.8-c), and the actual structure, which we are unable to represent, is said to be the **resonance hybrid.** The structures represented in Figure 15.2a are known as contributing forms of the resonance hybrid.

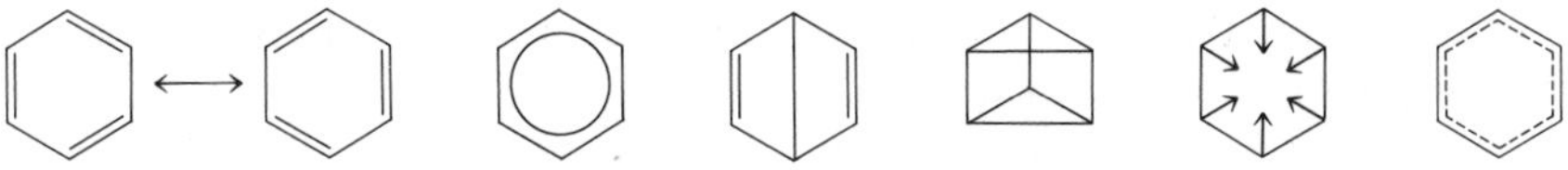
(a)
(b)
(c)
(d)
(e)
(f)

Figure 15.2 Various formulas for benzene: (a) Kekulé structures; (b) hexagon with inscribed circle; (c) Dewar's formula; (d) Ladenberg's prismatic formula; (e) Baeyer's formula; (f) Thiele's partial valence formula.

Some authors combine these two forms into a single structure as depicted in Figure 15.2b. The inner circle implies that the valence electrons are shared equally by all six carbon atoms (that is, the electrons are delocalized, or spread out, over all the carbon atoms). We have chosen to write either of the two contributing Kekulé structures. Each corner of the hexagon is understood to be occupied by one carbon atom. Attached to each carbon atom is one hydrogen atom, and all the atoms lie in the same plane. Whenever any other atom, or group of atoms, is substituted for a hydrogen atom, then it must be shown as bonded to a particular corner of the hexagon.

[2]When two or more structural formulas of a compound can be written that do not differ in the location of atomic nuclei but differ only in the position of valence electrons, we consider that the actual structure of the substance must lie somewhere between the structures represented by the individual formulas. Such a structure is said to exhibit resonance. This interpretation of the structure of benzene has been confirmed physically through experimentation. All the carbon–carbon bonds are of the same length, 1.39 Å, a value between the bond length of the carbon–carbon double bond (1.34 Å) and the carbon–carbon single bond (1.54 Å), but closer to the former.

15.2 Nomenclature of Aromatic Hydrocarbons

There is more than one system for the naming of aromatic hydrocarbons. Both common names and systematic names are encountered. The removal of a hydrogen atom from an aromatic hydrocarbon gives rise to an *aryl* group. (Often the letters **Ar** are used to represent any aryl group just as **R** is used to represent an alkyl group.) Two common aryl groups are *phenyl* and *benzyl.*

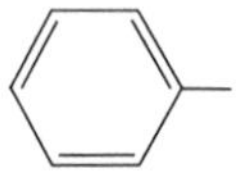

Phenyl

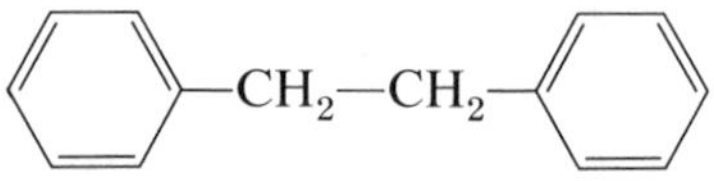

—CH2—CH2—

1,2-Diphenylethane

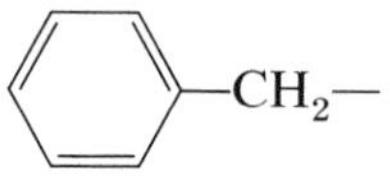

—CH2—

Benzyl

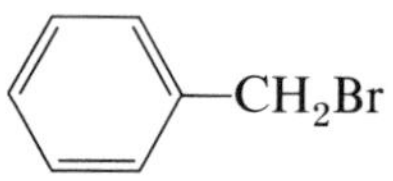

—CH2Br

Benzyl bromide

Some aromatic compounds are referred to exclusively as derivatives of benzene while others are more frequently denoted by their common names. Note that in the following structures it is immaterial whether the substituent is written at the top, side, or bottom of the ring. a hexagon is symmetrical, and all positions are equivalent.

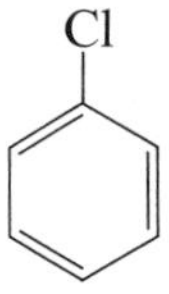

Cl

Chlorobenzene

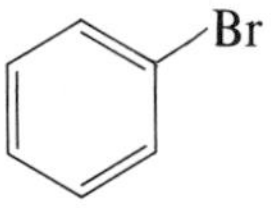

Br

Bromobenzene

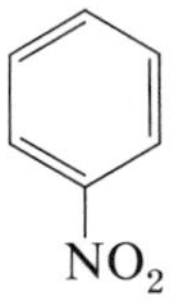

NO2

Nitrobenzene

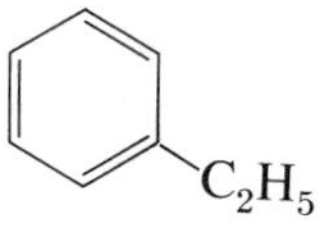

C2H5

Ethylbenzene

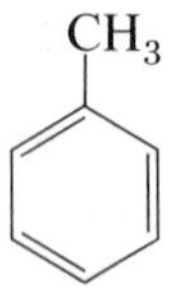

CH3

Toluene (Methylbenzene)

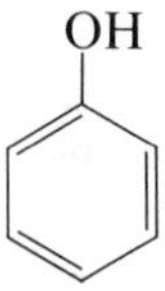

OH

Phenol (Hydroxybenzene)

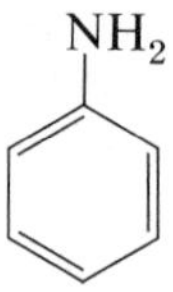

NH2

Aniline (Aminobenzene)

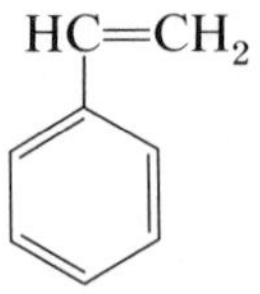

HC=CH2

Styrene (Vinylbenzene)

A further complication arises when there is more than one substituent attached to a benzene ring. When this occurs, all the positions on the hexagon are no longer equivalent, and the relative positions of the substituents must be designated. In the case of a disubstituted benzene, the prefixes *ortho* (*o*-), *meta* (*m*-), and *para* (*p*-) are used. These positions are indicated as follows.

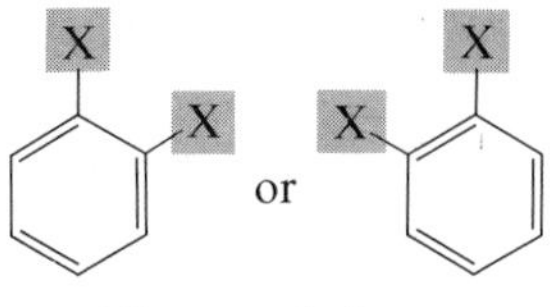
X
X
or
X
X

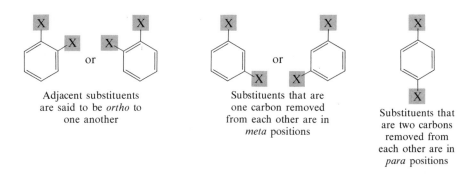

Adjacent substituents are said to be *ortho* to one another

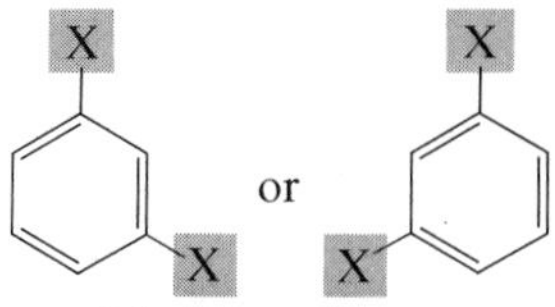
X
or
X
X
X

Substituents that are one carbon removed from each other are in *meta* positions

X
X

Substituents that are two carbons removed from each other are in *para* positions

A few examples will serve to illustrate this nomenclature. Notice that the convention is to indicate the substituents in alphabetical order.

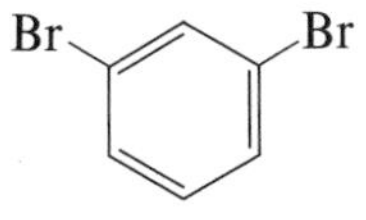
Br
Br

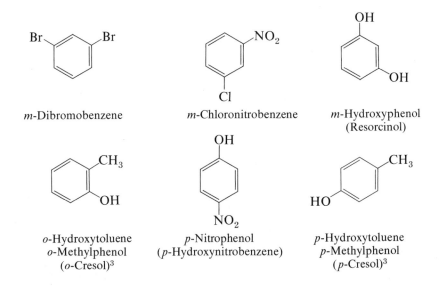

m-Dibromobenzene

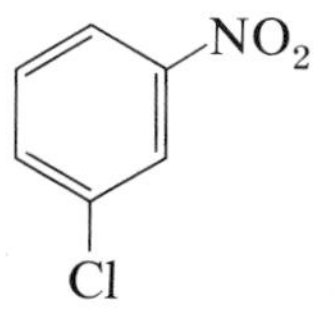
NO2
Cl

m-Chloronitrobenzene

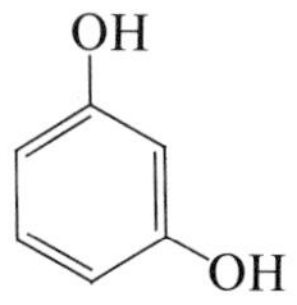
OH
OH

m-Hydroxyphenol
(Resorcinol)

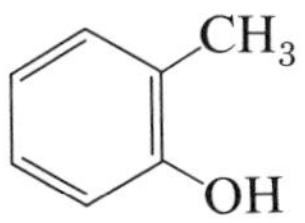
CH3
OH

o-Hydroxytoluene
o-Methylphenol
(*o*-Cresol)[3]

OH
NO2

p-Nitrophenol
(*p*-Hydroxynitrobenzene)

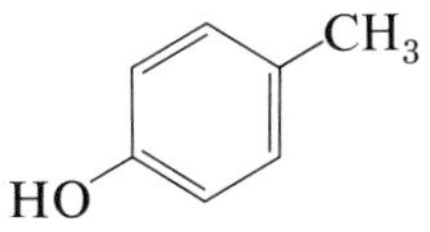
CH3
HO

p-Hydroxytoluene
p-Methylphenol
(*p*-Cresol)[3]

The common names of many aromatic hydrocarbons still persist in the chemical literature. There are three isomers corresponding to the formula $C_6H_4(CH_3)_2$. Along with toluene, they are homologs of benzene. The parent name of all three compounds is xylene.

Benzene
mp 6°
bp 80°

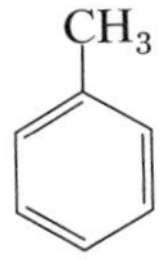
CH3

Toluene
mp −95°
bp 111°

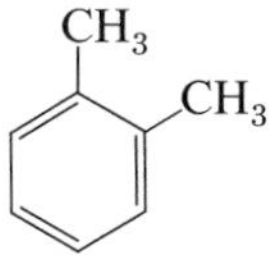
CH3
CH3

o-Xylene
mp −29°
bp 144°

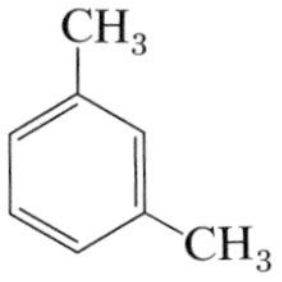
CH3
CH3

m-Xylene
mp −54°
bp 139°

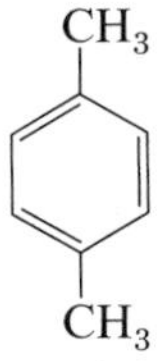
CH3
CH3

p-Xylene
mp 13°
bp 138°

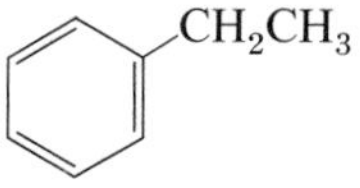
CH2CH3

Ethylbenzene
mp −94°
bp 136°

[3]The methyl derivatives of phenols are commonly called *cresols*.

The three xylenes are also isomeric with ethylbenzene, $C_6H_5(C_2H_5)$, since they all have the molecular formula C_8H_{10}. These four isomers again illustrate that the boiling points of compounds can be very nearly identical, while their melting points can differ over a wide range, with the symmetrical isomer (*p*-xylene) melting at the highest temperature.

When there are three or more substituents, the ring is numbered and the substituents are located by number. When a common name is used, the carbon atom that bears the group responsible for the name is considered to be carbon number one. Substituents are named in alphabetical order.

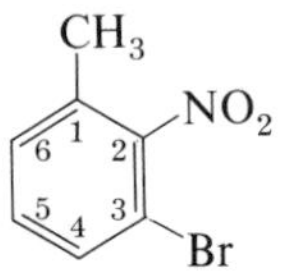

3-Bromo-2-nitrotoluene

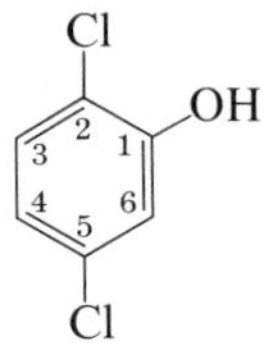

2,5-Dichlorophenol

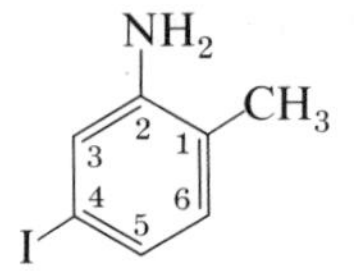

2-Amino-4-iodotoluene

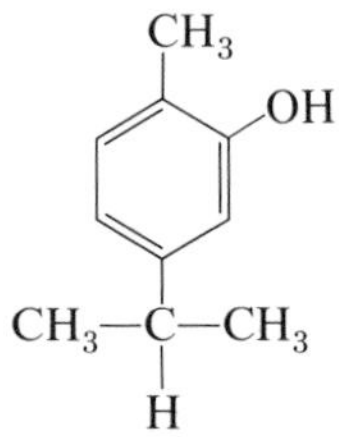

2-Hydroxy-4-isopropyltoluene
(Thymol—flavoring
constituent of thyme)

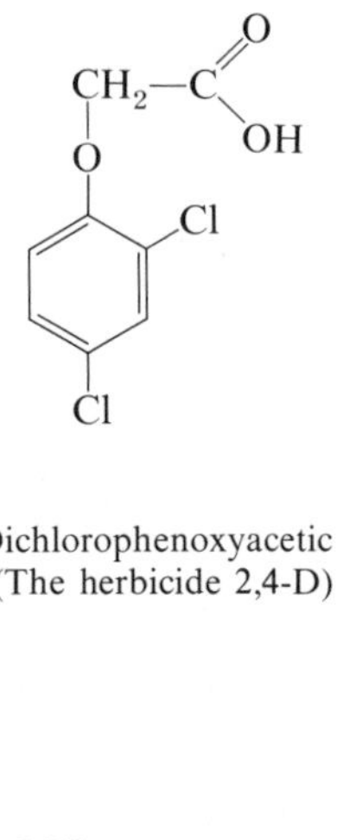

2,4-Dichlorophenoxyacetic acid
(The herbicide 2,4-D)

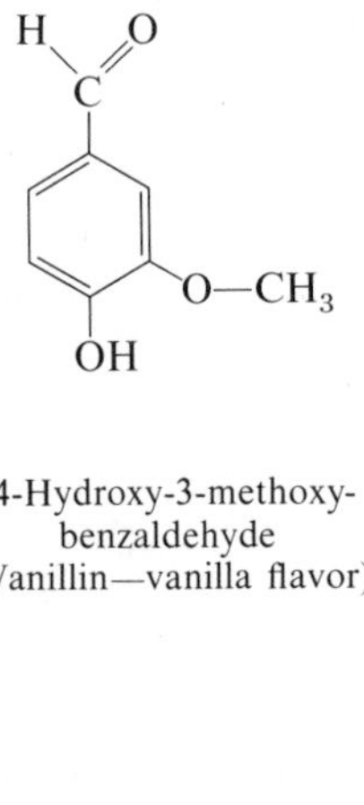

4-Hydroxy-3-methoxy-
benzaldehyde
(Vanillin—vanilla flavor)

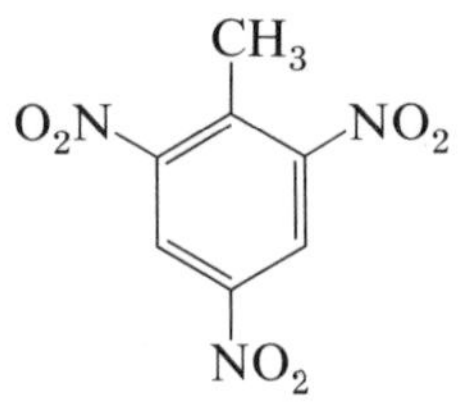

2,4,6-Trinitrotoluene (TNT)

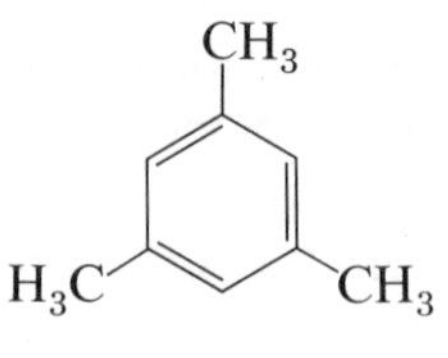

1,3,5-Trimethylbenzene
(Mesitylene)

Mention should be made of some common aromatic hydrocarbons that are not substituted benzenes but are condensed benzene rings:

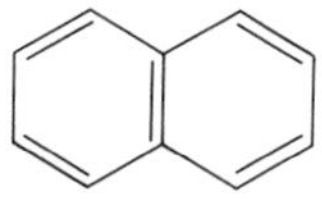

Naphthalene
mp 80°
bp 218°

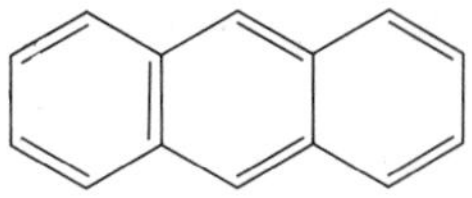

Anthracene
mp 218°
bp 342°

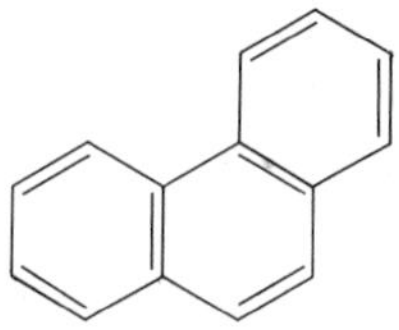

Phenanthrene
mp 101°
bp 340°

These three substances are obtained from coal tar. Naphthalene has a pungent odor and is commonly used as an ingredient in moth flakes. Anthracene is an important starting material in the manufacture of certain dyes. A large group of naturally occurring substances, the steroids, contain the hydrogenated phenanthrene structure (see Section 24.10).

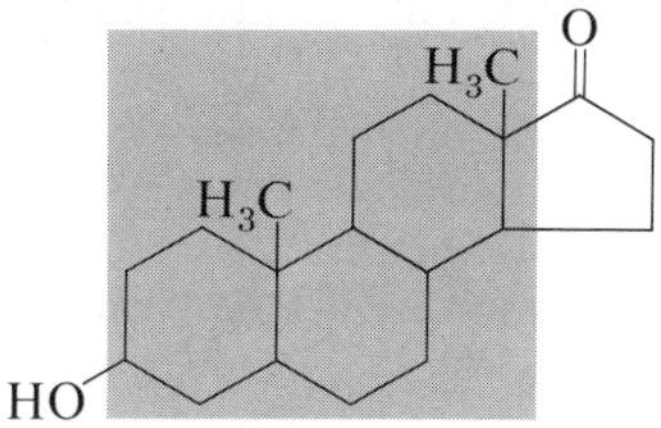
O
H3C
H3C
HO

Androsterone
(male sex hormone)

These polycyclic aromatic compounds do not exist in coal itself, but are formed due to the intense heating involved in the distillation of coal tar. For many years, it has been known that workers in coal-tar refineries are susceptible to a type of skin cancer known as tar cancer. Investigation has shown that a number of these aromatic polynuclear hydrocarbons have the ability to cause cancer when applied to the skin. Such compounds are called *carcinogens* (cancer-producers). One of the most active carcinogenic compounds, 1,2-benzpyrene (Figure 15.3), occurs in coal tar to the extent of 1.5%, and it has been isolated from cigarette smoke, automobile exhaust gases, and charcoal-broiled steaks. Only a few milligrams of 1,2-benzpyrene is required to induce cancer in experimental animals.

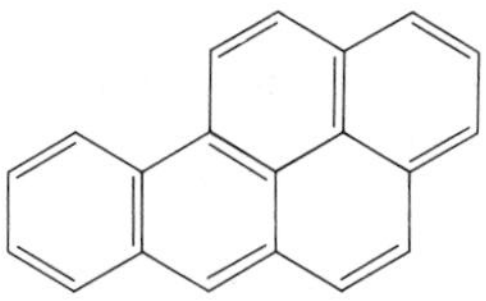

Figure 15.3 1,2-Benzpyrene.

The mechanism by which these compounds cause cancer has not yet been elucidated; therefore, this is a very active area of chemical and medical research. To a certain extent, fused polynuclear hydrocarbons are formed whenever organic molecules are heated to high temperatures. It is the current belief that lung cancer is caused by the formation of carcinogenic compounds in the burning of cigarettes.

The aromatic hydrocarbons containing only one benzene ring are generally liquids, while the polyring aromatics are generally solids. All of them are insoluble in water but are soluble in organic solvents.

15.3 Reactions of Aromatic Hydrocarbons

Although benzene can be made to undergo addition reactions under special conditions, its principal mode of reaction is substitution. Aromatic substitution reactions, unlike the substitution reactions of the alkanes, are rather easily controlled. The general equation for the substitution of a ring hydrogen is as follows.

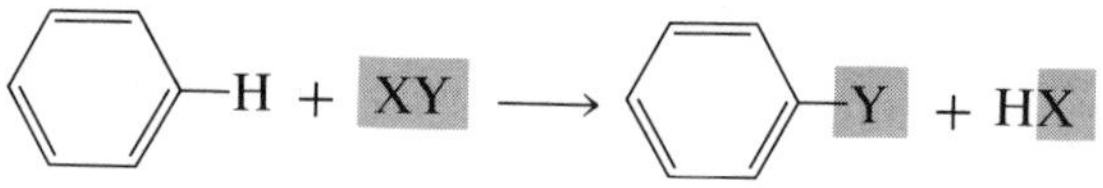
H + XY
Y + HX

Following are specific illustrative reactions.

Halogenation—the substitution of hydrogen by chlorine or bromine:

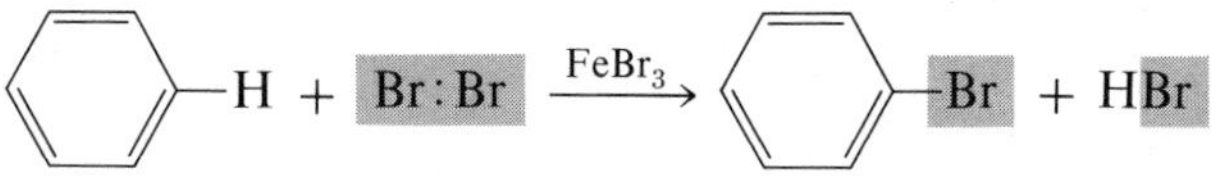
H + Br:Br
FeBr3
Br + HBr

Nitration—the substitution of hydrogen by the nitro group of nitric acid:

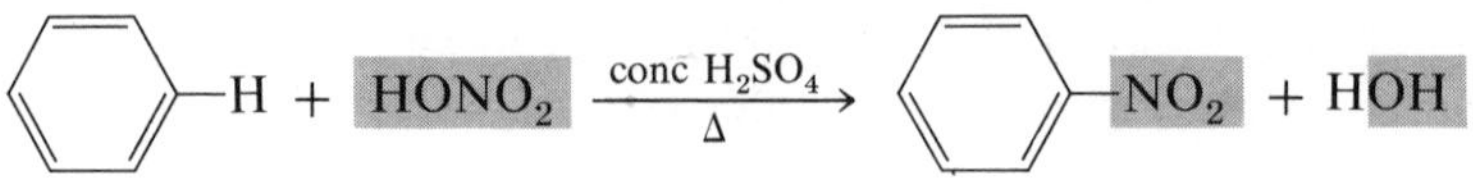
H + HONO2
conc H2SO4
Δ
NO2 + HOH

Sulfonation—the substitution of hydrogen by the sulfonic acid group:

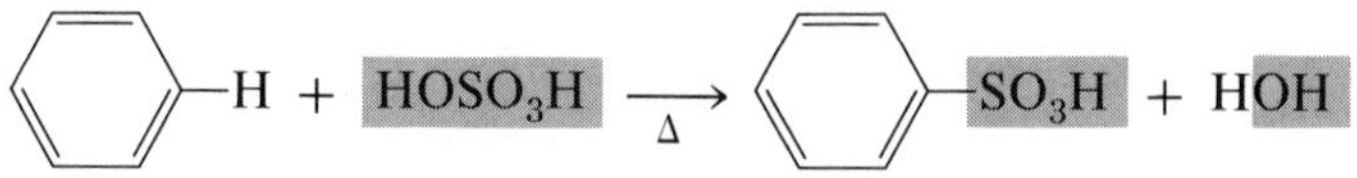
H + HOSO3H
Δ
SO3H + HOH

Alkylation—the substitution of hydrogen by an alkyl group:

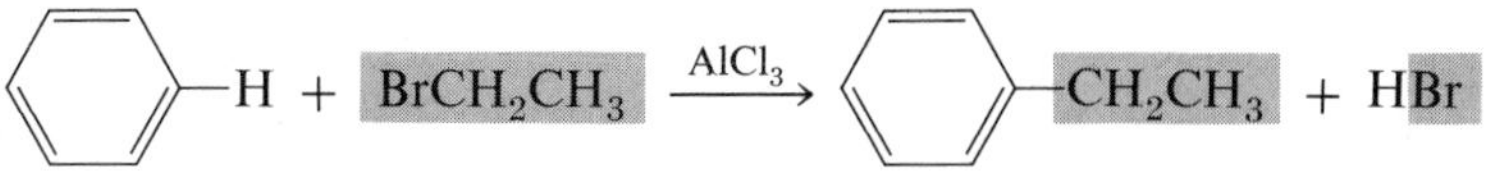

This last reaction is also known as the Friedel–Crafts reaction after its discoverers.

15.4 Mechanism of Aromatic Substitution

All of the aromatic substitution reactions mentioned in Section 15.3 proceed via an electrophilic attack on the electrons of the benzene ring. Catalysts are required in each case to promote the formation of an electron-deficient species (an electrophile): the bromonium ion, the nitronium ion, sulfur trioxide, and a carbonium ion, respectively.

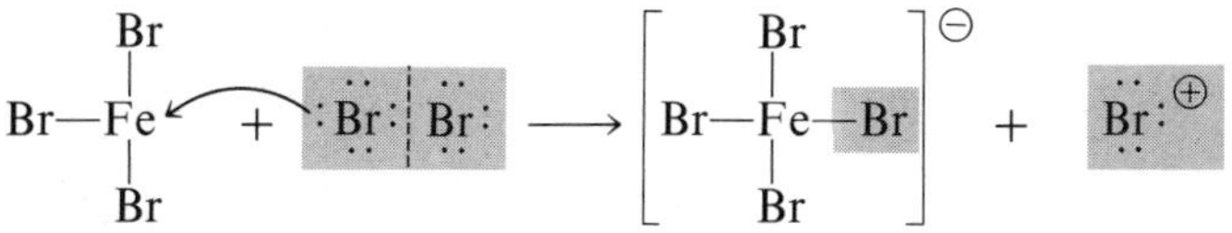

Ferric bromide

Bromonium ion

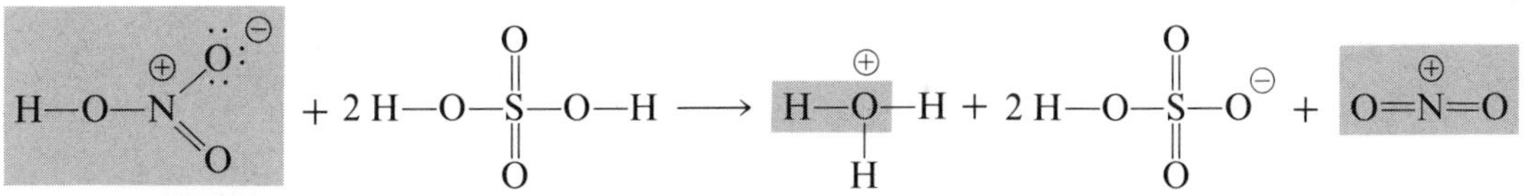

Nitric acid

Nitronium ion

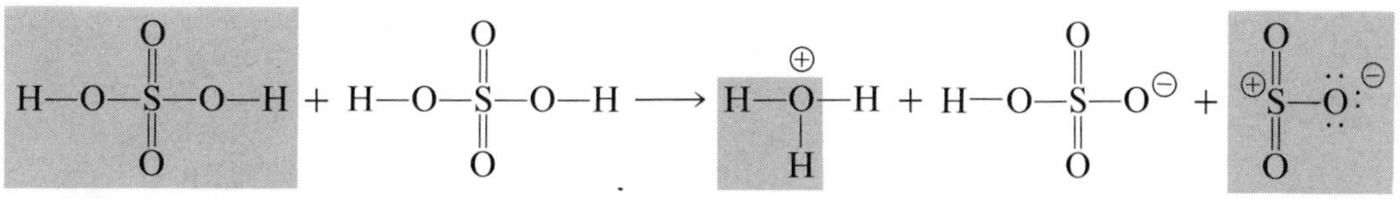

Sulfuric acid

Sulfur trioxide

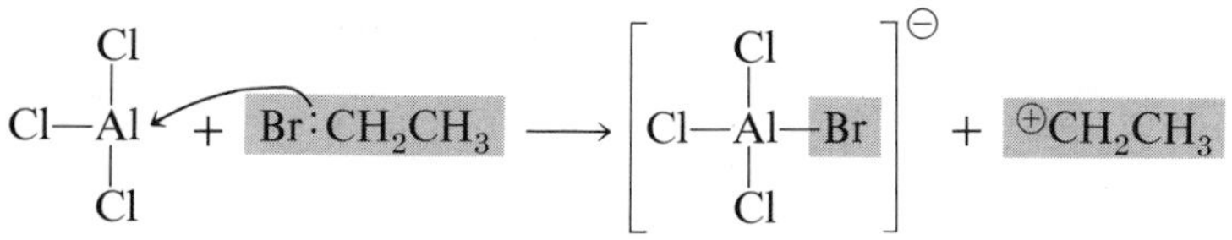

Aluminum chloride

Ethyl carbonium ion

As a result of the above reactions, highly reactive electrophilic agents are produced. These electrophiles attack the electrons of the ring to form intermediate resonance stabilized carbonium ions. One pair of electrons is used to form a single bond with the electrophile. The remaining four electrons are spread out over five carbon atoms and thus the resulting positive charge is delocalized throughout the benzene ring. The bromonium ion is illustrated here, but the mechanism is the same for all of the electrophilic species.

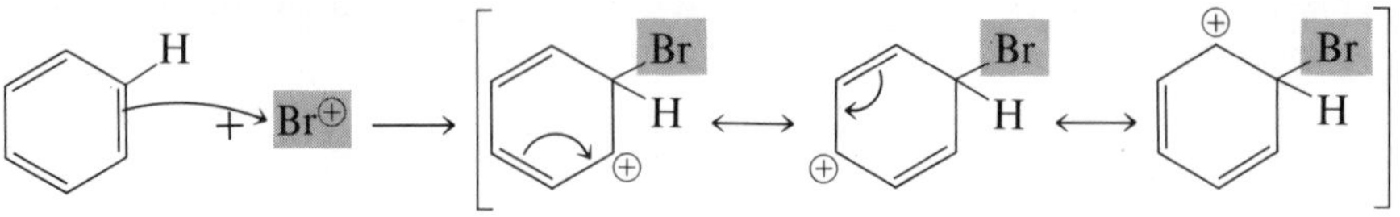

H
+
Br⊕
Br
H
⊕
Br
H
⊕
⊕
Br
H

Intermediate carbonium ion
stabilized by resonance

The reaction is then completed by attack of the anionic form of the catalyst upon the ring proton. Loss of this proton serves to re-establish the resonance-stabilized aromatic ring and to regenerate the catalyst.

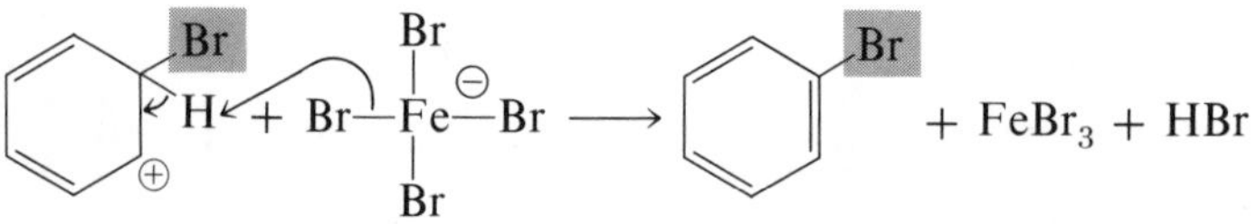

Br
H
⊕
+ Br—Fe—Br
Br
Br
⊖
Br
+ FeBr3 + HBr

Table 15.1 Directive Influence of Ring Substituents

Ortho-Para Directors	*Example*	*Meta Directors*	*Example*
Alkyl —R	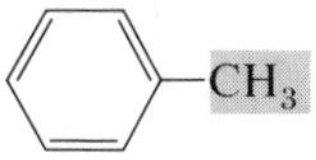	Nitro $-NO_2$	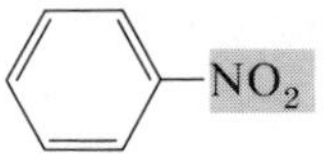
Hydroxyl —OH	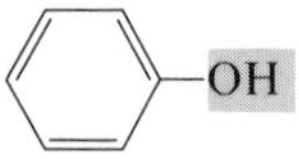	Carboxyl —COOH	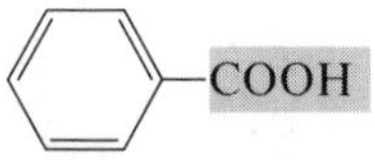
Alkoxyl —OR	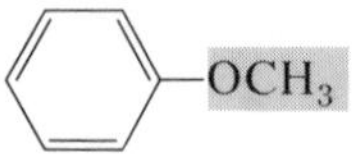	Cyano —CN	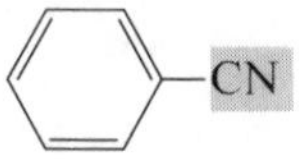
Halogen —X	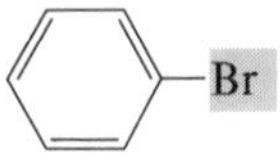	Sulfonic acid $-SO_3H$	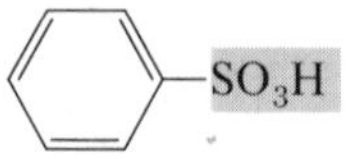
Amino $-NH_2$	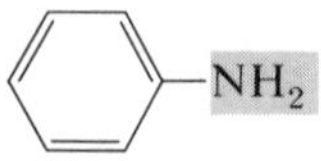		

15.5 Directive Influence of Substituents

A substituent positioned on the benzene ring affects both the facility of substitution of a second group and the position that the second group will occupy. The first substituent, then, is referred to as the *directing group*. Generally speaking, the position that a second substituent group will occupy is independent of the type of entering group involved and is entirely due to the influence of the directing group.

We identify two kinds of directing groups. Those directing the incoming substituent to *ortho* and *para* positions are termed **ortho-para directors,** and those directing the incoming group to the *meta* position are called **meta directors.** Among the ortho-para directors are alkyl, hydroxyl, alkoxyl, halogen, and amino groups. Some common meta directors are the nitro, carboxyl, cyano, and sulfonic acid groups (Table 15.1). Ortho-para directors generally donate electrons to the aromatic ring; meta directors withdraw electrons from the ring. The following examples illustrate these directive effects.

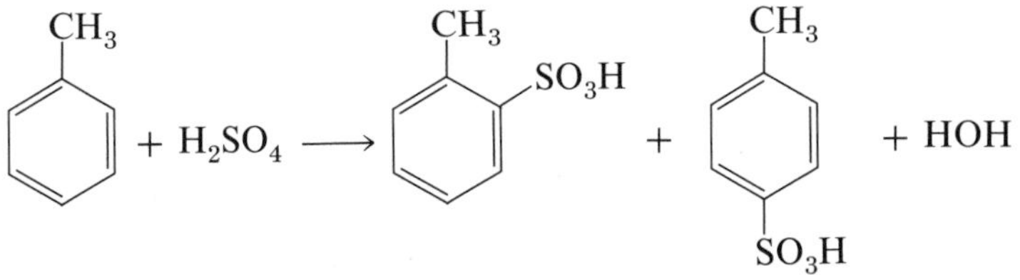

CH3
+ H2SO4
CH3
SO3H
+
CH3
SO3H
+ HOH

Toluene — *o*-Toluenesulfonic acid[4] — *p*-Toluenesulfonic acid[4]

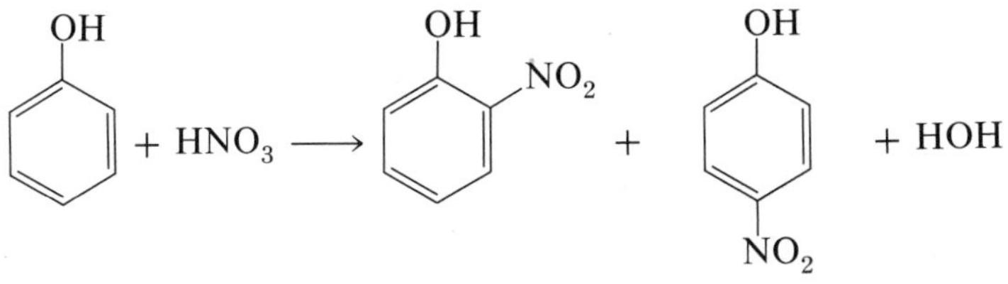

OH
+ HNO3
OH
NO2
+
OH
NO2
+ HOH

Phenol — *o*-Nitrophenol — *p*-Nitrophenol

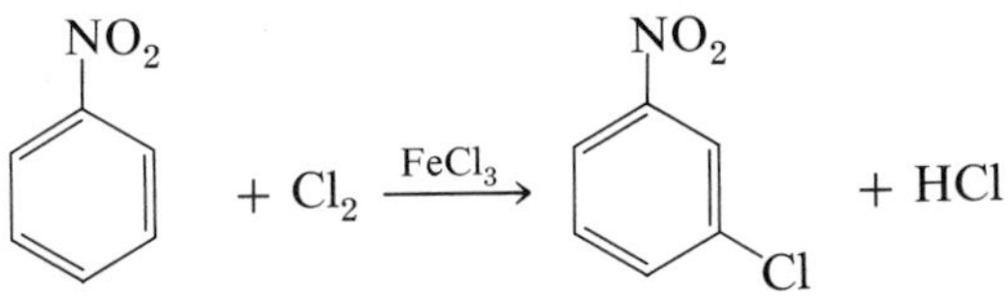

NO2
+ Cl2
FeCl3
NO2
Cl
+ HCl

Nitrobenzene — *m*-Chloronitrobenzene

[4]The relative amounts of the two isomers depend to a large extent upon the temperature of the reaction. The sulfonation of toluene at 0°C yields approximately equal amounts of the ortho and para isomers, whereas at 100°C the ratio of para to ortho isomer is four to one.

The directive effects of substituent groups must always be considered in the planning of a synthesis of a polysubstituted benzene derivative. For example, if one desires to prepare *m*-bromonitrobenzene, he should not nitrate bromobenzene, but rather should brominate nitrobenzene.

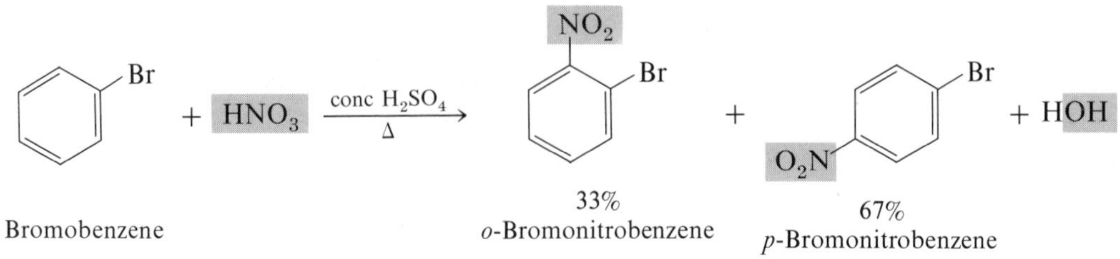
Br
+ HNO3
conc H2SO4
Δ
NO2
Br
+
Br
O2N
+ HOH
Bromobenzene
33%
o-Bromonitrobenzene
67%
p-Bromonitrobenzene

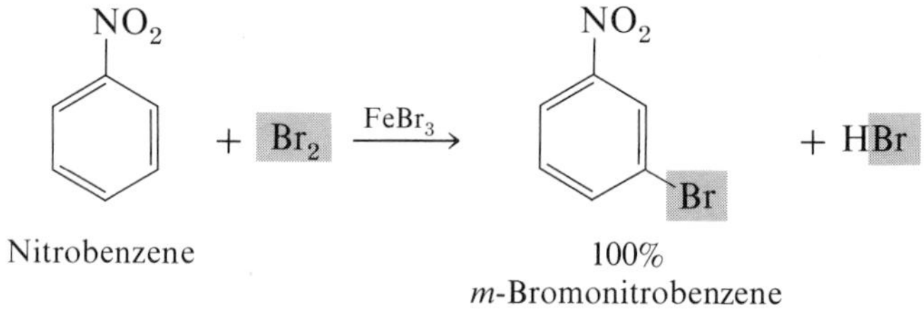
NO2
+ Br2
FeBr3
NO2
Br
+ HBr
Nitrobenzene
100%
m-Bromonitrobenzene

15.6 Uses of Benzene and Benzene Derivatives

Benzene is employed industrially as a solvent and as a starting material for the production of many other products. When using benzene in the laboratory, adequate ventilation is necessary because inhalation of large concentrations of benzene can cause nausea and even death due to respiratory or heart failure. Repeated exposure to benzene leads to a progressive disease in which the ability of the bone marrow to make new blood cells is eventually destroyed. This results in a condition called *aplastic anemia*. Because of its hazardous nature, many chemical laboratories are replacing benzene with toluene as a general solvent. Toluene is utilized as a solvent in lacquers and in the production of dyes, drugs, and explosives. Trinitrotoluene (TNT), unlike nitroglycerine, is not sensitive to shock on jarring and hence must be exploded by a detonator. The xylenes are good solvents for grease and oil and are used for cleaning slides and optical lenses of microscopes. *p*-Xylene is oxidized to terephthalic acid, which is then utilized in the production of Dacron (see Section 19.2-a).

Nitrobenzene is used extensively in the manufacture of aniline, the parent compound of many dyes and drugs. Phenol[5] is a good antiseptic and germicide, but its use is limited because of its toxicity (see Section 17.5-f). Moreover, contact with phenol causes severe blistering of the skin. It is used in the production of dyes, drugs, disinfectants, plastics such as Bakelite, and explosives such as picric acid.

[5] Phenol containing a small amount of water is a liquid, and in this form it is referred to as *carbolic acid*. Lysol is a dilute solution of cresol isomers and is used in hospitals as a disinfectant.

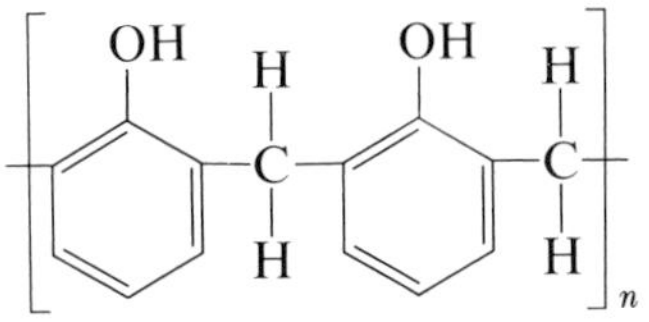

Bakelite

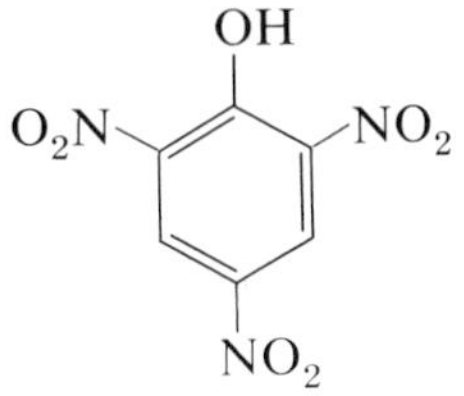

2,4,6-Trinitrophenol
(Picric acid—*note similarity to TNT*)

Substances containing the benzene ring are commonly found in both the animal and plant kingdoms, although they are more abundant in the latter. Plants have the ability to synthesize the benzene ring from carbon dioxide, water, and inorganic materials. Animals, on the other hand, are incapable of this synthesis but are dependent on benzenoid compounds for their survival. Therefore, the animal must obtain the compounds from the food that it ingests. Included among the aromatic compounds necessary for animal metabolism are the amino acids phenylalanine, tyrosine, and tryptophan and certain vitamins such as vitamin K, riboflavin, and folic acid (Figure 15.4). In addition, a great majority of drugs contain the benzene ring (see Section 20.13a–e).

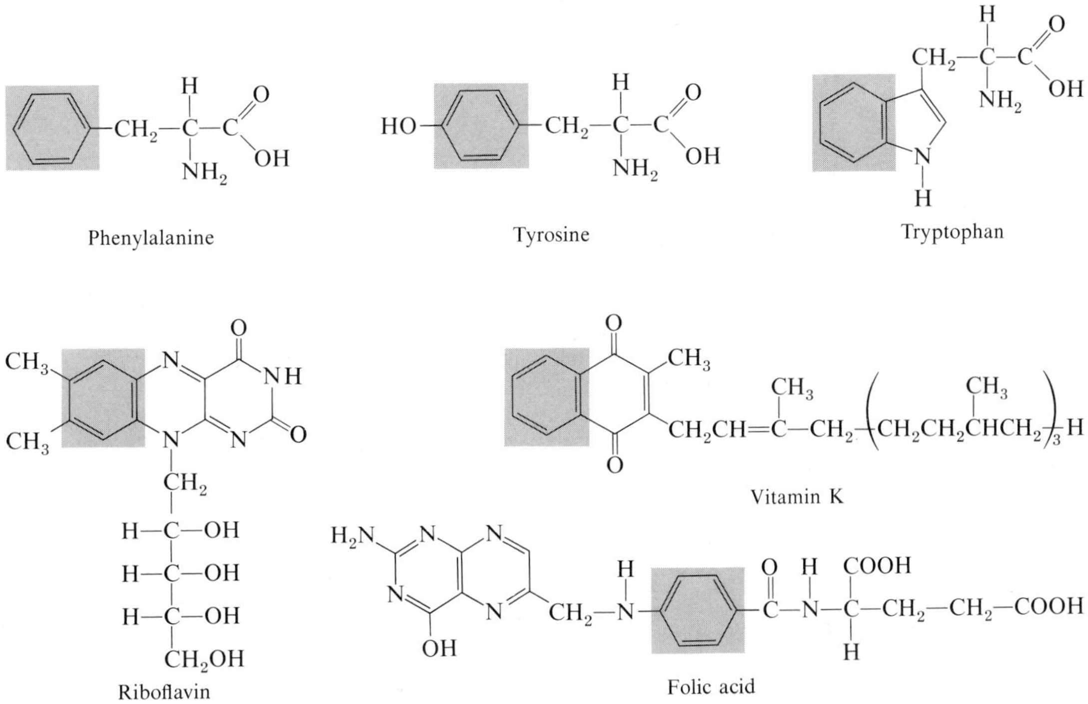
Phenylalanine
Tyrosine
Tryptophan
Vitamin K
Riboflavin
Folic acid

Figure 15.4 Some biologically important compounds that cannot be synthesized by animals.

Exercises

15.1 What characteristic features of aromatic hydrocarbons distinguish them from other hydrocarbons?

15.2 Briefly discuss the concept of resonance.

15.3 Name the following compounds.

(a) 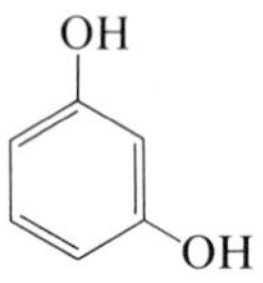

(e) 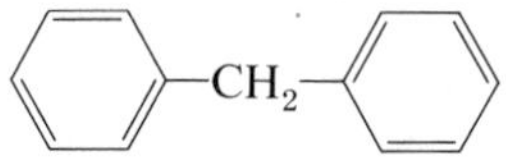

(b) 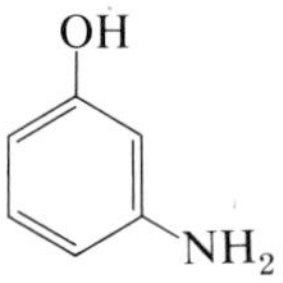

(f) 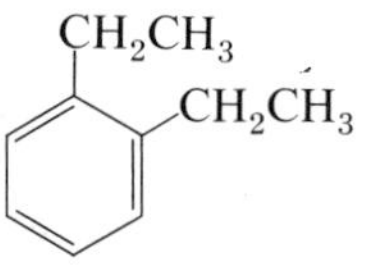

(c) 

(g) 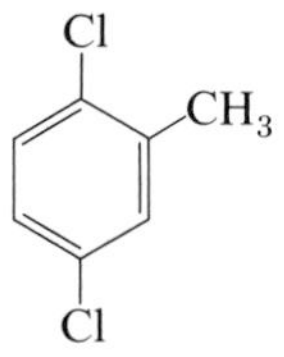

(d) 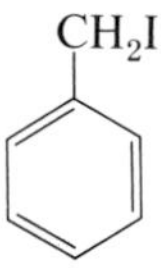

(h) 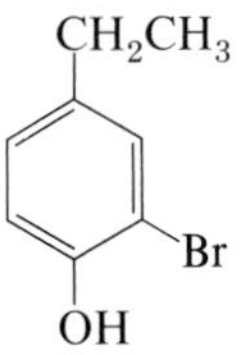

15.4 Write structural formulas for the following compounds.

(a) *m*-xylene
(b) *o*-chlorophenol
(c) *p*-dinitrobenzene
(d) 2-bromo-1,3-dimethylbenzene
(e) naphthalene
(f) TNT
(g) *p*-chloroiodobenzene
(h) 3,4-dimethyltoluene
(i) 2-phenyl-2-butene
(j) phenylacetylene

15.5 Draw and name all the C_9H_{12} aromatic isomers. (There are eight!)

15.6 Draw and name all the isomeric tetramethylbenzenes.

15.7 Draw and name all the possible tribromobenzenes obtained by bromination of each of the following.

(a) *o*-dibromobenzene
(b) *m*-dibromobenzene
(c) *p*-dibromobenzene

15.8 Distinguish between the terms *alkyl group* and *aryl group*. Give an example of each.

15.9 Write three Kekulé formulas for naphthalene, four for anthracene, and five for phenanthrene.

15.10 How many monochloro derivatives of naphthalene are possible? Draw their formulas.

15.11 What is meant by the terms *ortho-para director* and *meta director?*

15.12 Predict the principal product(s) of the following reactions.

(a) 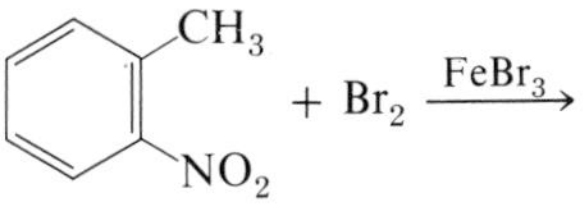

(b) 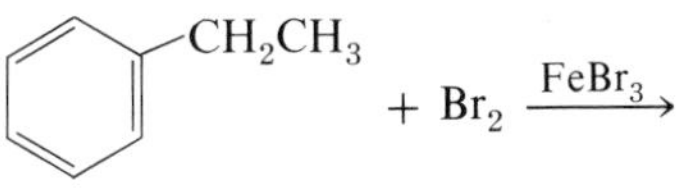

(c) 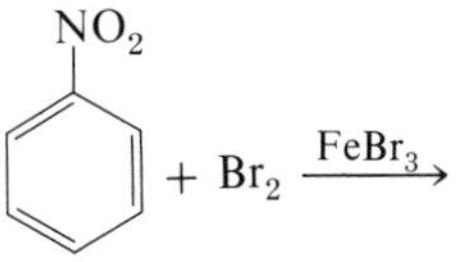

(d) 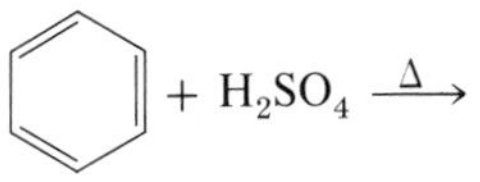

(e) 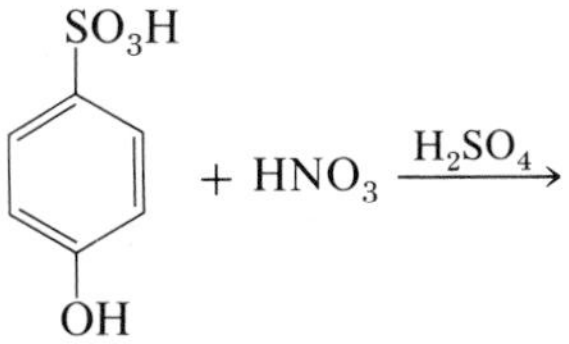

(f) 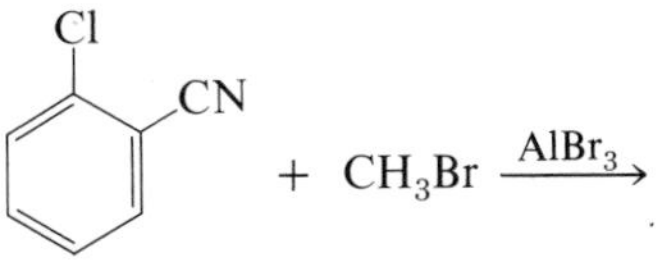

(g) 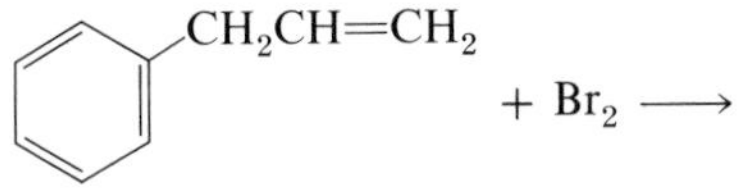

(h) 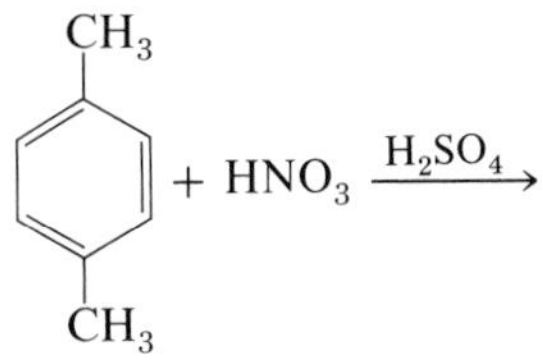

15.13 Contrast the bromination of benzene with that of methane.

15.14 Compare the chemical behavior of benzene and 1,3-cyclohexadiene.

15.15 What chemical tests would you perform to distinguish among cyclohexane, cyclohexene, and benzene?

15.16 List five substances found in your home that contain the benzene ring.

16

Halogen Derivatives of Hydrocarbons

The class of compounds we are about to discuss is of great importance to the industrial chemist, but only of limited value to the biochemist. The halogen derivatives of the hydrocarbons result from the replacement of one or more hydrogen atoms of a hydrocarbon with halogen atoms. If *the substituted hydrocarbon is aliphatic,* then the derivative is an **alkyl halide (R—X)**; if the *hydrocarbon is aromatic,* then the derivative is an **aryl halide (Ar—X).** Previously, the functional group of the different classes of hydrocarbons was a particular type of bond—single, double, or triple. In the case of the organic halides, the functional group is a single atom, the covalently bonded halogen atom.

16.1 Nomenclature of Alkyl Halides

The common names of the alkyl halides are quite prevalent in the chemical literature. The naming system in this case is quite similar to the naming of inorganic halide salts. For example, NaCl is sodium chloride and CH_3Cl is methyl chloride. The alkyl group is always named first and is followed by the name of the halide. Under the IUPAC system, the parent compound is the longest continuous chain containing the halogen atom. In designating the halogen substituent the *-ine* halogen ending is replaced by *-o*. Table 16.1 illustrates the nomenclature of the simple alkyl halides. Alkyl halides of more than five carbons are most conveniently named according to the IUPAC system. The following are examples of more complicated molecules.

$CH_3CH_2CH_2CH(CH_3)CHClCH_2Br$ or 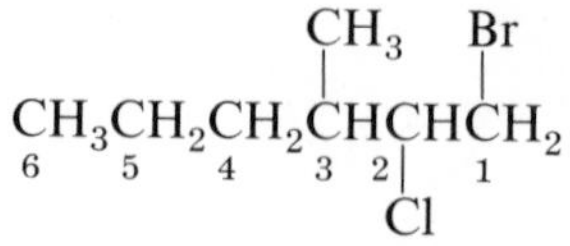

1-Bromo-2-chloro-3-methylhexane

$CH_3CCl_2CH_2CH_2CH(CH_3)CH(C_6H_5)CH_2CH_3$ or 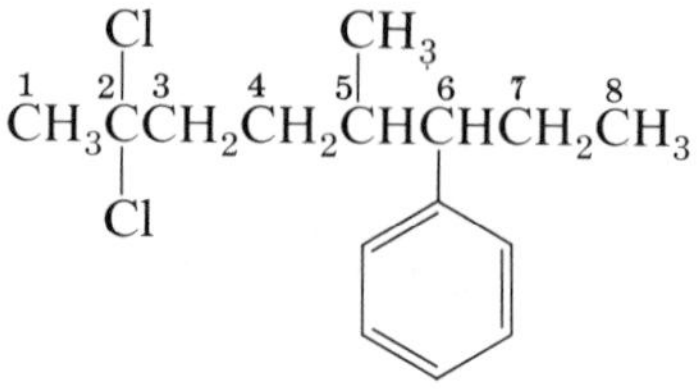

2,2-Dichloro-5-methyl-6-phenyloctane

$CH_3CH_2CI(C_2H_5)CH_2CHICH{=}CH_2$ or 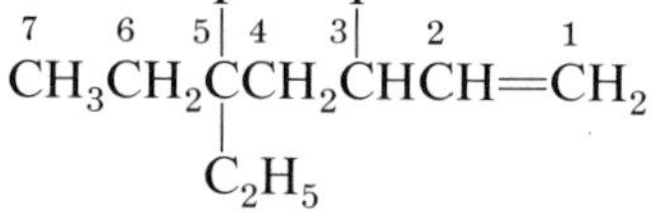

5-Ethyl-3,5-diiodo-1-heptene

Table 16.1 Nomenclature of the Alkyl Halides

Molecular Formula	*Structural Formula*	*Common Name*	*IUPAC Name*
CH_3I	CH_3I	Methyl iodide	Iodomethane
C_2H_5I	CH_3CH_2I	Ethyl iodide	Iodoethane
C_3H_7Cl	$CH_3CH_2CH_2Cl$	*n*-Propyl chloride	1-Chloropropane
C_3H_7Cl	$CH_3CHClCH_3$	Isopropyl chloride	2-Chloropropane
C_4H_9Br	$CH_3CH_2CH_2CH_2Br$	*n*-Butyl bromide	1-Bromobutane
C_4H_9Br	$CH_3CH_2CHBrCH_3$	*sec*-Butyl bromide	2-Bromobutane
C_4H_9Br	$CH_3CH(CH_3)CH_2Br$	Isobutyl bromide	1-Bromo-2-methylpropane
C_4H_9Br	$CH_3{-}CBr(CH_3){-}CH_3$	*t*-Butyl bromide	2-Bromo-2-methylpropane
C_3H_5Cl	$H_2C{=}CH{-}CH_2Cl$	Allyl chloride	3-Chloropropene
$C_2H_4Br_2$	$CH_2Br{-}CH_2Br$	Ethylene bromide	1,2-Dibromoethane

16.2 Preparation of Organic Halides

Naturally occurring organic halides are rare. They are, however, synthesized in great quantities because they are used extensively in industry as the starting materials for the preparation of a great variety of organic reagents. In addition, many organic halides are useful in their own right, as we shall see in Section 16.5.

a. From Alkenes

Dihalides are prepared by the addition of halogen molecules to the appropriate alkene, whereas monohalides are prepared by the addition of hydrogen halides to alkenes. Note that in equation (2) below the hydrogen bromide adds to the alkene in accordance with Markownikoff's rule.

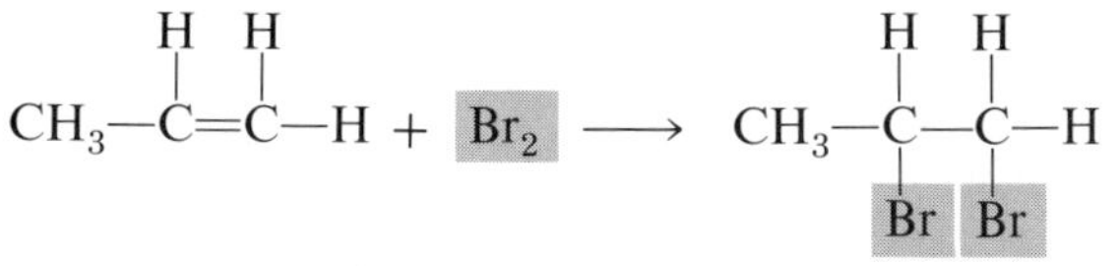
H H
CH3—C=C—H + Br2
H H
CH3—C—C—H
Br Br

(1)

Propene — 1,2-Dibromopropane

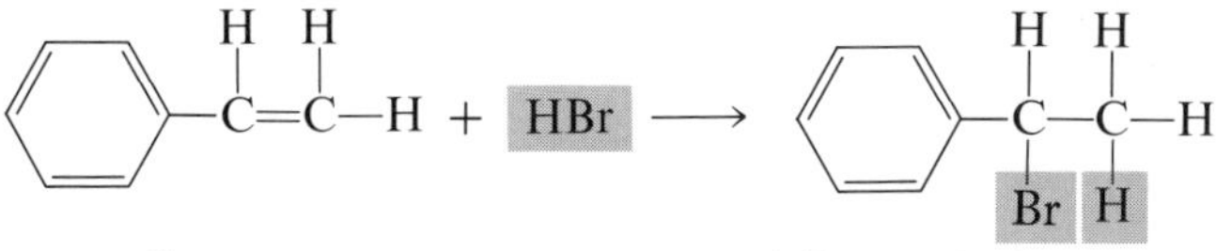
H H
C=C—H + HBr
H H
C—C—H
Br H

(2)

Styrene (Phenylethylene) — 1-Bromo-1-phenylethane

b. From Alcohols

The alcohols are the common laboratory reagents for the synthesis of the alkyl halides. There are several reagents which will effect the replacement of the alcoholic hydroxyl group by a halogen atom (see also Section 17.4-c).

$$3\ CH_3CH_2OH + PX_3 \longrightarrow 3\ CH_3CH_2X + H_3PO_3 \qquad (3)$$

$X = Cl, Br, I$

$$CH_3CH_2CH_2OH + \text{conc } HX \longrightarrow CH_3CH_2CH_2X + HOH \qquad (4)^1$$

$X = Br, I$; if $X = Cl$, $ZnCl_2$ is needed as a catalyst

[1]Conceptually, the reaction in which a hydrogen halide reacts with an alcohol to form an alkyl halide and water is similar to a neutralization reaction in which an acid and a base react to produce a salt and water. Care must be taken to note that while the reaction appears similar, it is *not* the same. Alcohols are not organic bases even though they do possess a hydroxyl group and alkyl halides are *not* organic salts.

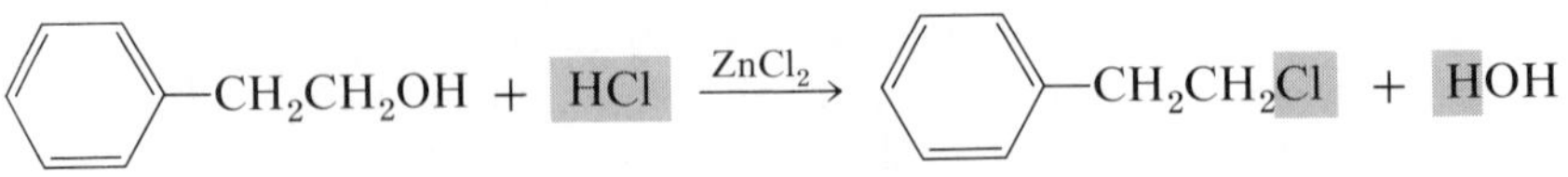 (5)

2-Phenyl-1-ethanol 1-Chloro-2-phenylethane

c. Halogenation of Aromatic Hydrocarbons

Direct halogenation is most frequently applied in the synthesis of aryl halides. Recall that the direct halogenation of alkanes results in the formation of polyhalogenated products (Section 13.4).

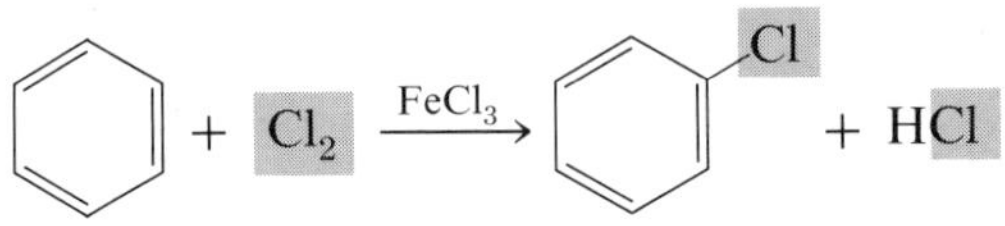

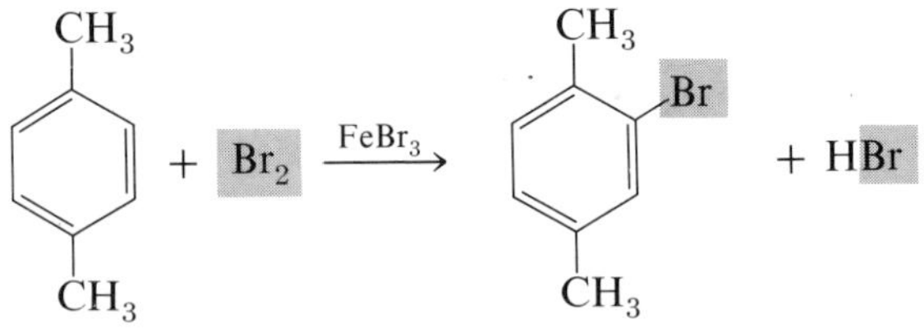

p-Xylene 2-Bromo-*p*-xylene

The unpleasant taste and odor of drinking water are sometimes caused by the chlorination process. In particular, chlorination of water containing phenol (a common contaminant of industrial waste water) leads to substitution of the ring hydrogens to produce mono- and dichlorophenols. The offensive taste of 2,6-dichlorophenol can be detected at a concentration of 0.1 ppb.

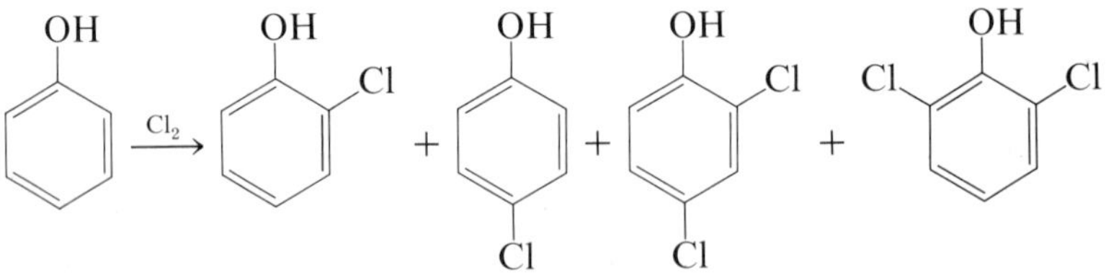

Photochemical halogenation of aromatic side chains results in the substitution of the hydrogen atom on the carbon atom adjacent to the

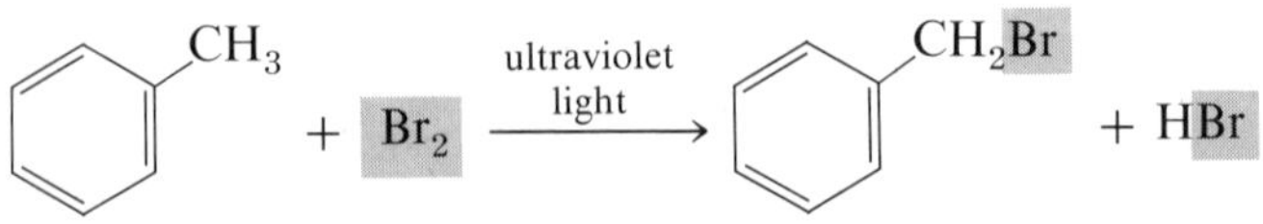

Toluene Benzyl bromide

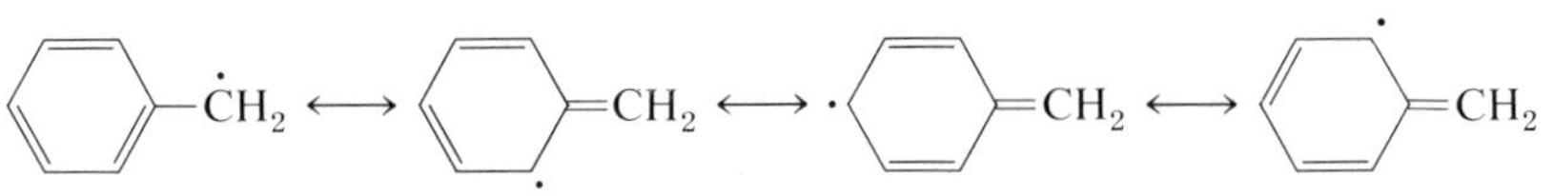

Figure 16.1 Resonance stabilization of a free radical on a carbon atom adjacent to the aromatic ring.

aromatic ring. This is because the free radical intermediate can be stabilized by resonance when the odd electron is adjacent to the aromatic ring (Figure 16.1).

d. Preparation of Alkyl Fluorides from Other Alkyl Halides

Fluorine gas and hydrogen fluoride are extremely reactive species that are difficult to handle safely. Therefore, special techniques are required to prepare alkyl fluorides. Use is made of inorganic fluorides in a process called halogen exchange.

$$2\,CH_3Br + Hg_2F_2 \longrightarrow 2\,CH_3F + Hg_2Br_2$$

$$CCl_4 + 2\,HF \xrightarrow{SbF_5} CF_2Cl_2 + 2\,HCl$$

The second equation represents the preparation of the fluorocarbon, freon-12 (freon-11 is $CFCl_3$). Because chlorofluormethanes are readily liquefied under pressure and are almost completely inert, they are ideal refrigerants and aerosol propellants. However, since these compounds are so inert, they do not react with substances with which they come in contact. As a result, they diffuse up through the atmosphere unchanged and eventually reach the stratosphere (altitude of 10 to 12 miles).

Ozone is an essential component of the stratosphere where it functions to absorb the high energy ultraviolet radiation from the sun. If this radiation were not filtered out, it would quickly burn exposed skin causing skin cancer and eye damage. The freons that reach the stratosphere absorb ultraviolet solar radiation and break down, liberating chlorine atoms ($Cl\cdot$). The chlorine atoms are believed to destroy the ozone layer by a sequence of reactions that yield oxygen.

$$CF_2Cl_2 \xrightarrow[\text{light}]{\text{ultraviolet}} Cl\cdot + \text{other products}$$

$$Cl\cdot + O_3 \longrightarrow ClO + O_2$$

This ozone depletion mechanism is based upon theoretical models and laboratory studies and not upon correlations of freon and ozone concentrations actually measured in the stratosphere. Nevertheless, the public has become aroused; the potential for increased incidences of skin cancer is too

Table 16.2 Boiling Points of Alkyl Halides (°C)

Alkyl Group	*Chloride*	*Bromide*	*Iodide*
Methyl	−24	5	43
Ethyl	12	38	72
n-Propyl	47	71	102
Isopropyl	37	60	90

high a price to pay for consumer comforts. Most manufacturers are now changing from aerosol sprays to spray pump dispensers.

16.3 Physical Properties of Alkyl Halides

Alkyl halides are insoluble in water. The chlorides are lighter than water, while the bromides and iodides are heavier. Table 16.2 illustrates once again the effect that increasing molar mass has upon the boiling points of a series of homologous compounds. It is convenient to remember that methyl iodide is the only methyl halide that is not a gas at room temperature. Aside from the rather uncommon ethyl fluoride, ethyl chloride is the only ethyl halide that is a gas at room temperature. Ethyl chloride has an extremely high vapor pressure, and use is made of this physical property in many ways. For example, it is employed in baseball games as a temporary anesthetic whenever a player is hit by a pitched ball. The affected area is sprayed with liquefied ethyl chloride. The halide then begins to evaporate, and as it does so it abstracts heat from the injured area. The tissue becomes temporarily frozen, and thus insensitive to pain. The same treatment is used in medicine when extremely painful injections must be administered. Caution must be taken in using alkyl halides because almost all of them are toxic. Exposure over prolonged periods may cause liver damage.

16.4 Chemical Properties of Organic Halides

As indicated earlier, the replacement of a hydrogen atom by a halogen atom effects an enormous difference in the reactivity of the molecule. Most of the reactions that the alkyl halides undergo are either substitution or elimination reactions. In some cases, both types of reaction are possible for the same alkyl halide and a given reagent. The mechanism of reaction in such an instance will depend upon the structure of the alkyl halide involved, the solvent utilized, and the nature and concentration of the participating reagent.

a. Elimination (Dehydrohalogenation)

The dehydrohalogenation reaction was mentioned in Section 14.2-a as a method for the preparation of alkenes.

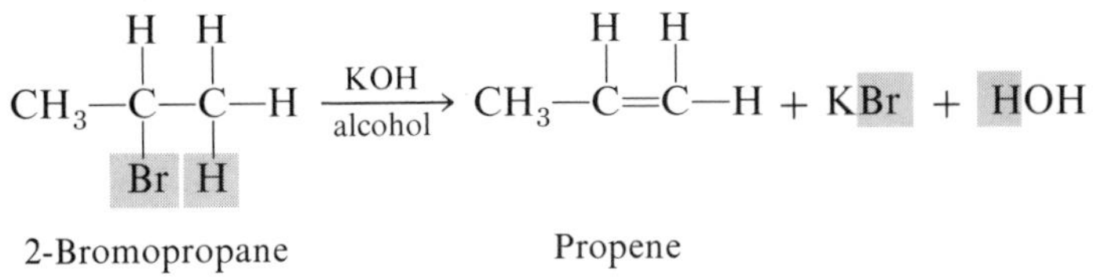
KOH
alcohol
2-Bromopropane
Propene

The reaction is effected by the use of a strong base in an alcoholic solution. A concerted mechanism has been proposed, that is, the removal of a proton occurs simultaneously with a shift of the electron pair and ionization of the halide. The overall result is the loss of hydrogen halide with formation of a carbon–carbon double bond. This type of reaction is termed an **elimination reaction.** It is defined as a *reaction that effects the loss of some simple molecule and the formation of a multiple bond.*

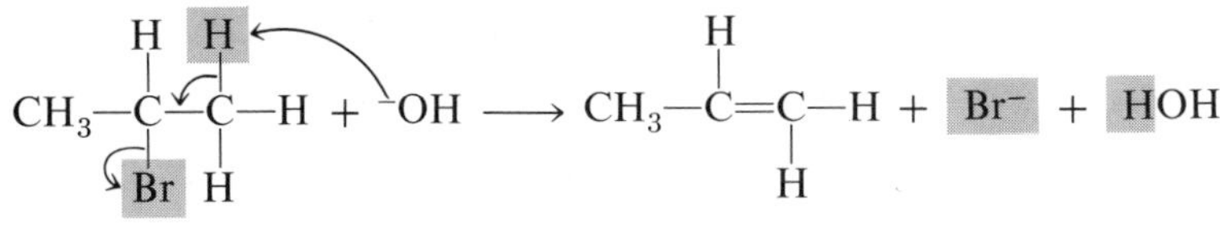

In the above example, all six of the hydrogen atoms adjacent to the halogen are equivalent, hence only one product is possible. When there is a case of nonequivalent hydrogens, more than one alkene may be formed. The predominant product will be the more highly substituted alkene (see also Section 17.4-b).

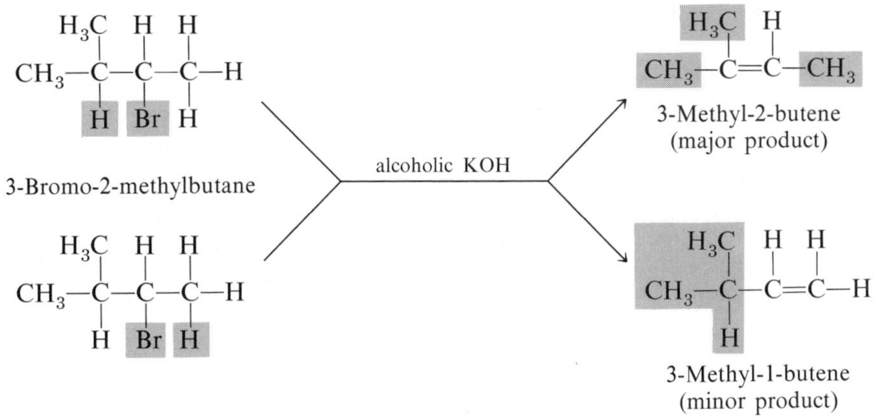
3-Bromo-2-methylbutane
alcoholic KOH
3-Methyl-2-butene
(major product)
3-Methyl-1-butene
(minor product)

The dehydrohalogenation reaction is useful for the preparation of some alkynes. For example, phenylacetylene may be synthesized by the dehydrobromination of styrene dibromide in the presence of fused potassium hydroxide.

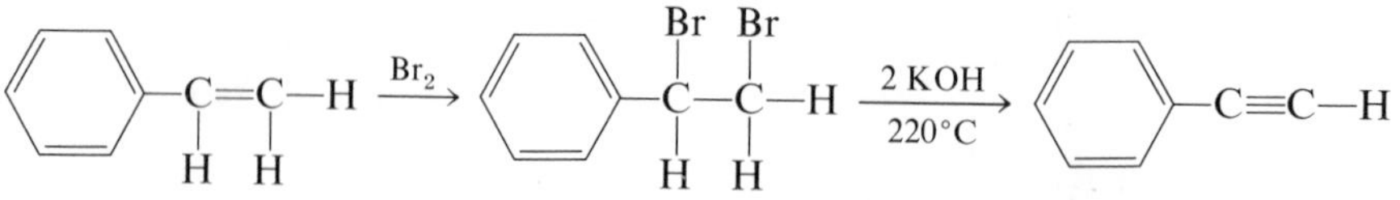

Phenylethene (Styrene) — 1,2-Dibromo-1-phenylethane (Styrene dibromide) — Phenylethyne (Phenylacetylene)

Several insect species have been found to be resistant to DDT (see Section 16.6). These insects have developed an enzyme, *DDT dehydrochlorinase,* that effects the dehydrochlorination of DDT and converts DDT into a noninsecticidal ethylene derivative (DDE).

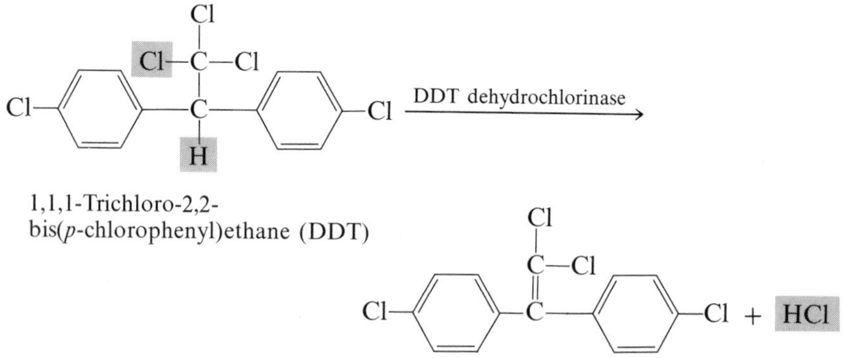
DDT dehydrochlorinase
+ HCl

1,1,1-Trichloro-2,2-bis(*p*-chlorophenyl)ethane (DDT)

1,1-Dichloro-2,2-bis(*p*-chlorophenyl)ethene (DDE)

b. Substitution

The halogen atom of alkyl halides may be displaced by a number of other negative groups to yield a wide variety of organic derivatives. These attacking groups are referred to as **nucleophiles** (nucleus-loving) or **nucleophilic reagents.** Nucleophilic reagents seek a center of positive charge, and therefore they are species that can donate a pair of electrons (that is, Lewis bases) in a chemical reaction. Common nucleophiles are negative ions or neutral molecules containing unshared electron pairs (for example, $\ddot{N}H_3$, $\ddot{P}(CH_3)_3$, $H_2\ddot{O}:$).

General Equation

$$R-CH_2-X + A^- \longrightarrow R-CH_2-A + X^-$$

where A = a nucleophile such as OH^-, OR^-, CN^-, NH_2^-, SH^-

Specific Equations

$$CH_3I + KCN \xrightarrow{\text{alcohol}} CH_3CN + KI$$

Methyl iodide — Methyl cyanide

$$CH_3CH_2CH_2Br + NaNH_2 \xrightarrow{NH_3} CH_3CH_2CH_2NH_2 + NaBr$$

n-Propyl bromide — *n*-Propylamine

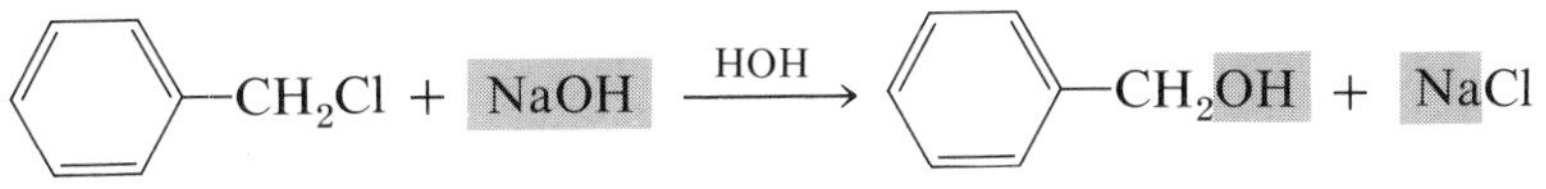

Benzyl chloride — Benzyl alcohol

Aromatic compounds containing halogen atoms in their side chain behave essentially like alkyl halides. On the other hand, the aryl halides, in which the halogen is directly bonded to a carbon of the aromatic ring, are very unreactive. They will not undergo the normal substitution reactions with nucleophilic reagents. However, if strongly electron-withdrawing groups (for example, the nitro group) are positioned ortho or para to the halogen atom, the halogen may be displaced. The following reaction was used by Frederick Sanger to determine the sequence of amino acids in proteins (see Chapter 25).

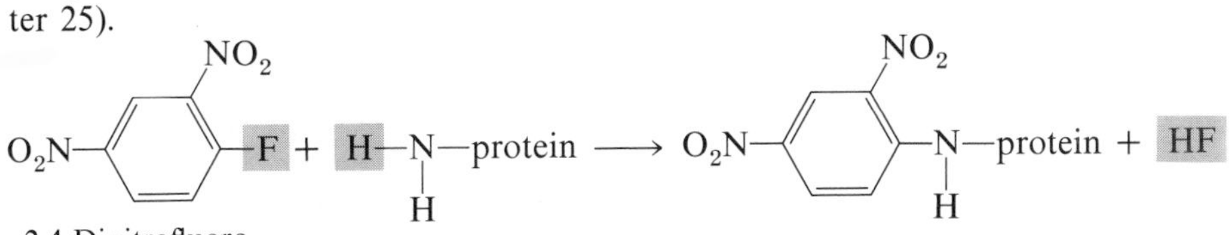

2,4-Dinitrofluoro-benzene

c. Reaction with Metals

Alkyl and aryl halides will react with lithium and magnesium to form extremely important organometallic compounds. These compounds are invaluable intermediates in the synthesis of many other organic compounds.

$$CH_3CH_2CH_2CH_2Br + 2\,Li \xrightarrow[\text{ether}]{\text{anhydrous}} CH_3CH_2CH_2CH_2Li + LiBr$$

n-Butyl bromide — *n*-Butyllithium

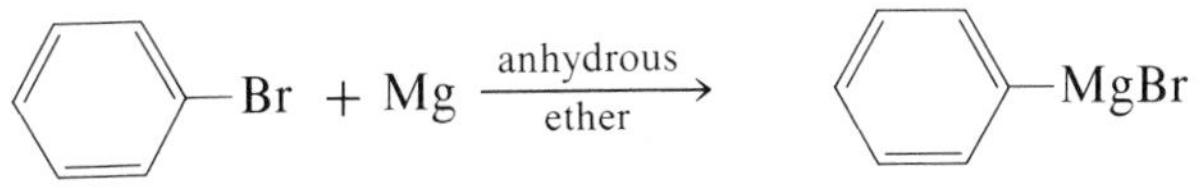

Bromobenzene — Phenylmagnesium bromide

The reaction of alkyl halides (or aryl bromides) with magnesium metal was discovered in 1901 by the French chemist Victor Grignard (Nobel Prize, 1912). The resulting alkyl or aryl halides are called *Grignard reagents* in honor of their discoverer. Grignard reagents and organic lithium compounds are not usually isolated, but rather are used in solution, and care must be taken to exclude water and oxygen. If traces of moisture are present, these reagents immediately react with the water to form the corresponding hydrocarbons. In fact, such a reaction is used in the preparation of small amounts of pure hydrocarbons. In subsequent chapters we shall see that these organometallic reagents are valuable synthetic tools for the preparation of many organic compounds.

$$CH_3CH_2CH_2CH_2Li + HOH \longrightarrow CH_3CH_2CH_2CH_3 + LiOH$$

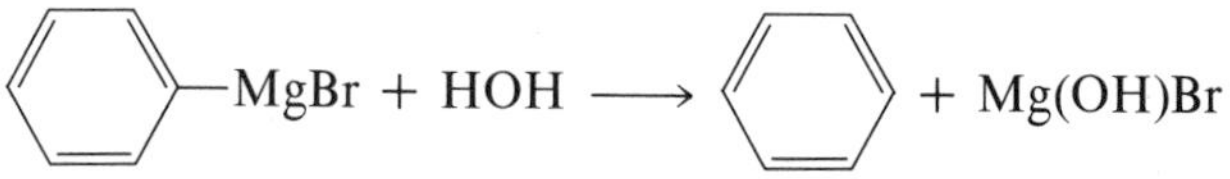
—MgBr + HOH ⟶
+ Mg(OH)Br

16.5 Uses of Halogenated Hydrocarbons

In addition to their value as intermediates in synthetic processes, organic halides have numerous commercial uses, for example, as solvents, refrigerants, aerosol sprays, antiseptics, and anesthetics. Table 16.3 lists some organic halides, their formulas, physical states, and chief uses.

Table 16.4, on the other hand, contains a listing of compounds that have been embroiled in controversy throughout the twentieth century. It includes toxic gases, metabolic poisons, plasticizers, aerosols, and herbicides. The one common feature of all of these compounds is that they contain carbon–halogen bonds. Organic halides, with but a few exceptions (e.g., the thyroid hormone, thyroxine; see Figure 25.2) are incompatible with living systems.

16.6 The Polychlorinated Pesticides

In 1962 Rachel Carson published the highly provocative book *Silent Spring* in which she provided dramatic examples of how the environment had been damaged by the indiscriminate use of chemicals.

> For the first time in the history of the world, every human being is now subjected to contact with dangerous chemicals, from the moment of conception until death. In the less than two decades of their use, the synthetic pesticides have been so thoroughly distributed throughout the animate and inanimate world that they occur virtually everywhere. They have been recovered from most of the river systems and even from streams of groundwater flowing unseen through the earth. Residues of these chemicals linger in soil to which they may have been applied a dozen years before. They have entered and lodged in the bodies of fish, birds, reptiles, and domestic and wild animals so universally that scientists carrying on animal experiments find it almost impossible to locate subjects free from such contamination. They have been found in fish in remote mountain lakes, in earthworms burrowing in soil, in the eggs of birds—and in man himself. For these chemicals are now stored in the bodies of the vast majority of human beings, regardless of age. They occur in the mother's milk, and probably in the tissues of the unborn child.

Table 16.3 Commercially Important Halogenated Hydrocarbons

Name	*Formula*	*Physical State*	*Remarks*
Carbon tetrachloride	CCl_4	Liquid	Solvent. CCl_4 is poisonous; should always be used with adequate ventilation.
Chloroform	$CHCl_3$	Liquid	Formerly used as inhalation anesthetic in surgery. It has a depressant action on the heart and on the respiratory center. Chloroform has been shown to cause cancer in laboratory mice and rats.
Iodoform	CHI_3	Yellow solid	Formerly used as an antiseptic for wounds; used in treatment of some skin diseases and skin ulcers.
Halothane	$CF_3CHClBr$	Gas	Inhalative anesthetic. It is effective and relatively nontoxic.
Freons	CF_2Cl_2 $CFCl_3$	Gas	Nontoxic, nonodorous, noncombustible; used as a refrigerant, and in aerosol spray propellants. Possible cause of some ozone depletion in the upper atmosphere.
Methyl chloride	CH_3Cl	Gas	Refrigerant.
Methyl bromide	CH_3Br	Gas	Toxic; used as a vermicide.
Ethyl chloride	C_2H_5Cl	Gas	External local anesthetic.
Teflon	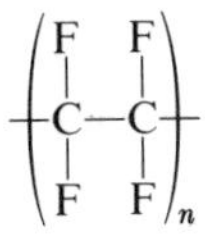	Waxy plastic	A polymer that is resistant to oxidation and corrosion. Used as an electrical insulator and as a liner in frying pans and other utensils to provide a nonsticking surface.
Trichloroethylene	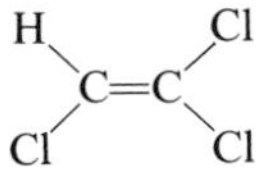	Liquid	Drycleaning solvents
Tetrachloroethylene	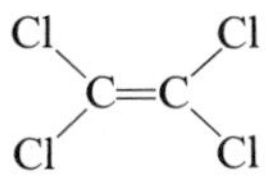	Liquid	Drycleaning solvents
p-Dichlorobenzene		Solid	Larvicide. It is extensively used in place of naphthalene to kill moths and caterpillars.
1,2-Dibromo-3-chloropropane (DBCP)	$CH_2BrCHBrCH_2Cl$	Liquid	Formerly used as a pre-emergence soil fumigant to kill nematodes. First shown to be a health hazard when workers in chemical plants became sterile due to exposure to DBCP. Fear that residue is retained in vegetables caused U.S. production to be halted.

Table 16.4 Toxic Organic Halides

Name	*Formula*	*Remarks*
Phosgene	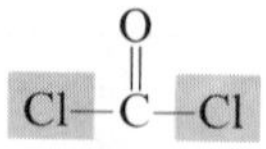	Lethal gas—developed during World War I. It attacks hydroxyl groups and amino groups of the proteins of the lung membranes, producing HCl and heat, which are responsible for its toxic effect.
Mustard gas	$Cl-CH_2CH_2-S-CH_2CH_2-Cl$	Skin inflammatory agent—its effect is to cause severe, slow-healing blistering on any part of the body that it contacts.
Diisopropylfluorophosphate (DFP)	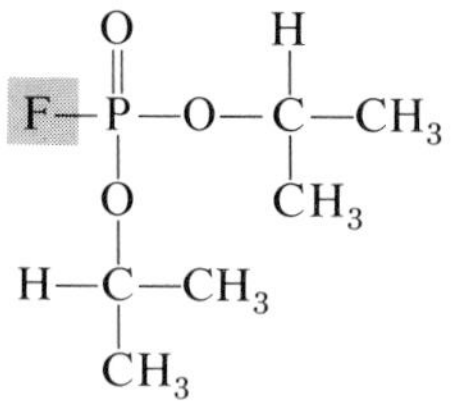	Nerve gas—combines with a specific hydroxyl group (of serine) situated at the active site of the enzyme *acetylcholine esterase,* the enzyme that functions in the transmittal of the nerve impulse. Deactivation of the enzyme by DFP disrupts coordination in the autonomic nervous system and in the muscles, causing tremors, convulsions, and death. The activity of DFP resembles that of the organophosphate pesticides (see page 504).
Phenacyl chloride	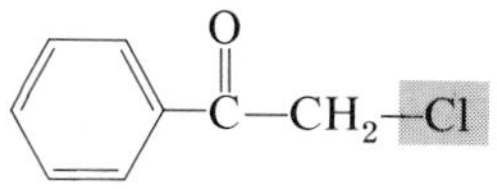	Lachrymator (tear gas)—used routinely in civilian police work and occasionally by the military.
Iodoacetate	$I-CH_2-COO^-$	Metabolic poison—combines with sulfhydryl groups (—SH) of enzymes and thereby deactivates the enzymes (see page 659).
Fluoroacetate	$F-CH_2-COO^-$	Metabolic poison—reacts with citric acid, an intermediate of the Krebs cycle (see Section 28.7), to form 2-fluorocitric acid, which is not further metabolized. Thus the principal effect is blockage of the Krebs cycle.

Table 16.4 (Continued)

Name	*Formula*	*Remarks*
Polychlorinated biphenyls (PCB's)	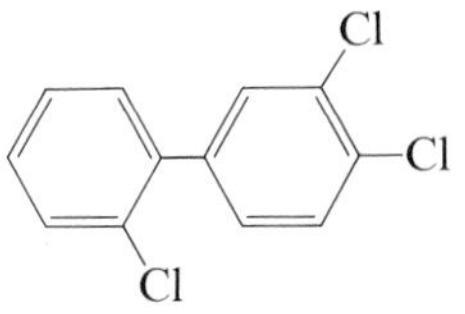 Partially chlorinated 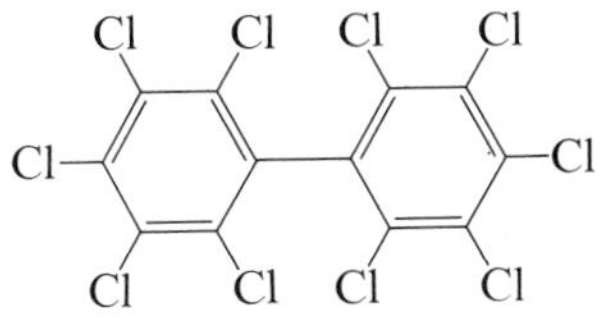 Completely chlorinated	There are 210 possible PCB's depending on the number of chlorine atoms bonded to the biphenyl structure. Commercial PCB's are various mixtures of these possibilities. PCB's have been used in lubricants, hydraulic fluids, waxes, paints, inks, and plastics. Their high heat capacities make them useful as heat transfer fluids and as insulators. PCB's are similar to DDT in that they are persistent in the environment, they accumulate in fatty tissue, and they are poisonous. In 1970 the principal American producer, the Monsanto Company, began restricting its sales of PCB's only for use in electrical equipment where there was not an acceptable substitute. In 1976 the company announced that it would phase out the production of PCB's.
Vinyl chloride	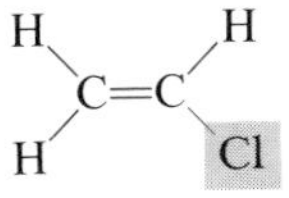	Formerly used as an aerosol propellant in household spray paints and finishes, protective and decorative coatings, paint removers, adhesives, and solvents. The compound has been linked to a form of liver cancer occurring among workers exposed to it in chemical plants. In 1974 vinyl chloride was banned from use in aerosol drugs, cosmetics, pesticides, and the above-mentioned household sprays. Vinyl chloride is still used in the production of polyvinyl chloride plastics.
Polyvinyl chloride (PVC)	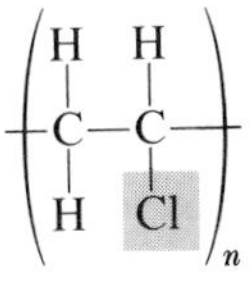	Plastic polymer made from vinyl chloride monomer units and used in phonograph records, plastic bottles, garden hoses, etc. Disposal of PVC products presents a problem because they undergo loss of HCl gas when heated in an incinerator. HCl is a corrosive irritant that significantly increases the air pollution

(Continued)

Table 16.4 (Continued)

Name	*Formula*	*Remarks*
		associated with municipal refuse incinerators.
2,4-Dichlorophenoxyacetic acid (2,4-D)	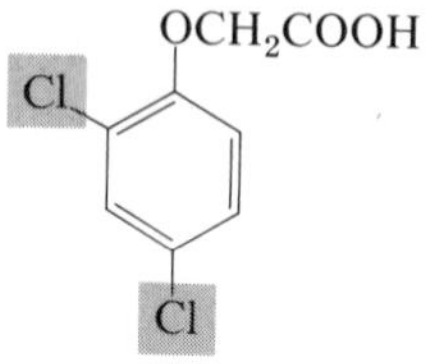	Both 2,4-D and 2,4,5-T are widely used as herbicides. They exert their effect by mimicking natural plant growth hormones. The cell walls take up water at a faster rate than normal, producing elongation in the stem, little or no root growth, and leaves that are deficient in chlorophyll. The cessation of normal physiological functions results in the death of the plant. The chief advantage of these compounds is a preferential attack on weeds without harm to grasses or trees. Both substances were used by the U.S. military in Vietnam to defoliate the jungle growth. Controversy has arisen over reports concerning increased birth defects in Vietnamese children in the sprayed areas. Manufacturers insist that these effects are due to an impurity in the production of 2,4,5-T. That compound is no longer applied on food crops, but is still used to control weeds in range lands.
2,4,5-Trichlorophenoxy acetic acid (2,4,5-T)	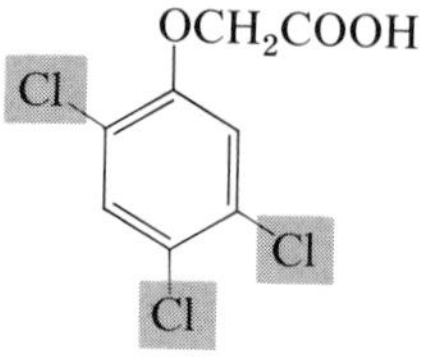	

Miss Carson focused her attention on two major groups of synthetic pesticides, the chlorinated hydrocarbons and the organic phosphates (see Section 21.3). We shall discuss the former group here with a view toward understanding their structure and the nature of their deadly action. Figure 16.2 contains the names and formulas of the common chlorinated hydrocarbon pesticides.

DDT (Dichlorodiphenyltrichloroethane[2])

DDT is formed by the reaction of chlorobenzene with chloral hydrate in the presence of concentrated sulfuric acid as a catalyst.

[2] Notice that this name is not a correct application of IUPAC nomenclature (see page 350). Commercial DDT is usually composed of 20–25% of the *o,p*-isomer and 75–80% of the *p,p*-isomer.

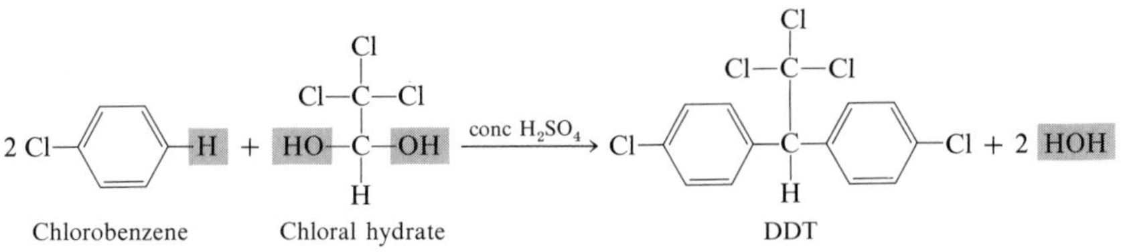
conc H2SO4
Chlorobenzene
Chloral hydrate
DDT

No other pesticide has been as assiduously studied as DDT. It was first synthesized in 1874 by Othmar Zeidler, a German chemist working on his doctoral thesis. Its properties as an insecticide went unnoticed until 1939 when Paul Mueller of

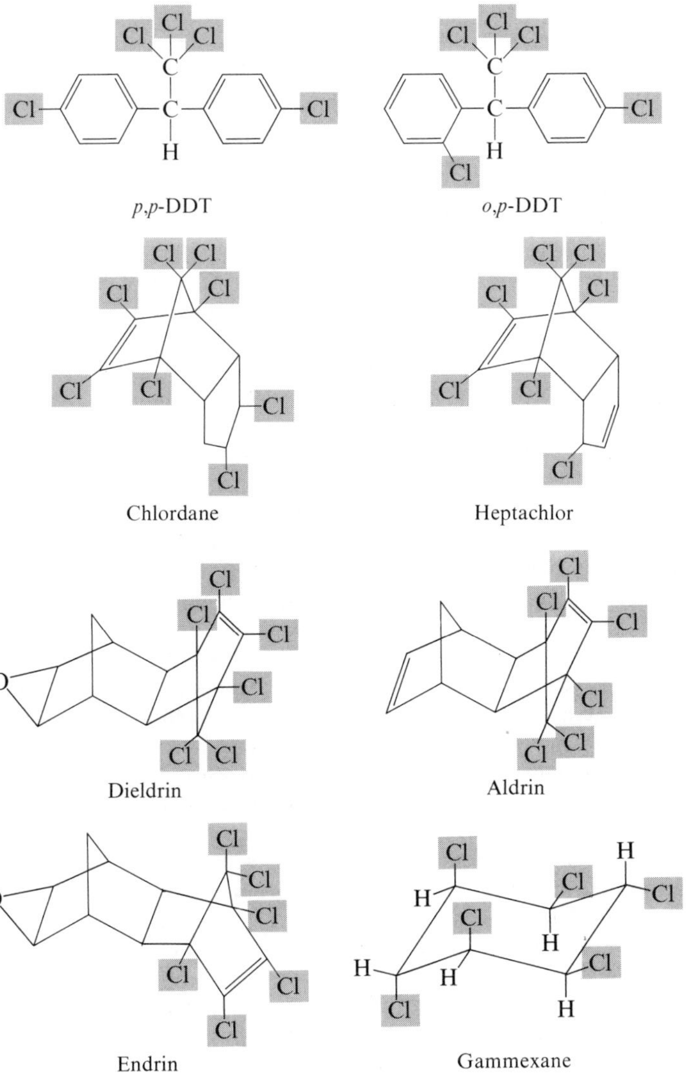
p,p-DDT
o,p-DDT
Chlordane
Heptachlor
Dieldrin
Aldrin
Endrin
Gammexane

Figure 16.2 The polychlorinated pesticides.

Switzerland observed its extreme toxicity to moths and houseflies. (Mueller was awarded the Nobel Prize in 1948 for discovering the insecticidal properties of DDT.) Almost immediately it was hailed as a life-saving protector against insects and insect-borne diseases. In 1943 U.S. military authorities aborted a typhus epidemic in Naples, Italy, by a massive dusting with DDT. The chemical has virtually eliminated the yellow fever mosquito in many South American countries. It has killed the black fly that transmitted a threadworm responsible for blindness in millions of Africans. Malaria had been accused of causing, either directly or indirectly, over one half of the world's human deaths.[3] Use of DDT on a world-wide basis resulted in a major control over the malarial mosquito, with a saving of millions of lives.

When Rachel Carson's book appeared, DDT was proclaimed to have permeated all living tissue. It has been found in eggs and in vegetables, especially the leafy varieties such as lettuce and spinach. The penguins in Antarctica, pheasants in California, salmon in Lake Michigan, and peregrine falcons in England have all been polluted with DDT and its analogs. There have been massive fish kills, the robin population has been severely depleted, and the bald eagle is in danger of extinction.

As yet, there has never been a proved instance of a human death having resulted from the proper use of DDT. Many people, therefore, regard the chemical as harmless to mankind. During World War II, thousands of soldiers, refugees and prisoners were dusted with DDT powder to kill the typhus-carrying louse. These individuals suffered no immediate ill effects of the chemical. The reason for this apparent harmlessness is that DDT, unlike other chlorinated hydrocarbons, is not readily absorbed through the skin when it is in *powder form*. However, it does find its way into the human system either through the nose or the mouth. When DDT is inhaled, it can be absorbed through the lungs; when it is ingested as a contaminant of food, it is slowly absorbed through the digestive tract. Once it has entered the body, it is stored in fatty tissue (probably because DDT is fat soluble) such as the adrenals, testes, or thyroid. Relatively large amounts are deposited in the brain, liver, and kidneys. Table 16.5 lists average DDT levels (in parts per million) in samples of human fat from various geographic locations.

The dangerous characteristics of DDT are: (1) The tremendous capability of DDT to spread throughout the world. Wind and water, migrating birds and fish carry it thousands of miles from the point of application. (2) DDT is one of the most persistent pesticides. It is insoluble in water and settles to the bottom of streams and lakes. When sprayed into the environment, it resists decomposition into its simpler constituents. The half-life of DDT in the soil is estimated at five to ten years. (3) DDT concentrates in living tissue. Each successive link in the food chain accumulates a much larger amount of DDT. Small fish absorb ten times more DDT than the plankton they feed upon and birds that eat these fish contain thirty times more DDT than the fish. (4) Numerous insects have developed a resistance to DDT. It is estimated that over two hundred insect species have become immune to the chemical. Some of these insects are found to contain an enzyme (DDT dehydrochlorinase) that converts DDT to a noninsecticidal ethylene derivative (Section 16.4-a). Other insects are able to form a nonlethal analog by replacing the tertiary hydrogen with a hydroxyl group. (In Section 21.3 we discuss the organophosphate pesticides that have been developed to combat insects that are resistant to DDT.)

[3] Louis L. Williams, "Pesticides: A Contribution to Public Health," in *Man—His Environment and Health,* a supplement to *American Journal of Public Health,* **54** (1), 34 (1964).

Table 16.5 Average DDT Levels in Human Fat Samples[a]

Geographic Area	*PPM*	*Geographic Area*	*PPM*
Arctic regions	3.0	Hungary	12.4
Canada	4.9	India (New Delhi)	26.0
Germany	2.3	Israel	19.2
England	3.9	United States	12.0
France	5.2		

[a] From Eugenia Keller, "The DDT Story," *Chemistry,* **43** (2), 12 (1970).

DDT and the other pesticides exert their effect by blocking the action of enzymes. Since the concentration of enzymes is relatively small, minute amounts of DDT can bring about vast bodily changes. In animal experiments, 3 parts per million has been found to inhibit an essential enzyme in heart muscle; only 5 parts per million has brought about necrosis or disintegration of liver cells. Animal experimentation has also indicated that the pesticides cause hormonal disturbances. In female rats, chlordane increased estradiol production by 385% and, in male pigeons, the rate of testosterone metabolism was doubled. DDT apparently interferes with the bicarbonate metabolism of birds. As a result they lay eggs that are too thin-shelled (about 20% thinner than normal) to survive the incubation period.

It has been suggested that thinning of eggshells and breakage of eggs are caused largely by the effects of DDT upon the enzyme *carbonic anhydrase* (see Section 31.5). This enzyme is found primarily in the red blood cells where it catalyzes the rapid formation and breakdown of bicarbonate ion at rates fast enough to meet requirements for CO_2 transport.

$$CO_2 + H_2O \rightleftharpoons H_2CO_3 \rightleftharpoons H^+ + HCO_3^-$$

Bicarbonate ions are the source of carbonate ions that combine with calcium ions in the oviduct to form the eggshell. Dr. Pocker (University of Washington) has found that DDT and other chlorinated compounds have the ability to coprecipitate minute amounts of the enzyme from solution, and hence reduce enzymic activity.[4]

The question is frequently asked: How reliable is extrapolation of animal data to man? We do not know the answer to this question. No animal species exists that handles pesticides in every respect like man does. Little is known about the chronic effects of continued and possibly increasing doses of pesticides over a long period of time. Until the cause and effect relationship between pesticide exposure and human disease is understood, it is necessary that we prevent their excessive use. At present, DDT and other chlorinated hydrocarbons are banned in the United States except for essential purposes, such as the control of disease of epidemic proportions or of a major infestation of crop insects. The reduction in DDT use in the United States has led to marked declines in DDT levels in fish, birds, food, and human fat tissue.[5]

[4] Other hypotheses for egg-shell thinning have also been postulated. Some researchers believe that the metabolite DDE exerts a more powerful inhibitory effect upon carbonic anhydrase than DDT. Others believe that decreased bird populations are a result of the induction by pesticides (and PCB's) of liver enzymes that lower the estrogen levels. See David B. Peakall, "Pesticides and the Reproduction of Birds," *Scientific American,* April 1970.

[5] David W. Johnston, "Decline of DDT Residues in Migratory Songbirds," *Science,* **186,** 841 (1974).

Other Chlorinated Hydrocarbons

Chlordane is a more potent pesticide than DDT. It has been used extensively in agricultural spraying, in gardens and in lawns. Like DDT, chlordane is not readily decomposed and persists for long periods of time in the soil, on foodstuffs, or on surfaces to which it may be applied. Unlike DDT, chlordane can enter the body through all available openings. It is absorbed through the skin, breathed in as a spray, and absorbed from the digestive tract. In common with all the other chlorinated hydrocarbons, its deposits build up in the body in a cumulative fashion. A diet containing only 2.5 parts per million of chlordane eventually leads to storage of 75 parts per million in the fat of experimental animals.

Heptachlor is a derivative of chlordane that is marketed as a separate agricultural formulation, often mixed with fertilizer to control pests in the soil. Heptachlor is also used against household pests, such as flies and mosquitoes. It has a particularly high capacity for storage in fatty tissue. A diet containing as little as 0.1 part per million will result in the deposit of measurable amounts of heptachlor in the body. In the soil and in the tissues of both plants and animals, it may be converted into another compound known as heptachlor epoxide. Tests on birds indicate that the epoxide is more toxic than the original compound.

Endrin is the most toxic of all the chlorinated hydrocarbons. It is fifteen times as poisonous as DDT to mammals, thirty times as poisonous to fish, and about three hundred times as poisonous to some birds. Endrine, like dieldrin and aldrin, is related structurally to chlorinated naphthalenes, a class of compounds that has been shown to cause hepatitis. Endrin had been used to combat soil and foliage insects.

Dieldrin[6] is a stereoisomer of endrin. It is about five times as toxic as DDT when swallowed but forty times as toxic when absorbed through the skin in solution. Dieldrin strikes quickly at the nervous system, sending the victim into convulsions. It had been substituted for DDT in malaria-control work in regions where the malaria mosquitoes had become resistant to DDT.

Aldrin[6] produces degenerative changes in the liver and kidneys. A quantity the size of an aspirin tablet can kill over 400 quail. Pheasants and rats that have been fed quantities too small to kill them produced fewer offspring and the majority of the offspring died within a week. Aldrin and dieldrin have been used in the control of forestry pests, where their long-lasting residual effect was considered advantageous. Recently, the use of dieldrin and aldrin as pesticides has been banned by the Environmental Protection Agency because of studies linking their use with increased liver cancers in humans.

Gammexane (Lindane) is prepared by the addition of three moles of chlorine to benzene. Nine different stereoisomeric 1,2,3,4,5,6-hexachlorocyclohexanes are pos-

[6] Dieldrin and aldrin are named for two German chemists, Otto Diels and Kurt Alder. Both men received the Nobel Prize in 1950 for discovering a very important organic reaction that bears their names. This reaction is utilized in the synthesis of these insecticides.

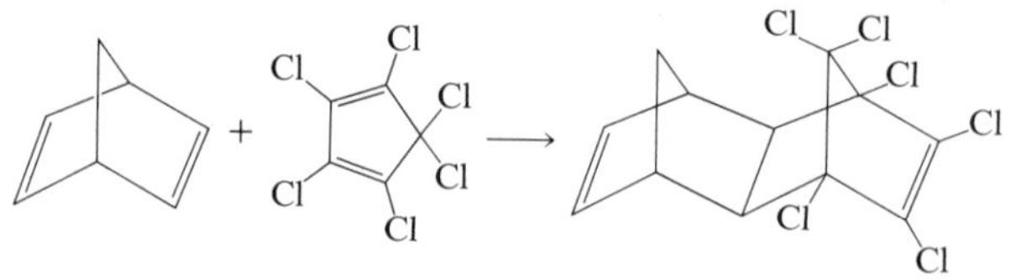
Cl
Cl
Cl
Cl
Cl
Cl
+
Cl
Cl
Cl
Cl
Cl
Cl

Aldrin

sible, and these are distinguished by the Greek letters α, β, γ, etc. It is remarkable that only the γ-isomer has insecticidal properties. Its stereochemical structure is given in Figure 16.2. Gammexane, like DDT, is a broad-spectrum insecticide, a poison active against many species of insects. It is used on fruits, vegetables, and cotton, in cattle sprays, and in the treatment of cut lumber. It is also used in many homes to eradicate flies. The most common use is in a vaporizer, which is essentially a small metal cup supported over a candle. A pellet of gammexane is placed in the cup and the heat of the burning candle melts the pellet, which then slowly evaporates. These devices have not been licensed by the federal government. Although the vapors keep rooms free from flies, the long-range effects upon humans have not been investigated.

Exercises

16.1 Draw and name all the isomers of $C_5H_{11}Br$.

16.2 Draw and name all the dichloro derivatives of *n*-butane and isobutane.

16.3 Name the following compounds.

(a) CHI_3

(b) CH_3CH_2MgBr

(c) $(CH_3)_3CBr$

(d) 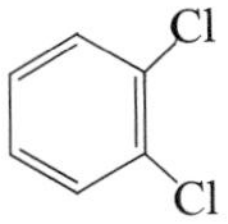
Cl
Cl

(e) $(CH_3)_2CFCH_2CH(C_2H_5)CH_2CHFCH_2CH_3$

(f) $CH_3CH_2CH{=}CHCHBrCH(CH_3)_2$

(g) $CH_3CH_2CHClCH(CH_3)CHBr_2$

(h) 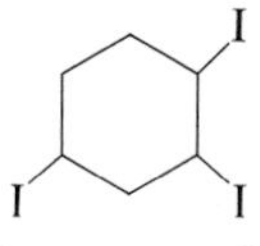
I
I
I

16.4 Write structural formulas for the following compounds.

(a) 3-chloropentane

(b) 1,2-dibromobutane

(c) *o*-iodotoluene

(d) 1,3,5-tribromocyclohexane

(e) chloroform

(f) DDT

(g) 1-chloro-2-methylpropane

(h) 2,3,5-triiodononane

(i) 2,5-difluoro-2,4,7-trimethyldecane

(j) 3-chlorocyclopentene

16.5 Using the appropriate reagents, write an equation for the preparation of each of the following compounds.

(a) 1-bromobutane
(b) 2-bromobutane
(c) 2,3-dichlorobutane
(d) 2-iodo-1-butene
(e) benzyl chloride
(f) *p*-chlorotoluene

16.6 Complete the following equations.

(a) $(CH_3)_3COH + HCl \longrightarrow \ ? \ \xrightarrow[\text{alcohol}]{\text{KOH}}$

(b) $(CH_3)_3CCl + Li \longrightarrow \ ? \ \xrightarrow{\text{HOH}}$

(c) 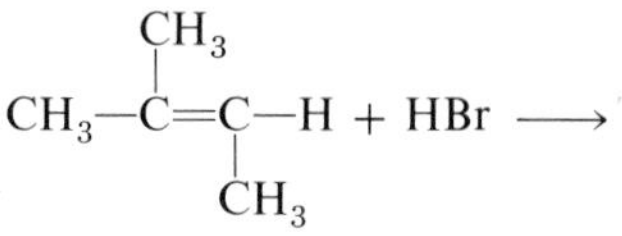
CH3
CH3—C=C—H + HBr →
CH3

(d) $CH_3CH_2I \xrightarrow[\text{HOH}]{\text{NaOH}}$

(e) 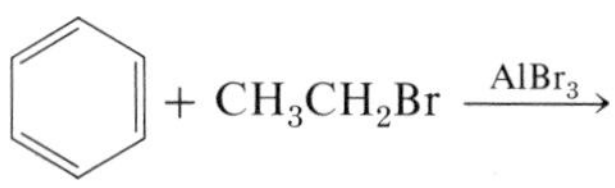
+ CH3CH2Br
AlBr3

(f) 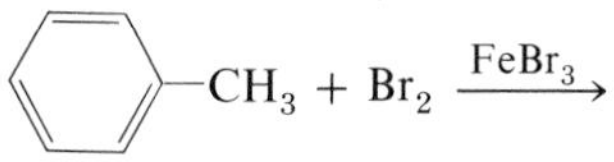
CH3 + Br2
FeBr3

(g) 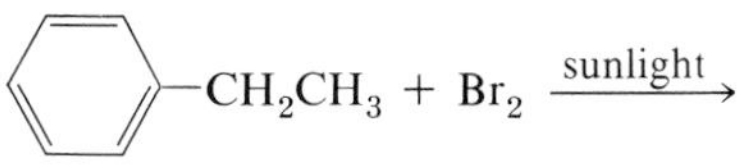
CH2CH3 + Br2
sunlight

(h) 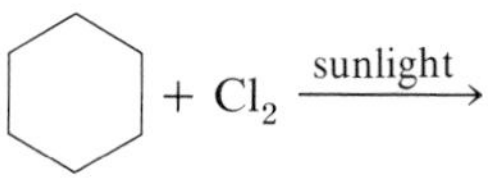
+ Cl2
sunlight

16.7 Show reaction sequences for the following conversions. Indicate reagents and reaction conditions. In some cases, more than one equation is required.

(a) $(CH_3)_2CHCH_2CH_2Br \longrightarrow (CH_3)_2CHCHBrCH_3$
(b) $CH_3CH_2CH_2Br \longrightarrow CH_3CH_2CH_3$
(c) $CH_2{=}CH_2 \longrightarrow CH_3CH_2NH_2$

(d) 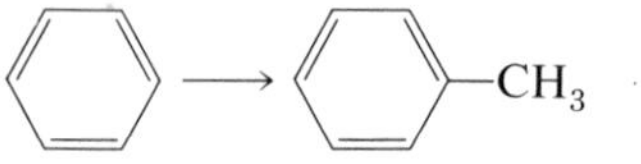
CH3

(e) 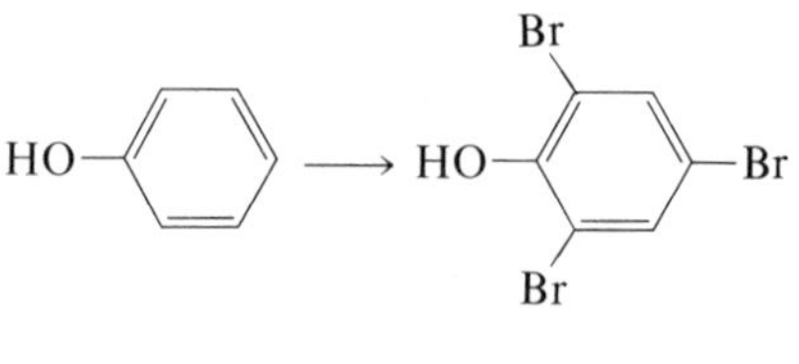
HO
HO
Br
Br
Br

16.8 Examine the labels of consumer products found in a supermarket (e.g., soaps, toothpaste, cleaning fluids, etc.) and list five that contain organic halogen compounds. Write the names of these compounds and their respective structural formulas.

16.9 Distinguish between the terms *antiseptic* and *anesthetic*. Give an example of each. What is the disadvantage of using chloroform as an anesthetic?

16.10 The analysis of an alkyl halide showed that it contained 55% chlorine. What is the empirical formula of this compound?

16.11 0.5 gram of an alkyl bromide was treated with lithium in anhydrous ether, followed by the addition of water to the reaction flask. Upon addition of the water, 91 ml (at STP) of a nonreactive gas was evolved. Propose two possible structures for the original alkyl bromide.

17

Alcohols and Ethers

Conceptually, both alcohols and ethers can be thought of as derivatives of water. When one of the hydrogen atoms of a water molecule is replaced by an alkyl group, an alcohol results.[1] The general formula for this class of compounds is **R—OH.** When both hydrogen atoms of water are replaced by alkyl or aryl groups, the product that is formed is an ether. The general formula for all ethers, then, is **R—O—R′**. The functional group of the alcohols is the hydroxyl group, and the functional group for the ethers is the [**—O—**] unit, or the ether group.

Alcohols

17.1 Classification and Nomenclature of Alcohols

The properties of alcohols depend upon the structural arrangement of the carbon atoms in the molecule. Alcohols may be divided into three classes which serve to distinguish these different structural arrangements from one another. The classes are known as primary, secondary, and tertiary, and alcohols are placed into these classes according to the type of carbon atom to which the hydroxyl group is attached.

1. A primary (1°) carbon atom is a carbon atom that is attached to only one other carbon.

[1]Substitution of an aryl group for one hydrogen atom of water yields a phenol.

Example

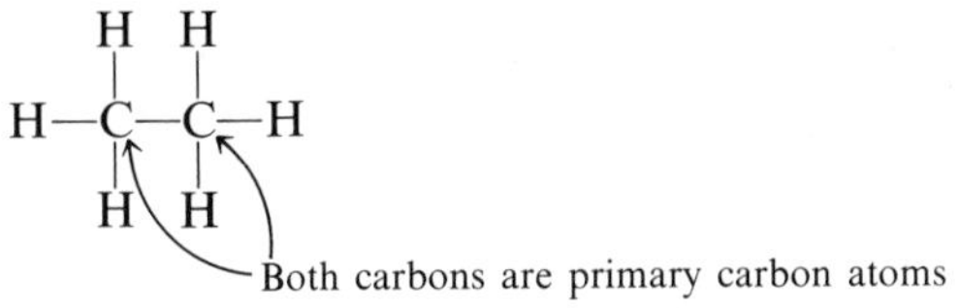
Both carbons are primary carbon atoms

2. A secondary (2°) carbon atom is one that is attached to two other carbon atoms.

Example

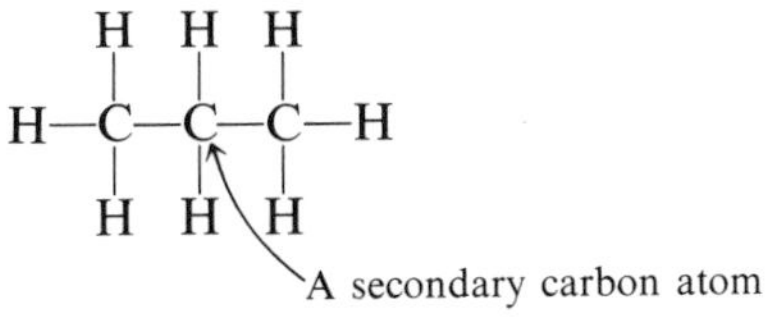
A secondary carbon atom

3. A tertiary (3°) carbon atom is one that is attached to three other carbon atoms.

Example

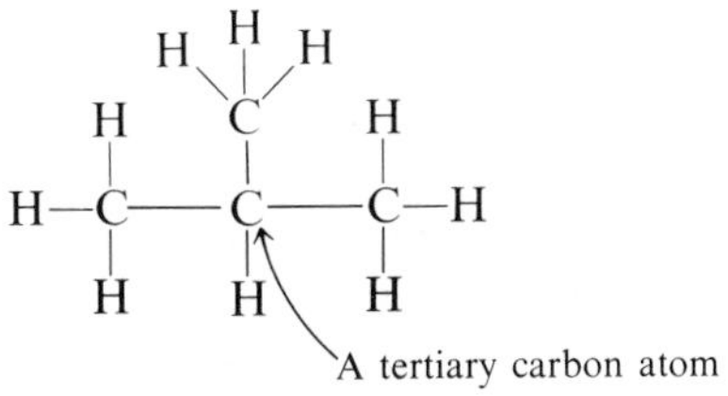
A tertiary carbon atom

Therefore, a *primary alcohol* is one in which the hydroxyl group is attached to a primary carbon atom, a *secondary alcohol* is one whose hydroxyl group is located on a secondary carbon atom, and a *tertiary alcohol* has its hydroxyl group bonded to a tertiary carbon. Table 17.1 lists the nomenclature and classification of some of the simpler alcohols.

As seen in the table, the common names of the lower members of the alcohol series are named in a manner analogous to that used for the naming of the alkyl halides. The name of the alkyl group is followed by the word alcohol. The IUPAC system is adapted for use in naming the higher homologs. In the IUPAC system, the designations primary, secondary, and tertiary are unnecessary and have no significance. Alcohols are named under the IUPAC system as follows.

1. The longest continuous chain of carbons containing the —OH group is taken as the parent compound.

Table 17.1 Classification and Nomenclature of Some Alcohols

Structural Formula	*Type of Alcohol*	*Common Name*	*IUPAC Name*
CH_3OH	Primary	Methyl alcohol	Methanol
CH_3CH_2OH	Primary	Ethyl alcohol	Ethanol
$CH_3CH_2CH_2OH$	Primary	*n*-Propyl alcohol	1-Propanol
$CH_3\overset{OH}{\overset{\mid}{C}}HCH_3$	Secondary	Isopropyl alcohol	2-Propanol
$CH_3CH_2CH_2CH_2OH$	Primary	*n*-Butyl alcohol	1-Butanol
$CH_3CH_2\overset{OH}{\overset{\mid}{C}}HCH_3$	Secondary	*sec*-Butyl alcohol	2-Butanol
$CH_3\overset{CH_3}{\overset{\mid}{C}}HCH_2OH$	Primary	Isobutyl alcohol	2-Methyl-1-propanol
$CH_3-\overset{OH}{\underset{CH_3}{C}}-CH_3$	Tertiary	*t*-Butyl alcohol	2-Methyl-2-propanol
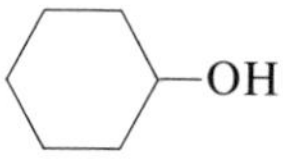	Secondary	Cyclohexyl alcohol	Cyclohexanol

2. The chain is numbered from the end closer to the hydroxyl group. The number which then appropriately indicates the position of the hydroxyl group is prefixed to the name of the parent hydrocarbon.
3. The *-e* ending of the parent alkane is replaced by the suffix *-ol*, the generic ending for the alcohols.
4. If more than one hydroxyl group appears in the same molecule (polyhydroxy alcohols) the suffixes diol, triol, etc., are used. In these cases the *-e* ending of the parent alkane is retained.

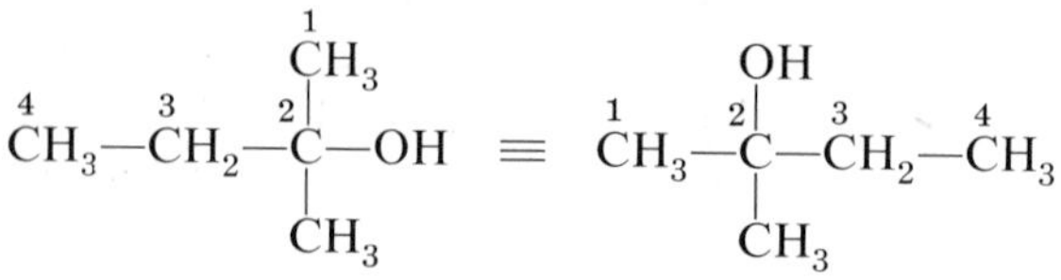

2-Methyl-2-butanol

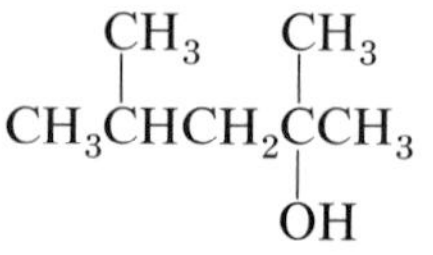

2,4-Dimethyl-2-pentanol

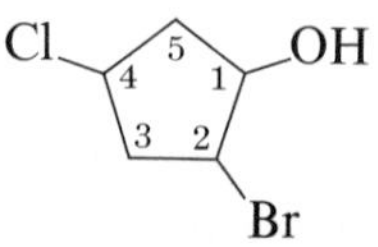

2-Bromo-4-chlorocyclopentanol

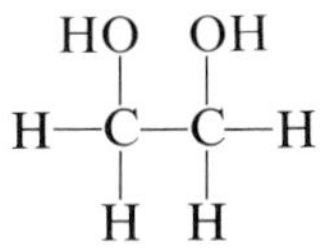

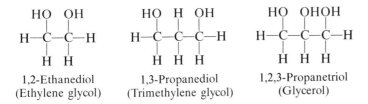

1,2-Ethanediol
(Ethylene glycol)

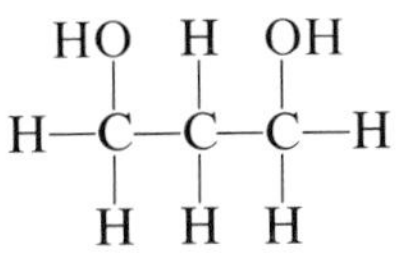

1,3-Propanediol
(Trimethylene glycol)

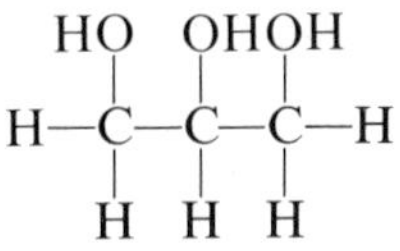

1,2,3-Propanetriol
(Glycerol)

The commercially important phenols are usually referred to by their common names rather than by their systematic names.

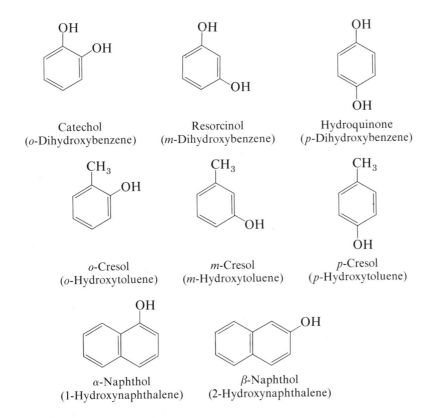

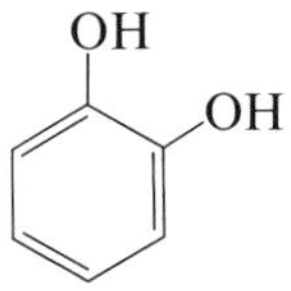

Catechol
(*o*-Dihydroxybenzene)

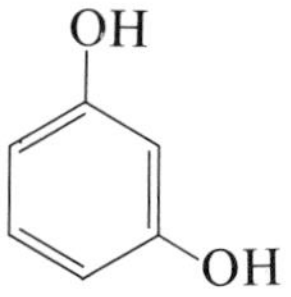

Resorcinol
(*m*-Dihydroxybenzene)

Hydroquinone
(*p*-Dihydroxybenzene)

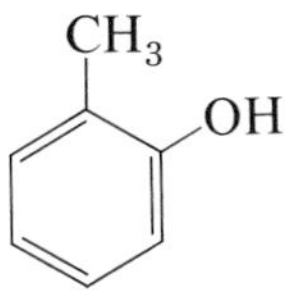

o-Cresol
(*o*-Hydroxytoluene)

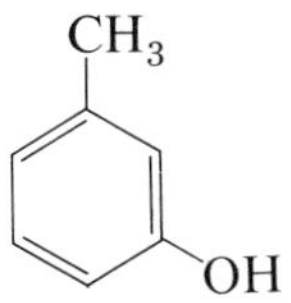

m-Cresol
(*m*-Hydroxytoluene)

p-Cresol
(*p*-Hydroxytoluene)

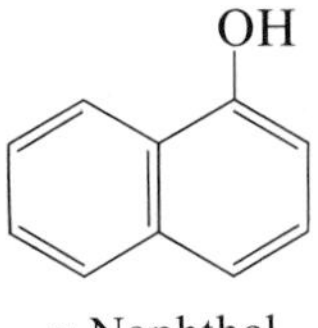

α-Naphthol
(1-Hydroxynaphthalene)

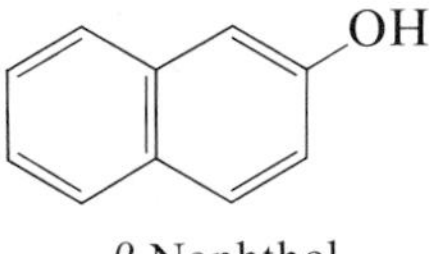

β-Naphthol
(2-Hydroxynaphthalene)

17.2 Preparation of Alcohols

In studying the reactions of the alkenes and the alkyl halides, we have already encountered reactions that are useful for the preparation of alcohols. The following examples, then, should serve as a review for the student. This is very often the case in organic chemistry. A typical reaction of one class of compounds is the preparative reaction of another class. In fact, a thorough knowledge of organic chemistry requires a knowledge of the interconversion of the various classes of organic compounds. In addition to the following methods of preparation, special reactions for the preparation of methyl and ethyl alcohols are discussed in Section 17.5.

a. Hydration of Alkenes

In Section 14.4-b the addition of sulfuric acid across the double bond of an alkene was illustrated. If the alkylsulfuric acid that results from this reaction is treated with water, an alcohol is formed. The $-OSO_3H$ group is replaced by the $-OH$ from the water.

$$R-\overset{\overset{H}{|}}{C}=\overset{\overset{H}{|}}{C}-H \;+\; \boxed{HOSO_3H} \;\longrightarrow\; R-\overset{\overset{H}{|}}{\underset{\underset{\boxed{HO_3SO}}{|}}{C}}-\overset{\overset{H}{|}}{\underset{\underset{\boxed{H}}{|}}{C}}-H$$

$$R-\overset{\overset{H}{|}}{\underset{\underset{HO_3SO}{|}}{C}}-\overset{\overset{H}{|}}{\underset{\underset{H}{|}}{C}}-H \;+\; \boxed{HOH} \;\longrightarrow\; R-\overset{\overset{H}{|}}{\underset{\underset{\boxed{HO}}{|}}{C}}-\overset{\overset{H}{|}}{\underset{\underset{H}{|}}{C}}-H \;+\; \boxed{HOSO_3H}$$

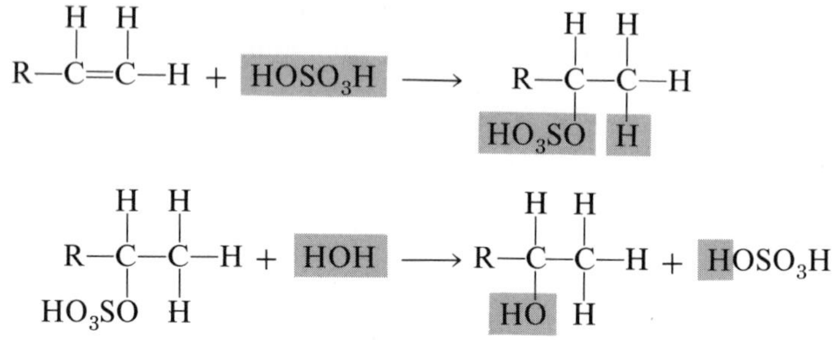

Since the sulfuric acid is regenerated, it acts as a catalyst, and the entire reaction can be represented by the following single equation.

$$R-\overset{\overset{H}{|}}{C}=\overset{\overset{H}{|}}{C}-H \;+\; \boxed{HOH} \;\xrightarrow{H_2SO_4}\; R-\overset{\overset{H}{|}}{\underset{\underset{\boxed{HO}}{|}}{C}}-\overset{\overset{H}{|}}{\underset{\underset{\boxed{H}}{|}}{C}}-H$$

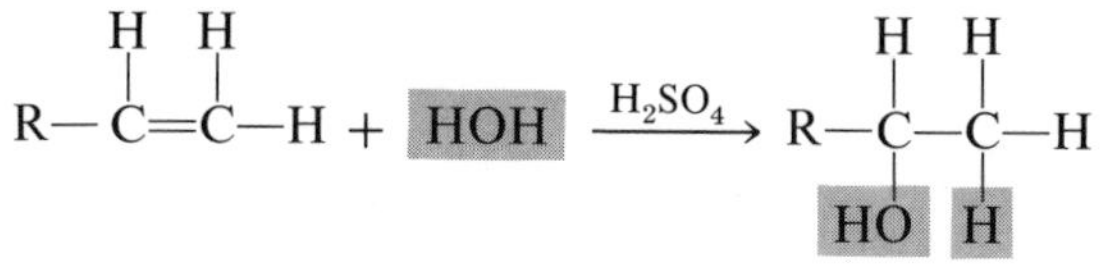

When the reaction is represented in this manner, it appears as though a direct addition of the H and OH from a molecule of water has taken place across the double bond of the alkene. This particular representation of the reaction is deceiving because the composition of the products does not give a correct indication of the mode (mechanism) of reaction. Notice that if the reacting alkene in the above reaction were ethylene (if R = H), the resulting alcohol would be ethyl alcohol. Ethyl alcohol is the only primary alcohol that can be prepared by this method. Why? (Hint—Recall Markownikoff's rule.) Because the alkenes are so readily available from petroleum, this reaction accounts for the major production of industrial ethyl alcohol and isopropyl alcohol (rubbing alcohol). Direct hydration of alkenes is an important industrial process and it occurs quite frequently in living systems (e.g., see pages 729 and 745). However, it is seldom utilized as a laboratory preparation of alcohols.

$$CH_3-\overset{\overset{H}{|}}{C}=CH_2 \;\xrightarrow[H_2SO_4]{HOH}\; CH_3-\overset{\overset{H}{|}}{\underset{\underset{OH}{|}}{C}}-CH_3$$

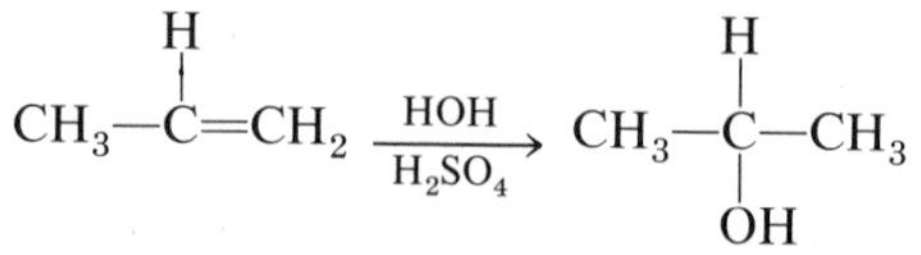

Propylene   Isopropyl alcohol

b. Hydrolysis of Alkyl Halides

It has been shown (Section 16.4-b) that the products of the hydrolysis of alkyl halides are dependent upon many factors, among them, reaction conditions. In order to prepare an alcohol by the hydrolysis of an alkyl halide, experimental conditions must be sought that will maximize the hydrolysis (substitution) reaction and minimize the elimination reaction. One method that accomplishes this involves the use of dilute aqueous solutions of sodium or potassium hydroxide.

$$\text{R—X} + \text{HOH} \xrightarrow[\text{NaOH}]{\text{dilute}} \text{R—OH} + \text{HX}$$

Another method involves the use of a finely divided suspension of silver hydroxide in moist ether.

$$\text{R—X} + \text{AgOH} \xrightarrow[\text{ether}]{\text{moist}} \text{R—OH} + \text{AgX}$$

The production of phenols from aryl halides requires the use of more drastic conditions because the aryl halides are very stable. The initial product, sodium phenoxide, is converted to phenol by acidification with hydrochloric acid.

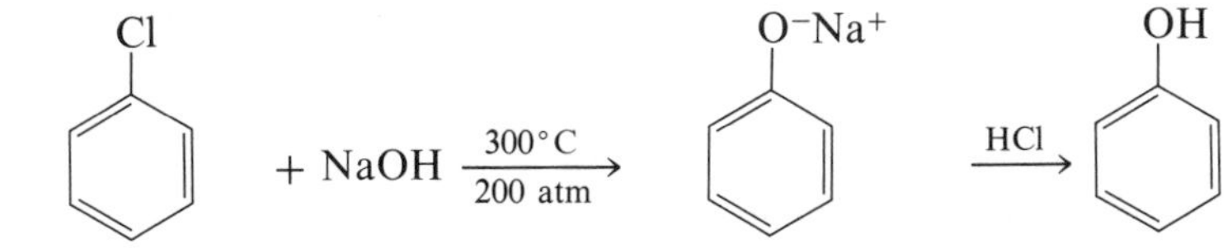
Cl
+ NaOH
300°C
200 atm
O⁻Na⁺
HCl
OH

Chlorobenzene   Sodium phenoxide   Phenol

17.3 Physical Properties of Alcohols

It was earlier stated that the alcohols can be considered to be derivatives of water. This relation becomes particularly apparent, especially for the lower homologs, in a discussion of the physical and chemical properties of the alcohols. Remember that methane, ethane, and propane are gases, and are insoluble in water. In contrast, methanol, ethanol, and propanol are liquids and are completely miscible with water. Therefore, replacement of a single hydrogen atom with a hydroxyl group brings about marked changes in solubility and physical state of the molecules. These differences result from the hydrogen-bonding capabilities of the alcohols. Hydrogen bonding is responsible for the intermolecular attractions between alcohol molecules, and hence even the lightest homolog of the series exists as a liquid at room temperature. Alcohols can form hydrogen bonds with water molecules as

Table 17.2 Comparison of Boiling Points and Molar Masses

Formula	*Name*	*M*	*BP* (°C)
CH_4	Methane	16	−161
HOH	Water	18	100
C_2H_6	Ethane	30	−89
CH_3OH	Methanol	32	65
C_3H_8	Propane	44	−45
CH_3CH_2OH	Ethanol	46	78
n-C_4H_{10}	n-Butane	58	−0.5
$CH_3CH_2CH_2OH$	1-Propanol	60	97

well, and so the lower homologs of the series are water soluble. The concept of hydrogen bonding is of extreme importance to the organic chemist as well as to the biochemist. Table 17.2 lists the molar masses and the boiling points of some common compounds. An examination of the table reveals that substances having similar molar masses do not always have similar boiling points.

In the case of the alcohols, the relatively high boiling points are a direct result of strong intermolecular attraction. Recall that a boiling point is a rough measure of the amount of energy necessary to separate a liquid molecule from its nearest neighbors. If the nearest neighbors of a molecule are associated to that molecule by means of hydrogen bonding, a considerable amount of energy must be supplied to break the hydrogen bonds. Only then can an individual molecule escape from the liquid into the gaseous state.

Figure 17.1 illustrates hydrogen bonding in water and in the alcohols. Reference to this schematic representation reveals the reasons why water boils at a higher temperature than methyl alcohol, even though water is a lighter molecule. The oxygen atom and both hydrogen atoms of the water molecule participate in hydrogen bonding to three, or even four, adjacent

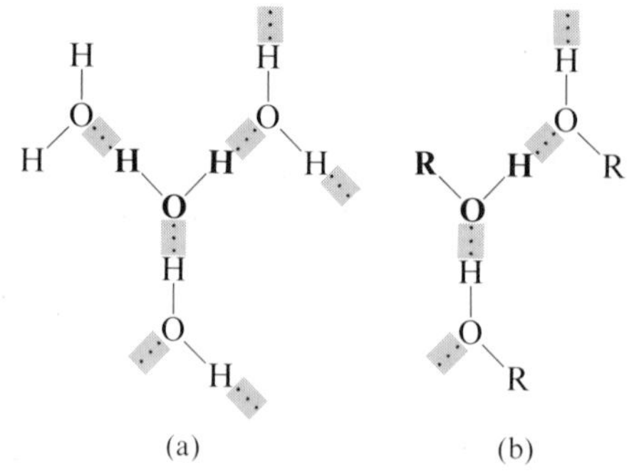
(a)
(b)

Figure 17.1 Intermolecular hydrogen bonding in water (a) and alcohol (b).

water molecules (see also Figure 6.3). Alcohols result from the replacement of one hydrogen atom of water with an alkyl group. The alkyl group does not participate in hydrogen bonding, and so the alcohol is associated to only two other alcohol molecules. More energy is required to disrupt three or four intermolecular bonds than two, and thus greater energy is needed to vaporize the water. (The energy required to break a hydrogen bond is about 5 kcal/mole. While this is distinctly less than the energy required to break any of the intramolecular bonds in water or alcohol, it is still an appreciable amount of energy and shows its significance in the boiling point differences.)

As mentioned previously (Section 7.3-b), polarity and hydrogen bonding are significant factors in the water solubility of alcohols. A common expression among chemists is "like dissolves like" (the word *like* refers to relative polarities). This rule of thumb implies the useful generalization that polar solvents will dissolve polar solutes, and nonpolar solvents will dissolve nonpolar solutes. Care must be taken, however, not to apply this generalization haphazardly to all cases. All alcohol molecules are polar, yet not all alcohols are water soluble. On the other hand, all alcohols are soluble in most of the common nonpolar solvents (carbon tetrachloride, ether, benzene). Only the lower homologs of the series have an appreciable solubility in water. As the length of the carbon chain increases, water solubility decreases.

The differences in water solubility can be explained in the following manner. The hydroxyl group confers polarity and water solubility upon the alcohol molecule. The alkyl group confers nonpolarity and water insolubility. Whenever the hydroxyl group comprises a substantial portion of a molecule, the molecule will be water soluble. (The hydroxyl group can be thought of as dragging the remainder of the molecule into the water structure.)

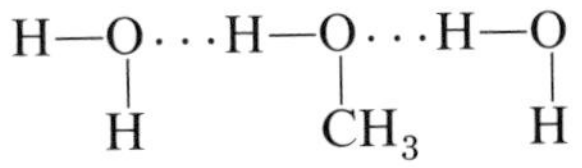

CH_3OH has more than 50% of its mass (17/32) located in the hydroxyl portion of the molecule. CH_3CH_2OH has more than 37% of its mass (17/46) located in the hydroxyl group. Both have infinite water solubility. As the carbon chain of monohydroxy alcohols increases, the contribution of the hydroxyl group to the total mass of the molecule decreases and soon becomes overwhelmed by the mass of the carbon chain. Consider *n*-decyl alcohol, in which the hydroxyl group is only 17/158 of the mass of the entire molecule. In this case the hydroxyl group is almost totally ineffective in dragging the large alkane portion of the molecule into the water structure (Figure 17.2). Therefore, *n*-decyl alcohol behaves more like an alkane than an alcohol in many of its physical properties.

$$\underbrace{CH_3CH_2CH_2CH_2CH_2CH_2CH_2CH_2CH_2CH_2}_{\text{Alkane portion}}OH$$

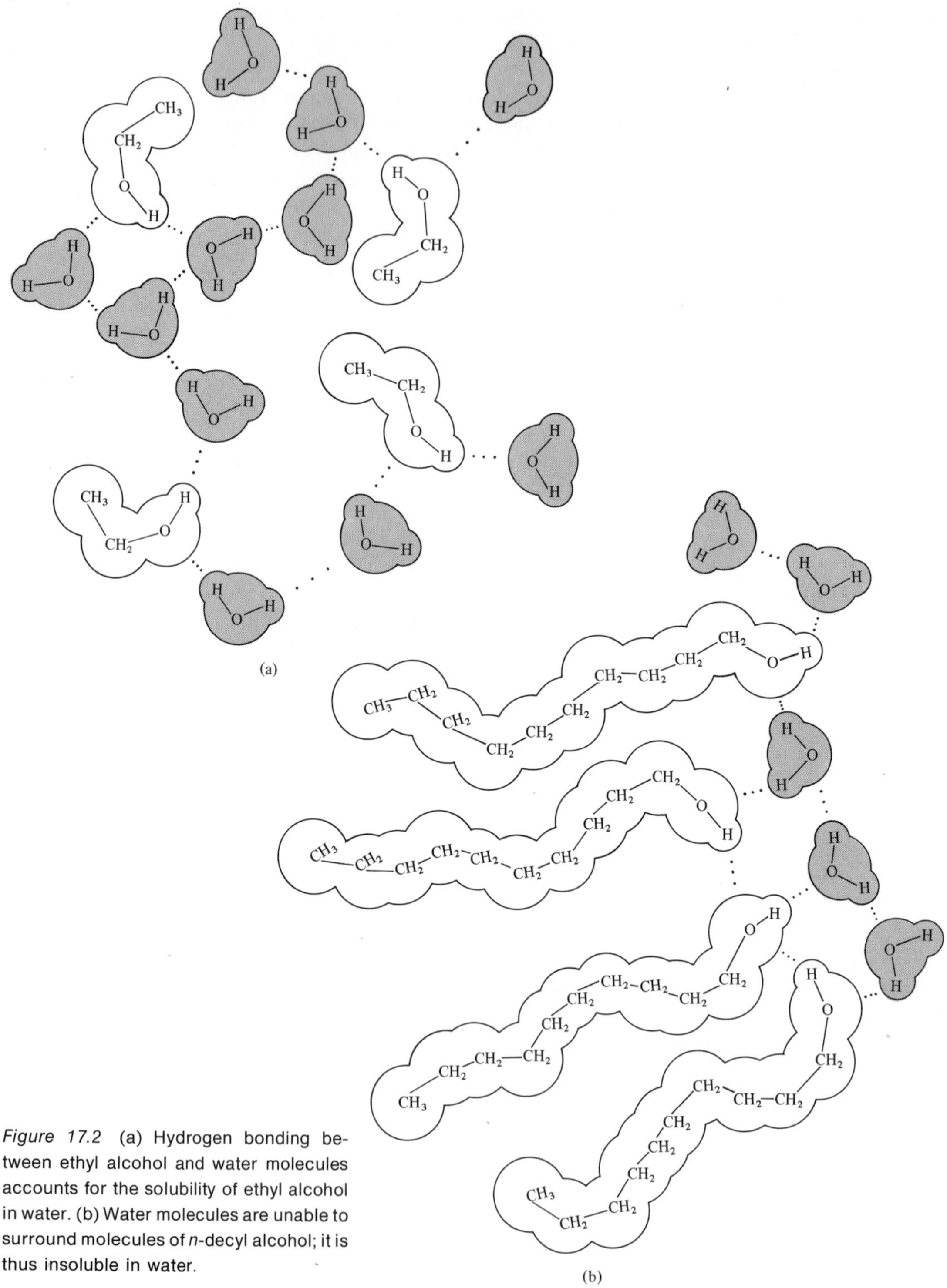

(a) (b)

Figure 17.2 (a) Hydrogen bonding between ethyl alcohol and water molecules accounts for the solubility of ethyl alcohol in water. (b) Water molecules are unable to surround molecules of *n*-decyl alcohol; it is thus insoluble in water.

Table 17.3 Solubilities of the Butyl Alcohols in Water

Alcohol	*Formula*	*Solubility grams/100 grams of water*
n-Butyl	$CH_3CH_2CH_2CH_2OH$	8
Isobutyl	$(CH_3)_2CHCH_2OH$	11
sec-Butyl	$CH_3CH_2CH(OH)CH_3$	12.5
t-Butyl	$(CH_3)_3COH$	Completely soluble

Consider Table 17.3, which contains the solubilities of the butyl alcohols in water. The large discrepancies in the water solubilities of these isomeric alcohols cannot be attributed to differences in molar mass or mass percent of the hydroxyl group. The differing solubilities are a result of the different geometrical shapes of the alcohols. The very compact *t*-butyl alcohol molecules experience weaker intermolecular van der Waals attractions and are more easily surrounded by water molecules. Hence, *t*-butyl alcohol has the highest solubility in water and lowest boiling point (83°C) of any of its isomers (all of which boil above 100°C).

In summary, solubility considerations involve the balance of polar and nonpolar groups within a molecule, as well as molecular shape. The more polar a molecule and the more compact its shape, the greater will be its water solubility.

17.4 Chemical Properties of Alcohols

a. Acidity and Basicity

The acid-base properties of the aliphatic alcohols are similar to those of water. That is, upon reaction with a strong acid these alcohols will accept protons (i.e., they act as bases) to form positively charged analogs of the hydronium ion. *Protonated alcohols are called* **oxonium ions.**

$$H-\ddot{\underset{\cdot\cdot}{O}}-H + H_2SO_4 \rightleftharpoons H-\underset{\underset{H}{|}}{\overset{\cdot\cdot\oplus}{O}}-H + HSO_4^-$$

Hydronium ion

$$CH_3-\ddot{\underset{\cdot\cdot}{O}}-H + H_2SO_4 \rightleftharpoons CH_3-\underset{\underset{H}{|}}{\overset{\cdot\cdot\oplus}{O}}-H + HSO_4^-$$

An oxonium ion

The acid dissociation constants (K_a) for methanol and ethanol are comparable to that of water, whereas the higher molar mass alcohols are less acidic. Phenols, on the other hand, are more acidic than water and alcohols because of the resonance stabilization of the phenoxide ion.

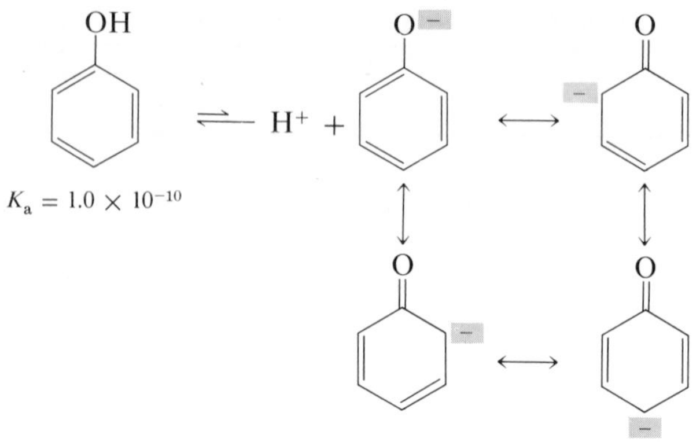

The acidity of phenols is a characteristic property that differentiates them from alcohols. Phenols can be neutralized by strong bases, whereas alcohols do not react with base.

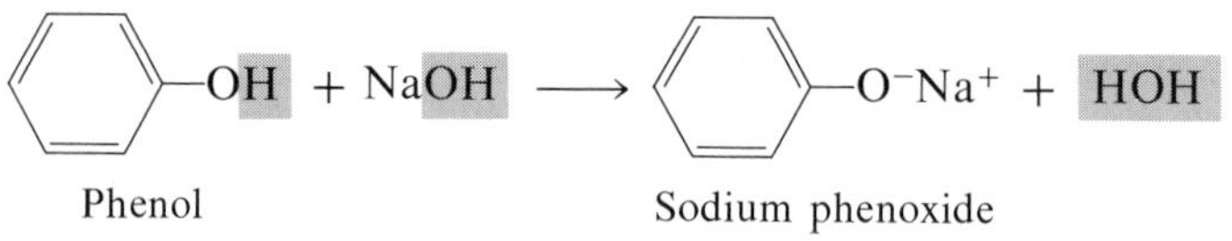

$$CH_3CH_2OH + NaOH \longrightarrow \text{No reaction}$$

Electron-withdrawing groups attached to the benzene ring enhance the acidity of phenols. 2,4,6-Trinitrophenol (picric acid), in which all three nitro groups help in delocalizing the negative charge, is almost as strong as hydrochloric acid. It readily reacts with weak bases to form salts.

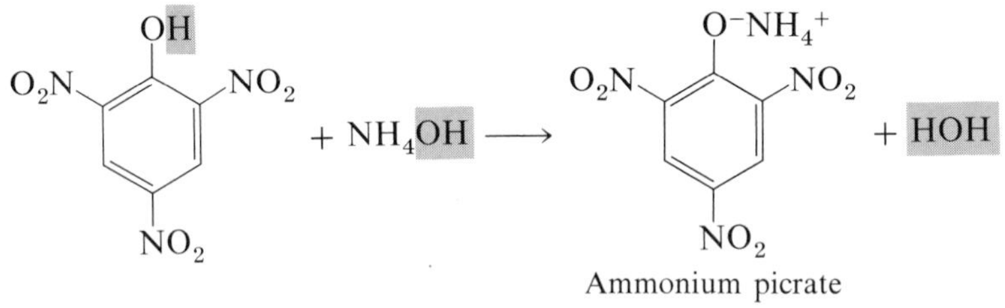

b. Dehydration of Alcohols (Preparation of Alkenes or Ethers)

The dehydration of alcohols was cited in Section 14.2 as a method for the preparation of alkenes. We shall have occasion to mention it again in our later discussions of carbohydrate and fatty acid metabolism. The reaction involves intramolecular dehydration, represented by the following general equation.

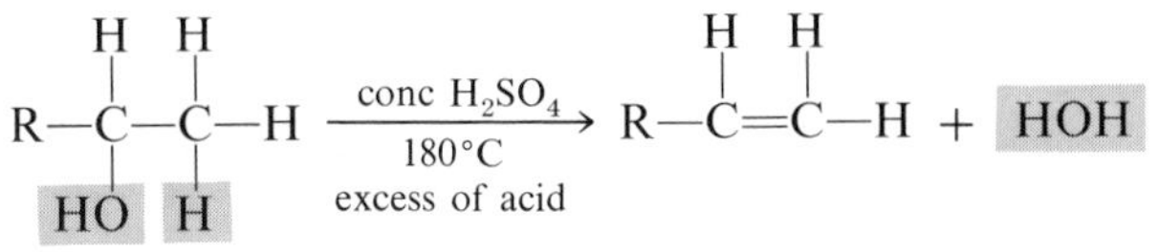

The first step in the reaction is the protonation of the alcohol to form an oxonium ion intermediate.

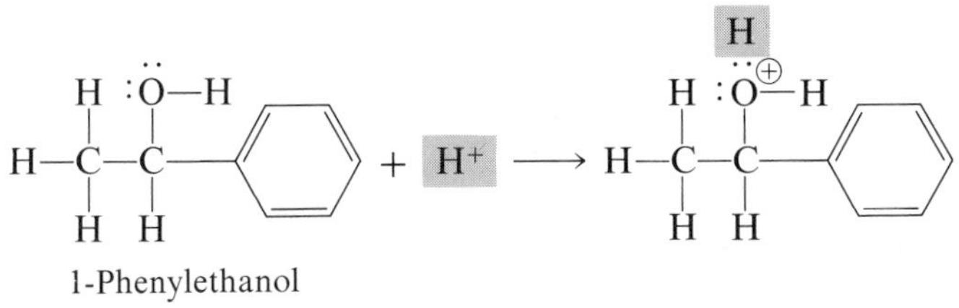

Next a molecule of water is eliminated, and the positive charge then resides on the carbon atom (i.e., a carbonium ion is formed).

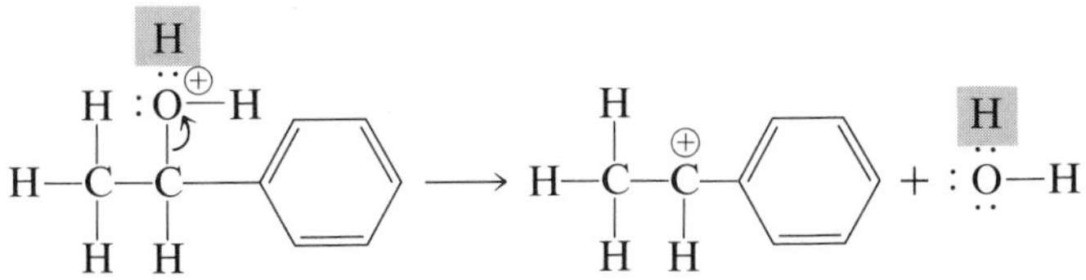

In the last step the acid catalyst is regenerated by loss of a hydrogen ion from the adjacent carbon atom.

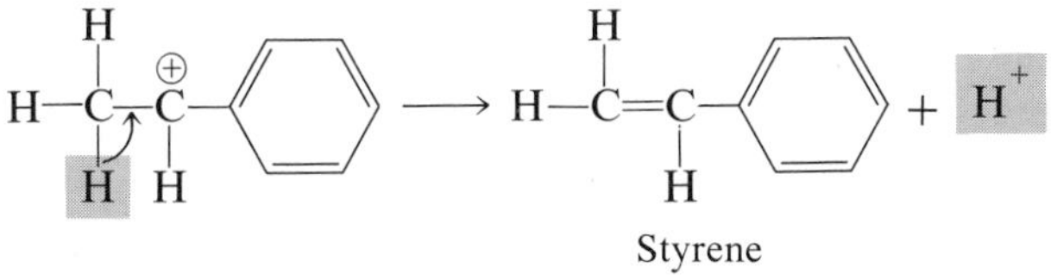

Dehydration of 1-phenylethanol can yield only one possible alkene. However, there are many alcohols that present a choice of possible products (e.g., $CH_3CH_2CHOHCH_3$), and we need to be able to predict the major product. This prediction is based upon a consideration of the relative stabilities of the prepared alkenes. Because alkyl groups are electron-donating, they tend to stabilize multiple carbon–carbon bonds, and thus *the more highly substituted alkene is usually the major product obtained from the acid-catalyzed dehydration of alcohols.*

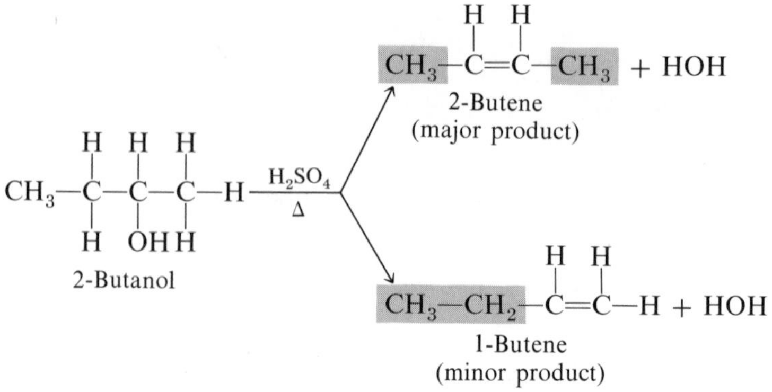
2-Butanol
2-Butene
(major product)
1-Butene
(minor product)

An alteration of reaction conditions, particularly reaction temperature, can bring about an intermolecular dehydration, which yields another class of organic compounds, the ethers (see Section 17.7).

General Equation

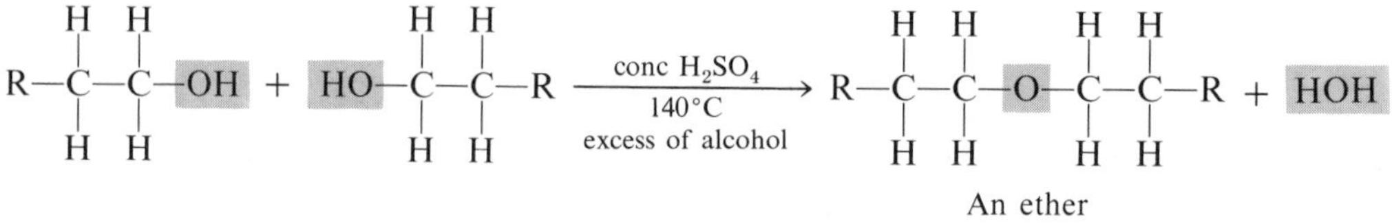
conc H_2SO_4
140°C
excess of alcohol
An ether

The first step of the mechanism again involves oxonium ion formation. However, at this lower temperature and in the presence of excess alcohol, displacement of a water molecule by another alcohol molecule takes precedence. The catalyst is then regenerated in the third step.

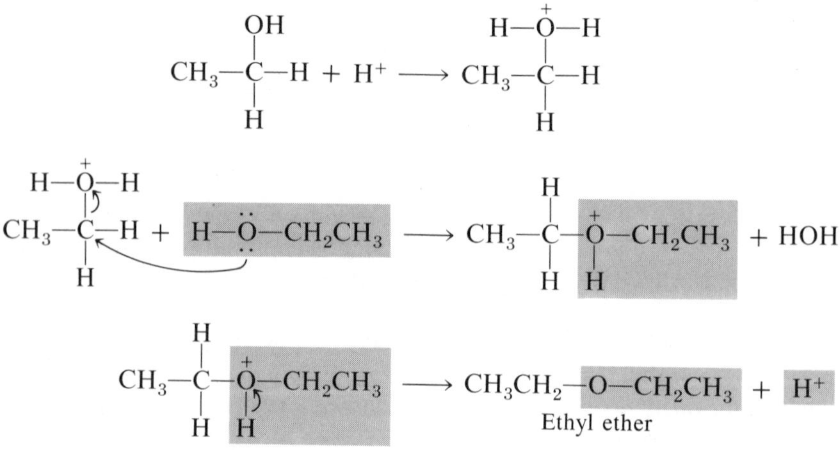
Ethyl ether

Dehydration (and its reverse, hydration) reactions occur continuously in cellular metabolism. In these biochemical dehydrations, enzymes serve as catalysts instead of acids and the reaction temperature is 37°C instead of the elevated temperatures required in the laboratory.

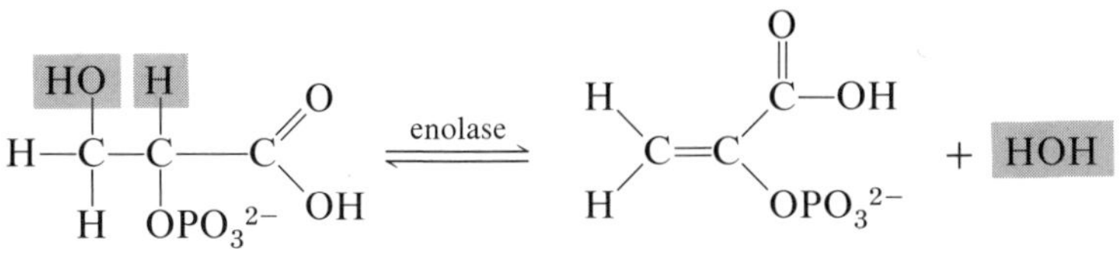

2-Phosphoglyceric acid   Phosphoenolpyruvic acid

c. Replacement of the Hydroxyl Group

The hydroxyl group of alcohols can be replaced by various acid anions (Section 16.2-b). Hydrobromic acid and hydriodic acid react readily with all alcohols. The less reactive hydrochloric acid requires the presence of zinc chloride for reaction with primary and secondary alcohols. A concentrated hydrochloric acid solution that has been saturated with zinc chloride is known as the *Lucas reagent.*

The phosphorus halides and thionyl chloride are also utilized to replace the hydroxyl group by halogen. Thionyl chloride is especially effective because the by-products of the reaction, SO_2 and HCl, are volatile and escape from the reaction mixture.

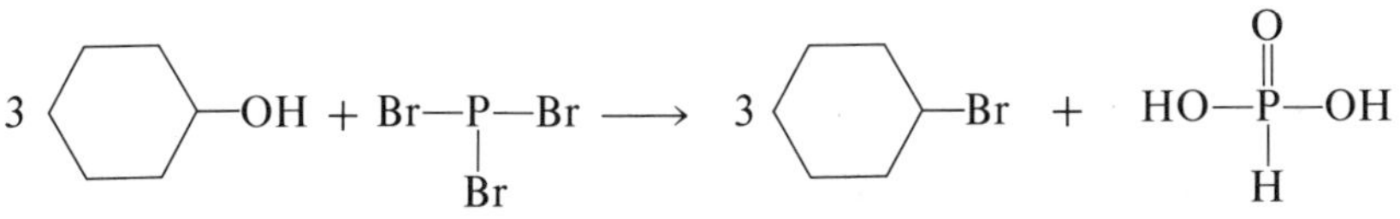

Cyclohexanol   Bromocyclohexane   Phosphorous acid

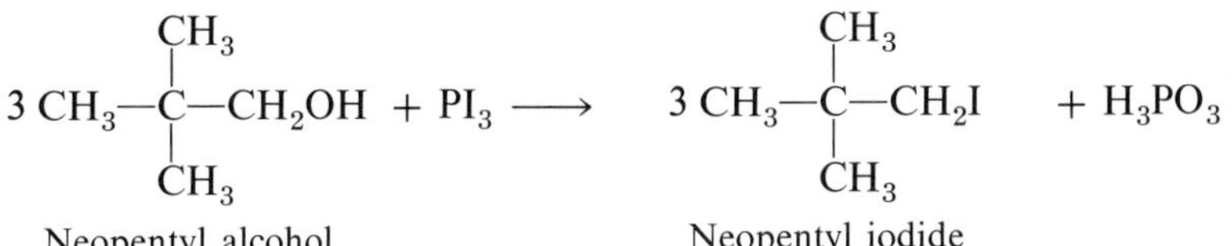

Neopentyl alcohol (2,2-Dimethylpropanol)   Neopentyl iodide (2,2-Dimethyliodopropane)

$$CH_2{=}CH{-}CH_2OH + SOCl_2 \longrightarrow CH_2{=}CH{-}CH_2Cl + SO_{2(g)} + HCl_{(g)}$$

Allyl alcohol (3-Hydroxypropene)   Thionyl chloride   Allyl chloride (3-Chloropropene)

d. Oxidation-Reduction Reactions

1. Dehydrogenation (Oxidation) to Form Aldehydes and Ketones. These reactions will be discussed in the next chapter and also in the chapters on carbohydrate metabolism and lipid metabolism. Primary alcohols are oxidized to aldehydes, while secondary alcohols are oxidized to ketones. Tertiary alcohols are stable to oxidation under the usual conditions. The most common oxidizing agents are potassium permanganate and potassium dichromate. The symbol [O] implies that an oxidizing agent is used without indicating the specific reagent.

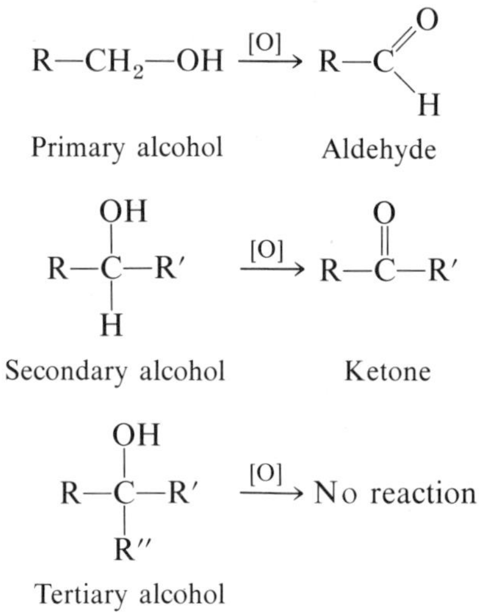
R—CH2—OH
[O]
R—C
O
H
Primary alcohol
Aldehyde
OH
R—C—R′
H
[O]
O
R—C—R′
Secondary alcohol
Ketone
OH
R—C—R′
R″
[O]
No reaction
Tertiary alcohol

2. Replacement of Hydrogen by Active Metals to Form Alkoxides (Reduction). This reaction can be considered to be a simple replacement reaction since the product contains a metal in place of a hydroxyl hydrogen. However, if we examine oxidation numbers or write two half-reactions to obtain the net equation, it is apparent that electron exchange must occur. The metal goes from an oxidation state of 0 to that of +1 while hydrogen undergoes an oxidation number change of +1 to 0. The reaction is strictly analogous to the reduction of water by an active metal.

$$\text{H—O—H} + \text{M} \longrightarrow \text{HO}^-\text{M}^+ + \tfrac{1}{2}\text{H}_2$$

A hydroxide

$$\text{R—O—H} + \text{M} \longrightarrow \text{RO}^-\text{M}^+ + \tfrac{1}{2}\text{H}_2$$

An alkoxide

$$\text{CH}_3\text{CH}_2\text{OH} + \text{Na} \longrightarrow \text{CH}_3\text{CH}_2\text{O}^-\text{Na}^+ + \tfrac{1}{2}\text{H}_2$$

Sodium ethoxide

Half-reactions

$$Na \longrightarrow Na^+ + e^-$$
$$CH_3CH_2OH + e^- \longrightarrow CH_3CH_2O^- + \tfrac{1}{2}H_2$$

The reactions of active metals with alcohols are not as vigorous as those with water and thus are more easily controlled. (Isopropyl alcohol is often employed in the laboratory to decompose excess pieces of sodium because this reaction is relatively slow and moderate.) The liberation of hydrogen gas from an unknown organic liquid upon treatment with sodium metal affords a quick test for the presence of the —OH group within a molecule.

e. Inorganic Ester Formation

The reaction of primary and secondary alcohols with oxy-acids rather than with halogen acids leads to the formation of compounds in which the —OH group of the acid is converted to —OR. Such compounds are called **esters,** and they are very important and useful substances. In Chapter 19 we shall discuss the organic esters prepared from alcohols and organic acids, and in Chapter 26 we discuss the esters formed from fatty acids and glycerol. The inorganic esters may be formed from alcohols and inorganic acids such as sulfuric, nitrous, nitric, and phosphoric.

The reaction of ethanol with sulfuric acid merits special mention because it illustrates the importance of specifying reaction conditions. At a temperature of 180 °C and in the presence of excess sulfuric acid, dehydration occurs to form ethylene (Section 17.4-b). At 140 °C and excess of alcohol, the product is ethyl ether (see Section 17.7). At room temperature the alcohol and acid react to yield the ester, ethyl hydrogen sulfate.

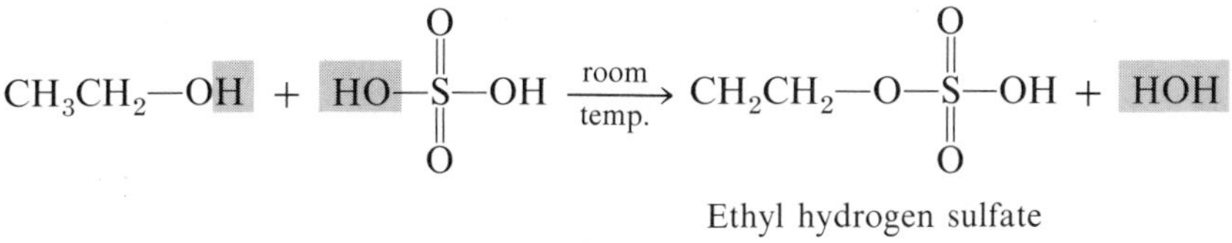
CH3CH2—OH + HO—S—OH
room
temp.
CH2CH2—O—S—OH + HOH

Ethyl hydrogen sulfate

Esters of nitrous acid form readily by the direct reaction of an alcohol with the acid. The valuable drug isopentyl nitrite functions by dilating the smaller blood vessels and by relaxing smooth muscles, thus reducing hypertension and relieving the severe pain of angina pectoris. Other drugs that are used clinically for the same purpose are the nitrates of glycerol (see Section 17.5-e) and pentaerythritol. These polynitrate compounds are very sensitive to shock and thus they are also very powerful explosives.

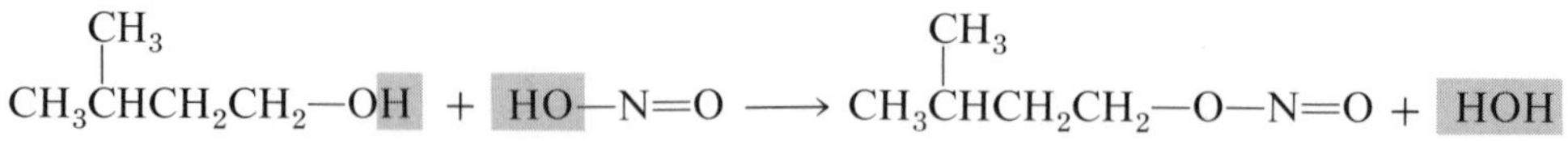

$$\underset{\text{Isopentyl alcohol}}{\overset{\overset{\displaystyle CH_3}{|}}{CH_3CHCH_2CH_2-OH}} + \underset{\text{Nitrous acid}}{HO-N{=}O} \longrightarrow \underset{\text{Isopentyl nitrite}}{\overset{\overset{\displaystyle CH_3}{|}}{CH_3CHCH_2CH_2-O-N{=}O}} + HOH$$

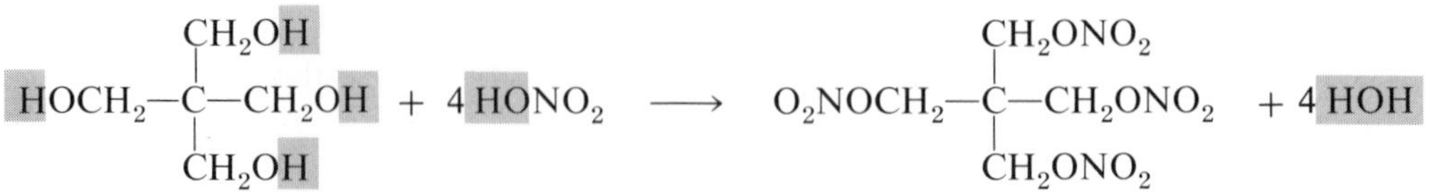

$$\underset{\text{Pentaerythritol}}{\overset{\overset{\displaystyle CH_2OH}{|}}{\underset{\underset{\displaystyle CH_2OH}{|}}{HOCH_2-C-CH_2OH}}} + 4\,HONO_2 \longrightarrow \underset{\text{Pentaerythritol tetranitrate (PETN)}}{\overset{\overset{\displaystyle CH_2ONO_2}{|}}{\underset{\underset{\displaystyle CH_2ONO_2}{|}}{O_2NOCH_2-C-CH_2ONO_2}}} + 4\,HOH$$

It is possible to replace all three hydroxyl groups of phosphoric acid by alkoxy groups. Phosphate esters play vital roles in all biological systems. We shall discuss these very important compounds in more detail in Chapters 21 and 28.

$$\underset{\underset{\displaystyle OH}{|}}{\overset{\overset{\displaystyle O}{\|}}{HO-P-OH}} \xrightarrow{CH_3OH} \underset{\underset{\displaystyle OH}{|}}{\overset{\overset{\displaystyle O}{\|}}{CH_3O-P-OH}} \xrightarrow{CH_3OH}$$

$$\underset{\underset{\displaystyle OH}{|}}{\overset{\overset{\displaystyle O}{\|}}{CH_3O-P-OCH_3}} \xrightarrow{CH_3OH} \underset{\underset{\displaystyle OCH_3}{|}}{\overset{\overset{\displaystyle O}{\|}}{CH_3O-P-OCH_3}}$$

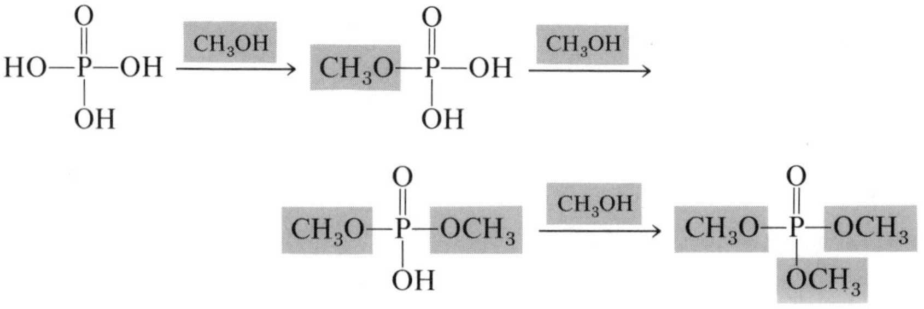

17.5 Important Alcohols

a. CH₃OH (Methanol, Methyl Alcohol, Wood Alcohol)

Prior to 1923, methanol was prepared by the destructive distillation of wood—hence its common name, wood alcohol. When wood is heated to a temperature of 250°C in the absence of air, it decomposes to charcoal and a volatile fraction. Two to three percent of this fraction is methanol, and it can be separated from the other components (acetic acid and acetone) by fractional distillation. On the average, a ton of wood produces about 35 pounds of the alcohol. Presently, methanol is more economically prepared by combining hydrogen and carbon monoxide under conditions of high temperature and pressure in the presence of a zinc oxide-chromium oxide catalyst.

$$2\,H_2 + CO \xrightarrow[\text{ZnO-Cr}_2\text{O}_3]{200\text{ atm, }350°C} CH_3OH$$

Methyl alcohol, unlike ethyl alcohol, is poisonous. Ingestion of as little as 15 ml of methanol can cause blindness; 30 ml can cause death. The poisoning effect is brought about by the body's oxidation of methanol to formic acid, and probably to some extent, to formaldehyde. Formic acid causes severe acidosis (Section 29.4) and other characteristic symptoms of methyl alcohol poisoning. The basis of the selective injury to retinal cells in methanol intoxication is not definitely known. Some believe that formaldehyde has a specific toxicity for the cells of the retina, while others believe that formic acid is the offending substance. Since the dominant feature of methanol poisoning is acidosis, the antidote is a solution of the weak base sodium bicarbonate (baking soda, $NaHCO_3$). Methyl alcohol should never be applied to the body, nor should its vapors be inhaled because methanol is readily absorbed through the skin and respiratory tract.

Methanol is employed commercially as a solvent for paint, gum, and shellac. It is used as the starting material for the commercial synthesis of formaldehyde, and it is used in some types of automobile antifreeze. It is the only alcohol that will dissolve cellulose nitrate. In recent years, attention has been given to the use of methanol (and ethanol) as a blend component of gasoline. Favorable results have been achieved with mixtures containing 80–90% gasoline and 10–20% alcohol. (Such a mixture is referred to as *gasohol.*) Since methanol can easily be synthesized from coal, wood chips, or municipal refuse (and ethanol from grain), this represents one method of augmenting our dwindling fuel supplies.

b. C_2H_5OH (Ethanol, Ethyl Alcohol, Grain Alcohol, Alcohol)

Ethyl alcohol is probably the best known and most important member of the alcohol series. In fact, when the layman uses the term alcohol, he is usually referring to ethyl alcohol.

The production of alcoholic spirits is one of the oldest known chemical reactions. Even in biblical times ethanol was prepared by the fermentation of sugars or starch from various sources (potatoes, corn, wheat, rice, etc.). Biochemical investigations have shown that the fermentation process is catalyzed by enzymes found in yeast, and that it proceeds by an elaborate multistep mechanism. These steps will be considered in detail in Chapter 28. The equation for the overall process can be written as follows.

$$\underset{\text{Starch}}{(C_6H_{10}O_5)_x} \xrightarrow{\text{enzymes}} \underset{\text{Glucose}}{C_6H_{12}O_6} \xrightarrow{\text{enzymes}} \underset{\text{Ethanol}}{2\,C_2H_5OH} + 2\,CO_2$$

The greatest use of ethanol is as a beverage. Wines contain about 12% ethanol by volume; beers and ciders contain about 4%; and whiskey, gin, and brandy contain 40–50%. The alcoholic content of a beverage is indicated by a measure known as *proof spirit.*[2] For some obscure reason, the proof value is

[2]This term has its origins in an early procedure used to test the alcoholic strength of a beverage. The whiskey to be tested was poured over a portion of gunpowder and the mixture was ignited. If the powder ignited, it was "proof" of spirit content in the whiskey. If the powder did not burn it was due to the presence of too much water in the whiskey.

twice the alcoholic content by volume. Thus whiskey that is 50% alcohol is said to be 100 proof.

Ethyl alcohol is potentially toxic. Rapid ingestion of one pint of pure alcohol would kill most people. Excessive ingestion over a long period of time leads to deterioration of the liver, loss of memory, and may lead to strong physiological addiction. If the alcohol is diluted (as in alcoholic beverages) and is consumed in small quantities, it is relatively safe. The body possesses enzymes that have the capacity to metabolize ethyl alcohol.[3]

Alcohol not intended for beverage purposes is commercially prepared from ethylene, which is a by-product of the petroleum industry. The alcohol so produced is 95% alcohol and 5% water. The water that remains in this mixture cannot be removed by ordinary distillation procedures, since 95% ethanol is a constant boiling mixture **(azeotrope).**[4] This 95% alcohol is used in chemical laboratories as a solvent. If 100% alcohol is needed, special procedures must be employed to prepare it. One method is to dry the alcohol over calcium oxide for several hours. The calcium oxide has a great affinity for water but will not combine with ethanol. The remaining alcohol is then distilled. Alcohol so prepared is known as *absolute* (100%) alcohol.

Ethanol is used as a solvent for perfumes, tinctures, lacquers, varnishes, and shellacs. It is also employed in the synthesis of other organic compounds. When ethanol is used for such industrial purposes it is not subject to a federal tax. To insure the legitimate use of tax-free alcohol, the government treats it with certain additives that make it unfit to drink. Such alcohol is known as *denatured* alcohol. Common denaturants are methanol and benzene. These compounds are toxic but do not interfere with the solvent properties of the alcohol.

c. $CH_3CHOHCH_3$ (Isopropyl Alcohol, 2-Propanol, Rubbing Alcohol)

Isopropyl alcohol is toxic when ingested but, unlike methanol, it is not absorbed through the skin. Its chief use is as rubbing alcohol, and it is often used as a solvent in place of ethanol. Since 2-propanol is unfit to drink, it escapes the legal restrictions placed on ethanol.

d. CH_2OHCH_2OH (Ethylene Glycol, 1,2-Ethanediol, Glycol)

Ethylene glycol is a colorless, viscous, sweet liquid that is water soluble

[3]Contrary to popular belief, alcohol is a depressant of the central nervous system, not a stimulant. The illusionary stimulation comes from its effects of depressing brain areas responsible for judgment. The resulting lack of inhibitions and restraints may cause one to feel "stimulated."

[4]An azeotropic mixture is a constant boiling mixture of two components. The two components are present in a fixed composition. A mixture of 95% ethanol and 5% water is an azeotropic mixture that boils at 78°C.

and denser than water. When taken internally, it is as toxic as methanol. The most important of the dihydroxy alcohols, it is used as a permanent radiator antifreeze because of its high boiling point (198°C) and its effectiveness in lowering the freezing point of water. A solution of 60% ethylene glycol in water will not freeze until the temperature falls to −49°C (−56°F).

e. $CH_2OHCHOHCH_2OH$ (Glycerol, 1,2,3-Propanetriol, Glycerin)

This important trihydroxy alcohol is also viscous, sweet, heavier than water, and water soluble. It is, however, nontoxic, and is an ingredient of all fats and oils. It is obtained as a by-product of the alkaline hydrolysis of fats during the soap-making process (see Section 24.8). Glycerol has widespread industrial use; the following list is not all-inclusive.

1. Preparation of hand lotions and cosmetics.
2. In inks, tobacco products, and plastic clays to prevent dehydration (glycerol is hygroscopic).
3. Constituent of glycerol suppositories.
4. Sweetening agent and solvent for medicines.
5. Lubricant, especially in chemical laboratories.
6. Production of plastics, surface coatings, and synthetic fibers.
7. Production of nitroglycerin.

The last glycerin derivative deserves special attention. Nitroglycerin was first prepared in 1846 by the Italian chemist Sobrero, who was lucky that he lived to tell of his discovery. Sobrero mixed nitric acid and glycerin, and the ensuing explosion nearly killed him. It was not until fifteen years later that the famed Swedish chemist and inventor Alfred Nobel discovered a method to prepare and transport the compound safely. He found that a type of diatomaceous earth, *Kieselguhr,* was capable of absorbing the nitroglycerin, thus rendering it insensitive to shock. The mixture was referred to as dynamite. Unless exploded by means of a percussion cap or by a detonator containing mercuric fulminate ($Hg[ONC]_2$), it was quite stable. The production of dynamite was a major breakthrough in the building industry. It made possible the construction of canals, dams, highways, mines and railroads. Its use as a weapon in warfare greatly disturbed the Nobel family, however, and after Alfred Nobel's death (1896) a trust fund was set up for an annual award for an outstanding contribution toward peace. (A trust fund was also set up to offer annual awards for contributions in the fields of chemistry, physics, economics, literature, and medicine or physiology.) It is surprising that a compound that is so sensitive to shock is also used as a drug. Nitroglycerin is administered either in tablet form (mixed with nonactive ingredients) or as an alcoholic solution (spirit of glyceryl trinitrate). Nitroglycerin functions in a manner similar to isopentyl nitrite (Section 17.4-e) in relieving the pains of angina pectoris.

The reaction for the preparation of nitroglycerin is analogous to the nitration of a monohydroxy alcohol, but three molecules of nitric acid are required for every molecule of glycerin.

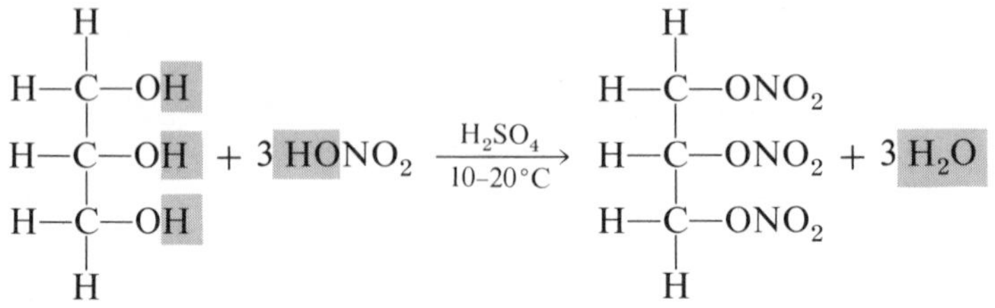

Glycerol trinitrate
(Nitroglycerin)

Nitroglycerin is a pale yellow, oily liquid that detonates upon slight impact.

$$4\ C_3H_5(ONO_2)_3 \longrightarrow 6\ N_{2(g)} + 12\ CO_{2(g)} + 10\ H_2O_{(g)} + O_{2(g)}$$

The reaction produces temperatures of over 3000°C and pressures above 2000 atmospheres. The explosion wave caused by such temperatures and pressures is enormous, accounting for the damaging effect of the detonation.

f. C_6H_5OH (Phenol, Hydroxybenzene, Carbolic Acid)

Phenol is a colorless crystalline solid, sparingly soluble in water and having a characteristic odor. Phenol is a powerful germicide, as are most of its derivatives. Phenol itself is rarely used as an antiseptic because of its extreme toxicity. It is highly caustic to the skin, and when taken internally it behaves as a violent poison. Like methanol, it can be absorbed through the skin. It is widely used as a standard of comparison for other germicides. The *phenol coefficient* is an arbitrary measure of the effectiveness of a given germicide. It is obtained by dividing the concentration of phenol needed to destroy a certain organism in a given time by the concentration of some other germicide needed to destroy the same organism in the same time. For example, a 1% germicide solution that destroys an organism in the same time as that required by a 5% phenol solution, is assigned a phenol coefficient of 5. Examine the labels on disinfectants, deodorants, soaps, throat lozenges, and muscle rubs that are in your home. You will find that most of them contain a phenolic derivative as an active ingredient.

Hexachlorophene is a polysubstituted phenol derivative that is a potent antiseptic (phenol coefficient of 125). It had been used in dilute form in the manufacture of toothpastes, deodorants, and germicidal soaps (such as PhisoHex). In 1972, studies showed that hexachlorophene is absorbed through the skin of rats, that it caused brain damage in monkeys, and that it may cause certain genetic defects. Consequently, hexachlorophene has been withdrawn from the commercial market pending further investigations. The FDA has recommended that its use be restricted to preventing the spread of staphylococcus infections, which cannot be dealt with in any other way.

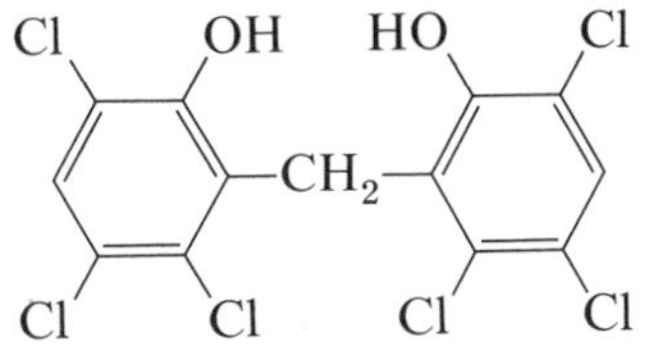

Hexachlorophene

Ethers

Ethers may be considered to be derivatives of water in which both hydrogen atoms have been replaced by alkyl or aryl groups. They may also be considered as derivatives of alcohol, in which the hydroxyl hydrogen has been replaced by an organic group.

$$\underset{\text{Water}}{\mathrm{H{-}O{-}H}} \xrightarrow[\text{hydrogens}]{\text{replace both}} \underset{\text{Ether}}{\mathrm{R{-}O{-}R'}} \xleftarrow[\text{hydrogen}]{\text{replace hydroxyl}} \underset{\text{Alcohol}}{\mathrm{R{-}O{-}H}}$$

The general formula for the ethers is **R—O—R′.** When both **R**— groups are the same, the compound is a *symmetrical* ether. When R and R′ are different, the ether is said to be a *mixed* ether.

R—O—R $H_3C{-}O{-}CH_3$	R—O—R′ $H_3C{-}O{-}CH_2CH_3$
Symmetrical ether	Mixed ether

17.6 Nomenclature of Ethers

Ethers are generally known by their common names rather than by their IUPAC names. The names of both organic groups are given, and are followed by the word ether. In naming the symmetrical ethers, the prefix *di-* is unnecessary, although it is often employed to avoid confusion.

Under the IUPAC system, one group is considered to have been derived from a parent hydrocarbon, and the other is considered part of an *alkoxy* substituent. As the name indicates, the substituent consists of both the organic group and the oxygen atom. By convention, the larger group is considered to

Table 17.4 Nomenclature of Ethers

Molecular Formula	*Structural Formula*	*Common Name*	*IUPAC Name*
C_2H_6O	$CH_3—O—CH_3$	(Di)methyl ether	Methoxymethane
C_3H_8O	$CH_3—O—CH_2CH_3$	Methyl ethyl ether	Methoxyethane
$C_4H_{10}O$	$CH_3CH_2—O—CH_2CH_3$	(Di)ethyl ether	Ethoxyethane
$C_4H_{10}O$	$CH_3—O—\overset{1}{C}H_2\overset{2}{C}H_2\overset{3}{C}H_3$	Methyl *n*-propyl ether	1-Methoxypropane
$C_4H_{10}O$	$CH_3—O—\overset{2}{C}H\overset{3}{C}H_3$ with $\overset{1}{C}H_3$ on C-2	Methyl isopropyl ether	2-Methoxypropane
C_7H_8O	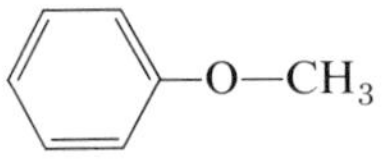	Methyl phenyl ether (Anisole)	Methoxybenzene
$C_8H_{10}O$	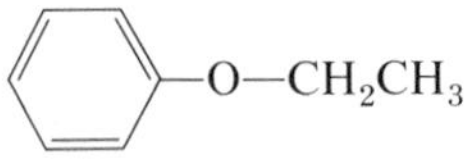	Ethyl phenyl ether (Phenetole)	Ethoxybenzene
$C_8H_{10}O$	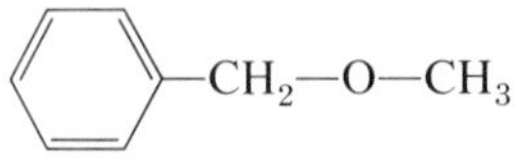	Methyl benzyl ether	Methoxyphenylmethane
$C_{12}H_{10}O$	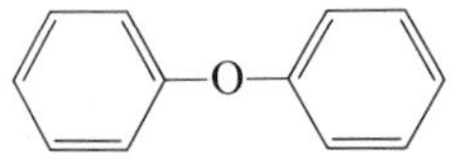	(Di)phenyl ether	Phenoxybenzene

have been derived from the parent hydrocarbon, and the smaller group is part of the alkoxy substituent. Thus, a methyl group attached to the oxygen (OCH_3) of an ether would be called a methoxy group, and an ethyl group bonded to an oxygen (OC_2H_5), an ethoxy group, and so on. Table 17.4 illustrates the rules of nomenclature for several simple and mixed ethers.

17.7 Preparation of Ethyl Ether

The aliphatic ethers, like the alkyl halides, are strictly synthetic compounds. By far the most common and most important ether to be synthesized is ethyl ether. Its preparation has already been mentioned in Section 17.4; a general scheme, including significant reaction conditions, is given below for the interactions of ethene, ethanol, and sulfuric acid.

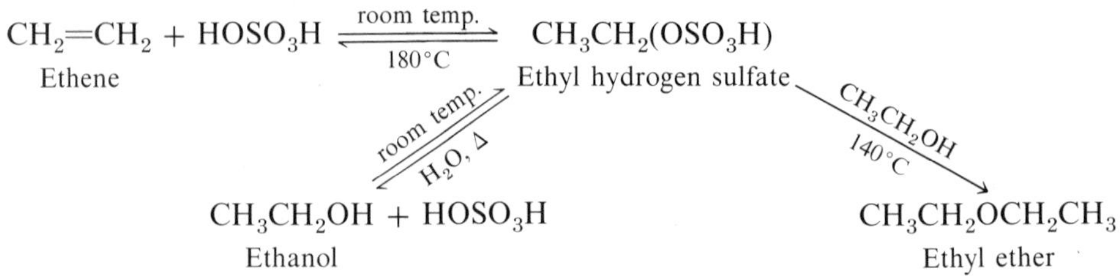
CH2=CH2 + HOSO3H
Ethene
room temp.
180°C
CH3CH2(OSO3H)
Ethyl hydrogen sulfate
room temp.
H2O, Δ
CH3CH2OH
140°C
CH3CH2OH + HOSO3H
Ethanol
CH3CH2OCH2CH3
Ethyl ether

The commercial manufacture of ethyl ether involves a continuous addition of ethanol into a mixture of sulfuric acid and ethanol that is maintained at a temperature of 140 °C. At this temperature, the ethyl ether distills out of the reaction mixture as it is formed. This is one reaction in which the maintenance of reaction temperature is critical. At temperatures below 130 °C, the reaction between acid and alcohol is so slow that unreacted alcohol will distill out of the mixture. At temperatures above 150 °C, the ethyl hydrogen sulfate intermediate will decompose to yield ethene and sulfuric acid.

17.8 Williamson Synthesis of Ethers

The symmetrical ethers can be produced by the elimination of a molecule of water from two molecules of alcohol.

$$\mathrm{R{-}O{-}H + H{-}O{-}R \xrightarrow{H_2SO_4} R{-}O{-}R + HOH}$$

This procedure, however, is not satisfactory for the preparation of mixed ethers.

In 1851, the British chemist Alexander Williamson discovered a general method for the synthesis of both symmetrical and mixed ethers. He reacted sodium alkoxides or sodium phenoxides with alkyl halides to produce a wide variety of ethers. For example, ethyl ether can be prepared from sodium ethoxide and ethyl bromide.

$$\mathrm{CH_3CH_2O^-Na^+ + CH_3CH_2Br \longrightarrow CH_3CH_2{-}O{-}CH_2CH_3 + NaBr}$$

The mechanism is similar to that for the preparation of alcohols from alkyl halides and sodium hydroxide (Section 17.2-b). In this case the strong nucleophilic alkoxide ion displaces the halide ion.

$$\mathrm{CH_3CH_2O^- + CH_3CH_2Br \longrightarrow \left[CH_3CH_2O\cdots\overset{\displaystyle CH_3}{\underset{\displaystyle H\ \ H}{C}}\cdots Br\right]^- \longrightarrow CH_3CH_2{-}O{-}CH_2CH_3 + Br^-}$$

The alkyl halides can be replaced by a dialkyl sulfate, particularly in the synthesis of aromatic ethers. Examples of the preparation of mixed ethers are:

$$\underset{\text{Sodium isopropoxide}}{\mathrm{(CH_3)_2CHO^-Na^+}} + \mathrm{CH_3Br} \longrightarrow \underset{\text{Methyl isopropyl ether}}{\mathrm{CH_3{-}\overset{\displaystyle CH_3}{\underset{\displaystyle H}{C}}{-}O{-}CH_3}} + \mathrm{NaBr}$$

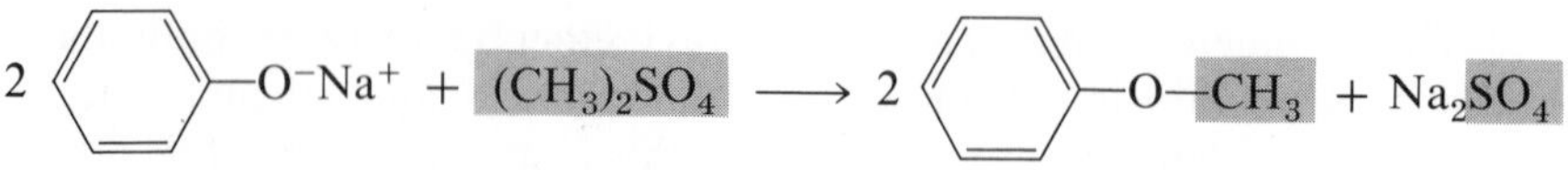

Sodium phenoxide — Methyl phenyl ether

17.9 Physical Properties of Ethers

The carbon–oxygen bond of the ethers is in many respects similar to the carbon–carbon single bond of the alkanes. The ethers, therefore, are colorless, highly volatile, very flammable, less dense than water, and practically water insoluble. The lack of a hydrogen–oxygen bond makes intermolecular hydrogen bonding impossible, and the ethers, like the alkanes, are unassociated. This is evidenced by their boiling points, which are similar to those of the alkanes of comparable molar mass and much lower than their isomeric alcohols. Table 17.5 lists comparative boiling points for some ethers, alkanes, and alcohols.

17.10 Chemical Properties of Ethers

The chemical behavior of the ethers and the alkanes is strikingly similar. Ethers are characterized by their inertness toward most of the common reagents under ordinary conditions. They are resistant to attack by metallic sodium, by oxidizing and reducing agents, and by strong bases. However, they are soluble in concentrated sulfuric acid, a characteristic that distinguishes ethers from alkanes. Solubility in H_2SO_4 is a result of the basic character of the ether molecule. Ethers are protonated by acids to form oxonium ions.

$$C_2H_5{-}\ddot{\underset{\cdot\cdot}{O}}{-}C_2H_5 + \text{conc } H_2SO_4 \longrightarrow C_2H_5{-}\overset{\cdot\cdot\oplus}{\underset{\underset{H}{|}}{O}}{-}C_2H_5 \ HSO_4^{\ominus}$$

Diethyloxonium hydrogen sulfate
(soluble in H_2SO_4)

Some of the strong acids, when heated with ethers, will effect a cleavage of the ether molecule. The acid commonly used for this reaction is hydriodic

Table 17.5 Comparison of Boiling Points of Alkanes, Alcohols, and Ethers

Formula	*Name*	*M*	*BP (°C)*
$CH_3CH_2CH_3$	Propane	44	−48
CH_3OCH_3	Methyl ether	46	−23
CH_3CH_2OH	Ethyl alcohol	46	78
$CH_3CH_2CH_2CH_2CH_3$	*n*-Pentane	72	36
$CH_3CH_2OCH_2CH_3$	Ethyl ether	74	35
$CH_3CH_2CH_2CH_2OH$	*n*-Butyl alcohol	74	118

acid. If an excess of the acid is employed, the alcohol formed initially is converted to an alkyl halide.

$$CH_3CH_2{-}O{-}CH_2CH_3 + HI \longrightarrow CH_3CH_2OH + CH_3CH_2I$$

$$CH_3CH_2{-}O{-}CH_2CH_3 \xrightarrow{\text{excess HI}} 2\ CH_3CH_2I + HOH$$

The reaction mechanism involves protonation of the ether to form an intermediate oxonium ion, followed by nucleophilic displacement of an alcohol molecule by iodide. With concentrated HBr and HCl, higher temperatures are required because the bromide and chloride ions are poorer nucleophiles than the iodide ion.

Aryl alkyl ethers are cleaved by acid to yield phenols. The reaction mechanism is the same as for aliphatic ethers.

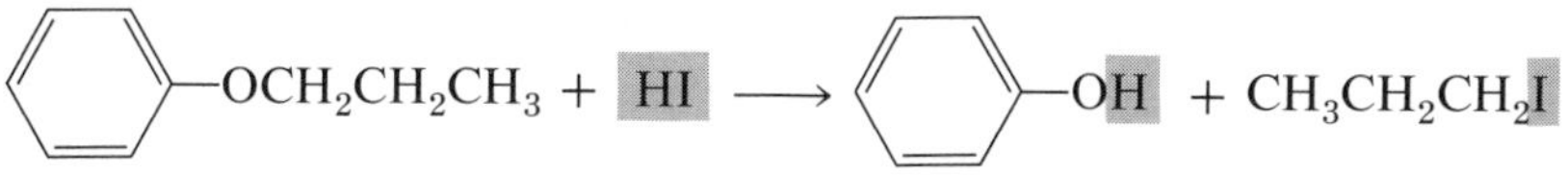

OCH2CH2CH3 + HI
OH + CH3CH2CH2I

n-Propyl phenyl ether — Phenol — *n*-Propyl iodide

Phenyl ethers are sufficiently stable to undergo the normal aromatic substitution reactions. Since the alkoxy group is an ortho-para director, a mixture of isomers can be expected.

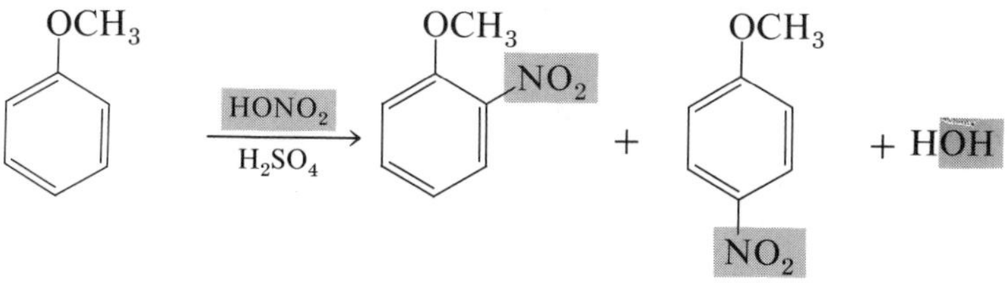

OCH3
HONO2
H2SO4
OCH3
NO2
+
OCH3
NO2
+ HOH

Methoxybenzene (Anisole) — *o*-Nitroanisole — *p*-Nitroanisole

17.11 Uses and Hazards of Ethyl Ether

Ethyl ether was, for many years, the most generally used of all the volatile anesthetics. Ether produces unconsciousness by depressing the activity of the central nervous system. The major disadvantages of ether are its irritating effects on the respiratory passages and the occurrence of postanesthetic nausea and vomiting. Many hospitals are now using methyl propyl ether (Neothyl), halothane (Fluothane, $CF_3CHClBr$), enflurane (Ethrane, CHF_2OCF_2CHFCl), and methoxyflurane (Penthrane, $CH_3OCF_2CHCl_2$) as general anesthetics. They are more potent than ethyl ether and less irritating to the respiratory passages. Caution must also be exercised in their use, since halothane has recently been shown to cause a severe liver toxicity in a small percentage of people given this anesthetic.

Modern surgical practice has moved away from a single anesthetic. Generally, the patient is given a strong sedative such as Pentothal by injection to produce unconsciousness. The gaseous anesthetic is then administered through a mask (in a stream of nitrous oxide and oxygen gases) to provide insensitivity to pain and to retain unconsciousness.

There are four stages associated with anesthesia. In the first stage there is a high degree of analgesia (incapacity to feel pain) without loss of consciousness. The second stage is characterized by loss of consciousness, by irregular deep respiration, and often by active and purposeless muscular activity. In the third stage the necessary muscular relaxation is achieved and surgery can be performed. The fourth stage is characterized by paralysis of the respiratory muscles. Death will follow unless corrective action is initiated. A principal function of the anesthesiologist is to determine the level of anesthesia by monitoring the alterations in respiration and the degree of muscular relaxation in the abdomen.

Ether is extremely hygroscopic. A freshly opened can of ether will immediately pick up about 1–2% water from the moisture in the air. Special techniques (dry box, nitrogen atmosphere) are employed in handling the compound when reaction conditions call for anhydrous ether (for example, as a solvent for Grignard reactions).

Ether serves as an excellent extraction medium because (1) it is a good solvent for many organic substances, (2) it is not miscible with water, (3) it is relatively unreactive, (4) it does not dissolve most inorganic substances, and (5) it has a low boiling point and can be easily removed from the extract by low temperature distillation. Ether is also employed in separating constituents of foodstuffs. Lipids are soluble in ether and thus can be separated from carbohydrates and proteins, which are ether insoluble.

Two hazards must be avoided when working with ether. First, since ether vapor is heavier than air, it remains at the bottom of a room and has the nasty property of "rolling" long distances across bench tops. Hence, the necessity of avoiding open flames in a laboratory in which ether is being used. Ether fires cannot be extinguished with water, since ether floats on top of water. A carbon dioxide extinguisher is most often used.

Secondly, by reaction with oxygen in the air, ethers may form peroxides when left standing in open containers. These peroxides remain with the residue when ether is distilled or evaporated to dryness. Solid peroxides are dangerous because they are readily explosive. For this reason, ether from a previously opened bottle should be treated with a reducing agent, such as ferrous sulfate, before it is used. Commercial absolute ether contains 0.05 ppm of sodium diethyldithiocarbamate, $(C_2H_5)_2NCS_2^{\ominus}Na^{\oplus}$, as an antioxidant.

17.12 Cyclic Ethers

All of the cyclic compounds that have been mentioned thus far have had only one element, carbon, in the ring. There is a large class of *organic compounds that contain two or more different elements in the same ring structure.* Such cyclic compounds are called **heterocyclic** (Greek, *heteros,* other). Heterocyclic compounds, especially those containing nitrogen as the hetero atom (see Section 20.12), are widely abundant in nature. The carbohydrates (Chapter 23) constitute the largest class of naturally occurring oxygen heterocycles.

Cyclic ethers with three atoms in the ring are called **1,2-epoxides.** 1,2-Epoxides are named as oxides of the corresponding alkenes; the most important member of this group is ethylene oxide.

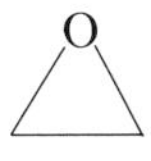
Ethylene oxide

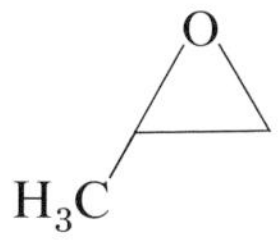
H3C
Propylene oxide

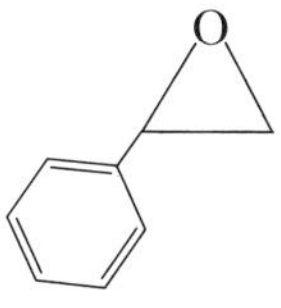
Styrene oxide

Ethylene oxide is prepared commercially by the reaction of ethylene with oxygen in the presence of a silver catalyst.

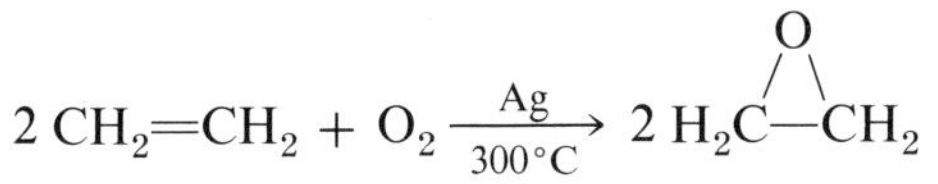
2 CH2=CH2 + O2
Ag
300°C
2 H2C—CH2
O

Because of the severe ring strain, ethylene oxide is a very reactive molecule. The strain is relieved by reactions which result in an opening of the ring by cleavage of a carbon–oxygen bond. (Recall the high reactivity of cyclopropane compared to the other cycloalkanes.) Ethylene oxide serves as an important industrial starting material for the synthesis of a wide variety of 1,2-disubstituted organic compounds. It reacts with water to yield ethylene glycol (Section 17.5-d), with alcohols to form ether alcohols, called Cellosolves, and with ammonia to produce amino alcohols. Cellosolves are solvents for quick-drying lacquers, varnishes and paints. Ethanolamines are used as emulsifying agents, in the preparation of soaps, and also as solvents.

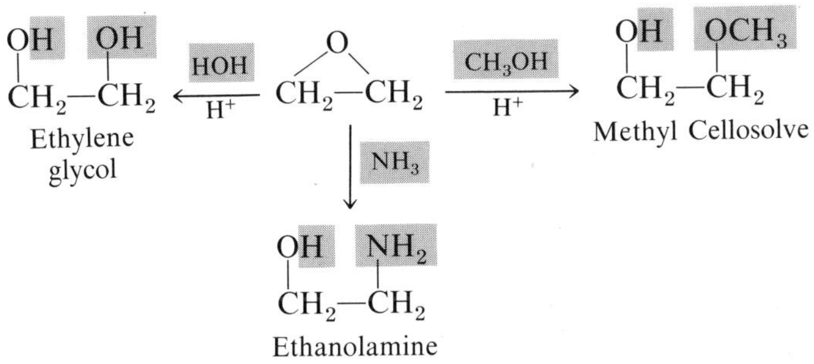
OH OH
CH2—CH2
Ethylene glycol
HOH
H+
O
CH2—CH2
CH3OH
H+
OH OCH3
CH2—CH2
Methyl Cellosolve
NH3
OH NH2
CH2—CH2
Ethanolamine

Other industrially important oxygen heterocycles are dioxane, tetrahydrofuran, and furan.

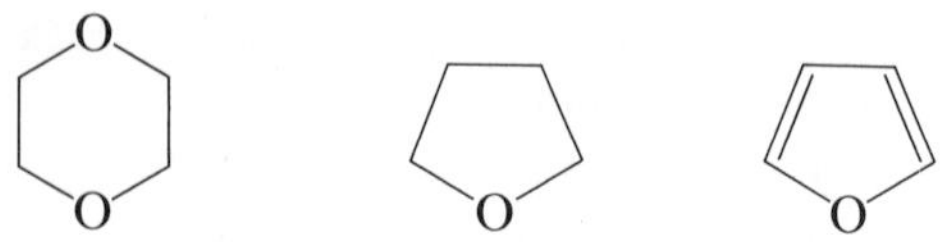

Dioxane (1,4-Dioxane) | Tetrahydrofuran (THF) | Furan

Dioxane and THF are very useful solvents because they are relatively unreactive to other reagents and they are miscible with both water and many organic compounds. Tetrahydrofuran is often used in place of ethyl ether as a solvent for Grignard reactions and for reduction reactions by metal hydrides. Furan is usually found combined in the structures of a number of natural products. In its reactions, furan most closely resembles its nitrogen analog, pyrrole (see Section 20.12-b). Furan displays many of the properties of a cyclic ether, of a 1,3-diene system, and of an aromatic compound. It is cleaved in aqueous acid to a 1,4-dicarbonyl compound, it adds two moles of hydrogen to form tetrahydrofuran, and it undergoes many electrophilic substitution reactions.

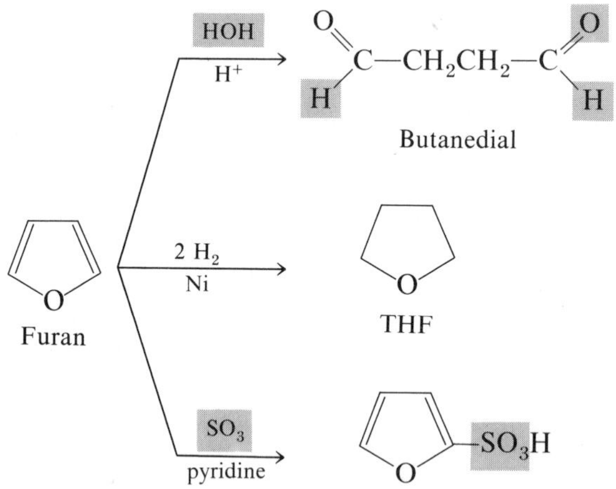

HOH
H+
C—CH2CH2—C
O
O
H
H
Butanedial
Furan
2 H2
Ni
THF
SO3
pyridine
SO3H
Furansulfonic acid
O

Exercises

17.1 Draw and name all the alcohols corresponding to the molecular formula $C_5H_{12}O$. Label each isomer as primary, secondary, or tertiary.

17.2 Draw and name all the isomeric ethers having the molecular formula $C_5H_{12}O$.

17.3 Draw and name all the aliphatic isomers of molecular formula $C_4H_{10}O$.

17.4 Name the following compounds.

(a) $(CH_3)_2CHCH_2CH_2OH$
(b) $CH_3CHOHCH(Br)_2$
(c) $CH_3CH_2CH(CH_3)CHOHC(CH_3)_3$
(d) 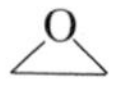
(e) $CH_3OCH_2CH_2CH_2CH_3$
(f) 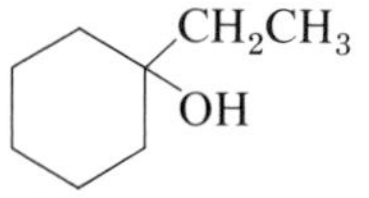
(g) $CH_3CH(CH_3)CH_2C(CH_3)_2OH$
(h) 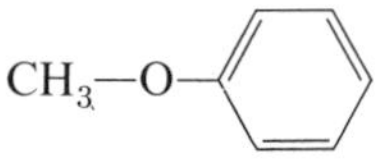
(i) $CH_3OCH_2CH_2OH$
(j) $CH_2{=}CHCHOHCH_3$
(k) $CH_3CH_2CH_2OCH_2CH_2CH_3$
(l) 

17.5 Write structural formulas for the following compounds.

(a) 4-methyl-2-pentanol
(b) *p*-nitrophenol
(c) 2-chloro-3-isopropylcyclohexanol
(d) 2,4,4-trimethyl-2-heptanol
(e) methyl isopropyl ether
(f) benzyl alcohol
(g) resorcinol
(h) 2,2,5,5-tetramethylcyclopentanol
(i) tetrahydrofuran
(j) ethoxypropane

17.6 What is wrong with each of the following names? Give the structure and correct name for each compound.

(a) 1,1-dimethyl-1-propanol
(b) 4-methyl-4-hexanol
(c) 1-ethyl-2-ethanol
(d) 2-*n*-propyl-2-pentanol
(e) 2,2-dimethyl-3-chloro-3-butanol
(f) 5-bromophenol

17.7 Write equations for the Williamson synthesis of methyl ethyl ether from **(a)** methyl alcohol and ethyl bromide and **(b)** ethyl alcohol and methyl bromide.

17.8 In the preparation of ethers from alcohols, why is it so critical to control the reaction temperature between 130°C and 150°C?

17.9 Write equations for the preparation of **(a)** 1-pentanol and **(b)** 2-pentanol.

17.10 Arrange the following compounds in order of decreasing solubility in water.

(a) $n\text{-}C_4H_9OH$, CH_3OH, $n\text{-}C_4H_9Br$
(b) $n\text{-}C_4H_9OH$, $CH_2OHCH_2CH_2CH_2OH$, $CH_2OHCHOHCH_2OH$
(c) $(CH_3)_3CCH_2OH$, $CH_3(CH_2)_4OH$, $(CH_3CH_2)_2CHOH$

17.11 Discuss the factors responsible for the differences in water solubility and in boiling points between methyl chloride and methyl alcohol.

17.12 1-Butanol and diethyl ether are isomers.

(a) Based on structure, which one would be expected to have the higher vapor pressure, higher boiling point, and greater solubility in water?

(b) The fact is that both isomers have comparable solubilities in water (8 g/100 ml). Offer an explanation.

17.13 Write equations summarizing the action of sulfuric acid on ethanol. Be sure to include all pertinent reaction conditions.

17.14 Complete the following equations.

(a) $CH_3CH_2CH_2CH_2OH + Na \longrightarrow$

(b) $CH_3CH_2CH_2OH + HI \longrightarrow$

(c)  $—CH_2CHOHCH_3 + H_2SO_4 \xrightarrow{\Delta}$

(d)  $—OH + NaOH$

(e) $CH_3CH_2C(CH_3){=}CH_2 + HOH \xrightarrow{H_2SO_4}$

(f) $CH_3CH_2CH_2OH + H_3PO_4 \longrightarrow$

(g)  $—O—CH(CH_3)_2 + HI \longrightarrow$

(h)  $+ HOH \xrightarrow{H^+}$

17.15 Devise simple chemical tests that would enable you to distinguish one compound from the other in each of the following pairs.

(a) ethyl alcohol and ethyl chloride

(b) ethyl alcohol and dimethyl ether

(c) *n*-butyl alcohol and *t*-butyl alcohol

(d) cyclohexanol and cyclohexane

(e) cyclohexanol and phenol

(f) furan and tetrahydrofuran

17.16 Give the name and one use for each of the following compounds.

(a) CH_3OH

(b) CH_3CH_2OH

(c) $CH_3CHOHCH_3$

(d) CH_2OHCH_2OH

(e) $CH_2OHCHOHCH_2OH$

(f) C_6H_5OH

17.17 What is meant by the following terms?

(a) absolute alcohol

(b) denatured alcohol

(c) 86 proof

(d) phenol coefficient

17.18 Dehydration of a butanol isomer with concentrated sulfuric acid at 180 °C gave 2-butene as the only product. What was the structural formula of the alcohol?

17.19 1.6 grams of an alcohol reacted with excess potassium to liberate 560 ml of hydrogen gas at STP. What is the molar mass of the alcohol?

17.20 4.6 grams of ethanol was treated with excess sulfuric acid. Assuming no side reactions, how many liters (at STP) of ethene could be produced?

17.21 Compound X has the molecular formula, C_4H_8. It adds hydrogen bromide to form compound Y which then reacts with AgOH to form a tertiary alcohol. What are the formulas of X and Y?

17.22 An ether containing 68.2% carbon, 13.6% hydrogen, and 18.2% oxygen reacts with hydrogen iodide to yield methanol and a compound containing 69.0% iodine. Treatment of this latter compound with a dilute potassium hydroxide solution yields a secondary alcohol. What is the structural formula of the original ether?

18

Aldehydes and Ketones

The study of aldehydes and ketones brings us for the first time to a consideration of an extremely important chemical grouping, the carbon–oxygen double bond, $-\overset{O}{\overset{\|}{C}}-$. This unit is referred to as the **carbonyl group.** It is common to both the aldehydes and ketones, as well as to several other classes of organic compounds. The carbonyl carbon of the aldehyde is a terminal carbon and is always bonded to a hydrogen, $\mathbf{-\overset{O}{\overset{\|}{C}}-H}$. The carbonyl carbon of the ketone, on the other hand, is never a terminal carbon since it must be bonded to two other carbon atoms. The general formula for an aldehyde is $\mathbf{R-\overset{O}{\overset{\|}{C}}-H}$, and the general formula for a ketone is $\mathbf{R-\overset{O}{\overset{\|}{C}}-R'}$. (Note that the general formula for an aldehyde may be abbreviated to RCHO but not to RCOH since the latter implies an alcohol.)

Aldehydes and ketones are thus two different classes of organic compounds. The rationale for their inclusion in the same chapter is based upon their similarity of properties. Most reagents that attack carbonyl groupings will react with both aldehydes and ketones. Any observable differences in their properties, with the exception of ease of oxidation, are differences of degree rather than kind.

18.1 Nomenclature of Aldehydes and Ketones

The generic name aldehyde has a significant meaning. The compound is obtained by the removal of hydrogen from an alcohol, and the name is derived from the two words, *al*cohol *dehyd*rogenation. Both the common and IUPAC names of the aldehydes are frequently used, however, usage of common names predominates for the lower homologs of the family. The common names are taken from the names of the acids into which the aldehydes are convertible by oxidation.

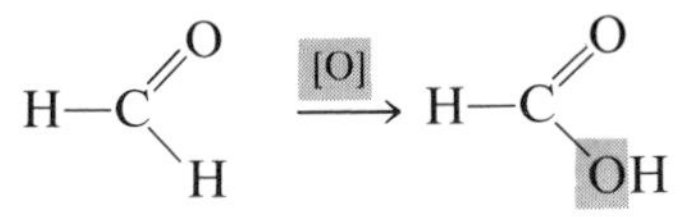

Formaldehyde Formic acid

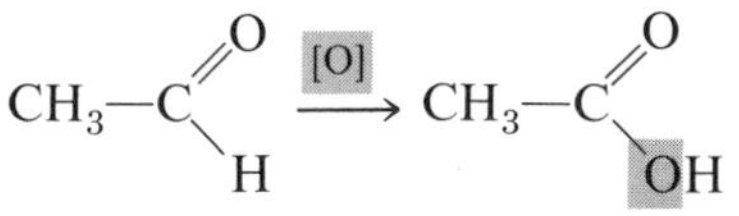

Acetaldehyde Acetic acid

Naming of the aldehydes according to the IUPAC system follows the previously established rules. The longest chain of carbons is taken to be the one containing the carbonyl group. The final *-e* of the alkane name is replaced by the suffix *-al,* which designates the functional group of the aldehydes, $\mathbf{-C(=O)H}$. The aldehyde functional group takes precedence over all the groups discussed in previous chapters. Thus, the carbonyl carbon is always considered to be carbon number one, and it is unnecessary to designate this group by number. Examples of nomenclature are provided in Table 18.1.

The general formula for the ketones is $\mathbf{R{-}\overset{\overset{\large O}{\|}}{C}{-}R'}$, where **R** and **R′** may be any alkyl or aryl groups. The common names of the ketones, like those of the ethers, are obtained by naming each of the attached organic groups and adding the word ketone. The lowest homolog of the ketone series is universally called acetone rather than dimethyl ketone. (It was first prepared from acetic acid.) According to the IUPAC system, the longest continuous chain containing the carbonyl group designates the parent hydrocarbon. The *-e* ending of the alkane name is replaced by the suffix *-one*, signifying that the compound is a ketone. Because the carbonyl carbon of a ketone can never be a terminal carbon, the position of the functional group must be indicated by the appropriate number. The chain is numbered so that the carbonyl carbon is assigned the lowest possible number. Table 18.2 illustrates the nomenclature for some of the ketones.

Table 18.1 Nomenclature of Aldehydes

Molecular Formula	*Condensed Structural Formula*	*Common Name*	*IUPAC Name*
CH_2O	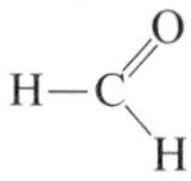	Formaldehyde	Methanal
C_2H_4O	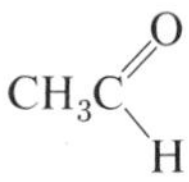	Acetaldehyde	Ethanal
C_3H_6O	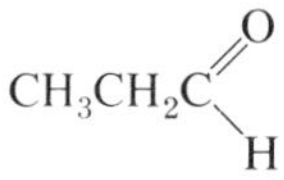	Propionaldehyde	Propanal
C_4H_8O	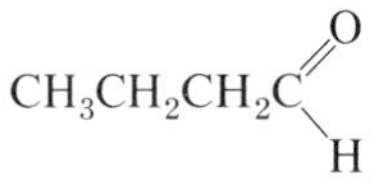	Butyraldehyde	Butanal
C_4H_8O	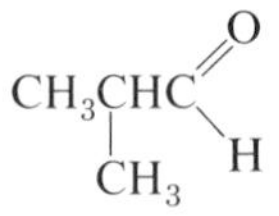	Isobutyraldehyde	2-Methylpropanal
$C_5H_{10}O$	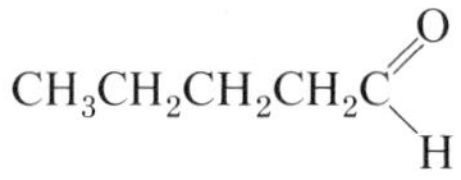	Valeraldehyde	Pentanal
$C_5H_{10}O$	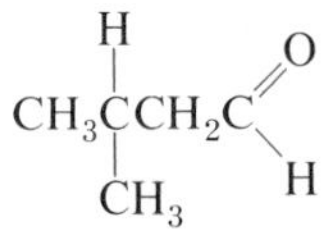	Isovaleraldehyde	3-Methylbutanal *not* 2-Methylbutanal
C_7H_6O	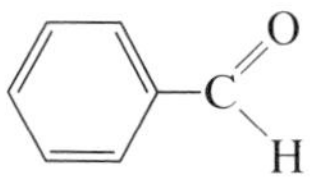	Benzaldehyde	Benzaldehyde (Phenylmethanal)
$C_8H_8O_3$	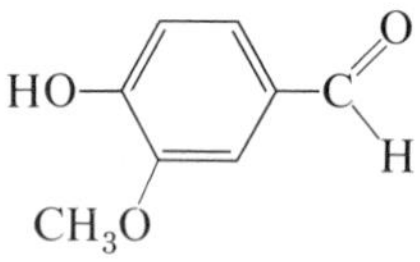	Vanillin (odor of vanilla)	4-Hydroxy-3-methoxybenzaldehyde
C_9H_8O	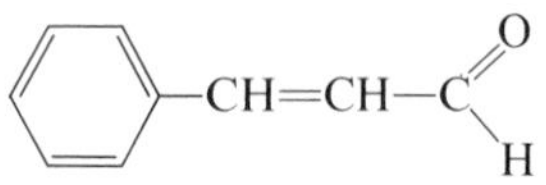	Cinnamaldehyde (odor of cinnamon)	3-Phenyl-2-propenal

Table 18.2 Nomenclature of Ketones

Molecular Formula	*Condensed Structural Formula*	*Common Name*	*IUPAC Name*
C_3H_6O	$CH_3\overset{O}{\overset{\|}{C}}CH_3$	Acetone (dimethyl ketone)	Propanone
C_4H_8O	$CH_3\overset{O}{\overset{\|}{C}}CH_2CH_3$	Methyl ethyl ketone	Butanone
$C_5H_{10}O$	$CH_3CH_2\overset{O}{\overset{\|}{C}}CH_2CH_3$	Diethyl ketone	3-Pentanone
$C_5H_{10}O$	$CH_3CH_2CH_2\overset{O}{\overset{\|}{C}}CH_3$	Methyl *n*-propyl ketone	2-Pentanone
$C_5H_{10}O$	$CH_3\underset{CH_3}{\underset{\|}{C}}H\overset{O}{\overset{\|}{C}}CH_3$	Methyl isopropyl ketone	3-Methyl-2-butanone *not* 2-Methyl-3-butanone
$C_6H_{10}O$	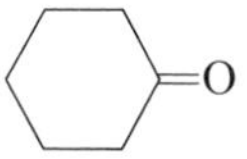	Cyclohexanone	Cyclohexanone
C_8H_8O	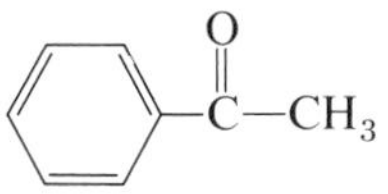	Acetophenone (Methyl phenyl ketone)	Phenylethanone
$C_{13}H_{10}O$	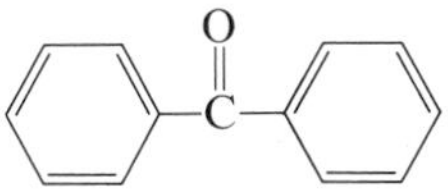	Benzophenone (Diphenyl ketone)	Diphenylmethanone

18.2 Preparation of Aldehydes and Ketones

When a primary alcohol is oxidized, it is sometimes possible to isolate an aldehyde as the chief product. The aldehyde that is formed is more susceptible to oxidation than was the original alcohol. Therefore, provision must be made to remove it from the reaction mixture as quickly as possible. Advantage is taken of the fact that the aldehyde has a greater volatility than its corresponding acid or alcohol. Only aldehydes that boil significantly below 100 °C can be conveniently prepared by the oxidation of primary alcohols. This method is therefore limited to the production of a few simple aldehydes. However, as we shall see in Chapter 28, the enzyme-catalyzed oxidation of alcohols to aldehydes is of great significance in biological systems.

Oxidation of a primary alcohol results in an aldehyde that contains two less hydrogen atoms than the original alcohol. Hence this reaction can also be considered as a dehydrogenation reaction. Recall from Section 8.4 that oxidation and dehydrogenation both involve the loss of electrons. This loss of electrons may be manifested by either the addition of oxygen or the removal of hydrogen. Similarly, reduction, which may appear to involve only a gain of hydrogen or a loss of oxygen, implies a gain of electrons. The great majority of oxidation-reduction reactions occur through a transfer of electrons, although the balanced net equation shows only a change of atoms within molecules. Since these reactions occur widely in organic chemistry, and because they are of utmost importance in biochemical metabolism, we shall write the pertinent half-reactions for many of the oxidation-reduction reactions that we study (Section 8.6). This will serve to emphasize the electron transfers that are involved. The general equation for the oxidation (dehydrogenation) of primary alcohols to produce aldehydes is

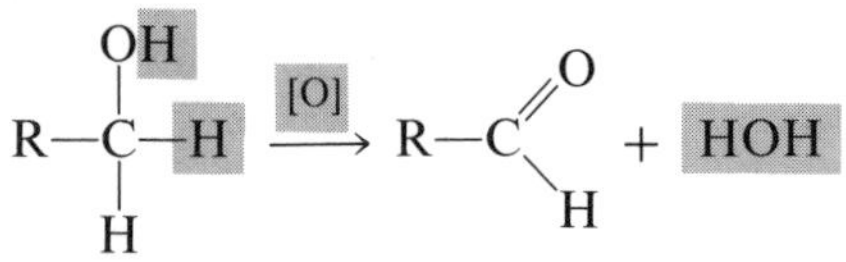

In the preparation of acetaldehyde, ethyl alcohol is slowly added to a warm mixture of potassium dichromate and dilute sulfuric acid, which is maintained at a temperature of 50 °C. The aldehyde boils at a temperature of 21 °C, and thus it distills off as quickly as it is formed. This prevents further oxidation to acetic acid.

$$\underset{\text{(bp 78 °C)}}{3\,CH_3CH_2OH} + \underset{\text{Orange}}{K_2Cr_2O_7} + 4\,H_2SO_4 \longrightarrow \underset{\text{(bp 21 °C)}}{3\,CH_3C(=O)H} + K_2SO_4 + \underset{\text{Green}}{Cr_2(SO_4)_3} + 7\,H_2O$$

Oxidation

$$CH_3CH_2OH \longrightarrow CH_3C(=O)H + 2\,H^+ + 2\,e^-$$

Reduction

$$14\,H^+ + Cr_2O_7^{2-} + 6\,e^- \longrightarrow 2\,Cr^{3+} + 7\,H_2O$$

Benzylic alcohols are conveniently oxidized to the corresponding aldehydes with freshly prepared manganese dioxide, MnO_2. This reagent is strong enough to oxidize the alcohol to the aldehyde but is unable to oxidize the aromatic aldehyde to the acid.

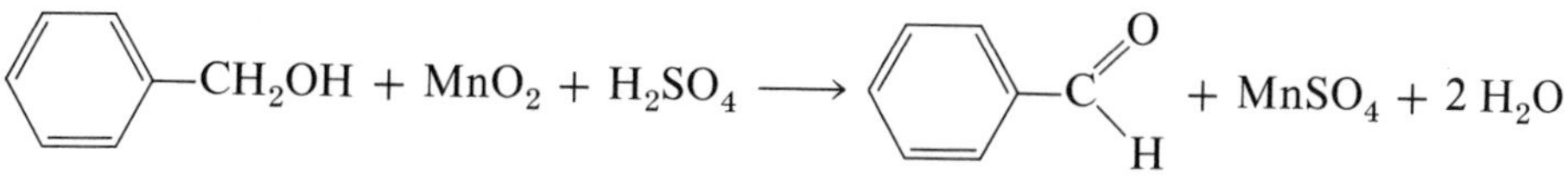

Benzyl alcohol Benzaldehyde

Oxidation

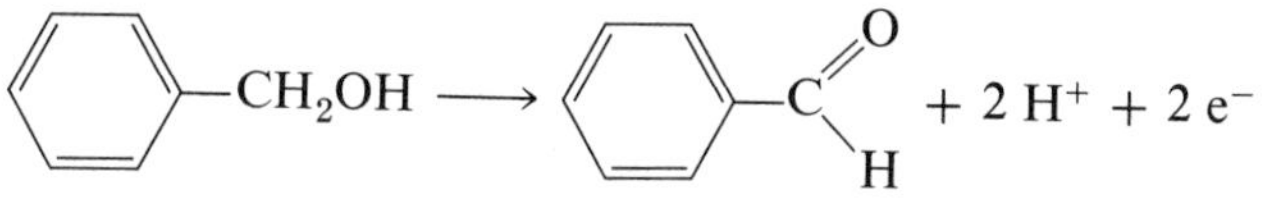

Reduction

$$4\,H^+ + MnO_2 + 2\,e^- \longrightarrow Mn^{2+} + 2\,H_2O$$

Like the aldehydes, ketones are obtained by the oxidation of an alcohol. However, the alcohol that is used must be a secondary alcohol and no special precautions need be taken to remove the ketone from the reaction mixture since it is not susceptible to futher oxidation. The oxidation is carried out in acid solution, and potassium dichromate or chromium(VI) oxide (CrO_3) is commonly used as the oxidizing agent.

General Equation

$$R{-}\overset{\displaystyle OH}{\underset{\displaystyle H}{C}}{-}R' \xrightarrow{[O]} R{-}\overset{\displaystyle O}{\overset{\|}{C}}{-}R' + HOH$$

Specific Equations

$$3\,CH_3\overset{\displaystyle OH}{\underset{\displaystyle H}{C}}CH_3 + K_2Cr_2O_7 + 4\,H_2SO_4 \longrightarrow 3\,CH_3\overset{\displaystyle O}{\overset{\|}{C}}CH_3 + K_2SO_4 + Cr_2(SO_4)_3 + 7\,H_2O$$

Isopropyl alcohol Acetone

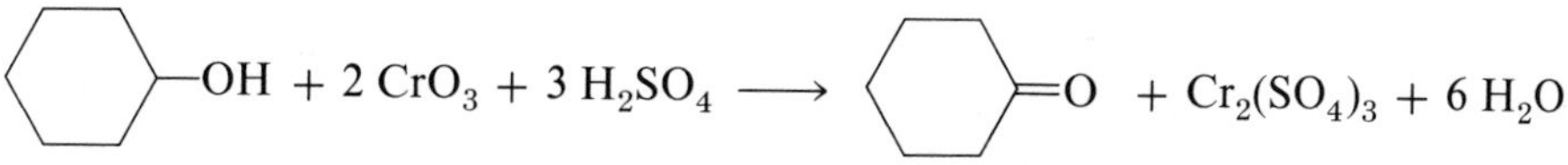

Cyclohexanol Cyclohexanone

A useful synthesis of aromatic or mixed aromatic-aliphatic ketones is the Friedel–Crafts reaction (Section 15.3). Lewis acids (such as $AlCl_3$, $ZnCl_2$, BF_3) catalyze the reaction of an aromatic hydrocarbon with an aryl or an alkyl acid chloride (see Section 19.5-b) to yield the desired ketone.

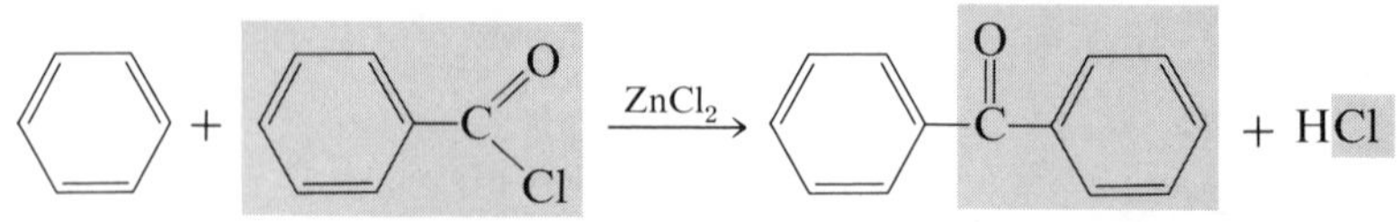

O
C
Cl
ZnCl2
O
C
+
+ HCl

Benzene Benzoyl chloride Benzophenone

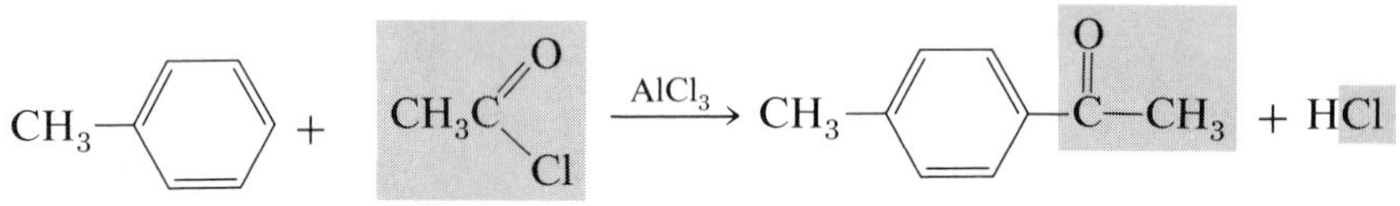

CH3
+
O
CH3C
Cl
AlCl3
CH3
O
C—CH3
+ HCl

Toluene Acetyl chloride *p*-Methylacetophenone

As we shall see in Chapters 28 and 29, the electrons that are released during the oxidation of alcohols to carbonyl compounds are converted into a form of energy that can be utilized by the cell. These biochemical oxidation reactions are carried out at body temperature (~37 °C) and are catalyzed by enzymes. Notice that in the following reaction the enzyme selectively catalyzes the oxidation of the secondary alcohol group to a ketone but it does not oxidize the primary alcohol group to an aldehyde. We shall discuss enzyme specificity in Section 26.6.

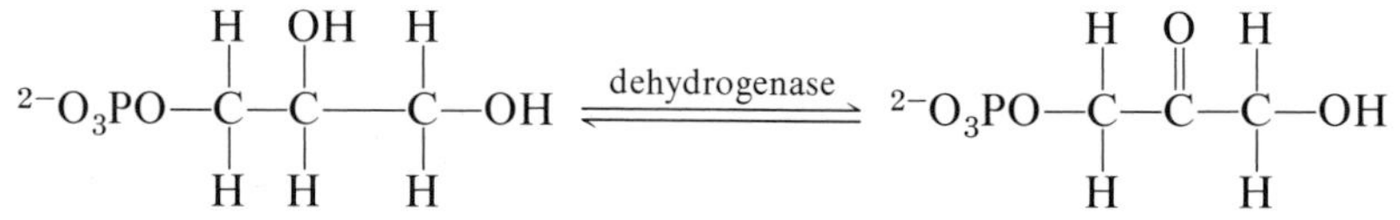

H OH H
2−O3PO—C—C—C—OH
H H H
dehydrogenase
H O H
2−O3PO—C—C—C—OH
H H

Glycerol 3-phosphate Dihydroxyacetone phosphate

18.3 Physical Properties of Aldehydes and Ketones

With the exception of the gaseous formaldehyde, the majority of the aldehydes are liquids (Table 18.3). (The physical state of acetaldehyde, bp 21 °C, depends upon the temperature of the laboratory; in warm rooms acetaldehyde exists as a gas.) Although the lower members of the series have pungent odors, many other aldehydes are used in making perfumes and artificial flavorings. Formaldehyde and acetaldehyde are infinitely soluble in water. The higher homologs are sparingly soluble in water, but generally are soluble in organic solvents. As a rule, they are lighter than water.

Since aldehydes lack hydroxyl groups, they are incapable of intermolecular hydrogen bonding; thus they have boiling points considerably lower than their corresponding alcohols. On the other hand, their boiling points are somewhat higher than the corresponding alkanes and ethers, indicating the presence of weak intermolecular attractions (dipole–dipole forces).

The physical properties of the ketones are almost identical to those of

Table 18.3 Physical Constants of Some Aldehydes

Formula	*Common Name*	*MP (°C)*	*BP (°C)*	*Solubility (grams/100 grams of water)*
$HCHO$	Formaldehyde	−92	−21	miscible
CH_3CHO	Acetaldehyde	−121	21	miscible
CH_3CH_2CHO	Propionaldehyde	−81	49	20
$CH_3(CH_2)_2CHO$	Butyraldehyde	−99	76	3.6
C_6H_5CHO	Benzaldehyde	−26	180	0.3

the corresponding aldehydes. Acetone is the only ketone that is completely soluble in water. The higher homologs are colorless liquids, are slightly soluble in water, and, unlike the aldehydes, have rather bland odors. As a general rule, ketones are less dense than water. Table 18.4 lists some physical constants for several of the ketones.

18.4 The Carbonyl Group

The chemical reactions of aldehydes and ketones are a function of the carbonyl group. Since the carbon–oxygen double bond is a site of unsaturation, it resembles the carbon–carbon double bond in many of its properties. The planar configuration around a double-bonded carbon (see Figure 14.2) occurs also for the carbonyl group (Figure 18.1a). The electrons of the carbon–oxygen double bond, like the electrons of the carbon–carbon double bond, are affected to a greater extent by approaching reagents than are the electrons in single bonds.

Table 18.4 Physical Constants of Some Ketones

Formula	*Name*	*MP (°C)*	*BP (°C)*	*Solubility (grams/100 grams of water)*
$CH_3\overset{O}{\overset{\parallel}{C}}CH_3$	Acetone	−95	56	Miscible
$CH_3\overset{O}{\overset{\parallel}{C}}CH_2CH_3$	Butanone	−86	80	25
$CH_3CH_2\overset{O}{\overset{\parallel}{C}}CH_2CH_3$	3-Pentanone	−40[a]	103	5
$CH_3\overset{O}{\overset{\parallel}{C}}CH_2CH_2CH_3$	2-Pentanone	−78	102	6
$(CH_3)_2CH\overset{O}{\overset{\parallel}{C}}CH_3$	3-Methyl-2-butanone	−92	94	—

[a]Again note that the symmetrical isomer has the highest melting point.

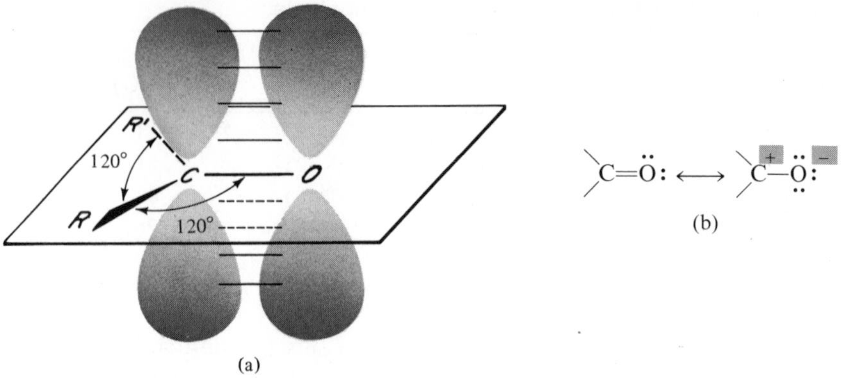
R'
120°
C
O
R
120°
(a)
(b)

Figure 18.1 (a) The carbonyl group and the two atoms directly bonded to the carbonyl carbon lie in a plane. (b) Resonance forms of the carbonyl group.

There is, however, one important difference between a C=C bond and a C=O bond. Recall that the electrons of a carbon–carbon double bond are shared equally between the two carbon atoms since they are identical and have identical electronegativities. The electronic distribution is symmetrical and the bond is strictly nonpolar, covalent. Oxygen, however, has an appreciably higher electronegativity than carbon, and so the oxygen atom has a tendency to draw the electron pair more closely to itself and away from the carbon atom. Consequently, the electronic distribution of the bond is not symmetrical; the oxygen end is slightly negative and the carbon end is slightly positive. The bond is said to be polarized and can be represented as the sum of two resonance contributors, a neutral form and a doubly charged form (Figure 18.1b). The doubly charged form is a valid way of representing the carbonyl group, and it accounts for the relatively high boiling points of the aldehydes and ketones. Physical measurements indicate that carbonyl compounds also have relatively high dipole moments. Thus, if aldehydes or ketones are placed between the poles of an electric field, the molecules will orient themselves so that the oxygen ends of the molecules face the positive pole. Most important is an examination of the products formed when reagents add across the carbon–oxygen double bond. The products always result from the positive fragment of the reagent adding to the oxygen, and the negative fragment adding to the carbon.

18.5 Chemical Properties of Aldehydes and Ketones

a. Addition Reactions

Aldehydes and ketones will react with a great variety of compounds by addition. The course of the addition reaction in each of the following examples is identical; that is, there is an initial attack by a nucleophilic species on the carbonyl carbon, followed by the addition of a positive species to the oxygen.

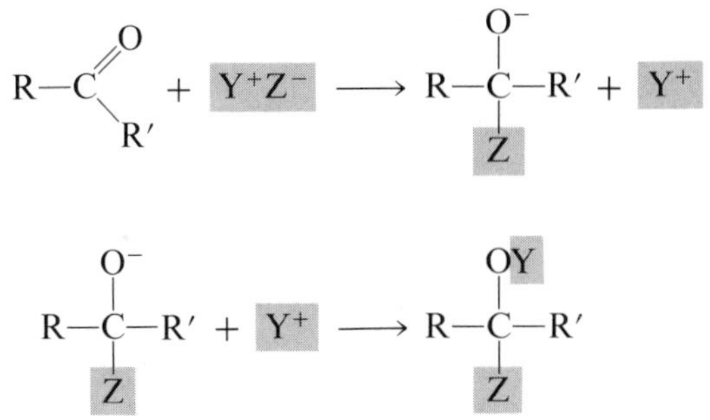

$$R-\overset{\displaystyle O}{\underset{\displaystyle R'}{\overset{\|}{C}}} + \boxed{Y^+Z^-} \longrightarrow R-\overset{\displaystyle O^-}{\underset{\displaystyle Z}{\overset{|}{C}}}-R' + \boxed{Y^+}$$

$$R-\overset{\displaystyle O^-}{\underset{\displaystyle Z}{\overset{|}{C}}}-R' + \boxed{Y^+} \longrightarrow R-\overset{\displaystyle OY}{\underset{\displaystyle Z}{\overset{|}{C}}}-R'$$

1. Addition of Water and Alcohol. The reaction of carbonyl compounds with water results initially in the formation of carbonyl hydrates. The addition of one mole of an alcohol to an aldehyde or ketone yields a *hemiacetal* or *hemiketal,* respectively. But with few exceptions, the products so formed are unstable and revert back to the original aldehyde or ketone.[1]

$$CH_3CH_2\overset{\displaystyle O}{\overset{\diagup}{\underset{\displaystyle H}{C}}} + \boxed{HOH} \rightleftharpoons \left[CH_3CH_2\overset{\displaystyle OH}{\underset{\displaystyle OH}{\overset{|}{C}}}-H \right]$$

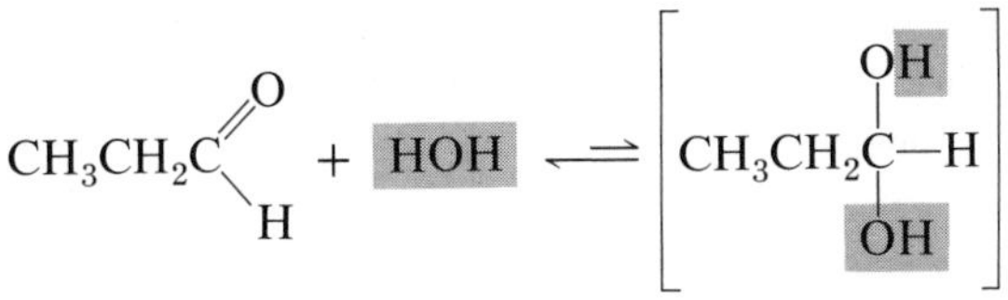

Unstable hydrate

$$CH_3-\overset{\displaystyle O}{\overset{\|}{C}}-CH_3 + \boxed{HOH} \rightleftharpoons \left[CH_3-\overset{\displaystyle OH}{\underset{\displaystyle OH}{\overset{|}{C}}}-CH_3 \right]$$

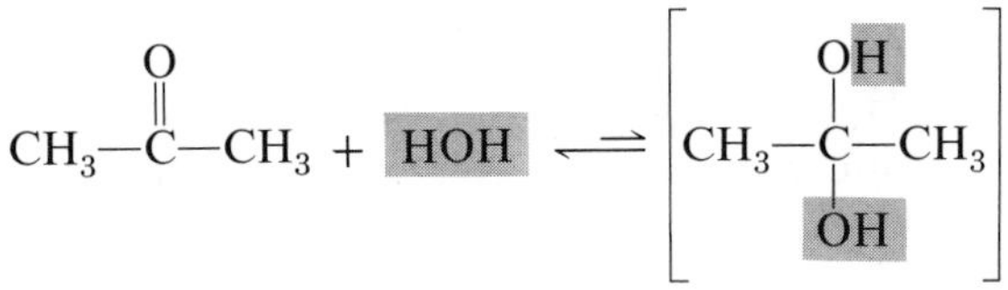

Acetone hydrate
(very unstable)

[1] However, if the carbonyl group is adjacent to electron-withdrawing groups, as in chloral, the hydrate can be isolated.

$$Cl-\overset{\displaystyle Cl}{\underset{\displaystyle Cl}{\overset{|}{C}}}-\overset{\displaystyle O}{\overset{\diagup}{\underset{\displaystyle H}{C}}} + \boxed{HOH} \longrightarrow Cl-\overset{\displaystyle Cl}{\underset{\displaystyle Cl}{\overset{|}{C}}}-\overset{\displaystyle OH}{\underset{\displaystyle OH}{\overset{|}{C}}}-H$$

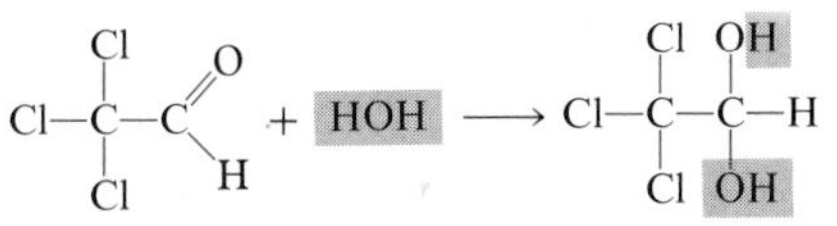

Chloral   Chloral hydrate

Chloral hydrate is stabilized by the presence of three electronegative chlorines that further increase the polarity of the carbon–oxygen bond. It is a solid, is very soluble in water, and is one of the very few organic compounds that possesses two hydroxyl groups on the same carbon atom. It is used in sedatives and is given to patients in hospitals as a nighttime sleeping pill (Noctec). Chloral hydrate dissolved in alcohol constitutes the so-called knock-out drops or Mickey Finns. Recall (Section 16.6) that chloral hydrate is one of the reagents for the synthesis of DDT.

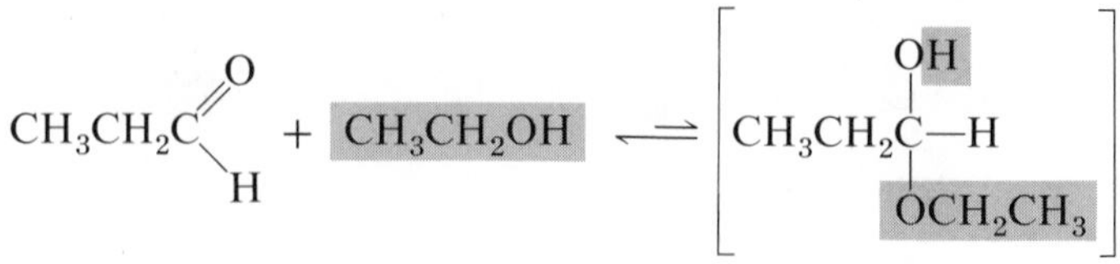

$$\underset{H}{\overset{O}{CH_3CH_2C}} + \boxed{CH_3CH_2OH} \rightleftharpoons \left[\underset{OCH_2CH_3}{\overset{OH}{CH_3CH_2C-H}} \right]$$

Unstable hemiacetal

$$\underset{}{\overset{O}{CH_3\overset{\parallel}{C}CH_3}} + \boxed{CH_3OH} \rightleftharpoons \left[\underset{\boxed{OCH_3}}{\overset{\boxed{OH}}{CH_3\overset{\vert}{\underset{\vert}{C}}CH_3}} \right]$$

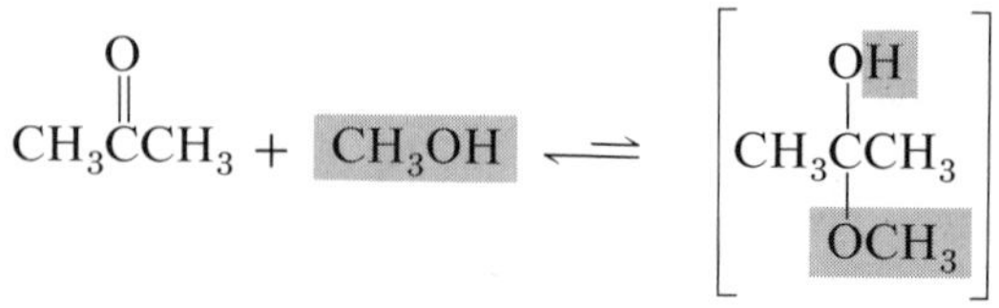

Unstable hemiketal

Examples of stable hemiacetals and hemiketals are found in the cyclic forms of four- and five-carbon hydroxy aldehydes and ketones. These cyclic compounds result from intramolecular interaction between an alcohol group and the carbonyl group. These cyclization reactions are of particular significance in our discussion of the structures of carbohydrates (see Section 23.4).

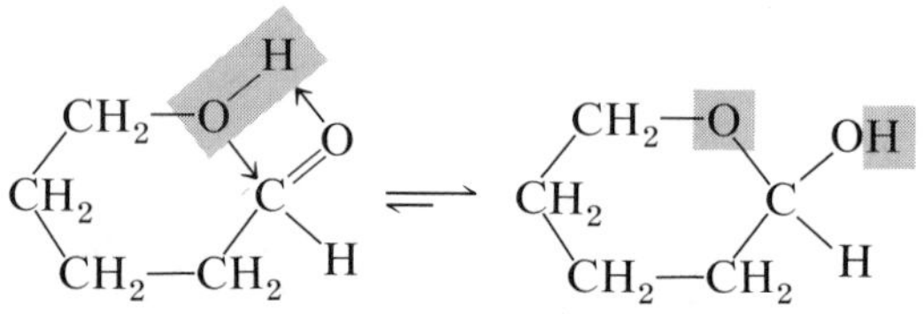

5-Hydroxypentanal   Stable cyclic hemiacetal

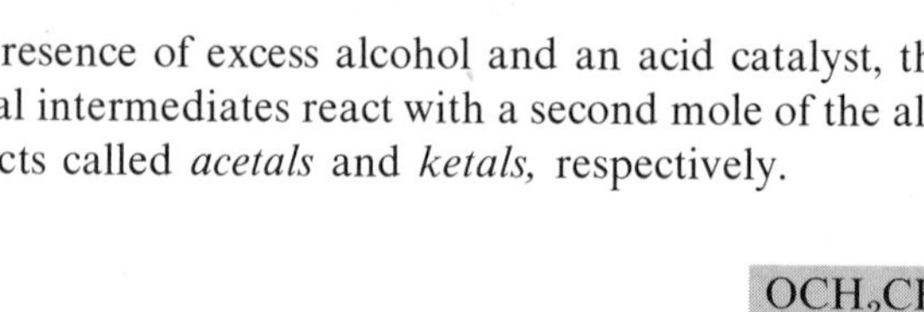

6-Hydroxy-2-hexanone   Stable cyclic hemiketal

In the presence of excess alcohol and an acid catalyst, the hemiacetal and hemiketal intermediates react with a second mole of the alcohol to form stable products called *acetals* and *ketals,* respectively.

$$\underset{OCH_2CH_3}{\overset{OH}{CH_3CH_2\overset{\vert}{\underset{\vert}{C}}-H}} + \boxed{CH_3CH_2OH} \overset{H^+}{\rightleftharpoons} \underset{OCH_2CH_3}{\overset{OCH_2CH_3}{CH_3CH_2\overset{\vert}{\underset{\vert}{C}}-H}} + \boxed{HOH}$$

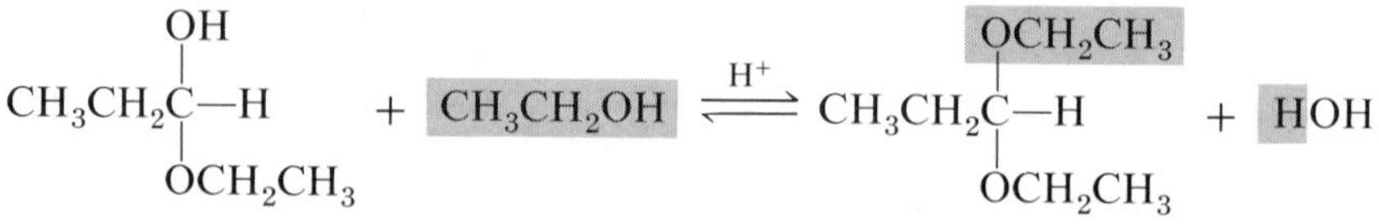

An acetal

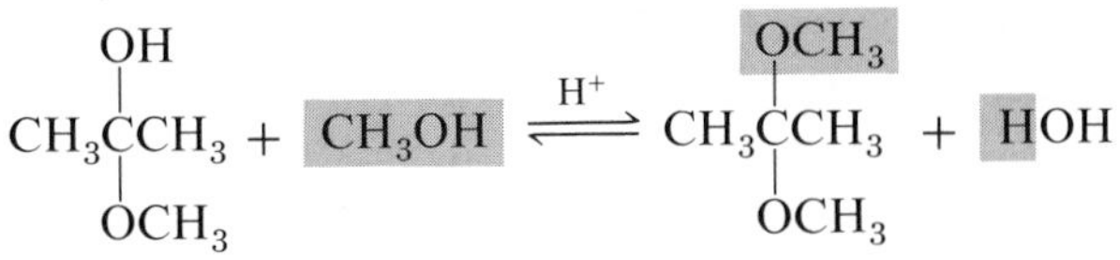

A ketal

*2. **Addition of Hydrogen Cyanide.*** Since HCN is a poisonous gas, it is usually prepared in the reaction mixture from potassium cyanide and sulfuric acid. The products of hydrogen cyanide addition to aldehydes and ketones are called *cyanohydrins* (or hydroxynitriles). They are important intermediates in the preparation of hydroxy acids and amino acids.

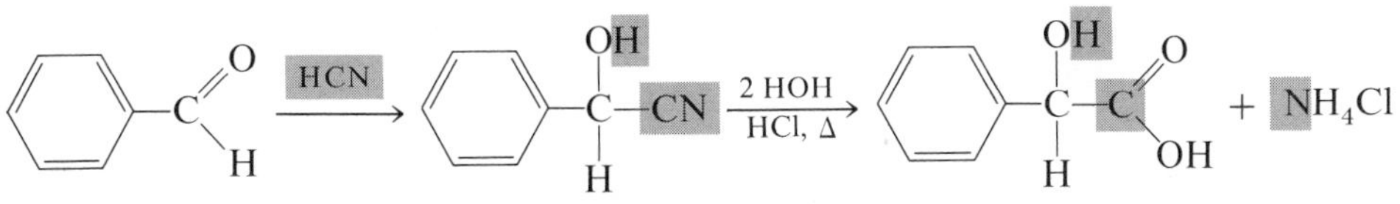

Benzaldehyde — Mandelonitrile — Mandelic acid

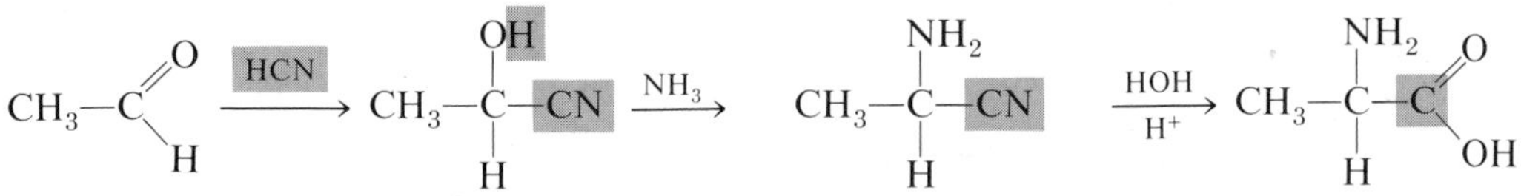

Acetaldehyde — Acetaldehyde cyanohydrin — 1-Amino-1-cyanoethane — Alanine (α-Aminopropionic acid)

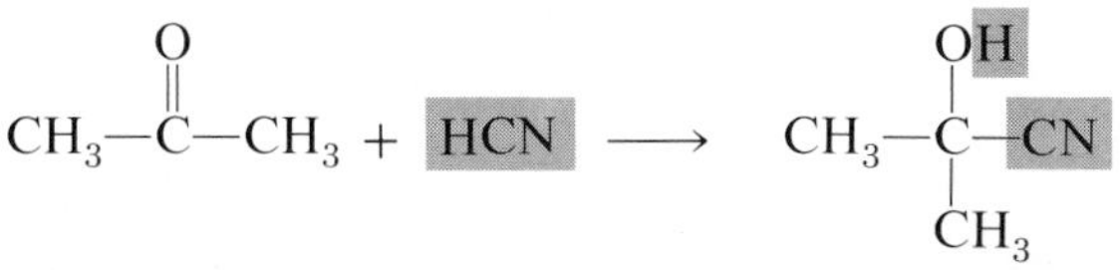

Acetone — Acetone cyanohydrin (2-Cyano-2-propanol)

In an important commercial reaction, acetone cyanohydrin is converted to methyl methacrylate by refluxing with methanol and a strong acid. The methyl methacrylate is then polymerized to form a transparent plastic known as Lucite and Plexiglas.

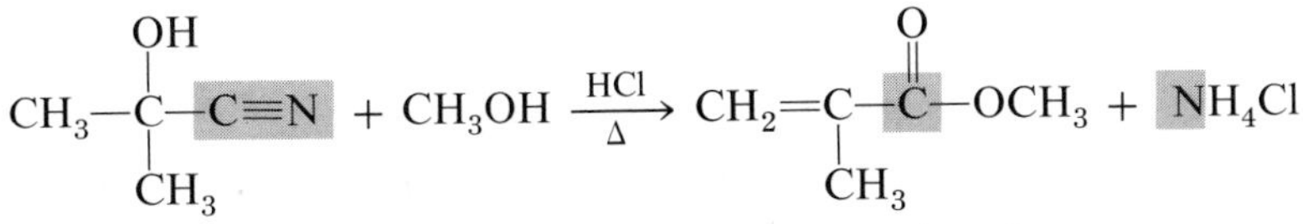

Methyl methacrylate

*3. **Condensation with Ammonia Derivatives.*** A condensation reaction is one in which two molecules combine chemically to form a larger molecule, usually with the concomitant elimination of some small molecule such as water. Certain derivatives of ammonia will react with aldehydes and ketones to form unstable intermediates that then lose water to form stable crystalline solids. Such products have very sharply defined melting points and so are often used for the identification of aldehydes and ketones. (Upon treatment with dilute acid, these compounds revert to the original carbonyl compounds.) The ammonia derivatives that are usually employed in this reaction are hydroxylamine (H_2NOH), semicarbazide ($H_2NNHCONH_2$), and phenylhydrazine ($H_2NNHC_6H_5$). Notice that all three reagents are of the form H_2NB. When B is —OH (from hydroxylamine), the product formed is an *oxime;* when B is $—NHCONH_2$ (from semicarbazide), the product formed is a *semicarbazone;* when B is $—NHC_6H_5$ (from phenylhydrazine), the product formed is a *phenylhydrazone.*

General Equation

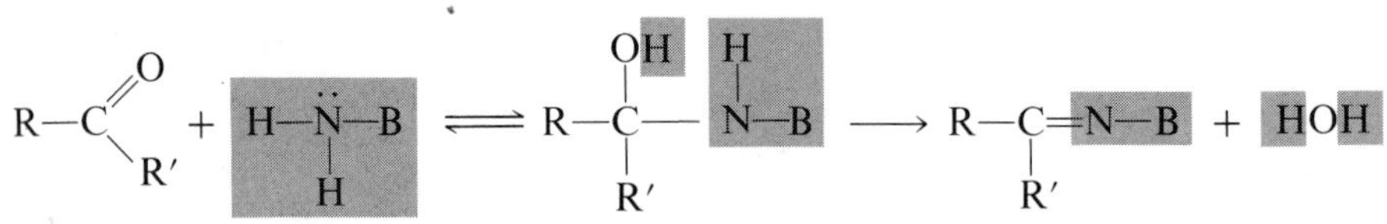

Unstable   An imine[2]

Specific Equations

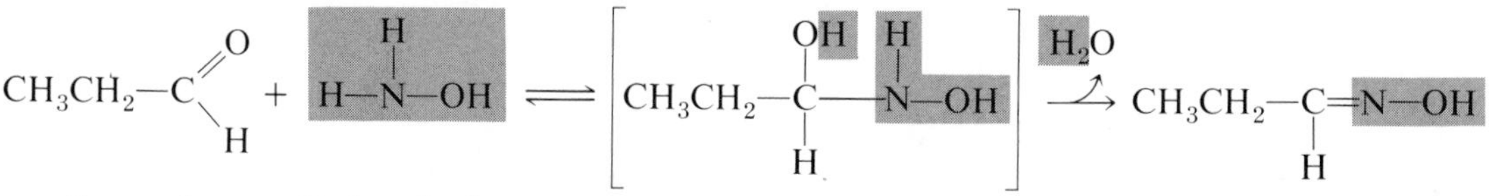

Propanal   Hydroxylamine   Propionaldehyde oxime

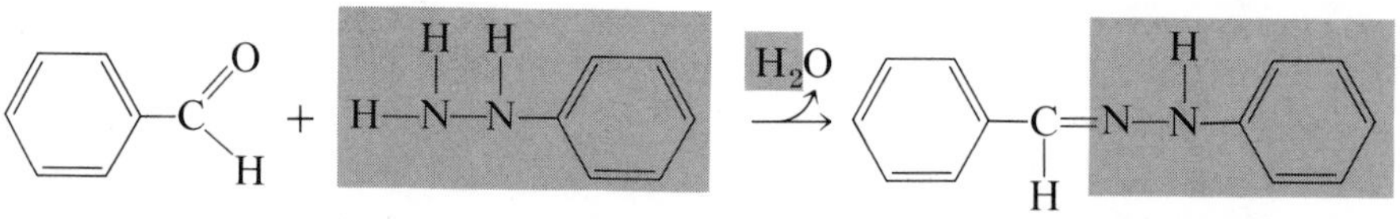

Benzaldehyde   Phenylhydrazine   Benzaldehyde phenylhydrazone

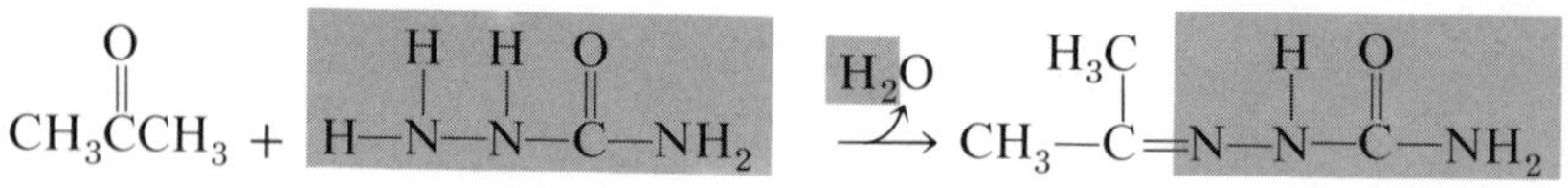

Acetone   Semicarbazide   Acetone semicarbazone

If B in the general equation above is an organic group, then the ammonia derivative is a primary amine (see Section 20.6). The condensation reaction between a carbonyl compound and a primary amine gives rise to an important type of imine, referred to as a *Schiff base*. Schiff bases are intermediates in several enzymic reactions involving interaction between amino groups and carbonyl groups.

[2]The carbon–nitrogen double bond (>C=N) is referred to as the *imino* group; compounds possessing this structure are called *imines*.

General Equation

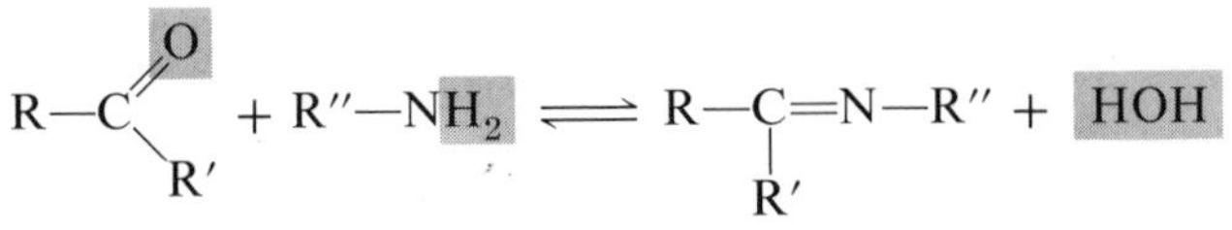

$$R-\overset{\overset{\displaystyle O}{\|}}{\underset{\underset{\displaystyle R'}{|}}{C}} + R''-NH_2 \rightleftharpoons R-\underset{\underset{\displaystyle R'}{|}}{C}=N-R'' + HOH$$

| Carbonyl | Primary | Schiff base |
| compound | amine | |

Specific Equation

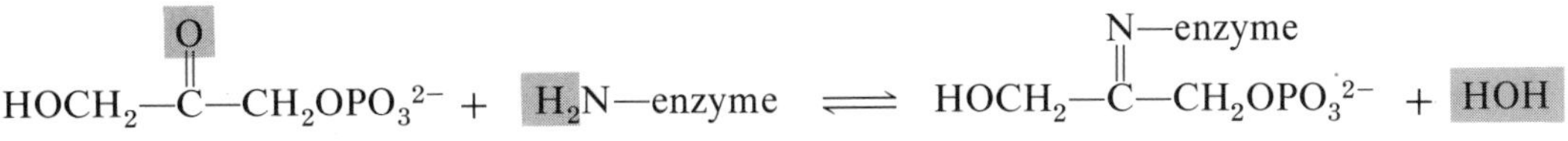

$$HOCH_2-\overset{\overset{\displaystyle O}{\|}}{C}-CH_2OPO_3^{2-} + H_2N-enzyme \rightleftharpoons HOCH_2-\overset{\overset{\displaystyle N-enzyme}{\|}}{C}-CH_2OPO_3^{2-} + HOH$$

| Dihydroxyacetone | Enzyme containing | Schiff base of dihydroxyacetone |
| phosphate | a free —NH$_2$ group | phosphate and enzyme |

4. Reaction with Organometallic Reagents. The reactions of carbonyl compounds with Grignard reagents or lithium alkyls (or aryls) afford convenient methods for the preparation of a wide variety of alcohols. These reactions are usually carried out by preparing the Grignard reagent or the lithium alkyl in an ether solution (Section 16.4-c) and then slowly adding the carbonyl compound. The reaction mixture is easily hydrolyzed by water to yield the desired alcohol. Notice that these reactions also proceed with addition across the carbon–oxygen double bond. The **R—** (or **Ar—**) group adds to the carbon and the magnesium halide or lithium adds to the oxygen.

Synthesis of a Primary Alcohol. Addition of formaldehyde to RMgX or RLi, followed by hydrolysis of the intermediate addition compound yields a primary alcohol.

General Equation

$$H-\overset{\overset{\displaystyle O}{\diagup}}{\underset{\underset{\displaystyle H}{\diagdown}}{C}} + RMgX \xrightarrow[\text{or THF}]{\text{ether}} H-\overset{\overset{\displaystyle OMgX}{|}}{\underset{\underset{\displaystyle R}{|}}{C}}-H \xrightarrow{HCl} H-\overset{\overset{\displaystyle OH}{|}}{\underset{\underset{\displaystyle R}{|}}{C}}-H + Mg\overset{\diagup Cl}{\diagdown X}$$

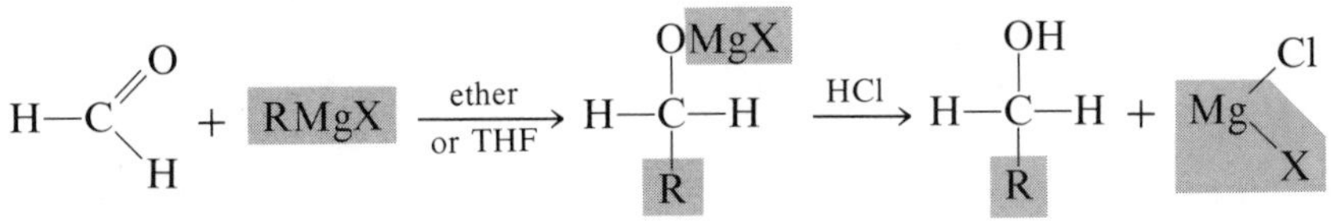

Specific Equation

$$H-\overset{\overset{\displaystyle O}{\diagup}}{\underset{\underset{\displaystyle H}{\diagdown}}{C}} + CH_3CH_2MgBr \xrightarrow[\text{or THF}]{\text{ether}} H-\overset{\overset{\displaystyle OMgBr}{|}}{\underset{\underset{\displaystyle CH_2CH_3}{|}}{C}}-H \xrightarrow{HCl} H-\overset{\overset{\displaystyle OH}{|}}{\underset{\underset{\displaystyle CH_2CH_3}{|}}{C}}-H + Mg\overset{\diagup Cl}{\diagdown Br}$$

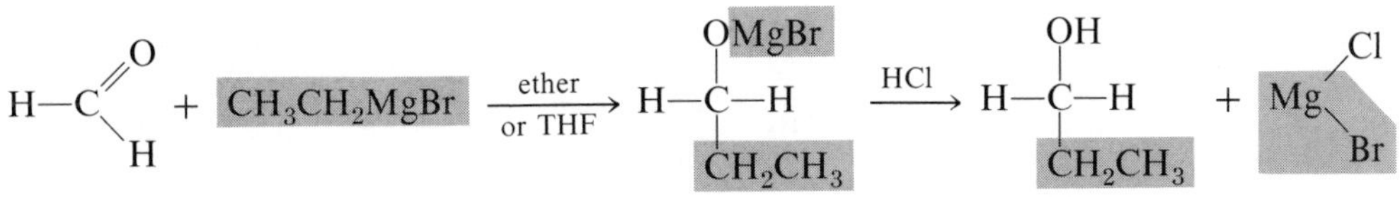

1-Propanol

Synthesis of a Secondary Alcohol. Formaldehyde is the only carbonyl compound that reacts with organometallic reagents to give a primary alcohol. All other aldehydes produce secondary alcohols exclusively.

General Example

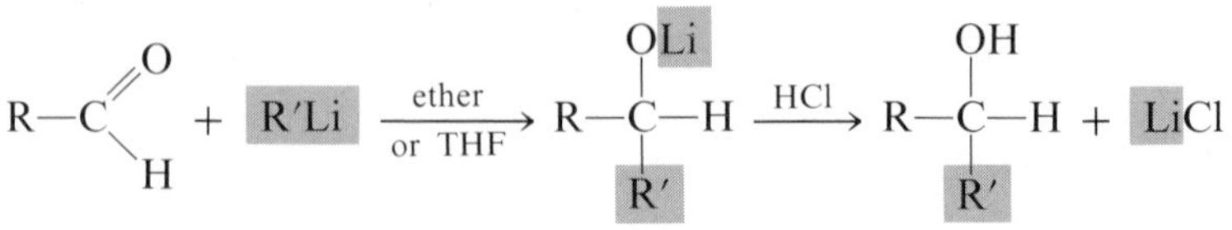
R—C(=O)H + R′Li
ether or THF
R—C(OLi)(R′)—H
HCl
R—C(OH)(R′)—H + LiCl

Specific Example

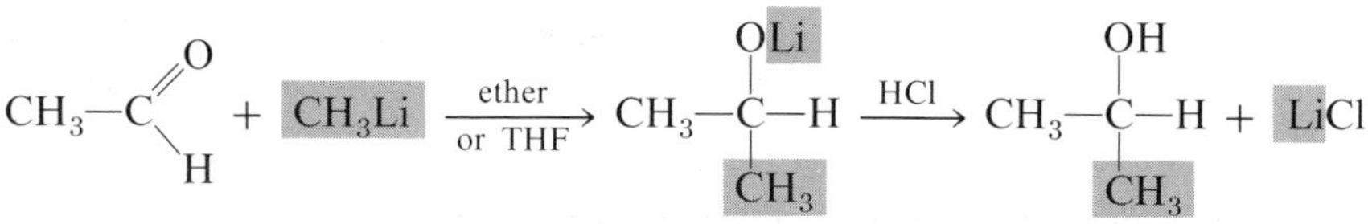
CH3—C(=O)H + CH3Li
ether or THF
CH3—C(OLi)(CH3)—H
HCl
CH3—C(OH)(CH3)—H + LiCl

Acetadehyde 2-Propanol

In the preparation of secondary alcohols by this method, there will always be a choice as to which aldehyde and which organic halide will be utilized. For example, 1-phenyl-1-ethanol can be synthesized from bromobenzene and acetaldehyde or from benzaldehyde and methyl bromide. The choice is usually made on the basis of cost and availability of the starting materials.

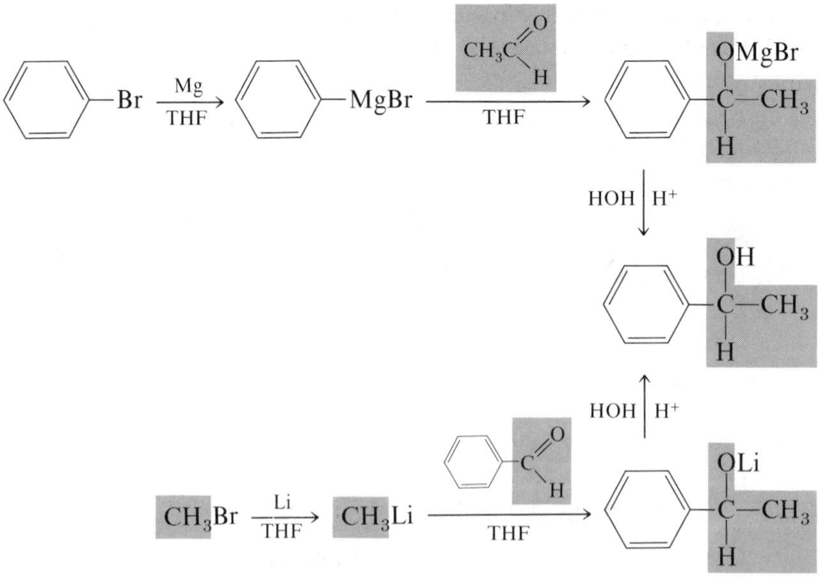
Br
Mg
THF
MgBr
THF
OMgBr
CH3
H
HOH
H+
OH
CH3
H
HOH
H+
OLi
CH3
H
CH3Br
Li
THF
CH3Li
THF

Synthesis of a Tertiary Alcohol. Ketones will react with organometallic reagents to produce tertiary alcohols.

General Example

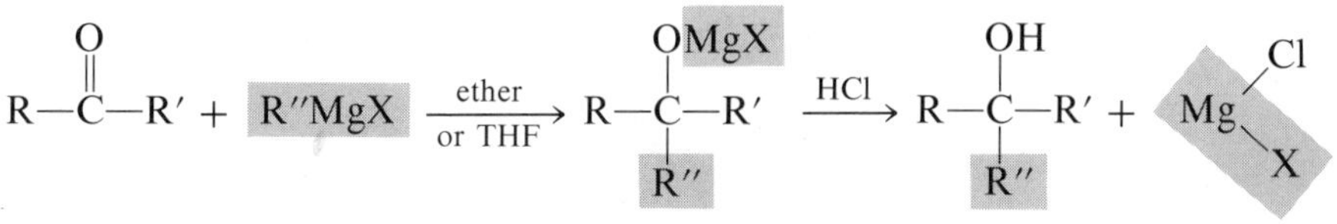

$$\underset{\substack{O \\ \| \\ R-C-R'}}{} + R''MgX \xrightarrow[\text{or THF}]{\text{ether}} \underset{\substack{OMgX \\ | \\ R-C-R' \\ | \\ R''}}{} \xrightarrow{\text{HCl}} \underset{\substack{OH \\ | \\ R-C-R' \\ | \\ R''}}{} + Mg\underset{X}{\overset{Cl}{<}}$$

Specific Examples

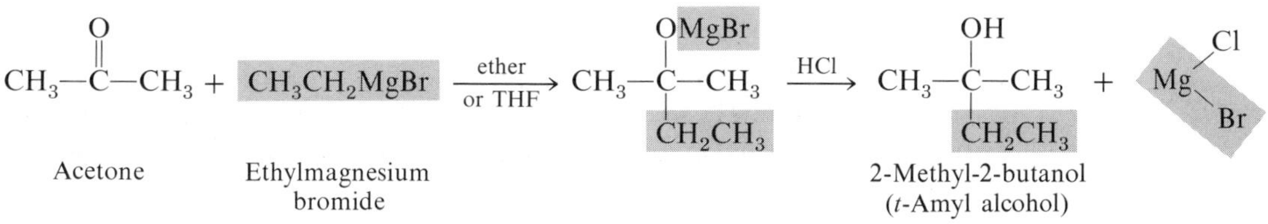

$$\underset{\substack{O \\ \| \\ CH_3-C-CH_3}}{} + CH_3CH_2MgBr \xrightarrow[\text{or THF}]{\text{ether}} \underset{\substack{OMgBr \\ | \\ CH_3-C-CH_3 \\ | \\ CH_2CH_3}}{} \xrightarrow{\text{HCl}} \underset{\substack{OH \\ | \\ CH_3-C-CH_3 \\ | \\ CH_2CH_3}}{} + Mg\underset{Br}{\overset{Cl}{<}}$$

Acetone   Ethylmagnesium bromide   2-Methyl-2-butanol (*t*-Amyl alcohol)

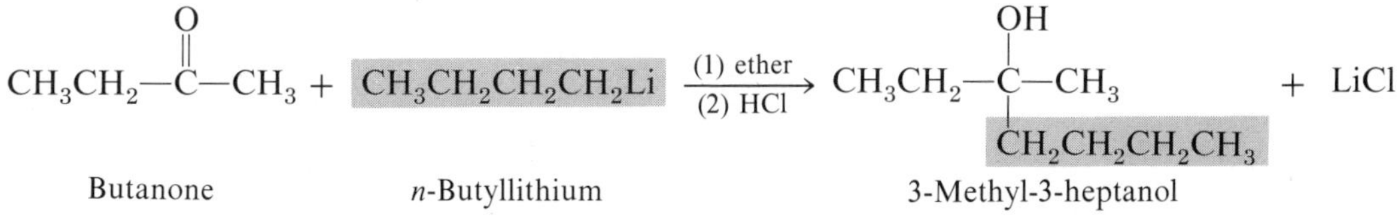

$$\underset{\substack{O \\ \| \\ CH_3CH_2-C-CH_3}}{} + CH_3CH_2CH_2CH_2Li \xrightarrow[\text{(2) HCl}]{\text{(1) ether}} \underset{\substack{OH \\ | \\ CH_3CH_2-C-CH_3 \\ | \\ CH_2CH_2CH_2CH_3}}{} + LiCl$$

Butanone   *n*-Butyllithium   3-Methyl-3-heptanol

5. Self-Addition (Aldol Condensation). The hydrogen atoms that are attached to the carbon atom adjacent to the carbonyl carbon have a particular significance because they are more easily released as protons than any of the other hydrogens in the molecule. The explanation for this is given in Section 18.5-b. Such hydrogens are referred to as *alpha hydrogens,*[3] because they are bonded to the alpha carbon.

Any aldehyde molecule that contains α-hydrogens may add to itself or to another aldehyde. The compound resulting from such a process contains both a carbonyl group and a hydroxyl group within the same molecule. The

---

[3] The Greek letters α, β, etc., are used to designate the positions of carbon atoms with respect to the carbonyl carbon. For example,

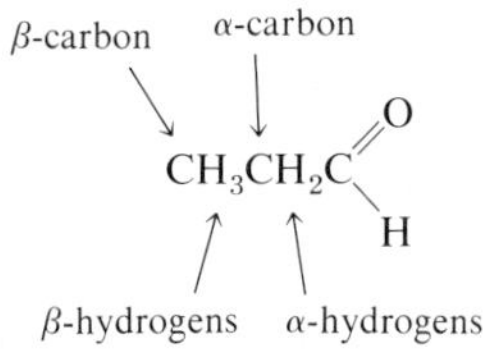

$$\underset{\underset{\substack{\uparrow \quad\quad \uparrow \\ \beta\text{-hydrogens} \quad \alpha\text{-hydrogens}}}{\overset{\overset{\substack{\beta\text{-carbon} \quad \alpha\text{-carbon} \\ \searrow \quad\quad \downarrow}}{}}{CH_3CH_2C}}}{} \overset{O}{\underset{H}{<}}$$

The hydrogens bonded to these carbons are then called α-hydrogens, β-hydrogens, etc.

reaction, which is catalyzed by a dilute base, is known as an *aldol condensation* (aldol is a composite word derived from the words *ald*ehyde and alco*hol*). The reaction proceeds in the same manner as all other carbonyl addition reactions. An α-hydrogen adds to a carbonyl oxygen; the remainder of the molecule adds to the carbonyl carbon.

$$CH_3-\overset{\displaystyle O}{\underset{\displaystyle H}{C}} \quad + \quad H-\overset{\displaystyle H}{\underset{\displaystyle H}{C}}-\overset{\displaystyle O}{\underset{\displaystyle H}{C}} \quad \overset{OH^-}{\rightleftharpoons} \quad CH_3-\overset{\displaystyle HO}{\underset{\displaystyle H}{C}}-\overset{\displaystyle H}{\underset{\displaystyle H}{C}}-\overset{\displaystyle O}{\underset{\displaystyle H}{C}}$$

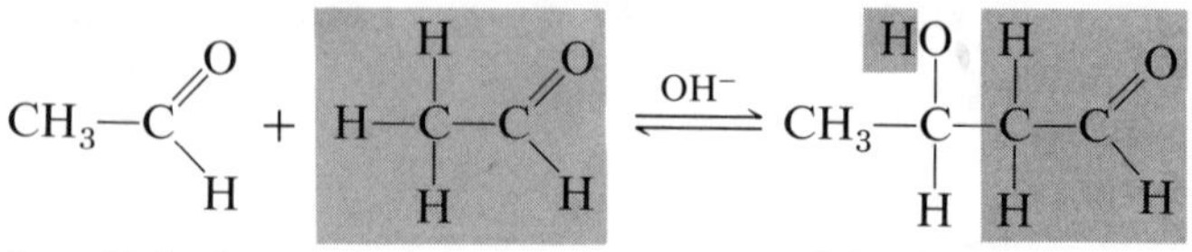

Acetaldehyde   3-Hydroxybutanal

The addition product is very susceptible to dehydration. Gentle warming of the solution results in the formation of an α,β-unsaturated aldehyde. (Such a molecule has a carbon–carbon double bond and a carbonyl group within the same molecule.)

$$CH_3-\overset{\displaystyle HO}{\underset{\displaystyle H}{C}}-\overset{\displaystyle H}{\underset{\displaystyle H}{C}}-\overset{\displaystyle O}{\underset{\displaystyle H}{C}} \quad \overset{\Delta}{\longrightarrow} \quad CH_3-\overset{\displaystyle}{\underset{\displaystyle H}{C}}=\overset{\displaystyle}{\underset{\displaystyle H}{C}}-\overset{\displaystyle O}{\underset{\displaystyle H}{C}} \quad + \quad HOH$$

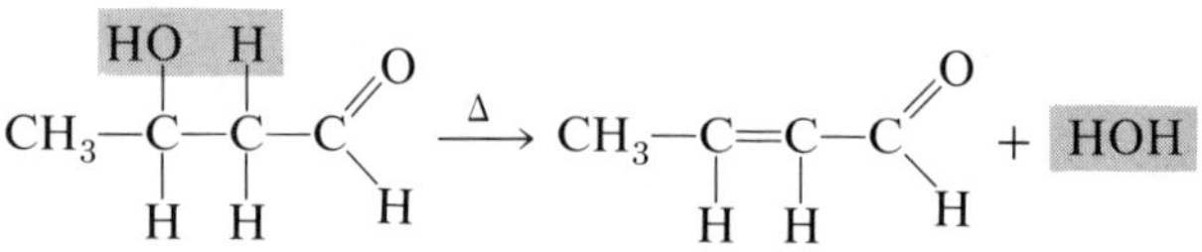

2-Butenal
(Crotonaldehyde)

$$CH_3CH_2\overset{\displaystyle O}{\underset{\displaystyle H}{C}} + CH_3\overset{\displaystyle O}{\underset{\displaystyle H}{\overset{\displaystyle}{\underset{\displaystyle}{C}}C}} \overset{OH^-}{\longrightarrow} CH_3CH_2\overset{\displaystyle HO}{\underset{\displaystyle H}{C}}-\overset{\displaystyle H}{\underset{\displaystyle CH_3}{C}}-\overset{\displaystyle O}{\underset{\displaystyle H}{C}} \overset{}{\underset{\Delta}{\longrightarrow}} CH_3CH_2\overset{\displaystyle}{\underset{\displaystyle H}{C}}=\overset{\displaystyle}{\underset{\displaystyle CH_3}{C}}-\overset{\displaystyle O}{\underset{\displaystyle H}{C}} + HOH$$

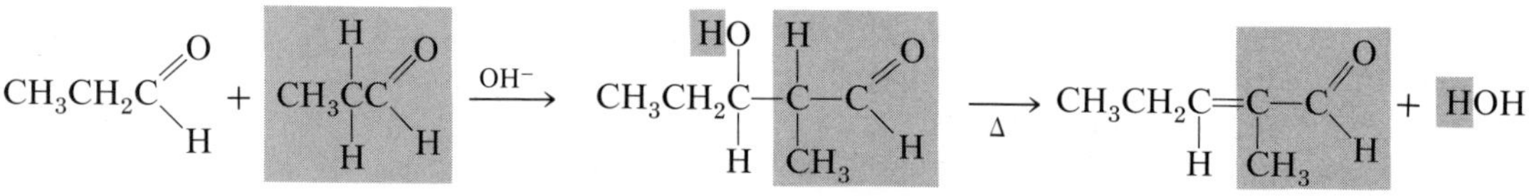

Propanal   3-Hydroxy-2-methylpentanal   2-Methyl-2-pentenal

$$CH_3-\overset{\displaystyle CH_3}{\underset{\displaystyle CH_3}{C}}-\overset{\displaystyle O}{\underset{\displaystyle H}{C}} \quad + \quad CH_3-\overset{\displaystyle CH_3}{\underset{\displaystyle CH_3}{C}}-\overset{\displaystyle O}{\underset{\displaystyle H}{C}} \quad \overset{OH^-}{\longrightarrow} \quad \text{No reaction}$$

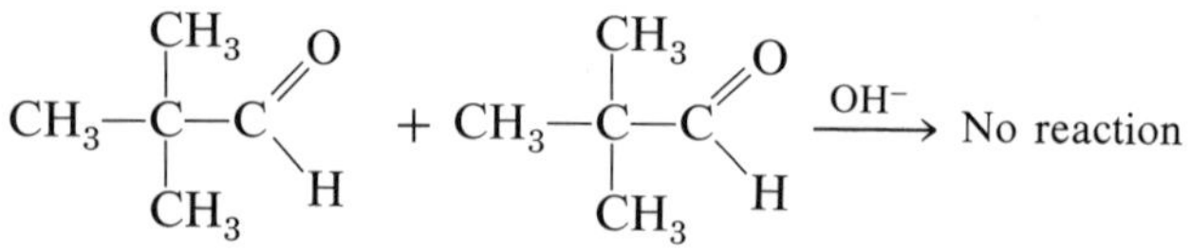

2,2-Dimethylpropanal

Aldehydes not containing α-hydrogens do not undergo self-addition reactions. They can, however, combine with other carbonyl compounds by acting as the acceptor molecule. One example is the preparation of cinnamaldehyde from the reaction of benzaldehyde and acetaldehyde.

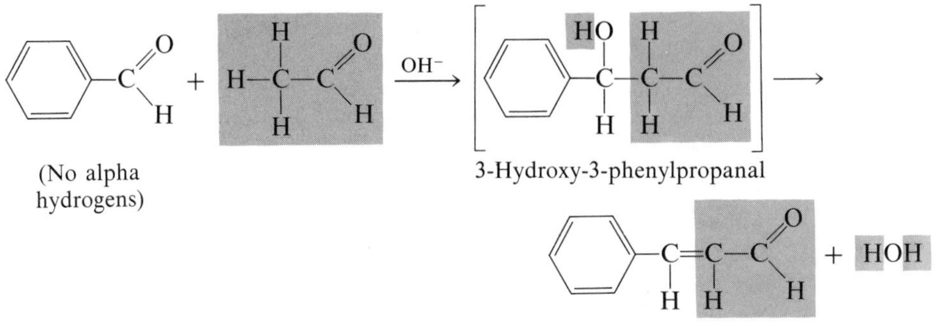

(No alpha hydrogens)

3-Hydroxy-3-phenylpropanal

Cinnamaldehyde

Ketones contain two alpha carbon atoms. If either of these carbons bears a hydrogen atom, the ketone will undergo a self-condensation. Upon heating, the condensation product loses a molecule of water to yield an α,β-unsaturated ketone. Ketones can also condense with other acceptor carbonyl compounds to effect the synthesis of a wide variety of organic compounds. A vital reaction in carbohydrate metabolism is the reversible aldol condensation between an aldehyde and a ketone (see page 716).

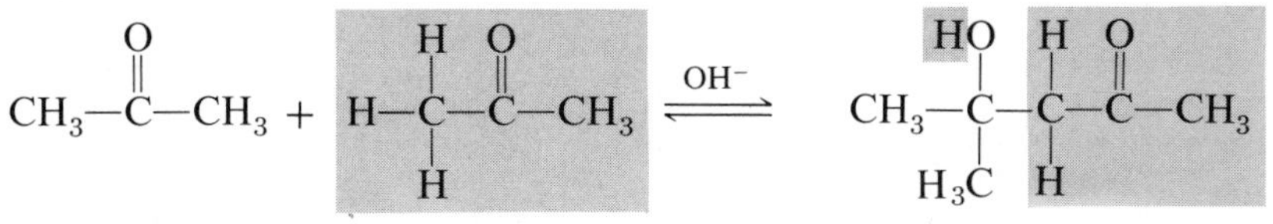

Acetone

4-Hydroxy-4-methyl-2-pentanone
(Diacetone alcohol)

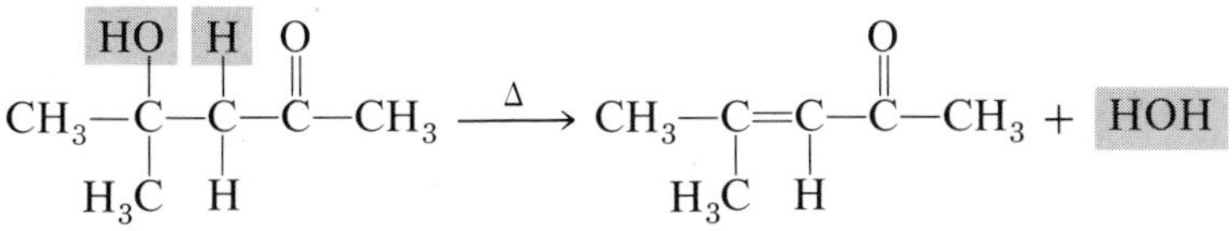

4-Methyl-3-penten-2-one
(Mesityl oxide)

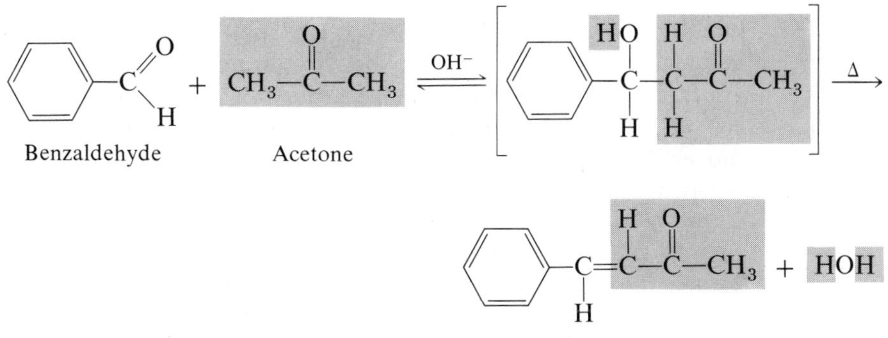

Benzaldehyde

Acetone

4-Phenyl-3-buten-2-one

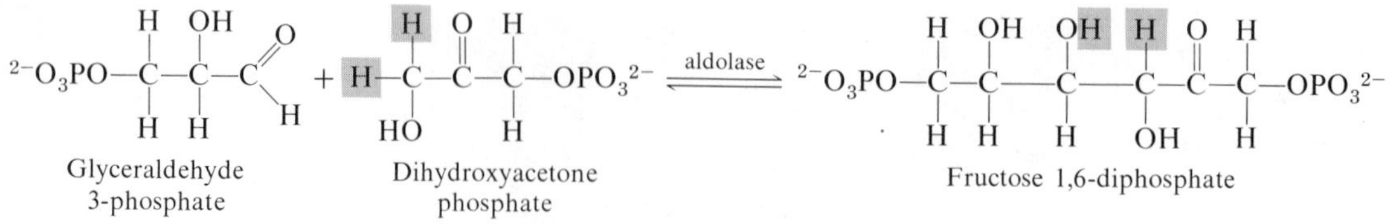
aldolase
Glyceraldehyde 3-phosphate
Dihydroxyacetone phosphate
Fructose 1,6-diphosphate

b. Substitution Reactions on the Alpha Carbon

It has been mentioned that only the alpha hydrogen atoms of a carbonyl compound are labile The phenomenon known as **tautomerism** is the basis for the explanation of this experimental observation. Tautomerism refers to certain rapid, reversible rearrangements that occur in nearly every carbonyl compound. Most often these rearrangements involve the simultaneous migration of a hydrogen atom with a resulting shift in double bonds. It is therefore possible for two distinct compounds, called **tautomers,** to coexist in a state of dynamic equilibrium. When a hydrogen atom is removed from either of the two tautomers, an anion is formed. This intermediate anion is stabilized by resonance and is referred to as the *enolate ion.* The equilibria between the two tautomers of acetaldehyde and the two tautomers of acetone, via the enolate ion, can be represented as follows.

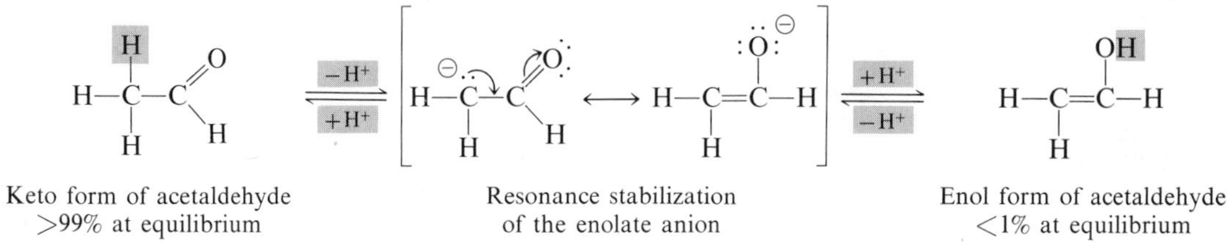
Keto form of acetaldehyde >99% at equilibrium
Resonance stabilization of the enolate anion
Enol form of acetaldehyde <1% at equilibrium

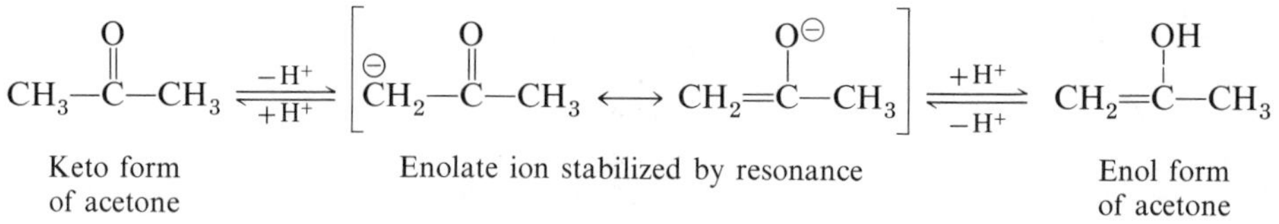
Keto form of acetone
Enolate ion stabilized by resonance
Enol form of acetone

The structure possessing the carbonyl group is known as the *keto form;* the structure containing the hydroxyl group bonded to an olefinic carbon is called the *enol form.* The term enol comes from a combination of the IUPAC ending for an alkene (*-ene*) and an alcohol (*-ol*). We shall have occasion to mention keto–enol tautomerism again in our discussions of carbohydrates and nucleic acids (Chapters 23 and 27).

As yet, no one has been able to isolate the enol form of the simple aldehydes and ketones. The keto form is considerably more stable for these compounds, and the equilibrium lies far in the direction of its formation. However, in the case of β-dicarbonyl compounds, both forms can be isolated.

The compound that has been extensively studied in this regard is *acetoacetic ester*. A solution of this ester at room temperature is a mixture containing about 93% of the keto form and 7% of the enol form.

$$\underset{\text{Keto form}}{CH_3-\overset{\overset{\displaystyle O}{\|}}{C}-CH_2-\overset{\overset{\displaystyle O}{\|}}{C}-OCH_2CH_3} \rightleftharpoons \underset{\text{Enol form}}{CH_3-\overset{\overset{\displaystyle OH}{|}}{C}=CH-\overset{\overset{\displaystyle O}{\|}}{C}-OCH_2CH_3}$$

By careful exclusion of alkali (ordinary glassware has enough alkali in it to catalyze the reaction—quartz does not), it is possible to distill the mixture in a quartz apparatus, into its two tautomeric forms. The enolic form is slightly more volatile and distills first, leaving the ketonic form behind in the distilling flask.

The student should not confuse the terms *resonance* and *tautomerism,* for there is a striking difference between them. Recall that resonance implies the ability to represent the one real structure of an organic molecule by writing the nonreal structures that are said to contribute to the resonance hybrid. At no time do the atoms of the contributing structures shift position, nor does any one of these structures at any time enjoy independent existence. In other words, there is not an equilibrium between the contributing structures of a resonance hybrid. To avoid confusion, we always represent two (or more) resonance structures by separating them with a two-headed arrow

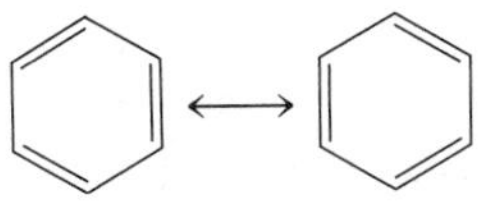

whereas the equilibrium between the two distinct tautomers is indicated by a pair of half-headed arrows, $\rightleftharpoons$.

The phenomenon of tautomerism helps explain the facile substitution of halogen atoms for the alpha hydrogen atoms of carbonyl compounds. The reaction is carried out in either an acidic or basic solution in order to facilitate enolate ion formation.

General Equation

$$R-CH_2-\overset{\overset{\displaystyle O}{\|}}{C}-R' + 2\,X_2 \xrightarrow{OH^-} R-\underset{\underset{\displaystyle X}{|}}{\overset{\overset{\displaystyle X}{|}}{C}}-\overset{\overset{\displaystyle O}{\|}}{C}-R' + 2\,X^- + 2\,HOH$$

$X = Cl, Br, I$

Specific Equation

$$CH_3-\underset{H}{\overset{H}{C}}-C\overset{O}{\underset{H}{\diagup\!\!\diagdown}} + 2\,Cl_2 \xrightarrow{NaOH} CH_3-\underset{Cl}{\overset{Cl}{C}}-C\overset{O}{\underset{H}{\diagup\!\!\diagdown}} + 2\,NaCl + 2\,HOH$$

Propanal 2,2-Dichloropropanal

Since acetone contains six alpha hydrogens, it is possible to obtain a variety of halogen-substituted products. The extent to which substitution occurs will be a function of the mole ratio of halogen and acetone in the reaction mixture and the pH of the solution. The enolization reaction is catalyzed by either acid or base.

$$CH_3\overset{O}{\overset{\|}{C}}CH_3 + 3\,Cl_2 \xrightarrow[\text{or base}]{\text{acid}} CCl_3\overset{O}{\overset{\|}{C}}CH_3 + 3\,HCl$$

1,1,1-Trichloroacetone

$$CH_3\overset{O}{\overset{\|}{C}}CH_3 + 6\,Cl_2 \xrightarrow[\text{or base}]{\text{acid}} CCl_3\overset{O}{\overset{\|}{C}}CCl_3 + 6\,HCl$$

Hexachloroacetone

c. Oxidation

Aldehydes may be characterized by the ease with which they are oxidized to carboxylic acids. Almost any oxidizing agent, including air, will bring about this oxidation. (Recall the experimental precautions employed in the preparation of aldehydes by the oxidation of a primary alcohol.) The general equation for the oxidation of an aldehyde is written as follows.

$$R-C\overset{O}{\underset{H}{\diagup\!\!\diagdown}} \xrightarrow{[O]} R-C\overset{O}{\underset{OH}{\diagup\!\!\diagdown}}$$

Aldehyde Acid

Ketones, on the other hand, are quite resistant to oxidation, and this difference in properties affords a method of distinguishing between the two classes of compounds. Three mild oxidizing solutions are used extensively as test reagents for the presence of the aldehyde group. They are incapable of oxidizing most other organic compounds. Named after the men who discovered their use, they are Benedict's solution, Fehling's solution, and Tollens' reagent. These reagents are commonly employed in testing for the presence of sugar in the blood or in the urine (see Section 23.9-b).

1. Benedict's Solution and Fehling's Solution. Both of these reagents are alkaline solutions of copper(II) (cupric) ion. The source of the ion is copper(II) sulfate. Because the cupric ion forms an insoluble hydroxide in basic solution, another reagent must be added to the solution to keep the copper ion from precipitating out as the hydroxide. In Benedict's solution, sodium citrate is employed for this purpose; copper remains in solution as the cupric citrate ion. The additional reagent in Fehling's solution is sodium potassium tartrate (Rochelle salt) with which copper forms the water soluble cupric tartrate ion. The blue color of these solutions is due to the presence of the cupric ion complexes. A positive test for the aldehyde group is evidenced by a color change to brick red, indicating the presence of the copper(I) (cuprous) oxide.[4]

The following equation is written for the reaction of an aldehyde with either Benedict's or Fehling's solution.

$$CH_3C\underset{\underset{\text{Blue}}{H}}{\overset{O}{\diagup}} + 2\ Cu^{2+} + 5\ OH^- \longrightarrow CH_3C\underset{\underset{\text{Red}}{O^-}}{\overset{O}{\diagup}} + Cu_2O_{(s)} + 3\ H_2O$$

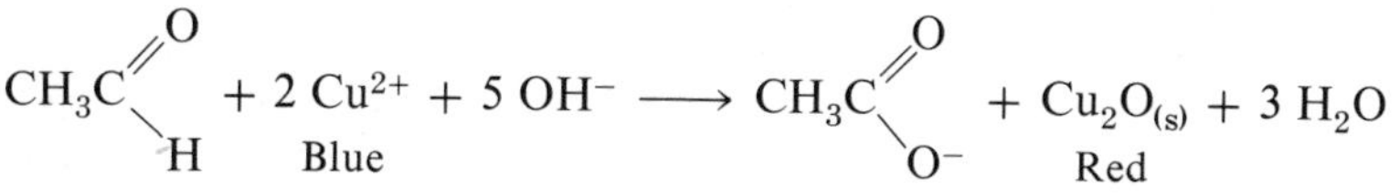

Oxidation

$$3\ OH^- + CH_3C\underset{H}{\overset{O}{\diagup}} \longrightarrow CH_3C\underset{O^-}{\overset{O}{\diagup}} + 2\ H_2O + 2\ e^-$$

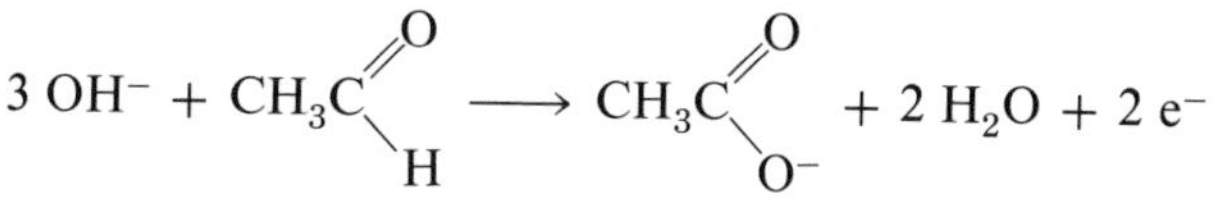

Reduction

$$2\ e^- + 2\ OH^- + 2\ Cu^{2+} \longrightarrow Cu_2O + H_2O$$

2. Tollens' Reagent. Tollens' reagent is an ammoniacal solution of silver nitrate. (In this case the silver is maintained in solution as the silver ammonia complex ion, $Ag(NH_3)_2^+$.) A positive test is indicated by the appearance of a silver mirror that forms on the inside surface of the reaction vessel due to the reduction of the silver ion to free silver. The deposition of silver on a glass surface by the oxidation of formaldehyde is used in the manufacture of mirrors.

$$H-C\underset{H}{\overset{O}{\diagup}} + 2\ Ag(NH_3)_2^+ + 3\ OH^- \longrightarrow H-C\underset{\underset{\underset{\text{mirror}}{\text{Silver}}}{O^-}}{\overset{O}{\diagup}} + 2\ Ag_{(s)} + 4\ NH_3 + 2\ H_2O$$

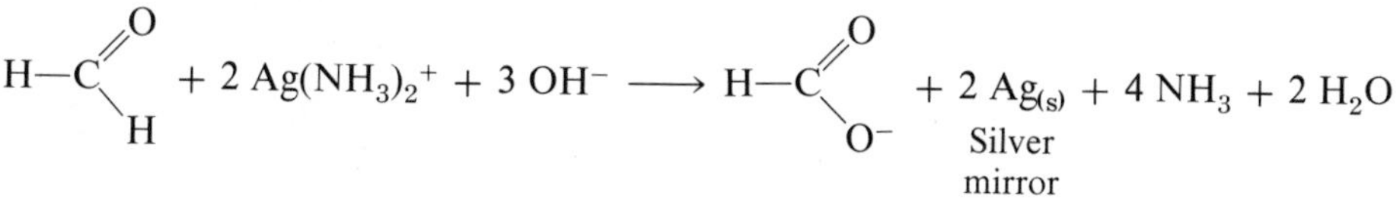

[4] The cupric ion ($+2$ oxidation state) is the oxidizing agent and therefore must be the substance that is reduced, in this case, to cuprous oxide ($+1$ oxidation state).

Oxidation

$$3\,OH^- + H{-}C({=}O){-}H \longrightarrow H{-}C({=}O){-}O^- + 2\,H_2O + 2\,e^-$$

Reduction

$$2\,Ag(NH_3)_2^+ + 2\,e^- \longrightarrow 2\,Ag + 4\,NH_3$$

d. Reduction

Aldehydes and ketones are readily reduced to the corresponding primary and secondary alcohols, respectively. A wide variety of reducing agents may be used. For laboratory applications, reduction with metal hydrides such as lithium aluminum hydride ($LiAlH_4$) or sodium borohydride ($NaBH_4$) has the great advantage of selectivity. Only carbon–oxygen double bonds (and not carbon–carbon double bonds) are reduced with these reagents.

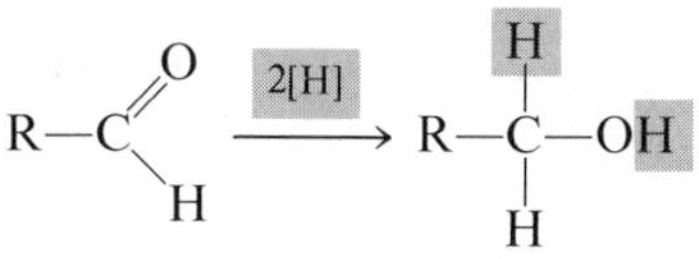

2[H]

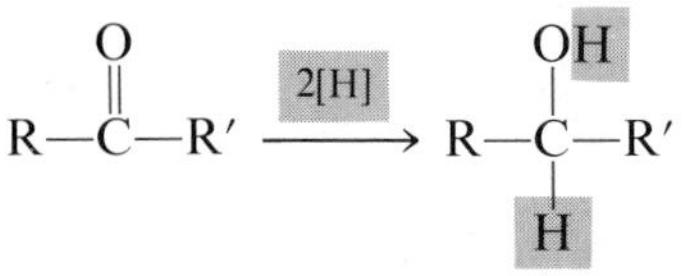

2[H]

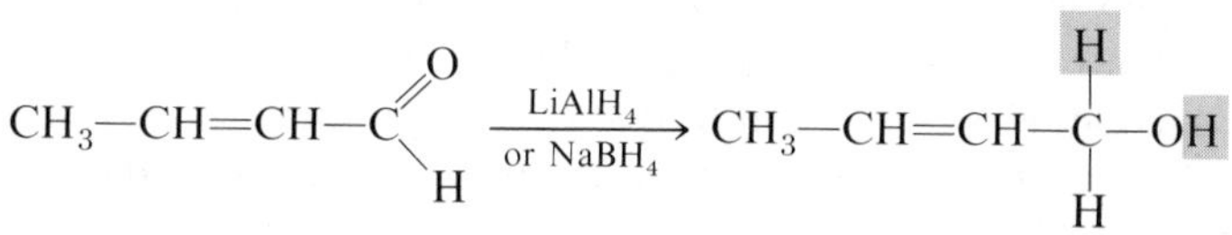

LiAlH4
or NaBH4

Crotonaldehyde (2-Butenal)

Crotyl alcohol (2-Buten-1-ol)

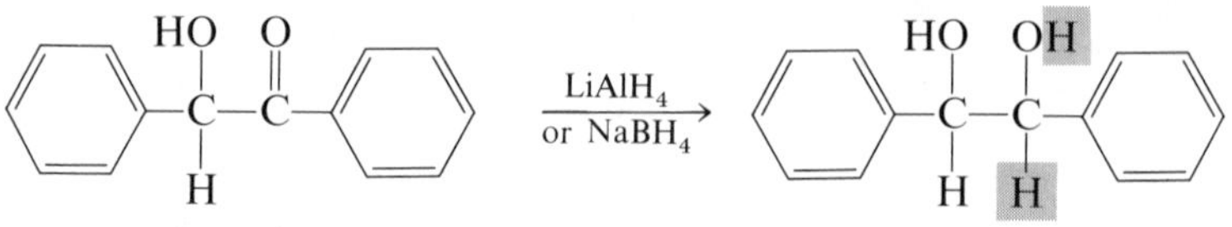

LiAlH4
or NaBH4

Benzoin (2-Hydroxy-1,2-diphenylethanone)

Hydrobenzoin (1,2-Diphenylethane-1,2-diol)

Carbonyl compounds may also be reduced to alcohols by hydrogen gas in the presence of a metal catalyst (catalytic hydrogenation). However, this method suffers from the disadvantages that many of the catalysts (Pt, Pd, Ru) are expensive and that other functional groups are also reduced.

non-specific reduction

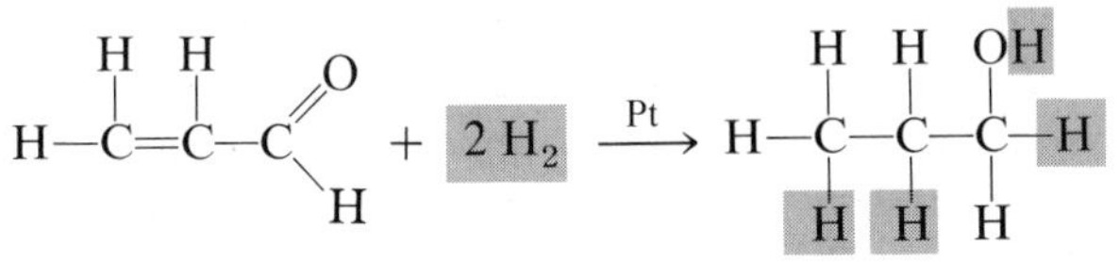

Acrolein (2-Propenal) — 1-Propanol

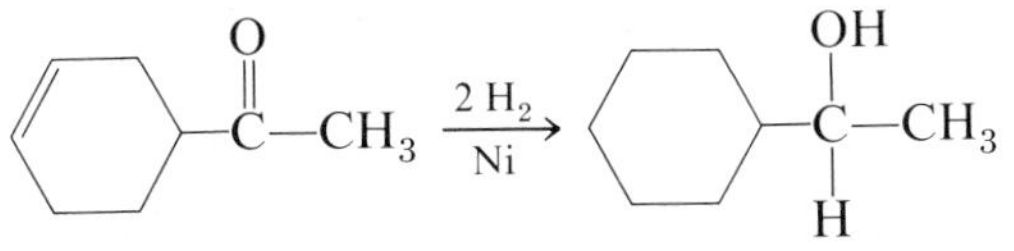

Two extremely important biochemical carbonyl reduction reactions will be discussed in Chapter 28. They are the reduction of acetaldehyde to ethyl alcohol and the reduction of pyruvic acid to lactic acid. In each case, the enzyme that catalyzes the reaction contains the coenzyme NADH (see Section 26.4), which is the reducing agent.

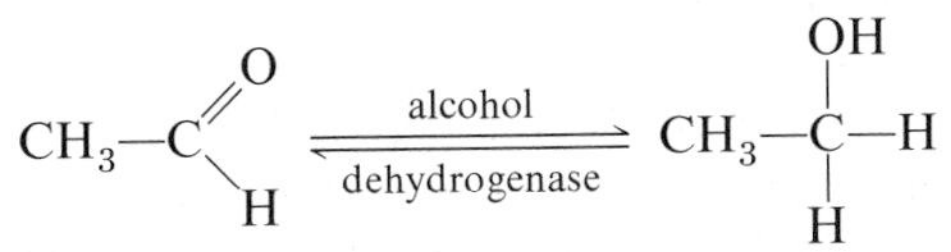

Acetaldehyde — Ethyl alcohol

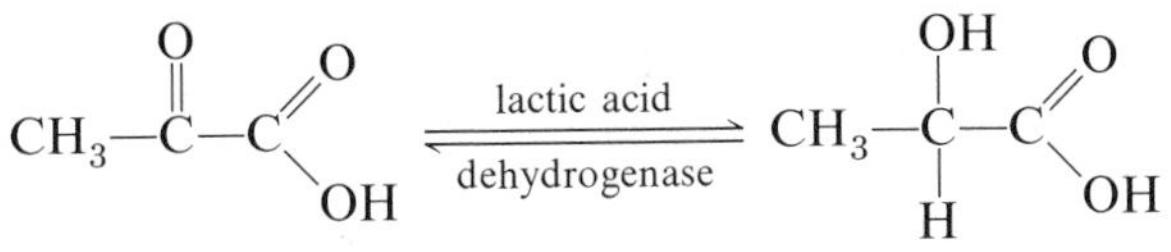

Pyruvic acid — Lactic acid

18.6 Important Aldehydes and Ketones

a. Formaldehyde (Methanal)

Formaldehyde is the simplest, and industrially, the most important member of the aldehyde family. Its commercial preparation involves the passage of methanol vapors over a copper or silver catalyst at temperatures of about 300 °C. Formaldehyde is a colorless gas with an extremely irritating odor. It cannot be easily handled in the gaseous state, and is therefore dissolved in water and sold as a 37–40% aqueous solution (such a solution is called *formalin*). If formalin is permitted to evaporate slowly under reduced pressure, a white solid polymer, known as *paraformaldehyde,* is formed. Because paraformaldehyde is readily reconverted to formaldehyde upon gentle heating, it serves as a convenient storage form of formaldehyde.

Formalin is used as a general antiseptic, as a reagent for the preparation of many other organic compounds, and for the synthetic preparation of resins such as Bakelite and Melmac. It possesses the ability to denature proteins, rendering them insoluble in water. For this reason it is used in embalming solutions and in the preservation of biological specimens. Candles made from paraformaldehyde are used in the fumigation of sickrooms. When a solution of formalin and ammonium hydroxide is evaporated to dryness, a crystalline solid, *hexamethylenetetraamine,* $(CH_2)_6N_4$, is formed. The compound is used as a urinary antiseptic, and has been given the trade name of Urotropin.

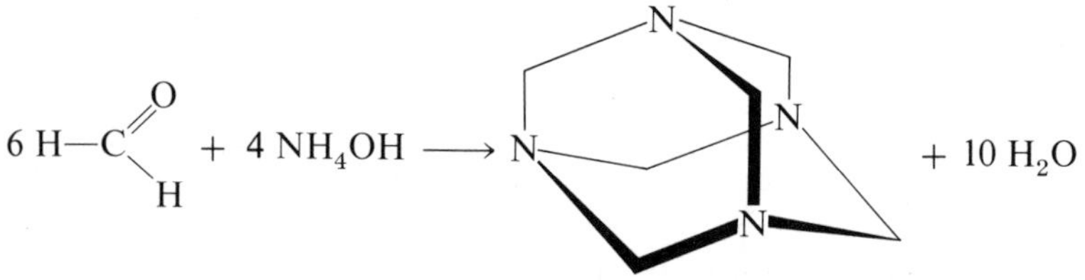

Hexamethylenetetraamine

b. Acetaldehyde (Ethanal)

Acetaldehyde is an extremely volatile, colorless liquid. It is prepared by the oxidation of ethyl alcohol, or by the hydration of acetylene (Section 14.9). It serves as a starting material for the preparation of many other organic compounds. When heated in the presence of an acid, it readily polymerizes into a stable liquid, called *paraldehyde.* Paraldehyde has been used as a hypnotic and as a sedative.

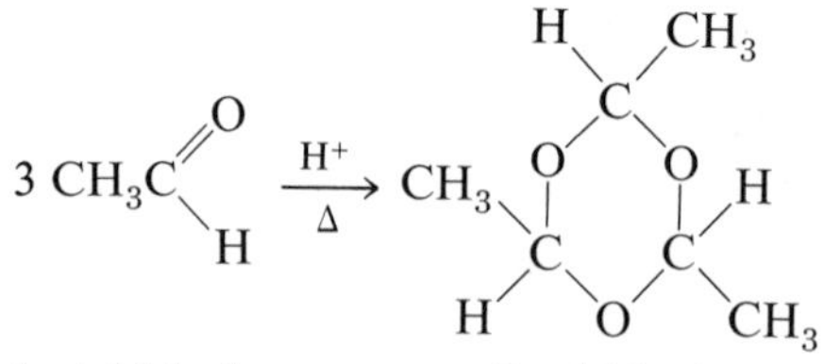

Acetaldehyde (bp 21°C) — Paraldehyde (bp 128°C)

c. Acetone (Propanone)

Acetone is by far the most important of the ketones. Until World War I, it was prepared by heating calcium acetate to high enough temperatures to cause its decomposition into calcium carbonate and acetone. However, during the war there was a great demand for acetone as a solvent for the production of the explosive cordite.[5] Chaim Weizmann, the British chemist who was later to become the first president of Israel, discovered that acetone could be obtained from the fermentation of corn or molasses. A bacterium is responsible for the process, and the products are a mixture of 30% butyl alcohol, 15% acetone, 5% ethyl alcohol, and 50% of the gases carbon dioxide and hydrogen. Acetone is also produced in large quantities by the oxidation of isopropyl alcohol.

Because it is miscible with water as well as most nonpolar organic solvents, acetone finds its chief use as a solvent. It is also an important intermediate for the preparation of chloroform, iodoform, dyestuffs, indigo, and many other complex organic compounds.

Acetone is formed in the human body as a by-product of lipid metabolism. Normally it does not accumulate to an appreciable extent because it is subsequently oxidized to carbon dioxide and water. The normal concentration of acetone in the human body is less than one milligram per 100 ml of blood. In the case of certain abnormalities, such as diabetes mellitus, the acetone concentration rises above this level. The acetone is then excreted in the urine where it can be easily detected. In severe cases, its odor can be noted on the breath (see Section 29.4).

Exercises

18.1 Draw and name all the isomeric carbonyl compounds having the molecular formula $C_5H_{10}O$.

18.2 Name the following compounds.

(a) $(CH_3)_2CHC(=O)H$

(b) $CH_3CHOHCH_2C(=O)H$

(c) $CH_3CH_2CH_2CH_2\overset{\overset{\displaystyle O}{\|}}{C}CH_2CH_3$

[5] Cordite is an explosive mixture containing 30% nitroglycerin, 65% guncotton (cellulose trinitrate), and 5% petroleum jelly. Acetone is necessary to make the mixture into a homogeneous gel.

(d) 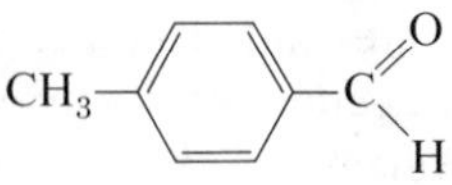

(e) $CH_3CH_2COH(CH_3)\overset{O}{\overset{\|}{C}}CH(CH_3)_2$

(f) 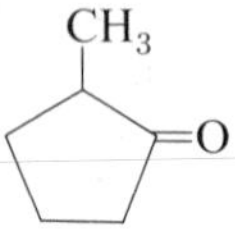

(g) $CH_3CHBrCHBr\overset{O}{\overset{\|}{C}}CH_3$

(h) $CH_3CH_2\overset{O}{\overset{\|}{C}}CH(CH_3)_2$

(i) $CH_3OCH_2\overset{O}{\overset{\|}{C}}CH_2CH_2CH_3$

(j) $CH_3CHClCH_2CH(CH_3)CH_2\overset{O}{\overset{\|}{C}}H$

18.3 Write structural formulas for the following compounds.

(a) 3,3-dimethylbutanal
(b) propionaldehyde
(c) 2-pentenal
(d) 4-hexen-2-one
(e) cyclopentanone
(f) 2-ethyl-3-iodoheptanal
(g) acetophenone
(h) 2-methoxy-3-heptanone
(i) 2,3-butanedione
(j) the oxime of 2-butanone

18.4 Write complete balanced equations for the oxidation of **(a)** 1-propanol by potassium permanganate and **(b)** 2-propanol by potassium dichromate.

18.5 Write equations for the preparation of:

(a) 3-methylbutanal from the appropriate alcohol
(b) cycloheptanone from the appropriate alcohol
(c) 1-propanol from ethylmagnesium bromide
(d) $CH_3CH_2CH_2\overset{OH}{\underset{CH_3}{C}}CH_2CH_3$ from a ketone and a Grignard reagent
(e) benzophenone via a Friedel–Crafts reaction

18.6 Complete the following equations.

(a) $CH_3CH_2\overset{O}{\overset{\|}{C}}H + KMnO_4 \longrightarrow$

(b) $CH_3CH_2\overset{O}{\overset{\|}{C}}H + 2\ CH_3CH_2CH_2OH \xrightarrow{H^+}$

(c) $CH_3\overset{\overset{O}{\|}}{C}CH_3 + H_2NNH_2 \longrightarrow$

(d) $CH_3\overset{\overset{O}{\|}}{C}CH_3 + H_2 \xrightarrow{Pt}$

(e) $CH_3\overset{\overset{O}{\|}}{C}CH_3 + CH_3\overset{\overset{O}{\|}}{C}CH_3 \xrightarrow{OH^-}$

18.7 Write equations for the preparation of each of the following compounds from a carbonyl compound.

(a) *t*-butyl alcohol
(b) 2-cyano-2-propanol
(c) 1-butanol
(d) 3-pentanol

18.8 What is the effective chemical reagent in **(a)** Fehling's solution, **(b)** Benedict's solution, and **(c)** Tollens' reagent?

18.9 Compare the reactions of potassium dichromate with **(a)** 1-propanol, **(b)** propanal, and **(c)** propanone.

18.10 What are the two principal sites of reactivity in propanal? Write an equation that illustrates each.

18.11 What is meant by the term *tautomerism?* Write the tautomeric structures of butanone. Indicate which are keto and enol forms.

18.12 Devise simple chemical tests that would enable you to distinguish each compound from the others in the following.

(a) 2-butanol, 2-methyl-1-propanol, and 2-methyl-2-propanol
(b) 1-butanol, butanal, and butanone

18.13 How are formaldehyde and acetone prepared commercially? What are their important uses?

18.14 What is a possible structure for the ketone that reacts with hydroxylamine to form an oxime which contains 13.9% nitrogen?

18.15 Compound A, whose molecular formula is C_5H_8O, decolorizes bromine water and gives a positive Benedict's test. Oxidation of A gave compound B ($C_5H_8O_2$) which, upon catalytic reduction with hydrogen, gave $(CH_3)_2CHCH_2COOH$. What are the possible formulas for A and B?

18.16 Compound A, whose molecular formula is $C_5H_{12}O$, is oxidized by a dilute solution of $K_2Cr_2O_7$ to compound B ($C_5H_{10}O$). Compound B reacts with phenylhydrazine to form compound C. Compound A can also be dehydrated with concentrated H_2SO_4 to form compound D. Addition of Br_2 to compound D yields 1,2-dibromo-3-methylbutane. Write structural formulas for A, B, C, and D.

19

Acids and Esters

Organic acids contain the functional group $-\overset{\overset{\displaystyle O}{\|}}{C}-OH$. This particular grouping of atoms is called the **carboxyl** group (derived from a combination of *carb*onyl and hydr*oxyl*) and thus organic acids are generally referred to as **carboxylic acids.** The general formula for the carboxylic acids is written **RCOOH,** where **R** may be either an aliphatic or an aromatic group.

Carboxylic acid derivatives are formally obtained by substituting another atom or group of atoms for the hydroxyl group. Replacement by a halogen atom, an acyloxy group, an alkoxy group, and an amino group gives rise respectively to *acid halides, anhydrides, esters,* and *amides.*

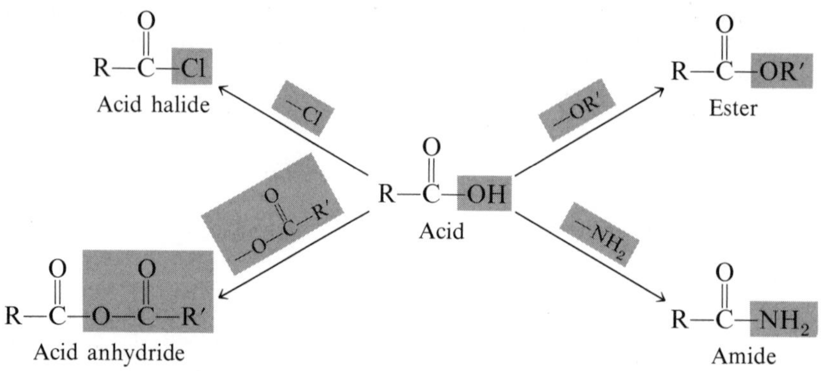
R—C(=O)—Cl
Acid halide
—Cl
R—C(=O)—OR′
Ester
—OR′
R—C(=O)—OH
Acid
—O—C(=O)—R′
—NH₂
R—C(=O)—O—C(=O)—R′
Acid anhydride
R—C(=O)—NH₂
Amide

Acids

19.1 Nomenclature of Carboxylic Acids

The acids most commonly encountered are known by their common names; many of these are based upon the source of the acid. Substituted acids are named by locating the position of the substituent group by means of the Greek letters α, β, γ, δ, etc. These letters refer to the position of the carbon atom in relation to the carboxyl carbon as illustrated.

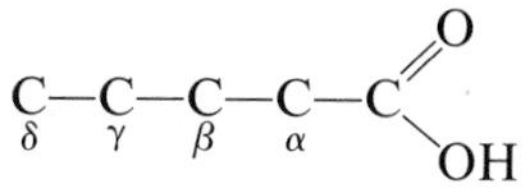

In the IUPAC system the parent hydrocarbon is taken to be the one that corresponds to the longest continuous chain containing the carboxyl group.

Table 19.1 Nomenclature of Some Aliphatic Carboxylic Acids

Formula	*Common Name*	*IUPAC Name*
$HCOOH$	Formic acid	Methanoic acid
CH_3COOH	Acetic acid	Ethanoic acid
CH_3CH_2COOH	Propionic acid	Propanoic acid
$CH_3(CH_2)_2COOH$	Butyric acid	Butanoic acid
$CH_3(CH_2)_3COOH$	Valeric acid	Pentanoic acid
$CH_3(CH_2)_4COOH$	Caproic acid	Hexanoic acid
$CH_3(CH_2)_6COOH$	Caprylic acid	Octanoic acid
$CH_3(CH_2)_8COOH$	Capric acid	Decanoic acid
$CH_3(CH_2)_{14}COOH$	Palmitic acid	Hexadecanoic acid
$CH_3(CH_2)_{16}COOH$	Stearic acid	Octadecanoic acid
$CH_3-\underset{H}{\overset{CH_3}{C}}-COOH$	Isobutyric acid (α-Methylpropionic acid)	2-Methylpropanoic acid
$Cl-\underset{Cl}{\overset{Cl}{C}}-COOH$	Trichloroacetic acid	2,2,2-Trichloroethanoic acid
$CH_3-\underset{OH}{CH}-CH_2COOH$	β-Hydroxybutyric acid	3-Hydroxybutanoic acid
$CH_2{=}CH-COOH$	Acrylic acid	Propenoic acid
$CH_3CH{=}CH-COOH$	Crotonic acid	2-Butenoic acid
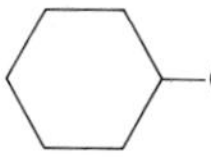—COOH	Cyclohexanecarboxylic acid	Cyclohexanecarboxylic acid

Table 19.2 Aliphatic Dicarboxylic Acids

Formula	*Common Name*	*Mnemonic*
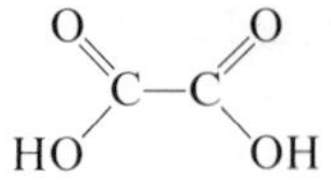	Oxalic acid	Oh
$HOOC{-}CH_2{-}COOH$	Malonic acid	My
$HOOC{-}(CH_2)_2{-}COOH$	Succinic acid	Such
$HOOC{-}(CH_2)_3{-}COOH$	Glutaric acid	Good
$HOOC{-}(CH_2)_4{-}COOH$	Adipic acid	Apple
$HOOC{-}(CH_2)_5{-}COOH$	Pimelic acid	Pie

The -*e* ending of the parent alkane is replaced by the suffix -*oic*, and the word acid follows. As in the case of the aldehydes, the carboxyl carbon is understood to be carbon number one. Table 19.1 contains the names and formulas of some of the most frequently encountered carboxylic acids.

Derivatives of the aliphatic dicarboxylic acids are very important in biological systems. These acids are almost always referred to by their common names; a mnemonic for remembering these names is given in Table 19.2.

The frequently encountered aromatic acids are known by their common names; others are named as derivatives of the parent acid, benzoic acid. If the aromatic group is not bonded directly to the carboxyl group, then the aromatic group is named as a substituent.

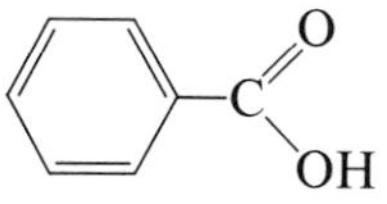

Benzoic acid
(Phenylmethanoic acid)

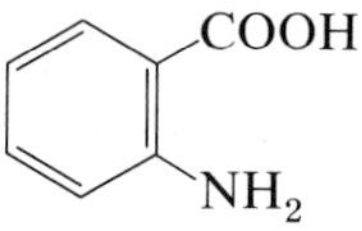

Anthranilic acid
(*o*-Aminobenzoic acid)

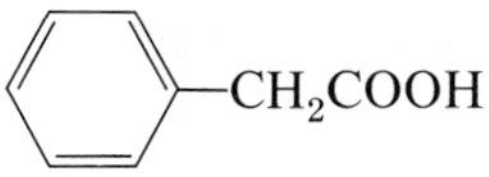

Phenylacetic acid
(Phenylethanoic acid)

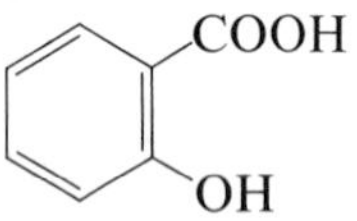

Salicylic acid
(*o*-Hydroxybenzoic acid)

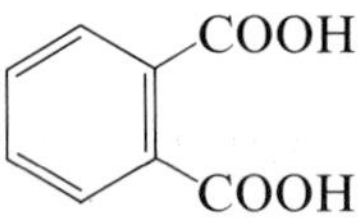

Phthalic acid
(Benzene-1,2-dicarboxylic acid)

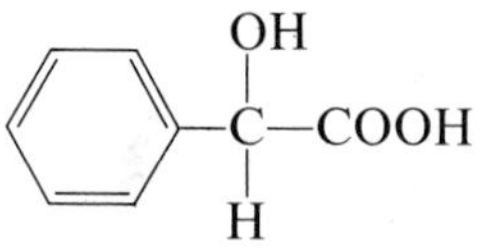

Mandelic acid
(α-Hydroxyphenylacetic acid)

19.2 Preparation of Carboxylic Acids

a. Oxidation

It has been mentioned before that acids are the final oxidation products of aldehydes and/or primary alcohols. The acid produced contains the same number of carbon atoms as did the precursor aldehyde or alcohol.

General Equations

$$RCH_2OH \xrightarrow{K_2Cr_2O_7/H^+} R{-}C(=O){-}OH$$

$$R{-}C(=O){-}H \xrightarrow{KMnO_4/H^+} R{-}C(=O){-}OH$$

Specific Equation

$$\underset{\text{1-Propanol}}{3\ CH_3CH_2CH_2OH} + 2\ K_2Cr_2O_7 + 8\ H_2SO_4 \longrightarrow \underset{\text{Propionic acid}}{3\ CH_3CH_2C(=O){-}OH} + 11\ H_2O + 2\ K_2SO_4 + 2\ Cr_2(SO_4)_3$$

Our bodies accomplish this same oxidation of alcohols to acids whenever we drink alcoholic beverages. The liver contains enzymes that convert the ethyl alcohol to acetic acid. Acetic acid is utilized to provide energy (Section 28.7), or it is converted into fat (Section 29.6). Excess alcohol that cannot be oxidized in the liver continues to circulate in the blood and eventually causes intoxication.

$$\underset{\text{Ethanol}}{CH_3CH_2OH} \xrightarrow[\text{dehydrogenase}]{\text{alcohol}} \underset{\text{Acetaldehyde}}{CH_3{-}C(=O){-}H} \xrightarrow[\text{dehydrogenase}]{\text{acetaldehyde}} \underset{\text{Acetic acid}}{CH_3{-}C(=O){-}OH}$$

Recall that in Section 17.5, we stated that methanol and ethylene glycol are poisonous to the body. It is not these compounds themselves, but rather their oxidation products, that cause the toxic effects. 1,2-Propylene glycol, on the other hand, is harmless because its oxidation product is pyruvic acid, which is a normal intermediate of carbohydrate metabolism.

$$\underset{\text{Methanol}}{CH_3OH} \xrightarrow{\text{enzyme}} \underset{\text{Formaldehyde}}{H{-}C(=O){-}H} \xrightarrow{\text{enzyme}} \underset{\text{Formic acid}}{H{-}C(=O){-}OH}$$

$$\underset{\text{Ethylene glycol}}{CH_2OHCH_2OH} \xrightarrow{\text{enzyme}} \underset{\text{Oxalic acid}}{HO{-}C(=O){-}C(=O){-}OH}$$

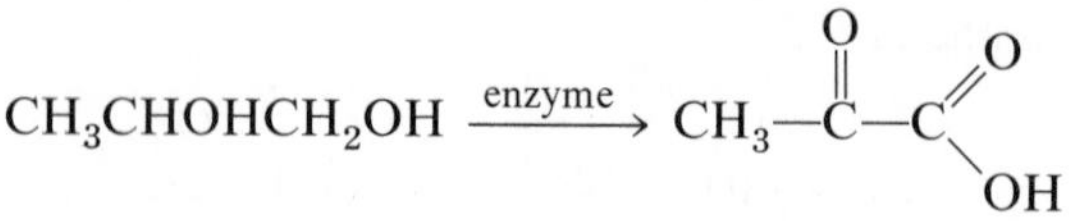
CH3CHOHCH2OH
enzyme
CH3—C—C
O
O
OH

1,2-Propylene glycol   Pyruvic acid

Aromatic acids are prepared by direct oxidation of alkylbenzenes. The length of the carbon chain is not a factor. Benzoic acid is the only acid formed; the remaining carbon atoms of the chain are oxidized to carbon dioxide and water.

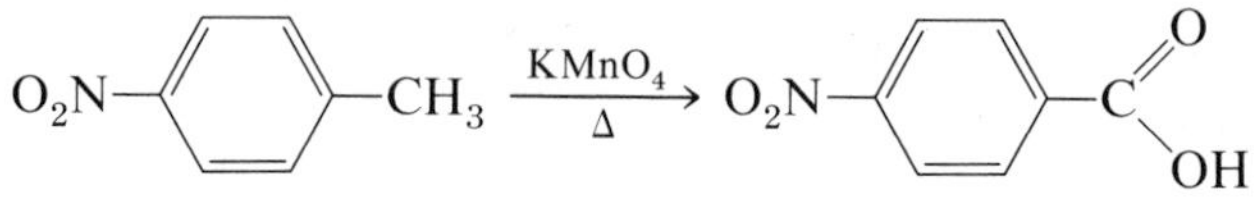
O2N
CH3
KMnO4
Δ
O2N
C
O
OH

p-Nitrotoluene   *p*-Nitrobenzoic acid

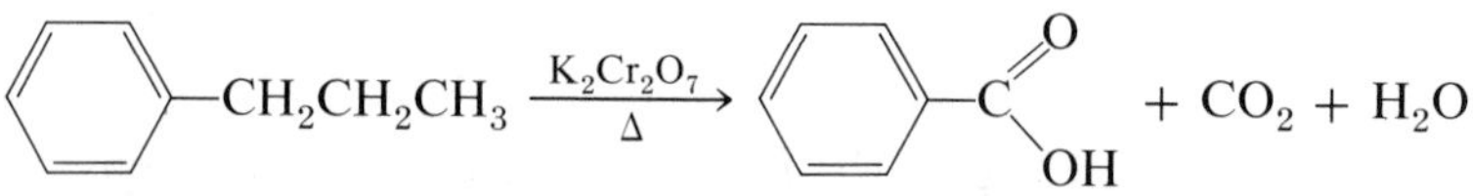
CH2CH2CH3
K2Cr2O7
Δ
C
O
OH
+ CO2 + H2O

n-Propylbenzene   Benzoic acid

There are two oxidation reactions that are of considerable commercial importance. The oxidation of cyclohexanol yields adipic acid, which is heated with hexamethylenediamine to form nylon 66. Oxidation of *p*-xylene produces terephthalic acid, which is combined with ethylene glycol in the preparation of a synthetic polyester fiber, Dacron. Many permanent press garments are made of Dacron polyester.

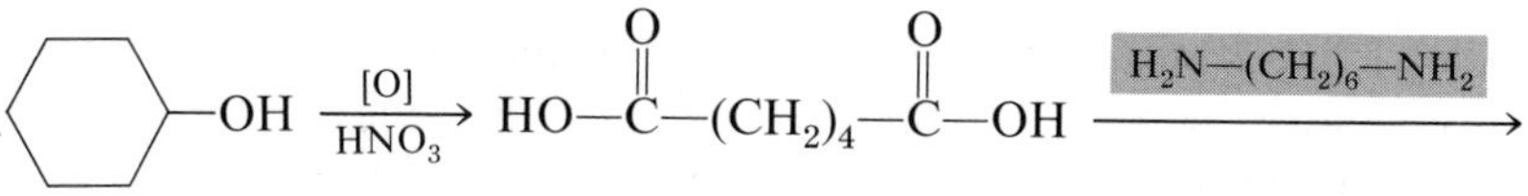
OH
[O]
HNO3
HO—C—(CH2)4—C—OH
O
O
H2N—(CH2)6—NH2

Cyclohexanol   Adipic acid

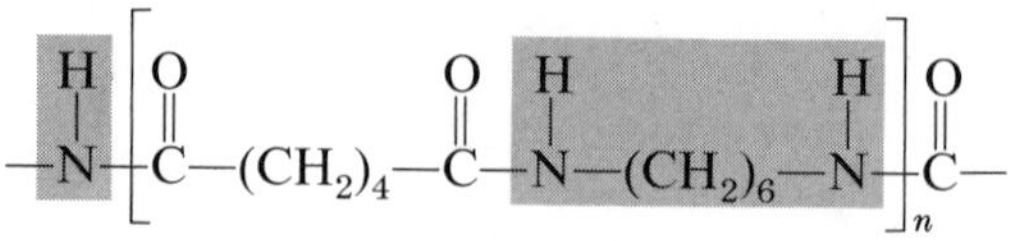
H
O
O
H
H
O
N
C—(CH2)4—C
N—(CH2)6—N
C
n

Nylon 66

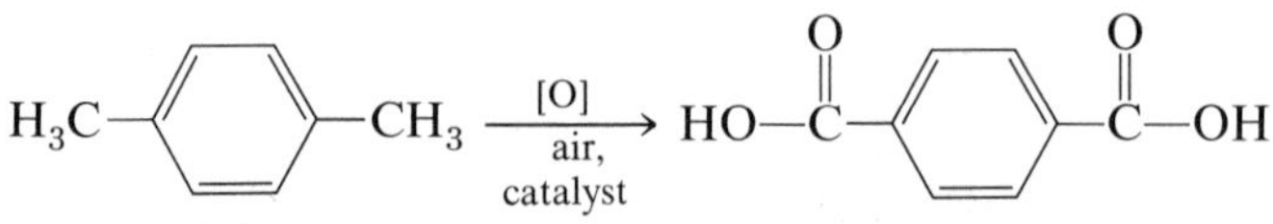
H3C
CH3
[O]
air,
catalyst
HO—C
C—OH
O
O

p-Xylene   Terephthalic acid

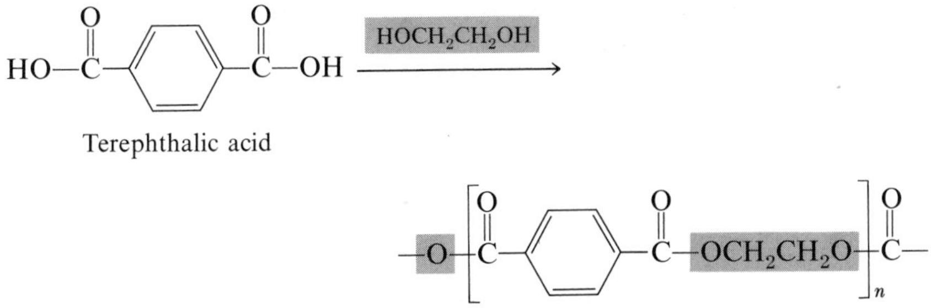

Terephthalic acid

Dacron

b. From the Corresponding Salts

This reaction is analogous to the preparation of many inorganic acids from their salts. Recall that hydrochloric acid can be prepared by heating sodium chloride and sulfuric acid (Section 8.2-e).

$$2\ NaCl + H_2SO_4 \longrightarrow 2\ HCl + Na_2SO_4$$

In the preparation of the lower boiling organic acids, sulfuric acid is added to the dry salt and the acid is distilled from the reaction mixture. The higher boiling organic acids are prepared by treating an aqueous solution of the salt with sulfuric acid. The acid so prepared is extracted into ether, and the ether is then separated from the acid by evaporation. Again note that the resultant acid contains the same number of carbon atoms as did the initial salt.

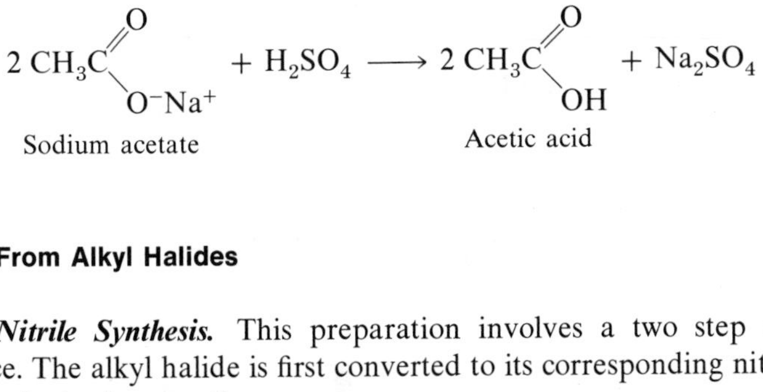

Sodium acetate — Acetic acid

c. From Alkyl Halides

1. Nitrile Synthesis. This preparation involves a two step reaction sequence. The alkyl halide is first converted to its corresponding nitrile in a simple substitution reaction.

$$CH_3CH_2Br + KCN \longrightarrow CH_3CH_2CN + KBr$$

Ethyl bromide — Ethyl cyanide (Propionitrile)[1]

[1] By naming the products as nitriles rather than cyanides, the relationship to the acids is stressed. Nitriles are named according to the acids which they yield upon hydrolysis. Thus CH_3CN is acetonitrile, and CH_3CH_2CN is propionitrile.

In a second step, the nitrile is hydrolyzed to the acid or to its salt. A strong acid or base is the catalyst.

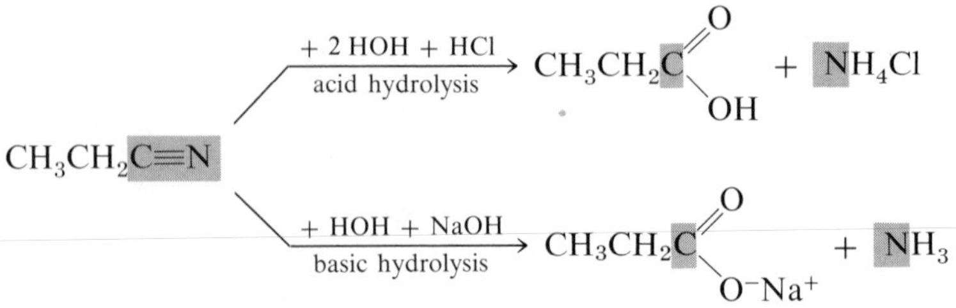
+ 2 HOH + HCl
acid hydrolysis
+ HOH + NaOH
basic hydrolysis

Notice that in all cases this reaction sequence yields an acid containing one more carbon atom than the initial alkyl halide. However, the Grignard synthesis, which is discussed next, is the preferred method for preparing carboxylic acids from the appropriate alkyl or aryl halide.

*2. **Grignard Synthesis.*** In this synthesis anhydrous carbon dioxide (dry ice) is added to an ether solution of the Grignard reagent (Section 16.4-c). The reaction mixture is then hydrolyzed by treatment with dilute acid, and the organic acid is liberated.

$$CH_3CH_2Br + Mg \xrightarrow{\text{ether}} CH_3CH_2MgBr$$

$$CH_3CH_2MgBr + O{=}C{=}O \xrightarrow{\text{ether}} CH_3CH_2{-}\overset{\overset{\displaystyle O}{\|}}{C}{-}O{-}MgBr$$

$$CH_3CH_2{-}\overset{\overset{\displaystyle O}{\|}}{C}{-}O{-}MgBr + HBr \longrightarrow CH_3CH_2C(=O)OH + MgBr_2$$

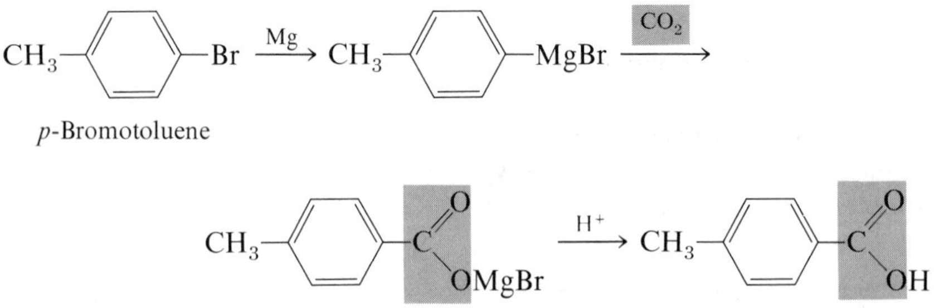
Mg
CO2
H+

p-Bromotoluene

p-Methylbenzoic acid

The Grignard synthesis is carried out in most college chemistry laboratories because it illustrates the preparation of the Grignard reagent as well as the preparation of a wide variety of organic acids.

19.3 Physical Properties of Carboxylic Acids

The first nine members of the carboxylic acid series are colorless liquids having very disagreeable odors. The odor of vinegar is that of acetic acid; the odor of rancid butter is primarily that of butyric acid. Caproic acid (Latin, *caper,* goat) is present in the hair and the secretions of goats. The acids from C_5 to C_{10} all have "goaty" odors (odor of Limburger cheese or an unaired locker room). The acids above C_{10} are waxlike solids, and because of their low volatility are practically odorless. As was the case with the alcohols, the lower acid homologs are soluble in water, but the solubility decreases sharply after butyric acid as the hydrocarbon character of the molecule increases. All the acids are soluble in organic solvents such as alcohol, benzene, carbon tetrachloride, and ether.

Table 19.3 contains physical constants for the first ten members of the aliphatic carboxylic acid family. Notice that the melting points show no regular increase with increasing molar mass. The boiling points of the acids are abnormally high, and increase approximately in 20°C increments. They are higher than the boiling points of alcohols or alkyl halides of comparable molecular weights. This low volatility is attributed to the strong association of acid molecules in the liquid state. In fact, molar mass measurements indicate that the acids exist primarily as dimers, or double molecules. For example, acetic acid is believed to have the following hydrogen-bonded structure.

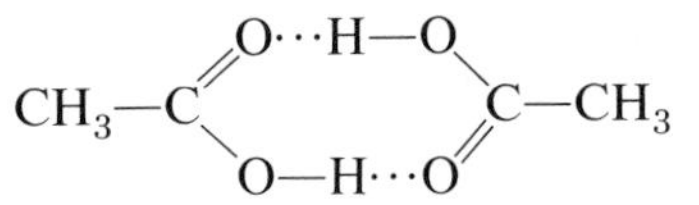

19.4 Effect of Structure upon Acidity

In Section 10.3 we mentioned that most organic acids are only slightly dissociated in solution and hence they are classified as weak acids. We ought now to ask: Why are carboxylic acids much more acidic (by a factor of about

Table 19.3 Physical Constants of Carboxylic Acids

Formula	Name of Acid	MP (°C)	BP (°C)	Solubility (grams/100 grams of water)	K_a 25°C
$HCOOH$	Formic	8.4	101	miscible	2.1×10^{-4}
CH_3COOH	Acetic	17	118	miscible	1.8×10^{-5}
CH_3CH_2COOH	Propionic	−22	141	miscible	
$CH_3(CH_2)_2COOH$	Butyric	−4.7	163	miscible	
$CH_3(CH_2)_3COOH$	Valeric	−35	187	5	
$CH_3(CH_2)_4COOH$	Caproic	−3	205		
$CH_3(CH_2)_5COOH$	Enanthic	−8	224		1.5×10^{-5} (propionic through capric)
$CH_3(CH_2)_6COOH$	Caprylic	16	238	less than 1 (caproic through capric)	
$CH_3(CH_2)_7COOH$	Pelargonic	14	254		
$CH_3(CH_2)_8COOH$	Capric	31	268		

10^{11}) than alcohols? The reason usually given is that the carboxylate anion formed upon dissociation is stabilized by resonance (two equivalent resonance structures) relative to the original acid molecule. In the alkoxide ion, RO^-, the negative charge is not delocalized, but is concentrated on the single oxygen atom. Although the tendency of organic acids to dissociate is slight, the stabilization of the anion through resonance facilitates the ionization process.

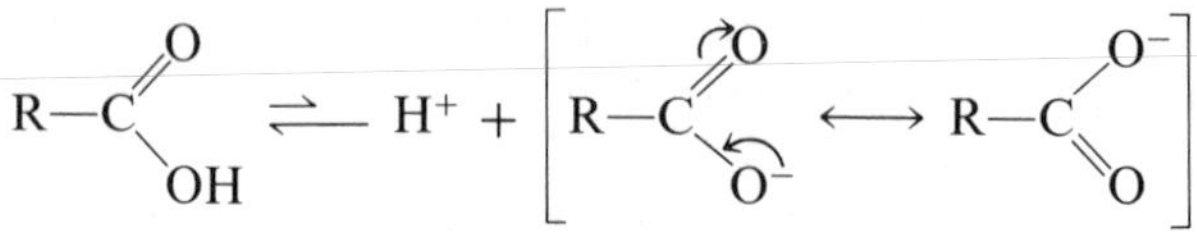

$$R-C\begin{smallmatrix}O\\\\OH\end{smallmatrix} \rightleftharpoons H^+ + \left[R-C\begin{smallmatrix}O\\\\O^-\end{smallmatrix} \longleftrightarrow R-C\begin{smallmatrix}O^-\\\\O\end{smallmatrix} \right] \qquad K_a = \frac{[RCOO^-][H^+]}{[RCOOH]}$$

Resonance-stabilized anion

If resonance were the only factor that accounted for the acidity, we would expect all carboxylic acids to be of comparable strength. We know that this is not so. Table 19.4 lists some of the acids and their respective dissociation constants. Notice that a wide range of values is possible. Carboxylic acids that contain strong electron-attracting groups (like halogens) on the alpha carbon are far more acidic than the unsubstituted acid. The methyl group, on the other hand, is an electron-releasing group. This accounts for the tenfold decrease in acidity in going from formic acid to acetic acid. These electrostatic factors, in which electrons are either attracted to or repelled from one atom (or group of atoms) with respect to another, are referred to as **inductive effects.** Electron-attracting groups facilitate ionization of the carboxylic acid by withdrawing electrons from the carboxylate group, weakening the oxygen–hydrogen bond, and thus aiding in the release of the proton. These groups also help to stabilize the acid anion, by delocalizing the negative charge from the carboxylate group to adjacent atoms. Inductive effects are additive, and the greater the number of electron-withdrawing groups on the alpha carbon, the stronger the acid. (Likewise, the more numerous the electron-releasing groups, the weaker the acid.)

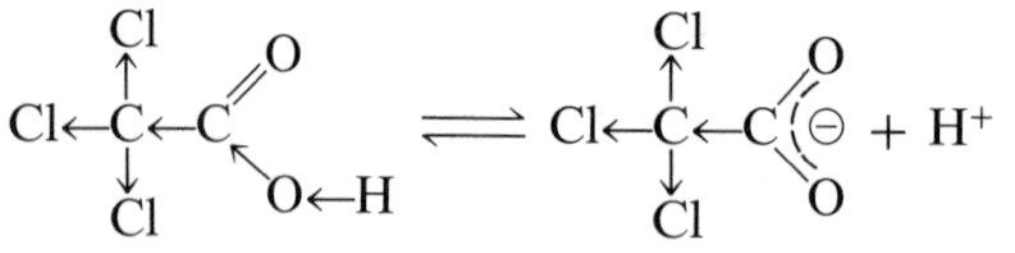

$$Cl\leftarrow\overset{\overset{Cl}{\uparrow}}{\underset{\underset{Cl}{\downarrow}}{C}}\leftarrow C\begin{smallmatrix}O\\\\O\leftarrow H\end{smallmatrix} \rightleftharpoons Cl\leftarrow\overset{\overset{Cl}{\uparrow}}{\underset{\underset{Cl}{\downarrow}}{C}}\leftarrow C\begin{smallmatrix}O^\ominus\\\\O\end{smallmatrix} + H^+ \qquad K_a = 2.2 \times 10^{-1}$$

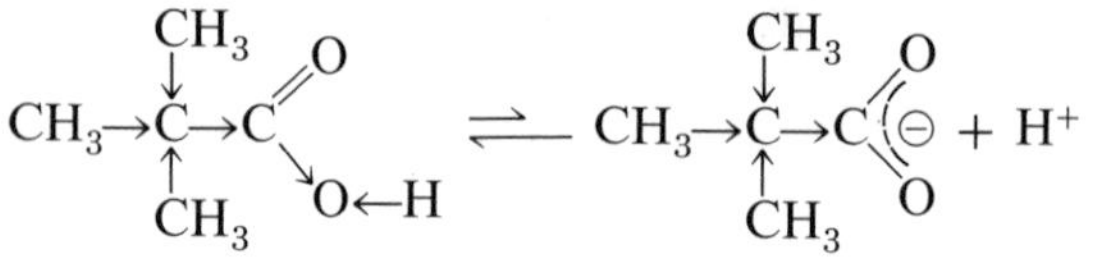

$$CH_3\rightarrow\overset{\overset{CH_3}{\downarrow}}{\underset{\underset{CH_3}{\uparrow}}{C}}\rightarrow C\begin{smallmatrix}O\\\\O\leftarrow H\end{smallmatrix} \rightleftharpoons CH_3\rightarrow\overset{\overset{CH_3}{\downarrow}}{\underset{\underset{CH_3}{\uparrow}}{C}}\rightarrow C\begin{smallmatrix}O^\ominus\\\\O\end{smallmatrix} + H^+ \qquad K_a = 9.5 \times 10^{-6}$$

Inductive effects diminish rapidly with distance; thus a halogen on a beta carbon is not nearly as effective as one on an alpha carbon (compare α- and β-chlorobutyric acids). Both induction and resonance operate in aromatic

Table 19.4 Factors Influencing Acid Strength

Factor	*Formula*	*Name of Acid*	K_a
Electronegativity of substituent	CH_2FCOOH	Fluoroacetic	2.6×10^{-3}
	$CH_2BrCOOH$	Bromoacetic	1.3×10^{-3}
	CH_2ICOOH	Iodoacetic	6.7×10^{-4}
	CH_3COOH	Acetic	1.8×10^{-5}
Number of electron-withdrawing groups on α-carbon	CCl_3COOH	Trichloroacetic	2.2×10^{-1}
	$CHCl_2COOH$	Dichloroacetic	5.5×10^{-2}
	$CH_2ClCOOH$	Chloroacetic	1.4×10^{-3}
Distance from carboxyl group	$CH_3CH_2CHClCOOH$	α-Chlorobutyric	1.4×10^{-3}
	$CH_3CHClCH_2COOH$	β-Chlorobutyric	8.9×10^{-5}
	$CH_2ClCH_2CH_2COOH$	γ-Chlorobutyric	3.0×10^{-5}
	$CH_3CH_2CH_2COOH$	Butyric	1.5×10^{-5}
Electron donors or acceptors	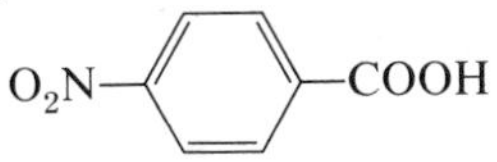O2N— —COOH	*p*-Nitrobenzoic	3.8×10^{-4}
	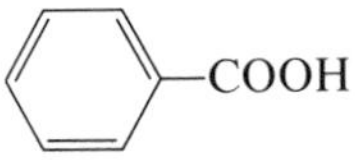—COOH	Benzoic	6.3×10^{-5}
	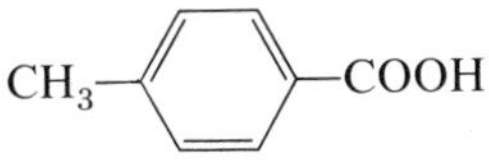CH3— —COOH	*p*-Toluic	4.2×10^{-5}
	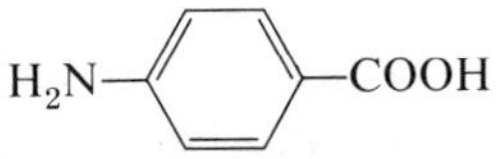H2N— —COOH	*p*-Aminobenzoic	1.2×10^{-5}

acids. Distance from the carboxyl group is not a factor here because the electrostatic effects are transmitted through the conjugated electron system. Compare the four aromatic acids in Table 19.4.

19.5 Chemical Properties of Carboxylic Acids

All of the reactions of the carboxylic acids can be considered to be substitution reactions. The nature of the other reacting species will determine whether substitution of the hydroxyl hydrogen, the hydroxyl group, or the alpha hydrogens will occur.

a. Substitution of Acidic Hydrogen—Neutralization

Characteristic of all acids is the ability to react with bases to form salts. Salts so formed are quite different from their parent acids. They are water-soluble solids and are completely dissociated in solution. They are insoluble in nonpolar solvents such as ether, benzene, and carbon tetrachloride. Aqueous solutions of most organic salts of sodium, potassium, calcium, and barium are basic owing to the hydrolysis of the organic anion (Section 8.2-a).

$$\text{HCl} + \text{NaOH} \longrightarrow \text{NaCl} + \text{HOH}$$

Acid   Base   Salt   Water

$$R-C\overset{O}{\underset{OH}{\big\langle}} + \text{NaOH} \longrightarrow R-C\overset{O}{\underset{O^-Na^+}{\big\langle}} + \text{HOH}$$

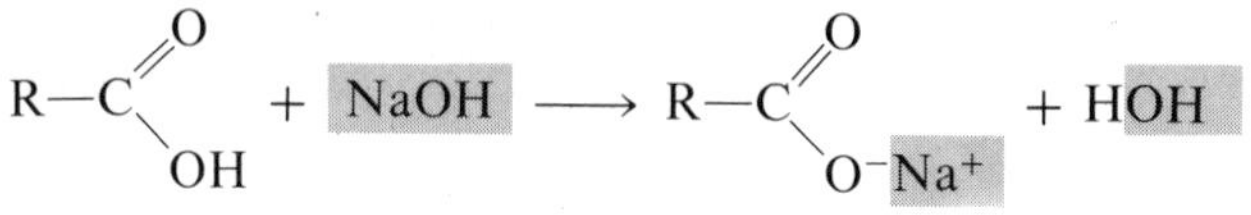

Neutralization of Organic Acids

$$CH_3C\overset{O}{\underset{OH}{\big\langle}} + \text{NaOH} \longrightarrow CH_3C\overset{O}{\underset{O^-Na^+}{\big\langle}} + \text{HOH}$$

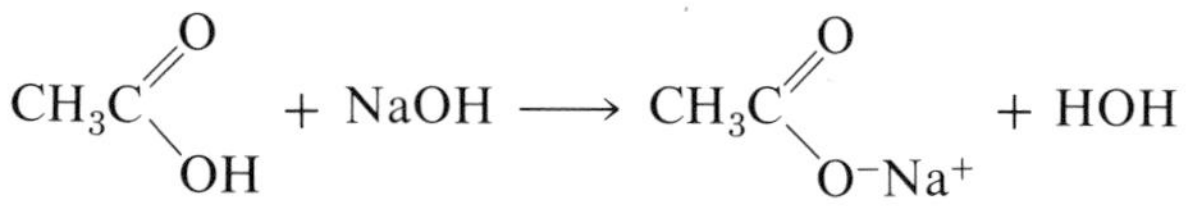

Acetic acid   Sodium acetate

Hydrolysis of Organic Salts

$$CH_3C\overset{O}{\underset{O^-}{\big\langle}} + \text{HOH} \longrightarrow CH_3C\overset{O}{\underset{OH}{\big\langle}} + \text{OH}^-$$

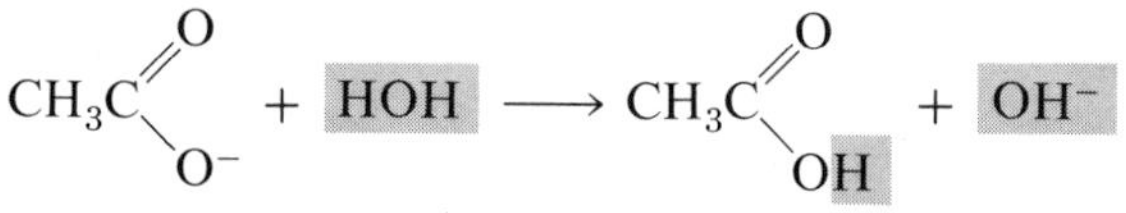

All carboxylic acids, whether or not they are water soluble, liberate carbon dioxide from a solution of sodium carbonate or sodium bicarbonate. This reaction may be used to distinguish the water-insoluble acids from all other classes of organic compounds. (The water-soluble acids are more conveniently distinguished by their ability to turn litmus red.)

$$2 R-C\overset{O}{\underset{OH}{\big\langle}} + \text{Na}_2\text{CO}_3 \longrightarrow 2 R-C\overset{O}{\underset{O^-Na^+}{\big\langle}} + \text{H}_2\text{O} + \text{CO}_{2(g)}$$

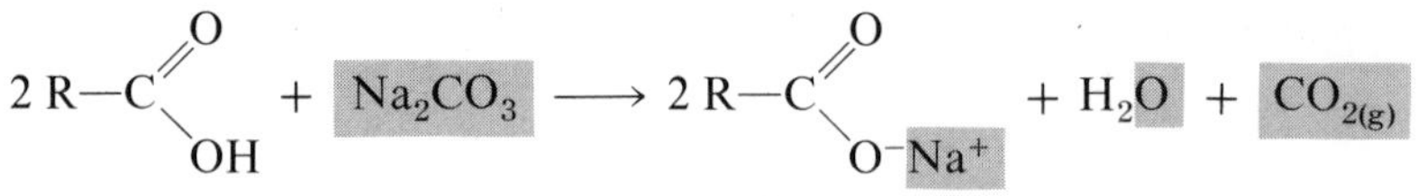

$$R-C\overset{O}{\underset{OH}{\big\langle}} + \text{NaHCO}_3 \longrightarrow R-C\overset{O}{\underset{O^-Na^+}{\big\langle}} + \text{HOH} + \text{CO}_{2(g)}$$

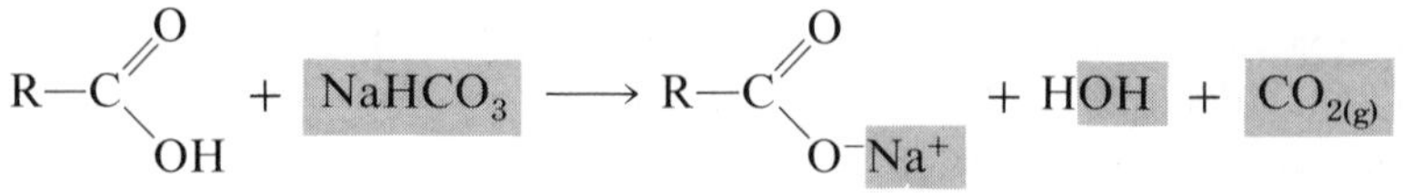

Organic salts are named in the same manner as are the inorganic salts. The name of the cation is followed by the name of the organic anion. The name of the anion is obtained by dropping the -*ic* ending of the acid name and replacing it with the suffix -*ate*. This applies whether we are using common names or IUPAC names. Following are some examples.

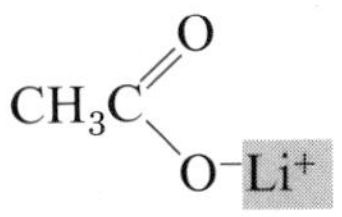

Lithium acetate (Lithium ethanoate)	Sodium butyrate (Sodium butanoate)	Potassium benzoate

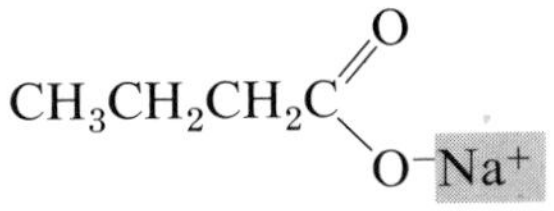

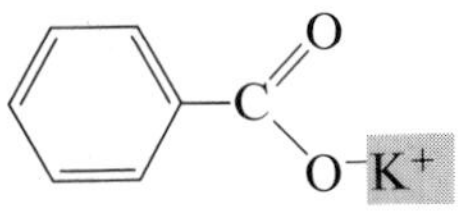

The sodium or potassium salts of long-chain carboxylic acids are called soaps. We shall discuss the chemistry of soaps in considerable detail in Section 24.8.

$$CH_3CH_2CH_2CH_2CH_2CH_2CH_2CH_2CH_2CH_2CH_2CH_2CH_2CH_2CH_2CH_2CH_2-C{\overset{O}{\underset{O^-Na^+}{}}}$$

18 C

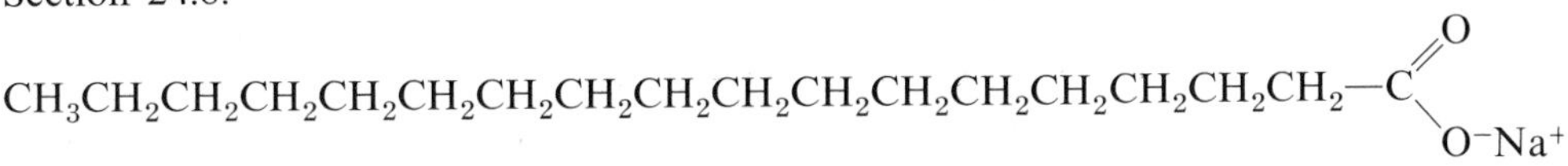

Sodium stearate (a soap)

Other salts that have commercial significance are calcium and sodium propionate, which are added to bakery goods to prevent spoilage (mold growth), sodium benzoate, which is used as a preservative (e.g., for cider), and zinc undecylenate, which is used in the treatment of athlete's foot.

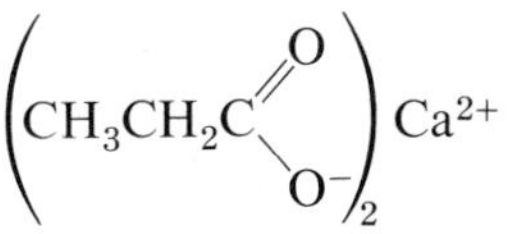

Calcium propionate	Zinc undecylenate

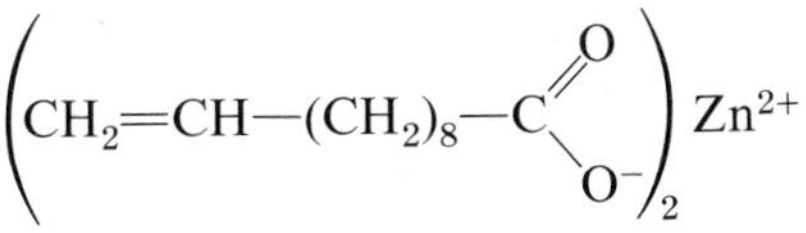

b. Substitution of the Hydroxyl Group

1. Formation of Acid Halides. The hydroxyl group of an acid may be replaced by a halide ion. The phosphorus halides and thionyl chloride are the reagents commonly employed in this reaction. Thionyl chloride, a low boiling liquid, is a particularly good reagent for the preparation of acid chlorides. Any excess reagent can be separated out by distillation and the by-products of the reaction are both gases and are easily removed.

$$3\ R-C{\overset{O}{\underset{OH}{}}} + PCl_3 \longrightarrow 3\ R-C{\overset{O}{\underset{Cl}{}}} + H_3PO_3$$

$$R-C{\overset{O}{\underset{OH}{}}} + PCl_5 \longrightarrow R-C{\overset{O}{\underset{Cl}{}}} + HCl_{(g)} + POCl_3$$

$$R-C{\overset{O}{\underset{OH}{}}} + SOCl_2 \longrightarrow R-C{\overset{O}{\underset{Cl}{}}} + SO_{2(g)} + HCl_{(g)}$$

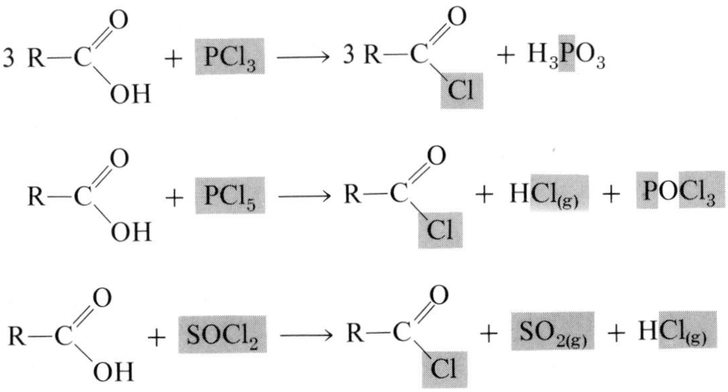

The acid halides are also referred to as acyl halides. Acyl bromides and iodides can be prepared by similar procedures, but acyl chlorides are made more easily and they are more reactive. Acid halides are the most reactive of the carboxylic acid derivatives. This increased reactivity is due to the presence of the electronegative halogen atom directly bonded to the carbonyl carbon. The inductive effect of the halogen is to withdraw electrons from the carbon atom making the carbon more susceptible to attack by nucleophilic reagents. The chief function of acid halides is to serve as intermediates in the preparation of many other organic compounds. The name of the acyl group is obtained from the parent acid by dropping the *-ic* ending and adding *-yl*. The acetyl group is an extremely important functional group in both organic chemistry and biochemistry (see Sections 28.7 and 29.2).

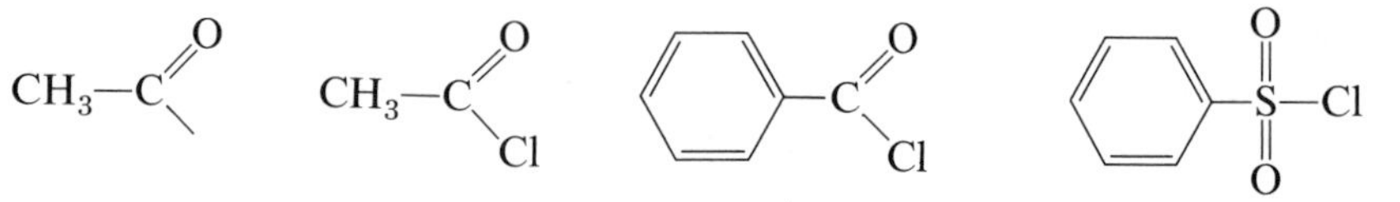

Acetyl group   Acetyl chloride   Benzoyl chloride   Benzenesulfonyl chloride

2. ***Acid Anhydrides.*** Organic acid anhydrides are analogous to inorganic anhydrides in that they are derived by the elimination of a molecule of water from the corresponding acid or base.

$$H_2SO_4 \longrightarrow SO_3 + H_2O$$
$$Ca(OH)_2 \longrightarrow CaO + H_2O$$

Structurally, acid anhydrides are considered to be formed by the removal of water between two molecules of acid. In practice, the anhydride is more easily prepared by the reaction of an acyl chloride and the sodium salt of an acid.

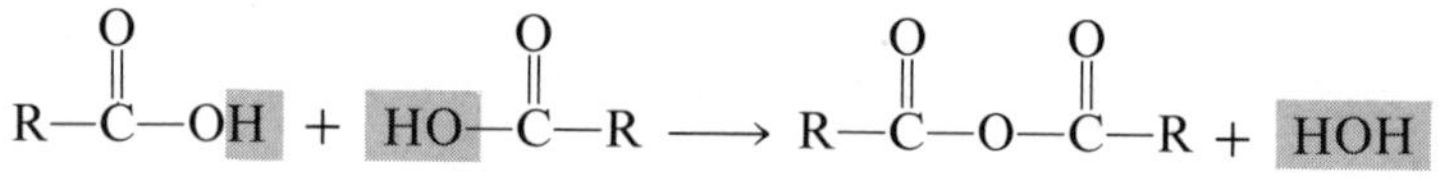

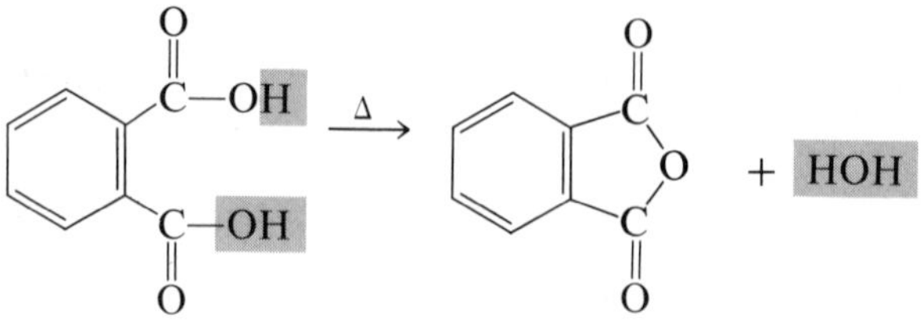

Phthalic acid   Phthalic anhydride

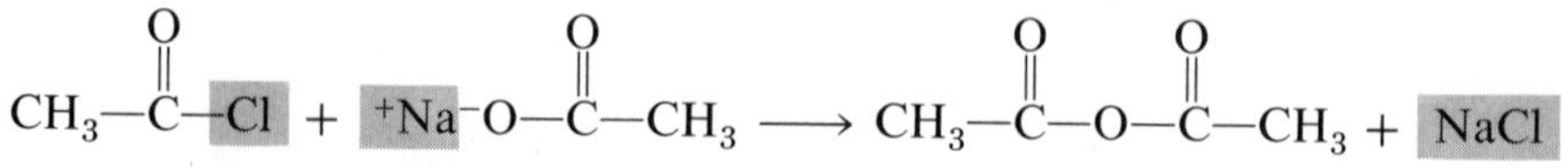

Acetyl chloride   Sodium acetate   Acetic anhydride

Both acetyl chloride and acetic anhydride are known as *acetylating agents* because they will react with many substances that have a replaceable hydrogen to form acetyl derivatives. The following reactions with water (hydrolysis), alcohol (alcoholysis), and ammonia (ammonolysis) illustrate the synthetic usefulness of these compounds.

Hydrolysis—Preparation of Acids. Acid halides and anhydrides are readily converted back to their parent acid by reaction with water.

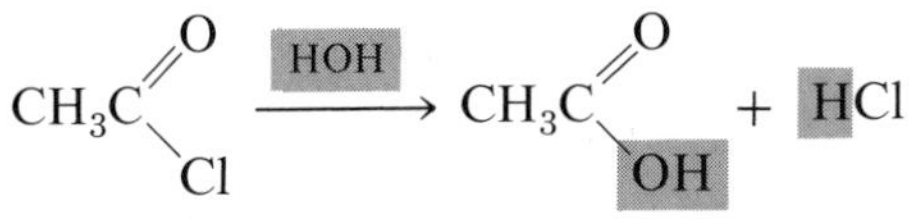

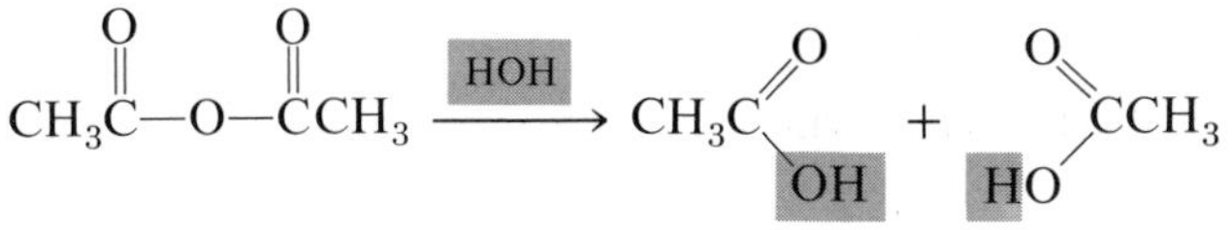

Alcoholysis—Preparation of Esters (*see Section 19.8*)

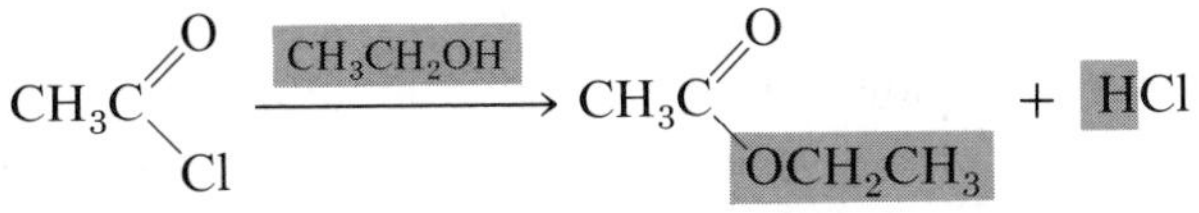

Ester

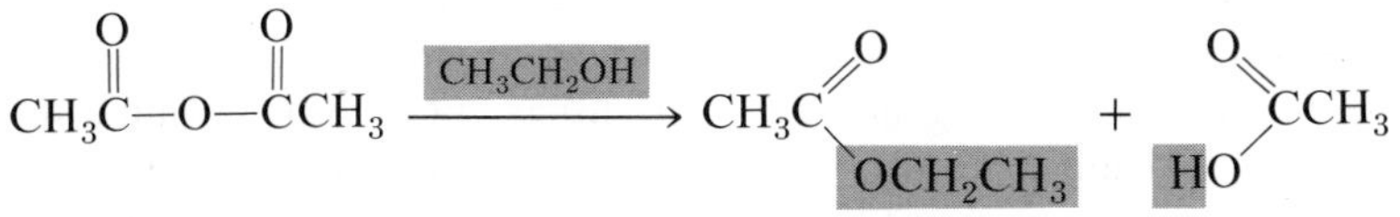

Ammonolysis—Preparation of Amides (*see Section 20.2*)

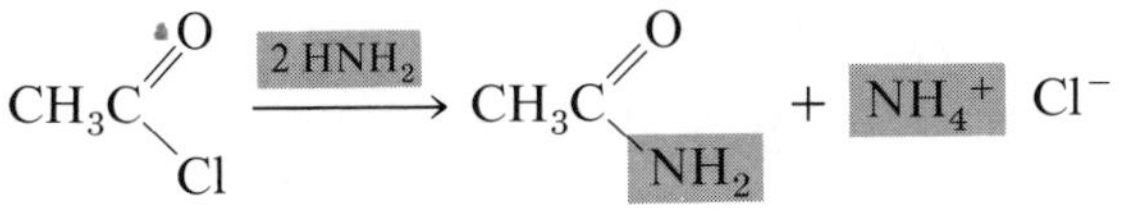

Amide

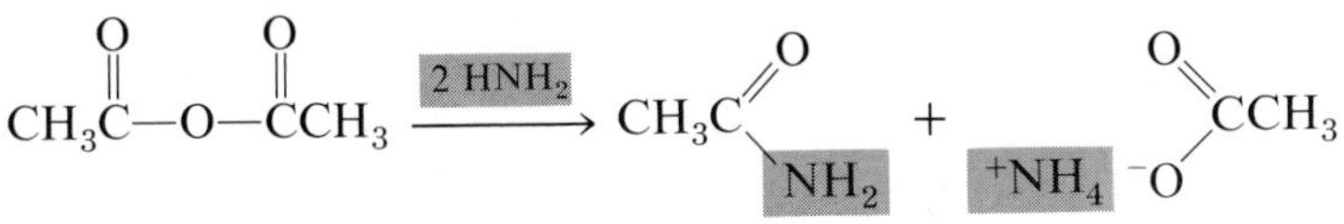

3. Formation of Esters (Esterification). When an acid and an alcohol are heated in the presence of an acid catalyst, a condensation reaction occurs that produces an ester and water.

General Equation

$$R'-C\overset{O}{\underset{OH}{\Big\langle}} + \boxed{ROH} \overset{H^+}{\rightleftharpoons} R'-C\overset{O}{\underset{OR}{\Big\langle}} + \boxed{HOH}$$

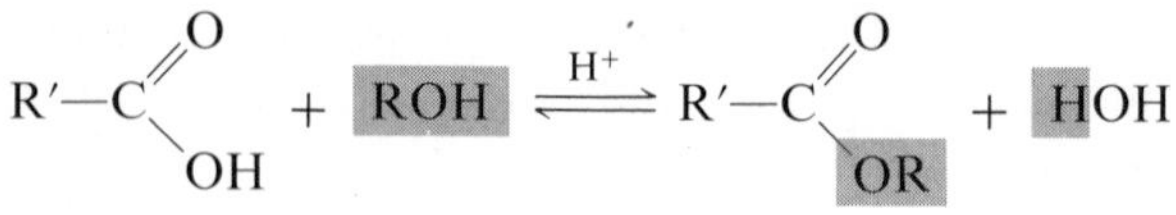

Ester

Specific Equation

$$CH_3-C\overset{O}{\underset{OH}{\Big\langle}} + \boxed{CH_3CH_2OH} \overset{H^+}{\rightleftharpoons} CH_3-C\overset{O}{\underset{OCH_2CH_3}{\Big\langle}} + \boxed{HOH}$$

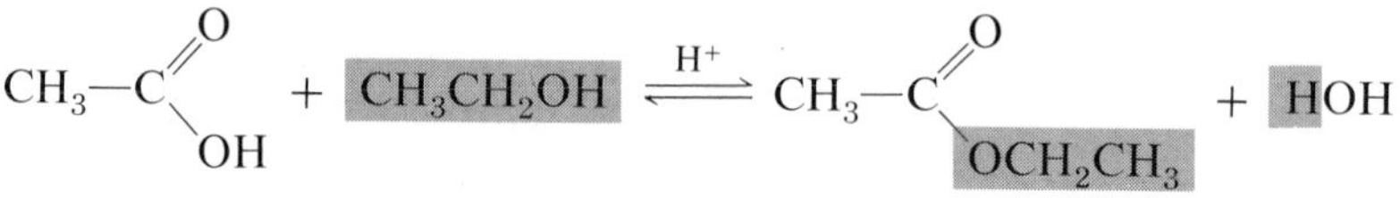

Ethyl acetate

The esterification reaction will be described in considerable detail in Section 19.8.

c. Replacement of Alpha Hydrogens

The carbonyl portion of the carboxyl group does not (1) add reagents such as hydrogen cyanide, (2) condense with the ammonia derivatives such as hydroxylamine, (3) self-condense with another molecule of acid as in the aldol condensation. However, the carbonyl group does influence the adjacent carbon atom, causing its hydrogens to be more labile than the hydrogens on carbon atoms further removed. The substitution of alpha hydrogens is brought about by treating the acid with either chlorine or bromine in the presence of sunlight or a suitable catalyst such as phosphorus or a phosphorus halide. (Iodine is too unreactive to undergo this reaction.) The reaction can be controlled with regard to the extent of substitution. The mono-, di-, or trihalo acids so produced are not to be confused with the acid halides.

$$CH_3-\overset{H}{\underset{H}{\overset{|}{C}}}-C\overset{O}{\underset{OH}{\Big\langle}} + \boxed{Br_2} \overset{P}{\longrightarrow} CH_3-\overset{H}{\underset{Br}{\overset{|}{C}}}-C\overset{O}{\underset{OH}{\Big\langle}} + H\boxed{Br}$$

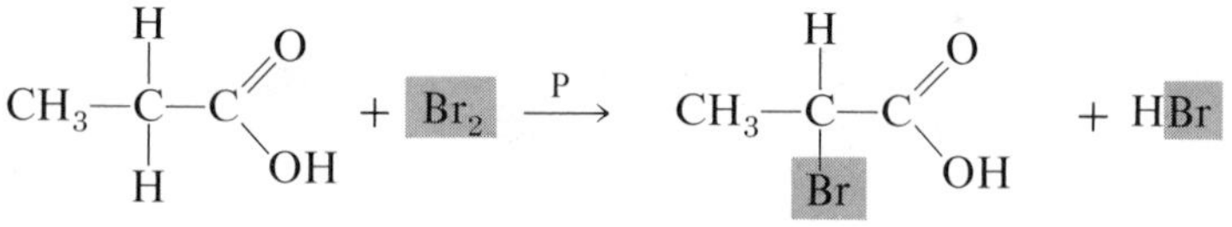

Propionic acid   2-Bromopropanoic acid
                        (α-Bromopropionic acid)

$$CH_3-\overset{H}{\underset{Br}{\overset{|}{C}}}-C\overset{O}{\underset{OH}{\Big\langle}} + \boxed{Br_2} \overset{P}{\longrightarrow} CH_3-\overset{Br}{\underset{Br}{\overset{|}{C}}}-C\overset{O}{\underset{OH}{\Big\langle}} + H\boxed{Br}$$

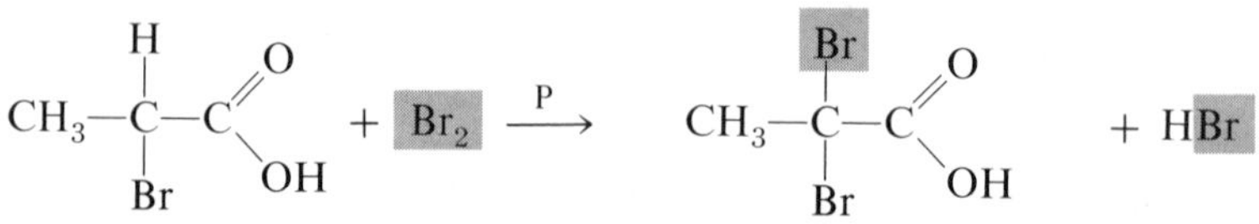

2,2-Dibromopropanoic acid
(α,α-Dibromopropionic acid)

Like the acid halides, haloacids are useful as intermediates in the preparation of other organic compounds. The alpha halogen atom (but not the alpha hydrogen) is easily replaced by the amino group and the hydroxyl group to yield amino acids and α-hydroxy acids, respectively.

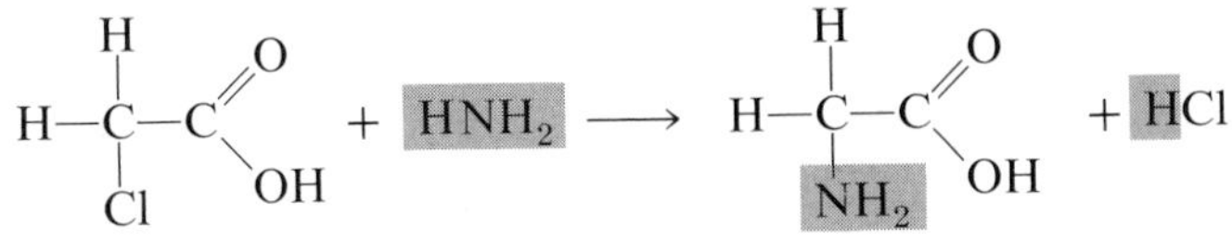

Chloroacetic acid — α-Aminoacetic acid (Glycine)

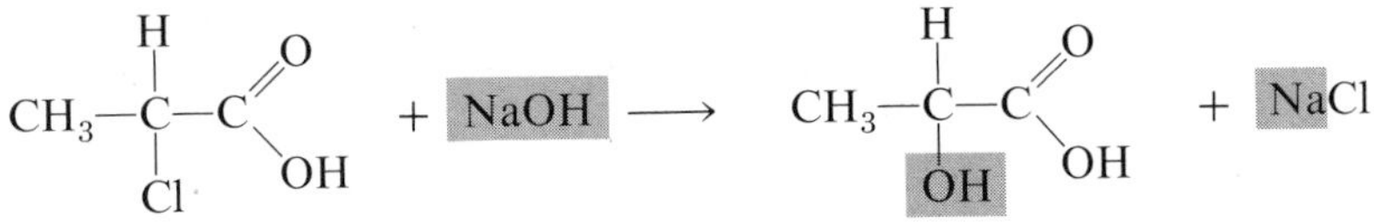

α-Chloropropionic acid — α-Hydroxypropionic acid (Lactic acid)

The weed killer 2,4-D is synthesized from chloroacetic acid and sodium 2,4-dichlorophenoxide.

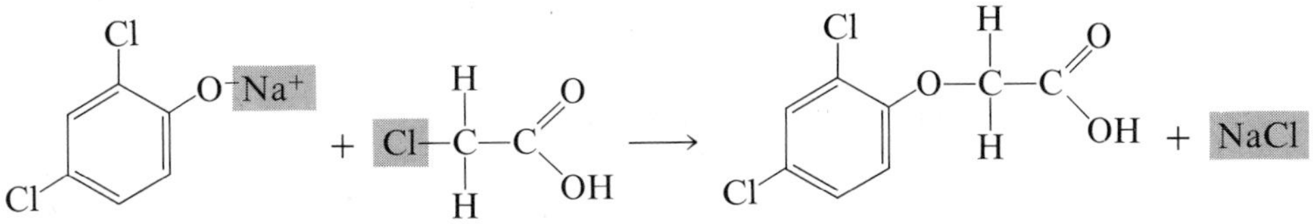

2,4-Dichlorophenoxyacetic acid

19.6 Important Carboxylic Acids

a. Formic Acid

Formic acid (Latin, *formica,* ant) is a principal component of the secretions of bees and ants. It is responsible for the blistering of the skin that follows a bee or an ant sting. Formic acid may be prepared commercially by heating carbon monoxide with sodium hydroxide. The resulting salt, sodium formate, is treated with sulfuric acid and formic acid is liberated.

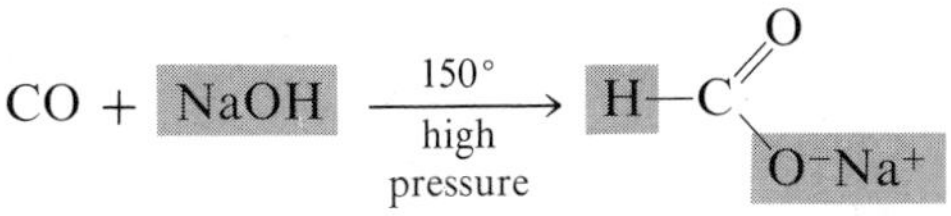

Sodium formate

Unlike all the other carboxylic acids, formic acid possesses an aldehydic group as well as a carboxyl group.

$$H-C\underset{OH}{\overset{O}{<}} \qquad H-C\underset{OH}{\overset{O}{<}}$$

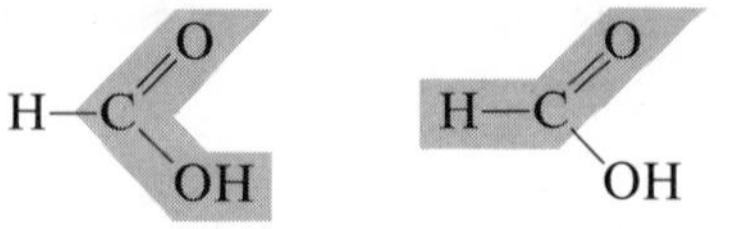

Carboxyl group   Aldehydic group

It is therefore not surprising to find that formic acid is capable of undergoing reactions that are not possible for the higher homologs of the acid series. Unlike most acids, it is a powerful reducing agent. It is easily oxidized to carbon dioxide and water, and thus gives a positive test with Tollens' reagent. It will decolorize potassium permanganate solution but will not reduce Fehling's or Benedict's solution.

$$H-C\underset{OH}{\overset{O}{<}} + Ag_2O \xrightarrow{\text{Tollens' reagent}} H_2O + CO_2 + 2\,Ag_{(s)}$$

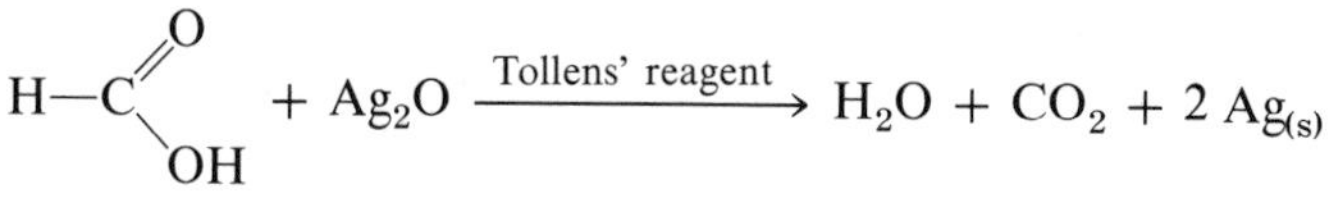

Furthermore, unlike all the other acids, formic acid can be both dehydrogenated and dehydrated. The latter reaction is employed in chemical laboratories to generate small amounts of pure carbon monoxide.

Dehydrogenation (Decarboxylation)

$$H-C\underset{OH}{\overset{O}{<}} \xrightarrow{160\,°C} H_2 + CO_2$$

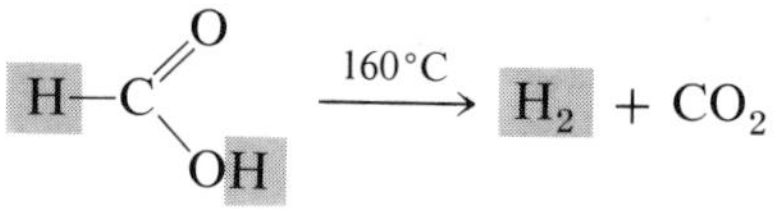

Dehydration

$$H-C\underset{OH}{\overset{O}{<}} \xrightarrow{H_2SO_4} H_2O + CO$$

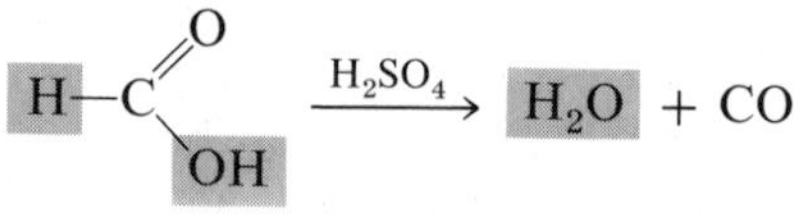

Another unique characteristic of formic acid is its inability to form a stable acid chloride. All attempts to prepare formyl chloride yield only hydrogen chloride and carbon monoxide.

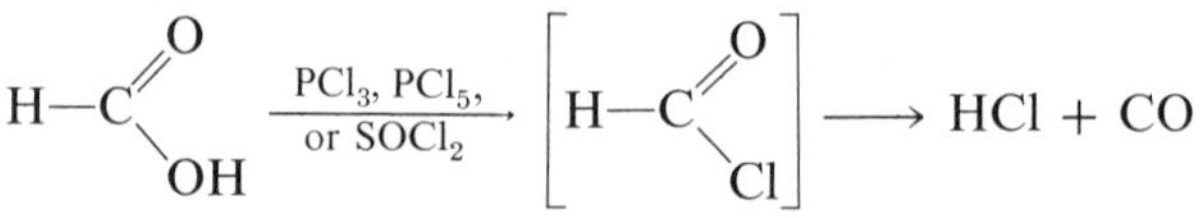

$$H-C\underset{OH}{\overset{O}{<}} \xrightarrow[\text{or SOCl}_2]{\text{PCl}_3,\,\text{PCl}_5,} \left[H-C\underset{Cl}{\overset{O}{<}} \right] \longrightarrow HCl + CO$$

Unstable

b. Acetic Acid

Acetic acid (Latin, *acetum,* vinegar) is by far the most important of the monocarboxylic acids. It is one of the earliest known organic compounds. Vinegar is a 4–5% solution of acetic acid in water. It is the acetic acid that imparts to the water the characteristic sour taste of vinegar. Pure acetic acid freezes at a temperature of 16.6°C (62°F). At normal room temperature, acetic acid is a colorless liquid. However in poorly heated laboratories, it may begin to crystallize, forming a glass-like solid. Hence the pure acid is often called *glacial* acetic acid.

Commercially, the acid is made by the oxidation of acetaldehyde or ethyl alcohol, or by the bacterial fermentation of wines, fruit juices, and ciders.

$$CH_3CH_2OH \xrightarrow[KMnO_4]{[O]} CH_3C\begin{matrix}{}^{\nearrow\!\!\!\!\!/}O \\ {}_{\searrow}OH\end{matrix}$$

$$C_6H_{12}O_6 \xrightarrow{\text{enzymes}} 2\ CH_3CH_2OH + 2\ CO_2$$

$$CH_3CH_2OH \xrightarrow[O_2\ (\text{air})]{\text{enzymes}} CH_3C\begin{matrix}{}^{\nearrow\!\!\!\!\!/}O \\ {}_{\searrow}OH\end{matrix}$$

Acetic acid is resistant to further oxidation and is frequently employed as a solvent for the oxidation of alkenes and alcohols.

c. Butyric Acid

The name of this acid is also indicative of its chief source (Latin, *butyrum,* butter) and butyric acid is liberated from butter that is left standing in the air. As previously mentioned, the disagreeable odor associated with rancid butter is that of butyric acid. Small quantities of this acid are added to oleomargarine to give it a butter-like taste.

d. Oxalic Acid

Oxalic acid is found as the monopotassium salt (HOOC—COOK) in many plants, especially spinach and rhubarb. It was formerly prepared by decomposing complex carbohydrates (in the form of sawdust) with sodium and potassium hydroxides. It is now obtained commercially by heating sodium formate. Lime is added to precipitate the insoluble calcium oxalate from which oxalic acid is obtained by treatment with sulfuric acid.

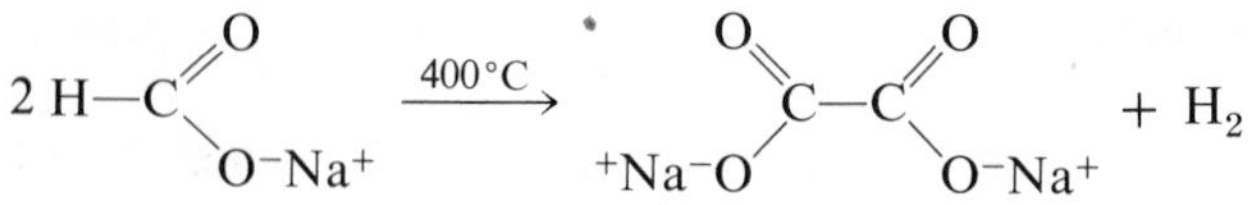

$$\begin{matrix} COO^-Na^+ \\ | \\ COO^-Na^+ \end{matrix} + CaO \longrightarrow Ca^{2+}\left[\begin{matrix} COO^- \\ | \\ COO^- \end{matrix}\right] \xrightarrow{H_2SO_4} \begin{matrix} COOH \\ | \\ COOH \end{matrix}$$

Like formic acid, and unlike all the other dicarboxylic acids, oxalic acid is readily oxidized to carbon dioxide and water by potassium permanganate. It is used in the quantitative analysis laboratory to standardize permanganate solutions.

$$5\,H_2C_2O_4 + 2\,KMnO_4 + 3\,H_2SO_4 \longrightarrow 2\,MnSO_4 + K_2SO_4 + 10\,CO_2 + 8\,H_2O$$

Oxalic acid is the strongest of the dicarboxylic acids. It is irritating to the skin and mucous membranes and is highly poisonous if ingested. (The cooking process destroys the acid in foods.) It is used as a bleaching agent and as a remover of rust and ink stains.

e. Lactic Acid

Lactic acid forms in sour milk as a result of the action of bacteria upon lactose (milk sugar).

$$\underset{\text{Lactose}}{C_{12}H_{22}O_{11}} + H_2O \xrightarrow[\text{in bacteria}]{\text{enzymes}} \underset{\text{Lactic acid}}{4\,CH_3CHOHCOOH}$$

Commercially it is prepared by the bacterial fermentation of cane sugar or corn starch. Its presence in certain sour foods (e.g., sauerkraut) prevents bacterial spoilage because most bacteria are unable to survive in an acidic environment. It is as strong an acid as formic acid. As we shall see in Section 28.3, lactic acid is an important compound biochemically. Some of the energy necessary for muscular activity is supplied by the breakdown of carbohydrates to lactic acid.

f. Tartaric Acid

Tartaric acid (HOOC—CHOH—CHOH—COOH) is obtained as a by-product of the wine industry. When the wine is aged in the casks, a white salt crystallizes out. This salt is potassium hydrogen tartrate (known as cream of tartar) and it is widely used in the manufacture of baking powder. Recall that Rochelle salt, used in Fehling's soluion, is the potassium sodium salt of tartaric acid ($^{+}Na^{-}OOC—CHOH—CHOH—COO^{-}K^{+}$). The free acid and its salts are used in medicine, dyeing and in some foods and soft drinks. We shall refer to tartaric acid in Section 22.4 in connection with its optical activity.

g. Citric Acid

Citric acid ($HOOC—CH_2—C(OH)COOH—CH_2—COOH$) is one of the most widely distributed naturally occurring acids. As its name implies, it is particularly abundant in citrus fruits. The acid is prepared commercially by the fermentation of glucose by a certain mold. Citric acid is produced by the mold as the mold consumes the glucose. This tricarboxylic acid is a normal constituent of blood serum and it is present in all living cells that obtain their energy from the aerobic metabolism of carbohydrates (see Section 28.7). Citric acid is widely used in the food industry and in the preparation of beverages because of its solubility in water and mildly sour taste. It forms a variety of salts, some of which are used in pharmaceutical preparations.

h. Salicylic Acid

Salicylic acid, $C_6H_4(OH)COOH$, occurs as the methyl ester in oil of wintergreen. It is prepared commercially from phenol and carbon dioxide. See Section 19.11 for a discussion of the important esters of salicylic acid.

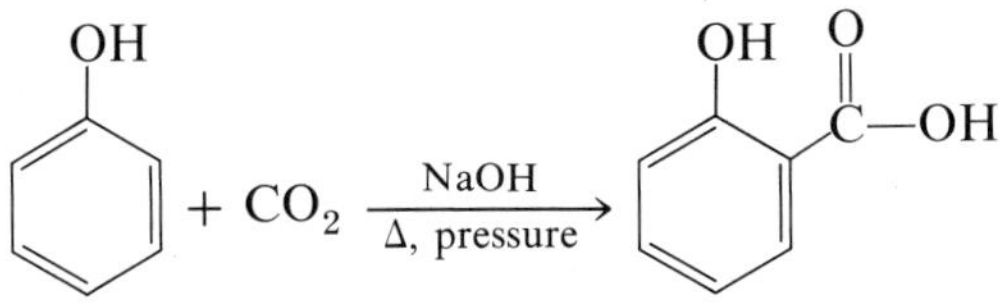

OH
+ CO2
NaOH
Δ, pressure
OH
O
C—OH

Esters

Esters are perhaps the most important class of derivatives of the carboxylic acids. The general formula for an ester is **RCOOR′**, where **R** may be a hydrogen, an alkyl group, or an aryl group, and where **R′** may be alkyl or aryl, but *not* hydrogen.

Esters are widely distributed in nature. Their occurrence is particularly important in fats and vegetable oils, which are esters of long-chain fatty acids and glycerol (Chapter 24). Esters of phosphoric acid are of the utmost importance to life, and they will be discussed in Chapters 21 and 28.

Table 19.5 Nomenclature of Esters

Formula	*Common Name*	*IUPAC Name*
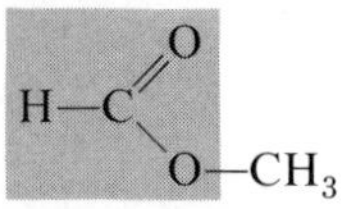	Methyl formate	Methyl methanoate
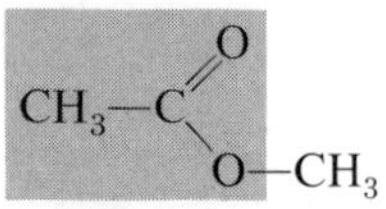	Methyl acetate	Methyl ethanoate
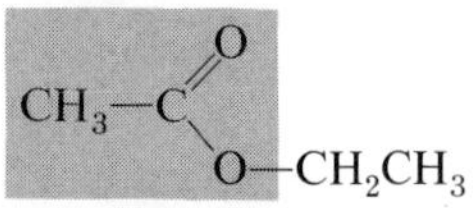	Ethyl acetate	Ethyl ethanoate
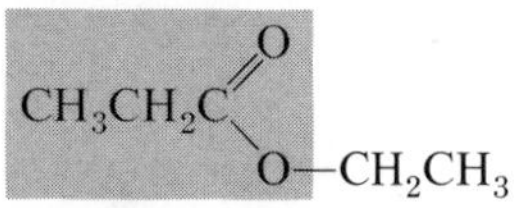	Ethyl propionate	Ethyl propanoate
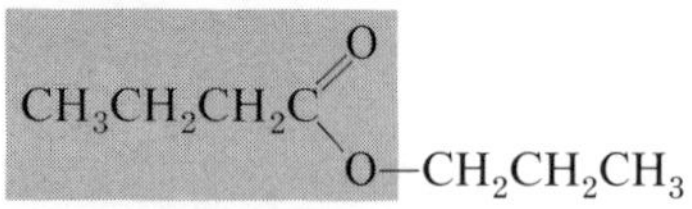	*n*-Propyl butyrate	*n*-Propyl butanoate
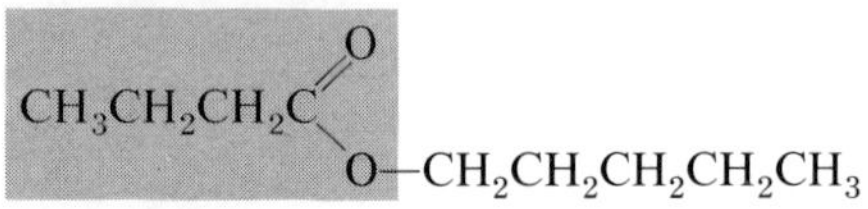	*n*-Pentyl butyrate (Amyl butyrate)	*n*-Pentyl butanoate
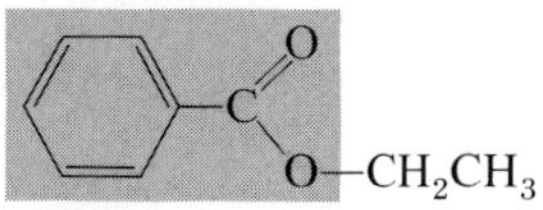	Ethyl benzoate	Ethyl benzoate
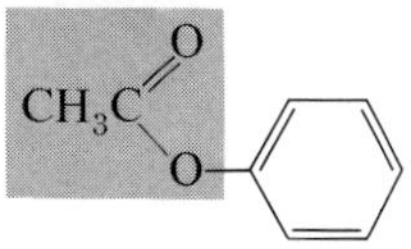	Phenyl acetate	Phenyl ethanoate

19.7 Nomenclature of Esters

Esters are named in a manner analogous to that used for naming organic salts. The group name of the alkyl or aryl portion is given first and is followed by the name of the acid portion. In both common and IUPAC nomenclature, the *-ic* ending of the parent acid is replaced by the suffix *-ate*. Table 19.5 illustrates the nomenclature of several esters.

19.8 Preparation of Esters

The preparation of esters from acid chlorides and anhydrides via alcoholysis has already been discussed (see Section 19.5-b). Mention has also been made of the esterification reaction between an acid and an alcohol. We will now examine this very important reaction in greater detail.

Recall that the equation for the reaction between acetic acid and ethyl alcohol is

$$CH_3COOH + CH_3CH_2OH \overset{H^+}{\rightleftharpoons} CH_3{-}C\overset{\displaystyle O}{\underset{\displaystyle OCH_2CH_3}{\lessgtr}} + HOH$$

Note the presence of the double arrows, indicating the reversibility of the reaction. When equilibrium is achieved, the reaction mixture contains approximately 66% of the ester and 34% of both the acid and the alcohol. An approximate value for the equilibrium constant for the esterification reaction can be calculated from the law of chemical equilibrium.

$$K = \frac{[CH_3COOCH_2CH_3][HOH]}{[CH_3COOH][CH_3CH_2OH]}$$

$$= \frac{(.66)(.66)}{(.34)(.34)} \simeq 4$$

Experimental data reveal that, for any esterification reaction, the composition of the equilibrium mixture will always be governed by the value of the equilibrium constant regardless of the relative amounts of initial alcohol and acid. Consideration of the law of chemical equilibrium and Le Châtelier's principle (Section 10.4-a) allows predetermination of reaction conditions that will produce a maximum yield of the desired ester. It is evident that the use of an excess of either the alcohol or the acid (whichever is less expensive) will produce an increase in the amount of the ester formed. In addition, if the water is removed from the reaction mixture as soon as it is formed, the equilibrium is shifted continuously to the right to conform to the law of chemical equilibrium. This can be accomplished if the acid, the alcohol, and the ester all boil at temperatures above 100°C. The water may then be distilled away from the reaction mixture as it is formed. A second method involves the addition of some inert substance, such as benzene, that forms an azeotropic mixture (boiling at 65°C) with water. In this instance, the water may be distilled off at a relatively low temperature. A third method involves the use of a dehydrating reagent that is inert toward the reaction but will absorb the water. Zinc chloride is sometimes used for this purpose.

In general, the reaction between an acid and an alcohol is extremely slow, and a catalyst must be employed to speed it up. Sulfuric acid is commonly used because it is both an acid and a good dehydrating agent. Thus it serves both to catalytically increase the rate of reaction and to shift the equilibrium to the right by reacting with water to form hydronium ions, H_3O^+, thus increasing the yield of the ester.

Theoretically, the esterification reaction can be considered as occurring by either of two pathways. The condensation of acid and alcohol could occur by the elimination of water through the loss of the hydrogen of the acid and the hydroxyl of the alcohol, or vice versa. The following equations illustrate the two possibilities. Note that both alternatives yield the *same* ester.

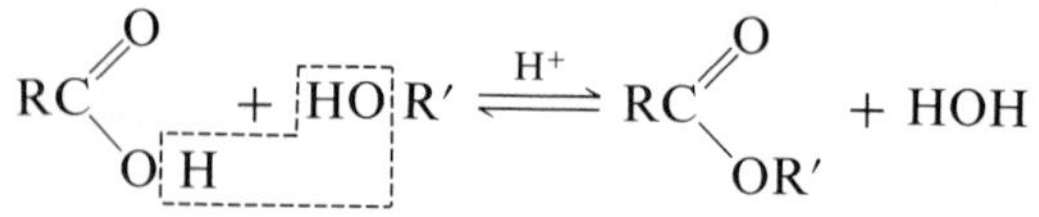

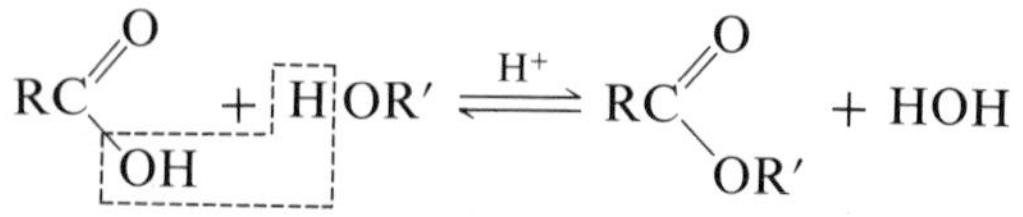

In Section 17.4-e, we indicated that the reaction proceeds according to the second equation. This knowledge was obtained by the use of isotopic labeling, a useful technique for the understanding of chemical mechanisms. Elucidation of this mechanism was accomplished by an esterification involving methyl alcohol that contained a heavy isotope of oxygen (^{18}O). When the alcohol was allowed to react with acetic acid, it was found that the resulting methyl acetate contained all of the isotopic oxygen; the water contained none of the labeled oxygen.

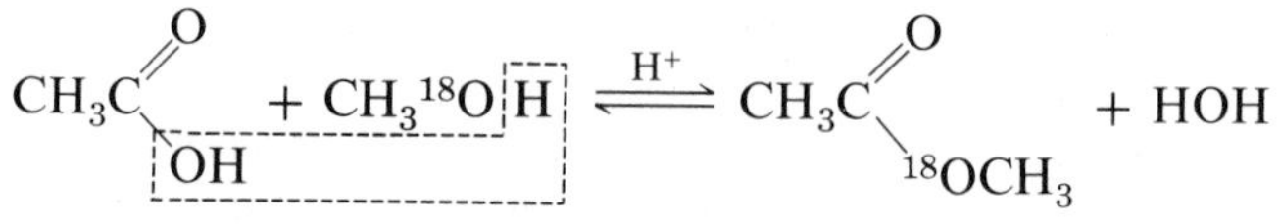

Similar experiments have been performed with a wide variety of acids and alcohols. In every case it was found that the hydroxyl group is lost from the acid and that the hydrogen is lost from the alcohol. It should be reemphasized that the ester formed in either case would be identical. We are concerned here only with a better understanding of the pathway of the reaction.

19.9 Physical Properties of Esters

Unlike the carboxylic acids from which they are derived, the esters have very pleasant odors. In fact, the specific aromas of many flowers and fruits are due to the presence of esters. Esters are utilized in the manufacture of perfumes and as flavoring agents in the confectionery and soft drink industries. (A mixture of nine esters is utilized to produce an artificial raspberry flavor.) Vapors of esters are nontoxic unless inhaled in large concentrations. (There is, therefore, danger in the fads of glue sniffing and smoking banana peels. The toxic ester in each case is amyl acetate.) Esters find their most important use as industrial solvents. Most esters are colorless liquids, insoluble in and lighter than water. They are neutral substances and usually have lower boiling points and melting points than the acids or alcohols of comparable

Table 19.6 Physical Properties of Esters

Formula	*Name*	*M*	*MP (°C)*	*BP (°C)*	*Aroma*
$HCOOCH_3$	Methyl formate	60	−99	32	
$HCOOCH_2CH_3$	Ethyl formate	74	−80	54	Rum
CH_3COOCH_3	Methyl acetate	74	−98	57	
$CH_3COOCH_2CH_3$	Ethyl acetate	88	−84	77	
$CH_3CH_2COOCH_3$	Methyl propionate	88	−88	80	
$CH_3CH_2COOCH_2CH_3$	Ethyl propionate	102	−74	99	
$CH_3CH_2CH_2COOCH_3$	Methyl butyrate	102	−95	102	Apple
$CH_3CH_2CH_2COOCH_2CH_3$	Ethyl butyrate	116	−93	121	Pineapple
$CH_3COO(CH_2)_4CH_3$	Amyl acetate	130	−71	148	Banana
$CH_3COOCH_2CH_2CH(CH_3)_2$	Isoamyl acetate	130	−79	142	Pear
$CH_3COOCH_2C_6H_5$	Benzyl acetate	150	−51	214	Jasmine
$CH_3CH_2CH_2COO(CH_2)_4CH_3$	Amyl butyrate	158	−73	185	Apricot
$CH_3COO(CH_2)_7CH_3$	Octyl acetate	172	−39	210	Orange

molar mass. Since they do not contain hydroxylic hydrogens, there is no possibility of intermolecular hydrogen bonding. Table 19.6 lists the physical properties of some common esters.

19.10 Chemical Properties of Esters

The most significant reaction of the esters is their hydrolysis (Greek, *hydro,* water; *lysis,* loosening). Hydrolysis, then, is the splitting of molecules by the addition of water. Most organic and biochemical compounds react very slowly, if at all, with water.[2] The chemist employs acids or bases to catalyze hydrolysis reactions in the laboratory. Living organisms utilize enzymes to catalyze these same reactions. In the laboratory, the process may take several hours; in living systems, it is complete in a fraction of a second.

a. Acid Hydrolysis of Esters

The acid hydrolysis of esters is simply the reverse of the esterification reaction (Section 19.8). The ester is refluxed with a large excess of water containing a strong acid. However, the equilibrium is unfavorable for ester hydrolysis ($K \simeq \frac{1}{4}$) and so the reaction never goes to completion.

General Equation

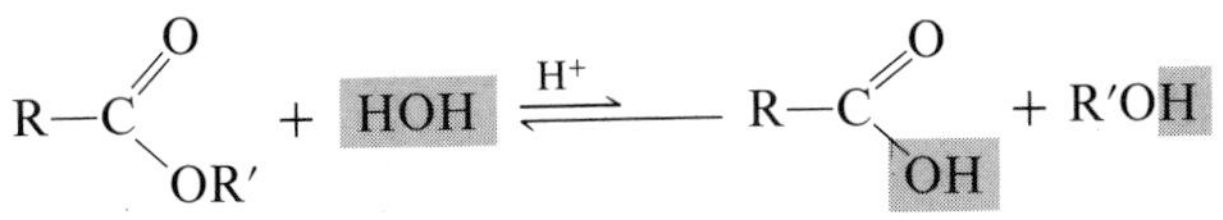

O
R—C
OR′
+ HOH
H+
O
R—C
OH
+ R′OH

[2]Two exceptionally rapid hydrolysis reactions were mentioned in connection with the reactions of acid chlorides and anhydrides with water (Section 19.5-b).

$$\text{(Methyl benzoate)} \quad + \quad \boxed{HOH} \quad \underset{}{\overset{H^+}{\rightleftharpoons}} \quad \text{(Benzoic acid)} \quad + \quad \boxed{CH_3OH}$$

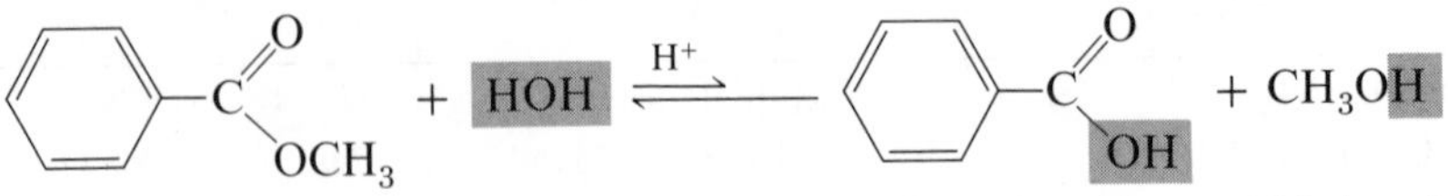

Methyl benzoate   Benzoic acid   Methanol

b. Alkaline Hydrolysis of Esters (Saponification)

When a base (such as sodium or potassium hydroxide) is used to catalyze an ester hydrolysis, the reaction goes to completion since the acid is removed from the equilibrium by its conversion to a salt. Organic salts do not react with alcohols, so the reaction is essentially irreversible. Accordingly, ester hydrolysis is usually carried out in basic solution. Because soaps are prepared by the alkaline hydrolysis of fats and oils, the term saponification (Latin, *sapon*, soap; *facere*, to make) is used to describe the alkaline hydrolysis of all esters (see also Section 24.5-a).

General Equation

$$R-C\overset{O}{\underset{OR'}{\diagup}} \quad + \quad \boxed{NaOH} \quad \longrightarrow \quad R-C\overset{O}{\underset{\boxed{O^-Na^+}}{\diagup}} \quad + \quad \boxed{R'OH}$$

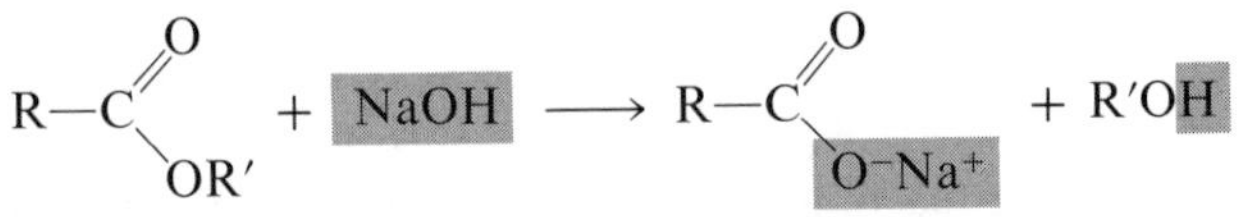

Specific Equation

$$CH_3-C\overset{O}{\underset{OCH_2CH_3}{\diagup}} \quad + \quad \boxed{NaOH} \quad \longrightarrow \quad CH_3-C\overset{O}{\underset{\boxed{O^-Na^+}}{\diagup}} \quad + \quad \boxed{CH_3CH_2OH}$$

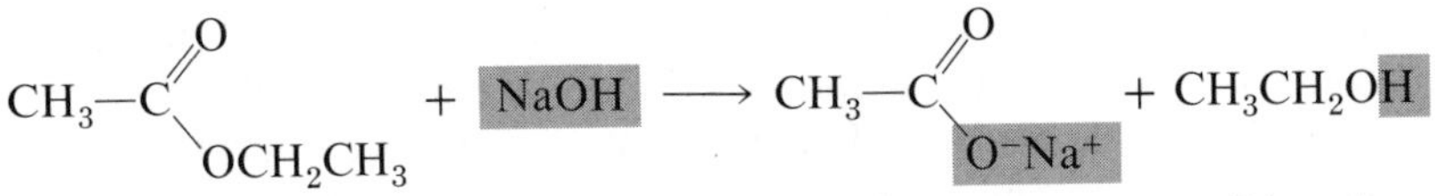

Ethyl acetate   Sodium acetate   Ethanol

c. Reduction of Esters (Formation of Alcohols)

Although carboxylic acids are obtained by the extensive oxidation of primary alcohols, they are not easily reduced to regenerate the initial alcohol. Esters, on the other hand, are converted to alcohols by the action of reducing agents such as lithium aluminum hydride ($LiAlH_4$), or sodium metal and an alcohol. Thus acids may be indirectly reduced by first converting them to their corresponding esters.

General Equation

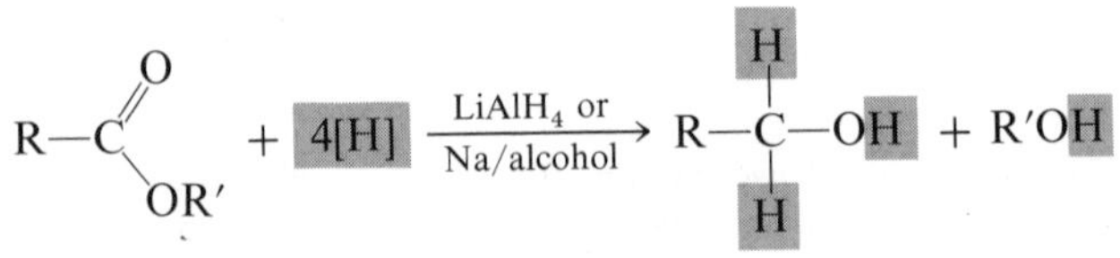

$$R-C\overset{O}{\underset{OR'}{\diagup}} \quad + \quad \boxed{4[H]} \quad \xrightarrow[Na/alcohol]{LiAlH_4 \text{ or}} \quad R-\overset{\boxed{H}}{\underset{\boxed{H}}{\overset{|}{\underset{|}{C}}}}-\boxed{OH} \quad + \quad \boxed{R'OH}$$

Specific Equation

$$\underset{\text{Methyl butyrate}}{CH_3CH_2CH_2C(=O)OCH_3} + 4\,[H] \longrightarrow \underset{n\text{-Butyl alcohol}}{CH_3CH_2CH_2CH_2OH} + \underset{\text{Methyl alcohol}}{CH_3OH}$$

d. Reaction with Ammonia (Ammonolysis)

Esters, like the other acid derivatives (acid halides and anhydrides), react with ammonia to form amides (Chapter 20).

General Equation

$$R{-}C(=O){-}OR' + HNH_2 \longrightarrow R{-}C(=O){-}NH_2 + R'OH$$

Specific Equation

$$\underset{\text{Methyl acetate}}{CH_3{-}C(=O){-}OCH_3} + NH_3 \longrightarrow \underset{\text{Acetamide}}{CH_3{-}C(=O){-}NH_2} + \underset{\text{Methanol}}{CH_3OH}$$

Notice that in all the reactions of esters thus far encountered, the carbonyl carbon–oxygen single bond is the one that is broken: $R{-}C(=O){\not-}OR'$. In all cases, an alcohol (R′OH) is one of the products formed.

19.11 Esters of Salicylic Acid

Salicylic acid is a bifunctional aromatic compound containing both a carboxyl group and a hydroxyl group. Thus, it may function as an acid or as an alcohol in an esterification reaction.

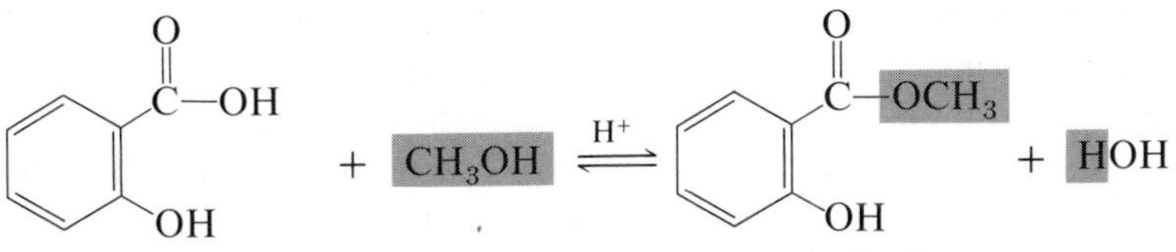

O
C—OH
OH
+ CH3OH
H+
O
C—OCH3
OH
+ HOH

o-Hydroxybenzoic acid (Salicylic acid) — Methanol — Methyl salicylate (Oil of wintergreen)

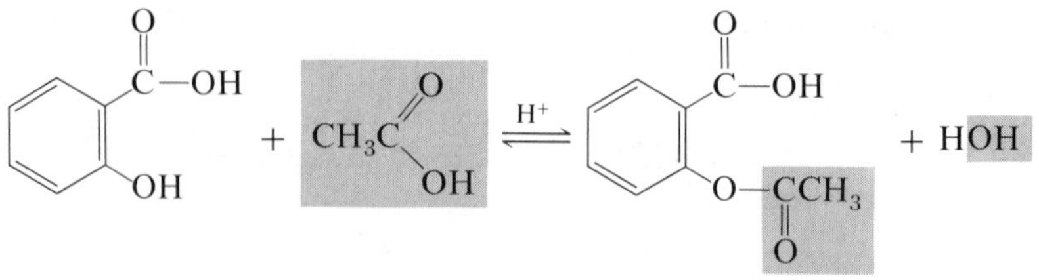

Acetic acid — Acetylsalicylic acid (Aspirin)

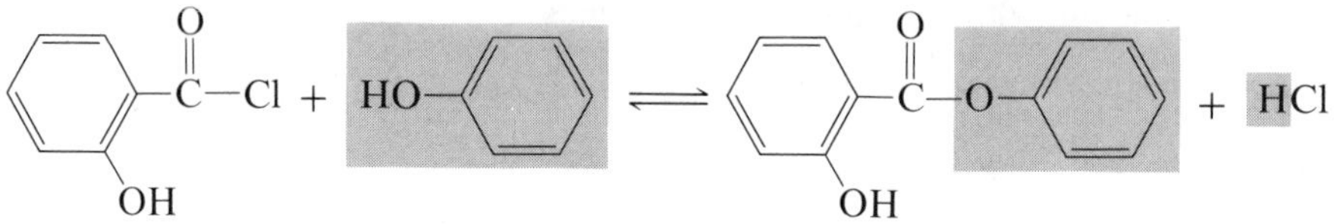

Acid chloride of salicylic acid — Phenol — Phenyl salicylate (Salol)

Methyl salicylate is an oil found in numerous plants and has a fragrance associated with wintergreen. Commercially, it is used in perfumes and for flavoring candy. It finds widespread use as the pain-relieving ingredient in liniments such as Ben-Gay. When rubbed on the skin, this ester has the unusual ability of penetrating the surface. Hydrolysis then occurs, liberating salicylic acid, which relieves the soreness.

Acetylsalicylic acid is a colorless solid. It is undoubtedly the most widely used drug in the world. Aspirin acts as a fever reducer (antipyretic), as a pain depresser (analgesic), and as an anti-inflammatory agent (reduces swelling in injured tissues and rheumatic joints). However, if taken in large quantities, it is also an effective poison! Thirty to forty grams constitutes a lethal dosage. An ordinary aspirin tablet contains about 0.32 gram (5 grains) of aspirin mixed with starch or some other inert binder to hold the tablet together. More than 35 million pounds of aspirin are produced annually in the United States. Of this amount, about half is converted into aspirin tablets and the rest is sold in the form of combination pain relievers and cold remedies.

We still do not understand just how aspirin accomplishes its extraordinary effects. There is some evidence to indicate that aspirin acts, in part, by inhibiting the synthesis of prostaglandins (see page 601). Among their many functions, prostaglandins are involved in inflammation, increased blood pressure, and the contraction of smooth muscle. Elevated concentrations of prostaglandins appear to activate pain receptors in the tissues, making the tissues more sensitive to any pain stimulus. There are certain hazards associated with the use of aspirin. Among these are a slight deterioration of the stomach lining and some intestinal bleeding. Prolonged use of aspirin can lead to gastrointestinal disorders. For a small percentage of the population, the effects of aspirin are more severe. Susceptible individuals experience skin rashes, asthmatic attacks, and even loss of consciousness. These people must be careful to avoid aspirin alone or in any combination, and the most satisfactory alternative available to them is acetaminophen (Tylenol, Tempra, Liquiprin). A panel of the Food and Drug Administration has recommended that aspirin and similar pain relievers should not be taken by women during the last three months of pregnancy except under a doctor's advice.

Salol is the phenyl ester of salicylic acid and is most easily prepared from the acid chloride rather than the free acid. Salol is a widely used intestinal antiseptic. It is not hydrolyzed by acids and therefore passes through the stomach unchanged. In the alkaline medium of the intestines, hydrolysis occurs to yield phenol and the salicylate ion. Salol is also employed as a coating for some medicinal pills in order to permit them to pass through the stomach intact, but to disintegrate in the intestines.

Exercises

19.1 Draw and name all the isomeric acids and esters having the molecular formula $C_5H_{10}O_2$.

19.2 Name the following compounds.

(a) $CH_3CH_2C(CH_3)_2C(=O)OH$

(b) $CH_3CH_2CH_2C(=O)O^-K^+$

(c) $CH_3CH_2CH(CH_3)C(=O)OH$

(d) $CH_3CH_2CH_2C(=O)OCH_3$

(e) $CH_3CH_2C(=O)Cl$

(f) $CH_3CHOHCH_2CH_2C(=O)OH$

(g) $CH_2(COOH)_2$

(h) 
NO2
COOH

19.3 Write structural formulas for the following compounds.

(a) citric acid
(b) β-hydroxybutyric acid
(c) *n*-butyl acetate
(d) *n*-amyl isobutyrate
(e) phthalic acid
(f) isopropyl propanoate
(g) adipic acid
(h) acetylsalicylic acid
(i) α,β-dibromobutyric acid
(j) methyl salicylate

19.4 Write equations showing the preparation of propanoic acid from **(a)** a sodium salt, **(b)** an alcohol, **(c)** a nitrile, **(d)** an alkyl halide, and **(e)** an ester.

19.5 Starting with toluene, devise a series of reactions for the preparation of **(a)** *m*-bromobenzoic acid, **(b)** *p*-bromobenzoic acid, and **(c)** benzyl bromide.

19.6 Give a brief description of the physical properties of the monocarboxylic aliphatic acids.

19.7 What is the explanation for the fact that formic acid melts about 30 degrees higher than propionic acid, even though its molar mass is considerably lower?

19.8 Write an equation to illustrate what occurs when an organic acid is placed in water. What determines the strength of an acid?

19.9 Arrange the following compounds in order of increasing acid strength: **(a)** acetic acid, **(b)** chloroacetic acid, **(c)** formic acid, **(d)** β-chloropropionic acid, **(e)** trichloroacetic acid.

19.10 A 0.10 *M* solution of the weak organic acid, RCOOH, dissociates to the extent of 3.4%. What is the dissociation constant of the acid?

19.11 Calculate the concentrations of H^+ and OH^- in a 1.0 *M* solution of acetic acid. What is the pH of the solution?

19.12 A 0.08 *M* solution of acetic acid is 1.5% dissociated. What is the hydrogen ion concentration and what is the pH of this solution?

19.13 What is the hydrogen ion concentration of a 0.01 *M* solution of chloroacetic acid ($K_a = 1.4 \times 10^{-3}$)? What is the pH of this solution?

19.14 A solution containing 1.0 *M* sodium acetate and 1.0 *M* acetic acid buffers at a pH of about 4. Show by equations what happens when acids or bases are added to this buffer solution.

19.15 Compare in a general way the pH of a 0.1 *M* acetic acid solution with that of a 0.1 *M* acetic acid solution that is also 0.1 *M* with sodium acetate. Account for these differences.

19.16 Assume that the equilibrium constant at 25°C for the esterification reaction between RCOOH and ROH is 3.5. Tell how the equilibrium will be affected by each of the following.

(a) rise in temperature
(b) addition of a small amount of sulfuric acid
(c) removal of water
(d) addition of NaOH.

19.17 Explain why basic hydrolysis of esters is preferable to acid hydrolysis.

19.18 Complete the following equations.

(a) $CH_3CH_2CH_2Br + KCN \longrightarrow\ ? \xrightarrow[H^+]{HOH}$

(b) 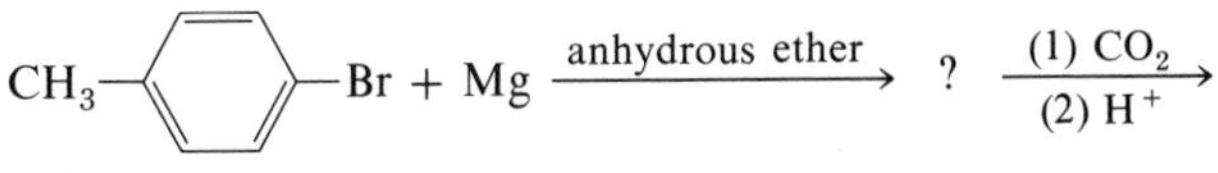

(c) $CH_3C(=O)OH + NaOH \longrightarrow$

(d) $CH_3C(=O)Cl + NH_3 \longrightarrow$

(e) $CH_3CH_2C(=O)OH + (CH_3)_2CHOH \longrightarrow$

(f) $CH_3CH_2CH_2C(=O)OH \xrightarrow[P]{Br_2}$

(g) $CH_3CH_2C(=O)OCH_2CH_2CH_3 + KOH \longrightarrow$

(h) 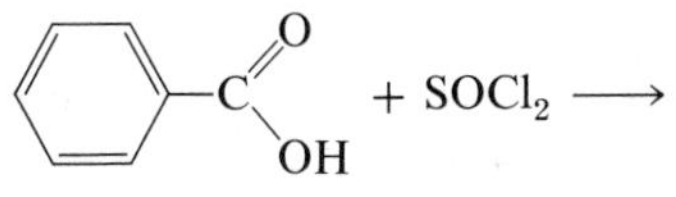

(i) $\left(CH_3C(=O)O^-\right)_2Ca + H_2SO_4 \longrightarrow$

(j) $CH_3CH_2C(=O)OCH_3 + LiAlH_4 \longrightarrow$

19.19 Starting with acetic acid, and any other organic or inorganic reagents, write equations for the preparation of each of the following.

(a) acetyl chloride
(b) acetic anhydride
(c) chloroacetic acid
(d) α-hydroxyacetic acid
(e) α-aminoacetic acid
(f) ethyl acetate

19.20 The following compounds contain a carbon-oxygen-carbon linkage. Compare their reactivity toward water: **(a)** ethyl ether; **(b)** acetic anhydride, and **(c)** ethyl acetate.

19.21 Devise simple chemical tests that would enable you to distinguish between the following pairs of isomers.

(a) $CH_3CH_2CH_2CH_2C\overset{O}{\underset{OH}{\diagup}}$ and $CH_3C(CH_3)_2C\overset{O}{\underset{OH}{\diagup}}$ 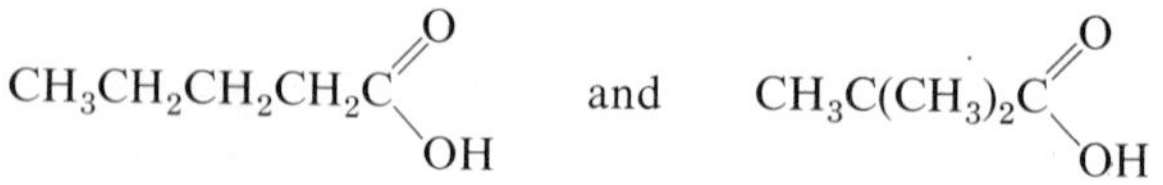

(b) $CH_3C\overset{O}{\underset{OCH_3}{\diagup}}$ and $H-C\overset{O}{\underset{OCH_2CH_3}{\diagup}}$ 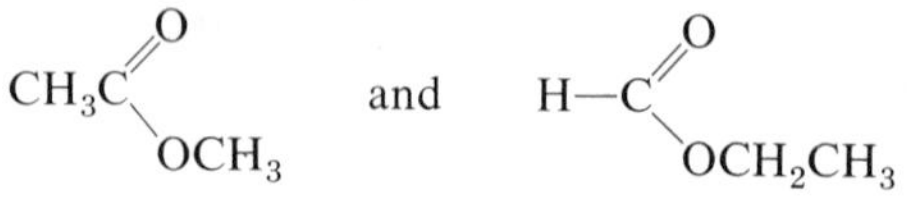

(c) $CH_3-\langle\!\bigcirc\!\rangle-C\overset{O}{\underset{OH}{\diagup}}$ and $CH_3-C\overset{O}{\underset{O}{\diagup}}-\langle\!\bigcirc\!\rangle$ 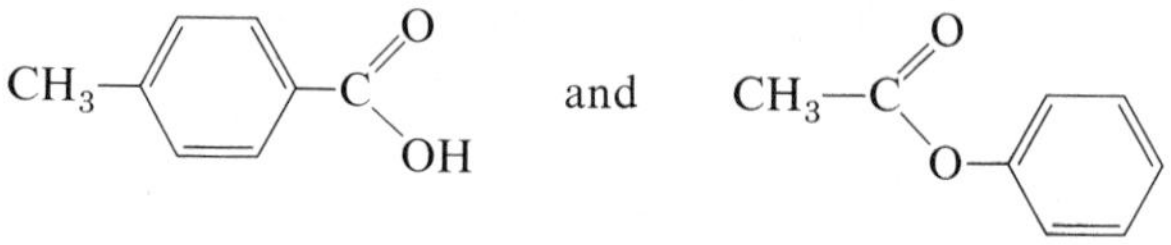

(d) $CH_3C\overset{O}{\underset{OCH_2CH_3}{\diagup}}$ and $CH_3OCH_2\overset{O}{\overset{\|}{C}}CH_3$ 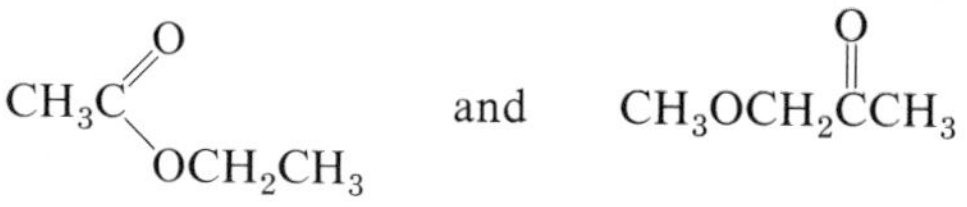

(e) $CH_3C\overset{O}{\underset{Cl}{\diagup}}$ and $CH_2ClC\overset{O}{\underset{H}{\diagup}}$ 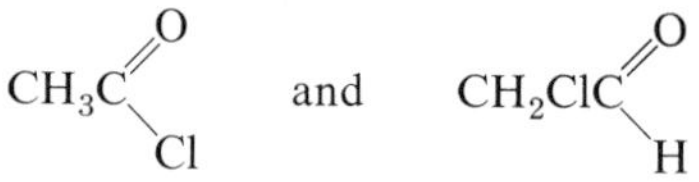

19.22 A compound, $C_7H_{14}O_2$, was treated with potassium hydroxide to yield ethanol and a potassium salt. Write four possible structural formulas of the original compound.

19.23 Hydrolysis of a neutral compound (A), $C_5H_{10}O_2$, gave an acid (B) and an alcohol (C). The alcohol reacted with PCl_3 to form an alkyl chloride that was then treated with Mg to form a Grignard reagent. Addition of CO_2, followed by hydrolysis, yielded an acid identical to (B). What are the structural formulas of A, B, and C?

19.24 5.1 grams of a monocarboxylic acid was required to neutralize 125 ml of a 0.4 M NaOH solution. Write all possible structural formulas for the acid.

19.25 If 3.0 grams of acetic acid reacted with excess methanol, how many grams of methyl acetate could be formed?

19.26 How many milliliters of a 0.10 M barium hydroxide solution would be required to neutralize 0.50 gram of dichloroacetic acid?

20

Amides and Amines

The compounds studied in the preceding chapters can be considered as derivatives of the alkanes or of water. In this chapter we shall deal with the organic nitrogen-containing compounds related to ammonia. Replacement of one or more hydrogen atoms of ammonia with alkyl or aryl groups yields **amines, RNH_2, R_2NH,** and **R_3N.** When a hydrogen atom of ammonia is replaced instead by an acyl group, RCO—, the resultant compound is an **amide, $RCONH_2$.**

Amides

Amides may also be considered to be derivatives of carboxylic acids. Replacement of the hydroxyl group by the $—NH_2$ group (the amino group), the —NHR group, or the $—NR_2$ group gives rise to an unsubstituted, a monosubstituted, or a disubstituted amide, respectively.

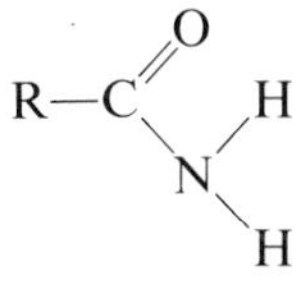

Unsubstituted amide

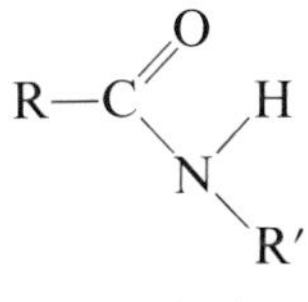

Monosubstituted amide

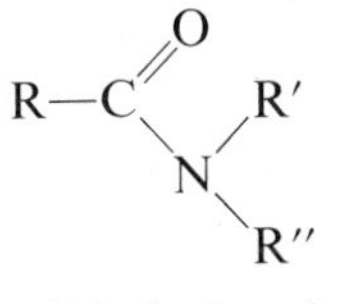

Disubstituted amide

The functional group of the amides is the 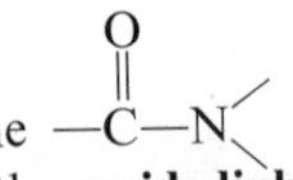 grouping. The *carbonyl carbon–nitrogen bond* is referred to as the **amide linkage.** This bond is very stable and is found in the repeating units of protein molecules (Chapter 25), in nylon (page 188), and in many other industrial polymers.

20.1 Nomenclature of Amides

Amides are named as derivatives of organic acids. The *-ic* ending of the common name or the *-oic* ending of the IUPAC name is replaced with the suffix *-amide.* Alkyl or aryl substituents on the nitrogen atom are denoted by

Table 20.1 Amide Nomenclature

Formula	*Common Name*	*IUPAC Name*
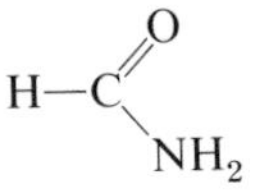	Formamide	Methanamide
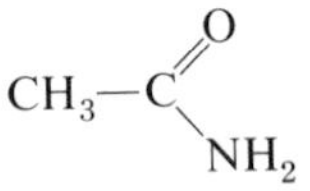	Acetamide	Ethanamide
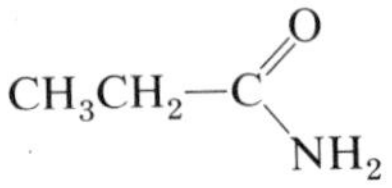	Propionamide	Propanamide
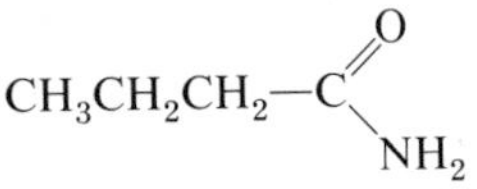	Butyramide	Butanamide
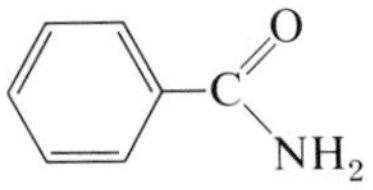	Benzamide	Benzenecarboxamide
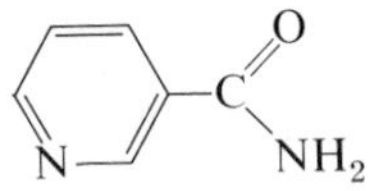	Nicotinamide (Niacin, a B vitamin)	Pyridine-3-carboxamide
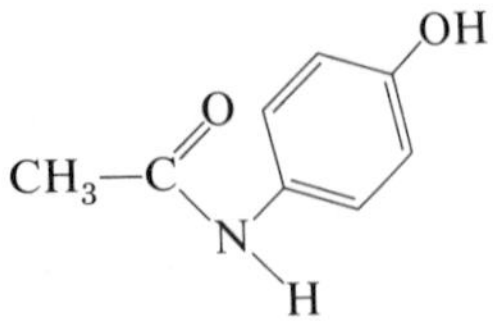	Acetaminophen (pain killer—major ingredient in Tylenol, Tempra, and Liquiprin)	*N-p*-Hydroxyphenylethanamide
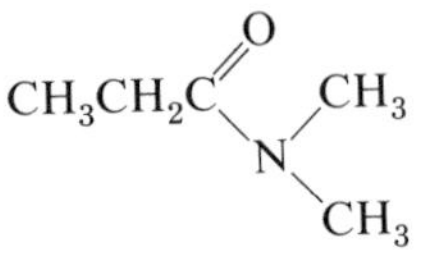	*N,N*-Dimethylpropionamide	*N,N*-Dimethylpropanamide

prefixing the name of the amide by *N*-, followed by the name of the substituent group. Table 20.1 gives some examples of amide nomenclature.

The unsubstituted amides are commonly prepared by the addition of ammonia to the carboxylic acid derivatives (acid chlorides, acid anhydrides, or esters). The route through the acid chloride is usually considered to be the most convenient. Each mole of acid chloride requires two moles of ammonia; the second mole neutralizes the liberated HCl.

20.2 Preparation of Amides

$$CH_3-\overset{\overset{\displaystyle O}{\|}}{C}-Cl \xrightarrow{NH_3} CH_3-\overset{\overset{\displaystyle O}{\|}}{C}-NH_2 \;+\; HCl \xrightarrow{NH_3} NH_4Cl$$

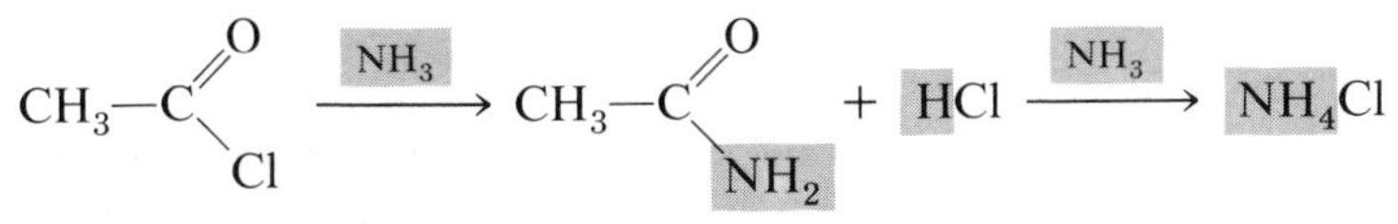

Acetyl chloride   Acetamide

$$\text{(Benzoyl chloride)} + 2\,NH_3 \longrightarrow \text{(Benzamide)} + NH_4Cl$$

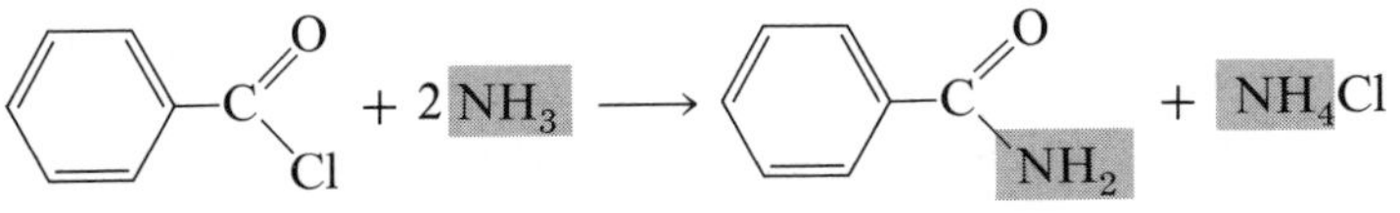

Benzoyl chloride   Benzamide

$$CH_3-\overset{\overset{\displaystyle O}{\|}}{C}-O-\overset{\overset{\displaystyle O}{\|}}{C}-CH_3 + 2\,NH_3 \longrightarrow CH_3-\overset{\overset{\displaystyle O}{\|}}{C}-NH_2 \;+\; \overset{\overset{\displaystyle O}{\|}}{C}-CH_3,\ {}^+NH_4{}^-O$$

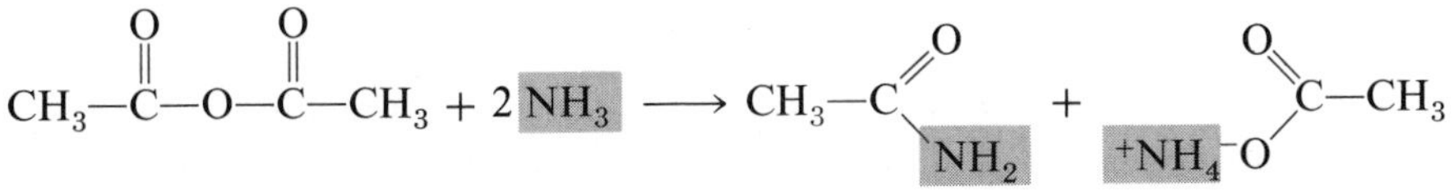

Acetic anhydride   Acetamide   Ammonium acetate

$$H-\overset{\overset{\displaystyle O}{\|}}{C}-OCH_3 \;+\; NH_3 \longrightarrow H-\overset{\overset{\displaystyle O}{\|}}{C}-NH_2 \;+\; HOCH_3$$

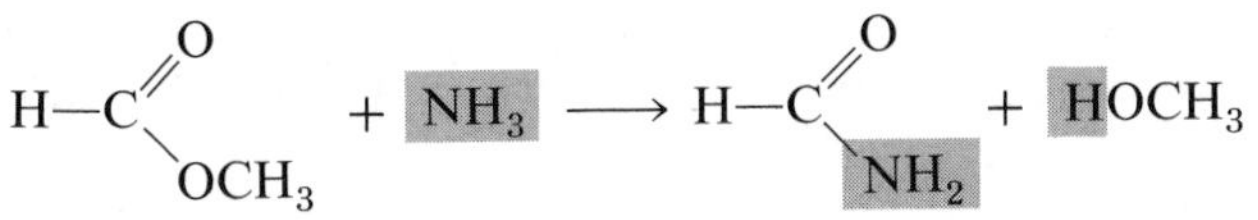

Methyl formate   Formamide   Methanol

Formation of amides by addition of ammonia to the free acid will occur, but the reaction is very slow at room temperature. The ammonium salt of the acid is formed first; then water can be split out if the reaction temperature is maintained above 100°C.

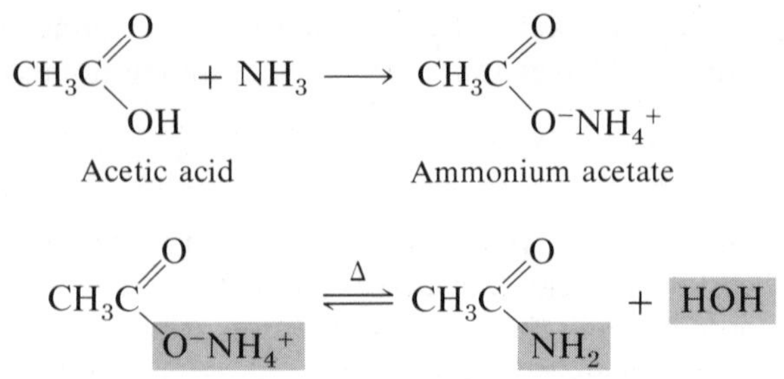

The second step is reversible; the equilibrium favors salt formation. Continuous removal of the water shifts the equilibrium to the right.

20.3 Physical Properties of Amides

With the exception of formamide, which is a liquid, all unsubstituted amides are sharply melting solids (Table 20.2). Most amides are colorless and odorless. The lower members of the series are soluble in both water and alcohol, water solubility decreasing as molar mass increases. The amide group is polar and, unlike the amines, amides are neutral molecules. The unshared electron pair is not localized on the nitrogen atom but is delocalized by resonance onto the oxygen atom of the carbonyl group. The dipolar ion structure restricts free rotation about the carbon–nitrogen bond. This geometric restriction has important consequences for protein structure (Section 25.4).

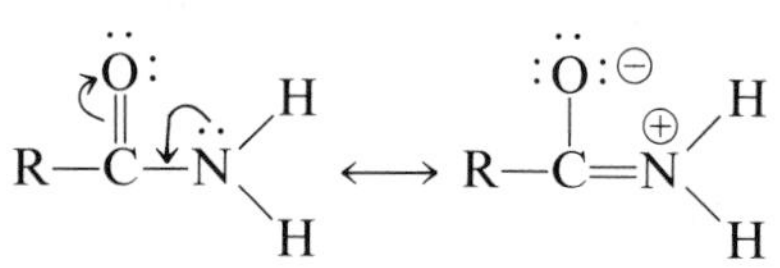

The amides have abnormally high boiling points and melting points. This phenomenon, as well as the water solubility of the amides, is a result of the polar nature of the amide group and the formation of hydrogen bonds (Figure 20.1). Thus electrostatic forces and hydrogen bonding combine to account for the very strong intermolecular attractions found in the amides. Note, however, that disubstituted amides have no hydrogens bonded to nitrogen and thus are incapable of hydrogen bonding. *N,N*-Dimethylacetamide has a melting point of −20 °C, which is about 100° lower than the melting point of acetamide.

Table 20.2 Physical Constants of Some Unsubstituted Amides

Formula	*Name*	*MP (°C)*	*BP (°C)*	
$HCONH_2$	Formamide	2	193	Soluble in water
CH_3CONH_2	Acetamide	82	222	Soluble in water
$CH_3CH_2CONH_2$	Propionamide	79	213	Soluble in water
$CH_3CH_2CH_2CONH_2$	Butyramide	116	216	Soluble in water
$C_6H_5CONH_2$	Benzamide	130	288	Insoluble

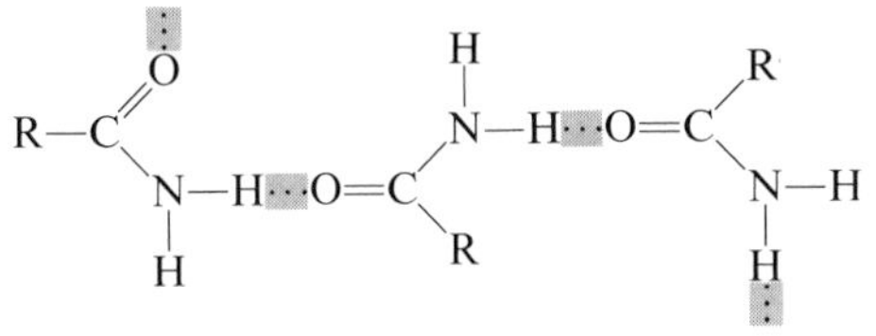

Figure 20.1 Intermolecular hydrogen bonding in amides.

20.4 Chemical Properties of Amides

a. Hydrolysis

The most important reaction of the amides is their hydrolysis. Amide hydrolysis reactions are strictly analogous to the hydrolysis of proteins, which will be discussed in Chapter 25. The hydrolysis may be effected in either acidic or basic solution.

General Equation

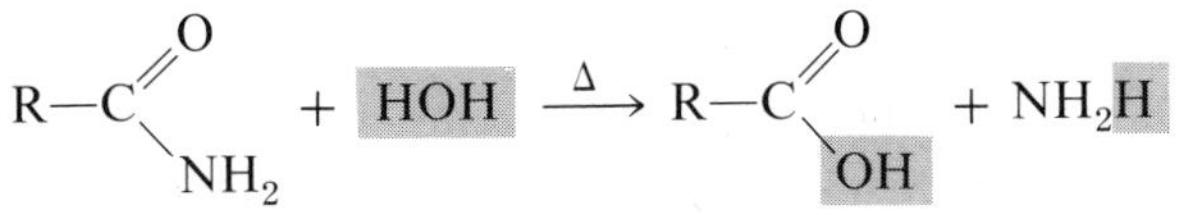

1. Acid Hydrolysis. Hydrolysis of unsubstituted amides in acid solution produces the free organic acid and an ammonium salt. Monosubstituted and disubstituted amides are hydrolyzed to their corresponding acid and amine salt.

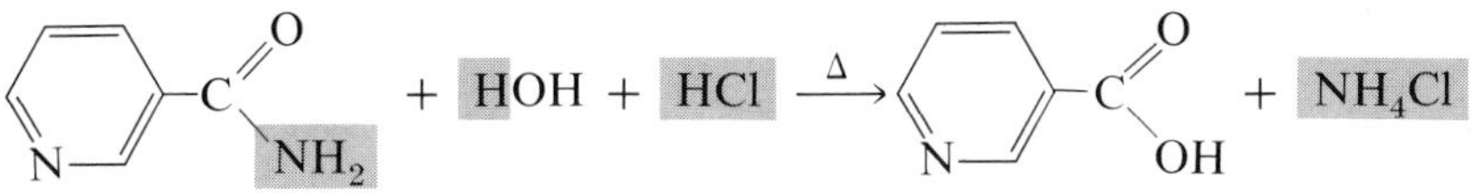

Nicotinamide — Nicotinic acid

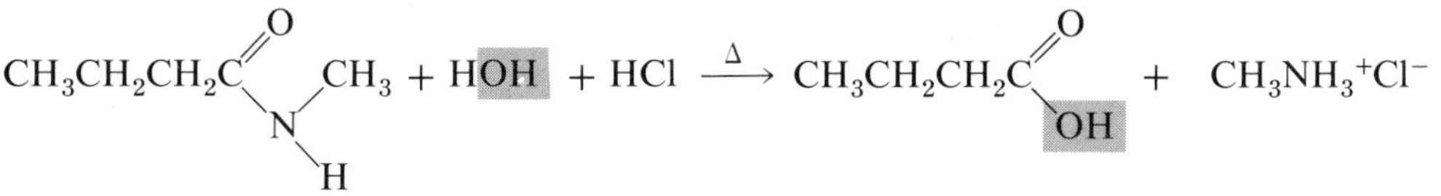

N-Methylbutyramide — Butyric acid — Methylammonium chloride

2. Basic Hydrolysis. Hydrolysis of unsubstituted amides in basic solution produces a salt of the organic acid and ammonia. The strong odor of ammonia signifies its formation as a reaction product.

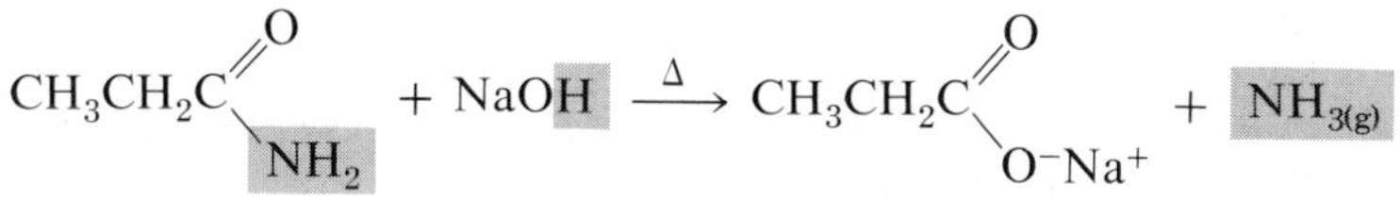

Propionamide — Sodium propionate

b. Dehydration (Formation of Nitriles)

When unsubstituted amides are heated in the presence of a strong dehydrating agent such as phosphorous pentoxide (P_4O_{10}) or thionyl chloride ($SOCl_2$), a molecule of water is lost and the corresponding nitrile is formed.

$$\underset{\text{Acetamide}}{CH_3{-}C(=O)NH_2} \xrightarrow[\Delta]{SOCl_2} \underset{\substack{\text{Acetonitrile}\\\text{(Methyl cyanide)}}}{CH_3{-}C{\equiv}N} + SO_2 + 2\,HCl$$

c. The Hofmann Reaction (Formation of Amines)

This reaction is of particular importance to the synthetic organic chemist. It affords him a method of preparing pure amines with one less carbon atom than the initial amide. An aqueous amide solution is treated with sodium hypochlorite or hypobromite and sodium hydroxide. The mixture is then distilled, and the volatile amine is collected in the distillate. The mechanism of the reaction is fully understood, but will not be presented here. Conceptually, it appears to involve the elimination of the carbonyl group.

General Equation

$$R{-}C(=O)NH_2 + NaOBr + 2\,NaOH \longrightarrow \underset{\text{An amine}}{RNH_2} + Na_2CO_3 + HOH + NaBr$$

Specific Equation

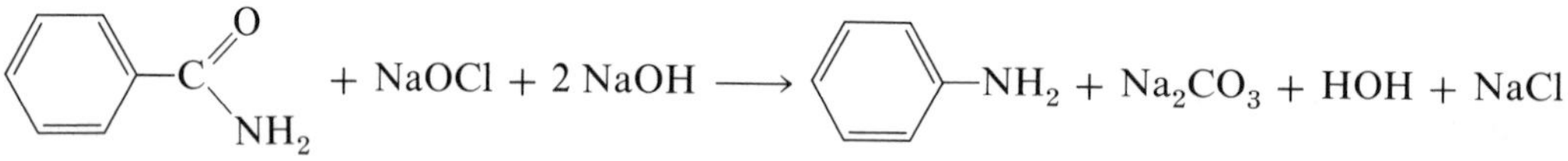

Benzamide — Aniline

d. Reaction with Nitrous Acid

Nitrous acid is generally utilized to convert amino groups into hydroxyl groups. Thus, when unsubstituted amides are treated with nitrous acid, the corresponding organic acid is produced and nitrogen gas is evolved. Nitrous acid is unstable and is generated in the reaction mixture by the combination of a nitrite salt and a strong acid.

General Equation

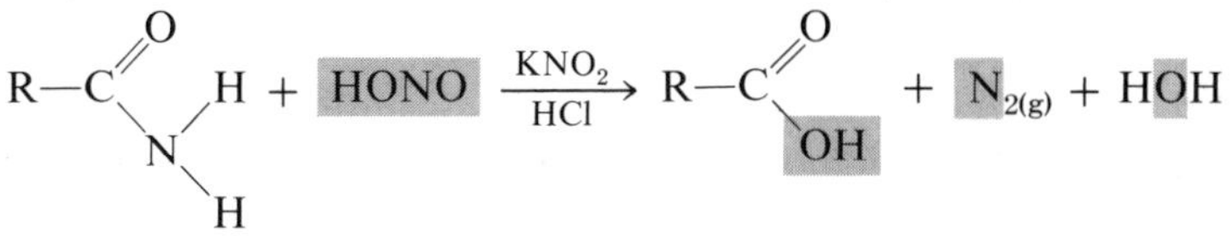
R—C
O
H
N
H
+ HONO
KNO2
HCl
R—C
O
OH
+ N2(g) + HOH

Specific Equation

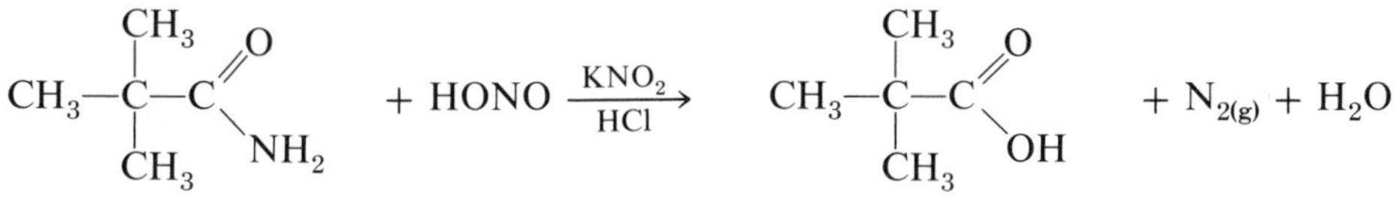
CH3
CH3—C—C
CH3
O
NH2
+ HONO
KNO2
HCl
CH3
CH3—C—C
CH3
O
OH
+ N2(g) + H2O

α,α-Dimethylpropionamide α,α-Dimethylpropionic acid

Amides that are highly substituted on the alpha carbon are difficult to hydrolyze by dilute acid or base. The substituents shield the carbonyl group and impede nucleophilic attack on the carbonyl carbon. Treatment of these shielded amides with nitrous acid is an effective method to accomplish their hydrolysis.

20.5 Urea

Urea is the diamide of carbonic acid ($HO-\overset{\overset{O}{\|}}{C}-OH$). Although the acid itself is unstable, many of its derivatives are known and are of great importance. You are familiar with salts of carbonic acid such as sodium bicarbonate (baking soda) and calcium carbonate (limestone). Recall that Wöhler's synthesis of urea from inorganic raw materials (potassium cyanate and ammonium chloride) was instrumental in the demise of the vital force theory and in the birth of organic chemistry as a separate science.

Urea is a white solid that is the end product of protein metabolism in mammals (Section 30.5). The normal adult excretes about 28–30 grams of urea daily in urine. Industrially, urea is prepared by heating carbon dioxide with ammonia under pressure.

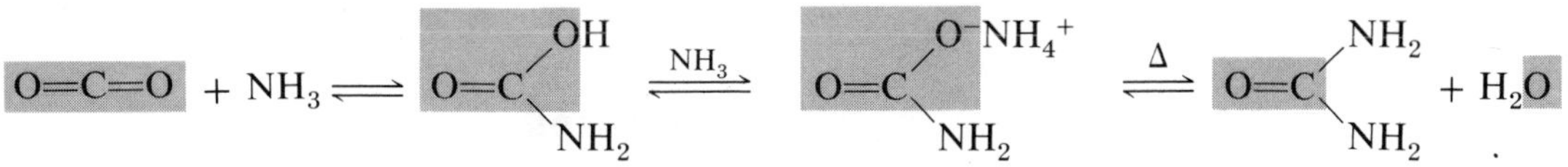
O=C=O
+ NH3
O=C
OH
NH2
NH3
O=C
O⁻NH4⁺
NH2
Δ
O=C
NH2
NH2
+ H2O

Ammonium carbamate Urea

There are several reactions unique to urea but we shall only mention one that is of clinical significance. The action of sodium hypobromite on urea yields nitrogen gas. A measurement of the volume of nitrogen released affords a quantitative determination of urea in various body fluids.

$$NH_2-\overset{\overset{\displaystyle O}{\|}}{C}-NH_2 + 3\,NaOBr + 2\,NaOH \longrightarrow N_{2(g)} + Na_2CO_3 + 3\,H_2O + 3\,NaBr$$

The chief commercial use of urea is as a fertilizer and as a starting material in the production of barbiturates and urea-formaldehyde plastics.

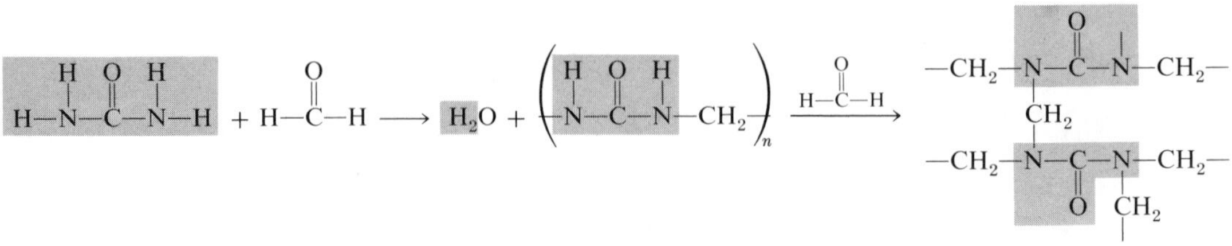

Urea condenses with carboxylic acid derivatives to form compounds called *ureides,* that have acyl groups bonded to one or both of the $-NH_2$ groups.

General Equation

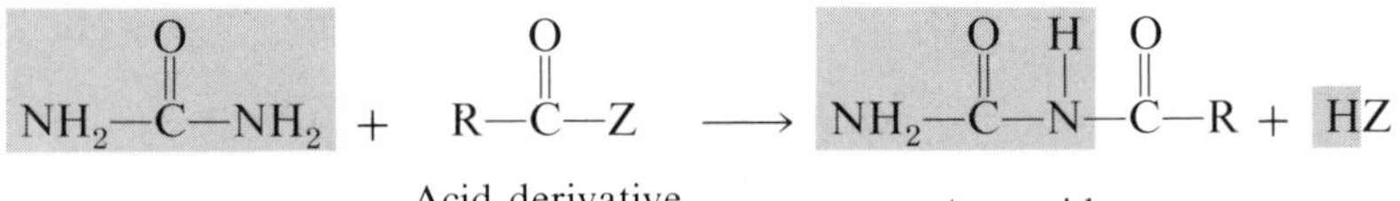

Acid derivative

An ureide

Particularly important ureides are the ones formed from dicarboxylic acids. Barbituric acid is synthesized by the reaction of urea with diethyl malonate in the presence of sodium ethoxide. It is the parent compound of the barbiturates (see Section 20.13-b).

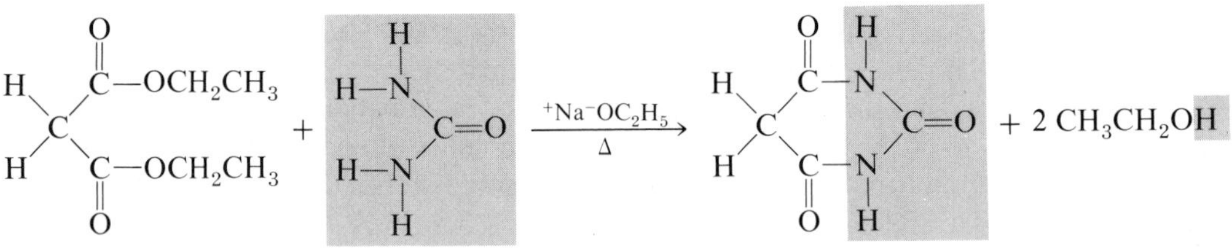

Diethyl malonate

Barbituric acid

Amines

20.6 Classification and Nomenclature of Amines

It will be recalled that the amines are the alkyl and aryl derivatives of ammonia. They are divided into three classes according to the number of hydrocarbon groups bonded to the nitrogen.

The simple aliphatic amines are commonly named by specifying the name(s) of the alkyl group(s) and adding the suffix *-amine.* (Note that the

H	H	R′	R′
H—N—H	R—N—H	R—N—H	R—N—R″
Ammonia	Primary amine	Secondary amine	Tertiary amine

name is written as one word in a manner analogous to the naming of the amides and in contrast to the naming of the alcohols.) The prefixes *sec*- and *tert*-, when part of the name of an amine, have no bearing on the classification of that amine. They refer instead to the nature of the carbon atom to which the nitrogen is attached (for example, *sec*-butylamine and *tert*-butylamine signify that the amino group is bonded to a secondary carbon and a tertiary carbon, respectively—see Table 20.3). When identical alkyl substituents occur in the same amine, the prefixes *di*- and *tri*- are employed. When the groups are dissimilar, they are named in order of increasing complexity. In the IUPAC system, the amino group ($-NH_2$) is usually treated as a substituent. Amine salts are named as derivatives of the ammonium ion. Aniline ($C_6H_5NH_2$) is the most important aromatic amine, and certain compounds are commonly named as derivatives of aniline. A capital *N* is placed before the name of the substituent to indicate that it is bonded to nitrogen. See Table 20.3 for examples of amine nomenclature.

a. The Hofmann Reaction (see Section 20.4-c)

20.7 Preparation of Primary Amines

$$R-C\overset{O}{\underset{NH_2}{\big\langle}} + NaOBr + 2\,NaOH \longrightarrow RNH_2 + NaBr + Na_2CO_3 + HOH$$

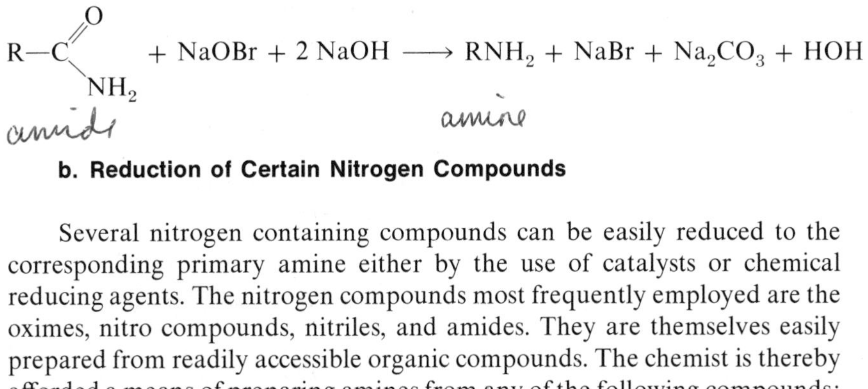

amide   *amine*

b. Reduction of Certain Nitrogen Compounds

Several nitrogen containing compounds can be easily reduced to the corresponding primary amine either by the use of catalysts or chemical reducing agents. The nitrogen compounds most frequently employed are the oximes, nitro compounds, nitriles, and amides. They are themselves easily prepared from readily accessible organic compounds. The chemist is thereby afforded a means of preparing amines from any of the following compounds: alcohols, alkyl halides, aldehydes, ketones, and acids.

1. Reduction of Oximes. Primary amines are formed by the chemical or catalytic reduction of oximes (page 408).

General Equations

$$R-\overset{H}{\underset{}{C}}=NOH + 4\,[H] \xrightarrow[\text{or } H_2/Ni]{\text{Na/ethanol}} R-\overset{H}{\underset{H}{C}}-NH_2 + HOH$$

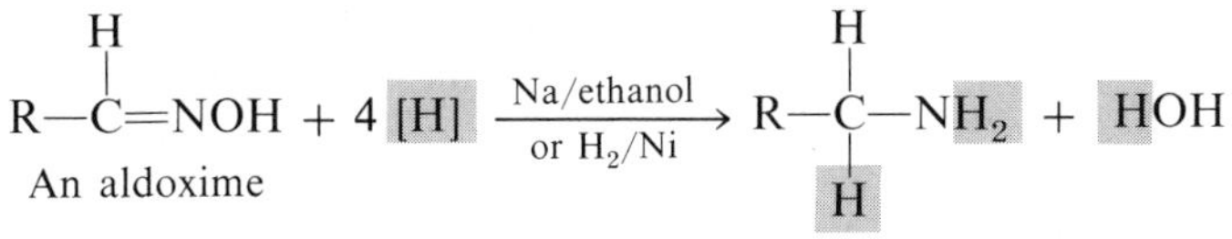

An aldoxime

Table 20.3 Nomenclature of the Amines

Amine Type	*Formula*	*Name*
Primary	CH_3NH_2	Methylamine (Aminomethane)
Primary	$CH_3CH_2NH_2$	Ethylamine (Aminoethane)
Secondary	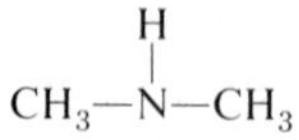	Dimethylamine
Secondary	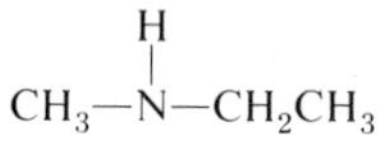	Methylethylamine
Tertiary	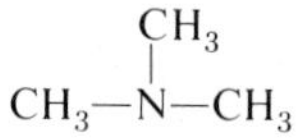	Trimethylamine
Tertiary	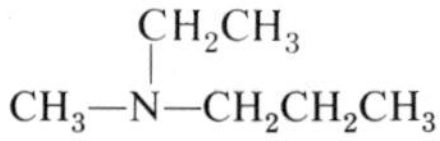	Methylethyl-*n*-propylamine
Primary	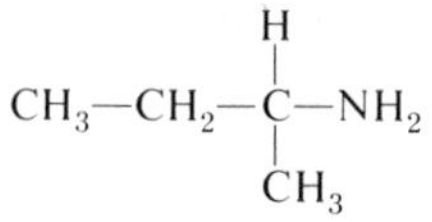	*sec*-Butylamine (2-Aminobutane)
Primary	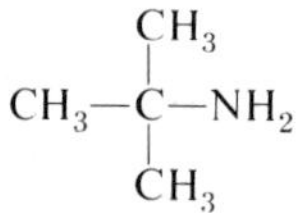	*tert*-Butylamine (2-Amino-2-methylpropane)
Primary	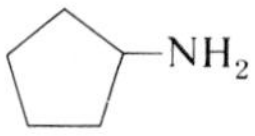	Cyclopentylamine
Primary	$H_2N{-}CH_2CH_2CH_2CH_2CH_2CH_2{-}NH_2$	Hexamethylenediamine (1,6-Diaminohexane)
Salt	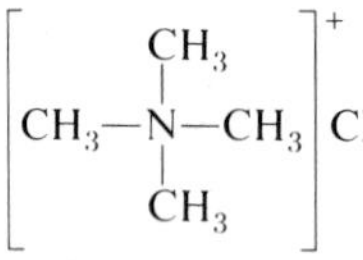	Tetramethylammonium chloride
Primary		*p*-Nitroaniline
Secondary	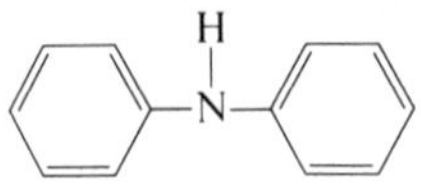	*N*-Phenylaniline (Diphenylamine)
Tertiary	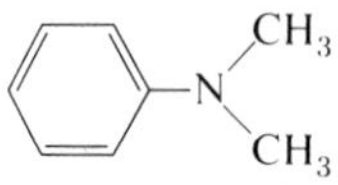	*N*,*N*-Dimethylaniline (Dimethylphenylamine)
Primary	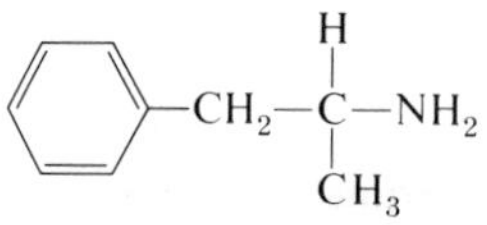	Benzedrine (2-Amino-1-phenylpropane) (Amphetamine, ingredient in "pep pills")

$$R{-}\overset{R'}{\overset{|}{C}}{=}NOH + 4[H] \xrightarrow[\text{or } H_2/Ni]{\text{Na/ethanol}} R{-}\overset{R'}{\overset{|}{\underset{H}{\underset{|}{C}}}}{-}NH_2 + HOH$$

A ketoxime

Specific Equations

$$CH_3CH_2OH \xrightarrow{[O]} CH_3C\overset{\displaystyle O}{\underset{\displaystyle H}{}} \xrightarrow{H_2NOH} CH_3{-}\overset{H}{\overset{|}{C}}{=}NOH \xrightarrow{Na/C_2H_5OH} CH_3{-}CH_2{-}NH_2$$

Ethanol — Acetaldehyde — Acetaldoxime — Ethylamine

$$CH_3CHOHCH_3 \xrightarrow{[O]} CH_3\overset{O}{\overset{\|}{C}}CH_3 \xrightarrow{H_2NOH} CH_3{-}\overset{CH_3}{\overset{|}{C}}{=}NOH \xrightarrow{H_2/Ni} CH_3{-}\overset{CH_3}{\overset{|}{\underset{H}{\underset{|}{C}}}}{-}NH_2$$

2-Propanol — Acetone — Dimethylketoxime — Isopropylamine

2. ***Reduction of Nitro Compounds.*** Reduction of aromatic nitro compounds is a particularly useful method for the formation of primary aromatic amines. Sodium hydroxide is added to convert the amine salt to the amine. Nitroalkanes, because they are less readily available, are seldom used for the preparation of aliphatic amines.

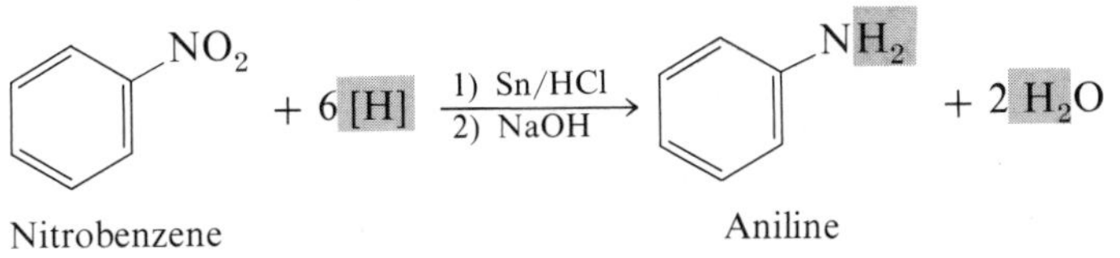
NO2
+ 6 [H]
1) Sn/HCl
2) NaOH
NH2
+ 2 H2O
Nitrobenzene
Aniline

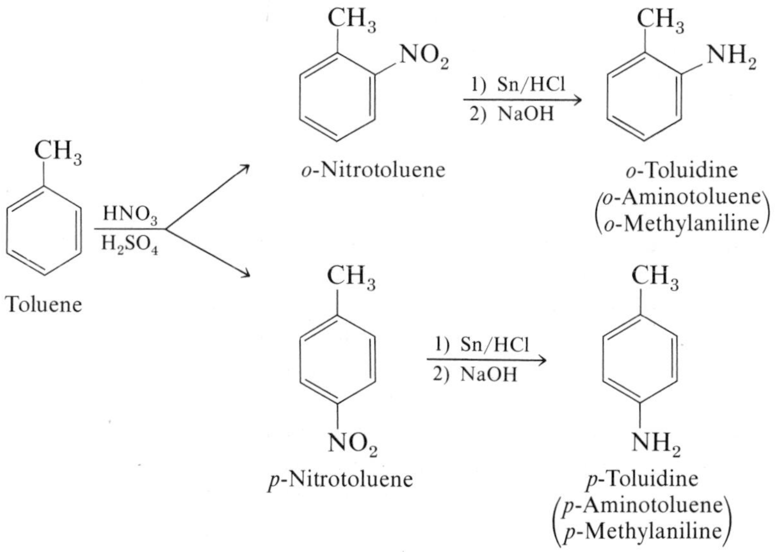
CH3
Toluene
HNO3
H2SO4
NO2
o-Nitrotoluene
1) Sn/HCl
2) NaOH
NH2
o-Toluidine
(o-Aminotoluene
o-Methylaniline)
p-Nitrotoluene
p-Toluidine
(p-Aminotoluene
p-Methylaniline)

*3. **Reduction of Nitriles (Cyanides).*** The resulting amine contains one more carbon than the original alcohol or alkyl halide.

General Equation

$$R{-}C{\equiv}N + 4\,[H] \xrightarrow{H_2/Ni} R{-}\underset{H}{\overset{H}{\underset{|}{\overset{|}{C}}}}{-}\underset{H}{\underset{|}{N}}{-}H$$

Specific Equation

$$\underset{\text{Methanol}}{CH_3OH} \xrightarrow{PBr_3} \underset{\text{Methyl bromide}}{CH_3Br} \xrightarrow{KCN} \underset{\text{Methyl cyanide (Acetonitrile)}}{CH_3CN} \xrightarrow{H_2/Ni} \underset{\text{Ethylamine}}{CH_3CH_2NH_2}$$

*4. **Reduction of Amides.*** Here the amine will have the same number of carbons as the amide (in contrast to the Hofmann reaction).

General Equation

$$R{-}C(=O){-}NH_2 + 4\,[H] \xrightarrow{LiAlH_4} RCH_2NH_2 + H_2O$$

Specific Equation

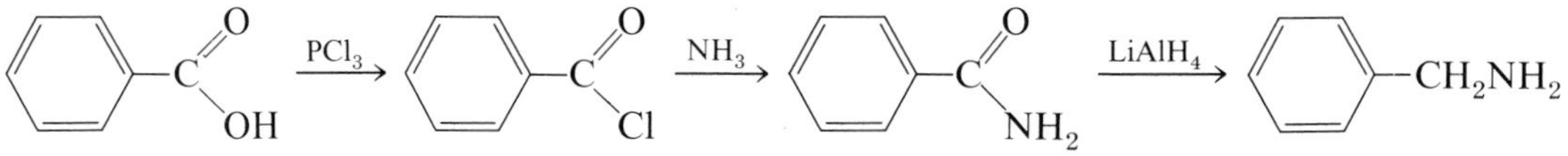

O
C
OH
PCl3
O
C
Cl
NH3
O
C
NH2
LiAlH4
CH2NH2

Benzoic acid — Benzoyl chloride — Benzamide — Benzylamine

20.8 Physical Properties of Amines

The lower members of the amine series resemble ammonia. They are colorless gases, soluble in water, and have pronounced odors that are somewhat similar to ammonia but are less pungent and more fish-like. The characteristic odor of fish is attributed to the presence of amines in the body fluid of the fish (for example, dimethylamine and trimethylamine are constituents of herring brine).

Some of the decomposition products found in decaying flesh are diaminoalkanes. They have foul odors, and their names attest either to their odor or their source (for example, $NH_2CH_2CH_2CH_2CH_2NH_2$ is putrescine and $NH_2CH_2CH_2CH_2CH_2CH_2NH_2$ is cadaverine). They arise from the decarboxylation of ornithine and lysine, respectively, amino acids found in animal cells (see Section 30.4-c).

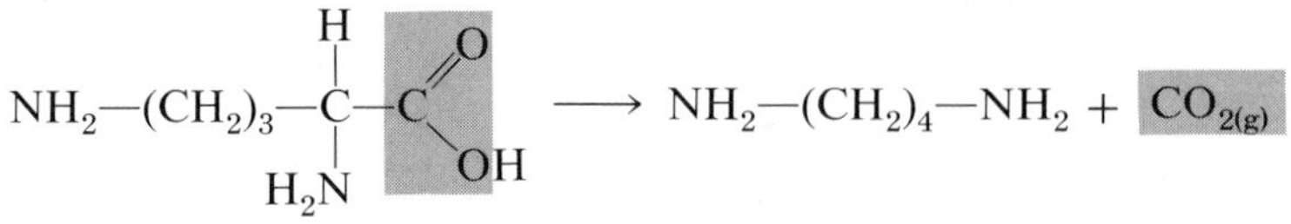

Ornithine Putrescine

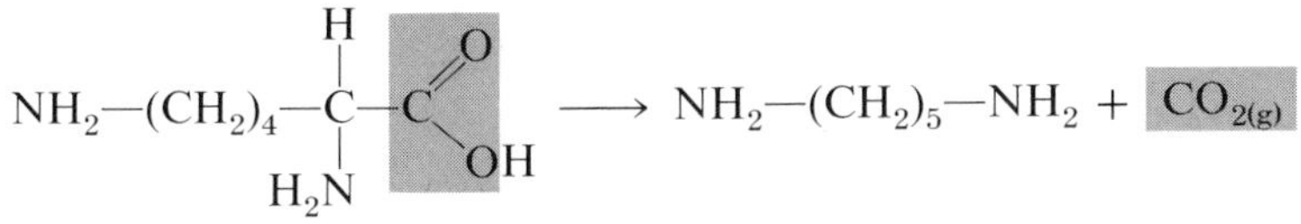

Lysine Cadaverine

Primary amines containing three to eleven carbon atoms are liquids; the higher homologs are solids. Dimethylamine is the only gaseous secondary amine and trimethylamine is the only gaseous tertiary amine. The larger the carbon skeleton of the amine, the lower is its water solubility. Aromatic amines are generally much less soluble in water than are aliphatic amines. There is some slight intermolecular association among amine molecules due to hydrogen bonding (Figure 20.2), but since nitrogen is less electronegative than oxygen, the bonding is less pronounced than in the alcohols. For example, compare the boiling points of CH_3NH_2 (−6°C) and CH_3OH (65°C). Table 20.4 lists the physical characteristics of some of the typical amines.

Table 20.4 Physical Properties of the Amines

Formula	*Name*	*MP (°C)*	*BP (°C)*	K_b
NH_3	Ammonia	−78	−33	1.8×10^{-5}
CH_3NH_2	Methylamine	−94	−6	4.4×10^{-4}
$CH_3CH_2NH_2$	Ethylamine	−82	17	5.2×10^{-4}
$(CH_3)_2NH$	Dimethylamine	−93	7	5.4×10^{-4}
$(CH_3)_3N$	Trimethylamine	−117	3	6.0×10^{-5}
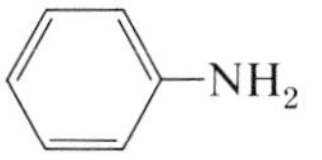	Aniline	−6	184	4.0×10^{-10}
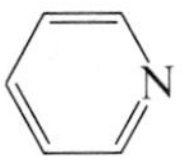	Pyridine	−42	115	2.0×10^{-9}
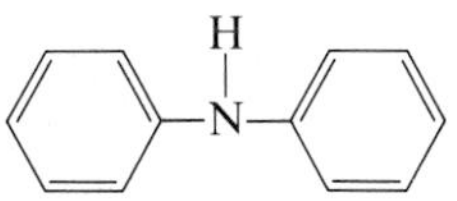	Diphenylamine	53	302	7.1×10^{-14}

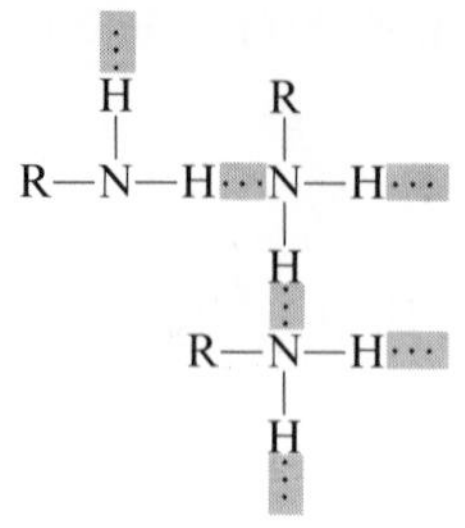

Figure 20.2 Intermolecular hydrogen bonding in amines.

20.9 Basicity of Amines

Whereas carboxylic acids are organic acids, amines are the organic bases. In Section 8.2 we discussed the properties of bases in general. Recall that the value of K_b, the basicity constant, is a measure of the strength of a base (i.e., a measure of its ability to act as a proton acceptor or electron pair donor). As in ammonia, the basicity of amines is a result of the unshared pair of electrons on the nitrogen atom. For amines we can write the following equation and equilibrium constant expression.

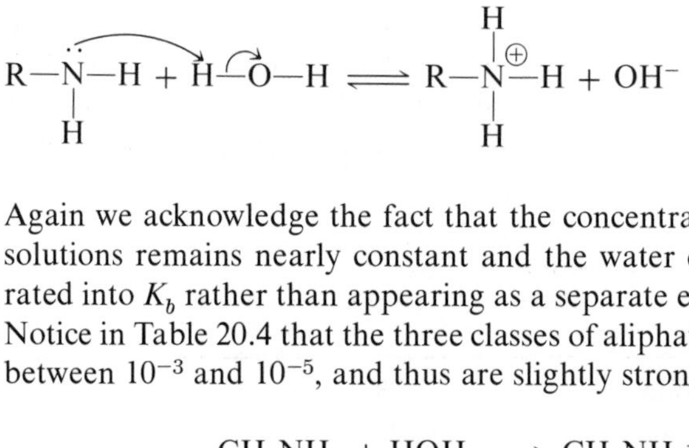

$$K_b = \frac{[RNH_3^+][OH^-]}{[RNH_2]}$$

Again we acknowledge the fact that the concentration of water in aqueous solutions remains nearly constant and the water concentration is incorporated into K_b rather than appearing as a separate entity in the denominator. Notice in Table 20.4 that the three classes of aliphatic amines have K_b values between 10^{-3} and 10^{-5}, and thus are slightly stronger bases than ammonia.

$$CH_3NH_2 + HOH \rightleftharpoons CH_3NH_3^+ + OH^-$$

$$K_b = 4.4 \times 10^{-4} = \frac{[CH_3NH_3^+][OH^-]}{[CH_3NH_2]}$$

The pH of an amine solution is calculated in a manner analogous to the calculation of the pH of an acid solution that has previously been described. The difference is that one first determines the pOH (the negative logarithm of the hydroxide ion concentration) and then subtracts it from 14 to obtain the desired pH.[1]

[1] Since $K_w = [H^+][OH^-] = 1.0 \times 10^{-14}$,

$$\log [H^+] + \log [OH^-] = \log 1.0 \times 10^{-14}$$

$$pH + pOH = 14$$

Example 20.1 What is the pH of a 0.30 *M* solution of trimethylamine ($K_b = 6.0 \times 10^{-5}$)?

$$(CH_3)_3N + HOH \rightleftharpoons (CH_3)_3NH^+ + OH^-$$

$$K_b = 6.0 \times 10^{-5} = \frac{[(CH_3)_3NH^+][OH^-]}{[(CH_3)_3N]}$$

Solving for $[OH^-]$ in a similar manner to that explained in Section 10.3:

$$6.0 \times 10^{-5} = \frac{x^2}{0.30}$$

$$x^2 = 1.8 \times 10^{-5}$$

$$x = [OH^-] = 4.3 \times 10^{-3}\ M$$

$$pOH = 2.4$$

$$pH = 11.6$$

Amines constitute the most important class of organic bases. By proper choice of aliphatic, aromatic, or heterocyclic amine, a broad spectrum of base strengths can be obtained. The degree of basicity of amines is dependent upon the availability of the unshared electron pair. An increase in the electron density about the nitrogen atom makes the electron pair more available; conversely, a decrease in electron density makes these electrons less available. The fact that the aliphatic amines are stronger bases than ammonia is explainable in terms of the inductive effect of the alkyl groups. Recall that alkyl groups are electron-releasing and thus they tend to increase the electron density at the nitrogen atom and also to disperse the positive charge on the cation that is formed.

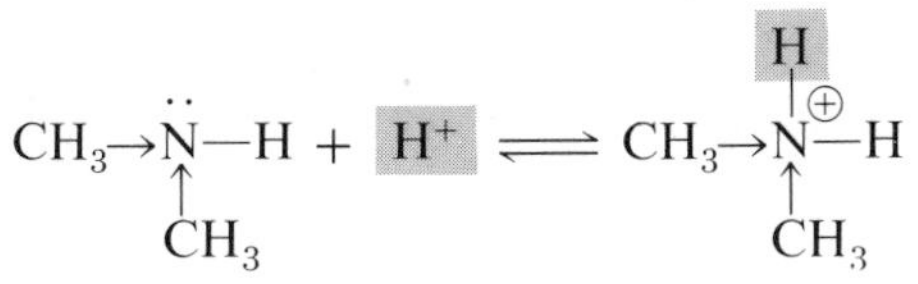

We find that tertiary amines are usually more basic than ammonia, yet less basic than the comparable secondary or primary amines. From this we must conclude that steric considerations outweigh inductive effects. The crowding of bulky groups about the nitrogen shields the electron pair from the attacking proton.

Aromatic amines in which the nitrogen is directly bonded to the aromatic ring are weaker bases than the aliphatic amines. This is attributed to the delocalization of the electron pair by resonance. The net effect of this resonance interaction is to decrease the electron density on the nitrogen.

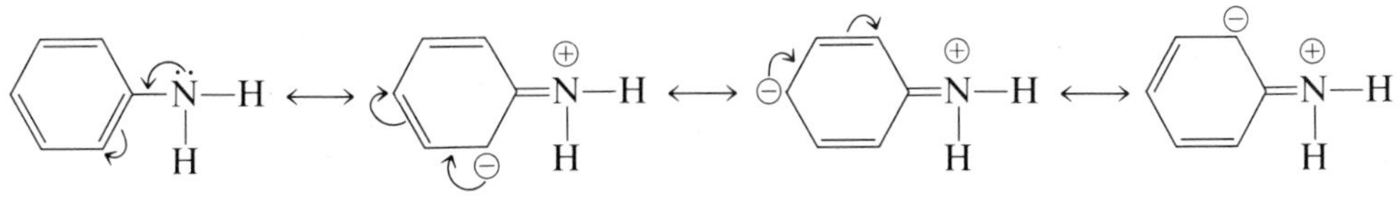

Substituents on the aromatic ring have predictable effects upon the basicity of aromatic amines. Electron-donating substituents increase the basicity whereas electron-withdrawing ones decrease it.

$$CH_3O-\langle\ \rangle-NH_2 > CH_3-\langle\ \rangle-NH_2 > \langle\ \rangle-NH_2 > O_2N-\langle\ \rangle-NH_2$$

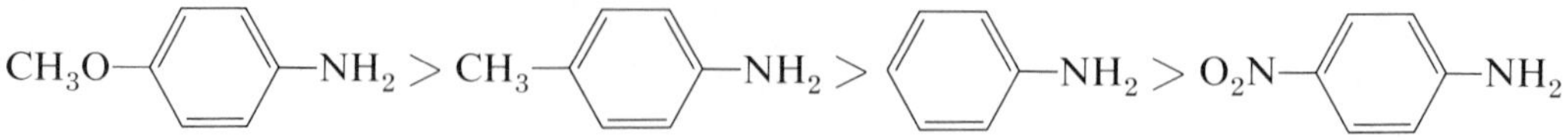

K_b: 2.2×10^{-9} 1.2×10^{-9} 4.0×10^{-10} 1.0×10^{-13}

20.10 Chemical Properties of Amines

a. Salt Formation

The most characteristic property of the amines is their ability to form salts with acids. These salts are similar to ammonium salts. They are formed either in aqueous solution or by the passage of HCl, HBr, HI, etc., through an etheral solution of the amine. In the latter case, the ether-insoluble amine salt separates out as a white solid.

$$CH_3CH_2-\overset{\overset{H}{|}}{\underset{\underset{CH_3}{|}}{N}}: \quad + \quad HCl \quad \longrightarrow \quad CH_3CH_2-\overset{\overset{H}{|}}{\underset{\underset{CH_3}{|}}{\overset{\oplus}{N}}}-H \ \overset{\ominus}{Cl}$$

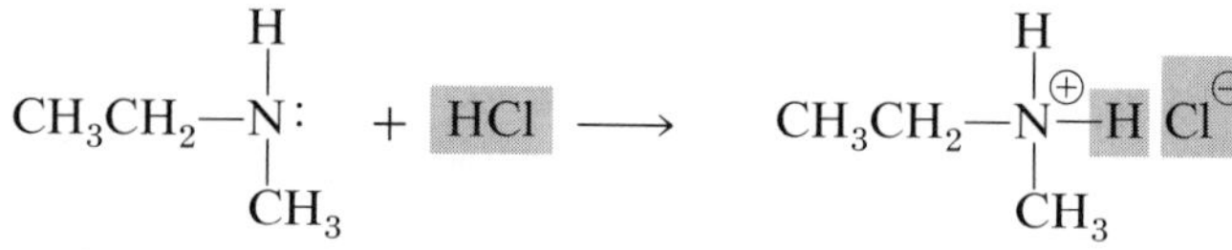

Methylethylamine   Methylethylammonium chloride

Amine salts are ionic and thus have properties considerably different from their parent amines. They are all odorless nonvolatile solids, soluble in water but insoluble in nonpolar solvents. The free amine is obtained from its salt by treatment with sodium hydroxide. The amine is recovered by distillation from the reaction mixture or by extraction into ether. (Recall that organic acids are obtained from their salts by treatment with a strong acid.)

$$CH_3CH_2-\overset{\overset{H}{|}}{\underset{\underset{CH_3}{|}}{\overset{\oplus}{N}}}-H \ Cl^- \xrightarrow{\text{NaOH}} CH_3CH_2-\overset{\overset{H}{|}}{\underset{\underset{CH_3}{|}}{\overset{\oplus}{N}}}-H \ OH^- \rightleftharpoons CH_3CH_2-\overset{\overset{H}{|}}{\underset{\underset{CH_3}{|}}{N}}: \quad + \quad HOH$$

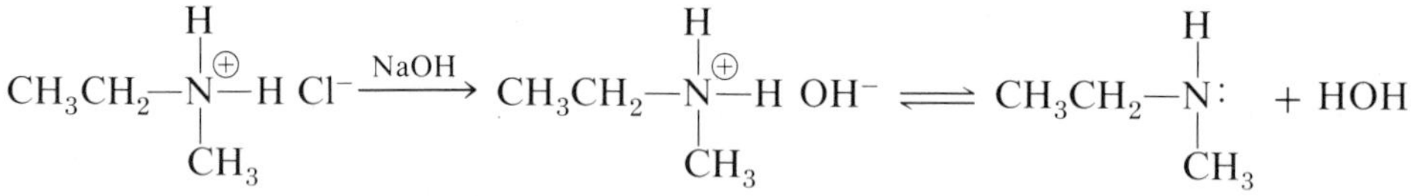

We shall see in Chapter 25 that an important characteristic of both amino acids and proteins is their ability to combine with protons (from acids) to form salts.

$$R-\overset{\overset{H}{|}}{\underset{\underset{H}{|}}{\underset{H-N:}{C}}}-C\overset{O}{\underset{OH}{\diagup}} \quad + \quad HCl \quad \longrightarrow \quad R-\overset{\overset{H}{|}}{\underset{\underset{H}{|}}{\underset{H-\overset{\oplus}{N}-H \ Cl^{\ominus}}{C}}}-C\overset{O}{\underset{OH}{\diagup}}$$

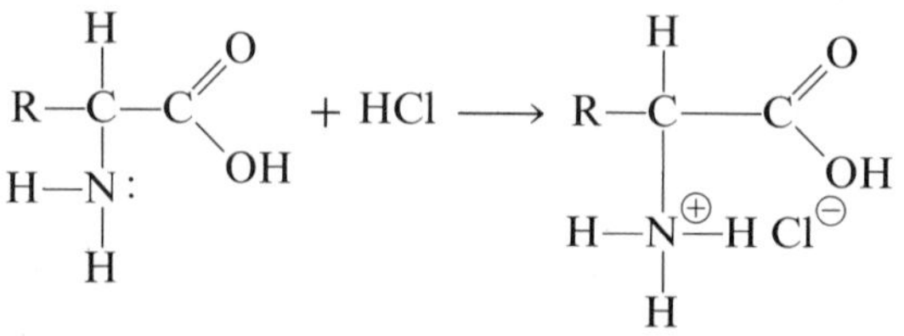

Amino acid   Amino acid salt

b. Alkylation

Like ammonia, amines react readily with alkyl halides to form more highly substituted amines. This reaction is employed to substitute alkyl groups for the hydrogens of primary and secondary amines. Unfortunately, the reaction has severe limitations from a synthetic standpoint. Like the chain reaction for the successive chlorination of the alkanes (Section 13.4), the reaction is difficult to control.[2] The amines that are products of the initial reaction may react further to yield a mixture of primary, secondary, and tertiary amines and substances known as *quaternary ammonium salts*.[3] The mechanism of the reaction is analogous to the displacement of halide ion from alkyl halides by hydroxide in the preparation of alcohols (Section 17.2-b).

$$H-\overset{\overset{\displaystyle H}{|}}{\underset{\underset{\displaystyle H}{|}}{N}}: + CH_3-I \longrightarrow H-\overset{\overset{\displaystyle H}{|}}{\underset{\underset{\displaystyle H}{|}}{\overset{\oplus}{N}}}-CH_3 \; I^{\ominus} \rightleftharpoons H-\overset{\overset{\displaystyle \cdot\cdot}{}}{\underset{\underset{\displaystyle H}{|}}{N}}-CH_3 + HI$$

$$CH_3NH_2 \; + \; CH_3I \longrightarrow CH_3-\overset{\overset{\displaystyle H}{|}}{\underset{\underset{\displaystyle CH_3}{|}}{N}}: + HI$$

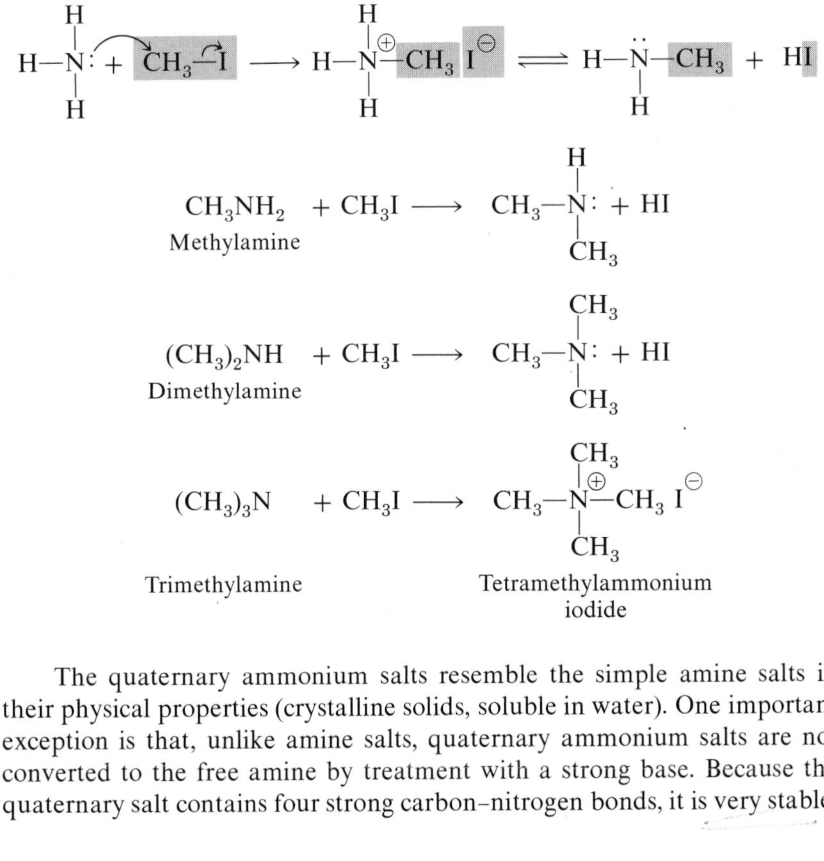

Methylamine

$$(CH_3)_2NH \; + \; CH_3I \longrightarrow CH_3-\overset{\overset{\displaystyle CH_3}{|}}{\underset{\underset{\displaystyle CH_3}{|}}{N}}: + HI$$

Dimethylamine

$$(CH_3)_3N \; + \; CH_3I \longrightarrow CH_3-\overset{\overset{\displaystyle CH_3}{|}}{\underset{\underset{\displaystyle CH_3}{|}}{\overset{\oplus}{N}}}-CH_3 \; I^{\ominus}$$

Trimethylamine   Tetramethylammonium iodide

The quaternary ammonium salts resemble the simple amine salts in their physical properties (crystalline solids, soluble in water). One important exception is that, unlike amine salts, quaternary ammonium salts are not converted to the free amine by treatment with a strong base. Because the quaternary salt contains four strong carbon–nitrogen bonds, it is very stable.

---

[2] In the commercial synthesis of amines, separation of the mixture is accomplished by fractional distillation.

[3] Some quaternary ammonium salts that contain one large alkyl group are useful detergents and bactericides (for example, dimethylethyloctadecylammonium bromide). Choline, $[(CH_3)_3NCH_2CH_2OH]^+OH^-$, and acetylcholine, $[CH_3COOCH_2CH_2N(CH_3)_3]^+OH^-$, are important biochemical compounds to be discussed in later sections.

Carbon–nitrogen cleavage does not occur; an equilibrium mixture is established between the quaternary ammonium salt and its hydroxide.

$$\underset{\text{Tetramethylammonium iodide}}{(CH_3)_4N^+I^-} + NaOH \rightleftharpoons \underset{\text{Tetramethylammonium hydroxide}}{(CH_3)_4N^+OH^-} + NaI$$

Unlike ammonium hydroxide, the tetraalkylammonium hydroxide is completely dissociated in aqueous solution and thus is of comparable base strength to sodium or potassium hydroxide. It may be prepared from its quaternary salt by treatment with silver oxide. This prevents the establishment of an equilibrium mixture because the silver halide precipitates out of solution.

$$2\,(CH_3)_4N^+I^- + Ag_2O + HOH \longrightarrow 2\,(CH_3)_4N^+OH^- + 2\,AgI_{(s)}$$

c. Acylation

The reaction between a primary or secondary amine and an acid chloride or acid anhydride yields an amide in which the acyl group has been substituted for a hydrogen bonded to the amine nitrogen. Tertiary amines do not undergo this reaction.

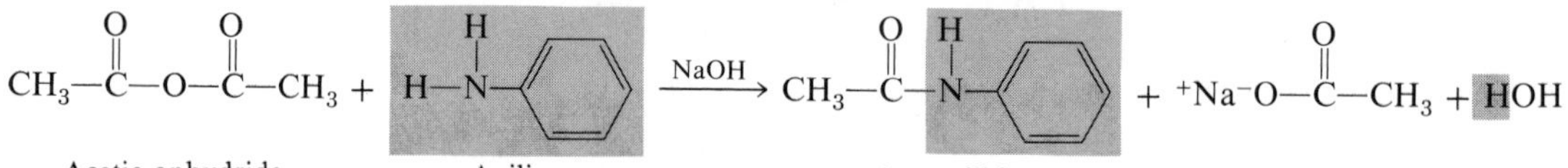
NaOH

Acetic anhydride   Aniline   Acetanilide

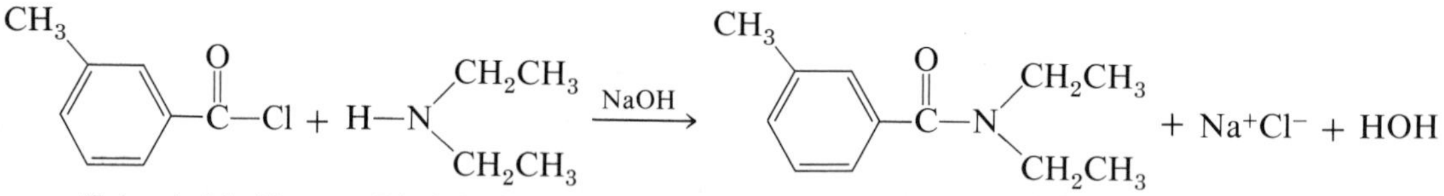
NaOH

m-Toluoyl chloride   Diethylamine   *N,N*-Diethyl-*m*-toluamide

The product *N,N*-diethyl-*m*-toluamide, also known as *m*-delphene, is the active ingredient in the popular insect repellent Off!. It is interesting that the two other isomers (*o*- and *p*-delphene) are ineffective as insect repellents. This remarkable specificity-of-structure requirement was mentioned previously in connection with pesticides (Section 16.6).

d. Reaction with Nitrous Acid (Nitrosation)

Nitrous acid is a unique reagent with respect to amino compounds in that it does not behave as a typical inorganic acid. That is, the majority of amines are not converted to their corresponding amine salts by nitrous acid. Instead, each class of amines reacts differently, and nitrous acid may be used to distinguish among them. Recall also that nitrous acid reacts readily with unsubstituted amides forming a carboxylic acid and liberating nitrogen gas (Section 20.4-d).

1. Reaction with Primary Amines. All primary amines react with nitrous acid forming an alcohol and liberating nitrogen gas. The evolution of nitrogen gas is quantitative, and this reaction forms the basis of the widely used van Slyke determination of free amino groups in proteins. For an understanding of the reaction, we will oversimplify and write the following equation showing the formation of an unstable, hypothetical intermediate which undergoes rearrangement to yield the observed products. Once again the nitrous acid is prepared in situ.

General Equation

$$R-N\begin{matrix}H\\\\H\end{matrix} + HO-N=O \xrightarrow[\text{HCl}]{\text{NaNO}_2} ROH + \left[\begin{matrix}N=O\\|\\H-N-H\end{matrix}\right] \longrightarrow N_{2(g)} + HOH$$

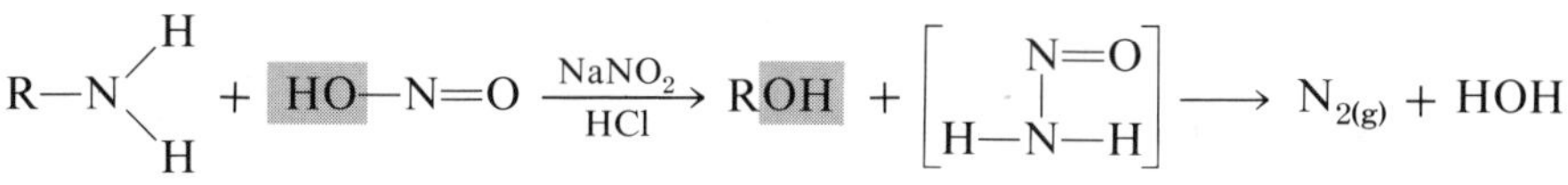

Specific Equation

$$CH_3CH_2NH_2 + HONO \longrightarrow CH_3CH_2OH + N_{2(g)} + H_2O$$
Ethylamine   Ethanol

If the primary amine contains more than two carbon atoms, the product formed is not exclusively the straight-chain alcohol. Rearrangement occurs to yield the branched-chain isomers as the principal products. For example, nitrosation of *n*-propylamine yields a mixture of *n*-propyl and isopropyl alcohols.

2. Reaction with Secondary Amines. Secondary amines react more slowly than the primary amines with nitrous acid, yielding compounds known as *nitrosoamines* (the —N=O group is the nitroso group).

General Equation

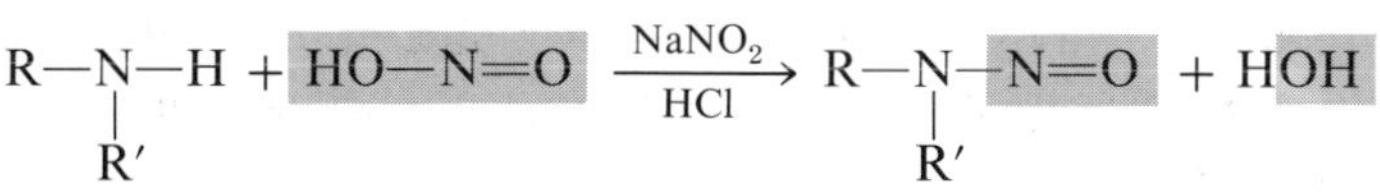

$$R-N-H + HO-N=O \xrightarrow[\text{HCl}]{\text{NaNO}_2} R-N-N=O + HOH$$
$$\begin{matrix}|\\R'\end{matrix} \qquad\qquad\qquad\qquad \begin{matrix}|\\R'\end{matrix}$$
A nitrosoamine

Specific Equation

$$CH_3CH_2-N-H + HONO \longrightarrow CH_3CH_2-N-N=O + H_2O$$
$$\begin{matrix}|\\CH_2CH_3\end{matrix} \qquad\qquad\qquad\qquad\qquad \begin{matrix}|\\CH_2CH_3\end{matrix}$$
Diethylamine   Nitrosodiethylamine

The nitrosoamines are not soluble in the aqueous medium, and they generally separate out as yellow oils. These nitroso compounds are known to be potent carcinogens, and there has been recent concern about the use of nitrites and nitrates as food preservatives. Bacteria in the stomach reduce nitrates to nitrites, and, in the presence of HCl in the stomach, the nitrites are converted into nitrous acid. The nitrous acid then reacts with certain secondary amines in the body to form nitrosoamines. A program has been initiated by the Department of Agriculture to greatly reduce the amount of sodium nitrate that is added to bacon and to substitute other preservatives.

*3. **Reaction with Tertiary Amines.*** Only the tertiary amines form amine salts when treated with nitrous acid. A reaction does occur in almost all cases but since the salt is soluble in aqueous solution, there is no visible sign of reaction.

General Equation

$$\underset{\displaystyle \mathrm{R'}}{\mathrm{R{-}\underset{|}{N}{-}R''}} + \mathrm{HONO} \xrightarrow[\mathrm{HCl}]{\mathrm{NaNO_2}} \left[\mathrm{R{-}\overset{\overset{\displaystyle H}{|}}{\underset{\underset{\displaystyle R'}{|}}{N}}{-}R''}\right]^{\oplus} \mathrm{ONO}^{\ominus}$$

An amine nitrite salt

Specific Equation

$$\underset{\text{Trimethyl-amine}}{(CH_3)_3N} + HONO \longrightarrow \underset{\text{Trimethylammonium nitrite}}{(CH_3)_3NH^{\oplus}NO_2^{\ominus}}$$

20.11 Diazonium Salts

The preceding reactions with nitrous acid are of little value except as diagnostic tests to distinguish among the three classes of amines. However, the reaction between nitrous acid and an aromatic primary amine is one of the most useful reactions in organic chemistry. Addition of sodium nitrite and hydrochloric acid to a primary aromatic amine results in the formation of a relatively stable cation, called a **diazonium ion.** The reaction is referred to as **diazotization.**

General Equation

$$\mathrm{Ar{-}NH_2} \xrightarrow[0\text{–}5\,^{\circ}\mathrm{C}]{\mathrm{NaNO_2/HCl}} \underset{\text{Diazonium ion}}{\mathrm{Ar{-}\overset{\oplus}{N}{=}N}}$$

Specific Equation

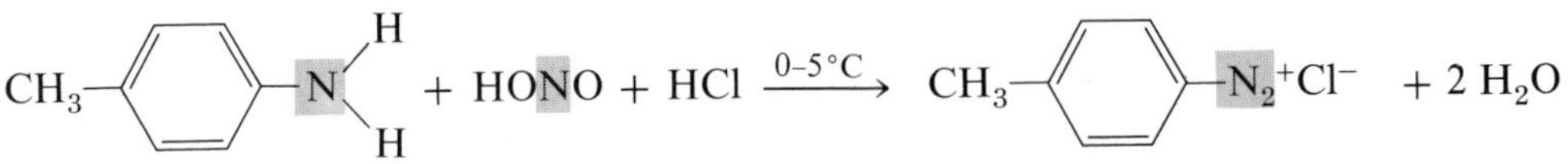

p-Toluidine

p-Toluenediazonium chloride (soluble salt)

Arenediazonium ions are resonance stabilized by delocalization of the electrons from the benzene ring to the diazonium group. This resonance stabilization accounts for the fact that arenediazonium salts are stable, whereas it is not possible to isolate the alkyldiazonium salts even at very low temperatures.

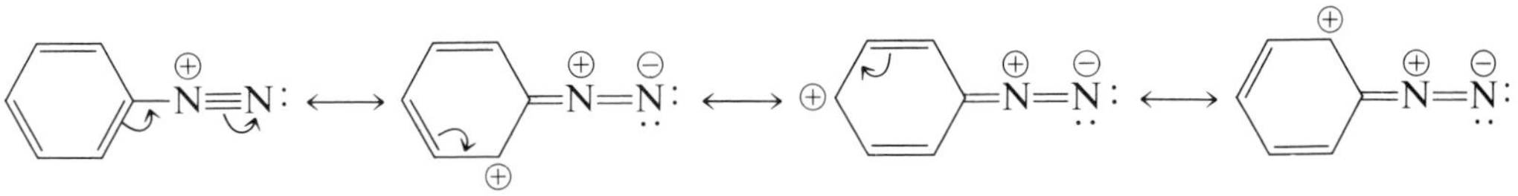

The diazonium salts are prepared as needed. Usually, no attempt is made to isolate them as solid materials because almost all of the dry salts are violently explosive. If the reaction mixture is heated, the diazonium salt decomposes yielding the corresponding phenol and nitrogen gas. This constitutes one of the best methods for the synthesis of substituted phenols.

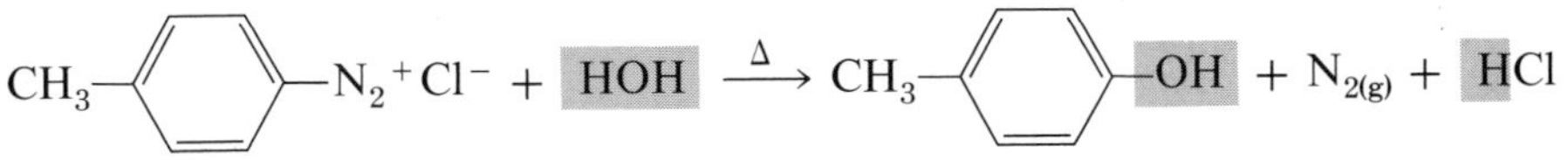

p-Toluenediazonium chloride

p-Cresol

Under suitable conditions, the diazonium group can be replaced by a wide variety of substituents to produce aromatic compounds that are difficult or impossible to prepare by other methods. The groups that will displace the nitrogen are nucleophilic agents. Some of the common nucleophilic displacement reactions that the diazonium salts undergo are illustrated in Figure 20.3.

The general mechanism for these reactions can be summarized as follows.

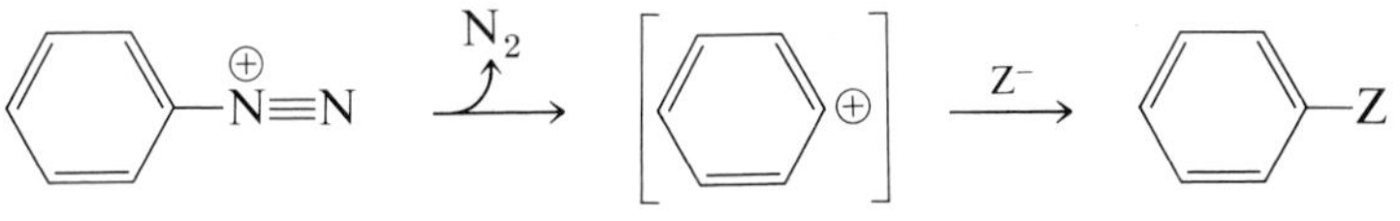

Unstable phenyl cation

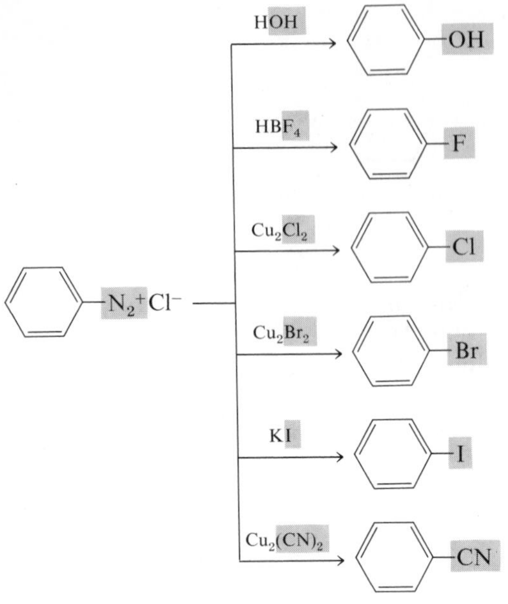

Figure 20.3 Summary of the displacement reactions of diazonium salts.

The diazonium salts can also undergo a *coupling* reaction with activated aromatic compounds such as phenols and aniline derivatives. Electron-donating substituents are required to facilitate attack by the positive diazonium ion on the benzene ring. Coupling always takes place in the position *para* to the activating group unless this position is already occupied, in which case the *ortho* compound results.

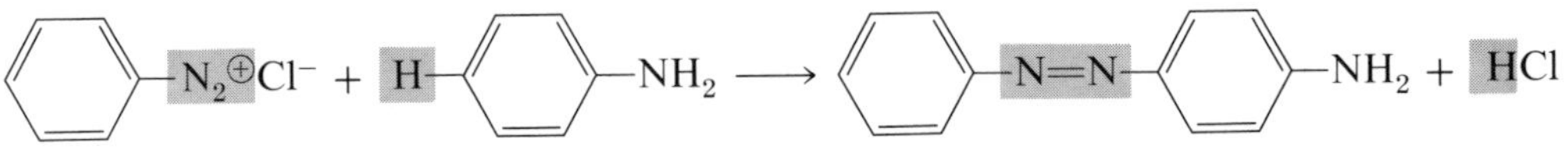

p-Aminoazobenzene

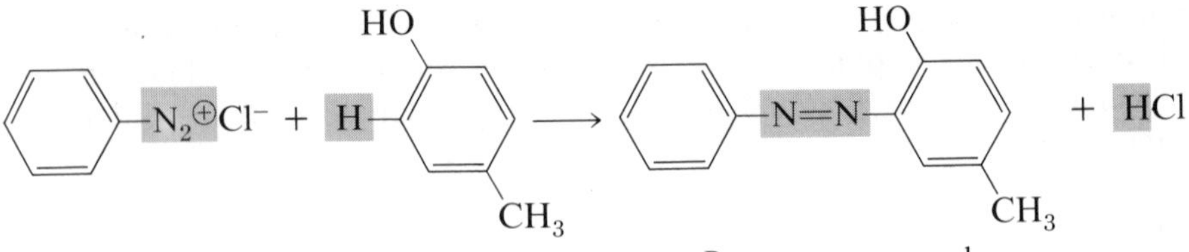

Benzeneazo-*p*-cresol

The products of the coupling reaction are substituted *azobenzenes;* they all contain the —N=N— or *azo group*. By varying the groups on the two rings, compounds having a variety of colors can be produced. A large number of

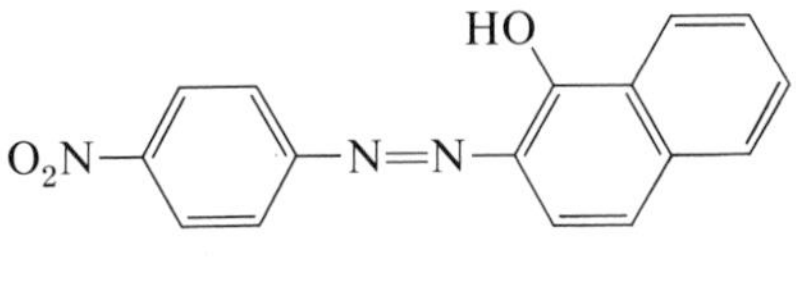

Para red

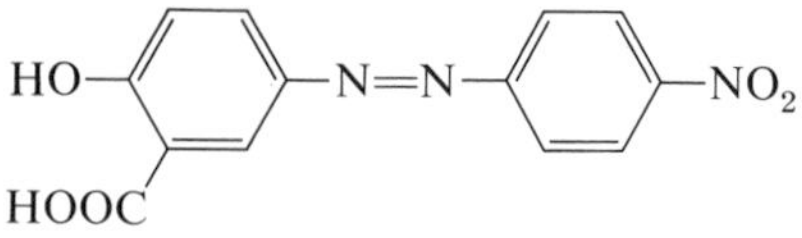

Alizarin yellow R

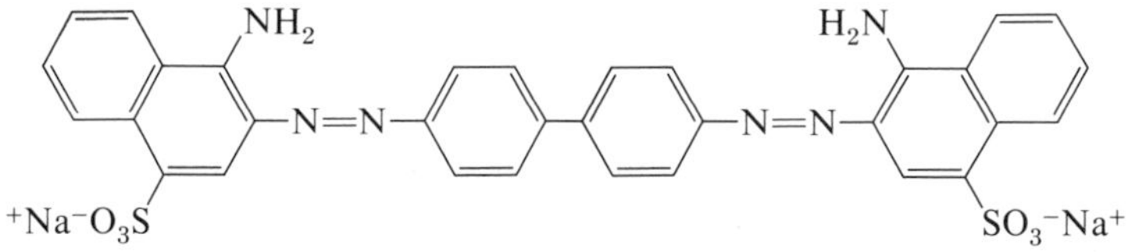

Congo red

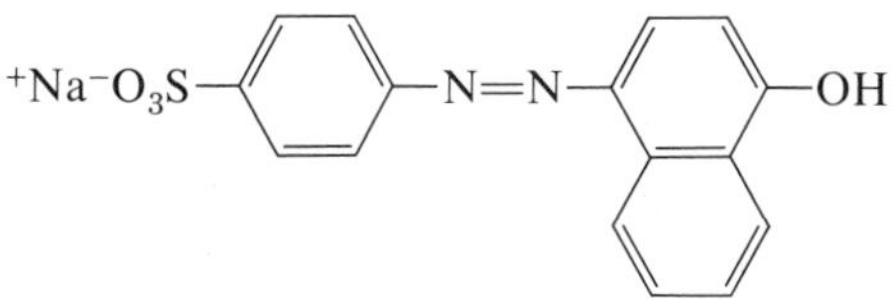

Orange I

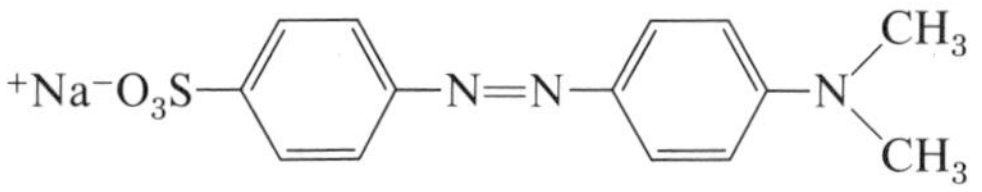

Methyl orange

Figure 20.4 Structures of some commercial azo dyes. Note the presence of polar groups to enhance water solubility and to aid in fixing the dye to the fabric.

dyes, known as *azo dyes,* are prepared by this reaction.[4] The ability of these compounds to absorb visible light is related to the presence of the conjugated system of double bonds in the dye molecules (Figure 20.4). The wavelength of light absorbed depends on the nature and position of the substituents and whether or not the substituents are ionized. The acid-base indicator, methyl red, can exist in two forms depending upon the pH of the solution. At pH values below 4.4 the red form predominates, while at pH values above 6.2 the yellow form predominates. At intermediate pH ranges (4.4–6.2), both forms are present and the color of the solution is orange.

[4] The azo dyes are a large and important class of synthetic dyes. A dye is defined as any compound that absorbs light in the visible region (i.e., is colored) and possesses the ability of attaching itself firmly to the fabric. Azobenzene is a highly colored (red) compound, yet it is not a dye because it lacks the requisite groups necessary to bond to fabrics.

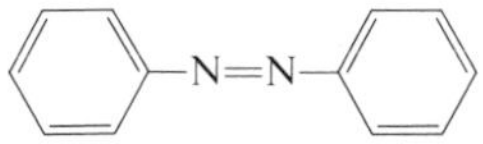

Azobenzene

Polar substituents, such as —SO_3Na, —OH, —COOH, and —NH_2, must be present to bind the dye molecule to the surface of the polar fibers (such as cotton or wool), which are carbohydrate or protein in nature. The major type of bonding that prevails in fixing the dye to the fabric is hydrogen bonding.

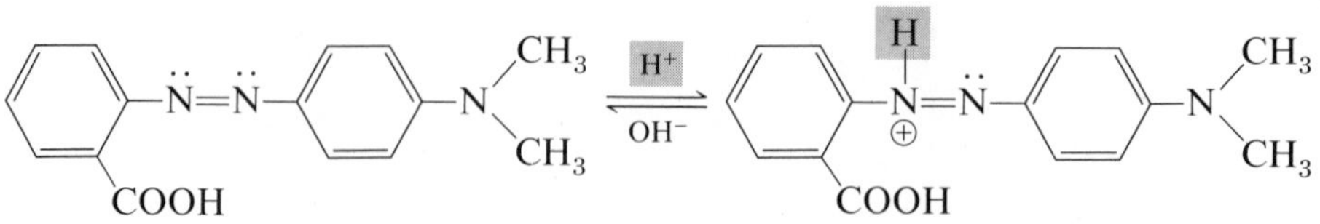

Methyl red (yellow form) — Methyl red (red form)

20.12 Nitrogen Heterocycles

Nitrogen-containing ring compounds comprise a large segment of organic chemistry and they are widely distributed in all living systems. The parent compounds of this class possess relatively simple structures. However, their derivatives, many of which exhibit physiological activity, are often quite complex. The common names and structures of the biologically significant heterocycles are presented in Table 20.5.

Many nitrogen heterocycles have aromatic properties similar to benzene. Pyridine and pyrrole are the chief examples of such compounds and they shall serve as representative systems about which we can make some useful generalizations. In Section 20.13 we shall discuss the important role that the heterocyclic compounds play in the chemistry of drugs.

a. Pyridine

Pyridine is the direct nitrogen analog of benzene. Like benzene, pyridine is an aromatic molecule. It is one of the most abundant heterocyclic compounds and is obtained commercially from coal tar. It has an obnoxious odor and is soluble in water as well as in most organic solvents. The unshared electron pair on nitrogen is not involved in the ring electron system. Therefore, these electrons are available to combine with hydrogen ions; thus, pyridine is a weak base.

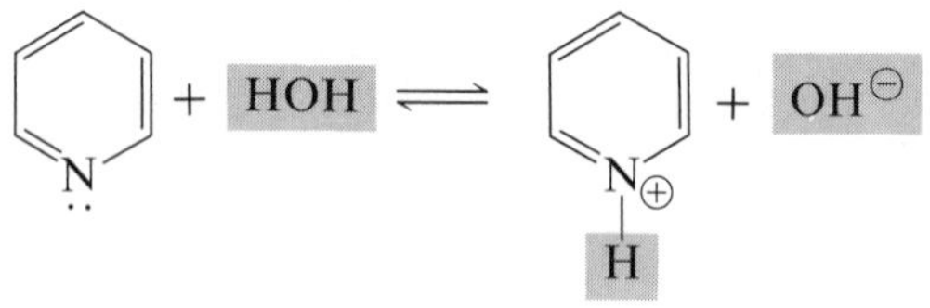

Pyridinium ion

$K_b = 2.0 \times 10^{-9}$

A greater number of contributing structures can be written to represent the resonance hybrid of pyridine than for benzene.

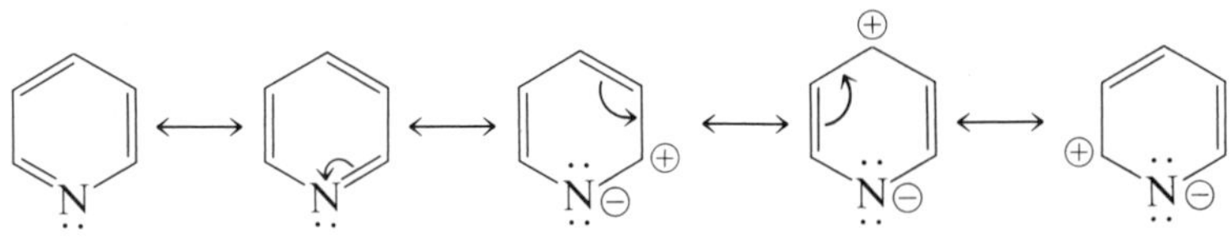

The nitrogen withdraws electron density from the ring, rendering carbons 2, 4, and 6 slightly electropositive. Consequently, the typical aromatic substitution reactions with electrophilic reagents occur much less readily with pyridine than with benzene. Reactions with electrophilic reagents require vigorous conditions and substitution takes place at the 3 positions. On the other hand, attack by nucleophilic reagents (such as sodamide) results in substitution at the 2 or 4 position.

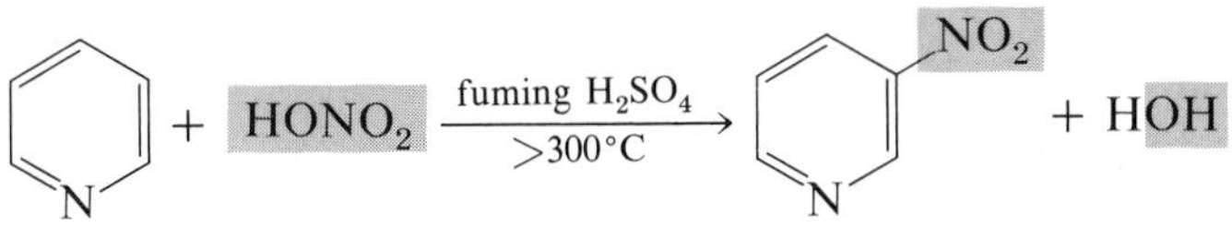

+ HONO$_2$
fuming H$_2$SO$_4$
>300°C
NO$_2$
+ HOH

3-Nitropyridine

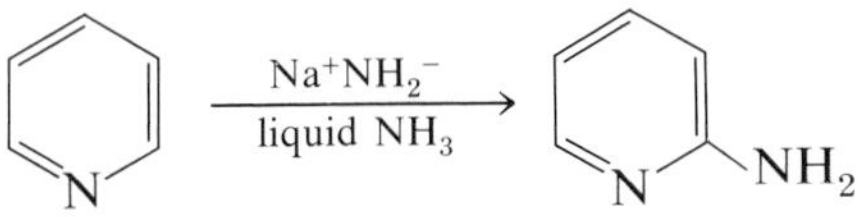

Na$^+$NH$_2^-$
liquid NH$_3$
NH$_2$

2-Aminopyridine

Partial reduction of pyridine yields a dihydropyridine; reduction occurring at the 4 position. This reaction can be reversed and it is the basis of the very important oxidation-reduction reactions that take place in living systems, catalyzed by the coenzyme NAD$^+$ (see Section 28.8). Complete reduction of pyridine results in the formation of piperidine.

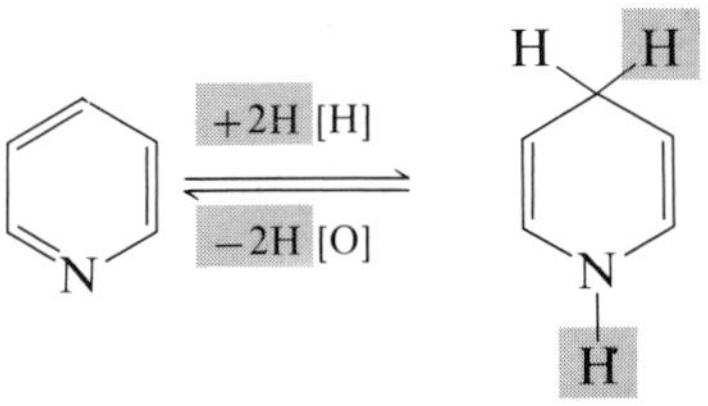

+2H [H]
−2H [O]

Dihydropyridine

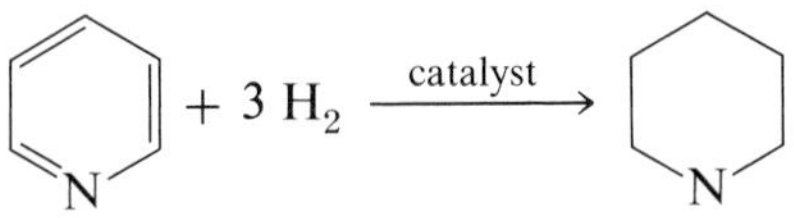

+ 3 H$_2$
catalyst

Piperidine

Table 20.5 Nitrogen Heterocycles

Formula	*Name*	*Occurrence*	*Example*
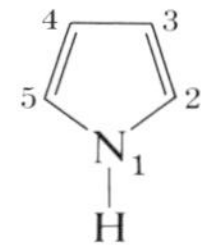	Pyrrole	Chlorophyll Hemoglobin Vitamin B_{12}	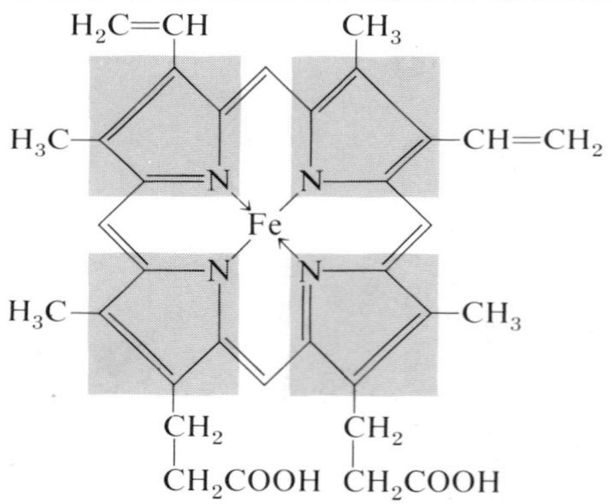 Heme portion of the hemoglobin molecule
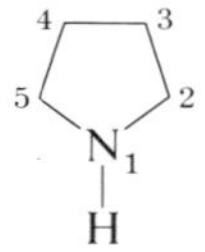	Pyrrolidine	Proline } amino acids Hydroxyproline } amino acids Nicotine	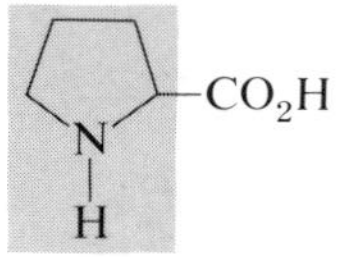 Proline
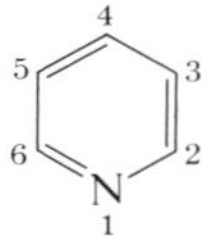	Pyridine	Nicotine Niacin } vitamins Pyridoxine } vitamins NAD^+ } coenzymes Pyridoxal phosphate } coenzymes	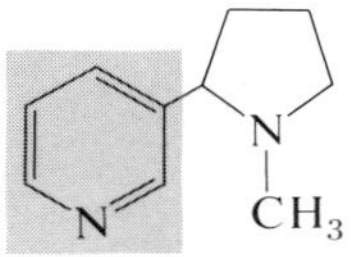 Nicotine
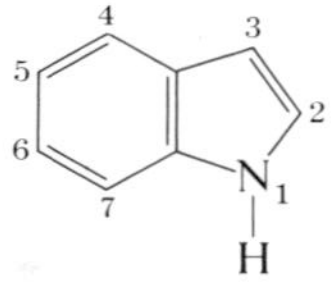	Indole	Tryptophan amino acid Serotonin } indole alkaloids Lysergic acid } indole alkaloids Strychnine } indole alkaloids Reserpine } indole alkaloids	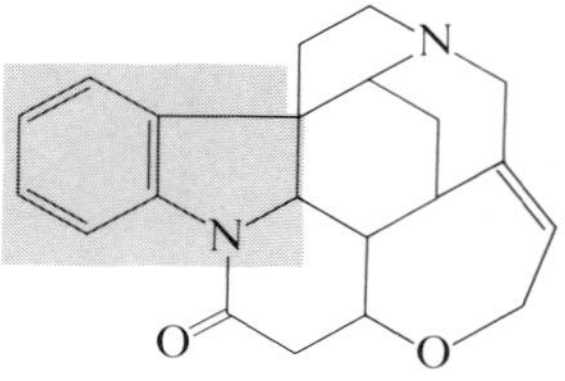 Strychnine

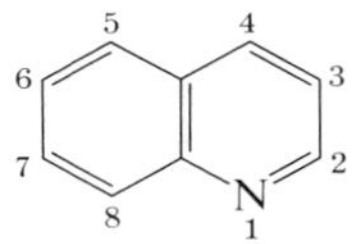	Quinoline	Quinine Curare	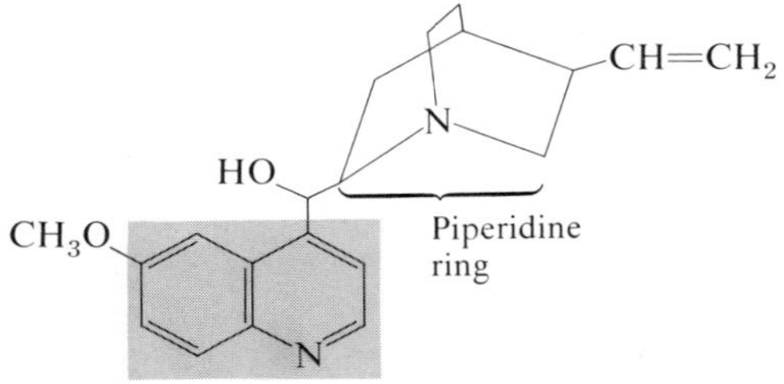 Quinine
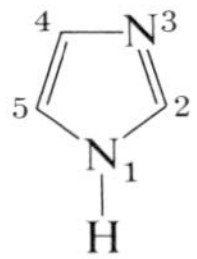	Imidazole	Histidine amino acid	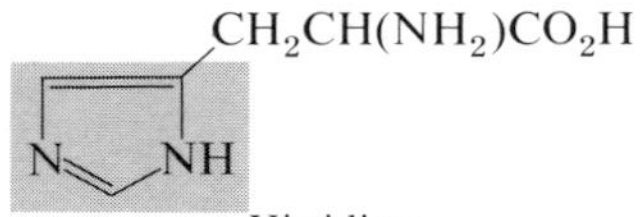 Histidine
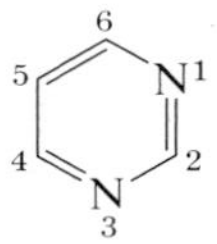	Pyrimidine	Vitamin B_1 Cytosine, Uracil, Thymine } in nucleic acids	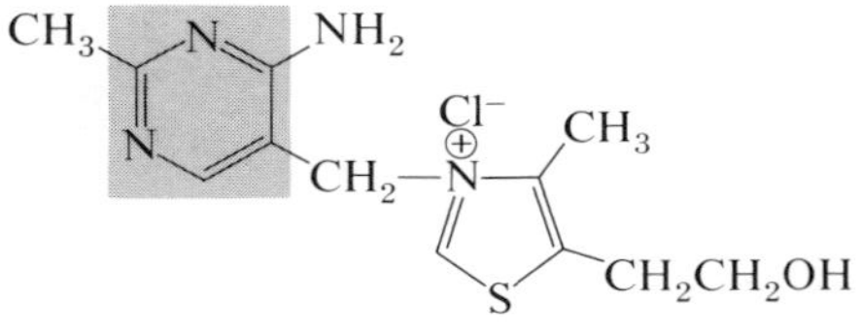 Vitamin B_1 (Thiamine)
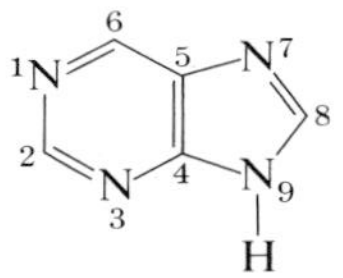	Purine	Adenine, Guanine } in nucleic acids NAD^+, FAD, Coenzyme A } coenzymes ATP, ADP, AMP	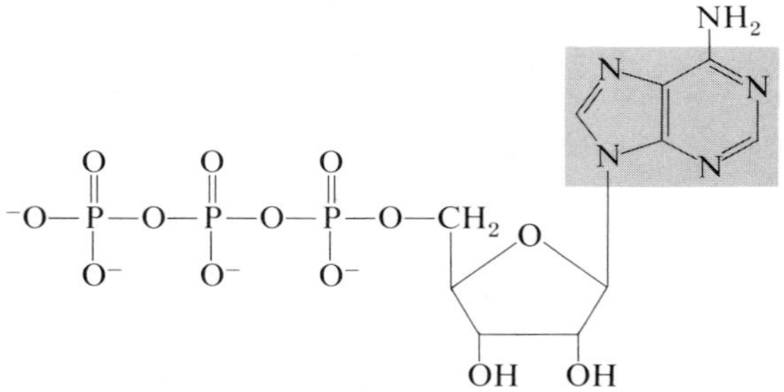 Adenosine triphosphate (ATP)

b. Pyrrole

Pyrrole is the nitrogen analog of furan (Section 17.12). Like furan, and unlike pyridine, the unshared pair of electrons is part of the aromatic system. Consequently, these electrons are not available to bond to a hydrogen ion and pyrrole is an example of an amine that does not behave as a base.

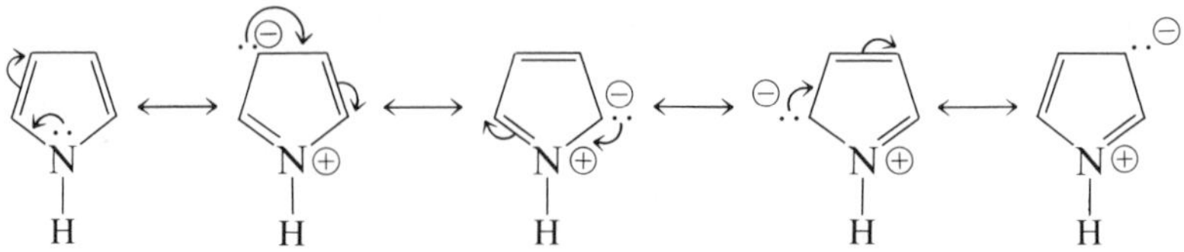

Pyrrole does undergo all of the normal aromatic substitution reactions (i.e., nitration, sulfonation, and halogenation). It can be alkylated with alkyl halides in the presence of a base to yield, primarily, the 2-alkyl derivatives, and it can be readily acylated by heating with acetic anhydride.

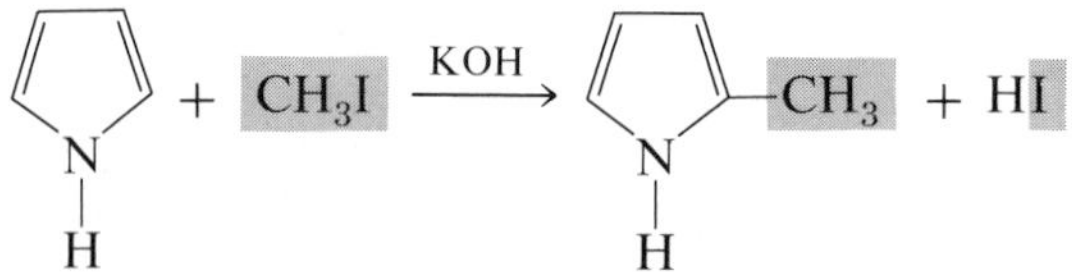
+ CH3I
KOH
CH3 + HI
N
H
N
H

2-Methylpyrrole

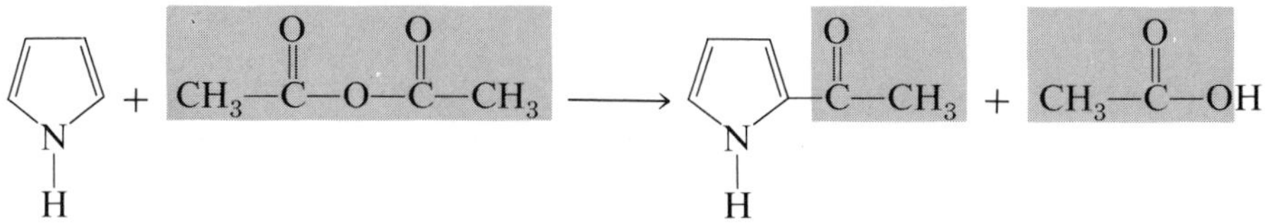
+ CH3—C—O—C—CH3
O
O
C—CH3 + CH3—C—OH
O
O
N
H
N
H

2-Acetylpyrrole

20.13 Drugs

According to the broadest definition, a **drug** is *any chemical substance that can bring about changes in the function of living tissue.* In the past few years the use of drugs, on a world-wide basis, has reached alarming proportions. Of particular concern are the so-called pop or mind drugs—the drugs that affect the brain and spinal cord. These organs make up the central nervous system (CNS), the control center not only of the body, but of everything we call the mind, that is, emotion, sensation, thought. Almost all drugs exhibit more than one type of response or activity depending on the dosage and the individual. The World Health Organization has classified seven different types of "mind" drugs according to the patterns of their action and the major responses they bring about. These are morphine, barbiturate–alcohol, cocaine, cannabis (marijuana), amphetamine, khat, and hallucinogen (LSD). Table 20.6 indicates some of the basic criteria by which part of this classification was made.

a. Morphine Narcotics

A narcotic (Greek, *narkotikos,* benumbing), when administered in medicinal doses, *relieves pain* and *produces sleep;* in poisonous doses it produces stupor, coma, convulsions, and ultimately death. The morphine narcotics are often referred to as opiates, since they are obtained from the opium poppy (*Papaver somniferum*). The dried juice from the unripened seed pod of this plant is commercial opium. Opium is a complex mixture containing more than twenty different compounds. Morphine, which constitutes about ten percent of opium, was first isolated in 1804, followed thereafter by codeine (methylmorphine) and other important medical agents, such as papaverine and noscapine which are antispasmodics. Heroin (diacetylmorphine), a semisynthetic derivative of morphine, was first prepared in 1874 by heating morphine and acetic acid (acetylation). Since then, a variety of synthetic drugs with morphine-like physiological activities (e.g., demerol, methadone) have been produced and have found widespread application (Figure 20.5).

The depressant action of the opiates is exerted entirely on the central nervous system. The exact mechanism by which these drugs act is unknown, although it is assumed that they block nerve transmission in some fashion. Among the principal effects are analgesia, sedation, hypnosis (drowsiness and lethargy), and euphoria. This elevation in mood and relief of apprehension is considered the major reason for drug abuse. As time passes, the euphoria wears off and the user becomes apathetic and gradually falls asleep. The opiates also cause pupillary constriction, and they

Table 20.6 Some Characteristics of Certain Types of Drugs

Type of Drug[a]	*Basic Action*	*Psychological (Psychic) Dependence*	*Physical Dependence*	*Withdrawal Symptoms (Abstinence Syndrome)*	*Development of Tolerance*
Morphine and morphine-like drugs	Depressant	Yes, strong	Yes, develops early	Severe, but rarely life-threatening	Yes
Barbiturate–alcohol	Depressant	Yes	Yes	Severe, even life-threatening	Yes, but only partial for alcohol
Cocaine	Stimulant	Yes, strong	No	None	No
Cannabis (marijuana)	Stimulant	Yes, moderate to strong	No	None	No
Amphetamine	Stimulant	Yes, variable	No	None	Yes, slowly
Hallucinogen (LSD)	Stimulant	Yes	No	None	Yes, rapid

From Benjamin A. Kogan, *Health: Man in a Changing Environment,* Harcourt Brace Jovanovich, Inc., New York, 1970. Reprinted with permission.

[a] Khat is not included here because the chewing of its leaves (producing an amphetamine-like drug effect) is not a problem in this country.

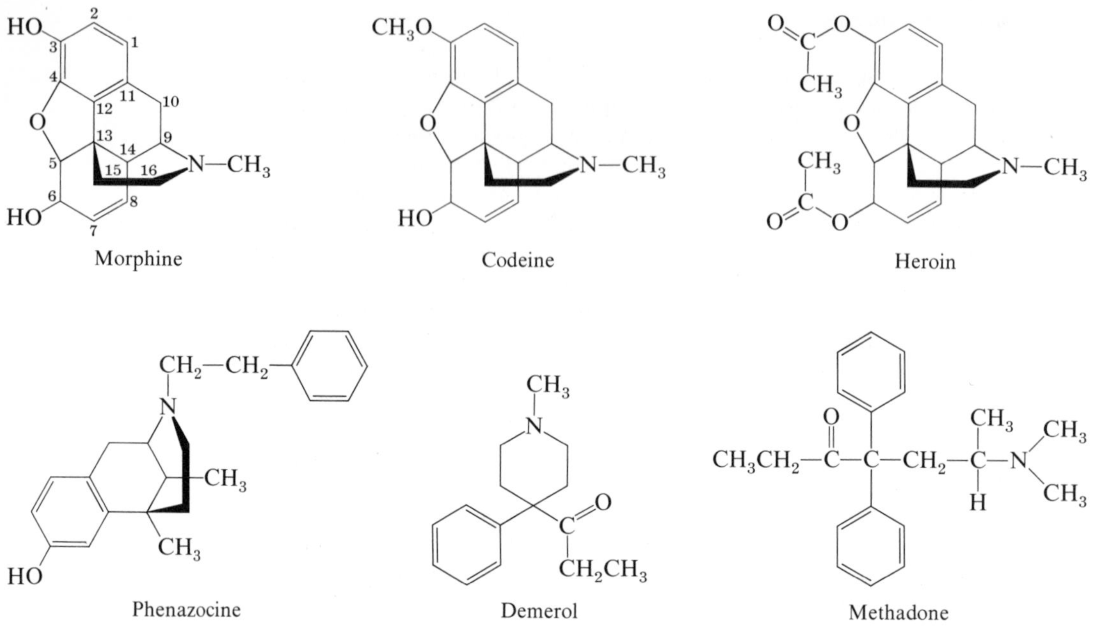
Morphine
Codeine
Heroin
Phenazocine
Demerol
Methadone

Figure 20.5 The chemical structures of some analgesic drugs.

depress the cough center (codeine being the most effective; thus its use in cough medicines). In the brain stem, these drugs will depress the respiratory center. Death from an overdose results from respiratory failure.

The chief medical function of the opiates is as analgesics for the relief of postoperative pain, cardiac pain, the pains of terminal cancer, and in childbirth. Morphine also causes constriction of the smooth muscles such as those found in the intestines and can, therefore, be used in the treatment of dysentery. Paregoric, formerly given to children suffering from diarrhea, is a mixture of compounds containing some morphine. The problem, however, is that addiction liability seems to parallel analgesic activity. Many of the synthetic compounds are more potent pain killers than morphine, but they are either dangerously addicting or hallucinogenic. There is a great deal of current research devoted to the synthesis of the ideal analgesic—one that is effective, yet nonaddicting and free of side effects such as respiratory depression and hallucinogenic activity.

Heroin is currently the most commonly abused narcotic; about half a million people in the U.S. use it daily. It is a bitter-tasting, white powder that is soluble in water and in alcohol. It is five to ten times more potent than morphine and because of the greater risk of addiction, heroin has been outlawed in the United States (except for research purposes). The major characteristics of heroin addiction are the development of both a tolerance to the drug and a physical dependence. *Tolerance* is a term used to signify the body's ability to adapt to a drug. If a tolerance is developed, increased quantities of the drug are required to produce the same effect as the original dose. Eventually, the abuser may require doses that would be lethal to a nonuser.

Physical dependence is related to dose and frequency of use. If the user is without the drug for 10–12 hours, he experiences vomiting, diarrhea, pains and tremors, restlessness, and mental disturbances (the withdrawal or abstinence syndrome). Opiate dependency can be passed from mother to fetus. A child born of an addict must spend the first days of its life withdrawing from the drug or death may result. Treatment can remove an addict's physical dependence upon a drug, but it is very difficult for him to abstain from any further use of the drug because of the psychological dependency that is produced.

The principal method of treating heroin addiction in this country is by administering minimal doses of methadone. About 90,000 heroin addicts are being treated in methadone maintenance clinics. Although methadone is also addictive, patients in the various maintenance programs have shown increased productivity and decreased antisocial behavior. Unlike the other narcotic drugs, methadone does not normally induce euphoria, but it eliminates the heroin hunger and withdrawal pains that would otherwise accompany the giving up of heroin. Moreover, at appropriate dosage levels methadone blocks the pleasurable sensations derived from heroin, thus further reducing heroin-seeking behavior. Another advantage of methadone is that it can be taken orally so that the dangers of infectious hepatitis (caused by a virus transmitted by unsanitary needles) are eliminated. Recently, however, several critics have voiced opposition to the methadone treatment. They question whether this approach achieves any lasting benefit for the addict, and they insist that methadone's long-term safety and effectiveness have not been adequately demonstrated. Withdrawal from methadone is generally regarded as being at least as difficult as withdrawal from heroin. If taken in excess by a person not tolerant of the drug, it can be lethal. In 1975, there were more methadone-related deaths in New York City than deaths attributed to heroin. Many of these poison victims were children who accidentally took a parent's methadone that had been premixed with fruit juice.

Many other drugs have been developed to treat heroin addicts. Chief among these are (*l*)-α-acetylmethadol (LAAM) and naltrexone. Both of these compounds are undergoing large-scale clinical testing and they are expected to be available in the 1980's.

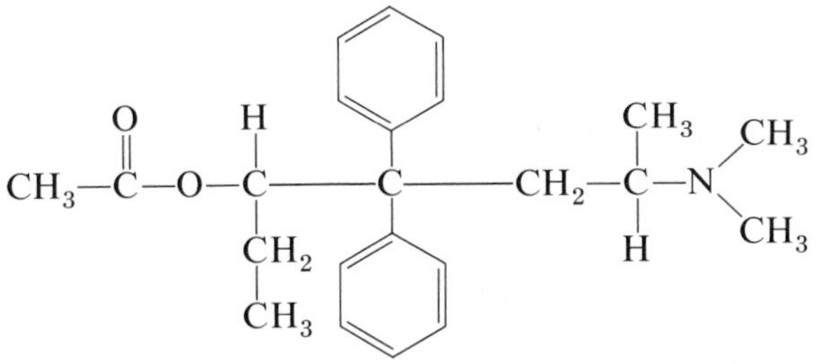
O
H
CH3
CH3
CH3—C—O—C——C——CH2—C—N
CH2
H
CH3
CH3

(*l*)-α-Acetylmethadol (LAAM)

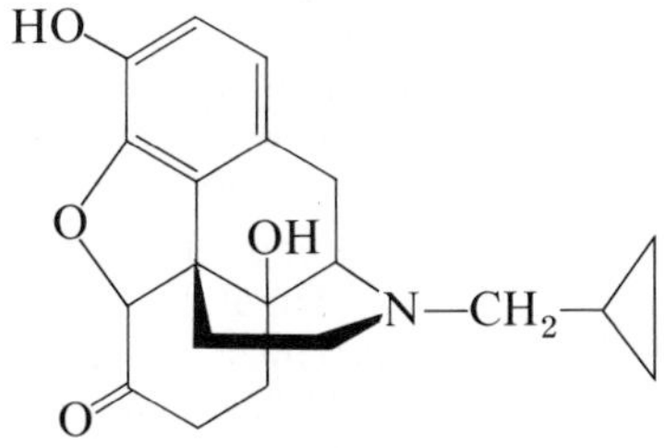
HO
O
OH
N—CH2
O

Naltrexone

LAAM is a methadone derivative, and like methadone it functions as a legal, orally active heroin substitute. The chief advantages of LAAM are (1) it is three times longer acting than methadone—it can be taken once every three days and therefore the addict can go to the clinic less frequently—and (2) it produces a smoother, more uniform effect than methadone and thus will make the addict less conscious of taking a drug. The major disadvantage of LAAM is that it is a narcotic and hence addictive.

Naltrexone, on the other hand, is a *narcotic antagonist,* i.e., it functions to prevent the narcotic (heroin) to have any effect. The narcotic antagonists that have been developed are all structurally similar to morphine. They exert their antagonistic effect by preferentially occupying the brain's opiate receptor sites. If a large enough dose of an antagonist is ingested before the addict uses heroin, the heroin will produce no euphoria, no sedation, and no analgesia because the opiate receptors are occupied by the antagonist. These narcotic antagonists are not narcotics and thus are nonaddicting.

b. Barbiturates

The barbiturates belong to a class of synthetic drugs that have *sedative* and *sleep-inducing* effects when administered in therapeutic doses. They are all prepared by heating urea with suitably substituted diethyl malonates. Figure 20.6 presents a synopsis of the preparation and structure of some of the well-known barbiturates.

Notice that the barbiturates differ only with respect to the substituent groups on carbon-5. The nature of the substituents plays an important role in establishing the pharmacological action of a particular barbiturate. Generally, the more highly lipid soluble a barbiturate is made by modifying the substituents, the more rapid is the onset of its action and the shorter is the duration of its pharmacological effects. Seconal and Nembutal are short-acting barbiturates (3–4 hours duration). They are used most often as sleeping pills and as preanesthetic sedatives, which are given to patients 30–60 minutes prior to surgery. Phenobarbital, on the other hand, is a long-acting barbiturate (10–12 hours duration). Once used widely as a tranquilizer, it is now employed chiefly as an anticonvulsant agent in the treatment of epilepsy and other brain disorders. The mechanism of this drug's action is unknown, but it is thought to suppress the spread of abnormal electrical discharges given off by the brain during a seizure.

Barbiturates, in the terminology of the drug culture, are "downers." The principal action of these drugs is on the central nervous system. It is believed that they depress the activity of the brain cells by interfering with oxygen consumption and the use and storage of energy. An average therapeutic dose to counteract insomnia or anxiety is 10–20 mg. An individual who makes repeated use of larger amounts of the drugs develops a tolerance, and soon many times the original dose becomes necessary. Habitual doses in excess of 800 mg per day depress the brain cells and effect a general depression of the central nervous system.

Of all the sedatives, the barbiturates are the most abused. (Ten billion barbiturate tablets are manufactured each year in the United States). It has definitely been established that barbiturates are habit-forming drugs. The symptoms of barbiturate addiction are similar to those of chronic alcoholism. A small dose can make the user rather sociable, relaxed, and good-humored, but his alertness and ability to react are decreased. Larger amounts result in blurred speech, mental sluggishness, confusion, loss of emotional control, and belligerence, and may induce a deep sleep or coma. Without immediate attention, the user may never wake.

Barbiturate dependency is becoming a more serious problem than opiate dependency. One reason is that, as tolerance to morphine develops, the user's body is able to handle what would be considered a lethal dosage for the nonuser without serious effects. However, with the barbiturates, even though a tolerance is developed, the lethal dose remains essentially constant. Thus, in a relatively short time the barbiturate addict's margin of safety decreases as his drug intake increases. Barbitu-

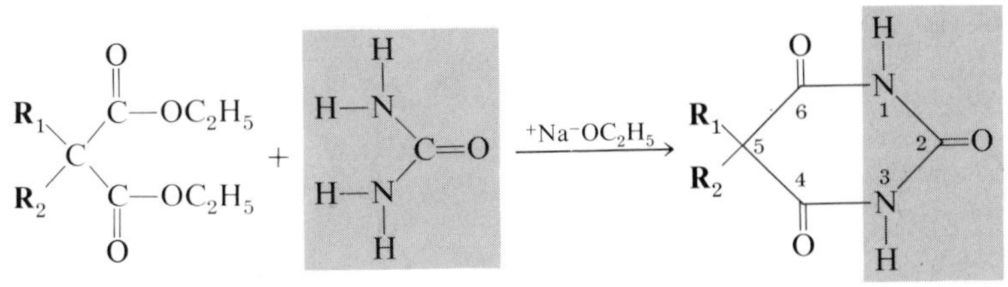

R_1	R_2	Name	Trade Name
C_2H_5-	C_2H_5-	Barbital	Veronal
C_2H_5-	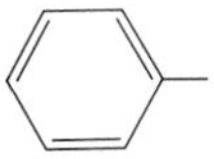	Phenobarbital	Luminal
C_2H_5-	$CH_3\overset{\overset{CH_3}{\vert}}{C}HCH_2CH_2-$	Amobarbital "Blue heaven"	Amytal
C_2H_5-	$CH_3CH_2CH_2\overset{\overset{CH_3}{\vert}}{C}H-$	Pentobarbital "Yellow jackets"	Nembutal
$CH_2{=}CHCH_2-$	$CH_3CH_2CH_2\overset{\overset{CH_3}{\vert}}{C}H-$	Secobarbital "Red devils"	Seconal

Figure 20.6 The preparation and structures of some barbiturates.

rates kill more people (mostly through deliberate overdoses) in the United States than any other drug, and they account for more than one fifth of all the poisonings reported. Since alcohol enhances the action of barbiturates, a particularly lethal combination is "booze" and sleeping pills.

c. Marijuana

Marijuana, colloquially known as "pot," "grass," or "Mary Jane," is found in the flowering tops of the female hemp plant (*Cannabis sativa*). (The longer male plant is used for fiber or hemp.) Hashish (*charas*) is a purer and more concentrated form of marijuana.[5] The major active constituent in marijuana is believed to be a derivative of tetrahydrocannabinol (Δ^9-THC) that is obtained from the resinous exudate of hemp plants. The potency of marijuana is related to its THC content. Typical marijuana available in the United States contains 2–3% THC, whereas much of the hashish has about 5% THC. It has been suggested that Δ^9-THC is rapidly metabolized to 11-hydroxy-Δ^9-tetrahydrocannabinol, and it is this latter compound

[5] Hashish was chewed by a band of assassins, active in 11th century Persia. The word *assassin* is derived from *Hashishin,* the name of the leader of these murderers.

that is responsible for the majority of the pharmacological effects.[6] This metabolic product of THC persists in the bloodstream for more than 72 hours, so that a frequent user of marijuana can get high on a smaller dose than that needed by the occasional user. By experimenting with synthetic THC derivatives of known purity, scientists hope to understand the pharmacological activity of marijuana.

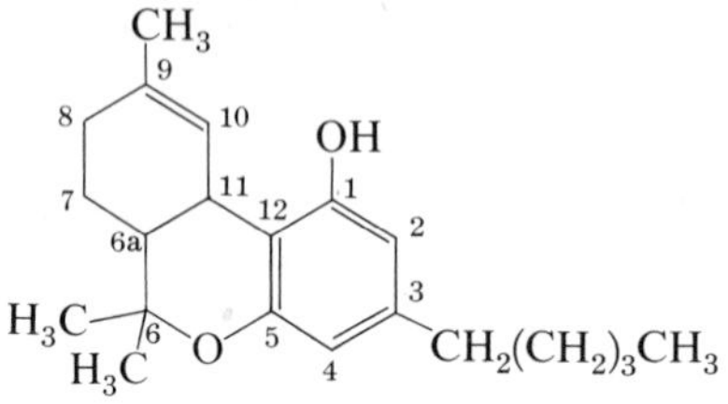
CH3
9
8
10
OH
1
7
11
12
2
6a
3
H3C
6
O
5
H3C
4
CH2(CH2)3CH3

Δ^9-Tetrahydrocannabinol

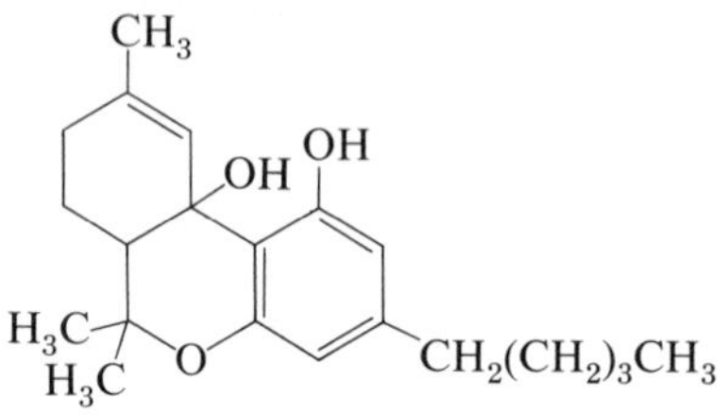
CH3
OH
OH
H3C
O
H3C
CH2(CH2)3CH3

11-Hydroxy-Δ^9-tetrahydrocannabinol

Marijuana acts on both the central nervous system and the cardiovascular system. When it is smoked, the effects of the drug can be felt within a few minutes and may last several hours. If marijuana is eaten with food, the onset of drug action is delayed for about an hour and the effects may persist from four to eight hours. Among the physiological effects of the drug are some or all of the following: increase in the pulse rate, bloodshot eyes, voracious appetite, nausea, and vomiting. The subjective effects include lethargy and hilarity, often at the same time. The user may experience an intensification of visual and auditory stimuli; time, space, and touch are usually distorted.

A great deal of misinformation, both pro and con, has developed about the smoking of pot. A publication by the National Clearinghouse for Mental Health Information (publication No. 5021, 1970, quoted below) deals with some of the fallacies associated with marijuana use.

Fable: Marijuana is a narcotic.
Fact: Marijuana is not a narcotic except by statute. Narcotics are opium or its derivatives (like heroin and morphine) and some synthetic chemicals with opiumlike activity.

Fable: Marijuana is addictive.
Fact: Marijuana does not cause physical addiction, since tolerance to its effects and symptoms on sudden withdrawal do not occur. It can produce habituation (psychological dependence), resulting in restlessness, anxiety, irritability, insomnia, or other symptoms.

Fable: Marijuana causes violence and crime.
Fact: Persons under the influence of marijuana tend to be passive. Sometimes a crime is committed by a person while under the influence of marijuana. The personality of the user is as important as the type of drug in determining whether chemical substances lead to criminal or violent behavior.

Fable: Marijuana leads to increase in sexual activity.
Fact: Marijuana has no aphrodisiac property.

[6] L. Lemberger et al., *Science* **177**, 62 (July 1972).

Fable: Marijuana is harmless.
Fact: Instances of acute panic, depression, and psychotic states are known, although they are infrequent. Certain kinds of individuals can also become overinvolved in marijuana use and center their lives around it. We do not know the effects of long-term use, although there is evidence that long-term users are typically apathetic and sluggish both physically and mentally.

(Note: A report on "Marijuana and Health" issued by the Department of Health, Education and Welfare expresses concern that marijuana's greatest danger involves operation of motor vehicles rather than the more widely publicized biological damage.)

Fable: Occasional use of marijuana is less harmful than occasional use of alcohol.
Fact: We do not know. Research on the effects of various amounts of each drug for various periods is under way.

Fable: Marijuana use leads to heroin.
Fact: We know nothing in the nature of marijuana that predisposes to heroin abuse. It is estimated that less than 5% of chronic users of marijuana will progress to experiment with heroin. (One study shows that more heroin addicts started out on alcohol than on marijuana.)

Fable: Marijuana enhances creativity.
Fact: Marijuana might bring fantasies of enhanced creativity but they are illusory, as are "instant insights" reported by marijuana users.

Fable: More severe penalties will solve the marijuana problem.
Fact: Marijuana use has increased enormously in spite of severely punitive laws.

d. Amphetamines

The amphetamines ("uppers" or "pep pills") represent a group of synthetic drugs that stimulate the central nervous system. The parent compound, amphetamine is not a single compound, but is a mixture of two optical isomers (see Exercise 22.22) that has been marketed under the name Benzedrine. It was synthesized in 1927 as a drug to simulate the actions of epinephrine (adrenalin). The latter is a hormone produced by the adrenal medulla; it plays a major role in making glucose rapidly available to tissues that require immediate energy. The amphetamines are often referred to as *sympathomimetic amines* since they mimic the action of epinephrine and norepinephrine on the central nervous system (Figure 20.7). Norepinephrine is released from storage sites at the end of the sympathetic nerves[7] and its stimulating effects are thought to be caused by a flooding of the brain synaptic sites with amines. Increased stimulation of the sympathetic nerves results in an elevation of the pulse rate and blood pressure, an increased rate of respiration, and a loss of appetite (anorexia).

Medically, the stimulating effects of the amphetamines make them useful for the treatment of a variety of conditions including mild depression, narcolepsy (an overwhelming compulsion to sleep), certain behavioral disorders (hyperkinetics) in

[7]The sympathetic nerves are a division of the autonomic nervous system. This system has the specialized task of regulating body functions over which an individual has no control such as circulation, digestion, heart rate, and respiration. Sympathetic nerves are also concerned with emotions.

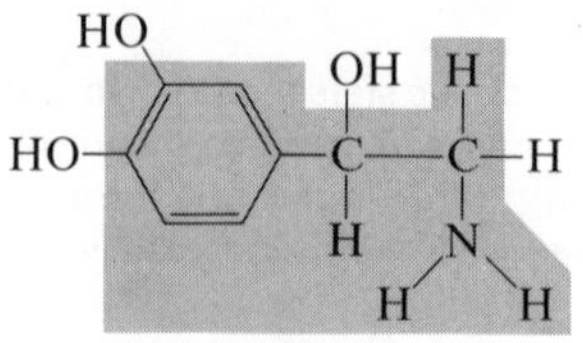

Norepinephrine

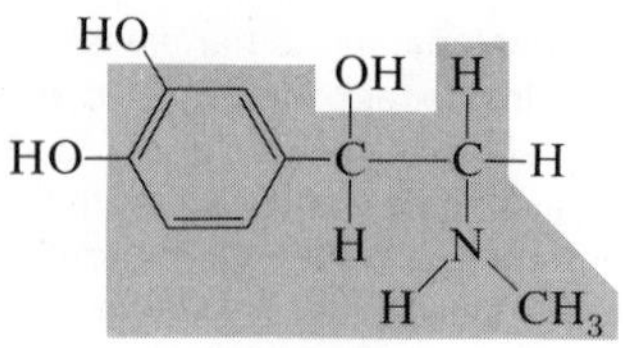

Epinephrine
(Adrenalin)

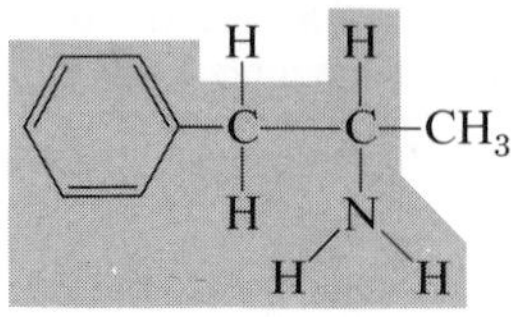

Amphetamine
(Benzedrine)

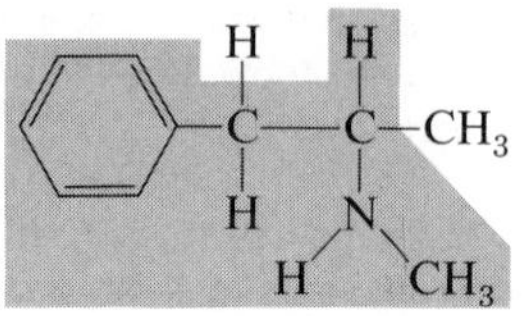

Methamphetamine
(Methedrine)

Figure 20.7 Amphetamines are synthetic drugs that mimic the actions of epinephrine and norepinephrine.

children, and obesity. (Most physicians do not recommend the use of amphetamines as appetite depressants for weight reduction because as soon as the drug is withdrawn, the patient's former appetite will return unless a new dietary pattern has been established.) Ordinary therapeutic doses of 10–30 mg per day provide a feeling of well-being, a decrease in fatigue and increased alertness. Overtired businessmen and truck drivers, students cramming for exams, and athletes who need to "get up" for a game are frequent misusers of these drugs.

Prolonged abuse of amphetamines can have ruinous effects upon a person's physical and mental health. Once ingested (or injected), they are not easily broken down, and by passing across the blood–brain barrier they tend to concentrate in the brain and cerebrospinal fluid. Extensive use of higher than normal doses can lead to long periods of sleeplessness, loss of weight, and severe paranoia. When excited, the person is liable to become violent to the point where he may hurt or kill somebody.

In recent years, the most widely abused amphetamine has been methamphetamine ("speed"). A user generally begins by taking the drug orally, but then progresses to intravenous injections ("shooting speed") to attain the desired effects. A tolerance is rapidly developed, and soon enormous amounts are required. With continued injections the "speed-freak" or "methhead" can stay awake for days. This constant wakefulness, which is accompanied with incessant babbling and tremendous nervous energy, is called a "run" or "speed-binge." Following a three or four day run, his body becomes exhausted and he falls into a coma-like sleep for 12–18 hours, usually with the aid of barbiturates or heroin. Upon wakening, depression sets in and the entire procedure is begun once again. Death due to serum hepatitis is quite common.

e. Hallucinogens

The hallucinogens are drugs that stimulate sensory perceptions (e.g., hearing strange voices and seeing alarming sights) that have no basis in physical reality. They

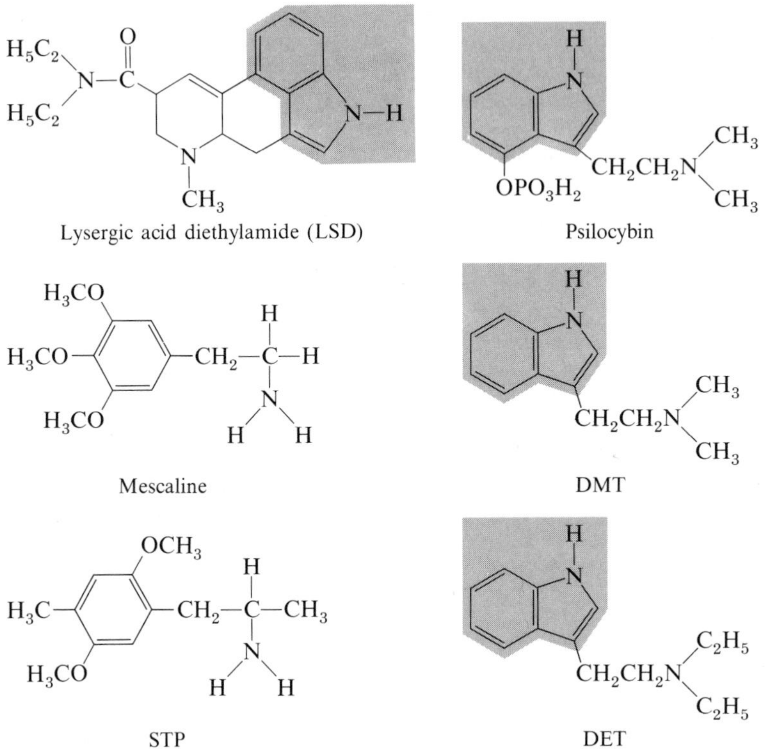
Lysergic acid diethylamide (LSD)
Psilocybin
Mescaline
DMT
STP
DET

Figure 20.8 Some hallucinogenic drugs.

are also referred to as *psychotomimetic drugs,* because in some cases their effects seem to mimic psychosis, and *psychedelic drugs,* because of the visions produced. Among the more abused hallucinogens are LSD (lysergic acid diethylamide), mescaline (3,4,5-trimethoxyphenylethylamine), psilocybin (4-phosphoryl-*N,N*-dimethyltryptamine), STP[8] or DOM (2,5-dimethoxy-4-methylamphetamine), DMT (*N,N*-dimethyltryptamine) and its diethyl analog DET (Figure 20.8).

The plants from which the major hallucinogens are obtained have been known for millenniums. Mescaline comes from the peyote cactus; psilocybin from certain species of mushrooms; and lysergic acid is derived from the parasitic fungus ergot, which grows on rye and wheat. The diethylamide of lysergic acid is then chemically synthesized in the laboratory. LSD is a white, crystalline powder that is odorless, tasteless, and readily soluble in water. It is, by far, the most powerful of the hallucinogens. In order to achieve the equivalent effects of 0.1 mg of LSD (an average "trip" dose), the user requires 5 mg of STP, 12 mg of psilocybin, and 400 mg of mescaline.

Legal use of LSD is limited to qualified investigators working under a government grant or for a state or federal agency. Its official distribution is strictly regulated by the National Institute of Mental Health. Many psychiatrists and psychologists

[8]The initials STP were originally popularized for a motor fuel additive, Scientifically Treated Petroleum. In today's drug culture the letters stand for Serenity, Tranquility, and Peace.

who have administered hallucinogens in a therapeutic setting claim specific benefits in the treatment of psychoneuroses, alcoholism, and social delinquency. LSD has also been used in the treatment of terminal cancer patients. It has lessened pain awareness and eased the suffering of many of these patients.

Oral doses of LSD as low as 20–25 μg (0.020–0.025 mg) are capable of producing psychological responses in sensitive individuals. The effects of the drug on the central nervous system are usually apparent within an hour following ingestion and most trips last from eight to twelve hours. The basic physiological effects are those typical of a mild stimulation of the sympathetic nervous system. These include elevated body temperature, dilated pupils, increased blood glucose, elevated blood pressure, and increased heart rate. The major subjective effect is a change in visual perception. "Many users feel a new awareness of the physical beauty of the world, particularly of visual harmonies, colors, the play of light and the exquisiteness of detail."[9]

Many attempts have been made to explain and localize the action of the hallucinogens. The structural resemblance of mescaline and STP to epinephrine and norepinephrine (see Figure 20.7) suggests a possible link between the drugs and the hormones. LSD (and the other substituted tryptamines) are structurally similar to serotonin, a hormone found in brain tissue that is believed to play an important role in the thought process. There is some evidence that the mental disease *schizophrenia* results from some abnormality in the metabolism of serotonin in the brain.

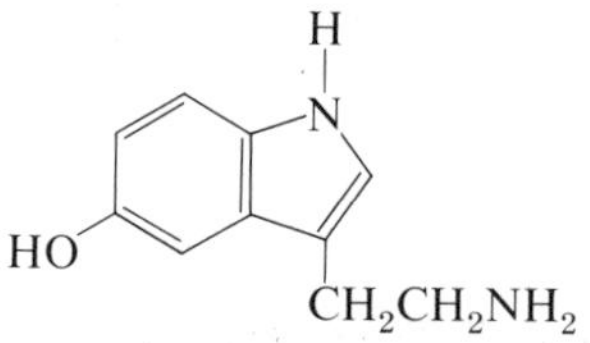
H
N
HO
$CH_2CH_2NH_2$

Serotonin
(5-Hydroxytryptamine)

It has been suggested that LSD may either block or replace serotonin in brain functions. While this particular hypothesis has not been substantiated, it does seem likely that LSD owes its activity to some interference with the transmission of impulses from cell to cell in the nervous system. Another theory that has been postulated is that LSD influences nerve cell activity by one of the following processes: (1) blocking energy production, (2) altering the permeability of the cell membrane, (3) increasing the permeability of the blood–brain tissue. If this permeability were increased, substances normally nontoxic because of their position on the blood side of the barrier could cross over and become toxic to the brain, thereby giving rise to bizarre symptoms.

There have been conflicting reports as to the teratogenic effects of LSD. (*Teratogenesis* refers to the production of physical defects in offspring while in the uterus.) Some investigators have reported an unusually high incidence of genetic damage to the chromosomes of the white blood cells of infants born to mothers who used LSD during their pregnancy. But other studies showed no such evidence, and these latter investigators conclude that "pure LSD, ingested in moderate doses, does not produce chromosome damage detectable by available methods."[10] In light of these contradictory findings, it is obvious that a great deal of further research is required.

[9] F. Barron et al., *Scientific American,* **210,** 32 (April 1964).
[10] N. Dishotsky et al., *Science,* **172,** 431 (April 1971).

Exercises

20.1 Draw and name all the isomeric amines having the molecular formula $C_4H_{11}N$.

20.2 Name the following compounds.

(a) $NH_2—C(=O)—NH_2$

(b) 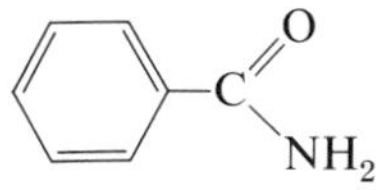

(c) $(C_2H_5)_2NCH(CH_3)_2$

(d) $C_2H_5N(CH_3)_2$

(e) 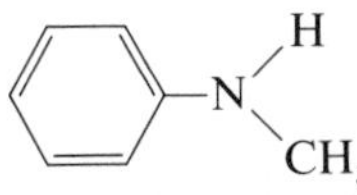

(f) $H_2NCH_2CH_2CH_2NH_2$

(g) $(CH_3)_2N(C_2H_5)_2{}^+I^-$

(h) $(CH_3)_2CHC(=O)NH_2$

(i) $CH_3CH(NH_2)CH_2CH(CH_3)CH_2OH$

(j) 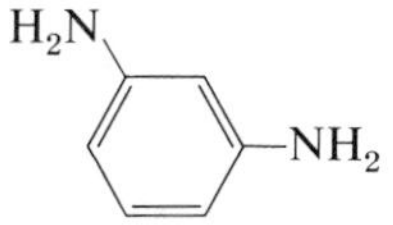

20.3 Write structural formulas for the following compounds.

(a) aniline
(b) acetanilide
(c) *n*-butylamine
(d) 2-amino-2-methyl-1-propanol
(e) formamide
(f) dimethyl *n*-propylamine
(g) cyclopentylamine
(h) nicotinamide
(i) di-*n*-butylamine
(j) tetraethylammonium chloride

20.4 Explain and illustrate the meaning of the terms *primary, secondary,* and *tertiary* when applied to amines.

20.5 Write equations for three preparations of acetamide.

20.6 Esters are usually hydrolyzed under alkaline conditions, but either acidic or basic hydrolysis is employed with amides. Why is this possible?

20.7 Contrast the physical properties of amides with those of amines.

20.8 Ethylamine and ethyl alcohol have comparable molar masses, yet ethyl alcohol has a much higher boiling point. Explain.

20.9 Write an equation to represent what occurs when dimethylamine is dissolved in water.

20.10 Arrange the following compounds in the order of decreasing basicity: **(a)** aniline, **(b)** ethylamine, **(c)** pyridine, **(d)** acetamide, **(e)** trimethylamine, and **(f)** sodium hydroxide.

20.11 Write equations for three different reactions of amides and amines.

20.12 Using RCOOH to represent the formula of lysergic acid, write an equation for the preparation of lysergic acid diethylamide.

20.13 Define, explain, or give an example of each of the following.

(a) diazotization
(b) diazonium ion
(c) azo dye
(d) acid-base indicator
(e) dihydropyridine
(f) drug
(g) narcotic
(h) morphine
(i) opium
(j) codeine
(k) heroin
(l) methadone
(m) barbiturate
(n) Benzedrine
(o) mescaline
(p) serotonin
(q) tolerance
(r) downer
(s) upper
(t) pot
(u) hashish
(v) speed
(w) LSD
(x) psychedelic drug

20.14 Complete the following equations.

(a) $CH_3C(=O)NH_2 + HCl \xrightarrow{HOH}$

(b) $CH_3C(=O)NH_2 + NaOH \xrightarrow{HOH}$

(c) 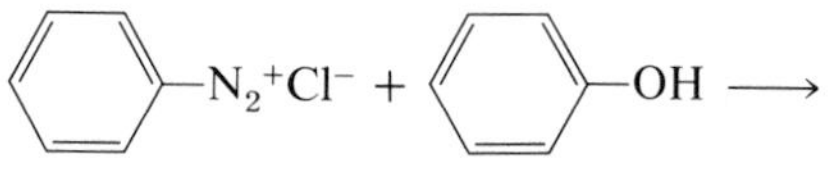

(d) $CH_3CH_2NH_2 + HCl \longrightarrow$

(e) 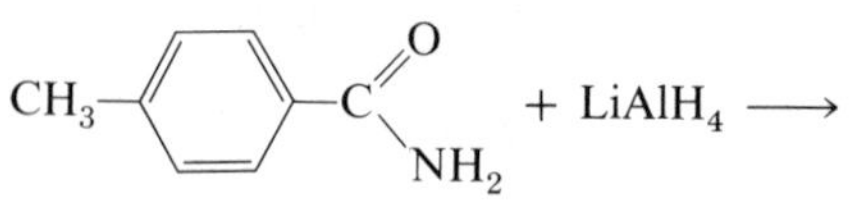

(f) $CH_3C(=O)NH_2 + NaOBr + NaOH \longrightarrow$

(g) 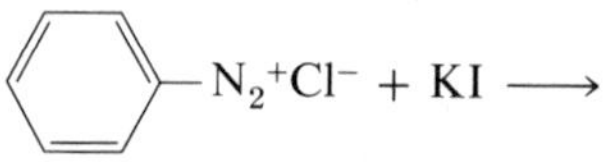

(h) 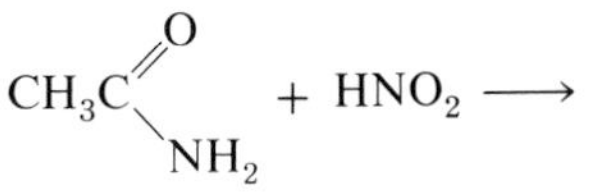

20.15 Devise simple chemical tests that would enable you to distinguish each compound from the other(s) in the following.

(a) $CH_3(CH_2)_4CH_2NH_2$ and 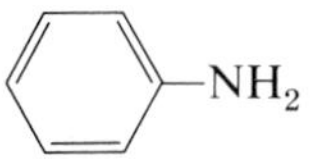

(b) 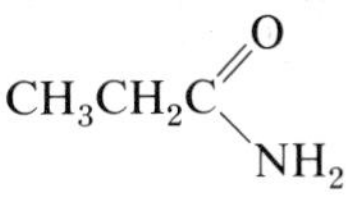 and $CH_3CH_2CH_2NH_2$

(c) $CH_3CH_2CH_2NH_2$, $(CH_3)_2NH$, and $(CH_3)_3N$

(d) 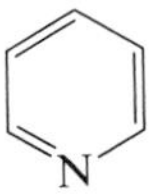 and 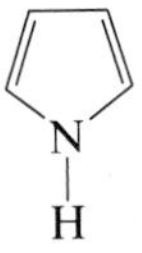

(e) morphine and heroin

(f) phenobarbital and Benzedrine

20.16 An organic compound, whose molecular formula is C_4H_9NO, is soluble in water. Addition of a NaOH solution produces ammonia and sodium butyrate. What is the structural formula of the compound?

20.17 What volume of nitrogen gas (at STP) is evolved when 9.0 grams of butanamide reacts with excess nitrous acid?

20.18 How many milliliters of 0.15 *M* hydrochloric acid is required to neutralize 0.25 gram of *n*-butylamine?

20.19 0.61 gram of an amine reacted with excess nitrous acid evolving 187 ml of nitrogen gas (at STP). Draw all of the possible structural formulas for the amine.

20.20 The dissociation constant for a primary amine, RNH_2, is 1.0×10^{-6}. Calculate the hydroxide ion concentration and the pH of a 0.30 *M* RNH_2 solution.

20.21 The pK_b of pyridine at 25°C is 8.7. Calculate the pH of a 0.10 *M* solution of pyridine in water.

21

Organic Compounds of Phosphorus and Sulfur

Organic compounds containing phosphorus and sulfur atoms are widely distributed in nature. Many of these substances are essential for the normal functioning of all living organisms. Phosphate or pyrophosphate esters are present in every plant and animal cell. They are biochemical intermediates in the transformation of food into usable energy (in the form of ATP). Organic phosphates are also important structural constituents of phospholipids (Section 24.9), nucleic acids (Section 27.1), coenzymes (Table 26.2), and insecticides. Several of the organic sulfur compounds are known primarily by their unpleasant odors such as the scent of skunks and the pungent odors and flavors of onions and garlic. Sulfur atoms are encountered in such diverse compounds as amino acids, coenzymes, detergents, dyes, mustard gas, penicillin, saccharin, and sulfa drugs.

Phosphorus Compounds

The great abundance of organophosphorus compounds is attributed to the existence of more than one oxidation state for the element. The three most important oxidation states of phosphorus are -3, $+3$, $+5$; that is, the organic compounds can be considered as derivatives of phosphine (PH_3), phosphorous acid (H_3PO_3), and phosphoric acid (H_3PO_4), respectively.

Phosphine is the phosphorus analog of ammonia. It resembles ammonia in being a pyramidal molecule with an unshared pair of electrons on phosphorus. Therefore, like ammonia, PH_3 can add a proton to form a phosphonium ion, PH_4^+.

$$H-\overset{\cdot\cdot}{\underset{\underset{H}{|}}{P}}-H + H^+ \longrightarrow H-\overset{\overset{H}{|}}{\underset{\underset{H}{|}}{P^\oplus}}-H$$

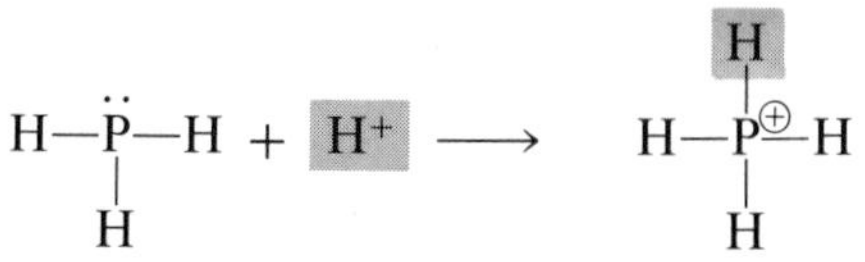

Phosphine   Phosphonium ion

Unlike ammonia, phosphine is practically insoluble in water (phosphorus is not sufficiently electronegative to participate in hydrogen bonding) and is much less basic. Phosphine is a poisonous gas, and it burns spontaneously in air because of the presence of impurities.[1]

The majority of the alkyl derivatives of phosphine are prepared directly from the alkyl halides or by the reaction of a Grignard reagent with phosphorus trichloride. Although alkylphosphines resemble amines in structure, they are less basic and more subject to oxidation.

$$3\ CH_3I + PH_3 + 3\ KOH \longrightarrow CH_3-\overset{\cdot\cdot}{\underset{\underset{CH_3}{|}}{P}}-CH_3 + 3\ KI + 3\ HOH$$

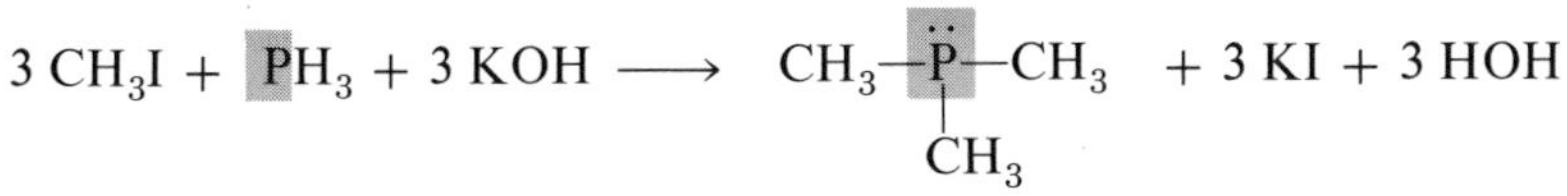

Trimethylphosphine

$$(CH_3)_3P + CH_3I \longrightarrow (CH_3)_4P^+I^-$$

Tetramethyl-
phosphonium
iodide

$$3\ \text{⬡}-MgBr + PCl_3 \longrightarrow \text{⬡}-\overset{\cdot\cdot}{\underset{\underset{\text{⬡}}{|}}{P}}-\text{⬡} + 3\ MgBrCl$$

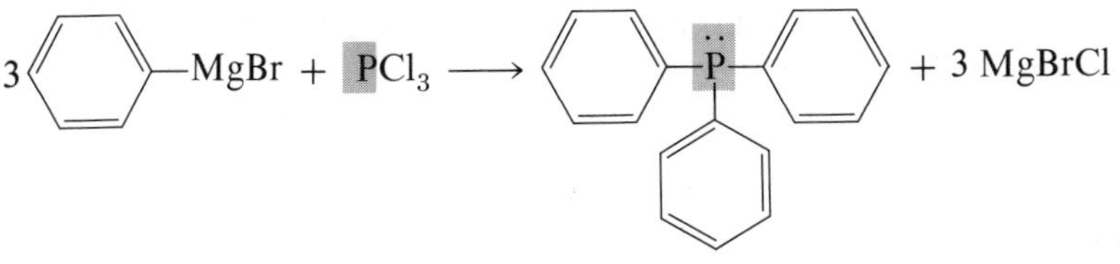

Triphenylphosphine

Esters of phosphorous acid are useful to the synthetic organic chemist but they do not occur to any appreciable extent in living systems. Trialkyl phosphites can be prepared by treating a phosphorus halide with an alcohol

[1] The faint flickering light sometimes observed in marshes may be due to spontaneous ignition of PH_3. The phosphine might be formed by reduction of naturally occurring phosphorus compounds.

in the presence of a tertiary amine. The tertiary amine is required to neutralize the HCl produced, otherwise the acid would hydrolyze the resultant phosphite to phosphorous acid and alkyl halide. Recall that the reaction of alcohols with PCl_3 (in the absence of an amine) was mentioned as a preparative method of alkyl halides (Section 16.2-b).

$$PCl_3 + 3\ CH_3CH_2OH + 3\ R_3N \longrightarrow CH_3CH_2O-\overset{\overset{\displaystyle ..}{|}}{\underset{\underset{\displaystyle OCH_2CH_3}{|}}{P}}-OCH_2CH_3 + 3\ R_3NH^+Cl^-$$

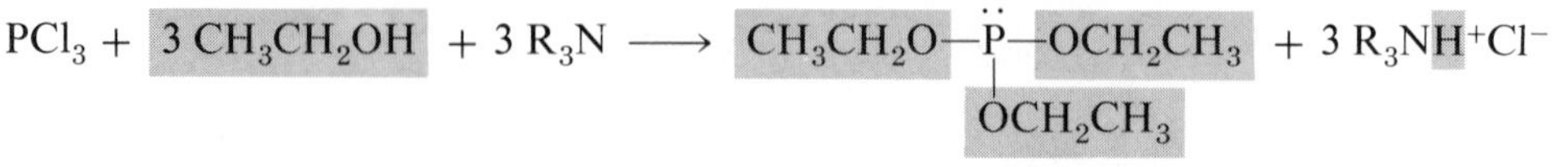

Triethyl phosphite

Trialkyl phosphites react with primary alkyl halides to produce phosphonate esters. This is the Arbuzov reaction, named after the Russian chemist who developed the synthetic technique.

$$CH_3CH_2O-\overset{\overset{\displaystyle ..}{|}}{\underset{\underset{\displaystyle OCH_2CH_3}{|}}{P}}-OCH_2CH_3 + CH_3CH_2CH_2CH_2I \xrightarrow{\Delta} CH_3CH_2CH_2CH_2-\overset{\overset{\displaystyle O}{\parallel}}{\underset{\underset{\displaystyle OCH_2CH_3}{|}}{P}}-OCH_2CH_3 + CH_3CH_2I$$

n-Butyl iodide

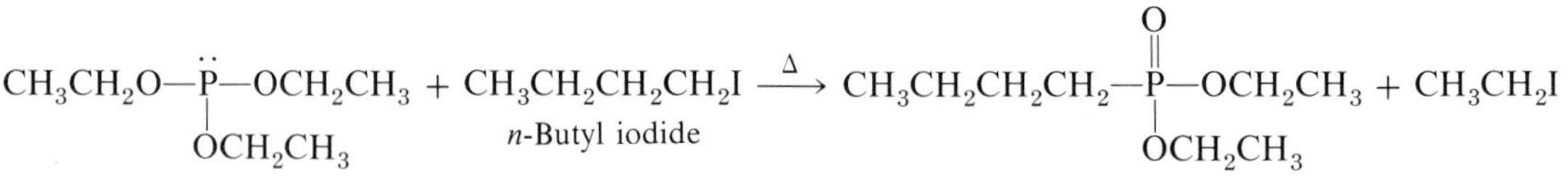

Diethyl butylphosphonate

21.3 Esters of Phosphoric Acid

Esters of phosphoric acid and pyrophosphoric acid are the most frequently encountered organophosphorus compounds. Pyrophosphoric acid ($H_4P_2O_7$) is formed by the elimination of a molecule of water from two moles of phosphoric acid; another phosphate can be added in a similar fashion to form a triphosphate (triphosphoric acid). Sodium salts of triphosphoric acids were utilized to a considerable extent in detergents to aid in the suspension of dirt particles (Section 24.8-b).

$$HO-\overset{\overset{\displaystyle O}{\parallel}}{\underset{\underset{\displaystyle OH}{|}}{P}}-OH \xrightarrow[-HOH]{\ HO-\overset{\overset{\displaystyle O}{\parallel}}{\underset{\underset{\displaystyle OH}{|}}{P}}-OH\ } HO-\overset{\overset{\displaystyle O}{\parallel}}{\underset{\underset{\displaystyle OH}{|}}{P}}-O-\overset{\overset{\displaystyle O}{\parallel}}{\underset{\underset{\displaystyle OH}{|}}{P}}-OH \xrightarrow[-HOH]{\ HO-\overset{\overset{\displaystyle O}{\parallel}}{\underset{\underset{\displaystyle OH}{|}}{P}}-OH\ } HO-\overset{\overset{\displaystyle O}{\parallel}}{\underset{\underset{\displaystyle OH}{|}}{P}}-O-\overset{\overset{\displaystyle O}{\parallel}}{\underset{\underset{\displaystyle OH}{|}}{P}}-O-\overset{\overset{\displaystyle O}{\parallel}}{\underset{\underset{\displaystyle OH}{|}}{P}}-OH$$

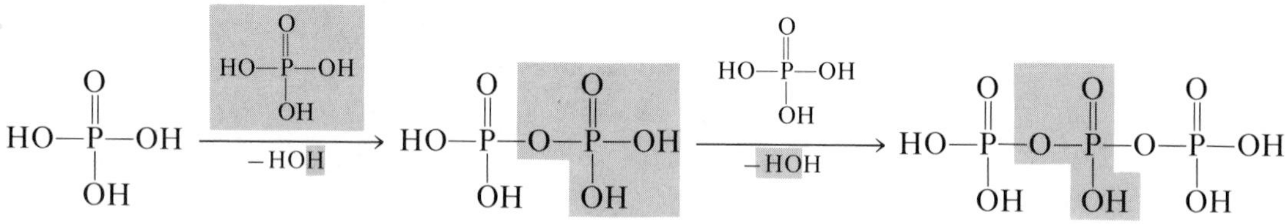

Phosphoric acid   Pyrophosphoric acid (Diphosphoric acid)   Triphosphoric acid

Replacement of one or more hydrogen atoms of phosphoric and pyrophosphoric acid by organic groups yields phosphate esters. Usually the name of the ester specifies the number of organic groups bound to the phosphate.

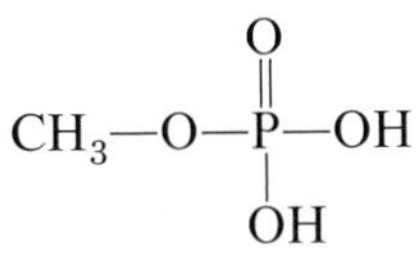

$$CH_3-O-\overset{\displaystyle O}{\underset{\displaystyle OH}{\overset{\|}{P}}}-OH \qquad CH_3CH_2-O-\overset{\displaystyle O}{\underset{\displaystyle O-CH_2CH_3}{\overset{\|}{P}}}-OH$$

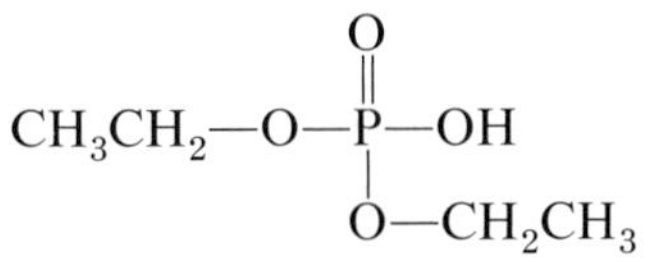

Methyl dihydrogen   Diethyl hydrogen
phosphate   phosphate

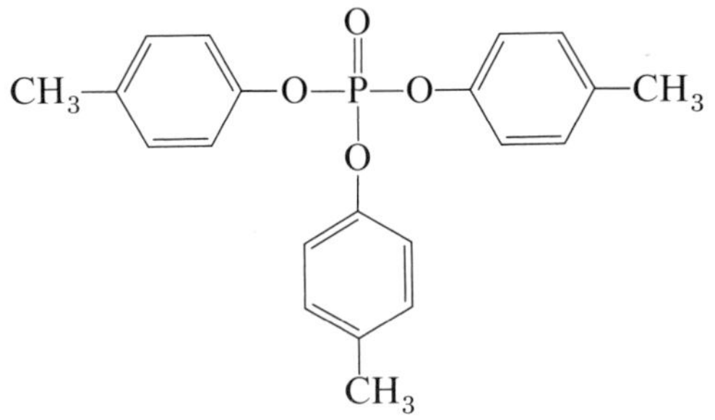

Tricresyl phosphate (TCP)   Tris(2,3-dibromopropyl) phosphate
(component of high octane fuels)   (Tris)

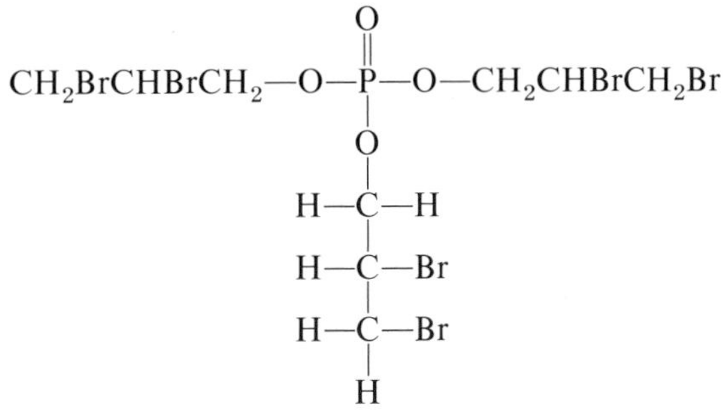

The combination of phosphorus and bromine atoms in the compound Tris imparts flame-retardant properties to garments. (The mechanism by which various flame retardants act to inhibit burning is still not completely known, although several theories have been postulated.) In 1977 the Consumer Product Safety Commission banned the sale of children's sleepwear treated with the compound Tris because the chemical was shown to cause kidney cancer in rats and mice. The fear was that Tris could be absorbed through the child's skin or swallowed if children chewed on their sleepwear. In the body Tris is hydrolyzed to 2,3-dibromopropanol, and this latter compound has been detected in the urine of children who have worn sleepwear treated with Tris.

Alkyl esters of phosphoric and pyrophosphoric acid are prepared in a similar fashion to those of carboxylic acids. One of the best methods involves the reaction of an alcohol with the acid chloride of phosphoric acid ($POCl_3$).

$$HO-\overset{\displaystyle O}{\underset{\displaystyle OH}{\overset{\|}{P}}}-OH \xrightarrow[SOCl_2]{PCl_5 \text{ or}} Cl-\overset{\displaystyle O}{\underset{\displaystyle Cl}{\overset{\|}{P}}}-Cl \xrightarrow{3\ CH_3CH_2CH_2CH_2OH} (CH_3CH_2CH_2CH_2O)_3-\overset{\displaystyle O}{\overset{\|}{P}} + 3HCl$$

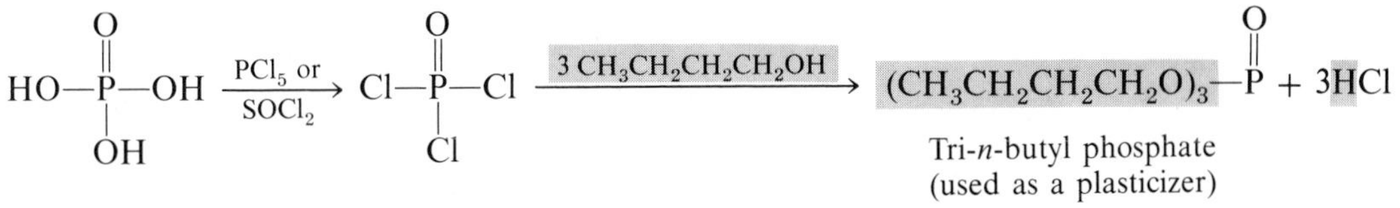

Tri-*n*-butyl phosphate
(used as a plasticizer)

A wide variety of phosphate esters occur naturally and are compounds of central importance in metabolism. Figure 21.1 presents only a few of these compounds. Notice that at physiological pH ($\sim$7), the phosphate groups are ionized. Phosphoric acid can also condense with carboxylic acid derivatives

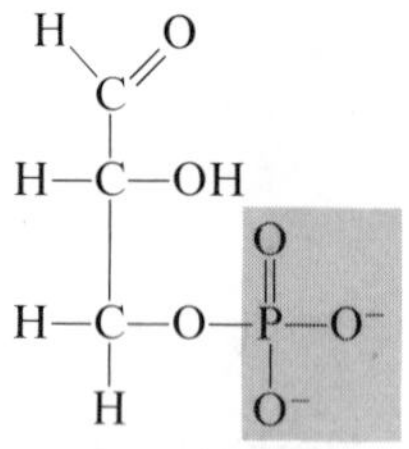

Glyceraldehyde 3-phosphate

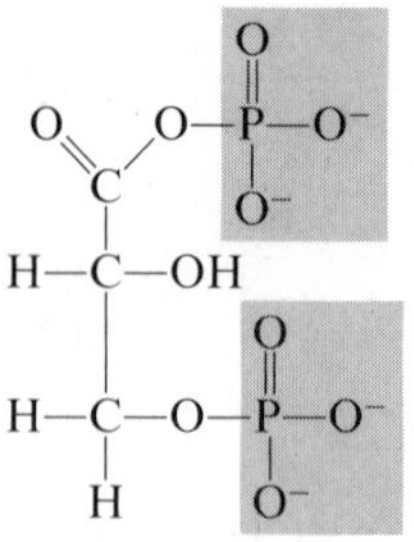

1,3-Diphosphoglyceric acid

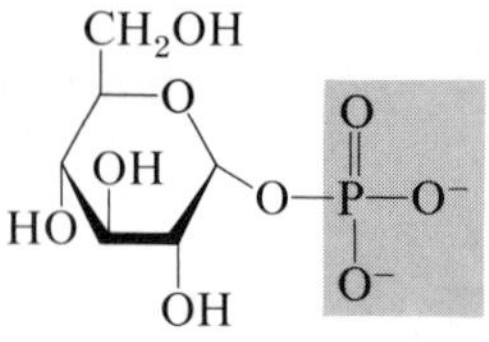

Glucose 1-phosphate

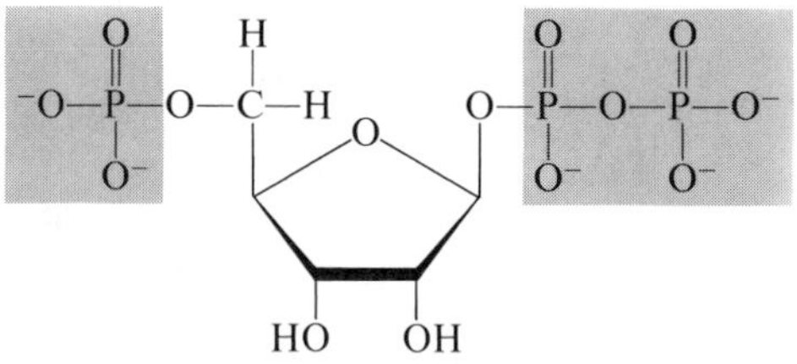

Phosphoribosyl pyrophosphate

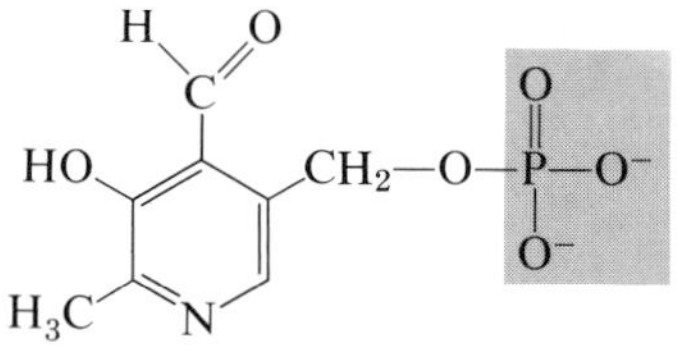

Pyridoxal phosphate

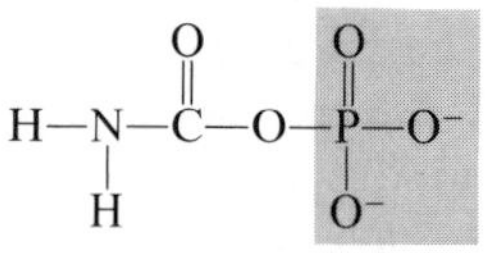

Carbamyl phosphate

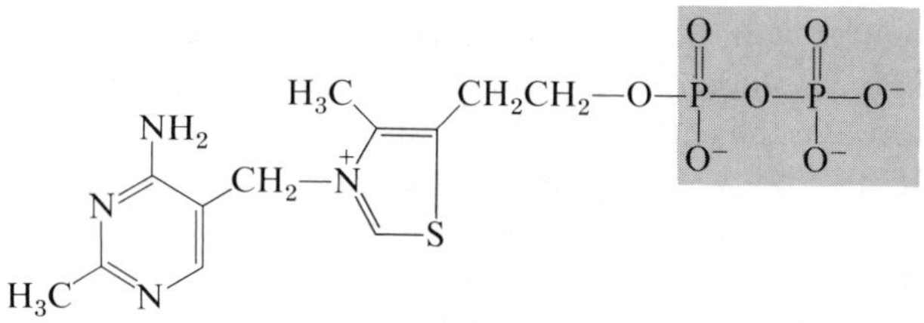

Thiamine pyrophosphate

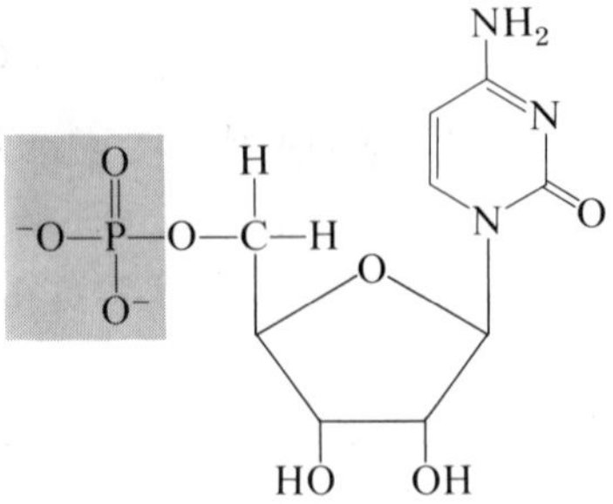

Cytidine nucleotide

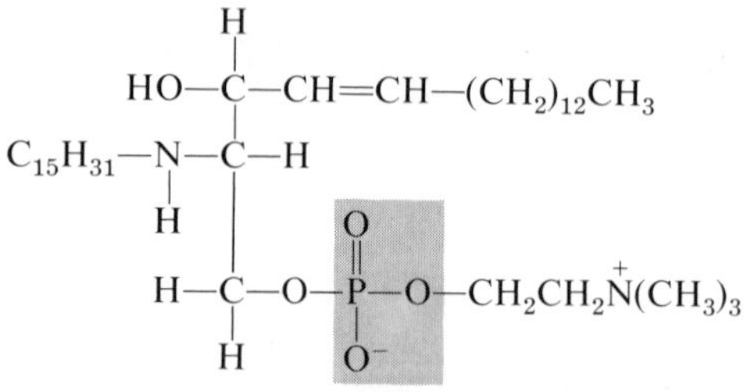

Sphingomyelin

Figure 21.1 Some phosphate compounds of biological importance.

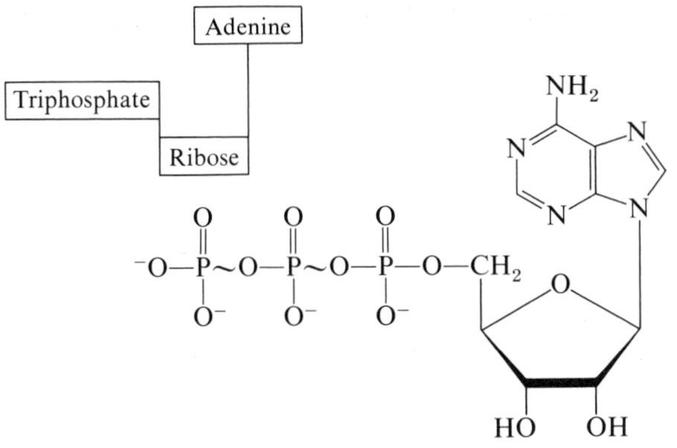
Adenine
Triphosphate
Ribose
NH2
HO
OH

Figure 21.2 The chemical structure of adenosine triphosphate (ATP).

to form mixed anhydrides. Some of these compounds, such as acetyl phosphate, have been isolated from biological sources where they probably function as acetylating agents.

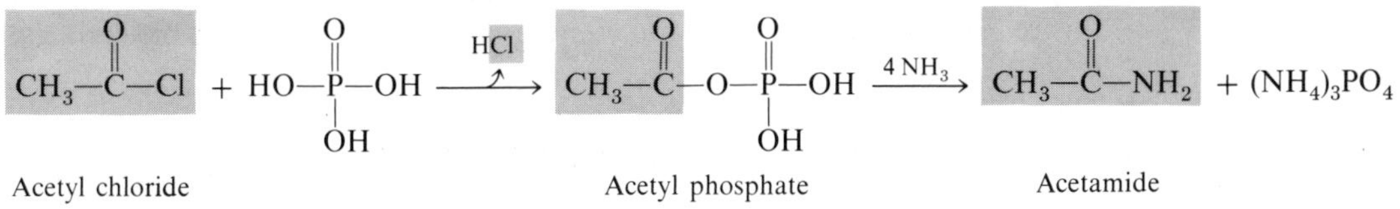
HCl
4 NH3
+ (NH4)3PO4
Acetyl chloride
Acetyl phosphate
Acetamide

Probably the single most important phosphate compound is *adenosine triphosphate* (ATP). Adenosine triphosphate was first isolated from skeletal muscle tissue and has since been shown to occur in all types of plant and animal cells. The concentration of ATP in the cell varies from 0.5 to 2.5 mg/ml of cell fluid. ATP is a nucleoside triphosphate composed of adenine, ribose, and three phosphate groups attached to the 5′ carbon of the ribose (Figure 21.2). As the names indicate, adenosine diphosphate (ADP) contains two phosphate groups, and adenosine monophosphate (AMP) contains one phosphate group (Figure 21.3).

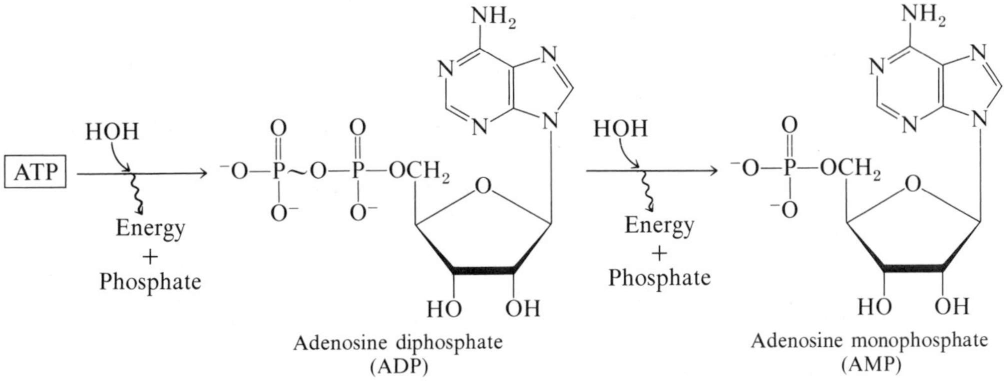
ATP
HOH
Energy
+
Phosphate
Adenosine diphosphate
(ADP)
HOH
Energy
+
Phosphate
Adenosine monophosphate
(AMP)

Figure 21.3 The relationships of ATP, ADP, and AMP.

The most significant feaure of the ATP molecule is the presence of the phosphoric acid anhydride, or pyrophosphate, linkage.

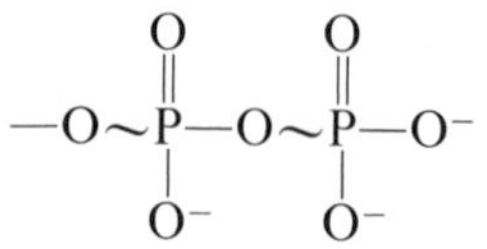

ATP is an *energy rich compound;* its pyrophosphate bonds are referred to as *high energy* bonds and are sometimes symbolized by a wiggle bond (~). A high energy bond is one that releases a relatively large amount of energy (>5000 cal/mol) when it is hydrolyzed. In this case, one of the driving forces for the reaction is to relieve the electron–electron repulsions associated with the negatively charged phosphate groups. The high energy phosphate bond may be likened to a coiled spring that is compressed when the phosphate group is attached. When the phosphate is released, the spring extends, thus releasing its energy. Energy rich compounds, then, are substances having particular structural features that yield high energies of hydrolysis, and for this reason they are able to supply energy for energy-requiring biochemical processes.

$$\text{ATP} \xrightleftharpoons{H_2O} \text{ADP} + P_i\text{[2]} + 7300\text{[3] cal/mole}$$

The important feature of this biochemical reaction is its reversibility. The hydrolysis of ATP releases energy, and its synthesis requires energy. Thus ATP is produced by those processes that supply energy to an organism (absorption of radiant energy of the sun in green plants and breakdown of foodstuffs in animals), and ATP is degraded by those processes that require energy (syntheses of carbohydrates, lipids, proteins; transmission of nerve impulses; muscle contraction, etc.). This coupling of the synthesis of ATP to processes that release energy and of the degradation of ATP to processes that require energy is one of the striking characteristics of living matter.

Although ATP is the principal medium of energy exchange in biological systems, it is not the only high energy compound. There are several other phosphate esters that are utilized to provide energy for certain energy-requiring reactions. Table 21.1 lists a number of phosphate compounds and

[2] P_i is the symbol for the inorganic phosphate anions $H_2PO_4^-$ and HPO_4^{2-} that are present in the intra- and extracellular fluids. About 1.2 grams of phosphate is needed in the daily diet to replace the amount excreted in the urine. (See also footnote a, Table 26.2).

[3] The values in the literature for the energy released when ATP is hydrolyzed vary somewhat. This situation is due in part to the fact that reaction conditions (concentration, temperature, pH) have not always been the same in different laboratories and in part to the difficulties in obtaining exact values for the equilibrium constants. Most texts and research articles report values between −7000 and −8000 cal/mole for the free energy of hydrolysis of ATP at 25 °C and a pH of 7.0. We shall use a value of −7300 throughout this text. It is certain, however, that under actual intracellular conditions the free energy of hydrolysis for ATP is considerably higher.

Table 21.1 Standard Free Energies of Hydrolysis of Some Phosphate Compounds

Type	*Example*	*ΔG′ (cal/mole)*	*See Page*
Enol phosphate 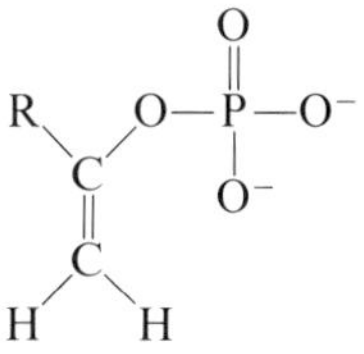	Phosphoenolpyruvic acid	−14,800	719
Acyl phosphate 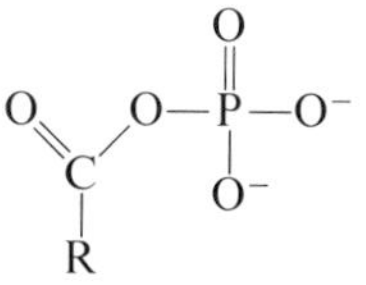	1,3-Diphosphoglyceric acid	−11,800	718
Guanidine phosphate 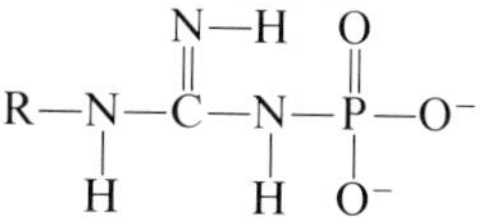	Creatine phosphate	−10,300	723
	Arginine phosphate	− 7,000	723
Pyrophosphate 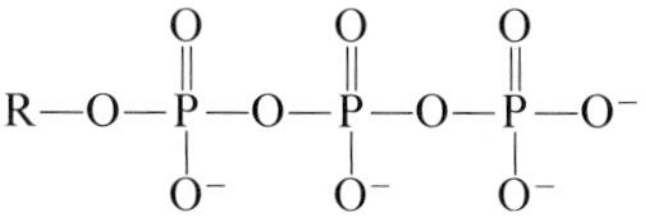	ATP $\longrightarrow$ AMP + P_i	− 8,600	767
	ATP $\longrightarrow$ ADP + P_i	− 7,300	722
	ADP $\longrightarrow$ AMP + P_i	− 6,500	
Sugar phosphate 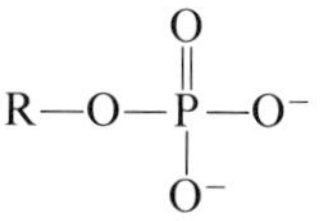	Glucose 1-phosphate	− 5,000	713
	Fructose 6-phosphate	− 3,800	716
	AMP $\longrightarrow$ Adenosine + P_i	− 3,400	
	Glucose 6-phosphate	− 3,300	714
	Glycerol 1-phosphate	− 2,200	

reference is given to subsequent pages in the text where the use of these compounds is illustrated.

Certain phosphate esters, in stark contrast to the above-mentioned metabolic intermediates, exert powerful inhibitory effects upon the transmission of the nerve impulse. The first such organophosphates were developed in Germany in the late 1930's in the course of research on antipersonnel nerve gases. An example of a nerve gas developed for warfare is diisopropyl fluorophosphate, DFP (Figure 21.4). The sheep kill near Dugway, Utah, in 1968 resulted from the accidental leakage of an organophosphate nerve gas from a test center several miles away.

The nerve gases affect all biological systems in a similar manner. They destroy the activity of a specific enzyme, *acetylcholine esterase,* that catalyzes the hydrolysis of acetylcholine. Acetylcholine is a quaternary ammonium salt present in an inactive, protein-bound form in nerve cells. The arrival of a nerve impulse leads to the release of the salt into the synaptic cleft (the gap between nerve cells). The acetylcholine molecules then diffuse across the synapse where they combine with specific receptor molecules on the adjacent nerve cell, causing the signal to be passed along. This process continues along the nerve fiber as the impulse is propagated. Once the impulse has been passed on, the acetylcholine must be immediately deactivated to leave the cell free to transmit the next impulse. The deactivation is accompanied by the hydrolysis of acetylcholine to choline and acetic acid, through the catalytic activity of *acetylcholine esterase*. It has been estimated that this enzyme-catalyzed hydrolysis reaction occurs in 40 μsec (40×10^{-6} sec).

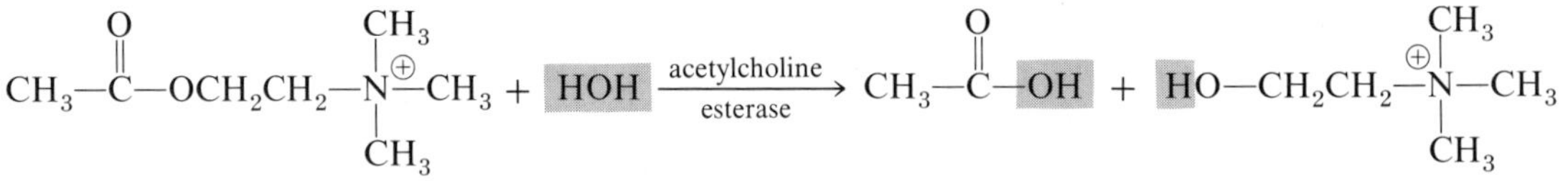

acetylcholine
esterase

Acetylcholine Acetic acid Choline

If the enzyme is inhibited and inactivation does not occur, acetylcholine accumulates on the cell, causing an overstimulation of the nerves and the glands and muscles that the nerves control. The movements of the entire body become uncoordinated; tremors, muscular spasms, convulsions, paralysis and death quickly follow. For this reason, the organic phosphate insecticides are among the most poisonous chemicals in the world. The structures of two of the common phosphorus-containing insecticides are given in Figure 21.4.

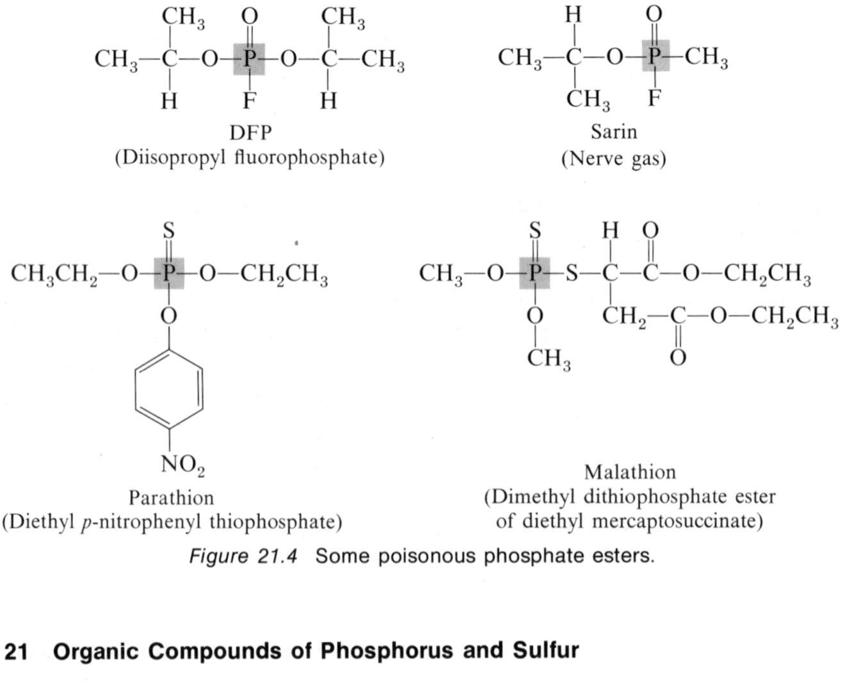

DFP
(Diisopropyl fluorophosphate)

Sarin
(Nerve gas)

Parathion
(Diethyl *p*-nitrophenyl thiophosphate)

Malathion
(Dimethyl dithiophosphate ester of diethyl mercaptosuccinate)

Figure 21.4 Some poisonous phosphate esters.

Parathion is one of the most powerful and dangerous of the organophosphate insecticides. Its only saving grace, albeit a small one, is that it decomposes rather rapidly. Unlike the chlorinated hydrocarbons, parathion will disappear from the environment within 15–20 days through biological degradation. (By contrast, at least half of the DDT may still be present five years after application.)

Since the banning of DDT, **malathion** has become one of the principal insecticides for use against household and garden insects. It is the least toxic of the organic phosphates to the higher animals because of the presence of an enzyme in the liver that quickly decomposes it. Malathion is widely employed to control fleas in dog kennels, and to protect dairy and other livestock.

Sulfur Compounds

Sulfur is located under oxygen in the same group of the periodic table and therefore we should expect to find a similarity among sulfur and oxygen compounds. A principal difference between the two series of compounds arises from the fact that sulfur exhibits a greater number of oxidation states than does oxygen. Thus, several important classes of organic compounds exist in which sulfur has a high positive oxidation number (+4 or +6). Table 21.2 presents a listing of the various types of organic sulfur compounds.

Table 21.2 Organic Sulfur Compounds

Class of Sulfur Compound	*General Formula*	*Specific Example*	*Nomenclature*
Thioalcohol (Mercaptan)	R—S—H	CH_3CH_2—S—H	Ethanethiol (Ethyl mercaptan)
Thiophenol	Ar—S—H	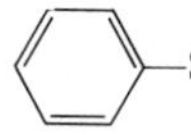—S—H	Thiophenol (Mercaptobenzene)
Thioether (Sulfide)	R—S—R′	CH_3—S—CH_3	Dimethyl thioether (Dimethyl sulfide)
Disulfide	R—S—S—R′	CH_3—S—S—C_2H_5	Methyl ethyl disulfide
Thioaldehyde	$R-\overset{\overset{\large S}{\|}}{C}-H$	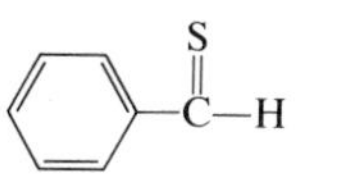	Thiobenzaldehyde
Thioketone	$R-\overset{\overset{\large S}{\|}}{C}-R'$	$CH_3-\overset{\overset{\large S}{\|}}{C}-CH_3$	Thioacetone

(*Continued*)

Table 21.2 (*Continued*)

Class of Sulfur Compound	*General Formula*	*Specific Example*	*Nomenclature*
Thioacid	R—C(=O)—S—H	CH_3—C(=O)—S—H	Monothioacetic acid
	R—C(=S)—S—H	CH_3—C(=S)—S—H	Dithioacetic acid
Thioester	R—C(=O)—S—R′	CH_3—C(=O)—S—C_2H_5	Ethyl thioacetate
Sulfoxide	R—S(=O)—R′	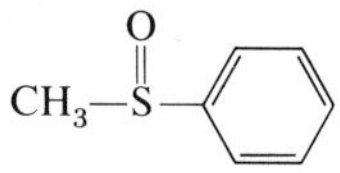	Methyl phenyl sulfoxide
Sulfone	R—S(=O)(=O)—R′	C_2H_5—S(=O)(=O)—C_2H_5	Diethyl sulfone
Sulfinic acid	R—S(=O)—OH	CH_3CH_2—S(=O)—OH	Ethanesulfinic acid
Sulfonic acid	R—S(=O)(=O)—O—H	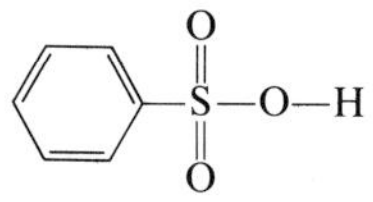	Benzenesulfonic acid
Sulfonate	R—S(=O)(=O)—O—R′	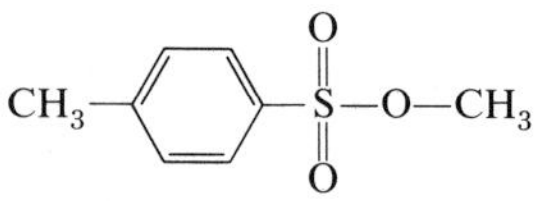	Methyl *p*-toluenesulfonate
Sulfonyl chloride	R—S(=O)(=O)—Cl	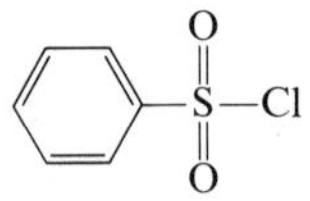	Benzenesulfonyl chloride
Sulfonamide	R—S(=O)(=O)—N(R)—R′	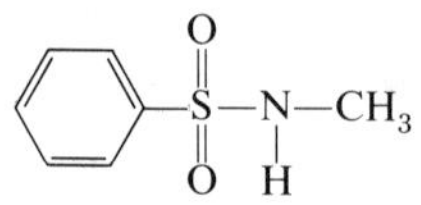	*N*-Methylbenzenesulfonamide
Sulfate	R—O—S(=O)(=O)—O—R′	CH_3—O—S(=O)(=O)—O—CH_3	Dimethyl sulfate

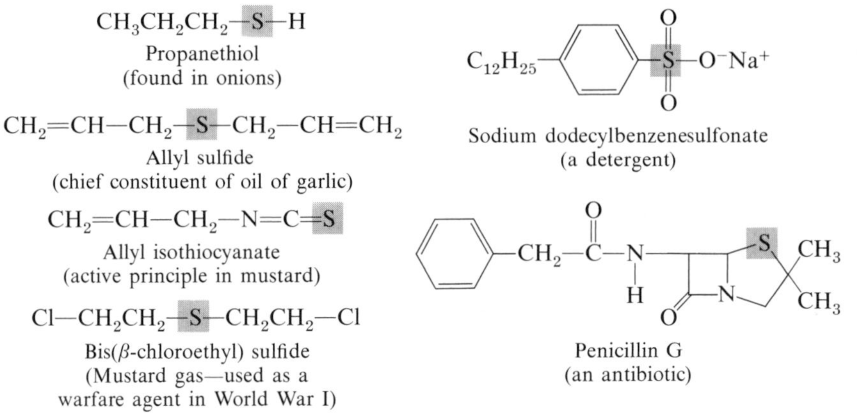
CH3CH2CH2—S—H
Propanethiol
(found in onions)
CH2=CH—CH2—S—CH2—CH=CH2
Allyl sulfide
(chief constituent of oil of garlic)
CH2=CH—CH2—N=C=S
Allyl isothiocyanate
(active principle in mustard)
Cl—CH2CH2—S—CH2CH2—Cl
Bis(β-chloroethyl) sulfide
(Mustard gas—used as a
warfare agent in World War I)
Sodium dodecylbenzenesulfonate
(a detergent)
Penicillin G
(an antibiotic)

Figure 21.5 Examples of sulfur-containing compounds.

(Notice that the term *thio* indicates the substitution of a sulfur atom for an oxygen atom.) Many of these compounds, such as the thioaldehydes, thioketones, and thioacids have only structural significance since they presently have no important uses. On the other hand, the thioalcohols, thioethers, and their oxidation products are of considerable importance to the organic chemist and biochemist. Examples of some sulfur-containing compounds are represented in Figure 21.5.

21.4 Derivatives of Hydrogen Sulfide

Thioalcohols, also known as thiols or mercaptans, are the monoalkyl derivatives of hydrogen sulfide, just as alcohols are the derivatives of water. The older term mercaptan (Latin, *mercurium captans,* mercury-seizing) derives from the ability of these compounds to form insoluble mercury salts, $Hg(SR)_2$. Many enzymes contain sulfhydryl groups (—SH), and the poisonous effects of mercuric salts are due to their ability to inactivate these enzymes by precipitation[4] (see Section 25.9-d).

The thioalcohols and thioethers are commonly prepared by the reaction of an alkyl halide with an alcoholic solution of an alkali salt of hydrogen

[4]The 1,2-dithiol derived from glycerin is called BAL (British antilewisite).

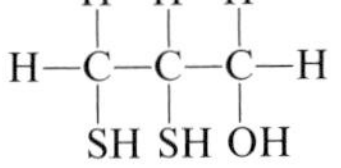

BAL

BAL is used as an antidote in heavy metal poisoning because the sulfhydryl groups form strong complexes with arsenic, mercury, silver, and lead, rendering these ions nontoxic.

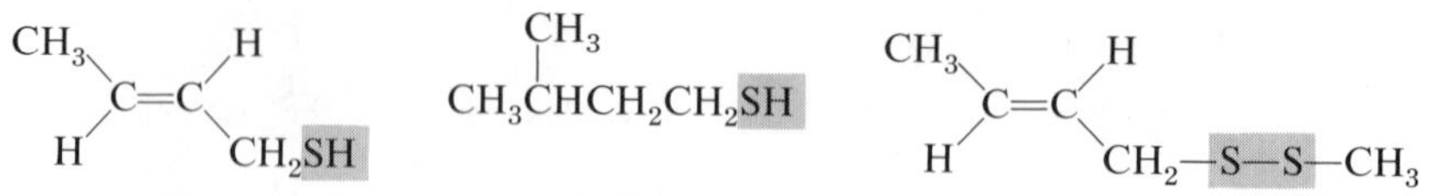

(a) (b) (c)

Figure 21.6 Skunk scent consists of three sulfur-containing compounds— (a) *trans*-2-butene-1-thiol, (b) 3-methyl-1-butanethiol, and (c) methyl-1-(*trans*-2-butenyl) disulfide—in a ratio of 4:3:3, respectively.

sulfide. These reactions are strictly analogous to the preparation of alcohols and ethers, that is, by nucleophilic displacement of an alkyl halide.

$$K^+SH^- + CH_3CH_2CH_2CH_2{-}Br \xrightarrow{\text{alcohol}} CH_3CH_2CH_2CH_2{-}SH + KBr$$

1-Butanethiol

$$CH_3CH_2CH_2{-}S^-Na^+ + CH_3{-}I \xrightarrow{\text{alcohol}} CH_3CH_2CH_2{-}S{-}CH_3 + NaI$$

Methyl *n*-propyl thioether
(Methyl *n*-propyl sulfide)

The thiols are characterized by their obnoxious odors. (Thioethers, on the other hand, do not have unpleasant odors.) Propanethiol is released from freshly chopped onions, and a mixture of three sulfur-containing compounds is responsible for the scent of skunks (Figure 21.6). The odor of ethanethiol can be detected at a dilution of 1 part per 50,000,000,000 parts air. Small amounts of thiols are added to natural gas so that the odor will make it obvious if a gas leak has occurred. The thiols boil at lower temperatures than the corresponding alcohols (e.g., ethanethiol boils at 35 °C compared to 78 °C for ethanol), and they are practically insoluble in water. This is indicative of the absence of hydrogen bonding between individual thiol molecules and between thiols and water molecules.

One of the chief distinguishing features of the thiols is that they are more acidic than their corresponding alcohols. (K_a of ethanethiol is 10^{-10} compared to a K_a of 10^{-16} for ethanol.) As a result, the sodium and potassium salts of thiols can be prepared from aqueous solutions of NaOH or KOH, whereas the corresponding salts of the alcohols require the use of sodium or potassium metal.

$$CH_3CH_2{-}S{-}H + NaOH \longrightarrow CH_3CH_2{-}S^-Na^+ + HOH$$

Sodium ethyl
mercaptide

This reaction is of practical use in the petroleum industry. Small amounts of organic sulfur compounds are often obtained in crude petroleum oils, and these are converted to thiols during the refining process. The thiols, of course, are objectionable because of their odor and the polluting nature of their combustion products (SO_2 and SO_3). They are removed from the petroleum products at the refineries by treatment with aqueous alkali. This process is referred to as *scrubbing*. The thiols are converted to nonvolatile, petroleum-insoluble salts that dissolve into the alkaline medium.

Another way that thiols and alcohols differ is in their behavior toward oxidizing agents. Recall that when alcohols are oxidized, it is the carbon bonded to the hydroxyl group, rather than the oxygen, that is oxidized (Section 17.4-d). When thiols are oxidized, the sulfur atom itself undergoes oxidation; the carbon chain remains unchanged. The product that is formed is dependent upon the type of oxidizing agent employed. Mild oxidizing agents such as air, ferric chloride, hydrogen peroxide, or iodine yield a disulfide. The disulfides can be considered to be the sulfur analogs of peroxides. They can be easily reduced to regenerate the thiols.

$$2\,CH_3{-}S{-}H + I_2 \longrightarrow \underset{\text{Dimethyl disulfide}}{CH_3{-}S{-}S{-}CH_3} + 2\,HI$$

$$CH_3{-}S{-}S{-}CH_3 \xrightarrow[\text{LiAlH}_4]{2\,[H]} 2\,CH_3{-}S{-}H$$

The interconversion of thiols and disulfides by oxidation-reduction reactions is utilized to a considerable extent in cellular metabolism. Two of the amino acids, cysteine and cystine, undergo such a transformation in many biochemical reactions (see Table 25.1). As we shall see in Section 25.6-c, disulfide bonds are found in proteins as a general aid to stabilization of the three-dimensional structure (tertiary structure).

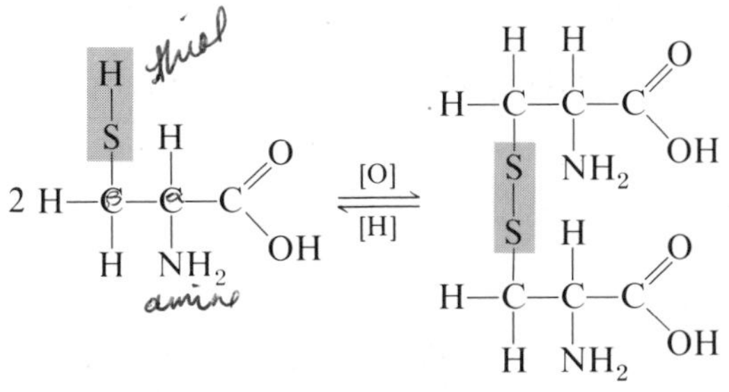

Cysteine Cystine

Another example is the oxidation and reduction of the coenzyme lipoic acid (see Table 26.2). Lipoic acid functions to catalyze certain oxidation reactions and in the process, it is reduced to the corresponding dithiol. Oxidation at a later stage by other components of the cell regenerates the lipoic acid.

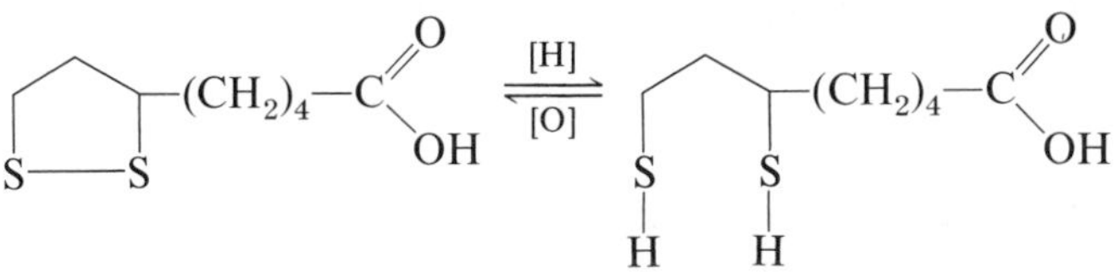

Lipoic acid Reduced form of lipoic acid

Thioethers, unlike ethers, are readily oxidizable. The oxidation of thioethers with hydrogen peroxide can yield either a *sulfoxide* or a *sulfone*.

$$CH_3-S-CH_3 \xrightarrow[H_2O_2]{H^+} CH_3-\overset{\overset{O}{\|}}{S}-CH_3 \xrightarrow[H_2O_2]{H^+} CH_3-\overset{\overset{O}{\|}}{\underset{\underset{O}{\|}}{S}}-CH_3$$

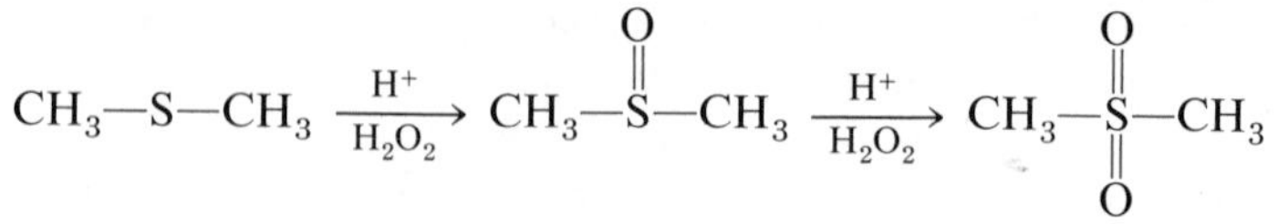

| Dimethyl sulfide | Dimethyl sulfoxide | Dimethyl sulfone |

Dimethyl sulfoxide (DMSO) is a highly polar solvent that is miscible with both polar and nonpolar compounds (i.e., water, alcohol, acetone, benzene, chloroform, ether). For this reason it has become a useful solvent for a multitude of organic reactions. Recently it has been utilized in place of water as a solvent in the study of certain enzymes. A commercially important disulfone is the drug Sulfonal. Sulfonal is a sedative and a soporific (agent that induces sleep).

$$C_2H_5-\overset{\overset{O}{\|}}{\underset{\underset{O}{\|}}{S}}-\overset{\overset{CH_3}{|}}{\underset{\underset{CH_3}{|}}{C}}-\overset{\overset{O}{\|}}{\underset{\underset{O}{\|}}{S}}-C_2H_5$$

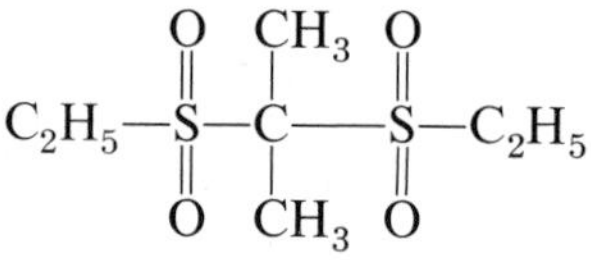

Sulfonal
(Diethylsulfondimethylmethane)

Thiols react with carboxylic acids to produce sulfur analogs of esters, called *thioesters*.

General Equation

$$R-\overset{\overset{O}{\|}}{C}-OH + \boxed{R'SH} \rightleftharpoons R-\overset{\overset{O}{\|}}{C}-\boxed{S-R'} + \boxed{HOH}$$

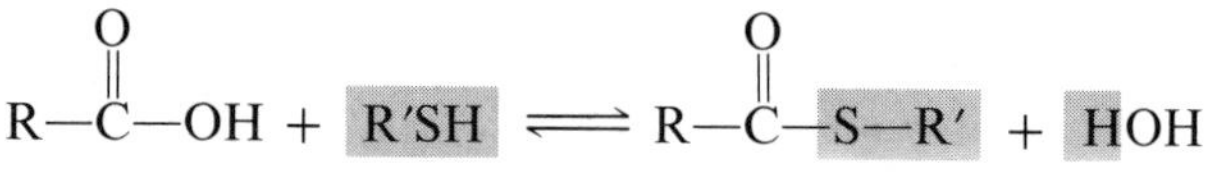

| Acid | Thiol | Thioester |

Specific Equation

$$CH_3-\overset{\overset{O}{\|}}{C}-OH + \boxed{CH_3-S-H} \underset{}{\overset{H^+}{\rightleftharpoons}} CH_3-\overset{\overset{O}{\|}}{C}-S-CH_3 + \boxed{HOH}$$

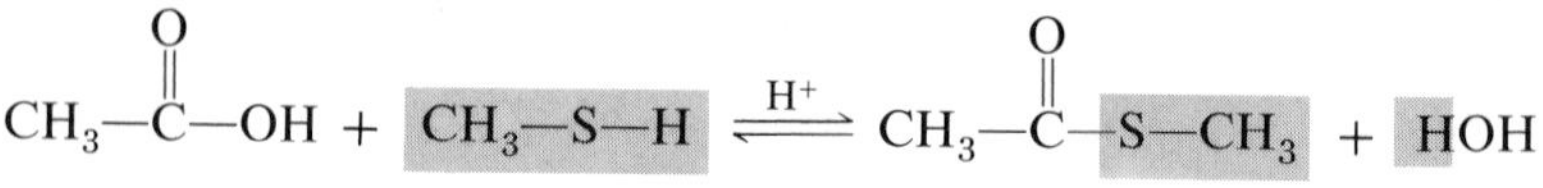

Methyl thioacetate

The fact that water is eliminated in this reaction, and not hydrogen sulfide, is consistent with the observation that in esterification reactions the hydroxyl group derives from the acid (Section 19.10).

Coenzyme A (abbreviated CoASH—see Table 26.2) is a complex sulfur-containing coenzyme that plays a central role in the metabolism of carbohydrates, fats, and proteins as well as in reactions associated with the transmission of nerve impulses and vision. The acetyl derivative of coenzyme A, called *acetyl CoA,* occurs in all living systems. The thioester linkage between the acetyl group and coenzyme A is a high energy bond. Because of its high degree of reactivity, acetyl CoA is a carrier of acetyl groups in the cell just as ATP is a carrier of high energy phosphate groups. Although the biochemical formation of acetyl CoA involves a series of reactions, we can conceptually visualize the process as a simple thioesterification (see Sections 28.7 and 29.2).

$$CH_3-\overset{\overset{\displaystyle O}{\|}}{C}-OH + \underset{\text{Coenzyme A}}{H-S-CoA} \longrightarrow \underset{\text{Acetyl CoA}}{CH_3-\overset{\overset{\displaystyle O}{\|}}{C}\sim S-CoA} + HOH$$

21.5 The Sulfonic Acids and Derivatives

The acid-catalyzed oxidation of thiols can lead to the formation of a series of acids. The final products of such an oxidation are compounds known as *sulfonic acids.*

$$\underset{\text{Ethanethiol}}{CH_3CH_2-SH} \xrightarrow{[O]} \underset{\text{Diethyl disulfide}}{CH_3CH_2-S-S-CH_2CH_3} \xrightarrow{[O]}$$

$$\underset{\text{Ethanesulfenic acid}}{CH_3CH_2-\ddot{\underset{..}{S}}-OH} \xrightarrow{[O]} \underset{\text{Ethanesulfinic acid}}{CH_3CH_2-\overset{\overset{\displaystyle O}{\|}}{S}-OH} \xrightarrow{[O]} \underset{\text{Ethanesulfonic acid}}{CH_3CH_2-\overset{\overset{\displaystyle O}{\|}}{\underset{\underset{\displaystyle O}{\|}}{S}}-OH}$$

On the other hand, you will recall that the aromatic sulfonic acids are prepared by the direct sulfonation of aromatic hydrocarbons (Section 15.3).

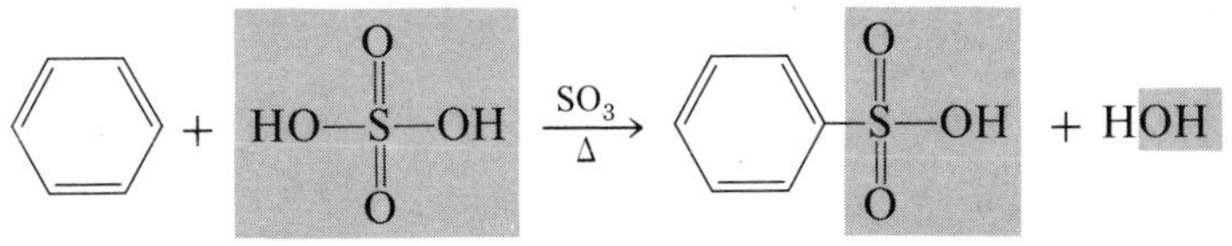

HO—S—OH
SO3
Δ
S—OH
+ HOH

Sulfuric acid   Benzenesulfonic acid

Sulfonic acids may be regarded as derivatives of sulfuric acid in which one hydroxyl group has been replaced by an alkyl or aryl group. In this respect they differ greatly from the organic esters of sulfuric acid (e.g., $CH_3CH_2OSO_3H$, in which a hydrogen atom is replaced by an organic group.

Like sulfuric acid, the sulfonic acids are extensively ionized, and are of comparable acid strength. Sulfonic acid groups are often introduced into organic molecules, such as dyes, to impart water solubility to these compounds (see Figure 20.4).

The sulfonic acids exhibit many of the chemical properties of the carboxylic acids. (The —SO_3H group of sulfonic acids is comparable to the —COOH group of carboxylic acids.) Thus, they are neutralized by bases, and are converted into the corresponding acid chlorides and esters.

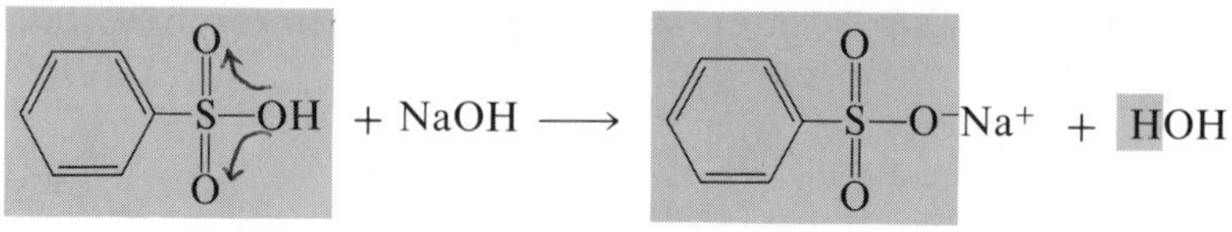

S—OH + NaOH ⟶ S—O⁻Na⁺ + HOH

Sodium benzenesulfonate

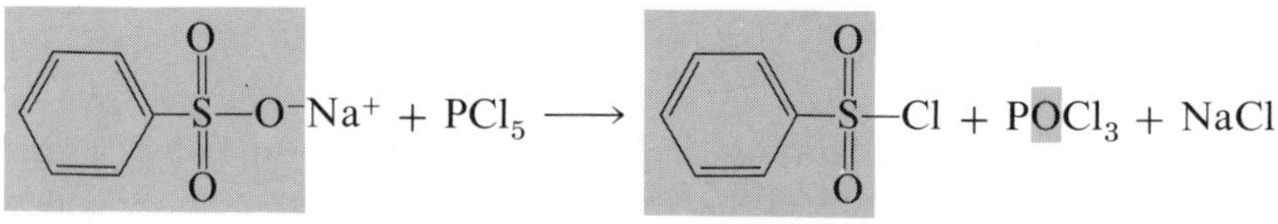

S—O⁻Na⁺ + PCl5 ⟶ S—Cl + POCl3 + NaCl

Benzenesulfonyl chloride

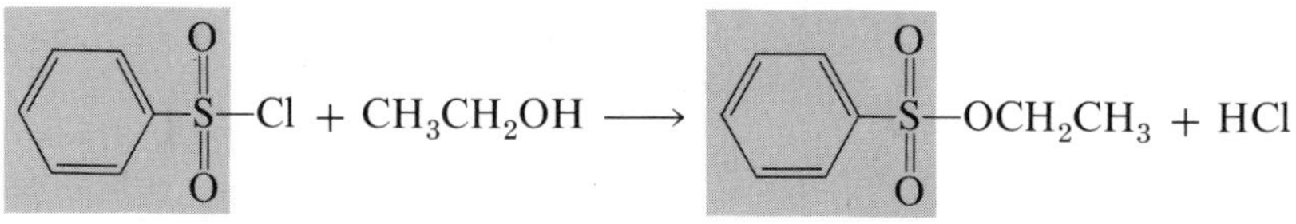

S—Cl + CH3CH2OH ⟶ S—OCH2CH3 + HCl

Ethyl benzenesulfonate

The amides and substituted amides of sulfanilic acid have achieved considerable importance in medicine as chemotherapeutic agents (see Section 26.9-a).

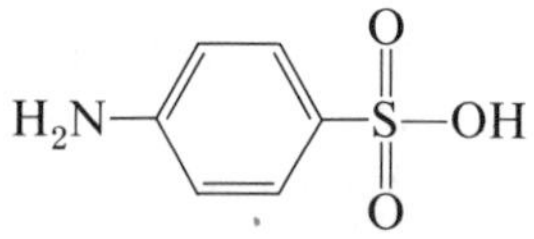

H2N—
S—OH

Sulfanilic acid
(*p*-Aminobenzene sulfonic acid)

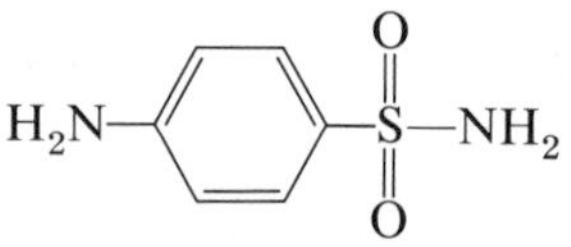

H2N—
S—NH2

Sulfanilamide
(*p*-Aminobenzene sulfonamide)

Exercises

21.1 Name the following compounds.

(a) $(C_2H_5)_3P$

(b) $(CH_3O)_3P$

(c) 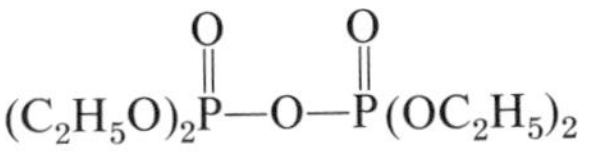

(d) $(C_3H_7O)_3PO$

(e) $CH_3{-}S{-}CH_2CH_3$

(f) 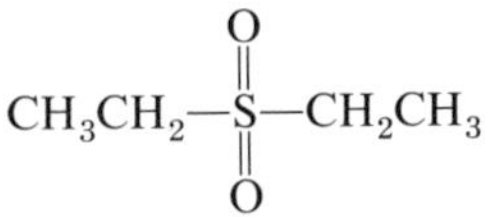

(g) 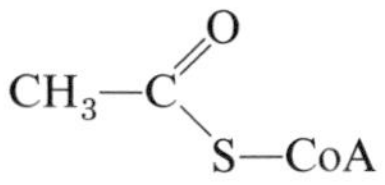

(h) 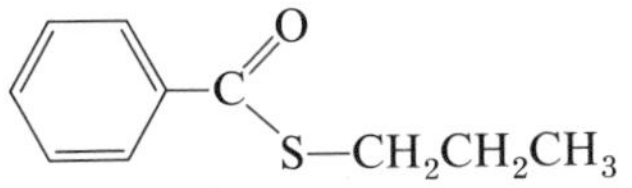

(i) 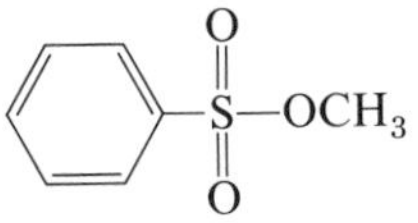

(j) 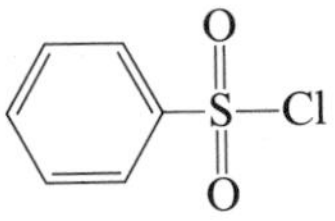

21.2 Write structural formulas for the following compounds.

(a) triisopropylphosphine
(b) dimethyl *n*-propylphosphonate
(c) ethyl dihydrogen phosphate
(d) diphenyl hydrogen phosphate
(e) ATP
(f) P_i
(g) 2-propanethiol
(h) DMSO
(i) *p*-toluenesulfonic acid
(j) acetyl phosphate
(k) potassium benzenesulfonate
(l) benzenesulfonamide

21.3 Define, explain, or give an example of each of the following.

(a) phosphate ester
(b) high energy bond
(c) nerve gas
(d) acetylcholine esterase
(e) malathion
(f) mercaptan
(g) scrubbing
(h) cysteine
(i) sulfonic acid
(j) coenzyme A

21.4 Which compound of each of the following pairs is the stronger base?

(a) NH_3 and PH_3
(b) $(CH_3)_3N$ and $(CH_3)_3P$
(c) $(C_6H_5)_3P$ and $(CH_3)_3P$
(d) $CH_3CH_2O^-$ and $CH_3CH_2S^-$
(e) 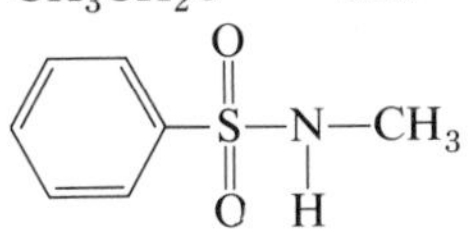 and 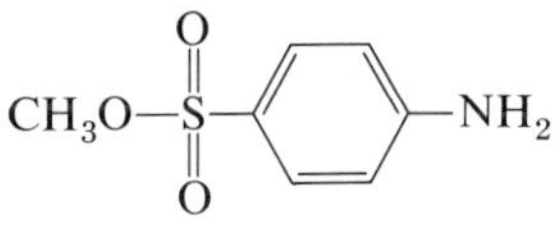

21.5 Which compound of each of the following pairs is the stronger acid?

(a) 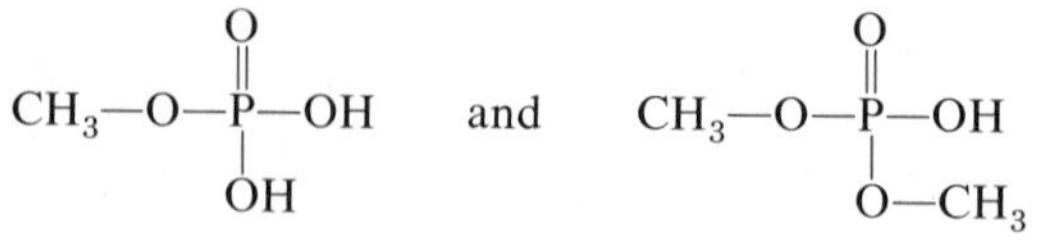

$$CH_3-O-\overset{\overset{\displaystyle O}{\|}}{\underset{\underset{\displaystyle OH}{|}}{P}}-OH \quad \text{and} \quad CH_3-O-\overset{\overset{\displaystyle O}{\|}}{\underset{\underset{\displaystyle O-CH_3}{|}}{P}}-OH$$

(b) CH_3SH and CH_3OH

(c) CH_3SH and ⬡—SH 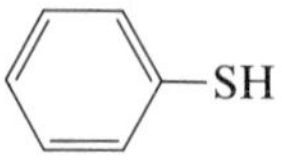

(d) 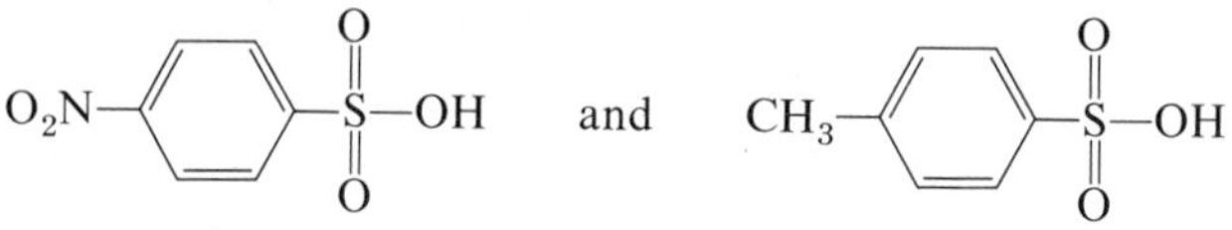

$$O_2N-⬡-\overset{\overset{\displaystyle O}{\|}}{\underset{\underset{\displaystyle O}{\|}}{S}}-OH \quad \text{and} \quad CH_3-⬡-\overset{\overset{\displaystyle O}{\|}}{\underset{\underset{\displaystyle O}{\|}}{S}}-OH$$

(e) 

$$⬡-\overset{\overset{\displaystyle O}{\|}}{\underset{\underset{\displaystyle O}{\|}}{S}}-OH \quad \text{and} \quad ⬡-\overset{\overset{\displaystyle O}{\|}}{C}\diagdown_{OH}$$

21.6 Arrange the compounds in each of the following groups in order of increasing solubility in water.

(a) $CH_3CH_2CH_2CH_2OH \quad CH_3CH_2CH_2CH_2SH$
$CH_3CH_2-S-CH_2CH_3 \quad CH_3CH_2CH_2CH_2S^-Na^+$

(b) $CH_3-\overset{\underset{\displaystyle CH_3}{|}}{P}-CH_3 \qquad CH_3-\overset{\underset{\displaystyle CH_3}{|}}{N}-CH_3 \qquad PH_3 \qquad (CH_3)_4P^+I^-$ 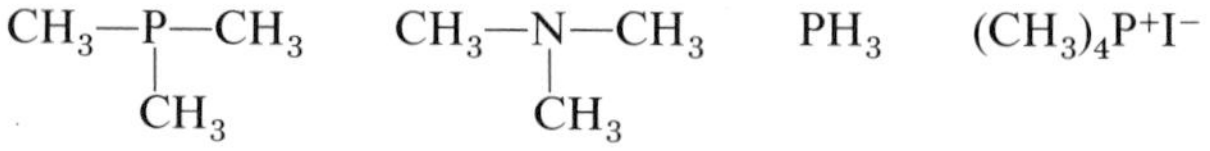

(c) $CH_3CH_2-\overset{\overset{\displaystyle O}{\|}}{S}-CH_2CH_3 \qquad CH_3SH$ 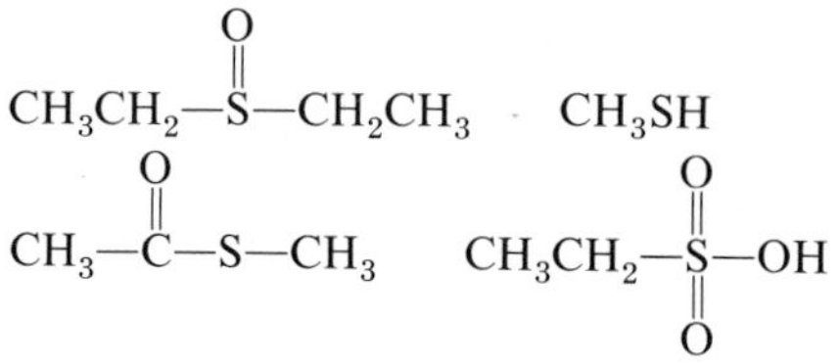

$CH_3-\overset{\overset{\displaystyle O}{\|}}{C}-S-CH_3 \qquad CH_3CH_2-\overset{\overset{\displaystyle O}{\|}}{\underset{\underset{\displaystyle O}{\|}}{S}}-OH$

21.7 Complete the following equations.

(a) $CH_3-⬡-MgCl + PCl_3 \longrightarrow$ 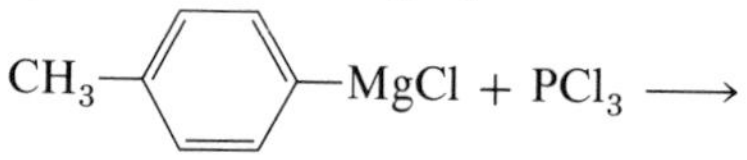

(b) $CH_3-\overset{\overset{\displaystyle H}{|}}{\underset{\underset{\displaystyle CH_3}{|}}{C}}-OH + POCl_3 \longrightarrow$ 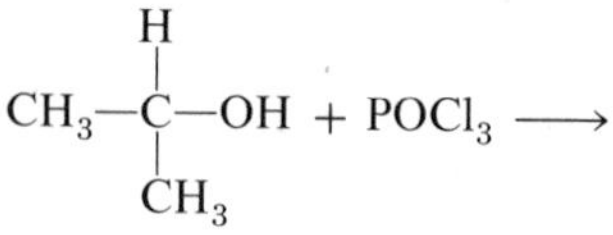

(c) $CH_3-\overset{\overset{\displaystyle O^{\ominus}}{\|}}{C}-O-\overset{\overset{\displaystyle O}{\|}}{\underset{\underset{\displaystyle OH}{|}}{P}}-OH + CH_3NH_2 \longrightarrow$ 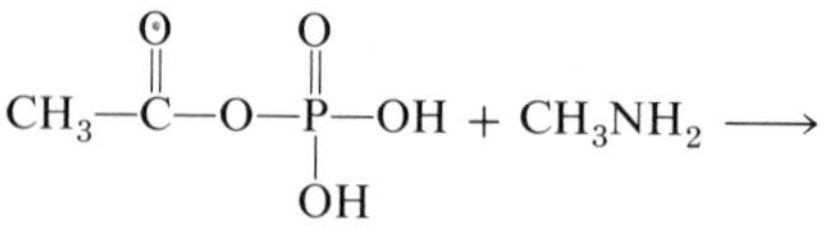

(d) $CH_3-\overset{\overset{\displaystyle O}{\|}}{C}-O-CH_2CH_2-\overset{\overset{\displaystyle \oplus}{\underset{\underset{\displaystyle CH_3}{|}}{N}}-CH_3}{\underset{}{\overset{\displaystyle CH_3}{|}}} + HOH \xrightarrow[\text{esterase}]{\text{acetylcholine}}$ 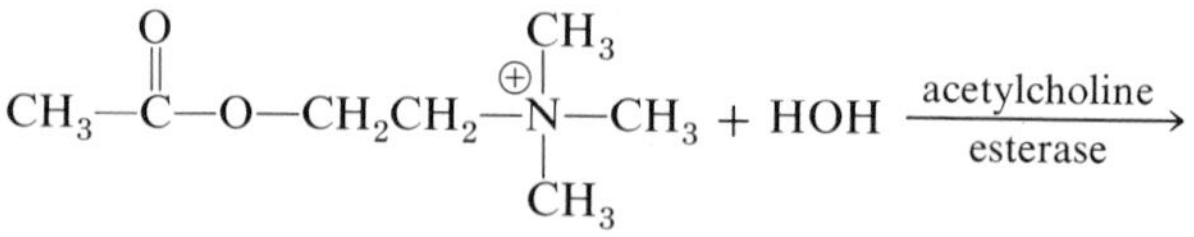

(e) $CH_3CH_2S^-Na^+$ + ⬡—CH_2Br $\xrightarrow{\text{alcohol}}$ 

(f) $CH_3SH + KOH \longrightarrow$

(g) $CH_3CH_2SH + H_2O_2 \longrightarrow$

(h) $CH_3CH_2-S-CH_2CH_3 + H_2O_2 \longrightarrow$

(i) ⬡—$\overset{\overset{\displaystyle O}{\|}}{C}$—OH + CH_3SH $\xrightarrow{H^+}$ $\rightleftharpoons$ 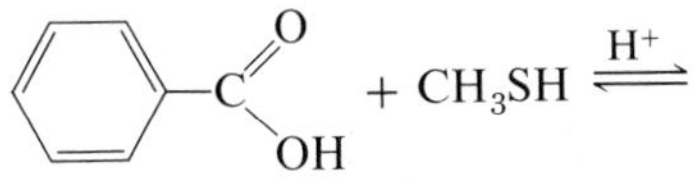

(j) ⬡ + H_2SO_4 $\xrightarrow[\Delta]{SO_3}$? $\xrightarrow{NaOH}$? $\xrightarrow{PCl_5}$ 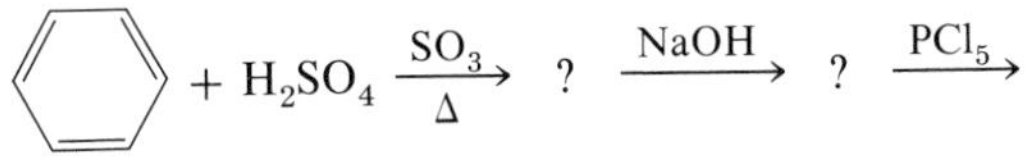

(k) $CH_3CH_2OH + H_2SO_4$ $\xrightarrow[\text{temperature}]{\text{room}}$

(l) CH_3—⬡—$\overset{\overset{\displaystyle O}{\|}}{\underset{\underset{\displaystyle O}{\|}}{S}}$—Cl + CH_3OH $\longrightarrow$ 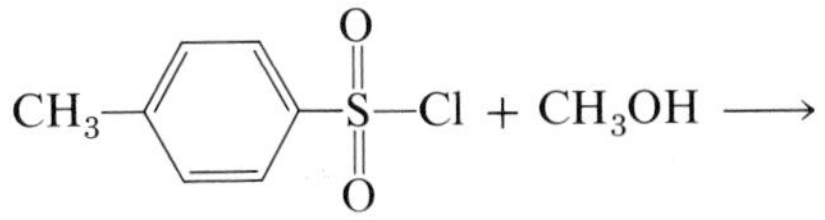

21.8 Devise a method that would enable you to distinguish each compound from the other in the following pairs.

(a) $CH_3-\underset{\underset{\displaystyle CH_3}{|}}{N}-CH_3$ and $CH_3-\underset{\underset{\displaystyle CH_3}{|}}{P}-CH_3$

(b) $CH_3-O-\underset{\underset{\displaystyle OH}{|}}{\overset{\overset{\displaystyle O}{\|}}{P}}-OH$ and $CH_3-O-\underset{\underset{\displaystyle O-CH_3}{|}}{\overset{\overset{\displaystyle O}{\|}}{P}}-O-CH_3$

(c) $CH_3-O-\overset{\overset{\displaystyle O}{\|}}{C}-CH_3$ and $CH_3-O-\underset{\underset{\displaystyle OH}{|}}{\overset{\overset{\displaystyle O}{\|}}{P}}-OH$

(d) CH_3CH_2OH and CH_3CH_2SH

(e) $CH_3-\overset{\overset{\displaystyle O}{\|}}{S}-CH_3$ and $CH_3-S-S-CH_3$

(f) $CH_3-\overset{\overset{\displaystyle O}{\|}}{C}-S-CH_3$ and $CH_3-\overset{\overset{\displaystyle O}{\|}}{C}-O-CH_3$

22

Stereoisomerism

Isomerism has been previously defined as the phenomenon whereby two or more *different* compounds are represented by *identical molecular* formulas. Isomeric molecules have different physical and chemical properties, and these differences are attributed to the existence of different structural formulas.

Two types of **structural isomers** have been mentioned—positional isomers and functional group isomers. *Positional isomers* result from the presence of an atom or a group of atoms at different positions on the carbon chain. These isomers were discussed in the chapters on alkanes, alkyl halides, and alcohols; examples are listed in Table 22.1.

Two molecules that have the same molecular formula but contain different functional groups are *functional group isomers*. Examples have been given for functional group isomers of alcohols and ethers, aldehydes and ketones, and acids and esters (Table 22.2).

The second major category of isomerism is **stereoisomerism,** or space isomerism. Unlike structural isomers, stereoisomers have the identical order

Table 22.1 Examples of Positional Isomers

Alkanes	*Alkyl Halides*	*Alcohols*
$CH_3CH_2CH_2CH_3$ *n*-Butane	CH_3CHCl_2 1,1-Dichloroethane	$CH_3CH_2CH_2OH$ 1-Propanol
$CH_3CH(CH_3)_2$ Isobutane	CH_2ClCH_2Cl 1,2-Dichloroethane	$CH_3CHOHCH_3$ 2-Propanol

Table 22.2 Examples of Functional Group Isomers

Alcohols and Ethers	*Aldehydes and Ketones*	*Acids and Esters*
CH_3CH_2OH	$CH_3CH_2C(=O)H$	$CH_3C(=O)OH$
Ethanol	Propanal	Acetic acid
and	and	and
CH_3OCH_3	$CH_3C(=O)CH_3$	$H-C(=O)OCH_3$
Dimethyl ether	Acetone	Methyl formate

of atoms and identical functional groups. *They differ only with respect to the spatial arrangement of atoms or groups of atoms within the molecule.* Therefore, **stereoisomers** are *isomers that have the same structural formulas, but different configurational formulas.* Stereoisomerism is possible whenever atoms or groups of atoms are in fixed positions relative to each other. There are two types of stereoisomerism—*optical isomerism* and *geometrical isomerism.*

22.1 Optical Isomerism

Optical isomers are *compounds that have identical physical properties* (melting points, boiling points, densities, etc.), *but differ with respect to their interaction with **plane-polarized** light.* Before continuing a discussion of optical isomerism, a review of the nature of plane-polarized light is necessary.

Ordinary light may be described as exhibiting an electromagnetic wave motion. As a light wave moves along in one direction, electromagnetic vibrations occur perpendicular to that direction (Figure 22.1a). If we ignore the electrical and magnetic components, we can simplify the situation and say that ordinary light consists of a multitude of light waves that can vibrate in all directions in the plane perpendicular to the direction of travel (Figure 22.1b).

If a beam of light is treated in such a manner as to allow only those waves which are traveling in one plane to be transmitted, the ensuing light beam is said to be *plane-polarized* (i.e., vibrating in a single plane, Figure 22.1c). Ordinary light is converted to plane-polarized light by passing through certain materials such as Polaroid (a plastic substance containing a finely divided crystalline substance) or calcite (a particular crystalline form of calcium carbonate). Both materials transmit only those light waves that are vibrating in a single plane, and deny passage to those waves vibrating in other planes.

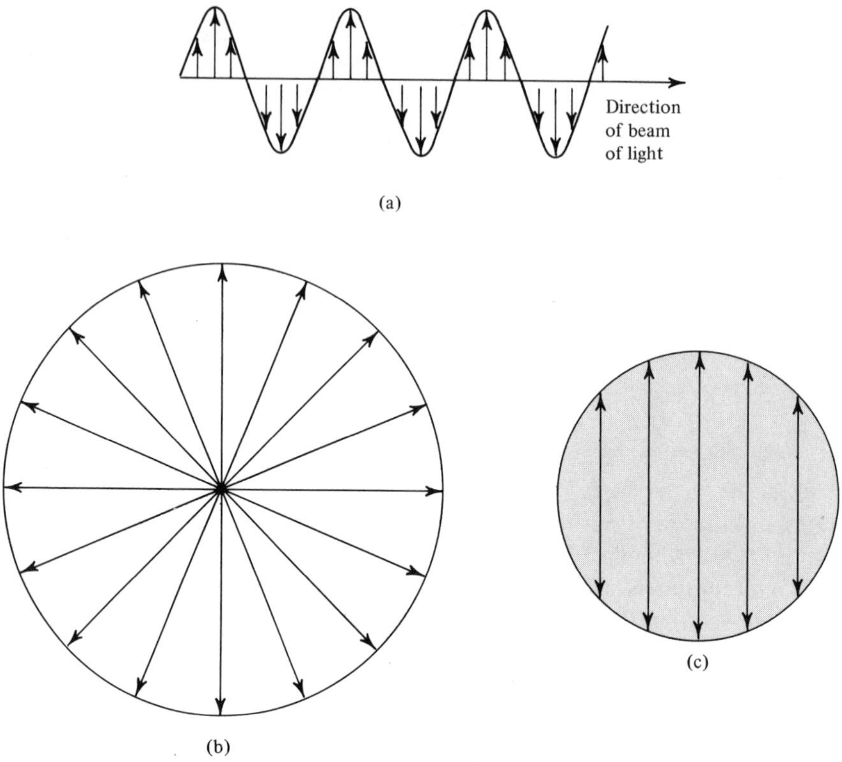
Direction
of beam
of light
(a)
(b)
(c)

Figure 22.1 (a) The waves in a beam of light vibrate in a direction perpendicular to the direction of propagation. (b) A cross-sectional view of a beam of monochromatic light vibrating in all planes. (c) A cross-sectional view of a beam of plane-polarized light.

A *polarimeter* is an instrument that detects and measures the effects of various substances upon plane-polarized light. A diagram of a polarimeter is given in Figure 22.2. The polarimeter consists of a monochromatic light source (i.e., single-wavelength light generated from a sodium or mercury lamp), two Nicol prisms (made of calcite), a tube to contain the sample,[1] and an eyepiece or phototube. They are arranged so that the light passes through the first prism, the polarizer, and is converted to plane-polarized light. The polarized light then passes into the sample tube. If the sample is *optically active,* it will rotate the polarized light a certain number of degrees either to the right or to the left. When it leaves the sample tube, the light will still be vibrating in one plane, but the plane will be at a different angle from the original plane. The rotated light next strikes the second, or analyzer, prism. The light will not pass through the analyzer prism until the prism is rotated the same number of degrees, in the same direction as the rotated light. A round dial, marked off in degrees, is attached to the analyzer. This permits

[1]The sample may be a pure gas, a pure liquid, a pure crystalline solid, or a solute dissolved in an appropriate solvent.

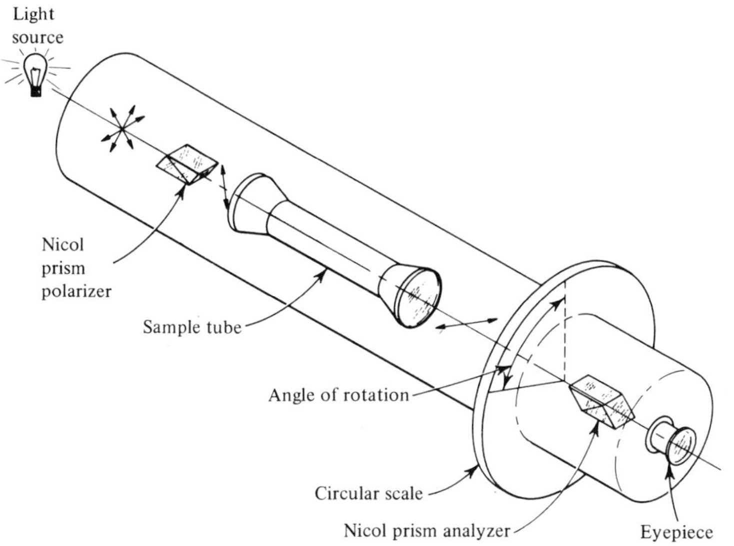
Light
source
Nicol
prism
polarizer
Sample tube
Angle of rotation
Circular scale
Nicol prism analyzer
Eyepiece

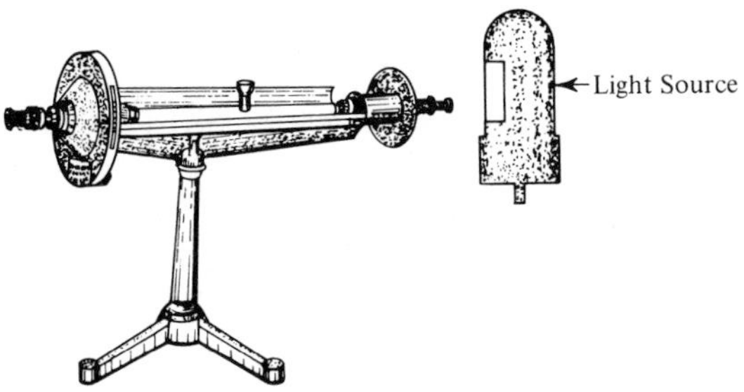
Light Source

Figure 22.2 A polarimeter

recording of the direction and number of degrees through which the analyzer must be rotated in order to align it properly with the polarized light. The *angle between the original and final planes of polarization* is known as the **optical rotation.**

The optical rotation of any sample depends upon the wavelength of the light source, the temperature, the length of the sample tube, the nature of the solvent (if any), and the structure of the compound. Therefore, chemists have established a set of standard conditions to provide for comparison of all optically active substances no matter what the concentration or phase of the substance. The **specific rotation,** $[\alpha]$, is defined as *the amount of rotation caused by 1 gram of substance per cubic centimeter in a sample tube 1 dm long.* The specific rotation is as important and just as characteristic a property of a compound as its melting point, boiling point, or density. For example, the specific rotation of an aqueous solution of sucrose is $[\alpha]_D^{20} = +66.5°$. Since specific rotation is affected by wavelength and temperature, these parameters should always be specified:

$$[\alpha]_{\mathrm{D}}^{t} = \frac{\alpha}{(l)(\text{gram/cm}^3)}$$

where $[\alpha]$ = specific rotation
D = sodium line of spectrum (5893 Å)
t = temperature
α = observed rotation
l = length of tube in decimeters
gram/cm^3 = concentration of solution or density of a liquid

22.2 Structure and Optical Activity

Optical activity has been defined as the ability of a substance to rotate the plane of polarized light. Conversely, if a substance does not deflect plane-polarized light, it is optically inactive. If light is rotated to the right (clockwise), the substance is **dextrorotatory** (Latin, *dexter,* right); those substances which rotate light to the left (counterclockwise) are **levorotatory** (Latin, *laevus,* left) (Figure 22.3). In denoting the direction of rotation, it is common practice to give a positive sign (+) to dextrorotatory substances and a negative sign (−) to levorotatory substances. Thus, sucrose is said to be dextrorotatory since it rotates plane-polarized light 66.5° in a clockwise direction.

So far, the properties of optical isomers have been described and certain terms have been defined for use in dealing with them. As yet, such fundamental questions as the following have not been answered. Why do some compounds exhibit optical isomerism? What is the spatial arrangement of these compounds? How do they differ from one another and from compounds that are not optically active?

The background to the discovery and explanation of optical isomerism is one of the most interesting sagas in the history of chemistry. The student is directed to a more comprehensive organic chemistry textbook, or, for full enlightenment, to the original papers of Pasteur (1848), Wislicenus (1873), van't Hoff, and Le Bel (both 1874). The key to their explanation of optical activity in organic compounds was the *tetrahedral carbon atom.*

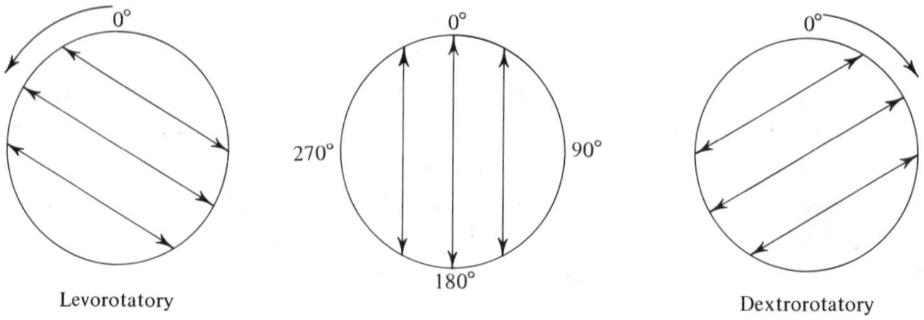
0°
0°
0°
270°
90°
180°
Levorotatory
Dextrorotatory

Figure 22.3 Direction of rotation of analyzer.

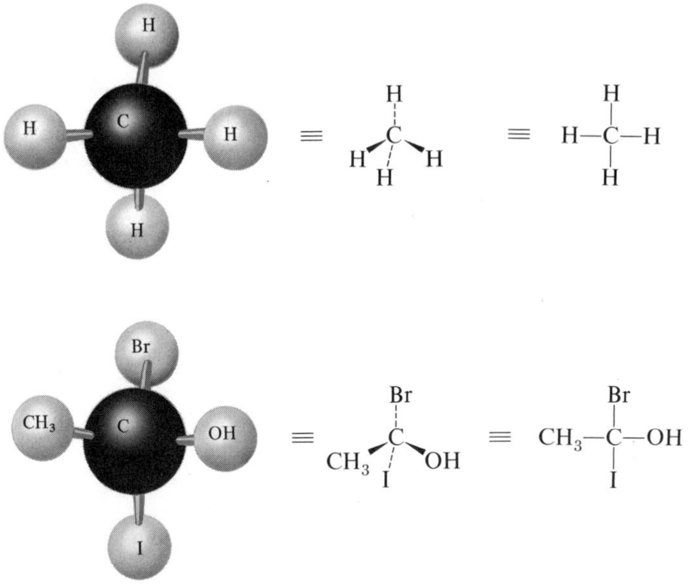
H
C
H
H
H
Br
CH3
C
OH
I

Figure 22.4 Tetrahedral configuration of methane and a tetrasubsituted compound.

Recall that the configuration of the methane molecule is tetrahedral. If the four atoms of hydrogen are substituted by other atoms or groups of atoms, the tetrahedral arrangement about the central carbon atom is still retained (Figure 22.4). If additional carbon atoms are added and if they all contain single bonds, the configuration about each of the carbon atoms will be tetrahedral.

The geometrical configuration that is imposed upon all singly bonded carbon atoms is the primary factor in optical isomerism. Consider the two generalized molecules shown in Figure 22.5. Compound I contains two identical substituents and two different substituents bonded to the central carbon, whereas compound II contains four dissimilar substituents. Compound II is an example of an *asymmetric* molecule. Any organic molecule in which there are four *different* atoms or groups of atoms bonded to a carbon atom is an asymmetric molecule.[2] Asymmetry (dissymmetry) is defined as nonsuperimposability upon a mirror image. To understand why compound I is classified as a symmetric molecule and compound II as an asymmetric molecule, we must place both compounds before a mirror and attempt to impose the mirror images upon the original molecules (Figure 22.6). The bonds may be twisted and turned but none of them may be broken. Because of inexperience in thinking in three-dimensional terms, the beginning student will fully understand this concept only after working with molecular models.

[2] Several texts use the term *chiral* (Greek, *cheir,* hand, hence chirality means "handedness") in place of dissymmetric and asymmetric. Molecules that are not superimposable on their mirror images are said to be chiral; that is, they possess chirality and thus are optically active. Molecules that are superimposable on their mirror images are *achiral,* and they are not optically active.

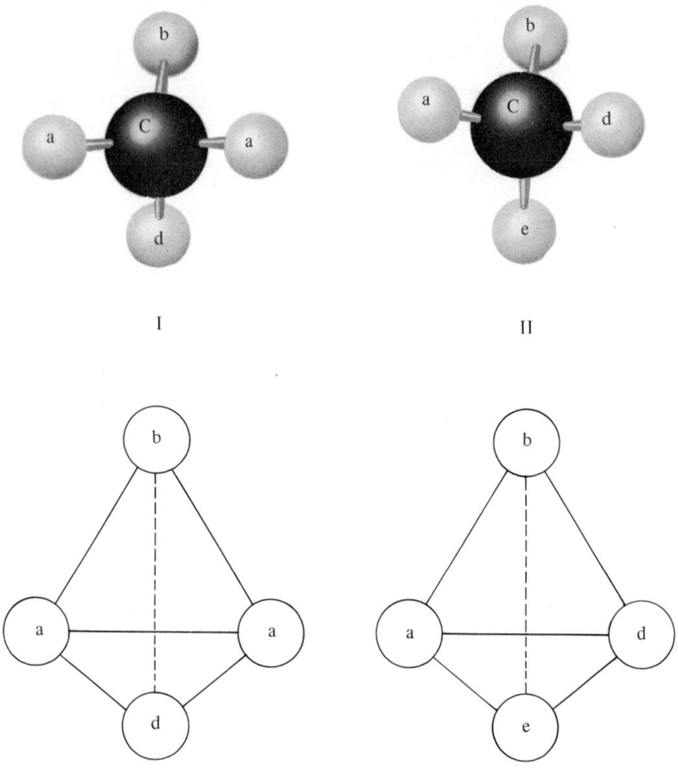
b
C
a
a
d
I
b
C
a
d
e
II
b
a
a
d
b
a
d
e

Figure 22.5 Compound I contains two similar and two dissimilar substituents. Compound II contains four dissimilar substituents.

A useful analogy can be drawn between nonsuperimposable mirror-image compounds and your right and left hands (Figure 22.7). Regardless of how you twist and turn them, you cannot superimpose your right hand upon your left, or vice versa. This is so because while your right and left hands are nearly perfect mirror images of each other, they are *not* identical. The difference becomes immediately apparent if you try to place your right hand in a left-hand glove.

Optical isomers that are nonsuperimposable mirror images are called **enantiomorphs** (Greek, *enantios,* opposite; *morph,* form), which is often abbreviated to **enantiomers.** Enantiomers have identical physical and chemical properties, and they rotate plane-polarized light the same number of degrees, *but in opposite directions.* For example, two forms of lactic acid, $CH_3CHOHCOOH$, are known to exist in nature. One form, isolated from muscle tissue, is identical in all respects to another form that is isolated from yeast, but the former is dextrorotatory ($[\alpha]_D^{20} = +2.3°$) and the latter is levorotatory ($[\alpha]_D^{20} = -2.3°$). The second carbon atom is the only one that is bonded to four different substituents, and it is therefore the one that is responsible for the asymmetry of the molecule.

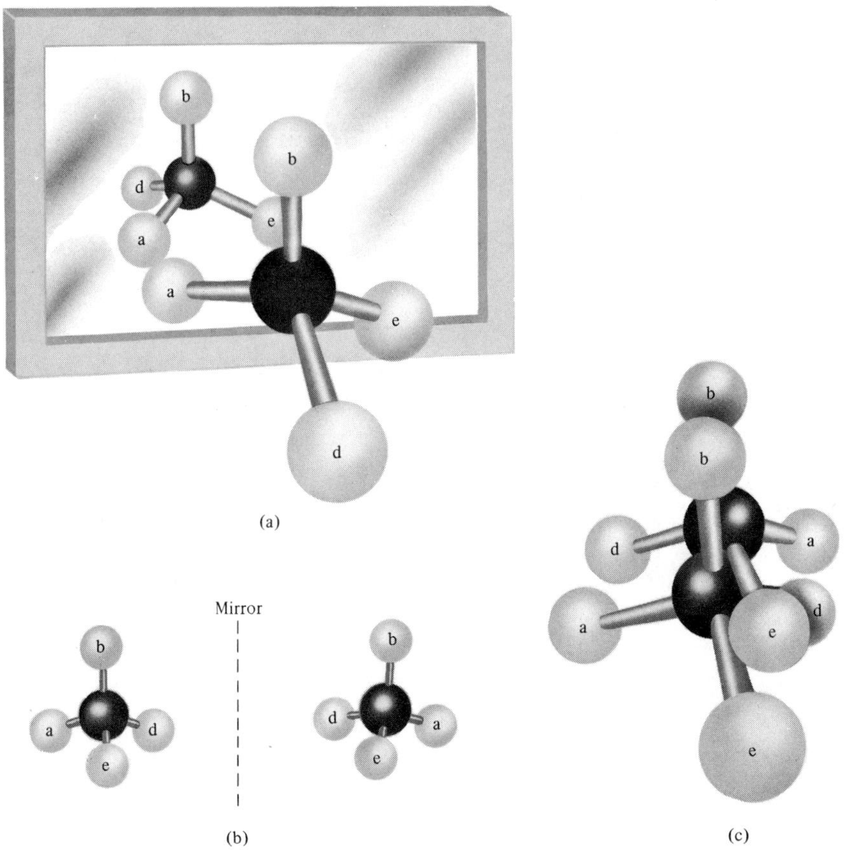
b
b
d
e
a
a
e
d
(a)
Mirror
b
a
d
e
b
d
a
e
(b)
b
b
d
a
a
e
d
e
(c)

Figure 22.6 (a) and (b) Mirror images of an asymmetric molecule. (c) Attempt to superimpose mirror images.

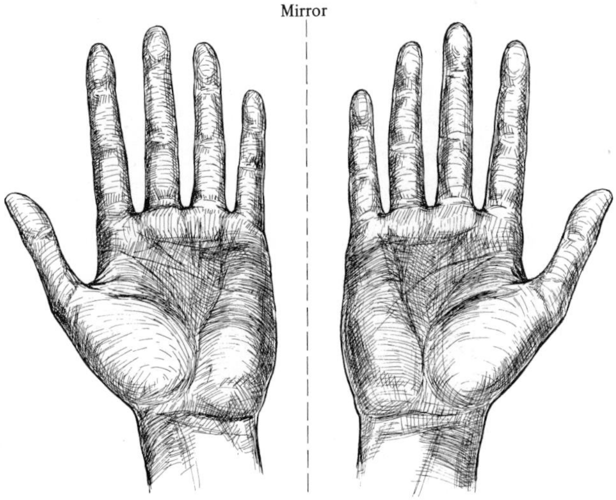
Mirror

Figure 22.7 The left and right hands are nonsuperimposable mirror images.

(a)

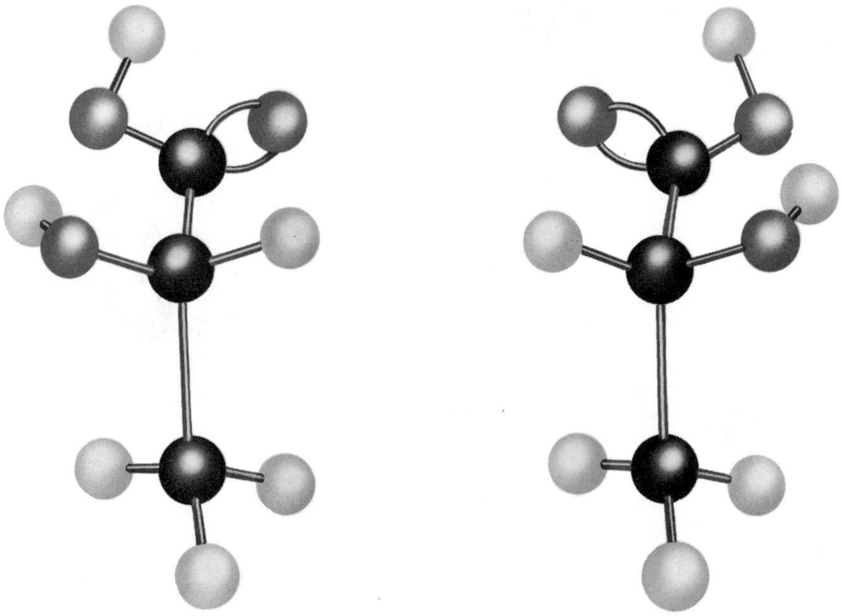

(b)

COOH
HO—C—H
CH3
≡
COOH
HO—C—H
CH3
COOH
H—C—OH
CH3
≡
COOH
H—C—OH
CH3

Dextro lactic acid

Levo lactic acid

(c)

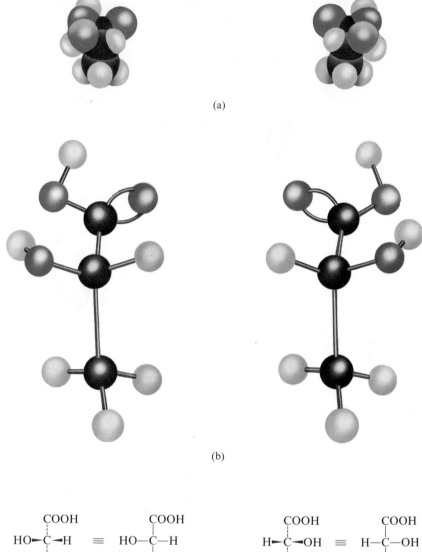

Figure 22.8 Enantiomers of lactic acid. (a) Space-filling models; (b) ball-and-stick models; (c) structural formulas.

It is preferable to draw perspective formulas when representing enantiomers since these formulas convey the three-dimensionality of the molecules. This is especially important because the very nature of stereoisomerism depends on the three-dimensional spatial arrangement of the atoms in a molecule. However, when we deal with most organic and biochemical molecules, writing perspective formulas is a cumbersome task. A convention has therefore been established to represent optical isomers meaningfully by the use of their structural formulas. It has been arbitrarily decided that substituents written at either side of the asymmetric carbon project forward toward the reader, while those positioned above and below the carbon extend backward into the page. The asymmetric carbon itself lies in the plane of the paper. The enantiomers of lactic acid are illustrated in Figure 22.8.

Thus far we have mentioned those compounds that contain only one asymmetric carbon atom. For such compounds there always exist two optical isomers, a dextrorotatory and a levorotatory form; these isomers are nonsuperimposable mirror images (enantiomers). A convenient method of drawing the enantiomer of an optically active compound is to maintain the position of two of the substituents (usually the larger ones) about the asymmetric center and invert the positions of the other two.

Example 22.1 2-Methyl-1-butanol exists in two optically active forms. The specific rotation of one is +5.756°, and the other −5.756°. Draw structural formulas of the enantiomers.

1. The structural formula for one enantiomer is written and the asymmetric center is identified.

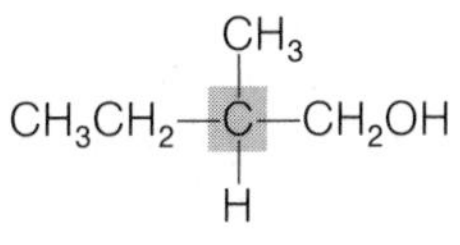

I

2. The positions of two of the substituents about the asymmetric center are inverted, while the positions of the other two are maintained.

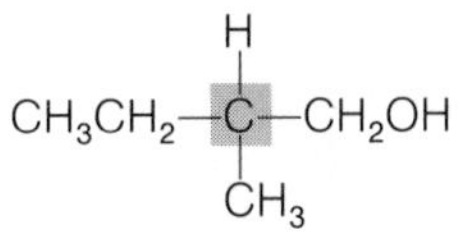

II

It is not possible to tell which isomer is dextrorotatory and which is levorotatory merely by inspection of the structural formulas. The distinction can only be made by measuring the specific rotation of each compound in a polarimeter.

A word of caution is required with this shorthand method of representing the three-dimensional arrangement of the atoms in an optical isomer. The unwary student must not be misled into thinking that he can superimpose the two formulas (I and II) simply by flipping one over on top of the other.

Figure 22.9 Enantiomers are nonsuperimposable mirror images.

"Mirror, mirror on the wall, who is the enantiomerest of them all?"

Identical two-dimensional formulas would then result. But molecules are three-dimensional; the substituents positioned above and below the flipped molecule would project *forward* and those positioned on either side would project to the *rear,* thus the molecules would *not* coincide. When molecules are compared to see whether or not they are optical isomers, convention dictates that two-dimensional structural formulas may only be rotated *in the plane* of the paper; they may never be flipped over or lifted out of the plane (Figure 22.9).

Table 22.3 lists some optically active and inactive compounds from several classes of organic compounds. A worthwhile exercise is the drawing of the enantiomers for each of the optically active compounds listed.

22.3 Diastereomers

Molecules that contain two or more asymmetric carbons can exist in more than two stereoisomeric forms. The first chemist to realize this was van't Hoff, who formulated a statement (van't Hoff's rule) allowing the prediction of the total number of possible optical isomers for a molecule with more than one asymmetric center. *The maximum number of different configurations is* 2^n, where n is the number of asymmetric carbon atoms. The rule is best illustrated with a specific example.

Table 22.3 Optically Active and Inactive Organic Compounds

Family	*Optically Active*	*Optically Inactive*
Alkane	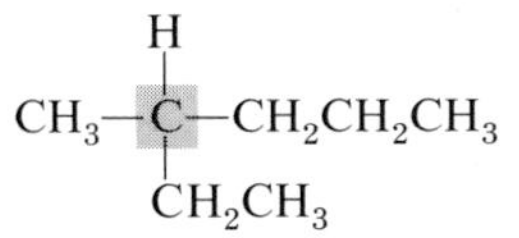	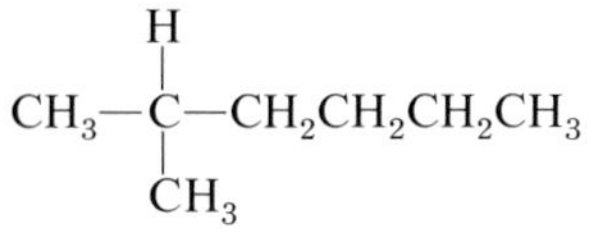
Alkyl halide	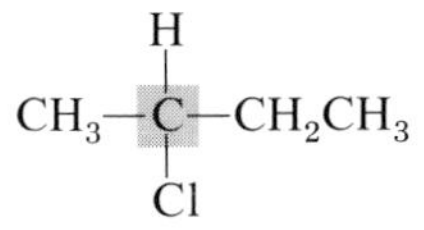	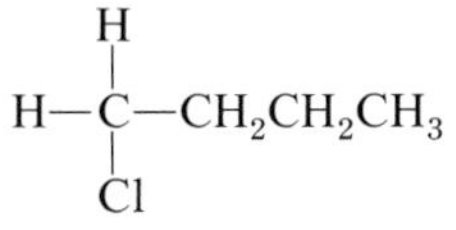
Alcohol	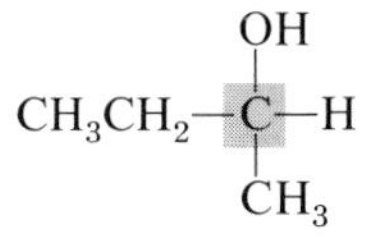	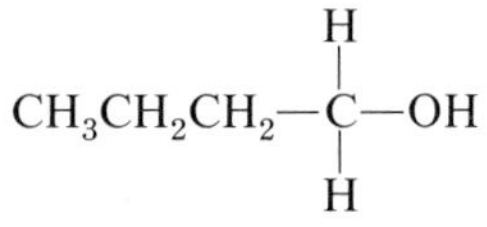
Aldehyde	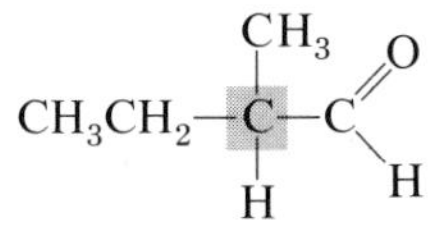	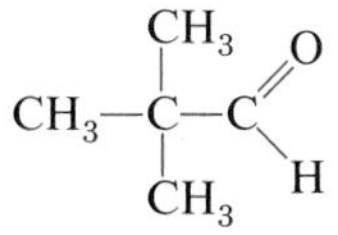
Ketone	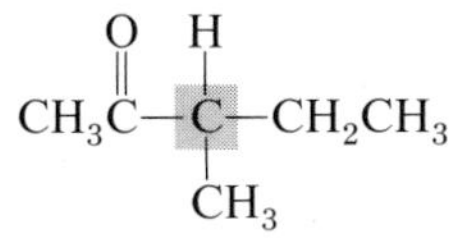	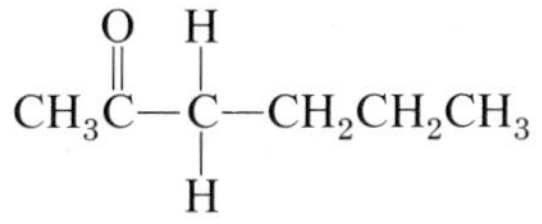
Amine	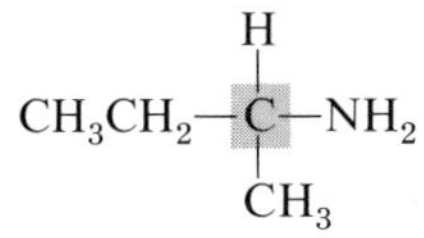	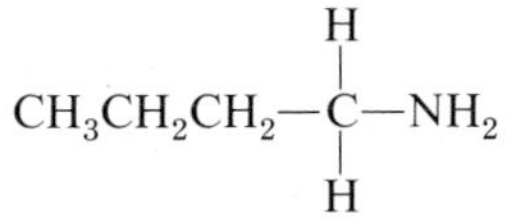
Hydroxy acid	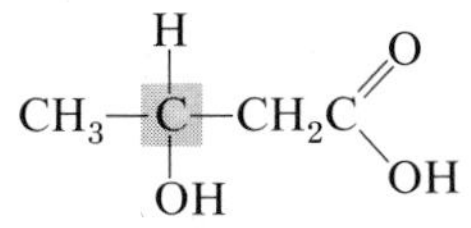	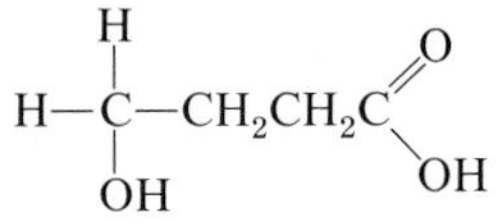
Amino acid	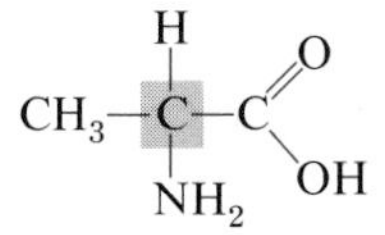	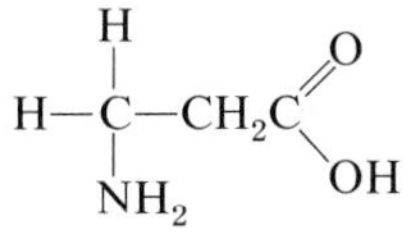
Ester	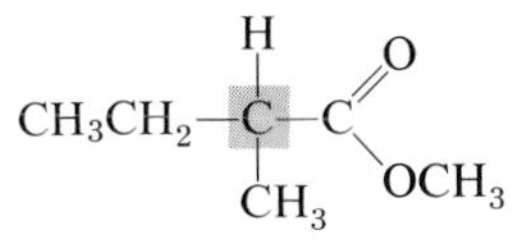	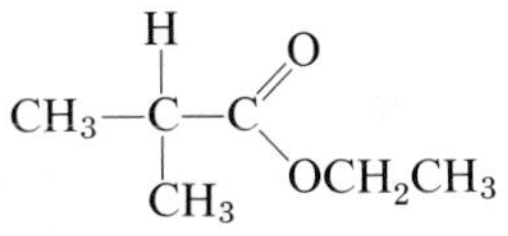

Example 22.2 Draw all the possible stereoisomers of 2-methyl-1,3-butanediol.

1. The structural formula is first drawn, and the number of asymmetric carbon atoms is noted. By use of the formula 2^n, the number of possible optical isomers is calculated.

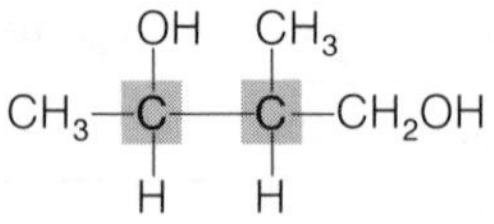

There are two asymmetric carbons and therefore there are 2^2 or 4 possible optical isomers.

2. Since two configurations are possible for each asymmetric carbon, there will be two sets of mirror images.

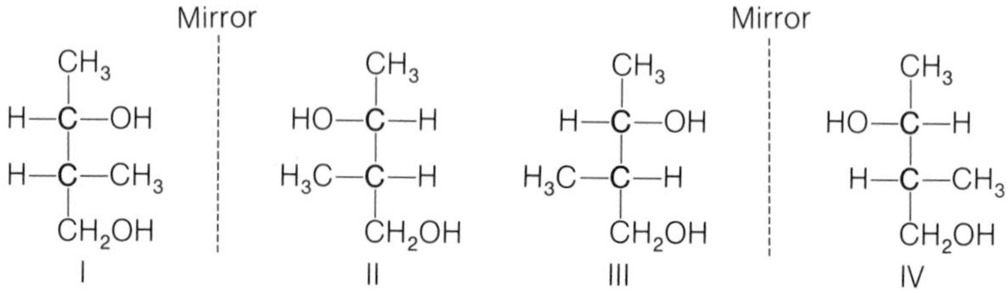
Mirror
Mirror

I II III IV

Structures I and II, and III and IV represent pairs of enantiomers[3] since they are nonsuperimposable mirror images. They have identical melting points, boiling points, densities, solubilities, etc. They rotate plane-polarized light to the same extent, but in opposite directions. This is not true of structures I and III, I and IV, II and III, or II and IV. They are all considered to be optical isomers (isomers because they have the same molecular formula, optical because they rotate the plane of polarized light), but they are not mirror images. Such pairs of optical isomers are **diastereomers.** Diastereomers, then, are *optical isomers that are not mirror images of each other.* Diastereomers do not have identical physical properties, nor do they rotate plane-polarized light to the same extent. They will have the same type of chemical properties, but the rate at which they react may be different.

22.4 Racemic Mixtures and Meso Compounds

We have mentioned that enantiomers rotate plane-polarized light the same number of degrees in different directions. One would predict, therefore, that a mixture containing equal amounts of dextrorotatory and levorotatory isomers would be optically inactive since the rotation caused by the molecules of one form would be exactly canceled by the rotation caused by the molecules of the other form. Experimental data bear out this prediction. A mixture containing equal amounts of a pair of enantiomers is called a **racemic mixture** or a **racemate;** it is symbolized by the notation (*dl*) or ($\pm$). A racemic mixture is optically inactive because of *external compensation.* There are just as many molecules rotating the plane-polarized light to the right as to the left; hence the net rotation observed in the polarimeter is zero.

Another phenomenon of stereoisomerism results when a molecule

[3]The number of enantiomeric pairs of optical isomers will always be equal to $2^n/2$.

contains more than one *similar* asymmetric carbon atom. Similar asymmetric carbon atoms are those in which the four unlike substituents bonded to one asymmetric carbon atom are identical to the four bonded to the other asymmetric carbon atom. Tartaric acid, HOOC—CHOH—CHOH—COOH, is the classic example of this phenomenon. The molecule contains two asymmetric carbon atoms, and four optical isomers might be predicted.

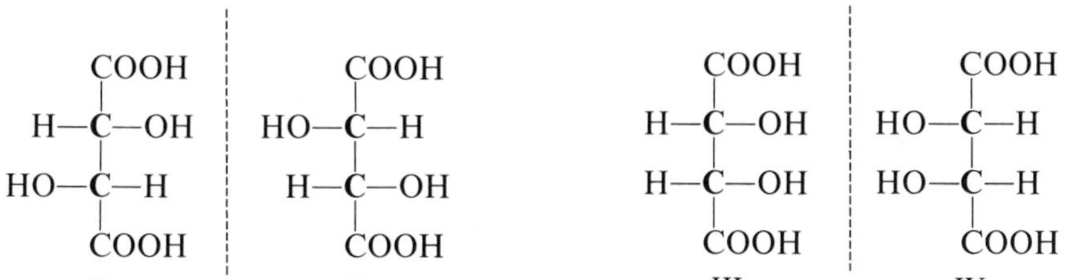

I II III IV

Enantiomers Identical

Structures I and II are indeed nonsuperimposable mirror images, and thus are both optically active. However, although structures III and IV are mirror images of one another, they are *superimposable* mirror images. This can be easily seen by rotating either structure 180° in the plane of the paper. Both structures are identical, and this one compound is an optically inactive diastereomer of both compound I and compound II. Furthermore, observe that if the bond between carbon number 2 and carbon number 3 of compounds III and IV is bisected by a mirror, the top half of the molecule is the mirror image of the bottom half. No similar plane of symmetry can be drawn through structures I and II.

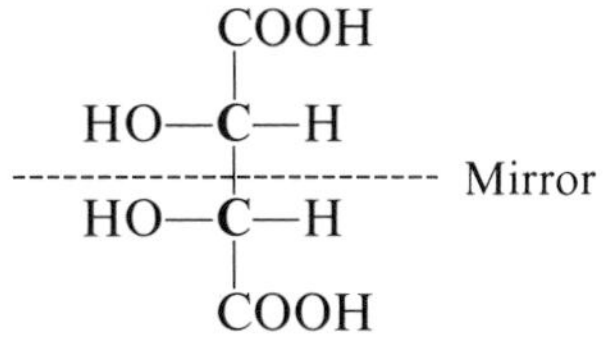
Mirror

The existence of a plane of symmetry indicates that a molecule is symmetric and cannot exist in an enantiomeric form. The symmetrical structure is referred to as *meso* tartaric acid.[4] It is optically inactive because

[4]A *meso* compound is characterized by an internal plane of symmetry. The molecule is superimposable on its mirror image even though it contains asymmetric carbon atoms. For this reason, it is incorrect to say that all molecules that contain an asymmetric carbon atom are asymmetric and thus optically active. Another example of a meso compound is ribitol, whose structural formula is

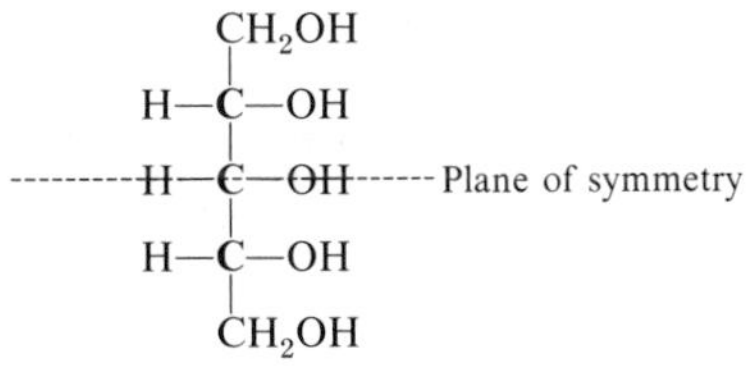
Plane of symmetry

Table 22.4 Physical Properties of Tartaric Acid Isomers

	MP (°*C*)	*Density* (*grams/cm*^{3})	*Solubility* (*grams/100 grams* H_2O)	$[\alpha]_D^{20}$ *water*
Dextro (*d*) tartaric acid	170	1.76	139	+12°
Levo (*l*) tartaric acid	170	1.76	139	−12°
Meso tartaric acid	140	1.67	125	0

of *internal compensation.* Any effect that the upper half of the molecule will have upon plane-polarized light will be exactly opposed to the effect of the lower half. As expected, the physical and chemical properties of meso tartaric acid are different from those of the optically active dextro and levo forms. Table 22.4 lists some properties of the isomeric tartaric acids.

22.5 The R-S Convention for Absolute Configuration

A problem arises in naming stereoisomers of compounds that contain more than one asymmetric carbon atom. That is, there is no evident correlation between the sign of optical rotation and the absolute configuration of the molecule. It therefore became necessary to devise a convention for the assignment of a specific designation to a given configuration. Such a system, referred to as the **R-S** convention, has been introduced by R. S. Cahn, C. Ingold, and V. Prelog. This system, which has been adopted by organic chemists, is not yet widely used in biochemistry, where the D- and L-notations still prevail (see Section 23.3).

The two configurations for an asymmetric molecule are designated **R** (Latin, *rectus,* right) and **S** (Latin, *sinister,* left) and are governed by a set of rules that establish a priority sequence for the four substituents.

1. Assign an order of priority, *a, b, c,* or *d,* to the atoms bonded directly to the asymmetric carbon with the atom of *highest atomic number* getting the highest priority such that $a > b > c > d$ (i.e., for the halogens the priority sequence is I > Br > Cl > F; other priorities are equally obvious, e.g., Cl > O > N > C).
2. Once priorities are assigned, visualize the orientation of the molecule in such a manner that it approximates the steering wheel of a car. The bond joining the asymmetric carbon atom to the substituent of lowest priority is considered to be the steering column. The other three substituents are considered to be the spokes of the steering wheel. Trace a path from *a* to *b* to *c*. If the direction of this path is clockwise, then the molecule is assigned the **R** configuration. If the path follows a counterclockwise direction, the molecule is assigned the **S** configuration.

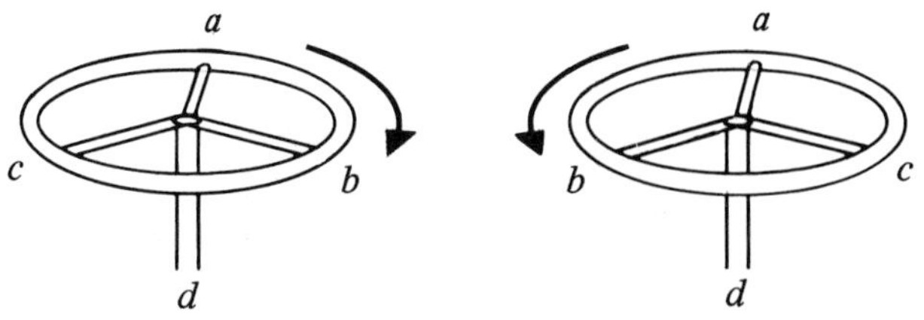

a
c
b
d
a
b
c
d

R　　　S

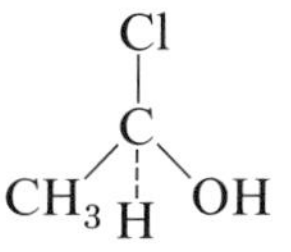

(R)-1-Chloroethanol

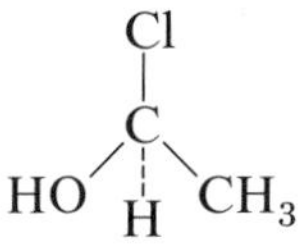

(S)-1-Chloroethanol

3. When two or more substituents on the asymmetric carbon atom contain the same atoms, priority is given to the substituent with the highest atomic number in the second atom from the asymmetric carbon. For example, the compound 2-aminobutane contains two carbons bonded to the asymmetric carbon atom. However, the C of the $—CH_2CH_3$ group is bonded to a carbon, whereas the C of the $—CH_3$ group is bonded to a hydrogen. The priority order for 2-aminobutane is $NH_2 > CH_2CH_3 > CH_3 > H$.

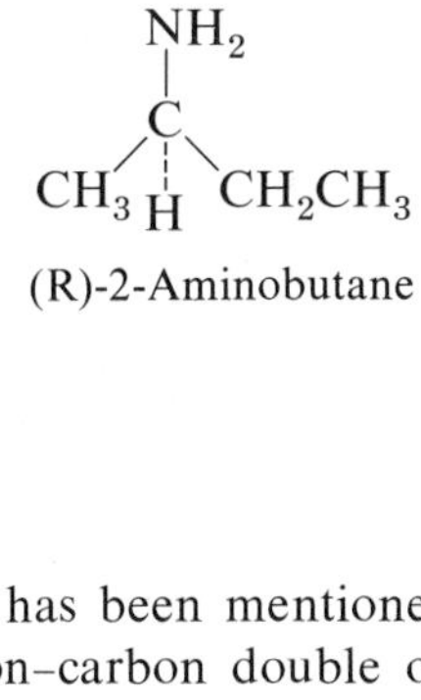

(R)-2-Aminobutane

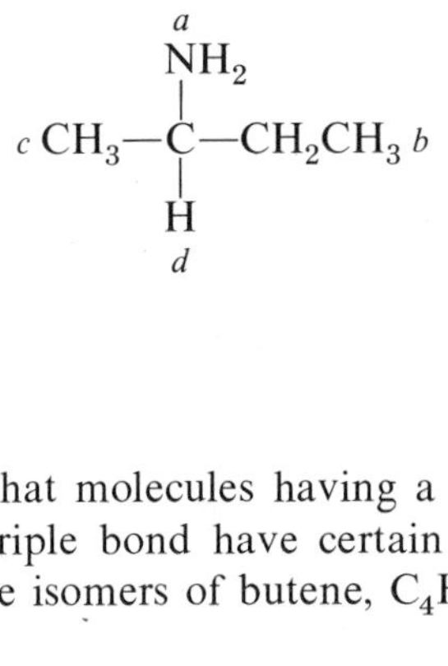

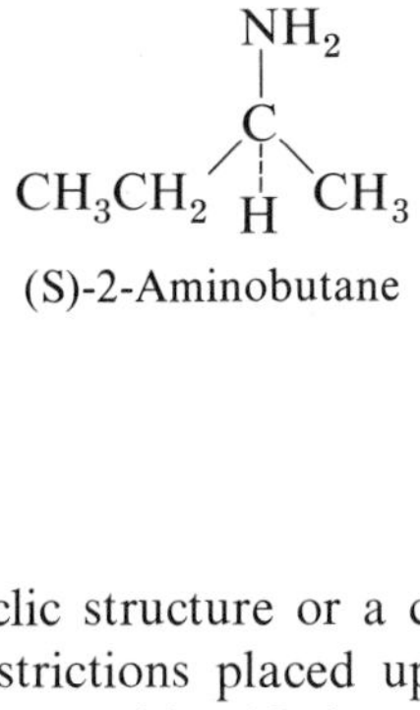

(S)-2-Aminobutane

22.6 Geometric Isomerism (*Cis-Trans* Isomerism)

It has been mentioned that molecules having a cyclic structure or a carbon–carbon double or triple bond have certain restrictions placed upon them. In Table 14.1 three isomers of butene, C_4H_8, were identified.

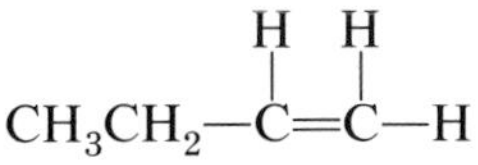

1-Butene
I

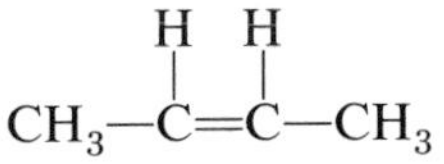

2-Butene
II

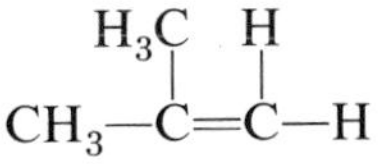

2-Methylpropene
III

Experimental evidence has shown, however, that there are four different butene molecules, all having distinctly different physical properties. The fourth isomer also has the structure that we have designated as 2-butene, and

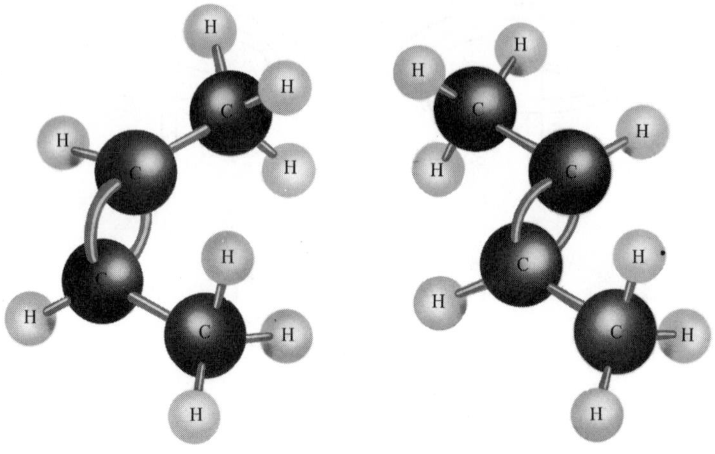

(a) (b)

Figure 22.10 (a) *cis*-2-Butene. (b) *trans*-2-Butene.

therefore these two isomeric 2-butenes, while they are structural isomers of compounds I and III, are not structural isomers of one another. Our knowledge of optical isomerism leads us to the conjecture that these two molecules might differ in the spatial configuration of their atoms. Because they have different physical properties, however, they cannot be optical isomers; a different type of configurational explanation must be sought to account for this phenomenon. The explanation is based upon the geometric arrangement of the carbon–carbon double bond.

Recall that the two carbon atoms of a C=C double bond and the four atoms that are attached to them are all in the same plane and that rotation around the double bond is prevented. (This is in sharp contrast to the free rotation enjoyed by carbon atoms that are linked to one another by means of single bonds.) Construction of ball-and-stick models indicates that there are two possible ways to arrange the atoms of 2-butene that are in keeping with its structural formulas (Figure 22.10). These three-dimensional ball-and-stick models are more simply represented as follows.

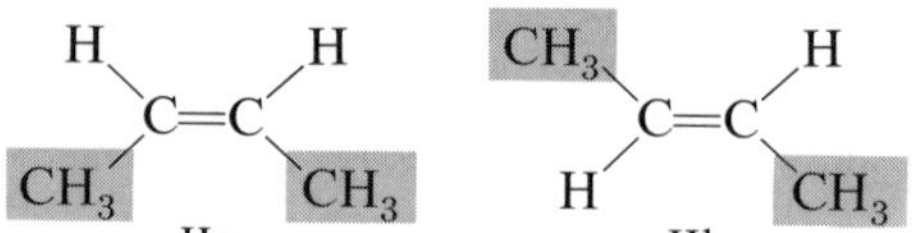

IIa	IIb
cis-2-Butene	*trans*-2-Butene
mp −139°C	mp −106°C
bp 4°C	bp 1°C

In structure IIa, both methyl groups lie on the same side of the molecule, and this compound is the *cis* isomer (Latin, *cis,* on this side). The methyl groups of structure IIb are on opposite sides of the molecule; it is the *trans* isomer (Latin, *trans,* across). Because of the restriction on free rotation about the double bond, the structures are clearly nonsuperimposable and hence not identical. *cis*-2-Butene and *trans*-2-butene are **geometric isomers** of one another.

Geometric isomers are compounds that have different configurations because of the presence of a rigid structure in the molecule. In contrast to the case of optical isomerism in which many isomers are possible for the same molecular formula, there are *only two* geometric isomers that correspond to the same formula (*cis* and *trans*). Because there are no geometric isomers of 1-butene or 2-methyl-1-propene, it is evident that the presence of a double bond is not the only criterion (nor is it a necessary one) for the occurrence of geometric isomerism. Geometric isomerism is not possible when there are identical substituents on either of the double-bonded carbons.

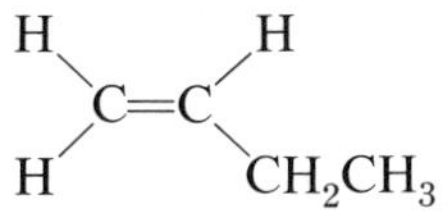

is identical to

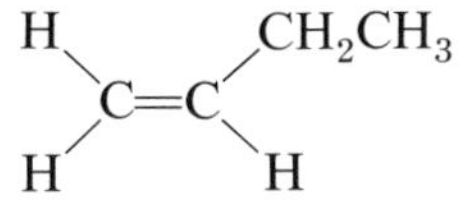

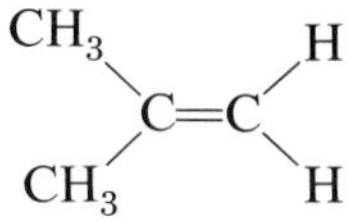

is identical to

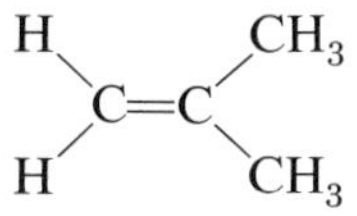

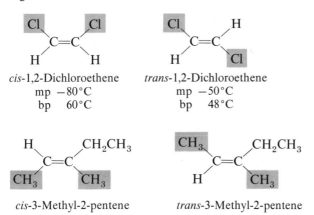

In general, geometric isomerism occurs in the alkenes when each carbon atom of a double bond bears two different substituents. When two identical (or nearly identical) substituents are on the same side of the double bond, the compound is the *cis* isomer; the *trans* isomer is the one in which similar groups are on opposite sides of the double bond. Following are some examples of geometric isomers.

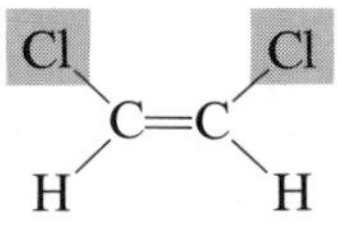

cis-1,2-Dichloroethene
mp −80°C
bp 60°C

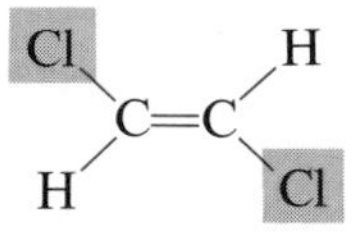

trans-1,2-Dichloroethene
mp −50°C
bp 48°C

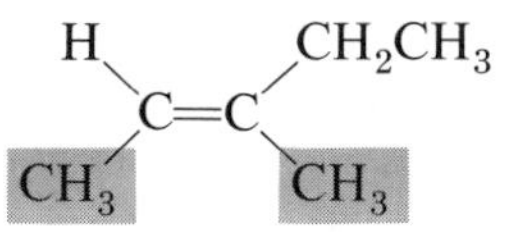

cis-3-Methyl-2-pentene

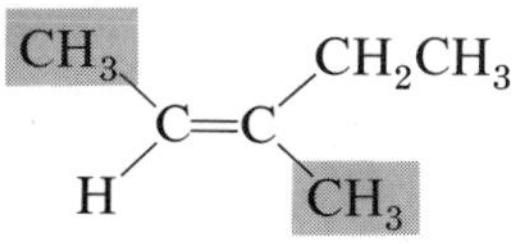

trans-3-Methyl-2-pentene

The *cis-trans* system for naming configurational isomers often leads to confusion. For example, is the isomer of 2-bromo-1-iodopropene *cis* or *trans?*

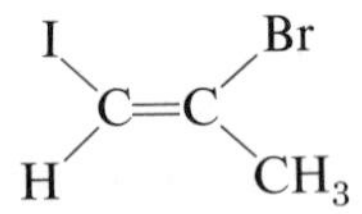

2-Bromo-1-iodopropene

To resolve this kind of ambiguity the **E, Z** nomenclature was devised. In this system the two atoms or groups of atoms bonded to each carbon of the double bond are assigned priorities as is done in naming enantiomers by the **R–S** convention (Section 22.5). The group of higher priority on one carbon is then compared with the group of higher priority on the other carbon. If the two groups of higher priority are on the same side of the molecule, the compound is the **Z** isomer (from the German *zusammen,* together). If the two groups of higher priority are on opposite sides of the molecule, the compound is the **E** isomer (from the German *entgegen,* in opposition to).

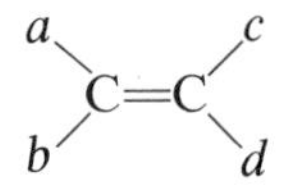

Z isomer

$a > b$

$c > d$

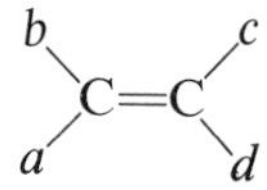

E isomer

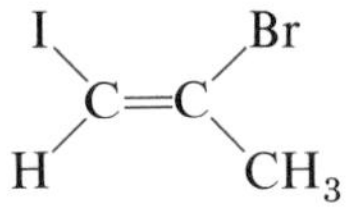

(Z)-2-Bromo-1-iodopropene

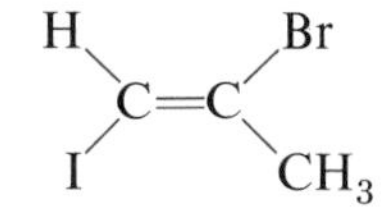

(E)-2-Bromo-1-iodopropene

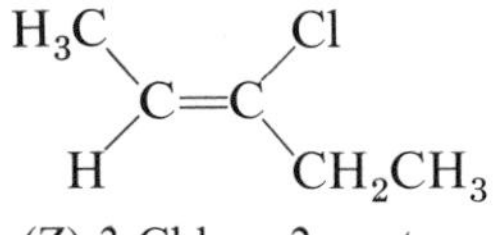

(Z)-3-Chloro-2-pentene

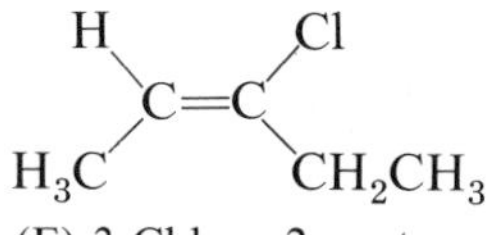

(E)-3-Chloro-2-pentene

Maleic and fumaric acids are classic examples of geometric isomers that have widely different chemical and physical properties (Figure 22.11). Because of the proximity of its carboxyl groups, maleic acid readily loses water to form an anhydride upon gentle heating.

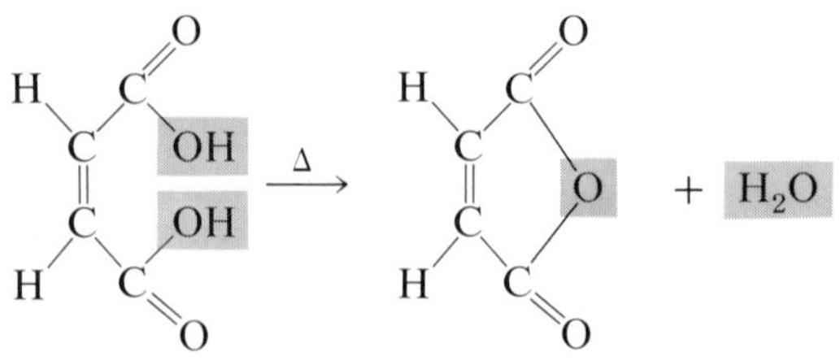

Maleic acid   Maleic anhydride

Fumaric acid is incapable of anhydride formation under the same reaction conditions. If it is heated to high temperatures (ca 300°C), fumaric acid

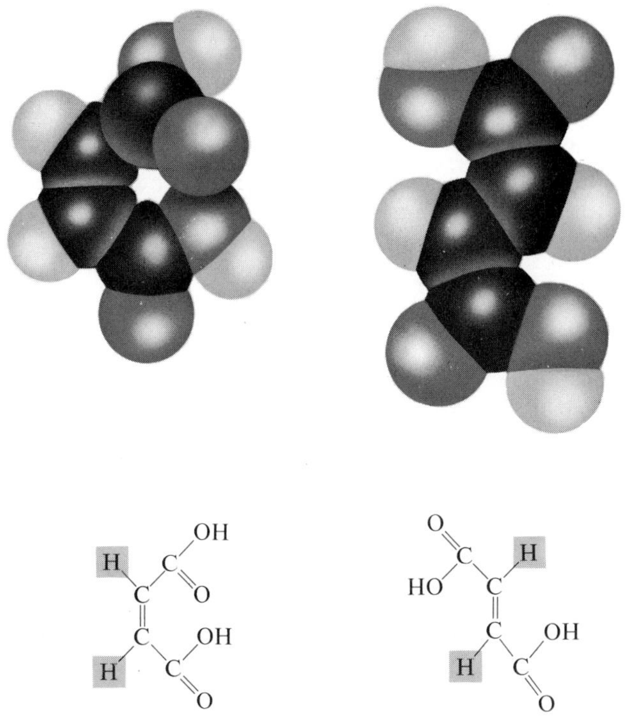

Maleic acid	Fumaric acid
(*cis*-Butenedioic acid)	(*trans*-Butenedioic acid)
mp 130°	mp 287°
Density 1.59 g/cm³	Density 1.64 g/cm³
Solubility in H_2O 78.8 g/100 ml	Solubility in H_2O 0.7 g/100 ml

Figure 22.11 Space-filling models, structural formulas, and properties of maleic acid and fumaric acid.

rearranges to form maleic acid, which then loses water to form the anhydride.

The nature of the bonding in the cycloalkanes also imposes geometric restraints upon the substituents that are bonded to the ring carbon atoms (see Section 13.8). Cyclopropane is a planar molecule, whereas cyclobutane and cyclopentane have slightly puckered rings. Recall that rings of six or more carbons are decidedly nonplanar. Common to all ring structures, however, is the inability of substituents to rotate about any of the ring carbon–carbon bonds. Therefore, substituents can either be on the same side of the ring (*cis*) or on opposite sides of the ring (*trans*). For our purposes, we will represent all cycloalkanes as planar structures; we will indicate substituents positioned above the plane by a solid line and substituents positioned below the plane by a dashed line.

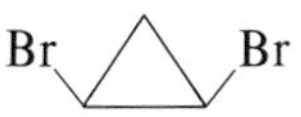

cis-1,2-Dibromocyclopropane

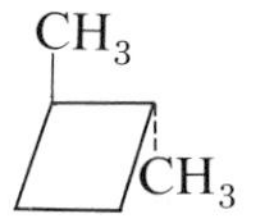

trans-1,2-Dimethylcyclobutane

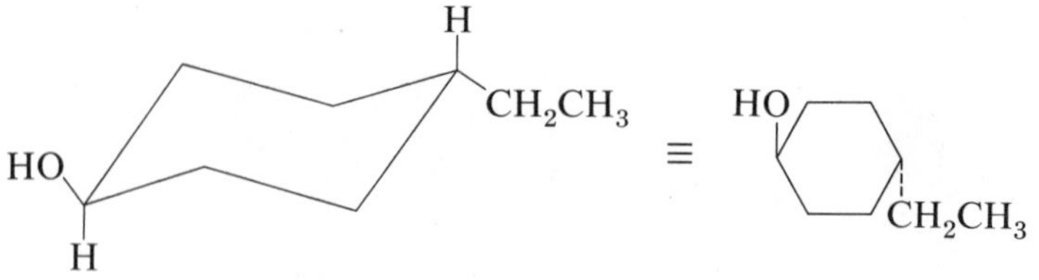
H
CH2CH3
HO
H
HO
CH2CH3

trans-4-Ethylcyclohexanol

22.7 Biochemical Significance

Molecular configurations are of the utmost importance in biochemistry. The example of the two enantiomers of lactic acid has already been cited. The dextrorotatory form is isolated from muscle tissue and the levorotatory isomer is found in yeast. Laboratory synthesis of lactic acid from acetaldehyde, propionic acid, or pyruvic acid produces a racemic mixture of (±)-lactic acid; it is impossible to synthesize chemically either of the optically pure forms. This is invariably the case. Whenever optically inactive reagents are utilized in a chemical synthesis, the resultant products will always be optically inactive even though one or more asymmetric centers may have been created in the new compounds. (A detailed explanation of this interesting phenomenon of organic synthesis may be found in a comprehensive text on organic chemistry.)

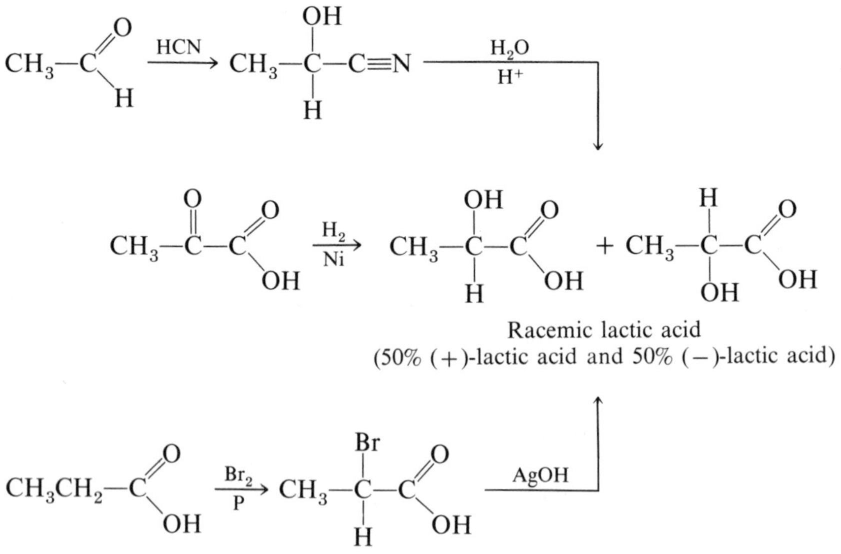
HCN
H2O
H+
H2
Ni
Br2
P
AgOH
Racemic lactic acid
(50% (+)-lactic acid and 50% (−)-lactic acid)

The obvious question, then, is how do muscle cells synthesize only (+)-lactic acid, and yeast only the (−)-isomer? The explanation here will arise many times during the study of biochemistry—enzymic control.

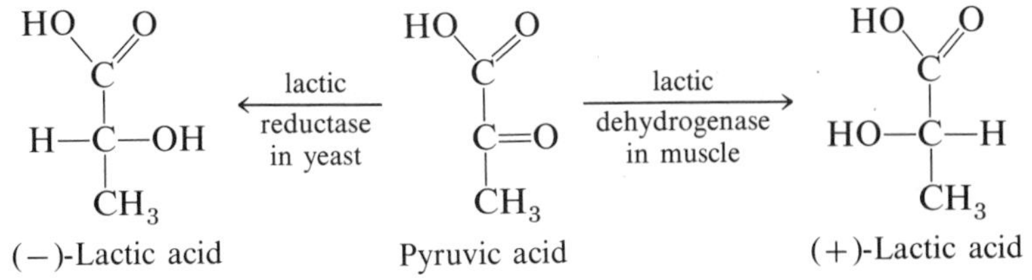

Enzymes are biological catalysts that are themselves optically active organic compounds. Similarly, almost every organic compound that occurs in living organisms is one enantiomer of a pair. Foods and medicines must have the proper molecular configurations if they are to be beneficial to the organism. For example, the popular meat-flavoring agent Accent is levorotatory monosodium glutamate. The dextrorotatory form of this salt would not enhance the flavor of meat because our taste buds could not recognize it. Similarly, the natural form of adrenalin is levorotatory. It has a physiological activity about fifteen to twenty times greater than that of its dextrorotatory enantiomer.

Because of the dangers of pesticides (Section 16.6), scientists have been searching for alternate methods to control insects. One method that has made a promising start involves the use of odorous chemicals to lure insects into lethal traps. An example of such a compound is trimedlure, which has been found to be strongly attractive to the male Mediterranean fruit fly. Trimedlure has eight possible stereoisomers, and they differ considerably in attraction for the insect. The fly is most strongly drawn to the isomer in which the methyl and ester groups are *trans* to each other.

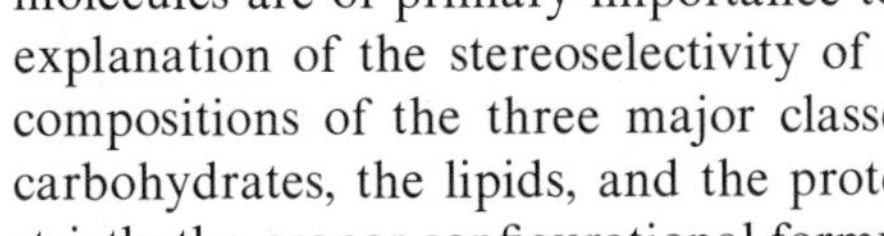

Methyl and ester group are *cis*.

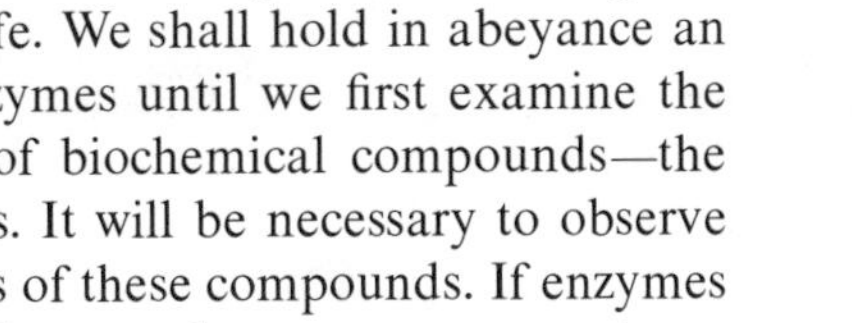

Methyl and ester group are *trans*.

The very subtle differences in structural configurations of organic molecules are of primary importance to life. We shall hold in abeyance an explanation of the stereoselectivity of enzymes until we first examine the compositions of the three major classes of biochemical compounds—the carbohydrates, the lipids, and the proteins. It will be necessary to observe strictly the proper configurational formulas of these compounds. If enzymes can recognize such subtle differences of shape and structure, so must we.

22.8 Molecules to See and to Smell

When light strikes the retina of the eye, a complex series of reactions is initiated by a *cis-trans* isomerization. This reaction is termed a **photochemical isomerization** (photoisomerization) because the energy of light causes the geometric change to occur. The only function of light in vision is to alter the

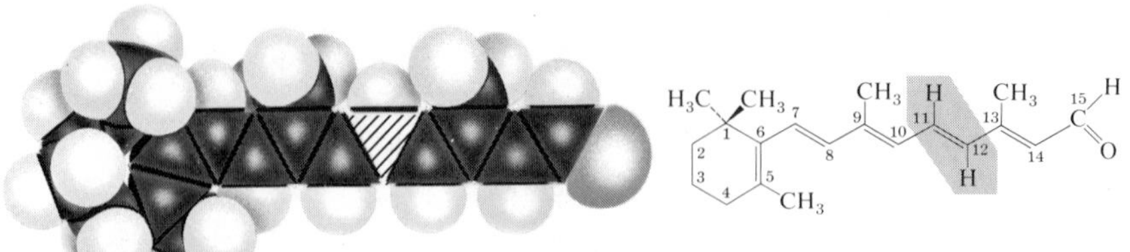

All-*trans* retinal

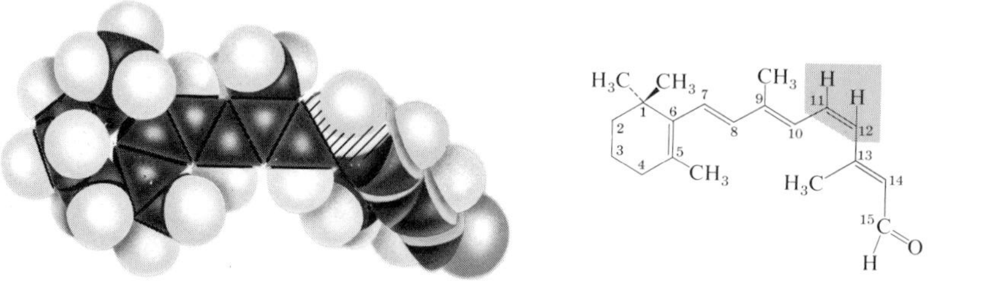

11-*cis*-Retinal

Figure 22.12 Fundamental molecule of vision is retinal ($C_{20}H_{28}O$), also known as retinene, which combines with proteins called opsins to form visual pigments. Because the nine-member carbon chain in retinal contains an alternating sequence of single and double bonds, it can assume a variety of bent forms. Two isomers of retinal are depicted here. In the space-filling models (*left*) carbon atoms are dark, except carbon-11, which is shown hatched; hydrogen atoms are light. The large atom attached to carbon-15 is oxygen. When tightly bound to opsin, retinal is in the bent and twisted form known as 11-*cis*. When struck by light, it straightens out into the all-*trans* configuration. This simple photochemical event provides the basis for vision. [From "Molecular Isomers in Vision" by Ruth Hubbard and Allen Kropf. Copyright © 1967 by Scientific American, Inc. All rights reserved.]

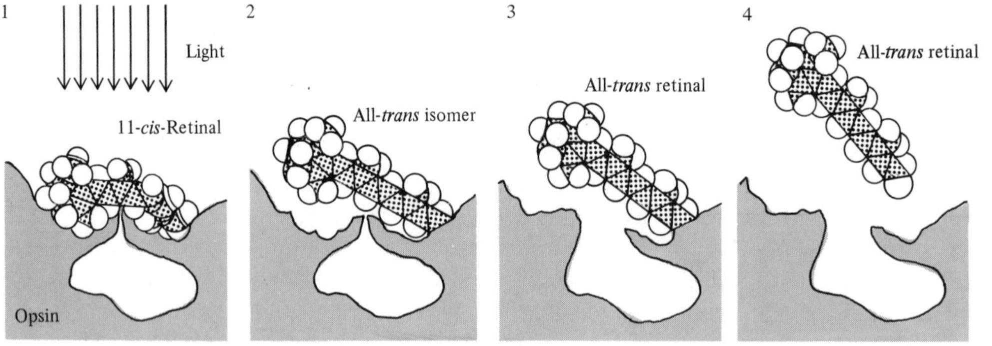
1
Light
11-*cis*-Retinal
Opsin
2
All-*trans* isomer
3
All-*trans* retinal
4
All-*trans* retinal

Figure 22.13 Molecular events in vision can be inferred from the known changes in the configuration of 11-*cis*-retinal after the absorption of light. In these schematic diagrams the twisted isomer is shown attached to its binding site on the protein molecule of opsin (1). After absorbing light the 11-*cis*-retinal straightens into the all-*trans* isomer (2). Presumably a change in the shape of opsin (3) facilitates the release of all-*trans* retinal (4). The configuration of the binding site in opsin is not yet known. [From "Molecular Isomers in Vision" by Ruth Hubbard and Allen Kropf. Copyright © 1967 by Scientific American, Inc. all rights reserved.]

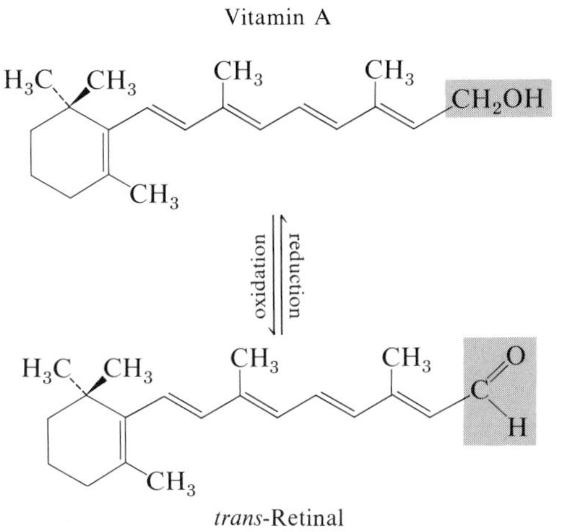
Vitamin A
oxidation
reduction
trans-Retinal

Figure 22.14 Vitamin A and *trans*-retinal are interconvertible via oxidation-reduction reactions.

shape of the absorbing molecule, 11-*cis*-retinal, to the *trans* configuration (Figure 22.12). Proteins in the eye are altered by this single photochemical act, the end result of which is a new impulse (Figure 22.13). The impulse is transmitted via the optic nerve to the brain. The method by which stimulation is transduced into nerve messages is as yet unknown. Notice that *trans*-retinal is analogous to vitamin A in all respects except that it contains a carbonyl group where the vitamin contains a primary alcohol group. Vitamin A is the reduced form of *trans*-retinal (or *trans*-retinal is the oxidized form of vitamin A), and the oxidation-reduction interconversions are essential to the chemical events of vision (Figure 22.14).

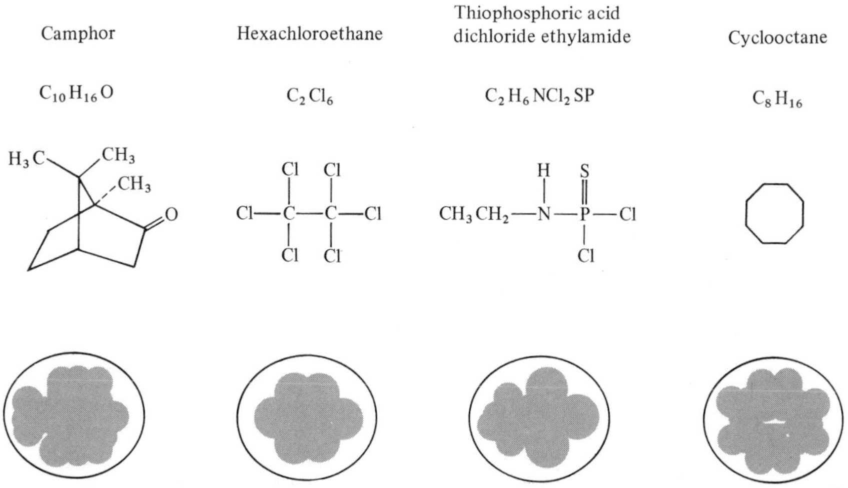
Camphor
Hexachloroethane
Thiophosphoric acid dichloride ethylamide
Cyclooctane
$C_{10}H_{16}O$
C_2Cl_6
$C_2H_6NCl_2SP$
C_8H_{16}

Figure 22.15 Unrelated chemicals with camphor-like odors show no resemblance in structural formulas. Yet, because the size and shape of their molecules are similar, they all fit the bowl-shaped receptor for camphoraceous molecules. [From "The Stereochemical Theory of Odor," by J. Amoore, J. Johnston, Jr., and M. Rubin. Copyright © 1964 by Scientific American, Inc. All rights reserved.]

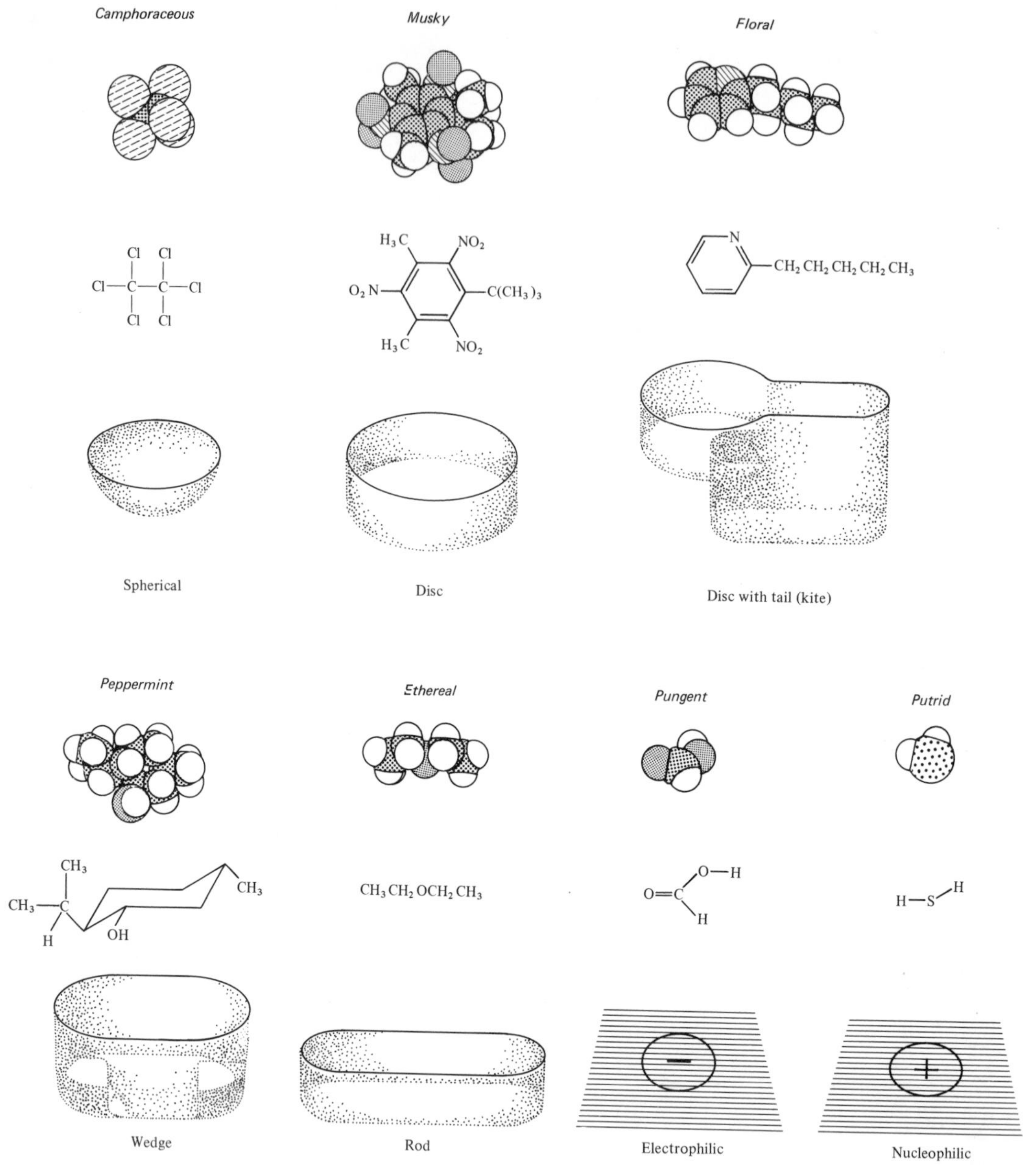
Camphoraceous
Musky
Floral
Spherical
Disc
Disc with tail (kite)
Peppermint
Ethereal
Pungent
Putrid
CH3CH2OCH2CH3
Wedge
Rod
Electrophilic
Nucleophilic

Figure 22.16 Olfactory receptor sites are shown for each of the primary odors, together with molecules representative of each odor. The first five respectively are hexachloroethane, xylene musk, α-amylpyridine, $(-)$-menthol, and diethyl ether. Pungent (formic acid) and putrid (hydrogen sulfide) molecules fit because of polarity rather than shape. [From "The Stereochemical Theory of Odor" by J. Amoore, J. Johnston, Jr., and M. Rubin. Copyright © 1964 by Scientific American, Inc. All rights reserved.]

Table 22.5 Primary Odors

Primary Odor	*Chemical Example*	*Familiar Substance*
Camphoraceous	Camphor	Moth repellent
Musky	Pentadecanolactone	Angelica root oil
Floral	3-Methyl-1-phenyl-3-pentanol	Roses
Peppermint	Menthone	Mint candy
Ethereal	Ethylene dichloride	Drycleaning fluid
Pungent	Acetic acid	Vinegar
Putrid	Butyl mercaptan	Rotten eggs

The current theory of olfaction was first postulated in 1949 by R. W. Moncrieff. He proposed that the olfactory system is composed of a limited number of receptor cells, each one representing a distinct primary odor. Furthermore, all odorous molecules produce their effects by fitting closely into receptor sites on these cells. (This hypothesis is essentially the same as the "lock and key" theory that we shall discuss in connection with enzymes in Section 26.5.) In essence, then, the major factor in determining the odor of a substance is the three-dimensional geometric shape of its molecules (Figure 22.15).

In 1952 John Amoore identified seven primary odors (Table 22.5). Every known odor can be made from these seven by mixing them in various proportions (in the same way that all the colors can be made from the three primary colors—red, yellow, and blue). In addition, Amoore described the size, shape, and chemical affinities of the seven different kinds of receptor sites that would recognize each of the primary odors. These receptor sites must necessarily have a distinctive shape, size, and chemical affinity so as to accept only those molecules with the correct geometric configuration or polarity. The shapes, sizes, and polarity of the molecules representing the seven primary odors are illustrated in Figure 22.16. If the molecules of a substance can fit into more than one receptor site, then the brain will perceive a more complex signal and the resultant odor will be a combination of the primary sites occupied by the molecules. Again, as in the case of vision, we still do not know the mode of stimulation of the olfactory nerve.

Exercises

22.1 Define or explain each of the following terms.

(a) plane-polarized light
(b) *d*-form
(c) optical isomerism
(d) asymmetric molecule
(e) polarimeter
(f) diastereomers

(g) levorotatory
(h) mirror image
(i) enantiomers
(j) meso compound
(k) specific rotation
(l) racemic mixture

22.2 Which of the heptane isomers are optically active? (Refer to Exercise 13.1.)

22.3 Draw all the optically active isomers for compounds having the following molecular formulas.
(a) C_4H_9Cl **(b)** $C_4H_8Br_2$

22.4 Draw all the optical isomers for the following compounds.
(a) $H_2NCH_2CHFCH_2OH$

(b) $(CH_3)_2COHCH_2C(=O)OCH_2CHOHCH_2OH$

(c) $(CH_3)_2CHCH(NH_2)CH_2CH(CH_3)_2$

(d) $CH_3C(=O)CHBrCH_3$

(e) $CH_3CHClCHOHC(=O)OH$

(f) $CH_3CH_2CH(CH_3)CHBrC(=O)H$

(g) $CH_3CHBrCHClC(=O)NH_2$

(h) $(CH_3)_2C{=}C(CH_3)CH_2CHICH_3$

22.5 Write formulas showing the configurations of all possible optical isomers of 2,3,4-tribromopentane.

22.6 Write formulas for all the stereoisomers of 3,4-dimethyl-1,5-hexadiene.

22.7 Write formulas showing the configurations of all the possible optical isomers of ribitol. Indicate which of these structures are enantiomers, which diastereomers, and which meso forms.

22.8 The active ingredient in a popular insect repellant is 2-ethyl-1,3-hexanediol. Draw its structural formula and all of its optical isomers.

22.9 Write structural formulas for the dextro, levo, and meso forms of 4,5-diethyloctane.

22.10 In what ways do enantiomers resemble each other? How do they differ?

22.11 How does a racemic mixture differ from a meso compound? In what ways are they similar?

22.12 What are geometric isomers? Write formulas for two pairs of noncyclic and two pairs of cyclic geometric isomers.

22.13 Write the formulas of the geometric isomers for each of the following compounds. Label them *cis* or *trans* (or **E** and **Z**). If there are no isomers, write *None.*

(a) 1-pentene

(b) 2-pentene

(c) 1-chloro-1-cyclopentene

(d) 1,3-dimethylcyclohexane

(e) 1-bromo-2-methylcyclobutane

(f) 3-penten-1-yne

(g) 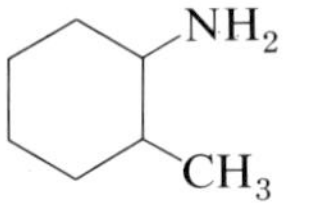

NH_2
CH_3

(h) $CH_3CH{=}CHCH(CH_3)_2$

(i) $CHI{=}CHCH_3$

(j) $(CH_3)_2C{=}CHCH_2CH{=}CHCH_3$

(k) $CH_3CH{=}CH\overset{\overset{\displaystyle O}{\|}}{C}CH_3$

(l) 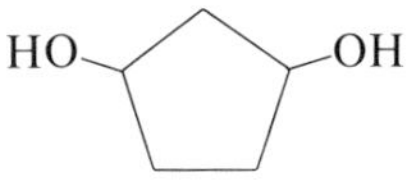

HO
OH

22.14 If the four bonds of carbon were directed toward the corners of a square, how many isomers of CH_2BrCl would exist? Draw them.

22.15 Draw and name all the geometric isomers (both noncyclic and cyclic) corresponding to the molecular formula C_5H_{10}.

22.16 Only one of the isomers of pentanal is optically active. What is its formula? Draw the two enantiomers of this compound and designate each as either **R** or **S.**

22.17 Give the formula of the smallest noncyclic alkane that could be optically active.

22.18 An alcohol has the molecular formula $C_4H_{10}O$.
(a) Write all the possible structural formulas for the alcohol.
(b) If the alcohol can be separated into two optically active forms, which of the formulas in (a) is correct?
(c) Assign an **R** or **S** designation to each of the enantiomers in (b).

22.19 The concentration of an optically active compound in water was 12 grams/liter. A 10 ml aliquot has an observed rotation of $+1.4°$ in a 10 cm polarimeter tube. Calculate the specific rotation of this compound.

22.20 2.00 grams of an optically active compound, dissolved in 10.0 ml of ethanol, rotated a beam of plane-polarized light 25° to the left in a 1 dm polarimeter tube. What is its specific rotation?

22.21 Examine the labels of some common household products (detergents, foods, drugs, sprays, cosmetic). Write structural formulas for the chemical compounds contained in these products.

22.22 Benzedrine is a racemic mixture of *d*- and *l*-amphetamine. The pure dextrorotatory enantiomer, Dexedrine, has a much greater physiological activity than either the racemic mixture or the pure levorotatory form. Draw the structural formulas for the two optical isomers of amphetamine (see Figure 20.7).

22.23 Lysergic acid diethylamide (LSD) contains two asymmetric carbon atoms (see Figure 20.8). Identify which carbon atoms are asymmetric and then draw structural formulas of the four optical isomers of LSD. (Only one of these four isomers, (*d*)-lysergic acid diethylamide is physiologically active.)

23

Carbohydrates

Carbohydrate literally means *hydrate of carbon*. This name is derived from the investigations of early chemists who found that when they heated sugars for a long period of time in an open test tube, they obtained a black residue, carbon, and droplets of water condensed on the sides of the tube. Furthermore, chemical analysis of sugars and other carbohydrates indicated that they contained only carbon, hydrogen, and oxygen, and many of them were found to have the general formula $C_x(H_2O)_y$.[1]

It is now known that some carbohydrates contain nitrogen and sulfur in addition to carbon, hydrogen, and oxygen. They definitely are not hydrated compounds, as are many inorganic salts (for example, $CuSO_4 \cdot 5\,H_2O$). Today the name **carbohydrate** is used to designate the *large class of compounds that are polyhydroxy aldehydes or ketones, or substances that yield such compounds upon acid hydrolysis.*

Carbohydrates are the major constituents of most plants, comprising from 60 to 90% of their dry mass. In contrast, animal tissue contains a comparatively small amount of carbohydrate (e.g., less than 1% in man). Plants utilize carbohydrates both as a source of energy and as supporting tissue in the same manner that proteins are used by animals. Plants are able to synthesize their own carbohydrates from the carbon dioxide of the air and from water taken from the soil. Animals are incapable of this synthesis, and therefore are dependent upon the plant kingdom as a source of these vital compounds. Man, the special animal, not only utilizes carbohydrates for his food (about 60–65% by mass of the average diet), but also for his clothing (cotton, linen, rayon), shelter (wood), fuel (wood), and paper (wood).

[1] Most carbohydrates do conform to the formula $C_x(H_2O)_y$. However, this formula is neither unique to the carbohydrates, nor is it a necessary criterion for membership in the carbohydrate class. For example, formaldehyde (CH_2O), acetic acid ($C_2H_4O_2$), and lactic acid ($C_3H_6O_3$) are certainly not carbohydrates, but the compound rhamnose ($C_6H_{12}O_5$) is.

23.1 Classification of Carbohydrates

Carbohydrates are classified according to their acid hydrolysis products. Three major categories are recognized:

1. The **monosaccharides,** or *simple sugars,* cannot be broken down into smaller molecules by hydrolysis.
2. The **disaccharides** yield two monosaccharide molecules upon hydrolysis.
3. The **polysaccharides** yield many monosaccharide molecules upon hydrolysis.

For the most part, the mono- and disaccharides are sweet-tasting,[2] crystalline solids that are readily soluble in water. Polysaccharides are frequently tasteless, insoluble, amorphous compounds with exceedingly high molar masses.

[2]A sweet taste is a physiological property that can be distinguished by the human tongue. Although sweetness is commonly associated with most mono- and disaccharides, it is not a specific property of carbohydrates. Many sugars are sweet to varying degrees, but several organic compounds have been synthesized that are far superior as sweetening agents. Sucaryl sodium, for example, is about 50 times as sweet as sucrose, while saccharin is about 450 times as sweet as sucrose. These synthetic compounds have no caloric value, and therefore they are useful for those persons (for example, diabetics) who must minimize their carbohydrate intake.

Sucaryl sodium and the analogous calcium salt, sucaryl calcium, are referred to collectively as the *cyclamates* (i.e., salts of cyclamic acid). Experimental findings have shown that large doses (50 times the daily maximum recommended for human consumption) of cyclamate induces cancer in rat bladders. In 1969 the Department of Health, Education, and Welfare ordered all beverages and other food products containing cyclamate off the market. It should be stated, however, that as yet there has been no connection between cyclamate and cancer or any other disease or disorder in humans. Many subsequent studies completed since 1969 by the FDA, by the National Cancer Institute, and by researchers in other countries have indicated that cyclamate is not carcinogenic.

During the period 1969–1977, saccharin was the only artificial sweetener permitted in the United States, although it too had been proclaimed by some to be carcinogenic. In 1977 the FDA disclosed that a three-year study carried out in Canada showed that rats fed a diet containing 5% saccharin developed bladder cancers. As a result, the FDA has proposed a ban on the sale of saccharin in the United States. This proposed ban was highly controversial. Opponents of the ban estimated that an individual would have to consume 800 12-ounce cans of diet soda each day for seven years to obtain the same amount of saccharin that was fed to the rats. Because of public (and diet food industry) opposition to the ban, Congress in November 1977 prohibited the FDA from acting against saccharin for 18 months and authorized a study on its safety by the National Academy of Sciences (NAS). One year later, a committee of the NAS unequivocally affirmed the earlier FDA decision to ban saccharin. The panel concluded that (1) saccharin is a carcinogen in animals, although of low potency, (2) it is a potential human carcinogen, (3) the compound itself and not impurities from manufacturing are responsible for carcinogenic activity, and (4) there is no evidence that saccharin offers any meaningful health benefits. At present, the 10 million Americans who suffer from diabetes have no alternative sugar substitute. Research efforts are currently directed toward the development of low calorie sweeteners composed of amino acids or peptide derivatives.

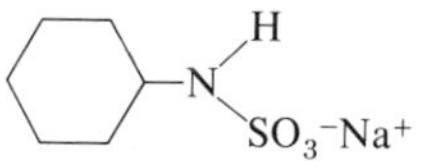
H
N
$SO_3^-Na^+$

Sucaryl sodium

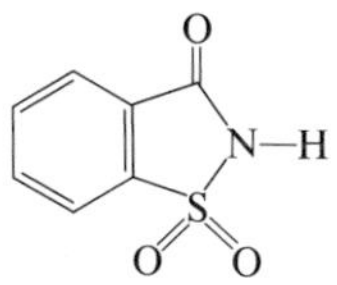
O
N—H
S
O
O

Saccharin

23.2 Monosaccharides—General Terminology

The general names for the monosaccharides are obtained in a manner analogous to the naming of organic compounds by the IUPAC system. The number of carbon atoms in the molecule is denoted by the appropriate prefix; the suffix *-ose* is the generic designation for any sugar. For example, the terms triose, tetrose, pentose, and hexose signify 3, 4, 5, and 6 carbon monosaccharides, respectively. In addition, those *monosaccharides that contain an aldehyde group* are called **aldoses;** *those containing a ketone group* are **ketoses.** (Some authors indicate the presence of a ketonic group in a monosaccharide by inserting the letters *-ul* into the general name of the compound.) By combining these terms, both the type of carbonyl group and the number of carbon atoms in the molecule are easily expressed. Thus, monosaccharides are generally referred to as aldotetroses, aldopentoses, ketopentoses (pentuloses), ketoheptoses (heptuloses), etc. The general formula for any aldose (I) and any ketose (II) may be represented as follows.

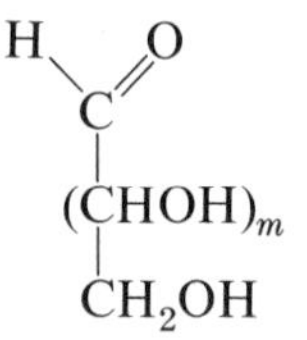

I
Aldoses
(m = 1, 2, 3, 4, 5)

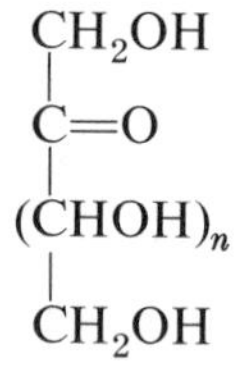

II
Ketoses
(n = 0, 1, 2, 3, 4)

Glucose (III) and fructose (IV) are specific examples of an aldose and a ketose.

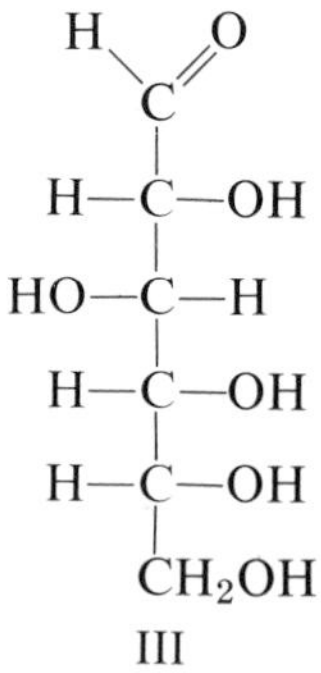

III
Glucose
(An aldohexose)

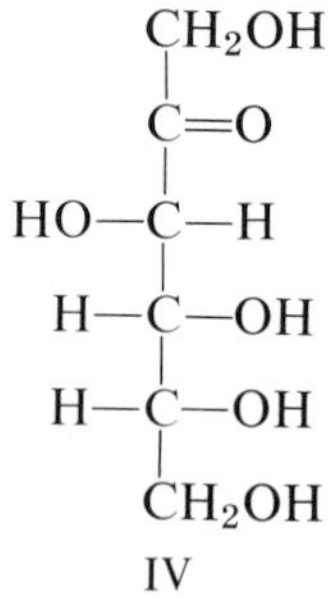

IV
Fructose
(A ketohexose—hexulose)

23.3 Stereochemistry

The simplest monosaccharides are the trioses, glyceraldehyde and dihydroxyacetone. Glyceraldehyde possesses an asymmetric carbon atom and thus may exist in two optically active forms. Dihydroxyacetone does not

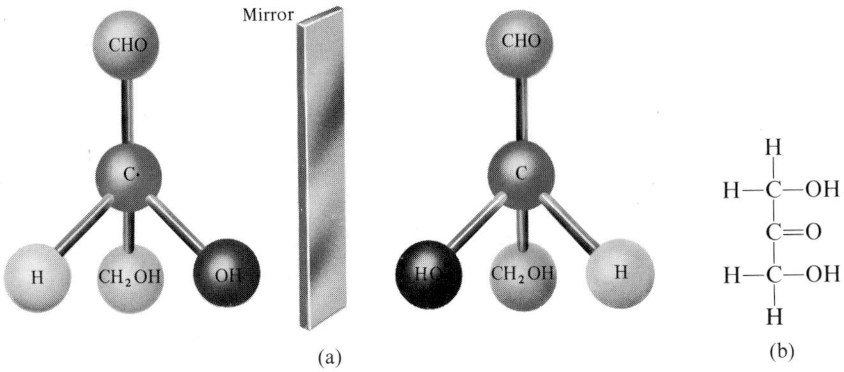
Mirror
CHO
C
H
CH₂OH
OH
CHO
C
HO
CH₂OH
H
(a)
(b)

Figure 23.1 (a) Enantiomers of glyceraldehyde. (b) Structural formula of dihydroxyacetone.

contain a center of asymmetry. Figure 23.1 shows ball-and-stick models of the two forms of glyceraldehyde and the structural formula of dihydroxyacetone. The two glyceraldehyde isomers are clearly mirror images. Except for the direction in which they rotate plane-polarized light, they have identical physical properties. One form has a specific rotation of $+13.5°$, the other a rotation of $-13.5°$. The great German chemist Emil Fischer (Nobel Prize, 1902) initiated the convention of projecting the formulas on a two-dimensional plane so that the aldehyde group is written at the top, with the hydrogen and hydroxyl written to the right and left. (Formulas of sugars represented in this manner are referred to as Fischer projections, Fischer models, or Fischer configurations.) Arbitrarily, Fischer then decided that the formula of glyceraldehyde in which the hydroxyl group is positioned to the right of the asymmetric carbon atom represents the dextrorotatory isomer. He assigned the letter D as its prefix. The levorotatory isomer, in which the —OH group is positioned to the left of the asymmetric carbon atom, was accordingly assigned the letter L as its prefix.[3]

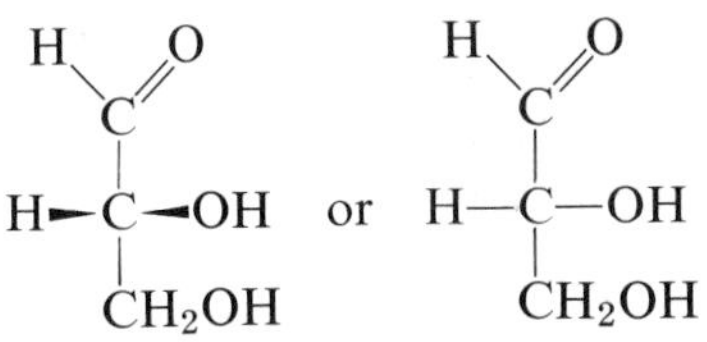

D-Glyceraldehyde

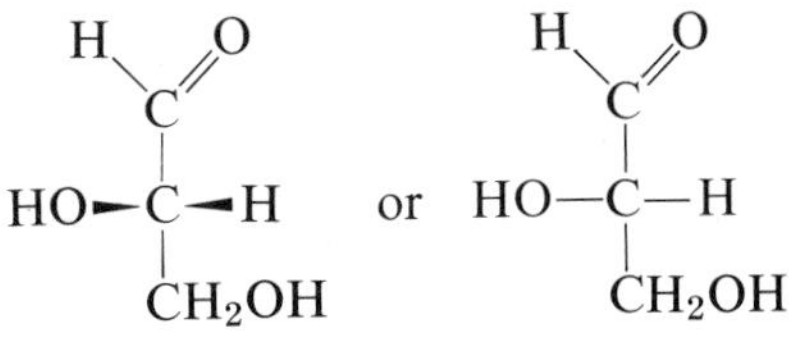

L-Glyceraldehyde

The two forms of glyceraldehyde are especially important because the more complex sugars may be considered to be derived from them. They serve as a reference point for designating and drawing all other monosaccharides.

[3] Fischer's arbitrary assignment proved to be correct. In 1951 chemists, with the aid of x-ray crystallography, determined the absolute configurations of the glyceraldehyde enantiomers and found that the D-isomer was indeed dextrorotatory.

Sugars whose Fischer projections terminate in the same configuration as D-glyceraldehyde are designated as **D-sugars;** those derived from L-glyceraldehyde are designated as **L-sugars.** The convention is illustrated by a consideration of the following four stereoisomeric aldotetroses.

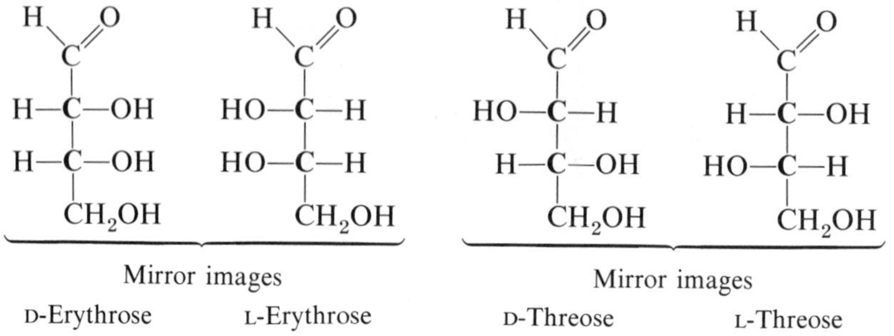

Notice that D- and L-erythrose and D- and L-threose are pairs of enantiomers. However, D-threose and L-threose are diastereomers of D-erythrose and L-erythrose. Thus, aside from having the same molecular formulas, the erythroses and the threoses are completely different sugars, having different chemical and physical properties.

The letters D and L are often very misleading for the beginning student. It must be emphasized that these prefixes serve only to signify the absolute configuration of a molecule. *A D-sugar is one that has the same configuration about the* **penultimate**[4] *carbon atom as does D-glyceraldehyde.* The letters do not in any way refer to the optical rotation of the molecule. D-Glyceraldehyde just happens to be dextrorotatory. It is not to be expected, however, that all compounds derived from it will have the same optical rotation, since such compounds will have additional asymmetric carbon atoms. In fact, D-erythrose and D-threose are both found to be levorotatory. The direction of rotation of plane-polarized light is a specific property of each optically active molecule. It is not at all dependent upon the configuration about the penultimate carbon. The symbols plus (+) and minus (−) are used to denote the optical rotation of the monosaccharides; (+) indicates a clockwise, or dextrorotatory, rotation, and (−) indicates a counterclockwise, or levorotatory, rotation. The correct designations for the isomers of glyceraldehyde, erythrose, and threose, that indicate both their configuration and their optical rotation, are given as follows: D(+)-glyceraldehyde, L(−)-glyceraldehyde, D(−)-erythrose, L(+)-erythrose, D(−)-threose, and L(+)-threose.

Figure 23.2 shows all the possible stereoisomeric aldopentoses and their complete names. Notice that the aldopentoses contain three asymmetric carbon atoms and that there are four enantiomeric pairs. Aldohexoses (Figure 23.3) contain four asymmetric carbon atoms and thus there are eight enantiomeric pairs, or sixteen isomers. Fortunately for students of biochemistry,

[4]The penultimate carbon is the asymmetric carbon atom that is farthest from the aldehyde or ketone group. The next-to-last carbon has been chosen, by convention, to be the reference carbon atom.

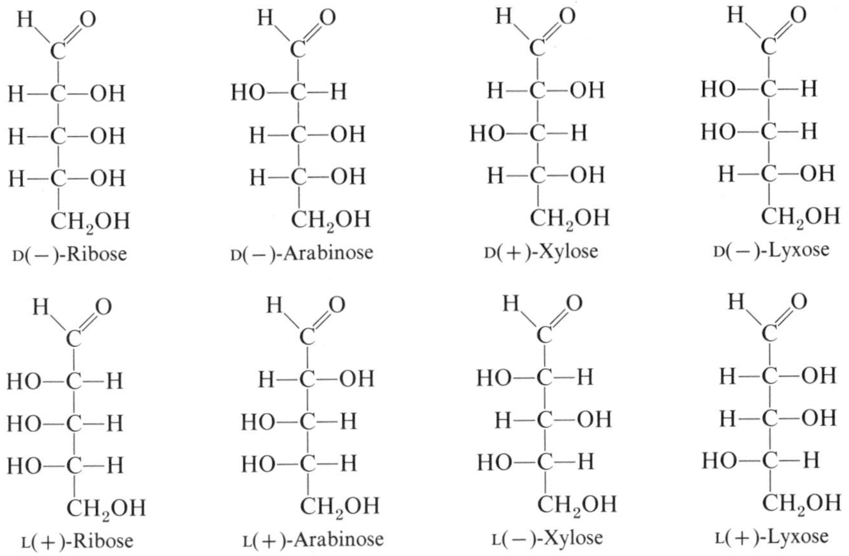

Figure 23.2 The eight stereoisomeric aldopentoses.

only three of the sixteen isomers are commonly found in nature, D(+)-glucose, D(+)-mannose, and D(+)-galactose. (All sixteen isomers have been prepared synthetically.) No one of these three stereoisomers is a mirror image of either of the others, so all three are diastereomers of each other. Furthermore, D(+)-glucose differs from D(+)-mannose and D(+)-galactose only in the configuration about one asymmetric carbon atom, carbon numbers 2 and 4, respectively. D(+)-Mannose and D(+)-galactose are said to be **epimers** of D(+)-glucose. A pair of diastereomers that differ only in the configuration about a *single* carbon atom are said to be epimers. No epimeric relationship exists between D(+)-mannose and D(+)-galactose. Finally, it should be mentioned that the aldohexoses are structural isomers of the only naturally occurring ketohexose, D(−)-fructose.

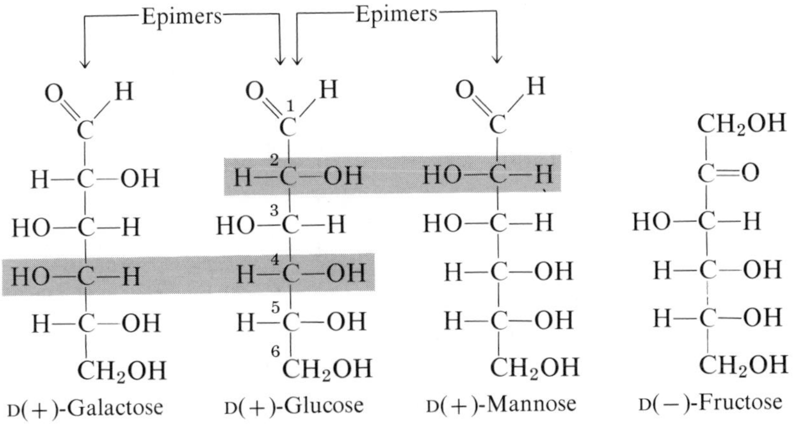
Epimers
Epimers
D(+)-Galactose
D(+)-Glucose
D(+)-Mannose
D(−)-Fructose

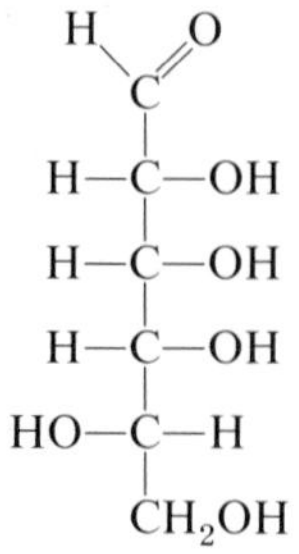

L(−)-Talose

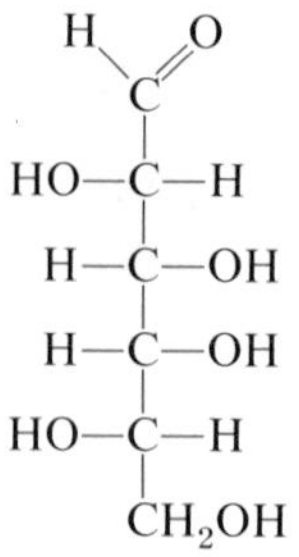

L(−)-Galactose

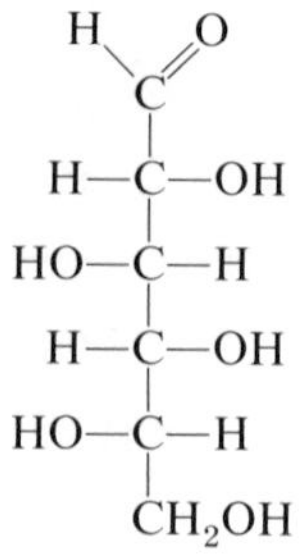

L(−)-Idose

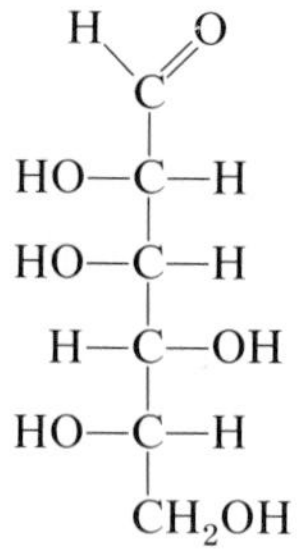

L(+)-Gulose

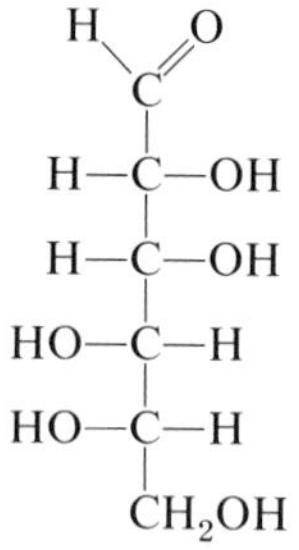

L(−)-Mannose

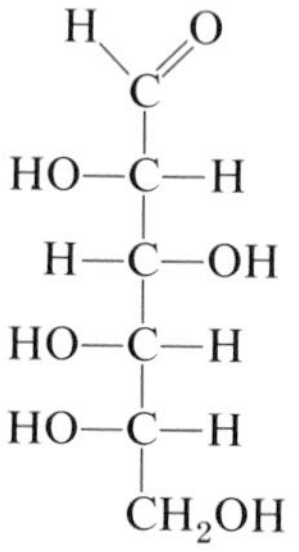

L(−)-Glucose

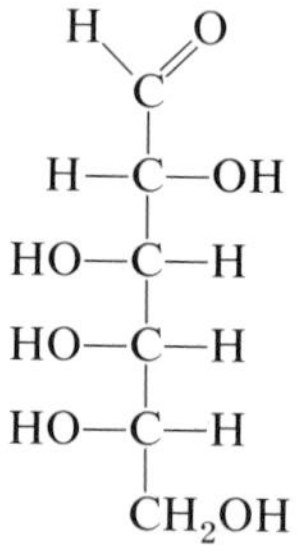

L(−)-Altrose

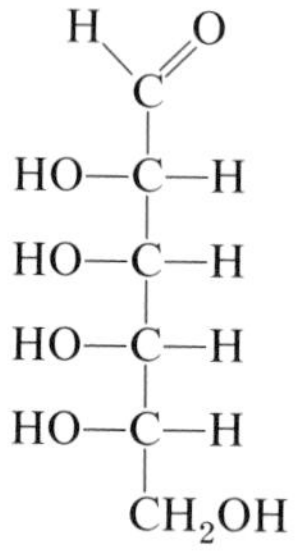

L(−)-Allose

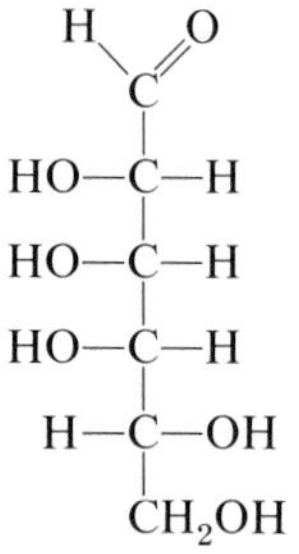

D(+)-Talose

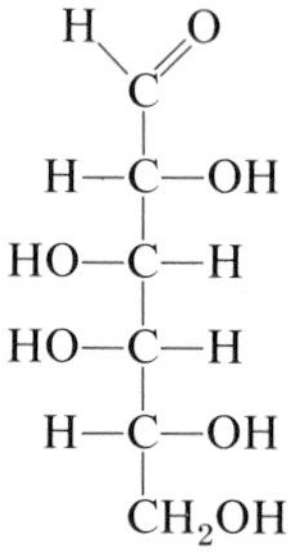

D(+)-Galactose

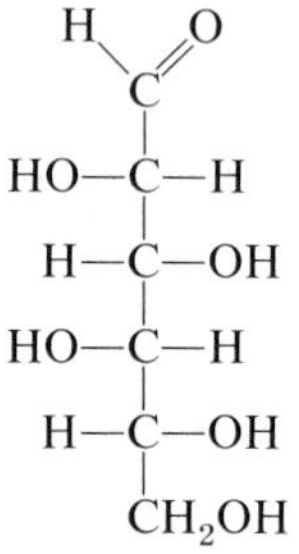

D(+)-Idose

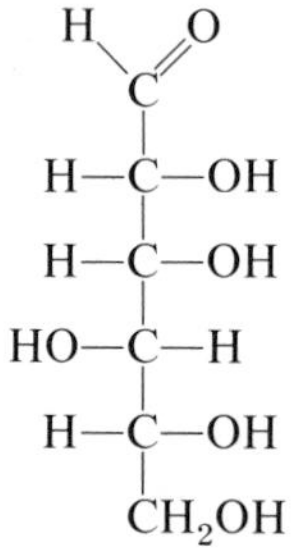

D(−)-Gulose

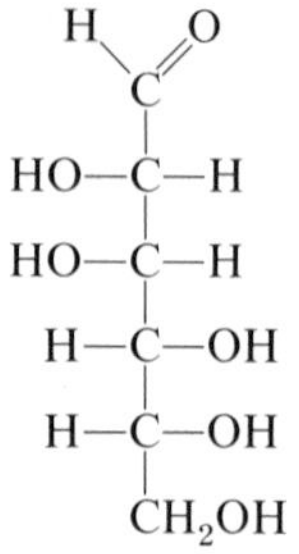

D(+)-Mannose

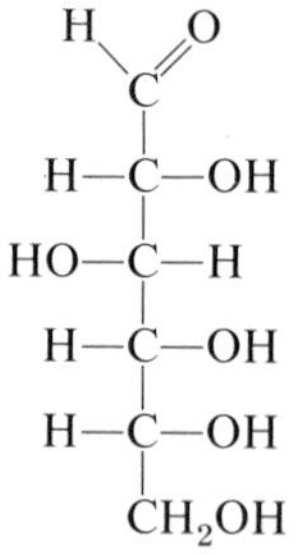

D(+)-Glucose

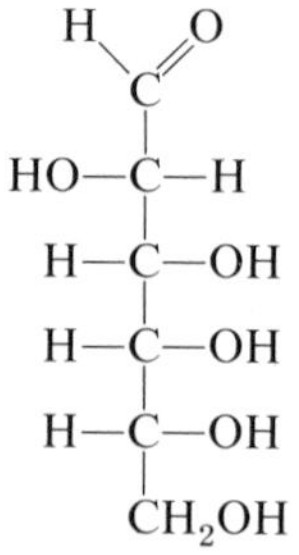

D(+)-Altrose

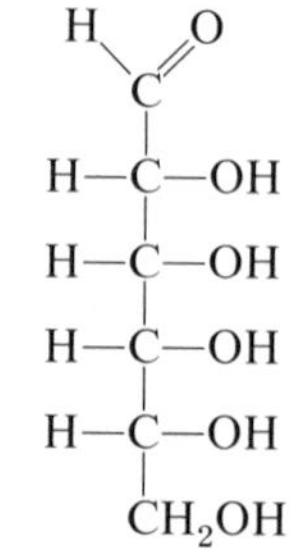

D(+)-Allose

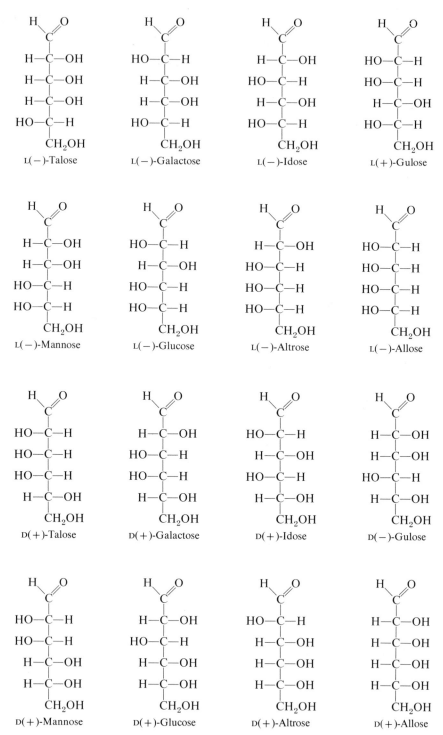

Figure 23.3 Configurations of the aldohexoses.

23.4 Cyclic Forms of Sugars

Thus far we have considered the sugars as existing only as open-chain structures containing a carbonyl group. Studies of the chemical and physical properties of these sugars indicate that other forms predominate, both in solution and in the solid state. The most convincing evidence was supplied by Charles Tanret (1895), a French pharmacist who first isolated two crystalline forms of D-glucose, each having distinctly different specific rotations. We shall consider this phenomenon further in Section 23.5.

The explanation of the existence of two forms of glucose can best be understood by considering the configuration of the molecule as represented by Figure 23.4. Observe that the hydroxyl group on carbon number five lies in proximity to the aldehyde group. In Section 18.5-a we observed that alcohols react intermolecularly with aldehydes to form unstable compounds called hemiacetals.

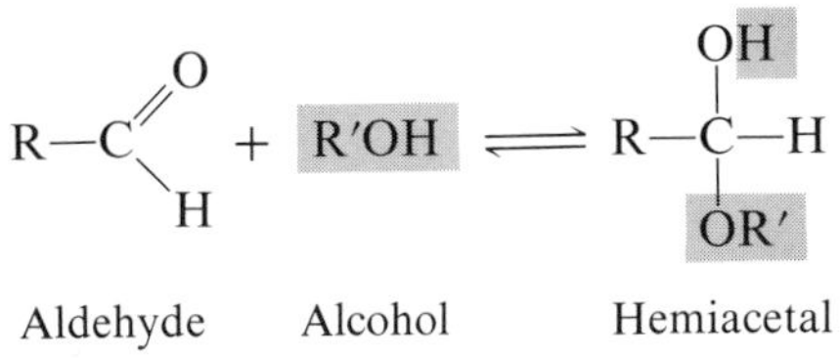
R—C
O
H
+
R′OH
R—C—H
OH
OR′
Aldehyde
Alcohol
Hemiacetal

Recall, however, that molecules containing both a hydroxyl group and a carbonyl group can react intramolecularly to form stable cyclic compounds. Because of the geometry of the glucose molecule, the hydroxyl group on

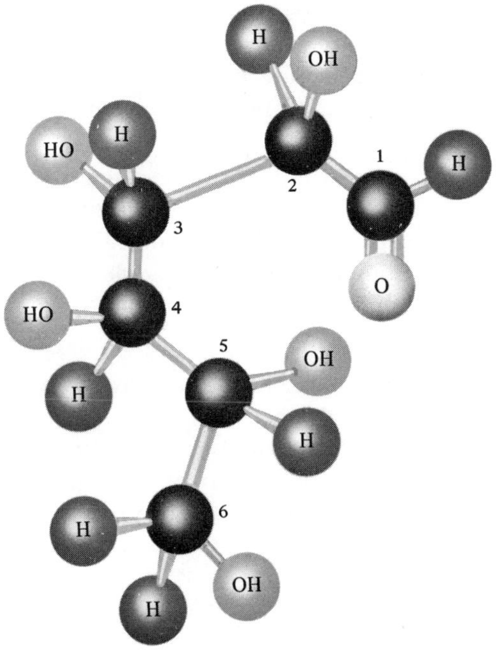
H
OH
HO
H
1
H
2
3
O
HO
4
5
OH
H
H
6
H
OH
H

Figure 23.4 A ball-and-stick model of the open-chain form of glucose.

carbon number five can react intramolecularly with the carbonyl group to form a stable cyclic hemiacetal.

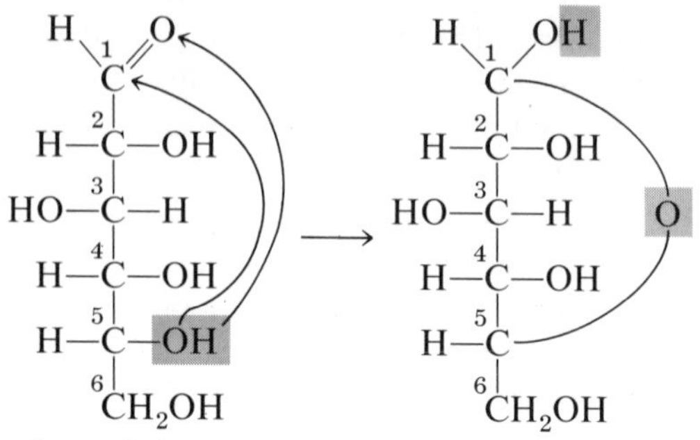
Open-chain form
Cyclic form

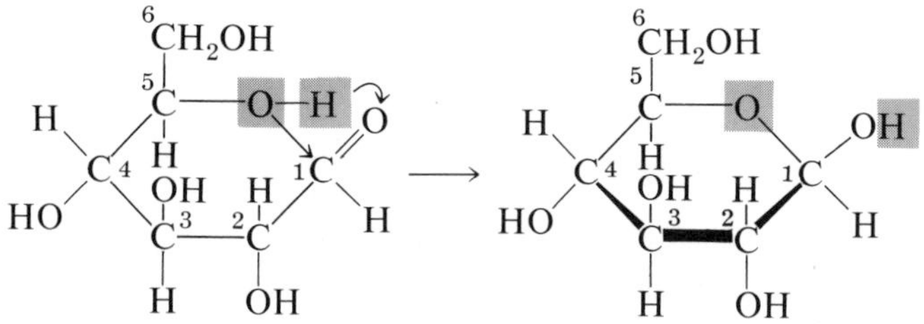

As a result of the cyclization reaction, two new problems of stereochemistry and nomenclature confront us.

1. A six-membered ring containing five carbon atoms and one oxygen atom has been formed. Since the reaction only involved an intramolecular transfer of atoms rather than the loss or gain of atoms, the new cyclic compound is a structural isomer of the open-chain compound.
2. Examination of the cyclic structure reveals that carbon number one has now become asymmetric, so that we would expect to find two stereoisomers of the ring compound. Additional terminology will be needed to distinguish among the three different structures of D-glucose—one open-chain form and two cyclic forms.

As we shall learn in the next section, a solution of glucose actually exists as an equilibrium mixture of the three forms. In this book the convention first suggested by an English chemist, Sir W. N. Haworth, for representing the formulas of the cyclic forms will be used. The molecules are drawn as planar hexagonal slabs with darkened edges toward the viewer. Ring carbon atoms and the hydrogen atoms directly attached to them are not shown. The disposition of the hydroxyl groups, positioned either above or below the plane of the ring, is sufficient to define the correct configuration of the molecule. Any group of atoms written to the right in the Fischer projection appears below the plane of the ring, and any group written to the left appears above the plane. The formulas for the three forms of D-glucose are shown.

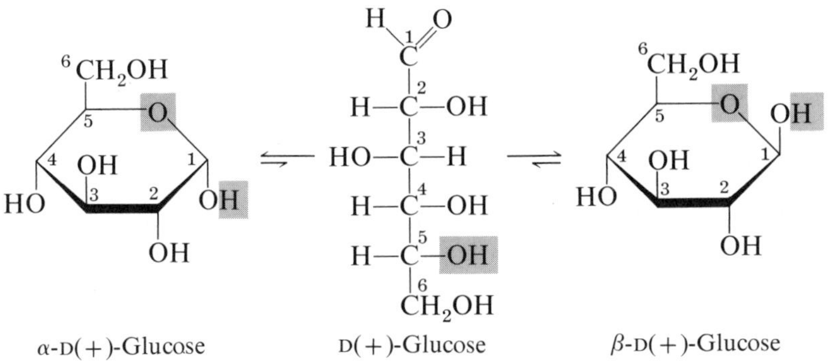

α-D(+)-Glucose
D(+)-Glucose
β-D(+)-Glucose

The two ring structures differ only in the relative configurations of the hydrogen and hydroxyl group with respect to carbon number one. Thus they are not enantiomers, but are related as epimeric diastereomers and have different chemical and physical properties. By convention, the *alpha* isomer is the one in which the hydroxyl group is positioned *below* the plane of the ring; if the hydroxyl group is positioned *above* the plane of the ring, it is the *beta* isomer.

Intramolecular hemiacetal formation is not unique to glucose. It occurs in galactose, mannose, fructose, and in the naturally occurring pentoses and heptoses. Figure 23.5 illustrates the equilibrium between the three forms of D-galactose, D-mannose, and D-fructose. Notice that galactose and mannose, like glucose, form a six-membered cyclic structure. Fructose can also exist in this form but is most commonly found in nature existing as a five-membered ring. Thus we are faced with another problem of nomenclature; we must have a way of distinguishing between sugars that form five-membered cyclic structures and those whose ring compounds contain six atoms. *Sugars with six-membered rings* are called **pyranoses** because they are related to the heterocyclic compound pyran; *those with five-membered rings* are **furanoses** and are related to the heterocyclic compound furan. The structures of pyran and furan are given below.

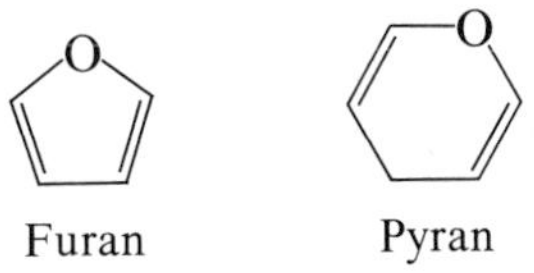

Furan
Pyran

One refers to the pyranose or furanose form of a monosaccharide to unambiguously designate the proper ring form of a sugar. For example, the following formulas both represent β-D-fructose.

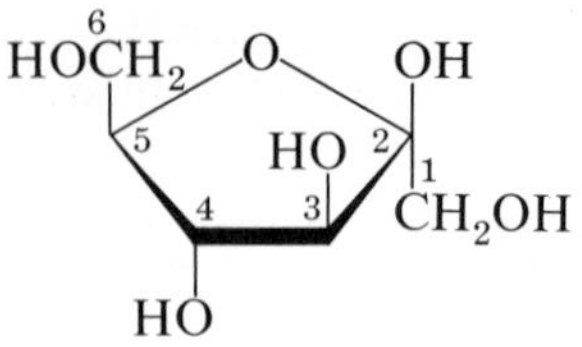

β-D-Fructose
(furanose form)

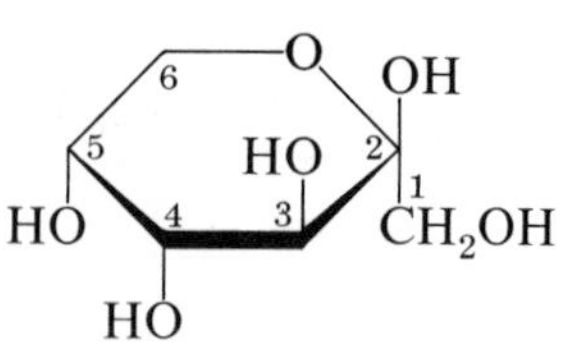

β-D-Fructose
(pyranose form)

Only by the use of the terms β-D-fructofuranose and β-D-fructopyranose can we differentiate between them.

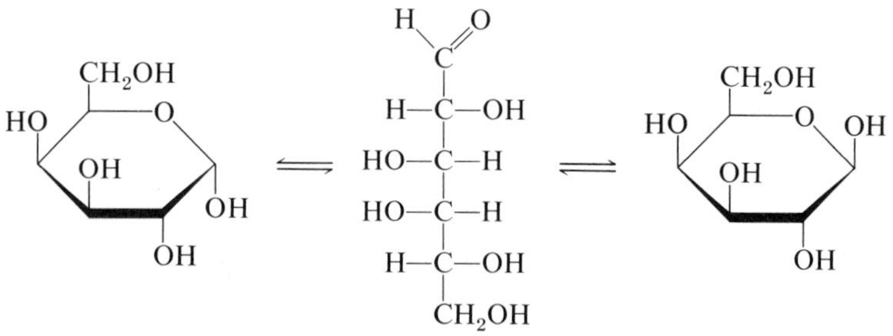

α-D(+)-Galactose D(+)-Galactose β-D(+)-Galactose

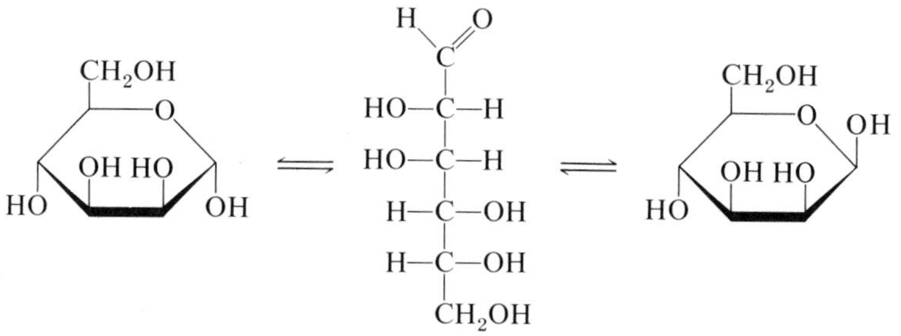

α-D(+)-Mannose D(+)-Mannose β-D(+)-Mannose

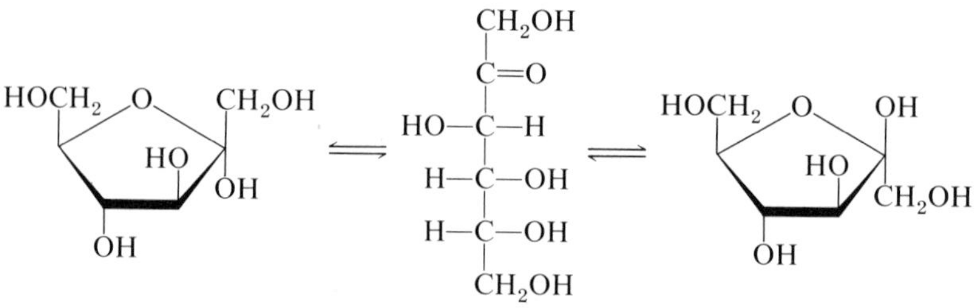

α-D(−)-Fructose D(−)-Fructose β-D(−)-Fructose

Figure 23.5 Mutarotation of D-galactose, D-mannose, and D-fructose.

23.5 Mutarotation

We have seen that α-D(+)-glucose and β-D(+)-glucose (α-D-glucopyranose and β-D-glucopyranose) are optical isomers but not mirror images. They are diastereomers and thus have different physical properties and can be separated by crystallization. If D-glucose is dissolved in water or in a 70% ethyl alcohol solution and is allowed to crystallize (by evaporation of the solvent or by lowering the temperature of the solution), the α-form of the sugar is obtained. Its melting point is 146 °C and the specific rotation is +112°. If glucose is crystallized from acetic acid or pyridine, the β-isomer is obtained. This form has a melting point of 150 °C and a specific rotation of +19°. Both α- and β-glucoses are stable in the crystalline form. When either of these isomers is dissolved in water, the optical rotation of each gradually changes with time and approaches a final equilibrium value of +53°. A quick calculation reveals that this value is not the average of the rotations of α- and β-glucoses. Instead of a 50–50 mixture of the two forms, the equilibrium condition must be such that one of the two isomers predominates. The percentages of α- and β-glucoses in an equilibrium mixture may be calculated by the use of simultaneous equations.

Let x = percent of β-isomer expressed as a decimal
Let y = percent of α-isomer expressed as a decimal

Therefore $$x(19°) + y(112°) = 53°$$

(The percentage of the β-isomer multiplied by its optical rotation plus the percentage of the α-isomer multiplied by its optical rotation must equal the optical rotation of the final solution.)

and $$x + y = 1.00$$

(Since both x and y are percentages expressed as a decimal, their sum must be unity.)

Solving for x and y in the two equations, we obtain

$$y = 0.366 = 37\%$$
$$x = 1.00 - y = 0.634 = 63\%$$

Thus, an equilibrium mixture of glucose contains 63% of the β-isomer, and 37% of the α-form. This *gradual change of optical rotation,* which continues until equilibrium is established, is known as **mutarotation.** It is a phenomenon that is common to all monosaccharides (and to some disaccharides) that can exist in α- and β-cyclic structures. There is sufficient evidence to believe that all monosaccharides that undergo mutarotation pass through the open-chain form, and a trace of this form is therefore present in the equilibrium mixture (less than 0.1%). It is for this reason that the monosaccharides

Table 23.1 Specific Rotations of Some Mono- and Disaccharides

Sugar	Specific Rotation (°)		
	α	β	Equilibrium Mixture
D-Glucose	+112	+19	+53
D-Fructose	−21	−133	−92
D-Galactose	+151	−53	+84
D-Mannose	+30	−17	+14
D-Lactose	+90	+35	+55
D-Maltose	+168	+112	+136

are said to have a *potential carbonyl group*. Table 23.1 lists specific rotation values for the common sugars.

23.6 Important Monosaccharides

a. Trioses

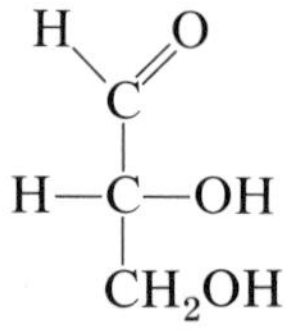

Glyceraldehyde

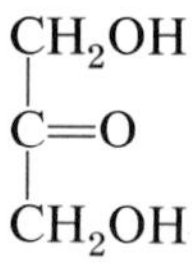

Dihydroxyacetone

Both of these compounds are found in plant and animal cells and play an important role in carbohydrate metabolism (Chapter 28).

b. Pentoses

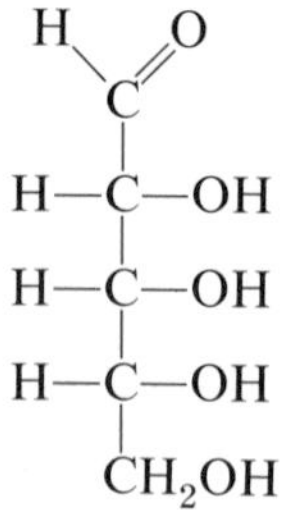

D(−)-Ribose

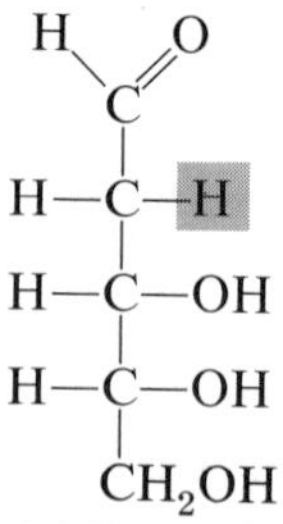

D(−)-2-Deoxyribose
(The prefix 2 is usually omitted from the name of the compound)

Ribose and deoxyribose are commonly found in nature. In the cyclic form they possess furanose structures.

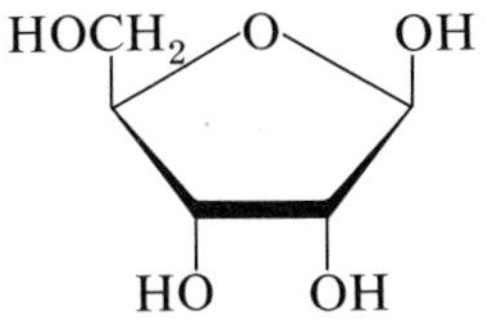

β-D-Ribose

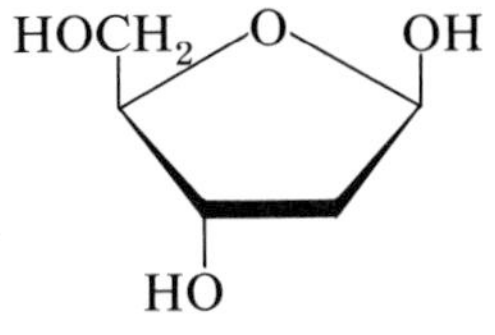

β-D-Deoxyribose

Both sugars are found in the furanose form in the nucleic acids of all living cells (Chapter 27). Ribose is also an intermediate in the pathway of carbohydrate metabolism and is a constituent of several of the coenzymes.

c. Hexoses

1. Glucose. Glucose is the most abundant sugar found in nature. It is commonly found in fruits, especially in ripe grapes, and for this reason it is often referred to as *grape sugar*. It is also known as *dextrose,* a name that derives from the fact that the predominant natural form of the sugar is dextrorotatory.

The majority of carbohydrates taken in by the body are eventually converted to glucose in a series of metabolic pathways. Glucose is the circulating carbohydrate of animals; the blood contains about 0.08% glucose and normal urine may contain anywhere from a trace to 0.2% glucose.

Commercially, glucose is made by the hydrolysis of starch. In the United States, corn starch is used in the process; in Europe the starch is obtained from potatoes. Glucose is only 75% as sweet as table sugar (sucrose) but it has the same caloric value.

2. Galactose. Galactose is formed by the hydrolysis of lactose, a disaccharide composed of a glucose unit and a galactose unit. It does not occur in nature in the uncombined state. The galactose needed by the human body for the synthesis of lactose (in the mammary glands) is obtained by the conversion of D-glucose into D-galactose. In addition, galactose is an important constituent of the glycolipids (Section 24.9-b) that occur in the brain and in the myelin sheath of nerve cells.

3. Fructose. Fructose is the only naturally occurring ketohexose. It exists in two structural forms, depending upon whether it is in the free state or combined. In the free state, fructose exists predominantly as a pyranose structure, whereas in the combined form (as in sucrose, inulin, and several phosphate esters), it exists as a furanose. Fructose is also referred to as *levulose* because it has an optical rotation that is strongly levorotatory (−92°). It is the sweetest sugar, and it is found, together with glucose and sucrose, in sweet fruits and honey.

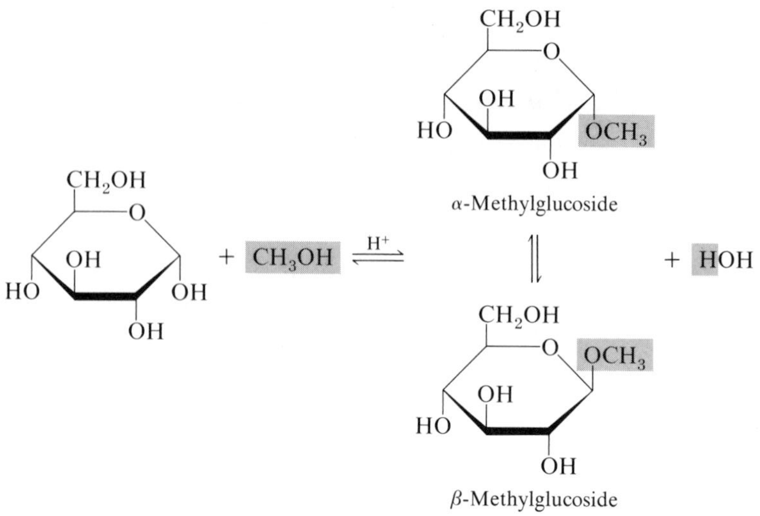
CH2OH
O
OH
HO
OH
OH
+ CH3OH
H+
CH2OH
O
OH
HO
OCH3
OH
α-Methylglucoside
+ HOH
CH2OH
O
OCH3
OH
HO
OH
β-Methylglucoside

Figure 23.6 The formation of an equilibrium mixture of α- and β-methylglucosides.

23.7 The Disaccharides

Most naturally occurring carbohydrates contain more than one monosaccharide unit. The manner in which two monosaccharide molecules are joined together is of particular interest. We have seen that the cyclic structures of the monosaccharides arise from intramolecular hemiacetal (or, in the case of fructose, hemiketal) formation. Recall that hemiacetals have a great tendency to react with a molecule of alcohol to form a stable compound known as an acetal (Section 18.5-a). If glucose is treated with methanol in the presence of an acid catalyst, an acetal, known as a *glucoside,* is formed (Figure 23.6). The more general term, **glycoside,** is used to refer to *the acetal that is formed when any carbohydrate reacts with a hydroxy compound.* Specific compounds may be named according to the sugars from which they are derived: hence, glucoside from glucose, galactoside from galactose, etc. *The carbon–oxygen–carbon linkage that joins the two components of the acetal* is called the **glycosidic linkage.** We shall see in the next sections that the biologically significant disaccharides and polysaccharides are composed of monosaccharide units that are joined by glycosidic linkages. The disaccharides all have the same molecular formula, $C_{12}H_{22}O_{11}$, and are therefore structural isomers of each other. They differ with respect to their constituent monosaccharide units and the type of glycosidic linkage connecting them.

a. Sucrose

Sucrose is known as beet sugar, cane sugar, table sugar, or simply as sugar. It is probably the largest selling pure organic compound in the world. As its names imply, sucrose can be obtained from sugar canes and sugar beets (whose juices contain 14–20% of the sugar) and is used chiefly as a food sweetener.

A molecule of sucrose may be envisioned to result from the combination of one molecule of α-D-glucopyranose and one molecule of β-D-fructofuranose; a molecule of water is eliminated in the process.

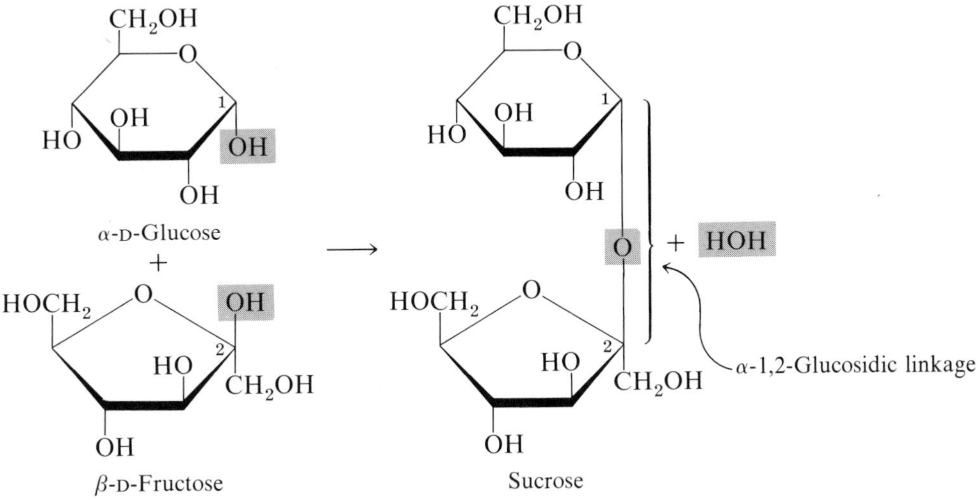

CH2OH
OH
HO
OH
OH
1
α-D-Glucose
+
HOCH2
O
OH
2
HO
CH2OH
OH
β-D-Fructose
Sucrose
+ HOH
α-1,2-Glucosidic linkage

The unique feature that characterizes the sucrose molecule is its glucosidic linkage; it involves the hydroxyl group on carbon number one of α-D-glucose and the hydroxyl group on carbon number two of β-D-fructose. By convention, sugars are read from left to right. This connecting linkage is therefore an α-1,2-glucosidic linkage. This bonding bestows certain properties upon sucrose that are quite different from those of the other disaccharides.

Sucrose, unlike the other disaccharides, is incapable of mutarotation. Thus, it exists in only one form both in the solid state and in solution. The presence of the 1,2-glucosidic linkage makes it impossible for sucrose to exist in the α- or β-configuration or in the open-chain form. This is a direct result of the fact that the potential aldehyde group of the glucose moiety and the potential ketone group of the fructose moiety have been tied up in the formation of the 1,2 (head-to-head) linkage. As long as the sucrose molecule remains intact, it cannot uncyclize to form the open-chain structure. Sucrose, therefore, does not undergo reactions that are typical of aldehydes and ketones, and it is said to be a *nonreducing* sugar (see Section 23.9-b).

The human body is unable to utilize sucrose or any other disaccharide directly because such molecules are too large to pass through cell membranes. Therefore, the disaccharide must first be broken down by hydrolysis into its two constituent monosaccharide units. In the body, this hydrolysis reaction is catalyzed by enzymes. The same hydrolysis reaction may be carried out in a test tube with dilute acid as a catalyst, but the reaction rate is much slower. The equation for the hydrolysis reaction is the reverse of the one given for the formation of sucrose.

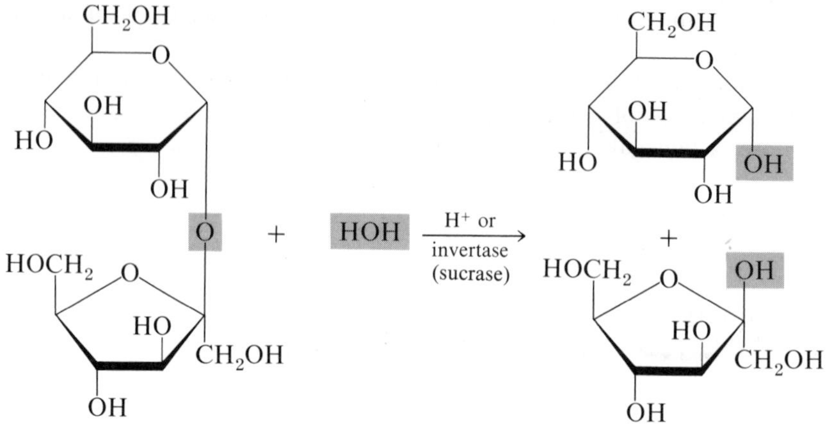

An interesting feature of this catalytic reaction involves the optical rotation of the hydrolyzate (the products of the hydrolysis reaction). Sucrose is dextrorotatory, $[\alpha]_D^{20} = +66.5°$, but upon hydrolysis the sign of the specific rotation changes (inverts) from positive to negative, $[\alpha]_D^{20} = -39°$. The product of the reaction is an equimolar mixture of D-glucose and D-fructose, both of which undergo mutarotation. The hydrolyzate actually contains an equilibrium mixture of α- and β-glucose, $[\alpha]_D^{20} = +53°$, and α- and β-fructose, $[\alpha]_D^{20} = -92°$. The specific rotation of the hydrolyzate is the sum of the specific rotations of each of the components.

$$\begin{array}{lccc} & \text{Sucrose} & \xrightarrow[\text{enzyme}]{\text{acid or}} & \underbrace{\text{D-Glucose} + \text{D-Fructose}}_{\text{Invert sugar}} \\ [\alpha]_D^{20} & +66.5° & & +53° \qquad -92° \end{array}$$

$$[\alpha]_D^{20} = (+53°) + (-92°) = -39°$$

The term **inversion** is often applied to the hydrolysis of sucrose since it is the only case in which the hydrolysis of a disaccharide effects a change in the sign of the specific rotation. *An equimolar mixture of* D-*glucose and* D-*fructose is called* **invert sugar.** The enzyme that catalyzes the hydrolysis is often referred to by its common name *invertase* rather than by its systematic name *sucrase.*

The inversion reaction has several practical applications. Since sucrose can exist in only one molecular configuration it is one of the most readily crystallizable sugars. Invert sugar has a much greater tendency to remain in solution. In the manufacture of jelly and candy and in the canning of fruit, crystallization of the sugar is undesirable; therefore, conditions leading to the hydrolysis of sucrose are employed in these processes. In addition, fructose is sweeter than sucrose, and hydrolysis adds to the sweetening effect.

b. Maltose

Maltose does not occur in the free state in nature to an appreciable extent. It occurs in animals as the principal sugar formed by the enzymic

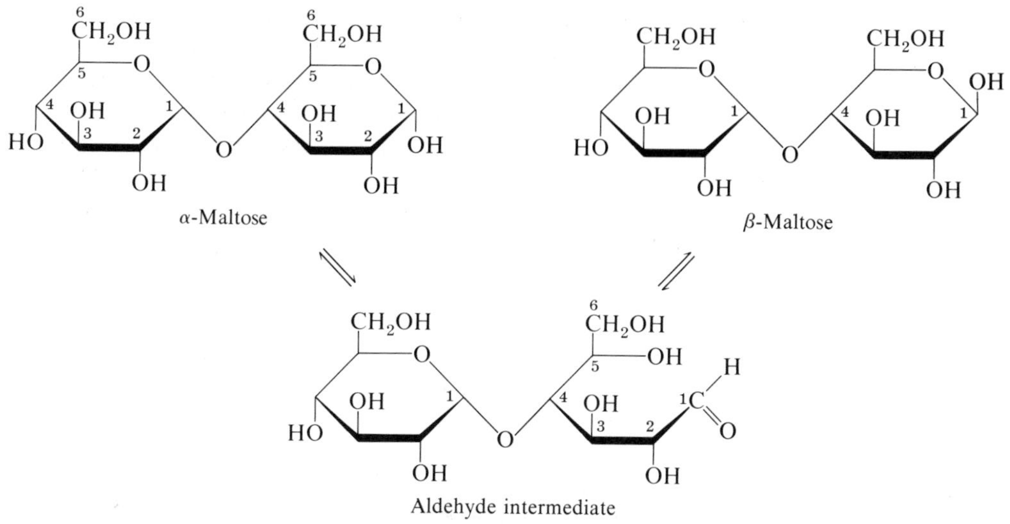
α-Maltose
β-Maltose
Aldehyde intermediate

Figure 23.7 Equilibrium mixture of maltose isomers.

(ptyalin) hydrolysis of starch. It is fairly abundant in germinating grain where it is formed by the enzymic (diastase) breakdown of starch. In the manufacture of beer, maltose is liberated by the action of malt (germinating barley) on starch, and for this reason it is often referred to as *malt sugar.*

Maltose is a reducing sugar, and it exhibits mutarotation. An equilibrium mixture of the maltose isomers is highly dextrorotatory, $[\alpha]_D^{20} = +136°$ (Figure 23.7). When maltose is hydrolyzed, either enzymically or by means of an acid catalyst, two molecules of D-glucose are produced. The formula of maltose must therefore incorporate two glucose molecules in such a way that the structure of the potential aldehyde group is retained. The glucose units in maltose are joined in a *head-to-tail* fashion through an alpha linkage from carbon number one of one glucose molecule to carbon number four of the second glucose molecule (i.e., an α-1,4-glucosidic linkage).

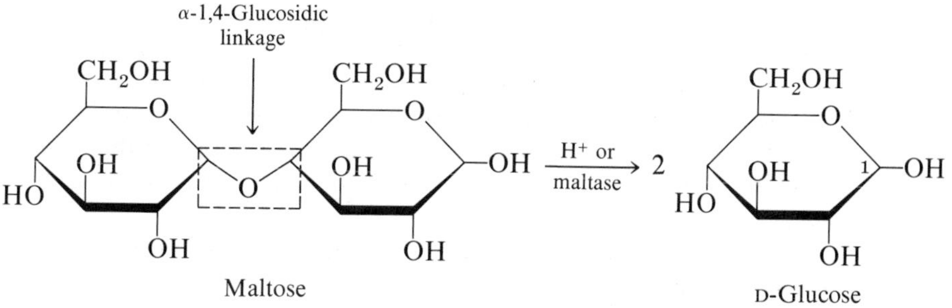
α-1,4-Glucosidic linkage
H+ or maltase
2
Maltose
D-Glucose

(We will use this convention for writing the hydroxyl group on carbon-1 when we don't wish to specify either the α-isomer or the β-isomer.)

Another disaccharide, **cellobiose,** is a stereoisomer of maltose. It is also composed of two glucose units, but in this case the two sugar moieties are joined by a β-1,4-glucosidic linkage. Like maltose, cellobiose is a reducing sugar and undergoes mutarotation. The enzyme maltase is stereospecific; it is capable only of hydrolyzing α-1,4-glucosidic linkages and for this reason does not act upon cellobiose. Cellobiose is obtained by the partial hydrolysis of cellulose. It may be further hydrolyzed to yield two molecules of D-glucose by the action of the enzyme *cellobiase* (which is specific for β-glucosidic linkages).

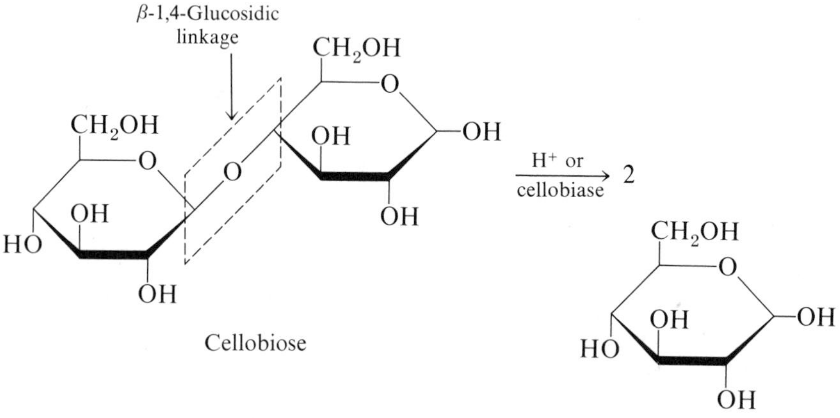
β-1,4-Glucosidic linkage
CH2OH
O
OH
HO
OH
O
CH2OH
O
OH
OH
OH
H+ or
cellobiase
2
CH2OH
O
OH
OH
HO
OH

Cellobiose

D-Glucose

c. Lactose

Lactose is known as *milk sugar* because it occurs in the milk of humans, cows, and other mammals. Human milk contains about 7.5% lactose, whereas cow's milk, which is not as sweet, contains about 4.5% lactose. Unlike most carbohydrates, which are plant products, lactose is one of the few carbohydrates associated exclusively with the animal kingdom. It is obtained commercially as a by-product in the manufacture of cheese. Lactose is one of the lowest ranking sugars in terms of sweetness (about one-sixth as sweet as sucrose). It is a reducing sugar and exhibits mutarotation. An equilibrium mixture of lactose has a specific rotation of $+55°$. The α-form of the sugar is of commercial importance as an infant food and in the production of penicillin.

Lactose is composed of one molecule of D-galactose and one molecule of D-glucose joined by a *β-1,4-galactosidic bond.* The two monosaccharides are obtained from lactose by acid hydrolysis or by the catalytic action of the enzyme *lactase.* About 70% of the world's population suffer from *lactose intolerance.* These people are unable to digest the sugar found in milk because they lack the enzyme *lactase.*

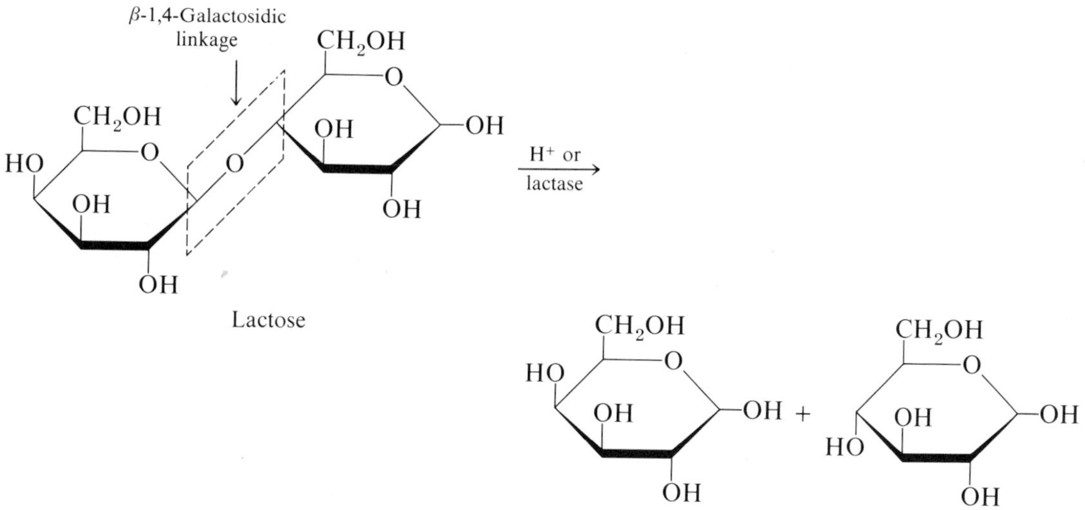
β-1,4-Galactosidic linkage
H+ or
lactase

Lactose

D-Galactose

D-Glucose

23.8 The Polysaccharides

The polysaccharides are the most abundant of the carbohydrates found in nature. They serve as reserve food substances and as structural components of plant cells. Polysaccharides are high molar mass (25,000–15,000,000) polymers of monosaccharides joined together by glycosidic linkages. Biochemically, the three most significant polysaccharides are starch, glycogen, and cellulose. They are also referred to as *homopolymers* since each of them yields only one type of monosaccharide (D-glucose) upon complete hydrolysis. (Inulin, another homopolymer, is found in the tubers of the Jerusalem artichoke and is composed entirely of fructofuranose units.) *Heteropolymers* may contain sugar acids, amino sugars, or noncarbohydrate substances. They are very common in nature (gums, pectins, hyaluronic acid) but will not be discussed in this text. The polysaccharides are nonreducing carbohydrates, are not sweet tasting (probably because of their limited solubility), and do not undergo mutarotation.

a. Starch

Starch is the most important source of carbohydrates in the human diet. (It is the chief caloric contributor to the diet.) It is found in most plants, particularly in the seeds, where it serves as a storage form of carbohydrate. The breakdown of starch to glucose serves to nourish the plant during periods of reduced photosynthetic activity. Starch is a mixture of two polymers, *amylose* and *amylopectin,* which can be separated from one another by physical and/or chemical methods. Natural starch consists of about 10–20% amylose, and 80–90% amylopectin.

Amylose is a straight-chain polysaccharide composed entirely of D-glucose units joined by an α-1,4-glucosidic linkage, as in maltose. Thus, amylose might be thought of as either *polymaltose* or *polyglucose.*[5]

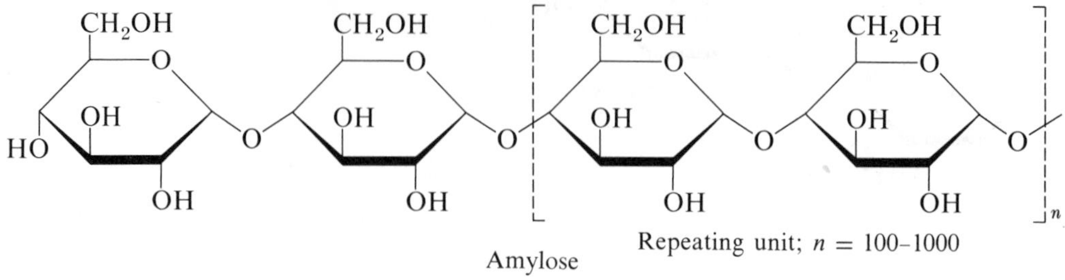

CH2OH
O
OH
HO
OH
Repeating unit; n = 100–1000

Amylose

Amylopectin is a branched-chain polysaccharide composed of glucose units that are linked primarily by α-1,4-glucosidic bonds, but have occasional α-1,6-glucosidic linkages, which are responsible for the branching. It has been estimated that there are over 1000 glucose units in amylopectin, and that branching occurs about once every 25–30 units (Figure 23.8b). The helical structure of amylopectin is disrupted by the branching of the chain, so instead of the deep blue color amylose gives with iodine, amylopectin produces a less intense red-brown color.

The complete hydrolysis of starch (amylose and amylopectin) yields, in three successive stages, dextrins, maltose, and glucose. Dextrins are glucose polysaccharides of intermediate size. The shine and stiffness imparted to clothing by starch is due to the presence of dextrins formed when the clothing is ironed. Because of their characteristic stickiness upon wetting, dextrins are used as adhesives on stamps, envelopes, and labels, and as pastes and mucilages. Since dextrins are more easily digested than starch, they are extensively used in the commercial preparation of infant foods (Dextrimaltose). A dried mixture of dextrins, maltose, and milk is used in the preparation of malted milk. Starch can be hydrolyzed by heating in the presence of dilute acid. In the human body it is degraded sequentially by several enzymes known collectively as *amylase.*

$$\text{Starch} \xrightarrow[\text{amylase}]{H^+,\ \Delta \text{ or}} \text{Dextrins} \xrightarrow[\text{amylase}]{H^+,\ \Delta \text{ or}} \text{Maltose} \xrightarrow[\text{maltase}]{H^+,\ \Delta \text{ or}} \text{Glucose}$$

b. Glycogen

Glycogen, often called *animal starch,* is the reserve carbohydrate of animals. Practically all mammalian cells contain some glycogen for the

[5] Experimental evidence indicates that the molecule is actually coiled like a spring and is not a straight chain of glucose units (Figure 23.8a). When coiled in this fashion the molecule has just enough room in its core to accommodate an iodine molecule. The characteristic blue-violet color that starch gives when treated with iodine is due to the formation of the amylose–iodine complex.

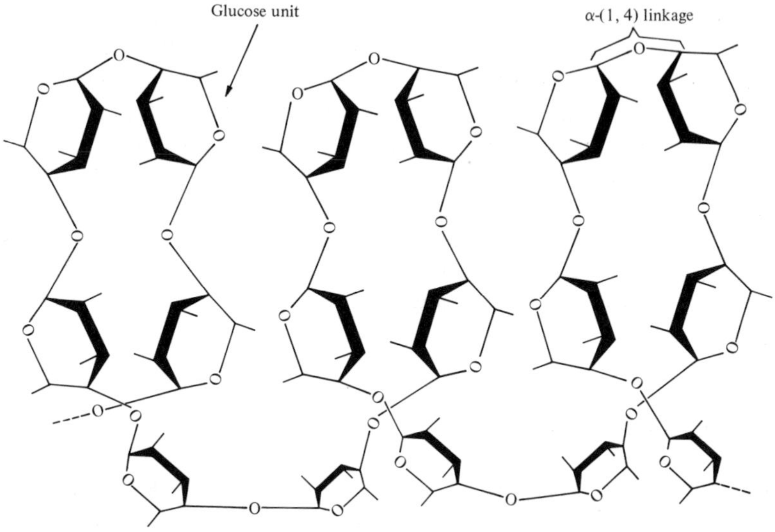

Glucose unit
α-(1, 4) linkage

(a)

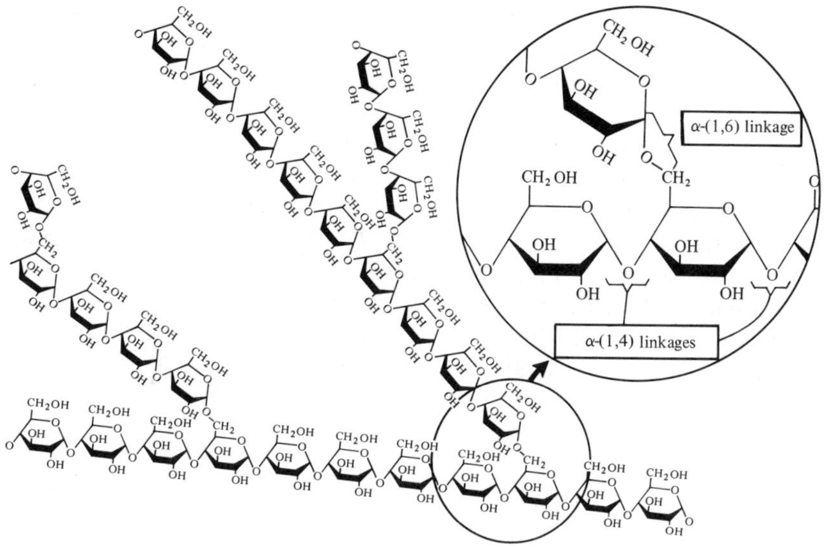

α-(1,6) linkage
α-(1,4) linkages

(b)

(c)

Figure 23.8 (a) The conformation of the amylose chain. (b) Branch points of amylopectin. (c) Branched array of glucose units in amylopectin or glycogen.

purpose of storing carbohydrate. However, it is especially abundant in the liver, 4–8%, and in muscle cells, 0.5–1.0%. When fasting or during periods of starvation, animals draw upon these glycogen reserves to obtain the glucose needed to maintain a proper state of metabolic balance.

In terms of structure, glycogen is quite similar to amylopectin, but it is more highly branched and its branches are shorter (8–12 glucose units in length). When treated with iodine, glycogen gives a red-brown color. Glycogen can be broken down into its D-glucose subunits by acid hydrolysis or by means of the same enzymes that attack starch. In animals, the enzyme *phosphorylase* catalyzes the breakdown of glycogen into phosphate esters of glucose (see Section 28.3).

c. Cellulose

Cellulose is a fibrous carbohydrate found in all plants where it serves as the structural component of the plant's cell wall. Cotton fiber and filter paper are almost entirely cellulose—wood is about 50% cellulose. From an industrial and economic standpoint, cellulose is the most important of all the carbohydrates. Since the earth is covered with vegetation, cellulose is the most abundant of all carbohydrates, accounting for over 50% of all the carbon found in the vegetable kingdom. Cellulose gives no color when treated with iodine.

Partial hydrolysis of cellulose yields only cellobiose. Thus the molecule must be composed of chains of D-glucose units (about 2000–3000) joined by β-1,4-glucosidic linkages. The chains are almost exclusively linear, unlike those of starch, which are highly branched. The linear nature of the cellulose chains allows a great deal of hydrogen bonding between hydroxyl groups on adjacent chains. As a result, the chains are closely packed into fibers, and there is little interaction with water or with any other solvent. Cotton, for example, is completely insoluble in water and of considerable mechanical strength.

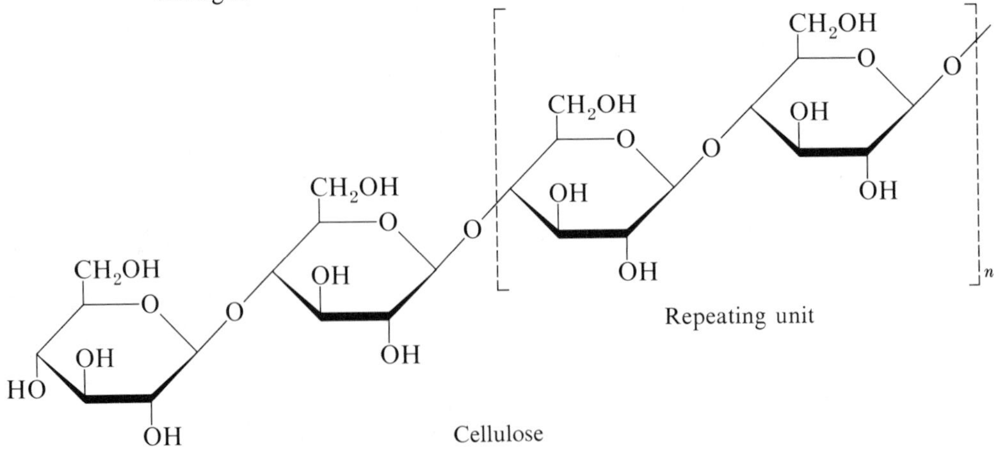
CH2OH
O
OH
HO
OH
CH2OH
O
OH
OH
O
O
CH2OH
O
OH
OH
O
CH2OH
O
OH
OH
O
n
Repeating unit

Cellulose

Cellulose yields D-glucose upon complete acid hydrolysis, yet man and the carnivorous animals cannot utilize cellulose as a source of glucose. Our digestive juices lack the enzymes that hydrolyze beta glucosidic linkages. Carnivorous animals need cellulose in the diet to provide bulk for the feces, thus preventing constipation. Many microorganisms and herbivorous animals (cow, horse, sheep) can digest cellulose. The higher animals can do so only because they contain microorganisms in their digestive tracts whose enzymes catalyze cellulose hydrolysis. The termite also contains such microorganisms, thus enabling it to subsist on a wood diet. This, once again, emphasizes the extreme stereospecificity of biochemical processes.

23.9 Tests for Carbohydrates

The chemical tests employed to detect carbohydrates and to distinguish among them may be divided into two broad categories: (1) tests based on the production of furfural derivatives and (2) tests based on the reducing property of sugars.

a. Tests Based on the Production of Furfural and Its Derivatives

When a monosaccharide is treated with a strong acid, dehydration of the monosaccharide occurs. If the monosaccharide is a pentose, the dehydrated product is *furfural;* a hexose gives rise to *hydroxymethylfurfural.* (Trioses and tetroses are incapable of undergoing this reaction since they do not possess the requisite minimum of five carbon atoms.) Under vigorous conditions, all carbohydrates containing five or more carbons can be made to react; under milder conditions only certain compounds undergo the dehydration reaction, thus affording us with more specific tests.

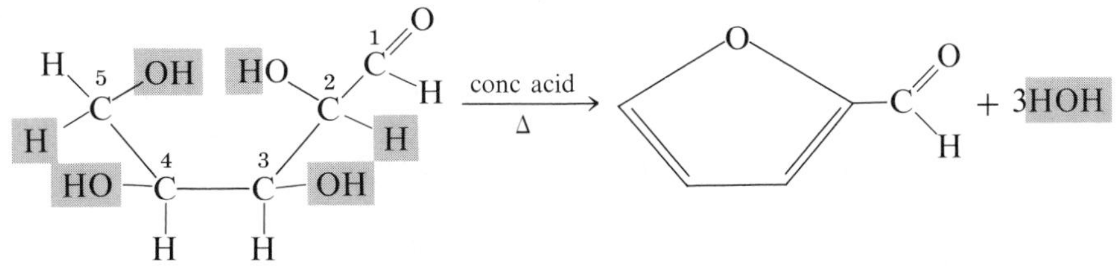
conc acid
Δ
+ 3HOH

A pentose

Furfural

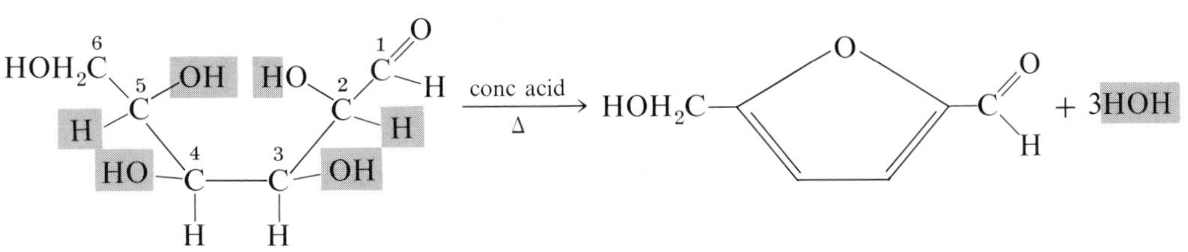
conc acid
Δ
+ 3HOH

A hexose

Hydroxymethylfurfural

In the presence of concentrated acid, various phenolic compounds (e.g., α-naphthol, orcinol, resorcinol) will condense with the furfural or the hydroxymethylfurfural to form colored dyes of the triphenylmethane type. The formation of these colored compounds is a positive test for the presence of carbohydrates.

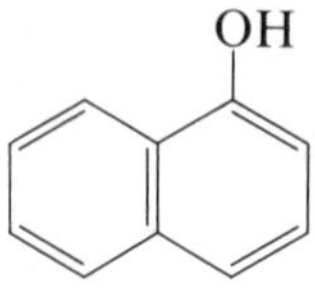
OH

α-Naphthol
(α-Hydroxynaphthalene)

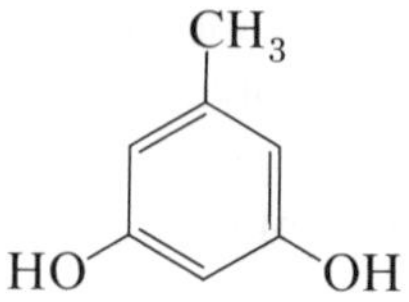
CH3
HO
OH

Orcinol
(3,5-Dihydroxytoluene)

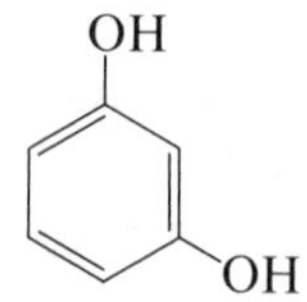
OH
OH

Resorcinol
(*m*-Hydroxyphenol)

1. Molisch Test. The Molisch test is the most general test for the presence of carbohydrates because it gives a positive test (indicated by the appearance of a purple ring) with all carbohydrates larger than tetroses. Concentrated H_2SO_4 will first hydrolyze all di- and polysaccharides to monosaccharides, the monosaccharides then form furfural or hydroxymethylfurfural, and these compounds condense with α-naphthol to form a purple condensation product.

2. Bial's Orcinol Test. Bial's orcinol test is important in the determination of pentoses and the pentose-containing subunits of nucleic acids. The reaction is *not* specific for pentoses, however, since other compounds such as trioses, uronic acids, and certain heptoses produce identical bright blue condensation products with orcinol. It is therefore important that this test be applied to reasonably pure samples. Bial's test is useful in differentiating between hexoses and pentoses because orcinol condenses with furfural (from pentoses) to form a blue-green compound and with hydroxymethylfurfural (from hexoses) to form a yellow-brown product.

3. Seliwanoff's Test. This test makes use of the fact that hot HCl dehydrates the ketohexoses to form hydroxymethylfurfural much faster than it acts upon the corresponding aldohexoses. Therefore, the test distinguishes aldohexoses from ketohexoses based on their differential rates of reaction. During a given time interval (usually chosen to be 60 seconds), fructose and those di- and polysaccharides that are hydrolyzed to fructose by HCl will react with the resorcinol in Seliwanoff's reagent to yield compounds having a deep red color. Aldohexoses yield substances that are just slightly pink.

b. Tests Based on the Reducing Property of Sugars

Because sugars are polyhydroxy aldehydes and ketones, they are usually able to undergo all of the normal reactions of the alcohols and of the carbonyl compounds. All monosaccharides and some disaccharides have the ability to reduce an alkaline solution of cupric ion. Any carbohydrate capable of this reduction without first undergoing hydrolysis is said to be a **reducing sugar.** It is not possible to write a complete oxidation-reduction equation for the process because many of the oxidized end-products have not yet been identified. (This reaction has been adopted as a simple and rapid diagnostic test for the presence of glucose in the blood or in the urine.)

$$\text{Reducing sugar} + Cu^{2+} \xrightarrow{OH^-} Cu_2O_{(s)} + \text{Oxidized products of the sugar}$$

The oxidation of sugars is a process that is best understood in terms of chemical equilibria. Recall that in solution only a very small fraction of most mono- and disaccharides exists in the open-chain, free-aldehyde form (Section 23.5). The cupric ions react with this small amount of free aldehyde, and in so doing, serve to continuously remove the open-chain form from the equilibrium mixture. This causes a continuous shifting of the equilibrium *away* from the cyclic forms *toward* the open-chain form. Gradually, all the sugar in the ring form is converted to the open-chain form and is subsequently oxidized. In Section 23.7-a it was mentioned that sucrose is not a reducing sugar because it does not contain a potentially free aldehyde group. This is equivalent to saying that sucrose cannot be converted to an open-chain form. Lactose and maltose are reducing sugars because they each possess a potentially free aldehyde group.

The alkaline reaction conditions facilitate the attainment of a tautomeric equilibrium between the enol and keto forms of the open-chain structure (Section 18.5-b). Enolization may result in the formation of glucose from fructose or vice versa. This fact is vital to an understanding of why fructose is a reducing sugar even though it contains a ketonic (rather than an aldehydic) carbonyl group. The following interconversion explains the phenomenon and will be dealt with again in the chapter on carbohydrate metabolism.

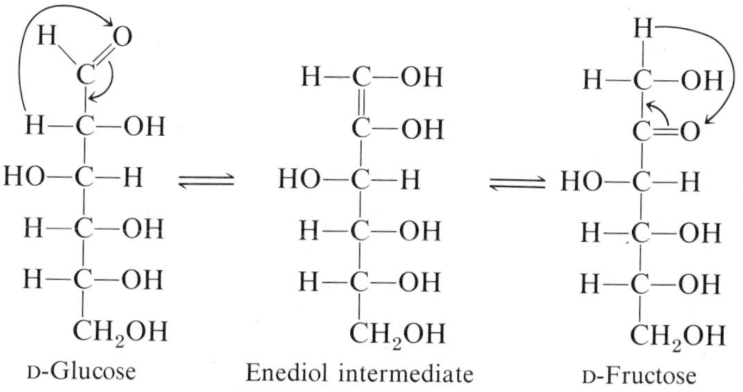
D-Glucose
Enediol intermediate
D-Fructose

1. Benedict's Test and Fehling's Test. (See Section 18.5-c.) These tests are performed under mildly alkaline conditions. They are extremely sensitive tests for the presence of carbohydrates but are by no means specific for any particular sugar. Aldehydes, as well as many other compounds, will give a positive test with these reagents. The formation of a red cuprous oxide precipitate is the criterion of a positive test. Clinitest tablets, which are used in clinical laboratories to test for sugar in the urine, contain cupric ions and are based on the same principles as the Benedict test. A green color indicates very little sugar, whereas a brick-red color indicates sugar in excess of 2 grams per 100 ml of urine.

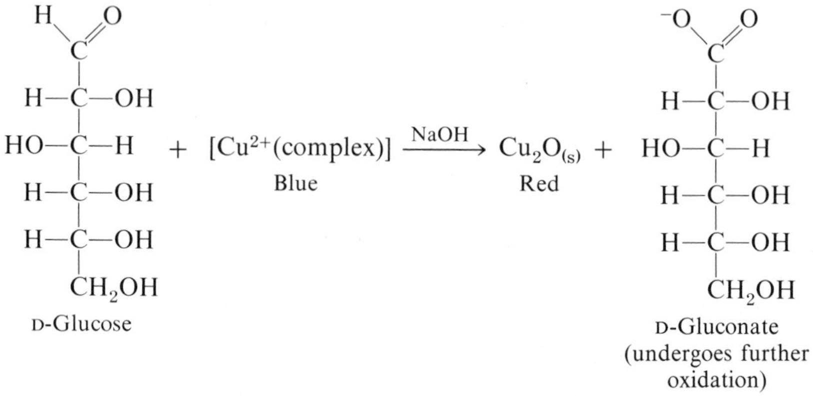

H
O
C
H—C—OH
HO—C—H
H—C—OH
H—C—OH
CH2OH
D-Glucose
+ [Cu2+(complex)]
Blue
NaOH
Cu2O(s)
Red
+
-O
O
C
H—C—OH
HO—C—H
H—C—OH
H—C—OH
CH2OH
D-Gluconate
(undergoes further
oxidation)

2. ***Barfoed's Test.*** The Barfoed test is used to distinguish between reducing monosaccharides and reducing disaccharides. A differential rate of reaction with cupric acetate in acetic acid is the basis of the test. Within a given time interval (usually 10 min), only monosaccharides will reduce the cupric ion, again forming the red insoluble cuprous oxide. Under acid conditions, cupric ion is a weaker oxidizing agent and is therefore only capable of oxidizing the monosaccharides within the assigned time. If heating is prolonged, the disaccharides may be hydrolyzed by the acid, and the resulting monosaccharides will give a positive test.

Exercises

23.1 Define, explain, or give an example of each of the following.

(a) carbohydrate
(b) triose
(c) hexose
(d) aldotetrose
(e) ketopentose
(f) monosaccharide
(g) disaccharide
(h) polysaccharide
(i) invert sugar
(j) reducing sugar
(k) glycosidic linkage
(l) glycoside
(m) glucoside
(n) sucrose
(o) α-lactose
(p) β-maltose
(q) α-D-glucopyranose
(r) $[\alpha]_D^{20} = +53°$
(s) epimers
(t) dextrose

23.2 Draw the structural formulas for three compounds that have the empirical formula $C_x(H_2O)_y$ but are not carbohydrates.

23.3 Draw structural formulas for the open-chain and cyclic forms of the four naturally occurring hexoses.

23.4 Draw all the possible stereoisomeric ketohexoses.

23.5 What structural relationship is indicated by the term D-sugar? Why are (+)-glucose and (−)-fructose both classified as D-sugars?

23.6 How does ribose differ from deoxyribose? Write a cyclic formula for each.

23.7 Explain the structural similarities and differences between each of the following.

(a) maltose and cellobiose
(b) starch and cellulose
(c) amylose and amylopectin
(d) amylopectin and glycogen

23.8 Give the common name of each of the following systematically named sugars.

(a) 4-(β-D-galactopyranosyl)-D-glucose
(b) 4-(α-D-glucopyranosyl)-D-glucose
(c) 4-(β-D-glucopyranosyl)-D-glucose
(d) α-D-glucopyranosyl-β-D-fructofuranoside

23.9 Draw formulas for each of the following.

(a) L-(−)-glucose
(b) L-(+)-fructose
(c) the enantiomer of D-(+)-mannose
(d) a ketopentose in the furanose form
(e) a ketohexose in the pyranose form
(f) an aldoheptose in the pyranose form
(g) α-methyl-D-fructoside
(h) β-methyl-L-glucopyranoside
(i) two mannose molecules joined by an α-1,4 glycosidic bond
(j) a trisaccharide composed of three different hexose units

23.10 Write formulas for the eight isomeric aldopentoses. Indicate which pairs of isomers are epimers.

23.11 Explain what is meant by the term mutarotation. How can it be shown that a solution of α-D-glucose exhibits mutarotation? Do any of the isomeric aldotetroses exhibit mutarotation? Explain.

23.12 What structural characteristics are necessary if a disaccharide is to be a reducing sugar? Draw the structure of a hypothetical nonreducing disaccharide composed of two aldopentoses.

23.13 Trehalose is a nonreducing disaccharide that yields glucose upon hydrolysis. Its systematic name is α-D-glucopyranosyl-α-D-glucopyranoside. Draw the structural formula of trehalose.

23.14 Ketones cannot be oxidized by mild oxidizing agents, yet fructose is oxidized by Benedict's reagent. Explain.

23.15 Dilute alkali also catalyzes the interconversion of fructose and mannose. Write structures to show this tautomeric equilibrium.

23.16 Which of the following will give a positive Benedict's test?

(a) L-galactose
(b) levulose
(c) D-mannose
(d) malt sugar
(e) cane sugar
(f) invert sugar
(g) milk sugar
(h) cellobiose
(i) inulin
(j) starch
(k) cellulose
(l) glycogen

23.17 Give the hydrolysis products of the compounds in (16) where possible.

23.18 What is the general test for a carbohydrate? Explain.

23.19 How could you distinguish between each of the following?

(a) glucose and starch
(b) glucose and fructose
(c) sucrose and invert sugar
(d) mannose and maltose
(e) galactose and ribose
(f) fructose and erythrose

23.20 Write equations for each of the following.

(a) hydrolysis of sucrose
(b) stepwise hydrolysis of starch
(c) oxidation of glucose with Benedict's reagent
(d) enolization of D-fructose to D-glucose
(e) reaction of D-galactose with HCN
(f) formation of a methyl glucoside
(g) production of furfural

23.21 α-D-Fructose has a specific rotation of $-21.0°$ and the specific rotation of β-D-fructose is $-133.0°$. An equilibrium mixture of the two forms exhibits a specific rotation of $-92.0°$. Calculate the percentage of α- and β-fructose present in the equilibrium mixture.

23.22 Show that D-glyceraldehyde is identical to R-glyceraldehyde and that L-glyceraldehyde is identical to S-glyceraldehyde (Section 22.5).

23.23 L-sugars occur in nature, but they are not nearly as abundant as D-sugars. L-Fucose (6-deoxy-L-galactose) and L-rhamnose (6-deoxy-L-mannose) are constituents of some bacterial cell walls. Draw structures for L-fucose and L-rhamnose.

24

Lipids

The lipids are a heterogeneous group of organic compounds that are important constituents of plant and animal tissues. They are arbitrarily classed together according to their solubility in organic solvents such as benzene, ether, chloroform, carbon tetrachloride (the so-called fat solvents) and their insolubility in water. Their solubility properties are a function of their alkane-like structures. Edible lipids constitute approximately 25–28% of the diet of the average American, and they serve as a starting material for the production of many important commodities such as soap products. Within the past two decades, the role of lipids in the diet has received a great deal of attention because of the apparent connection between saturated fats and blood cholesterol with arterial disease. As we shall learn in Chapter 29, the lipids are the most important energy storage compounds in the animal kingdom. In contrast, you will recall, plants store most of their energy in the form of carbohydrates, primarily as starch. In addition, lipids provide insulation for the vital organs, protecting them from mechanical shock and maintaining optimum body temperature. Lipids are an integral component of cell membrane structure (recall Figure 13.4) and, as such, are associated with transportation across cellular membranes.

24.1 Classification of Lipids

Unlike polysaccharides and proteins, lipids are not polymers—they lack a repeating monomeric unit. However, like carbohydrates, they can be classified according to their hydrolysis products and according to similarities in their molecular structures. Three major subclasses are recognized.

1. Simple lipids
   (a) Fats and oils,[1] which yield fatty acids and glycerol upon hydrolysis.
   (b) Waxes, which yield fatty acids and long-chain alcohols upon hydrolysis.
2. Compound lipids
   (a) Phospholipids, which yield fatty acids, glycerol, phosphoric acid, and a nitrogen-containing alcohol upon hydrolysis.
   (b) Glycolipids, which yield fatty acids, sphingosine or glycerol, and a carbohydrate upon hydrolysis.
   (c) Sphingolipids, which yield fatty acids, sphingosine, phosphoric acid, and an alcohol component upon hydrolysis.
3. Steroids—compounds containing a phenanthrene structure (Section 15.2) that are quite different from lipids made up of fatty acids.

24.2 Simple Lipids

Fats and oils are the most abundant lipids found in nature. Both types of compounds are called triacylglycerols[2] because they are *esters* composed of *three fatty acids* joined to *glycerol,* a *trihydroxy alcohol.*

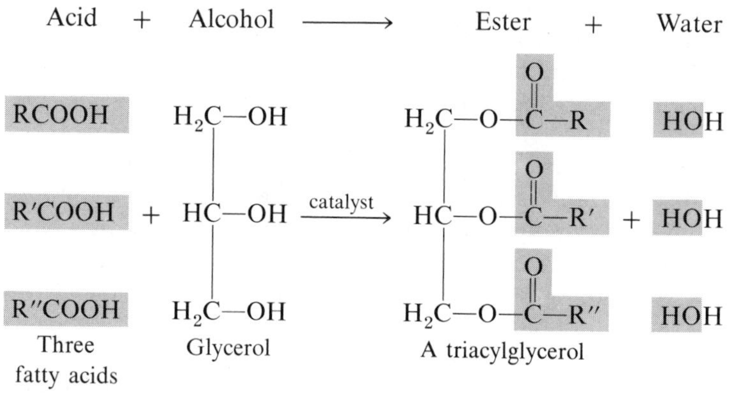

Acid + Alcohol ⟶ Ester + Water
RCOOH
R′COOH
R″COOH
H2C—OH
HC—OH
H2C—OH
catalyst
H2C—O—C—R
HC—O—C—R′
H2C—O—C—R″
O
HOH
Three fatty acids
Glycerol
A triacylglycerol

Further classification of triacylglycerols is made on the basis of their physical states at room temperature. It is customary to call a lipid a **fat** if it is a *solid at* 25 °C, and an **oil** if it is a *liquid at the same temperature.* (These differences in melting points reflect differences in the degree of unsaturation of the constituent fatty acids.) Furthermore, lipids obtained from animal sources are usually solids whereas oils are generally of plant origin. Therefore, we commonly speak of *animal fats* and *vegetable oils.*

[1]The student should not confuse the term oil, here used to refer to a particular group of lipids, with the hydrocarbon oils.

[2]Formerly these compounds were called triglycerides. An international nomenclature commission has recommended that this chemically inaccurate term no longer be used.

24.3 Fatty Acids

Since fats and oils both contain glycerol, the differences between them must be due to differences in their fatty acid components. It is therefore customary to describe triacylglycerols in terms of their fatty acids.

Fatty acids are long-chain, monocarboxylic acids. They may be either saturated or unsaturated, but they are invariably straight chains rather than branched or cyclic.[3] Furthermore, fatty acids that are obtained from lipids almost always contain an even number of carbon atoms.[4] The most abundant fatty acids are given in Table 24.1. Fatty acids are usually called by their common names. They are derived from Greek or Latin words that indicate the source of the compound.

Notice that stearic acid, a saturated fatty acid, contains the same number of carbon atoms as the three unsaturated acids listed in the table yet it has a much higher melting point. It is generally the case that the greater the degree of unsaturation of a fatty acid, the lower its melting point. This generalization enables us to explain the differences between fats and oils. *Fats contain a greater proportion of saturated fatty acids, whereas oils contain a greater percentage of unsaturated fatty acids.*

If all three hydroxyl groups of the glycerol molecule are esterified with the same fatty acid, the resulting ester is called a **simple triacylglycerol.** Although some simple triacylglycerols have been synthesized in the laboratory, they rarely occur in nature. All of the triacylglycerols obtained from naturally occurring fats and oils contain two or three different fatty acid components and are thus termed **mixed triacylglycerols.**

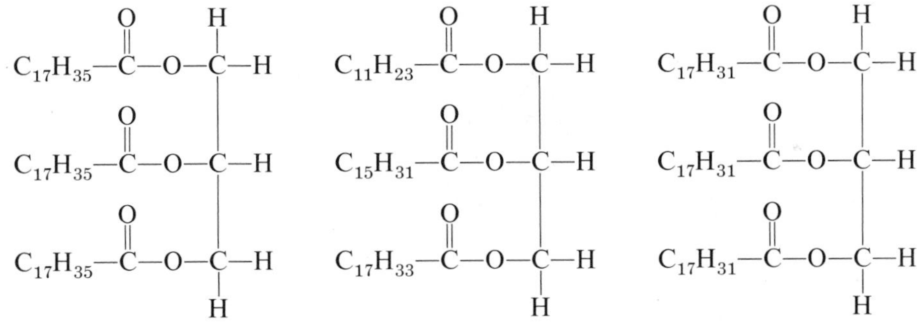

Glyceryl stearate (Tristearin) (a simple triacylglycerol) mp 71°C

Glyceryl lauropalmitooleate (a mixed triacylglycerol)

Glyceryl linoleate (Trilinolein) (a simple triacylglycerol) mp 9°C

[3]Two fatty acids are exceptions to this statement. Both malvalic acid and sterculic acid contain the very highly strained cyclopropene ring.

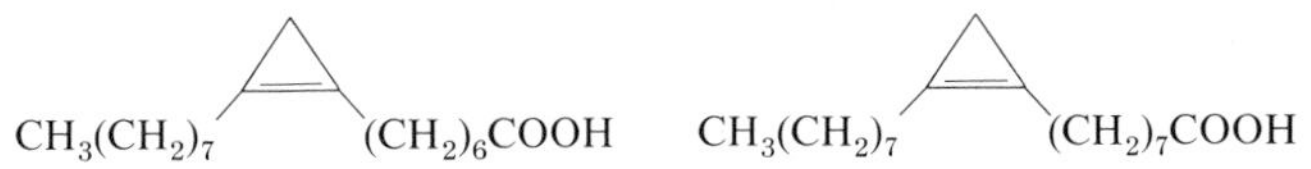

Malvalic acid — Sterculic acid

[4]The reason for this is that the hydrocarbon chain of a given fatty acid is biosynthesized two carbon units at a time (see Section 29.6).

Table 24.1 Common Fatty Acids

Name	MP (°C)	Condensed Structural Formula	Abbreviated[a] Formula	Source
Lauric acid	44	$CH_3(CH_2)_{10}COOH$	$C_{11}H_{23}COOH$	Coconut oil[b]
Myristic acid	54	$CH_3(CH_2)_{12}COOH$	$C_{13}H_{27}COOH$	Coconut oil
Palmitic acid	63	$CH_3(CH_2)_{14}COOH$	$C_{15}H_{31}COOH$	Most fats and oils
Stearic acid	70	$CH_3(CH_2)_{16}COOH$	$C_{17}H_{35}COOH$	Most fats and oils
Oleic acid	13	$CH_3(CH_2)_7CH{=}CH(CH_2)_7COOH$[c]	$C_{17}H_{33}COOH$	Most fats and oils
Linoleic acid	−5	$CH_3(CH_2)_4CH{=}CHCH_2CH{=}CH(CH_2)_7COOH$	$C_{17}H_{31}COOH$	Linseed oil
Linolenic acid	−11	$CH_3CH_2CH{=}CHCH_2CH{=}CHCH_2CH{=}CH(CH_2)_7COOH$	$C_{17}H_{29}COOH$	Linseed oil

[a]Saturated fatty acids have the general formula $C_nH_{2n+1}COOH$;unsaturated fatty acids are of the form $C_nH_{2n-1}COOH$, $C_nH_{2n-3}COOH$, $C_nH_{2n-5}COOH$, etc.

[b]Coconut oil is a liquid in the tropics, but at room temperature in the temperate zone it is a solid.

[c]Notice that because of the presence of the double bond in the molecule, there is the possibility of geometric isomers:

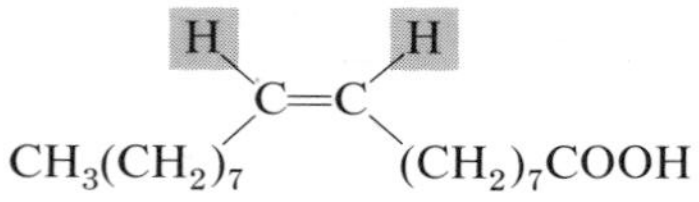
H
H
C=C
CH3(CH2)7
(CH2)7COOH

Oleic acid

and

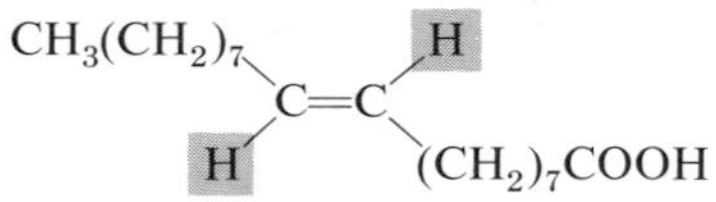
CH3(CH2)7
H
C=C
H
(CH2)7COOH

Elaidic acid

Elaidic acid is rarely found in nature. In fact, almost all of the naturally occurring unsaturated fatty acids have the *cis* configuration.

Table 24.2 Fatty Acid Components of Some Common Fats and Oils

	Component Acids (%)[a]								
	Myristic C_{14}	*Palmitic* C_{16}	*Stearic* C_{18}	*Oleic* C_{18}	*Linoleic* C_{18}	*Linolenic* C_{18}	*Eleostearic* C_{18}	*Saponification Values*	*Iodine Values*
Fats									
Butter	7–10	24–26	10–13	30–40	4–5			210–230	26–28
Lard	1–2	28–30	12–18	40–50	6–7			195–203	46–70
Tallow	3–6	24–32	20–25	37–43	2–3			190–200	30–48
Edible Oils									
Olive oil		9–10	2–3	83–84	3–5			187–196	79–90
Corn oil	1–2	8–12	2–5	19–49	34–62			187–196	109–133
Soybean oil		6–10	2–5	20–30	50–60	5–11		189–195	127–138
Cottonseed oil	0–2	20–25	1–2	23–35	40–50			190–198	105–114
Peanut oil		8–9	2–3	50–65				188–195	84–102
Safflower oil		6–7	2–3	12–14	75–80	0.5–0.15		188–194	140–156
Nonedible Oils									
Linseed oil		4–7	2–4	25–40	35–40	25–60		187–195	170–185
Tung oil		3–4	0–1	4–15			75–90	190–197	163–171

From *Organic Chemistry,* W. W. Linstromberg, D. C. Heath and Company, Boston, 1966. Reprinted by permission.
[a]Totals less than 100% indicate the presence of lower or higher acids in small amounts.

No single formula can be written to represent the naturally occurring fats and oils since they are highly complex mixtures of molecules in which many different fatty acids are represented. Table 24.2 shows the fatty acid composition of some common fats and oils. Notice that a fairly wide range of values occurs. The range is wide because the composition of lipids is variable and depends upon the plant or animal species involved as well as upon dietetic and climatic factors. For example, lard from corn-fed hogs is more highly saturated than lard from peanut-fed hogs. Linseed oil obtained from cold climates is more unsaturated than linseed oil from warm climates. Palmitic acid is the most abundant of the saturated fatty acids and oleic acid is the most abundant unsaturated fatty acid. Unsaturated fatty acids predominate over saturated ones for most plants and animals.

24.4 Physical Properties of Lipids

As previously mentioned, the lipids may be either liquids or noncrystalline solids at room temperature. Contrary to popular belief, *pure* fats and oils are colorless, odorless, and tasteless. The characteristic colors, odors, and flavors associated with lipids are imparted to them by foreign substances that have been absorbed by the lipid and are soluble in them. For example, the yellow color of butter is due to the presence of the pigment carotene; the taste of butter is a result of two compounds, diacetyl ($CH_3COCOCH_3$), and 3-hydroxy-2-butanone ($CH_3COCHOHCH_3$), that are produced by bacteria in the ripening of the cream. Fats and oils are lighter than water, having densities of about 0.8 gram/cm^3. They are poor conductors of heat and electricity and therefore serve as excellent insulators for the body.

24.5 Chemical Properties of Lipids

a. Saponification

Triacylglycerols may be hydrolyzed by several procedures, the most common of which utilizes alkali or enzymes called lipases. *Alkaline hydrolysis* is termed **saponification** because one of the products of the hydrolysis is a soap, generally sodium or potassium salts of fatty acids (Section 19.10-b). This hydrolysis reaction also provides a useful analytical method for the determination of a constant, the **saponification number,** which is characteristic of the simple lipids.[5] The saponification number of a lipid is defined as the *number of milligrams of potassium hydroxide required to saponify 1 gram of a fat or an oil.* Lipids containing shorter-chain fatty acids will have higher

[5] Within the past twenty years the analysis of lipids has been revolutionized. Use is now made of gas-liquid chromatography (glc) and thin-layer chromatography (tlc) for qualitative identification of lipids and fatty acid components and quantitative determination of the percentage of each component present. These techniques have largely replaced the older determinations of saponification number and iodine number.

saponification numbers than those containing longer-chain fatty acids. The following example illustrates the calculation of saponification numbers.

Example 24.1 Determine the saponification number of tristearin.

$$CH_3(CH_2)_{16}\overset{\overset{O}{\|}}{C}-O-\overset{\overset{H}{|}}{\underset{|}{C}}-H$$

$$CH_3(CH_2)_{16}\overset{\overset{O}{\|}}{C}-O-\overset{|}{\underset{|}{C}}-H + 3\ KOH \longrightarrow 3\ CH_3(CH_2)_{16}C\overset{\overset{O}{\diagup}}{\underset{O^-K^+}{\diagdown}} \quad + \begin{matrix} H \\ | \\ HO-C-H \\ | \\ HO-C-H \\ | \\ HO-C-H \\ | \\ H \end{matrix}$$

$$CH_3(CH_2)_{16}\overset{\overset{O}{\|}}{C}-O-\overset{|}{\underset{\underset{H}{|}}{C}}-H$$

Potassium stearate
(a soap)

Tristearin
(M = 890)

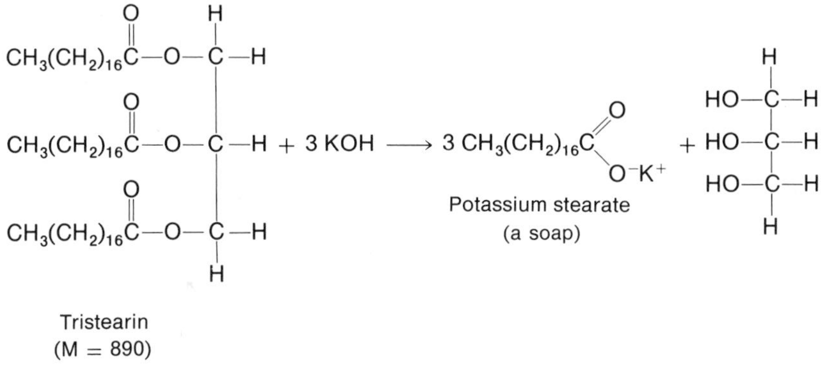

$$\text{Saponification no.} = \left(\frac{3\ \text{moles KOH}}{\text{mole lipid}}\right)\left(\frac{(56\ \text{g/mole KOH})(1000\ \text{mg/g})}{\text{g/mole lipid}}\right)$$

$$\text{Saponification no.} \atop \text{of tristearin} = \frac{168{,}000\ \text{mg KOH per mole lipid}}{890\ \text{g lipid per mole lipid}}$$

$$= 189\ \text{mg KOH per g lipid}$$

Experimentally, a weighed sample of a fat or an oil is saponified with a standard solution of alcoholic potassium hydroxide. Following saponification, the excess alkali is determined by titration with standard acid.

b. Halogenation

Unsaturated fatty acids, whether they are free or combined as esters in fats and oils, react with halogens by addition at the double bond(s). The reaction (halogenation) results in the decolorization of the halogen solution (Section 14.4-a). Since the degree of absorption by a fat or oil is proportional to the number of double bonds in the fatty acid moieties, the amount of halogen absorbed by a lipid can be used as an index of the degree of unsaturation. The index value is called the **iodine number** and is defined as the *number of grams of iodine (or iodine equivalent) that will add to 100 grams of fat or oil.* This value is influenced by a number of factors, such as percentage of unsaturated fatty acid in the triacylglycerol molecule and the degree of unsaturation of each fatty acid. As a general rule, a high iodine number indicates a high degree of unsaturation. Natural fats that have a preponderance of saturated fatty acids have iodine numbers of about 10–50; those that contain an abundance of polyunsaturated fatty acids have iodine numbers of 120–150.

Example 24.2 Determine the iodine number of triolein.

$$CH_3(CH_2)_7CH=CH(CH_2)_7\overset{\displaystyle O}{\overset{\displaystyle \|}{C}}-O-\overset{\displaystyle H}{\underset{\displaystyle |}{C}}-H$$

$$CH_3(CH_2)_7CH=CH(CH_2)_7\overset{\displaystyle O}{\overset{\displaystyle \|}{C}}-O-\overset{}{\underset{}{C}}-H \; + \; \underset{(M\,=\,254)}{3\,I_2} \longrightarrow$$

$$CH_3(CH_2)_7CH=CH(CH_2)_7\overset{\displaystyle O}{\overset{\displaystyle \|}{C}}-O-\overset{\displaystyle H}{\underset{\displaystyle |}{C}}-H$$

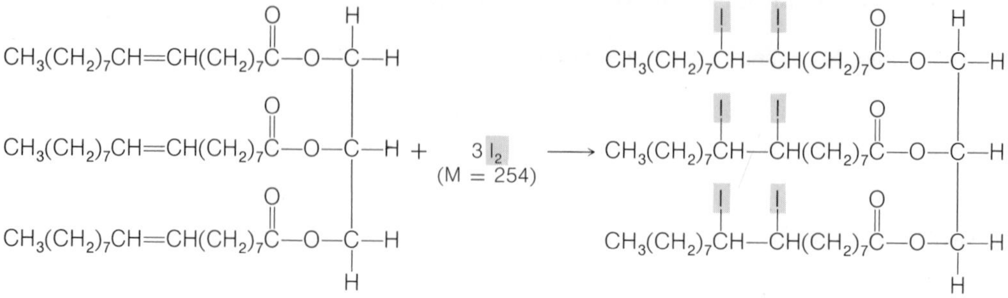

Triolein
(M = 884)

$$CH_3(CH_2)_7\overset{\displaystyle |}{\underset{}{C}H}-\overset{\displaystyle |}{\underset{}{C}H}(CH_2)_7\overset{\displaystyle O}{\overset{\displaystyle \|}{C}}-O-\overset{\displaystyle H}{\underset{\displaystyle |}{C}}-H$$

$$CH_3(CH_2)_7\overset{\displaystyle |}{\underset{}{C}H}-\overset{\displaystyle |}{\underset{}{C}H}(CH_2)_7\overset{\displaystyle O}{\overset{\displaystyle \|}{C}}-O-\overset{}{\underset{}{C}}-H$$

$$CH_3(CH_2)_7\overset{\displaystyle |}{\underset{}{C}H}-\overset{\displaystyle |}{\underset{}{C}H}(CH_2)_7\overset{\displaystyle O}{\overset{\displaystyle \|}{C}}-O-\overset{\displaystyle H}{\underset{\displaystyle |}{C}}-H$$

$$\text{Iodine no.} = \left(\frac{3 \text{ mole } I_2}{\text{mole lipid}}\right)\left(\frac{(254 \text{ g/mole } I_2)(100 \text{ g lipid})}{\text{g/mole lipid}}\right)$$

$$\begin{array}{c}\text{Iodine no.}\\ \text{of triolein}\end{array} = \frac{76200 \text{ g}}{884 \text{ g}} = 86$$

The above equation indicates the addition of molecular iodine. In actual practice, however, the reagents used are the interhalogens iodine monochloride (ICl), or iodine monobromide (IBr), both of which are more reactive than iodine alone. A weighed sample of lipid is treated with an excess of the iodine reagent. After the reaction is complete, the unused halogen is determined by titration with a standard solution of sodium thiosulfate.

c. Hydrogenation

A large-scale commercial industry has been developed for the purpose of transforming vegetable oils into solid fats. The chemistry of this conversion process is essentially identical to the catalytic hydrogenation reaction that has been described for the alkenes in Section 14.4-a. The process of converting oils to fats by means of hydrogenation is sometimes referred to as **hardening.** One method consists of bubbling hydrogen gas under pressure (25 lb/in.²) into a tank of hot oil (200°C) containing a finely dispersed nickel catalyst. An example is the conversion of triolein to tristearin.

$$CH_3(CH_2)_7-CH=CH-(CH_2)_7-\overset{\displaystyle O}{\overset{\displaystyle \|}{C}}-O-\overset{\displaystyle H}{\underset{\displaystyle |}{C}}-H$$

$$CH_3(CH_2)_7-CH=CH-(CH_2)_7-\overset{\displaystyle O}{\overset{\displaystyle \|}{C}}-O-\overset{}{\underset{}{C}}-H \; + \; 3\,H_2 \xrightarrow[\Delta]{Ni}$$

$$CH_3(CH_2)_7-CH=CH-(CH_2)_7-\overset{\displaystyle O}{\overset{\displaystyle \|}{C}}-O-\overset{\displaystyle H}{\underset{\displaystyle |}{C}}-H$$

Triolein

$$CH_3(CH_2)_{16}-\overset{\displaystyle O}{\overset{\displaystyle \|}{C}}-O-\overset{\displaystyle H}{\underset{\displaystyle |}{C}}-H$$

$$CH_3(CH_2)_{16}-\overset{\displaystyle O}{\overset{\displaystyle \|}{C}}-O-\overset{}{\underset{}{C}}-H$$

$$CH_3(CH_2)_{16}-\overset{\displaystyle O}{\overset{\displaystyle \|}{C}}-O-\overset{\displaystyle H}{\underset{\displaystyle |}{C}}-H$$

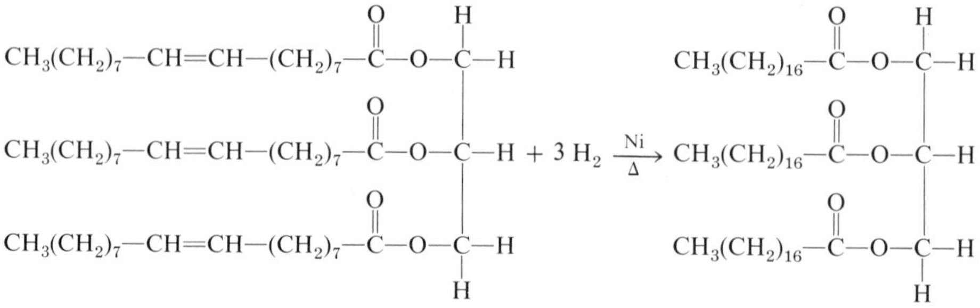

Tristearin

Figure 24.1 Hydrogenation of an oil. The beaker on the left contains a clear vegetable oil before hydrogenation; on the right is the same oil hardened by hydrogenation.

The equation represents the complete saturation of an unsaturated lipid. In the actual hardening process, the extent of hydrogenation is controlled so as to maintain a certain number of unsaturated linkages. (If all the bonds become hydrogenated, the product becomes hard and brittle like tallow.) If reaction conditions are properly controlled, it is possible to prepare a fat with a desirable physical consistency (soft and pliable). In this manner, inexpensive and abundant vegetable oils (cottonseed, corn, soybean) are converted into oleomargarine and cooking fats (Spry, Crisco, etc.—Figure 24.1). The peanut oil in peanut butter has been partially hydrogenated to prevent the oil from separating out. Today, because of the possible connection between saturated fats and arterial disease, many people are cooking with the vegetable oils (especially safflower seed oil) rather than with the hydrogenated products.

If the hydrogenation of an oil is allowed to continue for a long period of time, glycerol and long-chain alcohols are formed. This reaction is analogous to the previously mentioned reduction of esters (Section 19.10-c). These long-chain alcohols are employed in the manufacture of synthetic detergents (see Section 24.8-b).

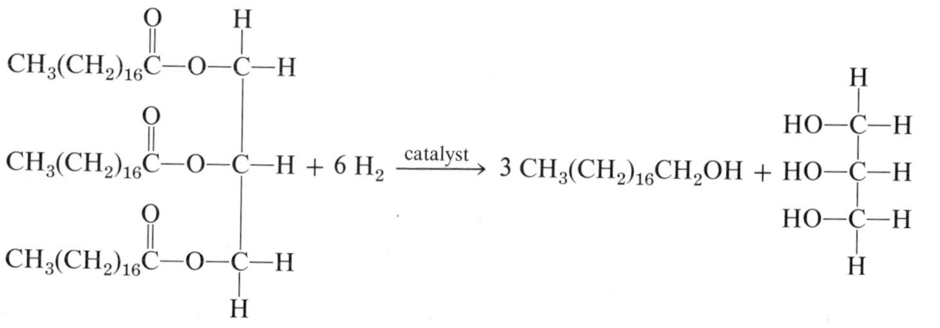

CH3(CH2)16C—O—C—H
+ 6 H2
catalyst
3 CH3(CH2)16CH2OH +
HO—C—H

Tristearin $\qquad$ *n*-Octadecanol $\qquad$ Glycerol

24.6 Special Features of Lipids

a. Rancidity

The term **rancid** is applied to *any fat or oil that develops a disagreeable odor*. Two principal chemical reactions are responsible for causing rancidity—*hydrolysis* and *oxidation*.

Butter is particularly susceptible to hydrolytic rancidity because it contains many of the lower molar mass acids (butyric, caproic) all of which have offensive odors. Under moist and warm conditions, hydrolysis of the ester linkages occurs, liberating the volatile acids. Microorganisms present in the air furnish the enzymes (lipases) that catalyze the process. Rancidity can easily be prevented by storing butter covered in a refrigerator.

Oxidative rancidity occurs in triacylglycerols containing polyunsaturated fatty acids. The reaction is quite complex, but it is believed that the first step involves the formation of a free radical, followed by production of hydroperoxides. Further oxidation reactions occur in which bonds are cleaved and the short-chain, offensive-smelling carboxylic acids are liberated.

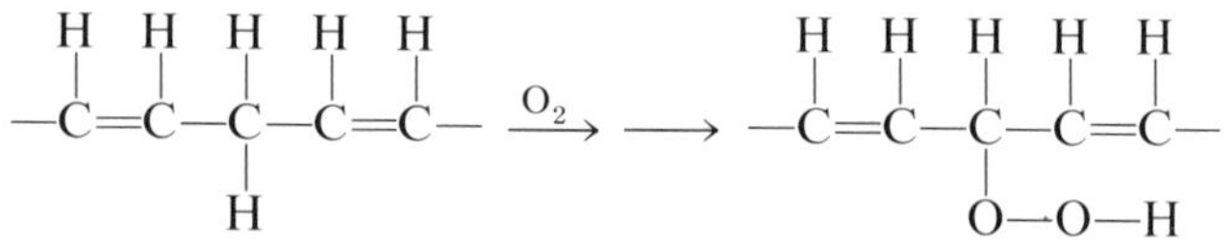

A hydroperoxide

Rancidity is a major concern of the food industry, and chemists involved in this area are continually seeking new and better substances to act as *antioxidants*. Such compounds are added in very small amounts (0.01–0.001%) to suppress rancidity. They have a greater affinity for oxygen than the lipid to which they are added and thus function by preferentially depleting the supply of adsorbed oxygen. We mentioned the synthetic antioxidant BHT in Section 13.5; two of the naturally occurring antioxidants are tocopherol (vitamin E—Figure 13.5) and ascorbic acid (vitamin C).

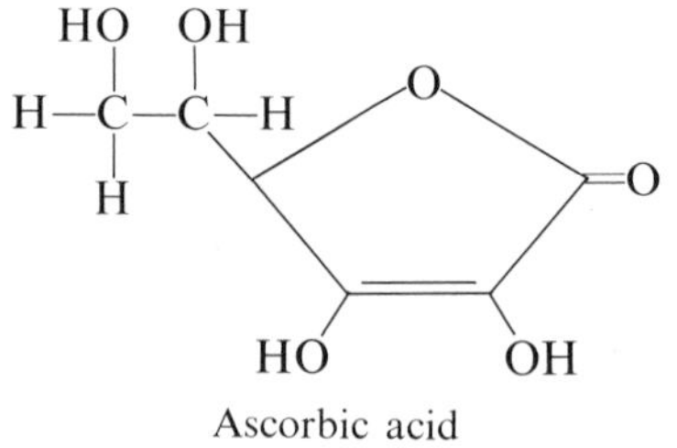

Ascorbic acid

The roles of vitamin E and vitamin C in human nutrition are still a subject of controversy (see Section 32.1-b, d). One of the functions of vitamin E may be to protect the unsaturated fatty acids in the lipids of cell membranes against the damaging effects of molecular oxygen. This vitamin has also received some notoriety as a "sex enhancer" probably because a deficiency of vitamin E has been shown to produce infertility in laboratory animals. Prolonged lack of vitamin C in the diet of humans causes the disease known as scurvy. Several years ago Linus Pauling suggested that massive doses of vitamin C ($>$1000 mg daily) should be taken for the prevention of the common cold. There is as yet no concrete evidence to substantiate or to invalidate this claim.

b. Drying Oils

A **drying oil** is *any substance that causes a paint or varnish to develop a hard, protective coating.* It is the susceptibility of highly unsaturated oils to react with oxygen that accounts for their usefulness in the paint industry. Linseed oil is especially reactive and is most commonly used. The term *drying* may be a misnomer because it implies that the protective coating is formed by the evaporation of the solvent. Instead, the drying process involves an oxidation followed by a polymerization reaction that results in the formation of a vast interlocking network of triacylglycerols joined by peroxide bridges. These oxidation-polymerization reactions are catalyzed by metal ions (lead, manganese, cobalt), and salts of these metals are included in paint to hasten the drying process.

Oil paints are suspensions of very finely divided pigments in linseed oil. *Oilcloth* is made by the application of several coats of linseed oil on woven fabric. *Linoleum* is a mixture of linseed oil, ground cork, and rosin that has been pressed together and "dried."

The reactions involved in the drying process are exothermic and heat is therefore given off to the surroundings. If the surroundings happen to be a poorly ventilated container, this heat will raise the oil to its kindling temperature and cause *spontaneous combustion.* Thus paint or oil rags should never be bundled together and stored in wood or cardboard containers.

24.7 Waxes

A wax is an ester of a long-chain alcohol (usually monohydroxy) and a fatty acid. The acids and alcohols normally found in waxes have chains of the order of 12–34 carbon atoms in length. Waxes are easily melted solids that are widely distributed in nature and are found in both plant and animal matter. They are not as easily hydrolyzed as the triacyglycerols and therefore are useful as protective coatings. Plant waxes are found on the surfaces of leaves and stems and serve to protect the plant from dehydration and from invasion by harmful organisms. Carnauba wax, largely myricyl cerotate, $C_{25}H_{51}COOC_{30}H_{61}$, is obtained from the leaves of certain Brazilian palm trees and is used as a floor and automobile wax and as a coating on carbon paper.

Animal waxes also serve as protective coatings. They are found on the surface of feathers, skin, and hair and help to keep these surfaces soft and pliable. Beeswax, which is mostly myricyl palmitate, $C_{15}H_{31}COOC_{30}H_{61}$, is secreted by the wax glands of the bee. Spermaceti wax, mainly cetyl palmitate, $C_{15}H_{31}COOC_{16}H_{33}$, is found in the head cavities and the blubber of the sperm whale. Spermaceti crystallizes in heavy white flakes when whale oil is exposed to air and chilled. It is used primarily in ointments, in cosmetics, and in the manufacture of candles. Lanolin, obtained from wool, is a mixture of fatty acid esters of the steroids lanosterol and agnosterol. It finds widespread medical applications as a base for creams, ointments, and salves.

24.8 Soaps and Synthetic Detergents

It has already been mentioned that soaps are salts of long-chain fatty acids. Alkaline hydrolysis of a fat or oil produces a soap and glycerol.[6] The older method of soap production consisted of treating molten tallow (the fat of livestock) with a slight excess of alkali in large open kettles. The mixture was heated and steam was bubbled through it. After the saponification process was completed, the soap was precipitated by the addition of sodium chloride, and then filtered and washed several times with water. It was then reprecipitated from the aqueous solution by the addition of more NaCl, The glycerol was recovered from the aqueous wash solutions. Currently, soap is prepared by a continuous process wherein lipids are hydrolyzed by water under high pressures and temperatures (700 lb/in.^{2} and 200°C). Na_2CO_3 is used instead of the more expensive NaOH to neutralize the acid.

$$\begin{array}{l} CH_2OOC(CH_2)_nCH_3 \\ | \\ CHOOC(CH_2)_nCH_3 \\ | \\ CH_2OOC(CH_2)_nCH_3 \end{array} \xrightarrow[\text{heat, pressure}]{H_2O} \underset{\text{}}{\text{Glycerol}} + \underset{\text{Fatty acid}}{3\ CH_3(CH_2)_nCOOH} \xrightarrow{Na_2CO_3} \underset{\text{Sodium salt of a fatty acid (a soap)}}{3\ CH_3(CH_2)_nCOO^- Na^+}$$

The crude soap is used as industrial soap without further processing. Pumice or sand may be added to produce scouring soap. Other ingredients, such as perfumes and antiseptics, are added to produce toilet soaps and deodorizing soaps, respectively. If air is blown through molten soap, a floating soap is produced. Such a soap is not necessarily purer than other soaps; it merely contains more air. Ordinary soap is a mixture of the sodium salts of various fatty acids. Potassium soaps (soft soap) are more expensive but produce a softer lather and are more soluble. They are used in liquid soaps and shaving creams.

a. The Cleansing Action of Soaps

A soap molecule can be considered to be composed of a large nonpolar hydrocarbon portion (*hydrophobic*—repelled by water) and a carboxylate salt end (*hydrophilic*—water soluble).

[6]Recall that the glycerol obtained as a by-product in the manufacture of soaps is used to make the explosive, nitroglycerin (Section 17.5-e). During World Wars I and II, people saved excess cooking fats and oils and turned them in for reclamation of the glycerol.

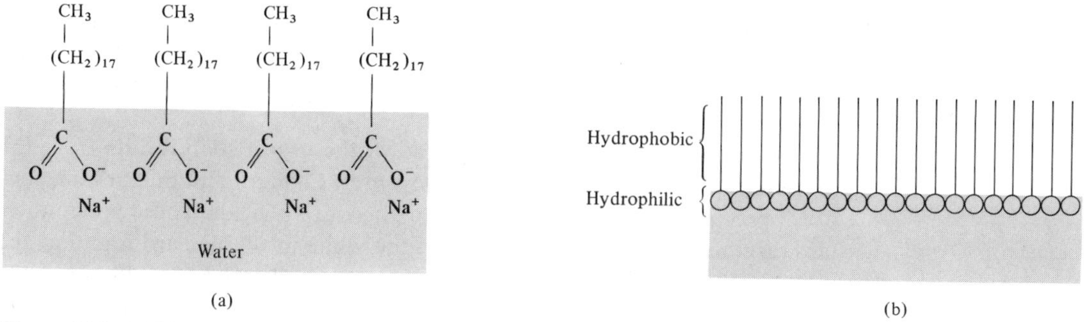
CH3
(CH2)17
C
O
O⁻
Na+
Water
(a)
Hydrophobic
Hydrophilic
(b)

Figure 24.2 (a) The formation of a thin film (a monolayer) on the surface of water. (b) A schematic representation.

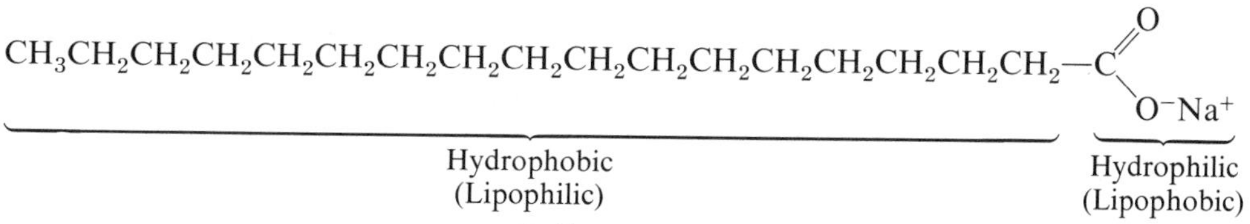
CH3CH2CH2CH2CH2CH2CH2CH2CH2CH2CH2CH2CH2CH2CH2CH2CH2—C
O
O⁻Na+
Hydrophobic
(Lipophilic)
Hydrophilic
(Lipophobic)

Sodium stearate

When soap is added to water, the hydrophilic ends of the molecules are attracted to the water and dissolved in it, but the hydrophobic ends are repelled by the water molecules. Consequently, a thin film (suds) forms on the surface of the water, drastically lowering its surface tension (Figure 24.2). When the soap solution is brought into contact with grease or oil (most dirt is held to clothes by a thin film of grease or oil), soap molecules become reoriented. The hydrophobic portions dissolve in the grease or the oil, and the hydrophilic ends remain dissolved in the aqueous phase. Mechanical action, such as scrubbing, causes the oil or grease to disperse into tiny droplets, and soap molecules arrange themselves around the surface of the globules. Oil or grease droplets surrounded by soap molecules are examples of *micelles*. (A micelle is an aggregate of dipolar species.) Because the carboxylate ends of the soap molecules project outward, the surface of each drop is negatively charged; therefore the drops repel one another and do not coalesce (Figure 24.3). The entire micelle becomes water soluble and is able

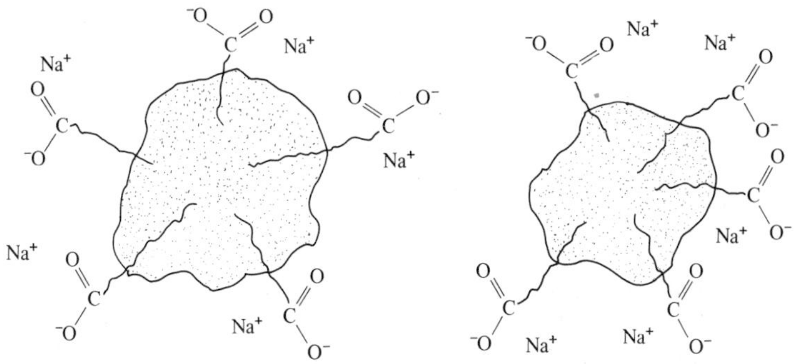

Figure 24.3 Two micelles will not coalesce because of the repulsions between their surrounding carboxylate groups.

to be washed away by a stream of water. The cleansing process involves both the lowering of the surface tension of water and *emulsification.* (As we shall see in Section 29.1, the bile salts serve as emulsifying agents in the digestion of dietary lipids.)

One major disadvantage involved in the use of soap results from the presence of certain metal ions in hard water. Calcium and magnesium ions form precipitates with the carboxylate ions of fatty acids. These precipitates are responsible for bathtub rings and the white insoluble curd found at the bottom of washing machines. The various methods used for softening hard water all involve the removal of the calcium and magnesium ions. The equation representing the action of hard water upon soap is

$$2 \text{ CH}_3(\text{CH}_2)_{16}\text{C} \overset{O}{\underset{O^-\text{Na}^+}{\diagup}} + \text{Ca}^{2+} \longrightarrow \left(\text{CH}_3(\text{CH}_2)_{16}\text{C} \overset{O}{\underset{O^-}{\diagup}} \right)_2 \text{Ca}^{2+} + 2 \text{ Na}^+$$

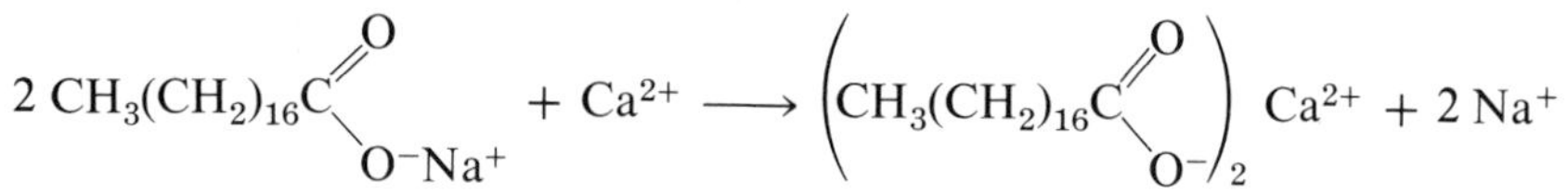

Sodium stearate
(soluble)

Calcium stearate
(insoluble)

b. Synthetic Detergents

The term *detergent* is a rather general one, used to denote any cleansing agent. Soaps would fall under such a broad definition. However, the popular use of the word generally refers to synthetic detergents also called *syndets.* Syndets have the desirable property of not forming precipitates with the ions of hard water. There are close to a thousand synthetic detergents commercially available in the United States. They may be classified as either anionic, cationic, or nonionic.

1. Anionic Detergents. Anionic detergents are sulfates of fatty acids or sulfonate salts of hydrocarbons.

                Hydrophobic   Hydrophilic

$$\text{CH}_3(\text{CH}_2)_{11}\text{O}{-}\overset{\textstyle O}{\underset{\textstyle O}{\overset{\|}{\underset{\|}{\text{S}}}}}{-}\text{O}^-\text{Na}^+ \qquad\qquad \text{CH}_3(\text{CH}_2)_{11}{-}\!\!\bigcirc\!\!{-}\overset{\textstyle O}{\underset{\textstyle O}{\overset{\|}{\underset{\|}{\text{S}}}}}{-}\text{O}^-\text{Na}^+$$

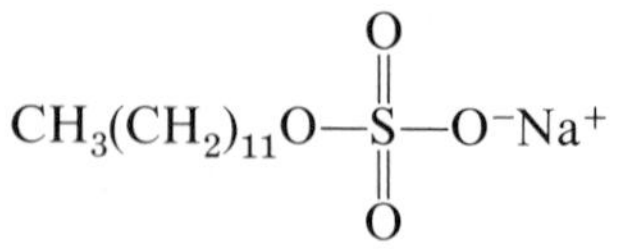

Sodium lauryl sulfate   Sodium *n*-dodecylbenzenesulfonate

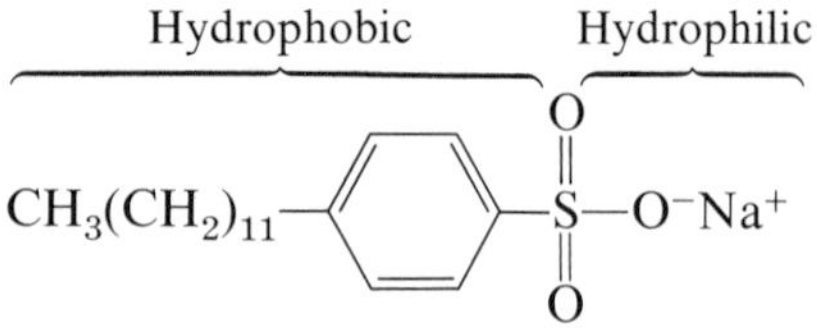

2. Cationic Detergents. These detergents are sometimes referred to as *invert soap* because their water-soluble end carries a positive, rather than a negative, charge. In addition to being good cleansing agents, they possess germicidal properties and are widely used in hospitals for this reason.

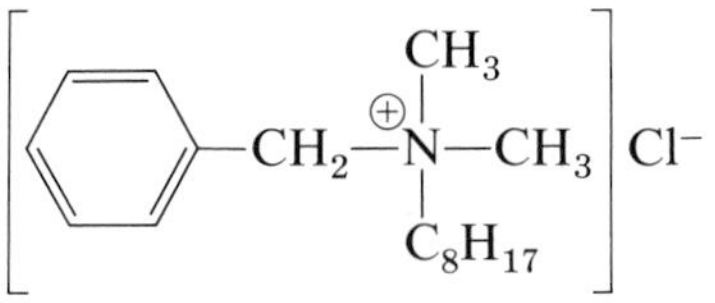

$$\left[\langle \bigcirc \rangle -CH_2 - \overset{\overset{\displaystyle CH_3}{|}}{\underset{\underset{\displaystyle C_8H_{17}}{|}}{\overset{\oplus}{N}}} - CH_3 \right] Cl^-$$

Benzyldimethyloctylammonium
chloride

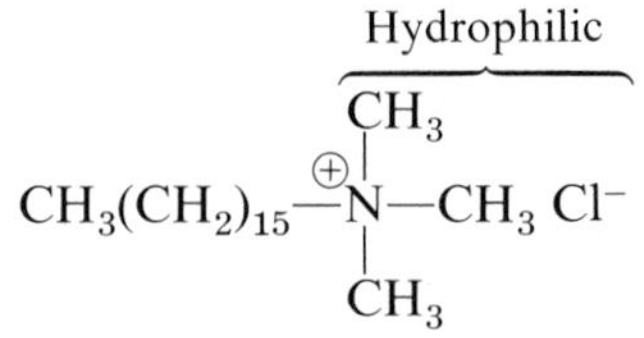

Hydrophilic

$$CH_3(CH_2)_{15} - \overset{\overset{\displaystyle CH_3}{|}}{\underset{\underset{\displaystyle CH_3}{|}}{\overset{\oplus}{N}}} - CH_3 \ Cl^-$$

Trimethylhexadecylammonium
chloride

3. Nonionic Detergents. These detergents contain polar covalent structures that provide the required water solubility. They are used extensively in dishwashing liquids and on all occasions that call for the absence of inorganic ions.

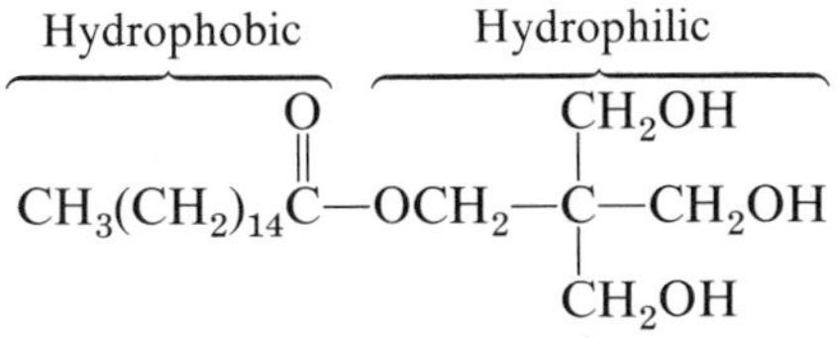

Hydrophobic   Hydrophilic

$$CH_3(CH_2)_{14}\overset{\overset{\displaystyle O}{\|}}{C} - OCH_2 - \overset{\overset{\displaystyle CH_2OH}{|}}{\underset{\underset{\displaystyle CH_2OH}{|}}{C}} - CH_2OH$$

Pentaerythrityl palmitate

4. Environmental Effects. The large-scale use of synthetic detergents during the twenty years following World War II created a serious disposal problem. Soaps, which contain straight-chain alkyl groups, can be removed from waste water through degradation by microorganisms in the soil (septic tanks) or in sewage treatment plants. (Soaps are therefore said to be *biodegradable*.) Many of the synthetic detergents could not be removed in this manner. The metabolism of microorganisms is adapted to the straight-chain alkyl groups found in soaps and natural fats; they are unable to break down the highly branched analogs used in the early syndets. The synthetic detergents continued to foam and make suds,[7] which clogged waste disposal plants, killed fish and wildlife by polluting streams, and even managed to make their way into city drinking water. Since 1966 all United States companies have been using straight-chain hydrocarbons in the production of syndets. Although this involves a greater expense, it seems to have alleviated some of the pollution problem.

A second environmental problem caused by detergents has not been solved. In their search for more effective cleansing agents, manufacturers have added "builders" to their detergents. Builders have little detergent effectiveness alone, but mainly function (1) to soften water, (2) to prevent soil from redepositing on clothes, and (3) to maintain a proper level of alkalinity in the wash water. Although a number of inorganic compounds have been

[7] It is interesting to note that many effective detergents do not foam in water. Experiments have proved that the degree of sudsing has very little to do with the efficiency of a detergent. However, the consumer has come to associate sudsing with cleaning ability, so manufacturers often add sudsing agents to their products.

used as builders (e.g., carbonates, bicarbonates, borates, silicates), the phosphates are the most effective. When phosphates were first added to detergents, a typical detergent might contain as much as 50% phosphate. Approximately half of the phosphate content of domestic sewage was contributed by detergents, the remainder being derived from human wastes.

As we shall see in subsequent chapters, phosphate is a nutrient required for plant (and animal) growth. It has thus been implicated as the principal cause of *eutrophication* of lakes and rivers. (Eutrophic derives from the Greek *eu*, well, and *trophos*, pertaining to nourishment.) Eutrophic lakes and rivers contain an overabundance of plants, especially algae (algae blooms), and decreased levels of oxygen. Only certain species of fish (carp, crappie, perch, bullhead) can live in such an environment. Moreover, the algae can produce chemicals that cause unpleasant tastes and smells in water and that, in some cases, are toxic to animals.

As a result of the environmental impact of eutrophication, communities have passed laws that either ban phosphate detergents or drastically reduce the amount of phosphates. One method of reducing the phosphate concentration in natural waters is to substitute other builders for phosphates in detergents. One such compound developed was *nitrilotriacetic acid,* NTA.

$$\begin{array}{c} HOOCCH_2{-}N{-}CH_2COOH \\ | \\ CH_2COOH \end{array}$$

In the late 1960s several detergents were marketed in Sweden using NTA as a builder. Preliminary testing uncovered no hazards to human or animal life, and NTA did not act as a significant nutrient source for algae. However, in 1970 tests performed in the United States showed that NTA, when administered in high doses to rats and mice, in combination with cadmium and mercury (two metals commonly found in waste water), caused a significant increase in birth defects in the test animals. NTA therefore appears to be teratogenic in combination with certain metals and is now considered to be a hazardous chemical. This lesson is illustrative of how little we know of environmental interactions. Extreme care must be exercised in replacing one objectionable compound with another.

24.9 Compound Lipids

a. Phospholipids

The phospholipids, also called phosphatides, are *compound lipids that are derivatives of glycerol phosphate.*

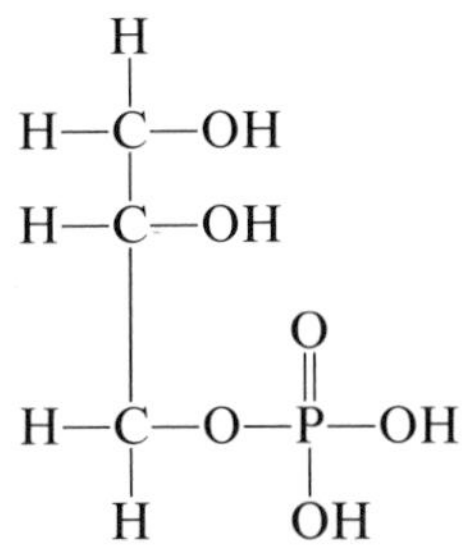

Glycerol phosphate

Phospholipids are found in all living organisms. Regardless of their source, they have quite consistent structures. Phospholipids are particularly abundant in liver, brain, and spinal tissue and are found in the outer membranes of most cells (see Figure 13.4). They appear to be essential components of cell structure since the amount of phospholipids present in animal tissues remains relatively constant, even during starvation when the cell's supply of simple lipids is depleted.

Phospholipids are large molecules containing both a polar and a nonpolar component. They are the most polar of all the lipids. It is believed that their primary function is to act as an emulsifying agent at cell membrane surfaces, where water-insoluble lipids and water-soluble materials (such as proteins) must be capable of intimate association. It is thought that phospholipids take part in fat metabolism by promoting the transportation of lipids in the bloodstream, primarily an aqueous medium. Phospholipids also play important roles in the electron transport system (the respiratory chain—Section 28.8), in secretory processes, and in the transport of ions across cell membranes. There is increased speculation regarding their functions in brain and nervous tissue, but at present their exact purpose is not known.

1. Lecithin.[8] Lecithin is probably the most common of the phospholipids. It contains the important quaternary ammonium salt choline, $HOCH_2CH_2N^+(CH_3)_3$, joined to a phosphoric acid residue by means of an ester linkage. The nitrogen in choline carries a positive charge and the phosphate a negative charge so that in solution at most pH values, lecithin exists as an **internal salt** or **zwitterion.** We shall have occasion to speak of zwitterions at greater length in the next chapter. The structure and hydrolysis products of lecithin are as follows.

[8]Since at least five different fatty acids are obtained on hydrolysis of highly purified lecithin, we are dealing with a mixture of compounds rather than with a single one. Technically, we should use the plural term lecithins instead of the singular name. However, most authors prefer to refer to the compounds collectively as lecithin.

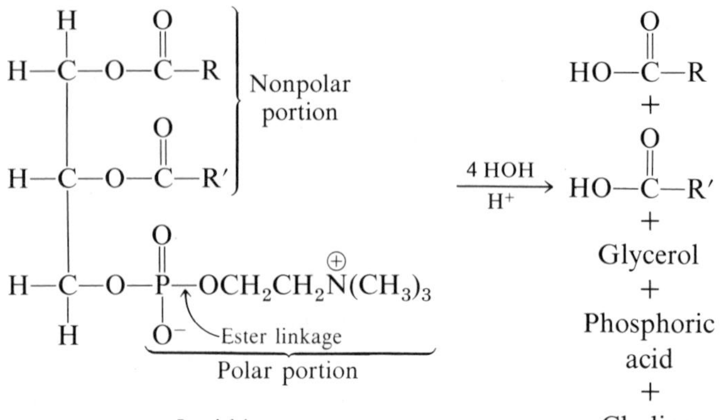
Nonpolar
portion
Ester linkage
Polar portion
4 HOH
H+
Glycerol
Phosphoric
acid
Choline

Lecithin
(Phosphatidyl choline)

Pure lecithin is a waxy white solid that quickly darkens when exposed to air. In contrast to fats and oils, it is colloidally dispersed in water and is insoluble in acetone. It is therefore possible to separate lecithin from an ether extract by the addition of acetone. Lecithin is especially abundant in egg yolk and soybeans. When obtained from the latter source it is used as an emulsifying agent in the dairy and confectionery industries.

2. ***Cephalin.*** The chief difference between the cephalins and the lecithins lies in the nitrogenous base component that is linked to the phosphate moiety. In the cephalins, the choline is replaced by *ethanolamine,* $HOCH_2CH_2NH_2$, or by the amino acid *serine,* $HOCH_2CH(NH_2)COOH$. The term cephalin is derived from its chief occurrence in the body, namely the head and spinal tissue (Greek, *kephalikos,* head). It is thought that cephalins play an important role in the process of blood clotting.

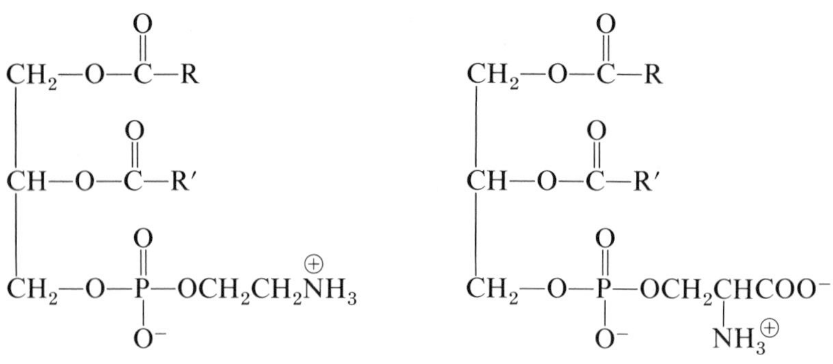

Phosphatidylethanolamine

Phosphatidylserine

Cephalins

b. Glycolipids

Several groups of compounds are found containing both lipid and carbohydrate structures. Those that are water soluble are termed liposaccharides and are thought of as derived carbohydrates. Those that retain solubility in nonpolar organic solvents are classed as glycolipids. One group of glycolipids contains fatty acids, glycerol, and various carbohydrates.

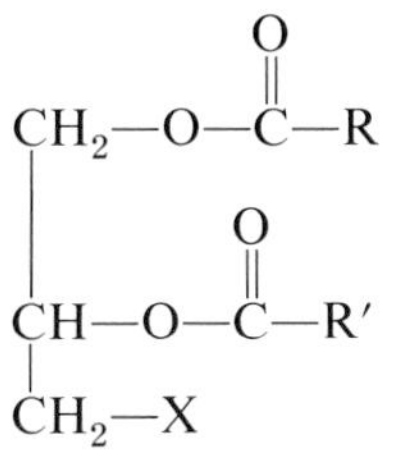

O
CH_2—O—C—R
O
CH—O—C—R′
CH_2—X

where X is —(galactose)$_{1\text{ or }2}$
—(glucose)$_{1\text{ or }2}$
—(mannose)$_{1\text{ or }2}$
—(glucose)(galactose)

Another group, the *cerebrosides,* are sphingosine derivatives (see Section 24.9-c) and thus may be classified either as glycolipids or sphingolipids. Cerebrosides occur primarily in the brain (7% of the solid matter) and in the myelin sheath of nerves. It has been suggested that they function in the transmission of nerve impulses across synapses. They are also believed to be present at receptor sites for acetylcholine (page 504) and other neurotransmitters. The fatty acids found in cerebrosides are unusual in that they contain 24 carbon atoms. Cerebrosides most often contain D-galactose attached by an acetal linkage at carbon-1 of sphingosine. Unlike most lipids, they are insoluble in ether but may be extracted into warm alcohol or pyridine.

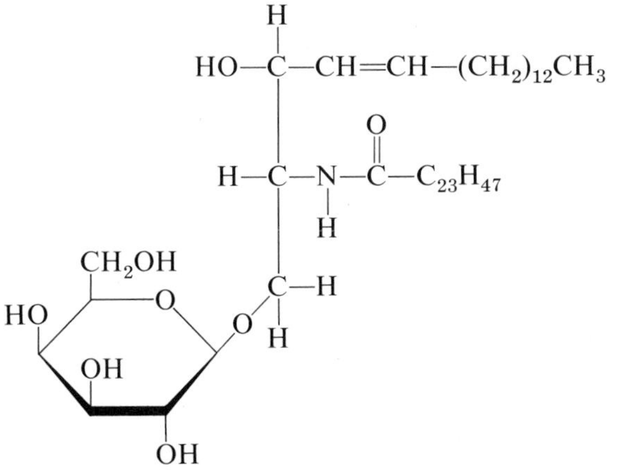

H
HO—C—CH=CH—$(CH_2)_{12}CH_3$
O
H—C—N—C—$C_{23}H_{47}$
H
CH_2OH
C—H
O
HO
O
H
OH
OH

A cerebroside

Two severe lipid storage diseases are caused by errors in the metabolism of the glycolipids. In Gaucher's disease, the glycolipids contain glucose instead of galactose. These abnormal glycolipids accumulate in the brain, spleen, and kidney cells. An infant with Tay–Sachs disease lacks an enzyme that breaks down glycolipids, so they accumulate in the tissues of the brain and the eyes (see Section 27.6).

c. Sphingolipids

Sphingolipids occur in the membranes of both plants and animals, with only a minor amount found in depot fat. They contain the long-chain unsaturated amino alcohol *sphingosine* instead of glycerol. Also present are fatty acids, phosphate, and an alcohol component. The most abundant sphingolipid is sphingomyelin, which contains choline as the alcohol group.

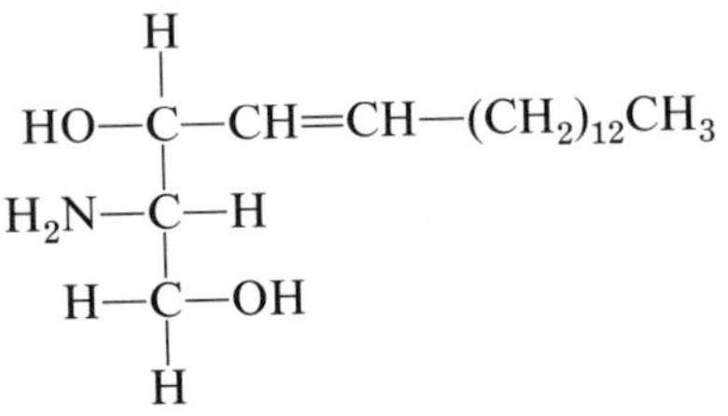
HO—C—CH=CH—(CH2)12CH3
H2N—C—H
H—C—OH

Sphingosine

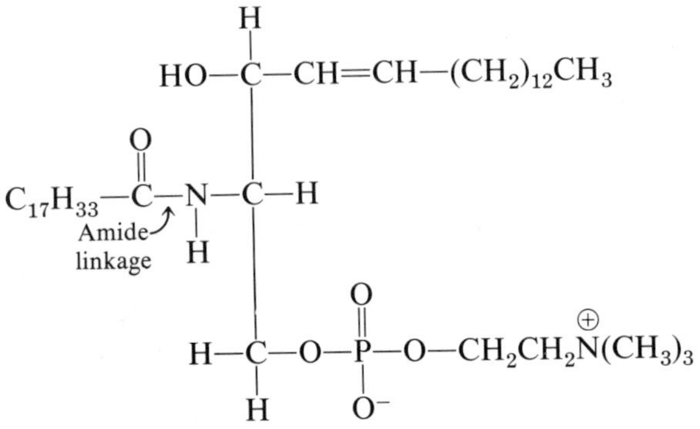
HO—C—CH=CH—(CH2)12CH3
C17H33—C—N—C—H
Amide
linkage
H—C—O—P—O—CH2CH2N(CH3)3

Sphingomyelin

Niemann–Pick disease is another lipid storage disease in which sphingomyelins build up in the brain, liver, and spleen, resulting in mental retardation and early death (see Section 27.6).

24.10 Steroids

Over thirty different steroidal compounds have been found in nature. They occur in plant and animal tissues, yeasts, and molds, but not in bacteria, and may exist free or combined with fatty acids or carbohydrates. All steroids have a perhydrocyclopentanophenanthrene skeleton, which consists of a completely saturated phenanthrene moiety fused to a cyclopentane ring. The rings are designated by capital letters, and the carbon atoms are numbered as follows.

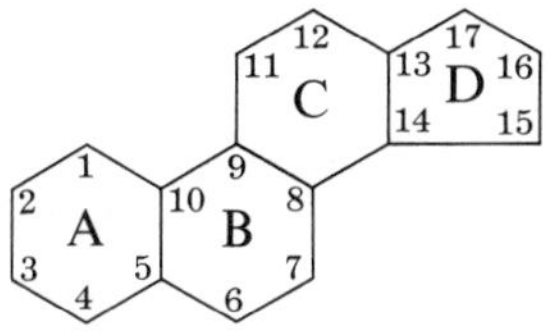
12
17
11
13
16
C
D
14
15
1
9
2
10
8
A
B
3
5
7
4
6

Perhydrocyclopentanophenanthrene

Although the parent ring structure is that of a cycloalkane, several steroids contain one, two, or three double bonds and many of them possess one or more hydroxyl groups. (These steroids are often termed *sterols,* signifying the presence of the hydroxyl group.) Most steroids contain methyl groups at carbon-10 and carbon-13 and a side chain at carbon-17. Since the steroid skeleton is a rigid cyclic system, any substituents will be situated either above or below the plane of the molecule. This admits the possibility of geometric isomerism. In addition, many of the ring carbon atoms are asymmetric and there are numerous optical isomers. Since the steroids are essentially high molar mass hydrocarbons, they are soluble only in the fat solvents. They differ from the other lipids in that they do not undergo saponification. The steroids are one of the most versatile classes of compounds found in nature. Very small amounts of steroids and slight variations in structure or in the nature of substituent groups effect profound changes in biological activity.

a. Cholesterol

Cholesterol does not occur in plants, but it is the best known and most abundant (about 240 grams) steroid in the human body and has a high occurrence in the brain and nervous tissue. It is the principal constituent of gallstones, from which it can be isolated as a white crystalline solid. Its name is derived from this source (Greek, *chole,* bile; *stereos,* solid).

Most meats and foods derived from animal products such as eggs, butter, cheese, and cream are particularly rich in cholesterol. In addition, cholesterol is synthesized in the liver from acetyl CoA. The human body manufactures about 3 grams of cholesterol each day. If cholesterol is present in the diet, then its biosynthesis in the liver is suppressed. Fasting also inhibits the biosynthesis of cholesterol, whereas high-fat diets (especially those containing saturated fatty acids) tend to accelerate cholesterol biosynthesis. Most of the cholesterol in the body is converted into cholic acid, which is used in the formation of bile salts. Cholesterol is also an important precursor in the biosynthesis of the sex hormones, adrenal hormones, and vitamin D.

Cholesterol has received much attention because of the suspected correlation between the cholesterol level in the blood and certain types of heart disease (see Section 32.4). About two-thirds of the cholesterol in the blood is esterified to unsaturated fatty acids. The cholesterol content of blood varies considerably with age, diet, and sex. Young adults average about 1.7 grams of cholesterol per liter of blood, whereas males at age 55 may have 2.5 g/liter or higher. Females tend to have lower blood cholesterol levels than males. About 85% of deaths due to cardiovascular disease are directly linked to *arteriosclerosis* (the hardening of the arteries that produces degenerative heart disease, stroke, and other arterial diseases). Atherosclerosis (a form of arteriosclerosis) results from the deposition of excess cholesterol. When cholesterol precipitates out of the blood and accumulates in the blood vessels, the resulting constriction reduces blood flow and leads to high blood pressure. Investigators are attempting to establish that reducing the blood cholesterol level by minimizing lipid intake through dietary restrictions, or by administering anticholesterol drugs, will help to prevent heart disease.

Cholesterol reacts with acetic anhydride and sulfuric acid causing a color change in the reaction mixture from pink to blue to green.[9] The reaction is the basis of the Liebermann–Burchard test for the quantitative determination of cholesterol.

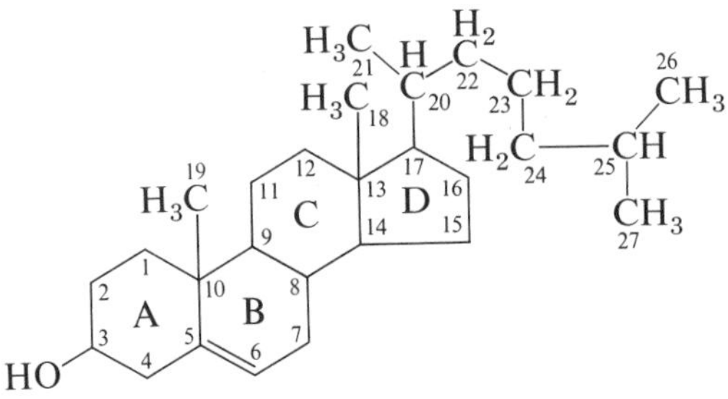

HO
A
B
C
D
H3C
19
H3C
18
H3C
21
H
C
20
H2
C
22
CH2
23
H2C
24
CH
25
CH3
26
CH3
27

Cholesterol

b. Bile Salts

The bile salts are sodium salts of peptide-like combinations of bile acids and amino acids. They are synthesized from cholesterol in the liver, stored in the gallbladder, and then secreted into the intestine. Bile salts are the most important constituents of bile, and their major function is to aid in the breakdown and absorption of dietary lipids, including cholesterol. These compounds are highly effective detergents because they contain both hydrophobic and hydrophilic groups. Thus, they perform their function by acting as emulsifying agents, that is, they break down large fat globules into smaller ones and keep these smaller globules suspended in the aqueous digestive environment (see Section 29.1).

[9] Owing to the difficulties in isolating the products, the mechanism of the reaction is not known. Acetic anhydride can condense with the hydroxyl group on carbon-3 of the cholesterol molecule to yield the corresponding ester.

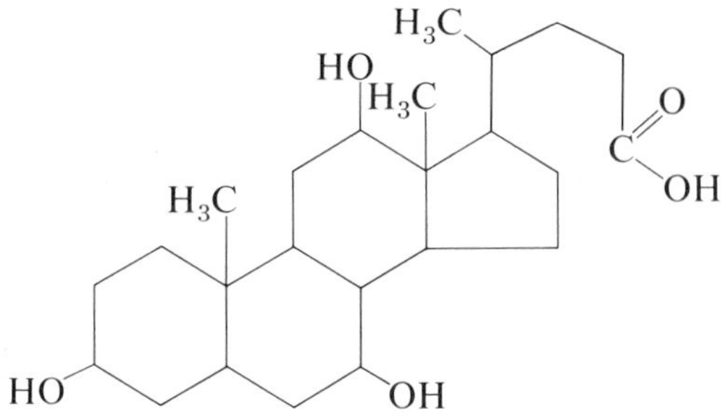
H3C
HO
H3C
O
C
OH
H3C
HO
OH

Cholic acid
(a bile acid)

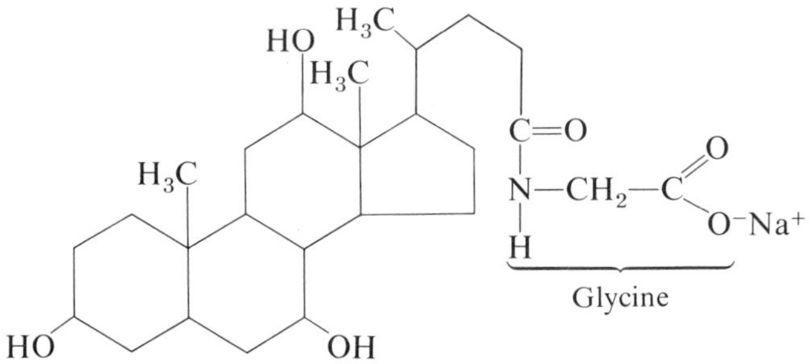
HO
H3C
H3C
C=O
O
H3C
N—CH2—C
O-Na+
H
Glycine
HO
OH

Sodium glycocholate
(a bile salt)

c. Adrenocortical Hormones

The outer part, or cortex, of the adrenal glands utilizes cholesterol to produce several hormones (about 30) that are essential to life. These compounds constitute a family of hormones of which *cortisone* can be considered to be the parent. These hormones effect the metabolism of foodstuffs, maintain the proper balance of electrolytes (sodium and potassium ions in particular), and control inflammations and allergies. Addison's disease, at one time a fatal illness caused by insufficient secretion of the adrenal hormones, is now successfully treated by oral administration of these hormones.

Adrenal hormones are used in the treatment of rheumatic fever and rheumatoid arthritis. Pharmaceutical companies have been able to obtain large quantities of synthetic cortisone by means of an involved process utilizing cholic acid isolated from the bile juices of cattle. Prolonged use of cortisone can have serious side effects including wasting of muscles and resorption of bone. In recent years cortisone has been supplemented with a synthetic analog, prednisolone, which is effective in much smaller doses, thereby greatly reducing the side effects.

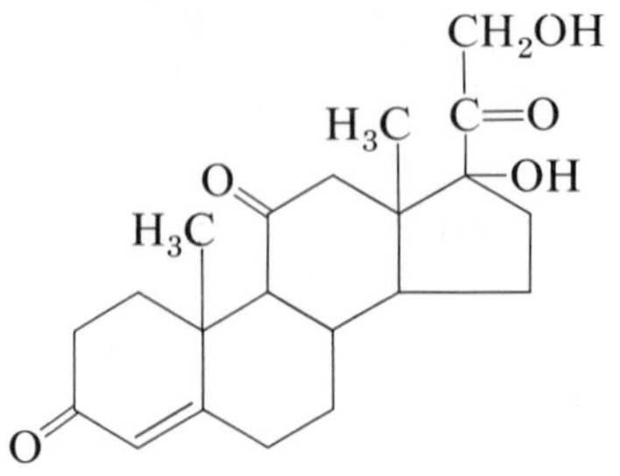
CH2OH
H3C
C=O
O
OH
H3C
O

Cortisone

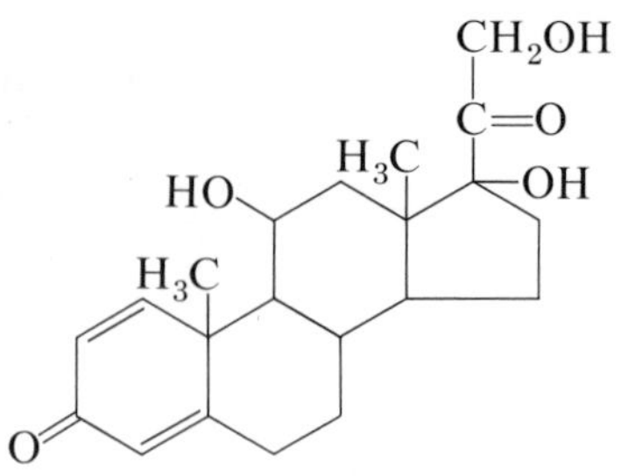
CH2OH
C=O
H3C
HO
OH
H3C
O

Prednisolone

d. Vitamin D

Vitamin D is the name given to a closely related group of compounds that are effective in preventing rickets, a disease characterized by skeletal abnormalities. The two most important members of the group are vitamin D_2 (ergocalciferol) and vitamin D_3 (cholecalciferol). Both vitamins are produced by ultraviolet irradiation of unsaturated steroids. Vitamin D_2 is derived from ergosterol, a steroid found in yeast, ergot, and other fungi. Vitamin D_3 is formed in the skin of animals by the action of sunlight on 7-dehydrocholesterol. Notice that these compounds differ only in the structure of the side chain at carbon-17, and in both cases the B ring of the steroid skeleton has been cleaved.

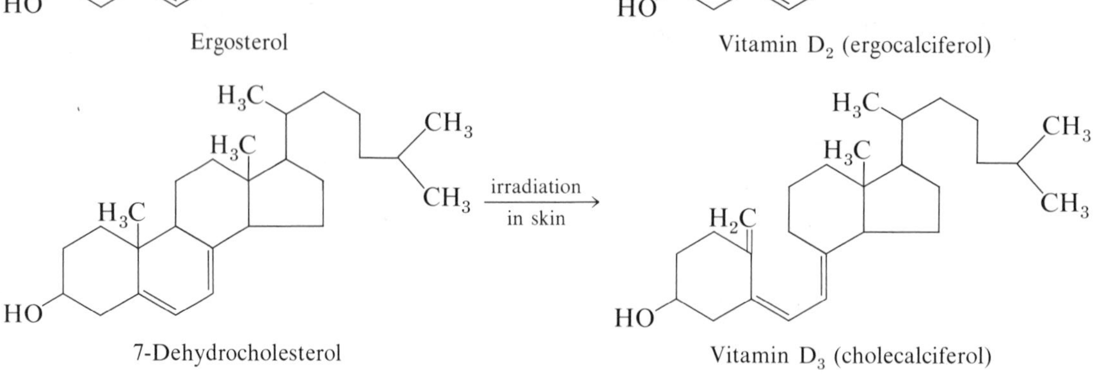
H3C
CH3
H3C
irradiation
CH3
CH3
H3C
HO
Ergosterol
H3C
CH3
H3C
H2C
CH3
CH3
HO
Vitamin D2 (ergocalciferol)
H3C
CH3
H3C
CH3
irradiation
in skin
H3C
HO
7-Dehydrocholesterol
H3C
CH3
H3C
CH3
H2C
HO
Vitamin D3 (cholecalciferol)

Most foods contain little or no vitamin D. The best natural sources of the vitamin are fish oils and egg yolks. Irradiated ergosterol (from yeast) is added to milk and margarine as a supplemental source of vitamin D. Individuals with a reasonable proportion of their skin exposed to sunlight rarely suffer from a vitamin D deficiency. The vitamin is activated in the skin (from 7-dehydrocholesterol) by exposure to solar radiation and is then transported in the body for utilization or storage in the liver.

The principal functions of vitamin D are to increase the absorption of calcium and phosphorus from the intestine and to increase the mobilization of these minerals from bone (see Section 32.1-C). The ingestion of excessive amounts of vitamin D is hazardous and produces abnormalities of bones and teeth, as well as calcification of the lungs and kidneys.

e. Sex Hormones

The primary male hormone *testosterone* is responsible for the development of the secondary sex characteristics peculiar to the male members of the species, such as facial hair, deep voice, and muscle strength. Two sex hormones are of particular importance in females. *Progesterone* is required for normal pregnancy, and the *estrogens* control the ovulation cycle. Notice that the male and female hormones exhibit only very slight structural differences, yet their physiological effects differ enormously.

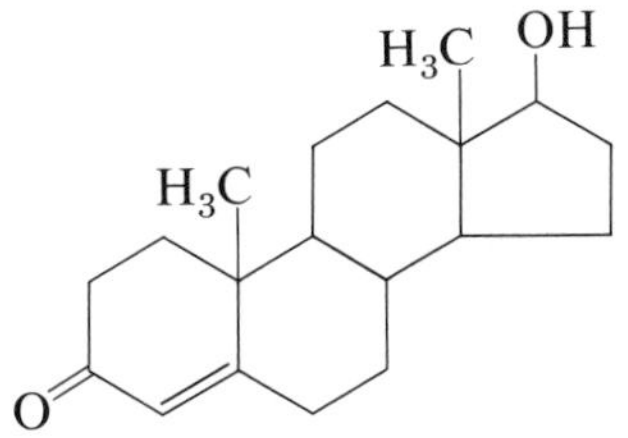
H3C
OH
H3C
O

Testosterone

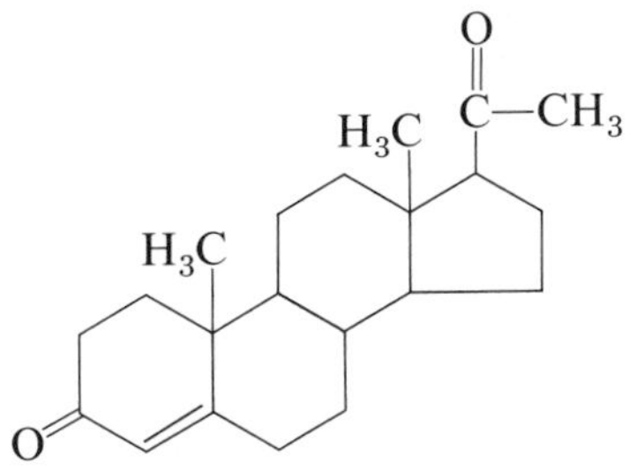
O
C—CH3
H3C
H3C
O

Progesterone

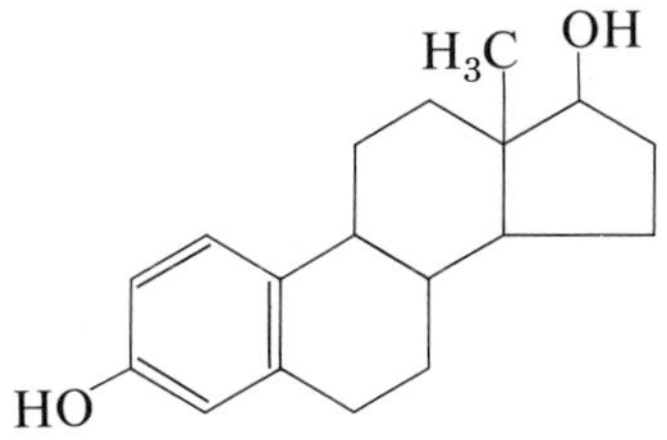
H3C
OH
HO

Estradiol

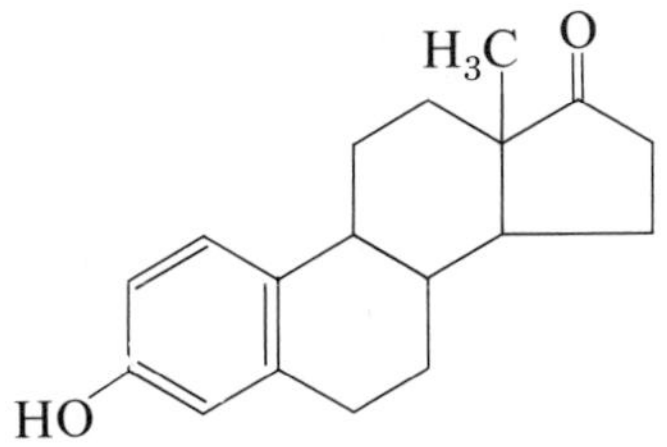
H3C
O
HO

Estrone

Estrogens

Synthetic derivatives of the female sex hormones have attracted widespread attention. When taken regularly, these drugs effectively function to prevent ovulation. The oral contraceptives are usually mixtures of two compounds that are analogs of progesterone and estradiol. For example, Enovid is a combination of norethynodrel and mestranol, whereas Ortho-Novum contains norethinodrone and mestranol. Most of the combination pills sold in the United States contain 2 mg of the progesterone analog and 0.1 mg of the estrogen analog (i.e., a mass ratio of 20:1).

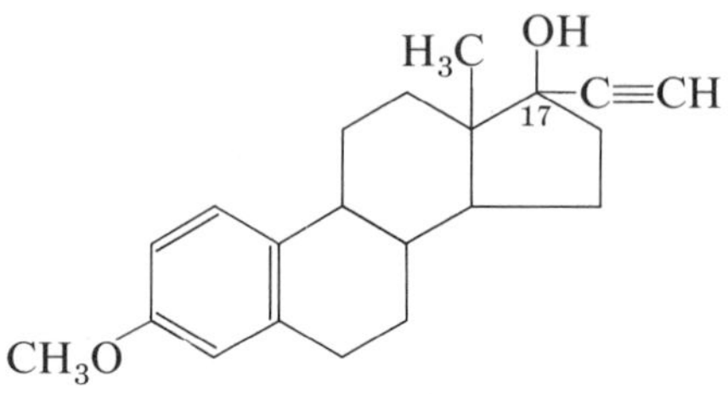

H3C
OH
C≡CH
17
CH3O

Mestranol
(analog of the estrogens)

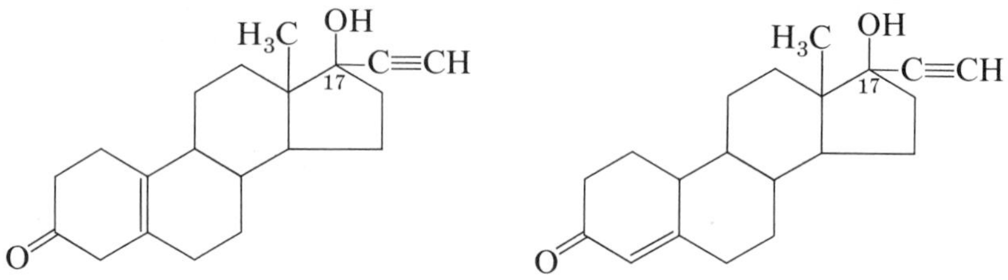

H3C
OH
C≡CH
17
O
H3C
OH
C≡CH
17
O

Norethynodrel Norethinodrone

analogs of progesterone

The prevention of ovulation is also effected by the administration of progesterone and estradiol, but these hormones must be injected into the body for maximal results. It is but a slight structural difference (incorporation of the acetylenic group at carbon-17) that confers upon the synthetic compounds the ability to be taken orally and to function in the same manner as the steroids produced by the body.

24.11 Birth Control

Three hundred years ago there were about 700 million people on Earth. The current population of the world now exceeds four billion. This population bomb has naturally made governments throughout the world acutely aware of the necessity for birth control as an obvious answer to overpopulation and all of its accompanying problems. Reliable contraceptives have been available to many men and women since the middle of the nineteenth century, but mankind had to wait another one hundred years for the method that would prove most effective. The first oral contraceptive—the Pill—did not appear until 1955 and it was not licensed by the Food and Drug Administration until 1960. Several billion pills are produced annually, and these are consumed by more than 100 million women all over the world.

In recent years there have been reports of increased incidence of breast and cervical cancers, thromboembolisms (blood clots), hypertension (high blood pressure), depression and psychiatric disturbances, benign liver tumors, gallbladder disease, urinary tract infections, and other risks among women who have taken the pill. In most cases, the increased risk of adverse side effects has been linked to the estrogen component in the pill. A 1975 report by the Rockefeller Foundation concluded that the pill is a "highly effective and generally safe method of birth control—but not for all women." The major points of the report, based on a review of oral contraceptive research from medical centers around the world, are

1. Oral contraceptives are highly effective and generally safe for most women.
2. The risk of developing serious illness as a consequence of taking the pill is small.
3. Death associated with the oral contraceptives is of a very low order of magnitude.
4. The pill's long-term side effects, unknown at this time, must continue to be monitored closely to safeguard users.
5. There is no evidence connecting use of oral contraceptives with cancer of the breast. For cervical cancer evidence is conflicting, with resolution dependent upon additional long-range data.
6. The increased risk of death from thromboembolism, established in early studies as approximately 3 per 100,000 women per year, is confirmed by more recent research. Increased risk of venous thrombosis of the legs and of cerebral thrombosis has also been established.

In order to understand how oral contraceptives work, it is first necessary to understand the normal physiology of the female menstrual cycle (Figure 24.4). This cycle is characterized by rhythmic monthly changes in the rates of secretion of the sex hormones (estrogens and progesterone) and corresponding changes in the sex organs themselves. Menstruation marks the terminal events of the normal ovarian cycle. The cycle begins when the pituitary gland releases *follicle stimulating hormone* (FSH) into the bloodstream. FSH is a glycoprotein that is primarily responsible for stimulating the growth of one of the eggs in the ovary.[10] Under the influence of FSH, the follicle or sack surrounding the egg secretes the estrogen hormones. The estrogens function to prepare the lining of the uterus to receive the fertilized egg. As the follicle ages, the secretion of estrogens increases and this rise in the estrogen level of the body has a negative feedback effect on the pituitary gland. As the amount of estrogens increases, the output of FSH decreases. At this stage another hormone, *luteinizing hormone* (LH), is secreted by the pituitary gland. LH oversees the final stages of egg-maturation and ensures that the egg is ejected from the ovary (ovulation).

[10] Recently, Dr. Cornelia Channing has identified a polypeptide (called *inhibin*) in mammalian ovaries that may function to inhibit the secretion of FSH from the pituitary gland. Research is being conducted to learn more about the biological activity of *inhibin* and how its effects are overcome so that only one egg matures in each fertility cycle.

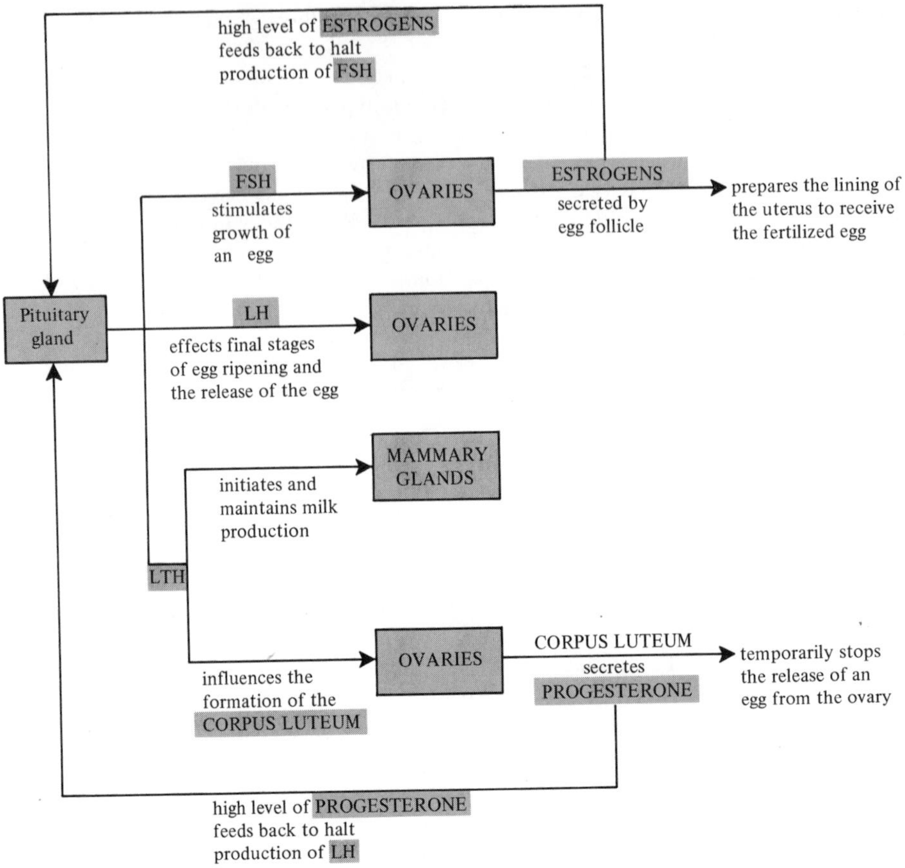
high level of ESTROGENS
feeds back to halt
production of FSH
FSH
stimulates
growth of
an egg
OVARIES
ESTROGENS
secreted by
egg follicle
prepares the lining of
the uterus to receive
the fertilized egg
Pituitary
gland
LH
effects final stages
of egg ripening and
the release of the egg
OVARIES
MAMMARY
GLANDS
initiates and
maintains milk
production
LTH
OVARIES
influences the
formation of the
CORPUS LUTEUM
CORPUS LUTEUM
secretes
PROGESTERONE
temporarily stops
the release of an
egg from the ovary
high level of PROGESTERONE
feeds back to halt
production of LH

Figure 24.4 Hormone interactions in the female reproductive system.

Following ovulation, a third hormone, *luteotrophic hormone* (LTH), is secreted by the pituitary gland. LTH influences the formation of a tissue called the *corpus luteum* (yellow tissue).[11] As it grows in size, the corpus luteum releases the hormone progesterone which temporarily stops the release of another egg from the ovary. Again there is negative feedback control whereby increased amounts of progesterone result in a diminution of LH secretion by the pituitary gland. Further egg-maturation is thus prevented until the fate of the ovulated egg has been decided.

If the egg is not fertilized, the life of the corpus luteum is short and it disintegrates. When this occurs, many of the small blood vessels that are present rupture and bleed. This small amount of blood, along with the fragments of glands and the mucin from these glands, forms the menstrual discharge. Also, if fertilization does not occur, there is a sharp decrease in the amounts of estrogens and progesterone. As a result, the pituitary gland secretes more FSH and LH and the cycle begins anew.

When a fertilized egg is implanted in the uterus, some mechanism prevents further ovulation until after the baby is born. In the last six months of pregnancy, the estrogens and progesterone are produced in large quantities by the placenta. The

[11]After childbirth, LTH initiates and maintain milk production.

high level of these hormones prevents the pituitary from secreting FSH and LH. In this way no additional mature eggs are released during pregnancy.

The two synthetic analogs of the female sex hormones that are combined in the Pill deceive the female body into thinking it is pregnant and acting accordingly. Finding an abundant supply of the sex hormones in the blood, the pituitary gland shuts down its supply of FSH and LH. The egg follicles are not developed nor are eggs released into the fallopian tube. The synthetic hormones also enhance the contraceptive effect by increasing the viscosity of the mucous around the cervix, and thus setting up a barrier that is difficult for the sperm to penetrate.

In recent years, there has been much publicity devoted to a "morning-after" pill. The subjects of this excitement are hormone-like substances that were originally isolated from semen and referred to as **prostaglandins.** At present, 16 prostaglandins have been discovered; they are found in extremely minute amounts in a variety of body tissues and biological fluids, including the menstrual fluid of women. Prostaglandins are a family of unsaturated fatty acids, each containing 20 carbon atoms and having the same basic skeleton as prostanoic acid (Figure 24.5). They are synthesized in the cell membranes from 20-carbon polyunsaturated fatty acids such as arachidonic acid. Slight structural changes are responsible for quite distinct biological effects.

The prostaglandins are among the most potent biological substances known. They are believed to have myriad potential therapeutic uses, including regulating menstruation and fertility, preventing conception, inducing labor, preventing and alleviating stomach ulcers, controlling inflammation and blood pressure, and relieving asthma and nasal congestion. Prostaglandins E_2 and $F_{2\alpha}$ have successfully been administered to induce abortions in humans. Their mechanism is uncertain but is different from that of the steroids. The prostaglandins cause uterine contractions and abortion of the embryo. Because they work in an abortive fashion, prostaglandins would only have to be taken once a month or only if a menstrual period

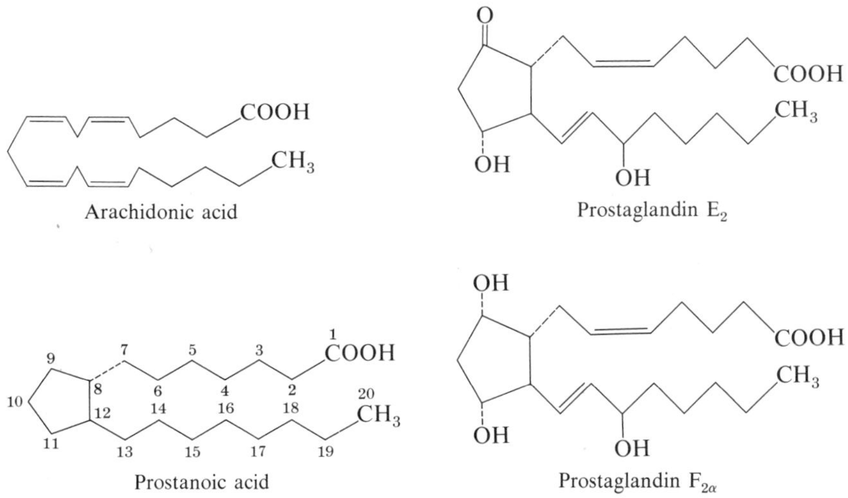

COOH
CH3
Arachidonic acid
O
COOH
CH3
OH
OH
Prostaglandin E2
1
COOH
9
7
5
3
10
8
6
4
2
20
12
14
16
18
CH3
11
13
15
17
19
Prostanoic acid
OH
COOH
CH3
OH
OH
Prostaglandin F2α

Figure 24.5 Arachidonic acid, prostanoic acid, and two of the naturally occurring prostaglandins. (Broken bonds are those that extend below the plane of the cyclopentane ring.)

were missed. Obviously, a great deal of controversy will be generated if and when these compounds become commercially available.[12] Practically every major pharmaceutical company now has active prostaglandin research programs under way to develop new syntheses and to discover new natural sources of prostaglandins. A rich source of prostaglandins exists in a sea plant (the sea whip or sea fan) that grows in coral reefs off the coast of Florida. Recently, chemists have developed several totally synthetic sequences for the preparation of prostaglandins.

Exercises

24.1 Define, explain, or give an example of each of the following.

(a) lipid
(b) triacylglycerol
(c) simple triacylglycerol
(d) mixed triacylglycerol
(e) saturated fatty acid
(f) unsaturated fatty acid
(g) compound lipid
(h) derived lipid
(i) rancidity
(j) soap
(k) syndet
(l) drying oil
(m) margarine
(n) sterol
(o) lecithin
(p) antioxidant
(q) spontaneous combustion
(r) micelle

24.2 Contrast the physical properties of fats and fatty acids. In what solvents are fats and oils soluble?

24.3 Define saponification number and iodine number. Of what significance are these numbers?

24.4 Name each of the following compounds.

(a)
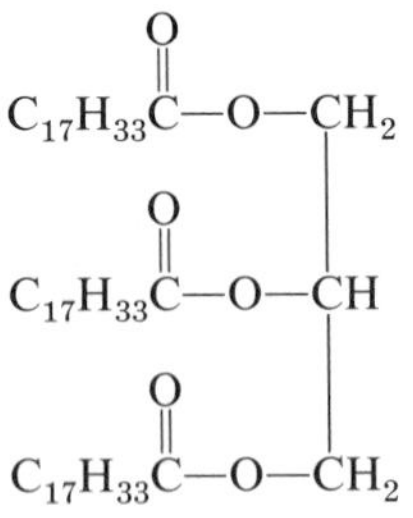

(b)
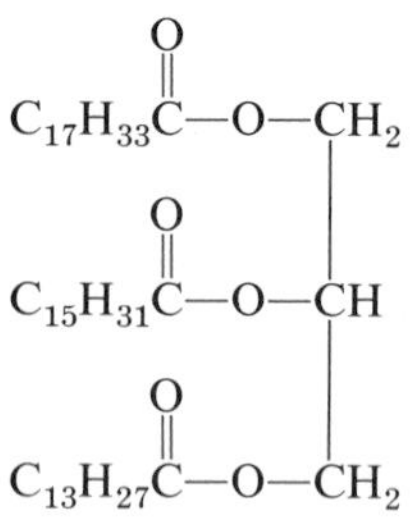

[12]At present, both prostaglandins must be administered by either intravenous or intrauterine injection. No orally active derivatives have been developed, and prospects are not hopeful because of the rapid metabolic degradation of the natural prostaglandins.

(c) 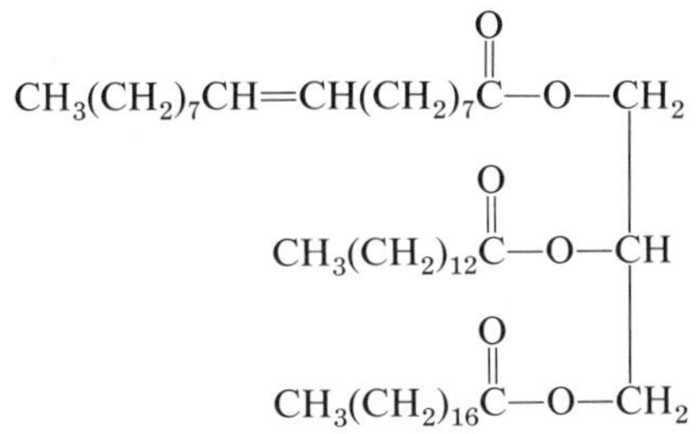

$$CH_3(CH_2)_7CH{=}CH(CH_2)_7\overset{\overset{\displaystyle O}{\|}}{C}{-}O{-}CH_2$$

$$CH_3(CH_2)_{12}\overset{\overset{\displaystyle O}{\|}}{C}{-}O{-}CH$$

$$CH_3(CH_2)_{16}\overset{\overset{\displaystyle O}{\|}}{C}{-}O{-}CH_2$$

(d) $C_{17}H_{35}C\overset{\displaystyle O}{\underset{\displaystyle O^-Na^+}{\diagup}}$ **(e)** $C_{15}H_{31}C\overset{\displaystyle O}{\underset{\displaystyle OC_{30}H_{61}}{\diagup}}$ 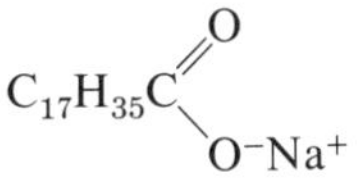

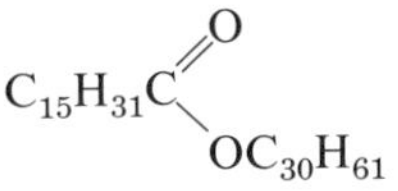

24.5 Write structural formulas for each of the following.
(a) glyceryl palmitate
(b) cetyl stearate
(c) a triacylglycerol of lauric, palmitic, and stearic acids
(d) a highly unsaturated oil
(e) sodium palmityl sulfate

24.6 Write a representative equation for each of the following reactions.
(a) hydrogenation of an oil
(b) halogenation of an oil
(c) manufacture of soap
(d) acid hydrolysis of triolein
(e) saponification of a mixed triacylglycerol
(f) hydrolysis of glycerol phosphate

24.7 What types of chemical reactions occur when fats become rancid?

24.8 Describe briefly (a) the manufacture of soap and (b) the cleansing action of soap.

24.9 Draw structural formulas for (a) an anionic detergent, (b) a cationic detergent, and (c) a nonionic detergent. Explain the significant differences in the structure of each. What structural features are necessary for a compound to be a good detergent? What advantages do detergents have over ordinary soaps? What are the disadvantages?

24.10 The active ingredient in the laundry detergent Cold Power is sodium tridecylbenzenesulfonate. Draw its structural formula.

24.11 Structurally, distinguish between each of the following.
(a) petroleum oil and vegetable oil   (e) a phosphatide and a glycolipid
(b) animal fat and vegetable oil   (f) a soap and a detergent
(c) a fat and a wax   (g) a hard and a soft soap
(d) a fat and a phosphatide   (h) butter and margarine

24.12 Contrast the structures of phosphatidyl serine, phosphatidyl ethanolamine, phosphatidyl choline, and sphingomyelin. To what subclass of phospholipids does each compound belong?

24.13 What is known of the function of the phospholipids? What is cephalin? What is its function in the body?

24.14 Phospholipids are considerably more soluble in water than triacylglycerols. Explain.

24.15 Draw the formula of the steroid nucleus and number the carbon atoms. Name three compounds that contain this structure and give their major functions in the body.

24.16 How many asymmetric carbon atoms are there in cholesterol? How many optical isomers could exist?

24.17 Write the equations for a possible reaction between cholesterol and **(a)** Br_2 and **(b)** acetic acid.

24.18 Which would require more potassium hydroxide for its saponification, lard or the same mass of linseed oil? Explain.

24.19 How many grams of soap can be made from 20 grams of glyceryl stearate?

24.20 How many milliliters of a 0.10 *M* KOH solution are required to saponify 10 grams of tristearin?

24.21 A 2.0 gram sample of a pure fat required the addition of 6.7 ml of 1.0 *M* KOH for complete hydrolysis. Calculate the saponification number of the fat.

24.22 A 20 gram sample of an oil was treated with iodine and it was observed that 17 grams of I_2 was required for complete reaction. Calculate the iodine number of the oil.

25

Proteins

The class of compounds known as **proteins** is an essential constituent of all cells. On the average, two thirds of the total dry weight of the cell is composed of protein. The Dutch chemist G. J. Mulder (1838) is credited as being one of the first scientists to recognize their importance. He used the term *protein* (Greek, *proteios,* first) to describe these vital compounds.

> There is present in plants and animals a substance which . . . is without a doubt the most important of the known components of living matter, and without it, life would be impossible. This substance has been named protein.

It is now apparent that the name was well chosen. Proteins are the major structural components of animal tissue just as cellulose provides for the structure of the plants. Proteins are components of skin, hair, wool, feathers, nails, horns, hoofs, muscles, tendons, connecting tissue, and supporting tissue such as cartilage. In addition, proteins are involved in communication (nerves), defense (antibodies), metabolic regulation (hormones), biochemical catalysis (enzymes), and oxygen transport (hemoglobin). They are utilized in the building of new tissue and in the maintenance of tissue that is already developed. Whereas the carbohydrates and lipids are used primarily as energy sources, the primary function of the proteins is body building and maintenance. Lipids and carbohydrates are stored by the body as energy reserves, but proteins are not stored to any appreciable extent. It is possible for animals to survive for a short period of time on a diet consisting of protein, vitamins, and minerals; an animal could not survive over the same period of time on a protein-free diet containing lipids, carbohydrates, vitamins, and minerals.

Proteins are giant polymeric molecules (linear polymers of amino acids) that vary greatly in molecular dimensions. Their molecular weights may range from several thousand to several million daltons.[1] In addition to carbon, hydrogen, and oxygen, all proteins contain nitrogen, and many contain sulfur, phosphorus, and traces of other elements. The elementary composition of most proteins is remarkably constant at about 51% carbon, 7% hydrogen, 23% oxygen, 16% nitrogen, 1–3% sulfur, and less than 1% phosphorus.

25.1 Classification of Proteins

Because of the great complexity of protein molecules (each one contains thousands of atoms), it is not possible to classify the proteins systematically in the same way as the carbohydrates and lipids are categorized, that is, on the basis of structural similarities. An older convention classified proteins under either of two descriptive headings. Stringy, elongated proteins such as silk fibroin and keratin were termed *fibrous;* spherical, compact proteins such as egg albumin, casein, and most enzymes were termed *globular*. Fibrous proteins tend to be insoluble in water and in most other solvents, whereas globular proteins are soluble in water and in solutions of salt and water.

Presently, there are two alternative methods for classifying proteins. One method classifies them according to composition as either **simple** or **conjugated.** A simple protein produces exclusively alpha amino acids upon hydrolysis. A conjugated protein produces a nonproteinaceous material, called a **prosthetic group,** upon hydrolysis in addition to the alpha amino acids. The simple proteins are further subdivided according to their solubility in various solvents. Subdivision of the conjugated proteins rests upon the nature of the prosthetic group. The second method of classification groups the proteins according to their function. Both systems are outlined here.

a. Classification of Proteins According to Composition

1. The Simple Proteins. **Albumins** constitute the most important and the most common group of simple proteins. They are present in egg white (egg albumin) and in blood (serum albumin). They are soluble in water and in dilute salt solutions.

[1] Presently, many biologists and biochemists are using the term "dalton" as a unit of mass synonymous with molcular weight (MW). A dalton is equivalent to the mass of one hydrogen atom (1.67×10^{-24} g). The molecular weight of a molecule expressed in daltons is numerically equivalent to its molar mass (M) expressed in units of grams per mole. Thus a protein whose molecular weight is 10,000 daltons has a molar mass of 10,000 g/mole.

Globulins are insoluble in water but are soluble in dilute salt solutions. They are a widely distributed group of proteins, for example, as antibodies in the blood serum and as blood fibrinogen.

Histones are basic proteins, since they contain a high proportion of basic amino acids (lysine and/or arginine). Histones are found associated with nucleic acids in the nucleoproteins of the cell. They are soluble in water and insoluble in dilute ammonium hydroxide.

Scleroproteins (albuminoids) have structural and protective functions and are characterized by their insolubility in water and in most other solvents. Examples are keratin (hair, skin, and nails), collagen (bone, tendon, and cartilage), and elastin (elastic fibers of connective tissue).

2. The Conjugated Proteins. **Phosphoproteins** are converted to phosphoric acid and amino acids upon complete hydrolysis. Casein, which is found in milk, is an important member of this class.

Glycoproteins contain a carbohydrate or a carbohydrate derivative as their prosthetic group covalently bonded to the protein chain. Mucin, a constituent of saliva, is a glycoprotein.

Chromoproteins consist of a pigmented prosthetic group combined with a simple protein. An example is hemoglobin. Hemoglobin possesses the iron-containing pigment **heme** (Figure 25.1) coordinated to the simple protein portion, the **globin.**

Nucleoproteins are complex substances that occur abundantly in the nuclei of plant and animal cells. The prosthetic groups are complex polymers of high molar mass, and are called nucleic acids (DNA and RNA; see Chapter 27).

Lipoproteins are often classified as compound lipids. They consist of triacyglycerols, cholesterol esters, and phospholipids attached to protein molecules. The nature of the linkage between the lipid and the protein molecule is obscure and is the subject of much current research. Most, if not all, of the lipid in mammalian blood is transported in the form of lipoprotein complexes. The electron transport system in the mitochondria is believed to contain large amounts of lipoprotein. Lipoproteins are also found in egg yolk, cell nuclei, ribosomes, and the myelin sheath of nerves.

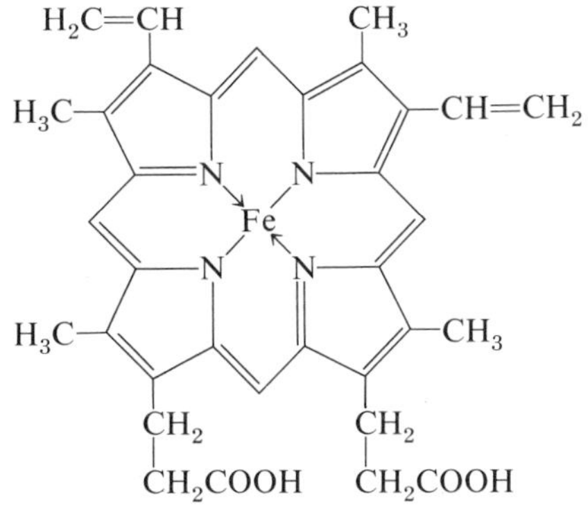

H2C=CH
CH3
H3C
CH=CH2
N
N
Fe
N
N
H3C
CH3
CH2
CH2
CH2COOH
CH2COOH

Figure 25.1 The structure of the heme portion of the hemoglobin and myoglobin molecules.

b. Classification of Proteins According to Function

1. **Structural Proteins.** More than half of the total protein of the mammalian body is collagen, found in skin, cartilage, and bone.
2. **Contractile Proteins.** Examples are myosin and actin isolated from skeletal muscle.
3. **Enzymes.** Enzymes represent the largest class of proteins. Nearly 2000 different kinds of enzymes are known. These biological catalysts are vitally important to all living systems and are discussed in detail in the next chapter.
4. **Hormones.** Several hormones have already been mentioned in connection with steroids. However, many hormones, such as insulin, are proteins.
5. **Antibodies.** The body produces antibodies to destroy any foreign materials (antigens) released into the body by an infectious agent. Gamma globulins are antibodies.
6. **Blood Proteins.** The albumins, globulins, and fibrinogen are the three major protein constituents of blood (see Chapter 31).

25.2 Amino Acids

Proteins may be defined as *high molar mass compounds consisting largely or entirely of chains of amino acids*. In this respect, proteins may be considered to be polymers analogous to the polysaccharides. In proteins, however, there are some twenty *different* structural monomeric units, and these are the amino acids. With the exception of proline and hydroxyproline (which contain an alpha secondary nitrogen atom) the amino acids that are the building blocks of the common proteins are characterized by a primary amino group *alpha* to the carboxyl group.

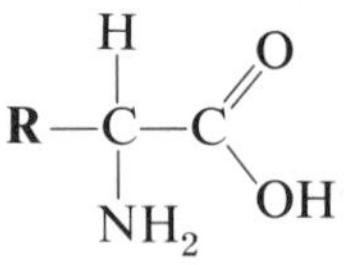
H
O
R—C—C
NH2
OH

Each individual amino acid has unique properties as a result of variations in the structures of the different **R** groups. (The **R** group is referred to as the *amino acid side chain.*)

Amino acids may be classified in several different ways. We choose to group them, according to the polarity of their side chains, into four different classes: (1) nonpolar, (2) polar and neutral, (3) acidic, and (4) basic. The structures of the common amino acids, their three-letter abbreviations, and certain of their distinctive features are given in Table 25.1. The amino acids are known exclusively by their common names, as the IUPAC names are too cumbersome. For example, asparagine was the first amino acid to be isolated

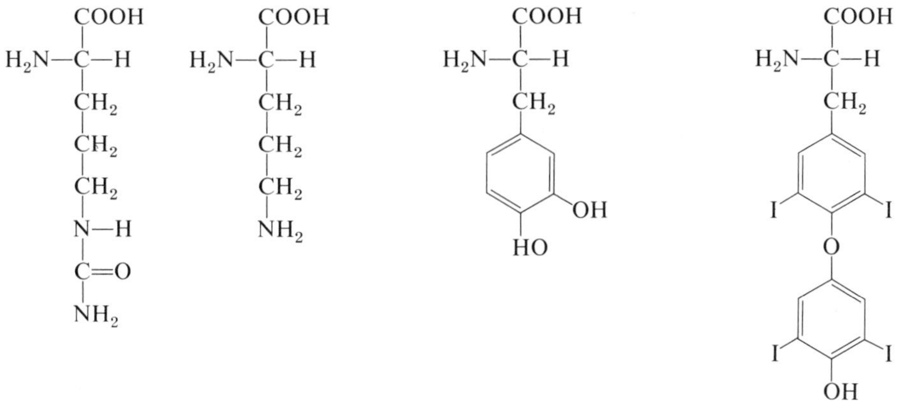

Citrulline — Ornithine
(intermediates in the urea cycle)

Dihydroxyphenylalanine (DOPA)
(intermediate in the formation of adrenalin)

Thyroxine
(thyroid hormone)

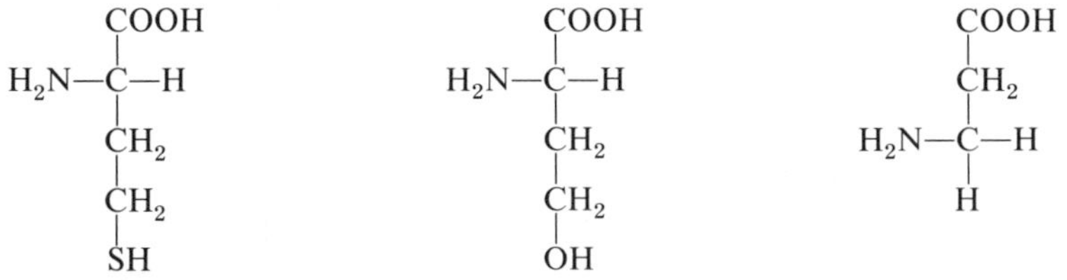

Homocysteine
(intermediate in the synthesis of methionine)

Homoserine
(intermediate in the synthesis of threonine)

β-Alanine
(component of coenzyme A)

Figure 25.2 Some biologically important nonprotein amino acids.

(1806) and was given its name because it was obtained from protein found in asparagus juice. Glycine, the major amino acid found in gelatin, received its name because of its sweet taste (Greek, *glykys,* sweet). There are more than 200 other amino acids that occur in nature (particularly in the plant kingdom), but they do not appear in proteins (Figure 25.2). These amino acids perform important biological functions (e.g., as intermediates in metabolic pathways) either as single molecules or combined in molecules of relatively small size.

25.3 General Properties of Amino Acids

a. Configuration

Notice, in Table 25.1, that glycine is the only amino acid whose α-carbon atom is not asymmetric. Therefore, with the exception of glycine, all the amino acids are optically active and may exist in either the D- or L-enantiomeric form (Figure 25.3). Once again, the reference compound for the

Table 25.1 Naturally Occurring Amino Acids

Name	*Abbreviation*	*Structural Formula*	*M*	*Distinctive Features*
1. Amino Acids with a Nonpolar **R**— Group				
Glycine	Gly	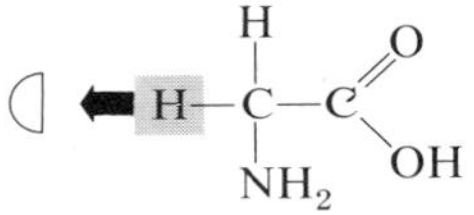	75	The only amino acid lacking an asymmetric carbon.
Alanine	Ala	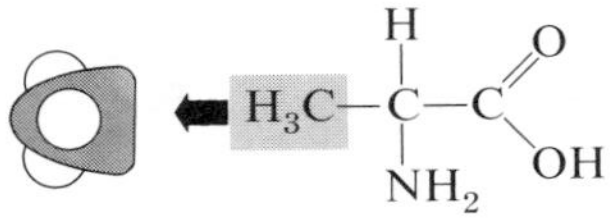	89	
Valine	Val	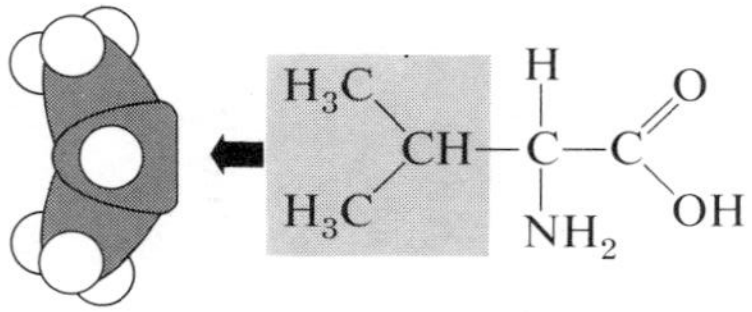	117	Most animals cannot synthesize branched-chain amino acids. They are, therefore, essential in the diet. (Valine, Leucine, Isoleucine)
Leucine	Leu	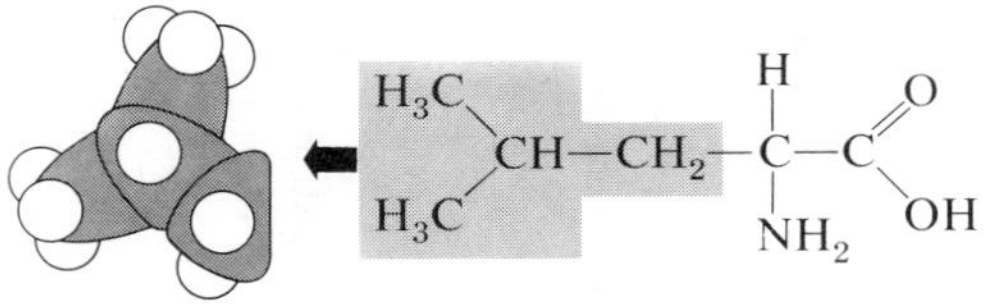	131	
Isoleucine	Ile	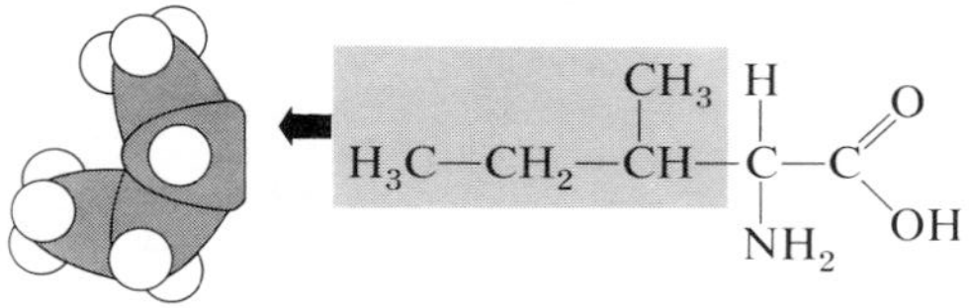	131	
Phenylalanine	Phe	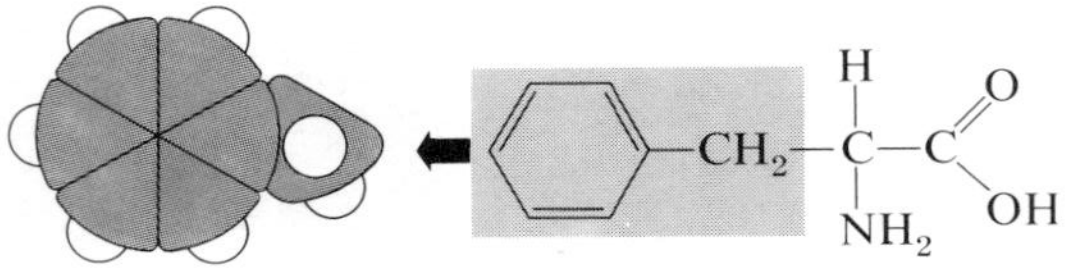	165	

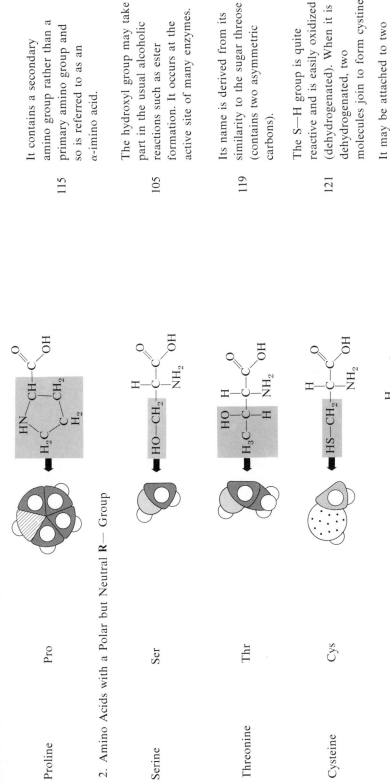

Proline	Pro	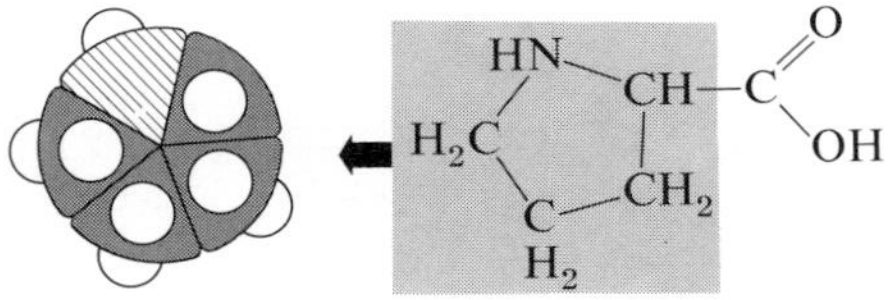	115	It contains a secondary amino group rather than a primary amino group and so is referred to as an α-imino acid.
2. Amino Acids with a Polar but Neutral **R**— Group				
Serine	Ser	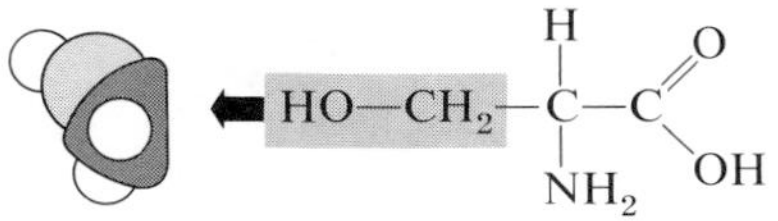	105	The hydroxyl group may take part in the usual alcoholic reactions such as ester formation. It occurs at the active site of many enzymes.
Threonine	Thr	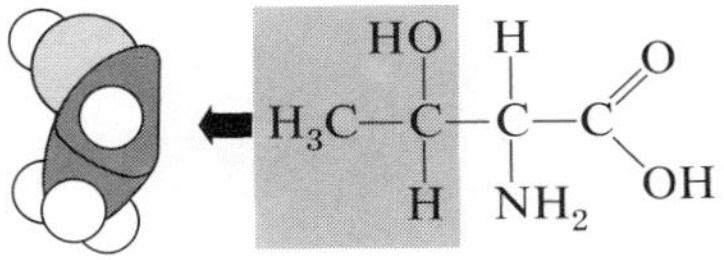	119	Its name is derived from its similarity to the sugar threose (contains two asymmetric carbons).
Cysteine	Cys	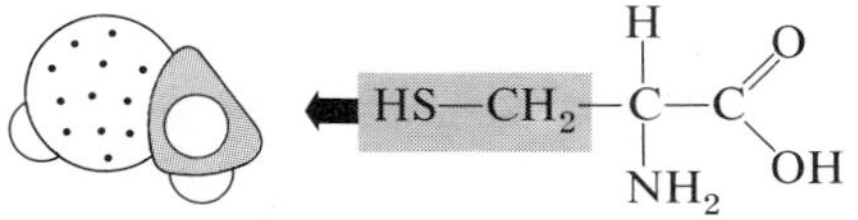	121	The S—H group is quite reactive and is easily oxidized (dehydrogenated). When it is dehydrogenated, two molecules join to form cystine.
Cystine	$(Cys)_2$	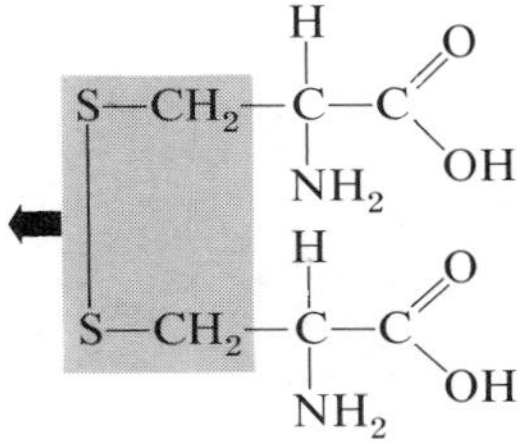	240	It may be attached to two different peptide chains at the same time or to the same chain at widely separated places. Such S—S disulfide linkages are common in most proteins and are important in establishing protein structure.
Methionine	Met	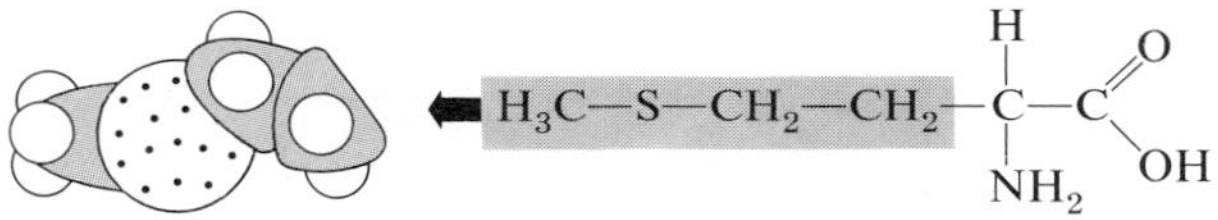	149	It is important as a donor of methyl groups.

(*Continued*)

Table 25.1 Naturally Occurring Amino Acids *(Continued)*

Name	*Abbreviation*	*Structural Formula*	*M*	*Distinctive Features*
2. Amino Acids with a Polar but Neutral **R**— Group *(Continued)*				
Tryptophan	Trp	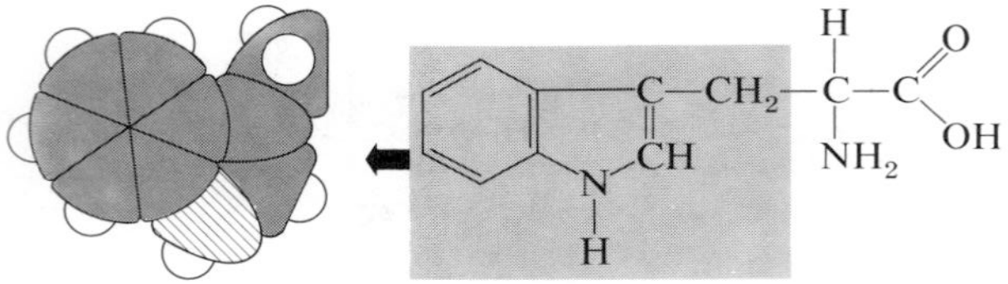	204	A heterocyclic amino acid (a derivative of indole).
Tyrosine	Tyr	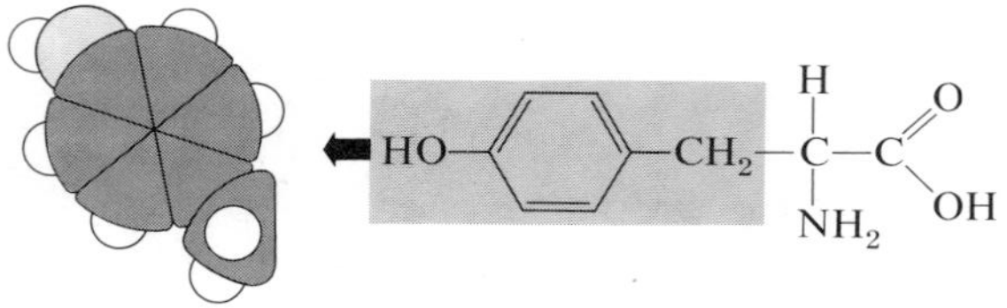	181	The phenolic group is weakly acidic. It loses its proton at pH values above 9.
Hydroxyproline	Hypro	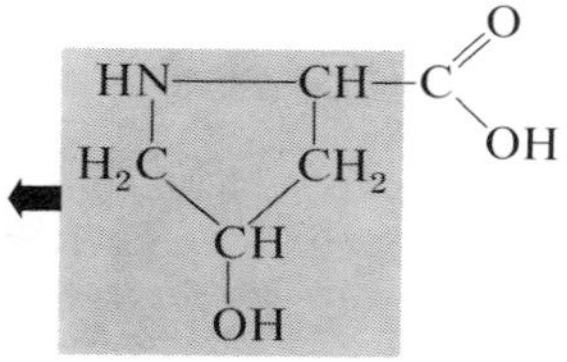	131	It is a major constituent of the structural protein collagen and is formed by the hydroxylation of proline already present in the protein molecule.
3. Acidic Amino Acids				
Aspartic acid	Asp	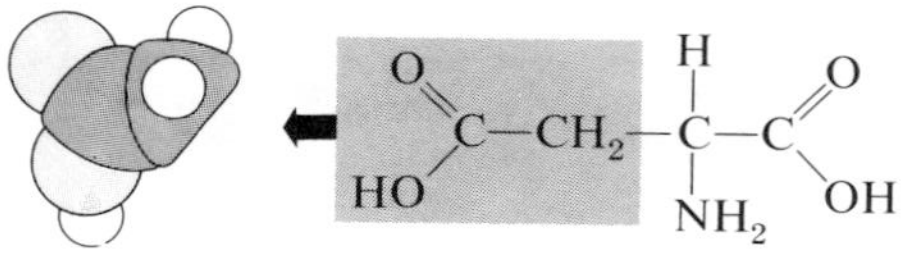	133	Also found in proteins as its amide derivative, asparagine (Asn).

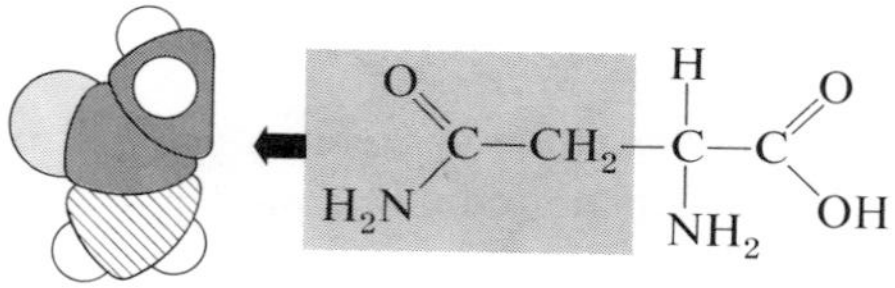

Glutamic acid	Glu	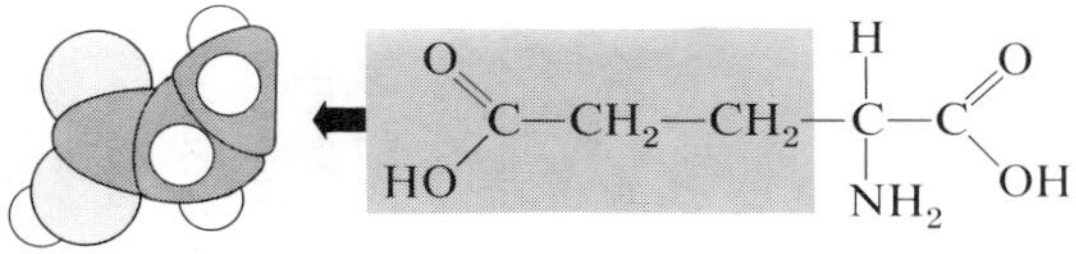	147	Also found in proteins as its amide derivative, glutamine (Gln).
		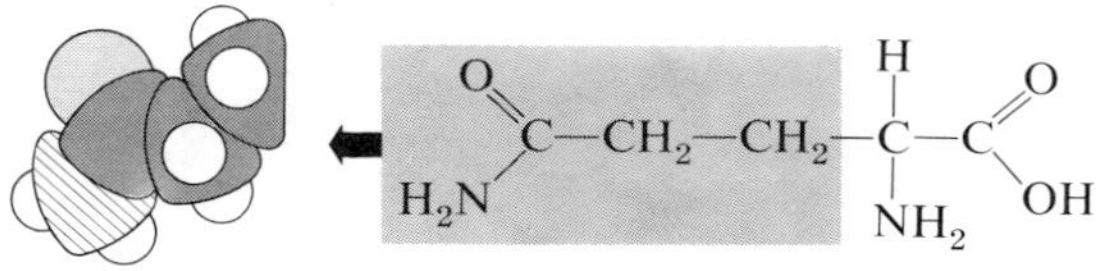		

4. Basic Amino Acids

Lysine	Lys	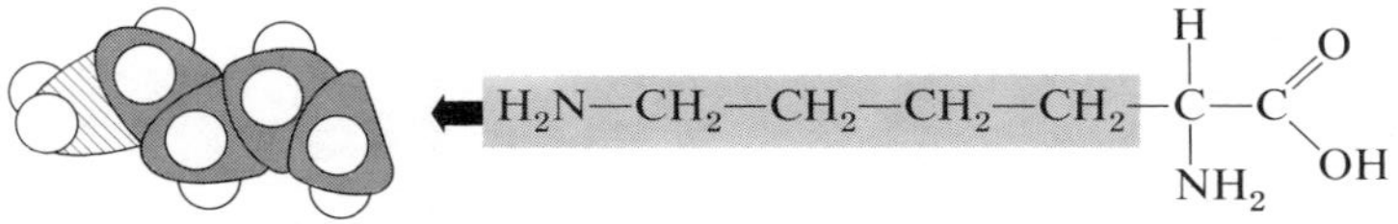	146	
Arginine	Arg	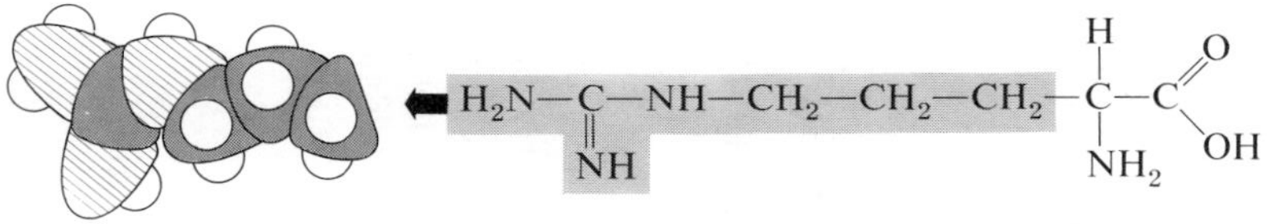	174	The very basic guanidyl group ($-\underset{H}{N}-\underset{NH}{\overset{\Vert}{C}}-NH_2$) makes arginine almot as strong a base as NaOH.
Histidine	His	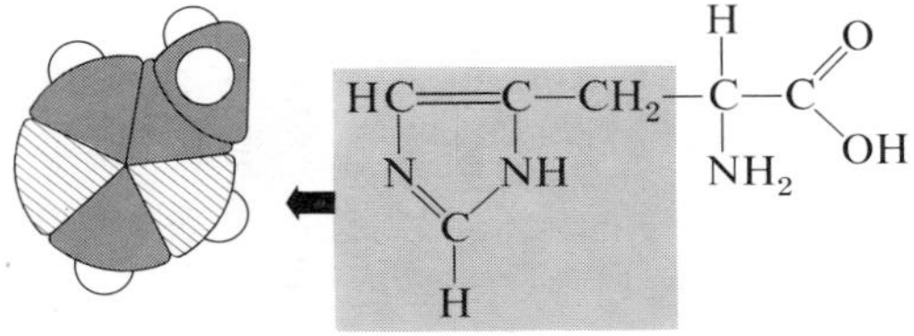	154	Histidine contains the weak heterocyclic base imidazole.

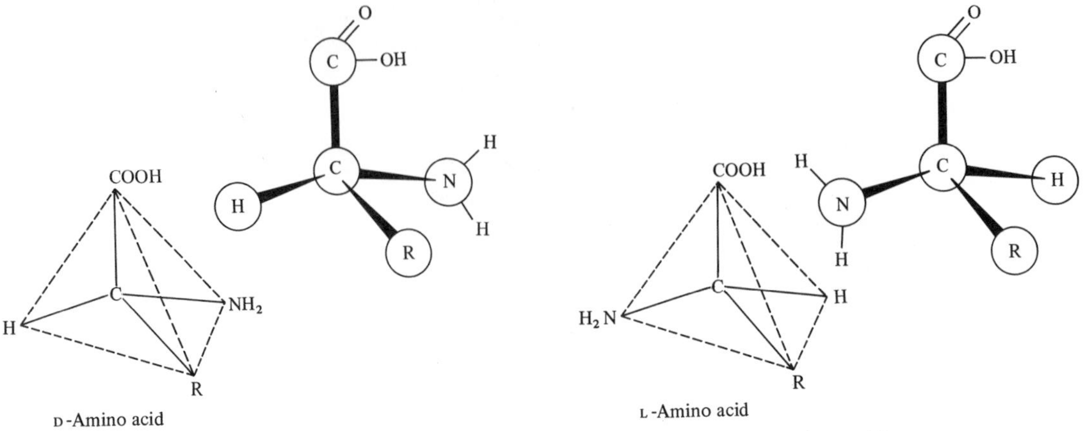
D-Amino acid
L-Amino acid

Figure 25.3 Optical isomers of an amino acid.

assignment of configuration is glyceraldehyde. (The amino group of the amino acid takes the place of the hydroxyl group of glyceraldehyde.)

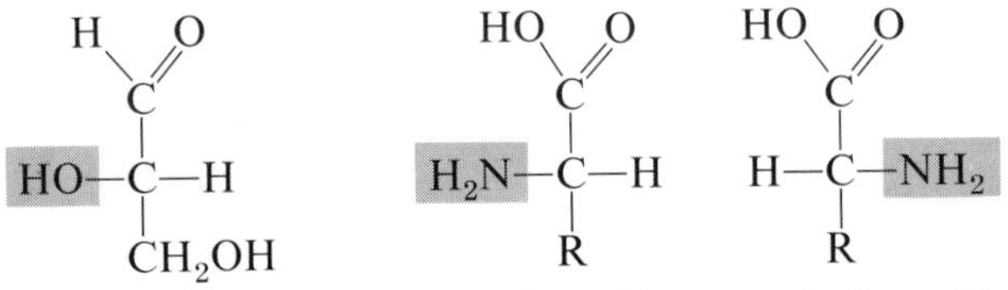

L(−)-Glyceraldehyde   L-Amino acid   D-Amino acid

It is interesting to note that the naturally occurring sugars belong to the D-series, whereas nearly all known plant and animal proteins are composed entirely of L-amino acids.[2] Recall that the letters D and L refer only to a specific configuration, not to the direction of optical rotation. For example, in a neutral solution L-alanine is dextrorotatory, whereas L-serine is levorotatory. The optical rotation of any amino acid is very much dependent upon the pH of the solution.

b. Dipolar Ion Structure

The amino acids are colorless, nonvolatile, crystalline solids, melting with decomposition at temperatures above 200°C. They are soluble in water in varying degrees and they are insoluble in nonpolar organic solvents. Their properties diverge widely from those of their unsubstituted carboxylic acid

[2]Certain bacteria are known to possess some D-amino acids as components of their cell walls. *Streptococcus faecalis* has a requirement for D-alanine; *Staphylococcus aureus* requires D-glutamic acid; several antibiotics (actinomycin for one) contain varying amounts of D-leucine, D-phenylalanine, and D-valine.

analogs. Organic acids of comparable molar mass are liquids or low melting solids that are soluble in organic solvents but have limited solubility in water. In fact, the properties of the amino acids are more similar to those of inorganic salts than to those of amines or organic acids.

The salt-like character of the amino acids is more readily accounted for if we assign a dipolar ion (also called **inner salt** or **zwitterion**) structure to amino acids in the solid state and in neutral solution. Since amino acids contain both acidic (—COOH) and basic (—NH_2) groups within the same molecule, we may postulate an intramolecular neutralization reaction, leading to salt formation.

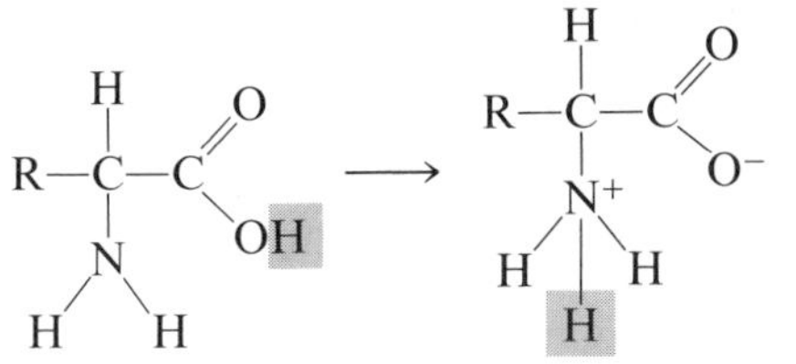

Zwitterion form
of an amino acid
(a dipolar ion)

Conductivity measurements confirm that at certain pH values, all amino acids exist in the neutral zwitterion form. When placed in an electric field, the dipolar ions migrate neither to the anode nor to the cathode, thus indicating a net charge of zero. *The pH at which an amino acid exists in solution as a zwitterion* is called the **isoelectric pH** (that is, the pH at which the amino acid is electrically neutral and has no tendency to migrate toward either electrode). Each amino acid has its own characteristic isoelectric pH. The isoelectric pH of the neutral amino acids ranges from 4.8 to 6.3, that of the acidic amino acids ranges from 2.8 to 3.2, and the isoelectric pH of the basic amino acids occurs between 7.6 and 10.8. Table 25.2 gives the isoelectric pH's for some representative amino acids.

Table 25.2 Isoelectric pH's of Some Representative Amino Acids

Amino Acid	*Type of Amino Acid*	*Isoelectric pH*
Alanine	Neutral, nonpolar	6.0
Valine	Neutral, nonpolar	6.0
Serine	Neutral, polar	5.7
Threonine	Neutral, polar	5.6
Aspartic acid	Acidic	2.8
Glutamic acid	Acidic	3.2
Lysine	Basic	9.7
Arginine	Basic	10.8

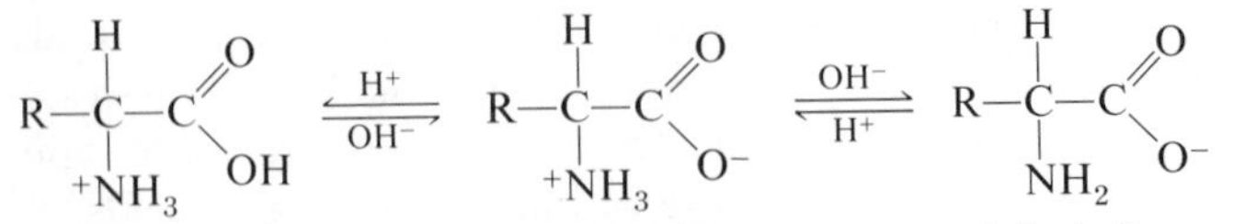
Cationic form
at pH's below
isoelectric pH
Net charge = +1
Zwitterion form
at isoelectric pH
Net charge = 0
Anionic form
at pH's above
isoelectric pH
Net charge = −1

Figure 25.4 Acid-base behavior of neutral amino acids.

Dipolar ions are **amphoteric** substances; they are *capable of donating or accepting protons* and thus may behave either as acids or as bases. When an acid is added to a solution of an amino acid that is at its isoelectric pH, hydrogen ions are accepted; the species becomes positively charged and will migrate to the negative electrode. Addition of a base causes the zwitterion to donate protons, thus becoming negatively charged; the species will now migrate toward the positive electrode. Figure 25.4 illustrates the acid-base behavior of the neutral amino acids.

For the neutral amino acids the pK_a of the α-carboxyl group (—COOH) is approximately 2–3, and the pK_a of the α-amino group ($—NH_3^{\oplus}$) is approximately 9–10. The isoelectric pH can be determined for these neutral amino acids by simply averaging the pK_a values for the two ionizable groups. Thus for alanine

$$\begin{aligned} pK_a(\text{—COOH}) &= 2.3 \\ pK_a(\text{—}\overset{\oplus}{N}H_3) &= \underline{9.7} \\ & \ 12.0 \end{aligned}$$

$$\text{isoelectric pH} = \frac{12.0}{2} = 6.0$$

We shall see shortly that the structure and properties of proteins depend largely upon the nature of the **R** groups of their constituent amino acids. Of particular importance are those amino acids that contain either acidic or basic side chains. Each of these groups has a characteristic pK_a; it is determined by titrating the amino acid with standard solutions of acid (e.g., HCl) and base (e.g., NaOH). The pK_a values of the acidic and basic amino acids are listed in Table 25.3. Figure 25.5 depicts the different species that exist in

Table 25.3 pK_a Values for the Acidic and Basic Amino Acids

	pK_a (—COOH)	pK_a ($—\overset{\oplus}{N}H_3$)	pK_a (**R** group)
Aspartic acid	2.0	9.7	3.7
Glutamic acid	2.2	9.7	4.3
Arginine	2.2	9.0	12.5
Lysine	2.2	9.0	10.5
Histidine	1.8	9.2	6.0

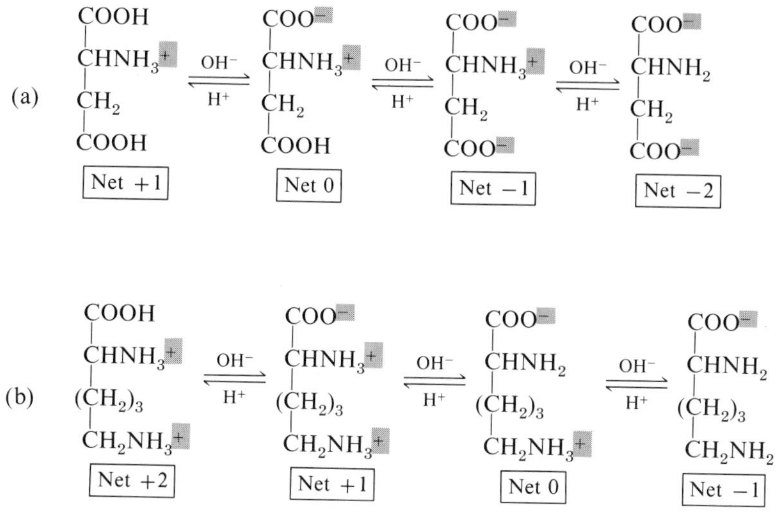
(a)
COOH
CHNH3+
CH2
COOH
Net +1
OH−
H+
COO−
CHNH3+
CH2
COOH
Net 0
COO−
CHNH3+
CH2
COO−
Net −1
COO−
CHNH2
CH2
COO−
Net −2
(b)
COOH
CHNH3+
(CH2)3
CH2NH3+
Net +2
COO−
CHNH3+
(CH2)3
CH2NH3+
Net +1
COO−
CHNH2
(CH2)3
CH2NH3+
Net 0
COO−
CHNH2
(CH2)3
CH2NH2
Net −1

Figure 25.5 Structures and charges for aspartic acid (a) and lysine (b) in solutions of varying hydrogen ion concentration.

solution for aspartic acid and lysine. Again, we can determine the isoelectric pH for these amino acids by averaging two pK_a values. For the acidic amino acids, the isoelectric pH is the average of the pK_a's of the two carboxyl groups, whereas the isoelectric pH for the basic amino acids is the average of the pK_a's of the two amino groups.

25.4 Primary Structure of Proteins—Peptides

Partial hydrolysis of proteins with dilute hydrochloric acid yields smaller molecules known as **peptides.** In 1902, Emil Fischer postulated that these peptides contain amino acids joined together by amide linkages between the amino group of one molecule and the carboxyl group of another (the union is accompanied by the elimination of a molecule of water). In protein chemistry such an amide linkage is given the name **peptide bond** to indicate its relevance to the combination of amino acids. Peptides containing two, three, and four amino acids are called dipeptides, tripeptides, and tetrapeptides, respectively. The term *polypeptide* is applied to relatively small molecules containing from five to thirty-five amino acids, and the term *protein* is arbitrarily used for longer polypeptides (i.e., having molecular weights above 5000 daltons). However, this is not meant to imply that a protein is composed of only one polypeptide chain. The enzymes lysozyme (Figure 25.6) and ribonuclease (Figure 25.7) contain 129 and 124 amino acids, respectively, in only one polypeptide chain, but the hormone insulin (Figure 25.8) contains two polypeptide chains (one chain contains 21 amino acids, the other 30) joined by disulfide linkages, and the hemoglobin molecule contains four amino acid chains. The **primary structure** of a protein refers to *the number and sequence of the amino acids in its polypeptide chain(s).*

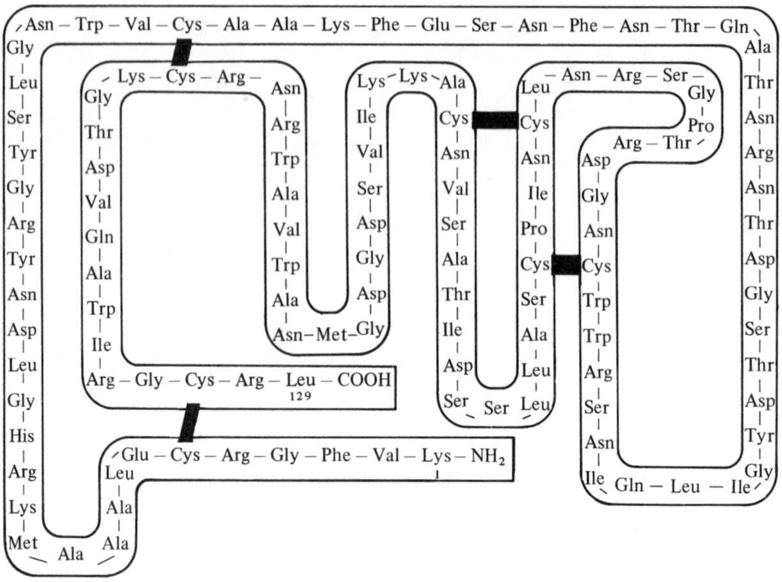
Asn – Trp – Val – Cys – Ala – Ala – Lys – Phe – Glu – Ser – Asn – Phe – Asn – Thr – Gln
Gly – Leu – Ser – Tyr – Gly – Arg – Tyr – Asn – Asp – Leu – Gly – His – Arg – Lys – Met – Ala – Ala – Ala – Leu
Glu – Cys – Arg – Gly – Phe – Val – Lys – NH2
1
Lys – Cys – Arg – Asn – Arg – Trp – Ala – Val – Trp – Ala – Asn – Met – Gly – Asp – Gly – Asp – Ser – Val – Ile – Lys – Lys – Ala – Cys – Asn – Val – Ser – Ala – Thr – Ile – Asp – Ser – Ser – Leu – Ala – Ser – Cys – Pro – Ile – Asn – Cys – Leu
Gly – Thr – Asp – Val – Gln – Ala – Trp – Ile – Arg – Gly – Cys – Arg – Leu – COOH
129
Asn – Arg – Ser – Gly – Pro – Thr – Arg – Asp – Gly – Asn – Cys – Trp – Trp – Arg – Ser – Asn – Ile – Gln – Leu – Ile – Gly – Tyr – Asp – Thr – Ser – Gly – Asp – Thr – Asn – Arg – Asn – Thr – Ala

Figure 25.6 The primary structure of lysozyme. Lysozyme is a protein enzyme that is found in many tissues and secretions of the human body, in plants, and in the whites of eggs. Its function is to destroy invading bacteria by catalyzing the breakdown of their cell walls.

By convention, we represent the structure of peptides beginning with the amino acid whose amino group is free (the so-called N-terminal end). The other end, therefore, contains a free carboxyl group and is referred to as the C-terminal end. Each amino acid in the peptide, with the exception of the C-terminal amino acid, is named as an acyl group in which the suffix *-ine* is

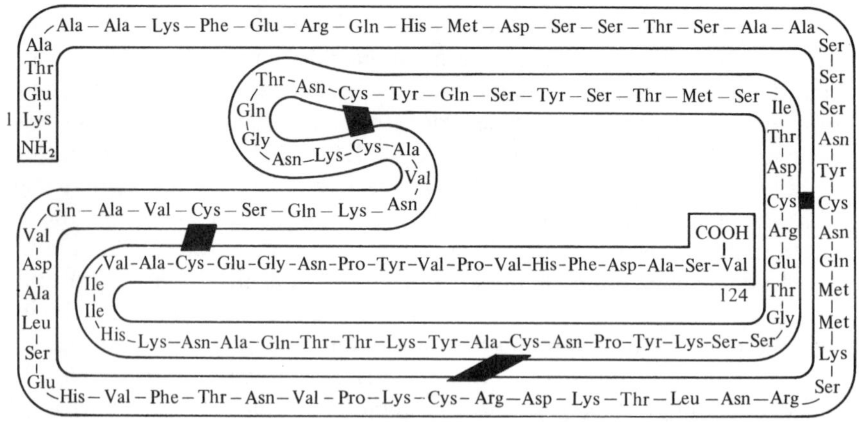
1
NH2 – Lys – Glu – Thr – Ala – Ala – Ala – Lys – Phe – Glu – Arg – Gln – His – Met – Asp – Ser – Ser – Thr – Ser – Ala – Ala – Ser – Ser – Ser – Asn – Tyr – Cys – Asn – Gln – Met – Met – Lys – Ser – Arg – Asn – Leu – Thr – Lys – Asp – Arg – Cys – Lys – Pro – Val – Asn – Thr – Phe – Val – His – Glu – Ser – Leu – Ala – Asp – Val – Gln – Ala – Val – Cys – Ser – Gln – Lys – Asn – Val – Ala – Cys – Lys – Asn – Gly – Gln – Thr – Asn – Cys – Tyr – Gln – Ser – Tyr – Ser – Thr – Met – Ser – Ile – Thr – Asp – Cys – Arg – Glu – Thr – Gly – Ser – Lys – Tyr – Pro – Asn – Cys – Ala – Tyr – Lys – Thr – Thr – Gln – Ala – Asn – Lys – His – Ile – Ile – Val – Ala – Cys – Glu – Gly – Asn – Pro – Tyr – Val – Pro – Val – His – Phe – Asp – Ala – Ser – Val – COOH
124

Figure 25.7 The primary structure of bovine ribonuclease. Ribonuclease is a protein enzyme that catalyzes the hydrolysis or ribonucleic acid (RNA).

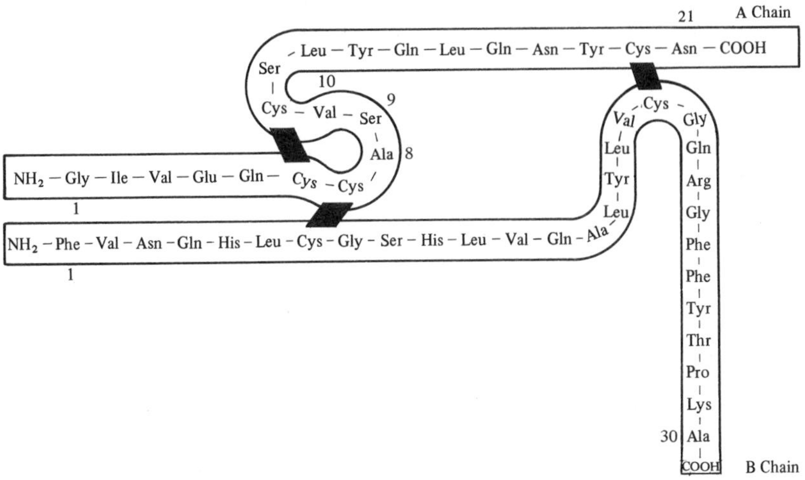
21 A Chain
Leu – Tyr – Gln – Leu – Gln – Asn – Tyr – Cys – Asn – COOH
Ser
10
9
Cys – Val – Ser
Ala 8
NH₂ – Gly – Ile – Val – Glu – Gln – Cys – Cys
1
NH₂ – Phe – Val – Asn – Gln – His – Leu – Cys – Gly – Ser – His – Leu – Val – Gln – Ala
1
Val Cys Gly Leu Gln Tyr Arg Leu Gly Phe Phe Tyr Thr Pro Lys
30 Ala
COOH B Chain

Figure 25.8 The primary structure of bovine insulin. Insulin is a protein hormone produced in the pancreas. It is essential for the regulation of carbohydrate metabolism. Insulin from other mammalian species has the identical structure except for amino acid units 8, 9, and 10 of the A Chain, which differ as shown below.

	Amino Acid Units		
Species	*8*	*9*	*10*
Cow	Ala	Ser	Val
Sheep	Ala	Gly	Val
Horse	Thr	Gly	Ile
Man	Thr	Ser	Ile
Pig	Thr	Ser	Ile
Whale	Thr	Ser	Ile

replaced by *-yl.* (In naming the individual units it is not necessary to include the prefix L since it is understood that the amino acids have the L-configuration.) Figure 25.9 illustrates the bonding and the naming of some peptides.

There are several naturally occurring peptides that possess significant biological activity. Oxytocin and vasopressin are nonapeptides produced by the pituitary gland (Figure 25.10). Notice that six of the nine amino acids are identical in both peptides, yet their physiological effects are markedly different. Oxytocin causes the contraction of smooth muscles. It is often administered at childbirth to induce delivery. Vasopressin causes a rise in blood pressure and it regulates the excretion of water by the kidneys. A deficiency of vasopressin results in diabetes insipidus, in which too much urine is excreted. The disease is treated by administering vasopressin.

One can begin to appreciate the enormous complexity of protein molecules from a consideration of these simple peptides. For example, if you take three different amino acids, A, B, and C, and use each one only once in forming a tripeptide, six possible isomers can be formed: ABC, ACB, BAC, BCA, CAB, and CBA. If there are *x* different amino acids in a polypeptide,

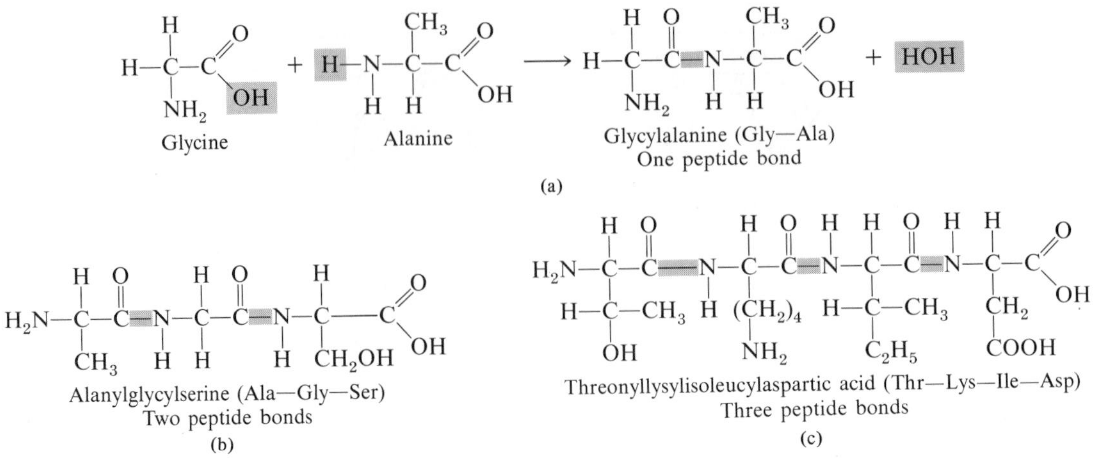
Glycine
Alanine
Glycylalanine (Gly—Ala)
One peptide bond
(a)
Alanylglycylserine (Ala—Gly—Ser)
Two peptide bonds
(b)
Threonyllysylisoleucylaspartic acid (Thr—Lys—Ile—Asp)
Three peptide bonds
(c)

Figure 25.9 (a) Formation of a dipeptide. (b) A tripeptide. (c) A tetrapeptide.

each one appearing only once, then the number of different possible structures is x factorial

$$x(x - 1)(x - 2) \cdots$$

Application of this formula to a peptide containing ten different amino acids reveals that there are over 3.6 million different possible structures.

Now consider further that in most proteins certain amino acids appear not once, but several times. A typical protein contains anywhere from 100 to 300 amino acids, and these units are joined together in a definite, characteristic sequence. Such a protein may have an enormous number of structural isomers based solely upon differences in amino acid sequence. It is estimated that the human body contains over 100,000 different protein molecules, all of which are characterized by different sequential arrangements of the twenty fundamental building blocks. The best analogy of the relationship between the twenty amino acids and the vast number of protein molecules they constitute is drawn between the letters of the alphabet and the number of words they form. All the words in the English language are composed of different sequential combinations of just twenty-six letters.

The truly remarkable phenomenon of living systems is that with so many possibilities for protein synthesis, generation after generation of cells produce identical protein molecules. This ability of the cell to produce the same proteins as its parent is vital to life because it is the sequence of amino acids within a protein that bestows biological function upon the molecule. In other words, if a protein molecule is to fulfill its physiological function, its amino acids must be ordered in a particular definite sequence. The substitution of even one amino acid in a large protein molecule can completely alter its biological function. A classic example of this phenomenon occurs in the

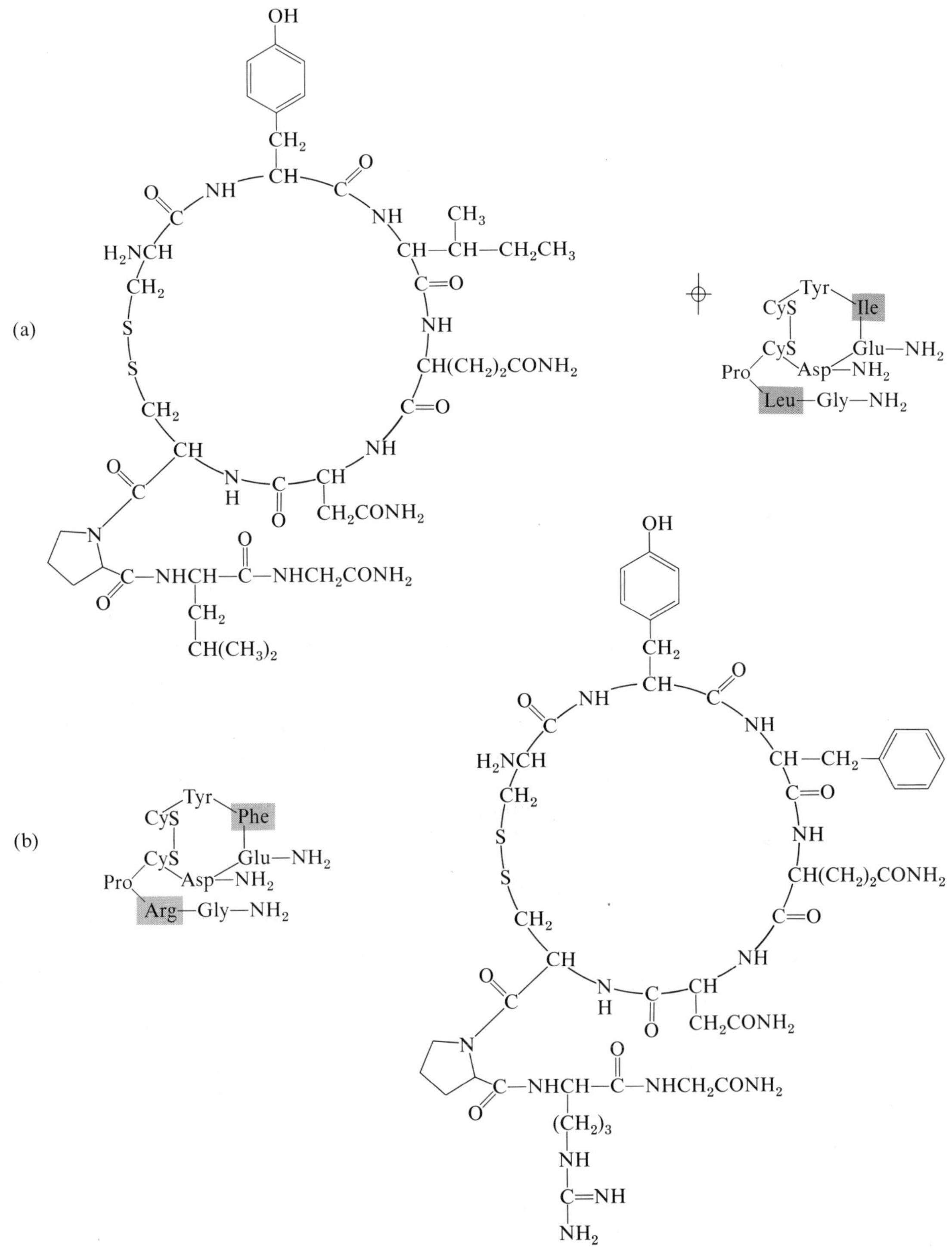
(a)
Tyr
CyS
Ile
CyS
Glu—NH2
Pro
Asp—NH2
Leu—Gly—NH2
(b)
Tyr
CyS
Phe
CyS
Glu—NH2
Pro
Asp—NH2
Arg—Gly—NH2

Figure 25.10 The peptide hormones oxytocin (a) and vasopressin (b). Vasopressin differs in structure from oxytocin only in the replacement of isoleucine by phenylalanine and leucine by arginine. Notice that the C-terminal glycine and the carboxyl groups in the aspartic acid and glutamic acid side chains are present as amides.

disease sickle cell anemia.[3] The hemoglobin of the afflicted individual contains valine in place of glutamic acid. The substitution occurs in two identical polypeptide chains (the β-chains) of the four contained in the hemoglobin molecule, and each β-chain is composed of 146 amino acids. This seemingly insignificant alteration in the molecule causes a drastic change in the properties of the molecule.

Normal hemoglobin:	Val-His-Leu-Thr-Pro-**Glu**-Lys. . .
	1 2 3 4 5 6 7
Sickle cell hemoglobin:	Val-His-Leu-Thr-Pro-**Val**-Lys. . .

Two of the most formidable research problems facing scientists today are to determine (1) the primary structure of proteins and (2) the relationship between amino acid sequence and protein function. The pioneer work on the elucidation of the primary structure of proteins was performed by Frederick Sanger at Cambridge, England. Sanger received the Nobel Prize for chemistry in 1958 for his work on the structure of proteins, especially that of insulin. He chose to study insulin because of its small size (MW = 5733 daltons) and because it was readily available in crystalline form from a variety of sources.[4] By employing methods of controlled hydrolysis of the protein to recognizable peptides, Sanger succeeded in determining the amino acid sequence in the two polypeptide chains of insulin. (It is beyond the scope of this book to elaborate on the techniques used in the analysis of protein structure. For detailed accounts of these methods, you are directed to more advanced texts.)

Sanger's extraordinary achievement opened the door to the elucidation of the structures of many additional proteins. Today, the primary structures are known for over 300 proteins, and partial sequences are known for many others (Table 25.4). Recent years have witnessed several breakthroughs in the areas of structure determination and protein synthesis. The following is only a partial listing of these accomplishments.

Two teams of chemists—one at Rockefeller University and the other at Merck Sharp & Dohme Research Laboratories—independently succeeded in synthesizing ribonucleases having the same enzymic activity as the naturally

[3]This is an inherited disease that may be fatal. The red blood cells of the victim assume a sickled shape instead of the normal round shape. Apparently the alteration of the hemoglobin polypeptide chain decreases the molecule's ability to function as an oxygen carrier.

[4]An interesting fact about living systems is that different ones contain, in many cases, almost identical protein molecules. Insulin is a striking example of this phenomenon (see Figure 25.8). Not only are the primary structures of a variety of mammalian insulin molecules similar but they have the same biochemical properties. This is fortunate for some diabetics who develop an allergy to a particular type of insulin; they can often be treated with insulin from another species that does not produce the allergic reaction. In this instance, alteration of some of the amino acids does not cause the protein to have an altered physiological effect, whereas in hemoglobin, as we have seen, it does. This explanation of the special physiological functions of proteins in terms of their structure will become apparent when we consider the concept of the *active site* in relation to enzymes (see Section 26.5).

Table 25.4 Some Proteins Whose Sequence of Amino Acids Is Known

Protein	*Number of Amino Acid Residues*
Enzymes	
Ribonuclease	124
Lysozyme	129
Papain	198
Trypsinogen	229
Chymotrypsinogen	246
Subtilisin (a bacterial protease)	274
Carboxypeptidase A	307
Others	
Nisin (an antibiotic)	29
Insulin	51
Trypsin inhibitor (bovine pancreas)	58
Cytochrome c (man, horse, pig, rabbit, chicken)	104
Cytochrome c (yeast)	108
Hemoglobin (human)	
α-chain	141
β-chain	146
Myoglobin	153
Tobacco mosaic virus protein subunit	158
Myelin (bovine)	170
Myelin (human)	172
Human growth hormone	188
Gamma globulin	1320

occurring ribonuclease. Ribonuclease is an enzyme that specifically cleaves RNA and degrades it into smaller fragments. This achievement represents the first total synthesis of an enzyme as well as the preparation of the largest known synthetic protein.

Dr. Gerald Edelman and his co-workers at Rockefeller University deciphered the complete amino acid sequence of the four polypeptide chains of a gamma globulin molecule. This protein, the largest ever to be analyzed, contains 1320 amino acids made up of 19,996 atoms and has a molecular weight of 150,000 daltons. Full knowledge of the three-dimensional shape of the molecule must await x-ray analysis of its crystals.

Nobel laureate Dorothy Hodgkin and co-workers at Oxford University have determined by x-ray crystallography the precise three-dimensional spatial arrangement of the two amino acid chains in the insulin molecule. This is the first protein hormone whose structure has been completely resolved.

Dr. Hans Neurath and co-workers at the University of Washington have defined the amino acid sequence of bovine carboxypeptidase A. In 1968, Dr.

William Lipscomb and co-workers at Harvard University completed a three-dimensional model of carboxypeptidase A based on results of their x-ray crystallography studies. The essential features of this model are borne out by the amino acid sequence studies at the University of Washington.

25.5 Secondary Structure of Proteins

The primary structure describes only the sequence of amino acids in the protein chain but tells nothing about the shape (conformation) of the molecule. Since most of the bonds in protein molecules are single bonds, one might assume that there is completely free rotation and that the molecules can assume an infinite number of shapes. However, it is known from a variety of physical measurements that each protein occurs in nature in a single, particular, three-dimensional conformation. *The fixed configuration of the polypeptide backbone* is referred to as the **secondary structure** of a protein.

Two major considerations are involved in the secondary structure of proteins. The first involves the manner in which the protein chain is folded and bent; the second involves the nature of the bonds that stabilize this structure.

Based upon x-ray studies, Linus Pauling (Nobel Prize for Chemistry, 1954, and for Peace, 1962) and Robert Corey postulated that some proteins

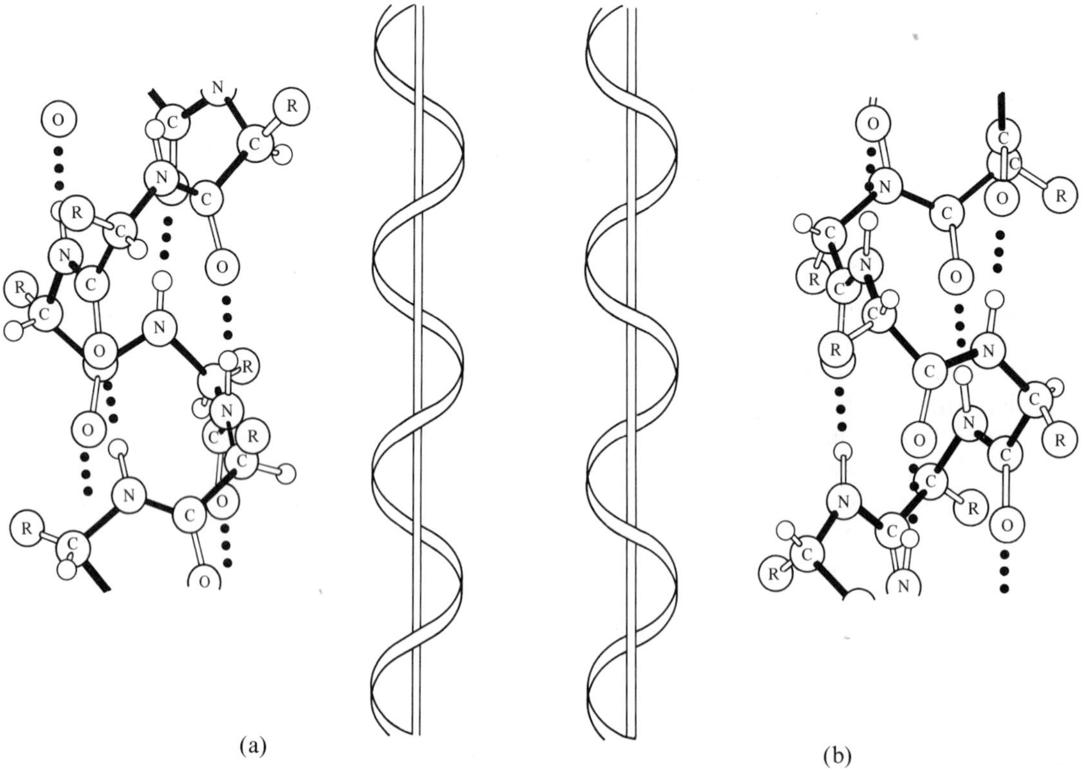
(a)
(b)

Figure 25.11 Alpha-helical conformation of protein. (a) Right-handed; (b) left-handed.

have a spiral shape (that is, they are shaped like a helix). This shape is best visualized as a spring coiled about an imaginary cylinder (Figure 25.11). The spring may be either *left-* or *right-handed.* The spiral structure assumed by proteins (which, you will recall, contain only L-amino acids) is invariably found to be right-handed. The spiral, or *helix,* is stabilized by hydrogen bond formation between the amide hydrogen of one peptide bond and the carboxyl oxygen above it which is located on the next turn of the helix. This *intra*polypeptide hydrogen-bonded structure is designated as α-helical. X-ray data indicate that each turn (gyre) of the helix consists of about four amino acid units and that the side chains of these units project outward from the coiled backbone.

Not all proteins assume a helical configuration. There are some proteins, such as gamma globulin, in which there is no helical structure. Other proteins, such as hemoglobin and myoglobin, are helical in certain regions of the polypeptide chain; the remaining portions assume random configurations. The polypeptide chains of other proteins such as silk fibroin and β-keratin are aligned side by side in a sheet-like arrangement. Adjacent chains are stabilized by *interchain* hydrogen bonds (contrasted to the *intrachain* hydrogen bonds of the α-helix). This configuration is less common than the α-helix because the R groups on adjacent chains sterically hinder one another, causing the chains to pucker or become pleated to minimize the R–R repulsions (Figure 25.12). This configuration, which involves *inter*polypeptide hydrogen bonds, is designated as the β-sheet configuration.

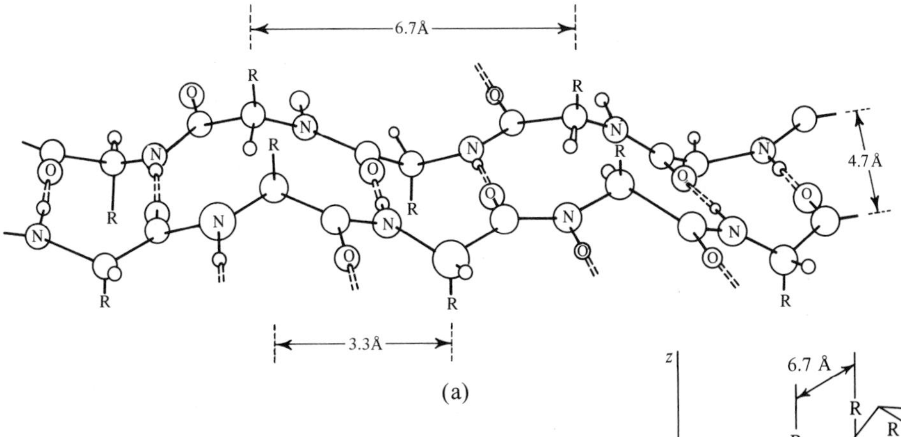

6.7Å
4.7Å
3.3Å

(a)

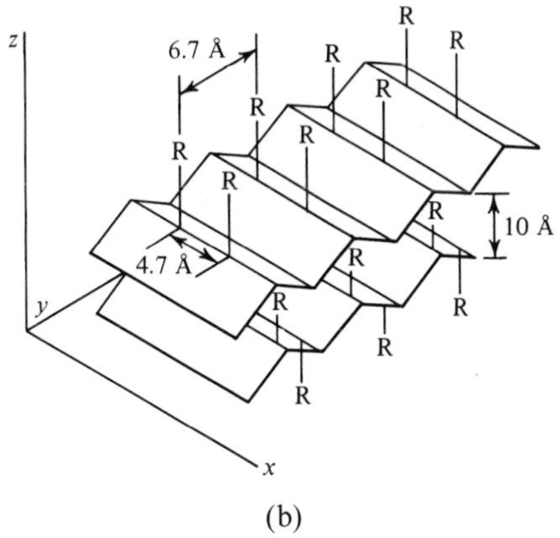

6.7 Å
4.7 Å
10 Å
z
y
x

(b)

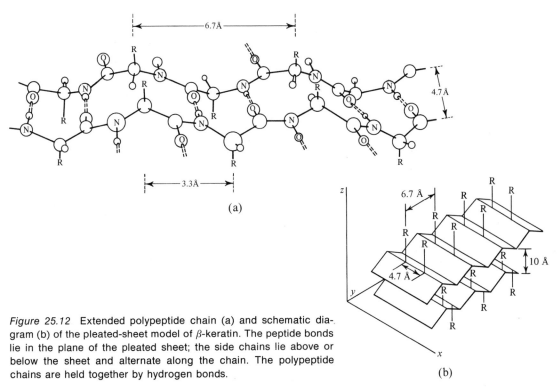

Figure 25.12 Extended polypeptide chain (a) and schematic diagram (b) of the pleated-sheet model of β-keratin. The peptide bonds lie in the plane of the pleated sheet; the side chains lie above or below the sheet and alternate along the chain. The polypeptide chains are held together by hydrogen bonds.

Table 25.5 Noncovalent Bonds and Interactions in Polypeptides[a]

Example	*Type of Bond*	*Approximate Stabilization Energy (kcal/mole)*
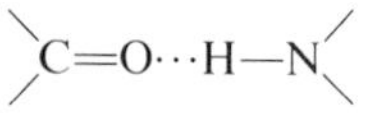	Hydrogen bond between peptides	2–5
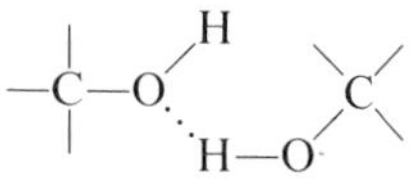	Hydrogen bond between neutral groups	2–5
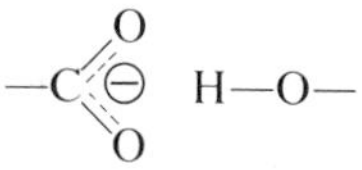	Hydrogen bond between neutral and charged groups	2–5
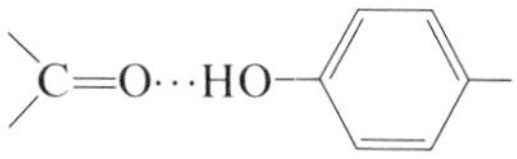	Hydrogen bond between peptide and **R** group	2–5
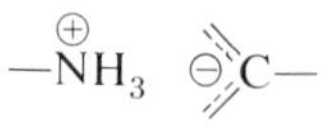	Salt linkage *or* ionic bond between charged groups (strongly dependent on distance)	<10
$—CH_3\ CH_3—$	Hydrophobic interaction	0.3
	Hydrophobic interaction —stacking of aromatic rings	1.5
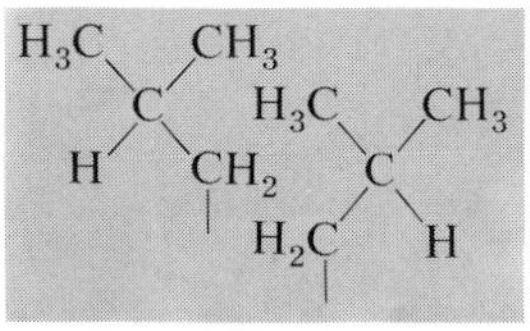	Hydrophobic interaction	1.5
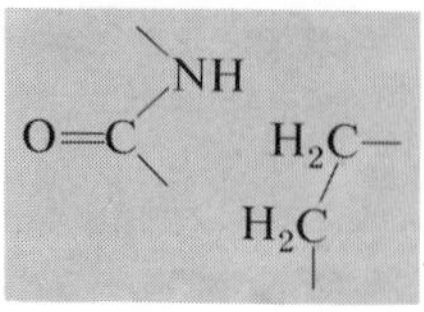	Hydrophobic interaction between **R** group (or part of **R** group, as in lysine) and the peptide bond. This interaction usually involves the hydrophobic **R** groups of the amino acid contributing the carboxyl group to the bond.	0.3 per CH_2
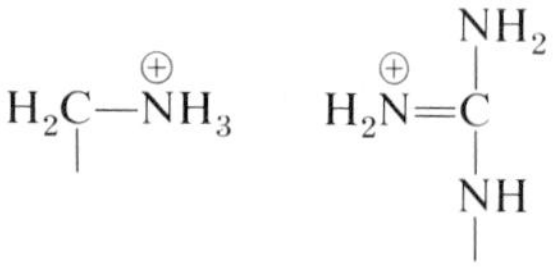	Repulsive interactions between similarly charged groups (strongly dependent on distance)	< -5

[a]From Robert Barker, *Organic Chemistry of Biological Compounds,* 1971, p. 103. Reprinted by permission of Prentice-Hall, Inc., Englewood Cliffs, N.J.

The physical characteristics of wool and silk are a result of their structural configurations. Wool is very flexible and extensible. It can stretch to twice its normal length without breaking, and the fiber will return to its original state upon release of tension. The stretching process involves breaking hydrogen bonds along turns of the α-helix (covalent bonds remain intact). The disulfide bonds between helices, together with re-formed hydrogen bonds, provide the forces that operate to restore the helix when tension is released. Silk, on the other hand, is already stretched out in the β-sheet configuration. The sheets are held together very tightly by hydrogen bonds. As a result, silk fiber is strong, but it is resistant to stretching.

25.6 Tertiary Structure of Proteins

The picture of a protein that we have thus far constructed is that of a loosely coiled spring lacking any definite geometrical arrangement. Intrachain hydrogen bonds are adequate for maintaining the α-helical configuration of a protein in the crystalline state, but they are not sufficiently strong to stabilize this conformation in aqueous solution. Recall that water is a good hydrogen-bonding substance, and therefore water molecules can successfully compete for the hydrogen-bonding sites on the α-helical backbone. This would cause disruption of the internal bonds, and the protein would assume a random configuration simultaneous with the disruption. Experimental evidence has shown, however, that the α-helical configuration of the protein is *not* destroyed when the molecule is dissolved in water. We must therefore conclude that there are other forces involved in compressing the long spiral chains into the definite geometrical structures characteristic of each different protein. *This unique three-dimensional shape, which results from the precise folding and bending of the helical coil,* is referred to as the **tertiary structure** of the protein. The tertiary structure of a protein is intimately involved with the proper biochemical functioning of that protein, as we shall see in the next chapter on enzymes.

The linkages responsible for the tertiary structure of a protein are a function of the nature of the amino acid side chains within the molecule. Many proteins are extremely compact, almost spherical in shape; such proteins have their nonpolar side chains directed toward the interior of the molecule (the hydrophobic or nonaqueous region) and their polar side chains project outward from the surface of the molecule toward the aqueous environment. The resulting picture is very similar to that of a micelle, which was discussed in connection with the properties of soap (Section 24.8). Some of the linkages that contribute to the tertiary structure of proteins are shown in Figure 25.13. Table 25.5 gives an indication of the relative strengths of interactions involving the noncovalent bonds found in proteins.

a. Salt Linkages

Salt linkages (ionic bonds) result from interactions between positively and negatively charged groups on the side chains of the basic and acidic

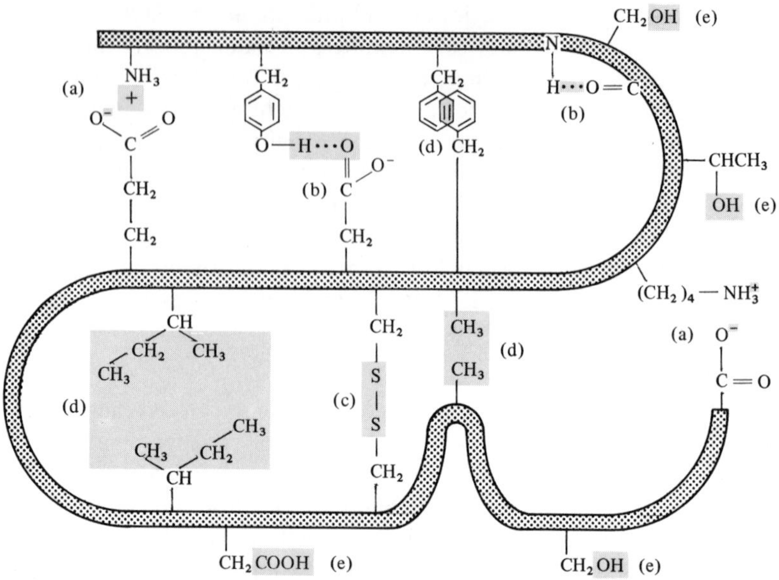

CH₂OH (e)
(a)
NH₃
+
O⁻
O
C
CH₂
CH₂
CH₂
O—H···O
O⁻
(b) C
CH₂
CH₂
(d) CH₂
N
H···O=C
(b)
CHCH₃
OH (e)
(CH₂)₄—NH₃⁺
(a) O⁻
C=O
CH
CH₂
CH₃
CH₃
(d)
CH₃
CH₃
CH₂
CH
CH₂
S
S
CH₂
(c)
CH₃
CH₃
(d)
CH₂COOH (e)
CH₂OH (e)

Figure 25.13 Bonds that stabilize the tertiary structure of proteins. (a) Salt linkages; (b) hydrogen bonds; (c) disulfide linkages; (d) hydrophobic interactions; (e) polar group interactions with water. [Adapted from T. P. Bennett, *Graphic Biochemistry,* Macmillan, 1968.]

amino acids. For example, the mutual attraction between an aspartic acid carboxylate ion and a lysine ammonium ion helps to maintain a particular folded area of the protein.

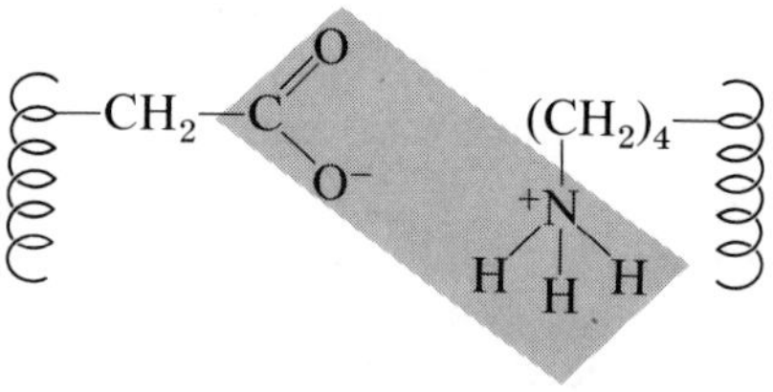

CH₂—C
O
O⁻
(CH₂)₄
+N
H H H

Aspartic acid Lysine

b. Hydrogen Bonding

Hydrogen bonds are formed principally between the side chains of the polar amino acids and between a carboxyl oxygen and a hydrogen donor group. The hydrogen-bonding capabilities of the terminal amino group of lysine and the terminal carboxyl groups of aspartic acid and glutamic acid are pH dependent. These groups can serve as both hydrogen-bond acceptors and hydrogen-bond donors only over a certain range of pH. Hydrogen bonds (as well as salt linkages) are extremely important in the interaction of proteins with other molecules.

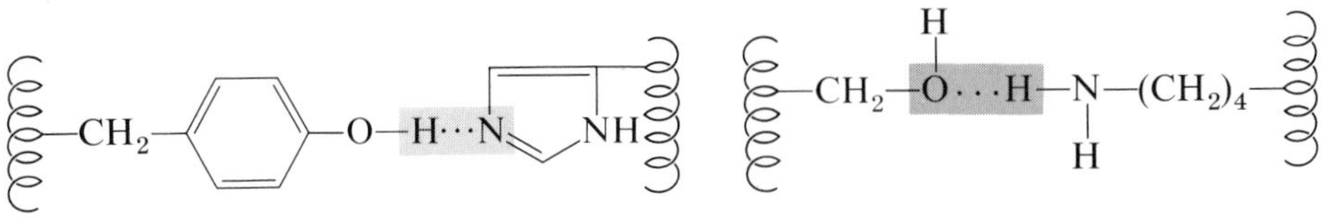

Tyrosine Histidine Serine Lysine

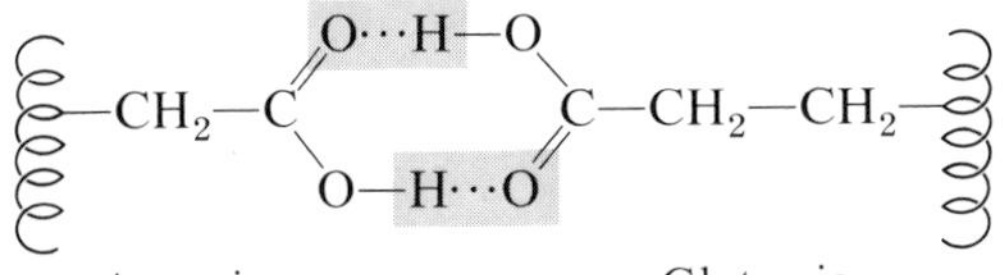

Aspartic acid Glutamic acid

A significant feature of hydrogen bonds is that they are highly directional. The strongest hydrogen bond results when the hydrogen donor and the acceptor atom are colinear. If the acceptor atom is at an angle to the covalently bonded hydrogen atom, the hydrogen bond is much weaker.

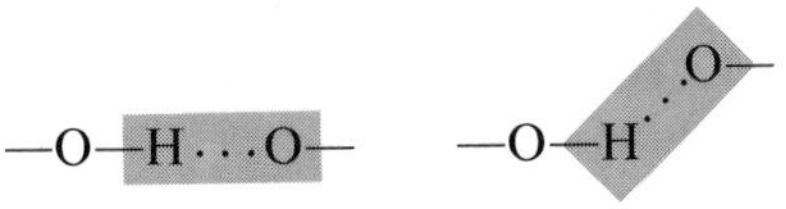

Strong H-bond Weak H-bond

c. Disulfide Linkages

Two cysteine residues may come in proximity as the protein molecule folds. The disulfide linkage results from the subsequent oxidation of the highly reactive sulfhydryl (—SH) groups to form cystine.

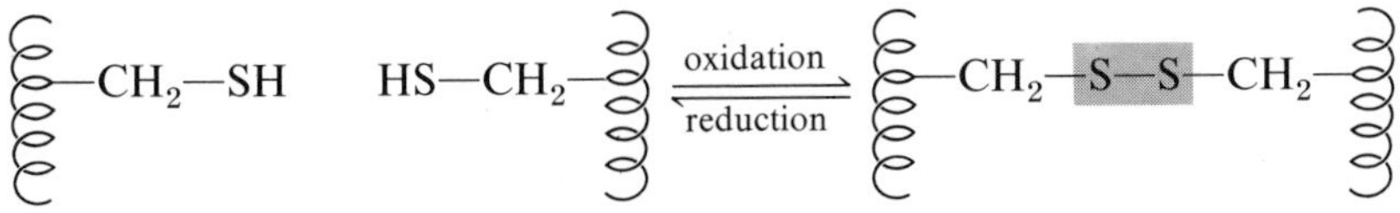
oxidation
reduction

Cysteine Cysteine Cystine

This disulfide bridge is the second most important covalent interaction involved in protein structure. (Recall that the peptide bond is the most important covalent interaction in the structure of proteins.) Disulfide linkages are frequently found in proteins as a general aid to the stabilization of the tertiary structure. Note, however, that one or more of these bonds may join one portion of a polypeptide chain covalently to another, thus interfering with the helical structure. Such linkages are clearly indicated in the protein structures given in Figures 25.6, 25.7, and 25.8.

d. Hydrophobic Interactions

Many investigators now believe that the noncovalent hydrophobic forces are the most significant in stabilizing the conformation of a polypeptide chain. It is not because they are so strong, but rather because there are so many of them. The majority of the nonpolar amino acid groups cluster together at the interior of the chain and the strength of all their hydrophobic interactions is considerable.

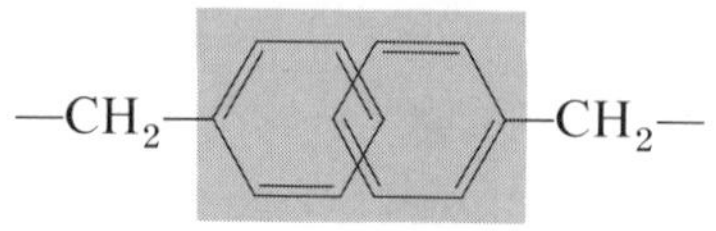
—CH2—
—CH2—

Phenylalanine Phenylalanine

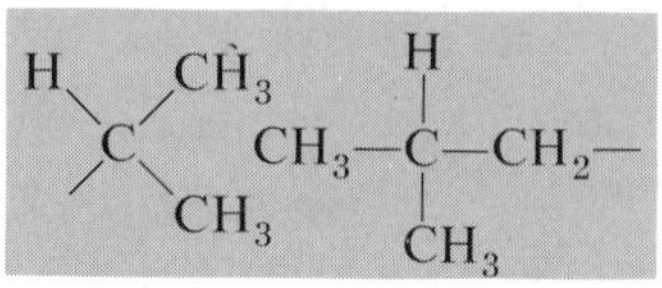
H
CH3
C
CH3
H
CH3—C—CH2—
CH3

Valine Leucine

25.7 Quaternary Structure of Proteins

Some proteins exist as aggregates of polypeptide subunits. *The manner in which separate polypeptide chains fit together* in those proteins containing more than one chain is referred to as the **quaternary structure** of the protein. The quaternary structure is stabilized by the same noncovalent forces (ionic, hydrogen, and hydrophobic bonds) that are involved in maintaining the tertiary structure.

J. C. Kendrew and M. F. Perutz (Nobel Prize winners in 1962) were able, through the use of x-ray diffraction studies, to elucidate completely the primary, secondary, and tertiary structures of myoglobin, a protein consisting of 153 amino acid units arranged in a single chain (Figure 25.14). Myoglobin functions to store oxygen in muscle cells. It is particularly abundant in marine animals, enabling them to remain under water for prolonged periods. The primary structure of myoglobin results from the formation of the single polypeptide chain. The secondary structure involves the coiling of this chain into an α-helix (about 70% of the protein strand has a spiral conformation). The tertiary structure results from the nonuniform folding of the chain to form a stable compact structure.

The hemoglobin molecule, the structure of which was also deduced by Kendrew and Perutz, consists of four separate polypeptide chains or subunits—two α-chains and two β-chains. Each of the four chains is very similar in structure to the single polypeptide chain of myoglobin (Figure 25.15). The four hemoglobin subunits are probably held together by noncovalent surface interactions between the polar side chains. Myoglobin has no quaternary structure (since it is composed of a single polypeptide chain), whereas collagen is made up of three strands wound together like a rope, and the protein coats of several viruses are composed almost entirely of protein subunits arranged in a highly ordered conformation (Figure 25.16). The polio virus contains 130 subunits, and the tobacco mosaic virus contains a grand total of 2130 subunits assembled around a central core of nucleic acids (see Section 27.7).

10
H_2N — Val — Leu — Ser — Glu — Gly — Glu — Trp — Gln — Leu — Val — Leu — His — Val — Tyr — Ala — Lys — Val —

20 30
Glu — Ala — Asp — Val — Ala — Gly — His — Gly — Gln — Asp — Ile — Leu — Ile — Arg — Leu — Phe — Lys —

40 50
Ser — His — Pro — Glu — Thr — Leu — Glu — Lys — Phe — Asp — Arg — Phe — Lys — His — Leu — Lys — Thr —

60
Glu — Ala — Glu — Met — Lys — Ala — Ser — Glu — Asp — Leu — Lys — Gly — His — His — Glu — Ala — Glu —

70 80
Leu — Thr — Ala — Leu — Gly — Ala — Ile — Leu — Lys — Lys — Gly — His — His — Glu — Ala — Glu —

90 100
Leu — Lys — Pro — Leu — Ala — Gln — Ser — His — Ala — Thr — Lys — His — Lys — Ile — Pro — Ile — Lys —

110
Tyr — Leu — Glu — Phe — Ile — Ser — Glu — Ala — Ile — Ile — His — Val — Leu — His — Ser — Arg — His —

120 130
Pro — Gly — Asn — Phe — Gly — Ala — Asp — Ala — Gln — Gly — Ala — Met — Asn — Lys — Ala — Leu — Glu —

140 150
Leu — Phe — Arg — Lys — Asp — Ile — Ala — Ala — Lys — Tyr — Lys — Glu — Leu — Gly — Tyr — Gln — Gly — COOH

(a)

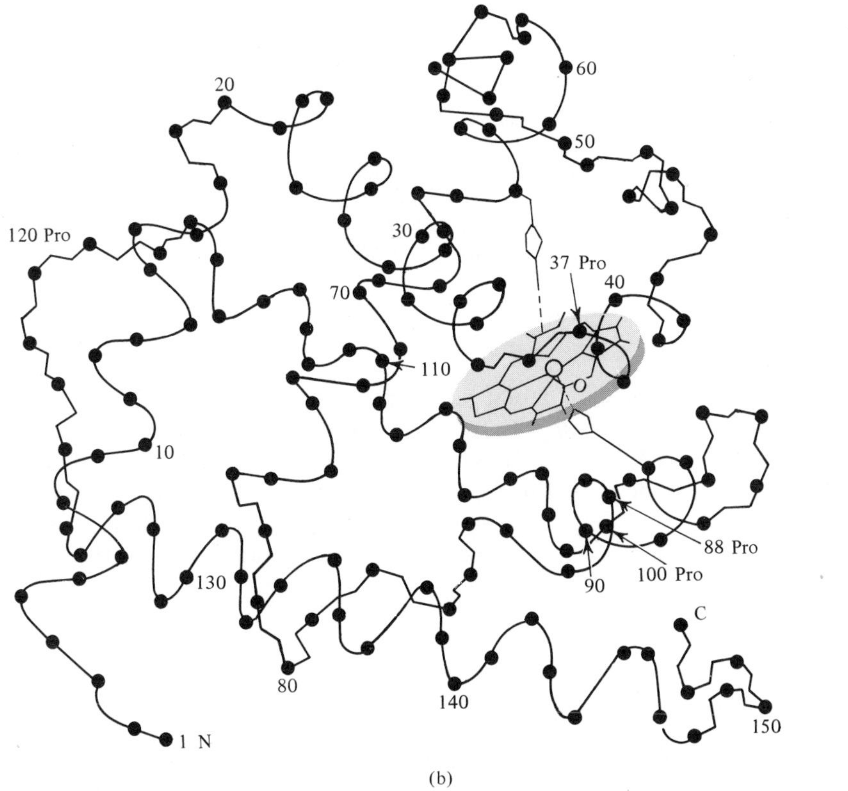

60
20
50
120 Pro
30
37 Pro
40
70
110
10
88 Pro
100 Pro
130
90
C
80
140
150
1 N

(b)

Figure 25.14 (a) The primary structure of myoglobin. (b) The conformation of myoglobin deduced from x-ray diffraction studies; the disk-like shape represents the heme group.

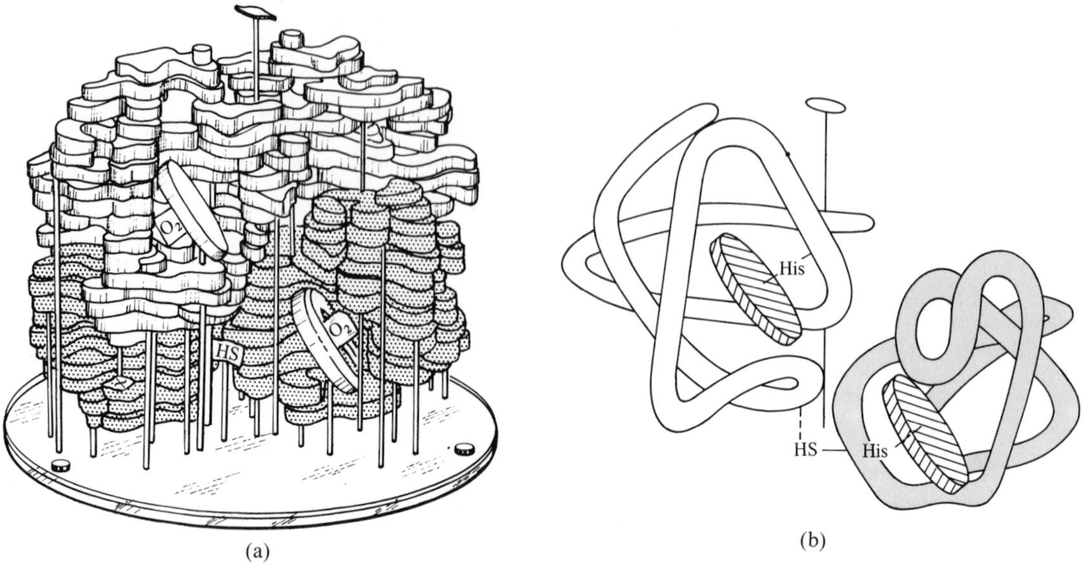
O2
O2
HS
His
HS
His
(a)
(b)

Figure 25.15 The hemoglobin molecule as deduced from x-ray diffraction studies showing tertiary and quaternary structure. The hemoglobin molecule is made up of two identical α-chains and two identical β-chains (shaded blocks). Each chain enfolds a heme prosthetic group, and the oxygen-binding site on this group is marked by O_2. Part (a) shows how closely these chains fit together in approximately tetrahedral arrangement. The model is built up from irregular blocks that represent electron density patterns at various levels in the molecule. (b) A schematic representation of an α-chain and a β-chain.

25.8 Electrochemical Properties of Proteins

When amino acids combine to form the polypeptide chain(s) of protein molecules, the majority of their amino and carboxyl groups are tied up in peptide bond formation. However, the side chains of aspartic and glutamic acids, and those of lysine, arginine, and histidine all retain their acidic and basic groups. In proteins, just as in the free amino acids, these groups will exist in solution as charged species, such as —COO^- and —NH_3^+. Accord-

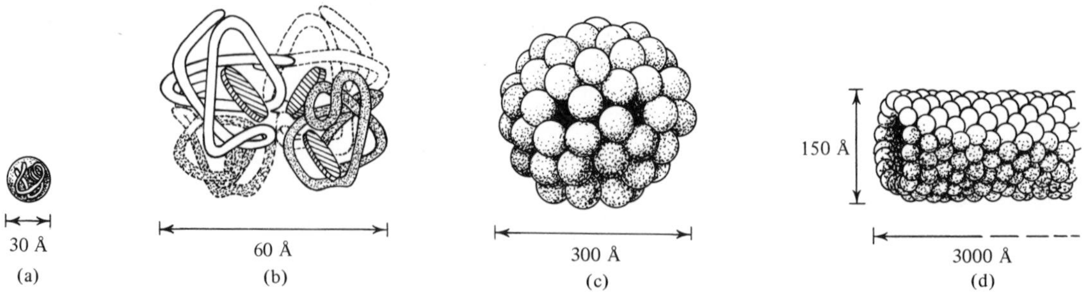
30 Å
(a)
60 Å
(b)
300 Å
(c)
150 Å
3000 Å
(d)

Figure 25.16 Schematic diagrams of the structures of several proteins (a) Myoglobin: (b) hemoglobin; (c) polio virus; (d) tobacco mosaic virus.

ingly, proteins are also amphoteric substances. Since all proteins contain some of the acidic and basic amino acids, positive and negative charges are found throughout the molecule.

When the net charge on a protein is zero, that is, when the number of negatively charged groups is equal to the number of positively charged groups, the protein will not migrate in an electric field. The pH value at which this charge cancellation occurs is the *isoelectric* pH of the protein under consideration. The isoelectric pH is characteristic of a given protein; it is dependent on the number, kind, and arrangement of the acidic and basic groups within the molecule. Those proteins having a high proportion of basic amino acids usually have relatively high isoelectric pH's, and those with a preponderance of acidic amino acids have relatively low isoelectric pH's. Table 25.6 lists the isoelectric pH's of several proteins.

The solubility of proteins in water is greatly dependent upon pH. As a general rule, proteins are least soluble at their isoelectric pH's. At pH values below its isoelectric pH, a protein will bear a net positive charge, and at pH values above its isoelectric pH, it bears a net negative charge. In either case, the net electrosatic effect is repulsion between adjacent protein molecules thus preventing coalescence. At the isoelectric pH, there is no *net* charge on the molecule. Individual molecules now have a greater tendency to approach one another due to the electrostatic attractions between the oppositely charged groups. They tend to clump together (coagulate) and precipitate out of solution.

The phosphoprotein casein is the major protein component of milk and it will precipitate from the milk in the form of white curds at its isoelectric pH of 4.6. The souring of milk results from the production of lactic acid by bacteria. The lactic acid lowers the pH value of milk from its normal value of about 6.6 to that of the isoelectric pH of casein.

Table 25.6 Isoelectric pH of Various Proteins

Protein	*Isoelectric pH*
Pepsin	<1.1
Pepsinogen	3.7
Casein	4.6
Egg albumin	4.7
Serum albumin	4.8
Urease	5.0
Insulin	5.3
Fibrinogen	5.5
Catalase	5.6
Hemoglobin	6.8
Ribonuclease	9.5
Cytochrome *c*	~ 10
Lysozyme	~ 11

25.9 Denaturation of Proteins

Denaturation may be defined as *any change that alters the unique three-dimensional configuration of a protein molecule without causing a concomitant cleavage of the peptide bonds* (Figure 25.17). Accompanying the disruption of the secondary and tertiary structures of proteins are dramatic changes in the physical and biochemical nature of the molecules. A wide variety of reagents and conditions can effect protein denaturation—some of them are outlined below.

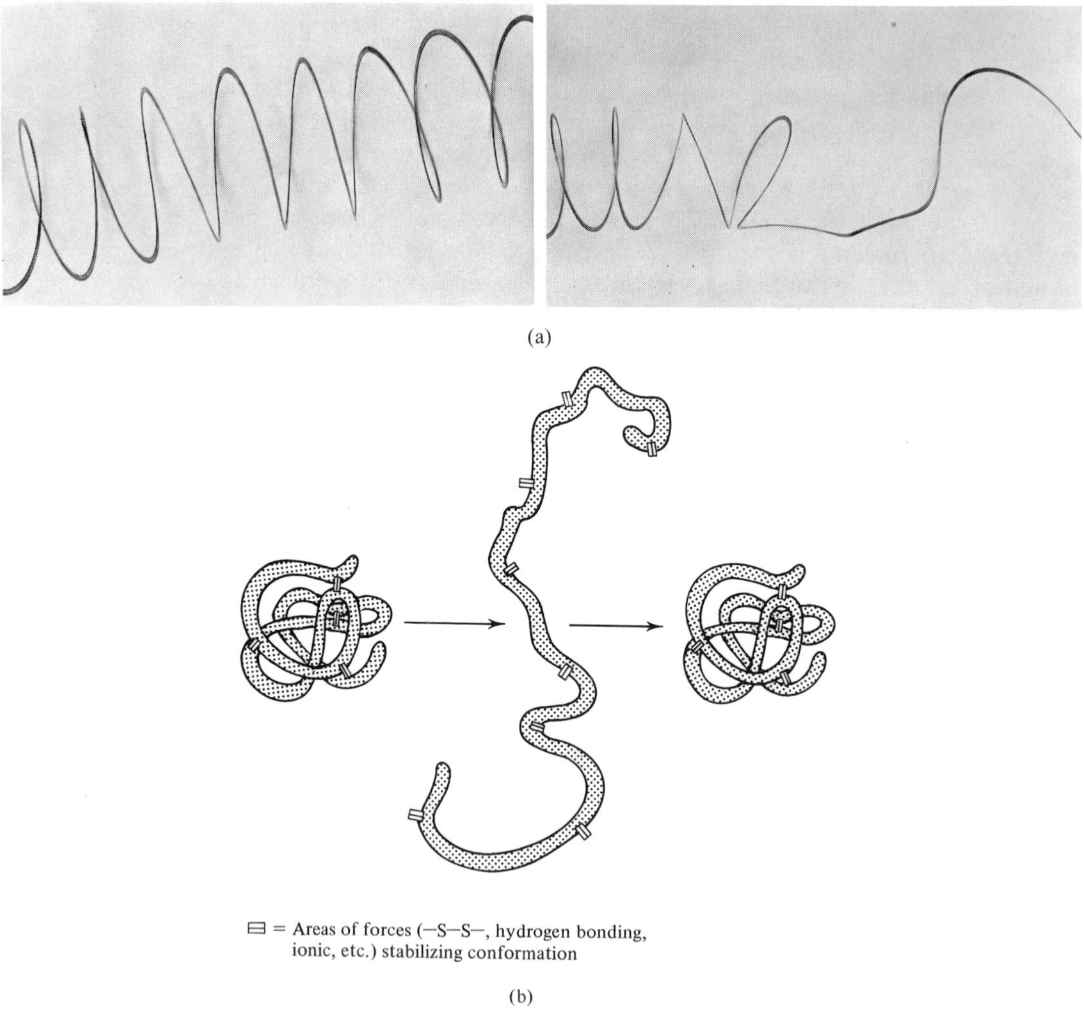
(a)
= Areas of forces (—S—S—, hydrogen bonding, ionic, etc.) stabilizing conformation
(b)

Figure 25.17 Denaturation of a protein. (a) Irreversible denaturation. The coiled spring represents the helical structure of a protein; when the elastic limit of the helix is exceeded, the shape is irreversibly altered. (b) A diagram illustrating reversible denaturation.

a. Heat and Ultraviolet Radiation

Heat and ultraviolet radiation supply kinetic energy to protein molecules causing their atoms to vibrate more rapidly, thus disrupting the relatively weak hydrogen bonds and hydrophobic bonds. This results in coagulation of the protein. The most common example is the frying or boiling of an egg. These methods are employed in sterilization techniques since they denature the enzymes in bacteria and, in so doing, destroy them. Denatured proteins are easier for enzymes to digest, and for this reason we cook our protein-containing food.

b. Treatment with Organic Solvents (Ethyl Alcohol, Rubbing Alcohol)

These reagents are capable of forming *intermolecular* hydrogen bonds with protein molecules and so disrupt the *intramolecular* hydrogen bonding within the molecule. A 70% alcohol solution is used as a disinfectant in cleansing the skin prior to an injection. The alcohol functions to denature the protein of any bacteria present in the area of the injection. A 70% alcohol solution effectively penetrates the bacterial cell wall, whereas 95% alcohol coagulates proteins at the surface, forming a crust that prevents the alcohol from entering into the cell.

c. Acids or Bases

Acidic and basic reagents disrupt salt linkages by altering the state of ionization of the carboxyl and amino groups and thus cause protein coagulation. If a protein is permitted to remain in prolonged contact with an acid or a base, cleavage of the peptide linkages will eventually occur.

d. Salts of Heavy Metal Ions (Hg^{2+}, Ag^{+}, Pb^{2+})

The cations of these heavy metals form very strong bonds with the carboxylate anions of the acidic amino acids and with the sulfhydryl groups of the sulfur-containing amino acids. Therefore they disrupt salt linkages and disulfide linkages and cause the protein to precipitate out of solution as insoluble metal–protein salts. This property makes some of the heavy metal salts suitable for use as antiseptics. For example, silver nitrate (also called lunar caustic), which is used to prevent gonorrhea infections in the eyes of newborn infants, and mercuric chloride, another antiseptic, act to precipitate the proteins in infectious bacteria. Most heavy metal salts are toxic when taken internally because they precipitate the proteins of all the cells with which they come in contact. Substances high in protein, such as egg whites and milk, are used as antidotes for heavy metal poisoning because their proteins readily combine with the heavy metal ions to form insoluble solids. The resulting insoluble matter is removed from the stomach by the use of an emetic, thus preventing the digestive juices from destroying the protein and once again liberating the poisonous heavy metal ions.

e. Alkaloid Reagents (Picric Acid, Tannic Acid)

These reagents are so named because they were originally used to study the structure of the alkaloids (morphine, cocaine, quinine). They function in a manner analogous to the heavy metal cations, but here the picrate and tannate anions combine with the positively charged amino groups to disrupt the salt linkages. In the manufacture of leather, tannic acid is used to precipitate the proteins in animal hides. This is the process called *tanning*. Tannic and picric acids are sometimes used in the treatment of burns. These acids combine with the protein in the exposed areas to form a leathery coating that excludes air and stops the loss of body fluids. The loss of water is the most significant cause of shock and the fatalities that result from burns. In an emergency, tea can serve as a source of tannic acid for the treatment of severe burns.

f. Denaturation of Hair (Permanent Waving)

In most cases, denaturation is an irreversible process. One important industry, however, owes its existence to a reversible denaturation procedure.

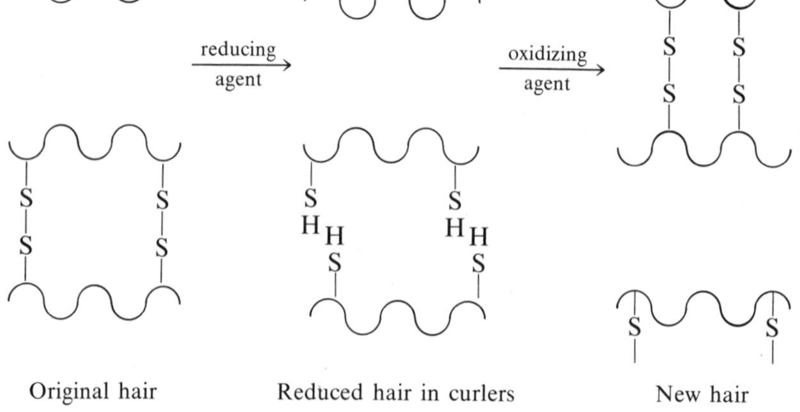
S
S
H
reducing agent
oxidizing agent
Original hair
Reduced hair in curlers
New hair

The protein of hair (keratin) has a high proportion of sulfur-containing amino acids, and it is the disulfide linkages between these amino acids that are largely responsible for the shape of the hair. The process of permanent waving involves the cleavage of these disulfide linkages by the addition of a reducing agent. The reduced, disordered hair is then placed on curlers and set in the desired pattern. An oxidizing agent is added to reform the disulfide linkages, this time between different amino acids. The effect of the process is the setting of the hair into a new pattern or hairdo.

25.10 Color Tests for Proteins

Several color tests for the qualitative detection of proteins are available. Two of them, the *ninhydrin* and *biuret* tests, are general in that they are positive for all proteins. The others are specialized and are positive only for those proteins which contain certain amino acids. These tests are outlined in Table 25.7.

Exercises

25.1 Define, explain, or give an example of each of the following.

(a) α-amino acid
(b) polypeptide
(c) protein
(d) peptide bond
(e) prosthetic group
(f) simple protein
(g) conjugated protein
(h) denaturation
(i) isoelectric pH
(j) zwitterion

25.2 What are the functions of proteins in the body?

25.3 How does the chemical composition of proteins differ from that of carbohydrates and lipids?

25.4 Compare globular and fibrous proteins. What property distinguishes the albumins from the globulins?

25.5 Give the names of four conjugated proteins and indicate the nature of their prosthetic groups. To what general class of conjugated protein does each belong?

25.6 Let the letters A, B, C, and D represent four different amino acids. Construct all the possible tetrapeptides using each letter only once.

25.7 Give IUPAC names for at least ten of the amino acids listed in Table 25.1.

25.8 Write the formula of a β-amino acid and give its IUPAC name.

25.9 Give the names and write structural formulas for two representatives from each of the four classes of amino acids.

25.10 Identify four amino acids that contain more than one asymmetric carbon atom. Draw all the optical isomers of **(a)** alanine, **(b)** serine, **(c)** threonine, and **(d)** isoleucine. Label each structure as D or L.

Table 25.7 Color Tests for Proteins

Name of Test	*Ingredients of Test Reagent*	*Criterion for Positive Test*	*Color Produced*	*Remarks*
Ninhydrin	Ninhydrin 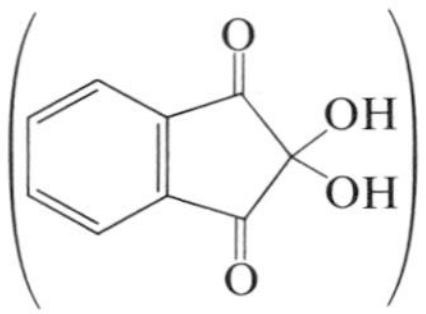	Free —NH_2 group	Blue	Positive test given by ammonia and primary amines as well as amino acids, peptides, and proteins. This test is generally used to detect the presence of amino acids. Proline and hydroxyproline give a characteristic yellow color.
Biuret	NaOH, dil $CuSO_4$	Two peptide bonds	Violet	Positive test given by tripeptides, polypeptides, and proteins but not by amino acids and dipeptides.
Xanthoproteic	Conc HNO_3	Amino acids containing a benzene ring	Yellow	Positive test given by proteins that contain tryptophan, tyrosine, or phenylalanine. (This test is performed inadvertently by students when they spill nitric acid on their skin.)
Millon	$Hg(NO_3)_2$, $Hg(NO_2)_2$	Tyrosine	Red	Positive test also given by any phenolic compound.

Hopkins–Cole	Glyoxylic acid 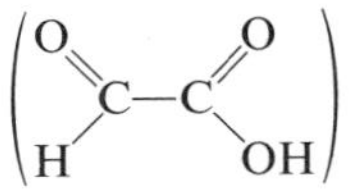 and conc H_2SO_4	Tryptophan	Violet ring	Positive test also given by any compound containing an indole ring.
Sakaguchi	α-Naphthol 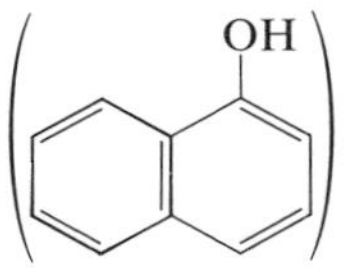 and sodium hypochlorite (NaOCl)	Arginine	Red	Positive test also given by any compound containing the guanidyl group.
Nitroprusside	Sodium nitroprusside $[Na_2Fe(NO)(CN)_5 \cdot 2H_2O]$	Cysteine	Red	The presence of sulfur in a protein can also be detected by hydrolyzing the protein with NaOH and then adding lead acetate. A black precipitate, PbS, is produced if sulfur is present.

25.11 Name and write all of the possible structural formulas for the tripeptides that upon hydrolysis yield one molecule each of glutamic acid, lysine, and serine.

25.12 How many different polypeptides containing two molecules of alanine and two molecules of valine could exist theoretically?

25.13 Write the formula of a hypothetical hexapeptide having an excess of basic groups over acidic. Name this hexapeptide.

25.14 Summarize the evidence in support of the zwitterion form of the amino acids.

25.15 Choose one amino acid from each class in Exercise 25.9, and write equations for its reaction with **(a)** a HCl solution and **(b)** a NaOH solution.

25.16 A direct current was passed through a solution containing alanine, histidine, and aspartic acid at a pH of 6.0. One amino acid migrated to the cathode, one migrated to the anode, and one remained stationary. Match the behavior with the correct amino acid.

25.17 What is meant by the primary, secondary, and tertiary structure of proteins?

25.18 What functional groups are found in the side chains of proteins?

25.19 Proteins help to maintain the pH of the organism. How can they perform this function?

25.20 Under what conditions does a protein have **(a)** a net positive charge, **(b)** a net negative charge, and **(c)** a net zero charge?

25.21 Two of the amino acids listed in Table 25.1 tend to disrupt the helical protein structure because they are incapable of forming the requisite hydrogen bonds. Identify these two amino acids.

25.22 Illustrate three types of cross-linkages found in proteins and indicate which amino acids are involved.

25.23 Briefly describe on the molecular level what occurs during each of the following processes.

(a) boiling an egg

(b) sterilization of surgical instruments

(c) permanent waving

(d) the use of a dilute solution of silver nitrate as a disinfectant in the eyes of newborn infants

25.24 In the precipitation of proteins what is the difference **(a)** between tannic acid and mercuric chloride and **(b)** between acids and alcohol?

25.25 Discuss the use of egg white as an antidote for heavy metal poisoning.

25.26 Name five color tests for proteins and indicate the specific group(s) in the protein molecule responsible for a positive test.

25.27 An octapeptide fragment from a hypothetical protein gave, upon complete hydrolysis, one molecule each of alanine (Ala), glutamic acid (Glu), glycine (Gly), leucine (Leu), methionine (Met), and, valine (Val) and two molecules of cysteine (Cys). The following fragments were isolated after partial hydrolysis: Gly-Cys, Leu-Val-Gly, Cys-Glu, Cys-Ala, Cys-Glu-Met and Met-Cys. Deduce the amino acid sequence in the octapeptide.

25.28 Each molecule of hemoglobin contains four iron atoms. The percentage of iron in the molecule is about 0.34%. What is the minimum molar mass of hemoglobin?

25.29 **(a)** The pK_a of the carboxyl group of leucine is 2.36 and the pK_a of its protonated amino group is 9.60. What is the isoelectric pH of leucine?

(b) Use Table 25.3 and calculate the isoelectric pH of histidine.

26

Enzymes

The various reactions that occur in biological systems are in many respects identical to those we have discussed previously (for example, hydration of unsaturated bonds, interconversions of alcohols, aldehydes, and acids, hydrolysis of esters and amides). In fact, scientists have been able to duplicate in the test tube (in vitro conditions) many of the reactions commonly carried out in living organisms (in vivo conditions). There is, however, one significant difference and that involves the rate at which the two types of reactions occur. The in vivo reactions take place about one hundred to one million times faster than the corresponding in vitro reactions. (It has been estimated that the reactions involved in the transmittal of a nerve impulse take between one and three millionths of a second.) If the in vitro reactions are to take place at an appreciable rate, drastic reaction conditions must be employed. Such conditions, which include high temperatures, the use of potent oxidizing or reducing agents and/or strong acids or bases, are all lethal to living systems. The in vivo reactions are carried out at body temperature (ca 37°C) and in the physiological pH range (pH ~7). The agents employed by the cell to effect the reactions under these conditions are the highly efficient, highly specific catalysts called enzymes. Life as we know it would be impossible without enzymes because nearly all functions of the cell depend directly or indirectly upon them.

Recall that a catalyst is any substance that increases the rate of a chemical reaction without being consumed in the reaction (Section 9.5). It does *not* affect the position of equilibrium in a reversible reaction; it only increases the rate of attainment of equilibrium. An **enzyme** may be defined as a *complex organic catalyst produced by living cells.* Most enzymes operate within the cell that produces them and are thus termed **intracellular.** If the enzyme's usual site of catalytic activity is outside the cell that produces it (as in the case of the gastric juices) the enzyme is designated as **extracellular.** The hydrolysis of sucrose affords a good example by which to distinguish enzyme action from the classical concept of a catalyst. If one were to exclude bacteria and molds, a solution of sucrose in water could be kept indefinitely without undergoing hydrolysis to any appreciable extent. If hydrochloric acid is added and the reaction mixture is heated, hydrolysis takes place producing glucose and fructose. If the enzyme invertase (sucrase) were added instead of the acid, the reaction would take place at a greater rate and the solution would not have to be heated at all. Furthermore, hydrochloric acid catalyzes the hydrolysis of lactose and maltose as well as that of sucrose; invertase is specific for sucrose alone and will not catalyze the hydrolysis of any other disaccharide.

26.1 Discovery of Enzymes

Since ancient times, man has been aware that yeast cells are capable of fermenting sugars. Louis Pasteur was one of the first scientists to study systematically such processes as alcoholic fermentation and the souring of milk. He observed that when glucose is stored in sealed, sterile containers, it remains indefinitely unchanged. Furthermore, he observed that the addition of yeast caused the glucose to be fermented to alcohol and carbon dioxide. Pasteur incorrectly believed that the presence of certain intact microorganisms was responsible for the observed catalytic activity and he used the term *ferment* to describe the entire group of catalysts. (The name still persists in the current German literature.) In 1878, Willy Kuhne proposed that the name enzyme (Greek, *en,* in; *zyme,* yeast) be used to describe these substances.

In 1897, Edward Buchner (Nobel Prize, 1907) prepared a cell-free extract by grinding yeast cells with sand and filtering the resultant material. The cell-free juice was found to be capable of promoting the fermentation of sugars in the same manner as the original intact yeast cells. This experiment is considered to mark the beginning of modern enzyme chemistry. Subsequent studies by many investigators elucidated all of the steps involved in this complicated but efficient process and showed that each step is controlled by a specific enzyme (Chapter 28). However, for thirty years, biochemists vainly attempted to isolate an enzyme in a pure form in order to study its properties. Finally, in 1926, James Sumner (Nobel Prize, 1946) succeeded in isolating a pure crystalline enzyme that he characterized as proteinaceous in nature. Shortly thereafter, other workers succeeded in isolating several different enzymes, all of which were shown to be proteins.

26.2 Nomenclature and Classification of Enzymes

The substance upon which an enzyme acts is known as its **substrate.** Enzymes are most commonly named by adding the suffix *-ase* to the root of the name of the substrate. Thus, for example, *urease* is the enzyme that acts upon urea and *sucrase* is the enzyme that acts upon sucrose. Sometimes an enzyme is named after the products that are formed as a result of its catalytic activity, as in the case of *invertase,* another name for sucrase. Still other enzymes have been given names indicating the type of reaction in which they are involved, such as *lactic dehydrogenase* and *pyruvic carboxylase.* Finally, there are those enzymes that still carry the common names given to them at the discretion of their discoverers. *Trypsin,* for example, comes from a Greek word meaning to rub and was selected because the enzyme was first obtained by grinding pancreatic tissue with glycerol. Many of the digestive enzymes have retained their older names, such as *pepsin, rennin,* and *ptyalin.*

About one thousand different enzymes have been identified; all are proteins of varying complexity, a small percentage of which have been obtained in pure form. To avoid the continued haphazard naming of enzymes, the International Union of Biochemistry, in 1961, recommended that enzymes be systematically classified according to the general type of reaction that they catalyze. Rules were established for the naming of an enzyme on the basis of both the precise chemical name of the substrate and the nature of the chemical reaction that is catalyzed. For example, the systematic name for sucrase is α-glucopyrano-β-fructofuranohydrolase. Because the resulting names are so cumbersome, this rational scheme has not been universally accepted, and the common names for enzymes persist in the literature. For this reason, we shall continue to use the common rather than the systematic names for the enzymes. Table 26.1 summarizes the six main divisions recognized in the classification of enzymes. A typical example is given of each type to illustrate the kinds of reactions that we shall deal with in later chapters on metabolism.

26.3 Characteristics of Enzymes

All enzymes are proteins. Some enzymes, such as pepsin, trypsin, and ribonuclease, are simple proteins since they consist entirely of amino acid units, but others are known to contain nonprotein portions and are therefore conjugated proteins. Several terms are commonly used in referring to the different components of a conjugated protein enzyme (also called a **holoenzyme**).

The *polypeptide segment of the enzyme* is known as the **apoenzyme.** The *nonprotein organic moiety,* which can frequently be separated from the apoenzyme, is called the **coenzyme.**[1] The apoenzyme is catalytically inactive

[1] Some authors reserve the term coenzyme for those substances that are readily dissociable from the enzyme. If the substance is firmly attached to the protein portion of the enzyme, it is referred to as a *prosthetic group.* We have already defined a prosthetic group as the nonprotein portion of any conjugated protein. In this sense, then, a coenzyme is a specific example of a prosthetic group.

Table 26.1 The International Union of Biochemistry Classification of Enzymes

Group Name of Enzyme	Reaction Type Catalyzed	Typical Reaction	Common Names of Enzymes
Oxidoreductase	Oxidation-reduction	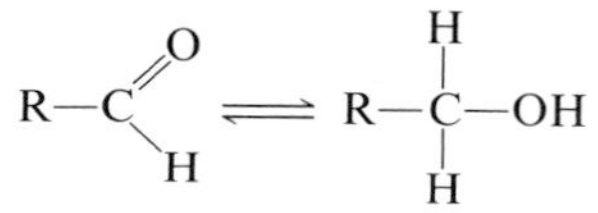	Dehydrogenase, oxidase, peroxidase
Transferase	Transfer of atoms or groups of atoms from one molecule to another	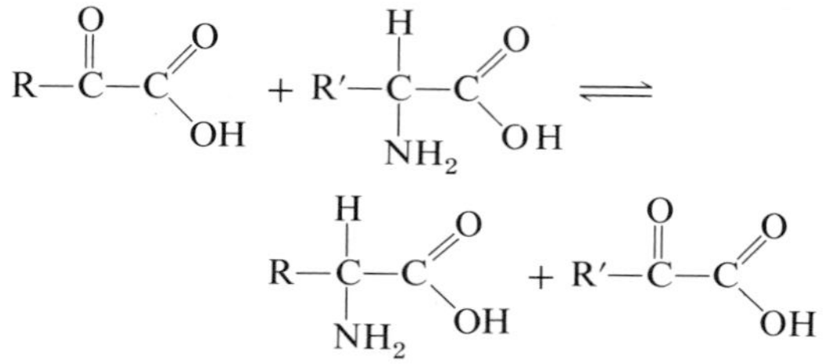	Transaminase, kinase, transacetylase
Hydrolase	Hydrolysis of a variety of substrates by water	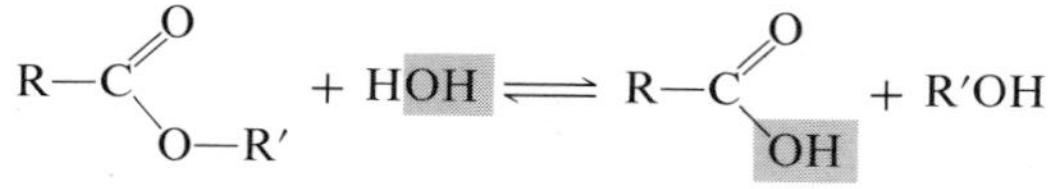	Lipase, phosphatase, amylase, amidase, peptidase
Lyase	Nonhydrolytic addition or removal of substituents from substrates	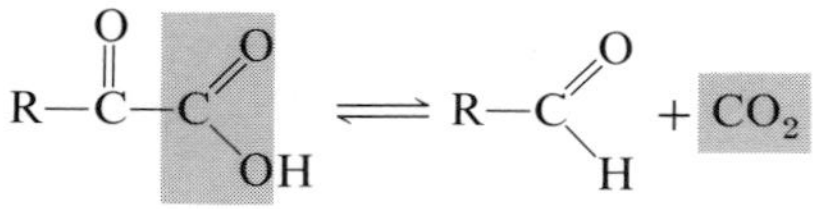	Decarboxylase, fumarase, aldolase
Isomerase	Internal rearrangement of certain substituents	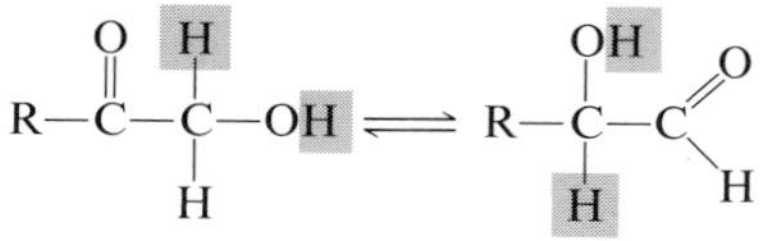	Isomerase, racemase, epimerase, mutase
Ligase	Linking together of two molecules concomitant with the breaking of a pyrophosphate bond	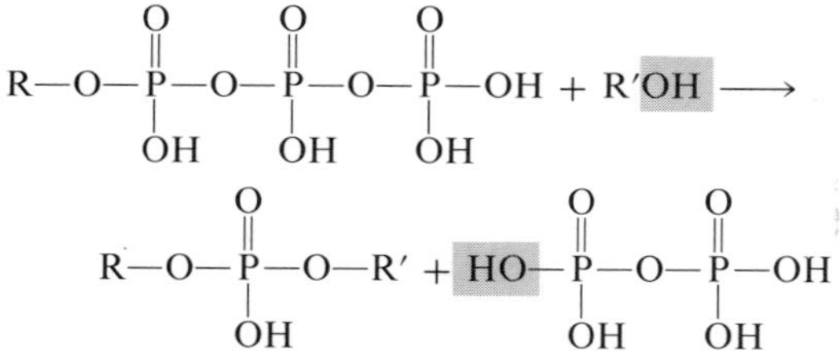	Synthetase, thiokinase

by itself, but its activity can be restored by the addition of the coenzyme. There are many metalloprotein enzymes in which the metal ion (e.g., Mg^{2+}, Mn^{2+}, Zn^{2+}) is bonded either to the apoenzyme or to the coenzyme. The metal ion is usually designated as the enzyme **activator.** It has been postulated that the metal ion acts to form a coordination complex between the enzyme and the substrate and to activate the substrate by promoting electronic shifts. In our later discussions of metabolic reactions, we shall have occasion to cite several examples of enzymes whose catalytic activity is dependent upon the presence of an activator.

The term **activation** takes into account the roles of both the coenzymes and the metal ion activators. In general, it refers to any *process in which an inactive protein is transformed into an active enzyme.* Many of the simple protein enzymes are secreted in an inactive form known as a **proenzyme** or a **zymogen.** Pepsinogen and trypsinogen are two such compounds that have been carefully studied. These proenzymes are converted into the active enzymes pepsin and trypsin by the action of other enzymes that remove an inhibitory peptide from the proenzyme molecule. In the activation of trypsinogen, a hexapeptide is cleaved from the molecule enabling the new protein trypsin, to attain the conformation essential to its catalytic activity (see Figure 30.1).

26.4 Coenzymes and Vitamins

The chemical nature of many of the coenzymes has been determined. Several of them have been found to be directly related to certain vitamins. **Vitamins** are *organic compounds that cannot be synthesized by an organism but nevertheless are essential for the maintenance of normal metabolism* (see Section 32.1). Since organisms differ in their synthetic abilities, a substance that is a vitamin for one species may not be so for another. It appears that the function of many of the vitamins (especially those of the **B** group) is to serve as structural units for the synthesis of coenzymes. This observation explains why vitamins play such a vital role in the normal functioning of an organism. Table 26.2 lists the names and structures of some of the coenzymes and their vitamin precursors. In our detailed study of metabolic reactions, we shall have occasion to refer back to this table.

26.5 Mode of Enzyme Action

In 1888, the Swedish chemist Svante Arrhenius proposed a scheme to account for catalytic activity. He suggested that a catalyst functions to combine with a reactant to form an intermediate compound. This intermediate is more reactive than the initial uncombined species.[2] This scheme

[2]The formation of an intermediate compound affords a lower energy reaction pathway. This, in effect, lowers the activation energy of the reaction, accounting for the increased rate of reaction. Enzymes reduce activation energies more effectively than other catalysts, thus enabling biochemical reactions to proceed at relatively low temperatures.

Table 26.2 Vitamins and Coenzymes

Vitamin	*Coenzyme*	*Function*
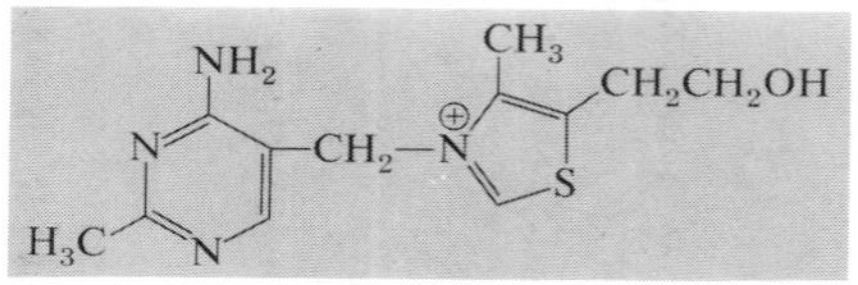 Thiamine (B_1)	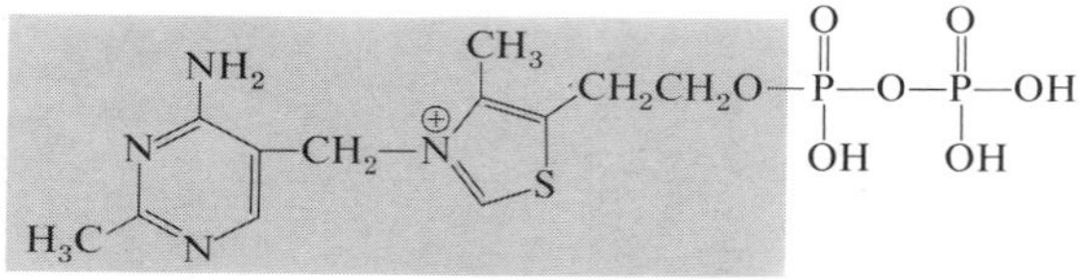 Thiamine pyrophosphate (TPP)	In reactions catalyzed by: α-Keto acid decarboxylases α-Keto acid oxidases Transketolase Phosphoketolase
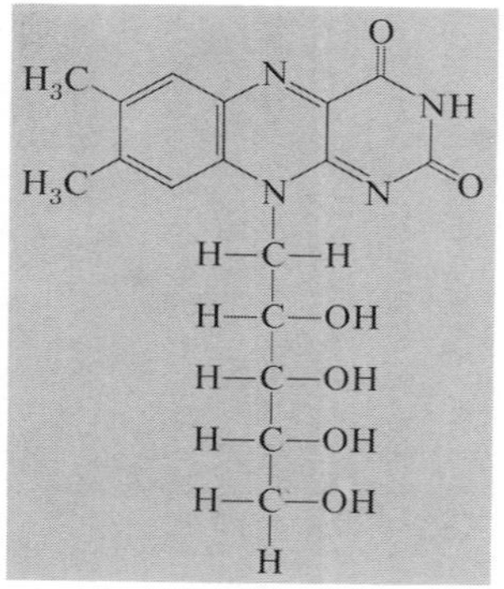 Flavin Ribotol Riboflavin (B_2)	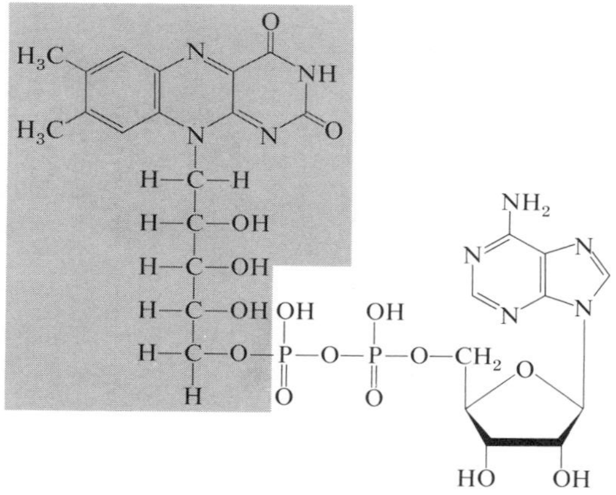 Flavin adenine dinucleotide (FAD)	In oxidation-reduction reactions
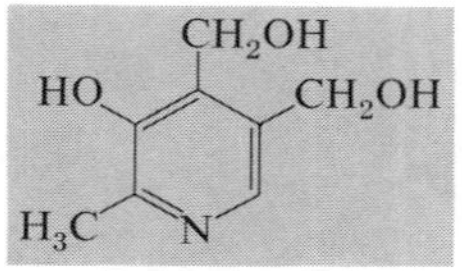 Pyridoxine (B_6)	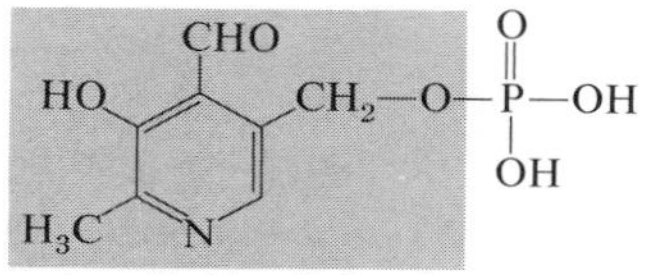 Pyridoxal phosphate	In several reactions of amino acid metabolism: Transamination Decarboxylation Racemization

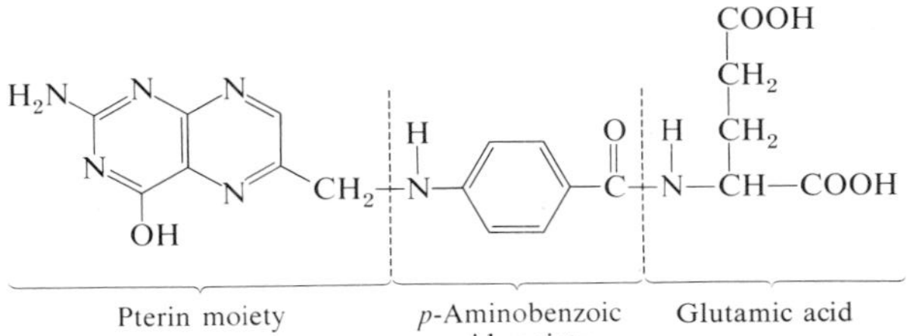
Pterin moiety
p-Aminobenzoic acid moiety
Glutamic acid

Folic acid (F)

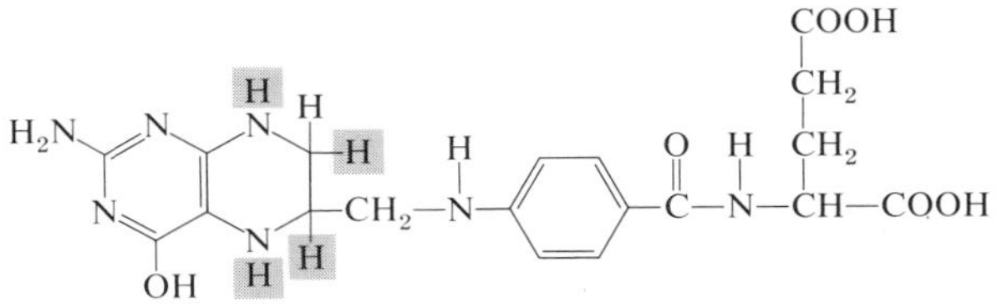

Tetrahydrofolic acid (FH_4)

In reactions that transfer single carbon units

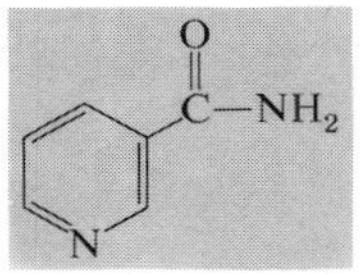

Nicotinamide

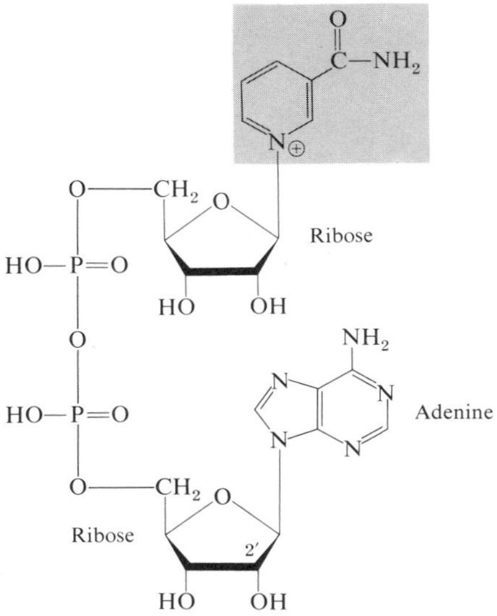
Ribose
Adenine
Ribose
2′

Nicotinamide adenine dinucleotide (NAD^+)[a]

In oxidation-reduction reactions

---

[a]When there is an additional phosphate group on the 2′ hydroxyl group of the ribose moiety, the compound is named nicotinamide adenine dinucleotide phosphate ($NADP^+$).

(Continued)

Table 26.2 Vitamins and Coenzymes (*Continued*)

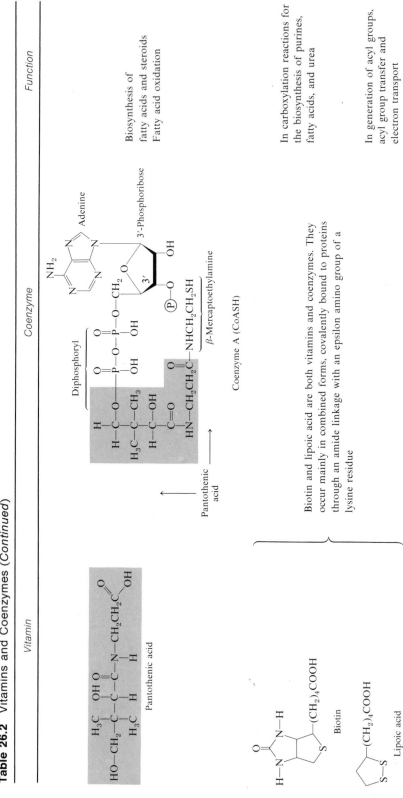

Vitamin	*Coenzyme*	*Function*
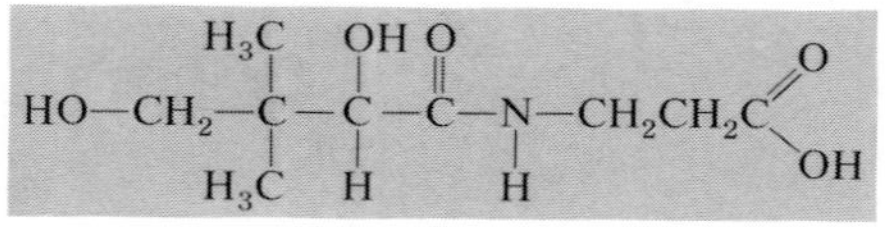 Pantothenic acid	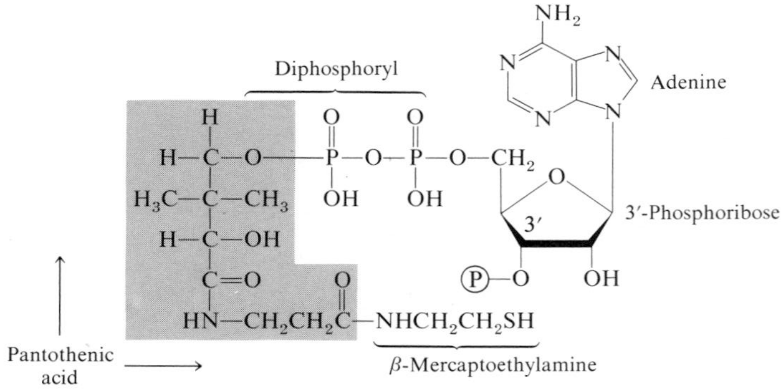 Diphosphoryl Adenine 3′-Phosphoribose Pantothenic acid β-Mercaptoethylamine Coenzyme A (CoASH)	Biosynthesis of fatty acids and steroids Fatty acid oxidation
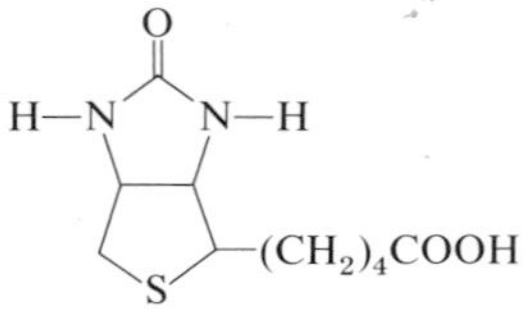 Biotin	Biotin and lipoic acid are both vitamins and coenzymes. They occur mainly in combined forms, covalently bound to proteins through an amide linkage with an epsilon amino group of a lysine residue	In carboxylation reactions for the biosynthesis of purines, fatty acids, and urea
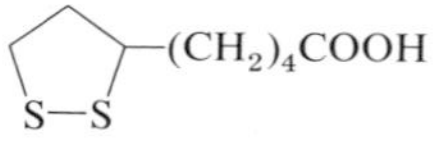 Lipoic acid		In generation of acyl groups, acyl group transfer and electron transport

applies to all catalytic reactions whether inorganic, organic, or biochemical. It is generally believed that catalytic reactions occur in at least two steps. In the first step, a molecule of the enzyme **(E)** and a molecule of the substrate **(S)** collide and react to form an intermediate compound, which is called the **enzyme–substrate complex (E—S).** (This step is reversible since the complex can break apart yielding the original substrate and the free enzyme.) The second step involves the formation of the products **(P),** followed by their release from the surface of the enzyme.

$$\mathbf{S + E \rightleftharpoons E{-}S} \qquad \text{Sucrose + Sucrase} \rightleftharpoons \text{Sucrose–sucrase complex}$$

$$\mathbf{E{-}S \longrightarrow P + E} \qquad \text{Sucrose–sucrase} + H_2O \longrightarrow \text{Glucose + Fructose + Sucrase}$$

The existence of an enzyme–substrate complex has been verified by spectroscopic and kinetic experiments. The bonds that hold the enzyme and the substrate together are probably the same forces involved in maintaining protein structure, that is, electrostatic, covalent, hydrogen bonds, etc. In addition, it has been demonstrated that the structural features or functional groups essential to the formation of the enzyme–substrate complex occur at a specific location on the surface of the enzyme. *This section of the enzyme, which combines with the substrate, and at which transformation from substrate to products occurs,* is called the **active site** of the enzyme. The active site possesses a unique conformation that is complementary to the structure of the substrate, thus enabling the two molecules to fit together in much the same manner as a key fits into a lock. The *lock and key theory* of enzyme action is illustrated in Figure 26.1.

The manner in which an enzyme transforms a substrate into product(s) has been extensively studied. As yet, however, the detailed mechanism by

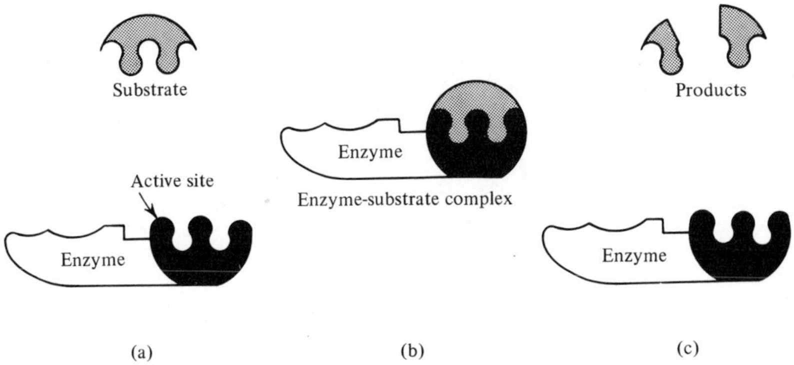

Substrate
Active site
Enzyme
Enzyme
Enzyme-substrate complex
Products
Enzyme
(a)
(b)
(c)

Figure 26.1 A schematic representation of the interaction of enzyme and substrate. (a) The active site on the enzyme and the substrate have complementary structures and hence fit together as a key fits a lock. (b) While they are bonded together in the enzyme–substrate complex, the catalytic reaction occurs. (c) The products of the reaction leave the surface of the enzyme, freeing the enzyme to combine with another molecule of substrate.

which enzymes increase the rate of reactions more efficiently than other catalysts is incompletely understood. We still do not know, for example, whether the full catalytic activity of an enzyme resides in its protein structure as a whole, or in a small region associated with the active site. Small peptides have been cleaved from some enzymes (such as ribonuclease) without appreciable loss of catalytic activity.

Many attempts have been made to implicate specific amino acid side chains (e.g., hydroxyl group of serine, sulfhydryl group of cysteine, imidazole group of histidine) as being part of the active sites of various enzymes. The active site probably consists of amino acids from several different positions along the protein chain; these amino acids can be brought into proximity as a result of the folding and bending of the polypeptide chain. The fact that enzymes are inactivated by denaturation, points to the importance of the secondary and tertiary structures in maintaining the active site in a precise three-dimensional arrangement. It is known that two amino acids, histidine-57 and serine-195, are involved in the active site of chymotrypsin (Figure 26.2). The remaining amino acids presumably function in maintaining the active site in the correct geometrical configuration to provide for maximum catalytic activity, as well as in imparting *specificity* to the molecule.

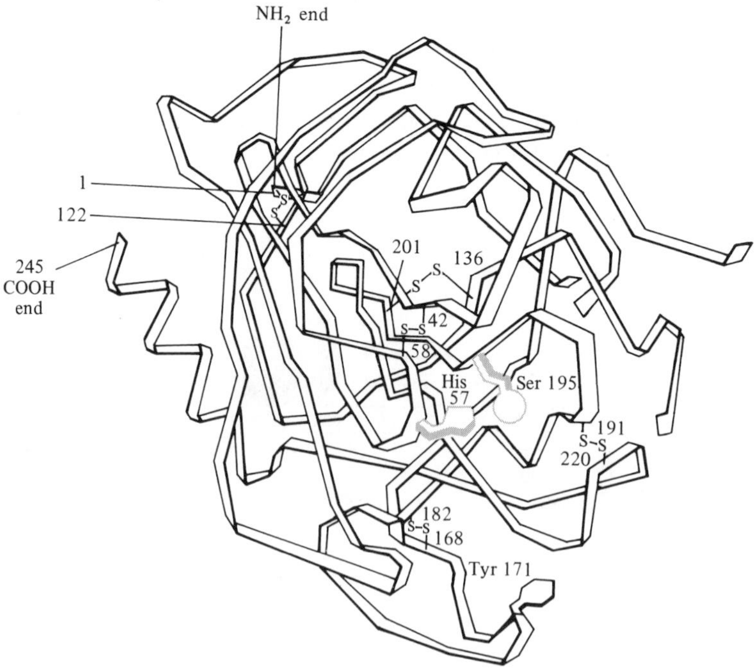

NH2 end
1
122
245
COOH
end
201
136
S–S
42
58
His
57
Ser 195
191
S–S
220
182
S–S
168
Tyr 171

Figure 26.2 Model of α-chymotrypsin. The enzyme has three polypeptide chains, which are interconnected by five disulfide bonds. The active site is in the region of histidine-57 and serine-195.

In 1963, D. E. Koshland, Jr., augmented the older lock and key theory by suggesting that an enzyme undergoes a change of conformation when it reacts with a substrate molecule to form the activated complex. After catalysis, the enzyme resumes its original structure.

> To explain the enzyme's ability to discriminate between similar compounds, a "fit" between the substrate and a portion of the enzyme surface seems essential. However, it appears possible that this fit is not a static one in which a rigid "positive" substrate fits on a rigid "negative" template, but rather, is a dynamic interaction in which the substrate induces a structural change in the enzyme molecule, as a hand changes the shape of a glove.[3]

This so-called *induced-fit theory* has not yet been completely verified, but it is an attractive proposal since it explains several experimental findings that are incompatible with the older theory. Koshland cites examples of compounds that bind to enzymes without undergoing further reaction as well as other compounds that are sterically not suited for the active site but that nevertheless react catalytically. According to Koshland, the active site of an enzyme consists of two components: one is responsible for substrate specificity (*contact groups*) and the other is responsible for catalysis (*catalytic groups*). The active site is a flexible region that can be induced to fit several structurally similar compounds. However, only the proper substrate is capable of correct alignment with the catalytic groups as well (Figures 26.3 and 26.4).

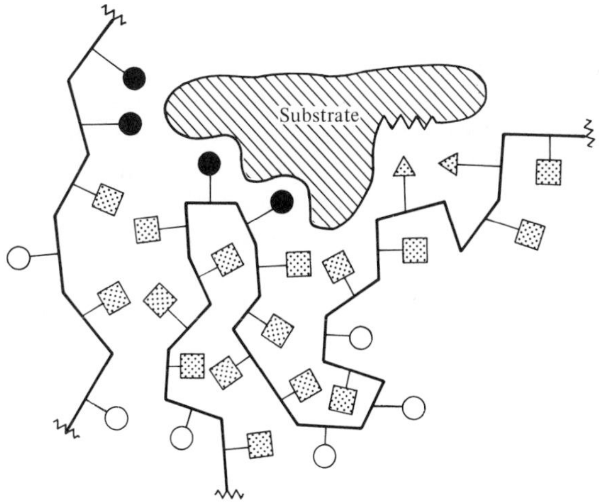
Substrate

Figure 26.3 A schematic representation of an active site. ●s = "contact" amino acid residues whose fit with substrate determines specificity; ▲s = catalytic residues acting on substrate bond, which is indicated by a jagged line; ○s = nonessential residues on the surface; and ▩s = residues whose interactions with each other maintain the three-dimensional structure of the enzyme. [Adapted from D. E. Koshland, Jr., *Science,* **142,** 1534 (1963).]

[3] D. E. Koshland, Jr., *Science*, **142,** 1533 (1963).

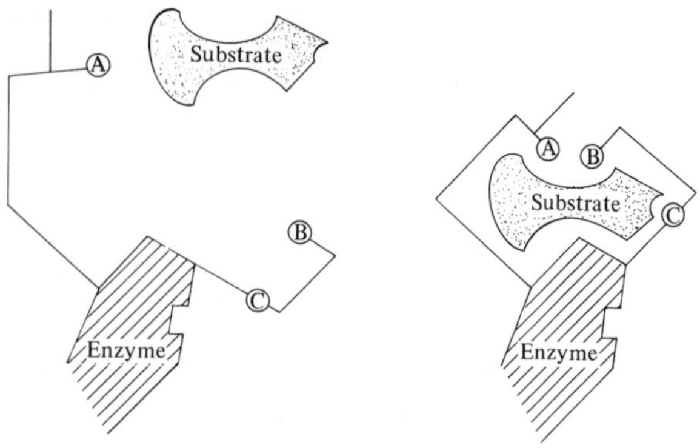
Substrate
A
B
C
Enzyme
Substrate
A
B
C
Enzyme
(a)

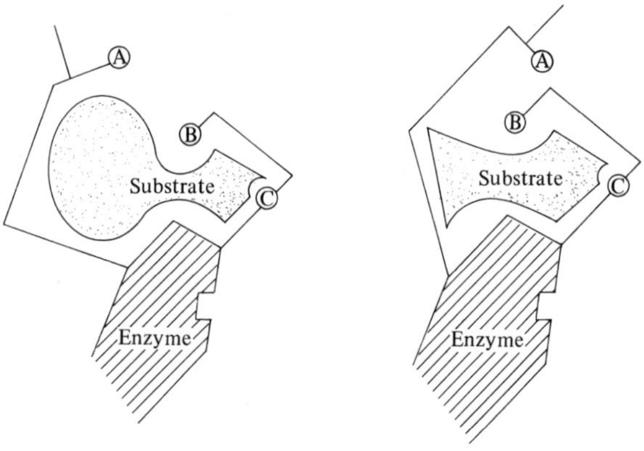
A
B
Substrate
C
Enzyme
A
B
Substrate
C
Enzyme
(b)

Figure 26.4 Schematic representation of a flexible active site. (a) Substrate binding induces proper alignment of catalytic groups A and B so that reaction ensues. (b) Compounds that are either too large or too small are bound, but fail to produce proper alignment of catalytic groups; hence, no reaction. [Adapted from D. E. Koshland, Jr., *Science,* **142,** 1539 (1963).]

26.6 Specificity of Enzymes

As we have already mentioned, one characteristic that distinguishes an enzyme from all other types of catalysts is its substrate specificity. Recall, for example, that hydrogen ions will catalyze the hydrolysis of disaccharides, polysaccharides, lipids, and proteins with complete impartiality, whereas different enzymes are required in all four cases. Enzyme specificity is a result of the uniqueness of the active site of each enzyme, and this uniqueness is a function of the chemical nature, electrical charge, and spatial arrangements of the groups located there. A wide range of enzyme specificities exist, and they are arbitrarily grouped as follows.

a. Absolute Specificity

Enzymes having absolute specificity will catalyze a particular reaction for one particular substrate only and will have no catalytic effect on substrates that are closely related. *Urease*, for example, will catalyze the hydrolysis of urea but not of methylurea, thiourea, or biuret.

$$\underset{\text{Urea}}{H_2N{-}\overset{\overset{\displaystyle O}{\|}}{C}{-}NH_2} + H_2O \xrightleftharpoons{\text{urease}} CO_2 + 2\,NH_3$$

$$\underset{\text{Methylurea}}{H_2N{-}\overset{\overset{\displaystyle O}{\|}}{C}{-}NH{-}CH_3} \qquad \underset{\text{Thiourea}}{H_2N{-}\overset{\overset{\displaystyle S}{\|}}{C}{-}NH_2} \qquad \underset{\text{Biuret}}{H_2N{-}\overset{\overset{\displaystyle O}{\|}}{C}{-}NH{-}\overset{\overset{\displaystyle O}{\|}}{C}{-}NH_2}$$

b. Stereochemical Specificity

Most enzymes show a markedly high degree of specificity toward one stereoisomeric form of the substrate. *Lactic acid dehydrogenase* catalyzes the oxidation of the L-lactic acid found in muscle cells but not the D-lactic acid found in certain microorganisms. *Fumarase* adds water to fumaric acid but not to its *cis* isomer, maleic acid.

c. Group Specificity

Enzymes having group specificity are less selective in that they will act upon structurally similar molecules having the same functional groups. Many of the *peptidases* fall into this category. Pepsin will hydrolyze all peptides having adjacent aromatic amino acids. *Carboxypeptidase* attacks peptides from the carboxyl end of the chain, cleaving the amino acids one at a time.

d. Linkage Specificity

Enzymes having linkage specificity are the least specific of all because they will attack a particular kind of chemical bond, irrespective of the structural features in the vicinity of the linkage. The *lipases,* which catalyze the hydrolysis of ester linkages in lipids, are an example of this type of enzyme.

26.7 Factors Influencing Enzyme Activity

Since enzymes are protein catalysts, they are affected by those factors that act upon proteins and upon catalysts in general. The activity of an enzyme may be measured by monitoring the reaction that it catalyzes at fixed time intervals. The rate of the reaction is determined by observing either the rate of disappearance of the substrate or the rate of formation of the product(s). In such experiments, the rate is the only variable; all other experimental conditions are held constant. (Recall the discussion of chemical kinetics in Chapter 9.)

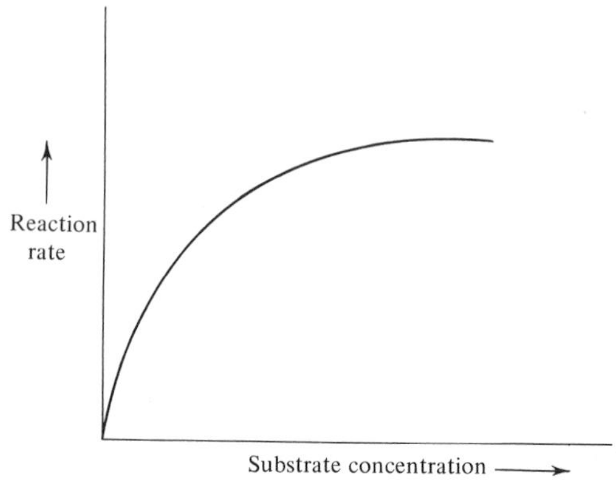

Reaction rate
Substrate concentration

Figure 26.5 Effect of substrate concentration on the rate of a reaction that is catalyzed by a fixed amount of enzyme.

a. Concentration of Substrate

The rate of an enzymic reaction increases as the substrate concentration increases until a limiting rate is reached. At this point, further increase in the substrate concentration produces no significant change in the reaction rate. Figure 26.5 is a characteristic plot of an enzyme-catalyzed reaction and it is

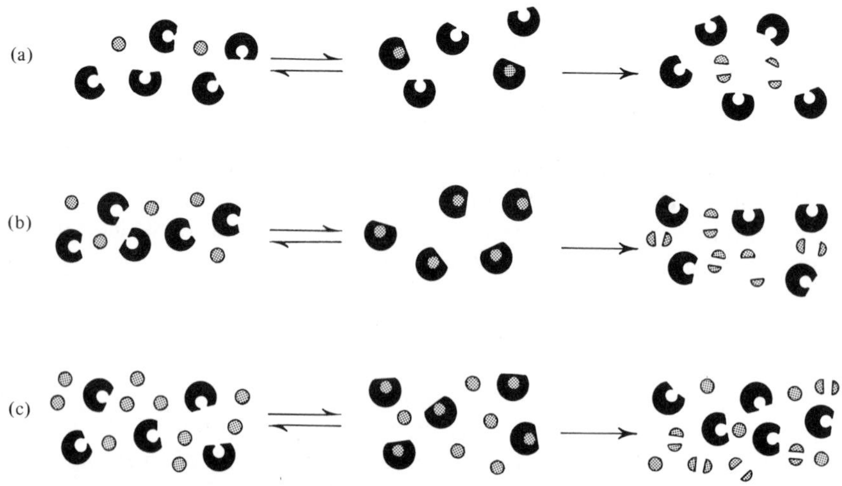

(a)
(b)
(c)

Figure 26.6 A schematic representation of relative concentrations of enzyme and substrate. (a) Low substrate concentration; (b) adequate substrate concentration; (c) excess substrate concentration.

taken as further evidence of the existence of the enzyme–substrate intermediate. At high substrate concentrations, practically all the enzyme molecules are saturated with the substrate at any given instant. Extra substrate molecules must wait until the enzyme–substrate complexes have dissociated to yield products and the free enzymes before they may undergo reaction. This relationship is illustrated in Figure 26.6 and can be summarized in terms of the two equations given in Section 26.5. At low substrate concentrations the formation of the **E—S** complex is the rate-determining step, while at high substrate concentrations, the slowest step is the dissociation of the **E—S** complex.

b. Concentration of Enzyme

Since in all practical cases the concentration of the enzyme is much lower than the concentration of the substrate (the substrate can never become saturated with the enzyme) the rate of an enzyme-catalyzed reaction is always directly dependent upon the concentration of the enzyme (Figure 26.7). This is not a new concept; the reaction rate of any catalytic reaction increases as the concentration of the catalyst is increased.

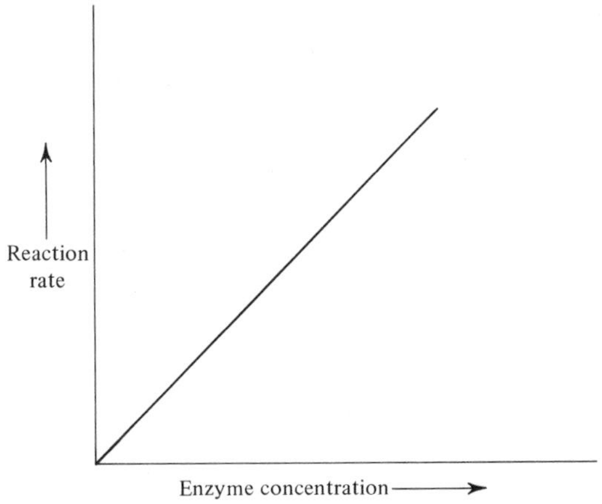
Reaction rate
Enzyme concentration

Figure 26.7 Effect of enzyme concentration on the rate of a reaction.

c. Temperature

A rule of thumb for most chemical reactions is that a 10 °C rise in temperature approximately doubles or triples the reaction rate. To a certain extent, this is true of all enzymic reactions. After a certain point, however, an increase in temperature causes a decrease in the rate of reaction (Figure 26.8). *The temperature that affords maximum activity* is known as the **optimum temperature** for the enzyme in question. Most enzymes of warm-blooded animals have optimum temperatures of about 37 °C (98 °F). The decrease in rate is a direct consequence of the fact that enzymes are proteins

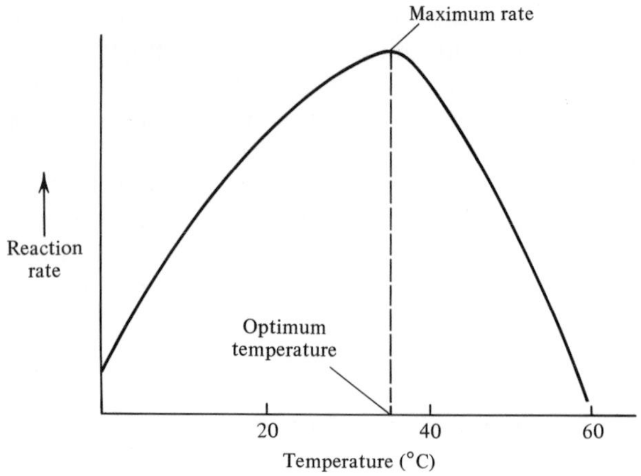
Maximum rate
Reaction rate
Optimum temperature
20
40
60
Temperature (°C)

Figure 26.8 Effect of temperature on the rate of an enzymic reaction.

and thus are denatured by heat. Heating disrupts the secondary and tertiary structures of the enzymes, effecting a disorientation of the active site. This disorientation renders the active site inaccessible to the substrate.

At temperatures of 0°C and 100°C the rate of enzyme-catalyzed reactions is nearly zero. This fact has several practical applications. We sterilize objects by placing them in boiling water so as to denature the enzymes of any bacteria that may be in contact with them. Animals go into hibernation because of a decrease in their body temperature (in the winter) and as a result, the rate of their metabolic processes decreases. The food required to maintain this lowered metabolic rate is provided by reserves stored in their tissues.

d. Hydrogen Ion Concentration

Again, as a consequence of their protein nature, enzymes are sensitive to changes in the pH of their environments. Extreme values of pH (whether high or low) can cause denaturation of the protein. However, any change in the hydrogen ion concentration alters the degree of ionization of acidic and

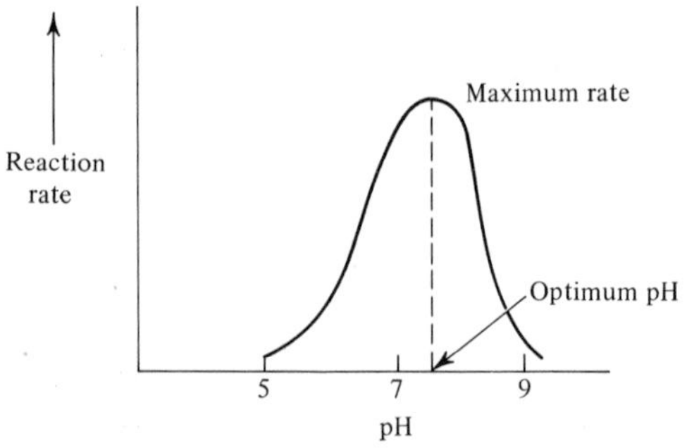
Reaction rate
Maximum rate
Optimum pH
5
7
9
pH

Figure 26.9 Effect of pH on the rate of an enzymic reaction.

basic groups both on the enzyme and on the substrate. If any ionizable groups are located at the active site, and if a certain charge is necessary in order for the enzyme to bind its substrate, then an enzyme molecule that has even one of these charges neutralized will lose its catalytic activity. An enzyme will exhibit maximum activity over a narrow pH range in which the molecule exists in its proper charged form. The median value of this range is known as the **optimum pH** of the enzyme (Figure 26.9). With the notable exception of gastric juice, most body fluids have pH values between 6 and 8. This is essential because most enzymes exhibit optimal activity in the physiological pH range of 7.0–7.5. However, each enzyme has a characteristic optimum pH and in a few cases this value is outside the usual physiological range. For example, the optimal pH for pepsin is 2.0 and that for trypsin is 8.0.

26.8 Enzyme Inhibition

In the preceding section, we noted that enzymes are inactivated by increased temperatures and by changes in pH. In a sense, then, temperature and hydrogen ion concentration may be considered to be factors that inhibit enzyme activity. In fact, any physical change or chemical reagent that denatures protein will adversely affect the rate of an enzymic reaction. This type of enzyme inhibition is referred to as **nonspecific inhibition** since it affects all enzymes in the same manner. In contrast, a **specific inhibitor** exerts its affect upon a single enzyme or a group of related enzymes. Most poisons act to inhibit specific enzymes (Table 26.3).

a. Competitive Inhibition

A **competitive inhibitor** is any *compound that bears a close structural resemblance to a particular substrate and competes with that substrate for*

Table 26.3 Poisons as Enzyme Inhibitors

Poison	*Formula*	*Enzyme Inhibited*	*Site of Action*
Cyanide	CN^-	Cytochrome oxidase, catalase	Binds Fe^{3+} activator
Fluoride	F^-	Enolase	Binds Mg^{2+} activator
Sulfide	S^{2-}	Phenolase	Binds Cu^{2+} activator
Arsenate	AsO_4^{3-}	Phosphotransacetylase	Substitutes for phosphate
Iodoacetate	ICH_2COO^-	Triose phosphate dehydrogenase	Binds to cysteine sulfhydryl group
Nerve gas	$F-\overset{\overset{\displaystyle O}{\|}}{P}(-OCH(CH_3)_2)-OCH(CH_3)_2$	Acetylcholine esterase	Binds to serine hydroxyl group

binding at the same active site on the enzyme. The inhibitor is not acted upon by the enzyme and so remains bound to the enzyme, preventing the substrate from approaching the active site. The degree of competitive inhibition depends upon the relative concentrations of substrate and inhibitor. If the inhibitor is present in relatively large quantities, it blocks the active sites on all the enzyme molecules and complete inhibition results. However, formation of the inhibitor–enzyme complex is reversible, and increased substrate concentration permits displacement of the inhibitor from the active site. Competitive inhibition can be completely reversed by addition of large excesses of substrate. The reversible nature of competitive inhibition has provided much information about the enzyme–substrate complex and about the specific groups involved at the active sites of various enzymes.

A classic example of competitive inhibition is the effect of malonic acid on the enzyme activity of *succinic acid dehydrogenase.* The former is a homolog of the enzyme's normal substrate, succinic acid. Malonic acid will bind to the active site, since the spacing of its carboxyl groups is not greatly different from that of succinic acid. No catalytic reaction occurs, and malonic acid remains bonded to the enzyme. We shall discuss this reaction again in connection with carbohydrate metabolism (Section 28.7).

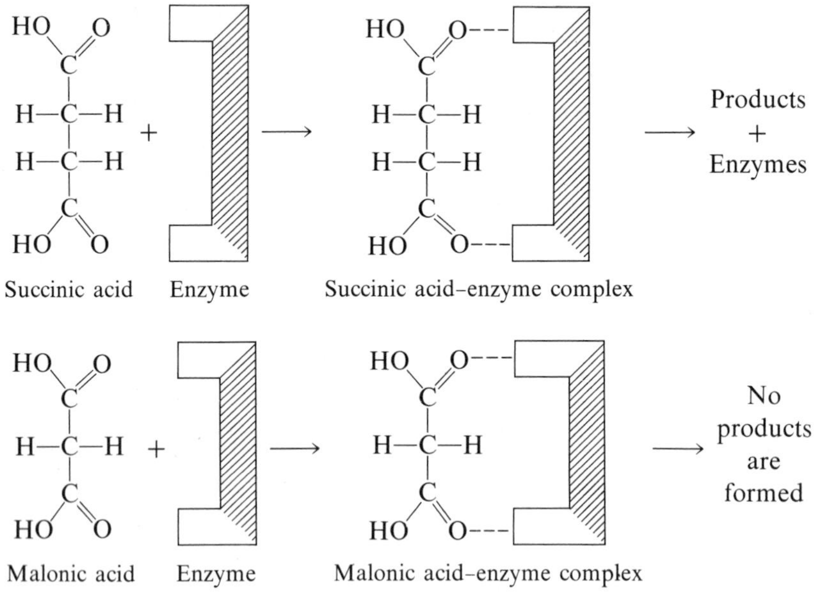
Products
+
Enzymes
Succinic acid
Enzyme
Succinic acid–enzyme complex
No
products
are
formed
Malonic acid
Enzyme
Malonic acid–enzyme complex

Another example of a competitive inhibitor is the compound chlorfenthol, which inhibits the enzyme *DDT dehydrochlorinase* that converts DDT to DDE. You will recall (page 350) that insects become resistant to DDT by a mutation resulting in their ability to biosynthesize the dehydrochlorinase enzyme. If chlorfenthol is applied to a field along with DDT, then the resistant insects will also be killed off because of the inhibition of their protecting enzyme.

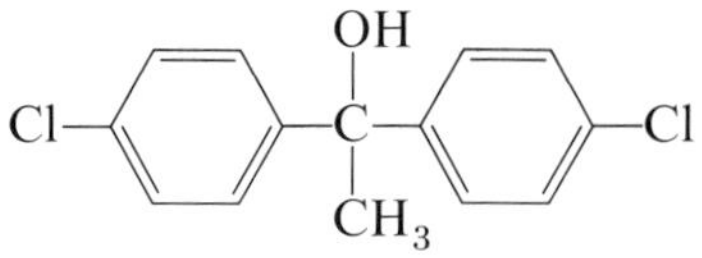

Chlorfenthol
1,1-Di(*p*-chlorophenyl)ethanol

b. Noncompetitive Inhibition

A **noncompetitive inhibitor** is a *substance that can combine either with the free enzyme or with the enzyme–substrate complex.* The noncompetitive inhibitor binds to the enzyme at a position relatively remote from the active site and in so doing alters the three-dimensional conformation of the enzyme. This effects a change in the configuration of the active site, so that either the **E—S** complex does not form at its normal rate, or once formed the **E—S** complex does not decompose at the normal rate to yield products. Since the inhibitor does not structurally resemble the substrate, the addition of excess substrate does not reverse the inhibitory effects.

Many enzymes contain reactive groups, such as $—COO^-$, $—NH_3^+$, —SH, or —OH, as functional groups that are located either at the active site or elsewhere in the enzyme and are essential for maintaining the proper three dimensional conformation of the enzyme. Any chemical reagent that is capable of combining with one or more of these groups will inhibit the enzyme. The heavy metal ions Ag(I), Hg(II), Pb(II) have strong affinities for carboxylate and sulfhydryl groups; we have already discussed their toxic effects with regard to protein denaturation. Iodoacetic acid is another inhibitor that combines specifically with —SH groups.

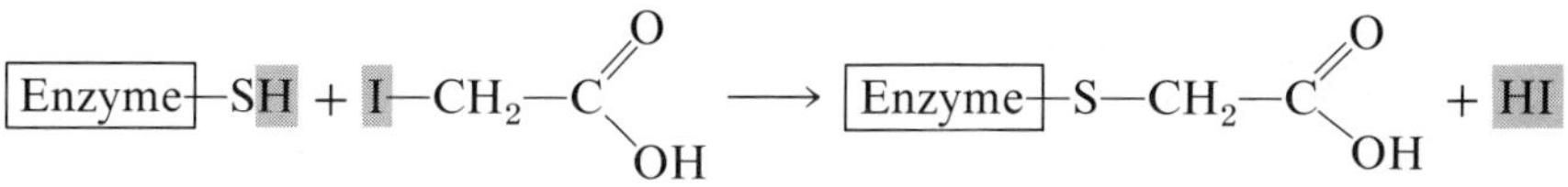

The nerve gases, especially diisopropylfluorophosphate (DFP), inhibit biological systems by forming an enzyme–inhibitor complex with a specific hydroxyl group (of serine) situated at the active site of the enzyme *acetylcholine esterase* (Section 21.3).

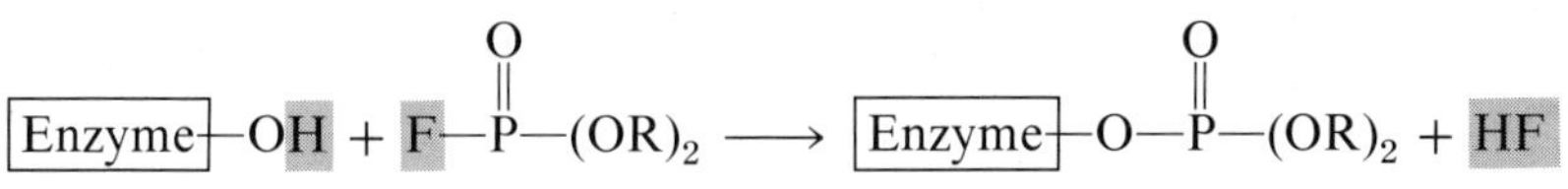

Metalloenzymes (those that require the presence of a metal ion for activation) are inhibited by substances that form strong complexes with the metal. Traces of hydrogen cyanide inactivate iron-containing enzymes such as catalase and cytochrome oxidase. Oxalic and citric acids inhibit blood clotting by forming complexes with calcium ions necessary for the activation of the enzyme *thromboplastin* (see Section 31.3).

c. End-Product Inhibition (Negative Feedback Control)

Some biosynthetic pathways are known in which the enzyme catalyzing the first reaction in the pathway is inhibited by the final product that the pathway produces. Consider the hypothetical series of reactions leading from starting compound A, through the series of intermediates B, C, and D, to the final product E. Each step is catalyzed by a different enzyme as indicated.

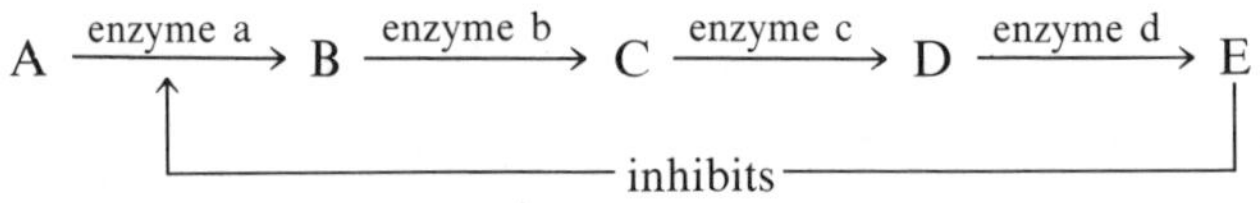
A
enzyme a
B
enzyme b
C
enzyme c
D
enzyme d
E
inhibits

The ultimate product of the sequence, E, serves to prevent formation of its own precursors by noncompetitively inhibiting the action of enzyme a. Thus, an organism is provided with a mechanism for controlling the rate of synthesis of a metabolic intermediate according to its need.

A specific example is the biosynthesis of cytidine triphosphate (CTP) from carbamoyl phosphate and aspartic acid in *E. coli*. The enzyme *aspartate*

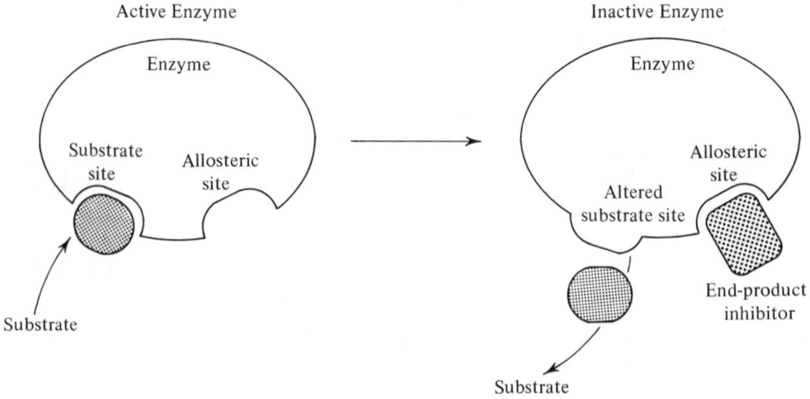
Active Enzyme
Enzyme
Substrate site
Allosteric site
Substrate
Inactive Enzyme
Enzyme
Allosteric site
Altered substrate site
End-product inhibitor
Substrate

Figure 26.10 A schematic representation of how the binding of an end-product inhibitor inhibits an enzyme by causing a change in structural conformation at the substrate binding site.

transcarbamoylase, which catalyzes the first step of this multistep reaction sequence, is inhibited by the end product, CTP, when a critical concentration of CTP is built up in the cell. When the concentration of CTP is sufficiently lowered by its metabolic utilization, its inhibitory effect decreases, and biosynthesis of the compound proceeds once again.

$$H_2N-\overset{\overset{\displaystyle O}{\|}}{C}-O-\overset{\overset{\displaystyle O}{\|}}{\underset{\underset{\displaystyle O^-}{|}}{P}}-O^- \;+\; HO-\overset{\overset{\displaystyle O}{\|}}{C}-CH_2-\overset{\overset{\displaystyle H}{|}}{\underset{\underset{\displaystyle NH_2}{|}}{C}}-\overset{\overset{\displaystyle O}{\|}}{C}-OH$$

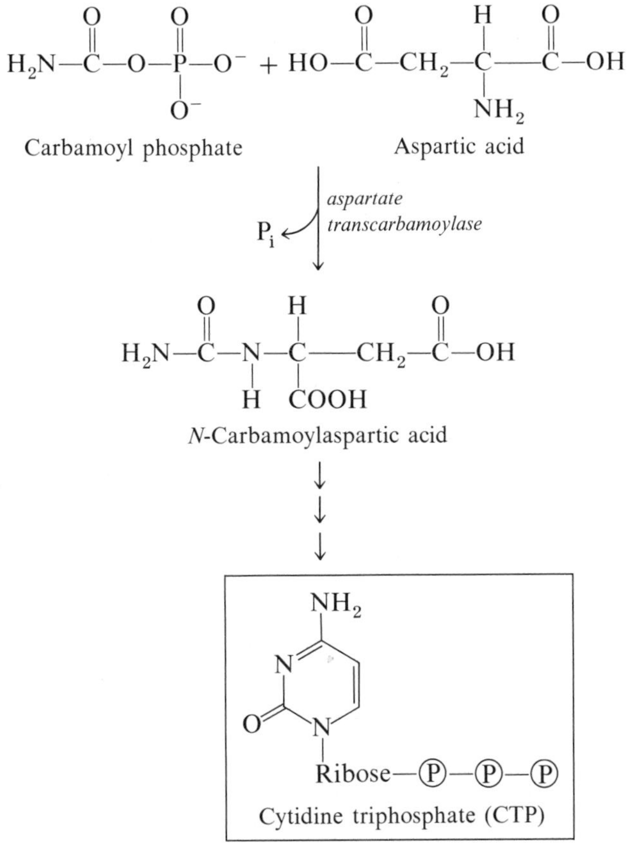

Carbamoyl phosphate   Aspartic acid

P_i ←   *aspartate transcarbamoylase*

$$H_2N-\overset{\overset{\displaystyle O}{\|}}{C}-\overset{\overset{\displaystyle H}{|}}{\underset{\underset{\displaystyle H}{|}}{N}}-\overset{\overset{\displaystyle H}{|}}{\underset{\underset{\displaystyle COOH}{|}}{C}}-CH_2-\overset{\overset{\displaystyle O}{\|}}{C}-OH$$

N-Carbamoylaspartic acid

↓
↓
↓

NH$_2$

Ribose—Ⓟ—Ⓟ—Ⓟ

Cytidine triphosphate (CTP)

It has been shown that in all cases of end-product inhibition, there are at least two distinct binding sites on the enzyme. One is the active site that binds to the substrate, and the other is the binding site for the inhibitory end product. This latter site is called the **allosteric site.** When cytidine triphosphate binds to the allosteric site of the enzyme it causes a change in its three-dimensional conformation, distorting the active site sufficiently to render it nonfunctional (Figure 26.10). In most cases of negative feedback control, the first step of a sequence of reactions is inhibited, thereby preventing the accumulation of useless intermediates.

26.9 Chemotherapy

Chemotherapy is the *use of chemicals* (drugs) *to destroy infectious microorganisms without damaging the cells of the host.*[4] From bacteria to man, the metabolic pathways of all living organisms are quite similar, and so the discovery of safe and effective chemotherapeutic agents is a formidable task. It is now well established that drugs function through their inhibitory effect on a critical enzyme in the cells of the invading organism.

a. Antimetabolites

An **antimetabolite** is a *substance that possesses a structure closely related to the normal substrate* (the metabolite) *of an enzyme* and ***competitively*** *inhibits a significant metabolic reaction.* One of the earliest (1930) and best understood antimetabolites is the synthetic antibacterial agent sulfanilamide. Its effectiveness rests on its structural similarity to *p*-aminobenzoic acid, a compound vital to the growth of many pathogenic bacteria. Bacteria require *p*-aminobenzoic acid for the synthesis of folic acid (a coenzyme precursor, Table 26.2) and sulfanilamide interferes with the enzyme-controlled step involving the incorporation of the *p*-aminobenzoic acid moiety into folic acid.

Sulfanilamide is not harmful to man (or to other mammals) because we cannot synthesize folic acid but must obtain it, preformed, from our diets.

[4]Chemotherapy is also widely used in the treatment of cancer patients. Drugs such as fluorouracil and 6-mercaptopurine interfere with the production of DNA and RNA in tumor cells by substituting for the pyrimidine and purine bases (see Section 27.1). Use of another drug, however, has engendered a great deal of controversy. Laetrile (amygdalin or "vitamin B_{17}"), a compound extracted from apricot pits (and produced synthetically), was first broadly promoted for cancer treatment in the early 1950s and rose to national prominence in the 1970s. It has been claimed to alleviate pain and prolong life for cancer victims. Laetrile is supposed to function by reacting with the enzyme *β-glucosidase* to release cyanide that kills cancer cells. Normal cells are said to be protected by an enzyme (*rhodanase*) that detoxifies cyanide by converting it into thiocyanate. The Food and Drug Administration, however, considers the drug unsafe and worthless. In a 1977 report, the commissioner of the FDA, Donald Kennedy, stated that "all available evidence indicates that Laetrile is a major health fraud in the United States and there is no documentation of its safety or effectiveness."

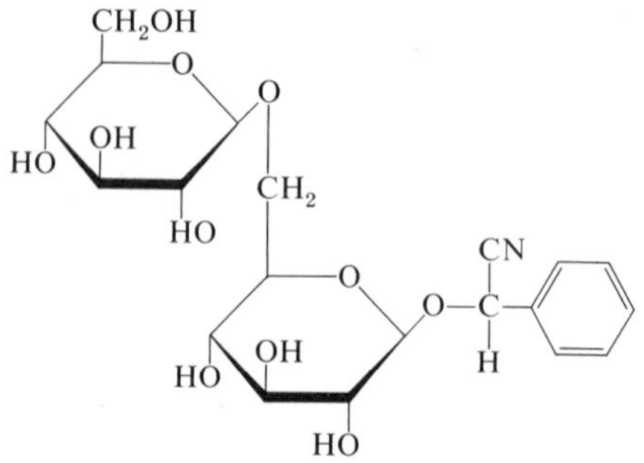
CH2OH
O
O
OH
HO
CH2
HO
CN
O
O—C—
H
OH
HO
HO

Laetrile
(Amygdalin)
(Mandelonitrile-β-gentobioside)

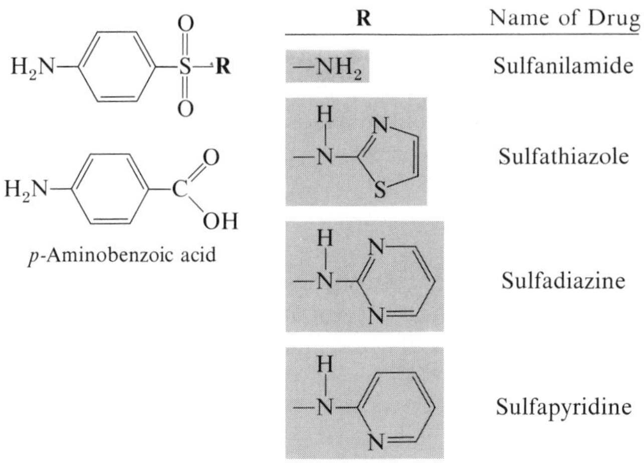
R
Name of Drug
Sulfanilamide
Sulfathiazole
Sulfadiazine
Sulfapyridine
p-Aminobenzoic acid

Figure 26.11 The sulfa drugs interfere with the normal metabolism of *p*-aminobenzoic acid.

After the drug was recognized as an antibacterial agent, many other sulfanilamide derivatives (sulfa drugs) were synthesized and found to be even more effective in this capacity. Many lives were saved during World War II as a result of these popularly named wonder drugs. Soldiers carried packages of powdered sulfa drugs to sprinkle on open wounds to prevent infection. Unfortunately, prolonged use of sulfa drugs causes a number of side effects, particularly kidney damage, so that they have been largely replaced by the penicillins and other antibiotics. Structures of several of the important sulfa drugs are given in Figure 26.11 along with the structure of the metabolite, *p*-aminobenzoic acid.

b. Antibiotics

Although some antibiotics are believed to function as antimetabolites, the terms are not identical. An **antibiotic** is a *compound produced by one microorganism that is toxic to another microorganism.* Antibiotics, many of which can now be synthesized in the laboratory, constitute no well-defined class of chemically related substances but, instead, possess the common property of effectively inhibiting a variety of enzymes essential to bacterial growth.

Penicillin, one of the most widely used antibiotics in the world, was fortuitously discovered by Alexander Fleming in 1929. In 1938, Ernst Chain and Howard Florey isolated it in pure form and proved its effectiveness as an antibiotic. (The three scientists received the Nobel Prize for Physiology and Medicine in 1945.) Penicillin was first introduced into medical practice in 1941. It is believed to prevent bacterial growth by interfering with the synthesis of essential components of the organisms' cell walls. Penicillin inhibits an enzyme (*transpeptidase*) that catalyzes the last step in bacterial cell-wall biosynthesis, namely, the cross-linking reaction. Several naturally occurring penicillins have been isolated. All have the empirical formula $C_9H_{11}O_4SN_2R$ and contain a four-membered ring fused to a five-membered ring (Figure 26.12). Penicillin G was the earliest penicillin to be used on a wide scale. However, it cannot be administered orally for it is quite unstable, and the acid pH of the stomach causes a rearrangement to an inactive derivative. Penicillin V and ampicillin, on the other hand, are acid-stable and they are the major oral penicillins. Some strains of bacteria become resistant to penicillin by developing an enzyme, *penicillinase,* that breaks down the antibiotic. To combat these strains, scientists have been able to synthesize penicillin analogs (e.g., methicillin) that are not inactivated by penicillinase.

Some people are allergic to penicillin and therefore must be treated with other antibiotics. (This allergic reaction is so severe that a fatal coma may occur if penicillin is inadvertently administered to a sensitive individual.) Fortunately, a number of antibiotics have been discovered (e.g., aureomycin, chloramphenicol, streptomycin, and tetracycline), and they have proved to be as effective as penicillin in destroying infectious microorganisms (Figure 26.13). Many of these antibiotics exert their effects by blocking protein synthesis in microorganisms.

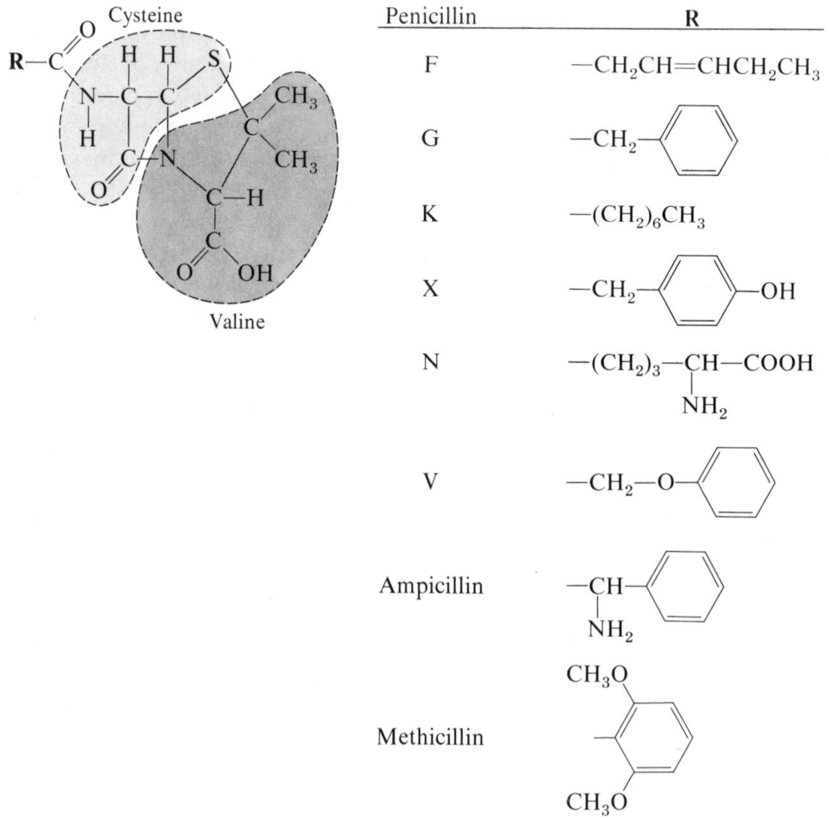
Cysteine
Valine
Penicillin
R
F
—CH2CH=CHCH2CH3
G
—CH2—
K
—(CH2)6CH3
X
—CH2—
—OH
N
—(CH2)3—CH—COOH
NH2
V
—CH2—O—
Ampicillin
—CH—
NH2
Methicillin
CH3O
CH3O

Figure 26.12 The penicillins differ only in the nature of their R groups. Notice that the amino acids valine and cysteine are incorporated into the penicillin structure.

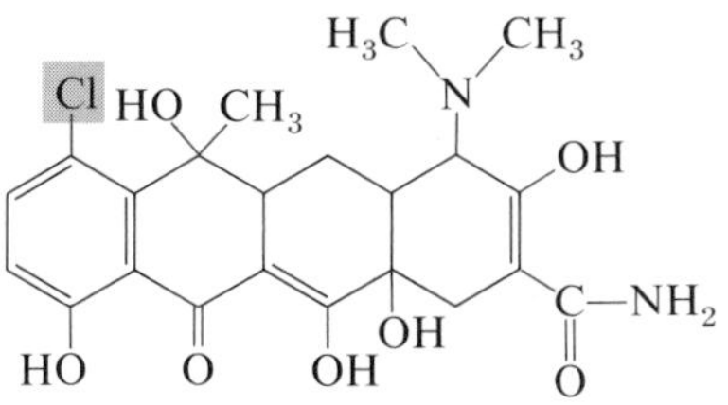

Aureomycin
(Chlorotetracycline)

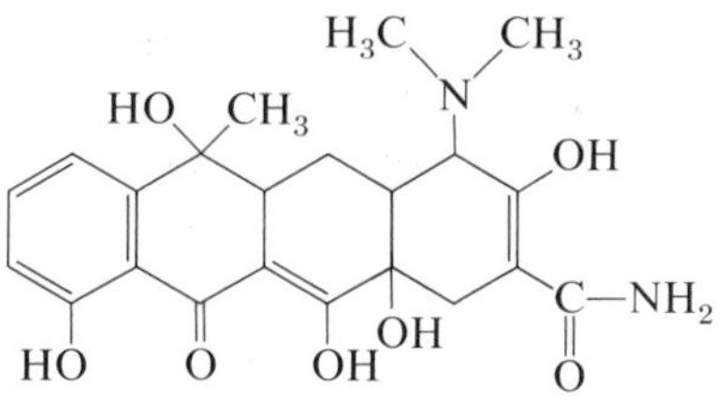

Tetracycline

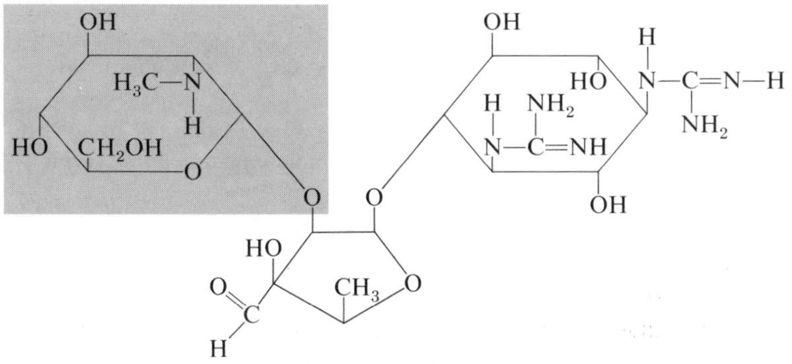

Streptomycin

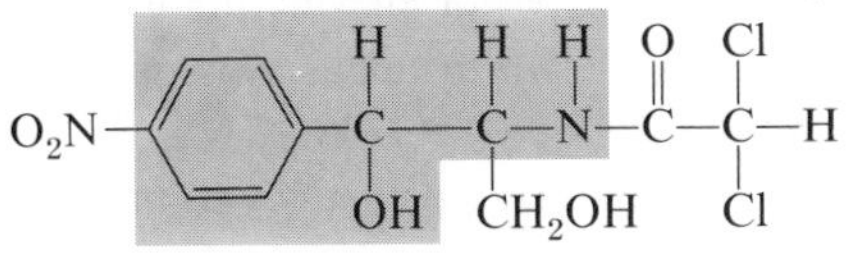

Chloramphenicol
(Chloromycetin)

Figure 26.13 Structures of some common antibiotics. Notice that aureomycin and tetracycline are similar in all respects except for the absence of a chlorine atom in the latter. Streptomycin is a glycoside containing an amino derivative of glucose, and chloramphenicol bears a resemblance to epinephrine.

Exercises

26.1 Define, explain, or give an example of each of the following.

(a) enzyme	**(k)** active site
(b) apoenzyme	**(l)** inhibitor
(c) coenzyme	**(m)** sulfa drug
(d) holoenzyme	**(n)** zymogen
(e) enzyme specificity	**(o)** activator
(f) energy of activation	**(p)** chemotherapy
(g) substrate	**(q)** antimetabolite
(h) optimum temperature	**(r)** antibiotic
(i) optimum pH	**(s)** vitamin
(j) allosteric site	**(t)** negative feedback control

26.2 Briefly recount the contributions of Pasteur, Buchner, and Sumner to our understanding of the nature of enzymes.

26.3 In what ways are enzymes similar to ordinary chemical catalysts? How do they differ? What is the effect of an enzyme on an equilibrium system?

26.4 Why are enzymes more specific than inorganic catalysts?

26.5 Animals can digest starch but not cellulose. Explain.

26.6 What is the relationship between coenzymes and vitamins?

26.7 What are the full chemical names for the coenzymes whose abbreviations are NAD^+ and FAD? Write structural formulas for these compounds.

26.8 What are the various types of specificity exhibited by enzymes? Give the name of an enzyme from each group, its substrate, and the products obtained from the enzymic reaction.

26.9 Briefly describe the nature of enzyme action. What functional groups are probably responsible for the formation and maintenance of the enzyme–substrate complex?

26.10 Compare the lock and key theory with the induced-fit theory of Koshland.

26.11 Only a small fraction of the enzyme molecule binds to the substrate. What function is served by the remainder of the enzyme molecule?

26.12 What factors influence the rate of an enzymic reaction? Discuss their effects?

26.13 Explain why enzymes become inactive above and below the **(a)** optimum temperature and **(b)** the optimum pH.

26.14 Briefly discuss the various types of enzyme inhibitors.

26.15 Experimentally, how could you distinguish a competitive inhibitor from a noncompetitive inhibitor?

26.16 Fluoroacetic acid, CH_2FCOOH, is a deadly poison. Suggest a possible mechanism for its toxic effect.

26.17 Explain why antimetabolites and antibiotics may both be classified as antiseptics.

27

The Nucleic Acids

In this chapter we shall consider the nucleic acids, the macromolecules whose structure determines the growth and development of all life forms. The nucleic acids are *informational* molecules; into their primary structure is encoded a set of directions that ultimately governs the metabolic activities of the living cell. We shall see how differences in primary nucleic acid structure account, on the molecular level, for the factors that differentiate organisms from one another.

27.1 Structure of Nucleic Acids

Nucleic acids are macromolecular polymers that occur in all living things. There are two types, deoxyribonucleic acid (DNA) and ribonucleic acid (RNA). Their chemical composition is best understood in terms of their hydrolysis products. Complete hydrolysis yields a mixture of heterocyclic amines (purines and pyrimidines), a five-carbon sugar (ribose or 2-deoxyribose), and phosphoric acid. Partial hydrolysis degrades the nucleic acid into somewhat larger subunits, **nucleotides** and **nucleosides.** The successive hydrolytic products of nucleic acids are shown in Figure 27.1.

a. Pyrimidine Bases

The pyrimidine bases that occur in nucleic acids are substituted derivatives of the parent compound pyrimidine. Pyrimidine is a heterocyclic

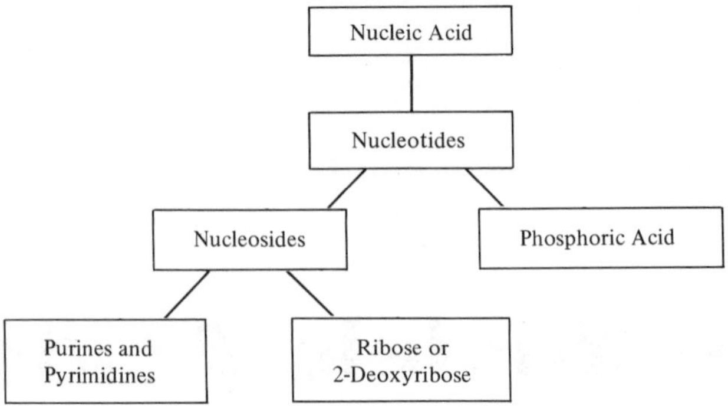
Nucleic Acid
Nucleotides
Nucleosides
Phosphoric Acid
Purines and Pyrimidines
Ribose or 2-Deoxyribose

Figure 27.1 Composition of nucleic acids.

six-membered ring compound containing two ring nitrogen atoms. It does not occur free in nature, but its derivatives *uracil, thymine,* and *cytosine* occur in nucleic acids. Figure 27.2 depicts the structures of the commonly occurring pyrimidine bases. In addition, several other pyrimidine derivatives are found in various nucleic acids. Among these are 5-methylcytosine and 5-hydroxymethylcytosine.

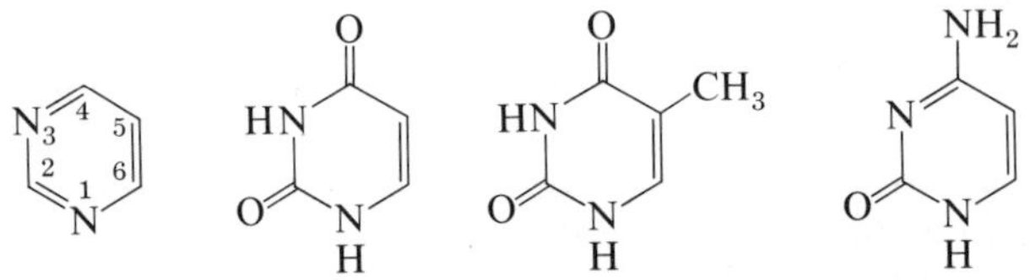

Pyrimidine | Uracil | Thymine | Cytosine

Figure 27.2 Pyrimidine and the major pyrimidine bases.

b. Purine Bases

The naturally occurring purine bases are derivatives of the parent compound purine, a heterocyclic amine consisting of a pyrimidine ring fused to an imidazole ring. Adenine and guanine are the major purine constituents of nucleic acids (Figure 27.3). 6-Methyladenine and 2-methylguanine are two of the minor purine bases that occur in certain nucleic acids.

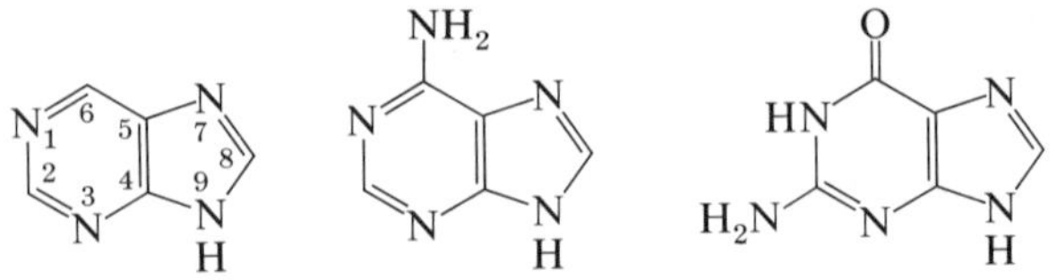

Purine | Adenine (6-Aminopurine) | Guanine (2-Amino-6-oxypurine)

Figure 27.3 Purine and the major purine bases.

c. Nucleosides

A purine or a pyrimidine base may be envisioned to combine with a pentose molecule, in a condensation-type reaction, to form a compound known as a **nucleoside.** The pentose may be ribose or 2-deoxyribose; both sugars exist in the furanose form.

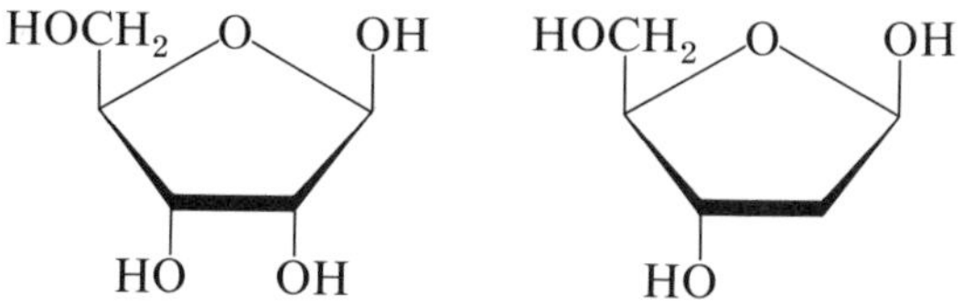

Ribose 2-Deoxyribose

The bond joining the pentose to the nitrogen base is termed an N-glycosyl linkage, and it is always beta in naturally occurring nucleosides. The N-glycosyl linkage is formed between carbon-1′ of the sugar and nitrogen-1 of the pyrimidine base or nitrogen-9 of the purine base.[1] A molecule of water is eliminated in the process.

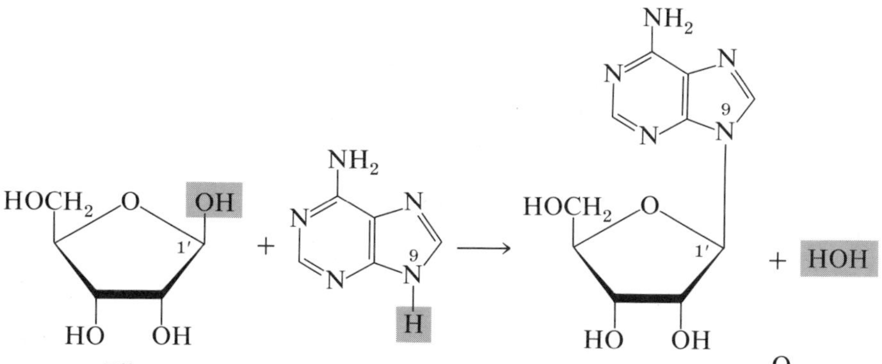

Ribose Adenine Adenosine

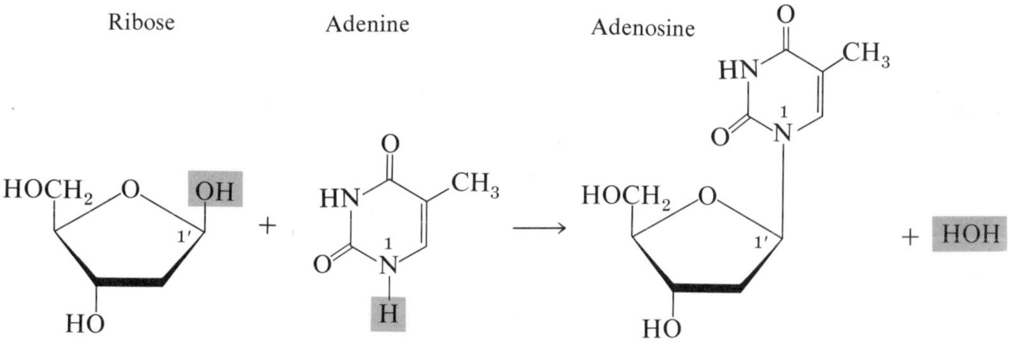

Deoxyribose Thymine Deoxythymidine

[1]The convention used in numbering is that atoms of the pentose ring are designated by primed numbers, whereas atoms of the purine or pyrimidine rings are designated by unprimed numbers.

The common names of the ribonucleosides are derived from the names of the nitrogenous bases. The prefix *deoxy-* is utilized if the base is combined with deoxyribose (deoxynucleosides): deoxyadenosine, deoxyguanosine, deoxycytidine, and deoxythymidine. Structures and names of the major ribonucleosides and one of the deoxyribonucleosides are given in Table 27.1.

d. Nucleotides

The nucleotides are phosphoesters of the nucleosides and are thought to result from the esterification of phosphoric acid with one of the free pentose hydroxyl groups.

Table 27.1 The Major Pyrimidine and Purine Nucleosides

Structure	*Name*	*Structure*	*Name*
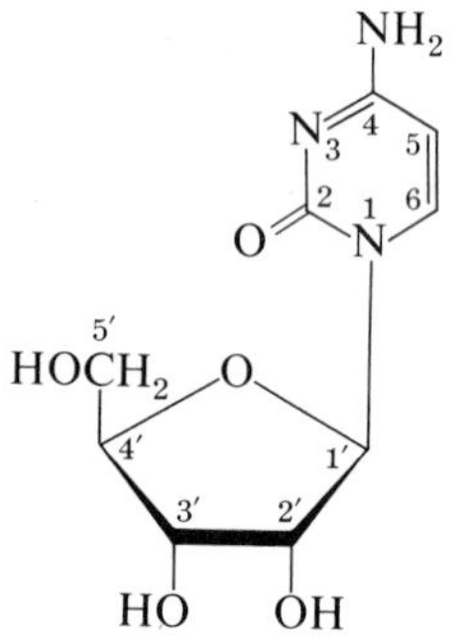	Cytidine	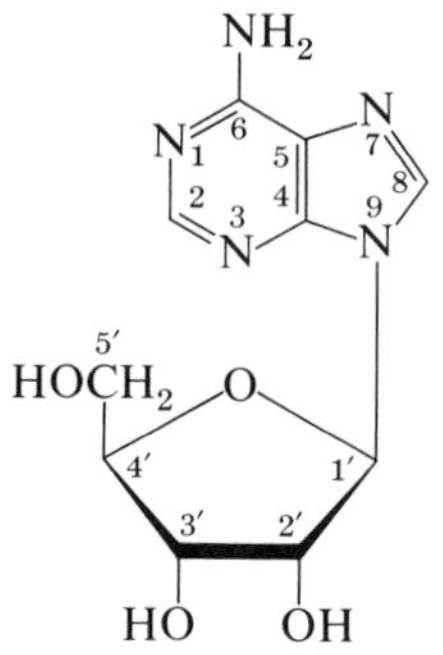	Adenosine
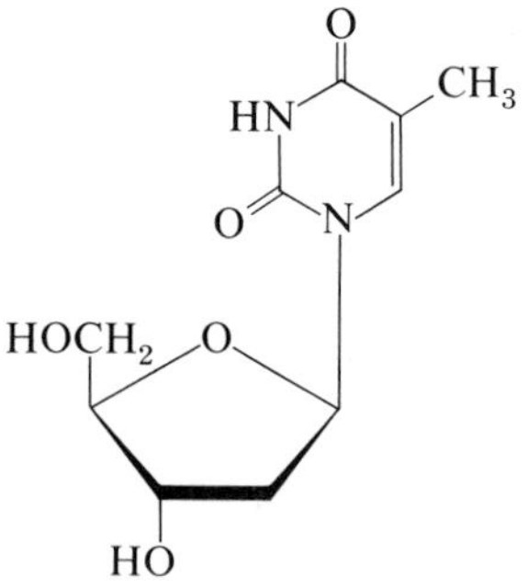	Deoxythymidine	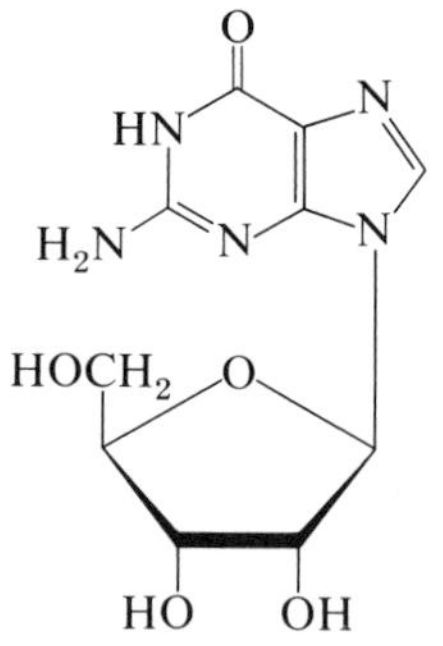	Guanosine
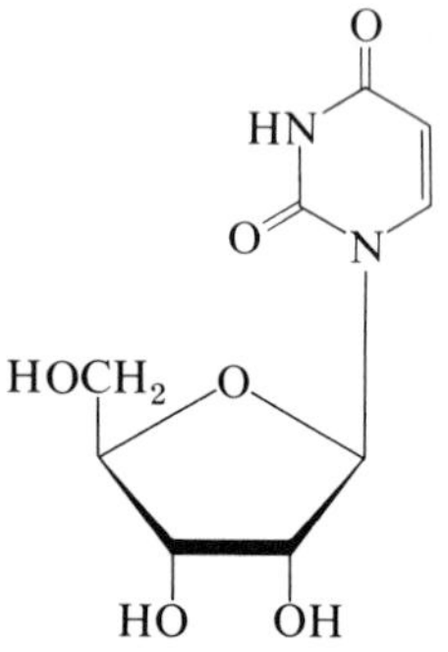	Uridine		

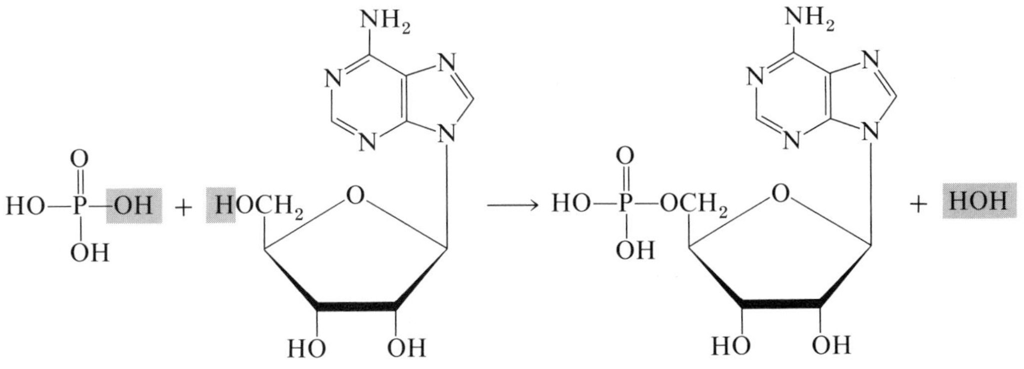
NH2
N
N
N
N
O
HO—P—OH
OH
HOCH2
O
HO
OH
HO—P—OCH2
HOH

Phosphoric acid | Adenosine | Adenylic acid (Adenosine monophosphate)

Nucleotides are the fundamental subunits of nucleic acids in the same manner that amino acids are the subunits of proteins. Nucleotides also occur in the free form in all cells. We have already mentioned adenosine monophosphate (AMP) and its derivatives ADP and ATP (see Figure 21.3). The nucleotides are named as nucleoside monophosphates, e.g., guanosine 5′-phosphate (GMP) and deoxyguanosine 5′-phosphate (dGMP), or as acids, e.g., cytidylic acid (CMP) and deoxycytidylic acid (dCMP). The names and structures of some nucleotides are given in Table 27.2.

Adenosine monophosphate may be further phosphorylated to yield adenosine diphosphate (ADP) and adenosine triphosphate (ATP). In addition, a cyclic 3′,5′-phosphate of adenosine occurs in which the phosphate group is bonded to two of the ribose carbons. The enzyme *adenyl cyclase* is bound to the membrane of the cell, where it catalyzes the formation of cyclic AMP from ATP. Cyclic AMP is an intracellular hormone that transmits messages from the cell membrane to enzymes within the cell.

$$\text{ATP} \xrightarrow{\text{adenyl cyclase}} \text{Cyclic AMP} + \text{PP}_i$$

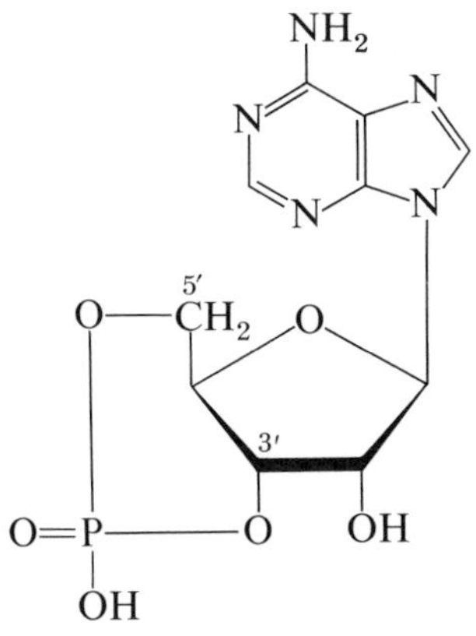
NH2
N
N
N
N
5′
O—CH2
O
3′
O=P—O
OH
OH

Adenosine 3′,5′-phosphate
(Cyclic AMP)

Table 27.2 The Pyrimidine and Purine Nucleotides

Structure	*Names*	*Structure*	*Names*
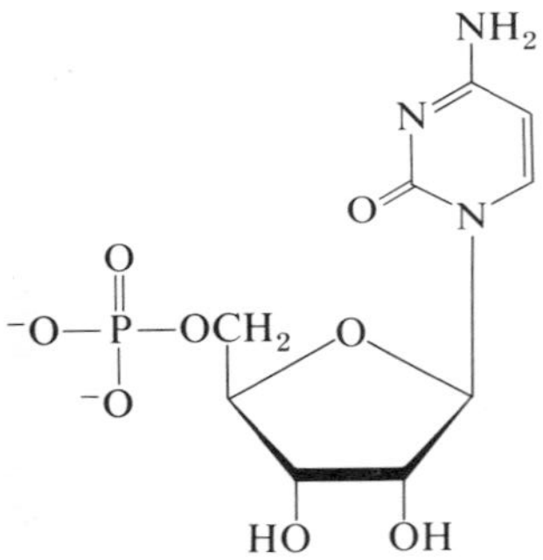	Cytidylic acid Cytidine 5′-phosphate	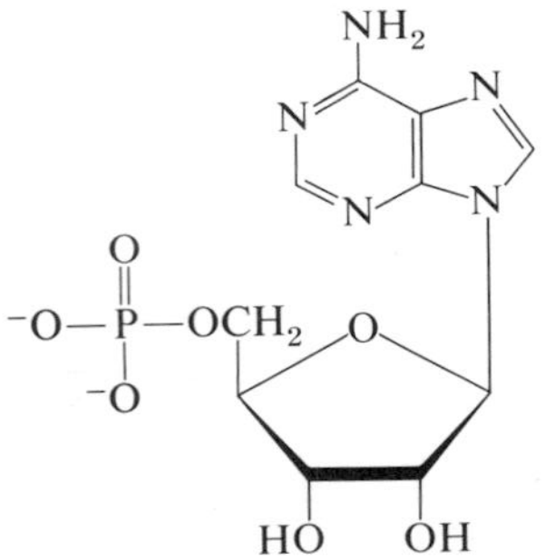	Adenylic acid Adenosine 5′-phosphate
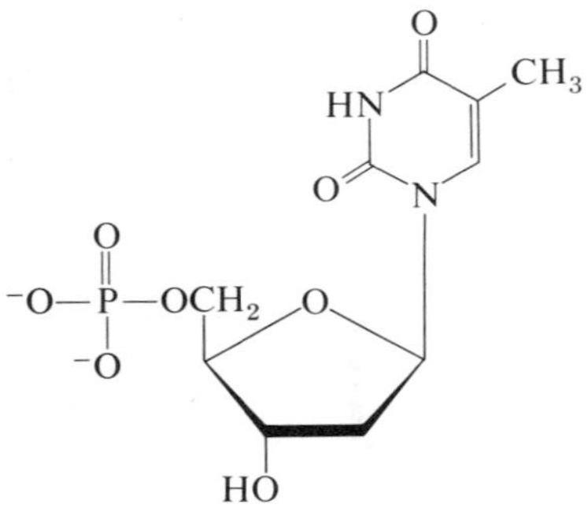	Deoxythymidylic acid Deoxythymidine 5′-phosphate	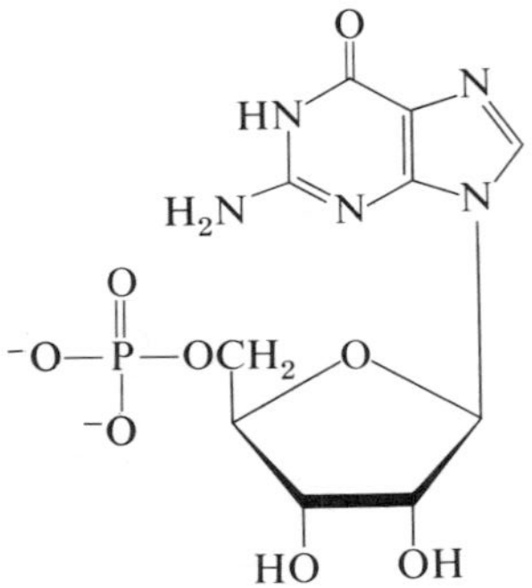	Guanylic acid Guanosine 5′-phosphate
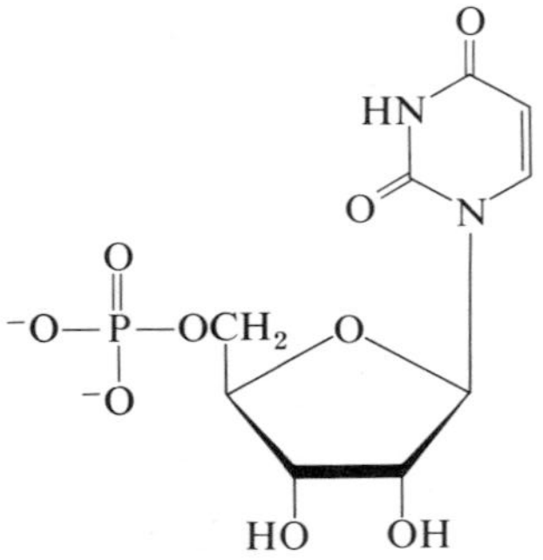	Uridylic acid Uridine 5′-phosphate		

This hormone is believed to influence membrane permeability, movement of ions, and the release of other hormones. Like ADP and ATP, cyclic AMP is not a constituent of nucleic acids.

Certain nucleotides are prosthetic groups of a number of important coenzymes. Among the nucleotides that function as coenzymes are FAD, NAD^+, and coenzyme A. Their structures were given in Table 26.2.

e. Nucleotide Polymers

Nucleic acids are polymers of the nucleotides in much the same way that proteins are polymers of amino acids. Both nucleic acids and proteins are irregular polymers since they are composed of several different types of basic repeating units (four distinct nucleotides in the case of the nucleic acids, and twenty distinct amino acids in the case of the proteins). Biosynthetically, nucleic acids are thought to arise from the polymerization of nucleoside triphosphates with the concomitant release of pyrophosphate. The polymerizing enzymes that are essential to this process are examples of enzymes that are classified as *ligases* (recall Table 26.1).

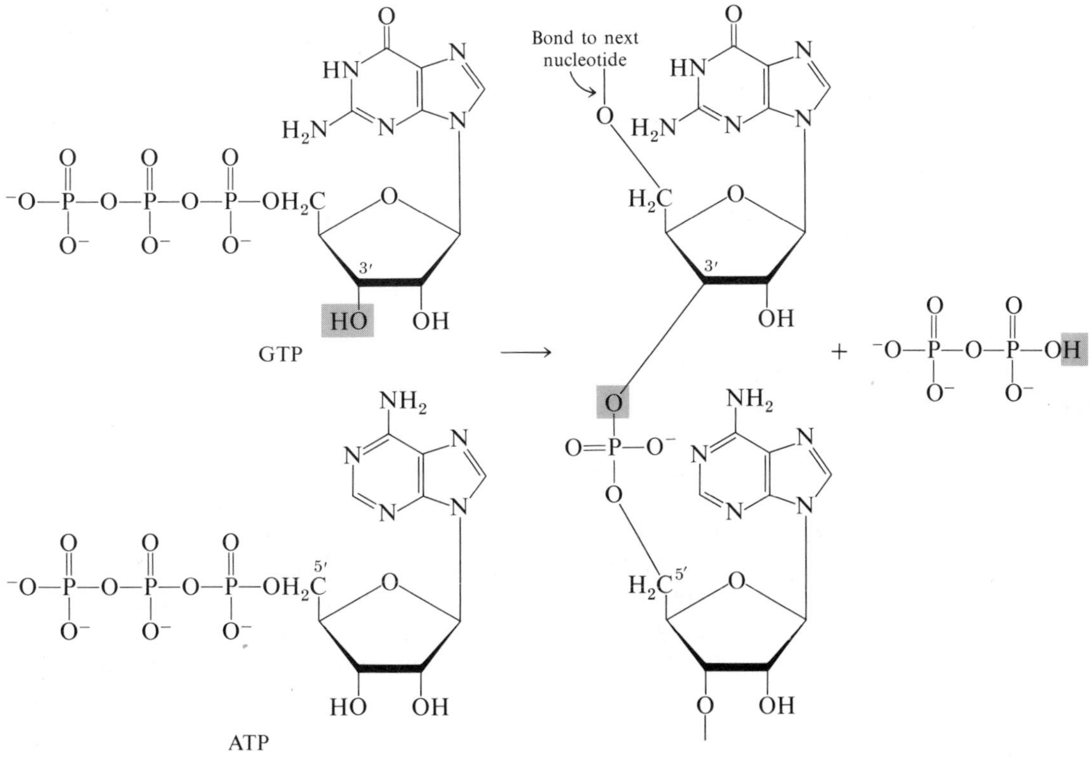
Bond to next nucleotide
GTP
ATP

When the nucleoside triphosphates condense to form nucleic acids, bonds are formed between the 5′-phosphate of one molecule and the 3′-

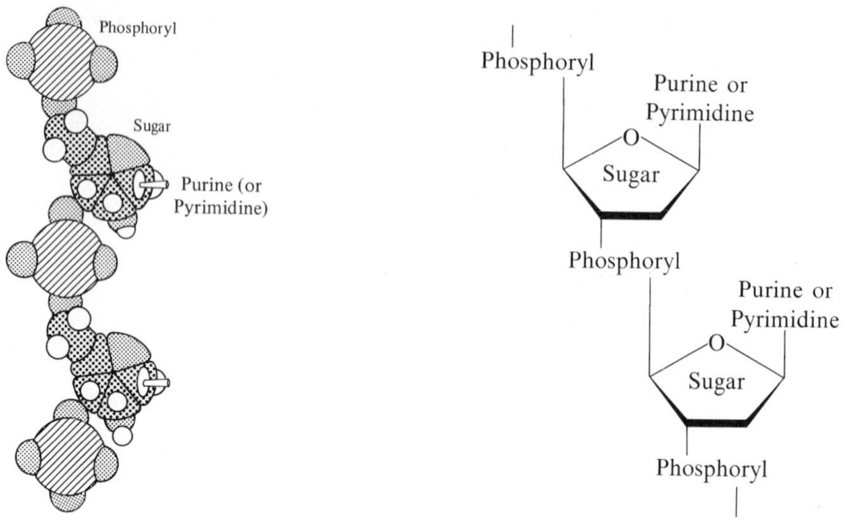
Phosphoryl
Sugar
Purine (or Pyrimidine)
Phosphoryl
Purine or Pyrimidine
O
Sugar
Phosphoryl
Purine or Pyrimidine
O
Sugar
Phosphoryl

Figure 27.4 General chemical structure of nucleic acids.

hydroxyl of an adjacent molecule. The phosphate group can be considered to be the connecting bridge between adjacent nucleosides. This gives rise to a molecule with an alternating sugar–phosphate backbone having a 3′,5′ phosphodiester linkage.

Nucleic acids, then, are polymeric chains with a backbone of repeating sugar units connected by phosphate bridges (Figure 27.4). The biosynthesis of nucleic acids has been greatly elucidated in the laboratories of A. Kornberg, S. Ochoa, and S. B. Weiss. Figure 27.5 illustrates partial structures for both DNA and RNA.

27.2 Deoxyribonucleic Acid (DNA)

DNA is a high molar mass nucleic acid that occurs almost exclusively in the nucleus of the cell.[2] It is found in almost every living organism, although some viruses contain RNA in its place (Section 27.7). During mitosis, DNA is apparent in the chromosomes, those structures directly responsible for the transmission of the genetic information. Molar mass determinations on various types of DNA yield values between 10^8 and 10^9 g/mole.

DNA is a double-stranded nucleic acid; within its structure two nucleic acid chains are intimately associated with one another by means of hydrogen bonding. The two strands of nucleic acid are oriented toward each other in such a way that the bases projecting from each are in proximity. The structures of the four bases permit hydrogen bonding between specific base

[2]Some DNA is found in the chloroplasts and in the mitochondria. For a discussion of the function of this extranuclear DNA, consult an advanced text such as Albert L. Lehninger, *Biochemistry,* 2nd ed., Worth Publishers, Inc., 1975, or J. Ramsey Bronk, *Chemical Biology,* Macmillan Publishing Co., Inc., 1973.

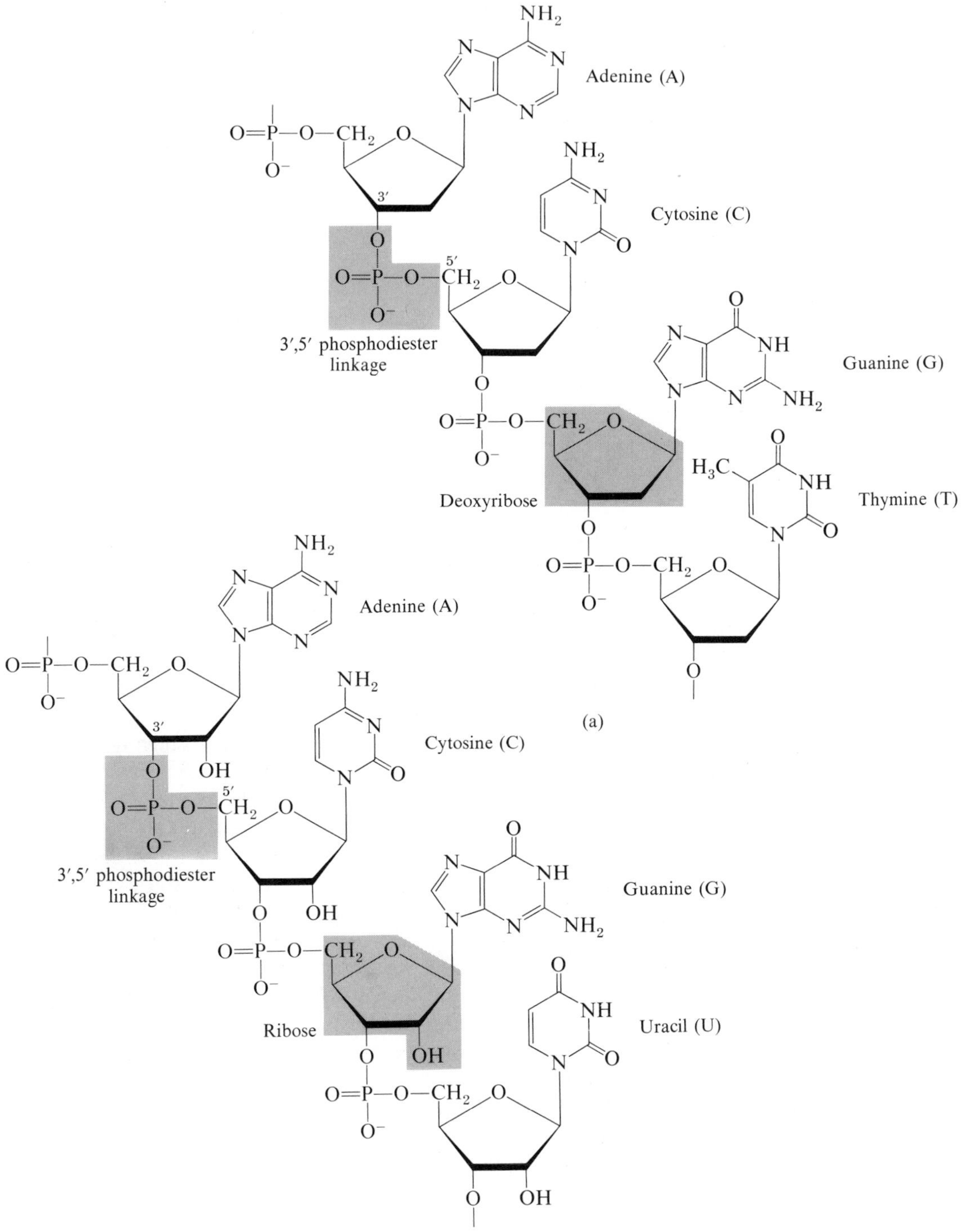
Adenine (A)
Cytosine (C)
Guanine (G)
Thymine (T)
Deoxyribose
3′,5′ phosphodiester linkage
(a)
Adenine (A)
Cytosine (C)
Guanine (G)
Uracil (U)
Ribose
3′,5′ phosphodiester linkage

(b)

Figure 27.5 Partial chemical structures of DNA (a) and RNA (b).

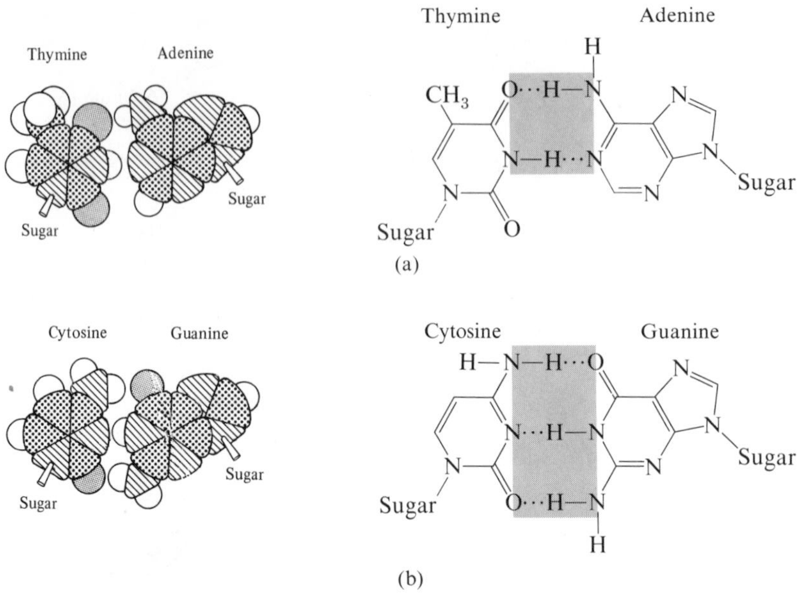
Thymine
Adenine
Sugar
Sugar
Thymine
Adenine
Sugar
Sugar
(a)
Cytosine
Guanine
Sugar
Sugar
Cytosine
Guanine
Sugar
Sugar
(b)

Figure 27.6 Pairings of thymine with adenine (a) and of cytosine with guanine (b) by means of hydrogen bonding as in DNA.

pairs (Figure 27.6). *The purines adenine and guanine always pair with the pyrimidines thymine and cytosine, respectively.* This intramolecular hydrogen bonding accounts for the secondary structure of DNA and is the major force holding the two strands of the molecule together. Figure 27.7 shows a schematic representation of a portion of a DNA molecule depicting hydrogen bonding between the bases of the two strands. Owing to the great length of the nucleic acid molecule, there are vast numbers of hydrogen bonds formed between the two chains. It is the additive contribution of all these bonds that imparts great stability to the DNA molecule. Notice that in order for the complementary bases to be adjacent, the two strands of the double helix must be aligned antiparallel; that is, their 3′,5′-phosphodiester bridges are in opposite directions.

The tertiary structure of DNA is better appreciated when the student understands some of the experimental data that led to its elucidation. A brief part of that experimental history follows.

a. Chargaff's Rules

The work of E. Chargaff and others has revealed a number of striking regularities in the "normal" DNA content of a variety of different species. It was Chargaff's insightful discoveries that led J. D. Watson and F. H. C. Crick to postulate their famous molecular model of DNA. Chargaff's findings, now known as **Chargaff's rules,** are summarized as follows.

1. The base composition of the DNA of an organism is constant throughout all the somatic cells of that organism and is characteristic for a given species. (Somatic cells include all body cells other than germ cells.)

2. Known base compositions vary greatly from one organism to another. This is clearly expressed by what is known as the *dissymmetry ratio,* (A + T)/(G + C). In other words, differing base compositions among organisms are reflected by variance in their dissymmetry ratios.

3. Closely related organisms often have similar base compositions and therefore have close values for their dissymmetry ratios.

4. The amount of adenine in the DNA of a given organism is always equal to the amount of thymine (A = T).

5. The amount of guanine in the DNA of a given organism is always equal to the amount of cytosine (G = C).

6. The total amount of purine bases in the DNA of a given organism is always equal to the total amount of pyrimidine bases (A + G = T + C).

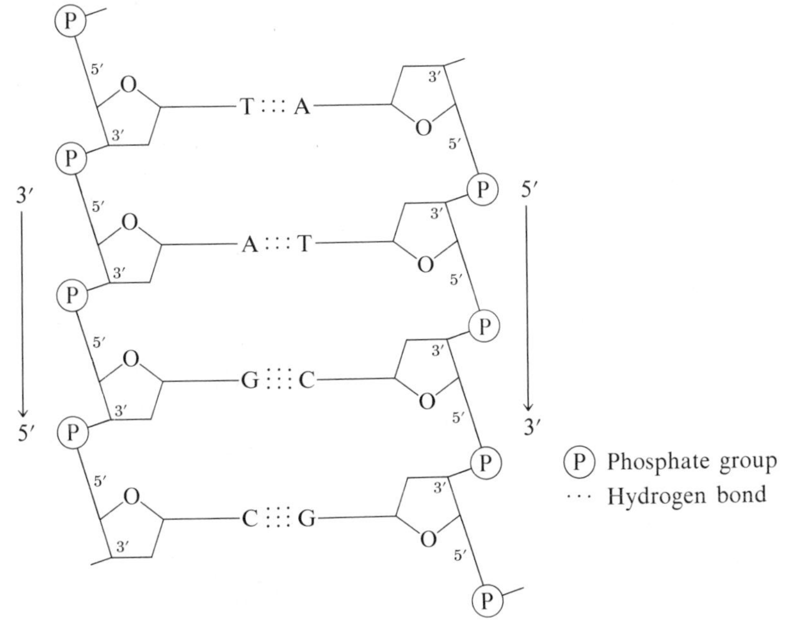
T ::: A
A ::: T
G ::: C
C ::: G
3′
5′
P
O
Ⓟ Phosphate group
··· Hydrogen bond

Figure 27.7 Secondary structure of a segment of a hypothetical DNA molecule. The two strands are complementary—*not* identical—and are held together through hydrogen bonding of complementary bases. The 3′,5′ phosphodiester linkages are in reverse order (antiparallel) in the two chains.

b. Watson–Crick Model for DNA

Chargaff's findings clearly suggested that the purine and pyrimidine bases of DNA were in some way linked to one another, adenine always pairing with thymine and guanine always pairing with cytosine. At about the same time (1950–53) that Chargaff and his colleagues published their findings, another experimenter, M. Wilkins, came forward with x-ray diffraction data on the lithium salt of DNA. Wilkins found that within the crystal there is a repeat distance of 34 Ångström units (Å) and there are ten subunits per turn. The repeat distance of the subunit is therefore 3.4 Å, which is the same order of magnitude that a single nucleotide might occupy in the chain. These data, coupled with Chargaff's findings, led to the Watson–Crick model for DNA (1953). Watson, Crick, and Wilkins received the Nobel Prize in 1962.

According to the model, DNA consists of two right-handed polynucleotide chains that are complementary and coiled about the same axis so that they form a **double helix.** The bases project inward in such a way that a purine of one strand always pairs with a pyrimidine of the other. Only certain base pairs can be spatially accommodated (again, according to Wilkins' x-ray data), and these are the pairs A—T and G—C. The bases are associated with one another by means of hydrogen bonding. Such pairing rules demand that the base sequence of one chain be completely determined by the base

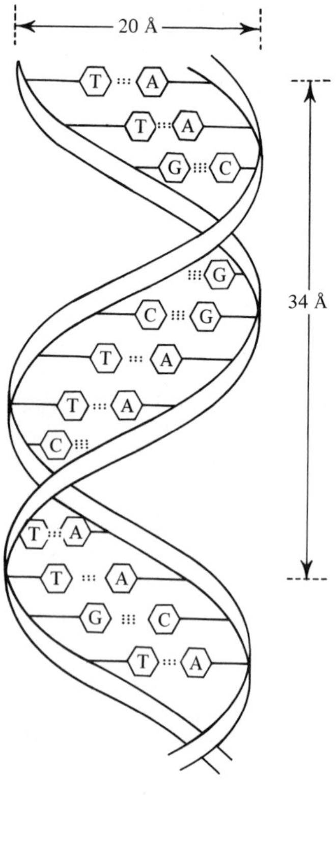
20 Å
T A
T A
G C
G
C G
T A
T A
C
T A
T A
G C
T A
34 Å

Figure 27.8 The Watson–Crick double-stranded helical model for DNA.

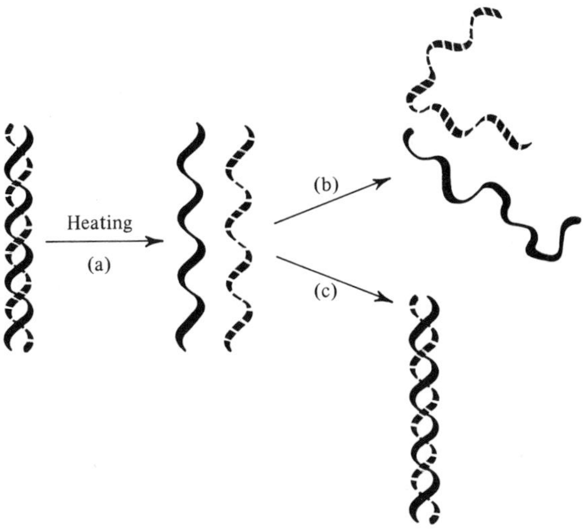
Heating
(a)
(b)
(c)

Figure 27.9 A schematic diagram showing the melting of DNA (a) and the effects of rapid (b) or slow (c) cooling.

sequence of the other. The two strands of the double helix are oriented with opposite (antiparallel) polarity. The negatively charged phosphoester backbone, therefore, projects outward from the helix, and the ionized phosphate groups are free to interact with the environment. (In particular, the DNA molecules form complexes with the histones—basic proteins whose protonated amino groups interact electrostatically with the negatively charged phosphate groups.) The helix has a diameter of about 20 Å and contains ten nucleotide pairs in each turn of the helix, which is 34 Å in length. The DNA in a typical mammalian cell contains about 30,000 nucleotides. Figure 27.8 depicts the Watson–Crick model for DNA.

It is not possible to separate the two strands of the DNA molecule without destroying its helical configuration. However, since the strands are complementary and therefore not identical, it should be possible to separate them either chemically or physically. This has, in fact, been done in DNA *melting* experiments (Figure 27.9). A solution of DNA is heated to a temperature sufficient to cleave the hydrogen bonds that hold the two strands together. (Cleavage and strand separation are indicated by a sharp change in the viscosity and optical density of the solution.) The solution is then cooled rapidly so that the complementary strands do not have the opportunity to realign themselves in their former configuration (*annealing*). The strands may then be separated by any number of chemical or physical separation techniques (chromatography, electrophoresis, density gradient centrifugation).

27.3 DNA as the Hereditary Material

At the time that Watson and Crick postulated the structure of DNA, biologists throughout the world were beginning to consider the possibility that the nucleic acids were the substances that carried hereditary information. Formerly, scientific speculation favored the proteins because it seemed that more information could be coded in the more diverse protein molecule. However, proteins were known to have a high turnover rate; that is, their metabolic synthesis and breakdown occurred at a high rate. The genetic material, it seemed, should have a relatively high degree of metabolic stability because of its responsibility for transmitting hereditary information from one generation to the next. DNA has the required metabolic stability. In addition, its structure suggests the possibility of **template replication,** a phenomenon that allows for the biosynthesis of two identical daughter molecules from the information contained in one parent molecule. Watson and Crick postulated the current theory of DNA replication according to their own model for the structure of the molecule.

a. DNA Replication—A Semiconservative Mechanism

If the sequence of bases along the DNA strand somehow contains the hereditary information, the biosynthesis of new molecules of DNA must proceed in some fashion that preserves this sequence of bases and hence conserves the information for distribution to progeny cells. Watson and Crick

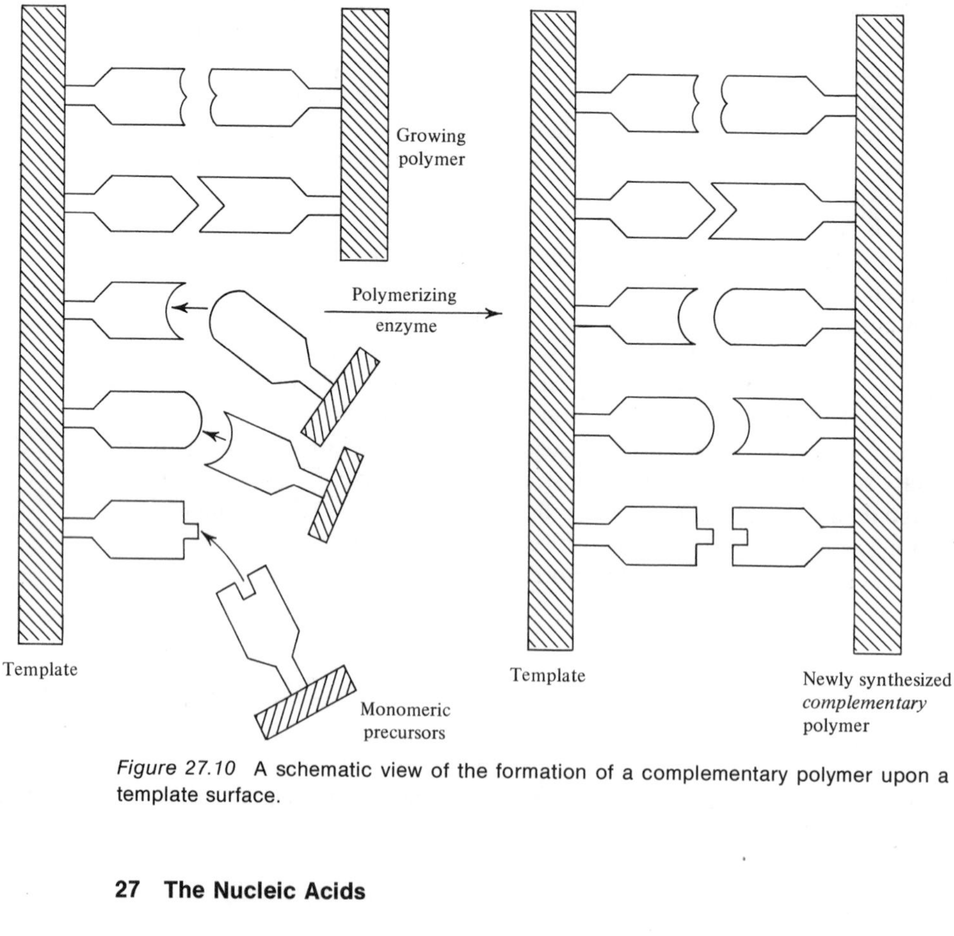
Growing
polymer
Polymerizing
enzyme
Template
Monomeric
precursors
Template
Newly synthesized
complementary
polymer

Figure 27.10 A schematic view of the formation of a complementary polymer upon a template surface.

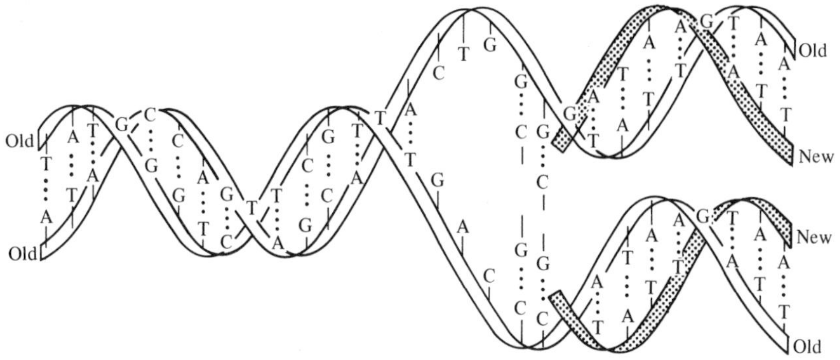
Old
Old
Old
New
New
Old

Figure 27.11 A schematic diagram of DNA replication. Replication is assumed to occur by sequential "unzipping" of the double helix. The new nucleotides are positioned (by an enzyme) and phosphate bridges are formed, thus restoring the original double-helical configuration. Each newly formed double helix consists of one old strand and one new strand.

proposed a semiconservative mode of replication for the molecule. They speculated that during DNA replication (which probably occurs during the interphase of mitosis), the two strands uncoil, and each strand acts as a **template** for the formation of a new complementary strand (Figure 27.10). In other words, the deoxynucleoside triphosphates that presumably exist in the free state in the nucleus are somehow attracted (according to the rules of base pairing) to the single-stranded, uncoiling, DNA. In this way, two new strands are formed that are *complementary* to the two parental strands. As the new daughter molecules begin to be formed, they assume the helical configuration of the parent molecule (Figure 27.11). It is because each molecule consists of one old parental strand and one newly synthesized strand that the mechanism is termed *semiconservative*. Thus, sometime prior to nuclear division, the hereditary material is duplicated and its information content is preserved for distribution to progeny cells.

The Watson–Crick hypothesis conformed to the specification that the base composition of the DNA of an organism is constant throughout the cells of the organism. It provided a simple, logical, and likely means by which the hereditary substance could be precisely replicated. It was a satisfactory hypothesis that needed only experimental verification for widespread acceptance.

The work of Meselson and Stahl (1957–58) offered some rather conclusive proof that DNA does indeed replicate by the Watson–Crick semiconservative mechanism. They grew the bacterium *Escherichia coli* on a medium containing ^{15}N-labeled ammonium chloride as a nitrogen source. Ordinary ^{14}N-ammonium chloride was then added to the cultures and samples of the cells were removed at various time intervals. The DNA from cells that had undergone one cell division had a density intermediate between the density of the DNA from cells grown entirely on ^{15}N or on ^{14}N culture mediums. DNA with this intermediate density was interpreted as being a *hybrid* molecule containing one heavy ^{15}N strand and one light ^{14}N strand. After two cell divisions, equal amounts of the hybrid DNA and the light DNA were detected (Figure 27.12). Initially, the double stranded molecule that contained the two heavy strands uncoiled, and the strands acted as templates for the formation of new strands containing the ordinary ^{14}N. This resulted in the first generation hybrid molecules. The replication process was then repeated, resulting in equal amounts of hybrid and light DNA.

b. DNA Synthesis in Vitro

Further evidence for template replication of DNA was provided by the work of A. Kornberg. He succeeded in isolating an enzyme from the bacte-

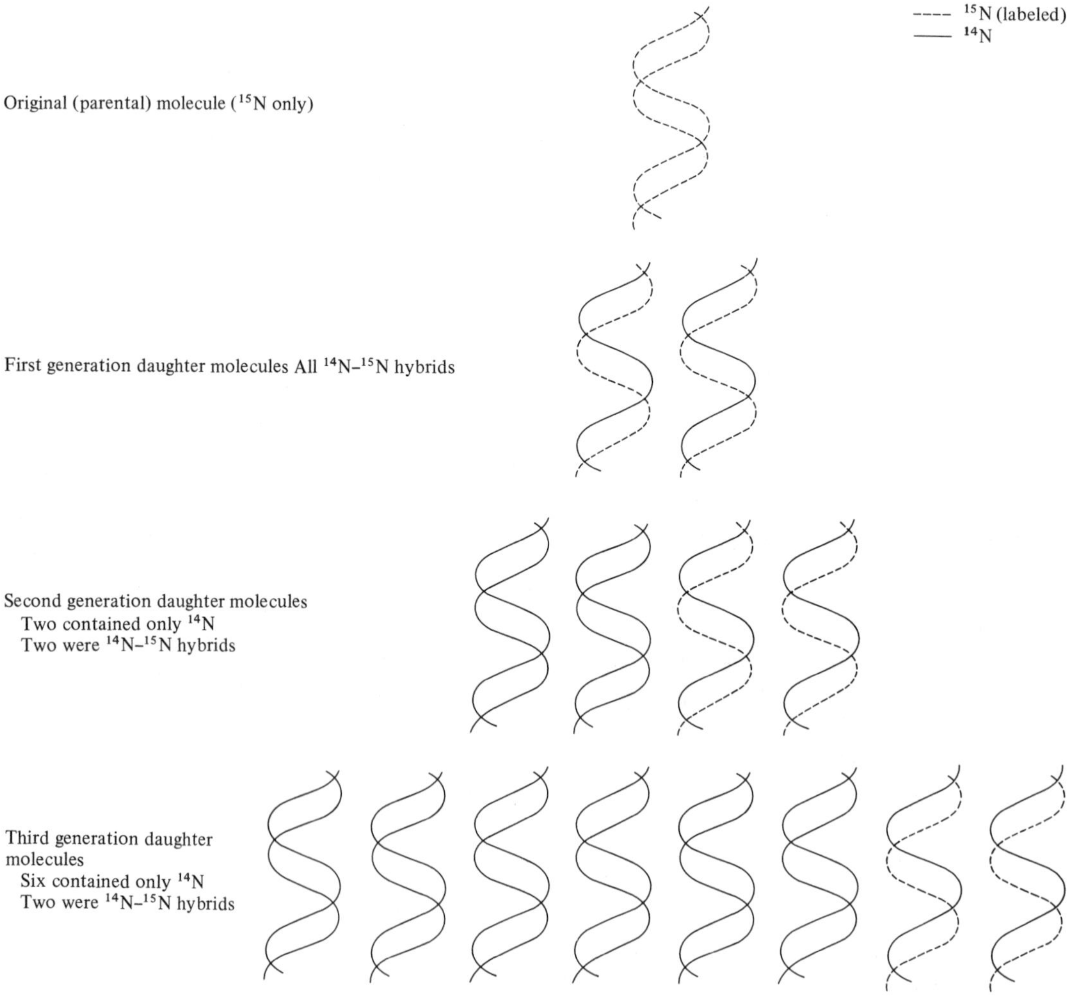

---- ¹⁵N (labeled)
—— ¹⁴N
Original (parental) molecule (¹⁵N only)
First generation daughter molecules All ¹⁴N–¹⁵N hybrids
Second generation daughter molecules
Two contained only ¹⁴N
Two were ¹⁴N–¹⁵N hybrids
Third generation daughter molecules
Six contained only ¹⁴N
Two were ¹⁴N–¹⁵N hybrids

Figure 27.12 The Meselson–Stahl experiment. [Adapted from Meselson and Stahl, *Proc. Natl. Acad. Sci. U.S.*, **44**, 675 (1958).]

rium *E. coli,* which he called *DNA polymerase.* This enzyme was capable of producing a DNA-like molecule having chemical and physical properties almost identical to those of native DNA. Kornberg's DNA replication system includes (1) all four deoxynucleoside triphosphates, (2) native DNA (in small amounts), which might serve both as a template and as a primer, and (3) the polymerase enzyme. The DNA that is formed by the system has a base composition nearly identical to the base composition of the native DNA primer, a fact lending support to the theory of template synthesis. In addition, Kornberg found that when the DNA primer was denatured so as to separate the two strands, synthesis proceeded much more efficiently. The equation that describes the system is illustrated.

$$\begin{matrix} n_1 \text{ dATP} \\ + \\ n_1 \text{ dTTP} \\ + \\ n_2 \text{ dGTP} \\ + \\ n_2 \text{ dCTP} \end{matrix} \xrightarrow[\text{DNA polymerase}]{\text{DNA template;}} \text{DNA} \begin{bmatrix} \text{dAMP} \\ \text{dTMP} \\ \text{dGMP} \\ \text{dCMP} \end{bmatrix}_{2n_1+2n_2} + 2(n_1 + n_2) \text{ PP}_i$$

Kornberg's enzyme is now called DNA polymerase I because two other DNA polymerases (II and III) have since been isolated. It now appears that DNA polymerase III is the major enzyme concerned in the replication process. DNA polymerase I is also believed to participate, but its main role appears to be in the repair of DNA. (The biological function of DNA polymerase II is as yet unknown.)

The polymerase enzyme joins nucleotides together *in one direction only,* linking the 5′-nucleoside triphosphates to the 3′-hydroxyl end of a growing polynucleotide chain. Since the two strands of DNA have opposite polarity, and since it is known from autoradiographic data that the two daughter molecules are synthesized side by side, we are left with the dilemma of postulating a mechanism whereby the strand having the terminal 5′-phosphate is replicated. That is, the antiparallel orientation of complementary DNA strands demands that one daughter strand elongate in the 5′⟶3′ direction and the other in the 3′⟶5′ direction. Synthesis in the latter direction would require a polymerizing enzyme with a quite different specificity from the one that operates in the 5′⟶3′ direction.

One theory about this replication came from the laboratory of R. Okazaki. He found that a large proportion of newly synthesized DNA exists as small fragments (sometimes called Okazaki fragments). As replication proceeds, these fragments become covalently joined into daughter strands by an enzyme termed *DNA ligase.* Dr. Okazaki postulates that DNA replication is a discontinuous process. Fragments of both daughter strands are synthesized in the same 5′⟶3′ direction, and the developing fragments are then joined by DNA ligase. The overall result is that one daughter strand grows in the 5′⟶3′ direction and the other begins at the "crotch" (or replication fork) of the partially uncoiled DNA strand and proceeds toward the loose end (Figure 27.13).

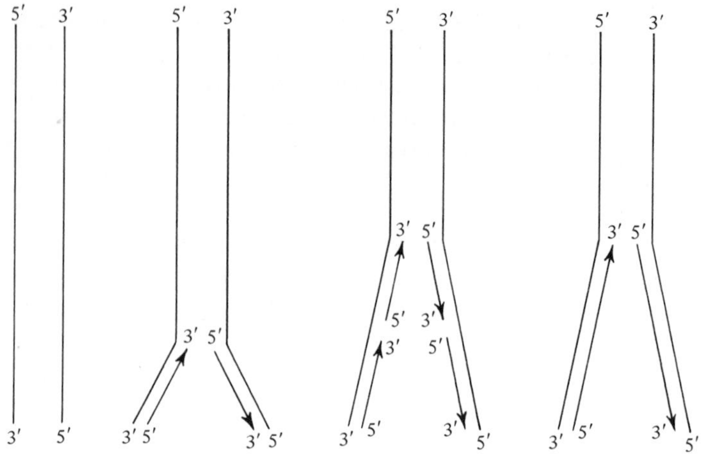
5′
3′
3′
5′
5′
3′
3′ 5′
3′ 5′
3′ 5′
5′
3′
3′ 5′
5′ 3′
3′ 5′
3′ 5′
3′ 5′
5′
3′
3′ 5′
3′ 5′
3′ 5′

Figure 27.13 Discontinuous replication of DNA. Both daughter strands are synthesized in the 5′ → 3′ direction. The enzyme *DNA ligase* ties the short fragments together.

27.4 Ribonucleic Acid (RNA)

RNA is a single-stranded nucleic acid having three important structural features that distinguish it from DNA, two of which have already been alluded to.

1. Ribonucleic acid contains the pentose ribose, in contrast to the 2-deoxyribose of DNA.
2. RNA contains the base uracil (Figure 27.2) rather than thymine.
3. Hydrolytic analysis of RNA indicates that its composition does not obey Chargaff's rules. In other words, the purine/pyrimidine ratio in RNA is not 1:1 as it is the case of DNA.

Recall that the 1:1 base ratio that was found to exist in the DNA molecule was the basis for the Watson–Crick postulate that DNA is a double-stranded molecule held together by base pairing. Accordingly, it has been theorized that such base pairing does not occur in RNA and the molecule, therefore, exists as a single strand (except in some viruses).

The absence of regular hydrogen bonding in RNA results in an irregular and relatively unpredictable structure for the molecule. X-ray diffraction data, however, indicate that portions of the molecule have a helical double-stranded structure. This is probably due to the presence of loops in the molecule formed when part of a strand folds back upon itself. Such regions would be characterized by A—U and G—C pairs, but they must have a limited occurrence since the purine/pyrimidine ratio usually does not approximate 1:1.

All of the RNAs appear to be synthesized, in vivo, by a template mechanism from a part of the DNA molecule. In other words, if the DNA of a particular organism is melted and the double helical structure is destroyed, the RNA of that organism will associate by means of base pairing with a particular section of one of the DNA strands. (Only one DNA strand serves as a template and is copied, but it is not always the same strand for all of the mRNA's.) Such a DNA–RNA complex is known as a **hybrid.**

Formation of DNA–RNA hybrids suggests that the biosynthesis of RNA proceeds in a manner highly analogous to the biosynthesis of DNA. The two strands of the DNA molecule begin to uncoil. The nucleoside triphosphates are attracted to the uncoiling region of the DNA molecule according to the rules of base pairing. They are then enzymically polymerized; a molecule of RNA is formed, with the concurrent elimination of pyrophosphate. The enzyme that catalyzes the reaction is known as *RNA polymerase.* It can initiate RNA synthesis in the middle of a DNA strand; chain growth proceeds in the 5′⟶3′ direction. The equation for the reaction is illustrated.

$$\begin{matrix} n_1 \text{ ATP} \\ + \\ n_1 \text{ UTP} \\ + \\ n_2 \text{ GTP} \\ + \\ n_2 \text{ CTP} \end{matrix} \xrightarrow[\text{RNA polymerase}]{\text{DNA template;}} \text{RNA} \begin{bmatrix} \text{AMP} \\ \text{UMP} \\ \text{GMP} \\ \text{CMP} \end{bmatrix}_{2n_1+2n_2} + 2(n_1 + n_2)\ \text{PP}_i$$

Because the RNA is a complementary copy of the information contained in the DNA, the process of RNA biosynthesis is referred to as **transcription.** We shall see later (Section 27.5-a) why this process is vital to all growth and development. Figure 27.14 contains a schematic representation of the process.

Three types of RNA are known to exist; the distinctions among them are made primarily on the basis of biochemical function. They do, however, differ also in molar mass and in secondary structure (Table 27.3).

a. Messenger RNA (mRNA)

Messenger RNA comprises only a few percent of the total amount of RNA within the cell. It has been shown to be complementary to a given segment of the DNA of the organism from which it is isolated. A molecule of

Table 27.3 RNA Molecules in *E. coli*

RNA Type	*Relative Amount (%)*	*Average Molar Mass*	*Approx. Number of Nucleotides*
Messenger RNA (mRNA)	5	4×10^5	1200
Transfer RNA (tRNA)	15	3×10^4	75
Ribosomal RNA (rRNA)	80	6×10^5	1800

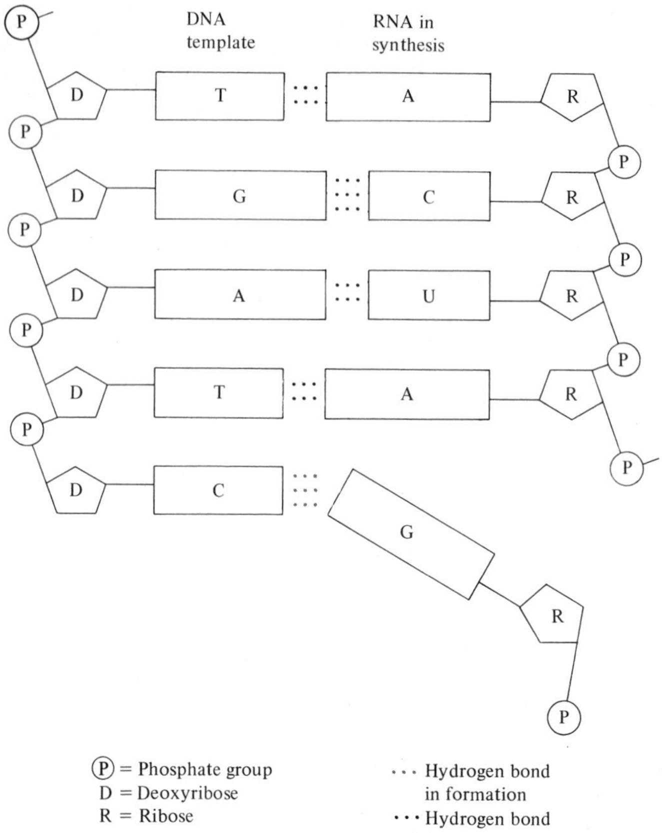
DNA
template
RNA in
synthesis
T
A
G
C
A
U
T
A
C
G
D
R
P
Ⓟ = Phosphate group
D = Deoxyribose
R = Ribose
··· Hydrogen bond
in formation
··· Hydrogen bond

Figure 27.14 Transcription—synthesis of RNA upon a DNA template. One strand of the DNA double helix is used as a template for alignment of RNA bases. Once a nucleotide is in position, a phosphate bridge is formed, resulting in a single strand of RNA.

mRNA exists for a relatively short time; like proteins, it is continuously being degraded and resynthesized. The rate of mRNA degradation is different from species to species and also from one type of cell to another. In bacteria, one-half of the total mRNA is degraded every two minutes, whereas in rat liver, the half-life is several days.

The molecular dimensions of the mRNA molecule vary according to the amount of genetic information that the molecule is meant to encode. It is known, however, that there is very little intramolecular hydrogen bonding in this type of RNA and that the molecule exists in a fairly random coil. After transcription the messenger RNA passes into the cytoplasm, carrying the genetic message from DNA to the ribosomes (Section 27.4-c), the sites of protein synthesis. In Section 27.5-a we shall see how mRNA directly governs that synthesis.

b. Transfer RNA (tRNA)

Transfer RNA (tRNA) is a relatively low molar mass nucleic acid, soluble in media commonly used to isolate the higher molar mass nucleic acids. It functions by attaching itself (with the aid of a specific enzyme—see Section 27.5-a) to a particular amino acid and carrying that amino acid to the site of protein synthesis at the precise moment specified by the genetic code. Each of the twenty amino acids found in proteins has at least one corresponding tRNA, and most amino acids have multiple tRNA molecules (Section 27.5-b). For example, there are two different tRNAs specific for the transfer of lysine, three for isoleucine, four for glycine, and six for serine. The existence of several tRNAs for the same amino acid is termed **multiplicity.**

A characteristic feature of tRNA molecules is that they contain a rather high percentage (~10%) of modified bases in addition to the common bases A, G, U, and C. Over forty different modified bases have been discovered in tRNAs, most of which are methylated forms of the common bases. Because of this feature, some regions of the cloverleaf diagram have been named for the modified bases that occur in them (Figure 27.15). For example, the T loop is so named because it includes thymine (T_r), and the D loop usually includes dihydrouracil (U_h). Other regions of the cloverleaf are the variable loop, which in different tRNAs has different numbers of nucleotides (ranging ing from 4 to 21), and the acceptor stem, which accepts the amino acid specific to that particular tRNA. All tRNA molecules have the same terminal sequence —C—C—A at the 3′ end of the acceptor stem. In addition, the anticodon loop contains a unique sequence of three nucleotides that is different in the tRNAs for different amino acids. This triplet is called the **anticodon** (Section 27.5-a). Finally notice that the stem regions contain intrachain hydrogen bonding between base pairs. The double-helical configuration occurs presumably by looping of the single-stranded tRNA chain (Figure 27.15-b).

In 1965 Robert Holley and his co-workers at Cornell University deduced the entire base sequence of an alanine tRNA isolated from bakers' yeast. The base sequences of one hundred other tRNA molecules have since been elucidated. Although these tRNAs have quite different base sequences, they all exist in the same cloverleaf conformation. Moreover, they all have a constant number of base pairs in the stem regions: seven in the acceptor stem, five in the T stem, five in the anticodon stem, and three or four in the D stem. These features are maintained in tRNA molecules from plants, animals, bacteria, and viruses. In the discussion of polypeptide synthesis (Section 27.5-a), we shall use the figure  to represent the tRNA molecule.

c. Ribosomal RNA (rRNA)

Ribosomal RNA (rRNA) comprises 80% of the total cellular complement of ribonucleic acid. The **ribosome** is *a cellular substructure that serves as the site for protein synthesis*. Its composition is about 60% rRNA and 40% protein. The ribonucleic acids and the proteins are bonded together by a large number of noncovalent forces such as hydrogen bonds and hydrophobic interactions. Structurally, a ribosome is composed of two spherical

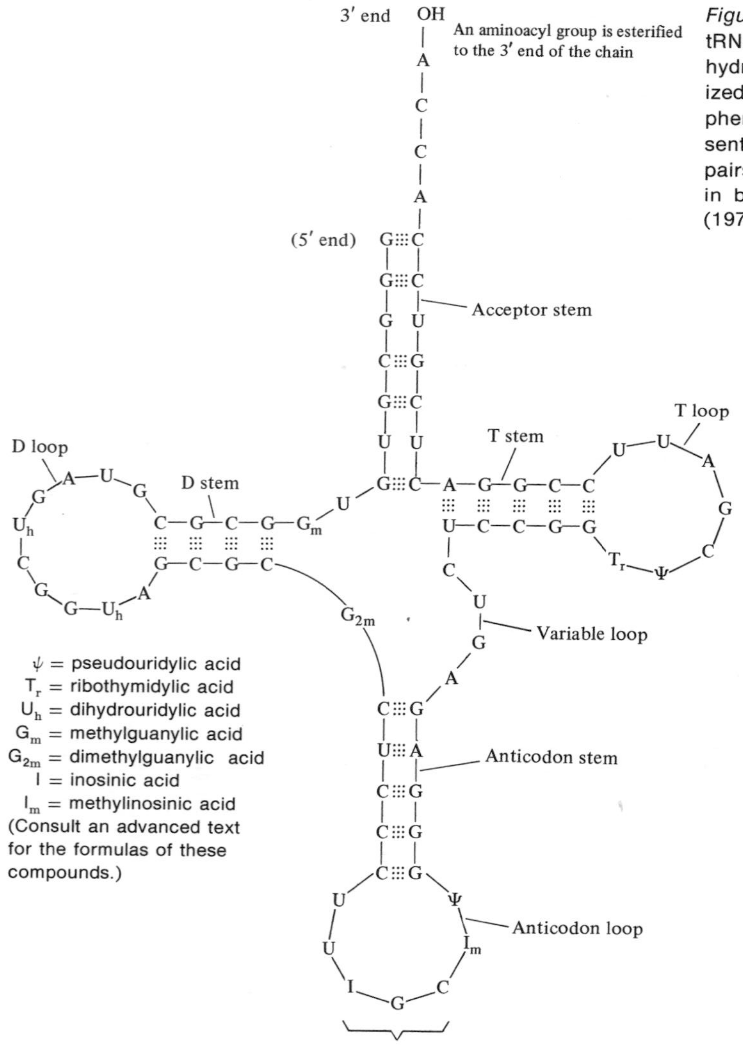
3′ end
OH
An aminoacyl group is esterified to the 3′ end of the chain
(5′ end)
Acceptor stem
T stem
T loop
D loop
D stem
Variable loop
Anticodon stem
Anticodon loop
Anticodon

ψ = pseudouridylic acid
T_r = ribothymidylic acid
U_h = dihydrouridylic acid
G_m = methylguanylic acid
G_{2m} = dimethylguanylic acid
I = inosinic acid
I_m = methylinosinic acid
(Consult an advanced text for the formulas of these compounds.)

(a)

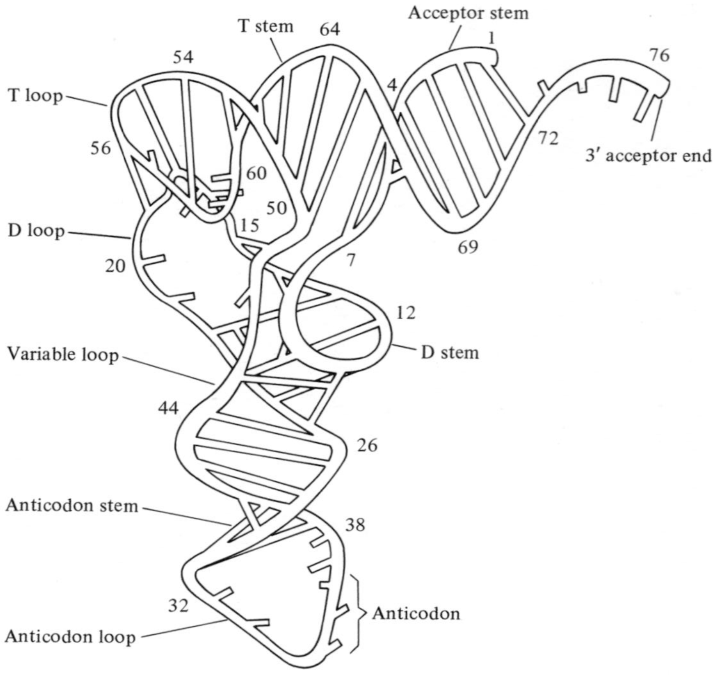
Acceptor stem
T stem
T loop
D loop
Variable loop
Anticodon stem
Anticodon loop
D stem
Anticodon
3′ acceptor end
1
4
7
12
15
20
26
32
38
44
50
54
56
60
64
69
72
76

(b)

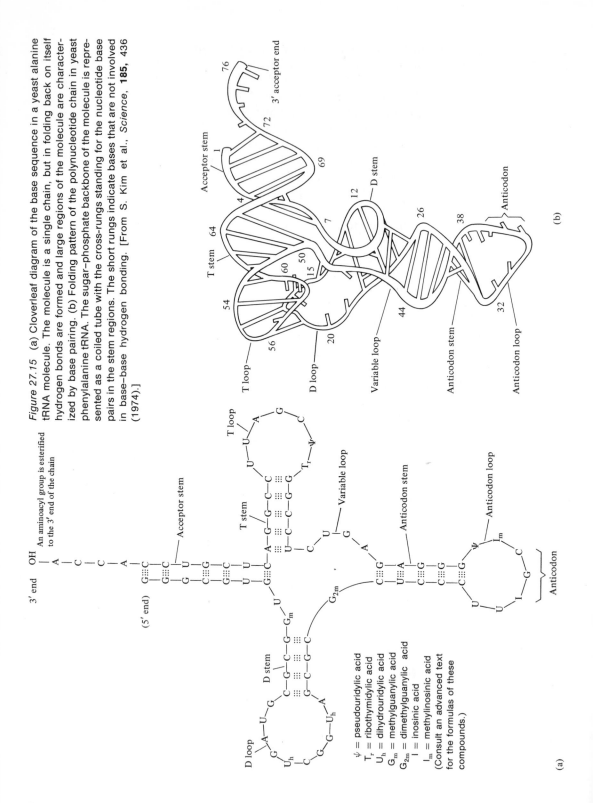

Figure 27.15 (a) Cloverleaf diagram of the base sequence in a yeast alanine tRNA molecule. The molecule is a single chain, but in folding back on itself hydrogen bonds are formed and large regions of the molecule are characterized by base pairing. (b) Folding pattern of the polynucleotide chain in yeast phenylalanine tRNA. The sugar–phosphate backbone of the molecule is represented as a coiled tube with the cross-rungs standing for the nucleotide base pairs in the stem regions. The short rungs indicate bases that are not involved in base–base hydrogen bonding. [From S. Kim et al., *Science*, **185,** 436 (1974).]

particles of unequal size. The smaller of them has a distinct affinity for mRNA; the larger has an attraction for tRNA. In terms of cellular structure, ribosomes are extremely small particles visible only with the aid of an electron microscope. More often than not, they are seen as clusters known as **polyribosomes,** or **polysomes.** When ribosomes occur in such aggregates, they are held together by strands of mRNA. On the average, five to eight ribosomes are simultaneously synthesizing the same polypeptide from the information in one mRNA strand, (large proteins require long strands of mRNA, and as many as 100 individual ribosomes may be attached). The time required for the synthesis of an average-size polypeptide (~300 amino acids) is about 15 seconds in a bacterial cell and two or three minutes in a mammalian cell.

Apart from its occurrence as a structural component of the ribosomes, the biological role of rRNA is yet unclear. Students of biochemistry often tend to take the chemical reactions that they study *out* of the cell, thereby taking them out of their proper frame of reference. The cellular substructures and organelles at which the various biochemical processes occur are often actively involved in these same processes. Such is definitely the case with the ribosomes and protein synthesis. We shall see that it is advantageous to regard the ribosome as a type of biological surface catalyst—an enzyme—rather than as some mysterious nonchemical cellular component (or passive workbench). A great deal remains to be discovered about the functions of ribosomal proteins and of ribosomal RNA.

27.5 The Genetic Code

Throughout the preceding sections, the role of the nucleic acids in *coding* information has been stressed. We have mentioned that the sequence of bases in the DNA molecule serves to direct protein synthesis. Now let us examine the nature of the code embodied in that sequence.

It is known that the amino acid sequence of any particular cellular protein is vital to the proper biochemical functioning of that protein in the overall scheme of cellular metabolism. For example, the conjugated protein hemoglobin contains some 280 amino acids in each of its two identical dimeric protein chains. Recall (Section 25.4) that the substitution of just one incorrect amino acid in 280 can result in the formation of nonfunctional hemoglobin molecules, as evidenced by the disease sickle cell anemia. Furthermore, there is a variety of other diseases (galactosemia, phenylketonuria—see Section 27.6) that appear to be related to the synthesis of nonfunctional, defective protein molecules, primarily enzymes. In many cases, there is just one incorrect amino acid in the molecule.

From such data, we may conclude that the primary structure, or amino acid sequence, of the protein molecule is of vital importance to life and that the cell must contain an intricate set of instructions wherein the proper sequence of amino acids for *all* its proteins is genetically contained. We now know that the code involves the sequence of bases in the DNA. *The segment of a DNA molecule that codes for the biosynthesis of one complete polypeptide chain* is called a **gene.** (If a given protein contains two or more polypeptide chains, each chain is coded by a different gene.)

How can a molecule with just four different monomeric units specify the sequence of the twenty different amino acids that occur in proteins? If each different nucleotide coded for one different amino acid, then obviously the nucleic acids could code for only four of the twenty amino acids. Suppose we consider the nucleotides in groups of two. There are 4^2 or sixteen different combinations of pairs of the four distinct nucleotides. Such a code is more extensive, but is still inadequate. If, however, the nucleotides are considered in groups of three, there are 4^3 or *sixty-four different combinations.* Here we have a code that is extensive enough to govern the primary structure of the protein molecule because it contains more than enough coding units to designate all twenty amino acids. Now we shall see how this code directs protein synthesis.

a. Building a Polypeptide

If the sequence of bases along the DNA strand determines the sequence of amino acids along the polypeptide chain, then the information contained in the DNA must be conveyed from the nucleus to the site of protein synthesis. This is accomplished by the orderly interactions of the nucleic acids with over 100 different enzymes. Recall that mRNA is made from a DNA template and so contains a base sequence that is complementary to that of the DNA upon whose surface it was synthesized. Once it is formed, the mRNA diffuses to the ribosomes carrying with it the genetic instructions. *Each group of three bases along the mRNA strand now specifies a particular amino acid, and the sequence of these triplet groups dictates the sequence of the amino acids in the protein.* Because the code involves three bases per coding unit, it is referred to as a **triplet code.** The coding unit is called a **codon.**

Now the cell faces the problem of lining up the amino acids according to the sequence called for by the mRNA, and of joining them together by means of peptide linkages. Because this process involves the transfer of the information encoded in the mRNA to the ultimate structure of the protein molecule, it is often referred to as **translation.**

Before the amino acids may be incorporated into a polypeptide chain, they must first be activated. Activation occurs prior to the reaction of the amino acid with its particular tRNA carrier molecule. The amino acid combines with a molecule of ATP, yielding a compound known as aminoacyl adenylate. Each of the twenty amino acids has its own specific activation enzyme system called *aminoacyl–tRNA synthetase.*

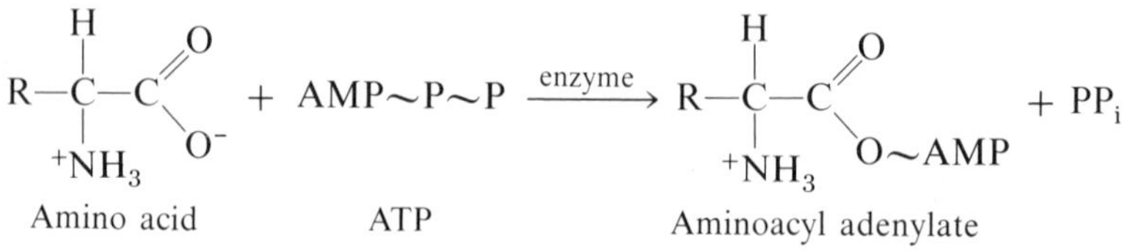
H
O
R—C—C
O⁻
⁺NH₃
+ AMP∼P∼P
enzyme
H
O
R—C—C
O∼AMP
⁺NH₃
+ PPᵢ
Amino acid
ATP
Aminoacyl adenylate

The aminoacyl adenylate remains on the surface of the enzyme and then undergoes reaction with the proper tRNA molecule to form the corresponding aminoacyl–tRNA complex. Both the enzymes (aminoacyl–tRNA synthetases) and the tRNAs are each highly specific for a particular amino acid.

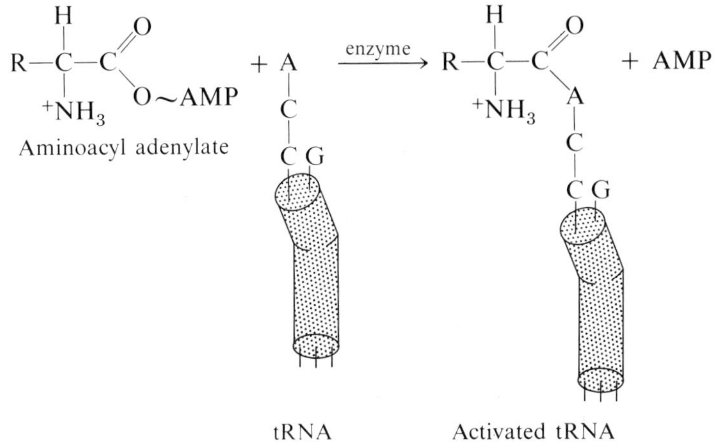

Aminoacyl adenylate
enzyme
+ AMP
tRNA
Activated tRNA

The aminoacyl–tRNA complex is often referred to as an *activated* tRNA molecule. Binding of the amino acid to its tRNA occurs at the 3′ position of the terminal adenosine nucleoside. Now that the amino acid molecules have been activated and have undergone reaction with their tRNA carriers, protein synthesis may take place. These events are summarized in Figure 27.16.

Figure 27.17 depicts a schematic stepwise representation of this all-important process. When a certain portion of the mRNA strand has been

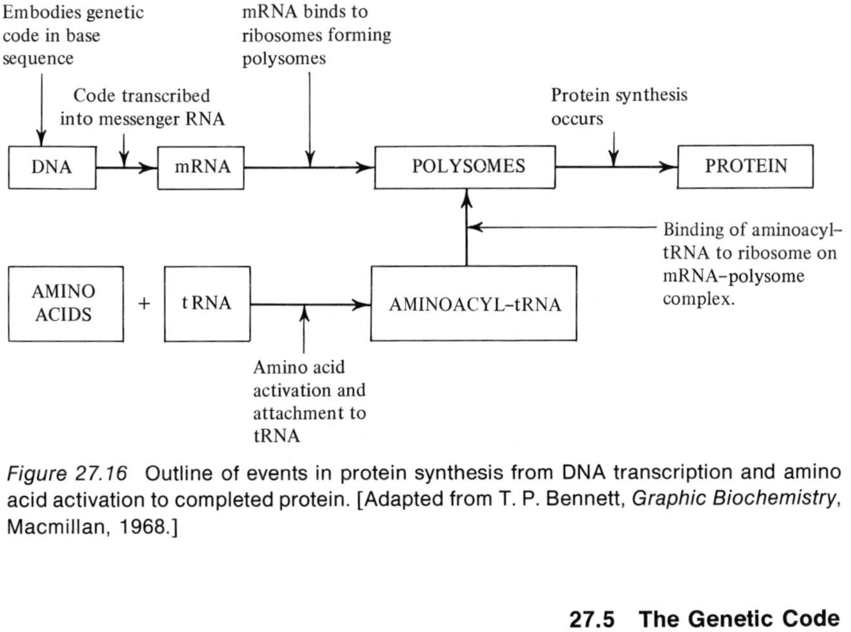

Embodies genetic code in base sequence
mRNA binds to ribosomes forming polysomes
Code transcribed into messenger RNA
Protein synthesis occurs
DNA
mRNA
POLYSOMES
PROTEIN
Binding of aminoacyl–tRNA to ribosome on mRNA–polysome complex.
AMINO ACIDS
+
tRNA
AMINOACYL–tRNA
Amino acid activation and attachment to tRNA

Figure 27.16 Outline of events in protein synthesis from DNA transcription and amino acid activation to completed protein. [Adapted from T. P. Bennett, *Graphic Biochemistry*, Macmillan, 1968.]

(a) Protein synthesis is already in progress at this ribosome. The growing polypeptide chain is bound to the peptidyl (**P**) site. At this point the aminoacyl (**A**) site is vacant. The codon UUU is lined up above the **A** site. An activated tRNA molecule whose anticodon is AAA approaches the ribosome.

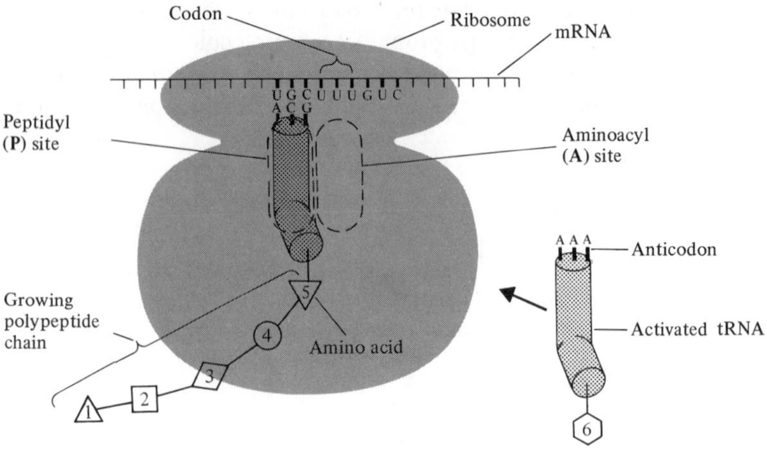
Codon
Ribosome
mRNA
U G C U U U G U C
A C G
Peptidyl (P) site
Aminoacyl (A) site
A A A
Anticodon
Activated tRNA
Growing polypeptide chain
Amino acid
1
2
3
4
5
6

(b) The activated tRNA molecule has become bound to the ribosome at the **A** site. It is also bound to the mRNA molecule by means of base pairing between codon and anticodon. Amino acid 6 is about to be incorporated into the polypeptide chain. The peptide linkage will be formed between the carboxyl group of amino acid 5 and the amino group of amino acid 6.

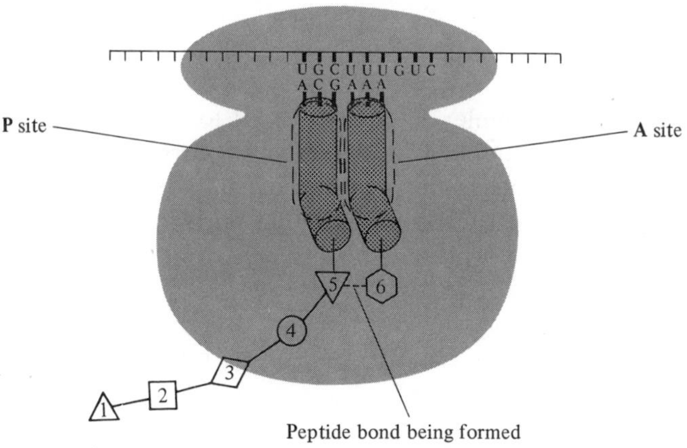
U G C U U U G U C
A C G A A A
P site
A site
1
2
3
4
5
6
Peptide bond being formed

(c) The peptide linkage has been formed, and the growing polypeptide chain is now attached to the **A** site. The tRNA molecule has dissociated from the **P** site and is about to move into the cytoplasm to pick up another amino acid.

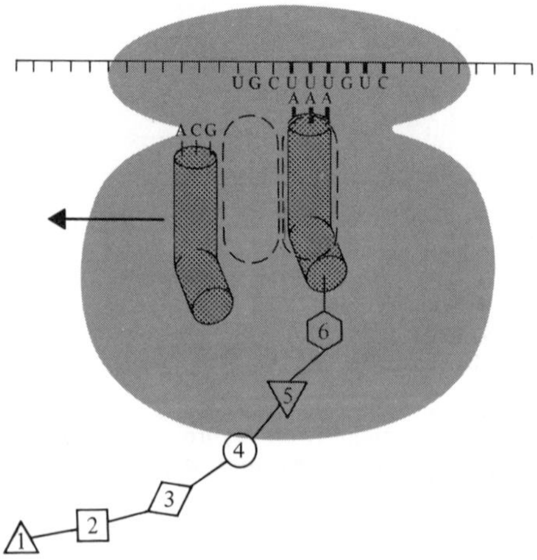
U G C U U U G U C
A A A
A C G
1
2
3
4
5
6

Figure 27.17 The elongation steps in protein synthesis.

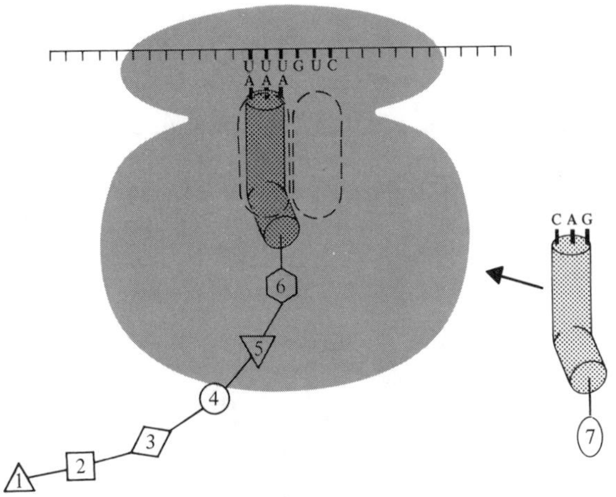
U U U G U C
A A A
C A G
6
5
4
3
2
1
7

(d) The ribosome moves from left to right along the mRNA strand. The polypeptide chain, along with the tRNA molecule to which it is bound, is simultaneously shifted from the **A** site to the **P** site. This brings the next codon, GUC, into line over the **A** site. Notice that an activated tRNA molecule (containing the next amino acid to be incorporated into the chain) is moving into position on the surface of the ribosome. Its anticodon is CAG.

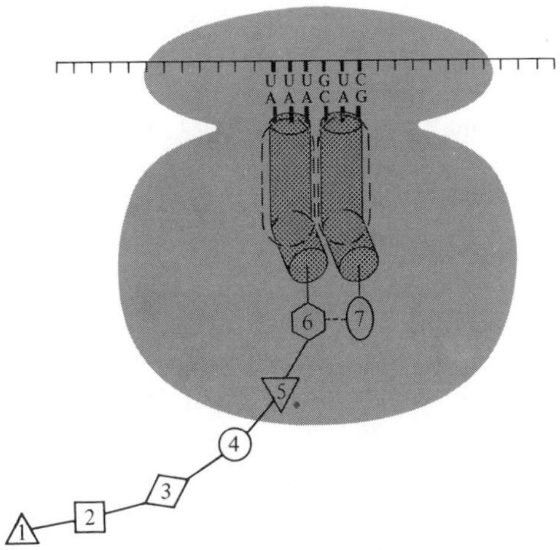
U U U G U C
A A A C A G
6
7
5
4
3
2
1

(e) The activated tRNA molecule carrying amino acid 7 is now in place on the ribosome. The peptide linkage between the carboxyl group of amino acid 6 and the amino group of amino acid 7 is about to be made.

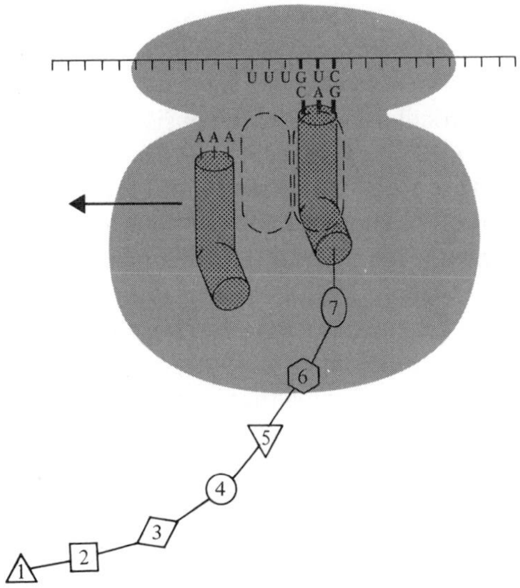
U U U G U C
C A G
A A A
7
6
5
4
3
2
1

(f) The peptide linkage has been formed, and the growing chain is attached, through a tRNA molecule, to the **A** site. The polypeptide chain is now seven amino acid units in length. The ribosome will move again, and the tRNA–polypeptide complex will be in position at the **P** site. This process will continue until the polypeptide chain is completed.

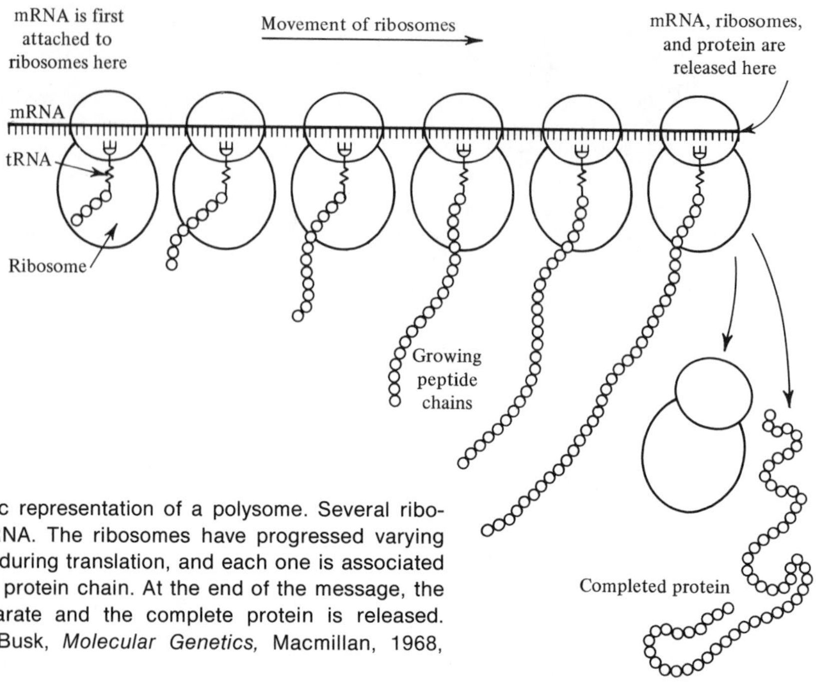
mRNA is first attached to ribosomes here
Movement of ribosomes
mRNA, ribosomes, and protein are released here
mRNA
tRNA
Ribosome
Growing peptide chains
Completed protein

Figure 27.18 Diagrammatic representation of a polysome. Several ribosomes are attached to mRNA. The ribosomes have progressed varying distances along the mRNA during translation, and each one is associated with a progressively longer protein chain. At the end of the message, the mRNA and ribosome separate and the complete protein is released. [Adapted from A. Gib DeBusk, *Molecular Genetics,* Macmillan, 1968, p. 34.]

read by a given ribosome, another ribosome may attach itself to the same strand and begin to read it simultaneously. Thus, in cells active in protein synthesis, we find clusters of ribosomes connected by a single strand of mRNA (Figure 27.18).

b. Cracking the Code

As we have stated, the sequence of bases on the mRNA directs the precise sequence of the amino acids for each protein. We have indicated that the codon, the unit that codes for a particular amino acid, consists of groups of three adjacent nucleotides on the mRNA. There are sixty-four possible triplet codons. Early experimenters were faced with the task of determining which codon (or perhaps codons) stood for each of the twenty amino acids. The cracking of the genetic code was the joint accomplishment of several well-known geneticists, notably H. Khorana, M. Nirenberg, P. Leder, and S. Ochoa (1961–64). A fairly extensive genetic dictionary has been compiled and is given in Table 27.4. Two important features of the code may be discerned from the table. First, the code is *degenerate;* that is several different codons may specify the same amino acid. Second, codons may be ambiguous; the specificity of one codon for one amino acid is not absolute. Further experimentation by Nirenberg has thrown much light on the nature of the genetic code. It now appears that

1. The code is universal—all animal, plant, and bacterial cells use the same codons to specify each amino acid.
2. The code is degenerate—in most cases more than one triplet will code for a given amino acid.
3. The first two bases of each codon are most significant; the third base often varies. This suggests that a change in the third base by a mutation may still permit the correct incorporation of a given amino acid into a protein.
4. The code is continuous and nonoverlapping—there are no special signals and adjacent codons do not overlap.
5. There are three codons that do not code for any amino acid. These codons had originally been called nonsense codons, but are now recognized as specific termination codons.
6. The codon AUG codes for methionine as well as being the initiation codon.

27.6 Genetic Diseases

We have seen that DNA directs the synthesis of proteins through the intermediary messenger RNA. On rare occasions, however, one of the bases in the DNA strand may be substituted, modified, or deleted. Such a change may lead to the insertion of the wrong amino acid(s) in a polypeptide chain.

Table 27.4 The Genetic Code

First Base	Second Base: *U*	*C*	*A*	*G*	Third Base
U	UUU Phe	UCU Ser	UAU Tyr	UGU Cys	*U*
	UUC Phe	UCC Ser	UAC Tyr	UGC Cys	*C*
	UUA Leu	UCA Ser	UAA Termination	UGA Termination	*A*
	UUG Leu	UCG Ser	UAG Termination	UGG Trp	*G*
C	CUU Leu	CCU Pro	CAU His	CGU Arg	*U*
	CUC Leu	CCC Pro	CAC His	CGC Arg	*C*
	CUA Leu	CCA Pro	CAA Gln	CGA Arg	*A*
	CUG Leu	CCG Pro	CAG Gln	CGG Arg	*G*
A	AUU Ile	ACU Thr	AAU Asn	AGU Ser	*U*
	AUC Ile	ACC Thr	AAC Asn	AGC Ser	*C*
	AUA Ile	ACA Thr	AAA Lys	AGA Arg	*A*
	AUG Met	ACG Thr	AAG Lys	AGG Arg	*G*
G	GUU Val	GCU Ala	GAU Asp	GGU Gly	*U*
	GUC Val	GCC Ala	GAC Asp	GGC Gly	*C*
	GUA Val	GCA Ala	GAA Glu	GGA Gly	*A*
	GUG Val	GCG Ala	GAG Glu	GGG Gly	*G*

If this change occurs at a crucial position, the resulting protein is unable to function (e.g., sickle cell hemoglobin). If the amino acid replacement occurs at a less important position, the protein's activity may be diminished or not affected at all. *Any chemical or physical change that alters the sequence of bases in the DNA molecule* is termed a **mutation.** Over 2000 diseases in humans result from mutations in genes; such diseases are called *inborn errors of metabolism* or *genetic diseases.* A partial listing of genetic diseases is presented in Table 27.5, and a few specific conditions are discussed below.

Phenylketonuria (PKU) results when the enzyme *phenylalanine hydroxylase* is absent. A person having PKU cannot convert phenylalanine to tyrosine, so the phenylalanine accumulates in the body, causes brain damage, and produces severe mental retardation. (About 1% of patients in mental institutions are phenylketonuric.) Some of the excess phenylalanine is transaminated (see Section 30.3-a) to phenylpyruvate.

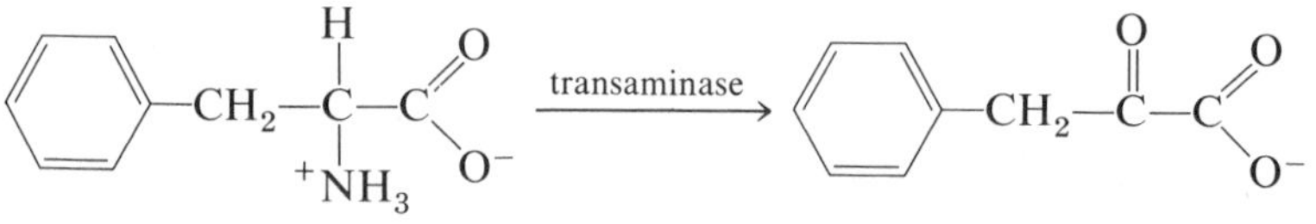
transaminase

Phenylalanine

Phenylpyruvate

The disease acquired its name from the high levels of this phenylketone in the urine. PKU may be diagnosed from a sample of blood or urine. If the condition is detected early enough (i.e., very soon after birth), mental retardation can be prevented by giving the patient a diet containing little or no phenylalanine. Because phenylalanine is so prevalent in natural foods, the low phenylalanine diet is composed of a synthetic protein substitute plus very small measured amounts of natural foods.

Galactosemia results from the lack of the enzyme that catalyzes the formation of glucose from galactose. The blood galactose level is markedly elevated, and galactose is found in the urine. This disease may result in damage to the central nervous system, mental retardation, and even death. The disease may be controlled by the administration of a diet free of galactose (i.e., a lactose-free diet).

In *Niemann–Pick disease,* a disease of infancy or early childhood, sphingomyelins accumulate in the brain, liver, and spleen because the enzyme *sphingomyelinase* is lacking. The accumulation of the sphingomyelins causes mental retardation and early death.

In *Gaucher's disease* glycolipids accumulate in the brain and cause severe mental retardation and death. Juvenile and adult forms of this disease are characterized by enlarged spleen and kidneys, hemorrhaging, mild anemia, and fragile bones. This disease is caused by the lack of a specific enzyme called *β-glucosidase.*

In the absence of another particular enzyme, *hexosaminidase A,* glycolipids accumulate in the tissues of the brain and eyes. This effect, called *Tay–Sachs disease,* is usually fatal to infants before they reach age two.

Table 27.5 A Partial Listing of Genetic Disorders in Man and the Malfunctional or Deficient Protein or Enzyme

Disease	*Protein or Enzyme*
Acatalasia	Catalase (red blood cells)
Albinism	Tyrosinase
Alkaptonuria	Homogentisic acid oxidase
Cystathionuria	Cystathionase
Fabry's disease	Ganglioside-hydrolyzing enzyme
Galactosemia	Galactose 1-phosphate uridyl transferase
Gaucher's disease	Glucocerebroside-hydrolyzing enzyme
Glycogen storage disease	Different types:
	α-Amylase
	Debranching enzyme
	Glucose 1-phosphatase
	Liver phosphorylase
	Muscle phosphofructokinase
	Muscle phosphorylase
Goiter	Iodotyrosine dehalogenase
Gout and Lesch–Nyhan syndrome	Hypoxanthine–guanine phosphoribosyl transferase
Hemolytic anemias	Different types:
	Glucose 6-phosphate dehydrogenase
	Glutathione reductase
	Phosphoglucoisomerase
	Pyruvate kinase
	Triose phosphate isomerase
Hemophilia	Clotting protein (antihemophilic factor) in blood
Histidinemia	Histidase
Homocystinuria	Cystathionine synthetase
Hyperammonemia	Ornithine transcarbamylase
Hypophosphatasia	Alkaline phosphatase
Isovalericacidemia	Isovaleryl SCoA dehydrogenase
Maple syrup urine disease	α-Keto acid decarboxylase
McArdle's syndrome	Muscle phosphorylase
Metachromatic leukodystrophy	Sphingolipid sulfatase
Methemoglobinemia	NADPH–methemoglobin reductase and NADH–methemoglobin reductase
Niemann–Pick disease	Sphingomyelin-hydrolyzing enzyme
Phenylketonuria	Phenylalanine hydroxylase
Pulmonary emphysema	Alpha globulin of blood
Sickle cell anemia	Hemoglobin
Tay–Sachs disease	Ganglioside-degrading enzyme
Tyrosinemia	Hydroxyphenylpyruvate oxidase
Von Gierke's disease	Glucose 6-phosphatase
Wilson's disease	Ceruloplasmin (blood protein)

From R. C. Bohinski, *Modern Concepts in Biochemistry,* Allyn and Bacon, Inc., Boston, 1973. Reprinted by permission.

Tay–Sachs disease can be diagnosed by assaying the amniotic fluid for the enzyme. The absence of *hexosaminidase A* in the amniotic fluid allows for a recommendation for a therapeutic abortion, since the disease is incurable.

27.7 The Nature of Viruses

A discussion of viruses seems particularly appropriate here, since viruses are composed almost entirely of proteins and nucleic acids. The field of virology is a rapidly expanding one, and recent research efforts to establish connections between viruses and cancer have yielded much important and exciting information.

Viruses are a unique group of infectious agents composed of a tightly packed central core of nucleic acids that is enclosed in one or more protein coats (see Figure 25.16). They are divided into two main classes on the basis of the nucleic acid content of the central core. Viruses contain either DNA or RNA *but never both.* (Recall that the cells of higher organisms, from bacteria to man, contain both kinds of nucleic acids.) Viruses differ from one another in their size[3] and shape. The influenza virus, for example, is about ten times bigger than polio virus. Viruses may be spherical, rod-shaped, or shaped like threads.

Characteristic of viruses is their infectivity. They are able to enter into the cells of specific hosts and there replicate or remain apparently latent. In order for viruses to replicate they must enter the cells of another organism, for they do not have the necessary biological machinery to replicate on their own. Outside the host cell a virus is an inert chemical complex; within the host cell it can use the cell's mechanisms for nucleic acid replication and protein synthesis to reproduce its own kind. The virus is therefore said to be an *obligatory intracellular parasite.* Viruses have a limited host range and they may be subclassed accordingly into animal, plant, and bacterial viruses. Within the subclasses, moreover, viruses are able to infect only certain species of organisms; for instance, a virus that can cause disease in a certain species of rodent may not be able to grow at all in a different but related species.

In order to understand the current ideas on how viruses may be responsible for *tumorigenesis,* it is necessary to know something of the nature of events that ensue when a virus infects a living cell.

a. Infective Processes

1. Productive Infection. Productive infection is said to occur when the virus utilizes the host's cellular apparatus for purposes of making more copies of itself. Somehow the virus is able to interfere with normal host metabolism, and to direct the synthesis of multiple copies of its own nucleic acids. At about the same time, the host's protein synthesis may halt as the synthetic machinery is taken over for the synthesis of multiple copies of virus protein coat. When the two major viral components are synthesized, the coats are assembled around the nucleic acid cores, and the newly formed virus particles are released from the cell either by budding or when the cell bursts. In either case, the host cell eventually dies.[4] The virus particles so released

[3]The size of the tobacco mosaic virus is approximately 60/5,000,000 of an inch by 3/5,000,000. Most viruses are visible only under the electron microscope.

[4]In certain instances, the host cell may continue to multiply for many generations, all the while releasing mature virus particles. The progeny of such cells are almost always infected with the virus and so they too release mature particles. The phenomenon is known as *endosymbiotic infection.*

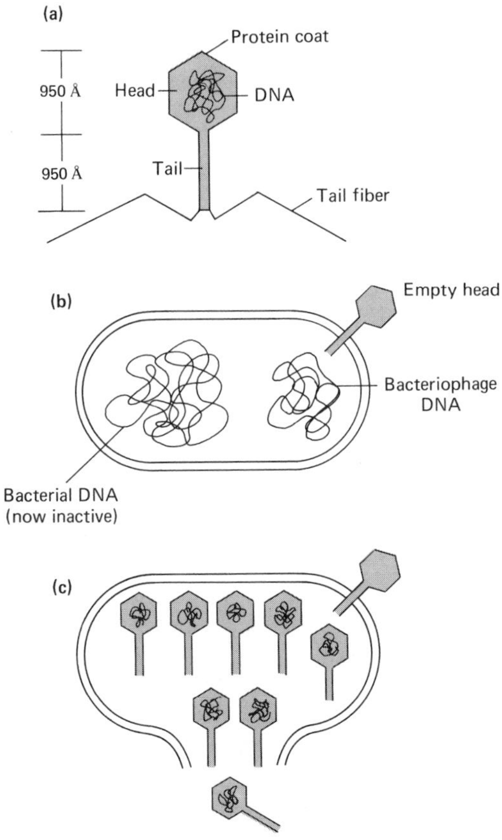
(a)
Protein coat
950 Å
Head
DNA
950 Å
Tail
Tail fiber
(b)
Empty head
Bacteriophage DNA
Bacterial DNA (now inactive)
(c)

Figure 27.19 Life cycle of bacteriophage. (a) Drawing of T2 bacteriophage. The bacteriophage DNA is located in the head of the particle, surrounded and protected by the protein coat. The end of the tail contains lysozyme, which can hydrolyze the polysaccharide of the bacterial cell wall. (b) When the bacteriophage infects a bacterium, the phage DNA is injected into the bacterial cell. Synthesis of bacterial nucleic acids and protein stops, and the synthetic machinery of the bacterial cell begins to produce phage DNA and protein. (c) As soon as all the phage constituents are synthesized, they begin to form new phage particles. When the bacterial cell becomes filled with phage particles, lysis occurs and the progeny are released. A total of 150 to 300 new phage are produced in a single bacterial cell, largely from bacterial constituents. [From J. R. Bronk, *Chemical Biology*, Macmillan, 1973, p. 190.]

are free to infect other cells and so the process is repeated. Cell death and the production of new viruses account for the symptoms of familiar viral infections such as flu and mumps. Figure 27.19 depicts the life cycle of a bacteriophage (a virus that infects bacteria).

*2. **Cell Transformation.*** Certain viruses (among them are the tumor viruses) may not cause the death of the host cell upon infection, but rather will alter the properties of the infected cell in such a way that it no longer behaves as it used to in controlling cell division. This type of infection is observed both in vivo and in vitro, and from studies in vitro has come a body of information about the altered properties of these virally *transformed cells*. In order to appreciate the significance of the changes it is necessary first to know something of the properties of normal (nontransformed) cells in tissue culture (in vitro).

Normal cells grown in a culture dish will continue to divide until their borders come in contact with each other. They usually align themselves in an orderly fashion with respect to one another. This phenomenon is known as *contact inhibition*. Cells infected with transforming viruses lose this characteristic and under the influence of the viral nucleic acid they multiply without restraint. They are seen to climb over one another in tissue culture and to divide so chaotically that three-dimensional clumps of cells are formed rather than single sheets. The normal mechanism that regulates cell division is lost due to some unknown influence exerted by the presence of the viral nucleic acids. This is significant in view of the fact that unregulated and chaotic cell division is typical of cancer in living animals.

Other evidence of altered cellular function is present in colonies of transformed cells. These evidences, often called *viral footprints,* include the appearance of new molecules (virus specific antigens) on the surface of and within the transformed cells, as well as certain enzymes, nonenzymic proteins, and mRNA, which is strongly suspected to be of viral origin. These latter footprints are not detected in uninfected cells of the same tissue culture line. The finding of footprints is often crucial in establishing that viral infection has taken place because in many instances, once the infecting virus has entered the cell it seems to disappear and further efforts to detect it (for example, by electron microscopy) will prove futile.

Transformation of cells in tissue culture and induction of cancer in animals are closely related phenomena, as we shall see, since often the inoculation of animals with transformed cells produces tumors.

b. RNA-Containing Tumor Viruses

The RNA-containing viruses are divided into subgroups on the basis of the kind of cancer they produce and the type of animal they infect. Included in this group are viruses that produce leukemias in birds, in mice, and in cats, and viruses that produce tumors of the mammary gland in mice. The latter virus is also referred to as the milk factor or Bittner virus since its main mode of transmission is from mother to offspring through the milk. The virus may be detected in breast tissue long before the appearance of cancer, and it is interesting to note that hormonal factors may play a role in activating the virus so that it may produce the malignant change. In addition, newborn and suckling mice are susceptible to the virus when it is given orally, or when injected under the skin or into the abdominal cavity. A viral etiology for human breast cancer is strongly suspected at present, and research efforts geared toward isolating the agent are underway.

Several years ago Dr. Howard Temin (University of Wisconsin) and Dr. David Baltimore (MIT), working independently, discovered that the usual order for the transmittal of genetic information can be reversed by certain RNA-cancer viruses.[5] Working with cultures of *Rous sarcoma* virus, an RNA-type that causes cancer in chicken cells, Dr. Temin found evidence that these transformed cells contained new DNA whose sequence of bases were complementary to those of the viral RNA. This could only happen if the viral RNA served as the template for the synthesis of DNA. (Recall from Section 27.4-a that this hypothesis is the reverse of the established dogma.) The discovery by Temin and Baltimore that these viruses contained an enzyme (*RNA-dependent DNA polymerase*) capable of transcribing the RNA virus message into DNA was a major breakthrough in the area of virus research. Shortly thereafter, other workers detected this enzyme in a variety of RNA-transformed cancer viruses, including human leukemia.[6] The significance of this research is that if the RNA-dependent DNA polymerase is required for the cancer-producing effect of

[5] Dr. Baltimore, Dr. Temin, and Dr. Renato Dulbecco of the Imperial Cancer Research Laboratory in London were awarded the 1975 Nobel Prize in Medicine for their discoveries "concerning the interaction between tumor viruses and the genetic material of the cell."

[6] Recent work has indicated that certain plant cells may also contain the enzyme that copies RNA from an RNA template. When the RNA-containing tobacco mosaic virus infects a tobacco plant, an enzyme in the plant (which is not found in the virus) catalyzes the transcription of viral RNA.

the RNA virus, a substance that inhibits this enzyme might prevent leukemia. Various drugs are presently being tested for treatment of leukemia and other forms of cancer in rats and mice.

c. DNA-Containing Tumor Viruses

Two of the best known viruses of the DNA-containing group are the polyoma virus and Simian virus-40 (SV-40). Both have been widely investigated as agents of cancer induction. The polyoma virus occurs naturally in mice both in the laboratory and in the wild. When introduced experimentally into various animals it produces a wide range of different types of tumors.

The virus is able to cause productive infection in certain tissue cultures, while in others it produces cell transformation. This latter event is detected by the finding of viral footprints as previously described. Cells of certain tissue culture lines, when so transformed, are capable of producing tumors when inoculated into various host animals. In addition, cells taken from the tumors so induced can be transplanted into animals of the same strain and tumor growth is again observed. Finally, preparations of the virus itself are tumorigenic in many species when inoculated under the skin.

Simian virus-40 (SV-40) is structurally similar to polyoma virus. It was originally discovered in kidney cultures of a certain species of monkey that were being used to propagate polio virus in the preparation of vaccine. Apparently SV-40 multiplies silently in the kidneys of these species without producing overt disease. However, the virus is tumorigenic in many other animals as well as in particular species of monkeys. Since its discovery as a contaminant in the vaccine preparations, much investigation has taken place, partially due to the finding that many children were observed to excrete the virus in urine for as long as five weeks after ingestion of live polio vaccine. SV-40 causes tumors at the site of inoculation into newborn hamsters. Hamster cells transformed in tissue culture also produce tumors when inoculated into the newborn animals. Cells of the tumors thus induced can be serially passed in animals of the same strain and the tumors to which they give rise are identical to the tumors produced by the virus itself.

From the foregoing discussion it can be seen that advances in animal models of virally induced tumorigenesis have been great. No comparable advances have been made in the field of human disease thus far. One of the biggest reasons is also the most obvious—experimental tumor induction in humans is not possible. Another stumbling block has been the elimination of contaminating viruses from tissue cultures of human cell lines. In addition, as we have seen, viruses often "disappear" after inducing malignant change so it is not surprising that efforts to identify viruses from human tumor tissue have not yet been very rewarding. However, research is actively in progress and it appears that before long, definitive results may be forthcoming in the area of certain human cancers for which a viral etiology is strongly suspected. Some of the areas presently under aggressive investigation include certain of the leukemias, lymphomas (tumors of the body's lymphoid tissues), mammary gland cancer, and various bone cancers.

27.8 Recombinant DNA

One of the most controversial issues in recent years to arouse the scientific community as well as government officials and the general public is that of gene splicing or recombinant DNA. The term **recombinant DNA** refers to *DNA molecules that have been created by splicing segments of DNA from one* *organism into pieces of DNA from another organism.* Almost any organism (e.g., plants, animals, viruses, bacteria) can serve as the DNA donor, but, to date, only the bacterium *Escherichia coli* has served as the DNA acceptor. *E. coli* is particularly suited for this work because its genetic identity has been so thoroughly studied. Most of *E. coli*'s genes (about 3000–4000) are contained within a single, large, ringed chromosome. In addition, there are much smaller, closed loops of DNA called **plasmids,** which consist of only a few genes. It is these plasmids that serve as the vehicles for the splicing technique. Figure 27.20 illustrates the production of recombinant DNA; the steps are outlined below.

1. *E. coli* bacteria are placed in a detergent solution to break open the cells.
2. The plasmids are separated from the chromosomal DNA by differential centrifugation.
3. Restriction enzymes are utilized to cleave the plasmid at a specific short sequence in a way that creates overlapping (so-called "sticky") ends.
4. In vitro combination of the same restriction enzyme with DNA from another organism (foreign DNA) produces segments of DNA having ends that are complementary to those of the plasmid. (Since different restriction enzymes have different cleavage sites, a given strand of DNA can be separated into many different segments of varying length. It is therefore possible to insert almost any foreign gene(s) into *E. coli.*)
5. The enzyme *DNA ligase* seals the foreign DNA segment into place in the plasmid.
6. The resealed plasmid is placed in a solution of calcium chloride containing *E. coli.* When the solution is heated, the bacterium cell becomes permeable, allowing the plasmid to enter.
7. *E. coli* reproduce by dividing (and thus doubling the population) at a rate of about once every 20–30 minutes. The new *E. coli* that are now created have characteristics dictated by their own genes as well as those that have been transplanted from a different species.

Proponents of recombinant DNA research are excited about its great potential benefits. Recombinant techniques are an enormous aid to scientists in mapping and sequencing the position of genes and in determining the functions of different segments of an organism's DNA. (An understanding of gene function and gene regulation is a primary goal of scientists working to cure cancer.) It is conceivable that recombinant DNA could lead to cures for genetic diseases. If appropriate genes could be successfully inserted into *E. coli,* then the bacteria would become miniature pharmaceutical factories, producing great quantities of insulin, clotting factor for hemophiliacs, missing enzymes, hormones, vitamins, antibodies, etc. In addition, it may be

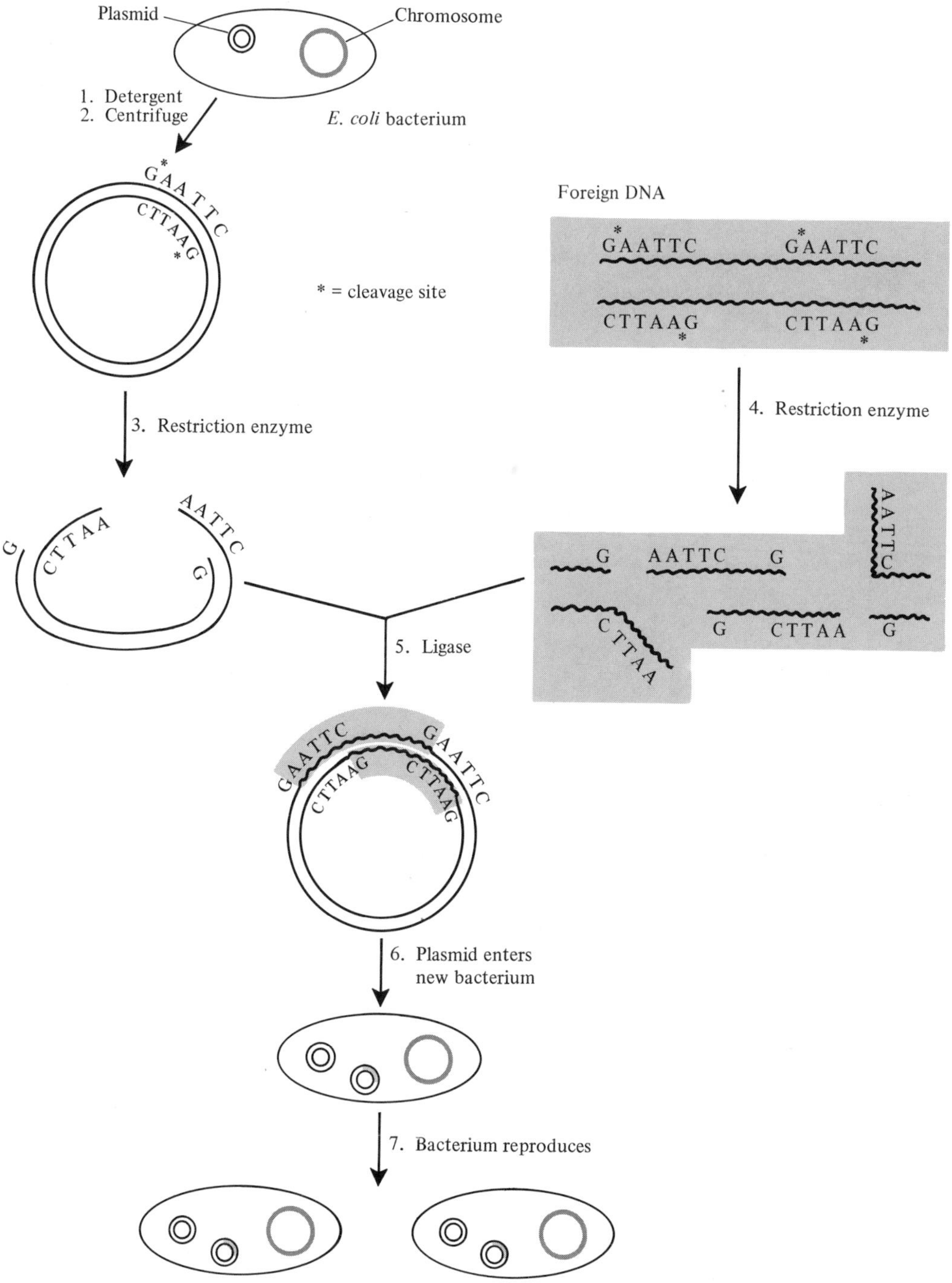
Plasmid
Chromosome
1. Detergent
2. Centrifuge
E. coli bacterium
GAATTC
CTTAAG
* = cleavage site
Foreign DNA
GAATTC GAATTC
CTTAAG CTTAAG
3. Restriction enzyme
4. Restriction enzyme
G CTTAA AATTC G
G AATTC G AATTC
CTTAA G CTTAA G
5. Ligase
GAATTC GAATTC
CTTAAG CTTAAG
6. Plasmid enters new bacterium
7. Bacterium reproduces

Figure 27.20 Recombinant DNA in *E. coli*.

possible to breed human intestinal bacteria that have the ability to digest cellulose, and to create new plants that can obtain their nitrogen directly from the air rather than from costly petroleum-based fertilizers.

Opponents of this research acknowledge the potential benefits, but they are concerned about the considerable risks. They fear that a man-made "Andromeda strain" could be unleashed, which would proliferate uncontrollably, causing mass disease and death.[7] Furthermore, they argue that this research can be misused for political and social purposes. These techniques might be exploited for genetically engineered control of human behavior. Finally, there are some who believe that we should not meddle with evolution by creating new forms of life different from any that exist on earth. Their slogan is "it's not nice to fool Mother Nature."

Concern about the potential biological hazards of recombinant DNA research first arose in 1973 among molecular biologists working in this area, and they voluntarily agreed to halt certain kinds of experiments. In 1976, the NIH issued research guidelines that apply to all federally supported work involving recombinant DNA. These guidelines establish a set of carefully controlled physical and biological containment conditions. The designations P1, P2, P3, and P4 represent increasing levels of physical containment, expressed in more elaborate laboratory facilities and procedures. P1 denotes standard laboratory precautions for experiments in the lowest-risk category (e.g., injecting harmless bacterial genes into *E. coli*). P4 denotes the ultrasecure laboratories for research of greatest risk (e.g., working with animal tumor viruses or primate cells). No escape of contaminated air, wastes, or untreated materials is permitted. As of 1977, there were only two P4 laboratories in existence in the United States.

Exercises

27.1 Give the names and formulas for all the compounds that can be obtained from the complete hydrolysis of DNA.

27.2 Name and draw the structural formulas of the pyrimidine and purine bases found most commonly in nucleic acids.

27.3 Draw structural formulas for the following minor pyrimidine and purine bases.
(a) 5-methylcytosine **(c)** 6-methyladenine
(b) 5-hydroxymethylcytosine **(d)** 2-methylguanine

27.4 Draw structural formulas for **(a)** deoxyuridine and **(b)** deoxyadenosine.

[7] A great breakthrough in the area of recombinant DNA research was made by geneticist Roy Curtiss. He was able to develop a strain of *E. coli* that could only live under laboratory conditions (i.e., they needed the nutrients supplied by the laboratory medium in order to survive). In 1976 the NIH certified this strain for use in recombinant DNA research.

27.5 Draw structural formulas for dinucleotides composed of **(a)** cytosine and adenine and **(b)** thymine and guanine. Show how these dinucleotides might link together by means of hydrogen bonds.

27.6 What is the relationship between each of the following?

(a) a purine base and a pyrimidine base

(b) ribose and deoxyribose

(c) a nucleoside and a nucleotide

(d) a nucleotide and a nucleic acid

(e) chromosomes, genes, and DNA

27.7 The possibilities of hydrogen bond formation are similar for uracil and thymine. Show that this is so.

27.8 Describe the Watson–Crick model for DNA.

27.9 Hereditary characteristics are encoded in the base sequence of DNA. Explain.

27.10 Briefly describe the processes involved in DNA replication.

27.11 How do DNA and RNA differ **(a)** structurally, **(b)** in base composition, and **(c)** in function?

27.12 Briefly outline the functional role of **(a)** DNA, **(b)** messenger RNA, **(c)** transfer RNA, and **(d)** ribosomal RNA in protein synthesis.

27.13 How many nucleotide units are present in a codon? Why is it generally assumed that a triplet of bases codes for each amino acid?

27.14 Describe the mechanism of protein synthesis in the cell.

27.15 If the sequence of bases in a section of mRNA is

AUGUACCACGGUACGCGGGUAUUGCUAGCCGAUGGGUAA

what would be the amino acid sequence in the peptide produced from this messenger RNA? (See Table 27.4.)

27.16 Suggest a possible nucleotide sequence in DNA that would code for the following polypeptide.

Met-Leu-Arg-Ser-Tyr-Glu-Gly-Phe-His-Lys-Thr

27.17 Recall (page 622) that sickle cell hemoglobin differs from normal hemoglobin as a result of the substitution of valine for glutamic acid as the sixth amino acid from the N-terminal end of the polypeptide chain. What alteration in the base sequence of the DNA could have caused this substitution? What would be the resulting base sequence in the mRNA? Can sickle cell anemia be considered a genetic disease? Why?

27.18 Viruses have been described as structures that are at the threshold of life. Comment on this description.

27.19 Differentiate the viral processes of productive infection and cell transformation.

28

Carbohydrate Metabolism

The term **metabolism** is used to describe the *various chemical processes by which food is utilized by a living organism to provide energy, growth substance, and cell repair.* Metabolic reactions include both the degradation of absorbed food substances into smaller molecules **(catabolism)** and the biosynthesis of complex molecules from simpler components **(anabolism).** Catabolic reactions are exothermic, and anabolic reactions are endothermic. Any chemical compound that is involved in a metabolic reaction is referred to as a **metabolite.**

In our earlier discussion of carbohydrates, it was mentioned that man's existence on this planet is directly dependent upon the plant kingdom. The animal world derives its foodstuffs, and hence its energy, from plant life. Plants obtain water, inorganic salts, and nitrogenous compounds from the soil, and carbon dioxide and oxygen from the atmosphere. With these raw materials, they are able to synthesize carbohydrates, lipids, and proteins. The energy required for these synthetic reactions is obtained from the sun; thus radiant energy (as sunlight) is the ultimate source of biological activity. *The overall process by which glucose is formed from carbon dioxide and water at the expense of solar energy* is termed **photosynthesis.** The general photosynthetic equation can be written as follows.

$$6\ CO_2 + 6\ H_2O + 686{,}000\ \text{cal} \xrightarrow{\text{sunlight}} \underset{\text{Glucose}}{C_6H_{12}O_6} + 6\ O_2$$

This synthesis is a distinguishing characteristic of green plants, and from the standpoint of man's survival it represents the most important series of reactions that occur on the surface of the earth. Notice that the formation of one mole of glucose requires the conservation of 686,000 calories. When compared to the energy requirement of most other endothermic chemical reactions, this is a very large sum. (It is roughly equivalent to the heat required to raise the temperature of 7 quarts of water from 0°C to 100°C.) On the solar scale, however, it is a drop in the bucket.[1]

Carbohydrates are the primary metabolites of the animal kingdom; recall that over half of the food that we ourselves consume is composed of carbohydrates. Carbohydrates serve as the chief fuel of biological systems, supplying living cells with usable energy. Like all fuels, carbohydrates must be burned or oxidized if energy is to be released. The combustion (oxidation) process ultimately results in the conversion of the carbohydrate into carbon dioxide and water; the energy stored in the carbohydrate molecule during photosynthesis is released in the reaction.[2]

$$C_6H_{12}O_6 + 6\ O_2 \longrightarrow 6\ CO_2 + 6\ H_2O + 686{,}000\ \text{calories}$$

This equation summarizes the biological combustion of foodstuff molecules by the cell (respiration). The term **respiration** is often used in a broader sense to include *all metabolic processes by which gaseous oxygen is used to oxidize organic matter to carbon dioxide, water, and energy.*

Both respiration and the combustion of the common fuels (wood, coal, gasoline) use oxygen from the air to break down complex organic substances to carbon dioxide and water. The energy released in the burning of wood, however, is manifested entirely in the form of heat, but excess heat energy is useless and even injurious to the living cell. Organisms conserve almost half of the 686,000 calories by a series of stepwise reactions that liberate small amounts of utilizable energy that is stored in the phosphate bond energy of ATP. The remainder of the energy is used to heat the body and thus to maintain proper body temperature.

28.1 Digestion and Absorption of Carbohydrates

Digestion may be defined as a *hydrolytic process whereby food molecules are broken down into simpler chemical units that can be absorbed by the body.* In man, digestion takes place in the digestive tract (Figure 28.1) and absorption occurs primarily in the small intestine.

Starch is the principal carbohydrate ingested by humans. Its digestion begins in the mouth where the enzyme *ptyalin* (an amylase) catalyzes its

[1] It has been estimated that on an average sunny day, each square centimeter of earth receives about 1 calorie of solar radiation every minute. Yet the energy intercepted by the earth is but a minute fraction $1/(5 \times 10^8)$ of the total energy given off by the sun, and only a small fraction of this intercepted energy (~1%) is utilized in photosynthesis.

[2] Plant and animal cells exist in a symbiotic cycle; each requires the products of the other for life. Although H_2O and O_2 are abundant in the atmosphere, CO_2 is only present to the extent of 0.02–0.05%. It has been estimated that if animals were removed from the earth, all the atmospheric CO_2 would be consumed in one to two years.

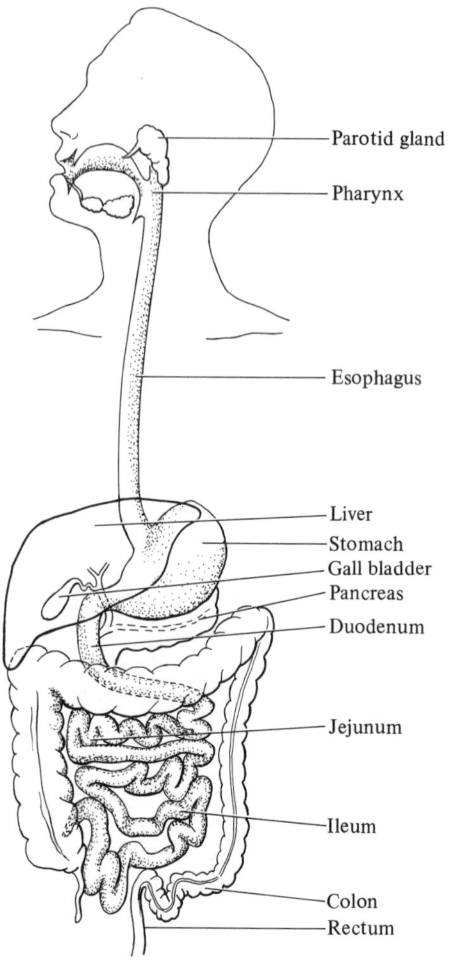
Parotid gland
Pharynx
Esophagus
Liver
Stomach
Gall bladder
Pancreas
Duodenum
Jejunum
Ileum
Colon
Rectum

Figure 28.1 The digestive tract.

hydrolysis to a mixture of dextrins. Ptyalin continues to function as food passes through the esophagus, but it is quickly inactivated when it comes in contact with the acidic environment of the stomach. Very little carbohydrate digestion occurs in the stomach; acid-catalyzed hydrolysis proceeds too slowly at body temperature to be effective. The primary site of carbohydrate digestion is the small intestine, where another amylase, *amylopsin,* converts the remaining starch molecules, along with the dextrins, to maltose. Maltose is then cleaved into two glucose molecules by the enzyme *maltase.* Disaccharides such as sucrose and lactose are not digested until they reach the small intestine where they are acted upon by the enzymes *sucrase* and *lactase.* Ultimately, the complete hydrolysis of disaccharides and polysaccharides produces three monosaccharide units—glucose, fructose, and galactose. These monosaccharides are then absorbed through the wall of the small intestine into the bloodstream. Absorption of most digested food takes place in the small intestine through the finger-like projections, called *villi,* that line the inner surface (Figure 28.2). Each villus is richly supplied with a fine network of blood vessels and a central lymph vessel. Absorption of monosaccharides occurs across the semipermeable membranous wall of each villus into the blood capillaries. However, the absorption does not occur by means of a simple process of osmosis or diffusion through an inert membrane. All cell membranes are selective in their action, a fact that implies that they play an active role in the absorption process. The monosaccharides differ in their rate of passage across the membrane, galactose > glucose > fructose, even though there is only a slight difference in their molecular size. It is believed that the passage of monosaccharides across the intestinal wall is an energy-requiring process, and so the term *active transport* is used to describe this type of absorption.

Following absorption, the monosaccharides are carried by the portal vein to the liver where galactose and fructose are enzymically converted to glucose. (Glucose is the only sugar that circulates in the blood, hence the name *blood sugar*.) The glucose may then pass into the general circulatory system to be transported to the tissues or it may be incorporated into glycogen to be stored in the liver as a reservoir for the maintenance of the normal level of blood glucose. The glucose in the tissues may be oxidized to CO_2 and H_2O, it may be converted to fat (see next chapter), or it may be

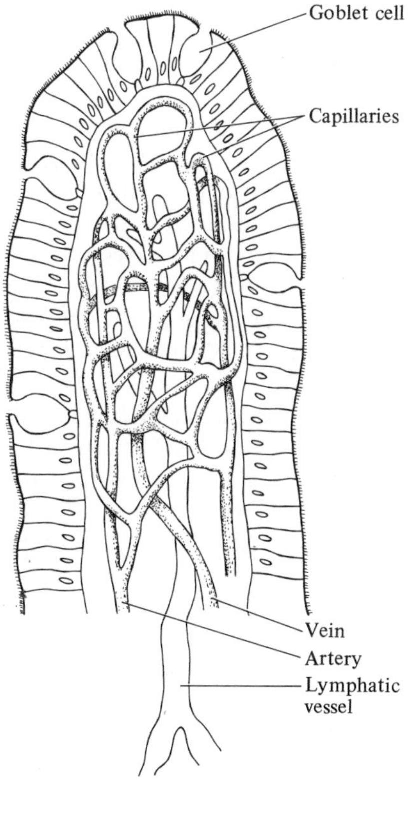
Goblet cell
Capillaries
Vein
Artery
Lymphatic vessel

Figure 28.2 Intestinal villus.

converted to muscle glycogen thus serving as a source of readily available energy for the muscle. The average person has a sufficient amount of stored glycogen (in liver and muscle cells) to supply energy for about 18 hours.

28.2 Blood Glucose

The concentation of glucose in the blood, referred to as the **blood-sugar level,** is vital to the proper functioning of the human body. Under normal circumstances the blood-sugar level remains remarkably constant at about 80 mg per 100 ml of blood. However, since individuals differ in their chemical makeup, a normal concentration of glucose may range from 70–100 mg per 100 ml. (This amounts to a total of about 5–6 grams of glucose, or one teaspoon, in the entire body.) Soon after eating a meal, the blood-sugar level may rise to 120 mg per 100 ml of blood, but this level returns to the normal level within two hours.

The condition resulting from a lower than normal concentration of glucose is called **hypoglycemia** and that resulting from a higher than normal concentration of glucose is **hyperglycemia.** In extreme cases of hyperglycemia, the *renal threshold* (160–170 mg per 100 ml) is reached, and excess glucose is excreted in the urine. Extreme hypoglycemia, which is usually due to the presence of excessive amounts of insulin, can cause unconsciousness, lowered blood pressure, and may result in death. Loss of consciousness is most likely due to the lack of glucose in the brain tissue, which has no capacity for glucose storage and thus is dependent upon this sugar in the blood for its energy requirements.

The most important contributing process in the maintenance of a constant glucose level is the synthesis and breakdown of glycogen in the liver. Several hormones are known to control the blood-sugar level. Insulin is secreted by the pancreas and functions to lower the blood-sugar level by making the cell membranes more permeable to the passage of glucose and by enhancing **glycogenesis,** the *formation of glycogen from glucose.* (The term *glyconeogenesis* is used to denote the formation of glycogen from noncarbohydrate sources.) If the pancreas does not secrete enough insulin, a diabetic condition may result.

The glucose tolerance test is the most important diagnostic test for diabetes mellitus. A patient's blood-sugar level is determined after a fast of several hours. Then a known amount (25–100 g) of glucose is administered, and the patient's blood is sampled for the level of circulating glucose. The initial rise in blood sugar falls rapidly in a normal individual. In a diabetic, the increase of the blood-sugar level is greater than the normal and will remain at the elevated levels for several hours (Figure 28.3).

All other hormones that affect sugar metabolism act to raise the concentration of sugar in the blood. Adrenalin (epinephrine) activates the enzyme *phosphorylase,* which is concerned with the breakdown of glycogen to glucose (*glycogenolysis*). During periods of emotional stress, such as anger or fright, adrenalin is secreted by the adrenal glands. It is transported by the

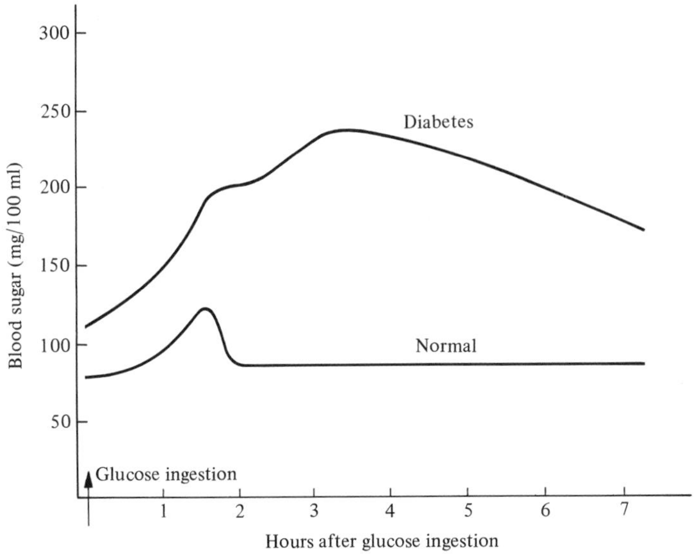
300
250
200
150
100
50
Blood sugar (mg/100 ml)
Diabetes
Normal
Glucose ingestion
1
2
3
4
5
6
7
Hours after glucose ingestion

Figure 28.3 Glucose tolerance test for a normal patient and for a diabetic.

blood to the muscles and liver where it promotes glycogen breakdown, thus increasing the amount of glucose in the blood. Glucagon, like insulin, is a polypeptide hormone that is produced by the pancreas. Unlike insulin, glucagon is secreted into the blood whenever the glucose level falls below the normal blood-sugar level. It functions by stimulating glycogenolysis in the liver (but not in muscle cells) to restore the blood glucose to its normal level. Glucagon thus counterbalances the effects of insulin.

$$\text{Blood glucose} \underset{\substack{\text{adrenalin,} \\ \text{glucagon} \\ \text{(glycogenolysis)}}}{\overset{\substack{\text{insulin} \\ \text{(glycogenesis)}}}{\rightleftharpoons}} \text{Glycogen}$$

28.3 Embden–Meyerhof Pathway/Anaerobic Glycolysis

To appreciate fully the role of carbohydrates in living systems, it is necessary to understand the biochemical nature of their *metabolic pathways.* A metabolic pathway is represented by a type of flow diagram that enables us to explain how an organism converts a certain reactant into a desired end-product. In the 1930s, the German biochemists G. Embden and O. Meyerhof elucidated the sequence of reactions by which glycogen and glucose are degraded in the absence of oxygen (*anaerobic* conditions) to pyruvic acid. It was discovered that the two apparently dissimilar processes of alcoholic fermentation in yeast and muscle contraction in animals proceed by the same pathway as far as pyruvic acid. In the absence of oxygen, muscle cells (and certain other animal tissues) convert pyruvic acid into lactic acid, but under similar conditions, the enzymes in yeast convert the pyruvic acid to ethyl alcohol and carbon dioxide. The former process is known as **glycolysis,** and the latter as **fermentation.**

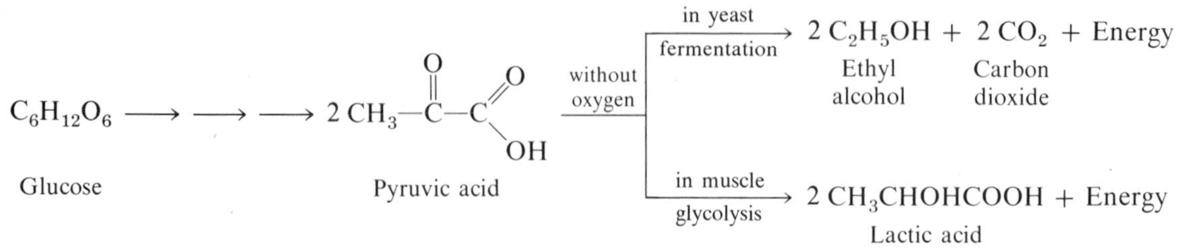

in yeast
fermentation
2 C2H5OH + 2 CO2 + Energy
Ethyl alcohol
Carbon dioxide
C6H12O6
Glucose
2 CH3–C(=O)–C(=O)OH
Pyruvic acid
without oxygen
in muscle
glycolysis
2 CH3CHOHCOOH + Energy
Lactic acid

It is important to notice that the conversion of glucose to pyruvic acid represents an oxidation reaction (that is, $C_6H_{12}O_6 \longrightarrow 2\,CH_3COCOOH + 4\,H^+ + 4\,e^-$), and yet oxygen is not required for the process to occur. We shall see that the outstanding characteristic of the Embden–Meyerhof pathway is the utilization of the coenzyme NAD^+ as the electron acceptor (oxidizing agent). Thus glycolysis and fermentation differ only in the eventual fate of the pyruvic acid, and hence in the means employed for the regeneration of the NAD^+.

A summary of the reactions and the metabolites involved in the Embden–Meyerhof pathway is given in Figure 28.4. This sequence of reactions is probably the best understood of all the metabolic pathways. Over a dozen different, specific enzyme molecules act in such a manner that the product of the first enzyme-catalyzed reaction becomes the substrate of the next. Since these enzymes are found in a soluble form in the cell, they were relatively easy to isolate and characterize. Each of the reactions given in Figure 28.4 has been assigned a number that corresponds to the discussion that follows. There is often the tendency to become lost in the complexity of this process and to lose an overall perspective. We shall be interested in seeing how the principles of organic chemistry apply to these biochemical reactions, but it must always be kept in mind that the central theme of metabolism is the extraction of chemical energy from foodstuff and not merely the degradation of these molecules into simpler substances. In the discussion that follows, the full names of the pertinent enzymes are given in the body of the text. Only the general names are written above or below the arrows of the biochemical equations.

Step i. Liver and muscle cells contain enzymes called *phosphorylases* that catalyze the phosphorylitic cleavage of the α-1,4-glucosidic bonds at the nonreducing end of the glycogen chain to produce glucose 1-phosphate. The reaction is analogous to a hydrolytic cleavage, except that here inorganic phosphate takes the place of a water molecule (i.e., a phosphorolysis). Although this reaction is reversible, a different enzyme and a different reaction sequence are employed for the biosynthesis of glycogen.

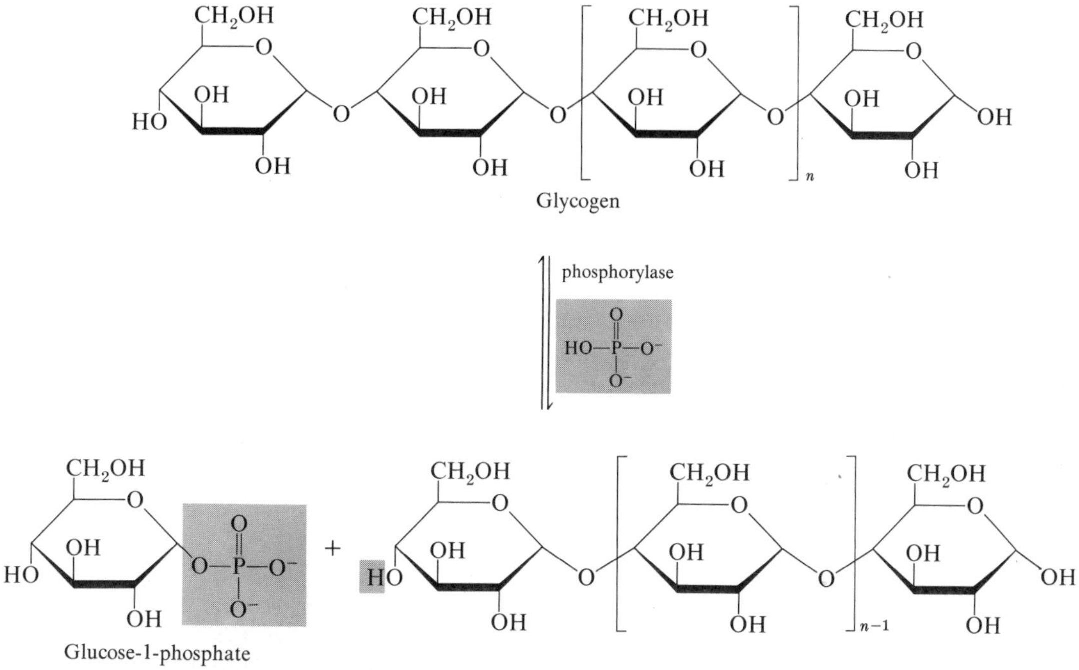

Glycogen
phosphorylase
Glucose-1-phosphate

Step ii. Glucose 1-phosphate and glucose 6-phosphate are readily interconvertible in the presence of Mg^{2+} ions and the enzyme *phosphoglucomutase.* For simplicity, the isomerization reaction may be thought of as an intramolecular transfer of the phosphate group from carbon-1 to carbon-6 and, although this process is reversible, glucose 6-phosphate formation is favored.

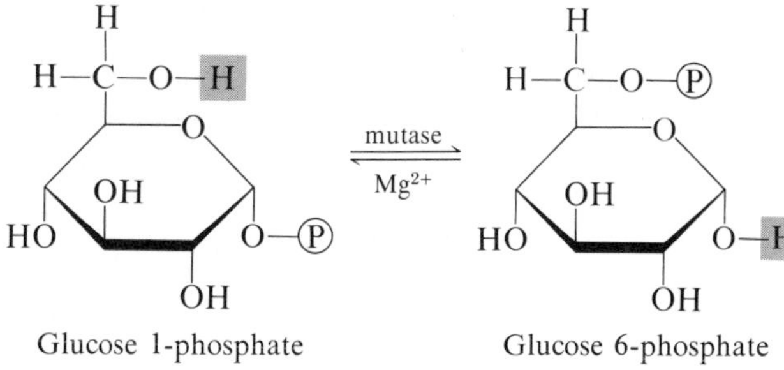

mutase
Mg2+
Glucose 1-phosphate
Glucose 6-phosphate

Step 1. The initial step in the utilization of glucose by yeast cells and by most animal cells is the phosphorylation to glucose 6-phosphate. The phosphate donor in this reaction is ATP, and the enzyme, which requires Mg^{2+} ions for its activity, is *hexokinase.* Hexokinase is not specific for glucose. It can also catalyze the transfer of phosphate from ATP to fructose, mannose, and 2-deoxyglucose.

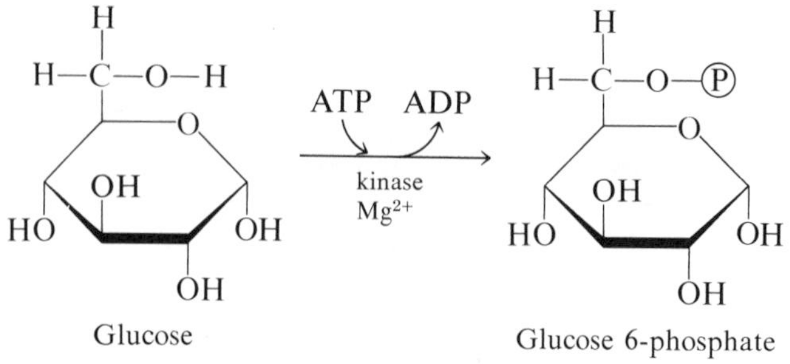
ATP
ADP
kinase
Mg2+
Glucose
Glucose 6-phosphate

The reaction is accompanied by an expenditure of energy, since a molecule of ATP is being utilized rather than synthesized. However, this step is necessary for the activation of the glucose molecule. (We shall see that a second molecule of ATP is expended in step 3. These two initiating molecules of ATP are recovered at a later stage of the pathway.) Reaction 1 is essentially irreversible in the cell; ATP is not formed to an appreciable extent by the reaction of ADP with a simple phosphate ester.

Step 2. Glucose 6-phosphate is isomerized to fructose 6-phosphate by the action of the enzyme *phosphoglucoisomerase*. This enzyme is highly specific for glucose 6-phosphate (e.g., another isomerase is required for the conversion of mannose 6-phosphate to fructose 6-phosphate).

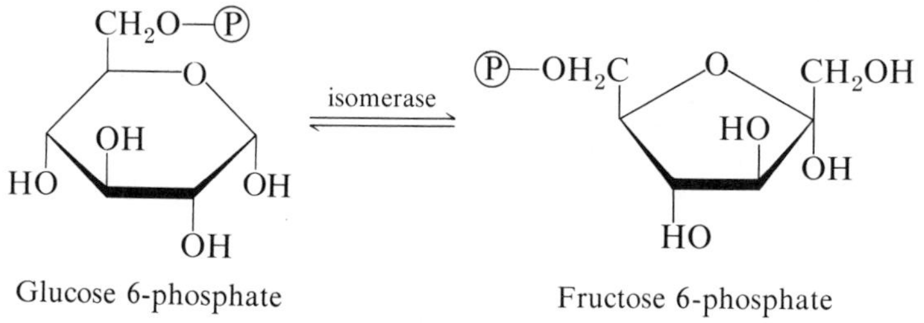
isomerase
Glucose 6-phosphate
Fructose 6-phosphate

A mechanism for this aldose–ketose transformation involves the formation of an enediol intermediate, and is best understood if the open-chain structures of the sugars are considered (Sections 18.5-b and 23.9-b).

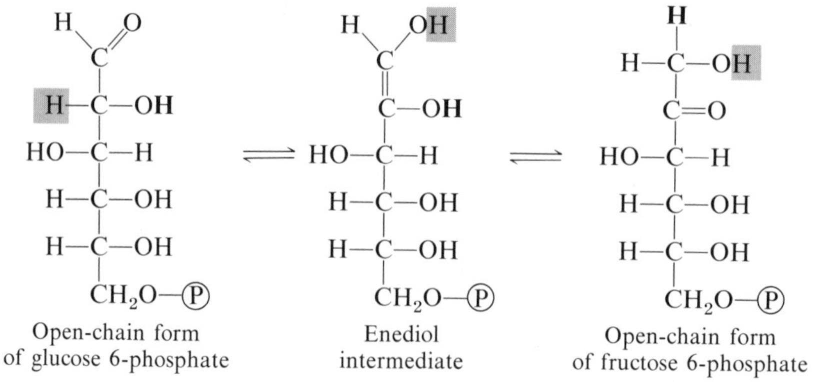
Open-chain form
of glucose 6-phosphate
Enediol
intermediate
Open-chain form
of fructose 6-phosphate

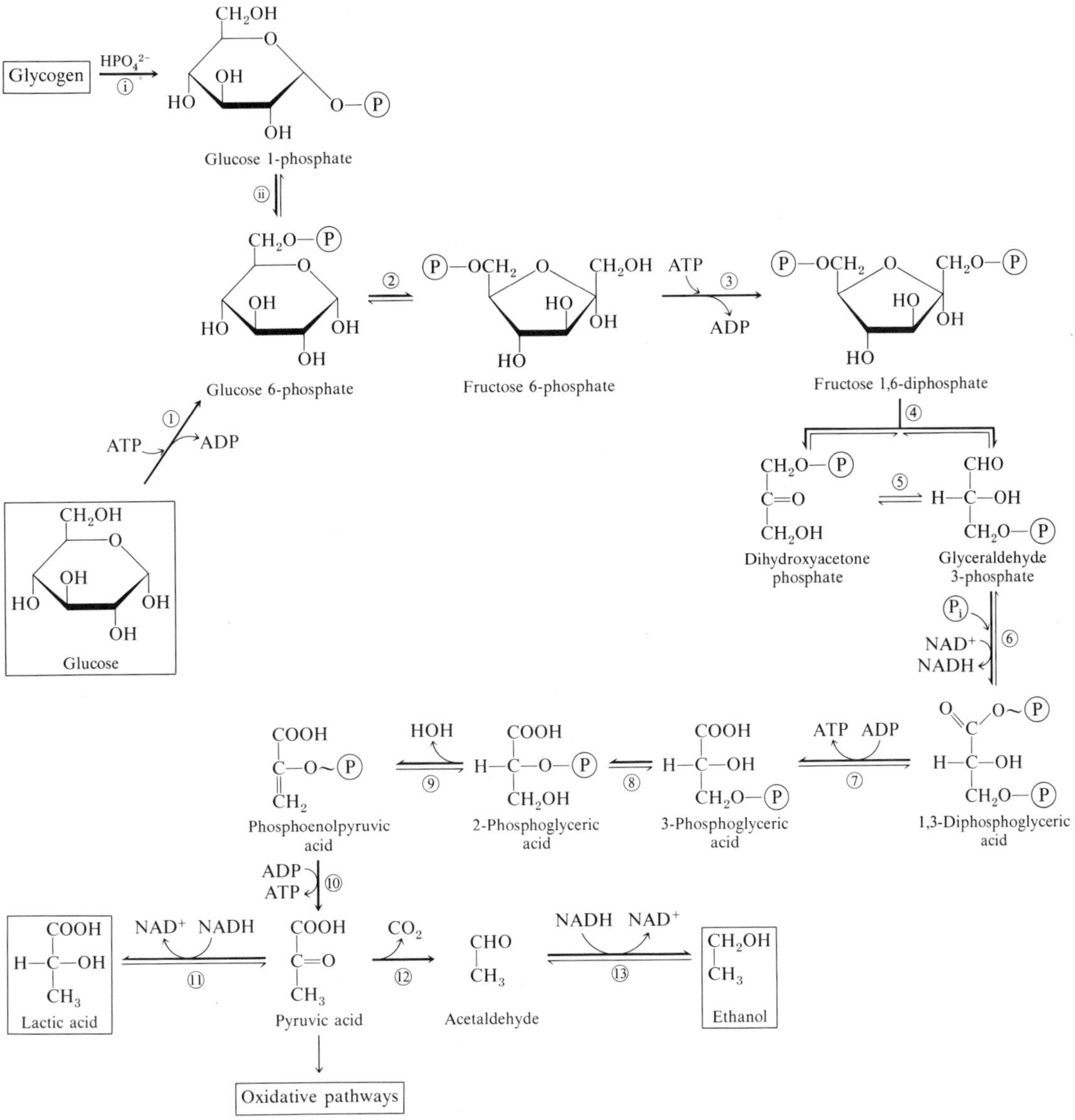
Glycogen
HPO_4^{2-}
(i)
Glucose 1-phosphate
(ii)
Glucose 6-phosphate
Fructose 6-phosphate
ATP
ADP
Fructose 1,6-diphosphate
Glucose
Dihydroxyacetone phosphate
Glyceraldehyde 3-phosphate
NAD^+
NADH
1,3-Diphosphoglyceric acid
3-Phosphoglyceric acid
2-Phosphoglyceric acid
HOH
Phosphoenolpyruvic acid
Pyruvic acid
Lactic acid
CO_2
Acetaldehyde
Ethanol
Oxidative pathways

Figure 28.4 Embden-Meyerhof pathway.

Step 3. Next follows another phosphorylation reaction, again involving the utilization of ATP as the phosphate-group donor. The enzyme *phosphofructokinase* is specific for fructose 6-phosphate and, like hexokinase, requires Mg^{2+} ions for activity. The reaction is irreversible, and represents the nonproductive utilization of a second molecule of ATP.

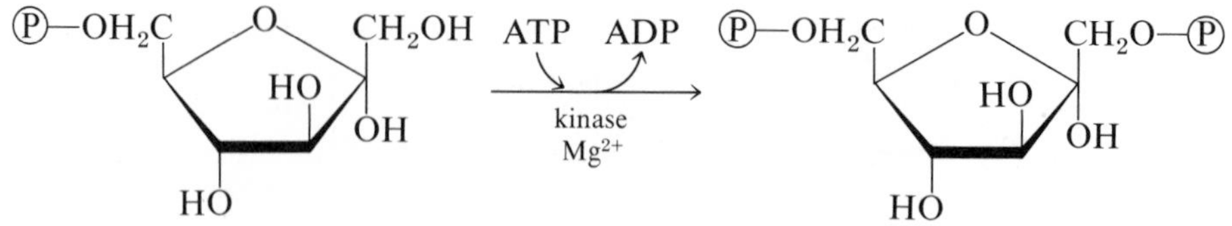

Fructose 6-phosphate → Fructose 1,6-diphosphate

Step 4. Fructose 1,6-diphosphate is then enzymically cleaved into two molecules of triose phosphate. This cleavage may be regarded as the reverse of an aldol condensation reaction (Section 18.5-a), and hence the enzyme has been named *aldolase*. Again, a better understanding of the reaction is achieved if the fructose 1,6-diphosphate is written in its open-chain form.

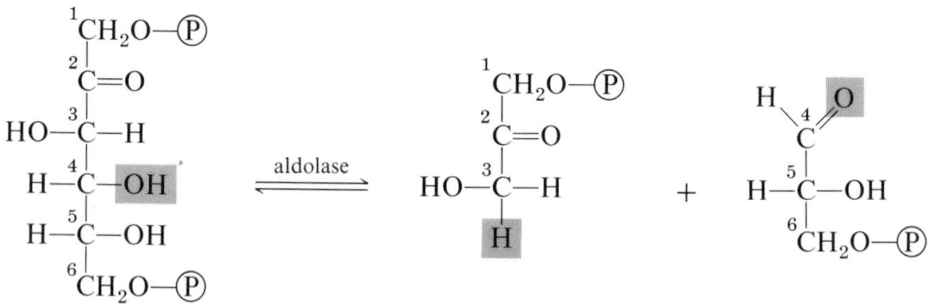

Fructose 1,6-diphosphate → Dihydroxyacetone phosphate + Glyceraldehyde 3-phosphate

The reaction catalyzed by the enzyme aldolase has been extensively investigated. Aldolase is not specific for fructose 1,6-diphosphate but will also cleave other ketose mono- and diphosphates (see exercise 28.6). In each case of cleavage, dihydroxyacetone phosphate is one of the products. This suggests that when aldolase catalyzes the reverse (condensation) reaction, it is absolutely specific for dihydroxyacetone phosphate, but a variety of aldehydes can serve as acceptor molecules.

Step 5. The next step is concerned with the interconversion of the triose phosphates. This is essential since only glyceraldehyde 3-phosphate can be further metabolized by the body. If cells were unable to convert dihydroxyacetone phosphate to glyceraldehyde 3-phosphate, then half of the energy stored in the original glucose molecule would be lost and would accumulate in the cell as the nonmetabolizable ketotriose phosphate. The enzyme for the isomerization reaction is *triose phosphate isomerase,* and the mechanism involves another enediol intermediate.

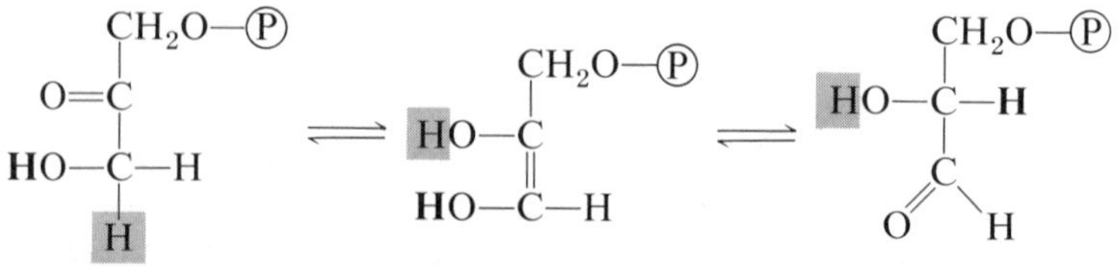

Dihydroxyacetone phosphate ⇌ Enediol intermediate ⇌ Glyceraldehyde 3-phosphate

A summation of steps 4 and 5 indicates that the enzymes aldolase and triose phosphate isomerase have effectively accomplished the conversion of one molecule of fructose 1,6-diphosphate into *two* molecules of glyceraldehyde 3-phosphate. Thus far, the glycolytic pathway has required an energy input in the form of two molecules of ATP but has not yet released any of the energy stored in the glucose.

Step 6. We now come to the first oxidation-reduction reaction in the glycolytic sequence. The conjugated enzyme *glyceraldehyde 3-phosphate dehydrogenase* contains the coenzyme NAD^+ as the prosthetic group. In the process of oxidizing the aldehyde to a carboxylic acid, the coenzyme is reduced to NADH. Notice that the same enzyme also catalyzes a phosphorylation reaction; inorganic phosphate is the phosphate donor. The energy obtained from the oxidation reaction is utilized for the phosphorylation reaction. The poisonous effect of iodoacetic acid (see Table 26.3) in blocking glycolysis occurs at this step. The dehydrogenase enzyme contains free sulfhydryl groups at its active site; iodoacetate noncompetitively inhibits the enzyme by covalently bonding to these catalytic groups.

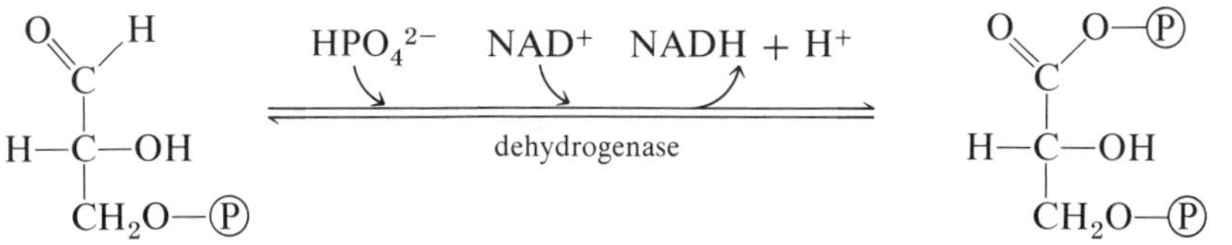

HPO_4^{2-}
NAD^+
$NADH + H^+$
dehydrogenase

Glyceraldehyde 3-phosphate

1,3-Diphosphoglyceric acid

The reaction is more easily understood if it is considered to occur in two steps: (a) oxidation and (b) phosphorylation. Note, however, that this presentation is an oversimplification that in no way implies the actual enzyme-catalyzed sequence of events.

(a) Oxidation

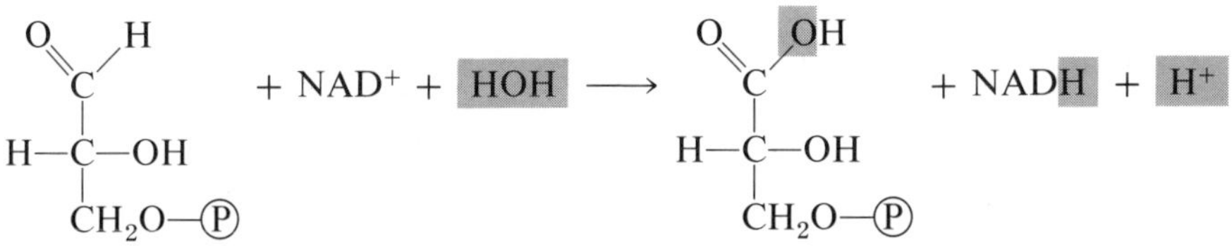

(b) Phosphorylation

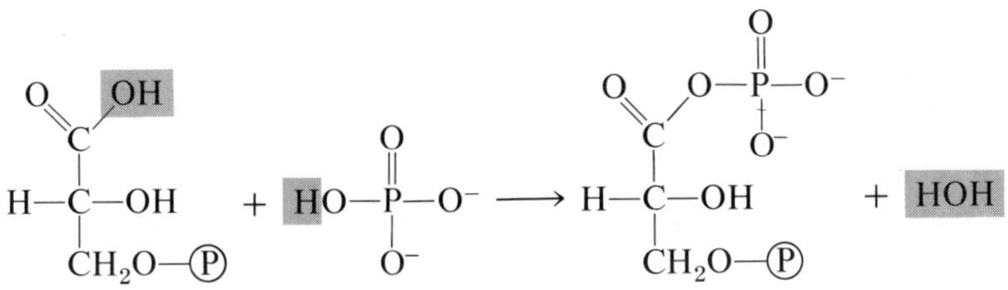

Step 7. Diphosphoglyceric acid, the product of the reaction in step 6, contains a high energy phosphate bond; it can transfer the phosphate group on carbon-1 directly to a molecule of ADP thus forming a molecule of ATP. The enzyme that catalyzes the reaction is *phosphoglycerokinase,* and, like all kinases, it requires Mg^{2+} ions for activity. It is in this reaction that ATP is first produced in the pathway. Since the ATP is formed by a direct transfer of a phosphate group from a metabolite to ADP, the process is referred to as **substrate level phosphorylation** to distinguish it from another ATP-synthesizing process to be considered shortly.

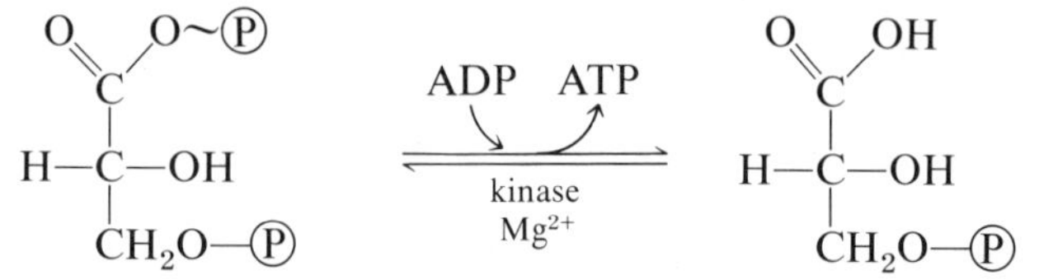

1,3-Diphosphoglyceric acid   3-Phosphoglyceric acid

Step 8. This step in the pathway is very similar to the intramolecular transfer of phosphate as seen in step ii. The enzyme *phosphoglyceromutase* catalyzes the exchange of a phosphate group from the hydroxyl group of carbon-3 to the hydroxyl group of carbon-2.

3-Phosphoglyceric acid   2-Phosphoglyceric acid

Step 9. The enzyme *enolase,* which also requires Mg^{2+} ions for activity, catalyzes an alcohol dehydration reaction to produce phosphoenolpyruvic acid, a compound with a high-energy enolic phosphate group. Fluoride is an effective poison of the glycolytic pathway because of its inhibition of the enzyme enolase (see Table 26.3). Fluoride ions bind to the activator magnesium ions to form a magnesium fluorophosphate complex.

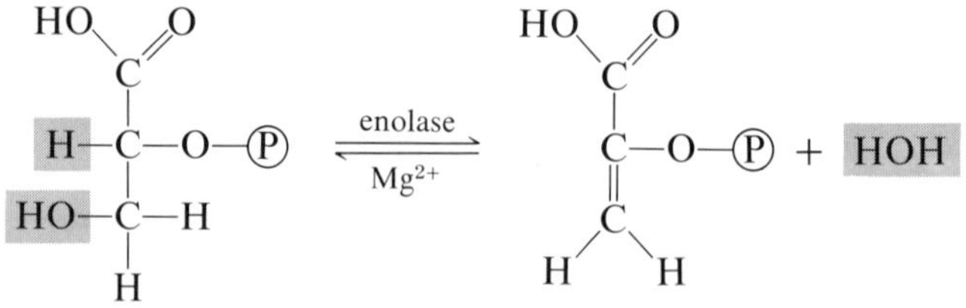

2-Phosphoglyceric acid   Phosphoenolpyruvic acid (PEP)

Step 10. This irreversible step provides a second example of substrate level phosphorylation. The phosphate group of phosphoenolpyruvic acid is transferred to ADP; one molecule of ATP is produced per molecule of PEP. It is likely that the reaction proceeds via the enol form of pyruvic acid, which is unstable with respect to rearrangement to pyruvic acid. The enzyme *pyruvate kinase* requires both Mg^{2+} and K^+ ions for activity.

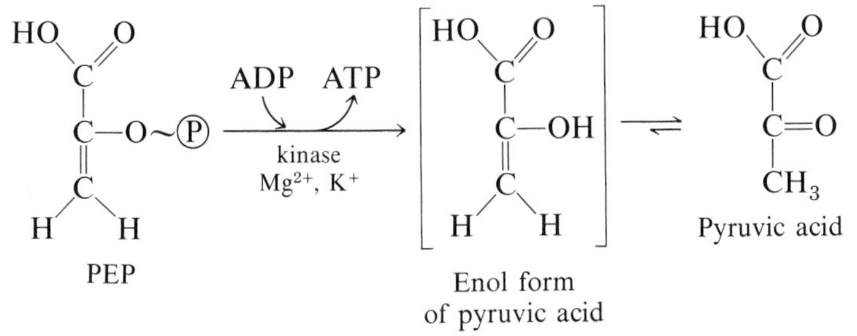
ADP ATP
kinase
Mg2+, K+
PEP
Enol form
of pyruvic acid
Pyruvic acid

Step 11. Steps 1 through 10 are identical for glycolysis and fermentation. Pyruvic acid, as we shall see, is the *crossroads* compound; its metabolic fate is dependent upon the conditions (anaerobic or aerobic) and upon the organism under consideration (Figure 28.5). Under usual conditions, muscle cells

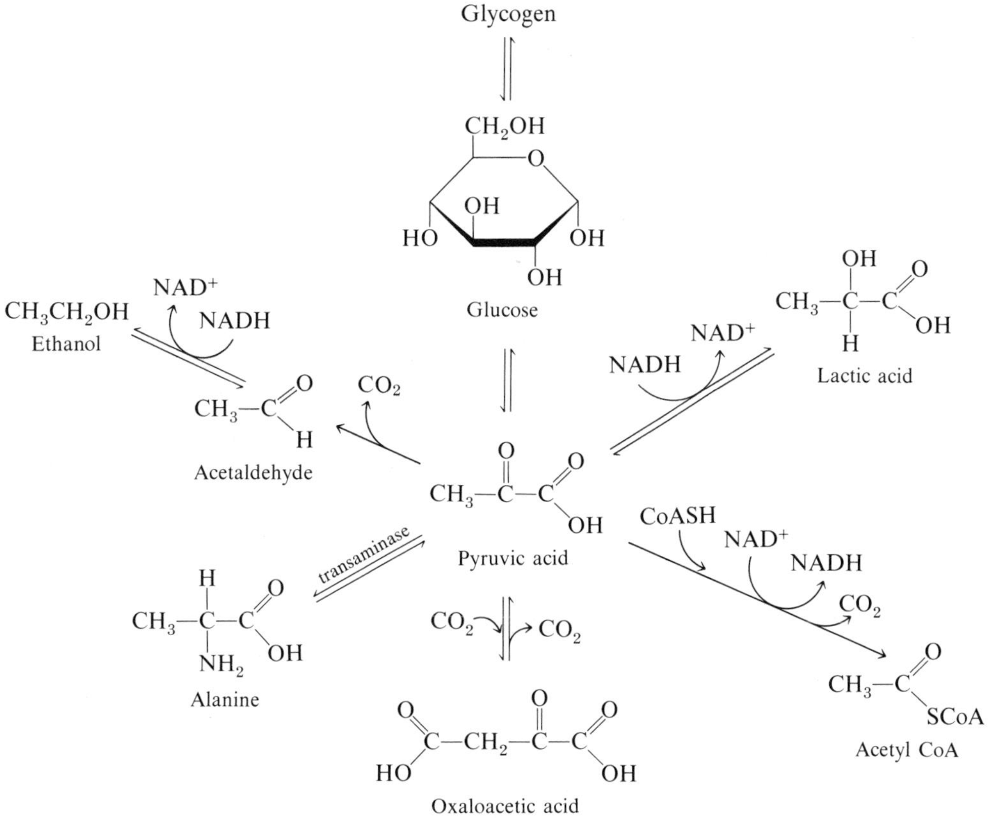
Glycogen
Glucose
Ethanol
NAD+
NADH
Acetaldehyde
CO2
Lactic acid
Pyruvic acid
CoASH
transaminase
Alanine
Acetyl CoA
Oxaloacetic acid

Figure 28.5 Importance of pyruvic acid in metabolism.

respire aerobically. However, during strenuous exercise the energy demand on the muscles is enormous and the respiratory system is unable to deliver oxygen to these cells in sufficient amounts to meet this demand. As a result, the muscle cells must obtain their required energy via the anaerobic pathway (i.e., glycolysis).

In the presence of NADH, the enzyme *lactic acid dehydrogenase*[3] catalyzes the reduction of pyruvic acid to lactic acid (ketone to secondary alcohol). This reaction is essential because it regenerates the NAD^+ that is needed in step 6. If NAD^+ were not replenished, the cell's supply of this coenzyme would be swiftly depleted, anaerobic glycolysis would cease, and glyceraldehyde 3-phosphate would accumulate in the cell.

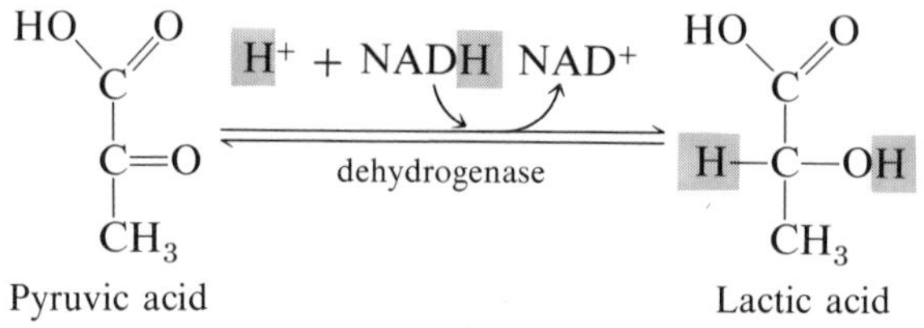

HO
O
C
C=O
CH3
Pyruvic acid
H+ + NADH NAD+
dehydrogenase
HO
O
C
H—C—OH
CH3
Lactic acid

Lactic acid, then, is the end-product of glycolysis, and if there were not some mechanism for its removal, it would accumulate in the muscle cells, raising the level of acidity in these cells. Increased acidity impedes muscle performance by causing muscle fatigue and exhaustion. Two processes act to maintain a proper level of lactic acid, both of which require oxygen. The heavy breathing that occurs after strenuous exercise helps to supply this oxygen.

1. 70–80% of the lactic acid diffuses into the bloodstream and is carried to the liver. There it may be oxidized to carbon dioxide and water (via the Krebs cycle) or converted back into glucose and/or glycogen.
2. The remaining 20–30% can be reoxidized to pyruvic acid, which then enters the Krebs cycle and is further oxidized to carbon dioxide and water during those periods when the muscle cells are afforded an ample supply of oxygen.

Step 12. Yeast and other microorganisms that can live in a limited supply of oxygen must also have some means of regenerating NAD^+. They do so by first decarboxylating pyruvic acid to acetaldehyde. This reaction, which is catalyzed by *pyruvic acid decarboxylase,* is responsible for CO_2 production in yeast (the reason yeast is used in certain baking processes). The enzyme requires the coenzyme thiamine pyrophosphate (TPP) and Mg^{2+} ions. The decarboxylation reaction is irreversible.

[3]This enzyme derives its name from its catalytic role in reversing this reaction. Recall that enzymes catalyze both the forward and the reverse reactions of an equilibrium mixture and that the enzymes may be named accordingly. An equally descriptive name for this enzyme is *pyruvic acid reductase* or *hydrogenase*.

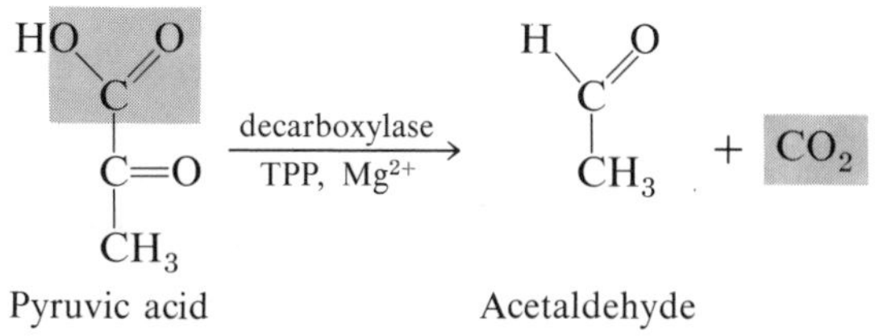
HO
O
C
C=O
CH3
decarboxylase
TPP, Mg2+
H
O
C
CH3
+ CO2
Pyruvic acid
Acetaldehyde

Step 13. The final step in the fermentation process is catalyzed by the enzyme *alcohol dehydrogenase.* Acetaldehyde is reduced by NADH to ethyl alcohol, thus regenerating NAD^+.

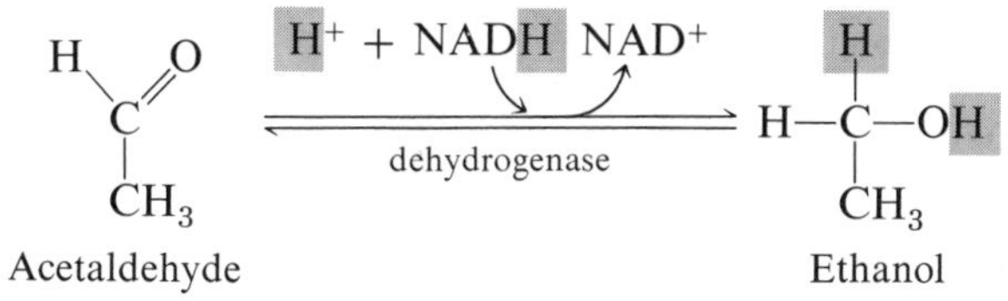
H
O
C
CH3
H+ + NADH NAD+
dehydrogenase
H
H—C—OH
CH3
Acetaldehyde
Ethanol

28.4 Reversal of Glycolysis and Fermentation

It has been mentioned that several of the degradative steps of the Embden–Meyerhof pathway are not reversible in vivo. (Single arrows are used in Figure 28.4 to indicate that reactions i, 1, 3, 10, and 12 are nonreversible.) This seems to be inconsistent with the fact that lactic acid and any of the other intermediates of the glycolytic sequence can be converted back to glycogen provided an energy source is available. Therefore, other reactions catalyzed by different enzymes must necessarily effect a reversal of steps i, 1, 3, and 10. The irreversibility of parts of a metabolic sequence is a common phenomenon; it will be encountered again in aerobic metabolism as well as in the metabolism of lipids and proteins. Catabolic and anabolic processes may have many of the same intermediates and many of the same enzymes, yet their pathways are not identical. Similarly, more than one pathway may exist for the same process. The separation of degradative and synthetic pathways allows an organism to control one series of reactions without affecting the other. Since we are concerned primarily with reactions that yield energy and produce ATP, we shall not take up the anabolism of carbohydrates. These reactions will be found in advanced biochemistry texts.

28.5 Energetics of Glycolysis and Fermentation

The net energy yield in the form of high energy phosphate bonds (moles of ATP) obtained from the anaerobic metabolism of each mole of glucose can be easily calculated from Figure 28.4.

One mole of ATP is expended in the initial phosphorylation of glucose (step 1).

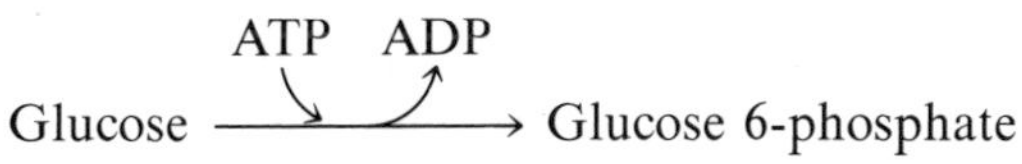
ATP ADP
Glucose
Glucose 6-phosphate

And a second mole is consumed in the phosphorylation of fructose 6-phosphate (step 3).

$$\text{Fructose 6-phosphate} \xrightarrow{\text{ATP} \searrow \nearrow \text{ADP}} \text{Fructose 1,6-diphosphate}$$

Steps 4 and 5 are very significant for it is in these reactions that each mole of six-carbon sugar is transformed into two moles of the triose phosphate, glyceraldehyde 3-phosphate. In step 7, each mole of glyceraldehyde 3-phosphate is converted into one mole of 3-phosphoglyceric acid; one mole of ATP is produced per mole of triose phosphate, or *two moles of ATP per mole of glucose.*

$$\text{1,3-Diphosphoglyceric acid} \xrightarrow{\text{ADP} \searrow \nearrow \text{ATP}} \text{3-Phosphoglyceric acid}$$

In reaction 10, one mole of ATP is generated for each mole of pyruvic acid that is formed from phosphoenolpyruvic acid; again two moles of ATP are produced for each mole of glucose that entered the pathway.

$$\text{Phosphoenolpyruvic acid} \xrightarrow{\text{ADP} \searrow \nearrow \text{ATP}} \text{Pyruvic acid}$$

A summation of all these steps reveals that for every mole of glucose degraded, two moles of ATP are initially consumed and four moles of ATP are ultimately produced. The net production of ATP is thus two moles per mole of glucose converted to lactic acid or to ethanol. If, however, glycogen is used as the source of glucose, steps i and ii are operative instead of step 1. It would then be necessary to expend only one mole of ATP (in step 3) in order to produce fructose 1,6-diphosphate, and the net yield of ATP would be three moles per mole of glucose 1-phosphate.

In Yeast

$$C_6H_{12}O_6 + 2\,\text{ADP} + 2\,P_i \longrightarrow 2\,C_2H_5OH + 2\,CO_2 + 2\,\text{ATP}$$

In Muscle—Starting with Glycogen

$$(C_6H_{12}O_6)_n + 3\,\text{ADP} + 3\,P_i \longrightarrow (C_6H_{12}O_6)_{n-1} + 2\,CH_3CHOHCOOH + 3\,\text{ATP}$$

Recall that about 7300 calories of free energy is conserved per mole of ATP produced (Section 21.3). Recall also that the total amount of energy that can theoretically be obtained from the complete oxidation of one mole of glucose is 686,000 calories. The energy conserved by anaerobic metabolism, then, is only a minute amount of the total energy that is available, that is, either 2.1% (14,600/686,000) or 3.2% (21,900/686,000). Thus, anaerobic cells extract only a small fraction of the total energy of the glucose molecule, yet this amount is sufficient for their survival. Since they are not nearly as efficient as aerobic cells, it is necessary that they utilize more glucose per unit of time to accomplish the same amount of cellular work.

28.6 Storage of Chemical Energy

Before continuing on to the aerobic phase of carbohydrate metabolism, we shall consider how muscle cells store and utilize the energy that they extract from carbohydrates (and other compounds). *The splitting of ATP to ADP and inorganic phosphate furnishes the energy for muscle contraction.*

In 1927, a compound called *creatine phosphate* was isolated from mammalian muscle and was subsequently demonstrated to be the storage form of energy in the muscles of vertebrates and in nerve tissue. *Arginine phosphate* serves the same purpose in the muscles of invertebrates. These high-energy phosphate compounds that serve as reservoirs of phosphate-bond energy are often called *phosphagens*. The high-energy bond of these particular phosphagens occurs between phosphorus and nitrogen rather than between phosphorus and oxygen.

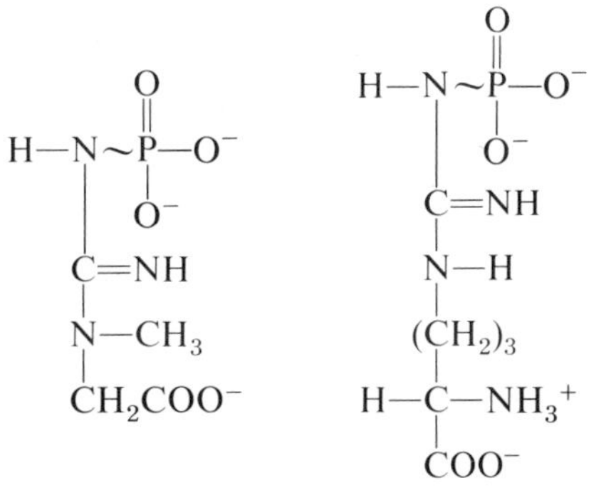

Creatine phosphate (Phosphocreatine) Arginine phosphate (Phosphoarginine)

The concentration of ATP in muscle is relatively low and cannot meet the demands of muscular exertion for more than a fraction of a second.[4] At rest, mammalian muscle contains 4–6 times as much creatine phosphate as ATP. As ATP is utilized, creatine phosphate, in the presence of *creatine phosphokinase*,[5] reacts with ADP to produce more ATP and creatine.

$$\text{Creatine phosphate} + \text{ADP} \underset{Mg^{2+}}{\overset{\text{phosphokinase}}{\rightleftharpoons}} \text{Creatine} + \text{ATP}$$

This reaction is readily reversible. When muscular activity is required, the reaction proceeds to the right; when there is an abundant amount of ATP

[4]The trained athlete can exert himself more strenuously because (1) he has a greater number of muscle cells and therefore produces ATP at a greater rate and (2) his muscles contain a greater amount of stored oxygen (bound to myoglobin) and will thus have prolonged aerobic activity.

[5]The measurement of creatine phosphokinase (CPK) activity in the blood is of value in the diagnosis of disorders affecting skeletal and cardiac muscle. A heart attack (myocardial infarction) damages heart cells, and the damaged cells "leak" their enzymes, including CPK, into the extracellular fluid and plasma, resulting in elevated serum CPK levels. The increase in serum CPK concentration is apparent 1–8 hours after a heart attack and reaches its highest level 12–24 hours after the attack. Thus the CPK test is a standard clinical test for the detection of a suspected myocardial infarction. (See also page 761 for the GOT test.)

(formed from the catabolism of foodstuffs) the reaction proceeds to the left, and creatine phosphate is stored in the muscle cells. Since the concentration of creatine phosphate in the muscle is limited, it is only useful in generating a quick source of utilizable energy. It has been estimated that the creatine phosphate can provide energy for about 20 seconds of strenuous exercise (e.g., long enough to sprint 200 meters). Energy for prolonged activity is obtained from the anaerobic breakdown of glycogen that is synthesized during long periods of muscular inactivity.

28.7 Aerobic Metabolism

We have mentioned that the anerobic phase of glucose catabolism ends with the production of lactic acid or ethanol and that it accounts for only a small fraction of the total available energy of the glucose molecule. Aerobic organisms (those that require oxygen to live) utilize the identical series of glycolytic reactions up to the production of pyruvic acid. Pyruvic acid, however, is not reduced to lactic acid but is oxidized in a number of discrete enzymic reactions to carbon dioxide and water.

$$2\,CH_3COCOOH + 5\,O_2 \longrightarrow 6\,CO_2 + 4\,H_2O + \sim 639{,}000 \text{ calories}$$

From the standpoint of energy production, the oxidation of pyruvic acid is of considerable significance because it liberates most of the energy (93%) stored in the glucose molecule. Much of this energy is stored in a chemical form in the high energy phosphate bonds of ATP.

A scheme for the complex series of reactions that effects the oxidation of glucose to carbon dioxide and water was first proposed by Sir Hans Krebs in 1937 (Nobel Prize for Medicine, 1953). Pyruvic acid is first decarboxylated to a two-carbon compound, which then enters a cyclic sequence of reactions known collectively as the *Krebs cycle,* the *tricarboxylic acid cycle,* or the *citric acid cycle.* The Krebs cycle functions to produce ATP and to provide metabolic intermediates for the biosynthesis of needed compounds. As we shall see in later chapters, the Krebs cycle is not restricted to the metabolism of carbohydrates. It also plays a vital role in the metabolism of lipids and proteins, and thus occurs in almost all cells of higher animals. Every enzyme that participates in the cycle has been identified, and the operation of this metabolic pathway in vivo has been completely verified by the use of isotopic tracers.

Figure 28.6 is a schematic outline of the Krebs cycle. Each reaction is numbered, and the individual steps of the sequence will now be considered in detail.

Step 1. The pyruvic acid formed in the Embden–Meyerhof pathway is not an intermediate in the Krebs cycle. It must first be enzymically decarboxylated and oxidized (oxidative decarboxylation) to yield the extremely important intermediate *acetyl coenzyme A.* (See discussion of acetyl CoA in

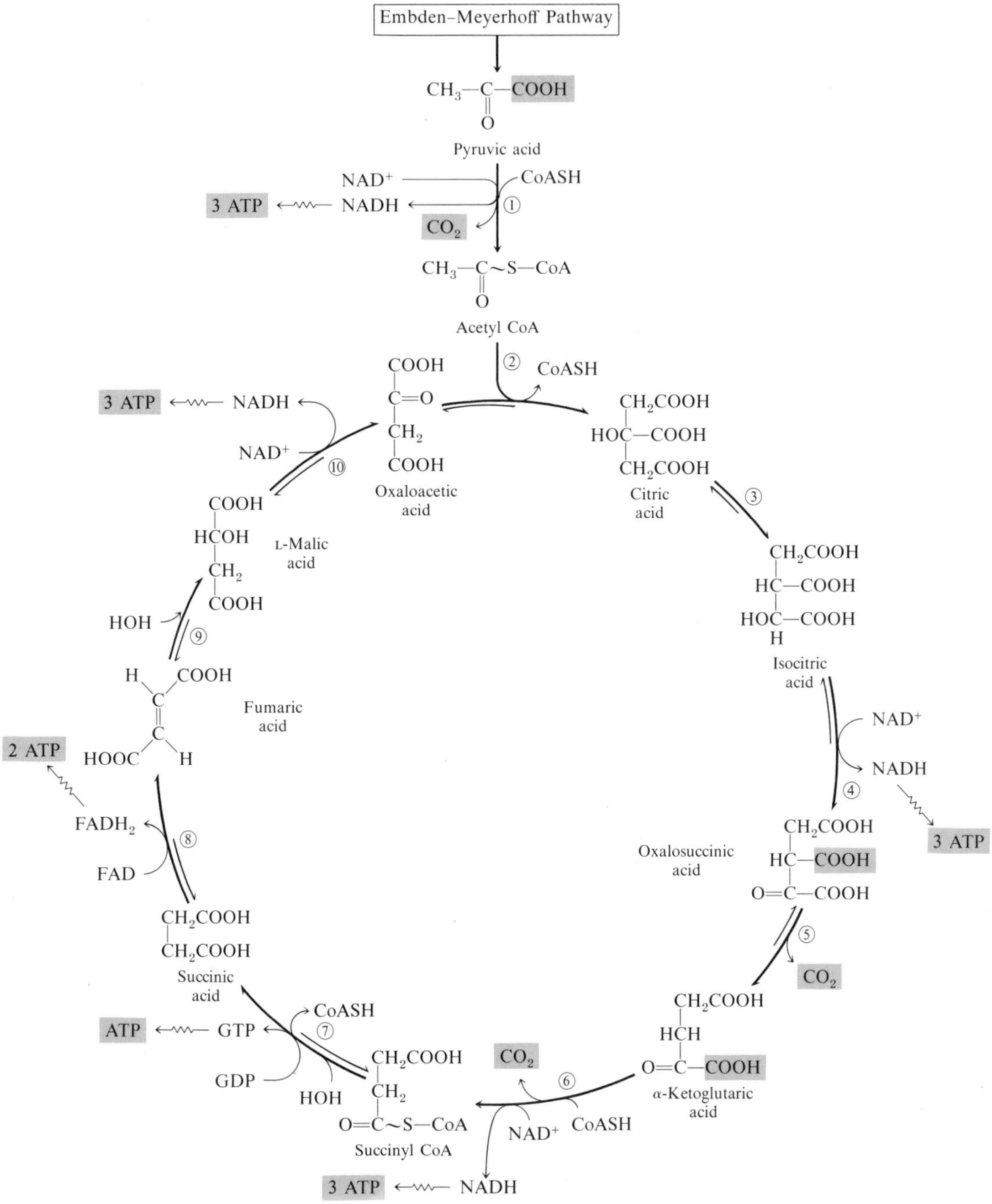
Embden–Meyerhoff Pathway
Pyruvic acid
NAD+
CoASH
3 ATP
NADH
CO2
Acetyl CoA
CoASH
3 ATP
NADH
NAD+
Oxaloacetic acid
Citric acid
L-Malic acid
HOH
Isocitric acid
Fumaric acid
NAD+
NADH
2 ATP
3 ATP
FADH2
FAD
Oxalosuccinic acid
CO2
Succinic acid
CoASH
ATP
GTP
GDP
HOH
CO2
Succinyl CoA
NAD+
CoASH
α-Ketoglutaric acid
3 ATP
NADH

Figure 28.6 The Krebs cycle.

Section 21.4 and formula of coenzyme A in Table 26.2.) The formation of acetyl CoA from pyruvic acid requires the participation of six cofactors: coenzyme A, thiamine pyrophosphate (TPP), lipoic acid, NAD^+, FAD, and Mg^{2+}.

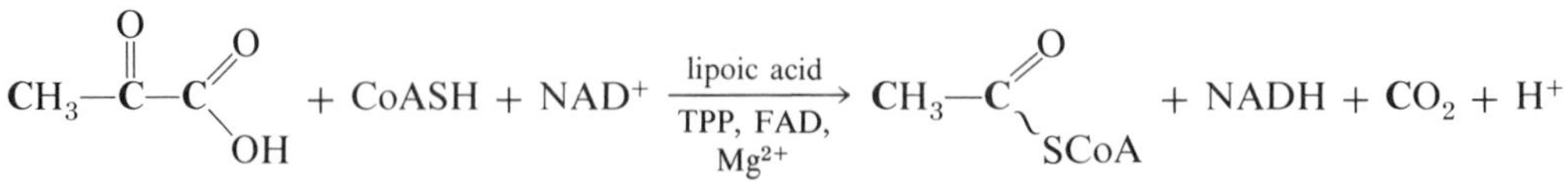

lipoic acid
TPP, FAD, Mg2+
+ CoASH + NAD+
+ NADH + CO2 + H+

Pyruvic acid

Acetyl CoA

The name given to this multienzyme system is *pyruvic acid dehydrogenase complex*. The mechanism of the reaction is believed to involve four distinct steps, decarboxylation, reductive acetylation, acetyl transfer, and electron transport. The initial decarboxylation step is irreversible; hence the overall conversion of pyruvic acid to acetyl CoA is irreversible.

Step 2. Acetyl CoA is likewise not a true intermediate in the Krebs cycle. It enters the cycle by condensing with the four-carbon dicarboxylic acid oxaloacetic acid, yielding citric acid, the six-carbon tricarboxylic acid for which the cycle was named. Notice that the reaction is very similar to an aldol condensation reaction (Section 18.5-a) and for this reason the enzyme was originally referred to as the *condensing enzyme*. This enzyme is now called *citric acid synthetase*. The two carbon atoms that originate from acetyl CoA are shown in boldface type here and in subsequent reactions. Note that this step regenerates coenzyme A.

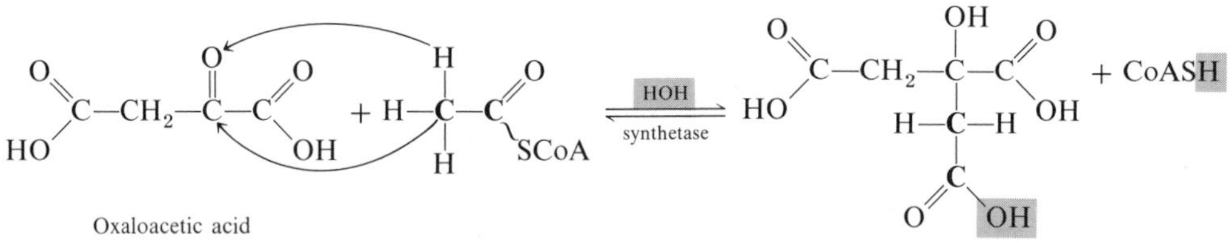

HOH
synthetase
+ CoASH

Oxaloacetic acid

Citric acid

Step 3. The next step is sometimes considered to be two separate reactions. The single enzyme *aconitase* catalyzes the successive elimination and reincorporation of a molecule of water. The net result is the isomerization of citric acid to its less symmetrical isomer, isocitric acid. The intermediate is *cis*-aconitic acid, and it normally remains bound to the enzyme. (Notice that the addition of water to the double bond of *cis*-aconitic acid is anti-Markownikoff; Section 14.4-b.)

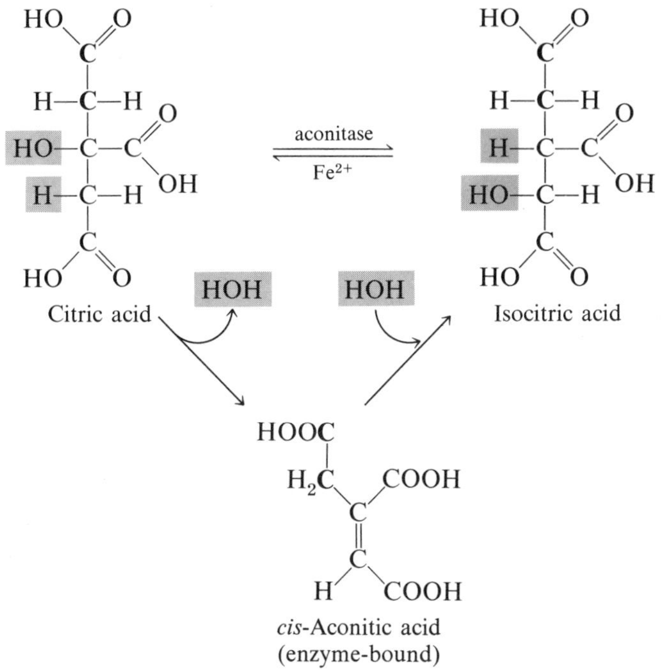
HO
O
C
H—C—H
O
HO—C—C
H—C—H
OH
C
HO
O
HOH
HOH
Citric acid
aconitase
Fe2+
Isocitric acid
H—C—C
HO—C—H
HOOC
H2C
COOH
C
C
H
COOH
cis-Aconitic acid
(enzyme-bound)

Steps 4 and 5. The next two steps are usually discussed together because experimental evidence indicates that the intermediate, oxalosuccinic acid, does not exist free but is firmly bound to the surface of the enzyme. Two enzymes, collectively called *isocitric dehydrogenase,* catalyze the oxidative decarboxylation of isocitric acid to α-ketoglutaric acid.

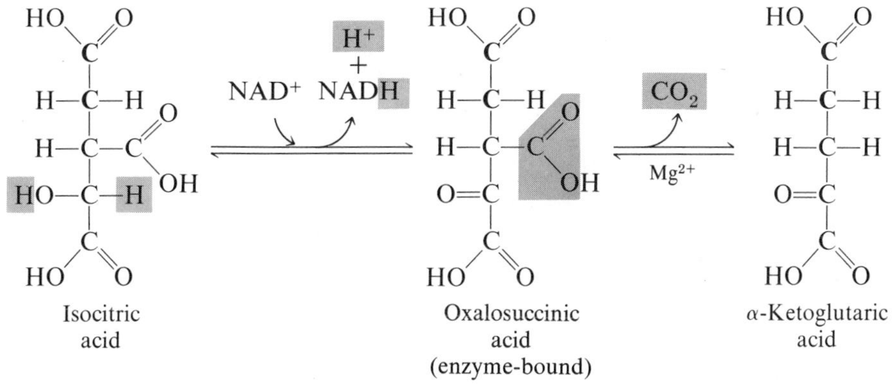
HO
O
C
H—C—H
H—C—C
HO—C—H
OH
C
HO
O
Isocitric acid
H+
+
NAD+
NADH
Oxalosuccinic acid
(enzyme-bound)
O=C
CO2
Mg2+
α-Ketoglutaric acid

Step 6. This step is practically identical to step 1; it is catalyzed by another multienzyme system known as *α-ketoglutaric acid dehydrogenase complex,* which requires the same six cofactors as pyruvic acid dehydrogenase. This is the only nonreversible reaction (because of the decarboxylation step) in the Krebs cycle and, as such, prevents the cycle from operating in the reverse direction.

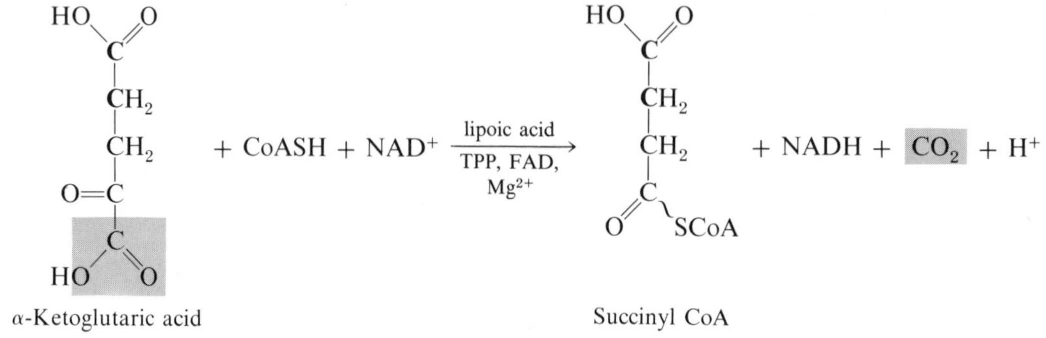
HO O C CH2 CH2 O=C C HO O
+ CoASH + NAD+
lipoic acid
TPP, FAD, Mg2+
HO O C CH2 CH2 C O SCoA
+ NADH + CO2 + H+
α-Ketoglutaric acid
Succinyl CoA

Step 7. In this reaction the chemical energy of the high-energy thioester bond is conserved by coupling the hydrolysis of succinyl CoA with the formation of guanosine triphosphate (GTP) from guanosine diphosphate (GDP) and inorganic phosphate. This reaction is significant because GTP can react with ADP to generate ATP. Here we have another example of substrate level phosphorylation.

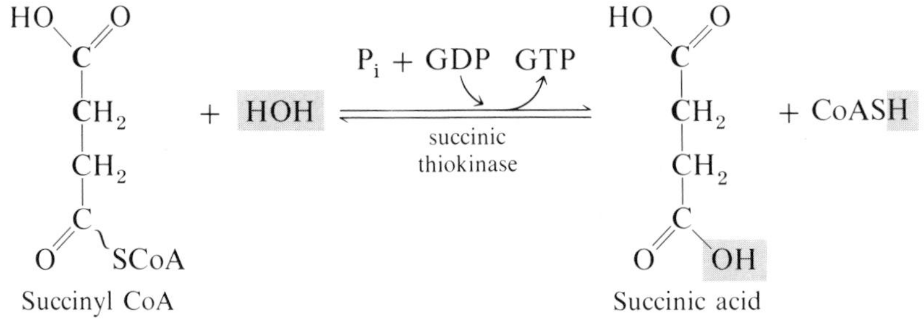
+ HOH
Pi + GDP GTP
succinic
thiokinase
+ CoASH
Succinyl CoA
Succinic acid

$$GTP + ADP \rightleftharpoons GDP + ATP$$

Step 8. The enzyme *succinic acid dehydrogenase* catalyzes the removal of two hydrogen atoms from succinic acid, thus forming fumaric acid. You will recall that this enzyme is competitively inhibited by malonic acid (Section 26.8). This dehydrogenation reaction is the only one in the cycle that utilizes the coenzyme FAD (Table 26.2) rather than NAD^+.

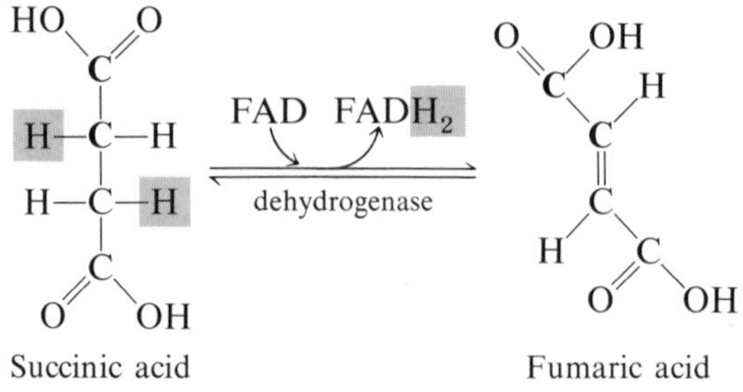
FAD FADH2
dehydrogenase
Succinic acid
Fumaric acid

Step 9. The addition of a molecule of water across the double bond of fumaric acid to form L-malic acid is catalyzed by the enzyme *fumarase.* This enzyme is highly stereospecific; only the L-isomer of malic acid is produced.

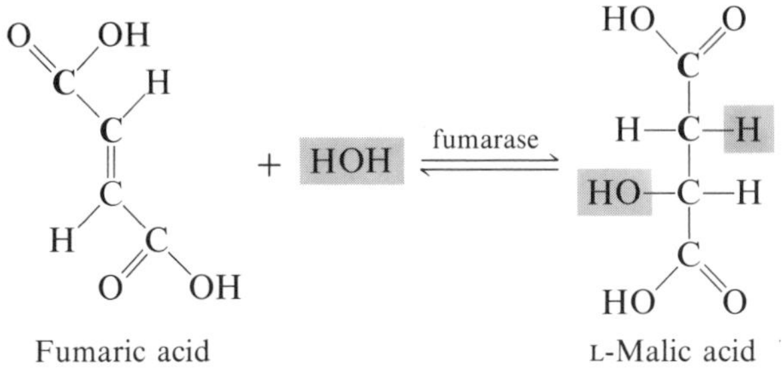

Fumaric acid
fumarase
L-Malic acid

Step 10. One revolution of the cycle is completed when the dehydrogenation of L-malic acid is brought about by the enzyme *malic acid dehydrogenase.* This is the fourth oxidation-reduction reaction that utilizes NAD^+ as the oxidizing agent.

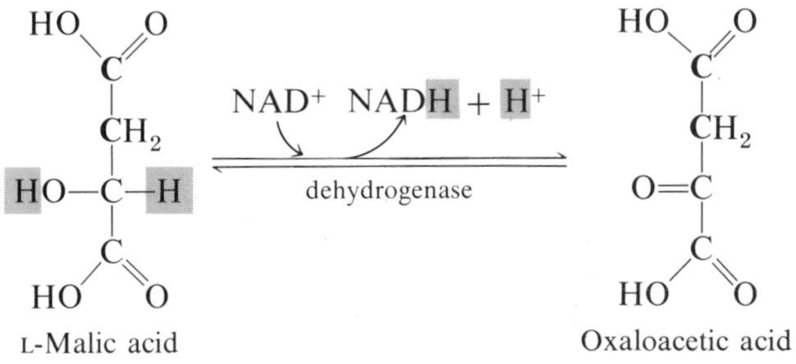

NAD+ NADH + H+
dehydrogenase
L-Malic acid
Oxaloacetic acid

28.8 Respiratory Chain (Electron Transport System)

We have stated that aerobic metabolism occurs only in the presence of molecular oxygen. It has also been mentioned that the major portion of solar energy stored in carbohydrates is conserved in the process. Yet nowhere in the Krebs cycle has the utilization of oxygen or the conservation of energy been indicated. None of the intermediates of the cycle has been shown to be linked to phosphate groups, and no direct synthesis of ATP took place from ADP and inorganic phosphate. The Krebs cycle deals primarily with the fate of the carbon skeleton of pyruvic acid, describing the metabolites involved in its conversion to carbon dioxide and water. The coenzymes NAD^+ and FAD are reduced to NADH and $FADH_2$; no mechanism has yet been indicated for their regeneration. The reduced coenzymes must be reoxidized if the aerobic phase of carbohydrate metabolism is to continue.

All the enzymes and cofactors that are necessary for the Krebs cycle and for the conservation of energy are localized in the **mitochondria,** small membrane-surrounded granules located throughout the cytoplasm that are often referred to as the "power plants" of the cell (Figure 28.7). The enzymes

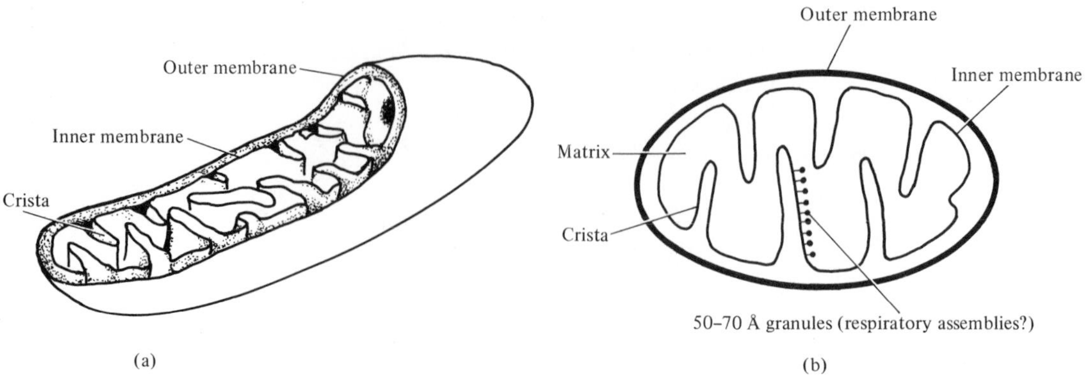
Outer membrane
Inner membrane
Crista
(a)
Outer membrane
Inner membrane
Matrix
Crista
50–70 Å granules (respiratory assemblies?)
(b)

Figure 28.7 Schematic three-dimensional (a) and two-dimensional (b) representations of a liver mitochondrion.

contained within the mitochondria are fixed in geometrically specific arrays such that they are capable of functioning as extremely efficient assembly lines. The *sequence of reactions whereby the reduced forms of the coenzymes are reoxidized* (ultimately by molecular oxygen) is known as the **respiratory chain** or the **electron transport system.**

Figure 28.8 illustrates the respiratory chain in the mitochondria. The sequence in which the electron carriers of the chain operate is determined by their respective reduction potentials (Section 8.7). The reduction potential is a measure of the tendency of a substance to gain electrons when compared to the standard hydrogen electrode (25 °C, 1 atm H_2, and 1 M H^+), which is arbitrarily assigned a value of 0.0 volt. Since in biological systems we are more interested in neutral solutions, a correction is made for changes in pH. At pH 7 the hydrogen electrode has a potential difference of −0.42 volt when measured against the standard hydrogen electrode. The reduction potentials of the intermediates of the respiratory chain are listed in Table 28.1.

For each molecule of pyruvic acid that is converted to carbon dioxide and water via the Krebs cycle, five dehydrogenation reactions occur. NAD^+ serves as the electron acceptor in four of these (steps 1, 4, 6, and 10), and FAD is the oxidizing agent in the fifth (step 8). In Figure 28.8, SH_2 symbolizes the reduced metabolites (i.e., pyruvic acid, isocitric acid, α-ketoglutaric acid, and malic acid) and S signifies the oxidized metabolites (i.e., acetyl CoA, oxalosuccinic acid, succinyl CoA, and oxaloacetic acid).

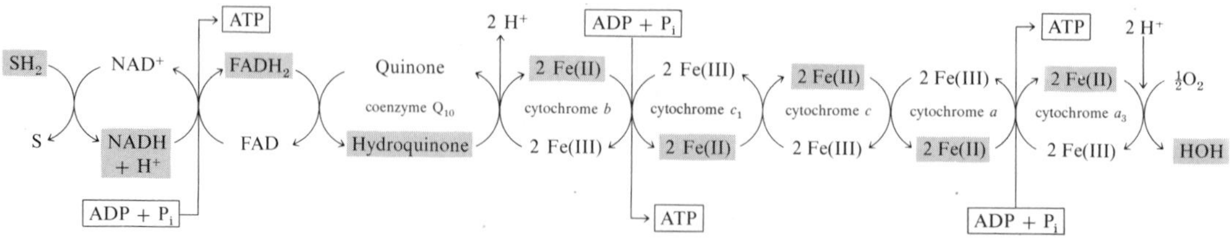
SH_2
S
NAD^+
NADH + H^+
ADP + P_i
ATP
$FADH_2$
FAD
Quinone
coenzyme Q_{10}
Hydroquinone
2 H^+
2 Fe(II)
cytochrome b
2 Fe(III)
ADP + P_i
2 Fe(III)
cytochrome c_1
2 Fe(II)
ATP
2 Fe(II)
cytochrome c
2 Fe(III)
2 Fe(III)
cytochrome a
2 Fe(II)
ATP
ADP + P_i
2 H^+
2 Fe(II)
cytochrome a_3
2 Fe(III)
$\frac{1}{2}O_2$
HOH

Figure 28.8 The respiratory chain and oxidative phosphorylation. (Reduced substances are indicated by shading.)

Table 28.1 Standard Reduction Potentials for Some of the Respiratory Chain Components

System (*Oxidant/Reductant*)	*E* (*pH 7*) (*volts*)
$NAD^+/NADH$	−0.32
$FAD/FADH_2$	−0.06
Cyt *b*-Fe(III)/cyt *b*-Fe(II)	+0.04
Cyt *c*-Fe(III)/cyt *c*-Fe(II)	+0.26
Cyt *a*-Fe(III)/cyt *a*-Fe(II)	+0.29
Cyt a_3-Fe(III)/cyt a_3-Fe(II)	+0.39
$\frac{1}{2}O_2/H_2O$	+0.82

We will now examine one of these conversions by writing a balanced half-reaction (Section 8.6).

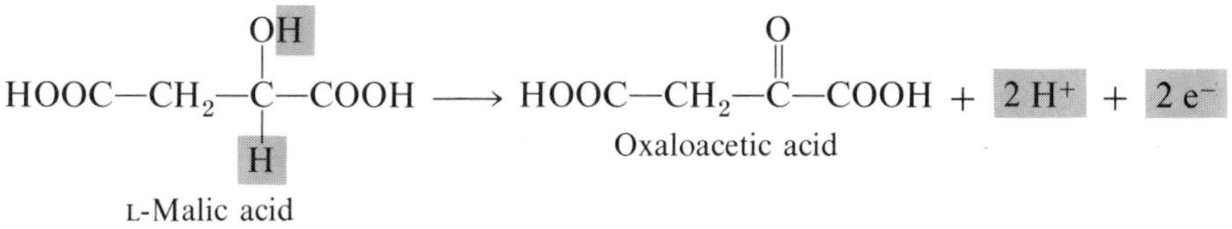
OH
HOOC—CH2—C—COOH
H
L-Malic acid
O
HOOC—CH2—C—COOH + 2 H+ + 2 e−
Oxaloacetic acid

It has been stated that the function of the coenzyme NAD^+ is to accept this pair of electrons. We can also write a balanced half-reaction for the reduction of the coenzyme; **R**— represents the remaining portion of the NAD^+ molecule (see Table 26.2).

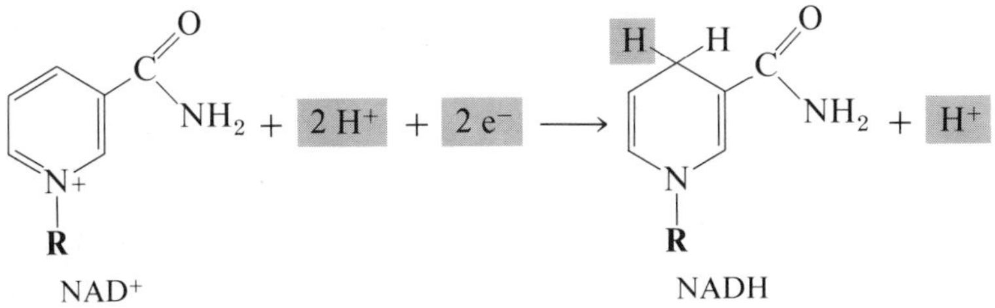
O
C
NH2 + 2 H+ + 2 e−
N+
R
NAD+
H
H
O
C
NH2 + H+
N
R
NADH

Two hydrogen ions and two electrons are removed from the substrate. The NAD^+ molecule accepts both electrons and one hydrogen ion; the other hydrogen ion is released into the medium.

In the next step of the respiratory chain, both of these hydrogen ions are passed on to the coenzyme FAD, causing its reduction to $FADH_2$. Simultaneously, the reduced form of the coenzyme NADH is reoxidized, and it is this step that accounts for the regeneration of NAD^+. Again we shall depict only the relevant portion of the FAD molecule.

$$NADH + H^+ \longrightarrow NAD^+ + 2\,H^+ + 2\,e^-$$

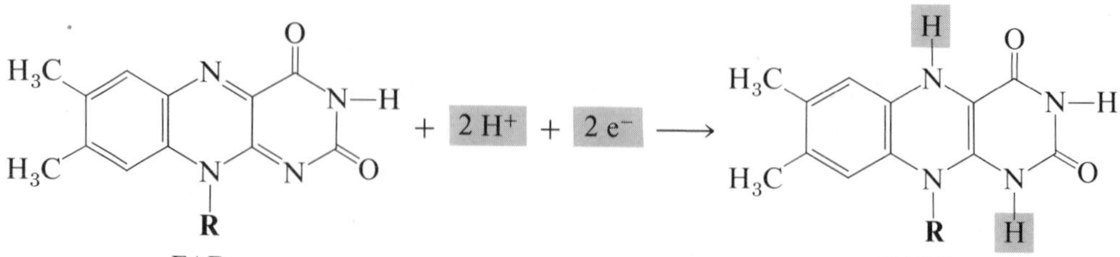

FAD FADH$_2$

(Note that in the oxidation of succinic acid to fumaric acid, FAD accepts two hydrogen ions and two electrons directly from the succinic acid.)

The electrons and the hydrogen ions are then transferred from FADH$_2$ to coenzyme Q (CoQ). This coenzyme is a quinone derivative with a long isoprenoid side chain. Coenzyme Q is also called ubiquinone because it is ubiquitous in all living systems. Several ubiquinones are known, differing only in the number of isoprene units in their side chain. The ubiquinone found in the mitochondria of animal tissues contains ten isoprene units and has been designated as CoQ$_{10}$. The quinone ring is reversibly reducible to a hydroquinone, and so CoQ serves as the electron carrier between the flavin coenzymes and the cytochromes. It is in this step that the oxidized form of the flavin coenzyme, FAD, is regenerated.

$$FADH_2 \longrightarrow FAD + 2\,H^+ + 2\,e^-$$

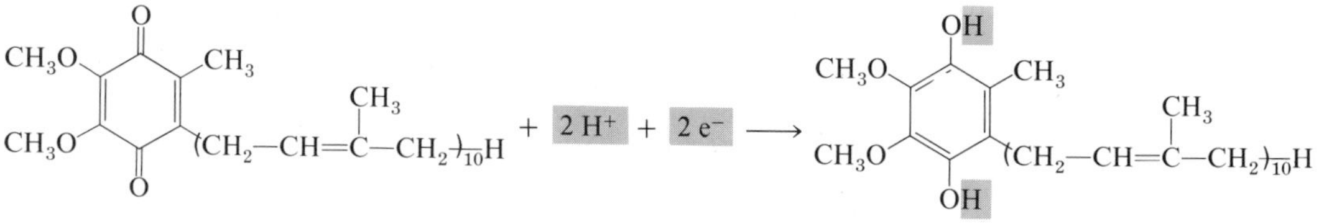

Coenzyme Q$_{10}$ Coenzyme Q$_{10}$H$_2$

A series of compounds called *cytochromes* were probably the first entities to be associated with electron transferring reactions. A number of these substances exist, and we have included five of them in Figure 28.8. The cytochromes are conjugated protein enzymes. Their coenzymes are iron porphyrins, which resemble heme, the pigment in hemoglobin and in myoglobin (Figure 25.1). The various cytochromes differ with respect to (1) their protein constituents, (2) the manner in which the porphyrin is bound to the protein, and (3) the substituents on the periphery of the porphyrin ring. Such slight differences in structure bestow differences in reduction potential upon the different cytochromes. The characteristic feature of the cytochromes is the ability of their iron atoms to exist in either the Fe(II) (ferrous) or Fe(III) (ferric) form. Thus each cytochrome in its oxidized form, Fe(III), can accept one electron and be reduced to the Fe(II)-containing form. This change in oxidation state is reversible, and the reduced form can donate its electron to the next cytochrome, and so on. Only the last cytochrome, *cytochrome* a_3, has the ability to transfer electrons to molecular oxygen.[6] (The combination of cyanide with the ferric ions of cytochrome a_3 completely inhibits the transfer of electrons to molecular oxygen. This accounts for the extreme toxicity of cyanide to living organisms.) The cytochromes are strictly electron carriers; the hydrogens of the reduced coenzyme Q_{10} are released as ions into the medium. (They are utilized in the last step in the reduction of oxygen to water.) Since the Fe(III)/Fe(II) system is only a one-electron exchange, two cytochrome *b* molecules are necessary to complete the oxidation of coenzyme $Q_{10}H_2$.

$$\text{Coenzyme } Q_{10}H_2 \longrightarrow \text{Coenzyme } Q_{10} + 2H^+ + 2\,e^-$$

$$2\text{ Cyt } b\,(Fe^{3+}) + 2\,e^- \longrightarrow 2\text{ Cyt } b\,(Fe^{2+})$$

Then

$$\text{Cyt } b \longrightarrow \text{Cyt } c_1 \longrightarrow \text{Cyt } c \longrightarrow \text{Cyt } a \longrightarrow \text{Cyt } a_3$$

In the final step, two molecules of the terminal electron carrier cytochrome a_3 pass their electrons to molecular oxygen, the ultimate electron acceptor. It has been estimated that about 95% of the oxygen utilized by cells reacts in this single process.

$$2\text{ Cyt } a_3(Fe^{2+}) \longrightarrow 2\text{ Cyt } a_3(Fe^{3+}) + 2\,e^-$$

$$\tfrac{1}{2}O_2 + 2\,H^+ + 2\,e^- \longrightarrow H_2O$$

Each intermediate compound in the respiratory chain is reduced by the addition of electrons and hydrogen ions in one reaction and is subsequently restored to its original form when it delivers the protons and electrons to the next compound. Thus each pair of hydrogen atoms that is removed from the substrates of the Krebs cycle ultimately reduces one atom of oxygen.

28.9 Oxidative Phosphorylation

The *process whereby ATP is synthesized as a result of the operation of the respiratory chain* is referred to as **oxidative phosphorylation** or **respiratory-chain phosphorylation.** The details of the mechanism that links the formation of ATP to the operation of the respiratory chain are largely unknown, although three theories have been proposed (see advanced biochemistry texts). It is known, however, that the energy required for the production of

[6] There is increased speculation that the terminal cytochromes (*a* and a_3) are not separate entities but rather exist as a complex within the same protein. This complex is called *cytochrome oxidase* and is designated as cytochrome aa_3.

ATP results from the passage of a pair of electrons from one carrier to the next. The electron transport system can be thought to function as a biochemical battery; that is, energy is obtained from oxidation-reduction reactions. The sites on the respiratory chain at which the oxidative phosphorylations are believed to occur are shown in Figure 28.8. From the data contained in Table 28.1, we can calculate the maximum amount of energy, E, that is made available when a single pair of electrons travels from NADH to oxygen along the chain.

$$
\begin{array}{rcll}
 & & & E \\
NADH & \longrightarrow & NAD^+ + H^+ + 2\,e^- & +0.32 \\
\frac{1}{2}\,O_2 + 2\,H^+ + 2\,e^- & \longrightarrow & H_2O & +0.82 \\
\hline
NADH + \frac{1}{2}\,O_2 + H^+ & \longrightarrow & NAD^+ + H_2O & +1.14 \text{ volts}
\end{array}
$$

The energy change for the reaction can be obtained by use of the following equation.[7]

$$\text{Energy change} = -nF\,\Delta E$$

where n = number of electrons transferred
F = Faraday's constant (23,000 cal/volt equivalent)

$$
\begin{aligned}
\text{Energy change} &= -(2)(23{,}000)(1.14) \\
&= -52{,}000 \text{ calories}
\end{aligned}
$$

This value of 52,000 calories represents a considerable amount of energy. If it were released all at once, much of it would be dissipated as heat and it might prove damaging to the cell. Therefore, the respiratory chain serves as a device for delivering this energy in small increments to be used to phosphorylate ADP. It has been experimentally observed that three molecules of ATP are formed for every molecule of NADH that is oxidized in the chain (but only two ATP's are formed if the primary acceptor is FAD, as in the case when succinic acid serves as a substrate). The net equation for the respiratory chain is

$$NADH + \tfrac{1}{2}\,O_2 + H^+ + 3\,ADP + 3\,P_i \longrightarrow NAD^+ + H_2O + 3\,ATP$$

Recall that 7300 calories are required for the conversion of one mole of ADP to ATP. It can be determined that almost half of the energy released in the electron transport is conserved in the formation of high energy phosphate bonds.

$$
\begin{aligned}
\text{Energy conserved by respiratory chain} &= \left(\frac{\text{energy conserved}}{\text{energy available}}\right)(100\%) \\
&= \frac{(3)(7300)}{(52{,}000)}(100\%) = 42\%
\end{aligned}
$$

[7]Since the actual concentrations of the various compounds are unknown, the use of reduction potentials can only yield a rough estimate of the actual energy change for the reaction.

28.10 Energy Yield of Carbohydrate Metabolism

It is now possible to summarize the energy conserved (in ATP production) from the complete oxidation of one molecule of glucose. It must be recalled that, under aerobic conditions, the glycolytic sequence terminates with the formation of two molecules of pyruvic acid. Two molecules of ATP are obtained from substrate level phosphorylation. In addition, since the pyruvic acid is not reduced to lactic acid, there are two molecules of NADH (from step 6) that remain in the cytoplasm. We know that NAD^+ must be regenerated from NADH for glycolysis to continue. The problem, however, is that NADH cannot pass across the membrane of the mitochondrion to be oxidized by the respiratory chain. A solution is achieved by a shuttle process involving glycerol 3-phosphate and dihydroxyacetone phosphate. These compounds can penetrate the mitochondrial barrier.

The first step in this shuttle (Figure 28.9) is the reduction of dihydroxyacetone phosphate by NADH to form glycerol 3-phosphate and to regenerate the NAD^+. Glycerol 3-phosphate then enters the mitochondrion where it is reoxidized to dihydroxyacetone phosphate, this time by a dehydrogenase that contains the coenzyme FAD instead of NAD^+. The dihydroxyacetone phosphate exits the mitochondrion and returns to the cytoplasm to complete the shuttle. The reduced form of the flavin coenzyme, $FADH_2$, is reoxidized by passing its electrons to coenzyme Q_{10} in the respiratory chain. As a result, two molecules of ATP (instead of three) are formed every time a cytoplasmic NADH molecule is reoxidized via the glycerol phosphate shuttle and the respiratory chain.

The aerobic continuation of glycolysis yields a total of 15 molecules of ATP from each molecule of pyruvic acid, or 30 molecules of ATP per molecule of glucose. Table 28.2 lists the various reactions that result in ATP synthesis.

$$C_6H_{12}O_6 + 6\ O_2 + 36\ ADP + 36\ P_i \longrightarrow 6\ CO_2 + 6\ H_2O + 36\ ATP$$

The above is the net equation for the oxidation of one mole of glucose in aerobic cells. The energy released (686,000 calories) is coupled to the syn-

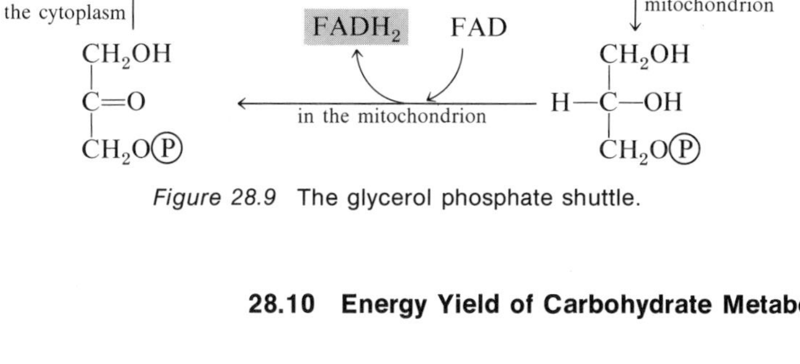
Dihydroxyacetone phosphate
Glycerol 3-phosphate
NADH + H+
NAD+
in the cytoplasm
diffuses into the cytoplasm
diffuses into the mitochondrion
FADH2
FAD
in the mitochondrion

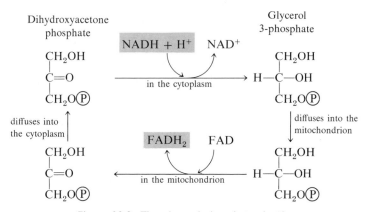

Figure 28.9 The glycerol phosphate shuttle.

Table 28.2 Production of ATP During the Oxidation of Glucose to Carbon Dioxide and Water

Reaction	*Type of Phosphorylation*	*Number of ATP Molecules Formed*
Glucose $\longrightarrow$ 2 Pyruvic acid	Substrate level	2
Glyceraldehyde 3-phosphate $\xrightarrow[FAD \searrow\ \nearrow FADH_2]{NAD^+ \searrow\ \nearrow NADH}$ 1,3-Diphosphoglyceric acid	Oxidative	2×2
Pyruvic acid $\xrightarrow{NAD^+ \searrow\ \nearrow NADH}$ Acetyl CoA	Oxidative	2×3
Isocitric acid $\xrightarrow{NAD^+ \searrow\ \nearrow NADH}$ Oxalosuccinic acid	Oxidative	2×3
α-Ketoglutaric acid $\xrightarrow{NAD^+ \searrow\ \nearrow NADH}$ Succinyl CoA	Oxidative	2×3
Succinyl CoA $\xrightarrow{GDP \searrow\ \nearrow GTP}$ Succinic acid	Substrate level	2×1
Succinic acid $\xrightarrow{FAD \searrow\ \nearrow FADH_2}$ Fumaric acid	Oxidative	2×2
Malic acid $\xrightarrow{NAD^+ \searrow\ \nearrow NADH}$ Oxaloacetic acid	Oxidative	2×3
	Sum	36

thesis of 36 moles of ATP from ADP and inorganic phosphate. Assuming that 7300 calories is required for the synthesis of each mole of ATP, then 36 × 7300 or 263,000 calories is conserved by the cell. The efficiency of conservation is thus:

$$\frac{(263{,}000)}{(686{,}000)}(100\%) = 38\%$$

This recovery of energy compares favorably with the efficiency of any man-made machine, and it represents a remarkable achievement on the part of the living organism.

Exercises

28.1 Define, or give an explanation for each of the following terms.

- **(a)** digestion
- **(b)** metabolism
- **(c)** anabolism
- **(d)** catabolism
- **(e)** metabolite
- **(f)** glycogenesis
- **(g)** blood-sugar level
- **(h)** hyperglycemia
- **(i)** hypoglycemia
- **(j)** renal threshold
- **(k)** anaerobic
- **(l)** aerobic

28.2 Briefly discuss the digestion and absorption of carbohydrates, enumerating the enzymes involved, and the sites along the gastrointestinal tract where catalysis occurs.

28.3 Carbohydrates are normally ingested at every meal, yet the body maintains a constant blood-sugar level. Summarize the factors that operate to maintain the blood-sugar level.

28.4 How does glycolysis differ from fermentation?

28.5 Distinguish between a *phosphatase* and a *phosphorylase*.

28.6 When *aldolase* catalyzes the cleavage of ketose mono- and diphosphates, dihydroxyacetone phosphate is always one of the products. Complete each of the following catalytic reactions.

(a)

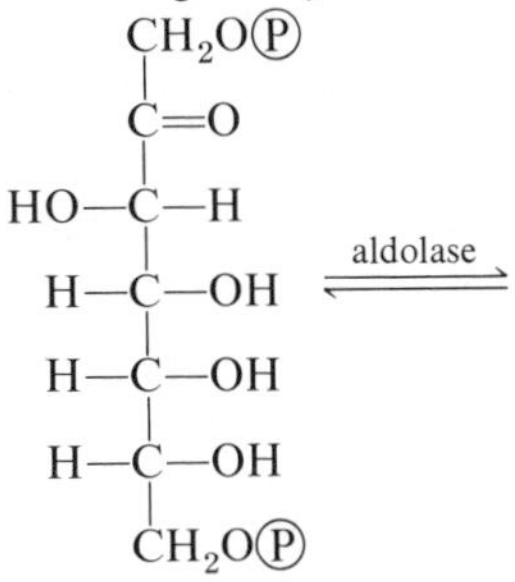
CH2O(P)
C=O
HO—C—H
H—C—OH
H—C—OH
H—C—OH
CH2O(P)
aldolase

Sedoheptulose
1,7-diphosphate

(b) fructose 1-phosphate $\xrightleftharpoons{\text{aldolase}}$

(c) erythrulose 1-phosphate $\xrightleftharpoons{\text{aldolase}}$

28.7 Outline, in detail, the sequence of reactions that occur in the Embden-Meyerhof pathway.

28.8 Label carbon numbers three and four in the glucose molecule.
(a) Which of these two carbon atoms will appear in lactic acid?
(b) Which of these two carbon atoms will appear in ethanol or in carbon dioxide?

28.9 Contrast the types of reactions catalyzed by mutases and isomerases.

28.10 Which is the oxidative step in the Embden–Meyerhof pathway? How is the reduced coenzyme reoxidized by aerobic cells? How is it reoxidized in muscle cells? How is it reoxidized in yeast cells?

28.11 The enzyme lactic acid dehydrogenase is not specific for pyruvic acid, but will catalyze the reduction of other keto acids. Write an equation for the enzymic reduction of phenylpyruvic acid. Does this surprise you in light of what was said about the stereochemical specificity of this enzyme in Section 26.6-b? Comment.

28.12 Write a balanced half-reaction for each of the following conversions.
(a) glucose $\longrightarrow$ pyruvic acid
(b) pyruvic acid $\longrightarrow$ lactic acid
(c) pyruvic acid $\longrightarrow$ ethanol + carbon dioxide

28.13 What is the fate of the lactic acid formed by muscular activity?

28.14 How much energy would be required to convert six moles of pyruvic acid to three moles of glucose?

28.15 List two reactions of glycolysis in which high energy phosphate bonds are formed.

28.16 Write the structural formula of creatine phosphate. What is its role in muscle contraction?

28.17 Outline, in detail, the sequence of reactions that occur in the Krebs cycle. What is the function of this cycle in carbohydrate metabolism?

28.18 Select the steps in the Krebs cycle that involve each of the following.
(a) addition of water
(b) addition of coenzyme A
(c) removal of coenzyme A
(d) removal of water
(e) oxidation reactions
(f) removal of carbon dioxide

28.19 Outline the electron transport system. What is its function?

28.20 Write balanced half-reactions for all of the oxidation-reduction reactions that occur in the Krebs cycle and in the respiratory chain.

28.21 How is the energy of carbohydrate metabolism made available to the body cells?

28.22 Only two molecules of ATP are produced by the aerobic conversion of **(a)** succinic acid to fumaric acid and **(b)** glyceraldehyde 3-phosphate to 1,3-diphosphoglyceric acid. Explain.

28.23 Write equations for all the substrate level phosphorylation reactions that occur during the metabolism of carbohydrates.

28.24 Name the direct precursor molecule of each of the following metabolites.

(a) fructose 1,6-diphosphate
(b) 1,3-diphosphoglyceric acid
(c) 2-phosphoglyceric acid
(d) ethanol
(e) acetyl CoA
(f) oxalosuccinic acid
(g) succinyl CoA
(h) L-malic acid

28.25 Give the names of eight enzymes that participate in carbohydrate metabolism and specify which reaction each one catalyzes.

28.26 Write the net equation for the oxidation of glucose in the body.

28.27 What purpose do carbohydrates fulfill in metabolism? In what respect does carbohydrate metabolism resemble the burning of table sugar in a pan? In what way does it differ?

29

Lipid Metabolism

Nearly all the energy required by the animal organism is generated by the oxidation of carbohydrates and lipids. Carbohydrates provide a readily available source of energy, and lipids function as the principal energy reserve. The amount of energy stored in fats far exceeds that stored in the form of glycogen.

Of all major nutrients, triacylglycerols are the richest energy source. The oxidation of one gram of a typical lipid liberates about 9000 calories; the oxidation of an equal mass of carbohydrate liberates only about 4000 calories. A lipid molecule has a much higher proportion of carbon and hydrogen than a carbohydrate molecule. Therefore, lipids have a greater capacity to combine with oxygen, and consequently have a higher heat content.[1]

The nutritional aspects of lipids are still not completely understood. Lipids are not dietary necessities; an organism can survive on a lipid-free diet if carbohydrates and proteins are supplied as a source of metabolic energy. Certain lipids, however, are required for normal growth and development. These lipids supply certain fatty acids which the organism cannot synthesize. Prominent among these *essential* fatty acids are the unsaturated acids containing more than one double bond, linoleic, linolenic and arachidonic acids. (Recall from Section 24.11 that these fatty acids are required as

[1] A large percentage of the lipid molecule is of the nature of a saturated hydrocarbon (i.e., highly reduced). Thus, fats may be thought to be analogous to combustible petroleum products, whereas carbohydrates may be considered to be analogous to alcohols (more oxidized), which are not nearly as effective as fuels.

precursors in the biosynthesis of prostaglandins.) The essential fatty acids are necessary for maintaining the structure of cell membranes and for the efficient transport and metabolism of cholesterol.

29.1 Digestion and Absorption of Lipids

Lipids are not digested by the body until they reach the upper portion of the small intestine. In this region, a hormone is secreted that stimulates the gall bladder to discharge bile into the intestine. The bile acts as an emulsifier, and serves to disrupt some of the hydrophobic bonds that tend to hold the lipid molecules together. This emulsification process is essential to lipid digestion. It converts the water-insoluble lipids into smaller globules, which are more susceptible to attack by the water-soluble, fat-splitting enzymes. Another hormone then promotes the secretion of the pancreatic juice. This juice contains the lipases that catalyze the digestion of lipids into diacylglycerols, monoacylglycerols, fatty acids, and glycerol. The hydrolysis of lipids probably occurs in a stepwise fashion, each step requiring a different lipase.

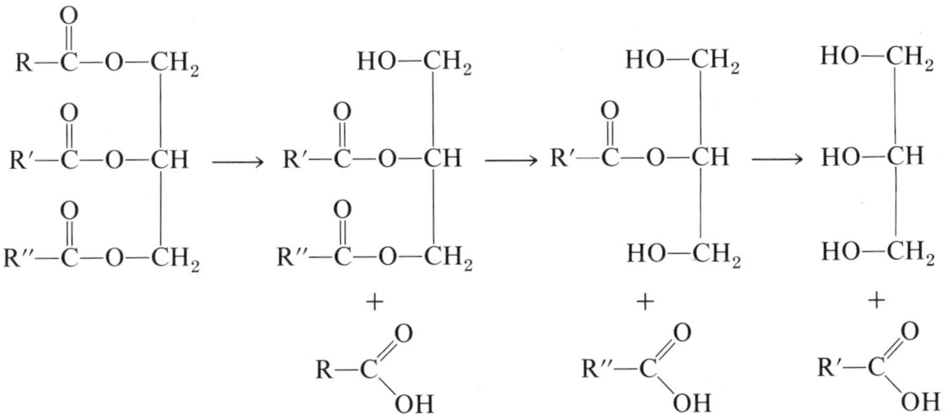

Biochemists do not agree upon the extent to which a lipid must be hydrolyzed in order to be absorbed. It is probable that some lipids after emulsification are absorbed directly through the intestinal membrane before being hydrolyzed. Once the diacylglycerols, monoacylglycerols, and fatty acids pass into the cells of the intestinal epithelium, they are immediately resynthesized into triacylglycerols or phospholipids. It has been shown experimentally that the glycerol needed to form these compounds is supplied by the metabolic pool in the intestinal cells and not from the glycerol obtained by lipid hydrolysis. As we shall see, these latter glycerol molecules are absorbed and incorporated into the glycolytic sequence. The acylglycerols then become associated with proteins and are transported into the bloodstream via the lymphatic system.[2] Some of the fat is carried to the liver

[2] Lymph is a clear fluid, similar to blood but without red corpuscles. The blood system and the lymphatic system are separate, but there is a crossover point near the thoracic duct.

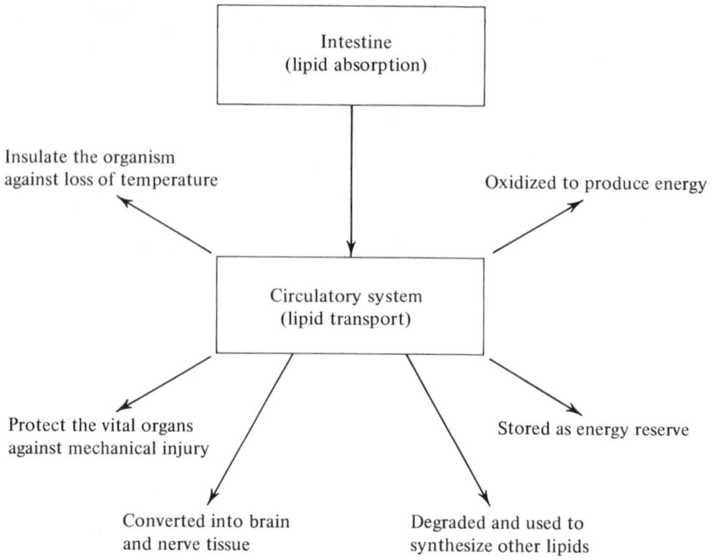
Intestine
(lipid absorption)
Insulate the organism
against loss of temperature
Oxidized to produce energy
Circulatory system
(lipid transport)
Protect the vital organs
against mechanical injury
Stored as energy reserve
Converted into brain
and nerve tissue
Degraded and used to
synthesize other lipids

Figure 29.1 Metabolic fate of lipids.

where it is utilized to provide energy for cellular functions. The liver is the organ responsible for maintaining the proper physiological lipoprotein concentration.

An early theory of lipid metabolism held that a portion of ingested food was utilized and that the remainder was stored in adipose tissue as *fat depots*. The stored fat was believed to be metabolically inactive, excess lipid that remained undisturbed until needed by the body. Through the use of tracer experiments, however, we now know that lipids are in a continuous state of dynamic equilibrium; they are constantly being transported, degraded, utilized, and resynthesized.

The human body has a large capacity for the storage of fat, a fact all too evident to many people. Fat depots, such as those directly beneath the skin and in the abdominal region, are specialized tissues in which a relatively large percentage of cytoplasm is replaced by lipid. Adipose tissue is the only tissue in which free triacylglycerols occur in appreciable amounts. Elsewhere, in cells or in the blood plasma, lipids are bound to proteins as lipoprotein complexes. The fate of lipids in the animal body is outlined in Figure 29.1.

29.2 Fatty Acid Oxidation

Before triacylglycerols are used for the production of energy, they are cleaved to fatty acids and glycerol by the lipases. The glycerol is converted to a carbohydrate derivative (see Section 29.5) and is metabolized via the Embden–Meyerhof pathway. The fatty acids, which contain the bulk of the energy of the lipids, are broken down in a series of sequential reactions accompanied by the gradual release of utilizable energy. Some of these reactions are oxidative and require the same coenzymes (NAD^+, FAD) as those which take part in the oxidation of carbohydrates. The enzymes involved in fatty acid catabolism are localized in the mitochondria along with the enzymes of the Krebs cycle, respiratory chain, and oxidative phosphorylation. This factor (the localization within the mitochondria) is of the utmost importance since it provides for the efficient utilization of the energy stored in the fatty acid molecules.

In the beginning of the twentieth century, the German biochemist Franz Knoop showed that the breakdown of fatty acids occurs in a stepwise fashion with the removal of two carbon atoms at a time. Subsequent investigations have resulted in the separation and purification of the enzymes involved. The details of the degradation sequence are now fully understood. The reaction scheme is shown in Figure 29.2.

Step 1. Fatty acids, like carbohydrates and amino acids, are relatively inert and must first be activated by conversion to an energy-rich fatty acid derivative of coenzyme A (called *fatty acyl CoA*). This activation process is a two-step reaction that is catalyzed by the enzyme *acyl CoA synthetase.* For each molecule of fatty acid that is activated, one molecule of coenzyme A and one molecule of ATP are used. First the fatty acid reacts with ATP to form a fatty acyl adenylate. (Note the similarity to the activation of amino acids prior to protein synthesis—Section 27.5a.) Then the sulfhydryl group of coenzyme A attacks the acyl adenylate to yield fatty acyl CoA and AMP. Finally, the pyrophosphate formed in the initial reaction is hydrolyzed to phosphate ions by the enzyme *pyrophosphatase.* The net effect is the utilization of two high-energy bonds of ATP to activate each molecule of a fatty acid.

$$\underset{\text{Fatty acid}}{R{-}CH_2CH_2CH_2CH_2CH_2C(=O)OH} + \underset{\text{ATP}}{AMP{\sim}P{\sim}P} \xrightarrow{\text{synthetase}} \underset{\text{Fatty acyl adenylate}}{R{-}CH_2CH_2CH_2CH_2CH_2C(=O)O{\sim}AMP} + PP_i$$

$$\underset{\text{Fatty acyl adenylate}}{R{-}CH_2CH_2CH_2CH_2CH_2C(=O)O{\sim}AMP} + \underset{\text{Coenzyme A}}{HSCoA} \xrightarrow{\text{synthetase}} \underset{\text{Fatty acyl CoA}}{R{-}CH_2CH_2CH_2CH_2CH_2C(=O)SCoA} + AMP$$

$$PP_i + HOH \xrightarrow{\text{pyrophosphatase}} 2\ P_i$$

The fatty acyl CoA now enters into a sequence of reactions known as *β-oxidation* (since the carbon beta to the carboxyl group undergoes successive oxidations), or the *fatty acid spiral* (because the fatty acid goes through the cycle again and again until it is finally degraded to acetyl CoA molecules). The reactions involved in steps 2 through 4, dehydrogenation, hydration, and dehydrogenation, are analogous to those involved in the conversion

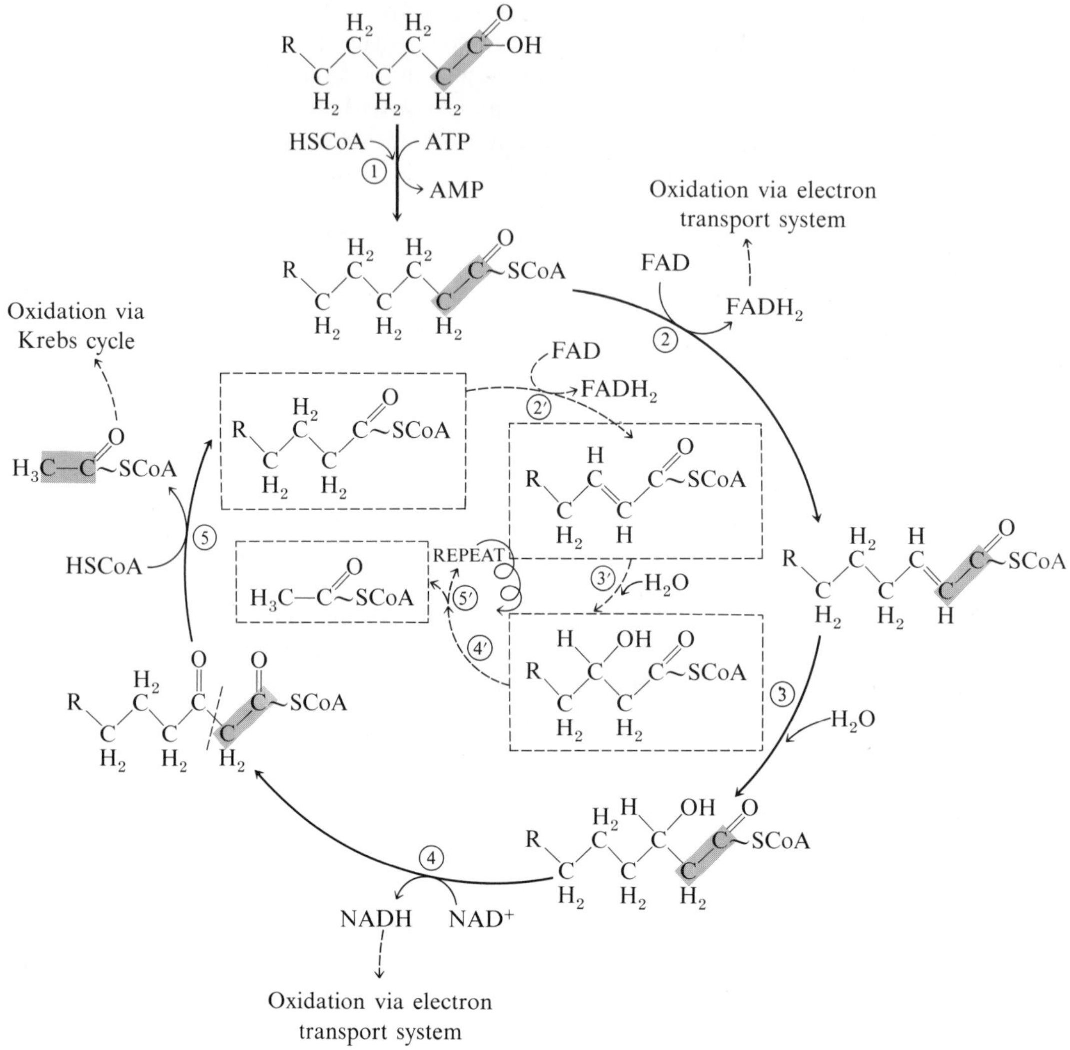

HSCoA
ATP
AMP
Oxidation via electron transport system
FAD
FADH$_2$
Oxidation via Krebs cycle
REPEAT
HSCoA
H_2O
NADH
NAD$^+$
Oxidation via electron transport system

Figure 29.2 Fatty acid oxidation (β-oxidation).

of succinic acid to oxaloacetic acid (succinic ⟶ fumaric ⟶ malic ⟶ oxaloacetic). The functional group at the beta carbon can be visualized as undergoing the following changes: alkane ⟶ alkene ⟶ secondary alcohol ⟶ ketone. The net effect is the introduction of a keto group beta to a carboxyl group.

Step 2. This first oxidation reaction is catalyzed by the enzyme *acyl dehydrogenase*. The coenzyme FAD accepts two hydrogen atoms from adjacent carbons, one from the α-carbon and one from the β-carbon. The enzyme is stereospecific in that only the *trans* alkene is obtained.

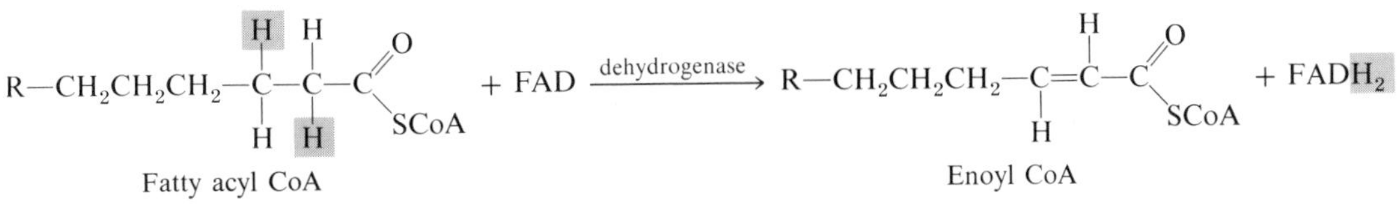
dehydrogenase
Fatty acyl CoA
Enoyl CoA

Each molecule of $FADH_2$ that is reoxidized via the electron transport chain supplies energy to form two molecules of ATP.

Step 3. Again, as in the Krebs cycle, only the L-isomer is formed when the stereospecific enzyme *enoyl CoA hydrase* adds water across the *trans* double bond.

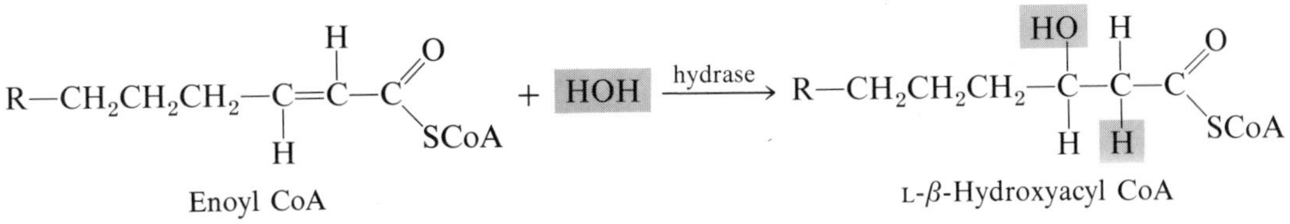
hydrase
Enoyl CoA
L-β-Hydroxyacyl CoA

Step 4. The second oxidation reaction is catalyzed by *L-β-hydroxyacyl dehydrogenase,* which exhibits an absolute stereospecificity for the L-isomer. Here the coenzyme NAD^+ is the hydrogen acceptor.

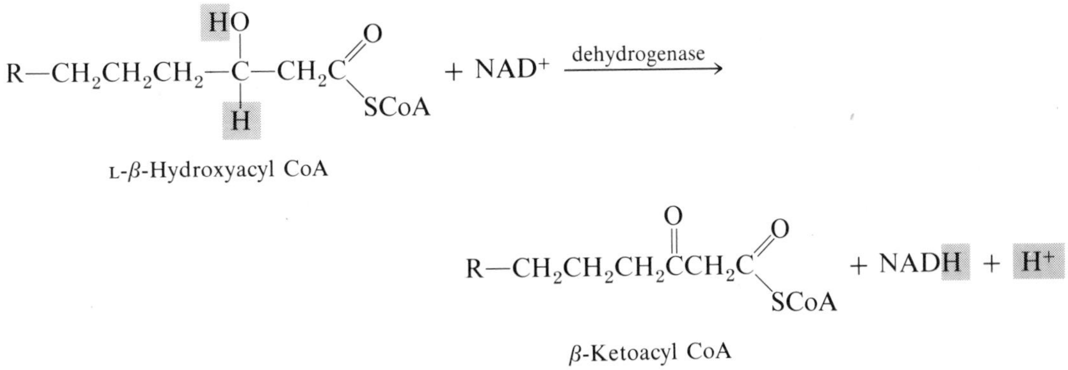
dehydrogenase
L-β-Hydroxyacyl CoA
β-Ketoacyl CoA

The reoxidation of NADH and the transport of hydrogen ions and electrons through the respiratory chain furnishes three molecules of ATP.

Step 5. The final reaction is a cleavage of the β-ketoacyl CoA by a molecule of coenzyme A. The products of the reaction are acetyl CoA and a fatty acyl CoA whose chain length is shortened by two carbon atoms. The enzyme is *β-ketothiolase.*

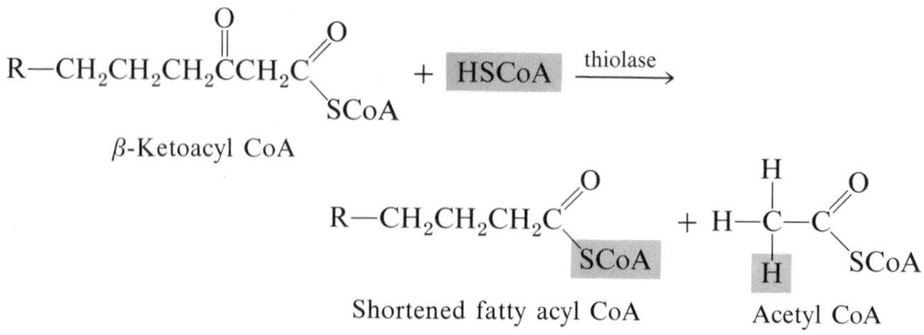

$$R-CH_2CH_2CH_2\overset{\overset{\displaystyle O}{\|}}{C}CH_2\overset{\overset{\displaystyle O}{\diagup}}{C}{\diagdown}_{SCoA} + \boxed{HSCoA} \xrightarrow{\text{thiolase}}$$

β-Ketoacyl CoA

$$R-CH_2CH_2CH_2\overset{\overset{\displaystyle O}{\diagup}}{C}{\diagdown}_{\boxed{SCoA}} + \overset{\displaystyle H}{\underset{\displaystyle H}{H-C-}}\overset{\overset{\displaystyle O}{\diagup}}{C}{\diagdown}_{SCoA}$$

Shortened fatty acyl CoA   Acetyl CoA

The newly formed fatty acyl CoA can then be degraded further by repetition of steps 2 through 5; a molecule of acetyl CoA is liberated at each turn of the spiral. Normally, the spiral is repeated as many times as is necessary to break down a fatty acid containing an even number of carbon atoms, n, into $n/2$ molecules of acetyl CoA. It should be noted that no further addition of ATP is necessary since the shortened fatty acids already contain the thiol ester. One molecule of ATP is sufficient to activate any fatty acid regardless of the number of carbon atoms in its hydrocarbon chain. The unsaturated fatty acids are also incorporated into the β-oxidation sequence. Two ancillary enzymes are required to convert their degradation products into normal substrates of the fatty acid spiral.

The acetyl CoA formed in fatty acid oxidation may enter the Krebs cycle or may be utilized in a variety of synthetic reactions (e.g., formation of triacylglycerols, phospholipids, cholesterol). The overall equation for one turn of the fatty acid degradation spiral is as follows.

$$CH_3(CH_2)_n\overset{\overset{\displaystyle O}{\diagup}}{C}{\diagdown}_{SCoA} + FAD + NAD^+ + HSCoA + H_2O \longrightarrow$$

$$CH_3(CH_2)_{n-2}\overset{\overset{\displaystyle O}{\diagup}}{C}{\diagdown}_{SCoA} + CH_3\overset{\overset{\displaystyle O}{\diagup}}{C}{\diagdown}_{SCoA} + FADH_2 + NADH + H^+$$

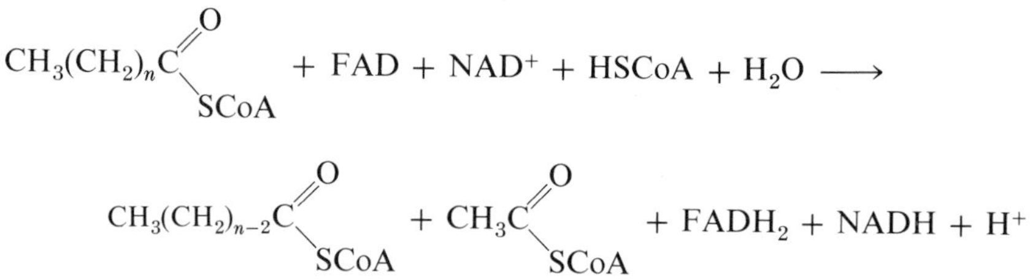

29.3 Bioenergetics of Fatty Acid Oxidation

The combustion of one mole of palmitic acid releases a considerable amount of energy, and the precise amount can be determined by conducting the experiment in a calorimeter.

$$C_{16}H_{32}O_2 + 23\ O_2 \longrightarrow 16\ CO_2 + 16\ H_2O + 2{,}340{,}000\ \text{calories}$$

The amount of energy that is made available to the cell and conserved in the form of ATP is readily calculable.

The breakdown by an organism of one mole of palmitic acid, necessitates the utilization of one mole of ATP, and eight moles of acetyl CoA are formed. You will recall that each mole of acetyl CoA metabolized by the Krebs cycle yields twelve moles of ATP. Each turn of the fatty acid spiral produces one mole of NADH and one mole of $FADH_2$. Reoxidation of these compounds by the respiratory chain yields three and two moles of ATP, respectively. The complete degradation of palmitic acid requires seven turns of the spiral, and a total of seven moles of $FADH_2$ and NADH is therefore formed. The energy calculations may be summarized as follows.

	Yield of ATP
1 mole of ATP is split to AMP and 2 P_i	−2
8 moles of acetyl CoA formed (8 × 12)	96
7 moles of $FADH_2$ formed (7 × 2)	14
7 moles of NADH formed (7 × 3)	21
Total moles of ATP	129

The percentage of available energy that can theoretically be conserved by the cell in the form of ATP is calculated as follows:

$$\frac{\text{(energy conserved)}}{\text{(total energy available)}}(100\%) = \frac{(129)\ (7300)}{(2{,}340{,}000)}(100\%) = 40\%$$

The efficiency with which fatty acids are metabolized is seen to be comparable to that of the carbohydrate metabolism system.

29.4 Ketosis

In general, all of the acetyl CoA formed in the breakdown of fatty acids is utilized by the body. Any abnormal physiological condition that causes cells to metabolize their stored fats at an accelerated rate will result in the production of more acetyl CoA than can be oxidized by the Krebs cycle or utilized in synthetic reactions. Such a condition generally occurs in conjunction with an impairment of carbohydrate metabolism, for example, fasting, diabetes mellitus, or certain types of cellular poisoning. In these cases, very little carbohydrate is available to the cell, and therefore the production of oxaloacetic acid is drastically reduced.[3] Recall that acetyl CoA enters the Krebs cycle by condensing with oxaloacetic acid. If this latter compound is not present in sufficient quantity, acetyl CoA will accumulate. Furthermore, in the absence of carbohydrate metabolism, the rate of fatty acid oxidation increases to meet the energy demands of the cell, and the buildup of acetyl

[3]This deficiency is probably caused by a lack of pyruvic acid, which can be converted into oxaloacetic acid by a carboxylation reaction. The energy required for this reaction is supplied by ATP.

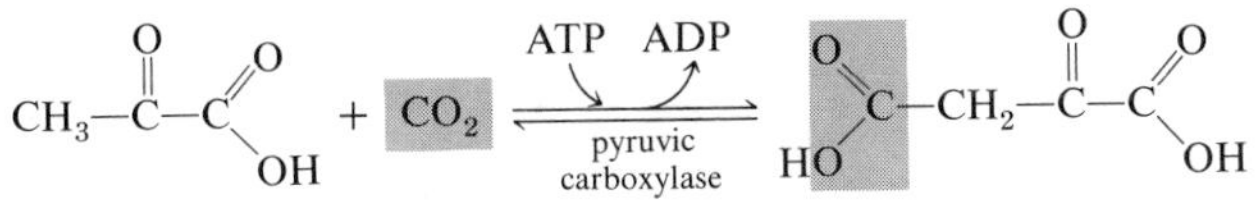
ATP ADP
pyruvic carboxylase

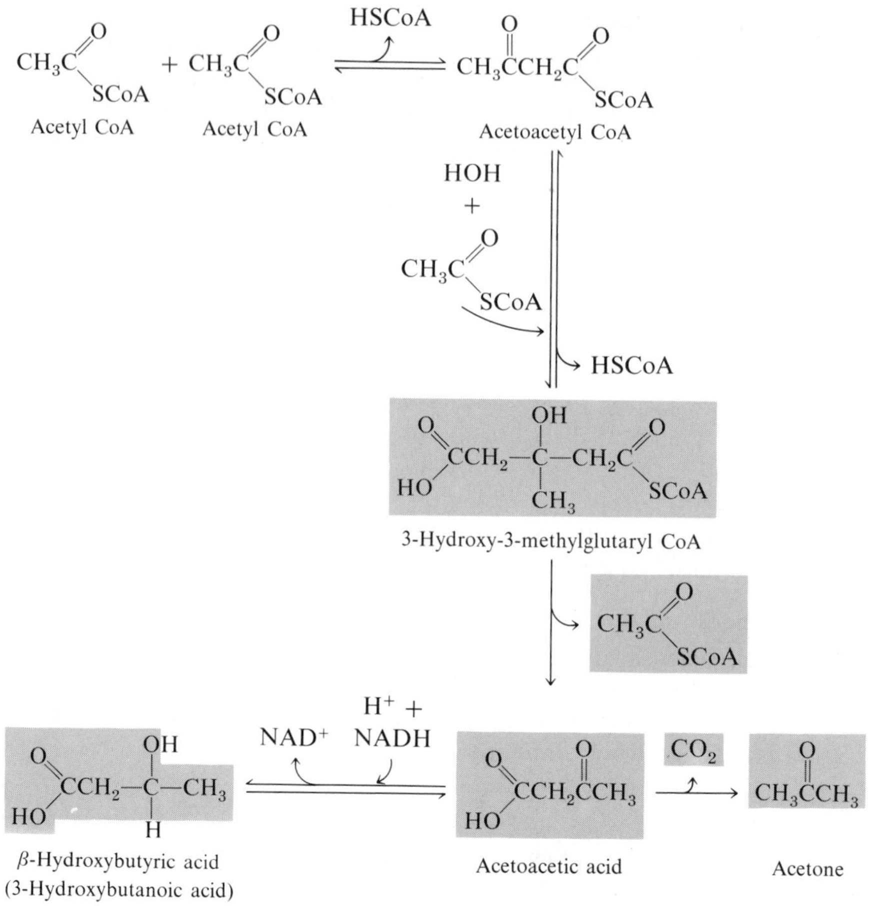

HSCoA
Acetyl CoA
Acetyl CoA
Acetoacetyl CoA
HOH
HSCoA
3-Hydroxy-3-methylglutaryl CoA
H+ +
NAD+
NADH
CO2
β-Hydroxybutyric acid
(3-Hydroxybutanoic acid)
Acetoacetic acid
Acetone

Figure 29.3 The formation of ketone bodies.

CoA increases accordingly. When the concentration of acetyl CoA in the cell reaches a certain level, a reversal of the last step of the fatty acid spiral occurs to produce acetoacetyl CoA. The acetoacetyl CoA reacts with another molecule of acetyl CoA and with water to form 3-hydroxy-3-methylglutaryl CoA, which is then cleaved to acetoacetic acid and acetyl CoA. Since some acetoacetic acid is normally produced in fatty acid metabolism, cells possess a mechanism for its oxidation. (It is believed that acetoacetic acid may serve as an energy source when the cell's supply of glucose is diminished.) When, however, the rate of formation of acetoacetic acid exceeds that of its disposal, the excess amount may either be reduced to form β-hydroxybutyric acid or be decarboxylated to CO_2 and acetone.[4]

Acetoacetic acid, β-hydroxybutyric acid, and acetone are collectively referred to as *ketone bodies*, and their accumulation in the blood (at concentrations greater than 1 mg/100 ml) and urine leads to a condition known as

[4] Since acetone is a fairly volatile compound, it is often expelled in the breath. Thus *acetone breath* is a symptom of ketosis and is frequently noticed in diabetics.

ketosis. A general scheme for the formation of ketone bodies is shown in Figure 29.3. Since two of the three ketone bodies are acids, their presence in the blood in excessive amounts causes a marked decrease in its pH. This decrease in the pH of the blood leads to a more serious condition known as *acidosis*. The body tends to lose fluids (become dehydrated) as the kidneys discharge large quantities of water in an attempt to eliminate these excess acids. Acidosis results in interference with the transport of oxygen by the hemoglobin molecule (see Section 31.5), and a fatal coma may result. Any treatment that promotes the utilization of carbohydrates (for example, an injection of insulin) can alleviate both ketosis and acidosis.

29.5 Glycerol Metabolism

One molecule of glycerol is obtained from each molecule of triacylglycerol or phospholipid that is hydrolyzed. Glycerol is readily incorporated into the scheme of carbohydrate metabolism by conversion to dihydroxyacetone phosphate, a key intermediate in the glycolytic pathway. This conversion includes the phosphorylation of a primary hydroxyl group followed by the oxidation of a secondary hydroxyl group to a ketone. This last reaction is readily reversed whenever glycerol phosphate is required for the synthesis of triacylglycerols (see Section 29.7).

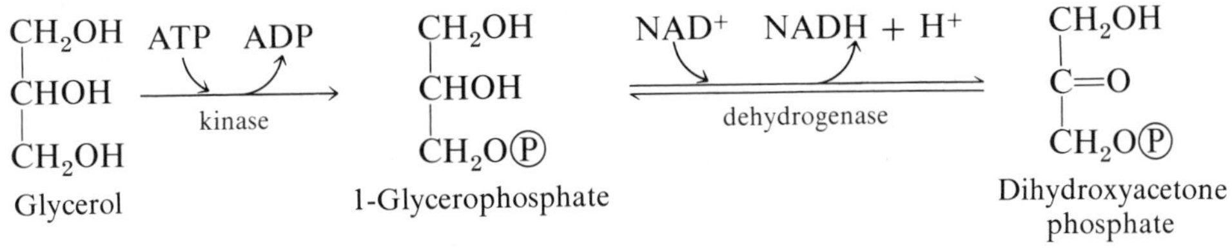

CH₂OH
CHOH
CH₂OH
Glycerol
ATP
ADP
kinase
CH₂OH
CHOH
CH₂O(P)
1-Glycerophosphate
NAD+
NADH + H+
dehydrogenase
CH₂OH
C=O
CH₂O(P)
Dihydroxyacetone phosphate

29.6 Synthesis of Fatty Acids

Both plant and animal organisms can convert their excess carbohydrates into lipids. Animal cells are unable to accomplish the reverse process, whereas plant seeds rapidly convert their fat deposits to sucrose upon germination. Apparently animal tissues lack essential enzymes that are present in the seeds.

We have seen that acetyl CoA is a common intermediate of both fatty acid and carbohydrate degradation; therefore it is logical to assume that this compound is the starting material for the synthesis of fatty acids. The original hypothesis was that the reactions of the fatty acid spiral were reversible, and that fatty acid biosynthesis proceeded by a simple reversal of the degradation reactions. However, in the early 1950s, the enzymes of the β-oxidation sequence were isolated, purified, and shown to be incapable of catalyzing the biosynthesis of fatty acids from acetyl CoA. These crucial experiments launched many investigators on an active search for a suitable pathway for the synthetic process. Thanks largely to the work of Feodor Lynen (Nobel Prize for Medicine, 1964), we now know that the enzymes which catalyze fatty acid synthesis are found in the cytoplasm rather than in the mitochondria where the enzymes for fatty acid degradation are located. Furthermore, all of these enzymes are associated in a single enzyme complex known as *fatty acid synthetase.* The various intermediates produced during the synthetic process are covalently bound to the synthetase complex by sulfhydryl groups (—SH). No free intermediates are released, and the correct sequence of reactions is governed by the position of each enzyme in the complex. In most organisms the end product of the fatty acid synthetase system is palmitic acid, which serves as the precursor of all other higher saturated fatty acids as well as all of the unsaturated fatty acids.

In 1959, several researchers observed that the biosynthesis of fatty acids requires ATP, CO_2, and the coenzyme *biotin* (Table 26.2). It was postulated that the first step is the carboxylation of acetyl CoA to malonyl CoA in the presence of the enzyme *acetyl CoA carboxylase.*

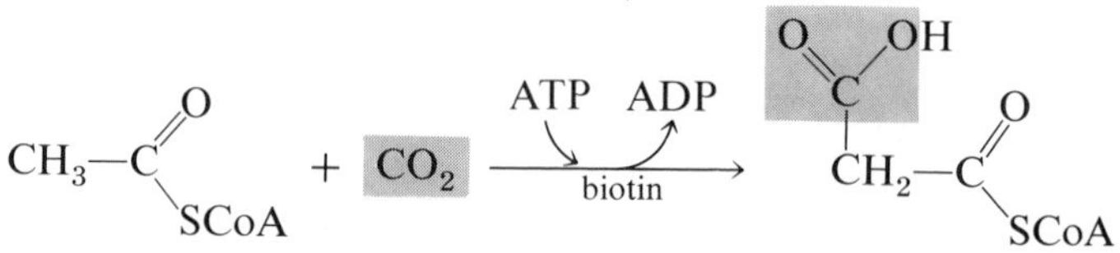
ATP ADP
biotin

Malonyl CoA

The mechanism of the reaction is quite complex. Present evidence indicates that carbon dioxide covalently complexes with biotin, and the CO_2 is then transferred to acetyl CoA in a subsequent step. The hydrolysis of ATP provides the energy for the formation of the CO_2–biotin complex.

$$CO_2 + \text{Biotin} + \text{ATP} \longrightarrow CO_2\text{–biotin} + \text{ADP} + P_i$$

$$CO_2\text{–biotin} + \text{Acetyl CoA} \longrightarrow \text{Malonyl CoA} + \text{Biotin}$$

A relatively low molar mass protein (~10,000 daltons) has been isolated from *E. coli* and has been identified as a key component in fatty acid biosynthesis. This protein, called *acyl carrier protein* (ACP–SH), contains a sulfhydryl group that covalently bonds to the acyl intermediates. The function of ACP–SH in fatty acid biosynthesis is to serve as a sort of anchor to which the acyl intermediates are attached. This process is described as *transacylation*—an acyl group is transferred from a coenzyme A thioester to the sulfhydryl group of the acyl carrier protein. The acetyl group is subsequently transferred to the sulfhydryl group of a second enzyme (a *synthase*), of the fatty acid synthetase system.

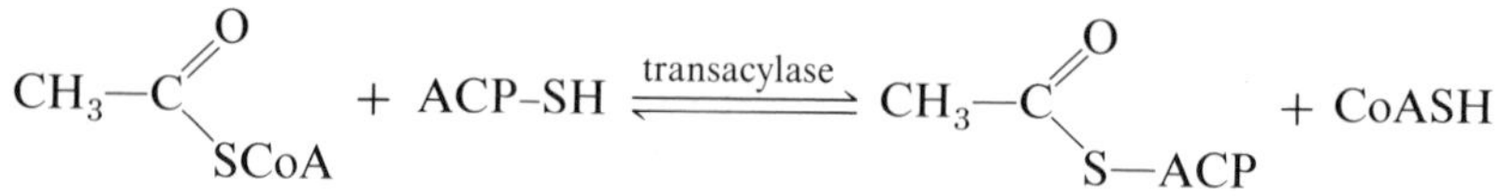
+ ACP–SH
transacylase
+ CoASH

Acetyl CoA — Acetyl-S-ACP

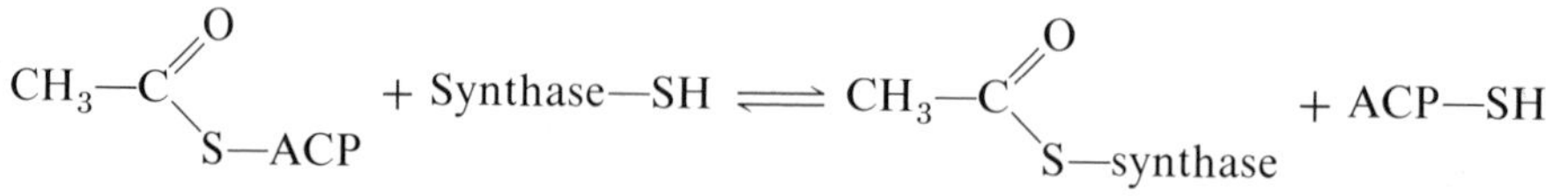
+ Synthase—SH
+ ACP—SH

In the next reaction, the malonyl CoA formed initially reacts with ACP–SH to yield malonyl–S–ACP. Malonyl–S–ACP and acetyl–S–synthase then combine to form acetoacetyl–S–ACP. Notice that the carbon dioxide released in this reaction is the same CO_2 that was required for the formation of malonyl CoA from acetyl CoA. Carbon dioxide can thus be considered to play a catalytic role in fatty acid biosynthesis. It is utilized in one step and regenerated in a later step.

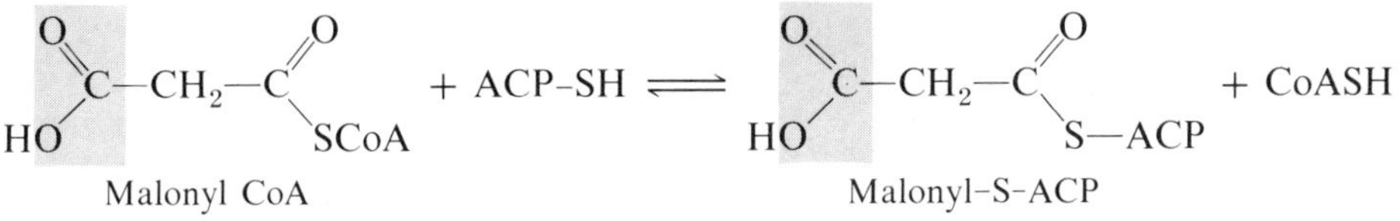
+ ACP-SH
+ CoASH
Malonyl CoA
Malonyl-S-ACP

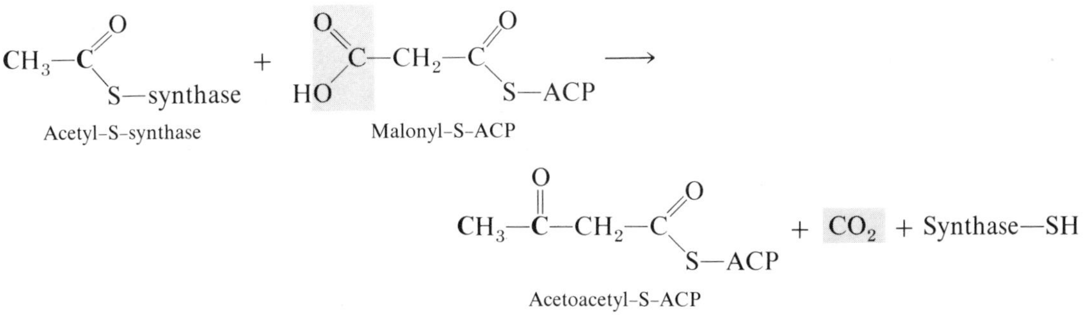
Acetyl-S-synthase
Malonyl-S-ACP
+ CO2 + Synthase—SH
Acetoacetyl-S-ACP

The next three reactions in the anabolic sequence are reduction, dehydration, and a second reduction. These processes are essentially the reverse of the catabolic reactions, except that here both reduction reactions are catalyzed by NADPH rather than by NADH and $FADH_2$.

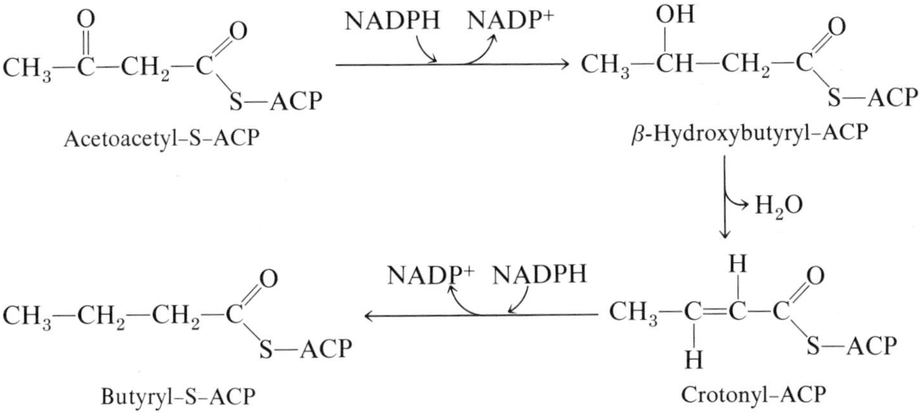
NADPH NADP+
Acetoacetyl-S-ACP
β-Hydroxybutyryl-ACP
H2O
NADP+ NADPH
Butyryl-S-ACP
Crotonyl-ACP

The butyryl group is then transferred to the synthase enzyme; this complex reacts in turn with another molecule of malonyl–S–ACP, carbon dioxide is given off, and the entire series of reactions is repeated. In this manner, the hydrocarbon chain of a given fatty acid is synthesized two carbon units at a time. This explains why the naturally occurring fatty acids contain an even number of carbon atoms. We have said that plants are able to synthesize fatty acids containing several double bonds. Animals are only capable of introducing one double bond into a fatty acid. They can convert stearic acid to oleic acid but lack the appropriate enzymes to carry out further dehydrogenations. The overall equation for the synthesis of palmitic acid is written as follows.

$$8\ CH_3C(=O)SCoA + 14\ NADPH + 14\ H^+ + 7\ ATP + HOH \longrightarrow$$

$$CH_3(CH_2)_{14}C(=O)OH + 7\ ADP + 7\ P_i + 14\ NADP^+ + 8\ CoASH$$

It should be noted that the existence of these separate pathways for the synthesis and degradation of fatty acids is analogous to the situation that exists for the anabolism and catabolism of glycogen. Again, the presence of separate pathways permits the two operations to proceed independently of each other. The rates of oxidation and biosynthesis are only governed by the general metabolic state of the organism.

29.7 Biosynthesis of Triacylglycerols

The coenzyme A derivatives of the fatty acids condense with glycerophosphate (Section 29.5) to form triacylglycerols. The composition of the resulting simple lipids depends upon which fatty acids and which specific enzymes participate in the synthetic reactions.

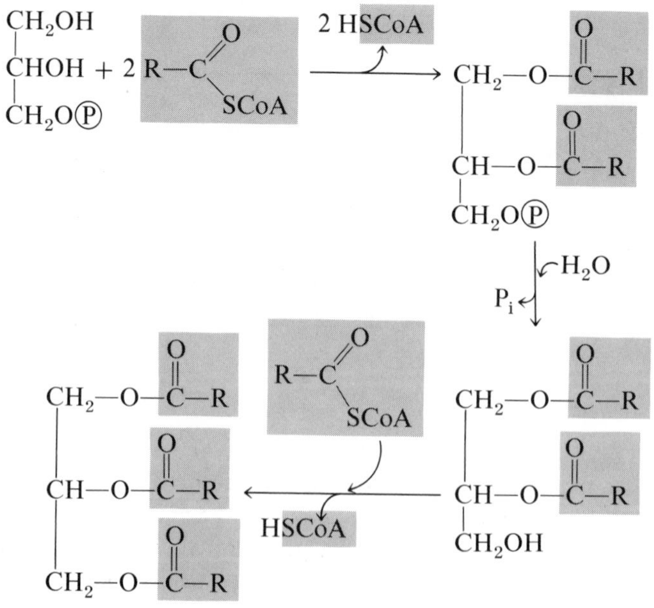
CH2OH
CHOH + 2 R—C
O
SCoA
CH2OⓅ
2 HSCoA
CH2—O—C—R
O
CH—O—C—R
CH2OⓅ
H2O
Pi
CH2—O—C—R
CH—O—C—R
CH2OH
R—C
SCoA
HSCoA
CH2—O—C—R
CH—O—C—R
CH2—O—C—R

Exercises

29.1 Compare the energy released when one gram of carbohydrate and one gram of lipid are oxidized completely in the body.

29.2 What is a characteristic structural feature of the essential fatty acids?

29.3 Briefly discuss the digestion and absorption of lipids.

29.4 Show, with equations, the chemical changes that lipids undergo during digestion. What is the function of bile?

29.5 Write the formulas for four fatty acids that would be formed during the digestion of the lipids found in soybeans. (See Table 24.2.)

29.6 What are the functions of depot fat?

29.7 What is meant by the term β-oxidation?

29.8 Why is the system for the oxidation of fatty acids called a spiral instead of a cycle?

29.9 Outline, in detail, the sequence of reactions that occur in the complete oxidation of lauric acid.

29.10 In which segment of fatty acid catabolism is the most energy made available?

29.11 Calculate the total number of moles of ATP that can be produced from the oxidation of one mole of glyceryl stearate to CO_2 and H_2O in an aerobic organism.

29.12 What are the ketone bodies? Write equations for their formation from acetyl CoA.

29.13 Why do diabetics accumulate relatively large amounts of ketone bodies?

29.14 What is the relationship between ketosis and acidosis?

29.15 Show, with equations, the fate of the glycerol obtained from the hydrolysis of lipids.

29.16 Start with acetyl CoA, and show the sequence of reactions for its incorporation into palmitic acid.

29.17 If ^{14}C carbon dioxide were incorporated into the pathway for the synthesis of stearic acid, which carbon atoms of the acid would be labeled?

29.18 Why do the majority of naturally occurring fatty acids contain an even number of carbon atoms?

29.19 The ingestion of excess carbohydrates results in the deposition of fats in adipose tissue. Explain.

29.20 What is the significance of acetyl CoA in metabolism?

30

Protein Metabolism

In Chapter 25 we indicated the essential nature of protein molecules and their vital functions in all living organisms. In Chapter 27 the biosynthesis of proteins was discussed in connection with the study of nucleic acids. In this chapter we shall deal primarily with the catabolic aspects of protein metabolism, with particular reference to the metabolic role of the amino acids.

30.1 Digestion and Absorption of Proteins

Like polypeptides and lipids, intact proteins cannot normally be absorbed across intestinal membranes. They must first be hydrolyzed into their constituent amino acids. Protein digestion begins in the stomach where the action of the gastric juice results in the hydrolysis of about 10% of the peptide bonds. This bond rupture results in the formation of smaller polypeptides (known as *proteoses* or *peptones*) having molecular weights of the order of 600–3000 daltons. Recall that gastric juice has a pH between 1 and 2, owing to the presence of hydrochloric acid. This high concentration of acid serves to denature the dietary proteins, making them more susceptible to attack by the proteolytic enzymes. The principal digestive component of gastric juice is *pepsinogen,* a zymogen that is catalytically converted by hydrogen ions into its active form, *pepsin* (Section 26.3).[1] Once some pepsin is formed, it also assists in the activation of the remaining pepsinogen. This phenomenon is known as **autocatalysis,** the catalysis of a reaction by one of the products of

[1] All of the enzymes that digest proteins (*peptidases*) are secreted in the form of their inactive precursors, probably to protect the tissues from digestion by their own enzymes.

that reaction. Pepsin is an *endopeptidase,* catalyzing the hydrolysis of peptide linkages within the protein molecule. It is believed to act preferentially on linkages involving the aromatic amino acids tryptophan, tyrosine, and phenylalanine as well as methionine and leucine. The gastric juice of infants contains *rennin,* an enzyme having a specificity very similar to that of pepsin.

Protein digestion is completed in the small intestine by the action of the digestive juices of the pancreas and of the intestinal mucosal cells. These juices are sufficiently alkaline to neutralize the acidic material passed on from the stomach. The pancreatic juice contains the zymogens *trypsinogen* and *chymotrypsinogen.* The former is activated to trypsin by the enzyme *enterokinase* (Figure 30.1); trypsin then activates chymotrypsinogen to chymotrypsin. Both of these active enzymes are endopeptidases. It has been postulated that chymotrypsin preferentially attacks peptide bonds involving the carboxyl groups of phenylalanine, tryptophan, and tyrosine, whereas trypsin attacks peptide bonds involving the carboxyl groups of the basic

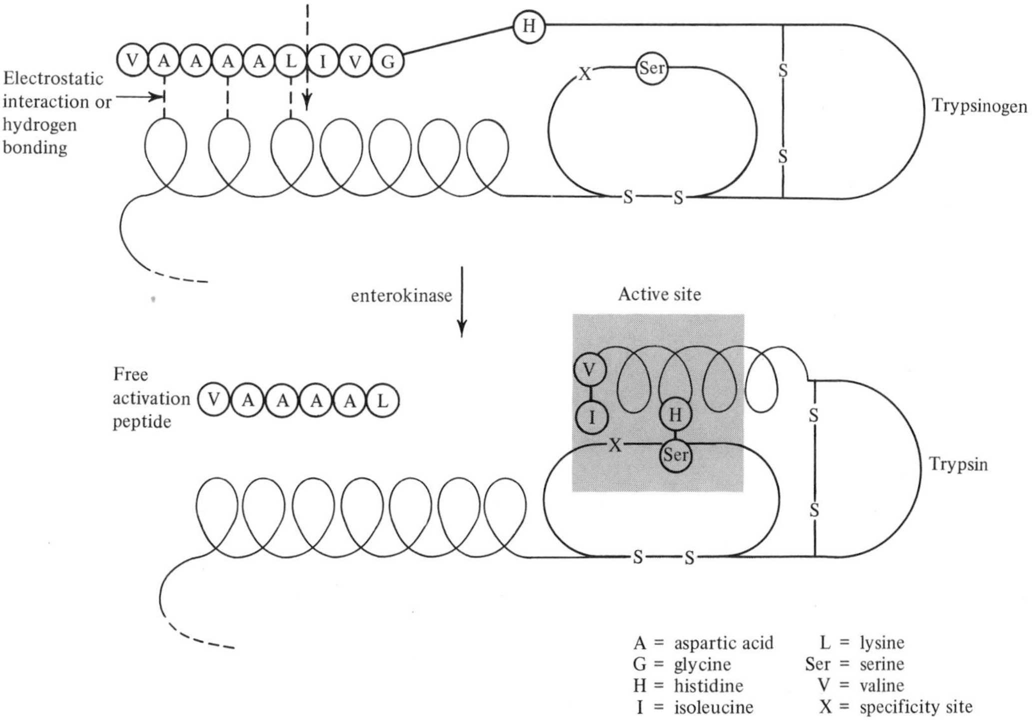

Electrostatic interaction or hydrogen bonding
V A A A A L I V G
H
X
Ser
S
S
S—S
Trypsinogen
enterokinase
Active site
Free activation peptide
V A A A A L
V
I
H
X
Ser
S
S
S—S
Trypsin
A = aspartic acid
G = glycine
H = histidine
I = isoleucine
L = lysine
Ser = serine
V = valine
X = specificity site

Figure 30.1 Schematic representation of the structural changes involved in the activation of trypsinogen. Rupture of lysyl–isoleucine bond in the N-terminal region (dashed arrow) leads to the liberation of the activation peptide and causes the newly formed N-terminal region of the polypeptide chain to assume a more nearly helical configuration. This in turn permits a histidine and serine side chain to come into juxtaposition so that these catalytic groups are properly aligned. The specificity site of the protein (X) is believed to be pre-existent in the zymogen molecule. [Adapted from H. Neurath, "Protein Structure and Enzyme Action," in J. L. Oncley (ed.), *Biophysical Science*, Wiley, 1959.]

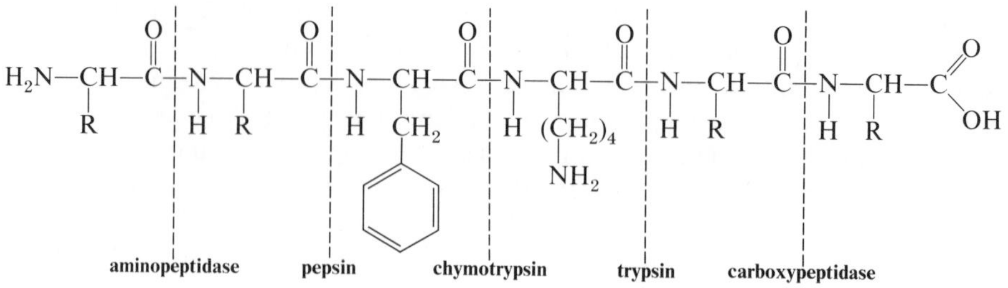

aminopeptidase
pepsin
chymotrypsin
trypsin
carboxypeptidase

Figure 30.2 Specificity of peptidase hydrolysis.

amino acids lysine and arginine. Pancreatic juice also contains an *exopeptidase, carboxypeptidase,* that catalyzes the hydrolysis of the peptide linkages at the free carboxyl end of the peptide chain, resulting in the stepwise liberation of free amino acids from the carboxyl end of the polypeptide.

Two types of peptidases are secreted in the intestinal juice: (1) an exopeptidase, *aminopeptidase,* which acts upon the peptide linkages of terminal amino acids possessing a free amino group, and (2) *dipeptidase* and *tripeptidase,* which cleave dipeptides and tripeptides. Figure 30.2 illustrates the specificity of the protein-digesting enzymes.

The combined action of the proteolytic enzymes of the gastrointestinal tract results in the complete hydrolysis of the dietary proteins into free amino acids.[2] These amino acids are actively transported (an energy-requiring process) across the intestinal wall into the portal circulation, and are carried to the liver. The liver is the principal organ responsible for the degradation and synthesis of amino acids. It is also the site of synthesis of several blood proteins (e.g., albumins, globulins, fibrinogen, prothrombin—see Chapter 31). The amino acids synthesized in the liver and the amino acids obtained from protein digestion, along with the amino acids derived from the turnover of tissue proteins, are transported by the blood to all the tissues of the body.

30.2 Nitrogen Balance

The cellular proteins are in a continual state of flux; they are constantly being degraded and resynthesized from their constituent amino acids. This theory was first validated by the classic study of R. Schoenheimer and D. Rittenberg in 1939. They fed isotopically labeled amino acids to healthy adult rats who had no pressing need for dietary amino acids for growth of tissue protein. They found that almost 60% of the ingested amino acids (labeled with the heavy isotope ^{15}N) were incorporated into tissue proteins. The apparent stability of the adult nongrowing organism is not due, therefore, to metabolic inertness, but rather to a delicate balance between the rates

[2]Occasionally, small peptides and polypeptides may be absorbed into the bloodstream. These foreign polypeptides act as *antigens,* that is, they stimulate the formation of specific *antibodies* and are probably responsible for the allergic reactions that some individuals develop for certain foods.

of synthesis and degradation of its constituent compounds. The cell may be visualized as containing a grand mixture, or metabolic pool, of amino acids derived either from the diet or from the degradation of tissue protein. The liver is metabolizing its proteins at such a rate that the equivalent of half the liver tissue is being replaced with fresh protein every six days (i.e., the half-life of liver protein is 6 days). Protein of muscle and connective tissue may have half-lives of from 180 to 1000 days. Hair has no half-life because only synthesis (and not degradation) of hair protein occurs within the ectodermal (skin) cells.

A dietary intake of nitrogen is required to provide for the biosynthesis of the various nitrogenous compounds. Nitrogen is lost in the continuous degradation of tissue protein and in the excretion of certain nitrogen-containing waste materials. Under normal conditions, an *individual's intake of dietary nitrogen is equal to the amount of nitrogen in the excrement.* Such a condition is referred to as **nitrogen balance** or **nitrogen equilibrium.** Organisms are said to be in positive nitrogen balance (intake exceeds excretion) whenever tissue is being synthesized, as, for example, during periods of growth and convalescence from disease. Negative nitrogen balance results from (1) an inadequate intake of protein (for example, fasting), (2) fevers, infections, wasting diseases, or (3) a diet that lacks, or is deficient in, any one of the essential amino acids. These factors cause an accelerated breakdown of tissue protein and nitrogen excretion exceeds intake.

Higher plants and certain microorganisms are capable of synthesizing all of their proteins from carbon dioxide, water, and inorganic salts. They obtain their required nitrogen either from soil nitrates or from atmospheric nitrogen (via nitrogen-fixing bacteria). Thus these organisms can grow on a medium that does not contain any preformed amino acids. Animals, however, can synthesize only about half of their naturally occurring amino acids; the remainder must be supplied in the diet. All of the amino acids required for the synthesis of a particular protein must be available to the cell at the time of protein synthesis. If just one amino acid is absent, or is present in insufficient quantity, the protein will not be synthesized. An **essential** or *indispensable* amino acid is one that cannot be synthesized by an organism from the substrates ordinarily present in its diet at a rate rapid enough to meet the demands of natural growth and repair. A list of essential amino acids is given in Table 30.1. Notice that, as a rule, these compounds contain carbon chains, or aromatic rings, that are not present as intermediates of carbohydrate or lipid metabolism. The inability to synthesize these amino acids does not arise from a lack of the necessary nitrogen but results rather from the animal's inability to manufacture the correct carbon skeleton. If, for example, an animal is supplied with phenylpyruvic acid, it can readily synthesize the amino analog phenylalanine. Lysine appears to be an exception since the entire preformed amino acid must be supplied. The general effect of a deficiency of one or more essential amino acids is to restrict growth and protein synthesis and produce a negative nitrogen balance.

Essential amino acids are best provided by animal protein. (Casein, the

Table 30.1 Essential and Nonessential Amino Acids for Man

Essential	*Nonessential*
Lysine	Glycine
Leucine	Alanine
Isoleucine	Serine
Methionine	Tyrosine[a]
Threonine	Cysteine
Tryptophan	Aspartic acid
Valine	Glutamic acid
Phenylalanine	Proline
Histidine[b]	Hydroxyproline
Arginine[c]	

[a]In the presence of adequate amounts of phenylalanine.

[b]Histidine is not a dietary essential for the adult male.

[c]Arginine can be synthesized but not at the rate required for normal growth.

protein from milk, is especially beneficial since it is well balanced in its amino acid distribution.) Proteins that lack an adequate amount of the essential amino acids are termed *incomplete proteins.* Gelatin is an example of an incomplete animal protein because it is deficient in tryptophan. Most plant proteins are deficient in lysine and/or one other essential amino acid (e.g., corn is deficient in lysine and tryptophan, rice is low in lysine and threonine, wheat is low in lysine, and the legumes are low in methionine). The average individual requires about 50 grams of protein in his daily diet. Prolonged deprivation of protein leads to exhaustion of all carbohydrate and fat reserves; the organism is ultimately left with no source of energy except its own tissue proteins. The Nigerian civil war (1967–68) made the world brutally aware of the serious condition of protein deprivation called *kwashiorkor,* a disease resulting in extreme emaciation, bloated abdomen, discoloration of the hair, and eventual death. This disease is prevalent in Latin America, Asia, and Africa and can be best treated by the administration of adequate amounts of well-balanced protein. The problem, of course, is that animal protein is a scarce commodity in many of these areas.

30.3 Metabolism of Amino Acids

Most of the amino acids in the metabolic pool are utilized for the synthesis of any of the myriad of proteins necessary to the living organism. This is the major function of amino acids in the body; about 75% of the metabolized amino acids in the normal adult are used for protein synthesis. This is necessary because of the constant destruction of body proteins by wear and tear. It should be emphasized, however, that amino acids also play an essential role in the metabolism of all nitrogenous compounds. In addition, amino acids may be catabolized to yield glycogen, fat, or energy, or they may participate in a specific biochemical pathway. Generally, the first step in the breakdown of amino acids is the separation of the amino group from the carbon skeleton. The amino groups are then incorporated into almost all the other nonprotein nitrogen-containing compounds, for example, hormones, nucleic acids, porphyrins, and creatine (Figure 30.3). A discussion of these specific metabolic pathways is beyond the scope of this text.

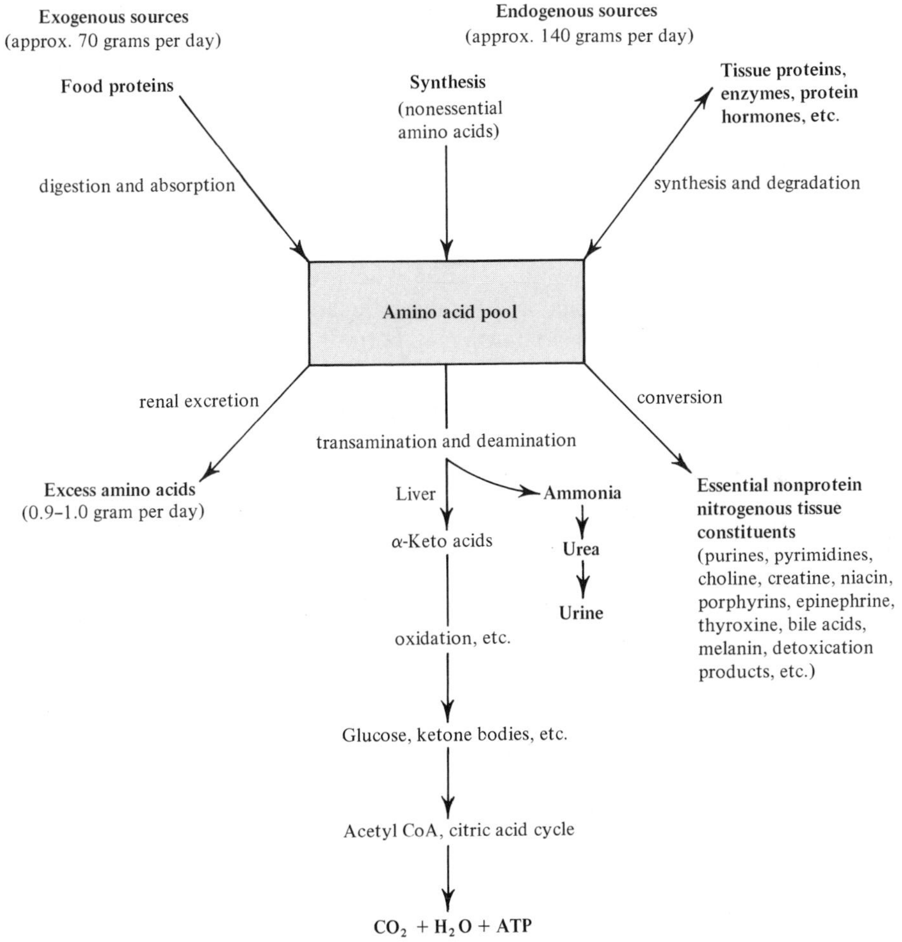

Exogenous sources
(approx. 70 grams per day)
Endogenous sources
(approx. 140 grams per day)
Food proteins
Synthesis
(nonessential
amino acids)
Tissue proteins,
enzymes, protein
hormones, etc.
digestion and absorption
synthesis and degradation
Amino acid pool
renal excretion
conversion
transamination and deamination
Excess amino acids
(0.9-1.0 gram per day)
Liver
Ammonia
α-Keto acids
Urea
Urine
oxidation, etc.
Essential nonprotein
nitrogenous tissue
constituents
(purines, pyrimidines,
choline, creatine, niacin,
porphyrins, epinephrine,
thyroxine, bile acids,
melanin, detoxication
products, etc.)
Glucose, ketone bodies, etc.
Acetyl CoA, citric acid cycle
$CO_2 + H_2O + ATP$

Figure 30.3 Scheme of general paths of amino acids in metabolism (average human). [From James M. Orten and Otto W. Neuhaus, *Human Biochemistry*, 9th ed., The C. V. Mosby Co., St. Louis, 1975.]

a. Transamination

In this reaction, the amino group is transferred from an amino acid to an α-keto acid, resulting in the formation of a new keto acid and a new amino acid.

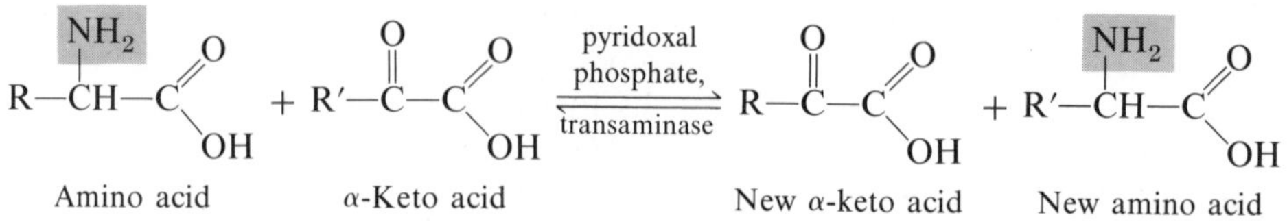

pyridoxal phosphate,
transaminase
Amino acid
α-Keto acid
New α-keto acid
New amino acid

Practically all of the naturally occurring amino acids have been found to undergo this reaction. It is probable that there are a large number of transaminases, each one catalyzing a reaction between an α-amino acid and some α-keto acid. The transaminases are widely distributed in nature, and are known to function in conjunction with the coenzyme pyridoxal phosphate (Table 26.2). It is believed that the first step involves the attachment of the amino acid to an enzyme, which then transfers the amino group to the appropriate α-keto acid (usually pyruvic acid, oxaloacetic acid, or α-ketoglutaric acid).

The production of α-keto acids by this process yields molecules that can enter the glycolytic pathway or the Krebs cycle. A single reaction is necessary for the conversion of alanine, glutamic acid, and aspartic acid to pyruvic acid, α-ketoglutaric acid, and oxaloacetic acid, respectively. Additional reactions are necessary to form metabolic intermediates from the other amino acids.

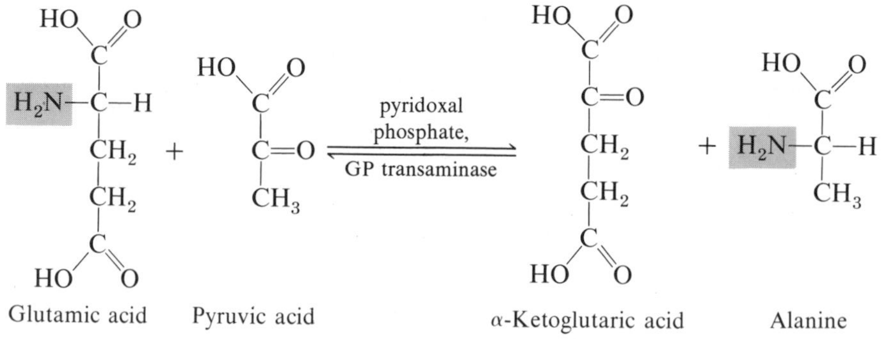

pyridoxal phosphate,
GP transaminase
Glutamic acid
Pyruvic acid
α-Ketoglutaric acid
Alanine

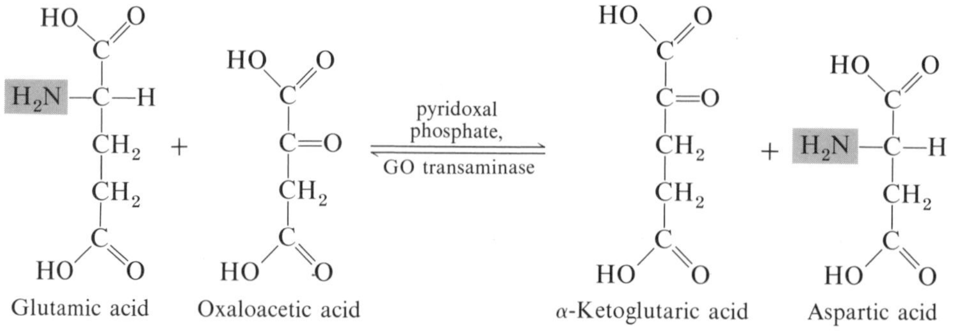

pyridoxal phosphate,
GO transaminase
Glutamic acid
Oxaloacetic acid
α-Ketoglutaric acid
Aspartic acid

Table 30.2 Glucogenic and Ketogenic Amino Acids

Glucogenic		*Ketogenic*	*Glucogenic and Ketogenic*
Alanine	Hydroxyproline	Leucine	Isoleucine
Arginine	Methionine		Lysine
Aspartic acid	Proline		Phenylalanine
Cysteine	Serine		Tyrosine
Glutamic acid	Threonine		Tryptophan
Glycine	Valine		
Histidine			

These two particular transamination reactions are of special clinical interest. Normally the blood contains a low concentration of transaminase enzymes. However, when extensive tissue destruction occurs, it is accompanied by rapid and striking increases in the blood transaminase levels. Liver tissue is particularly rich in *glutamic pyruvic transaminase* (GPT) and an elevated serum level of this enzyme is indicative of liver damage. Heart muscle contains an abundance of *glutamic oxaloacetic transaminase* (GOT). In the diagnosis of myocardial infarction (heart attack), it is common practice to test blood samples for both GOT and CPK (*creatine phosphokinase*—see footnote, page 723).

It is especially significant that the transamination reactions are reversible, since this reversibility links protein metabolism with the metabolism of carbohydrates and lipids. Those amino acids that can form any of the metabolites of carbohydrate metabolism can be converted to glucose or glycogen (or oxidized to CO_2, H_2O, and energy) and are referred to as *glucogenic* (Section 28.2).[3] Amino acids that give rise to acetyl CoA or acetoacetyl CoA are said to be *ketogenic*. Certain amino acids fall into both categories; leucine is the only amino acid that is exclusively ketogenic. Table 30.2 classifies the amino acids with regard to the metabolic fate of their carbon skeletons.

b. Oxidative Deamination

Oxidative deamination effects the replacement of the amino group by oxygen with the release of ammonia. This reaction is catalyzed by *amino acid oxidase,* an enzyme containing the oxidizing coenzyme FAD. The reduced coenzyme is reoxidized by molecular oxygen (and not by the respiratory chain) to form H_2O_2.

[3]During fasting or starvation, the liver glycogen is depleted very rapidly and blood glucose must be manufactured from other sources to supply energy for the brain cells. The major source of this blood glucose is these glucogenic amino acids obtained from the degradation of tissue protein.

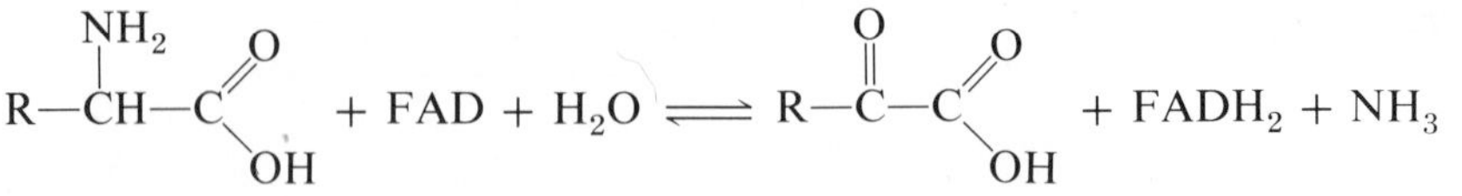

$$FADH_2 + O_2 \longrightarrow FAD + H_2O_2$$

General Equation

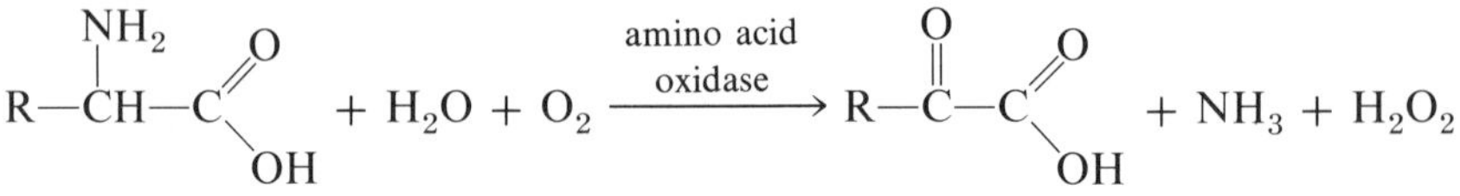
amino acid
oxidase

Specific Equation

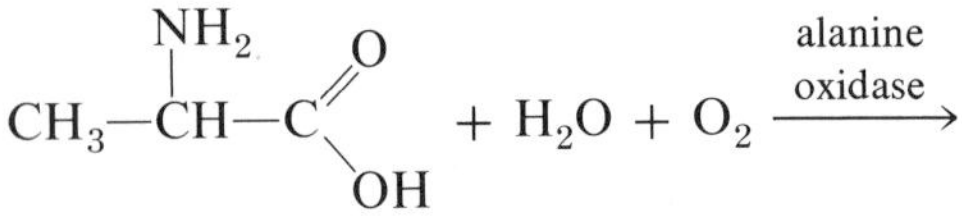
alanine
oxidase

Alanine

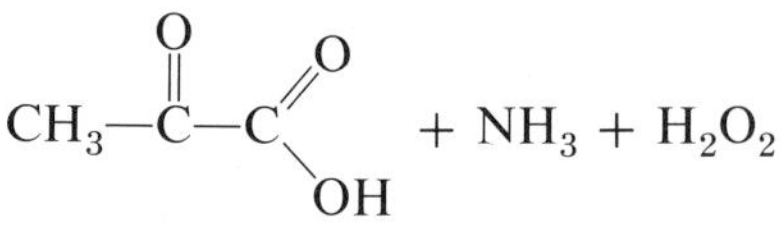

Pyruvic acid

Once again, the α-keto acids that are formed can be incorporated into the schemes of carbohydrate or lipid metabolism. The other products of the reaction, NH_3 and H_2O_2, are highly toxic and must be removed from the body. Ammonia is converted to urea and excreted in the urine (see Section 30.5). Under the influence of *catalase*, hydrogen peroxide is decomposed to water and oxygen. Catalase is extremely efficient; a single enzyme molecule catalyzes the decomposition of about 5×10^4 molecules of H_2O_2 per second.

$$H_2O_2 \xrightarrow{\text{catalase}} H_2O + \tfrac{1}{2}O_2$$

Although amino acid oxidases are widely distributed in nature, they are limited to the kidney and the liver of mammals. Furthermore, their activity in these organs is extremely low, and removal of amino groups by oxidases is of less physiological significance than removal by transaminases.

There are some specific amino acid oxidases that require pyridine nucleotides; in these cases, the reduced coenzymes (NADH and NADPH) are oxidized by way of the respiratory chain. The most important specific oxidase is *glutamic acid oxidase*. It has been crystallized from beef liver and shown to contain divalent zinc ions. It catalyzes the following reaction.

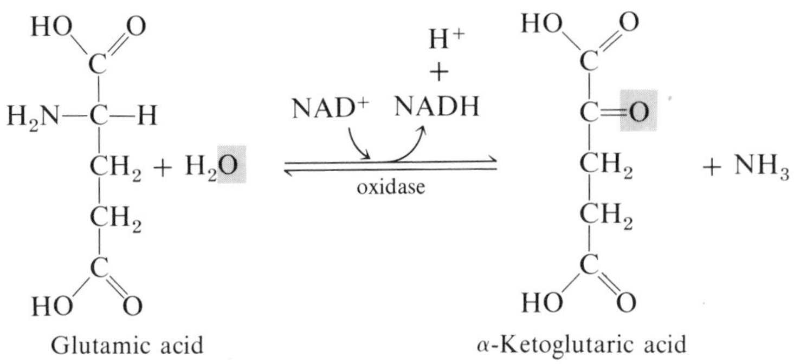

The reverse of this reaction provides a means for the synthesis of amino acids from ammonia and an intermediate of carbohydrate metabolism. The amino group in the glutamic acid can be passed on through transamination reactions, producing all the other cellular amino acids, providing the appropriate α-keto acids are available.

c. Decarboxylation

Several amino acid decarboxylases exist, which eliminate carbon dioxide from amino acids to form primary amines. Pyridoxyl phosphate is the necessary coenzyme.

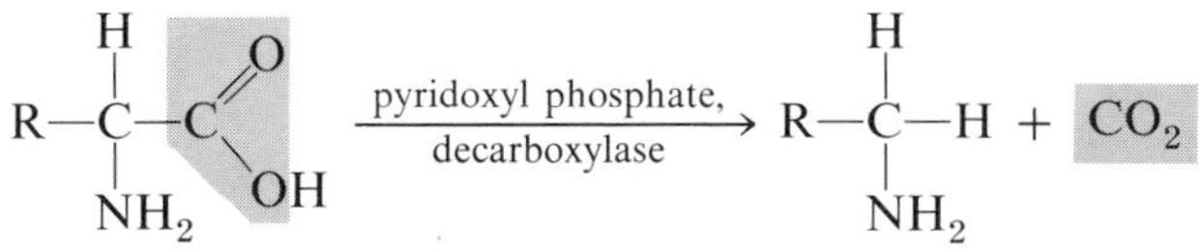

These enzymes are found primarily in microorganisms but they are also found in some animal tissues. Intestinal bacteria are responsible for amino acid decarboxylation, and the foul smell of the feces is due in part to the resulting amines. (Recall that cadaverine and putrescine are formed upon decarboxylation of lysine and ornithine—Section 20.8.)

Some of the amines produced as a result of decarboxylation have important physiological effects. Thus *histamine,* formed by decarboxylation of histidine, causes a decrease in blood pressure, swelling, and increased secretions from the eyes and nose. Release of small amounts of histamine is believed to be associated with allergic reactions to foreign proteins in ragweed (hay fever), certain foods, and insect stings. Many nasal decongestants and hay fever preparations contain antihistamines that reduce or eliminate the effects of histamine. The neurotransmitter serotonin (Section 20.13-e) is formed by the action of a specific decarboxylase on 5-hydroxytryptophan. Another decarboxylase catalyzes the decarboxylation of 3,4-dihydroxyphenylalanine (L-dopa) to form dopamine. A deficiency of dopamine in the brain cells is a primary cause of Parkinson's disease, a disorder of the central nervous system. Dopamine itself cannot be administered because it does not pass across the blood-brain barrier. A major breakthrough in the treatment of Parkinson's disease has been the use of L-dopa. Large doses of L-dopa are administered orally; the drug is able to pass from the digestive system into the blood and then cross the blood-brain barrier. In the brain cells L-dopa is converted into the necessary compound, dopamine.

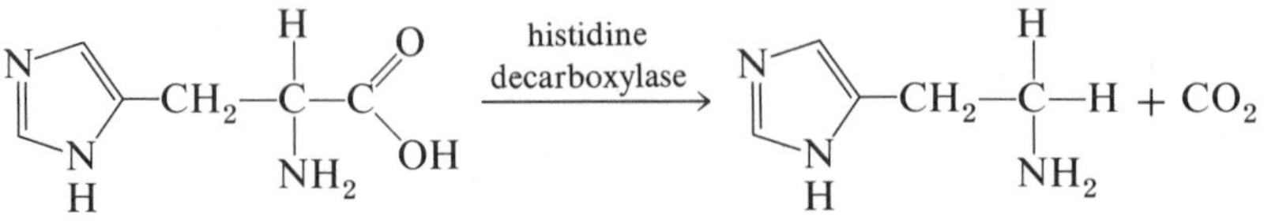
histidine
decarboxylase

Histidine Histamine

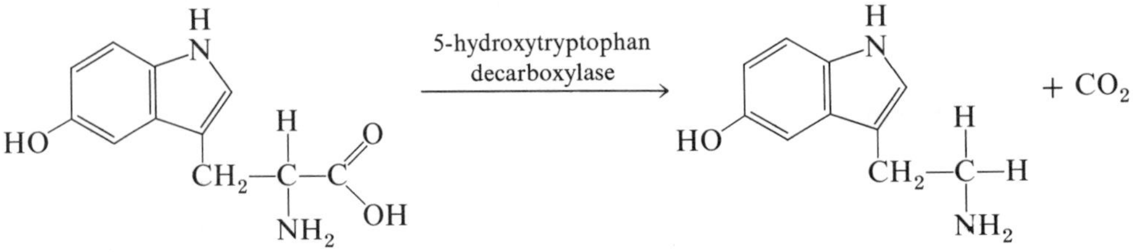
5-hydroxytryptophan
decarboxylase

5-Hydroxytryptophan

Serotonin
(5-Hydroxytryptamine)

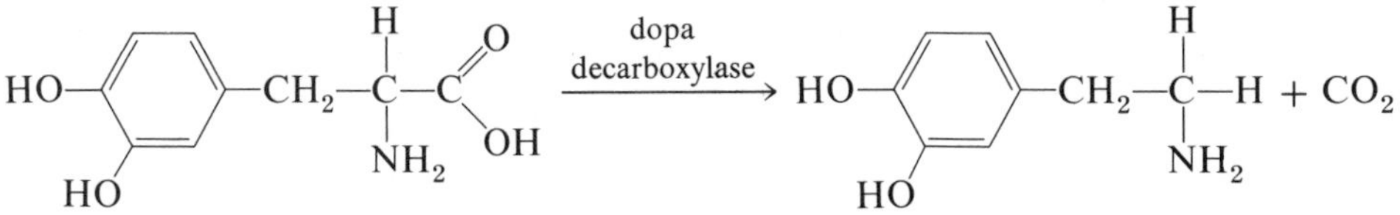
dopa
decarboxylase

3,4-Dihydroxyphenylalanine
(L-Dopa)

3,4-Dihydroxyphenylethylamine
(Dopamine)

30.4 Storage of Nitrogen

In contrast to carbohydrates and lipids, proteins are not stored to an appreciable extent in living organisms. Small amounts of amino acids are excreted by humans, but this is essentially a wasteful process since amino acids contain utilizable energy. Accordingly, most excess amino acids are deaminated; the resulting carbon skeleton is oxidized to produce energy or is stored as glycogen or fat. Most of the ammonia that is liberated is converted to urea. A relatively small proportion combines with glutamic acid in the presence of *glutamine synthetase* and ATP to form glutamine (the amide of glutamic acid).

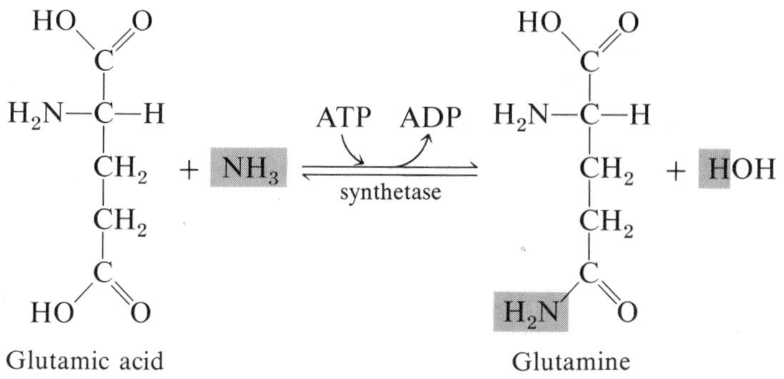

Glutamine is present in many tissues and in the blood and serves as a temporary storage and transport form of nitrogen. The amide amino group can be donated in specific reactions to appropriate acceptor molecules to form a variety of nitrogenous compounds (e.g., purines, pyrimidines, NAD^+). Asparagine, the amide of aspartic acid, is a common constituent of many plant proteins and is probably the storage form of nitrogen in plants.

30.5 Excretion of Nitrogen—the Urea Cycle

Whenever amino acids are utilized for energy production or for the synthesis of glycogen or fat, an amino group is liberated in the form of ammonia. Living organisms must have some mechanism of removing this ammonia from the cell environment because even low concentrations of ammonia are poisonous. (Levels of only 5 mg ammonia per 100 ml blood are toxic to humans. The normal concentration of this compound is about 1–3 μg per 100 ml.) Organisms differ biochemically in the manner in which they excrete excess nitrogen. Most vertebrates excrete nitrogen as urea in the urine. Birds, reptiles, and insects convert nitrogen to uric acid (Figure 30.4). All marine organisms, from unicellular organisms to fish, excrete free ammonia.

Ninety-five percent of the nitrogenous waste products in humans results from amino acid catabolism; the remaining 5% comes from the metabolism of other nitrogen-containing compounds. It is interesting to note that the breakdown of purine bases in the human body results in the production of uric acid, and very small concentrations of this acid are found in the urine and in the body fluids. Under certain pathological conditions, large quantities of uric acid are produced and deposited as the sparingly soluble mono-

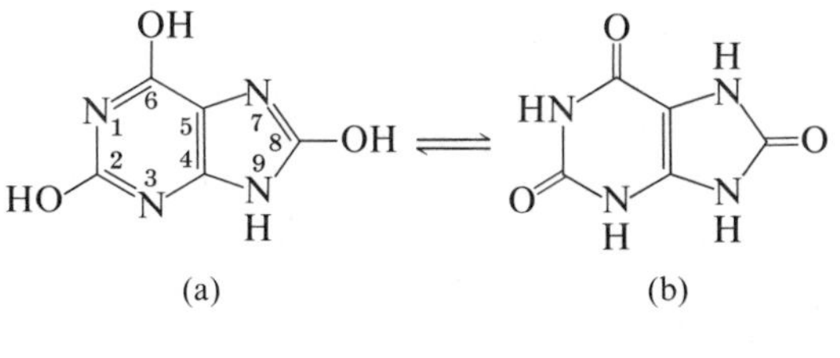

Figure 30.4 Uric acid is a purine base and exists in solution as an equilibrium mixture of enol (a) and keto (b) tautomers. Nitrogens-1, -3, and -9 derive from ammonia; nitrogen-7 derives from glycine.

sodium salt. Such deposits occur in the joints and produce the painful symptoms characteristic of gout.

The principal organ concerned with the formation of urea in mammals is the liver. Extensive liver damage results in the accumulation of ammonia and may lead to death. Once produced, urea is transported by the bloodstream to the kidneys and is eliminated in the urine. The normal individual excretes about 25 grams of urea daily, although this value varies greatly from day to day and is dependent upon protein intake.

H. Krebs and K. Hensleit (1932) were the first to propose a *cyclic sequence of reactions for the synthesis of urea*. The essential features of this cycle, called the **urea cycle,** or the **ornithine cycle,** have remained unchanged and are depicted in Figure 30.5. Each reaction has been numbered and will be discussed in the following sections.

Step 1. The formation of carbamoyl phosphate from NH_3 and CO_2 is the initial step in urea synthesis. The reaction is catalyzed by *carbamoyl phosphate synthetase*. Energy for the process is supplied by two molecules of ATP. The reaction mechanism has not yet been elucidated.

$$H_2O + NH_3 + CO_2 + 2\,ATP \longrightarrow H_2N{-}\overset{\overset{\displaystyle O}{\|}}{C}{-}O{-}Ⓟ + 2\,ADP + P_i$$

Carbamoyl phosphate

Step 2. The high energy compound, carbamoyl phosphate, then reacts with ornithine in the presence of *ornithine carbamoyltransferase* to form citrulline. The high energy phosphate bond is cleaved in the process.

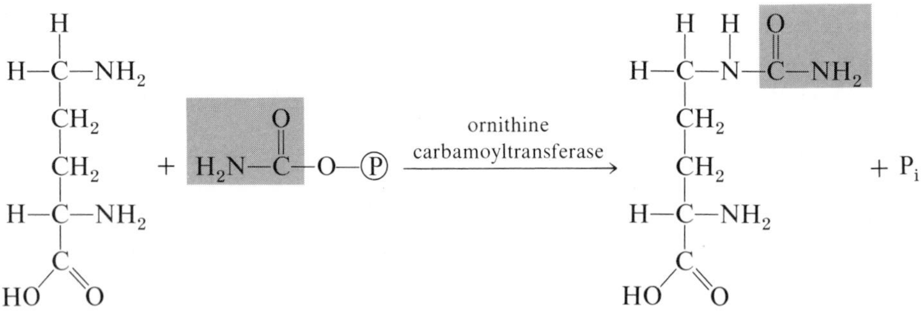
ornithine
carbamoyltransferase

Ornithine Citrulline

Step 3. The conversion of citrulline to arginine involves two enzymic steps. The first is the reversible condensation of citrulline with aspartic acid in the presence of *argininosuccinic synthetase*. The reaction produces argininosuccinic acid, a stable intermediate. It is cleaved in a subsequent step to yield a molecule of arginine and a molecule of fumaric acid. The synthesis of fumaric acid is important since it can be converted via the Krebs cycle to oxaloacetic acid. The latter can be transaminated to yield aspartic acid, or it can condense with acetyl CoA to form citric acid.

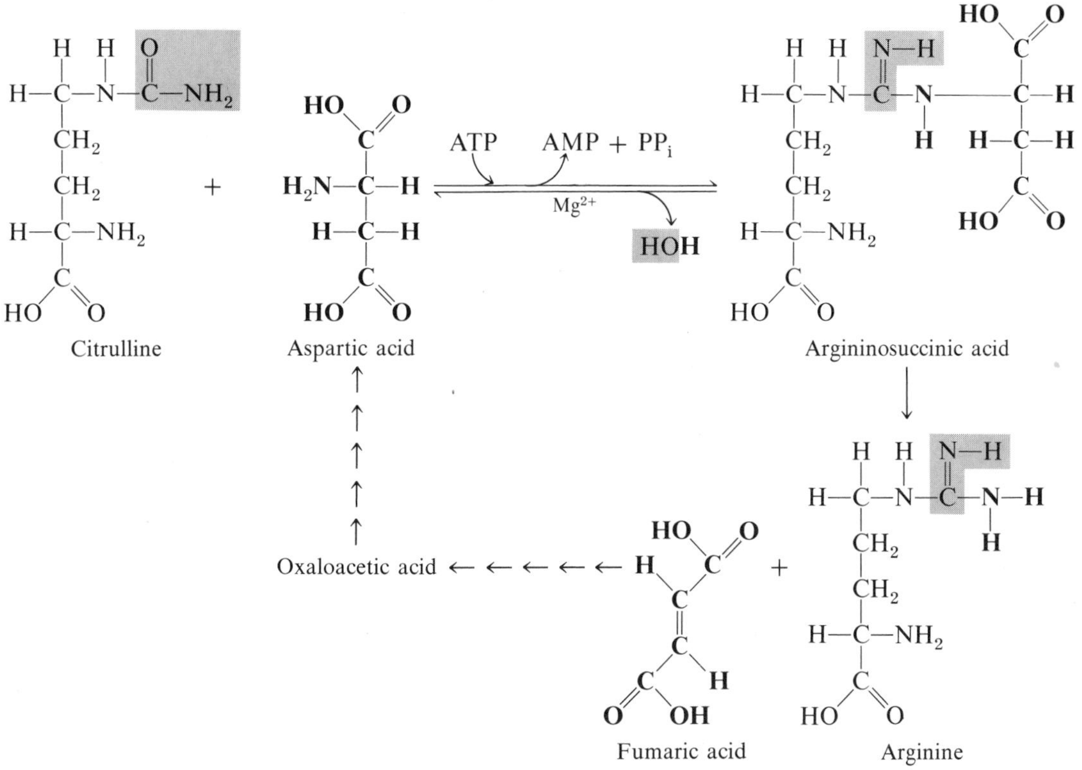

Citrulline
Aspartic acid
ATP
AMP + PP$_i$
Mg^{2+}
HOH
Argininosuccinic acid
Oxaloacetic acid
Fumaric acid
Arginine

Step 4. In the final step of the cycle, arginine is hydrolytically cleaved by *arginase,*[4] yielding urea and regenerating ornithine. The latter is now ready to accept another molecule of carbamoyl phosphate, thus completing the cycle.

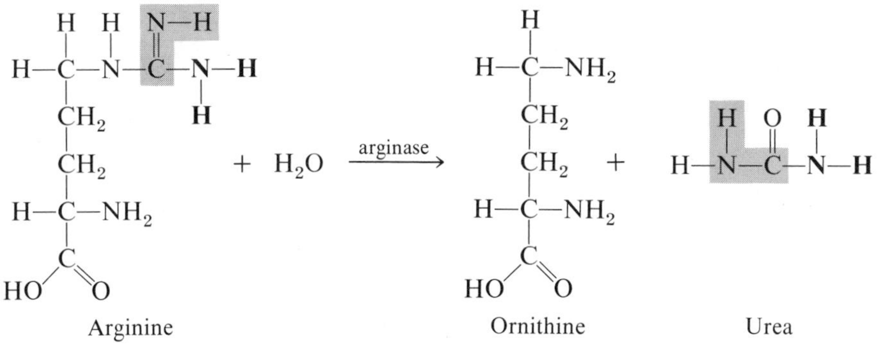

arginase
Arginine
Ornithine
Urea

[4]There is a direct correlation between arginase activity and the ability of a tissue to synthesize urea. In mammals, arginase is found in the liver, whereas birds, reptiles, and insects lack this enzyme and thus excrete their surplus nitrogen in the form of uric acid.

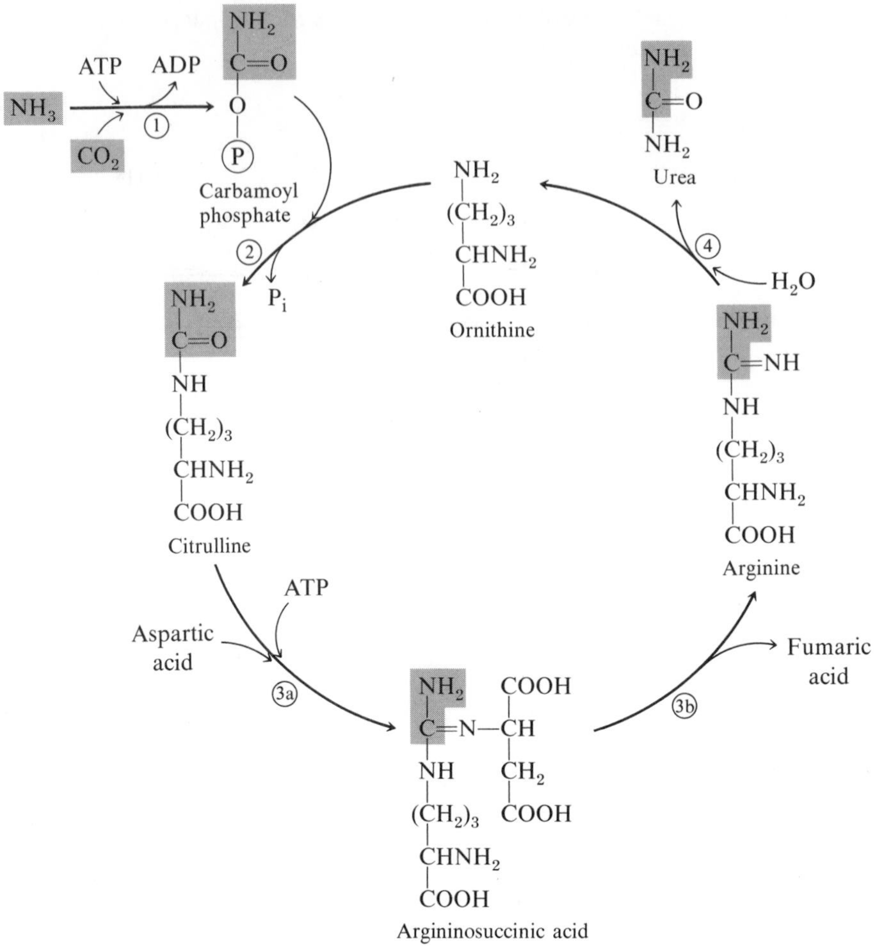
NH2
C=O
ATP
ADP
NH3
CO2
1
O
P
Carbamoyl phosphate
2
Pi
NH2
C=O
NH
(CH2)3
CHNH2
COOH
Citrulline
NH2
(CH2)3
CHNH2
COOH
Ornithine
NH2
C=O
NH2
Urea
4
H2O
NH2
C=NH
NH
(CH2)3
CHNH2
COOH
Arginine
ATP
Aspartic acid
3a
Fumaric acid
3b
NH2
COOH
C=N—CH
NH
CH2
(CH2)3
COOH
CHNH2
COOH
Argininosuccinic acid

Figure 30.5 The urea cycle.

Notice that one of the nitrogen atoms that is eventually converted to urea is contributed by aspartic acid and not by ammonia. However, NH_3 may have been incorporated into the aspartic acid in the following manner.

$$NH_3 + \alpha\text{-Ketoglutaric acid} \underset{\text{dehydrogenase}}{\overset{\text{glutamic acid}}{\rightleftharpoons}} \text{Glutamic acid}$$

$$\text{Glutamic acid} + \text{Oxaloacetic acid} \underset{\text{transaminase}}{\rightleftharpoons} \alpha\text{-Ketoglutaric acid} + \text{Aspartic acid}$$

The overall equation for the enzymic synthesis of urea from ammonia, carbon dioxide, and the amino group of aspartic acid may be written as follows.

$$NH_3 + CO_2 + H_2O + 3\ ATP + \text{Aspartic acid} \longrightarrow \text{Urea} + \text{Fumaric acid} + 2\ ADP + AMP + 2\ P_i + PP_i$$

30.6 Relationships Among the Metabolic Pathways

A great variety of organic compounds can be derived from carbohydrates, lipids, and proteins. All organisms utilize the three major foodstuffs to form acetyl CoA or the metabolites of the Krebs cycle. These, in turn, supply energy upon subsequent oxidation by the cycle. A brief summary of the interrelationships of the metabolic pathways is given in Figure 30.6. Notice

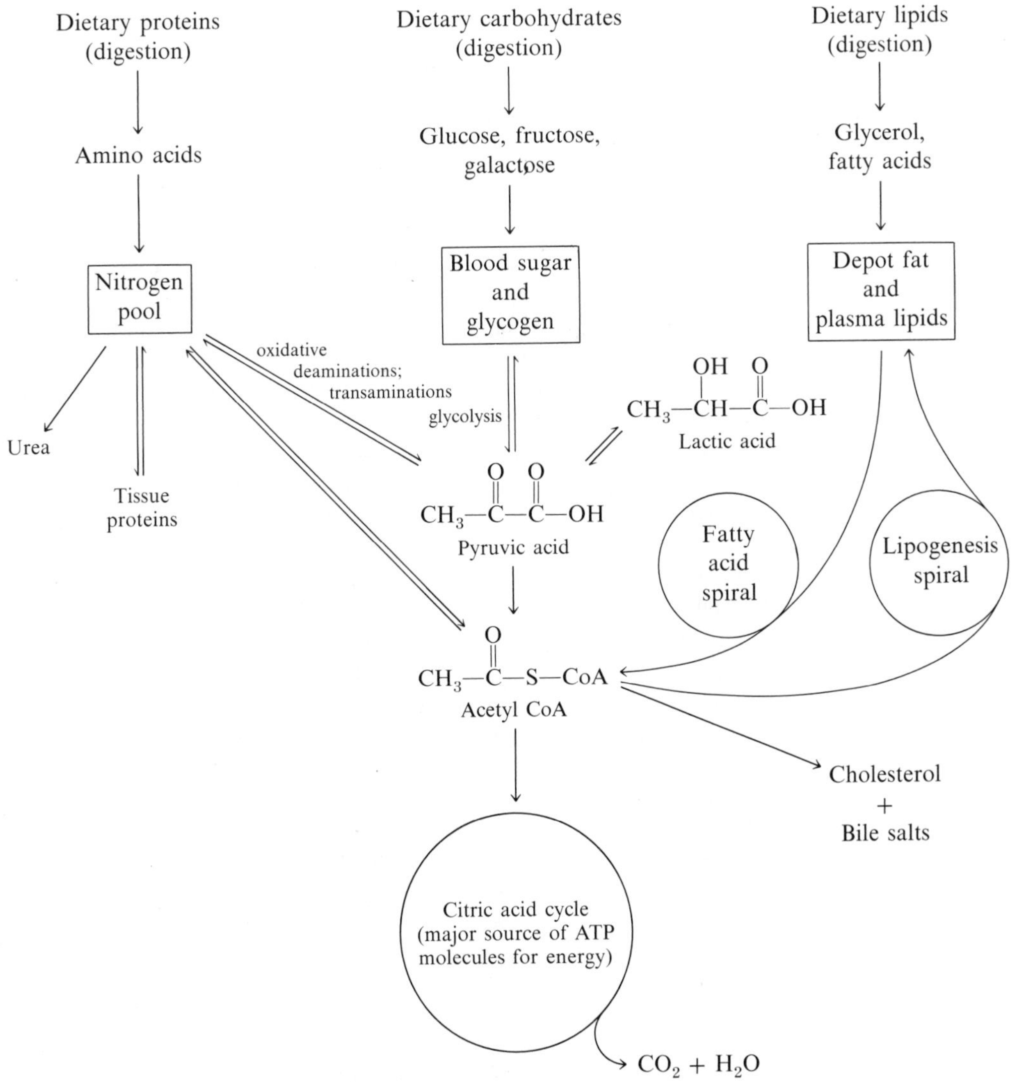

Dietary proteins (digestion)
Dietary carbohydrates (digestion)
Dietary lipids (digestion)
Amino acids
Glucose, fructose, galactose
Glycerol, fatty acids
Nitrogen pool
Blood sugar and glycogen
Depot fat and plasma lipids
oxidative deaminations; transaminations
glycolysis
Urea
Tissue proteins
Lactic acid
Pyruvic acid
Fatty acid spiral
Lipogenesis spiral
Acetyl CoA
Cholesterol + Bile salts
Citric acid cycle (major source of ATP molecules for energy)
CO2 + H2O

Figure 30.6 Interrelationships of metabolic pathways. [Adapted from J. R. Holum, *Elements of General and Biological Chemistry*, Wiley, 1968.]

that one nonreversible reaction is the conversion of pyruvic acid to acetyl CoA. Thus an acetyl CoA molecule derived from fatty acid degradation cannot be directly transformed into pyruvic acid. There is, therefore, no direct route for the synthesis of carbohydrates from fatty acids. Plants and some microorganisms that can convert fatty acids to carbohydrates and proteins utilize a series of enzymic reactions not found in animal tissues, referred to as the *glycolate cycle*. The student is referred to a more advanced text for the details of this pathway.

Exercises

30.1 Briefly discuss the digestion and absorption of proteins.

30.2 Vegetable protein is converted in the animal body into animal protein. Explain.

30.3 In the treatment of diabetes, insulin is administered by injection rather than given orally. Explain.

30.4 What is the defense mechanism employed by the body to protect its tissues from the digestive action of the proteolytic enzymes?

30.5 Indicate the location in a polypeptide chain that is cleaved by **(a)** an aminopeptidase and **(b)** a carboxypeptidase.

30.6 Indicate the expected products from the enzyme action of pepsin, chymotrypsin, and trypsin on each of the following tripeptides.
(a) Ala-Phe-Tyr
(b) Ile-Tyr-Ser
(c) Phe-Arg-Leu
(d) Thr-Glu-Lys

30.7 What was the significance of Schoenheimer's and Rittenberg's experiments with labeled amino acids?

30.8 What is meant by the term *metabolic pool?*

30.9 Amino acids unlike carbohydrates and lipids, are not stored by the body. What does this imply about a proper diet?

30.10 Distinguish between the following terms. Give a suitable example for each.
(a) endopeptidase and exopeptidase
(b) positive nitrogen balance and negative nitrogen balance
(c) essential amino acid and nonessential amino acid
(d) glucogenic amino acid and ketogenic amino acid
(e) transamination and deamination

30.11 What are the possible metabolic fates of the amino acids?

30.12 Which metabolic reactions serve to link together the interconversion of carbohydrates and proteins?

30.13 Write equations for each of the following. Include names of enzymes and names or initials of coenzymes.
(a) oxidative deamination of an amino acid
(b) transamination of an amino acid
(c) decarboxylation of an amino acid

30.14 Write equations for the formation of each of the following compounds from an amino acid.

(a) pyruvic acid
(b) histamine
(c) α-ketoglutaric acid
(d) cadaverine
(e) oxaloacetic acid
(f) ethanolamine

30.15 Write an equation for the synthesis of glutamine. What is the function of this compound in the body?

30.16 What is meant by the term detoxication?

30.17 Outline in detail the sequence of reactions that occur in the synthesis of urea.

30.18 How many moles of an amino acid are catabolized to produce one mole of urea? What compound is the source of the carbon atom in urea?

30.19 Indicate the relationship among the metabolisms of carbohydrates, lipids, and proteins.

31

The Blood

The three circulating fluids in the animal organism are *blood, interstitial fluid* (fluid within the tissue spaces), and *lymph* (fluid within the lymph vessels). Blood is by far the most important of these. It is the connecting link between all of the tissue cells and the external environment. The circulating blood transports oxygen and nutritive materials to all of the cells and metabolic waste products to the excretory organs. It transports hormones from the site of their secretion to the organs that require them. In addition to its transport functions, the blood serves to (1) help maintain osmotic relationships between the tissues (water balance and fluid distribution), (2) protect the organism against infection, (3) regulate the body temperature, and (4) control the pH of the body. The blood contains a number of different types of cells and chemical compounds that enable it to carry out these varied functions.

31.1 General Characteristics of Mammalian Blood

Circulatory blood consists of a viscous fluid portion (the plasma) and the formed elements (blood cells). It accounts for about one twelfth of the body weight of the average individual. A 150 pound (68 kg) man has about 13 pounds (6 kg) or 6 liters of blood; roughly 55–60% of it is **plasma.** Plasma is an extremely complex solution, containing all of the biochemically significant compounds (carbohydrates, amino acids, proteins, lipoproteins, enzymes, hormones, vitamins and inorganic ions). Although these compounds are continuously entering and leaving the circulatory system, the overall composition of the plasma remains remarkably constant (a state of dynamic equilibrium exists). Approximately 90–92% of plasma is water; the remaining 8–10% consists of dissolved solids, the greatest component of which are the plasma proteins. Dispersed throughout the plasma are the **formed elements,** which account for about 40–45% of the total blood volume. The formed elements of the blood are the **erythrocytes** (red blood cells), **leukocytes** (white blood cells), and **thrombocytes** (platelets).

a. Plasma Proteins

The protein content of the blood is about the same as that of the muscle and other tissues. More than 100 proteins have been identified in the blood; most of them are synthesized in the liver. At times of protein deprivation the plasma protein concentration is maintained at the expense of tissue protein. Plasma proteins are grouped into three main classes on the basis of their solubility properties and methods of isolation.

*1. **Albumins.*** Albumins are the most abundant proteins (~55%) in the plasma. Their major function is the maintenance of osmotic pressure (see Section 31.2), but they are also important for their buffering capacity and in the transport of certain drugs. Normal concentrations of albumins range from 4 to 5 grams per 100 ml of plasma.

*2. **Globulins.*** Globulins account for about 40% of the total plasma protein. They have higher molecular weights than the albumins (100,000 daltons as compared to 70,000 daltons). Three subclasses of globulins are recognized, and they differ from one another with respect to their rates of movement in an electric field; α-globulins (~1.0 gram per 100 ml of plasma), β-globulins (1.0–1.5 grams per 100 ml), and γ-globulins (0.5–1.0 gram per 100 ml). α-Globulins and β-globulins form complexes with the water-insoluble lipids and thus serve to transport these compounds through the aqueous media. γ-Globulins contain antibodies that combat certain infectious diseases (such as diphtheria, influenza, measles, mumps, typhoid) and are referred to as *immunoglobulins.*

*3. **Fibrinogen.*** Fibrinogen constitutes about 5% (~0.30 gram per 100 ml of plasma) of the total blood protein. It functions in blood coagulation (see Section 31.3). The fibrinogen content of plasma increases when inflammatory or infectious processes exist and during menstruation and pregnancy.

b. Formed Elements

Apart from the respiratory function of the hemoglobin molecules in the erythrocytes, the blood cells have specific roles that are not directly concerned with the general metabolic processes. This is in contrast to the plasma, which serves as the metabolic transport medium of the organism.

*1. **Erythrocytes.*** The red blood cells are the most numerous. The blood of the average adult female contains about 5 million of these cells, and that of the average male about 5.5 million, in every cubic millimeter of blood. (One drop of blood is the equivalent of 20–30 mm^3.) Any condition that tends to lower the oxygen content of the blood causes an increase in the number of erythrocytes. Persons living at high altitudes generally have a higher erythrocyte count than do those living at sea level. Conversely, increased barometric pressure results in a decrease in the erythrocyte count. The term *hematocrit value* is applied to the volume (in percent) of packed red blood cells in a sample (usually 10 ml) that has been centrifuged under standard conditions. The cells are spun to the bottom of a centrifuge tube and the supernatant liquid (the plasma) is drawn off the top. Any variation from the normal value may be indicative of the existence of certain pathological conditions. When *anemia* occurs, for instance, the percentage of erythrocytes is abnormally low; *polycythemia* is the condition arising from an abnormally high percentage.

Erythrocytes, unlike most other cells, contain neither mitochondria nor a nucleus. They cannot reproduce, have no aerobic metabolism, and are unable to synthesize carbohydrates, lipids, or proteins. They obtain all of their energy from the pentose phosphate shunt, and from substrate level phosphorylation in the Embden–Meyerhof pathway. The most significant components of the red blood cell are the hemoglobin molecules.

The major function of the erythrocyte is the transportation of oxygen to the cells and of carbon dioxide to the lungs (see Section 31.5). Human red blood cells have a life span of about four months. During this period, there is no degradation or resynthesis of hemoglobin molecules in the erythrocyte. New cells are formed in the bone marrow at the same rate that old cells are eliminated by special tissues in the liver and the spleen, in order that a constant level of erythrocytes be maintained. It has been estimated that there are approximately 30 trillion red blood cells in an average adult male and about 3 million of these are destroyed every second. Assuming that there are 300 million hemoglobin molecules in each erythrocyte, then 900 trillion molecules of hemoglobin must be synthesized every second (by cells in bone marrow) in order to maintain a constant supply.

*2. **Leukocytes.*** Leukocytes number about 7000 per cubic millimeter of blood, but this value is subject to considerable variation. The fatal disease *leukemia* is accompanied by an abnormally high leukocyte count, whereas in *typhoid fever* it is abnormally low. The chemical composition of white blood cells resembles that of other tissue cells (i.e., they contain glucose, lipids, proteins and other soluble organic substances and inorganic salts). The different varieties of leukocytes have specialized functions. *Polymorphonuclear* leukocytes destroy invading bacteria and other foreign substances. *Lymphocytes* serve as storage sites for the γ-globulins, and are probably involved in the synthesis of antibodies.

3. Thrombocytes. There are about 250,000 platelets in every cubic millimeter of blood. These small cells contain proteins and relatively large amounts of phospholipids, much of which is cephalin. The blood platelets liberate species that are instrumental in the mechanism of blood clotting.

31.2 Osmotic Pressure

When a living cell is placed in water, water gradually flows through the semipermeable membrane into the cell. The environment outside the cell is 100% water, while the percentage of water in the internal environment is considerably lower due to the presence of dissolved substances. Since the semipermeable membrane prevents these substances from leaving the cell, equalization of concentration can only be attained by the passage of the small water molecules into the cell. The diffusion of water from a dilute solution (high concentration of water), through a semipermeable membrane, into a more concentrated solution (low concentration of water) is known as the process of osmosis. **Osmotic pressure** is defined as *the pressure required to prevent the occurrence of osmosis when two solutions of unequal concentrations are separated by a semipermeable membrane.* The osmotic pressure depends solely upon the concentration of solute particles (either ions or molecules) in the solutions involved (recall Section 7.5-d).

The concentration of protein in the plasma far exceeds the concentration of protein in the interstitial fluid outside the blood vessels. This concentration gradient results in an osmotic pressure of about 18 torr. Therefore, if no external force were applied, fluids would be expected to diffuse from the interstitial fluid into the bloodstream. However, the pumping action of the heart creates the so-called **blood pressure,** and this pressure is greater at the arterial end of a capillary (~32 torr) than at the venous end (~12 torr). Since the blood pressure at the arterial end is higher than the osmotic pressure, the natural tendency is reversed and there is a net flow *from* the capillary *into* the interstitial fluid. The fluid which leaves the capillary contains the dissolved nutrients, oxygen, hormones, and vitamins needed by the tissue cells. As the blood moves along the capillary branches, the blood pressure decreases until at the venous end the osmotic pressure is greater than the blood pressure and there is a net flow of fluid *from* the interstitial fluid *into* the capillary. The incoming fluid contains the metabolic waste products such as carbon dioxide and excess water (Figure 31.1).

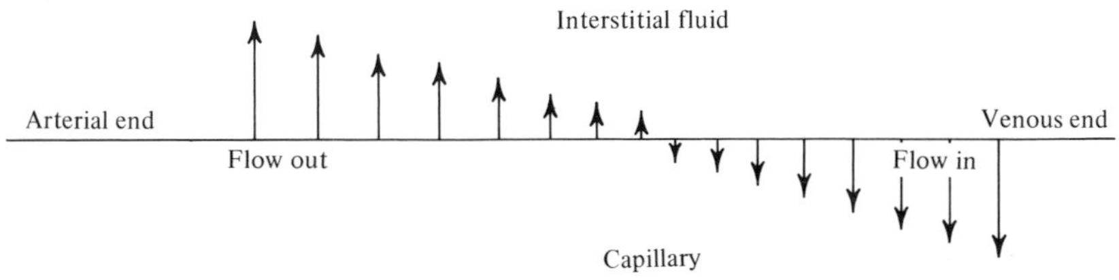
Interstitial fluid
Arterial end
Venous end
Flow out
Flow in
Capillary

Figure 31.1 Fluid flow between capillary and interstitial fluid.

Osmotic pressure, and hence the delicate balance of fluid exchange, is directly related to the concentration of albumins in the plasma. If the albumin level is low, as might be the case from (1) malnutrition (low protein intake), (2) abnormal protein synthesis, or (3) the loss of protein in the urine as a result of kidney disease (nephrosis), the osmotic pressure decreases. This results in a net efflux of fluids from the capillaries into the interstitial and cellular regions. This abnormal accumulation of fluids produces noticeable swelling in some tissues, particularly in the lower extremities. The condition so characterized is known as *edema*.

31.3 Clotting of Blood

The phenomenon of blood clotting or coagulation is of the utmost importance to the organism. If such a mechanism did not exist, loss of blood would occur whenever a blood vessel was injured. The details of the coagulation process, which is quite complex, have been clarified considerably by research in recent years. We shall restrict our discussion to a general description of the fundamental reactions involved. (The student is directed to an advanced biochemistry text for a more detailed explanation.)

When blood clots, the soluble protein fibrinogen is converted to the insoluble protein *fibrin*. Fibrin undergoes a polymerization reaction that results in the formation of needle-like threads. These threads enmesh the blood cells and effectively seal off the area where the blood vessel has been damaged. Once the fibrinogen has been removed from the plasma, the straw-colored fluid that remains is called the **serum.** The only distinction, therefore, between blood plasma and blood serum is the presence of fibrinogen in the plasma. Because blood serum lacks fibrinogen, it is unable to clot.

The clotting mechanism becomes operative only when a tissue is cut or injured. Blood platelets and damaged tissue cells are somehow activated and release a group of compounds collectively referred to as *thromboplastin*. In the presence of calcium ions and other cofactors, thromboplastin autocatalytically converts *autoprothrombin III* to *autoprothrombin C*. Autoprothrombin C is the enzyme that acts with calcium ions, phospholipids, and other cofactors to form *thrombin* from *prothrombin*. Autoprothrombin III and prothrombin are zymogens; their activation is analogous to the activation of the various digestive enzymes. Thrombin is the actual clotting enzyme. It is believed to effect the conversion of fibrinogen to fibrin by hydrolysis of a peptide fragment from the former. The process involved in blood coagulation may be summarized as follows.

$$\text{Autoprothrombin III} \xrightarrow[\text{Ca}^{2+}]{\text{thromboplastin}} \text{Autoprothrombin C}$$

$$\text{Prothrombin} \xrightarrow[\substack{\text{Ca}^{2+}\text{, phospholipids,} \\ \text{other cofactors}}]{\text{autoprothrombin C}} \text{Thrombin}$$

$$\text{Fibrinogen} \xrightarrow{\text{thrombin}} \text{Fibrin}$$

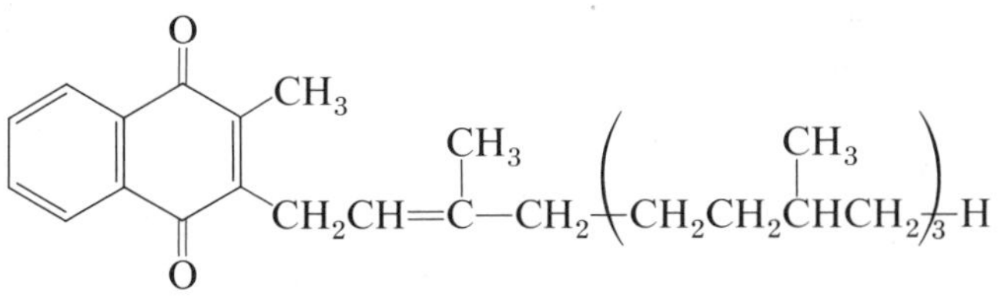

Vitamin K

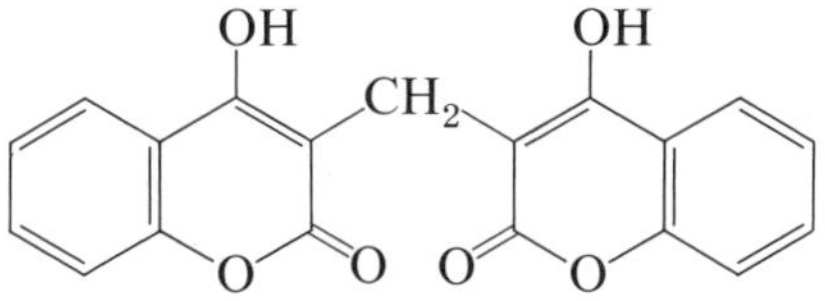

Dicumarol

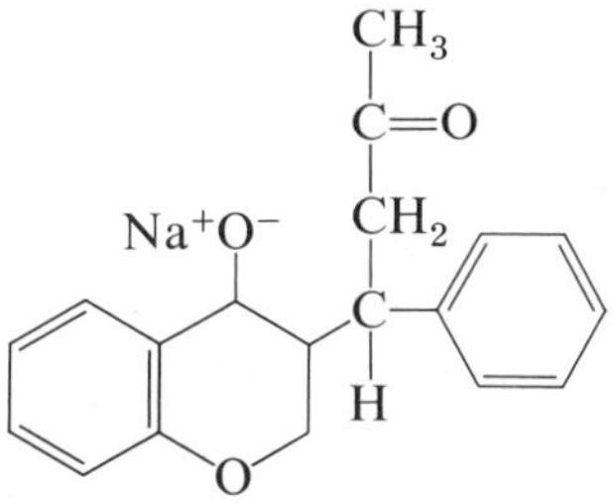

Warfarin sodium

Figure 31.2 The drugs dicumarol and warfarin sodium prevent thrombosis by acting as structural analogs of vitamin K and blocking the effects of the vitamin.

A number of substances, the *anticoagulants,* inhibit the clotting of blood by interfering with one or another of the reactions. *Heparin* is one of the principal anticoagulating agents. It is a polysaccharide, rich in sulfate–ester groups, and is believed to block the catalytic activity of both thromboplastin and thrombin. Low concentrations of heparin are normally present in the circulatory system to prevent *thrombosis,* the formation of a clot within a blood vessel. Certain sodium salts are employed as anticoagulants when blood is collected for clinical purposes. The anions of these salts, citrate, oxalate, and fluoride, form strong complexes with calcium ions, thus preventing them from existing in the free ionic form. If calcium ions are absent in the plasma, blood will not clot.

Two other compounds, vitamin K and dicumarol, are known to affect the clotting process. The blood from animals deficient in vitamin K has a prolonged coagulation time because of a lack of prothrombin in the plasma. Vitamin K is presumably involved in the biosynthesis of this protein. Dicumarol is believed to act as a metabolic antagonist of vitamin K. It functions either by repressing prothrombin formation, or by inhibiting an enzyme for which vitamin K is a cofactor. In recent years chemists have succeeded in synthesizing new anticoagulants that have a greater potency than dicumarol. One of these is warfarin sodium (Coumadin),[1] which is unique in that it can be administered orally, intravenously, intramuscularly, or rectally. These drugs are usually administered before an operation or after heart attacks to minimize thrombosis (Figure 31.2).

[1]Warfarin is also employed as a rat poison. It is safe for use as a rodenticide because regular ingestion of massive doses is fatal to rodents, whereas a single, accidental ingestion by children or pets is harmless.

31.4 Hemoglobin

The ability of mammalian blood to transport large quantities of oxygen depends upon the presence of the respiratory pigment hemoglobin (15 grams per 100 ml of blood). You will recall that hemoglobin is a conjugated protein having a molecular weight of about 68,000 daltons. Upon hydrolysis, it yields the simple protein, globin, and four *heme* groups (iron porphyrins—see Figure 25.1). The heme groups account for about 4% of the total molar mass. The characteristic red color of blood is due entirely to the presence of hemoglobin or, more precisely, to the presence of the heme groups, which absorb strongly in the blue region of the spectrum (~4000 Å). It is also the heme moiety that combines with molecular oxygen. However, the entire hemoglobin molecule is necessary for oxygen transport. If the globin is removed, or replaced by another protein, respiration is inhibited.

We have mentioned other heme proteins, such as catalase and the cytochromes, in which the central iron atom undergoes reversible oxidation-reduction; that is, $Fe^{2+} \rightleftharpoons Fe^{3+}$. The iron of the hemoglobin molecule, however, is in the ferrous state, and does not change to the ferric state at any time during normal oxygen transport. When oxygen is bound to the ferrous ion of hemoglobin, the compound is termed *oxyhemoglobin.* The deoxygenated form is sometimes referred to as reduced hemoglobin, but this term is misleading because it implies that the iron in the molecule is reduced. Therefore, we shall refer to this unoxygenated form simply as hemoglobin. The addition of oxidizing agents such as potassium ferricyanide can oxidize the iron to the ferric state. The same result is achieved in vivo by the action of nitrates, nitrites, and certain organic compounds (for example, acetanilide, nitrobenzene, the sulfa drugs). The resulting compound, which contains iron in the +3 oxidation state, is called *methemoglobin,* and is incapable of oxygen transport. Small amounts of methemoglobin are normally present (about 0.3 gram per 100 ml blood) in the erythrocytes, but appreciable amounts of this substance result in the pathological condition *methemoglobinemia.*[2]

Carbon monoxide hemoglobin, or *carboxyhemoglobin,* is formed by the combination of carbon monoxide with hemoglobin. The ferrous ions in hemoglobin have a much greater affinity for carbon monoxide than they do for oxygen (by a factor of 200) and thus will preferentially combine with any carbon monoxide that is in the blood. Carboxyhemoglobin will not transport oxygen, because all of the iron-binding sites are tied up by the carbon monoxide molecules. If sufficiently large numbers of hemoglobin molecules become saturated with carbon monoxide, death occurs as a result of a failure of the blood to supply the brain with oxygen.

When red blood cells are destroyed, their hemoglobin molecules are completely catabolized. The porphyrin ring is first cleaved; the globin and the iron are subsequently removed. The globin is then digested, and its amino acids join the others in the metabolic pool. The iron is set free, and

[2] Salami and other preserved meat products contain nitrite salts as preservatives. (Recall that nitrites have been implicated in nitrosoamine formation—page 474.) People who consume relatively large quantities of these foods will have a tendency to develop methemoglobinemia.

incorporated into the iron-storage protein *ferritin*. This iron will be reutilized for the synthesis of new hemoglobin in the bone marrow. The porphyrin skeleton is of no further use to the body. It undergoes a series of degradation reactions that lead to the production of the bile pigments, chiefly *biliverdin* and *bilirubin*. These are colored substances that give the bile its yellow color. The degraded pigments are stored in the gall bladder and released into the large intestine. As they travel down the intestinal track, they undergo additional transformations that result in a darkening of their color, thus accounting for the characteristic color of feces and urine.

31.5 Respiratory Functions of the Blood

The human body requires an enormous amount of oxygen to satisfy the demands of the energy-yielding oxidative phosphorylation reactions. The hemoglobin molecule is well suited to meet these demands because of its affinity for oxygen and because the attachment of oxygen to heme is readily reversible. In the alveoli of the lungs, hemoglobin comes into direct contact with a rich supply of oxygen (100 torr) and is converted to oxyhemoglobin. The oxyhemoglobin is carried by the arterial circulation to the cells in which there is a low oxygen concentration (40 torr) and a relatively high concentration of carbon dioxide (60 torr). The oxygen is given up to the cells; the resulting hemoglobin carries some of the carbon dioxide back to the lungs to be expelled, and more oxyhemoglobin is formed.

Oxyhemoglobin does not transfer all of its oxygen to the tissue cells. Normally, every 100 ml of arterial blood combines with about 19 ml of oxygen. In the resting individual, the venous blood carries about 12 ml of oxygen per 100 ml of blood. Therefore, more than 60% of the hemoglobin in venous blood is still combined with oxygen. If a person is engaged in strenuous exercise, his oxygen demand is high and the percentage of oxyhemoglobin in the venous blood may fall as low as 25%. Arterial blood is crimson in color; venous blood is a darker red, but it is not purple or blue.

The actual mechanism of the respiratory process is more complex than we have indicated above. Figure 31.3 and the following discussion represent an attempt to indicate the essential events that take place as blood traverses the capillaries of the alveoli and the tissues.

Atmospheric oxygen is taken into the lungs where a difference in pressure exists between the alveoli and the capillaries. Oxygen diffuses into the red blood cell (1) where it combines with hemoglobin, HHb, to form oxyhemoglobin, HbO_2^- (2).

$$\mathrm{HHb} + \mathrm{O_2} \rightleftharpoons \mathrm{HbO_2^-} + \mathrm{H^+}$$

Two factors tend to shift this equilibrium to the right. They are the high oxygen concentration in the lungs and the neutralization of hydrogen ions by the bicarbonate ions present in the red blood cells (3). The oxyhemoglobin is then carried to the tissues where carbon dioxide is being produced as a result

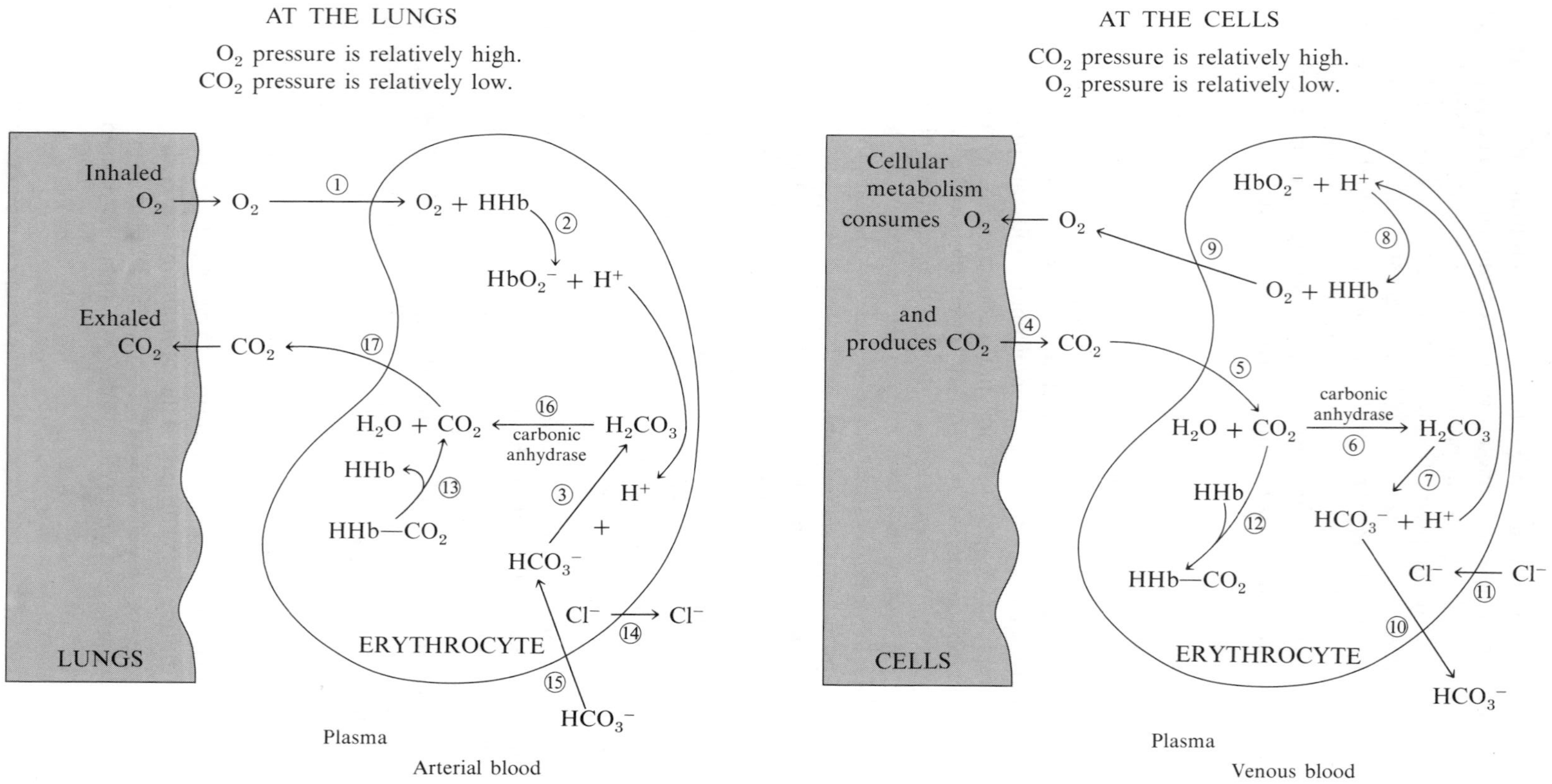
AT THE LUNGS
O2 pressure is relatively high.
CO2 pressure is relatively low.
Inhaled O2
O2
O2 + HHb
HbO2⁻ + H⁺
Exhaled CO2
CO2
H2O + CO2
carbonic anhydrase
H2CO3
HHb
HHb—CO2
H⁺
+
HCO3⁻
Cl⁻
Cl⁻
ERYTHROCYTE
LUNGS
HCO3⁻
Plasma
Arterial blood
AT THE CELLS
CO2 pressure is relatively high.
O2 pressure is relatively low.
Cellular metabolism consumes O2
O2
HbO2⁻ + H⁺
O2 + HHb
and produces CO2
CO2
carbonic anhydrase
H2O + CO2
H2CO3
HHb
HCO3⁻ + H⁺
HHb—CO2
Cl⁻
Cl⁻
ERYTHROCYTE
CELLS
HCO3⁻
Plasma
Venous blood

Figure 31.3 Oxygen–carbon dioxide transport by the blood. [Adapted from J. R. Holum, *Elements of General and Biological Chemistry*, Wiley, 1968.]

of cellular metabolism (4). The carbon dioxide diffuses into the erythrocyte (5) where the enzyme *carbonic anhydrase* catalyzes its combination with water to form carbonic acid (6). Carbonic acid subsequently dissociates into bicarbonate ions and hydrogen ions (7).

$$H_2CO_3 \rightleftharpoons H^+ + HCO_3^-$$

The major factor that shifts this equilibrium to the right is the neutralization of the hydrogen ions by oxyhemoglobin (8), which is concomitantly split into hemoglobin and oxygen. Because the oxygen pressure is greater in the capillaries than in the tissues, the oxygen diffuses into the tissue cells (9) to be utilized in the oxidative reactions of metabolism. As the bicarbonate ion concentration increases, the ions diffuse out of the red blood cell into the plasma (10). The loss of negative ions by the cell is then balanced by the migration of chloride ions into the cell from the plasma. The process, referred to as the *chloride shift* (11), brings about a re-establishment of electrolyte equilibrium. Most of the carbon dioxide is taken to the lungs as bicarbonate ion (~90%), but some combines with hemoglobin (probably with free amino groups in the globin portion) (12). This species is referred to as *carbamino-hemoglobin.*

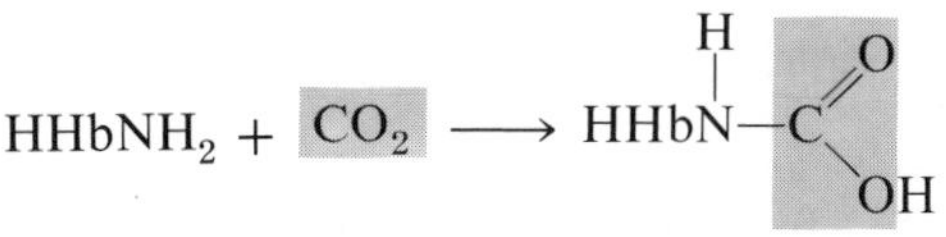
HHbNH2 + CO2
HHbN
H
C
O
OH

When the red blood cells return to the lungs, a reversal of events (12), (13), (14), (3), (15), (16), (17) sends carbon dioxide into the lungs to be expelled and the cyclic process repeats itself. It should be emphasized that hemoglobin plays the major role in the respiratory process. In addition to its involvement in oxygen transport, it is also involved in the ability of the blood to carry carbon dioxide without appreciable changes in pH.

31.6 Blood Buffers

Recall that the maintenance of pH within narrow limits is vital to the well-being of an organism. Any appreciable change in hydrogen ion concentration inhibits oxygen transport and alters the rates of the metabolic processes by decreasing the catalytic efficiency of enzymes. The blood plasma and the erythrocytes contain three buffering systems that maintain the pH of the blood between the values of 7.3 and 7.5. In Section 10.6 we mentioned the buffering action of the bicarbonate pair H_2CO_3/HCO_3^- and of the phosphate pair $H_2PO_4^-/HPO_4^{2-}$. Recall that there is 1.6 times as much monohydrogen phosphate in the blood as dihydrogen phosphate, and 10 times as much bicarbonate ion as carbonic acid. In Section 25.3, we indicated that the amino acids, in their zwitterionic forms, can neutralize small concentrations of either acids or bases. Since the plasma proteins and the globin of hemoglobin contain both acidic and basic amino acids, they tend to minimize changes in pH by combining with, or liberating, hydrogen ions and thus serve as excellent buffering agents over a wide range of pH values.

Of the three, the bicarbonate pair is the most important buffer system due to its intimate connection with the respiration processes. Under normal conditions, the primary factor that tends to change the pH of an organism is the continuous production of carbon dioxide and the acidic metabolites (acetoacetic acid, pyruvic acid, α-ketoglutaric acid, etc.). When acids enter the blood, they are neutralized by bicarbonate ions, and the slightly dissociable acid, carbonic acid, is formed.

$$H^+ + HCO_3^- \rightleftharpoons H_2CO_3$$

It would seem that this reaction would alter the buffer ratio by decreasing the bicarbonate ion concentration and increasing the concentration of carbonic acid. The excess carbonic acid, however, is readily decomposed to water and carbon dioxide by the enzyme *carbonic anhydrase*. The respiration rate is increased, and the carbon dioxide is eliminated at the lungs, thus preserving the proper buffer ratio.

$$H_2CO_3 \xrightarrow[\text{anhydrase}]{\text{carbonic}} H_2O + CO_2$$

Because of their function in neutralizing acids, the bicarbonate ions in the blood have been referred to as the *alkaline reserve* of the body.

Exercises

31.1 Define or explain the meaning of each of the following terms.

(a) anemia
(b) polycythemia
(c) edema
(d) thrombosis
(e) leukemia
(f) chloride shift
(g) osmosis
(h) osmotic pressure
(i) lymph
(j) formed elements

31.2 List five functions of the blood.

31.3 The blood accounts for about 8% of the total body weight. Calculate the volume of blood that your body contains. (Density of blood = 1.06 g/ml.)

31.4 Classify the plasma proteins and give a function for each class.

31.5 Explain the processes involved when fluid is exchanged between the plasma and the cells at the arterial end and the venous end of a capillary.

31.6 Briefly explain the process of blood coagulation. Include in your discussion the function of each of the following.

(a) prothrombin
(b) thrombin
(c) fibrinogen
(d) thromboplastin
(e) calcium ions

31.7 What is the role of vitamin K in blood coagulation? What is the effect of **(a)** heparin and **(b)** dicumarol in preventing thrombosis? How do anions such as oxalates, fluorides, and citrates prevent blood coagulation?

31.8 What is the difference between plasma and serum?

31.9 Describe the chemical structure of hemoglobin.

31.10 Name three heme-containing compounds.

31.11 Contrast the structural features of each of the following compounds.

(a) globin and hemoglobin
(b) heme and hemoglobin
(c) hemoglobin and myoglobin
(d) hemoglobin and oxyhemoglobin
(e) oxyhemoglobin and carboxyhemoglobin
(f) oxyhemoglobin and methemoglobin

31.12 What is the oxidation state of iron in **(a)** hemoglobin, **(b)** oxyhemoglobin, **(c)** methemoglobin, and **(d)** carboxyhemoglobin?

31.13 Why is carbon monoxide such a deadly poison?

31.14 Contrast the contents of arterial blood with that of venous blood.

31.15 Briefly explain the process of oxygen transport from lungs to cells, and carbon dioxide transport from cells to lungs. Write equations for the reactions that occur at the cells and at the lungs.

31.16 What is the function of carbonic anhydrase?

31.17 What is the relationship between acidosis and oxygen transport?

31.18 Illustrate, with equations, how the blood buffers prevent the change in pH of the blood when small amounts of acid or base are produced during metabolic reactions.

31.19 Name the two blood buffer pairs. Why is the bicarbonate buffer pair effective in spite of the fact that the ratio of acid to anion is 1:10?

32

Nutrition and Health

by Howard J. Sanders[1]

The focus of public interest in nutrition has changed radically in recent decades. In the first 30 or 40 years of this century, the main nutritional interest of many Americans was in getting enough food to eat and in avoiding such vitamin-deficiency diseases as pellagra, rickets, scurvy, and beriberi. By now, these diseases have been largely conquered in the United States, thanks to improved foods and a better standard of living. As D. Mark Hegsted, administrator of USDA's Human Nutrition Center, states, "In the past, the message was, in essence, to eat more of everything. Now, we are faced with the more difficult problem of teaching the public to be more discriminating. Increasingly, the message will be to eat less [of many foods]."

The reason for the turnabout is that many foods now are believed to be factors in causing or promoting such degenerative diseases as heart disease, stroke, hypertension, some forms of cancer, diabetes, osteoporosis, tooth decay, kidney disease, and cirrhosis of the liver. Diet also is involved in an especially prevalent disability—obesity.

Large percentages of the American public avoid foods containing cholesterol and saturated fat to reduce the risk of coronary heart disease. Many people curtail their intake of salt to reduce the risk of high blood pressure. They eat less sugar to decrease tooth decay or eat more dietary fiber in the hope of preventing colon cancer, diverticulosis, hemorrhoids, and other disorders.

[1]Reprinted with permission from *Chemical and Engineering News,* Volume 57, pp. 27–46 (March 26, 1979).

To function normally, the human body needs from 45 to 50 essential nutrients that can be obtained only from the diet. When the diet supplies all of these nutrients, as well as enough calories, the body can synthesize the many other compounds it also requires for proper health. The body needs macronutrients—proteins (amino acids), carbohydrates (starches and sugars), and fats (fatty acids). It also needs micronutrients—vitamins and minerals (trace elements). In addition, the body, of course, needs water. Recent progress in understanding the role of some of these nutrients in human health is reviewed in this chapter.

32.1 Vitamins

Doubtless, one of the biggest nutritional advances of the century was the discovery in 1913 of the first vitamin—vitamin A—by Elmer V. McCollum, a biochemist at the University of Wisconsin. In 1922, he went on to discover vitamin D. In the 1930's, vitamins captured the public's imagination. Probably nothing did more to arouse wide interest in nutrition during this period than the public's awareness of the importance of vitamins. Gulping down vitamin pills or eating vitamin-enriched foods (milk, bread, breakfast cereals, and so on) became almost a national obsession.

Over the years, scientists have isolated 13 vitamins needed by humans. Researchers today are in general agreement that no more vitamins are likely to be found. One reason is that more than 30 years have elapsed since the last one (vitamin B_{12}) was discovered, despite an intensive search since then for others. Another reason is that many people have lived quite well for years on intravenous solutions containing only the known vitamins and other nutrients.

Vitamins often are classified as either water-soluble (vitamin C and the B-complex vitamins) or fat-soluble (vitamins A, D, E, and K). Most water-soluble vitamins act as coenzymes by promoting the metabolic reactions of other compounds (see Table 26.2). The fat-soluble vitamins have more varied functions (Table 32.1).

Fat-soluble vitamins, if taken in high doses, can accumulate in hazardously large amounts because they are stored in body fat. People who consume too much vitamin D, for example, can develop bone pain, bone-like deposits in the kidneys, and mental retardation. Water-soluble vitamins, on the other hand, generally are rapidly excreted in the urine. Thus, even when taken in relatively large amounts, they usually are not toxic.

Vitamins have been the subject of more fads and more misrepresentations than any other group of nutrients. Claims have been made that, among other things, vitamins cure cancer, arthritis, and mental illness, increase sexual potency, prevent colds, and overcome muscular weakness. It is small wonder that the public is baffled by such claims, which many nutritionists reject.

Table 32.1 Deficiencies of Vitamins Can Cause a Wide Range of Health Problems

Vitamin	*Structural Formula (See Page)*	*Dietary Sources*	*Deficiency Symptoms*
Fat-Soluble Vitamins			
Vitamin A	539	Fish liver oils, liver, eggs, fish, butter, cheese, milk. A precursor, β-carotene, is present in green vegetables, carrots, tomatoes, squash	Night blindness, eye inflammation
Vitamin D	596	Fish liver oils, butter, vitamin-fortified milk, sardines, salmon. The body also obtains this compound when ultraviolet light converts 7-dehydrocholesterol in the skin to vitamin D	Rickets, osteomalacia
Vitamin E	295	Vegetable oils, margarine, green leafy vegetables, grains, fish, meat, eggs, milk	Anemia in premature babies fed inadequate infant formulas
Vitamin K	777	Spinach and other green leafy vegetables, tomatoes, vegetable oils	Increased clotting time of blood, bleeding under the skin and in muscles
Water-Soluble Vitamins			
Thiamine (vitamin B_1)	646	Cereal grains, legumes, nuts, milk, beef, pork	Beriberi
Niacin (nicotinamide)	647	Red meat, liver, turnip greens, fish, eggs, peanuts	Pellagra
Riboflavin (vitamin B_2)	646	Milk, red meat, liver, green vegetables whole wheat flour, fish, eggs	Dermatitis, glossitis (tongue inflammation), anemia
Pyridoxine (vitamin B_6)	646	Eggs, meat, liver, peas, beans, milk	Dermatitis, glossitis, increased susceptibility to infections, irritability, convulsions in infants
Pantothenic acid	648	Liver, beef, milk, eggs, molasses, peas, cabbage	Gastrointestinal disturbances, depression, mental confusion
Folic acid	647	Liver, mushrooms, green leafy vegetables, wheat bran	Anemias, gastrointestinal disturbances
Vitamin B_{12} (cyanocobalamine)		Liver, meat, fish eggs, milk, oysters, clams	Pernicious anemia, retarded growth, glossitis, spinal cord degeneration
Biotin	648	Beef liver, kidney, peanuts, eggs, milk, molasses	Dermatitis
Ascorbic acid (vitamin C)	582	Citrus fruit, tomatoes, green peppers, strawberries, potatoes	Scurvy

a. Vitamin A

As is well known, vitamin A plays a major role in night vision (Section 22.8). It also is important to the growth and maintenance of skin. In the past, some physicians used large doses of vitamin A for treating acne. Not only were the doses potentially toxic but they were not really effective. Recently, however, chemists at Hoffman-La Roche in Switzerland have found that a synthetic derivative of vitamin A (13-*cis*-retinoic acid) appears promising as a drug for treating severe acne.

In 1975, Erik Bjelke of the Norwegian Cancer Registry in Oslo reported that the more vitamin A that cigarette smokers consumed, the less apt they were to develop lung cancer. Their source of the vitamin was chiefly vegetables (particularly carrots) and milk. Bjelke comments, "In view of the difficulties of influencing smoking behavior, the suggestion that ingested agents such as vitamin A may be potent preventives against the cancer-causing effects of smoking should be of more than theoretical interest."

Michael B. Sporn of the National Cancer Institute, on the other hand, believes that naturally occurring forms of vitamin A are ineffective in preventing human cancer. In high concentrations, vitamin A, he says, would be too toxic for this purpose. He does believe, however, that this vitamin's synthetic analogs, known as retinoids, show promise in preventing various forms of human cancer, such as cancer of the lung, breast, and bladder.

b. Vitamin C

In 1747, James Lind, while studying medicine at the University of Edinburgh, discovered that citrus fruit was effective in treating sailors suffering from scurvy. Not until 1932, however, was the active ingredient in citrus fruit, vitamin C (ascorbic acid), isolated by Charles G. King at the University of Pittsburgh and by Albert Szent-Györgyi in Hungary. Vitamin C was the first dietary component to be recognized as essential for preventing a human disease. Most animals can synthesize vitamin C, but humans, monkeys, and guinea pigs cannot.

In recent years, a subject of fierce controversy in the nutrition field has been the claim that vitamin C is effective in preventing or curing the common cold. Since at least the 1930's, people have been debating whether this vitamin actually is useful in combating colds.

The issue did not flare up into a major public debate, however, until 1970, when Linus C. Pauling, emeritus professor of chemistry at Stanford University, published his best-selling book *Vitamin C and the Common Cold*. He stated that vitamin C in doses ranging from 1 to 5 grams a day could prevent colds and that as much as 15 grams a day could cure a cold. Scientists investigating Pauling's claims have obtained conflicting results.

Some scientists report that taking 1.5 grams of vitamin C every half hour over a period of two hours can cure a cold. Others concede that this may be true for a small group of the population. However, as one researcher puts it: "For the remainder of the population, vitamin C is relatively or completely ineffective in curing the common cold, since large-scale controlled studies have been quite unimpressive."

After studying the value of vitamin C in preventing or treating the common cold, Terence W. Anderson of the medical faculty at the University of Toronto concluded, "We found little effect from the purely prophylactic regimens we used. There was no evidence of a gradient in effect across the dosage range of 250, 1000, or 2000 mg a day . . . . [The vitamin C treatment] had little or no effect on the frequency of colds, upon their total duration, or upon nasal symptoms."

Victor Herbert of the Bronx Veterans Administration Hospital says, "There is some evidence that large doses of vitamin C may have a mild antihistaminic effect. However, taking a mild antihistamine for relief [of the common cold] rather than large doses of vitamin C would clearly be preferable."

Some scientists point to the possible dangers of taking massive doses of this vitamin. Such doses, for example, may raise the uric acid level in body fluids and thus cause gout in people predisposed to this disease.

Pauling and other researchers also have been exploring the use of large doses of vitamin C to treat cancer and schizophrenia and to retard aging. Others have been studying its use in treating heart disease.

James D. Cook and Elaine R. Monsen at the University of Washington have reported that 75 mg a day of vitamin C can increase significantly the absorption of iron from the intestine into the bloodstream. The vitamin could be quite useful for this purpose, they emphasize, because many people suffer from iron deficiency caused not by an inadequate intake of iron but by its poor absorption. Walter Mertz, chairman of USDA's Nutrition Institute, comments, "Although cereal products in the U.S. have been enriched with iron for about 40 years, a disappointingly high incidence of iron deficiency still exists in some parts of the population. A treatment that would promote the absorption of iron into the bloodstream would be highly beneficial."

c. Vitamin D

Vitamin D is important mainly because it promotes the absorption of calcium ions from the intestine into the bloodstream, so that the concentration of calcium ions reaches a level high enough to permit proper bone mineralization (Section 24.10-d). A shortage of this vitamin in infants and children can cause the bone disease rickets. In adults, such a deficiency can produce the bone disease osteomalacia.

In the years following McCollum's discovery of vitamin D (actually two compounds, D_2 and D_3, there being no D_1), the substance was believed to function directly in the body without chemical change. In 1967, however, Hector F. DeLuca and co-workers at the University of Wisconsin found that

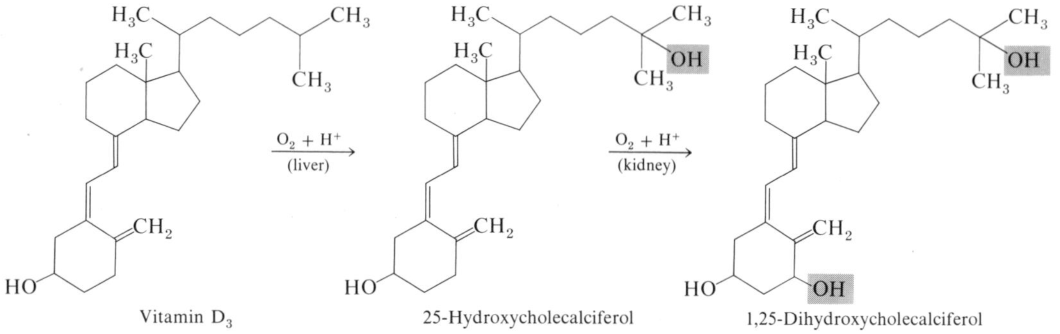
H3C
CH3
H3C
CH3
O2 + H+
(liver)
H3C
CH3
H3C
OH
CH3
O2 + H+
(kidney)
H3C
CH3
H3C
OH
CH3
CH2
CH2
CH2
HO
HO
HO
OH
Vitamin D3
25-Hydroxycholecalciferol
1,25-Dihydroxycholecalciferol

Figure 32.1 Vitamin D must acquire two hydroxyl groups to be active.

vitamin D must be converted to other compounds before it can perform its necessary functions in the body. DeLuca and co-workers discovered that vitamin D_3 is metabolized in the liver to 25-hydroxycholecalciferol, which in turn is further metabolized in the kidney to 1,25-dihydroxycholecalciferol (Figure 32.1). This latter compound is believed to be the form of the vitamin that is active in calcium transport.

These discoveries have led to methods for helping patients who, because of liver or kidney diseases, cannot convert vitamin D into the active form. Lacking the needed form of the vitamin, these patients often develop osteoporosis, osteomalacia, vitamin D-resistant rickets, or other bone diseases. Now, a patient with a kidney disease can be given the synthetic compound 1-hydroxycholecalciferol, which the liver can convert to 1,25-dihydroxycholecalciferol, or can be given the dihydroxy compound directly to obtain vitamin D in a form that can combat the bone disorder. Says DeLuca, "If current efforts prove successful, as seems likely, 1,25-dihydroxycholecalciferol or a related compound could have a sizable impact on world health. It could have a particular effect on the aged, who so often are afflicted with both kidney and bone diseases."

d. Vitamin E

One of the most vigorously promoted and highly controversial vitamins is vitamin E. Discovered in 1923 by Herbert M. Evans and Katharine S. Bishop at the University of California, it is actually a group of compounds called tocopherols. Deficiencies of this vitamin in humans occur almost exclusively in premature babies who are fed improper infant formulas. In infants, a shortage of this vitamin can cause anemia, produced by the accelerated destruction of red blood cells.

Among the sensational claims made for vitamin E are that it increases virility and sexual endurance in humans. It also is alleged to overcome problems of sterility. But a panel of the Institute of Food Technologists says, "Leading scientific and medical opinion does not support the view that supplemental doses of vitamin E have any value in preventing male impotence or sterility or in affecting human reproduction."

Some of the medical claims made for vitamin E have resulted, at least in part, from the erroneous extrapolation to humans of tests carried out in laboratory animals. Some years ago, scientists found that vitamin E is among the factors required to prevent sterility in male rats and to permit normal pregnancy in female rats. From these findings, some people concluded that vitamin E in humans is effective in preventing male impotence or sterility and in promoting successful pregnancy. This view, however, is not supported by scientific or medical evidence gathered over the past 40 years or more.

Various people have alleged that large doses of vitamin E are useful in treating such disorders as coronary heart disease, muscular weakness, ulcers, arthritis, diabetes, and skin disorders. The vitamin also is said to improve athletic ability, retard aging, and increase the lungs' resistance to air pollution. Scientists have found that rats deprived of vitamin E do develop muscular weakness and scaly skin, but this has not been shown to occur in humans.

Some investigators have reported that, when taken in large amounts, vitamin E, which is an antioxidant, may help to protect the lungs from oxidative damage caused by ozone and other air pollutants (Section 13.5). For some time, researchers have known that vitamin E, as an antioxidant, protects the fatty acids in cell membranes from oxidative attack and thus prevents damage to these membranes. On the other hand, vitamin E can cause problems. Some people who take large daily doses develop nausea, intestinal disorders, and blurred vision.

e. Other Vitamins

In 1935, Henrik Dam in Copenhagen discovered vitamin K, which he showed could combat hemorrhaging in chicks. Later, other researchers found that this vitamin is necessary for the formation of prothrombin and thus permits the desired clotting of blood (Section 31.3).

For many years, this clot-promoting ability was thought to be the only function of vitamin K in humans and animals. (The vitamin, incidentally, got its name from the word *Koagulation,* in German.) Peter V. Hauschka of Children's Hospital in Boston has found that this vitamin also is involved in the formation of bone.

The B vitamin niacin (nicotinic acid) is best known for its ability to prevent pellagra. The vitamin also serves as a component of enzyme systems involved in the body's oxidation-reduction reactions (Section 28.8).

In recent years, a number of investigators have alleged that large doses of niacin are useful in treating schizophrenia. After exploring this claim in detail, a task force of the American Psychiatric Association concluded that no reliable evidence exists that niacin is of any value for this purpose. Moreover, niacin in large amounts can produce such adverse side effects as impaired liver function and abnormal glucose metabolism.

Although research on vitamins is not as active today as it was in the 1930's and 1940's, studies of these compounds continue. Research is in progress to determine where vitamins operate in the human body and exactly how they function. Considerable work also is being done to determine how vitamins might be used to prevent or cure disease.

Says John G. Bieri of NIH,

> In the vitamin field, the easy problems already have been solved. The more difficult ones remain. Scientists would like to find out, for example, how vitamin A functions at the cellular level to maintain normal epithelial tissue throughout the body. They would like to know how vitamin D regulates calcium metabolism and bone formation. They would like to have a better understanding of why a deficiency of vitamin E increases an animal's need for the essential trace element selenium. Obviously, in the vitamin field, a great deal more needs to be learned.

32.2 Trace Elements

In recent decades, growing research has centered on the trace elements in the human diet. These elements, although present in living tissue in concentrations of only micrograms per gram or less, can have a profound effect on human health.

Research on dietary trace elements, which has been under way sporadically for 300 years, accelerated markedly about 20 years ago. Scientists realized that, if such elements as iron, copper, selenium, and cobalt are essential to the human diet, other elements also might be essential. To date, 15 trace elements have been judged to be essential to the health of humans and/or animals. The exact human requirements of only three—iron, iodine, and zinc—are known, however.

The essential trace elements in the body may be a part of or complexed with hormones, enzymes, and vitamins, as well as bone and tissue structures. They also may serve a wide range of other functions. Iron is needed to make hemoglobin and myoglobin. Iodine is a constituent of thyroid hormones and prevents iodine-deficiency goiter. Chromium is an activator of insulin. Cobalt is a component of the vitamin B_{12} molecule.

Deficiencies of trace elements in the human body may occur because of excessive loss of these elements by way of the intestine, urinary tract, or skin. Shortages also may occur because of inadequate dietary intake of these elements. Refined flour, for example, provides the body with much less iron and zinc than does whole grain. Fabricated foods such as substitutes for orange juice or whipped cream may contain smaller amounts of trace elements than their conventional counterparts.

Walter Mertz of USDA's Nutrition Institute observes, "In the years ahead, the trend toward the increased consumption of preprocessed, refined, or even synthetic foods can be expected to continue and to decrease our intake of trace elements. Thus, research concerned with the essential functions of new trace elements . . . appears to be of the greatest importance."

A major problem in studying the trace elements is that they are often difficult to measure because of their low concentration. For example, the vanadium content of fresh peas is usually less than 0.4 ng/gram. To measure trace elements, scientists may use atomic absorption spectrometry, activation analysis, emission spectrometry, and other techniques. Development of still more sensitive analytical methods would be useful.

The first trace element found to be essential in the human diet was iron. In 1681, English physician Thomas Sydenham wrote of using "iron and steel filings steeped in cold Rhenish wine" to treat chlorosis, an iron-deficiency anemia.

The second trace element shown to be essential in humans was iodine. In about 1850, French chemist Jean Baptiste Boussingault discovered that the salt deposits that some South American Indians believed could cure goiter were able to do so because the deposits contained iodine compounds.

More recently, seven more trace elements—copper, manganese, zinc, cobalt, molybdenum, selenium, and chromium—have been found to be essential to human nutrition. An additional six—tin, vanadium, fluorine, silicon, nickel, and arsenic—are required by various animals. The dietary trace elements can be divided into three broad categories, as indicated in Table 32.2.

The most common trace element deficiency is a dietary shortage of iron, the first of the Group 1 dietary trace elements. An abnormally low intake of iron causes anemia, which is characterized either by a low concentration of hemoglobin in the blood or by a low volume of packed red blood cells.

Iron deficiency is especially prevalent among women who, because of menstrual bleeding, need more of it than men. The Recommended Dietary

Table 32.2 Number of Dietary Trace Elements Shown to Be Essential in Humans/Animals Is Increasing

Element	*Date Discovered to Be an Essential Dietary Trace Element in Animals*	*Dietary Sources*	*Functions in Humans and/or Animals*
Group 1 (Need in Humans Established and Quantified)			
Iron	17th century	Meat, liver, fish, poultry, beans, peas, raisins, prunes	Component of hemoglobin and myoglobin; oxidative enzymes
Iodine	1850	Iodized table salt, shellfish, kelp	Needed to make the thyroid hormones thyroxine and triiodothyronine; prevents iodine-deficiency goiter
Zinc	1934	Meat, liver, eggs, shellfish	Present in at least 90 enzymes and in the hormone insulin

Table 32.2 Number of Dietary Trace Elements Shown to Be Essential in Humans/Animals Is Increasing (*Continued*)

Element	*Date Discovered to Be an Essential Dietary Trace Element in Animals*	*Dietary Sources*	*Functions in Humans and/or Animals*
Group 2 (Need in Humans Established But Not Yet Quantified)			
Copper	1928	Nuts, liver, shellfish	Component of oxidative enzymes; involved in absorption and mobilization of iron needed for making hemoglobin
Manganese	1931	Nuts, fruits, vegetables, whole-grain cereals	Involved in formation of enzymes, bone
Cobalt	1935	Meat, dairy products	Component of vitamin B_{12}
Molybdenum	1953	Organ meats, green leafy vegetables, legumes	Involved in formation of enzymes, proteins
Selenium	1957	Meat, seafood	Involved in enzyme formation, fat metabolism
Chromium	1959	Meat, beer, unrefined wheat flour	Required for glucose metabolism
Group 3 (Need Established in Animals But Not Yet in Humans)			
Tin	1970	[a]	Essential for normal growth of rats
Vanadium	1971	[a]	Needed for optimum growth of chicks and rats
Fluorine	1972	Fluoridated water	Essential for normal growth of rats
Silicon	1972	[a]	Needed for growth and bone development in chicks; needed for growth of rats
Nickel	1974	[a]	Needed for normal growth and for formation of red blood cells in rats; needed by chicks and swine
Arsenic	1975	[a]	Required for normal growth of rats, goats, minipigs; shortage of element can impair reproductive ability of goats, minipigs

[a] Present in trace quantities in many foods and in the environment.

Allowance for iron among women in the age range of 23–50 is 18 mg/day, whereas it is only 10 mg/day for men in the same age group.

Iodine is needed to prevent iodine-deficiency goiter. A small percentage of goiter patients (about 4%) do not have the disease because of a shortage of iodine but because of the action of antithyroid compounds in some foods, such as turnips, cabbage, and kale. Iodine is present in two thyroid hormones—thyroxine (Figure 25.2) and triiodothyronine. These compounds increase the metabolic rate and oxygen consumption of cells.

a. Zinc

Zinc is present in at least 90 enzymes, as well as in the hormone insulin. It plays an important role in the processes of growth, including cell division and protein synthesis. It is involved in the functioning of the pituitary, adrenal glands, pancreas, and gonads.

Zinc was first reported to be essential to animals in 1934 by Wilbert R. Todd, Conrad A. Elvehjem, and co-workers at the University of Wisconsin. The importance of this element in humans was not appreciated until 1963, when Ananda S. Prasad of Wayne State University reported the effects of zinc deficiency among adolescent boys in Egypt. These boys showed retarded growth, delayed sexual development, enlarged liver, and other disorders. Their zinc deficiency was found to be caused partly by their large consumption of bread containing a high concentration of dietary fiber and phytate, which inhibit the absorption of zinc from the intestine. These boys also ate little meat, which contains zinc in a form that is readily absorbed.

Later, other scientists discovered zinc deficiency among children in Turkey and Iran. In 1972, it was found, surprisingly, in Denver, among some preschool children whose growth was noticeably retarded, probably because they ate very little meat and obtained little zinc from other sources. Zinc deficiency also can cause an impaired sense of taste and smell, a delayed healing of wounds and burns, and impaired immune responses. These disorders can be corrected by feeding zinc supplements.

Says Harold H. Sandstead, director of USDA's Human Nutrition Laboratory, "We do not know how often either marginal or deficient dietary intakes of zinc occur in the United States. However, limited knowledge leads to the suspicion that they may not be so rare as previously believed, especially in populations whose diets are high in cereals and low in meat and other animal products."

b. Copper

Copper, the first of the Group 2 dietary trace elements, serves an important function in the body's oxidative processes as a component of such oxidative enzymes as tyrosinase, cytochrome oxidase, monoamine oxidase, and ceruloplasmin. In some animals, a deficiency of copper can cause loss of hair pigmentation because tyrosinase is needed to convert tyrosine to the pigment melanin.

A deficiency of copper can produce other disorders, such as anemia, a low level of white cells in the blood, and impaired bone formation. According to the most widely accepted theory, a shortage of copper can cause anemia because this element is needed for the absorption and mobilization of iron required to make hemoglobin.

c. Manganese

Manganese, shown to be essential in animals in 1931, functions partly by activating a number of enzymes, among them glycosyltransferases. In 1969, Roland M. Leach Jr., professor of poultry science at Pennsylvania State University, reported that manganese deficiency in chicks causes some of its adverse effects by decreasing the activity of these enzymes and thus interfering with the synthesis of glycosaminoglycans. These compounds play an important role in the body by maintaining the strength and stability of connective tissue, such as cartilage associated with bone. Leach's studies in chicks have shown that reduced synthesis of this compound is associated with bone abnormalities, congenital ataxia (a nervous disorder), and defective eggshell formation.

In 1972, Edward A. Doisy Jr. of St. Louis University's school of medicine described what is believed to be the first reported case of manganese deficiency in man. During a study of the human requirement for vitamin K, one of Doisy's patients was fed a diet from which manganese was accidentally omitted. The reported effects of manganese deficiency included a low level of serum cholesterol, weight loss, slowed growth of hair and nails, and a reddening of the hair. These symptoms were relieved when the patient was returned to a normal diet.

In 1978, Yukio Tanaka of St. Mary's Hospital in Montreal reported the case of a 12-year-old boy who suffered from ataxia and occasional convulsions that were traced to the unusually low level of manganese in his blood. Previously, scientists had found that manganese deficiency causes nervous disorders in chicks, rats, mink, and guinea pigs, but not in humans.

d. Other Elements

In 1935, scientists at Australia's Department of Agriculture reported that a shortage of the trace element cobalt causes "wasting disease" in sheep and cattle. The researchers found that this disease could be prevented or cured by using cobalt as a dietary supplement. In human nutrition, cobalt is needed because it is a component of vitamin B_{12}, which can prevent pernicious anemia.

Molybdenum, which was found to be essential in animals in 1953, is needed partly because it is a component of several enzymes, such as xanthine oxidase, aldehyde oxidase, and sulfite oxidase. Researchers have found that when chicks, lambs, and young turkeys fail to receive enough molybdenum their growth is stunted. Although molybdenum is known to be needed in man, a deficiency of this element has never been reported in humans.

The essential dietary role of selenium in animals was discovered in 1957. A deficiency of selenium was shown to produce muscular disorders in sheep and cattle and to cause liver and kidney ailments in rats and swine. An inadequate intake of selenium also may produce diseases in humans. Studies in Guatemala and Thailand indicate that some children with kwashiorkor (severe protein malnutrition) are deficient in this element.

In 1959, it was shown that chromium is essential for the normal metabolism of glucose in rats. Later, this need for chromium also was demonstrated in humans. According to present theory, chromium facilitates the action of insulin in promoting the utilization of glucose and amino acids. Only chromium complexed with an unidentified organic molecule, a so-called "glucose tolerance factor," has been shown to be metabolically active in the body. Chromium deficiency has been observed among children with severe protein-energy malnutrition in Nigeria, Jordan, and Turkey. It also has been found among some elderly people in the United States.

During the years 1970–1975, six Group 3 trace elements—tin, vanadium, fluorine, silicon, nickel, and arsenic—were found to be essential to various animals. Thus far, however, none of these newer trace elements has been shown to be essential to humans. R. Leach of Pennsylvania State University comments, "Nutritional scientists generally believe that, if the shortage of a dietary trace element causes a disease in experimental animals, it also can cause a disease in humans—even if no such human disease has yet been discovered." W. Mertz adds,

> I believe that there is a very good chance that, within the coming decade, every one of the newer trace elements will be shown to be essential in humans. The question of essentiality is, of course, completely independent of the question of whether a deficiency of a newer trace element occurs in man, because humans may consume entirely sufficient amounts of that element. Because of obvious ethical considerations, we cannot deliberately produce trace element deficiencies in humans and thus cause bodily harm. The only way we can discover such deficiencies in man is by studying deprived peoples in underdeveloped parts of the world or by studying peoples in developed countries who eat unconventional or bizarre diets.

Various reasons explain the rapid increase since 1970 in the number of dietary trace elements shown to be essential in animals. One is the availability of highly sensitive analytical methods, such as activation analysis and electrothermal atomic absorption spectrometry. Another is the development of special isolation chambers that enable researchers to maintain animals under closely controlled conditions that keep out all unwanted contaminants. If a scientist wants to determine the effect of a diet containing less than a specified minuscule amount of tin, for example, he must be sure that his animals do not pick up any tin from the test chamber (now usually made of plastic) or from the air the animals breathe. Moreover, the diets fed these animals must be highly purified and contain a precisely controlled amount of tin.

In recent years, scientists have speculated about what new trace elements might be found in the future to be necessary to the health of animals and possibly also of humans. Among the elements that seem to be the most likely candidates are cadmium, lead, boron, lithium, and bromine.

32.3 Fiber

Few topics in human nutrition have stimulated public interest as greatly in recent years as has the subject of fiber in the diet. Advertisements and TV commercials proclaim the merits of various breakfast foods, breads, and other products on the basis of their fiber content. Many nutritionists are concerned that high-fiber foods are being publicized as if they were miracle drugs.

Because of changing American eating habits, the consumption of crude fiber from cereals and potatoes has dropped about 50% since the turn of the century. In the same period, the intake of crude fiber from fruits and vegetables has declined about 20%. On the other hand, the public has been eating more meat, milk, and other foods that contain little or no fiber.

All this is of considerably more than academic interest because some scientists believe that a low-fiber diet is associated with the increased incidence of colon cancer, diverticulosis (an intestinal disorder), coronary heart disease, atherosclerosis, gallstones, and hemorrhoids. Some people also claim that a low-fiber diet is associated with the increased incidence of appendicitis, hiatus hernia, varicose veins, and other ailments.

Although many scientists question these views, large segments of the general public seem convinced that a high-fiber diet is the key to solving a great many health problems. About one thing, however, there is general agreement. An increased intake of dietary fiber is definitely beneficial in treating constipation.

a. Determining Fiber Content

Americans obtain most of the fiber in their diet from wheat and other grains, potatoes, fruits, and vegetables. But what exactly is fiber (also known in common parlance as roughage)? Peter J. Van Soest, professor of animal nutrition at Cornell University, says, "Every research group around the world seems to have a somewhat different definition. But basically there is increasing agreement that fiber is the cell walls of plants." Others prefer to define fiber as a mixture of carbohydrates and related substances (such as lignin) that are obtained from plants and that, in humans, undergo little or no digestion by enzymes.

The fiber in foods usually is designated as either crude fiber or dietary fiber, depending on the laboratory procedure used to measure it. Generally, the percentage of crude fiber in any given food is much less than its percentage of dietary fiber. The reason is that the test for determining crude fiber fails to measure a large percentage of various fiber components present in foods.

The method for determining the crude fiber in plant foods was developed in Europe in about 1800 and came into general use between 1820 and 1840. In this procedure, a food sample is first treated with an organic solvent, such as diethyl ether or ethyl alcohol. It then is treated with hot, dilute sulfuric acid and finally with hot, dilute sodium hydroxide. The remaining solid material obtained after the residue is heated to a high temperature is the crude fiber.

This vigorous treatment of the food sample removes not only the lipids, proteins, and readily digestible carbohydrates such as starch (which should be removed) but a sizable percentage of the lignin, hemicellulose, and cellulose. According to some estimates, the results of such an analysis generally include only 10–50% of the lignin, 20% of the hemicellulose, and 50–80% of the cellulose, which should be measured in full in any accurate determination of plant fiber.

Although the results of the crude fiber test are poor, scientists continue to perform the test because it has become traditional and deeply ingrained. Moreover, the procedure is officially approved by the Association of Official Analytical Chemists and is required by the food regulatory laws of many states. In addition, the test is fairly easy to carry out and gives a rough measure of cellulose content.

The measurement of dietary fiber, on the other hand, gives a much more accurate indication of a food's fiber content. It accounts, for example, for all of the cellulose, hemicellulose, and lignin.

Over the past 25 years, scientists have attempted to develop a variety of improved methods for determining the dietary fiber content of foods. In 1963, Van Soest devised a method using a boiling solution containing the detergent sodium lauryl sulfate, along with sodium borate, sodium phosphate, and ethylenediaminetetraacetic acid (EDTA), to isolate the plant cell wall. The detergent removes the proteins and all of the other nonfiber components. The sodium borate and phosphate act as a buffer to maintain the pH at 7. EDTA, a chelating agent, prevents calcium, magnesium, iron, and other cations from interfering with the test.

In 1969, David A. T. Southgate, then at Dunn Nutrition Laboratory in Cambridge, England, reported a procedure for measuring fiber in which he first extracts the lipids with an organic solvent such as ethyl alcohol. He then digests the proteins and starches with enzymes.

Both the Van Soest and Southgate methods are widely used today in food research laboratories. In the years ahead, measurements of crude fiber are expected to be replaced more and more by measurements of dietary fiber because of their far greater accuracy.

b. Dietary Fiber

Dietary fiber consists of cellulose, hemicellulose, and lignin. It also includes pectins, gums, and mucilages, although these substances are not fibrous. The dietary fiber in food undergoes little or no hydrolysis by digestive enzymes in the small intestine. It then passes through the large intestine essentially unchanged. Actually, some components of dietary fiber are degraded by bacteria in the large intestine to form volatile carboxylic acids (propionic, butyric, and others) and gases (carbon dioxide and methane).

In the past, nutritionists were concerned primarily with the total quantity of dietary fiber present in typical servings of foods. Now, they are becoming more and more concerned about the percentages of the specific components of dietary fiber in foods. Part of the reason for this changing concern is the growing recognition that the various components of dietary fiber act in different ways in the body.

Cellulose, hemicellulose, and pectin, for example, have a high capacity to absorb water. Lignin, on the other hand, has a very low capacity for water absorption. Hemicellulose and pectin can be digested fairly readily by bacteria in the large intestine, whereas cellulose is digested by these bacteria to only a limited extent.

The most abundant component of dietary fiber is usually cellulose. A linear polymer of glucose, cellulose has a molecular weight ranging from about 600,000 to 2 million daltons (Section 23.8-c).

Hemicellulose is likewise a polysaccharide. It consists of pentoses and hexoses, many of which are branched. It has a relatively low molecular weight of about 10,000–20,000 daltons.

Lignin, which is not a carbohydrate, is a polymer based on phenylpropane units. Its molecular weight ranges from about 1000 to 10,000 daltons. Lignin has a much greater ability to bind bile acids than does cellulose. Thus, lignin increases the excretion of these acids from the intestine. This causes the body to make more of these acids, which are synthesized in the liver from cholesterol (Section 24.10-b). This tends to lower the serum cholesterol level.

Pectin is a polygalacturonic acid with a molecular weight of about 60,000–90,000 daltons. Pectin tends to promote the excretion of bile acids. Thus, like lignin, it tends to lower the blood's level of cholesterol.

Much of the public's lively interest in dietary fiber was triggered by the findings of Denis P. Burkitt, a physician at St. Thomas Hospital medical school in London. In the early 1970's, he reported that rural people in middle and southern Africa who eat large amounts of high-fiber foods have an exceptionally low incidence of colon cancer. This disease is a major cause of cancer deaths in the United States and Western Europe. It also is common in

African cities where more affluent people do not eat a largely vegetarian diet, as do their rural countrymen.

Also in the early 1970's, Hugh C. Trowell, then a consulting physician at Makerere University in Kampala, Uganda, independently reported similar results among peoples in Africa and Asia. As far back as 1954, a few scientists suggested that the Bantus of Africa almost never developed colon cancer and coronary heart disease because their diets contained a high percentage of fiber.

Burkitt and Trowell have reported that rural African populations eating a high-fiber diet not only show a very low incidence of colon cancer and coronary heart disease but also of diverticulosis, diabetes, appendicitis, hemorrhoids, and hiatus hernia. These disorders are, of course, extremely common in the western world.

Repeated studies have shown that the typical diet of westerners contains much less fiber than that of rural Africans. The average American, for example, consumes only about 4 grams of crude fiber a day, whereas the rural Africans studied by Burkitt and Trowell eat about 25 grams of crude fiber daily, obtained mainly from corn, legumes, and root crops. Burkitt and Trowell believe that the prevalence of various major diseases in the United States and Europe is increased markedly because of the low-fiber diets usually consumed by westerners.

Burkitt attempts to explain the rarity of colon cancer among rural Africans who eat a high-fiber diet by pointing out that dietary fiber increases fecal bulk. The bulk is enhanced by the absorption of water by the fiber and by the increase in the amount of indigestible material in the feces. Because of this greater bulk, the average fecal transit time through the intestine for Ugandan villagers is only 35 hours, compared to 71 hours for British Royal Navy men, who eat a low-fiber diet. The decreased transit time, says Burkitt, speeds the elimination of possible carcinogens from the intestine. The greater bulk of the fecal matter also reduces the concentration of possible carcinogens and makes them less hazardous.

What these carcinogens really are has been a matter of continuing speculation. Some scientists propose that they are compounds formed from bile acids and sterols by bacteria in the large intestine. Others suggest that bile acids promote colon cancer when they react with other dietary components in the large intestine to form carcinogens. It has been proposed that one of the carcinogens might be 1,2-benzpyrene (Figure 15.3), which is present in smoked meats and fish.

Many scientists, on the other hand, seriously doubt whether colon cancer actually is related to a low intake of dietary fiber. Rather, they believe it is associated with a high intake of meat and thus of saturated fat and protein. Rural Ugandans, for example, eat a largely vegetarian diet and thus consume relatively little meat.

A high-fiber diet also is reported to decrease the level of serum cholesterol and thus may reduce the risk of developing atherosclerosis and coronary heart disease. Some types of dietary fiber increase the excretion of cholesterol from the intestine. Thus, the fiber reduces the absorption of cholesterol into the bloodstream and lowers the serum cholesterol level.

Intense disagreement exists over whether dietary fiber is capable of reducing the risk of developing coronary heart disease or other cardiovascular diseases. Donald B. Zilversmit, a nutritional scientist at Cornell University, says, "The evidence that dietary fiber has a beneficial effect on the mortality or morbidity caused by cardiovascular disease is not convincing. The experimental evidence that one or more components of dietary fiber lower serum cholesterol concentrations is conflicting."

As he and others have pointed out, the Bantus, for example, eat not only a diet high in fiber but one low in meat and saturated fat. Their low incidence of coronary heart disease could well be caused by their low consumption of saturated fat, rather than their high intake of fiber.

Burkitt and various other scientists also believe that a high-fiber diet reduces the chance of developing diverticulosis, hemorrhoids, and other intestinal disorders. By absorbing water, fiber promotes the formation of a relatively soft, bulky stool. This reduces the need for straining during bowel movements and is believed to decrease the likelihood that these intestinal ailments will occur.

Some people have concluded that a high-fiber diet is the answer to a vast array of medical problems. However, a fiber-rich diet can cause problems of its own. Since dietary fiber can bind metals like calcium, magnesium, iron, zinc, copper, and chromium, which are important in human nutrition, a high-fiber diet can promote the elimination of these elements and thus reduce their absorption into the bloodstream, with serious consequences.

Phytic acid[2] (also known as phytate) is found in many high-fiber foods such as cereals. Phytic acid, in large amounts, is especially noted for binding calcium, iron, and zinc. Says Kristen W. McNutt, executive officer of the National Nutrition Consortium, "We should be cautious about recommending that people radically increase their fiber intake by eating large quantities of foods that also will increase their phytate consumption. Iron-deficiency anemia, for example, is already a serious health problem in many women and children in the United States."

Scientists see an urgent need to learn more about the composition of dietary fiber. Better methods are needed for measuring each of its constituents. More needs to be known about the properties of these components and the effects of each component on the human body. More needs to be learned about how, if at all, specific intake levels of each component affect the incidence of disease.

[2] Phytic acid is a hexaphosphate of inositol,

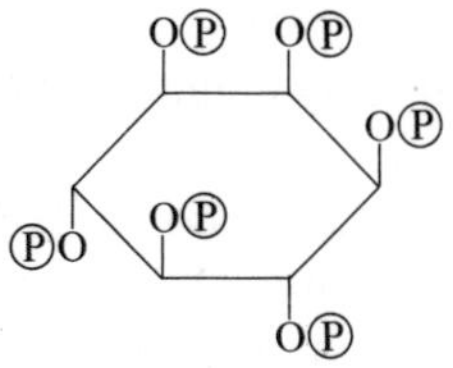

32.4 Cholesterol

In the past 20 years, few subjects in the nutrition field have attracted as much public attention as cholesterol. Today, everything from margarine and vegetable oils to egg substitutes and meat analogs is advertised on the basis that it contains little or no cholesterol.

a. Cholesterol and Cardiovascular Disease

Cholesterol is believed to be a primary factor in the development of atherosclerosis, coronary heart disease, and stroke. As the leading cause of death in the United States, heart attack and stroke together take about 840,000 lives a year.

The disorder underlying both heart attack and stroke is atherosclerosis, characterized by the buildup of deposits (plaques) on the inner surfaces of arteries. If a blood clot forms in such a constricted artery leading to the heart or brain and causes a complete stoppage of blood flow, a heart attack or stroke occurs almost instantly. In atherosclerotic plaques, the lipid in highest concentration is cholesterol. Present in lower concentrations are two other types of lipids—phospholipids and triacylglycerols.

Scientists generally agree that elevated cholesterol levels in the blood, as well as high blood pressure and cigarette smoking, are associated in humans with an increased risk of heart attack. A long-term investigation by the NIH has shown that, among men aged 30–49, the incidence of coronary heart disease was five times greater if their cholesterol level was 260 mg/100 ml of serum or more, compared to the men whose level was 200 mg/100 ml of serum or less.

What has not been entirely proved, say many scientists, is that a lowering of serum cholesterol levels in humans by the use of diet or drugs will reduce the incidence of coronary heart disease. However, such a reduced incidence seems likely on the basis of animal studies.

Some scientists, on the other hand, believe that research in humans already has demonstrated that a lowered cholesterol level does reduce the risk of coronary heart disease. Research reported in 1972 by Osmo Turpeinen and co-workers at the University of Helsinki has shown that people who reduce their cholesterol level by eating diets low in saturated fat and low in cholesterol have a noticeably decreased incidence of coronary heart disease. The gathering of proof to support this contention is exceedingly difficult because large numbers of people must be studied over long periods of time. Useful data can be obtained only by a slow, laborious process, partly because only a small percentage of any group will die of coronary heart disease in any given year.

Cholesterol, a wax-like substance found in all body cells, is essential to normal body function (Section 24.10-a). This compound, a steroid alcohol (sterol), is a precursor of adrenal and sex hormones. It also is a precursor of bile acids and vitamin D. In the human body, cholesterol is synthesized primarily in the liver. The body also obtains it from food of animal origin, such as eggs, dairy products, and meat. Generally, the more cholesterol a person eats, the less his body synthesizes.

Because lipids such as cholesterol are not soluble in water, they cannot be transported in the blood (an aqueous medium) unless complexed with water-soluble proteins. These complexes, which have a wide range of properties, are known as lipoproteins.

Lipoproteins generally are classified according to their density and origin. There are four broad categories: chylomicrons (density of less than 1.006 g/ml and made in the intestine), very low-density lipoproteins (density of less than 1.006 and made mainly in the liver), low-density lipoproteins (1.006–1.063), and high-density lipoproteins (1.063–1.21). The chylomicrons contain up to [illegible]% by weight of lipids, and the very low-density lipoproteins (VLDL's) contain up to about 95% lipids. The low-density lipoproteins (LDL's) contain about 75% lipids, and the high-density lipoproteins (HDL's) only about 50%.

In research on cholesterol and its role in heart disease, the types of lipoproteins that have received the greatest attention in recent years have been the LDL's and HDL's. The reason is that they almost always contain a higher percentage of cholesterol than do the chylomicrons or the VLDL's.

b. HDL, the Good Cholesterol

One of the most fascinating discoveries in this field is that, contrary to what might be assumed, cholesterol in the form of HDL reduces (not increases) a person's risk of developing coronary heart disease. For this reason, HDL has come to be known as "good cholesterol." On the other hand, cholesterol in the form of LDL increases a person's risk of developing coronary heart disease and thus sometimes is referred to as "bad cholesterol."

Back in 1954, John W. Gofman and co-workers at the University of California, Berkeley, reported that atherosclerosis and coronary heart disease are associated with elevated levels of serum LDL's, rather than with serum lipoproteins in general. They also reported that the level of serum LDL's was a better predictor of coronary heart disease risk than was the level of serum cholesterol.

Initially, the findings of Gofman and associates were not taken very seriously by many scientists studying coronary heart disease because they tended to think solely in terms of serum cholesterol levels rather than of whether the cholesterol was present as LDL's or HDL's. Now, scientists recognize that, even if the total cholesterol level is high, the risk of coronary heart disease may be only moderate if the level of LDL is low and that of HDL is high. Low-density lipoproteins contain an unusually high proportion of cholesterol—40–45%. High-density lipoproteins contain only about 15–20% cholesterol.

LDL's are believed to promote coronary heart disease by first penetrating the coronary artery wall, where they are broken down enzymically to cholesterol, cholesterol ester, and protein. The cholesterol and cholesterol ester are then deposited in the artery wall, becoming major parts of the atherosclerotic plaque.

Robert W. Mahley, head of the comparative atherosclerosis and arterial metabolism section of the National Heart, Lung & Blood Institute, points out that patients who, because of hereditary factors, have high levels of LDL in their blood have a high incidence of coronary heart disease. Often, they die from this disease in their teens.

How do HDL's reduce the risk of developing coronary heart disease? No one knows for sure—but there are some fairly compelling theories.

Daniel Steinberg and co-workers at the University of California, San Diego, have suggested that HDL competitively inhibits the uptake of LDL by cells by occupying the binding sites that otherwise might be occupied by LDL. Research by Yechezkiel and Olga Stein, a husband-and-wife team at Hadassah University Hospital in Jerusalem, also supports the view that HDL helps to block the attachment of LDL to the cells. The independent studies of Steinberg and the Steins also indicate that HDL may help to combat coronary heart disease by transporting cholesterol out of the cells.

John A. Glomset of the University of Washington was the first to propose that HDL promotes the removal of cholesterol from cells. He showed that the enzyme lecithin cholesterol acyl transferase (LCAT) converts the cholesterol in HDL to cholesterol ester, which leaves the surface of the molecule and is transferred to the core of the HDL particle. The HDL then can pick up additional free cholesterol from the cells and carry it to the liver, where it is converted to bile acids. These acids pass into the intestine and are excreted.

According to Robert J. Nicolosi, a biochemist at the New England Regional Primate Research Center, HDL's ability to remove cholesterol from tissue, at least in vitro, may depend on the nature of the specific protein-phospholipid complexes present in the HDL. These complexes act as substrate for the LCAT enzyme.

Assuming that HDL helps to protect the body against coronary heart disease, what can be done to increase serum levels of this lipoprotein?

One way, for reasons not now known, is by exercise. Last year G. Harley Hartung of Houston's Methodist Hospital reported that the group of marathon runners he studied had a decidedly higher average HDL level than did a group of sedentary men. He believes that the difference in HDL levels of the two groups was related to their degree of physical activity and was not significantly related to their diet.

Robert S. Lees, director of MIT's Arteriosclerosis Center has shown that HDL levels also can be increased by losing weight. He recommends that a person's weight be held to within a few pounds of his or her so-called "ideal weight."

Christian Gulbrandsen and co-workers at NIH's Hawaii Heart Study have reported that HDL levels in the blood can be increased by drinking alcohol in moderation. The amount of alcohol used in their research project per person was equivalent to about three 12 oz. bottles of beer a day.

Actually, not everyone agrees that raising HDL levels in the blood is an effective way to combat coronary heart disease. Robert Mahley declares,

> It is premature to say that all HDL-cholesterol is beneficial. We really do not know enough about the different types of lipoproteins that, in general terms, are referred to as HDL. Contrary to what other investigators have implied, our research group finds that more than 90% of the HDL does not interact with the cell receptor sites and thus does not block the attachment of LDL. However, a minor subclass of HDL is capable of binding to the receptor. From these findings, we have concluded that HDL represents a mixed group of lipoproteins with at least two distinctly different metabolic roles.

A reduction of low-density lipoprotein levels is believed by many scientists to be desirable. This may be accomplished by eating a diet low in cholesterol and low in saturated fat. And some physicians recommend that overweight people go on reducing diets to achieve their recommended weight.

32.5 Diet and Behavior

For many years, scientists have known that diet affects the behavior and brain development of laboratory animals and humans. Studies in rats and other animals show that malnutrition can reduce exploratory behavior and problem-solving ability. It also can cause apathy, irritability, and fearfulness.

In the 1950's, investigators began focusing their attention on the fact that a low intake of proteins and/or calories in children can retard their behavioral development, leading to poor muscular coordination and undesirable personality traits (lethargy, irritability, lack of curiosity, and so on). In addition, it can cause children to get low scores on standard intelligence tests.

Studies in laboratory animals show that diet has a profound effect on brain development. In rats, a severely limited intake of proteins and/or calories early in life can cause a permanent reduction in brain size and weight. It also can reduce the brain's content of DNA (a measure of the total number of brain cells) and its content of myelin (a complex lipid whose concentration is a measure of the brain's growth).

Similar results also have been reported among children who have suffered severe protein-calorie malnutrition. In 1969, Dr. Myron Winick and co-workers at Columbia University's Institute of Human Nutrition compared the brains of nine children who died of malnutrition in the first year of life with the brains of well-nourished children who died of other causes. The brains of the malnourished infants were unusually small, and their total number of brain cells was markedly below normal. In the case of three of the infants, the DNA content of their brains was only about 40% of normal.

In MIT's department of nutrition and food science, Dr. Richard J. Wurtman, Dr. John D. Fernstrom, and co-workers have been studying the role of short-term and long-term diet in changing the concentration of various neurotransmitters in the brain of rats. Neurotransmitters are compounds that chemically relay the electrical impulses between neurons. The concentration of these compounds in the brain is believed to have a major effect on behavior.

In 1971, Wurtman, Fernstrom, and associates observed that the amount of the amino acid tryptophan present in the rat diet can influence its subsequent conversion in the brain to the neurotransmitter serotonin (page 764). The amount of the amino acid tyrosine in the diet, they later found, can influence its conversion in the brain to the neurotransmitter norepinephrine (Figure 20.7). Most recently, Wurtman and associates found that the amount of the nonamino acid choline in the diet can affect the formation of the neurotransmitter acetylcholine (page 504). Hence, the functioning of neurons in the brain and thus behavior might be influenced by diet.

Loy D. Lytle, then at MIT, has reported that rats fed a diet containing corn as the only source of protein have an ususually high sensitivity to pain—specifically, the pain of an electrical shock. Corn contains a relatively low concentration of tryptophan, which is the precursor of serotonin. Low concentrations of serotonin in the neurons may be associated with high responsiveness to pain and to stimuli in general.

As the events of recent decades have shown, eating habits can be changed. People today are eating more polyunsaturated fat and decreasing their consumption of eggs, and thus of cholesterol, to reduce the risk of heart disease. They are eating less salt to lower the risk of high blood pressure. They are drinking more orange juice to increase their intake of vitamin C. They are watching their total food intake to avoid becoming overweight.

The major problem facing nutritionists, biochemists, and other scientists in the field will be to determine what changes in the human diet will be most effective in promoting health and preventing disease. Although the controversies are certain to persist, the experts will be called upon to give reliable answers to the all-important question: What, indeed, is proper nutrition?

Index